The Brooks and Olmo Register of Fruit & Nut Varieties

third edition

ISBN 0-9615027-4-6

Printed in the United States of America

ASHS Press
600 Cameron Street
Alexandria, VA 22314-2562
telephone 703.836.4606 • fax 703.836.2024 • e-mail ashspres@ashs.org

Contents

Contributing Specialists

Sandra L. Anagnostakis
—chestnut
The Connecticut Agricultural Experiment Station
123 Huntington St.
Box 1106
New Haven, CT 06504
phone 203.789.7253
fax 203.789.7232; e-mail slanagno@caes.state.ct.us

Robert L. Andersen
—beach plum, cherry, cherry plum, Chinese bush cherry, plum, sandcherry
Dept. of Horticultural Sciences
Hedrick Hall
New York State Agricultural Experiment Station
Geneva, NY 14456-0462
phone 315.787.2235
fax 315.787.2216
e-mail rla2@nyaes.cornell.edu

Carl W. Campbell
—loquat
15301 S.W. 269th Terrace
Homestead, FL 33032
phone 305.247.3597
fax 305.246.7003

Richard J. Campbell
—mango
Fairchild Tropical Garden
11935 Old Cutler Rd.
Miami, FL 33156-4299
phone 305.665.2844
fax 305.665.8032

John R. Clark
—grape
Dept. of Horticulture
Univ. of Arkansas
Fayetteville, AR 72701
phone 501.575.2603
fax 501.575.7273
e-mail jrclark@comp.uark.edu

James Cummins
—rootstocks
Dept. of Horticultural Sciences
Hedrick Hall
New York State Agricultural Experiment Station
Geneva, NY 14456-0462
phone 315.789.7083
fax 315.787.2216
e-mail jn3c@nyaes.cornell.edu

Hugh Daubeny
—raspberry
3558 West 15th St.
Vancouver, B.C. V6R 2Z4, Canada

Louise Ferguson
—fig, olive
Univ. of California
Kearney Agricultural Research Ctr.
9240 Riverbend Ave.
Parlier, CA 93648
phone 209.891.2541
fax 209.891.2593
e-mail louise@uckac.edu

Ross Ferguson
—kiwifruit (Chinese gooseberry)
HortResearch
120 Mt. Albert Research Centre
Mt. Albert Rd.
Private Bag 92 169
Auckland, New Zealand
phone 64.9.849.3660
fax 64.9.815.4201; e-mail rferguson@marccri.marc.cri.nz

Gene Galletta
—strawberry
U.S. Dept. of Agriculture
Agricultural Research Service
Fruit Laboratory
Beltsville Agricultural Research Center–West
Beltsville, MD 20705
phone 301.504.5652
fax 301.504.5062

Fred G. Gmitter, Jr.
—citrus
Citrus Research and Education Ctr.
700 Experiment Station Rd.
Univ. of Florida
Lake Alfred, FL 33850-2299
phone 941.956.1151
fax 941.956.4631
e-mail fgg@gnv.ifas.ufl.edu

Thomas M. Gradziel
—almond, processing peach
Dept. of Pomology
Univ. of California
Davis, CA 95616
phone 916.752.1575
fax 916.752.8502
e-mail tmgradziel@ucdavis.edu

L.J. Grauke
—*hickory, pecan*
U.S. Dept. of Agriculture
Agricultural Research Service
Pecan Genetics and Breeding Research
Rte. 2, Box 133
Highway 50 South
Somerville, TX 77879
phone 409.272.1402
fax 409.272.1401
e-mail ljg@tamu.edu

Duane W. Greene
—*apple*
Dept. of Plant and Soil Sciences
Bowditch Hall
Univ. of Massachusetts
Amherst, MA 01003-0910
phone 413.545.5219
fax 413.545.0260
e-mail dgreene@plsoil.umass.edu

Kim E. Hummer
—*currant, elderberry, gooseberry, jostaberry, pear*
U.S. Dept. of Agriculture
Agricultural Research Service
National Clonal Germplasm Repository
33447 Peoria Rd.
Corvallis, OR 97333-2521
phone 541.750.8712
fax 541.750.8717
e-mail hummerk@bcc.orst.edu

R. Scott Johnson
—*pomegranate*
Univ. of California
Kearney Agricultural Research Center
9240 Riverbend Ave.
Parlier, CA 93648
phone 209.891.2500
fax 209.891.2593
e-mail sjohnson@uckac.edu

Robert J. Knight
—*atemoya, avocado, cherimoya, sapodilla, white sapote*
Univ. of Florida
18905 S.W. 280th St.
Homestead, FL 33031
phone 305.242.4795
fax 305.246.7003

Robert R. Krueger
—*date*
U.S. Dept. of Agriculture
Agricultural Research Service
National Germplasm Repository for Citrus and Dates
1060 Martin Luther King, Jr. Blvd.
Riverside, CA 92507
phone 909.787.4399
fax 909.787.4398

Desmond R. Layne
—*pawpaw*
129 Atwood Research Facility
400 E. Main St.
Kentucky State Univ.
Frankfort, KY 40601-2355
phone 502.227.5942
fax 502.227.6381
e-mail dlayne@ca.uky.edu

Richard E.C. Layne
—*apricot*
Agriculture and Agri-Food Canada
Harrow Research Centre
Harrow, Ont. N0R 1G0, Canada
phone 519.738.2251
fax 519.738.2929

Paul Lyrene
—*blueberry*
Horticultural Sciences Dept.
1143 Fifield Hall, Univ. of Florida
Gainesville, FL 32611
phone 352.392.4711, ext. 307
fax 352.392.5653
e-mail pml@gnv.ifas.ufl.edu

Gale H. McGranahan
—*Persian walnut*
Dept. of Pomology
Wickson Hall
Rm. 1055
Univ. of California
Davis, CA 95616
phone 916.752.0113
fax 916.752.8502

Shawn A. Mehlenbacher
—*hazelnut (filbert)*
Dept. of Horticulture
4017 Agricultural and Life Sciences Bldg.
Oregon State Univ.
Corvallis, OR 97331-7304
phone 541.737.5467
fax 541.737.3479
e-mail mehlenbs@bcc.orst.edu

James N. Moore
—blackberry
Dept. of Horticulture
316 Plant Science Bldg.
Univ. of Arkansas
Fayetteville, AR 72701
phone 501.575.2603
fax 501.575.8619

Robert A. Norton
—apple
700 13th St. N.E.
East Wenatchee, WA 98802
phone & fax 509.886.9204

Dick Okie
—plum
U.S. Dept. of Agriculture
Agricultural Research Service
Southeast Fruit and Tree Nut
Research Laboratory
111 Dunbar Rd.
Byron, GA 31008
phone 912.956.6405
fax 912.956.2929
e-mail a03dokie@attmail.com

Dan E. Parfitt
—pistachio
Dept. of Pomology
Univ. of California
Davis, CA 95616
phone 916.752.7031
fax 916.752.8502
e-mail deparfitt@ucdavis.edu

Joseph D. Postman
—mayhaw, pear, quince
U.S. Dept. of Agriculture
Agricultural Research Service
National Clonal Germplasm
Repository
33447 Peoria Rd.
Corvallis, OR 97333-2521
phone 541.750.8712
fax 541.750.8717
e-mail postmanj@bcc.orst.edu

Bill Reid
—black walnut
Pecan Experiment Field
Kansas State Univ.
P.O. Box 247
Chetopa, KS 67336-0247
phone 316.597.2972
fax 316.597.2758

Curt R. Rom
—apple
Dept. of Horticulture
316 Plant Science Bldg.
Univ. of Arkansas
Fayetteville, AR 72701
phone 501.575.2603
fax 501.575.8619
e-mail crom@comp.uark.edu

Phillip Rowe
—banana
P.O. Box 1776–FHIA
Gulfport, MS 39502
phone 504.68.2078
fax 504.68.2313

Kay Ryugo
—persimmon
Dept. of Pomology
Univ. of California
Davis, CA 95616
phone 916.752.0122
fax 916.752.8502

Wayne B. Sherman
—nectarine, peach
Horticultural Sciences Dept.
1143 Fifield Hall
P.O. Box 110690
Univ. of Florida
Gainesville, FL 32611
phone 352.392.4711, ext. 306
fax 352.392.5653
e-mail wbs@gnv.ifas.ufl.edu

Archie Sparks
—butternut, heartnut
Box 98
Beaver, IA 50031
phone 515.275.4277

Robert L. Stebbins
—apple
1525 Southwest Brooklane Dr.
Corvallis, OR 97333

Richard G. St. Pierre
—saskatoon (serviceberry)
Dept. of Horticultural Science
Univ. of Saskatchewan
51 Campus Dr.
Saskatoon, SK S7N 5A8,
Canada
phone 306.966.5867
fax 306.966.8106
e-mail stpierre@sask.usask.ca

Thomas E. Thompson
—chestnut, hican, hickory, pecan
U.S. Dept. of Agriculture
Agricultural Research Service
Pecan Genetics and Breeding
Research
Rte. 2, Box 133
Highway 50 South
Somerville, TX 77879
phone 409.272.1402
fax 409.272.1401

Nicholi Vorsa
—*cranberry*
Blueberry and Cranberry Research and Extension Center
Rutgers Univ.
Chatsworth, NJ 08019
phone 609.726.1590
fax 609.726.1593
e-mail vorsa@aesop.rutgers.edu

Roger Way
—*apple*
Dept. of Horticultural Sciences
Hedrick Hall
New York State Agricultural Experiment Station
Geneva, NY 14456-0462
phone 315.787.2217
fax 315.787.2397

Francis Zee
—*Barbados cherry, carambola, guava, longan, lychee, macadamia, papaya, passion fruit, pineapple*
U.S. Dept. of Agriculture
Agricultural Research Service
P.O. Box 4487
Hilo, HI 96720
phone 808.959.5833
fax 808.959.3539
e-mail hilof2@ars-grin.gov

ALMOND

Adarcias.—See peach rootstocks.

Adafuel.—Orig. in 1970 as an almond x peach selection from an o.p. population of Marcona at E.E. Aula Dei, Zaragoza, Spain. Introd. by R. Cambra in 1990 as an easy-to-root, vigorous hybrid clone for almond and peach. Tests show higher rooting by hardwood cuttings than GF 677. Resistant to *Sphaeroteca pannosa, Tranzchelia pruni-spinosa*, and *Coryneum beijerinckii* but susceptible to *Agrobacterium tumefaciens* and *Meloidogyne* species. One-year-old shoots are intensely red; shoots long with little lateral shoot growth. Fruit is intermediate between almond and peach.

Aldrich.—Discovered in 1973 by Everett and Randall Aldrich as a chance seedling adjacent to a commercial orchard near Hughson, Calif. Plant patent 5320; 6 Nov. 1984; assigned to Maryanne, Clarence, and Randall Aldrich. Introd. in 1984. Nut: soft to paper shell; well-sealed. Kernel: ovate; width-to-length ratio is 0.55; slightly plump; medium to small size; somewhat wrinkled; few doubles; resembles Thompson. Matures 2 weeks after Nonpareil. Easy to harvest and to hull. Tree: large; upright. Blooms with Nonpareil; cross-pollinates with Nonpareil, Carmel, Fritz, Monterey, and Price.

Alenik.—Orig. by A. Yadrov, N. Popov, and A. Rhikter at State Nikitski Botanical Gardens, Yalta, Crimea, Ukraine. Parentage: Mission x (*Prunus dulcis* 128/2 x *Amygdalus bucharica* 2714). Introd. in 1988. Nut: size medium to small; soft shelled with average shelling percentage of 55%; outer shell sparsely pored. Kernel: size intermediate, oval-shaped; color intermediate; flavor sweet; about 15% double kernels. Maturity is early. Bloom time late to very late. Pollen cross-compatible with Nikitskyi 62 and Nikitskyi 2240. Tree: medium size; branching intermediate. Precocity similar to Nonpareil. Flower buds have high heat requirement and are extremely resistant to winter cold. Some resistance to *Rhizopus stolonifera* and *Monilinia* spp.

Arboleda.—Orig. in Merced, Calif., by Guido Bandoni. Introd. in 1974. Plant patent 3490; 19 Feb. 1974; assigned to Bright's Nursery, Le Grand, Calif. Parentage unknown, discovered in 1965. Nut: 1 3/8 inches in length, elongated, plump; shell soft, well-sealed, straw colored; kernel 13/16 inch in length, 9/17 inch in width, 5/16 inch average thickness; harvest late September, easy to harvest and hull; kernels uniform, free of defects, sweet, 100% singles, kernel averages 63.2% of the nut. Tree: large, stocky, dense, compact, vigorous, hardy, heavy and consistent productivity, no tendency to drop blossoms, no limb breakage. Very subject to noninfectious bud-failure.

Ayles.—A self-fertile, late-blooming, o.p. seedling of Tuono. Introd. in 1987 in Zaragoza, Spain, by Antonio Felipe and Rafael Socias i Company, Unidad de Fruticulture. Naturally autogamous due to coincidence of anthers and stigma. Blooms late, 10 to 14 days after Nonpareil and 1 day after Ferragnes. Nut is very hard shelled (30% to 34% kernel). Kernel: heart-shaped (width-to-length ratio is 0.64), few doubles, medium to large, good commercial quality. Matures in late September. Low sensitivity to late frosts. Tree: medium in size, spreading, relatively compact growth, fruit on spurs.

Ballico.—Orig. in Ballico, Merced County, Calif., by Vincent B. Rodgers. Introd. in 1957. Plant patent 1568; 19 Feb. 1957. Parentage unknown; discovered in 1947; propagated in 1949. Nut: size uniform, 28.3 x 19.5 mm; shell soft, well-sealed; flavor good; no doubles; resembles Mission (Texas); kernel cracks out about 50% of nut; kernel averages 13 x 21 mm. Tree: large, upright, vigorous, productive, blooms several days later than Mission. Cross-incompatible with Mission.

Bell.—Orig. in Durham, Calif., by James C. Bell. Plant patent 4378; 13 Feb. 1979. Introd. in 1979. Whole tree mutation, discovered in 1960 in a row of Ne Plus Ultra. Nut large, elongated; shell paper-thin, well-sealed; percentage of kernel to nut 68.6%. Tree: medium size, foliage dense; total absence of brown rot; good pollenizer, especially for Nonpareil; flowers pink; blooms early, more upright than Ne Plus Ultra, making it easier to knock.

Bonanza.—Orig. near Yuba City, Calif., by LeRoy Koch. Introd. in 1975. Plant patent 3699; 1 Apr. 1975. Chance seedling in Peerless orchard planted in 1930, discovered in 1962. Nut: medium; 222 per lb; shaped like Ne Plus Ultra; ovate; shell soft, medium thick, smooth; outer shell crumbly; inner shell hard, thin, fairly well-sealed; kernel: small to medium, 55% of nut, elongated, averages 25 per oz, medium light brown to reddish brown; flavor sweet; quality good; borne in clusters on short spurs, hangs well on tree; harvest early, with Nonpareil or Peerless; easy to harvest and process. Tree: large; vase shaped; spreading; vigorous; regular and heavy bearer; full bloom with or 3 to 5 days after Mission; light pink; foliage glossy green, darker than known varieties.

Bonita.—Orig. near Chico Butte County, Calif., by Joseph Pavcik. Plant patent 3951; 14 Sept. 1976. Discovered growing in an orchard of Nonpareil almond trees. Nut: medium; soft shell; kernal: medium size, dry, sweet flavor; good quality; much heavier crop than Nonpareil; harvest period same as Nonpareil. Tree: large; vigorous; frost hardy.

Britz.—Orig. in Kortfontein, Riversdale, Cape Province, South Africa, on the farm of a Mr. Britz. Introd. in 1925 by Nel Bros., Montagu, Cape Province, South Africa, and discovered by them in 1924. Parentage unknown. Nut: small, round; shell: semisoft. Tree: willowy growth habit; regular and prolific bearer.

Butte.—Orig. in Le Grand, Calif., by F.W. Anderson, Merced, Calif. Introd. in 1963 by Fowler Nurseries, Newcastle, Calif. Plant patent 2380; 17 Mar. 1964; assigned to Fowler Nurseries. Mission x Nonpareil; cross made about 1953; selected in 1961; tested as Anderson 2T16A. Nut: medium to small, averaging 16 per oz; shell: soft to hard, thick, solid, well-sealed; cracks out about 52% of kernel; nut and kernel intermediate between Mission and Nonpareil; kernel: medium to small, averaging 30 to 33 per oz, wedge-shaped, smooth, few doubles; quality good; pellicle thin; harvested after Nonpareil and before Mission; resembles Drake, Ruby, and Mission. Tree: vigorous; produces well; grows well on almond and peach rootstocks; blooms late with Mission; interfruitful with Nonpareil, Mission, Emerald, and Ruby.

Carmel.—Orig. in Merced, Calif., by Gene Y. Arakaki. Introd. in 1966. Plant patent 2641; 7 June 1966. Nut: medium; thick; color light; kernel 60% of nut, averaging 20–24 per oz, lighter color than Nonpareil, quality and flavor good; harvest 3 weeks later than Nonpareil, 1 week earlier than Mission (Texas). Tree: medium, more rounded than Nonpareil, but smaller; blooms with Nonpareil; productive; bears regularly and in clusters; interfertile with Nonpareil and Mission. Susceptible to noninfectious bud-failure.

Carrion.—Orig. in Arbuckle, Calif., by Alfonso R. Carrion. Introd. in 1972. Plant patent 3467; 5 Feb. 1974. Unknown seedling, probably Nonpareil x Mission, discovered in 1965. Nut: size medium; elongated; shell soft, well-sealed; kernel: pleasant flavor; 65% of nut; harvest 1 Sept. in Arbuckle, about 10 Sept. in Modesto. Tree: stocky and dense; vigor excellent; hardy; heavy bearer; blooms 2 days before Mission at Modesto, 2 days after Nonpareil at Arbuckle; good pollenizer for both varieties.

Craven.—Orig. in Dallas, Ore. Introd. about 1935. Parentage unknown; discovered about 1930. Nut: shell thin; quality good, equaling the Hatch varieties. May be a renamed old variety.

Cressey.—Orig. in Ballico, Calif., by Vincent B. Rodgers. Introd. in 1963. Plant patent 2330; 24 Dec. 1963; assigned to Burchell Nursery, Modesto, Calif. Thought to be a bud mutation of Nonpareil; discovered in 1954. Nut: small; elongated, averaging 26.0 mm long; papershell, light brown, outer part crumbling; inner part well-sealed; kernel: small, averaging 20.3 mm long, 12.0 mm wide, 33 kernels per oz, elongated, plump, pellicle thin, light brown, quality very good; kernel averages 71% of nut; ripens with Nonpareil, which it resembles.

Tree: large; vigorous; rounded, spreading; fruit spurs numerous; productivity very heavy; leaves smaller than those of Nonpareil; interfruitful with Ne Plus Ultra, Jordanolo, Mission (Texas), Ballico, and Merced; incompatible with Nonpareil; blooms with Nonpareil and Merced.

Cruz.—Discovered in 1970 by Irwin H. Bonds as a chance seedling adjacent to a commercial almond orchard near Cressey, Calif. First tests established in 1978. Plant patent 7497, 16 Apr. 1991; assigned to Ruben A. Cruz, Modesto, Calif.; introd. in 1991. Nut: medium size; ovate; soft shell (60% kernel); well-sealed. Kernel: medium size; somewhat rounded; width-to-length ratio is 0.71; moderately plump; somewhat wrinkled; few doubles. Ripens 7 to 10 days after Nonpareil; easy to harvest and to hull. Tree: medium size; upright, open; heavy, consistent production. Bloom begins about 3 days before Nonpareil and extends for longer period.

Davey.—Orig. in Davis, Calif., by Arthur E. Davey, California Agr. Expt. Sta., and Milo N. Wood U.S. Dept. of Agriculture. Introd. in 1952. Nonpareil x Sans Faute; cross made in 1936 by Wood; selected in 1944 by Davey and Wood; tested as Calif. 3-3. Nut: size medium, similar to Nonpareil; broadly ovate; shell soft, smooth, thin, well-sealed, resembles I.X.L.; kernel: attractive; size medium, ovate, smooth; good texture and flavor; doubles few; pellicle light brown; blanching quality good; ripens early, about with Nonpareil; difficult to knock and hull; borne on spurs, usually scattered, not in clusters. Tree: vigorous, very upright; blooms with Nonpareil and can be used as a pollenizer for it, but also interfruitful with Ne Plus Ultra, Peerless, Mission (Texas), and Jordanolo; somewhat tolerant of mild salinity. Tends to come into bearing later than other varieties.

Dehn.—Orig. in Ogden, Utah, by Emil Dehn. Introd. in 1952. Northland o.p.; discovered in 1945. Nut: Resembles Northland and I.X.L. Tree: bears more heavily than Northland or I.X.L.

Emerald.—Orig. in Le Grand, Calif., by F.W. Anderson, Merced, Calif., Introd. in 1962. Plant patent 2208; 8 Jan. 1963; assigned to Reedley Nursery, Reedley, Calif. Mission x (Nonpareil x Mission); tested as Anderson 17P450. Nut averages about 200 per lb; kernel averages 28 per oz; quality good; shell: hard, well-sealed; harvest period slightly later than that of Mission; resembles Nonpareil. Tree: blooms late, about 5 days after Mission and 10 days after Nonpareil; very productive; precocious bearer; interfruitful with Mission and Ruby.

Empire.—Orig. in Le Grand, Calif., by F.W. Anderson, Merced, Calif. Introd. in 1964. Plant patent 2367; 25 Feb. 1964; assigned to Dave Wilson Nursery, Modesto, Calif. Mission x peach-almond seedling; selected in 1960. Nut: large, averaging about 1 5/8 inch long and 200 per lb; elongated; shell paper, thin, ragged, well-sealed, tan; kernel averages 62% of nut, large, elongated, averaging 20 per oz; pellicle thin, brown; very few doubles, averaging less than 1%; quality good; flavor normal; resembles Ne Plus Ultra, but larger nut and softer shell. Tree: large; vigorous; blooms about 2 weeks after Ne Plus Ultra, about 1 week after Nonpareil, and about with Mission; production heavy.

Ferciana.—See Ishtara®.

Ferraduel.—Introd. in 1969 by Charles Grasselly, Sta. Rech. de Arb. Fruit. de la Grande Ferrade, to provide cross-pollenizer for Ferragnes. Cristomorto x Ai. Tested at INRA, La Grande Ferrade, Mandeuel, and Nimes, France. Similar to Ferragnes. Nut: hardshell (shelling percentage is 28); kernel: large, flat, very thin integuments.

Ferragnes.—A late-blooming, precocious introduction from Sta. Rech. de Arb. Fruit. de la Grande Ferrade, INRA, France. Tested at La Grande Ferrade, Manduel, and Nimes. Introd. in 1969 by Charles Grasselly. Cristomorto x Ai. Blooms very late (after Mission). Nut: hardshell (shelling percentage is 30 to 43); kernels: large, elongated (width-to-length ratio is 42), brown, somewhat wrinkled, and pubescent. Nuts mature in mid-September, 2 to 3 days after Nonpareil. Self-incompatible. Tree is vigorous, precocious, and productive. An important commercial variety in the Mediterranean area.

Foros.—Orig. by A. Yadrov, N. Popov, and A. Rhikter at State Nikitski Botanical Gardens, Yalta, Crimea, Ukraine. Hatf-159 x Takida badem. Introd. in 1990. Nut: size large; shell: semi-hard with 47% shell; outer surface sparsely pored. Kernel: large; ovate; intermediate color intensity; wrinkling intermediate; flavor sweet. Low tendency to doubles. Nut maturity early. Bloom time very late. Pollen cross-compatible with Nikitskyi 62 and Desertyi. Flower buds have high heat requirement and high resistance to winter cold. Tree: medium in shape and size; tends toward sparse branching and precocious bearing; some tendency for biennial bearing. Reported to have resistance to *Rhizopus stolonifera* and high resistance to *Monilinia*.

Francoli.—Originated by F.J. Vargas Garcia and M.A. Romero Romero, Centre de Mas Bove, IRTA, Reus (Tarragona), Spain. Cristomorto x Gabaix. Cross made in 1976; introd. in 1992. Nut: very hard shell; shelling percentage about 30%. Kernel: large; semi-elliptical and pointed; few double kernels; slightly wrinkled; light brown pellicle of medium thickness. Bloom very late. Self-incompatible; pollinates Glorietta and Ferragnes. Maturity is early. Tree: medium to strong vigor; medium upright; branching medium-sparse; foliage dense; easy to prune. Precocious; spur-bearer; very high production.

Fritz.—Orig. in Manteca, Calif., by John A. Fritz. Introd. in 1969. Plant patent 3005; 24 Nov. 1970, assigned to Burchell Nursery, Modesto, Calif. Chance seedling. Nut: small; ovate; shell: soft, medium thin; straw color; kernel: small; ovate, medium brown; flavor slightly bitter; percentage of kernel 49.9%; harvested after Mission; easily hulled. Tree: size medium; vigorous heavy producer; tends to produce double kernels; good pollenizer for Nonpareil.

GF 677.—Rootstock. See under Peaches.

GF 8-1.—See under Peaches. Almond/GF 8-1 trees are of stature similar to almond/peach.

Glorietta.—Developed by F.J. Vargas Garcia and M.A. Romero Romero, Centre de Mas Bove, IRTA, Reus (Tarragona), Spain. Primorsky x Cristomorto. Cross made in 1975; introd. in 1992. Nut: large; very hard-shelled; shelling percentage slightly greater than 30%. Kernel: large, elliptical, and pointed; few double kernels; pellicle smooth, light brown, medium thick; high quality. Maturity medium late, early September. Bloom: late to very late. Self-incompatible; cross-fertile with Masbovera, Francoli, and Ferragnes. Tree: strong vigor; upright; slightly open structure; dense foliage; very easy to train and prune. Bears primarily on spurs; precocious. Easy to harvest and hull.

Guara.—A frost-tolerant, self-fertile introduction by Antonio Felipe and Rafael Socias i Company, Unidad de Fruticultura, Zaragoza, Spain. Introd. in 1987. Unknown origin; discovered in 1974 in a variety collection under an incorrect name, freed from known viruses, and tested with Ayles and Moncayo. Fruit matures early (first of September). Nut: very hardshelled (32% to 35% kernel); kernel is ovate (width-to-length ratio is 0.59); medium to large; few double kernels, good commercial quality. Naturally autogamous due to coincidence of anthers and stigma. Blooms in mid-March in central Spain, approximately with Ayles; high density of medium-sized blossoms on spurs. Blossoms have low sensitivity to frost. Tree: medium size and spreading; somewhat compact.

Hall (*Hall's Hardy*).—Orig. in Coffeyville, Kan., by a Mr. Hall. Introd. about 1925 by Willis Nursery Co., Ottawa, Kan. Parentage unknown; may be a peach x almond hybrid. Nut: large. Tree: small; hardy; can be grown where peaches are successful. Apparently similar to or identical with Ridenhower. Has a very strong almond flavor.

Hall's Hardy.—See Hall.

Hansen 536.—See rootstocks for almonds.

Hansen 2168.—See rootstocks for almonds.

Harpareil.—Orig. in Davis, Calif., by M.N. Wood (1883 to 1953), U.S. Dept. of Agriculture, and W.P. Tufts, California Agr. Expt. Sta. Introd. in 1938. Nonpareil x Harriott; cross made in 1926; first crop in 1930; selected in 1932; tested as Calif. 8/31; sibling of Jordanolo. Nut: large; generally elliptical; hulled nut plump; shell soft, outer shell very thin, often spongy or crumbly, medium brown; inner shell very thin, hard; kernel: large, sometimes produces doubles, moderately plump, wide ovate, blanches well; pellicle: rather light to medium brown, rather thick and tough; easily harvested and hulled; quality not as good as Jordanolo. Tree: vigorous; bears heavily but not as regularly as Jordanolo; not as susceptible to brown rot as Drake; blooms early after Jordanolo; pollen compatible with Ne Plus Ultra, I.X.L., Peerless, Nonpareil, Drake, and Mission; incompatible with Jordanolo. The name Harpareil is coined from its parents' names. Very susceptible to noninfectious bud failure.

Harvey.—Orig. in Yuba City, Calif., by Harvey Schmidt. Introd. in 1957 by George R. McFeely, Oswald Nursery Co., Yuba City. Unknown seedling, probably Mission x Nonpareil. Discovered in 1956 growing beneath a walnut tree in the Schmidt's orchard. Plant patent 2813; 28 May 1968, assigned to George R. McFeely. Nut: size medium; ovate; shell paper, thin, ragged at times; outer shell hard; inner shell smooth, hard, brittle, readily broken with fingers; kernels: 66.60% of nut, size medium, 20 to 25 per oz, medium light brown, smooth, occasional wrinkling, flavor sweet, quality very good; borne in clusters on short spurs; harvest 5 to 7 days after Nonpareil. Tree: size medium; vigor medium; very productive and regular bearer; in comparison to Nonpareil, is more upright, foliage smaller and darker green, more pronounced tendency to develop spurs and fruit buds on current season's growth; interfertile with Nonpareil and compatible with Marianna 2624 rootstock. Very susceptible to noninfectious bud failure.

Hashem #1.—Orig. in Denair, Calif., by Hashem Naraghi, H. Naraghi Farms, Modesto, Calif. Introd. in 1979. Plant patent 4466. Unknown seedling. Nut: resembles Nonpareil, kernels being slightly lighter in color and larger. Tree: large, vigorous, robust, bushy, retains full green leaf color; productive.

Hashem II.—Discovered in 1976 by Hashem Naroghi as a limb sport in a commercial almond orchard near DeNair, Calif. Plant patent 4845; 11 May 1982; introd. in 1980. Nut: soft shell. Kernel: large, long, flat, similar to Nonpareil; few doubles. Blooms with Nonpareil but extends later. Ripens just after Nonpareil. Tree: slightly smaller than Nonpareil; upright; productive. Bears in clusters.

Heleodor.—Developed by A. Yadrov, N. Popov, and A. Rhikter at State Nikitski Botanical Gardens, Yalta, Crimea, Ukraine. Nitikskyi 62 x IXL. Introd. in 1988. Nut: size small, similar to Mission; shell light, soft, and easily broken by hand; shelling percentage about 55%; outer shell densely pored. Kernel: small (like Mission); oval; sweet; slight wrinkling; color intensity intermediate; low tendency to double. Nut matures early (with Nonpareil). Bloom late to very late; cross-compatible with Desertnyi and Nitikskyi 2240. Heat requirement for flower bud emergence very high (like Primorski). Tree: medium size; tends to be precocious, bearing on 4- to 5-year-old trees; some tendency for biennial bearing. Resistant to winter cold. Shows resistance to *Rhizopus stolonifera* and *Monilinia* spp.

Ishtara®.—Rootstock. See under Plums.

Jordanolo.—Orig. in Davis, Calif., by M.N. Wood (1883 to 1953), U.S. Dept. of Agriculture, and W.P. Tufts, California Agr. Expt. Sta. Introd. in 1938. Nonpareil x Harriott; cross made in 1926; first crop in 1930; selected in 1932; tested as Calif. 8-32; sibling of Harpareil. Nut: large; roughly elliptical; shell: thin, soft, smooth, attractive light yellow, large, round or oval, shelling out about 60%; pellicle color light, rather thick, tough, smooth; kernel: large, rather plump, irregular oval or ovate, attractive, nearly smooth; quality high; blanches very well; hulls easily; resembles Jordan. Tree: upright, vigorous; bears early and heavily; blooms early, usually with or somewhat before Ne Plus Ultra; considerable resistance to mites (red spider) because of dense foliage; very susceptible to noninfectious bud failure; pollen compatible with Ne Plus Ultra, I.X.L., Nonpareil, and Peerless; incompatible with Harpareil. Jordanolo is a coined name from its similarity to Jordan and from its origin in Yolo County, Calif.

Jubilee.—Orig. in Templeton, Calif., by Frank T. Peterson, nurseryman. Introd. in 1929 by the Frank C. Runitz Ranch, Templeton. Parentage unknown; discovered about 1925. Nut: shell hard; kernel tends to be small; resembles Drake. Tree: prolonged, late blooming period, with Mission, escaping early frosts; pollinates Nonpareil and Mission; exhibits noninfectious bud failure.

Kapareil.—Orig. in Davis, Calif., by Dale E. Kester, California Agr. Expt. Sta. Introd. in 1962. Nonpareil x Calif. 24-6 [Eureka x seedling A5-25 (Nonpareil x Eureka)]; cross made in 1951 by E.F. Serr and H.I. Forde of above station; planted in 1952; selected in 1955 by Kester; tested as Calif. 18-24. Nut: shell thin, being a papershell type, similar to Nonpareil; kernel small, averaging 30 or more per oz in most years, suitable for the candy bar market; kernel: thin, rated as good as Nonpareil in eating quality; matures about 1 week before Nonpareil; resembles Nonpareil. Tree: size and shape resemble Nonpareil; productive; blooms with Nonpareil; cross compatible with Nonpareil, Mission, and Davey.

Kern Royal.—Orig. in Kern County, Calif., by Leland O. Mugg, Wasco, Calif. Introd. in 1967. Bud mutation of Nonpareil. Plant patent 2771; 17 Oct. 1967; assigned to Kern County Land Co., San Francisco. Nut: shape similar to Nonpareil, but smaller; paper-shelled, well-sealed; kernel 20 per oz, skin of immature kernel red; hull slightly rougher than Nonpareil; ripens 2 days later than Nonpareil. Tree: vigorous; similar to Nonpareil in size and shape, distinguished by red blossoms.

Kochba.—A high-yielding seedling from the cross Greek x Nonpareil made by P. Spiegel-Roy and J. Kochba in 1967 at Volcani Center, Bet Dagan, Israel. Introd. in 1985. Nut: semi-hard; completely sealed with retention of outer cork; shelling percentage about 51%; kernel: oblong-ovate (width-to-length ratio is 60), relatively flat, smooth, light-colored; few double kernels; good flavor. Flowering early, before Ne Plus Ultra. Tree: medium size, larger than Ne Plus Ultra; drooping branches; somewhat difficult to train. Bears predominantly on spurs. Consistently higher yielding that other selections of the same series.

Lauranne.—Developed by Charles Grasselly, INRA, Avignon, France. Ferragnes x Tuono. Introd. in 1989. Shell: hard; shelling percentage about 38%; smooth surface. Kernel tends to be small and ovate; slightly wrinkled; color light; tendency to double (5% to 15%). Maturity medium early (mid-September). Bloom late to very late. Tree: spreading to drooping; medium vigor; precocious.

Le Grand.—Orig. in Le Grand, Calif., by Frederic W. Anderson, Merced, Calif. Introd. in 1972. Plant patent 3130; 25 Apr. 1972; assigned to Arthur Bright, Le Grand. Selected in 1966 as most promising among a large number of crosses and backcrosses between almonds and peaches, tested as Anderson 34P865. Nut: most nearly resembles Mission; small; ovate; shell soft to hard, well-sealed; kernel size medium, oval to ovate, smooth to wrinkled; harvest before Mission. Tree: size medium; dense; vigorous; bears regularly; productivity heavy; self-compatible but not highly self-pollinating. Difficult to remove from tree.

Livingston.—Discovered by P.P. Hill, Modesto, Calif., as chance seedling. Plant patent 4116; 27 Sept. 1977. Nut: paper shell, well-sealed; kernel medium size. Blooms 3 days after Nonpareil and matures 8 days after Nonpareil. Tree: medium size, semi-upright, resembles Nonpareil. Easy to harvest.

Lodi.—Discovered by Irwin H. Bonds in 1976 as a chance seedling adjacent to a cultivated orchard near Manteca, Calif. Tests established in 1977. Plant patent 7068; 12 Dec. 1989. Nut: ovate; soft shell; well-sealed; shelling percentage about 60%. Kernel: medium to large; somewhat broad; width-to-length ratio is 0.62; medium thick; few doubles; slightly bitter flavor. Matures about 1 week after Nonpareil. Tree: average size; medium shape. Compatible with Marianna 2624 rootstock. Blooms about with Nonpareil.

Mangup.—Developed by A. Yadrov, N. Popov, and A. Rhikter at State Nikitski Botanical Gardens, Yalta, Crimea, Ukraine. Nonpareil x (Nikitskyi 62 x Nikitskyi 53). Introd. in 1989. Nut: large (like Ne Plus Ultra); shell semi-hard with shelling percentage of 49% (like Mission); surface sparsely pored. Kernel: size large (like Ne Plus Ultra); ovate; slightly bitter taste (like Mission); wrinkling intermediate; color intensity intermediate; some tendency to wrinkle; low doubling percentage. Matures early (with Nonpareil). Bloom late (with Mission). Pollen cross-compatible with Desertnyi, Nikitskyi 2240. Heat requirement of flower buds very high (like Primorski). Flower buds resistant to winter cold. Tree: medium size and shape; branching intermediate; some tendency to biennial bearing. Resistant to *Rhizopus stolonifera* and to *Monilinia* spp.

Marianna 2626.—Rootstock. See under Peaches. Poor compatibility with Nonpareil.

Masbovera.—Developed by F.J. Vargas Garcia and M.A. Romero Romero, Centre de Mas Bove, IRTA, Reus (Tarragona), Spain. Primorski x Cristomorto. Cross made in 1975; introd. in 1992. Nut: very hard shell; shelling percentage about 30% or slightly less. Kernel: size medium to large; few doubles; elliptic and pointed; pellicle smooth, light brown, medium thick; quality high. Matures medium-late (early September). Bloom late to very late. Self-incompatible; pollenizers include Glorietta, Francoli, and Ferragnes. Tree: vigorous; medium upright; dense foliage. Precocious; very productive; strong spur-bearer.

Merced.—Orig. in Ballico, Merced County, Calif., by Vincent B. Rodgers. Introd. in 1958. Plant patent 1730; 15 July 1958; assigned to Burchell Nursery, Modesto, Calif. Mission x Nonpareil; selected in 1948. Nut: shell thin, well-sealed, lighter in color than Nonpareil; kernel averaging 58% to 63% of nut; ripens with Nonpareil. Tree: size medium; upright; blooming period just after Nonpareil; productivity very heavy; cross-compatible with Nonpareil.

Mission.—(*Texas*). Originated as a chance seedling in Texas, about 1891. Nut: small, averaging about 25–28 kernels per oz, shell: medium hard and well-sealed. Harvest: 4 weeks after Nonpareil. Tree: blooms late, productive, with upright growth. Susceptible to *Ceratocystis* and salt injury.

Monarch.—Introd. by Theodore J. Moschitto, Modesto, Calif., in 1982. Plant patent 4852; 15 May 1982; assigned to Burchell Nursery, Modesto. Nut: semi-hardshell with good seal; kernel large, plump. Blooms 1 day after Nonpareil; matures 7 days after Nonpareil. Tree upright, similar to Mission.

Moncayo.—A frost-tolerant selection from the Tardive de la Verdiere x Tuono cross, introd. in 1987 by Antonio Felipe and Rafael Socias i Company, Unidad de Fruticultura, Zaragoza. Cross made in 1974. Shell is very hard; shelling percentage is 25 to 28; kernel oblong (width-to-length ratio is 0.56), medium to large; few doubles; good commercial quality. Blooms in late March, 2 or 3 days after Ayles; profuse blooming on spurs. Low sensitivity to late frosts. Self-compatible; reported to be naturally autogamous due to coincidence of anthers and stigma. Tree: medium to large, spreading to drooping, but easy to train. Fruit mature in mid-September.

Money Tree.—Orig. in Chico, Calif., by Frederick R. Wells. Introd. in 1971. Plant patent 3046; 13 Apr. 1971, assigned to Sierra Gold Nursery, Yuba City, Calif. Unknown seedling. Discovered in 1967. Nut: 1 1/2 inches in length; elongated; paper shell; kernel light tan, flavor sweet, 72% of nut; date of harvest midseason. Tree: size medium, vigorous, dense in growth with abundant foliage; leaves more glossy and larger than Nonpareil; blooms 2 days before Nonpareil. Tree subject to low-productivity syndrome.

Mono.—Orig. in Le Grand, Calif., by F.W. Anderson, Merced, Calif. Introd. in 1972. Plant patent 3125; 18 Apr. 1972; assigned to Arthur Bright, Le Grand. Seedling of a late-blooming unnamed variety; discovered in 1966, tested as Anderson SP570. Nut: resembles Nonpareil; size medium; oval; shell thin; outer shell hard; inner shell soft, well-sealed; kernel: size medium, oval, smooth, light brown, flavor sweet; harvest early, with Nonpareil. Tree: large; open; vigorous; bears regularly, productivity heavy; blooms 1 week after Nonpareil.

Monterey.—Orig. in Merced, Calif., by Robert and Kenneth Stretch. Introd. in 1974. Plant patent 3483; 19 Feb. 1974. Discovered and selected in 1962. Nut 1/2 x 3/4 inch; elongate, slightly greater width than thickness; shell hard, well-sealed; kernel: 1 x 1/2 inch; harvested with or just after Mission. Tree: large; vigorous, resilient limbs; more tolerant of frost than Nonpareil; heavy and regular bearer; tolerant of brown rot and shot hole fungus; foliage retains rich green color through harvest.

Myran®.—See under Peach Rootstocks.

Nonpareil.—Originated in Suisum, Calif., in 1879 by A.T. Hatch. Selected from o.p. seed thought to be from Lanquedoc region of France. Nut: very high quality; medium thick, color light; kernel: 65% to 70% of nut, averaging 22–25 per oz, with poorly sealed paper shell. Harvests early, 4 weeks before Mission. Tree: vigorous, medium size with consistently high yields. Susceptible to noninfectious bud failure.

Norman.—Orig. in Le Grand, Calif., by F.W. Anderson, Merced, Calif. Introd. in 1963 by Fowler Nurseries, Newcastle, Calif. Plant patent 2426; 14 July 1964; assigned to Fowler Nurseries. Chance seedling; discovered as sucker shoot below the graft union of a Nonpareil tree; selected in 1958. Nut: medium to small, averaging 22 per oz; shell: thin, paper, smooth to ragged; percentage of kernel to nut averages 66%; kernel: medium to small, averaging 32 per oz; ovate; smooth; quality good; few doubles; pellicle thin; resembles Nonpareil but averages 10% to 20% smaller; ripens very early; about with Nonpareil. Tree: vigorous, upright; crops heavy to very heavy, bears regularly; blooms about 3 days after Nonpareil, about midway between Nonpareil and Mission, being interfruitful with both varieties.

Northland.—Orig. in Ogden, Utah, by Arthur Combe, Valley Nursery. Introd. in 1938. O.p. seedling of I.X.L.; discovered in 1935. Nut: similar to I.X.L. Tree: hardy.

Padre.—Orig. in Davis, Calif., by Dale E. Kester, Agr. Expt. Sta., Univ. of California. Introd. in 1983. Mission (Texas) x Swanson; cross made in 1946; selected in 1954; tested as CP 5-58. Nut: kernel plump, broad, similar to Mission, averages 25 to 30 per oz; shell 45% to 50% of nut, hard, resists worm damage; matures moderately late but before Mission. Tree: upright, resembles Mission; productive and consistent bearer; blooms late near Mission; pollen compatible with Mission, Butte, Nonpareil, Ripon, Ruby, Thompson and Fritz; susceptible to brown rot (*Monilinia* spp.).

Paxman Midget.—Orig. in Ripon, Calif., by Edna L. Paxman. Introd. in 1963 by Stuart's Nursery, French Camp, Calif. Plant patent 2276; 27 Aug. 1963. Parentage unknown; discovered in 1954. Nut: small; rounded; paper or soft shell, well-sealed; kernel small, averaging 35 to 43 per oz, averaging 60% to 70% of nut; ripens with or later than Nonpareil. Tree: large, upright; vigor moderate; production medium to light.

Plateau.—Discovered by Leonard D. James, Modesto, Calif., as a chance seedling. Plant patent 4739; 9 June 1981; assigned to Burchell Nursery, Modesto. Introd. in 1981. Blooms 4 days after Nonpareil and matures in mid-September, 8 days after Nonpareil. Nut is soft-shelled, well-sealed; kernel large. Tree: medium size, semi-upright.

Price Cluster.—Orig. in Durham, Calif., by Joseph L. Price of Redding, Calif. Introd. in 1964. Plant patent 2350; 4 Feb. 1964. Discovered in 1953 in a row of Nonpareil trees. Nut: shell soft-kernel averages 26 per oz; borne in cluster-like formation along branches; resembles Nonpareil in shape but more plump; some double kernels. Tree: very heavy bearing, with alternate bearing; hardy; leaf deeper green than Nonpareil; blooms with Nonpareil, pollinating that and other varieties.

Profuse.—Orig. in Delhi, Calif., by Herbert L. Corcorran. Introd. in 1957. Plant patent 1622; 16 July 1957. Nonpareil o.p.; discovered in 1952. Nut: shell thickness medium, soft, well-sealed; kernel size averages 18 to 25 per oz, crack-out averages 48%, flavor sweet, quality good; matures about 1 week later than Nonpareil. Tree: vigorous; productivity heavy. Cross-incompatible with Nonpareil.

Reinero.—Orig. in Winton, Merced County, Calif., by Peter K. Reinero and Adolph J. Reinero, Merced, Calif. Introd. in 1968. Plant patent 2788; 9 Jan. 1968; assigned to Reinero Bros., Winton. Chance seedling, thought to be Nonpareil o.p. Nut: size medium; shell soft; kernel 58% of nut, 12.6 per oz, flavor: sweet; resembles Davey; shell characterized by larger pits, streak shorter, broader and darker, and wing broader; ripens 12 days later than Davey. Tree: upright; vigorous; resembles Davey but has darker foliage.

Rhikter.—Developed by A. Yadrov, N. Popov, and A. Rhikter at State Nikitski

Botanical Gardens, Yalta, Crimea, Ukraine. (*P. dulcis* 1281/2 x *P. bucharica*) x Vinoslivyi. Introd. in 1984. Nut: medium size; medium shell color; semi-soft; 52% shell; surface densely pored. Kernel: size medium; oval; slight wrinkling; color intermediate. High tendency to double (25%). Maturity date medium (like Ne Plus Ultra). Bloom very late. Flower buds have very high heat requirement. Pollen cross-compatible with Nikitskyi 62 and Desertnyi. Tree: medium shape; medium size; branching dense (like Marcona); precocious. Flower buds resistant to winter cold. Tends to have thin shoots.

Ripon.—Orig. in Le Grand, Calif., by F.W. Anderson, Merced, Calif. Introd. in 1962. Plant patent 2379; 17 Mar. 1964; assigned to Irvin M. Burchell, Burchell Nursery, Modesto, Calif. O.p. seedling of an unnamed late-blooming mutation of Nonpareil; selected in 1958; tested as Anderson 14P1160. Nut: medium to small, averaging 240 per lb; ovate; shell: hard, well-sealed, light beige; kernel: medium to small, averaging 29 to 39 per oz; pellicle thin, light tan; quality good; doubles none to few; harvested about with Ne Plus Ultra. Tree: blooms late, after Mission and about 2 weeks after I.X.L.; heavy bearer; vigorous; interfruitful with Drake, Emerald, Nonpareil, Ruby, and Mission.

Robson.—Orig. in Durham, Calif., by S.J. Toy, Agscience Serv., Chico, Calif. Introd. in 1972. Plant patent 3085; 28 Mar. 1972. Chance seedling, discovered in 1964. Nut: size medium; hull thin; shell: soft, well-sealed, shaped like Nonpareil but thicker with pronounced neck at point of attachment to pedicel; kernel: slightly darker than Nonpareil but not as dark as Mission, averages 25 per oz, flavor sweet, 55% to 60% of nut. Tree: resembles Nonpareil; very productive; flower blooms 1 to 2 days earlier and 1 to 2 days later than Nonpareil; cross-pollinates with Nonpareil, NePlus Ultra, Peerless, Mission. and Drake.

Roy.—Orig. in Stayton, Ore., on the property of E. Roy. Introd. about 1930. Parentage unknown; discovered about 1925. Nut shell fairly soft.

Ruby.—Orig. in Le Grand, Calif., by F.W. Anderson. Introd. in 1958. Plant patent 1698; 6 May 1958; assigned to Reedley Nursery, Reedley, Calif. Parentage unknown; selected in 1946. Nut: shell thicker, harder, and better sealed than Nonpareil; kernel averages 53% of nut; kernel averages 28 per oz; matures with Mission. Tree: blooms late, 2 to 3 days after Mission; production high; bears regularly.

Samish.—Orig. as a seedling selection from Marcona x Greek cross made in 1966 at the Volcani Center, Bet Dagan, Israel by P. Spiegel-Roy, Joshua Kochba, and R. Iris. Selected in 1970, introd. in 1982. Nut: semi-hard (41% kernel), completely sealed, light-colored, and smooth. Kernel: uniform, smooth, elongated to somewhat rounded; size medium, but small with heavy crop. Low number of doubles. Adapted to in-shell, kernel, and confectionary trade. Blooms early, before Ne Plus Ultra. Cross-compatible with Ne Plus Ultra and Um el Fahm. Bears predominantly on spurs but some on other shoots. Well adapted to mechanical harvest. Yield has been high. Tree: growth vigorous and large.

Sandall.—See Tardy Nonpareil.

Sauret #1.—Orig. in the Oak Flat area of San Luis Obispo County, Calif., by Clarence H. Sauret. Introd. in 1965. Plant patent 3293; 16 Jan. 1973; assignor of fractional part interest to Richard K. Sauret. Chance seedling, discovered in 1952. Nut: small; averages. 370 per lb; ovate; hull thin; shell thin, ragged, outer shell crumbly, inner shell medium to soft, well-sealed; kernel: brown, averages 35 1/2 per oz, flavor good, 66% of nut; harvested with Nonpareil; easily harvested and processed. Tree: large; upright; vigorous; medium dense growth with abundant foliage; heavy and regular producer; blooms and pollinates with Nonpareil.

Sauret #2.—Orig. in the Oak Flat area of San Luis Obispo County, Calif., by Clarence H. Sauret. Introd. in 1965. Plant patent 3294; 16 Jan. 1973; assignor of fractional part interest to Richard K. Sauret. Chance seedling, discovered in 1950. Nut: large; averages 300 per lb; elongated; hull medium thick; shell: medium thin, outer shell crumbly, inner shell medium hard, ragged, well-sealed; kernel: brown, averages 29.3 per oz, flavor good, 60% of nut; harvested 10 days before Mission; easily harvested and processed. Tree: large; upright to spreading; vigorous; dense growth with abundant foliage; heavily clustered producer, regular bearer; very few defectives; blooms 3 days after Nonpareil.

Solano.—Orig. in Davis, Calif., by Dale E. Kester, Agr. Expt. Sta., Univ. of California. Introd. in 1983. 21-19A [Nonpareil x A1-30 (Nonpareil x Eureka)] x 22-20 (Nonpareil x A1-30); cross made in 1946; selected in 1954; tested as S. 5A-3. Nut: kernel high quality, smooth, attractive, flat, resembles Nonpareil, 20 to 25 per oz, 60% kernel to nut; shell soft, better sealed than Nonpareil, susceptible to worm damage; matures slightly later (7 to 10 days) than Nonpareil. Tree: medium to upright, similar to Nonpareil but more slender and longer branches; bears on spurs with some tendency to alternate bearing; yield similar to Nonpareil; blooms near Nonpareil; pollen compatible with Nonpareil, Mission, Merced, NePlus Ultra, Harvey, Norman, Robson, and Jordanolo.

Sonora.—Orig. in Davis, Calif., by Dale E. Kester, Agr. Expt. Sta., Univ. of California. Introd. in 1983. 21-19A [Nonpareil x A1-30 (Nonpareil x Eureka)] x 22-20 (Nonpareil x A1-30); cross made in 1946; selected in 1954; tested as S. 5A-20. Nut kernel large, 15 to 20 per oz 65% to 70% kernel to nut; shell thin, not well-sealed, susceptible to worm damage; matures slightly after Nonpareil. Tree: medium to round, slightly smaller than Nonpareil; produces much of crop laterally on long shoots; bears early and productive; susceptible to alternate bearing; blooms early between NePlus Ultra and Peerless; susceptible to brown rot (*Monilinia* spp.); pollinates Nonpareil, NePlus Ultra, Peerless, Mission, Merced, and Thompson; incompatible with Solano and Vesta.

Steliette.—Orig. by Chas. Grasselly, INRA, Avignon, France. Ferragnes x Tuono. Introd. in 1989. Shell: semi-hard with a shelling percentage about 45%; outer surface smooth. Kernel: size large; ovate; moderately wrinkled; light color. Some double kernels, ranging from 5% to 10%. Early maturing. Bloom late to very late; self-compatible. Tends toward early bearing. Tree: medium vigor; somewhat difficult to train.

Supernova.—A self-compatible radiation mutant produced at Fruit Res. Inst., Rome, Italy, by F. Monastra, G. Della Strada, C. Fidighelli, and R. Quarta. A plant of the self-incompatible cultivar Fascinello was subjected to ionizing radiation (Co60, at 3 kR) in 1970. Introd. in 1987. Nut: semi-hardshell (40% kernel), 15% doubles; kernel elliptical-elongated (width-to-length ratio is 0.54), large, veined. Blooms late. Maturity is moderately early. Tree: average vigor, medium erect, precocious. Susceptible to *Monilia*. High regular production in all areas of central Italy.

Tardy Nonpareil (*Sandall*).—Orig. in Escalon, Calif., by Ernest T. Sandall. Introd. in 1955. Plant patent 1389; 3 May 1955. Bud mutation of Nonpareil; discovered about 1942. Nut: matures 5 days later than Nonpareil; shell: soft, smooth, well-sealed, cracks easily; kernel: averages about 25 per oz, cracks out about 55% of nut by weight; resembles Nonpareil. Tree: blooms 2 weeks later than parent, thus having more chance to escape frosts. Not grown commercially.

Texas.—See Mission.

Thompson.—Orig. in Clovis, Calif., by L.M. Thompson. Introd. in 1957. Plant patent 1526; 6 Nov. 1956. Parentage unknown; discovered in 1946. Nut: small and difficult to knock; subject to nut gumming; shell soft, well-sealed; kernel averages 60% of nut, sweet, quality good. Tree: blooms late, usually with Mission; bears consistently; vigorous.

Tioga.—Orig. in Le Grand, Calif., by F.W. Anderson, Merced, Calif. Introd. in 1966. Plant patent 2695; 27 Dec. 1966; assigned to Arthur Bright, Le Grand, Calif. Open-pollinated seedling of a late-blooming, double-flowered mutation of Nonpareil, probably pollinated by Mission. Nut: medium to small, average length 1 1/4 inch, average thickness 7/16 inch; ovate; shell: soft, smooth; outer coat hard; inner shell hard to soft, well-sealed; kernel 50% of nut, 33 to 39 per oz, medium light brown, thin, flat, few doubles; harvest between Nonpareil and Mission; for confectionery uses. Tree: medium; vigorous; open; blooms 2 to 5 days after Mission.

Titan.—See rootstocks for almond.

Tokyo.—Orig. in Le Grand, Calif., by Federic W. Anderson, Merced, Calif. Introd. in 1972. Plant patent 3242; 25 July 1972; assigned to Arthur Bright, Le Grand. O.p. seedling of unknown parentage, discovered in 1965, tested as Anderson 5P300. Nut: size medium to large; oval; shell hard, smooth, well-sealed; kernel medium to large, oval, smooth to slightly wrinkled, flavor sweet; harvest after Nonpareil, before Mission; less worm damage than Nonpareil. Tree: large; open to dense; vigorous; bears regularly; productivity medium to heavy.

Utah.—Orig. in Ogden, Utah, by Arthur Combe, Valley Nursery. Introd. in 1938. I.X.L. o.p.; discovered in 1935. Nut: long, narrow. Tree: hardy.

Vesta.—Orig. in Ripon, Calif., by U.S. Dept. of Agriculture at the John Weststeyn Ranch. Introd. in 1968. Later flowering sport of Nonpareil x 5A-3 [(Nonpareil x Jordan) x Jordanolo] x [Nonpareil x (Nonpareil x Eureka)]; cross made in 1956; seedling grown in Fresno, Calif., in plots cooperative with the Fresno State College Foundation; first fruited in 1960; selected by Robert W. Jones, U.S. Dept. of Agriculture in 1960. Nut: shell soft, easily cracked; suture closed and tight, making the kernels less subject to worm damage than Nonpareil; kernel: very similar to Nonpareil, but larger, few doubles, flavor good; ripens 10 to 14 days after Nonpareil. Tree: blooms with Nonpareil and pollinates that variety; vigorous; leaves more resistant to red spider mites than Nonpareil.

Wawona.—Orig. in Le Grand, Calif., by F.W. Anderson, Merced, Calif. Introd. in 1965. Plant patent 2548; 24 Aug. 1965; assigned to Arthur Bright, Bright's Nursery, Le Grand. Ruby x Mission; tested as Anderson 8P525. Nut: size medium, averaging 2.2 g, ovate, width medium to flat; shell: hard, smooth; percentage of kernel to nut, 48%; kernel size medium, averaging 28 per oz; ovate, thin, smooth, medium brown; flavor sweet; quality good; very few doubles (less than 2%); desirably longer and thinner nuts and kernels than Mission; ripens late, about 1 week after Mission; resembles Ruby. Tree: medium to large; vigorous; blooms late, about 1 week after Mission; bears regularly and heavily.

Wood Colony.—Discovered by David E. Blickenstaff, Modesto, Calif., as a chance seedling. Plant patent 5583; 3 Dec. 1985; assigned to Burchell Nursery, Modesto. Introd. in 1985. Nut: semi-soft with good seal; kernel medium size, plump. Blooms 2 days after Nonpareil, matures 7 days after Nonpareil. High yield potential on small to medium-sized spreading tree.

Yosemite.—Orig. in Le Grand, Calif. by F.W. Anderson, Merced, Calif. Introd. in 1965. Plant patent 2534; 8 June 1965; by F.W. Anderson, assignor to Arthur Bright, Bright's Nursery, Le Grand. Mission x (probably) Nonpareil; selected in 1957; tested as Anderson 3T14. Nut: hull thin, easy to harvest, easily removed; shell ovate, hard, smooth, heavier, harder, and better sealed than Nonpareil; uniformly light straw color, attractive, medium to small, average 1 1/8 inches long, 13/16 inch thick, 250 to 300 per lb, smaller than Nonpareil, suitable for in-shell purposes; kernel: medium to small, average 30 to 35 per oz, ovate; smooth; pellicle thin; sweet; quality very good; very few doubles produced; ripens early to midseason, with or shortly after Nonpareil. Tree: large, vigorous; blooms late, about with Mission; interfruitful with Nonpareil, Mission, Ruby, Emerald, Butte, Norman, and Ripon.

Rootstocks for Almond

Adarcias.—Described under peach rootstocks. Compatible with most commercial almond varieties.

Adafuel.—Orig. in 1970 as an almond x peach selection from an open-pollinated population of Marcona at E.E. Aula Dei, Zaragoza, Spain. Introd. by R. Cambra in 1990 as an easy-to-root, vigorous hybrid clone for almond and peach. Tests show higher rooting by hardwood cuttings than GF 677. Resistant to *Sphaeroteca pannosa, Tranzchelia pruni-spinosa,* and *Coryneum beijerinckii* but susceptible to *Agrobacterium tumefaciens* and *Meloidogyne* species. One-year-old shoots are intensely red; shoots long with little lateral shoot growth. Fruit is intermediate between almond and peach.

Ferciana.—See Ishtara®.

GF 677.—See under Peaches. Compatible

with very broad range of almond varieties. Induces very rapid growth, trees within the planting are very uniform. Almond/GF 677 trees are very vigorous, uniform, and productive and tolerant of drought and calcareous soils.

GF 8-1.—See under Peaches. Almond/GF 8-1 trees are of stature similar to almond/peach.

Hansen 536.—Orig. in Davis, Calif., by C.J. Hansen, Dept. of Pomology, Univ. of California. Selected and introd. in 1986 by D.E. Kester and R.N. Asay. Almond B x Peach Selection 1-8-2; cross made in 1964; pollen parent has *Prunus davidiana*, Chinese peach PI6582, and Okinawa in pedigree. Very vigorous clonal rootstock for almonds, peaches, and nectarines; also compatible with prunes and Japanese plums, but not with apricots. In California, propagated by hardwood cuttings taken in late fall and planted directly in the nursery row, where survival is slightly lower than that of Hansen 2168. Root system of mature tree is intermediate between almond seedling and peach; large branch roots, large mass of medium to small fibrous roots; more deeply rooted than Marianna 2624. Anchorage excellent. No suckers. Not cold hardy. Immune to root-knot nematodes; susceptible to root-lesion nematodes; tolerant to lime-induced chlorosis. Selected because of its slightly better tolerance of *Phytophthora syringae* than Hansen 2168, but has shown less field survival than peach. Relatively low chilling, about 400 to 500 h, similar to Nonpareil almond.

Hansen 2168.—Orig. in Davis, Calif., by C.J. Hansen, Dept. of Pomology, Univ. of California. Introd. in 1986 by D.E. Kester and R.N. Asay. Very vigorous rootstock for almonds, peaches, nectarines, and Japanese plums; not compatible with apricots. Almond B x Peach Selection 1-8-2; cross made in 1962; pollen parent has *Prunus davidiana*, Chinese peach PI 6582, and Okinawa in pedigree. Propagated by hardwood cuttings and by micropropagation. Root system of mature tree intermediate between almond seedling and peach; more deeply rooted than Marianna 2624. Anchorage excellent; no suckers. Not cold hardy. Resistant to root-knot nematodes; immune to both *M. javanica* and *M. incognita acrita*; susceptible to root-lesion nematodes; tolerant to lime-induced chlorosis. Slightly less tolerant of *Phytophthora syringae* than Hansen 536 in original tests. Relatively low chilling, about 400 to 500 h, similar to Nonpareil almond.

Ishtara®.—See under Plums. Compatible under Ai, Ferragnes, Ardechoise, and Ne Plus Ultra; no incompatibilities reported. Vigor similar to that of peach and GF 8-1. Excellent anchorage expedites shaking.

Myran®.—See under Peach Rootstocks. Compatible under all cultivars of almond tested. Very vigorous, similar to GF 677. Young almond trees on Myran show no symptoms of verticillium wilt when planted in contaminated soil. More tolerant of root asphyxia than are peach x almond rootstocks.

Titan.—Orig. in Fresno, Calif., by the U.S. Dept. of Agriculture in cooperation with Fresno State College. Introd. in 1971. O.p. seedling of a late-flowering mutant of Nonpareil, selected by Robert W. Jones, U.S. Hort. Field Sta., Fresno, tested as 92-54. For the production of seed for rootstock: blooms with Nemaguard peach, and when planted adjacent to it, will produce a very high proportion of F_1 hybrid almond-peach seed; such plantings should be separated from other almond trees to prevent contamination. Seedlings from these F_1 hybrid seed are very vigorous, making excellent rootstock for almond in peach replant areas, in sandy areas where nematode stunting is severe on Nemaguard, and where moisture is inadequate for peach rootstock. Nut: soft but sealed shell; soil should be well-drained for seed germination.

APPLE

A.2.—See Alnarp 2 in apple rootstocks.

Abbott.—Orig. in Saskatoon, Saskatchewan, Canada, by C.F. Patterson, Univ. of Sask. Introd. in 1960 for home gardens. Elsa x Melba; selected in 1945; tested as A59-383. Fruit: up to 2 1/4 inches in diam.; skin blushed over yellow; flesh very mildly acid; quality good.

Ace® Spur Delicious.—Orig. in Quincy, Wash., by Carl Perleberg. Limb sport of Oregon Spur Delicious. Discovered in 1970. Plant patent 4587; 12 Aug. 1980; assigned to Columbia Basin Nursery. Similar to Oregon Spur Delicious, except earlier coloring and longer type.

Acey Mac (*Buhr McIntosh*).—Orig. in Peru, N.Y., by Arthur Burrill. Discovered before 1969. Assigned to Columbia Basin Nursery, Quincy, Wash. Fruit: very similar to McIntosh in appearance, flavor, shape, flesh color, and texture, but larger, firmer, and less inclined to preharvest drop. Ripens 5 days after Rogers McIntosh. Tree: similar to McIntosh.

Acheson.—Orig. in Edmonton, Alberta, Canada, by Mrs. J.H. Acheson. Introd. in 1941. Delicious o.p. Fruit: large; ovate; skin greenish-yellow; flesh fine-grained, subacid, quality above medium; season mid-August into September. Tree: hardy; productive.

Adams Delicious.—Orig. in Hood River, Ore., by L.A. Adams. Introd. in 1956. Bud mutation of Delicious; discovered in 1954 on a 40-year-old tree. Fruit: skin cherry-red, darker than parent, color appearing early; shape typically Delicious.

Adams Jonathan (*Giant Jonathan*).—Orig. in Pittsfield, Ill., by Bessie D. Adams. Propagation materials obtained by Stark Bros. Nurseries & Orchards Co., Louisiana, Mo., but never introd. Considered to be a bud mutation of Jonathan; discovered in 1939, when tree was apparently 14 years old. Fruit: large; rough; quality poor. A periclinal cytochimera 2-2-4-4. Chimeral nature determined at the New York State Agr. Expt. Sta., Geneva; accessioned in 1949 by this station.

Adanac.—Orig. in Unity, Saskatchewan, Canada, by John Lloyd. Introd. in 1960. Battleford o.p.; selected in 1960; tested as Jack's No. 1. Fruit: up to 2 1/4 inches in diam.; good for cooking; ripens early; keeps in storage for 2 to 3 months.; resembles Battleford. Tree: hardy, to –50F, being hardier than Battleford.

Adel.—Orig. in Ames, Iowa, by the Pomology Section, Iowa Agr. Expt. Sta. Introd. in 1921. Parentage unknown; from mixed seed; selected in 1918; tested as Ames 419. Fruit: size medium; skin greenish-yellow or striped yellow with lively red, resembling Gravenstein; flesh crisp, tender, sprightly subacid, pleasant, good to very good; season October to January. Tree: vigorous; productive. Apparently no longer being propagated.

Adina™ (*Frankad, Stark® Adina™*).—Orig. in Stanthorpe, Queensland, Australia, by H. Franklin. Plant patent 6361; 1 Nov. 1988. Assigned to Stark Bros. Nursery, Louisiana, Mo. Fruit: large; color red to purplish red, occasional stripe; shape round conic; flesh creamy white, firm, very juicy; sweet, with distinctive hint of cinnamon; superior eating quality; harvest season very early, mid-June. Tree: vigorous; precocious; low chilling requirement (350 h); grows in hardiness zones 6 to 9; Dorsett Golden is a suitable pollenizer.

Advance.—Orig. in Saskatoon, Saskatchewan, Canada, by C.F. Patterson, Univ. of Sask. Introd. in 1959 for home gardens. Columbia x Wealthy; selected in 1944; tested as A58-308. Fruit: up to 2 1/2 inches in diam., skin green with some blushing and striping; flesh mildly subacid, quality good; keeps well; ripens from midseason on. Tree: hardy; very susceptible to fire blight.

Akane (*Tohoku No.3, Primerouge*).—Orig. in Fujisaki, Aomori, Japan, by the Fruit Tree Res. Sta., Ministry of Agr. & Forestry, Morioka, Japan. Introd. in 1970. Jonathan x Worcester Pearmain; cross made in 1939; selected in 1953; tested as

Tohoku No. 3. Fruit: medium; round oblate; skin bright red, slight scarfskin, thick, tough, smooth; flesh white, firm, crisp, juicy, slightly coarse; flavor subacid, resembles Jonathan; quality good; recommended as a dessert apple; ripens early to mid-September at McIntosh season in Morioka; keeps 3 to 4 weeks. Tree: slightly upright; vigor rather weak; crops light, annual. Less susceptible to scab than most varieties.

Alaska.—Orig. in Ettersburg, Calif., by Albert F. Etter. Introd. in 1944. Plant patent 699; 18 June 1946; assigned to George C. Roeding, Jr., California Nursery Co., Fremont, Calif. Bedfordshire Foundling o.p.; selected in 1944. Fruit: size medium; stem short; calyx closed; skin yellow, shiny; flesh snow white, tender, juicy; resembles Bedfordshire Foundling. Tree: heavy producer.

Alice.—Orig. in Fjälkestad, Sweden, by Balsgård Fruit Breeding Inst. Introd. in 1951. Parentage unknown. Fruit: medium; round to oblate; skin bright yellow with solid red blush on exposed side, attractive; flesh white, juicy, crisp when not too ripe, quality good, insipid; ripens between Stark Earliest and James Grieve; resembles Ingrid Marie. Tree: vigorous; hardy; spreading; regular cropper; blooms midseason.

All Red Jonathan.—Orig. in Allentown, Lehigh County, Pa., by Homer S. Kemp, Bountiful Ridge Nurseries, Princess Anne, Md. Introd. in 1937. Apparently a bud mutation of Jonathan; discovered in 1934. Fruit: skin attractive, full bright red, not too dark under eastern U.S. conditions; otherwise resembles Jonathan. Tree: slightly more vigorous than Jonathan.

Almeda.—Orig. in Hendersonville, Sumner County, Tenn., by James W. Savely. Introd. in 1939. Plant patent 327; 18 Apr. 1939. Considered to be a bud mutation of Rhode Island Greening; discovered in 1927. Fruit: large, 2 3/4 to 3 inches in diam.; skin green, sometimes with a slight blush, thin, slightly waxy; flesh greenish-white, crisp, juicy, acid to subacid; ripens early in late July and early August; keeps exceptionally well; good for cooking; suitable for shipping; resembles Rhode Island Greening, but with more acid and a different ripening season. Tree: size medium; strong, vigorous; bears annually; very productive; blooms in midseason.

Alnarp 2 (*A.2*).—See rootstocks for apple.

Alnarps Favorit.—Orig. in Alnarp, Sweden, by Dept. of Pomology, Agr. College of Sweden. Introd. in 1967. McIntosh x Alfa 68; cross made in 1944. Fruit: large; conical; skin smooth with red stripes; flesh creamy white, firm, juicy; good cooking and storage qualities; ripens early October. Tree: size medium; vigor medium; productivity good; triploid; recommended for southern Sweden.

Alton.—Orig. in Geneva, N.Y., by Richard Wellington, New York State Agr. Expt. Sta. Introd. in 1938. Early McIntosh x New York 845 (Red Canada x Yellow Transparent); cross made in 1923; first full crop in 1935. Fruit: similar to Crimson Beauty; flavor mild, subacid; ripens just after Crimson Beauty; eating and cooking apple for home and roadside markets. Tree: resembles Crimson Beauty.

Alvah.—See Flamespur Red Rome.

Ambassy™.—Red sport of Delcorf, introd. by Davodeau-Ligonniere Nursery, Angers, France. Fruit: striped red-orange; flavor acid; ripens 2 weeks before Gala; 2-month storage life at –0.5C.

Ambrosia.—A good-quality red apple ripening in Delicious season. Chance seedling discovered in early 1980s by Wilfred and Robert Mennell, Cawston, British Columbia, Canada. Assigned to Okanagan Plant Improvement Co., Canada. Fruit: medium to large size; attractive; crisp; sweet, low acid; very juicy; distinct, pleasant aroma; mild flavor. Tree: spur type, upright growth, not a tip bearer.

Ames.—Orig. in Ames, Iowa, by H.L. Lantz, Iowa Agr. Expt. Sta. Introd. in 1921. Allen Choice x Perry Russett; first fruited in 1920. Fruit: size medium; skin deep red, medium thick, smooth; flesh firm, crisp, juicy, yellow, subacid; quality good to very good; season January to May; adapted in particular to southern Iowa. Tree: vigorous; productive and regular bearer; strong framework in the orchard and nursery.

Andersen.—Orig. in Saskatoon, Saskatchewan, Canada, by C.F. Patterson, Univ. of Sask. Introd. in 1960 for home gardens. Columbia x Melba; selected in 1945; tested as A58-301. Fruit: up to 2 inches in diam.; somewhat conical; skin light red with some splashing; flesh white, mildly subacid, quality good; keeps extremely well; ripens in mid-September. Tree: extremely hardy.

Anderson Jonathan.—Orig. in the orchard of Robert Anderson, Covert, Mich. Introd. in 1927. Bud mutation of Jonathan; discovered in 1923; selected by Robert Anderson and by Roy Gibson, Greening Nursery Co., Monroe, Mich.; propagated by this nursery. Fruit: medium to large; skin bright red over a yellow ground color; flesh white, firm, crisp, juicy, subacid; stores well; resembles Jonathan, except more red color. Tree: bears well annually; standard type.

Anna.—Orig. in Doar Na Shomron, Israel, by Abba Stein. Introd. there in 1963; in the U.S. as P.I. 280400 in 1965. Red Hadassiya (a local variety) x Golden Delicious; cross made in 1959. Fruit: large; skin red cheeked; flesh subacid to sweet, flavor mild; ripens in Israel end of June to first of July. Tree: annually productive; low chilling requirement, about 400 hrs.

Annalee (*Blount Golden*).—Orig. in Irving, Texas, by Floyd Blount, Jr., Dallas, Texas Introd. in 1974. Plant patent 3496; 26 Feb. 1974. Golden Delicious o.p., discovered in 1962. Fruit: about 3 inches in diam.; conical; skin yellow with some dots, some russeting, waxy, smooth, thick, tough; flesh satiny white with yellow tint, firm, juicy, flavor good; a dessert variety; ripens early August in Dallas. Tree: size medium; upright, inverted vase-shape; very vigorous; productivity very good; bears regularly. Superior heat resistance for Texas area.

Anoka (*Champion Supreme*).—Orig. in Brookings, S.D., by N.E. Hansen, South Dakota Agr. Expt. Sta. Introd. in 1920. Mercer x Duchess; selected in 1918. Fruit: quality poor. Tree: very hardy; bears very young.

Applegate Red Rome.—Orig. in Monmouth County, N.J., by Leslie N. Applegate, Freehold, N.J. Introd. in 1951. Parentage unknown; original tree selected by members of the New Jersey Apple Inst. Fruit: skin better colored than Red Rome, which it resembles.

Applethorpe Earlidel.—See Earlidel.

Applethorpe Summerdel.—See Summerdel.

Arlet.—See Swiss Gourmet.™

Ashworth.—See Autumn Arctic.

Atha.—Orig. in Cullman County, Ala., by Atha Warnick. Introd. about 1930. Red Astrachan o.p. (perhaps crossed with Yellow Transparent); selected about 1915. Fruit: resembles Yellow Transparent, but more vigorous, less tendency to bear biennially.

Atlas.—Orig. in Ottawa, Ontario, Canada, by Canada Dept. Agr., Res. Sta. Introd. about 1924. Winter St. Lawrence o.p.; selected in 1912. Fruit: skin striped red, highly colored; good for dessert and culinary purposes; keeps well. Tree: very hardy; upright, strong; productive but requires no thinning.

Atwood Delicious.—Orig. in Wenatchee, Wash., by Charles Atwood. Introd. in 1958. Bud mutation of Delicious; discovered in 1952. Fruit: skin solid red, coloring darker and earlier than Delicious. Tree: standard type. No longer propagated.

Atwood Spur Delicious.—Orig. in Wenatchee, Wash., by Charles H. Atwood. Introd. in 1967. Assigned to Yakima Valley Nursery, Wenatchee, Wash. Bud mutation of the standard Delicious; discovered in 1956. Fruit: resembles Delicious; colors earlier and darker than Starking Delicious, blush or solid type, brighter red. Tree: vigorous; medium size; productive; spur-type.

August Delicious.—Orig. in Monticello, Minn. by Marion Hagerstrom. Introd. in 1977 by Herschel L. Boll, Champaign, Ill. Jersey Sweet x Sweet Delicious. Cross made about 1943. Fruit: resembles Delicious, but slightly larger, ripens late August in Minnesota. Tree: highly productive, vigorous, hardy.

Austin.—Orig. in Natal, South Africa, by a Mr. Smith. Introd. in 1934. Believed to be a bud mutation of Lady Carrington (*Alma*). Fruit: small to medium; roundish; skin light green to creamy ground color, blushed with red; flesh white, rather acid, easily becomes over-ripe and insipid; resembles Alma; ripens in January. Tree: vigorous; spreading; very productive.

Autumn Arctic (*Ashworth*).—Orig. in Barnard, Vt., on the Dorothy Thompson farm by Fred M. Chase. Introd. in 1956 by

Fred L. Ashworth, nurseryman, Heuvelton, N.Y. Considered to be a seedling of Arctic; discovered in 1952. Fruit: medium to large; shape, skin color, and flavor of Baldwin; better quality than Arctic; ripens in September; resistant to apple scab; no thinning required.

Autumn Gold (*Hein cv.*).—Orig. in Hein Orchards, Tieton, Wash., as a chance seedling in 1985; U.S. plant patent applied for. Introd. in 1994; assigned to Columbia Basin Nursery, Quincy, Wash. Fruit: large, conical, similar to Golden Delicious in shape; pink blush over yellow-green ground color; highly resistant to russet; quality similar to Golden Delicious but with maturity two weeks later; storage life better than that of Golden Delicious, with much less tendency to shrivel. Tree: medium vigor, spreading habit, midseason blooming, diploid.

Azwell.—See Wellspur Delicious.

B.9.—See Red-Leaved Paradise in apple rootstocks.

Baldwin Double Red.—See Sawyer Baldwin.

Baldwin Scarlet.—See Sawyer Baldwin.

Ballou Grimes.—See McClintock Grimes.

Bancroft.—Orig. in Ottawa, Ontario, Canada, by Canada Dept. Agr., Res. Sta. Introd. about 1935. Forest x McIntosh; selected in 1930. Fruit: sizes well; skin highly colored red; keeps well in cold storage until April; quality moderately good. Tree: strong grower of good form; bears at a young age; appears to be as hardy as McIntosh.

Banks Crimson Beauty.—Orig. in Waterville, Kings County, N.S., Canada, by H.P. Banks. Introd. about 1940. Bud mutation of Crimson Beauty; discovered in 1910. Fruit: skin solid red, very attractive; otherwise very similar to parent. Tree: identical to parent; semi-vigorous; stout, well spaced branches and a moderate number of heavy fruit spurs.

Bardwell Red Duchess.—See Daniels Red Duchess.

Barkley Red Rome.—Orig. in Manson, Wash., by George L. Barkley. Introd. in 1953. Plant patent 1470; 17 Apr. 1956; assigned to Van Well Nursery, Wenatchee, Wash. Bud mutation of Rome Beauty; discovered in 1944. Fruit: medium to large; roundish conic, tends to be longer than most Rome Beauty fruit; skin thick, tough, smooth, with more red than Rome Beauty or Red Rome, distinctive striping; flesh nearly white, firm, rather crisp, juicy, mildly subacid; keeping quality very good; resembles Rome Beauty. Tree: a strong grower; size medium; upright at first but later roundish to somewhat spreading and drooping. No longer propagated.

Barrie.—Orig. in Saskatoon, Saskatchewan, Canada, by C.F. Patterson, Univ. of Sask. Introd. in 1960 for home gardens. Columbia x Wealthy; selected in 1950; tested as A60-202. Fruit: up to 2 1/2 inches long and 2 inches in diam.; skin cream overlaid with light red; flesh mildly subacid, quality good; ripens in midseason. Tree: hardy; productive.

Barry.—Orig. in Geneva, N.Y., by the New York State Agr. Expt. Sta. Introd. in 1957. McIntosh x Cox's Orange Pippin; selected in 1936. Fruit: medium to large; skin with a solid, dark red blush, appearance fine; quality good to very good for both dessert and culinary uses; matures in the fall, 2 weeks before McIntosh; cold storage life of 3 months; resembles McIntosh.

Bates Lobo.—Orig. in Stevensville, Pa., by W.L. Bates. Not introd. Considered to be a bud mutation of Lobo, arising as a whole tree. Fruit: large; irregular. A periclinal cytochimera 2-2-2-4; tends to revert to diploid; chimeral nature determined at the New York State Agr. Expt. Sta., Geneva; accessioned in 1952 and 1953 by this station. Tree: vigorous; somewhat spreading; stout branches; somewhat alternate bearing habit. No commercial plantings, academic interest only.

Battleford.—Orig. in North Battleford, Saskatchewan, Canada, by Murray Somerville. Introd. about 1945. Parentage unknown; discovered in 1934. Fruit: size medium; skin pale yellow, striped, washed dull red; quality only fair, with flesh becoming mealy in a short time; season late August and early September. Tree: not very hardy.

Baxters Black Winesap.—Orig. in Nauvoo, Ill., by Cecil J. Baxter, Sr., and Frederick K. Baxter. Introd. in 1944. Plant patent 619; 21

Mar. 1944; Frederick K. Baxter, assignor to Gem City Vineland Co., Nauvoo. Parentage unknown; selected in 1930. Fruit: large, round; skin deep red, approaching black, smooth, waxy; flesh whitish with yellow tinge, crisp, firm, juicy; ripens in late October, 10 days after Winesap; holds well on tree; excellent keeper in common storage; ships well. Tree: grows rapidly; very hardy; early bearer.

Beacon (*Early Delicious, Fenton, Miller Red*).—Orig. in Excelsior, Minn., by the Univ. of Minnesota Fruit Breeding Farm. Introd. in 1936. Malinda o.p.; seed planted in 1908; tested as Minn. 423. Fruit: large; skin solid red, covering almost entire surface, tough; flesh juicy, mildly acid, quality good, superior to Maiden Blush, Duchess, and Wealthy; dessert rating good; matures about 1 Aug., 10 to 14 days before Wealthy. Tree: productive; quite susceptible to cedar rust. This is not the Fenton variety described in 1873.

Beauty Blush.—Orig. in Saskatoon, Saskatchewan, Canada, by C.F. Patterson, Univ. of Sask. Introd. in 1960 for home gardens. Elsa x Wealthy; selected in 1947; tested as A59-342. Fruit: up to 2 inches in diam.; somewhat oblong; skin yellow, slightly blushed; flesh mildly subacid, flavor pleasant, fair quality; ripens in midseason. Tree: upright grower.

Belchard.—See Chantecler.

Belle de Quebec.—See Quebec Belle.

Bemali.—See rootstocks for apple.

Best Ever (*Bestever*).—Orig. in Sebastopol, Calif., by Wm. E. Silva. Introd. in 1963. Plant patent 2486; 9 Mar. 1965. Gravenstein x (Red Rome x Delicious); cross made in 1945; selected in 1958. Fruit: averages 2 3/4 inches in diam.; oblong; skin mottled dark red, striped, smooth, thin, some waxiness; flesh yellow, juicy, crisp, flavor very good; aroma heavy and pleasing; keeps very well in common storage until January; ripens 20 Oct. to 10 Nov. Tree: medium large; upright; very productive.

Bestever.—See Best Ever.

Beverly Hills.—Orig. in Los Angeles, by William H. Chandler, Univ. Calif. Introd. in 1945. Melba x Early McIntosh; seed planted in 1939; first fruited in 1942; tested as UCLA 302. Fruit: skin pale yellow with red stripes, splashed with red; flesh white, quality very good; resembles McIntosh. Tree: moderately well suited to southern California coastal climate because of low chilling requirement.

Bisbee Delicious.—See Starkrimson Delicious.

Black Baldwin.—See Sawyer Baldwin.

Black Jonathan.—See Blackjon.

Black McIntosh.—See Blackmack.

Black Mickey McIntosh.—Orig. in Dansville, N.Y., by Isaac C. Rogers. Introd. about 1930. Bud mutation of McIntosh; discovered in 1929. Fruit: skin with solid blush, but not as deeply colored as Rogers McIntosh; resembles McIntosh.

Black Rome Beauty.—See Ruby Rome Beauty.

Black Wine.—Orig. in Wayne County, Ill., by J.E. Markham, Markham's Nursery, Flora, Ill. Introd. in 1931. Considered to be Winesap o.p.; selected in 1924. Fruit: large; skin bright red.

Black Winesap.—Orig. in Tieton, Wash., by Bert Brooks, Lafayette Nursery, Lafayette, Ore. Introd. about 1936. Winesap o.p.; discovered in the early 1930s. Fruit: skin solid red; resembles Winesap. Now obsolete.

Blackjon (*Black Jonathan, Double Red Jonathan, Red Jon, Red Jonathan*).—Orig. in Wenatchee, Wash., by A.T. Gossman, Columbia & Okanogan Nursery Co. Introd. in 1931. Bud mutation of Jonathan; selected in 1929. Fruit: skin earlier coloring and brighter red than Jonathan, otherwise, identical to parent.

Blackmack (*Black McIntosh*).—Orig. in Oliver, British Columbia, Canada, by H. Simpson. Introd. in 1930 by Columbia & Okanogan Nursery Co., Wenatchee, Wash. Bud mutation of McIntosh; selected in 1928. Fruit: skin colors red early, permitting earlier picking, thereby eliminating loss by windfalls; resembles McIntosh. Tree: standard type.

Blair.—Orig. in Ottawa, Ontario, Canada, by the late D.S. Blair and Paul Roy, Canada Dept. Agr. Introd. in 1973. McIntosh x Fameuse; cross made in 1929, selected in 1944, tested as 0-294. Fruit: 2 1/2 to 2 3/4 inch in diam., fairly uniform; round, more pointed than McIntosh; skin solid dark red, occasionally some light green areas, thin, tough; colors better than McIntosh, otherwise very similar; flesh white, rarely pale green, juicy and aromatic, similar to McIntosh, but ripening 10 days earlier; storage life almost as long as that of Lobo. Tree: spreading like McIntosh with fewer suckers after pruning and hardier; blooms with McIntosh; moderately productive.

Blairmont.—Orig. in Blairsville, Ga., by J.M. Thompson, South East Fruit & Tree Nut Res. Lab, Byron, Ga., and U.S. Dept. of Agriculture. Introd. in 1982. QH 11-75 (York x Minjon) x Grove. Cross made in 1961 by Jerry B. Hardigree; selected in 1971 by J.M. Thompson; tested as 6132-106 and BL 32-106. Fruit: 3 inches in diam., subconic; skin attractive with 90% washed solid dark and lively red over pale yellow; flesh firm, cream-colored, fine texture, juicy; ripens just after Mollie's Delicious, 3 weeks before Delicious. Tree: moderately vigorous, spreading; annual bearer, self-infertile, pollenizers required (Golden Delicious recommended); resistant to fire blight, scab, and bitter rot.

Blaxtayman.—Orig. in Wenatchee, Wash., by J.H. Dickey, Columbia & Okanogan Nursery Co. Introd. in 1930. Bud mutation of Stayman Winesap; discovered about 1926. Fruit: skin overall red; much higher percentage of extra fancy fruit than parent.

Blaze.—Orig. in Urbana, Ill., by the Illinois Agr. Expt. Sta. Introd. in 1958. Collins x Fanny; cross made in 1939 under the direction of M.J. Dorsey; selected in 1951; tested as Ill. 17. Fruit: size medium; round oblate in vertical section, round in transverse section; skin nearly covered with medium red over a bright-yellow ground color, thin, tough, with a very attractive smooth, glossy finish, lenticels inconspicuous; flesh yellowish, crisp, firm, fine-grained, juicy, mildly subacid, quality good; matures 3 1/2 weeks before Jonathan, or about 1 Aug. in southern Illinois; flavor somewhat less tart than Jonathan, which it resembles. Tree: upright-spreading; vigorous; productive; blooms 2 days after Jonathan; moderately susceptible to apple scab and cedar apple rust; susceptible to black rot; moderately resistant to natural infections of apple blotch, powdery mildew, and fire blight in the orchard. Recommended to extend the fresh-market season of Jonathan-type apples in commercial apple regions with a climate similar to that of southern Illinois.

Bledisloe Cox.—Orig. in Gisborne, New Zealand, by W.M. Hall, orchardist. Introd. there in 1934. Bud mutation of Cox's Orange Pippin; discovered in 1932. Fruit: skin color improved over parent; not fully red but very heavily striped with a characteristic dappled appearance in the same way that Beauty of Bath is dappled. Named in honor of Lord Bledisloe, Governor-General during the 1934 period. Tree: identical to Cox's Orange Pippin.

Blount Golden.—See Annalee.

Blushing Golden.—Orig. in Coben, Ill., by Ralph Griffith. Introd. in 1968. Parentage unknown; chance seedling discovered in 1959. Plant patent 2835; 1 Oct. 1968; assigned to Stark Bros. Nurseries & Orchards Co., Louisiana, Mo. Fruit: resembles Golden Delicious with red blush; round conic shape; skin deep yellow ground cover with up to 50% light red blush, small dots, smooth, very attractive; flesh very firm; very good quality, flavor between Jonathan and Golden Delicious; long storage life; ripens 2 weeks after Golden Delicious. Tree: vigorous; spreading; semi-spur type; productive.

Bolding.—See Red Delight.

Boller McIntosh.—Orig. in Sodus, N.Y., on the farm of Arthur Boller. Introd. in 1954. Bud mutation of McIntosh; discovered about 1950. Fruit: resembles McIntosh but firmer, somewhat smaller, and more highly colored red than McIntosh. Tree: standard type.

Bonan Red June.—See Bonum Red June.

Bonanza.—Orig. in Sebastopol, Calif., by Wm. E. Silva. Introd. in 1965. King x Rome; selected in 1965; tested as Silva 5. Fruit: 2 1/2 to 3 inches in diam.; stem long; elongate and pointed like Delicious; skin bright red, striped; flesh firm, juicy, aroma prominent; keeps well in common storage; ripens in mid-September. Tree: strong; limbs droop on end where fruit develops heavily, does not fruit well in center.

Bonfire.—Orig. in Sebastopol, Calif., by Wm. E. Silva. Introd. in 1965. Rome x Delicious; selected in 1965; tested as Silva 4. Fruit:

shape and size approaching Delicious; skin bright red, smooth finish; flesh firm, white, fine-grained, aromatic; season beginning of September for fresh market and storage. Tree: large; upright.

Bonser.—Orig. in Batlow, New South Wales, Australia. Introd. in 1960. Chance seedling discovered in 1950. Fruit: medium to large; skin a bright carmine, attractive; excellent cool-storage qualities; very highly resistant to apple scab; no thinning required; resembles Jonathan. Tree: vigorous; bears well and consistently.

Bonum Red June (*Bonan Red June*).—Orig. in Auburn, Ala., by the Alabama Polytechnic Inst. Introd. in 1940. Parentage unknown; received as a mislabeled tree; discovered in 1919. Fruit: small; roundish oblong, somewhat flattened at ends; quality good; keeps well; skin solid bright red; ripens 3 June to 30 Aug.; resembles Hunt Red June and Hunts Everbearing.

Bonza.—Orig in Batlow, New South Wales, Australia, by B. Athenson. Introd. in 1962. Chance seedling, parentage unknown, discovered in 1953. Fruit size medium to large; conical oblate, even; skin 80% to 100% bright crimson, lightly streaked, coarse and unattractive first few years, subsequently will fine-up and, in some localities, resemble Jonathan; flesh pale, creamy white, firm texture crisp, fine-grained, flavor pleasant; good storage quality, about 8 months at 30 to 32F; ripens 25 Mar in N.S.W. Tree: vigor moderate; hardiness unknown; very productive, tip bearing; moderately tolerant to scab.

Bountiful (*Malling Bountiful*).—Orig. in Kent, England, by F.H. Alston, East Malling Res. Sta. Introd. in 1986. Cox's Orange Pippin o.p. Fruit: large, 75 to 80 mm in diam.; skin green with occasional orange/red patches or stripes; acidity lower than Bramley; a culinary apple, cooks to give a deliciously flavored sauce; best-suited for home orchards. Ripens in late September; at 3.5C, stores in marketable condition until January. Tree: compact, about half the size of Bramley trees; precocious; productive; little pruning required; few pollination problems.

Bountiful Winesap.—Orig. in Martins Mountain Orchard, Paw Paw, W.Va., by Harry W. Miller, Consolidated Orchard. Introd. in 1939 by Bountiful Ridge Nurseries, Princess Anne, Md. Parentage unknown; discovered in 1928. Fruit: very much larger than Winesap; skin as highly colored as Winesap, which it resembles. Tree: produces well and regularly; bears earlier than Winesap.

Bowyer.—Orig. in Prince George, British Columbia, Canada, by Mrs. W.R. Bowyer. Introd. in 1950 by Skinner's Nursery, Dropmore, Manitoba, Canada. Chance seedling; seed planted in 1930. Fruit: size medium; quality good. Tree: not hardy.

Boyd Duchess (*Boyd Red Duchess*).—Orig. in Anna, Ill., by Claud J. Boyd, J.J. Boyd & Sons, Cobden, Ill. Introd. in 1930 by Davis Nursery, Anna. Chance seedling found in an orchard of Duchess of Oldenburg; discovered in 1930. Fruit: large; flesh firm; matures in mid-July. Tree: comes into bearing very early; rather susceptible to *Botryosphaeria* canker.

Boyd Red Duchess.—See Boyd Duchess.

Boyer Early.—Orig. in Black River, N.Y., by Robert Boyer. Introd. in 1950 by Fred L. Ashworth, nurseryman, Heuvelton, N.Y. Considered to be Fameuse o.p.; discovered in 1949. Fruit: size and appearance similar to Fameuse, but not quite as juicy or as good quality; ripens early; resembles Fameuse.

Boys Joy.—Orig. in Hammond, N.Y. Introd. in 1957 by Fred L. Ashworth, nurseryman, Heuvelton, N.Y. Parentage unknown; tree about 100 years old, perhaps a seedling of Baxter; selected in 1953. Fruit: large; skin dark red; flesh mild, sweet, soft; ripens in late September.

Braeburn.—Orig. in Waiwhero, Upper Moutere, Nelson, New Zealand, by O. Moran. Introd. in 1952 by Williams Bros., Braeburn, Upper Moutere, Nelson. Parentage unknown, probably Lady Hamilton x open-pollinated seedling. Discovered in 1952. Fruit: medium to large; oval with a definite crown, shape is taller and crown more pronounced in colder climatic regions; skin glossy, covered with short stripes of dark crimson, three-fourths overlaid with dark scarlet blush, bright, attractive, considered a striped variety; flesh pale cream, very firm, crisp, juicy, flavor subacid, quality excellent; ripens very late, just before Fuji and with Red Dougherty; keeps well in storage, not susceptible to bitter pit; hangs well on tree without preharvest drop. Tree: spreading; vigor moderate; productivity heavy and precocious, tendency toward biennial bearing but easily controlled by pruning.

Breakey.—Orig. in Morden, Manitoba, Canada, by Canada Dept. Agr., Res. Sta. Introd. in 1935. Blushed Calville o.p.; selected in 1929. Fruit: size medium; round oblate; skin a well-colored red and scarlet; flesh white, fine-textured, melting, juicy, flavor mild, spicy; quality good as dessert or sauce; season September to November. Tree: vigorous; upright, spreading; hardy.

Brechsteinia.—Orig. in Lander, Wyo., by J.M. Steinbrech, Wyo. Agr. Expt. Sta. Introd. in 1931. Fruit: large; skin light green, blushed on one side; flesh fine-grained, juicy, subacid; ripens first of October. Tree: head markedly cone-shaped, formed by multiple leaders; upright, somewhat open; moderately vigorous; very hardy; twigs stout; internodes short; buds obtuse and free; buds heavily pubescent; lenticels round; bark light red. Named by Aven Nelson in honor of J.M. Steinbrech, Superintendent of the State Expt. Farm at Lander.

Bridgham Delicious.—Orig. in First Creek, Wash. Introd. in the 1950s. Bud mutation of Delicious; discovered in 1930. Fruit: large; skin bright red; ripens about with Delicious, which it resembles. Tree: resembles Delicious.

Bright 'N Early.—Orig. in Chelan, Wash., by Don Craig. Introd. in 1976. Plant patent 4372. Bud mutation of Oregon Spur Delicious. Fruit: flesh white and juicy; skin is brightly colored 124 days after bloom, 1 week earlier than other red Delicious varieties. Tree: vigorous, with wide crotch angles.

Brightness.—Orig. in Saskatoon, Saskatchewan, Canada, by C.F. Patterson, Univ. of Sask. Introd. in 1959 for home gardens. Columbia x Wealthy; selected in 1942; tested as A58-320. Fruit: up to 2 1/2 inches in diam.; broadly conical; skin bright, glossy red; flesh very mildly subacid, flavor pleasing; keeps fairly well; ripens in early September; resembles Prolific. Tree: yields heavily; extremely hardy.

Britegold.—Orig. in Trenton, Ontario, by L.P.S. Spangelo, S.J. Leuty, and H.B. Heeney, Agr. Canada, Ottawa, and Smithfield Experimental Farm, Trenton, Ontario. Introd. in 1980. Sandel (Delicious x Sandow) x Ottawa 522 {Platt Melba x R6T68 [Jonathan x (Rome Beauty x *Malus floribunda* 821 sib.)]}. Tested as Ottawa 652. Fruit: medium to large; skin bright yellow; flesh cream-colored to yellow slightly coarse, tender, juicy; flavor low acid; only medium quality; processed fruit rated good for slices and fair for sauce; harvest date after McIntosh; storage life short, 1 to 2 months at 2C. Tree: low to medium vigor; spreading; tendency to droop; yields low; some tendency to become biennial; some fruit drop before ripe; blooms with McIntosh; resistant to apple scab, very susceptible to cedar apple and quince rusts.

Britemac.—Orig. in New Brunswick, N.J., by M.A. Blake, New Jersey Agr. Expt. Sta. Introd. in 1964. Melba x Kildare; cross made in 1934; selected in 1942; first distributed for trial as NJ 88634 by Blake; later tested as NJ 3. Fruit: medium to large; slightly oblate; skin striped and blushed, colors better than McIntosh, very attractive, resembles Cortland in appearance and shape, sometimes becoming russeted; flesh white, juicy, crisp, similar to McIntosh and Cortland; flavor mildly subacid, slightly aromatic; apparently not well-adapted for processing; high dessert quality, as good as or better than McIntosh; hangs better on tree than McIntosh in central New Jersey; primarily of local value (New Jersey and eastern Pennsylvania); ripens just after McIntosh, but of pleasant eating quality more than 2 weeks before it is ripe. Tree: considered to be very hardy; strong, spreading; somewhat smaller than McIntosh; bears well; blooms in midseason.

Brooks 27.—Orig. in Brooks, Alberta, Canada, by D. Bark, Provincial Hort. Sta. Introd. in 1937. Parentage unknown. Fruit: average 2 1/2 inches in diam.; skin greenish, overlaid with attractive red; pleasant dessert flavor and good for culinary purposes; keeps well in common storage; quality very good. Tree: hardy, comparing favorably with Heyer 12.

Brooks Beauty.—Orig. in Washington by Bert Brooks, Lafayette Nursery, Lafayette, Ore. Introd. about 1934. Reported to be Gravenstein x Tompkins King; selected about 1930. Fruit: large, about the size of Tompkins King; flesh firm, fine-grained, juicy, good for culinary purposes but flavor flat; keeping quality good; ripens about 1 Oct.; resembles Tompkins King. Now obsolete; no longer being propagated.

Browning Beauty.—See Potomac.

Buckley Giant™.—Orig. in Sumner, Wash., by Buckley Nursery Co., Buckley, Wash. Introd. in 1963. Unknown seedling formerly known as Olsen. Discovered in 1954. Fruit: large; oval; skin ground color golden green, 40% covered by red blush, thin; flesh white, crisp; ripens 1 to 10 Sept.; highly susceptible to bitter pit; recommended for fresh fruit and cooking. Tree: medium size; round; highly productive.

Bud.9.—See Red-Leaved Paradise in apple rootstocks.

Budagovsky 9.—See Red-Leaved Paradise in apple rootstocks.

Budagovsky 57-490.—See rootstocks for apple.

Budagovsky 57-491.—See rootstocks for apple.

Buhr McIntosh.—See Acey Mac.

Burgundy.—Orig. in Geneva, N.Y., by R.D. Way and R.C. Lamb, New York Agr. Expt. Sta. Introd. in 1974. Monroe x N.Y. 18491 (Macoun x Antonovka), cross made by R.C. Lamb in 1953, first fruited in 1960, selected by R.D. Way in 1961, tested as NY 161. Fruit: medium size, 2 3/4 inches in diam.; oblate; skin very dark red, little striping, glossy; flesh creamy, semi-firm to firm, crisp, juicy, flavor subacid, eating quality good, somewhat resembles Jonathan; sunburn and watercore can be problems; does not drop badly after ripening; stores longer than most other early fall varieties; ripens about 11 Sept. in Geneva, 2 1/2 weeks before McIntosh; a dessert variety. Tree: vigorous; moderately productive, fruit borne mainly on spurs, bears annually, no overcropping; blooms early midseason, about 1 day after McIntosh, good pollenizer for other varieties; susceptible to apple scab and cedar apple rust, highly susceptible to fire blight.

C6.—See rootstocks for apple.

C.P. Close.—See Close.

C & O Blaxtayman.—See Nured Stayman.

C and O Red Rome 262.—See Nured Rome.

C and O Winesap.—Orig. in Wenatchee, Wash., by J.A. Snyder, Columbia & Okanogan Nursery Co. Introd. in 1930. Bud mutation of Winesap; discovered in 1928. Colors earlier than Winesap.

Cal-King.—Orig. in Van Nuys, Calif., by George King. Introd. in 1947. Parentage unknown; discovered in 1942. Fruit: skin colors well; resembles Beverly Hills. Tree: grows well in southern California.

Cameo™ (*Caudle cv.*).—A chance seedling discovered in the Smith-Caudle orchard near Dryden, Wash., in the 1980s. U.S. plant patent applied for. Fruit: attractive bright red stripe over yellow-green ground; round to conical, lacking prominent lobes at apex; size medium to large; flesh firm and creamy-white; flavor sweet-tart; eating quality very good. Matures with Rome Beauty. Stores extremely well; can be stored more than 12 months in CA; not subject to scald. Tree: vigorous, many spurs produced; hardy; precocious and productive.

Campbell Spur.—See Red Chief.

Caravel (*Portia*).—Orig. in Ottawa, Ontario, Canada, by Canada Dept. Agr., Res. Sta. Named and introd. in 1964. Melba x Crimson Beauty; selected by D.S. Blair (1909 to 1959); tested as Ottawa 277 (released in 1942 under this number) early testing by Blair, tested later and released by Lloyd P.S. Spangelo. Fruit: roundish conic; skin yellow, heavily washed with bright red (up to 65% of surface), attractive; flesh cream, tender; good flavor and dessert quality, superior to Crimson Beauty, nearly equals Melba; ships well; ripens 3 to 4 days before Melba; requires thinning to secure commercial fruit size; developed primarily for eastern Canada. Tree: hardy; bears biennially; sets heavily; comes into bearing early.

Cardinal Stayman.—Orig. in Mount Jackson, Va., in the F.H. Wissler orchard in the Shenandoah Valley. Introd. in 1950. Bud mutation of Stayman Winesap; discovered in 1945. Fruit: skin bright solid red; otherwise, similar to parent.

Carlson Red Rome.—Orig. in Wiley Heights, Yakima, Wash., by John Carlson. Introd. in 1961; assigned to Morrison Nursery, Zillah, Wash. Bud mutation of Rome Beauty; discovered in 1954. Fruit: Rome Beauty type; skin solid red, appearing early; in other characteristics resembles parent. Tree: standard type. No longer propagated.

Carlton.—Orig. in Geneva, N.Y., by Richard Wellington, New York State Agr. Expt. Sta. Introd. in 1923. Montgomery x Red Astrachan; cross made in 1912; tested as Astrachan 2391. Fruit: large, round conic; skin an attractive dark red; flesh white, tender, juicy, subacid, of Red Astrachan flavor; home use and roadside markets; ripens 1 month later than Red Astrachan. Tree: vigorous; annual bearer.

Carroll.—Orig. in Morden, Manitoba, Canada, by C.R. Ure, Canada Dept. Agr., Res. Sta. Introd. in 1961. Morden 5029-E152 (Moscow Pear apple seedling) x Melba; selected in 1947; tested as Morden 366. Fruit: medium large; skin mostly solid red; ripens in late August to early September; keeps well into December; rated excellent as an early dessert variety, fair to good cooked; resembles a highly-colored Melba. Tree: dwarfish to medium height; rounded; hardy; scaffold branches structurally good; productivity medium; tends toward annual bearing.

Cascade Compact Red Delicious.—See Starkspur Compact Red Delicious.

Case Red Wealthy.—See Case Wealthy.

Case Wealthy (*Case Red Wealthy, Red Wealthy, Wealthy Double Red*).—Orig. in Sodus, N.Y., by James G. Case. Introd. in 1940. Bud mutation of Wealthy; discovered in 1933. Fruit: resembles parent except for darker red skin.

Caudle.—See Cameo.

Cauley.—Orig. in Grenada, Miss., by J.W. Willis, Delta Expt. Sta., Mississippi Agr. Exp. Sta., Stoneville. Introd. in 1942 by McNair Nurseries, Learned, Miss. Parentage unknown. Fruit: extremely large, weighing as much as 1 lb. each; skin with faint red stripes; flesh yellow, crisp, juicy; culinary and jelly uses; resembles Yellow Newtown. Tree: yields heavily, averaging 26 bushels per tree.

Champion Supreme.—See Anoka.

Chantecler (*Belchard®*).—Orig. in Angers, France, by Y. Lespinasse, J.M. Oliver, J. Lespinasse, and M. Le Lezec, Station d'Arborculture Frutiere, INRA. Assigned to Agri Obtentions, 78283 Guyancourt, France. Introd. to French growers in 1977. Golden Delicious x Reinette Clochard. Fruit: most nearly resembles Reinette Clochard; medium; conical globose; skin golden yellow with slight russeting; flesh cream-colored, medium firm; exceptional flavor. Ripens 3 weeks after Golden Delicious. Tree: vigor medium to strong; precocious; productive.

Charden.—A triploid yellow-skinned apple introd. in France about 1975. Fruit: size large, 70 to 80 mm; skin yellow-green, sometimes with 10% faint pink blush; less russet than Golden Delicious; shape round-conic, less conic than Golden Delicious. Flesh very firm, juicy, yellow; flavor slightly acid, more tart than Golden Delicious, sometimes slightly astringent; eating quality fair to medium good; some resemblance to Mutsu. Tree: vigorous; very productive; large leaves.

Cheddar Cross.—Orig. in Bristol, England, by G.T. Spinks, Univ. Bristol Res. Sta., Long Ashton. Introd. in 1949; brought into the U.S. in 1952 as P.I. 199100. Allington Pippin x Star of Devon; selected in 1946. Fruit: large; flavor good; very attractive; ripens in early midseason; resembles Star of Devon. Tree: medium vigor; fairly upright; dense growth; spurs freely.

Chehalis.—Orig. in Oakville, Wash., by Lloyd Lonborg. Introd. in 1965. Parentage unknown; discovered in 1955; selected in 1962. Fruit: medium length, stout; basin medium narrow, deep, abrupt, smooth; calyx medium large, usually open with stamens visible; skin thick, smooth, greenish yellow to full yellow with occasional pinkish blush on exposed side; dots fairly large, some russeted; flesh white to cream colored, medium-fine textured, crisp, breaking and moderately juicy at prime maturity, becoming medium soft; flavor pleasantly mild, subacid, not noticeably aromatic; ripe 3 to 4 weeks before Golden Delicious, slightly earlier than McIntosh, keeps well in cold storage. Resembles

Golden Delicious. Tree: moderate vigor, upright, spreading, moderately productive, fruiting on twigs and spurs. Resistant to apple scab.

Chelan Red.—Orig. in Manson, Wash., by J.D. Hamilton, J.D. Hamilton Fruit Co., Wenatchee, Wash. Introd. in 1953. Bud mutation of Delicious. Fruit: skin an allover red ("Hay's Maroon"), similar to Richared but lacking the stripes of Starking Delicious; skin reaches full red at least 3 weeks before Starking Delicious; at maturity, skin medium mahogany to dark red, rather heavy, tough, comparable to Winesap; flavor similar to Delicious; maintains very good quality in cold storage; resembles Richared. Tree: spur type.

Cherry Cox.—Orig. in Lindved, Hjallese, Denmark, at Kjer's orchard. Introd. in 1951. Bud mutation of Cox's Orange Pippin; discovered in 1948. Fruit: skin a bit more intense red than Cox's Orange Pippin, which it resembles.

Cherry Red Rome Beauty.—See Ruby Rome Beauty.

Chesapeake.—Orig. in Wallingford, Ky., by Frank Browning. Introd. in 1967 by Bountiful Ridge Nurseries, Inc., Princess Anne, Md. Red Rome o.p.; discovered in 1958. Fruit: medium large; shaped similar to Rome; skin highly colored, very smooth; flesh white, juicy, high dessert quality; storage qualities good, keeps color and does not scald; ripens slightly ahead of Rome. Tree: size medium; vigorous; hardy; productive; recommended for commercial plantings.

Chick-A-Dee McIntosh.—Orig. in Monmouth, Maine, as a whole-tree spur-type mutation of Rogers McIntosh in the Chick Orchard. Patented in 1984; assigned to Roaring Brook Nurseries, Monmouth. Fruit: indistinguishable from McIntosh. Tree: compact, very spurry growth habit; crotch angles wider than Rogers McIntosh; very slow growing; has little or no need for pruning due to horizontal branching; original tree about half the size of adjacent Rogers McIntosh trees of the same age.

Chief.—Orig. in Sebastopol, Calif., by Wm. E. Silva. Introd. in 1965. Moore Sweet x Williams; selected in 1964; tested as Silva 1. Fruit: 2 1/2 to 3 in. in diam.; rounded to oblate; skin red; flesh firm, crisp, juicy; ripens early in last half of July. Tree: large; long branching, heavy scaffold limbs, up to 3 fruits per spur. For local, early markets, canning and sauce.

Chief Washakie.—See Washakie.

Chieftain.—Orig. in Ames, Iowa, by S.A. Beach, Hort. Dept., Iowa State Univ. Introd. in 1966. Jonathan x Delicious; cross made in 1917; tested as A-603. Fruit: averages between Jonathan and Delicious in size; symmetrical, slightly conic; skin bright red, solid coloring, smooth; flesh firm, juicy, subacid; flavor good to excellent, milder than Jonathan, more sprightly than Delicious; excellent storage qualities, free of Jonathan spot. Tree: hardy; strong framework; moderately vigorous; resembles Jonathan; productive; resistant to apple scab and fire blight.

Childs Delicious.—Orig. in Dee, Ore., by Leroy Childs, Oregon Agr. Expt. Sta., Hood River. Introd. in 1960. Bud mutation of Starking Delicious; discovered in 1953. Fruit: a Starking Delicious type, except red color of skin appears earlier than in parent; resembles Starking Delicious. No longer being propagated commercially.

Chipman.—Orig. in Saskatoon, Saskatchewan, Canada, by C.F. Patterson, Univ. of Sask. Introd. in 1960 for home gardens. Columbia x Melba; selected in 1944; tested as A58-328. Fruit: up to 2 1/2 inches in diam.; shape typically Delicious; skin greenish-yellow with blushing and striping of dull red, very attractive; flesh very firm; quality good; ripens from midseason to late September. Tree: extremely hardy.

Christmas Red.—Orig. in Unity, Saskatchewan, Canada, by John Lloyd, John Lloyd's Fruit Nursery. Introd. in 1960. Parentage unknown; seedling discovered in 1950. Fruit: 2 to 2 1/4 inches in diam.; skin has red blush; flesh has good keeping qualities, keeps until spring; ripens 20 Sept. to 1 Oct. Tree: small, round; hardy to –40F; recommended for Northern Prairie district only; will not tolerate lime soil.

Clarkrich.—Orig. in Peshastin, Wash., by Ed S. Clarke. Introd. in 1958; assigned to Rich & Sons, Hillsboro, Ore. Bud mutation of Starking Delicious; discovered in 1955. Fruit: skin bright red, coloring 10 days before parent; resembles Starking Delicious. Tree: standard type; similar to Delicious.

Classic.—Orig. in Hood River, Ore., by Clifford S. Misuyo and George Nakamura. Introd. in 1973. Plant patent 3182; 23 May 1972; assigned to Carlton Nursery, Forest Grove, Ore., and C & O Nursery, Wenatchee, Wash. Limb sport of Starking Delicious, discovered in 1963. Fruit: large; uniform; conical, angular; skin colors earlier than Starking Delicious, distinct stripes by mid-July, with solid red appearance about 1 month before maturity; flesh satiny white with greenish tint, juicy, firm; ripens 25 Sept. to 5 Oct. at Hood River. Tree: large, spreading, vigorous, hardy; regularly productive.

Clawson.—Orig. in Chelan, Wash., by O.T. Clawson. Introd. in 1952. Bud mutation of Starking Delicious; discovered in 1945. Fruit: large; typical of parent, which it resembles. Tree: bears regularly; about two-thirds the size of typical Starking Delicious.

Clawson Spur.—Orig. in Chelan, Wash., by O.T. Clawson. Introd. in 1958. A tree froze to the ground in 1935; in 1936 a sprout arose from the winter-injured trunk, which developed into the present tree. Fruit: skin red over yellow; typical of Starking Delicious, which it resembles. Tree: small; spur-type growth.

Clear Gold (*Stauffer*)—Orig. in Myerstown, Pa., by Orville Stauffer, Mt. Aetna Orchards. Introd. in 1965 by Worley's Nurseries, York Springs, Pa. Limb sport of Golden Delicious. Discovered in 1962. Fruit: size medium; conical; skin clear gold, almost no russeting even under adverse conditions; flesh same as Golden Delicious; ripens 1 Oct. Tree: medium large; vigor, hardiness and productivity good.

Clifton Rome.—Orig. in Memphis, N.Y., by L.J. Clifton. Introd. in 1957. Bud mutation of Rome Beauty; discovered in 1920. Fruit: skin brilliant, bright, solid red blush with no striping. Tree: standard type.

Close (*C. P. Close*, *Red Transparent*).—Orig. in Arlington, Va., by C.P. Close, Arlington Expt. Farm, U.S. Dept. of Agriculture. Introd. in 1938. Parentage uncertain; selected about 1925; tested as USDA 57. Fruit: fairly large, 2 1/2 inches or more in diam.; quality fairly good for both dessert and cooking; ripens with or slightly ahead of Yellow Transparent; stands high summer temperatures well; for home gardens and commercial purposes. Tree: widely adapted.

Coffman (*Coffman June*, *Summer Red*).—Origin unknown. Introd. about 1930. Fruit: medium large; roundish oblong; skin greenish-yellow, striped and splashed with red, with white dots; flesh white, juicy, quality good; ripens last of June in Georgia. Tree: upright grower. Now obsolete.

Coffman June.—See Coffman.

Collet.—Orig. in Notre Dame de Lourdes, Manitoba, Canada, by Victor G. Collet. Parentage unknown; discovered in 1948. Tested at the Expt. Farm, Canada Dept. Agr., Morden, Manitoba. Fruit: size medium, averaging 2 1/2 to 3 inches in diam.; oblate to conic oblate; skin attractive, 80% medium red; flesh white, firm, crisp, texture moderately fine; dessert quality fair to good, excellent as sauce or pie; resembles Godfrey. Tree: medium to tall; rounded to pyramidal; structurally strong; very hardy; annually productive. Promising in southern Manitoba.

Colora Red York (*Red York Imperial*).—Orig. in Colora, Cecil County, Md., by Lloyd Balderson III. Introd. in 1935. Plant patent 168; 10 Mar. 1936; assigned to Bountiful Ridge Nurseries, Princess Anne, Md. Bud mutation of York Imperial; discovered in 1933. Fruit: skin a full bright red without any stripe, coloring uniformly in all parts of the tree; other characteristics identical with York Imperial. Tree: identical with York Imperial.

Colvis Spice.—Orig. in Chester, Ill., by Bernie Colvis. Introd. in 1979 by Southmeadow Fruit Gardens, Birmingham, Mich. Bud mutation of Golden Delicious; discovered in 1974. Fruit: large; skin shiny, fawn russet; flesh a distinct spicy flavor.

Conard.—Orig. in Mountain Grove, Mo., by Paul H. Shepard, Missouri State Fruit Expt. Sta. Introd. in 1935. Ben Davis x Jonathan; selected in 1920; tested as A-756. Fruit: large; roundish to conic; skin almost entirely red, resembles Jonathan in appearance; flesh white, crisp, firm, moderately juicy, quality good, flavor tart, fine-grained; ripens about 1 week after Jonathan; stores well; not subject to preharvest drop. Tree: vigorous; bears large crops annually; less susceptible to scab than Jonathan.

Connell Fireside.—See Connell Red.

Connell Red (*Connell Fireside*).—Orig. in Dunn County, Wis., by Wm. F. Connell of Menomonie, Wis. Introd. in 1957. Plant patent 1602; 14 May 1957. Bud mutation of Fireside; discovered in 1949. Fruit: large, about 3 inches in diam.; oblate to slightly conic; skin dark red, usually smooth, thin, glossy; flesh creamy white, juicy, firm, tender, crisp, mildly acid; quality good; ships and stores well until April and May. Tree: very hardy; very productive.

Co-op 30.—See Enterprise.

Co-op 32.—See Pristine.

Co-op 38.—See GoldRush.

Cooper.—See TRECO Red Gala.

Cooper Granny Smith.—See Earlee Grannee.

Cooper Winesap.—Orig. in Pateros, Wash., by Calvin Cooper. Intro. in 1958. Bud mutation of Black Winesap; discovered in 1957. Fruit: overall red of skin with pin stripe of darker red, not becoming too dark at maturity; resembles Mosebar Winesap. Tree: similar to Winesap.

Cornell McIntosh.—Orig. in Ithaca, N.Y., by L.G. Klein, New York State Agr. Expt. Sta. Introd. in 1956. Bud mutation of McIntosh; discovered in 1954. Fruit: apparently identical with regular McIntosh, except for much improved color in a solid blush pattern and less tendency than other sports to revert to striped color pattern. Tree: standard type.

Corodel (*Coromandel Red*).—Discovered in 1972 in New Zealand as a chance seedling of unknown parentage by R. Knottenbelt on the Coromandel Peninsula. U.S. plant patent assigned to Brandt Fruit Trees, Parker, Wash. Fruit: medium; 60% to 80% solid cherry red; flesh white, nonbrowning, very firm, crisp, juicy, flavor mildly subacid; starchy at harvest time; after ripening, quality good; harvest 3 to 4 days before Granny Smith. Tree: vigor moderate; spreading; heavy cropper; annual bearer; little or no preharvest drop; flowers late.

Coromandel Red.—See Corodel.

Cortland—Orig. in Geneva, N.Y., by S.A. Beach, New York State Agr. Expt. Sta. Introd. in 1915. Ben Davis x McIntosh; cross made in 1898. Fruit: large; roundish-oblate; ribbed uniform; skin attractive red, darkly and obscurely striped; bloom heavy; flesh white, slow to discolor on exposure to air, fine-grained, crisp, tender, juicy, subacid, quality good; hangs well on tree; stands handling well; season November to February; dessert, cooking, and market usage; resembles McIntosh. Tree: early and annual bearer; very hardy; heavy cropper, susceptible to scab and mildew.

Cowin Rome Beauty (*Double Red Rome, Rome Beauty Double Red, Stark Double-Red Rome Beauty*).—Orig. in Wapato, Wash., by E.E. Cowin. Introd. in 1927. Bud mutation of Rome Beauty; discovered in 1917. Fruit: skin all red; blush type; colors early; keeps very well in storage. Tree: standard type; similar to Rome Beauty; bears annually and heavily.

Cox Rome Beauty (*Dark Red Rome Beauty*).—Orig. in Proctorville, Ohio, by U.T. Cox. Introd. about 1925 by Stark Bros. Nurseries & Orchards Co., Louisiana, Mo. Considered to be a bud mutation of Rome Beauty. Fruit: skin dark red; otherwise same as parent. Tree: identical to Rome Beauty.

Crab C.—See rootstocks for apple.

Crandall.—Orig. in Urbana, Ill., by Charles S. Crandall, Illinois Agr. Expt. Sta. Introd. in 1952. Rome Beauty x Jonathan; cross made in 1914; selected in 1925; tested as Ill. 1. Fruit: medium to small; roundish oblate; skin red, thick; dessert and cooking usage; ripens before Winesap; keeps well in storage through April. Tree: productive; comes into bearing early; blooms in midseason; relatively disease resistant. Named in memory of the late Professor Crandall.

Creswell.—Orig. in Eugene, Ore. Introd. about 1936. Parentage unknown; discovered in 1927. Fruit: quality very good, superior to Spitzenburg; matures 3 to 4 weeks earlier than Spitzenburg, which it resembles. Tree: vigorous; upright, spreading, large, hardy; blossoms midseason with Golden Delicious.

Crimson Cox.—Orig. in Carey, Herefordshire, England, by John Harris of Haynes Farm. Introd. about 1928; introd. in Canada about 1934. Bud mutation of Cox's Orange Pippin; discovered about 1913. Fruit: resembles parent but skin practically covered with crimson, which tends to dull on keeping; generally regarded as slightly inferior to parent in flavor. Tree: similar to parent.

Crimson Gold.—Orig. in Ettersburg, Calif., by Albert F. Etter. Introd. in 1944. Parentage unknown. Fruit: size medium; round; skin red; flesh white.

Crimson Spy.—Orig. in Coburg, Ontario, Canada, by S.B. Hathaway. Introd. in 1931. Considered to be a bud mutation of Northern Spy. Fruit: skin almost solid crimson; otherwise, very similar to Northern Spy.

Crimson Winesap.—See Mosebar Winesap.

Criterion.—Orig. in Lower Yakima Valley near Parker, Wash., by Francis M. Crites, Wapato, Wash. Introd. in 1972. Plant patent 3505, assigned to Carlton Nursery Co., Forest Grove, Ore. Unknown seedling. Fruit: large, uniform; conic, ribbed; skin yellow with occasional pink blush, medium thick, tough, smooth, glossy, waxy; flesh cream color, flavor mildly sweet, firm, juicy, aromatic; good shelf life; for dessert and processing; ripens about 9 to 14 Oct. in central Washington. Tree: large, standard growth; spreading; vigorous; hardy; productive.

Currie.—Orig. in Ottawa, Ontario, Canada, by Canada Dept. Agr., Res. Sta. Seed planted in 1903; first fruited in 1916. Fruit: large, conical; Introd. in 1921. Northern Spy o.p. Flesh straw-colored, soft; flavor spicy, unlike that of Northern Spy, but resembling it in color and shape; ripens in early fall.

Custer.—Orig. in Mandan, N.D., by W.P. Baird, U.S. Northern Great Plains Field Sta. Introd. in 1952. Red Siberian crab x Wealthy; selected in 1939. Fruit: small; skin striped red on greenish-yellow ground color; sweet, pleasant to eat; ripens latter part of August or first of September, resembles Roselin. Tree: very hardy; medium upright; prolific bearer.

D.B.—See Viking.

Dakota.—Orig. in Mandan, N.D., by William A. Oitto, Northern Great Plains Res. Center. Introd. in 1965. Wealthy x Whitney crab apple; tested as Mandan 42-13. Fruit: large, up to 3 inches in diam.; round-oblate; stem slender, long; skin an attractive deep red over yellow, sometimes with very dark red splotches; flesh creamy yellow, crisp, juicy, firm, fine-grained, sprightly subacid; excellent for dessert and culinary uses; ripens in mid-September; keeps in common storage for several weeks. Tree: spreading; vigorous; productive; hardy. Primarily for the home garden.

Daliguy Jonagold.—Orig. in France; discovered by the Davodeau-Ligonniere Nursery. Plant patent 6148; 12 Apr. 1988. A red fruit mutation of Jonagold.

Daliter.—See Elton.

Dana Red Delicious.—Orig. in Dana, N.C., by Ben Lytle. Introd. in 1982 by Van Well Nursery, Wenatchee, Wash. Bud mutation of Oregon Spur Red Delicious. Fruit: in North Carolina is larger than parent and colors earlier, blush-type color; ripens 10 days before parent. No longer propagated.

Daniels Red Duchess (*Bardwell Red Duchess, Minnesota Red Duchess, Stark's Red Duchess*).—Orig. in Excelsior, Minn., by William Bardwell. Assigned to and introd. in 1925 by Franc P. Daniels, Daniels Nursery, Long Lake, Minn. Bud mutation of Duchess of Oldenburg; discovered in 1902. Fruit: skin color attractive, being solid red with faint striping; otherwise, resembles parent. Tree: less vigorous than regular Duchess of Oldenburg; low, wide, crotched, and spreading.

Daniels Redstreak.—Orig. in Oswegatchie, N.Y., by Robert Daniels. Introd. in 1964 by Fred L. Ashworth, nurseryman, Heuvelton, N.Y. Parentage unknown; discovered about 1960. Fruit: skin red; flesh red; good quality; ripens early; resembles Early McIntosh. Tree: hardy.

Dark Red Rome Beauty.—See Cox Rome Beauty.

Dark-Red Staymared.—See Staymared.

Davey.—Orig. in North Grafton, Mass., by S. Lothrop Davenport. Introd. in 1950. Plant patent 906; 20 Dec. 1949; assigned to Kelly Brothers Nurseries, Dansville, N.Y. McIntosh o.p.; discovered in 1928. Fruit: large; skin red, highly colored; hangs well on tree; ripens after McIntosh; keeps all winter;

flesh quality and flavor equal to Baldwin, flavor distinctive; resembles Wealthy in shape, but skin more highly colored. Tree: bears earlier and more regularly than Baldwin; resistant to scab. Awarded a first-class certificate in 1945 by the Massachusetts Hort. Soc.

Davies.—Orig. in Norfold, St. Lawrence County, N.Y., by George Davies. Introd. in 1961. Yellow Transparent o.p.; selected in 1936. Fruit: flavor pear-like; ripens very early; resembles Yellow Transparent. Tree: hardy.

Davis.—See GalaSupreme.

Dawn.—Orig. in Saskatoon, Saskatchewan, Canada, by C.F. Patterson, Univ. of Sask. Introd. in 1959. Columbia x Melba; selected in 1944; tested as Sask. A55-101. Fruit: up to 1 3/4 inches in diam. at Saskatoon under non-irrigated field conditions; skin well covered with light crimson; mildly acid; quality good for dessert and cooking; ripens 10 days before Rescue. Tree: production heavy; extremely hardy.

Dayton.—Orig. in Urbana, Ill., by the Purdue, Rutgers, and Illinois (PRI) cooperative breeding program. Introd. in 1987 by S.S. Korban, J. Janick, E.B. Williams, and F.H. Emerson. Plant patent 5584; 3 Dec. 1985. Assigned to Newark Nurseries, Hartford, Mich. NJ 123249 x PRI1235-100; cross made in 1967. Pedigree includes Melba, Red Rome, Jonathan, Wealthy, Starr, Rome Beauty, and *Malus floribunda* 821. Selected in 1976; tested as Co-op 21; named in honor of retired breeder D.F. Dayton. Fruit: medium-large, 75 mm in diam.; roundish to slightly oblate; 90% of surface bright medium red; smooth, very glossy and attractive; flesh pale yellow, fine-grained, firm, crisp, juicy; flavor mildly subacid; quality very good; a summer dessert apple ripening 4 weeks before Delicious; in 1C storage, retains quality up to 4 weeks. Tree: moderately upright and vigorous; field immune to apple scab; moderately resistant to powdery mildew and fire blight; fruit hang on the tree 2 weeks after ripe without losing firmness or dessert quality.

Delawine.—Orig. in Port Washington, Ohio, by W.F. Hines. Introd. in 1948 by the Henry Field Seed and Nursery Co., Shenandoah, Iowa. Delicious x Stayman Winesap. Fruit: shape, size, and color of Delicious; flesh juicy, flavor of Stayman Winesap. Tree: strong grower; spreading; foliage heavy; annual bearing.

Delbardestivale®.—A red mutation of Delcorf, discovered in France and Netherlands; typical color of Delcorf but with more extended blush, allowing harvest in a single pick. All other characteristics same as Delcorf.

Delco.—See Kidd's Orange.

Delcon.—Orig. in Mountain Grove, Mo., by Paul H. Shepard, Missouri State Fruit Expt. Sta. Introd. in 1948. Conard x Delicious; cross made in 1916; selected in 1929; tested as A-1286. Fruit: size medium when not thinned; shaped like Chenango, elongated, conical; skin light to dark red over yellow; flesh fine-grained, juicy, good flavor; ripens about 10 days before Delicious; keeps well in storage; resembles Delicious. Tree: dwarf to semi-dwarf, branches drooping; for close planting and home orchards.

Delcorf.—Bred and introd. by Georges Delbard Nursery, Malicorne, France. Golden Delicious x Stark Jongrimes. Fruit: size medium to medium-large; skin very smooth with few lenticels, half covered with bright red stripes over bright yellow ground; shape upright cylindrical; flesh yellow, crisp, juicy; aromatic; slightly astringent; fruit firm, not subject to bruising. Harvest in late August, 2 weeks before Gala; sensitive to premature fruit drop. Storage life 1 month; does not become mealy. Tree: medium vigor; branches split off easily in young trees; productive, tendency to biennial bearing. Sets fruit on 1-year-old wood.

Delistein.—Orig. in Newton County, Ark., by Herman Smith. Introd. about 1940. Parentage unknown, but thought to be Delicious x Gravenstein; however, the original tree was received as a seedling from a nursery located in Farmington, Ark.; discovered in the late 1920s. Fruit: skin red over green. Apparently obsolete.

Delkistar.—See Regali.

Dewar.—See Starkspur Ultramac.

Dexter Winesap.—See Mosebar Winesap.

Dieterich Starking Delicious.—Orig. in Watsonville, Calif., by L.G. Dieterich. Introd. in 1960. Bud mutation; discovered in 1950 as a whole tree in an orchard of Starking Delicious. Fruit: skin solid red; flesh firm, flavor sweet; keeps well in cold storage. Tree: standard type of growth, now obsolete.

Discovery.—Orig. in Suffolk, Essex, England, by John F.P. Matthews, Matthews Fruit Trees. Introd. in 1962. Worcester Pearmain o.p., discovered in 1955. Fruit: size medium; usually round, but sometimes flat; skin orange-red, attractive, thick; flesh white, crisp, quality very good; fruit has tendency to crack; ripens August to September. Tree: size medium; round headed, vigor moderate; very hardy; slow to come into production; tip bearer but also crops on spurs; tolerant to mildew.

Dodd.—Orig. in De Peyster Township, St. Lawrence County, N.Y., by Howard Dodd. Introd. in 1962 by Fred L. Ashworth, nurseryman, Heuvelton, N.Y. Tolman Sweet o.p.; Fameuse may have been the pollen parent; selected about 1958. Fruit: banana-flavored; resembles Winter Banana; keeps well into the winter months. Tree: hardy.

Dominant.—Orig. in Sebastopol, Calif., by Wm. E. Silva. Introd. in 1965. Reported to be Delicious x King, selected in 1965; tested as Silva 3. Fruit: large; skin pink blushed on cream background; flesh firm, juicy, aromatic; stem long, slender; matures in late August; for fresh market. Tree: large, upright.

Dorsett Golden.—Orig. in Nassau, New Providence Islands, Bahamas, by Mrs. I. Dorsett. Introd. in 1964 by Newcomb Nursery, Homestead, Fla. Chance seedling of Golden Delicious; seed planted in 1953; selected in 1959; imported into the U.S. in 1961. Fruit: resembles Golden Delicious, but poorer coloring; often seedless. Tree: produces crops in a tropical climate where no temperatures occur below 45F; tendency to leaf out late in southern U.S. following very warm winters.

Double Early Red Delicious.—See Imperial Delicious.

Double-Red Baldwin.—See Sawyer Baldwin.

Double Red Delicious.—See Starking Delicious.

Double-Red Duchess.—See Red Van Buren.

Double Red Jonathan.—See Blackjon.

Double Red McIntosh.—See Rogers McIntosh.

Double Red Rome.—See Cowin Rome Beauty.

Double Red Stayman.—See Nured Stayman.

Double Red Winesap.—Orig. in Adams County, Pa., by Adams County Nursery & Fruit Farms, Aspers, Pa. Bud mutation of Winesap; discovered in 1948. Fruit: skin with a superior red color; otherwise, with same characteristics as Winesap. Tree: like Winesap.

Double Red York.—See Yorking.

Double Red York Imperial.—See York-A-Red.

Douglas Wormless.—Orig. in Norfold, N.Y., by George Davies. Introd. in 1963 by Fred L. Ashworth, nurseryman, Heuvelton, N.Y. Parentage unknown; discovered in 1957. Fruit: medium; skin red; ripens late; keeps well in common cold storage; of the Blue Pearmain type. Tree: leaves apparently resistant to attack by the tent caterpillar insect; sometimes difficult to propagate, as grows on some rootstocks and not on others. Primary value is for use in breeding.

Downingland.—Orig. in New Madison, Ohio, by Ernest J. Downing. Introd. in 1968. Golden Delicious x Cox Rome Beauty; cross made in 1934; first fruited in 1938; tested as Downing 2; named in 1968. Fruit: size medium, usually averaging 2 1/2 inches diam. or less, may not size well under unfavorable conditions; shape resembles Golden Delicious; skin yellow, nearly covered deep red to crimson, attractive, must be left on tree to develop satisfactory color; flesh quality better than Rome Beauty; ripens after both parents; tends not to drop; late keeper in common storage, season mid-winter. Tree: bears annually; very susceptible to fire blight; develops a strong central leader, scaffold branches produce wide angles; blooms late; little or no thinning required.

Dr. Bill.—Orig. in Lethbridge, Alberta, Canada, by the Canada Dept. Agr., Res. Sta. Introd. in 1941. Patten Duchess o.p. Fruit: size medium, up to 3 inches in diam.; oblate to round oblate; skin medium red; flesh white, medium coarse, flavor mild to subacid, very good cooked, fair as dessert; ripens 1 week after Transparent. Tree: medium to tall; rounded, spreading; hardy; annually productive.

Dulcet.—Discovered by L. James Bailey, Wenatchee, Wash., about 1975; introd. in 1988; plant patent 8187; assigned to Van Well Nursery. Parentage unknown. Fruit: medium, 70 to 75 mm; skin deep dull red with faint stripe and pronounced lenticels; shape round-conic; flesh creamy white, firm, crisp; flavor mildly sweet and aromatic; harvest date early September, 4 weeks before Delicious. Tree: vigor precocious, annual bearer, appears to be resistant to powdery mildew.

Dunning.—Orig. in Geneva, N.Y., by Richard Wellington, New York State Agr. Expt. Sta. Introd. in 1938. Early McIntosh x Cox's Orange Pippin; cross made in 1923; first full crop in 1934. Fruit: size medium; skin striped red, very attractive; flesh sweet, quality good; ripens very early, first part of August; for local and roadside markets.

Earlee Grannee™ (*Early Granny, Cooper Granny Smith*).—Orig. near Brewster, Wash., by C.L. Cooper. Discovered as whole-tree sport of Granny Smith in 1979. Introd. in 1986. Plant patent 5761; 8 July 1986; assigned to TRECO®, Woodburn, Ore. Fruit: ripens 7 to 14 days before Granny Smith; requires 165 frost-free days to mature. Tree: more compact than Granny Smith. Similar to Granny Smith in other particulars.

Earlibrite Delicious.—Orig. in Zillah, Wash., by Harold Matson. Introd. in 1969. Plant patent 3025; 2 Feb. 1971, assigned to Stark Bros. Nurseries & Orchards Company, Louisiana, Mo. Bud mutation of Ryan Red. Fruit: large; conic; skin develops solid bright red blush very early; flesh characteristics same as Ryan Red; ripens mid-September in Zillah. Tree: large; spreading; vigorous; hardiness similar to Red Delicious; productivity heavy.

Earlichief™.—A whole-tree, red mutation of Redchief® Delicious, discovered in the orchard of Arden Winkel, Hartford, Mich. Introd. in 1989. Plant patent 7928; assigned to Inter-Plant. Fruit: develops full red color 2 weeks before Redchief® Delicious; otherwise identical to Redchief® Delicious.

Earlidel (*Applethorpe Earlidel*)—Orig. at Applethorpe, Queensland, Australia, by Granite Belt Hort. Res. Sta. Released in 1988. Delicious x Early McIntosh. Tested as 35-155 and GB 155. Fruit: medium, 60 to 70 mm; skin 90% to 100% cherry red with darker red overstripes; round to slightly conical; flesh creamy white, firm, medium texture; subacid, sweet, pleasant flavor. Ripens 1 week before Gravenstein, 4 to 6 weeks before Delicious; shelf life about 10 days. Tree: more vigorous than Delicious; precocious; regular bearing; flowers early, with Granny Smith.

Earligold.—Orig. in Selah, Wash., by Richard A. Harvey. Introduced in 1979. Plant patent 4820; assigned to Columbia & Okanogan Nursery, Wenatchee, Wash. Discovered in 1971 as a chance seedling in a Golden Delicious orchard. Fruit: size medium; uniform roundish-conical; skin smooth, yellowish-green; flesh white, smooth, high acid level; good storage life; excellent for processing; ripens 1–8 Aug. Tree: similar to Golden Delicious in shape and hardiness; vigor moderate; productive, self-pollenizing.

Earlired Delicious (*Ulrich Delicious*).—Orig. in Pateros, Wash., by Edmund W. Heath, Heath's Nursery. Introd. in 1954. Bud mutation of Starking Delicious; discovered in orchard of Edward Ulrich. Fruit: red skin color develops early; flesh also early whitening; otherwise resembles Starking Delicious. Tree: standard type; identical to Delicious.

Earlistripe Red Delicious.—Orig. in Pateros, Wash., by Calvin L. Cooper. Introd. in 1966. Plant patent 2606; 15 Mar. 1966; assigned to Columbia Basin Nursery, Quincy, Wash., Heath Nursery, Brewster, Wash., and May Nursery, Yakima, Wash. Parentage uncertain, possible bud mutation of Wellspur Delicious; discovered in 1960. Fruit: resembles Top Red; 3 3/8 inches in diam.; skin smooth, thick, tough, waxy, red with pin stripes, becoming darker as fruit ripens with blushed effect; colors 10 to 14 days earlier than Starkspur and has more intense blush; fruit that drop about thinning time are already partly pigmented; flesh creamy white, tinged with carmine and sea-foam yellow, juicy, fine-grained and crisp. Tree: semi-dwarf; spur-type;

upright habit; very productive; early bearing.

Early Cortland (*Mari*).—Orig. in Geneva, N.Y., by New York State Agr. Expt. Sta., Cornell Univ. Introd. in 1982. Cortland x Lodi. Cross made in 1938; selected in 1949: tested as NY 49-19 in Geneva and Lofthus, Norway, where it was named Mari by Olaf Einset, formerly of the Geneva Station. Not a mutation of Cortland, but a distinct variety. Fruit: similar to Cortland, except has considerable distinct striping and lighter green ground color, flavor more tart and ripens a month earlier than Cortland; fruit hangs well and has longer storage life than most early varieties. Tree: similar to Cortland; vigor medium; productive; susceptible to fire blight (*Erwinia amylovara*); can serve as a pollen source for other early blooming varieties. Introd. to fill a need between very early and late varieties.

Early Crimson.—Orig. in Bedford, England, by Laxton Bros. Introd. in 1931. Worcester Pearmain x Gladstone. Cross made in 1908. Fruit: small; round-conical; skin crimson; flesh white, soft, rather dry; ripens July to August. A parent of Merton Ace.

Early Delicious.—See Beacon.

Early Golden.—See Lodi.

Early Granny.—See Earlee Grannee and Goldsmith.

Early McIntosh (*Rob Roy*).—Orig. in Geneva, N.Y., by Richard Wellington, New York State Agr. Expt. Sta. Introd. in 1923. Yellow Transparent x McIntosh; seed planted in 1909; first crop produced in 1918. Fruit: resembles McIntosh but is less aromatic and more sprightly; ripens in mid-August; flavor excellent; good for home and local markets. Tree: vigorous; productive; biennial in habit, tendency toward small size unless thinned very early.

Early Red One.—Orig. in Wenatchee, Wash., by A.M. Ward. Plant patent 3556; 24 May 1974; assigned to Van Well Nursery, Wenatchee. Introd. in 1972. Bud mutation of Red King Delicious, discovered in 1966, tested as Ward apple. Fruit: size medium, conic; skin completely covered with red, colors early; flesh white, very firm, moderately juicy; ripens 10 to 12 Sept. Tree: size medium; upright; moderately vigorous and hardy; productive. Recommended for poor coloring areas only.

Early Red Rich (*Tonasket Red Delicious*).—Orig. in Tonasket, Wash., by Guy Scacco. Introd. in 1958. Considered to be a bud mutation of Starking Delicious; discovered in 1954. Fruit: skin colors early; matures early; resembles Delicious. Tree: standard type; identical to Delicious.

Early Red Stayman.—See Neipling Red Stayman.

Early Spur Rome.—A bud mutation of Lawspur Rome, discovered by Wilfred Berger, Quincy, Wash. Plant patent 7328; 18 Sept. 1990; assigned to C&O Nursery, Wenatchee, Wash. Fruit: color deep red by mid-August just as Lawspur Rome is beginning to color; flavor of Rome Beauty; handles and stores well; does not bruise easily. Tree: spur habit of growth.

Early Thompson.—A good-quality summer apple orig. at the U.S. Dept. of Agriculture (USDA) Georgia Mountain Branch Expt. Sta., Blairsville, by J.M. Thompson. Introd. in 1993 jointly by USDA and Univ. of Tennessee. NJ 125355 x NJ 6055; cross made in 1963 by L.F. Hough and Catherine Bailey at Rutgers Univ., New Brunswick, N.J.; selected at Blairsville by J.M. Thompson in 1971. Tested as 634011-96. Not patented. Fruit: firm, attractive; skin 80% bright red on pale-yellow ground; flesh white, fine, juicy. Fruit quality good to very good; flavor mild with good balance. Size medium; heavy thinning required to achieve 68 to 70 mm in diam. and to avoid biennial bearing. Ripens in early summer for pick-your-own and local markets; adapted to southern highlands. Tree: vigorous; spreading; similar to Rome Beauty in susceptibility to fire blight. Chilling requirement more than 1200 h.

Early Vance Spur.—Orig. in Auburn, Ala., by J.D. Norton, Auburn Nursery. Introd. in 1971. Plant patent 3040; 13 Apr. 1971. Limb mutation of Vance Delicious. Fruit: 3 1/4 inches in diam.; oblong, inclined to conic; skin pale yellow undercolor, highly covered with bright dark red, thin, moderately tough; flesh white, tinged with pale yellow, firm, moderately fine and juicy, sweet, highly flavored; ripens 15 July to 1 Aug., 2 to 3 weeks before parent; borne on spurs; recommended for home and commercial plantings. Tree: semi-dwarf; abundant spurs; moderate vigor; roundish or spreading; tolerant to fire blight and cedar rust.

Early Worcester.—See Tydeman's Early Worcester.

Eastman Sweet.—Orig. in De Peyster, N.Y., by Benjamin Eastman. Introd. in 1953 by Fred L. Ashworth, nurseryman, Heuvelton, N.Y. Parentage unknown; discovered in 1860. Fruit: medium; oblong; skin yellow, scarlet striping over an orange blush; cavity green, no russeting; flesh creamy white, sweet, dessert quality good; winter season; resembles Lady Sweet. Tree: vigorous; hardy; long-lived, with trees known to be 100 years old in 1962.

Edgar.—Orig. in Ottawa, Ontario, Canada, by Canada Dept. Agr., Res. Sta. Not introd. commercially; only on trial basis. McIntosh x Forest; selected in 1929. Fruit: skin dull black-red; flesh white, sometimes tinged red; quality and texture very good; storage season longer than that of McIntosh. Tree: upright; light cropper. Not recommended, as too slow to come into bearing and not as productive as McIntosh.

Edgewood.—Orig. in Ames, Iowa, by the Iowa Agr. Expt. Sta. Introd. in 1921. Salome x Jonathan; selected in 1920. Fruit: size good; flesh firm, juicy, sprightly, quality fair; ripens in mid-October; keeps well in cold storage until May or later; hangs well on tree; free of Jonathan spot and internal browning. Tree: grows slowly; should be topworked on vigorous, hardy, understock; good yields; blooms in midseason.

Edith Smith.—Orig. in Lucky Lake, Saskatchewan, Canada, by Leonard Smith. Introd. in 1964 by Lakeshore Nurseries, Saskatoon, Saskatchewan. Parentage unknown; discovered in 1940. Fruit: 2 1/2 to 3 inches in diam.; ovate, basin and cavity very shallow; skin yellow, washed on orange-red; flesh firm, texture fine; keeps well in storage. Tree: quite hardy.

Egri piros.—See Red of Eger.

Ein Shemer.—Orig. in Doar Na Shomron, Israel, by Abba Stein. Introd. in Israel in 1963; in the U.S. as P.I. 280401 in 1964. Zabidani (local variety) x Golden Delicious. Fruit: resembles Golden Delicious; medium; conical to roundish; skin greenish-yellow; flesh crisp, melting, tender, moderately juicy, flavor subacid, quality good; ripens in Israel beginning of July. Tree: very precocious; productive; low chilling requirement.

Elan.—Orig. from breeding program at Inst. for Hort. Plant Breeding, Wageningen, The Netherlands. Introd. in 1984; plant breeders rights granted 20 Nov. 1989. Golden Delicious x James Grieve. Fruit most resembles Elstar, but larger, redder, smoother-skinned, and less acid; round conical; flesh medium firm, fine-grained. Harvest 1–5 days after Elstar, requiring two or more pickings; short shelf life; can be stored only for a few weeks. Tree: medium vigor; precocious; tip bearing; very productive; regular bearing; upright spreading.

Elbee (*Galagolden*).—Orig. at Untervaz, Switzerland, by Leo Basser. Introd. in 1976. U.S. plant patent 5947. A partially red mutation of Golden Delicious. Fruit: skin greenish-yellow with reddish-brown blush. Russets less than Golden Delicious. Resembles Golden Delicious in all other respects.

Elista™.—A natural mutation of Elstar found in the orchard of M. Peter in France in 1983. Differs from the parent variety only in having much more and brighter red-striped skin surface, which permits complete harvest in two pickings. Storage until February if picked early.

Elliott Spur Golden.—Orig. in Wapato, Wash., by M.W. Elliott. Introd. in 1964. Bud mutation of Golden Delicious; discovered in 1962. Fruit: characteristics of Golden Delicious; Tree: spur type of growth, internode length being about 1/2 to 2/3 that of a standard tree.

Elstar (*Lustre Elstar™*)—Orig. in Wageningen, The Netherlands, by T. Visser, Inst. for Hort. Plant Breeding. Introd. in 1972. U.S. plant patent 6450; 6 Dec. 1988; assigned to Carlton Plants, Dayton, Ore. Golden Delicious x Ingrid Marie; cross made in 1955. Fruit: medium to large, axial diam. 60 to 70 mm, transverse diam. 70 to 80 mm; shape round-conic; good color, yellow with very attractive orange-red stripe; more susceptible to sunburn than Gala; performs best in cool districts; flesh firm, white, rather coarse; very good flavor but acid at harvest time; harvest date 3 1/2 weeks before Golden Delicious; requires 2 weeks storage to develop acceptable flavor. Tree: very vigorous; precocious cropping; less productive than Gala, with some tendency to biennial cropping; no preharvest drop; diploid, produces viable pollen; foliage not shed until January. More susceptible to powdery mildew than Granny Smith and Jonagold; scab susceptible; fruit very susceptible to *Phytophthora* rot.

Elta.—Orig. in Brookings, S.D., by N.E. Hansen, South Dakota Agr. Expt. Sta. Introd. in 1927. Wealthy o.p. Fruit: small, averaging about 2 1/2 inches in diam.; round, conical; skin light orange-yellow, striped with red; sweet; season late fall. No longer propagated.

Elton (*Daliter*).—A red-fruited sport of Elstar, discovered by M. Peter, Loire Atlantique, France. Fruit: bright red-orange, striped. Otherwise fruit and tree are identical with Elstar. Heavily planted in France.

Empire.—Orig. in Geneva, N.Y., by Roger D. Way, New York State Agr. Expt. Sta. Introd. in 1966. McIntosh x Delicious. Cross made in 1945; first fruited in 1954; tested as N.Y. 45500-5 and sometimes designated as M-5. Fruit: medium; round to somewhat blocky; skin dark red, up to 100% colored, heavy waxy bloom, attractive; flesh cream, crisp, juicy, subacid, aromatic, quality excellent; strictly a fresh-market variety and not suited for processing; storage quality good; ripening 2 weeks after McIntosh. Tree: vigorous, upright spreading; crotch angles especially broad and strong; easily pruned; begins to bear at 5 to 6 years of age; annual bearer; more productive than Delicious; fruits are evenly spaced all along branches on the 2- to 3-year-old wood; not self-fruitful; can serve as a cross-pollenizing variety with other varieties that bloom early and in midseason.

Empire Red.—Orig. in Grand Forks, British Columbia, Canada, by Robert Campbell. Introd. in 1942. Plant patent 608; 23 Nov. 1943; assigned to Stark Bros. Nurseries & Orchards Co., Louisiana, Mo. Parentage unknown; selected in 1928. Fruit: skin almost solid bright red, thin, tough, glossy, dots whitish; flesh white, streaked with red, very firm, crisp, juicy, mildly subacid; ships well; cooking and dessert uses; ripens early; resembles McIntosh. Tree: vigorous; upright, tall; rapid growing; hardy; productive; regular bearing, no longer propagated.

Empress.—Orig. in Geneva, N.Y., by R.D. Way, K.G. Livermore, and R.C. Lamb, New York State Agr. Expt. Sta., Cornell Univ., and by L.F. Hough, Rutgers Univ, New Brunswick, N.J. Introd. in 1988. Plant patent 7118. Jonamac x Vista Bella; cross made in 1969; selected in 1977; tested as NY 651. Named for its similarity in size, appearance, and quality to Empire, but ripens 6 weeks earlier. Fruit: medium, mostly about 60 to 65 mm in diam.; shape round and symmetrical; color attractive, nearly totally blushed red and without stripes; skin covered with a heavy waxy bloom; flesh light yellow; texture semi-firm; softens rapidly; flavor subacid, with a pleasing aroma; a late summer apple, ripens 6 weeks before Delicious; at –0.5C, stores 1 month. Tree: only 70% as large as many vigorous varieties; easy to manage in the orchard; heavily cropping; annual bearing; no known resistances to diseases and insects. Bloom: late midseason, between Delicious and Golden Delicious; diploid, produces viable pollen.

Empress Spur Golden.—Orig. in Manson, Wash., by Graydon Templin. Introd. in 1965; assigned to Columbia Basin Nursery, Quincy, Wash. Bud mutation of Golden Delicious; discovered in 1962. Fruit: similar to parent. Tree: heavy spur type growth habit; highly productive.

Enterprise (*Co-op 30*).—A high-quality winter apple with multiple disease resistances, orig. in West Lafayette, Ind., by a cooperative breeding program consisting of the Agricultural Expt. Stations of Indiana, New Jersey, and Illinois. Introd. in 1993 by J.A. Crosby, J. Janick, P.C. Pecknold, Joseph Goffreda, and S.S. Korban. U.S. plant patent 9193; assigned to Purdue Res. Foundation, West Lafayettte. Parentage: PRI 1661-2 x PRI 1661-1; pedigree includes McIntosh and

Delicious. Tested as PRI 2693-1. Fruit: size medium to large, 70 to 75 mm; skin glossy, very bright, dark red on green-yellow to deep yellow ground; flesh pale yellow to cream-colored; texture medium fine, crisp, breaking; flavor moderately acid. Harvest season 2.5 weeks after Delicious; fruit hangs well on tree. After 6 months in refrigerated storage, excellent texture, mildly acid, slightly spicy, full flavored. Immune to apple scab (*Vf* gene); resistant to fire blight and cedar apple rust; moderately tolerant to powdery mildew. Tree: moderate to high vigor; spreading; cropping moderate, annual; bloom late; fruit borne singly on moderate-length spurs.

Erickson.—Orig. in Aitkin, Minn., by John and Charles Erickson. Introd. in 1923. Parentage unknown; seed planted in 1905; selected in 1910. Fruit: large; skin green; ripens early; resembles Hibernal. Tree: very hardy; annual bearer.

Erlijon.—Orig. in Coin township, Carroll County, Ark., by Ralph Banta. Introd. in 1968 by Merle J. Lucas, Green Forest Nursery, Green Forest, Ark. Parentage unknown. Fruit: 2 1/2 to 3 inches in diam.; skin yellow ground color covered with dark red, attractive; flesh very white, juicy, crisp, firm, flavor slightly resembling Jonathan; keeps fairly well, better than Transparent or Lodi but not as well as Jonathan; ripens 2 to 3 weeks earlier than Jonathan. Tree: semi-dwarf; similar to Jonathan in early growth, but with more vigor and more upright, slowing down and spreading at maturity.

Etter's Gold.—Orig. in Ettersburg, Calif., by Albert F. Etter. Introd. in 1944. Plant patent 659; 28 Aug. 1945; assigned to George C. Roeding, Jr., California Nursery Co., Fremont, Calif. Parentage unknown. Fruit: large; round, slightly ribbed; skin golden yellow, glossy; basin shallow; calyx tightly closed; stem medium slender; flesh light yellow, crisp, mildly subacid, sprightly; core small; keeps well; ripens in October; resembles Wagener.

Evarts Delicious.—Orig. in the orchard of Russell Evarts, Matamora, Mich., by Roy E. Gibson, Greening Nursery Co., Monroe, Mich. Introd. about 1950. Bud mutation of Delicious; discovered by Gibson. Fruit: skin striped, dark red, coloring early; flesh crisp, fine-grained, juicy; resembles Starking Delicious. Tree: vigorous; upright; requires early training to produce well-formed scaffolds.

Excells Red (*Nickell*).—Orig. in Pateros, Wash., by Nickell Bros. Introd. in 1956. Bud mutation of Starking Delicious; discovered in 1952. Fruit: slightly larger than Starking Delicious; red skin color appears early; resembles Starking Delicious. Apparently no longer being propagated.

Exeter.—Orig. in Saskatoon, Saskatchewan, Canada, by C.F. Patterson, Univ. of Sask. Introd. in 1959 for home gardens. Columbia x Melba; selected in 1945; tested as A58-305. Fruit: up to 2 1/2 inches in diam.; slightly conical; skin cream overlaid with bright red, some striping, attractive; flavor pleasing, mildly subacid, quality good; ripens in early September. Tree: hardy.

Exeter Cross.—Orig. in Bristol, England, by G.T. Spinks, Univ. of Bristol, Res. Sta., Long Ashton. Introd. in 1952. Worcester Pearmain x Beauty of Bath; selected in 1952. Fruit: medium to small; skin color attractive; ripens early; characteristics intermediate between parents. Tree: moderately vigorous; upright; tip-bearing.

Extra Red Delicious.—See Starking Delicious.

Fairglo.—Orig. in Mason, Mich., by Sara Tena Fair. Introd. in 1963. Plant patent 2178; 2 Oct. 1962. Chance seedling; seed planted about 1921 by John William Bennett, father of Fair; selected about 1929 by Fair. Fruit: averages about 3 1/8 inches in diam.; globose; skin moderately thin, glossy, waxy, attractive yellow with a bright blush; flesh white, not juicy, texture mellow; quality good; ripens in August; keeping quality good to best for a summer maturing variety. Tree: produces moderate crops.

Fall Red.—Orig. in Morden, Manitoba, Canada, from cross made by C.R. Ure. Selected at Univ. of Alberta; introd. by Prairie Fruit Breeding Coop. Committee. Duchess of Oldenburg x Haralson. Fruit: medium large, 70 to 80 mm in diam.; oblong-round; skin dark red on yellow-green ground color; flesh white, medium-high acidity, good flavor. Harvest mid-September in the southern prairies of Canada;

stores well. Fruit quality suitable for juice as well as dessert. Tree: vigor moderate; tendency to biennial bearing. Moderately susceptible to fire blight, but recovers quickly. Hardy to zone 2A.

Falstaff.—Orig. in East Malling, Kent, England, by F.H. Alston. Introd. in 1989. James Grieve x Golden Delicious. Owned by Plant Breeding Intl. Fruit: medium, 65 to 75 mm; skin red stripes on yellow-green background, smooth; flesh crisp, juicy; good sugar–acid balance, very good flavor. Most resembles Jonagold, but smaller size. Matures 2 weeks after Cox's Orange Pippin. Fruit stores well in refrigerated storage. Tree: medium vigor; drooping habit, with many drooping leaves presenting lower surfaces to the light; very productive. Diploid; spur flowers open 3 days before Cox's Orange Pippin; many flowers on l-year-old wood.

Fantazja (*Phantasy*).—Orig. in Skierniewice, Poland, by Aleksander Rejman, Warsaw Agr. Univ., Dept. of Pomology, Warszawa, Nowoursynowska 166, Poland. Introd. in 1960. McIntosh x Linda; cross made in 1944; selected in 1954, tested as D.II.38. Fruit: similar to Empire in appearance; medium to large; roundish; skin very thin, smooth, with deep red blush; flesh yellow-cream, fine, very juicy, tart, quality good, flavor good; ripens after Gala in eastern U.S., just before Empire, keeps well in cold storage; recommended as a home garden and commercial variety for nearby markets. Tree: large, round; vigorous; hardy; very productive on MM 106; susceptible to scab; may be valuable for breeding purposes.

Farley McIntosh.—Orig. in the orchard of Farley Bros. Orchard & Nursery Co., Albion, Mich., by Roy E. Gibson, Greening Nursery Co., Monroe, Mich. Introd. about 1940. Bud mutation of McIntosh; discovered in 1930. Fruit: medium; similar to McIntosh except skin is blushed rather than striped. Tree: low yields.

Farley Northern Spy.—Orig. in the orchard of Farley Bros. Orchard & Nursery Co., Albion, Mich., by Roy E. Gibson, Greening Nursery Co., Monroe, Mich. Introd. about 1924. Bud mutation of Northern Spy; discovered about 1920; tested as Farley 9. Fruit: large; skin well colored with red, with indistinct striping; flesh firm, crisp, juicy, flavor excellent; typically Northern Spy. Tree: comes into bearing earlier than parent.

Faurot.—Orig. in Mountain Grove, Mo., by Missouri State Fruit Expt. Sta. Introd. in 1935. Ben Davis x Jonathan; selected in 1920; tested as A-986. Fruit: size medium, resembling Jonathan; skin color like Jonathan, dots large and very conspicuous; quality fair; keeps late; ripens with Winesap; less likely to drop than Jonathan, which it resembles. Tree: vigorous; less susceptible to fire blight than Jonathan, resistant to canker; produces large crops.

Fechtig.—See Par-Fect Spur Criterion.

Fenton.—See Beacon.

Fiesta™ (*Malling Fiesta*).—Orig. in East Malling, Kent, England, by F.H. Alston, J.R. Stow, J.W. Palmer, L.H. Clark, R.F.V. Williams, and H.F. Ermen, East Malling Res. Sta. Introd. in 1986. Cox's Orange Pippin x Idared, crossed in 1971; selected in 1979. U.S plant patent 7956; assigned to TRECO®, Woodburn, Ore. Fruit: attractive but somewhat dull red blush and stripes on a yellow-green background; round and slightly flattened; texture very crisp and juicy; flavor pleasantly aromatic, similar to Cox's Orange Pippin; harvest in mid-September with Cox's Orange Pippin; at 1C common storage, stores until April; at 3.5C in controlled atmosphere storage, keeps crisp and juicy until May; resistant to most storage disorders. Tree: moderately vigorous, larger than Cox; wide branch angles, drooping growth habit; productive, yield is double that of Cox; thinning is necessary; fruit hangs well after it ripens. Blooms 2 days later than Cox; styles are very receptive, even at low temperatures; slightly self-fertile; diploid, pollen viable.

Fireside.—Orig. in Excelsior, Minn., by the Univ. of Minnesota Fruit Breeding Farm. Introd. in 1943. Parentage unknown; selected in 1917; selected for further testing in 1927; tested as Minn. 993. Fruit: large, 3 inches in diam.; skin medium red, lightly striped with darker red; flesh yellowish, medium coarse, medium tender, medium juicy; flavor mildly subacid; quality excellent for dessert purposes; season of use

November to April. Tree: hardy; vigorous; resistant to cedar rust but only slightly less susceptible than Wealthy to apple scab and fire blight.

Firmgold.—Orig. in Zillah, Wash., by Z.B. Howell. Plant patent 4166; 20 Dec. 1977; assigned to Claud Callahan and Marjorie McCormick, Yakima, Wash. Chance seedling growing among Starkspur Golden Delicious and Starkrimson Red Delicious. Fruit: uniform, shape like Starkrimson Red Delicious, very long stem; skin somewhat greenish, thick, smooth; flesh juicy, firm, crisp; flavor sweet, keeps 300 to 330 days in ordinary storage; harvest 2 weeks after Starkspur. Tree: medium; vigorous; spreading; only moderately productive; resistant to mildew; buds and blossoms freeze resistant; larger and darker pink flowers than Starkspur Golden.

Flamespur Red Rome (*Alvah, Taylor*).—Orig. in Selah, Wash., by Alvin Taylor. Introd. in 1982. Plant patent 4974; 18 Jan. 1983; assigned to McCormick Fruit Tree Co., Yakima, Wash. Mutation of Starkspur Rome Beauty; discovered in 1971. Fruit: solid red overall, in contrast to usual coloring of Red Rome; harvest 20 days before Starkspur Rome Beauty; good keeping quality. Tree: less compact than Starkspur Rome Beauty.

Flaming Rome.—See Nero Rome.

Flanagan Delicious.—See Starkspur UltraRed Delicious.

Florina (*Querina®*).—Orig. in Beaucouze, France, by Y. Lespinasse, J.M. Oliver, J. Lespinasse, and M. Le Lezec, Station d'Arborculture Frutiere, INRA. Patented in France. Made available to French growers in 1977. 612-1 [14-126(Golden Delicious x F226829-2-2) x Simpson's Giant Starking] x Jonathan. Assigned to Agri Obtentions, 78283 Guyancourt, France. Fruit: most nearly resembles Empire; medium to large; color three-quarters purplish red on yellow background, shiny waxy bloom, attractive; truncate conical; flesh cream-colored; medium firm, resistant to bruising; rather sweet, good eating quality; open calyx; harvest 2 weeks later than Golden Delicious; stores well, market season October to January, softens quickly out of storage. Tree: medium to strong vigor; early bearing; productive; harvest is easy; resistant to apple scab (*Vf* gene), moderately susceptible to powdery mildew, tolerant to rosy apple aphid and to fire blight.

Folwell.—Orig. in Excelsior, Minn., by the Univ. of Minnesota Fruit Breeding Farm. Introd. in 1922. Malinda o.p., grown by T.E. Perkins, Red Wing, Minn.; seed planted by Wyman Elliot, Minneapolis; seedling moved to the Fruit Breeding Farm in 1908; selected in 1913; tested as Minn. 237. Fruit: large; roundish oblate, irregular; skin greenish-yellow, heavily blushed, attractive; flesh firm, tender, juicy, slightly subacid, fine-grained, quality good; season similar to Wealthy. Tree: large, spreading; very vigorous; produces regularly; hardy; resembles Hibernal.

Fortune.—Orig. in Geneva, N.Y., from the Cornell Univ. breeding program (New York Agr. Expt. Sta.). Introd. in 1995. Red Spy (Schoharie cv.) x Empire; U.S. plant patent applied for. Tested as NY 429. Fruit: large to very large, round to oblong, overall red color; flesh cream, crisp, firm, subacid, and slightly aromatic, as good as or better than Empire; ripens early October, with Empire; storage 4–5 months in common storage, better than that of Empire. Tree: vigorous and upright, precocious, annual bearing, midseason bloom, diploid, hardiness and disease susceptibility comparable to Empire.

Francis.—Orig. in Esterhazy, Saskatchewan, Canada, by Francis Hermansky. Introd. in 1965 by Lad. Martinovsky, Gerald, Saskatchewan. McIntosh o.p.; selected in 1945. Fruit: large, up to 3 inches in diam.; ripens early; keeps well in storage.

Frankad.—See Adina.

Franklin.—Orig. in Wooster, Ohio, by the Ohio Agr. Expt. Sta. Introd. in 1937. McIntosh x Delicious; selected in 1936. Fruit: size above medium; resembles Delicious in shape; skin nearly covered with red, very attractive; combines flesh characteristics of each parent, flavor mild, dessert quality high; season late fall and early winter, extends McIntosh season 1 week, ripening in late September; suggested for roadside and local markets. Trees: susceptible to scab.

Frazier Goldenspur.—Orig. in Wapato, Wash., by Loren Frazier. Introd. in 1963 by

Carlton Nursery Co., Forest Grove, Ore. Bud mutation of Golden Delicious; discovered in 1961. Fruit: typically Golden Delicious. Tree: spur-type growth, with heavy spurring tendencies.

Frecon.—See Starkspur Winter Banana.

Freedom.—Orig. in Geneva, N.Y., by Robert C. Lamb, New York State Agr. Expt. Sta. Introd. in 1983. NY 18492 (Macoun x Antonovka) x NY 49821-46 [Golden Delicious x F_2 26829-2-2 (Rome x *M. floribunda* 821 x Rome x *M. floribunda* 821)]. Cross made in 1958; selected in 1965. Plant patent 5723. Fruit: large; oblate; skin covered with 80% bright red stripes on yellow background; flesh cream-colored, firm, medium-fine, tender, and juicy; flavor sprightly subacid; quality good; will keep in refrigerated storage until January; for fresh market, sauce, and juice. Tree: vigorous, spreading; very productive, precocious; hardy; blooms 15 May at Geneva or 3 days after McIntosh, good pollenizer; resistant to apple scab, cedar apple rust, powdery mildew, and fire blight, can be grown without fungicide or bacterial sprays, named to indicate this resistance.

Fremont.—Orig. in Lander, Wyo., by J.M. Steinbrech, Wyoming Agr. Expt. Sta. Introd. in 1933. Fruit: skin pale yellow with red stripes; flesh whitish, subacid; resembles Wealthy in shape, size, color, flavor, and keeping quality, but less juicy; ripens about 15 Sept.

Freyberg (*Kidd's 21*).—Orig. in Greytown, Wairarapa, New Zealand, by J.H. Kidd. Introd. there in 1958; by Georges Delbard in France in 1959. Golden Delicious x Cox's Orange Pippin. Discovered in 1939. Fruit: small to medium; round; skin golden yellow, blushed with mahogany red; resembles Golden Delicious; skin very thick, bruises slightly but firmer than Golden Delicious; flesh yellow, fine, firm, crisp, juicy, sweet, excellent quality; storage quality is good but tends to scald and shrivel; named for Lord Freyberg, commander of the 2nd New Zealand Expeditionary Force in World War II and Governor-General of New Zealand from 1946 to 1952. Tree: dwarf; vigorous; upright branches; spur-bearing; productive; connoisseur variety for home gardens; matures same season as Golden Delicious; fruit hangs well on tree.

Frimley Beauty.—Orig. in Frimley Park, Hastings, New Zealand, by Thos. Horton, Premier Nurseries. Introd. in 1916; in the U.S. in 1920 as P.I. 51171; in 1939 as P.I. 133515; and in 1940 as P.I. 139159. Bud mutation of Rome Beauty. Fruit: large, 2 3/4 x 3 1/4 inches in diam.; uniform; round; transverse outline round, slightly crowned, regular; basin smooth, shallow; stem long; skin fully red with a brilliant scarlet tone, thick; flesh pale tinged yellow, coarse, tender, juicy, subacid, bland, fair quality; hangs well on tree, allowing for long harvest period; storage excellent; ripens with Rome Beauty; resembles Rome Beauty in every respect except color. Tree: moderately vigorous; pendulous; long pliant branches tend to run to blind wood; bad crotch angles cause breakage of fruiting arms; strong tip-bearing growth; difficult tree to train; susceptible to mildew, scab; produces heavily; slow to commence bearing if not properly pruned.

Frostproof.—Orig. in Mineral, Va., by Max Bazzanella, Max Nursery. Introd. in 1947. Plant patent 722; 7 Jan. 1947. Parentage unknown; discovered in 1930. Fruit: size medium, 2 1/4 to 3 inches in diam.; round; skin bronze, tough; flesh white, firm, spicy, subacid, flavor quince-like; core small; quality fair; ripens with Winesap; resembles Rustycoat. Tree: bears regularly; blooms 30 days after Winesap; upright, spreading.

Fruitland Delicious.—Orig. in Fruitland, Idaho, by Fruitland Nursery. Not introduced commercially. Bud mutation of Delicious, discovered previous to 1950. Fruit: 3 1/2 to 4 inches in diam.; less conic and more blocky than Delicious; skin blush, faintly striped; more bitter pit than Delicious; 2-4-4 periclinal cytochimera, chimeral nature determined by H. Dermen, confirmed by J. Einset and C. Pratt at the New York State Agr. Expt. Sta.; ripens 10 Oct. in Geneva, N.Y. Tree: size and shape same as Delicious; less productive than Delicious. Recommended for possible tetraploid parent.

Fuji.—Orig. in Fujisaki, Minami-Tsugaru, Aomori Prefecture, Japan, by Tohoku Hort. Res. Sta., Morioka City, Iwate Prefecture, Japan. Introd. there in 1962; in the U.S. as

P.I. 282898 in 1962. Ralls Janet x Delicious; selected in 1957; tested as Tohoku 7. Fruit: resembles Delicious in color; similar to Ralls in flesh characteristics; mild and sweet flavor; very high sugar, low acid; liable to produce water core; keeping quality good compared to Ralls; ripens 10 Nov. in Japan. Tree: tall form, spreading growth, willowy; leaves susceptible to Alternaria spot (*Alternaria mali*), apple scab, and fire blight.

Fukunishiki.—Orig. in Fukutami, Kuroishi-City, Aomori Prefecture, Japan, by Aomori Apple Expt. Sta. Introd. in U.S. in 1952 as PI 199526. Ralls x Delicious. Cross made in 1933, selected in 1947. Fruit: medium size; round or roundish-oblate; skin light red over green ground color with broken red stripes (in Japan brownish-red); flesh crisp, pale yellow, tender, juicy and mild flavored; requires a long warm season to ripen (early November); dessert variety; flavor and keeping quality superior to Ralls. Tree: large, spreading; vigorous, productive; susceptible to root rot. No longer grown commercially in Japan or elsewhere.

Fulford Gala.—See Regal Gala.

Fyan.—Orig. in Mountain Grove, Mo., by the Missouri State Fruit Expt. Sta. Introd. in 1935. Ben Davis x Jonathan; selected in 1920; tested as A-422. Fruit: large to very large; nearly round; skin fairly well colored red over greenish-yellow, attractive, tough, coarse, light-blue bloom; flesh firm, fine-grained, juicy, tart, flavor good; fine keeper; ripens 2 weeks after Jonathan, which it resembles. Tree: spreading; very vigorous; regular bearer of large crops; somewhat tolerant to disease.

Gala (*Kidd's D.8*).—Orig. in Greytown, Wairarapa, New Zealand, by J.H. Kidd. Introd. in 1960. Kidd's Orange x Golden Delicious. Selected in 1939. Introduced in U.S. in 1973. Plant patent 3637; 15 Oct. 1974; assigned to Stark Bros. Nurseries and Orchards Co., Louisiana, MO. Named and introduced in 1973. Fruit: medium, 2 1/4 x 2 3/4 inches, uniform; oval, round in transverse outline, crowned to slightly crowned; basin of medium width and depth; stem slender, very long; cavity wide and deep; skin pale yellow to golden yellow, heavily striped with red, glossy, thin, does not bruise readily, completely free from blemish, no russet; flesh yellow, fine, firm, crisp, juicy, sweet, low acid, quality excellent; long stem good for spray penetration; free from bitter pit; better storage quality and longer shelf life after removal from storage than Cox's Orange Pippin; susceptible to apple mosaic, ripens two weeks before Delicious. Tree: vigorous; long pliant branches; heavy spur-bearing habit similar to Delicious; foliage is healthy with tendency to leaf burn; yields are precocious, heavy, and regular.

Gala Must.—See Regal Gala.

Galagolden.—See Elbee.

GalaSupreme™ (*Davis cv.*).—A chance seedling (not a Gala sport) found growing from the root of a Delicious tree in the orchard of Nick Davis, Wenatchee, Wash. Introd. about 1992. U.S. plant patent pending. Fruit: somewhat similar in appearance to Gala with yellow ground covered with a red-orange stripe; size medium to large; flesh firm and crisp, cream colored in contrast to more yellowish flesh of Gala; resists browning when exposed to air; flavor sweet; maturity four weeks after Gala. Tree: vigorous, upright, spreading, resembles Gala.

Galaxy.—Orig. in Havelock North, Hawkes Bay, New Zealand, by K.W. Kiddle. Introd. in 1988. U.S. plant patent 6955; 1 Aug. 1989; assigned to Stark Bros. Nurseries, Louisiana, Mo. Limb mutation of Royal™, which is a mutation of Kidd's D-8 Gala™. Discovered in 1985 as a single mutated spur. Fruit: solid cherry red color over the entire surface with frequent, indistinct darker red overstripes. Color develops earlier than Tenroy Gala™. Galaxy is the same as Gala™ in the following characteristics: tree vigor, blossom period, crop yield, disease resistance, and fruit shape, size, and taste.

Galbraith Baldwin.—Orig. in Amherst, Mass., by Floyd Galbraith, Massachusetts Agr. Expt. Sta. Introd. in 1948. Bud mutation of Baldwin; selected in 1934. Fruit: skin highly colored red; otherwise resembles Baldwin. Tree: similar to Baldwin.

Galer.—Orig. in Hood River, Ore., by Walter Galer. Introd. in 1951 by Carlton Nursery Co., Forest Grove, Ore. Open-pollinated seedling, possibly Delicious or

Spitzenburg; selected in 1945. Fruit: skin colors well in interior valleys, partially colored in coastal areas; ripens 20 to 30 days before Delicious; flesh juicy, tart, excellent for eating fresh and for culinary purposes. Tree: moderately vigorous; similar to Esopus Spitzenburg; bark on limbs often conspicuously rough and scaly; blossoms open late, with Northern Spy but before Rome Beauty.

Gallen.—Orig. in Jaatsi, Tyrvää, Finland, by Matilda Gallen. Introd. in Finland in 1930. Introd. into the U.S. in 1955 as P.I. 224609. Parentage unknown. Fruit: medium to large, 3 inches in diam.; skin light green, but when fully ripe, pale yellow with white markings, tough; flavor good; good combination of sugar and acid; better eating quality than Lodi; susceptible to water core. Tree: vigorous; leafs out in early spring; blooms early.

Garden Delicious.—Orig. in Modesto, Calif., by C.F. Zaiger. Plant patent 3808; 25 Nov. 1975. Fruit: medium size, oblong with ends flattened; flesh is firm, juicy, sub-acid; skin medium thick, light yellow with slight blush of light red; full red color in cooler climates; ripens 15 Sept. Tree: semi-dwarf, about 6 to 8 feet; upright; nonvigorous; very productive. Very good for cooking and canning.

Gardner Delicious.—Orig. in Yakima, Wash., in the orchard of R.L. Gardner by J.T. Bregger, Ithaca, N.Y. Introd. about 1940. Bud mutation of Delicious; discovered about 1932. Fruit: skin with solid blush, no striping; resembles Richared. No longer being propagated by commercial nurseries.

Garland.—Orig. in Morden, Manitoba, Canada, by C.R. Ure, Canada Dept. Agr., Res. Sta. Introd. in 1961. Melba x Haralson; selected in 1949; tested as Morden 364. Fruit: medium to large; skin solid red, appearance good; ripens in late September, keeping to early January; rated good as dessert and very good for sauce, pie, and baking; resembles Bancroft, McIntosh, Lobo. Tree: moderately vigorous; healthy; hardy; scaffold branches strong; very productive.

Garrison.—Orig. in Mandan, N.D., by W.P. Baird, Northern Great Plains Field Sta. Introd. in 1957. Duchess of Oldenburg x Starking Delicious; selected in 1949; tested as Mandan 49-5; sibling of Thorberg and Mandan. Fruit: medium, averaging about four per lb; round; stem long; skin attractive red with carmine stripes; flesh medium firm, subacid, juicy, pleasant tasting with some of the flavor of Delicious; excellent dessert quality, good for culinary use; matures in early September. Tree: hardy; growth moderate; spreading.

Gavin.—Orig. in Bayfordbury, England, by the breeding program at the John Innes Inst. DA 20-2 x Merton Worcester, 1965. Named in honor of A. Gavin Brown. Fruit: medium, 60 to 65 mm in diam.; color red flush on yellow-green ground; shape oblong; flesh yellow, crisp, juicy; flavor sweet. Tree: less productive than Cox 's Orange Pippin; immune to apple scab, V_f resistance genes.

Geneva 11 (*G.11*).—See rootstocks for apple.

Geneva 65.(*G.65*).—See rootstocks for apple.

Geneva Early.—Orig. in Geneva, N.Y., by New York State Agr. Expt. Sta., Cornell Univ. Introd. in 1982. Quinte x Julyred. Cross made in 1964; selected in 1973; tested as NY 444. Fruit: large for its season, round-oblate; flesh cream-colored; subacid and aromatic, eating quality excellent, semifirm to soft; skin 60% to 100% red, mostly solid blush, sometimes striped; storage life less than 1 week. One of earliest ripening of all varieties, 1 Aug. in Geneva, sometimes as early as 13 July, requires two to three pickings. Tree: vigor good; effective pollenizer for early and midseason varieties; highly susceptible to fire blight and cedar-apple rust (*Gymnosporangium juniperi-viriginianae*) and apple scab (*Venturia inaequalis*). For home orchards, pick-your-own, or roadside markets.

Geneva McIntosh.—Orig. in Geneva, N.Y., by L.G. Klein, New York State Agr. Expt. Sta. Introd. in 1959. Bud mutation of McIntosh; discovered in 1956. Fruit: skin solid bright red blush with no striping; color considered to be brighter, and fruit appears to be more uniform in shape, than Rogers McIntosh; resembles Rogers McIntosh.

Geneva Ontario.—Orig. in orchard of the New York State Agr. Expt. Sta., Geneva;

accessioned in 1910 from T.C. Maxwell Nursery, Ontario, N.Y. Apparently not introd.; included here because of its primary interest to fruit breeders. Considered to be a bud mutation of Ontario. Fruit: large. Produces diploid gametes; a periclinal cytochimera 2–4–4–4. Chimeral nature determined at the New York State Agr. Expt. Sta.

Geneva Red Spy.—See Greene Spy.

Gensinger Winesap.—See Mosebar Winesap.

George.—Orig. in Edmonton, Alberta, Canada, by R. George. Introd in 1948. McIntosh o.p. Fruit: medium to large, up to 3 inches in diam.; oblong conic; skin solid red; flesh yellowish-white, juicy, mild, subacid, pleasant, firm; quality good for dessert and culinary uses; season midwinter to late winter. Tree: vigorous; hardy; well formed.

Gertie.—Orig. in East Bay, Manitoba, Canada, by Ed Robinson, Gaybird Nursery, Wawanesa, Manitoba. Introd. in 1945. Gertrude o.p.; discovered in 1940. Fruit: 2 1/4 x 2 inches in diam.; skin bright red with darker red streaks, attractive; flesh white, firm, tart; ripens 15 Sept. Tree: spreading; hardy; vigorous.

Giant Jonathan.—See Adams Jonathan.

Giant Rome.—See Loop Rome.

Giant Spy.—See Loop Spy.

Giant Transparent.—See Perrine Yellow Transparent.

Giant Wealthy.—See Loop Wealthy.

Giant Yellow Transparent.—See Perrine Yellow Transparent.

Gibbs A.—Orig. in Tift County, Ga., by Major C. Collins, Tifton, Ga. Introd. in 1977. Unknown seedling; discovered in early 1960s. Fruit: 3 x 2 3/4 inches; irregularly crowned and subconic, with one carpel prominently ridged into a hump; skin 80% to 90% dark red, undercover greenish-yellow; flesh cream-colored, fine, and crisp, quality very good; ripens August. Tree: vigorous: a dooryard apple for the Southeast.

Gibson Golden Delicious.—See Smoothee.

Gilbert 2.—See Imperial Delicious.

Gilmore Everbearing.—Orig. in Chatham County, N.C., by Glenn G. Gilmore, North State Nursery Co., Julian, N.C. Introd. in 1946. Parentage unknown. Fruit: small-medium; skin red; flavor mild; ripens over a long time, beginning in July. Tree: moderately vigorous; low spreading; bears at early age.

Gilmore Sweetening.—Orig. in Chatham County, N.C., by Glenn G. Gilmore, North State Nursery Co., Julian, N.C. Introd. in 1945. Parentage unknown. Fruit: medium-small; skin white ground color with 25% to 50% of surface covered with bright red stripes; flavor mild and sweet; ripens mid-to late summer. Tree: moderately vigorous; open, spreading; bears at early age.

Ginger Gold™.—Orig. in Nelson Co., Va., as a seedling growing in a Winesap orchard planted in 1980 by Clyde H. and Ginger Harvey, Arlington, Va. Introd. in 1989. First crop in 1984. Plant patent 7063; 5 Dec. 1989. Assigned to Adams County Nursery, Aspers, Pa. Fruit: resembles Golden Delicious; medium-large; color green to gold, nonrusseting; shape round to oblate; flesh white, slow oxidizing, firm, juicy, good flavor, pressure test 18 to 19 lb; soluble solids, 11.5%. Harvest 5 weeks before Golden Delicious; fruit hangs well. Keeps exceptionally well in storage for an early cultivar. Tree: medium vigorous; spreading; productive; precocious; hardy; blooms late with Golden Delicious. Susceptible to fire blight and powdery mildew.

Glendale.—Orig. in Miami, Manitoba, Canada, by William Oakes, Glenelm Nursery. Introd. in 1956. McIntosh o.p.; discovered in 1948. Fruit: flesh white, sweeter than McIntosh, which it resembles.

Glenelm.—Orig. in Miami, Manitoba, Canada, by William Oakes, Glenelm Nursery. Introd. in 1952 by Portage Plains Nursery, Portage La Prairie, Manitoba. McIntosh o.p.; discovered in 1945. Fruit: large; shape of Golden Delicious, which it resembles; quality good for dessert and culinary purposes. Tree: bears annually.

Glenmary.—Orig. in Miami, Manitoba, Canada, by William Oakes, Glenelm Nursery. Introd. in 1948 by Portage Plains Nurseries, Portage La Prairie, Manitoba McIntosh o.p.; discovered in 1940. Fruit: very attractive; quality good for dessert and culinary uses; ripens in the fall; stores well; resembles McIntosh. Tree: very hardy.

Glenn Red.—Orig. in Yakima, Wash., by Roger F. Williams, Williams Fruit Co. Introd. in 1957. Bud mutation of Delicious; discovered in 1955. Fruit: skin solid red; claimed to mature about 3 weeks before other red-colored Delicious types. Tree: produces very well.

Glenora.—Orig. in Miami, Manitoba, Canada, by William Oakes, Glenelm Nursery. Introd. in 1960 by Portage Plains Nursery, Portage La Prairie, Manitoba. Parentage unknown; discovered in 1958. Fruit: round, flat; skin red, tough; flesh white, fine-grained, flavor good; stores well.

Glenton.—Orig. in Ottawa, Ontario, Canada, by Canada Dept. Agr., Res. Sta. Introd. in 1946. Northern Spy o.p.; selected in 1909; named in 1911. Fruit: large; skin with a beautiful blush; ripens just after Wealthy; flavor pleasing, semi-tart, good for eating fresh as well as for culinary uses; resembles Northern Spy but quality not as good as that variety. Tree: resembles Northern Spy; vigorous; symmetrical; bears annually.

Glenwal.—Orig. in Miami, Manitoba, Canada, by William Oakes, Glenelm Nursery. Introd. in 1958. McIntosh o.p.; selected in 1940. Fruit: very large; sweet, dessert quality good; skin greenish-yellow; resembles Rhode Island Greening. Tree: hardy.

Gloster (*Gloster 69*).—Orig. in the breeding program, Jork Fruit Res. Sta., Jork, Germany. Introd. in 1969. Glockenapfel x Richared Delicious. Fruit: similar to Delicious; large, 70 to 75 mm in diam.; color and shape similar to Richared Delicious; flesh greenish-yellow, no bitter pit; sweet-tart flavor, more acid than Delicious, fair quality; harvest late with Golden Delicious and Jonagold; calyx open, subject to moldy core; long storage life. Tree: vigorous, upright; precocious; very heavy, annual cropping, more productive than Delicious; moderate preharvest drop.

Gloster 69.—See Gloster.

Godfrey.—Orig. in Morden, Manitoba, Canada, by the Canada Dept. Agr., Res. Sta. Introd. in 1931. Patten Greening o.p.; selected in 1927. Fruit: medium to small; skin well-washed dark red; flesh white, tinged greenish, crisp, texture fine, sweet, sprightly, quality good; season October to February. Tree: vigorous; spreading; productive.

Gold-N-Rose.—See Maigold.

Golden.—See South Dakota Golden.

Golden Beauty.—Orig. in Sebastopol, Calif., by Wm. E. Silva. Introd. in 1966. Golden Delicious x Red Sheet Pippin; selected in 1965. Fruit: large; not pointed but full at stem end; skin golden with a pink blush; flesh very sweet, low acid, flavor good; keeping qualities good. Tree: blossoms late; self-fruitful; heavy scaffold limbs; recommended for California foothills and colder climates.

Golden Glory™.—A semi-spur limb mutation of Smoothee®, discovered by Dan Simmons in Ohio. Tested as DS-165. Plant patent 7851, Apr. 7, 1992; assigned to Inter-Plant. Fruit: similar to Golden Delicious, except semi-spur and very smooth skin.

Golden Hackworth.—See Yellow Hackworth.

Golden Lodi.—See Lodi.

Golden Nugget.—Orig. in Kentville, Nova Scotia, Canada, by C.J. Bishop, Canada Dept. Agr., Res. Sta. Introd. in 1964. Golden Russet x Cox's Orange Pippin; selected in 1943. Fruit: small, which limits its commercial value but not its home gardening appeal; resembles Cox's Orange Pippin. Tree: moderately vigorous with thin branches with many slender fruit spurs.

Golden Sunset.—Orig. in Winlaw, British Columbia, Canada, by H.J. Cutler. Parentage unknown. Introd. about 1932. Fruit: about the size of McIntosh; skin pale yellow with some blush, like Duchess; flesh juicy, flavor sweet; ripens in early August; keeps well in common storage. Tree: bears well annually.

Golden Supreme™.—Discovered 1960 by Warren Carnefix, Fruitland, Idaho; assigned to Walter Krause, Reedley, Calif. Tested as WK-181. Licensed to Hilltop Nurseries, Hartford, Mich., and to Willow Drive Nursery, Ephrata, Wash. Fruit: medium to large; similar to Golden Delicious; yellow with pink blush on exposed fruit; typey shape; russet resistant; firm, juicy, mild flavored; harvest 10 to 12 days earlier than Golden Delicious; ripens unevenly, two pickings required; tendency to preharvest drop; storage life up to 4 months without shrivel. Tree: vigorous, nonspur, tendency to blind wood, biennial bearing.

Golden Treasure.—Orig. in Des Moines, Iowa, by C.J. Dykstra. Introd. in 1959. Plant patent 1838; 2 June 1959. Chance seedling; discovered in 1941. Fruit: large, average 2 3/4 inches in diam.; skin yellow-green with faint pink stripes; flesh crisp, juicy, very firm; matures with Yellow Transparent, which it resembles; keeps well. Tree: medium; vigorous; bears annually with distinct characteristics of fruit substantially evenly spaced apart along the longitudinal length of its branches.

Goldenir (*Lysgolden*®).—Orig. at Angers, France, by Natl. Inst. Agr. Res. A radiation mutant of Golden Delicious; selected by L. Decourtye. Introd. in 1977; assigned to Agr. Obtentions, Guyancourt, France. Trademarked in the U.S. by Hilltop Nurseries, Hartford, Mich. Fruit: most resembles Golden Delicious; medium; strong yellow color, sometimes with red blush; conical, typier than Golden Delicious; little or no russet; skin tougher, smoother, waxier than Golden Delicious; flesh and flavor same as Golden Delicious. Better resistance to bruising than Golden Delicious. Harvest 10 to 15 days after Golden Delicious; longer storage life. Tree: more vigorous than Golden Delicious; growth habit more leggy, with some barewood; less productive; less tendency to overcrop; susceptible to powdery mildew.

Goldensheen #2.—See Stark Puregold.

Goldo.—Orig. in Brookings, S.D., by N.E. Hansen, South Dakota Agr. Expt. Sta. Introd. in 1922. Duchess x Grimes Golden. Fruit: skin yellow; quality fair; does not store well; resembles Grimes Golden. Tree: somewhat lacking in hardiness; susceptible to fire blight. No longer available from nurseries.

GoldRush (*Co-op 38*).—A high-quality, yellow, winter apple resistant to apple scab and powdery mildew, orig. in West Lafayette, Ind., by a cooperative breeding program consisting of the Agricultural Expt. Stations of Indiana, New Jersey, and Illinois. Introd. in 1993 by J.A. Crosby, J. Janick, P.C. Pecknold, Joseph Goffreda, and S.S. Korban. U.S. plant patent applied for; assigned to Purdue Res. Foundation, West Lafayettte, Ind. Parentage: Golden Delicious x Co-op 17 (Co-op 17 has complex parentage, including Winesap). Tested as PRI 2750-6. Fruit: size medium, typically 65 to 75 mm (requires thinning); larger in warmer areas; skin deep yellow, nonwaxy, thin, conspicuous russeted lenticels. Flesh pale yellow, very firm, medium coarse, crisp, breaking, juicy; flavor tart at harvest, developing very rich, well-balanced flavor after a few weeks in storage. Harvest season 25 days after Delicious; outstanding storage life. Tree: moderately vigorous, semi-spur with strong central leader. Immune to apple scab (*Vf* gene); moderately resistant to fire blight; susceptible to cedar apple rust; moderately tolerant to powdery mildew.

Goldsmith (*Early Granny, Summer Granny*).—Orig. in Elgin, South Africa. Introd. in 1975. Thought to be chance seedling of Granny Smith x Golden Delicious. Fruit: similar to Granny Smith, except ripens 2 weeks earlier; large; conical; skin color green to buff yellow; texture fine, firm; flesh color cream; quality fair; storage life 6 to 8 months, somewhat shorter than Granny Smith. Tree: very vigorous, very precocious; bloom late.

Goldspur (*Sundale, Sundale SturdySpur*).—Orig. in Roosevelt, Wash., by Grady Auvil, Sundale Orchards. Introd. in 1963. Plant patent 2334; 24 Dec. 1963; assigned to Sundale Orchards. Bud mutation of Golden Delicious; discovered in 1960. Fruit: resembles Golden Delicious, but more prone to russet. Tree: medium small, semi-dwarf, with heavy thick limbs, having buds much closer together than on Golden Delicious; precocious bearer; large crops.

Goodenough Spitzenburg.—Orig. in Hood River, Ore., by W.H. Goodenough, Jr., now of Sumner, Wash. Introd. in the 1930s. Bud mutation of Esopus Spitzenburg; discovered in 1922 or 1923. Fruit: skin exceptionally dark red to black; otherwise, similar to parent. Tree: similar to Esopus Spitzenburg.

Goodland.—Orig. in Morden, Manitoba, Canada, by Canada Dept. Agr., Res. Sta. Introd. in 1955. Patten Greening o.p.; selected in 1925; tested as Morden 354. Fruit: medium to large; roundish oblate, shape uniform; one-half of skin an attractive red; flesh crisp, juicy, tender, sweetly subacid, aromatic; excellent as dessert or sauce; season mid-September to mid-January. Tree: medium tall; very hardy; annually productive.

Goodner Winesap.—Orig. in Wenatchee, Wash., by J.C. Goodner. Introd. in 1955. Bud mutation of Winesap; discovered in 1942. Fruit: skin striped, appearing early, no russet; resembles parent. Tree: same as Winesap.

Gordon.—Orig. in Whittier, Calif., by R.B. Gordon. Plant patent 4144; 8 Nov. 1977. Chance seedling of unknown parentage. Fruit: medium to large, nearly globose; skin green, blushed red, striped, medium thick, smooth, medium tender; flesh crisp, juicy; texture firm; color near white; good for eating and cooking. Tree: medium large; vigorous; upright; very productive; regular bearer. Low chilling requirement, suitable for growing in mild climatic coastal regions of southern California; prolonged season of flowering and fruiting.

Goro.—Orig. in Wadenswil, Switzerland, by E. Schaer and B. Krapf, Swiss Federal Res. Sta. for Fruit Growing. Introd. in 1973. Golden Delicious x Swiss Orange; cross made in 1951, tested as 308/M. Fruit: medium to large; spherical to flat spherical; skin whitish yellow, striped light red, shiny, slightly greasy, moderately easily bruised, no russet, dots inconspicuous, attractive; flesh yellowish, crisp to mellow, juicy, aroma strong, high sugar content, acid slight, quality good, slightly below that of Gravenstein; ripens 3 weeks after Gravenstein; ready to market straight from tree; optimum ripeness for eating in October, just before McIntosh, fills gap between Gravenstein and current autumn apples. Tree: vigorous; productive; regular; hangs very well; susceptibility to scab low.

Gourmet Golden™ (*Keystone*).—A chance seedling discovered by Randy Wick in Keystone Orchards, Riverside, Wash. U.S. plant patent 7209; 3 Apr. 1990; propagation rights assigned to North American Tree Co., Portland, Ore. Parentage: probably Delicious x Golden Delicious. Fruit: medium to large; typical Delicious shape; skin smooth in Pacific Northwest, sometimes slightly russeted in the East; attractive pink blush in most areas; flesh cream-colored, firm, crisp. Harvest season just before Rome Beauty; much less subject to bruising. Highly aromatic; very low acid; flavor excellent from controlled-atmosphere storage. Shelf life after storage similar to Delicious. Tree: vigorous; nonspur; upright-spreading.

Graham.—See Graham Spy.

Graham Red.—See Graham Spy.

Graham Special.—See Graham Spy.

Graham Spy (*Graham, Graham Red, Graham Special, Red Graham*).—Orig. in Manistee, Mich., by Thomas Elwood Graham, Manistee Orchard Co. Introd. in 1936. Plant patent 278; 17 May 1938; assigned to Greening Nursery Co., Monroe, Mich. Bud mutation of Northern Spy; discovered in 1926. Fruit: large, up to 3 1/2 inches in diam.; skin bright solid red; size, shape, and quality of Northern Spy; ripens earlier than Northern Spy, just before McIntosh; keeps in cold storage until spring; good pie and general purpose variety. Tree: growth similar to Northern Spy; bears moderately well.

Grandparent.—See Perrine Yellow Transparent.

Granspur (*Spur Granny Smith*).—Orig. in Brewster, Wash., by Cal Cooper and Grady Auvil, Orondo, Wash. Introd. in 1980 by C&O Nursery, Wenatchee, Wash. Plant patent 4676. Bud mutation of Hannaford Granny Smith. Fruit: larger and more uniform than Granny Smith; flesh white, fine-textured; skin has slight striping with darker shades of green on a lighter green background, slight red blush; ripens with standard Granny Smith; excellent keeping quality in cold storage. Tree: compact: needs pollenizer, no longer propagated.

Greendale.—Orig. in Geneva, N.Y., by Richard Wellington, New York State Agr. Expt. Sta. Introd. in 1938. McIntosh x Lodi; cross made in 1924; first full crop in 1935. Fruit: well-shaped; skin an attractive green, develops a faint pinkish cheek in most seasons; flavor mild, quality good; for dessert and cooking purposes; extends season of Lodi for local and roadside markets; resembles McIntosh in shape and Lodi in color.

Greene Spy (*Geneva Red Spy, Greene's Red Spy, Northern Spy Double Red, Red Spy*).—Orig. in Victor, N.Y., by William S. Greene. Introd. in 1923; discovered in 1895. Bud mutation of Northern Spy; sent to the New York State Agr. Expt. Sta. in 1910 by C.E. Greene (son of Wm. S. Greene). Fruit: identical with Northern Spy, except more red pigment in skin; originally solid bright red without either stripes or splashes; in 1966, reverted, distinct striping. Tree: vigorous; hardy; identical with Northern Spy.

***Greene's Red Spy.*—**See Greene Spy.

Greensleeves (*Malling Greensleeves*).—Orig. in East Malling, Kent, England, by F.H. Alston, R. Watkins, and J.W. Bates, East Malling Res. Sta. Introd. in 1977. Protected by Plant Breeders' Rights. James Grieve x Golden Delicious; seed germinated in 1966; tested as A120/6. Fruit: medium, 65 mm or more; skin green-yellow with a trace of russet in the stem cavity; juicy; flavor is pleasantly subacid; ripens in early midseason. Tree: semi-dwarf, ideal for high-density orchards; very precocious; productive (yielded 100% more fruit than Cox's Orange Pippin in the first four fruiting years); when stored at 3.3C, fruit is marketable until early December; flowers profusely 1 week before Cox's Orange Pippin, long flowering period, flowers showed some resistance to severe frosts.

Greenspur Granny Smith.—Orig. in Brewster, Wash., by Calvin L. Cooper. Introd. in 1980. Plant patent 4741; 18 Apr. 1980; assigned to Auvil Fruit Co., Orondo, Wash. Bud mutation of Hannaford Granny Smith; selected in 1972. Fruit: large; skin dark green with numerous and distinctive white spots: flesh firm with whiter flesh than parent; outstanding storage capabilities. Tree: spur-type, semi-dwarf; resembles parent, but has increased bloom set, with higher setting of double and triple clusters, sharply increased productivity, and reduced internodal distances on the terminal growths of branches; ripens with Granny Smith.

***Griffspur.*—**See Spur Goldblush.

Grove.—Orig. in Mountain Grove, Mo., by Paul H. Shepard, Missouri State Fruit Expt. Sta. Introd. in 1935. Ingram x Delicious; cross made in 1915; selected in 1930; tested as A-1150. Fruit: medium to large; conical, slightly pinched on one side; skin attractive, coloring a bright red well before ripening time; quality good; excellent keeper and shipper; ripens a little later than Winesap; resembles Delicious. Tree: blooms late, about 10 days after Delicious; resistant to scab, blotch, blight; vigorous.

***Gurney Viking.*—**See Viking.

H.T.O. Red Rome.—Orig. in Bangor, Mich., by H.T. Bigelow. Introd. about 1930 by Hill Top Nurseries and Orchards, Hartford, Mich. Parentage unknown; discovered about 1926. Fruit: skin bright red; shape uniform; similar to Red Rome. Tree: standard type; bears well. Apparently no longer being propagated.

Hadlock Reinette.—Orig. in Hammond, N.Y., on the Hadlock farm by Mrs. Edward Biondi of Ogdensburg, N.Y. Introd. in 1964 by Fred L. Ashworth, Heuvelton, N.Y. Parentage unknown; discovered in 1963. Fruit: skin blushed bronze-orange, resembling Cox's Orange Pippin in appearance; flesh firm, fine-grained, moderately subacid, quality good; keeps late in cold storage; three seeds per carpel; resembles Golden Russet. Tree: very hardy, more so than Golden Russet.

Hagey Sweet.—Orig. in Carrollton, Ohio, by John B. Birch, J.B. Birch Nursery Co. Introd. in 1921. Chance seedling discovered about 1909 by Birch on farm belonging to a Mr. Hagey. Fruit: very sweet; excellent for fresh use, very poor for culinary purposes and for storage; poor for commercial usage; resembles McIntosh in size and skin color. No longer being propagated.

Hampshire.—Chance seedling discovered in the Gould Hill Red Delicious orchard, Contocook, New Hampshire, in the late 80s. Plant patent 8519, 1994. Fruit: attractive, large, round, 100% dark red blush maturing with Red Delicious; eating quality good, especially after cold storage. Tree: semi-spur, upright, precocious, moderately vigorous, hardy.

Haralred™ (*Lantz Haralson, Red Haralson*).—Orig. in La Crescent, Minn., by Louis Lantz. A red-fruited mutation of Haralson. Plant patent 4824, 16 Feb. 1982;

assigned to TRECO®, Woodburn, Ore. Fruit: redder, earlier, sweeter than Haralson; early coloring without russet; sweet, juicy; good storage. Tree: same as Haralson.

Haralson.—Orig. in Excelsior, Minn., by the Univ. of Minnesota Fruit Breeding Farm. Named and introd. in 1923. Malinda o.p.; selected in 1913; tested as Minn. 90. Fruit: size medium, like Wealthy; roundish conic; skin an attractive red, dots greenish-russet, conspicuous; core small, closed; flesh white, firm, moderately tender, fine-grained, juicy, mild, pleasant, subacid; good cooking quality; stores well; season January to April. Tree: productive; vigorous; hardy; upright.

Hardibrite Red Delicious.—Orig. in Oregon by Carlton Nursery Co., Dayton, Ore. Introd. in 1975. Whole tree mutation of Hardi-Spur Red Delicious; selected in 1973; discovered in 1970. Fruit: medium-large; elongated; skin brighter red than Hardi-Spur, solid blush color, flesh white, crisp; ripens 155 days after full bloom, same as Hardi-Spur Red Delicious. Tree: spur-type, half size on seedling root; upright; compact; vigorous; average hardiness; excellent productivity. Recommended for fresh-fruit marketing and processing.

Hardi-Spur Delicious (*Lonely*).—Orig. in Dryden, Wash., by J.E. Griffith of Cashmere, Wash. Introd. in 1962 by Morrison Nursery, Zillah, Wash. Whole tree bud mutation; discovered in 1958 in an orchard of Starking Delicious. Fruit: skin solid red, which appeared consistently earlier than Starking Delicious; ripens with Starking Delicious. Tree: spur-type; produces heavily; one-half to two-thirds size of a standard apple tree.

Hardy Cumberland.—Orig. at the U.S. Dept. of Agriculture (USDA) Georgia Mountain Branch Expt. Sta., Blairsville. Introd. jointly by the USDA and the Univ. of Tennessee Plateau Expt. Sta., Crossville. Parentage: Lyons x Detroit Red; cross made at Blacksburg, Va., in 1961. Fruit: large, over 70 mm; shape round-oblate; color 80% washed and striped carmine; flesh cream-colored; flavor balanced aroma and acidity; eating quality excellent; no bitter pit. Harvest season slightly later than Rome Beauty. Tree: 90% the size of Golden Delicious tree; hardy, survived –25F without injury; no serious disease problems.

Hared.—See Starkspur Dixiered Delicious.

Harper Red Spy.—See Schoharie Spy.

Harrold Red.—Orig. in Zillah, Wash., by Wayne Harrold. Canadian propagation rights sold in 1957 to Fruit Tree Industries, British Columbia. Bud mutation of Starking Delicious; discovered in 1953 on a tree 5 or more years old. Fruit: skin nearly solid red with some striping, coloring extremely early; Delicious type. Tree: standard type.

Harvest Special.—Orig. in Saskatoon, Saskatchewan, Canada, by C.F. Patterson, Univ. of Sask. Introd. in 1960 for home gardens. Columbia x Melba; selected in 1946; tested as Saskatoon A60-203. Fruit: up to 2 inches in diam.; slightly oblong, up to 2 1/2 inches long; skin predominately red, with striping, attractive; flesh of good quality when sound, but tends to break down prematurely; ripens early to midseason. Tree: hardy.

Harvester.—Orig. in Saskatoon, Saskatchewan, Canada, by C.F. Patterson, Univ. of Sask. Introd. in 1960 for home gardens. Silvia x Melba; selected in 1946; tested as Sask. A 59-379. Fruit: up to 2 1/4 x 1 3/4 inches under non-irrigated field conditions at Saskatoon; skin cream, blushed, striped; quality good for its season, mildly acid; tendency to early breakdown; ripens early to midseason. Tree: extremely hardy.

Hatsuaki.—Orig. at Morioka, Iwate, Japan, by Y. Yoshida, S. Tsuchiya, T. Haniuda, T. Sanada, and S. Sadamori, Morioka Branch, Fruit Tree Res. Sta. Introd. in 1976. Jonathan x Golden Delicious; cross made in 1939; selected in 1967. Tested as Tohoku #8. Fruit: medium to large, 250 to 300 g; short round-conical to oblong conical, slightly uneven in shape; skin light rose-red on yellow ground, rather thick, dull finish, numerous small russet dots; some rough russet in and around basin; flesh yellowish white, crisp, rather coarse, very juicy, slightly acid, similar to Jonagold. Dessert quality excellent. Harvest with Gala, 17 days before Delicious. Tree: vigorous; spreading; regularly productive. Diploid; blossoms with Delicious.

Hawaii.—Orig. in Sebastopol, Calif., by Wm. E. Silva. Introd. in 1963. Gravenstein x

Golden Delicious; cross made in 1945; selected in 1959. Plant patent 2512; 11 May 1965. Fruit: averages 2 1/2 inches in diam. and 3 inches long, oblong angular; skin golden; flesh yellow, firm, juicy, strong pineapple-like flavor; aroma pleasant; ripens first of October; keeps well; resembles Golden Delicious. Tree: size medium; bears alternately, requiring heavy thinning in heavy producing years; fruit borne on small spurs; vigorous, rapid, and upright grower.

Hawkes Bay Red Delicious (*Law Red Delicious*).—Orig. in Twyford, Hastings, Hawkes Bay, New Zealand, by R.L. Law. Introd. in 1929. Bud mutation of Delicious; selected in 1927. Fruit: resembles Delicious, having improved color; skin very heavily striped bright red, superior to the block-red strains which become too dark and dull in Hawkes Bay.

Hawkeye Greening.—Orig. in Ames, Iowa, by H.L. Lantz, Iowa Agr. Expt. Sta. Introd. in 1921. Parentage unknown; seed produced in 1900, seedling planted in 1907; selected in 1917; named in 1921. Fruit: skin greenish-yellow; Rhode Island Greening type; medium to large; flesh fine-grained, juicy, good for dessert purposes but especially for sauce and pies; keeps exceptionally well in cold storage until March. Tree: very hardy; heavy and regular bearer. Recommended primarily for home gardens.

Hazen.—Orig. at Mandan, N.D., by Neal S. Holland. Dept. of Hort., North Dakota State Univ., Fargo, and U.S. Dept. of Agriculture. Introd. in 1980. Named to commemorate the town Hazen and the late Dean Arlon G. Hazen, Dean of College of Agr., and Director of North Dakota Expt. Sta. while cultivar was tested. Duchess x Starking Delicious. Tested as Mandan 41-27. Fruit: large, up to 3 inches in diam., round to round oblate; basin at blossom end quite broad and shows tendency toward the five points typical of Starking Delicious; skin attractive dark red with 80% colored; flesh greenish-yellow, medium firm, juicy, slightly coarse; flavor mild, subacid; ripens late August at Fargo. Tree: semi-dwarf; bears at 3 to 4 years; hardy, free of fire blight, annual cropper.

Heaver.—Orig. in Baljennie, Saskatchewan, Canada, by Kenneth N. Heaver. Introd. in 1964. Parentage unknown; selected about 1945; tested as Heaver 5. Fruit: averages 2 1/2 to 3 inches in diam.; skin yellow with a slight blush; flesh white, sweet, firm, juicy; keeps very well in cold storage; ripens during October. Tree: vigor medium; not too hardy in early years but seemingly hardy when established.

Hein.—See Autumn Gold.

Herman Smith.—Orig. in Riverside, Calif., by Herman Smith. Introd. in 1958. Considered to be Orleans x Gravenstein; cross made in 1945. Fruit: shape similar to Orleans; skin colors well; flesh mildly subacid; keeps well in storage; hangs well on tree; resembles Orleans.

Hermansky.—Orig. in Esterhazy, Saskatchewan, Canada, by Francis Hermansky. Introd. in 1965 by Lad. Martinovsky, Gerald, Saskatchewan. McIntosh o.p.; selected in 1945. Fruit: size medium; skin not too attractive; ripens fairly late; keeping quality fair. Tree: can be trained to bush form.

Heyer 12.—Orig. in Neville, Saskatchewan, Canada, by A. Heyer. Introd. about 1940. Seedling of a Russian apple received from A.P. Stevenson, Morden, Manitoba. Fruit: 2 1/2 inches in diam.; skin greenish-yellow; flesh juicy, moderately coarse, acid; quality fair for cooking; season mid-August to October. Tree: extremely hardy; productive; late blooming. Chief value is a hardy, early apple for northern climate.

Heyer 20.—Orig. in Neville, Saskatchewan, Canada, by A. Heyer. Introd. about 1936. Parentage unknown. Fruit: large, about 3 inches in diam.; flesh crisp, juicy, yellow. Tree: vigorous.

Hidala.—See Hillwell.

Hi-Early (*Hi-Early Delicious, Hi-Early Red Delicious, Neff Delicious*).—Orig. in Pateros, Wash., by George W. Neff. Introd. in 1955 by Heath's Nursery, Brewster, Wash. Bud mutation of Starking Delicious; discovered in 1945. Fruit: skin colors about 3 to 4 weeks earlier with a more intense red than parent, which it resembles. Tree: standard; identical to Delicious.

Hi-Early Delicious.—See Hi-Early.

Hi-Early Red Delicious.—See Hi-Early.

Hi-Red (*Hi-Red Delicious, Watson Delicious*).—Orig. in Cashmere, Wash., by

Everett Watson. Introd. in 1954 by the Chas. Morrison Nursery, Zillah, Wash. Bud mutation of Starking Delicious; discovered in 1949. Fruit: skin develops a beautiful, cherry red about 10 to 14 days before parent; resembles parent. Tree: identical in appearance to Delicious.

Hi-Red Delicious.—See Hi-Red.

Hillwell (*Hidala, Red Braeburn*).—Orig. at Hastings, New Zealand, by Mr. and Mrs. John Hill. Discovered as a limb mutation of Braeburn in 1981. Introd. in 1990. Plant patent 7526; assigned to TRECO®, Woodburn, Ore. Identical with Braeburn, except the apples of Hillwell are much redder.

Himekami.—Orig. at Morioka, Iwate, Japan, by Y. Yoshida, T. Haniuda, S. Tsuchiya, T. Sanada, T. Masuda, H. Bessho, and S. Sadamori, Morioka Branch, Fruit Tree Res. Sta. Introd 1984. Fuji x Jonathan; cross made in 1963; selected in 1976; tested as Morioka #37. Japanese Reg. No. P931, 18 July 1985. Fruit: medium to large, 250 to 300 g; shape oblong to oblong-conic; skin bright red over yellow-green ground, smooth, some russet in cavity; flesh cream-colored, juicy, sweet, water core; good sugar/acid balance; mild flavor. Harvest just after Gala; cold storage 2 months; shelf life 4 to 5 weeks. Tree: vigorous; upright; precocious; regularly productive. Diploid; blooms with Delicious. No longer propagated commercially.

Hinchliffe.—Orig. in Mexia, Texas, by N.N. Hinchliffe. Introd. about 1945. Parentage unknown; discovered about 1940. Fruit: large, about 3 inches in diam.; skin yellow, deeply blushed; flavor good; matures first of August; stores well into November. Tree: starts bearing at an early age; annual bearer.

Hokuto (*Northern Star*).—Orig. in 1983 from the breeding program at the Aomori Apple Expt. Sta., Kuroishi, Aomori Prefecture, Japan. Fuji x Mutsu. Japanese Reg. No. P 368. Fruit: medium, 200 to 250 g; shape round; color purplish-red stripe over yellow ground; flesh cream-colored, dense; sweet with moderate acidity, excellent quality; harvest late, with Mutsu; long storage, but susceptible to moldy core.

Holdren's Early Red.—Orig. in Little Hocking, Ohio, by Robert E. Holdren. Introd. in 1973. Plant patent 3558; 28 May 1974. Wealthy x Rome Beauty, cross made in 1956, selected in 1971. Fruit: 2 3/4 inches axial diam., 3 inches transverse diam.; skin red, thick, tough, smooth, glossy, waxy; flesh cream color, texture firm, fine crisp, quality good; ripens 28 July to 10 Aug. in Little Hocking area; ships better than other varieties in its season. Tree: upright, dense; vigorous; hardy; bears regularly; less affected by scab than other varieties grown in same area.

Holland.—See Summer Champion.

Holiday.—Orig. in Wooster, Ohio, by Freeman S. Howlett, Ohio Agr. Expt. Sta. Introd. in 1964. Macoun x Jonathan; cross made in 1940; tested as Ohio 6187. Fruit: skin deep red; flesh white, crisp; high dessert quality; keeps well in cold storage, being suitable fresh from 20 Dec. through 15 Jan.; harvest date of 10 Oct. over 17 years; resembles Macoun. Tree: moderately vigorous; upright; dense; somewhat hard to prune, thereby having a tendency to grow too tall.

Hollin.—Orig. in Fairburn, Ga., by Major C. Collins. Introd. in 1977. Unknown seedling. Fruit: 3 x 2 1/2 inches in diam.; conic oblate; skin 80% to 90% washed with a medium red, with hint of a stripe; flesh fine-grained, cream-colored, quality good; ripens early August. Tree: vigorous; similar to Red Delicious in leaf color and type. Recommended as a dooryard variety.

Holly.—Orig. in Wooster, Ohio, by F.S. Howlett, Ohio Agr. Expt. Sta. Introd. in 1970. Jonathan x Delicious. Cross made in 1952, selected in 1961, tested as 8408. Fruit: medium-large, oblong conic; skin completely covered with cherry red, smooth, thick; flesh subacid, texture excellent; less tendency to become mealy in storage than Delicious; harvest about 8 days after Delicious, 20 Oct. in Wooster. Tree: medium-large, spreading; vigorous; hardy; moderately productive. Named after F.S. Howlett, Chairman of Dept. of Hort., previously mentioned station, from 1947 to 1970.

Holmberg.—See Starkspur Golden Russet.

Holstein (*Holsteiner Cox, Vahldiks Cox Sämling No. III*).—Orig. in Eutin, Holstein, Germany, by a Mr. Vahldik. Introd. about

1950. Presumed to be Cox's Orange Pippin o.p., discovered about 1918. Fruit: medium large (larger than Cox), heavy; tallish truncate-conic, ribbed at apex; skin greenish to golden yellow, variable orange-red blush and slight russet; flesh firm, coarse, yellowish-white, flavor resembles Cox; ripens November to January. Tree: large, wide-spreading, moderately dense, vigorous, spurs freely; productive; very susceptible to mildew, a little to canker; flower subject to frost damage, triploid, incompatible with Cox, difficult to pollinate.

Holsteiner Cox.—See Holstein.

Honeycrisp.—Orig. near Excelsior, Minn., by breeders of Univ. Minnesota Hort. Res. Center. Introd. in 1991. Macoun x Honeygold; selected in 1974; tested as Minn. 1711. Plant patent 7197, assigned to TRECO, Woodburn, Ore. Fruit: medium, 60 to 70 mm in diam. without thinning; oblate to roundly oblate; skin 50% to 70% blushed scarlet over lemon yellow, dull finish, dimpled; numerous small conspicuous lenticels; flesh cream-colored, coarse, very crisp, juicy; flavor mildly aromatic, subacid, excellent dessert quality. Optimum harvest 5 days after McIntosh, but can be picked 10 days earlier or later; fruit tends not to drop; 5 months storage at 1C. Tree: moderately vigorous; upright-spreading; regular bearing; hardy in northern Minnesota.

Honeygold.—Orig. in Excelsior, Minn., by Hort. Res. Center, Univ. of Minn. Introd. in 1969. Golden Delicious x Haralson. Cross made in 1935 and selected in 1947 by W.H. Alderman, tested as Minn. 1595, named and released by Cecil Stushnoff, both of the previously mentioned station. Fruit: medium to large; round-conic, irregular; skin yellow with a bronze blush, smooth, thick, tender, dots corky, stellate; flesh crisp, yellow, juicy, Golden Delicious flavor, excellent dessert quality; ripens 1 Oct., Haralson season; quality of Golden Delicious, but storage characteristics superior. Tree: size medium, moderately open; vigor moderate; hardiness equal to Haralson; productivity moderate; susceptible to fire blight. Recommended for Minnesota and South Dakota where Golden Delicious is marginal.

Hoople's Antique Gold.—Orig. in Otway, Ohio, by Harry Hoople, Hoople Fruit Farm. Introd. in 1976 by Southmeadow Fruit Gardens, Birmingham, Mich. Bud mutation of Golden Delicious; discovered in 1970. Fruit: resembles Hudson's Golden Gem; skin antique gold russet; flavor richer and more intense than parent.

Hoosier Seedling.—See Stark JonGrimes

Horace.—Orig. in Ottawa, Ontario, Canada, by Canada Dept. Agr., Res. Sta. Has not been introd. commercially. Langford Beauty o.p.; selected in 1912. Fruit: medium; skin highly colored, striped; good quality; season September to November; resembles Fameuse.

Houser Delicious.—Orig. in Selah, Wash., by Lawrence E. Houser. Introd. in 1958 by Morrison Nursery, Zillah, Wash. Bud mutation of Starking Delicious; discovered in 1953. Fruit: skin colors very early and evenly; all characteristics of a fine type Delicious. Tree: of same growth habit as Delicious.

Houser Winesap.—Orig. in Selah, Wash., by Lawrence E. Houser. Introd. in 1940. Bud mutation of Winesap; discovered in 1925. Fruit: skin darker red than parent, appearing earlier in season; otherwise, resembles Winesap. Tree: similar to Winesap.

Hubbs Delicious.—Orig. in Milton-Freewater, Ore., by O.F. Hubbs, O.F. Hubbs Nursery. Introd. in 1950. Whole tree bud mutation in a mature orchard of Starking Delicious; discovered in 1950. Fruit: skin colors earlier and darker than parent, with some stripes; Delicious type; resembles Red King, but with more solid color. Tree: standard type.

Hudson's Golden Gem.—Orig. in Oregon by A.D. Hudson. Introd. about 1931 by Hudson's Wholesale Nurseries, Tangent, Ore. Discovered in a fence-row thicket. Fruit: large; conical; stem very long; skin russeted, very attractive, but susceptible to cracking; flesh rather tart, excellent for dessert purposes; hangs on tree a long time. Tree: productive; vigorous; bears annually.

Huebner Starking.—Orig. in Kelowna, British Columbia, Canada, by T. Huebner. Introd. in 1962 by Carlton Nursery Co., Forest Grove, Ore. Bud mutation of Starking Delicious; discovered in 1957 on a 9-year-old tree. Fruit: skin striped, dark red appearing early; resembles Delicious. Tree: standard growth habit.

Hume.—Orig. in Ottawa, Ontario, Canada, by Canada Dept. Agr., Res. Sta. Introd. about 1924. McIntosh o.p.; selected in 1916. Fruit: skin with high color; quality good; ripens a little earlier than McIntosh. Tree: hardier than McIntosh.

Hunts Everbearing.—Origin unknown. Introd. before 1931. Parentage unknown. Fruit: small, about 2 inches in diam.; roundish with unequal halves; skin dark red; stem long and slender; flesh white but darker than Red June; quality excellent, being sweet and pleasing to eat out of hand; ripens in Stillwater, Okla., about mid-July. Tree: small to dwarf habit; productive; bears regular crops. Primarily for home gardens.

Hurlburt's Golden Pippin.—Orig. in Indiana. Introd. in 1920 by Fairbury Nurseries, Fairbury, Neb. Chance seedling; propagating material sent to this nursery by person or persons now unknown. Fruit: appears to ripen somewhat earlier than Golden Delicious; in all other respects, resembles Golden Delicious. Tree: excellent producer; resembles Golden Delicious. Propagation discontinued about 1942 as so similar to Golden Delicious.

Hybl.—Orig. in Fremont, Neb., by James Z. Hybl, Sr. Introd. about 1945 by Plumfield Nurseries, Fremont. Delicious o.p.; discovered about 1939 or 1940. Fruit: skin yellow with a red cheek; quality very good; keeps well in cold storage; ripens late; resembles Delicious in size and shape.

Idagold.—Orig. in Moscow, Idaho, by Leif Verner, Idaho Agr. Expt. Sta. Introd. in 1944. Spitzenburg x Wagener; selected in 1939; tested as Idaho A 2-23. Fruit: medium to large; oblong conic to oblate conic, irregularly ribbed, rather variable; skin rich, clear yellow sometimes with a faint blush and often characterized by dark green in the stem cavity and calyx basin, smooth, waxy, moderately thick; flesh rich cream color, fairly coarse, often very firm, moderately juicy, sweet, aromatic; ripens with or slightly ahead of Rome Beauty; high dessert and keeping qualities. Tree: moderately vigorous, wide spreading, open, with strong crotches.

Idajon.—Orig. in Moscow, Idaho, by Leif Verner, Idaho Agr. Expt. Sta. Introd. in 1949. Wagener x Jonathan; selected in 1936. Fruit: medium to large; round to round conic, regular or slightly inconspicuously streaked with carmine, smooth, moderately thick; flesh nearly white, juicy, crisp, fine-grained, very mild subacid to somewhat sweet; quality good; ripens 10 days before Jonathan. Tree: moderately vigorous, upright, spreading; productive; less susceptible than Jonathan to mildew and fire blight; tends to drop fruit before mature.

Idared.—Orig. in Moscow, Idaho, by Leif Verner, Idaho Agr. Expt. Sta. Introd. in 1942. Jonathan x Wagener; selected in 1935. Fruit: skin nearly solid red, attractive; core small; texture good; exceptional keeping quality; excellent dessert and cooking quality, mild; ripens in early October; resembles Wagener. Tree: strong, vigorous, upright; precocious and very productive; susceptible to fire blight, very susceptible to powdery mildew.

Imperial Delicious (*Double Early Red Delicious, Gilbert 2, Imperial Double Early Red Delicious, Imperial Double Red, Imperial Double Red Delicious*).—Orig. in Parker, Wash., by Elon J. Gilbert, Richey & Gilbert Co., Yakima, Wash. Introd. in 1957. Bud mutation (whole tree) discovered in 1954 in a Starking Delicious orchard. Fruit: large; skin bright crimson red, which appears early; flavor very good; keeps well; resembles Royal Red Delicious. Tree: standard type.

Imperial Double Early Red Delicious.—See Imperial Delicious.

Imperial Double Red.—See Imperial Delicious.

Imperial Double Red Delicious.—See Imperial Delicious.

Imperial Lady Clarendon (*Improved York, York Imperial Sport*).—Orig. in Misgund East, Cape Province, South Africa, by Ford Kritzinger & Sons, growers. Introd. there in 1948. Bud mutation of York Imperial; discovered in 1942. Fruit: larger, but flatter than York Imperial, 2 3/4 to 3 inches in diam.; resembles parent in all other respects; inclined to preharvest drop. Tree: weaker, shoots more sparse, leaves larger, darker and fruit buds larger than York Imperial; blossoms and ripens simultaneously. Named by Ford Kritzinger in honor of the wife of a one-time Governor-General of South Africa.

Improved Early Transparent.—See Lodi.

Improved Red Winesap.—See Swanson Winesap.

Improved Royal Red Delicious.—See Nured Royal Delicious.

Improved York.—See Imperial Lady Clarendon.

Into.—See Maga.

Iwakami.—Orig. at Morioka, Japan, by Y. Yoshida, T. Haniuda, S. Tsuchiya, T. Sanada, T. Masuda, H. Bessho, and S. Sadamori, Morioka Branch, Fruit Tree Res. Sta. Fuji x Jonathan. Introduced in 1984. Japanese Reg. no. 932, 18 July 1985. Tested as Morioka #39. Fruit: medium, 250 g; shape round; skin bright red stripes on yellow green ground, smooth, medium thick, russet in cavity, lenticels small; flesh whitish-yellow, crisp, juicy; flavor sprightly, aromatic. Ripens with Gala. Tree: vigorous; spreading; precocious; regularly productive. Diploid; blossoms with Jonathan, before Delicious.

J9.—See Jork 9 in apple rootstocks.

Jenner Sweet.—Orig. in Colton, N.Y., by Howard Jenner. Introd. in 1964 by Fred L. Ashworth, nurseryman, Heuvelton, N.Y. Fameuse o.p.; discovered in 1924. Fruit: medium; skin blushed, more so than Fameuse; flavor fine, mildly sweet, typically Fameuse; ripens in mid-autumn; resembles Fameuse. Tree: extremely hardy.

Jenred.—See Starkspur UltraStripe Delicious.

Jerseymac.—Orig. in New Brunswick, N.J., by L.F. Hough and Catherine H. Bailey, New Jersey Agr. Exp. Sta. Introd. in 1971. NJ24 = [Melba x (Wealthy x Starr) x (Red Rome x Melba)] x Julyred. Cross made in 1956; selected in 1961; tested as NJ38. Fruit: size 2 3/4 to 2 7/8 inches in diam.; round, uniform; skin light green ground color with 50% to 75% blushed with bright red, thin, tough; colors better than McIntosh, otherwise very similar; flesh juicy and aromatic like McIntosh, white; resembles McIntosh; ripens 1 month before McIntosh. Tree: upright, spreading like McIntosh. Early budbreak. Very susceptible to scab.

Jerseyred.—Orig. in New Brunswick, N.J., by M.A. Blake, New Jersey Agr. Expt. Sta. Introd. in 1954. Gallia Beauty x White Winter Pearmain; selected in 1947; tested as N.J. 1. Fruit: large; round oblate; skin colors better than Rome Beauty in New Jersey; quality better than Rome Beauty and may be picked later than that variety; keeps well in storage; resistant to breakdown; excellent for eating fresh and for cooking; resembles Rome Beauty. Tree: vigorous; dense; resistant to blight.

Jester (*Malling Jester*).—Orig. in East Malling, Kent, England by F.H. Alston, East Malling Res. Sta. Introduced in 1981. Worcester Pearmain x Starkspur Golden Delicious. Fruit: medium, 65 to 70 mm; color distinctive bright red, predominantly red on a yellow/green background; skin finish good; skin texture good; flesh crisp and juice quality good; best suited for home orchards; ripe in early October; storage life short. Tree: regular, heavy cropping; precocious; compact growth habit; requires little pruning; flowers 2 days after Cox's Orange Pippin; styles and pollen resistant to low temperatures (5C).

Joan.—Orig. in Ames, Iowa, by the Dept. Hort., Iowa Agr. Expt. Sta. Introd. in 1932. Anisim x Jonathan; cross made in 1906; selected in 1918. Fruit: large; round; skin a solid bright red, very attractive; flesh white, subacid, juicy, quality fair; a good culinary type. Tree: annual bearer; very productive; subject to fire blight.

Joel's Red Delicious.—Orig. in Grass Valley, Calif., by Joel Bierwagen. Introd. in 1960. Bud mutation of Delicious. Fruit: large, 12 to 14 oz; resembles Delicious; skin solid red, no striping; flesh white, slight greenish tint; texture finer grained than Starking Delicious, 2 to 7 Sept. in Grass Valley. Tree: vigorous; upright; spreading if trained; hardy; productive, bears regularly; resistant to bacterial bark disease.

Joey.—Orig. in Wyndmere, N.D., by R.L. Wodarz. Introd. in 1955. McIntosh o.p. Fruit: size averages 3 1/4 inches in diam.; rounded oblate; skin solid deep red, shiny; flesh white, firm, fine textured, a little tart, very juicy, excellent for cooking; ripens in late September; 4 to 5 months storage. Tree: hardy; annual bearer; very productive; medium spreading.

John Roberts.—See Roberts.
Jon-A-Red.—See Jonared.
Jonadel.—Orig. in Ames, Iowa, by H.L. Lantz, Iowa Agr. Expt. Sta. Introd. in 1958. Jonathan x Delicious; selected in 1928. Fruit: larger and more conic than Jonathan; milder in flavor than that variety but with more acid than Delicious; equal in red skin color to Jonathan; dessert qualities excellent; flesh juicy, firm; ripens with Jonathan and directly after McIntosh; keeps well in storage until March or later; exhibits no evidence of Jonathan spot or internal browning; skin durable and withstands handling well; tends to drop from tree unless treated with NAA. Tree: less affected by fire blight than Jonathan; hardy; spreading.
Jonafree.—Orig. in Urbana, Ill., by a cooperative breeding program consisting of the Agr. Expt. Sta. of Illinois (D.F. Dayton and J.B. Mowry), Purdue Univ. (E.B. Williams, J. Janick, and F.H. Emerson), and Rutgers Univ. (L.F. Hough and C. Bailey). Introd. in 1979. Plant patent 4633; 27 Jan. 1981. 855-102 x NJ31; original seedling planted in 1965; a fifth-generation derivative of C.S. Crandall's Rome x *M. floribunda* 821 cross. Ancestry also includes Jonathan, Gallia Beauty, Red Spy, and Golden Delicious. Selected in 1972; tested as Co-op 22. Fruit: medium, axial diam. 55 to 70 mm, transverse diam. 63 to 76 mm; roundish to slightly oblate; color 95% medium red without stripes; undercolor yellow-green; smooth with dry wax, generally no russet, glossy, attractive; appearance resembles Jonathan; flesh light yellow, fine-grained, crisp; flavor subacid, moderately rich, pleasant aroma, good dessert apple. Season: with Jonathan. Storage: at 1C retains flavor 10 weeks, no Jonathan spot, no shrivel in storage. Tree: semi-spreading, vigorous, nonspur type; field immune to apple scab; slightly susceptible to powdery mildew and fire blight.
Jonagold.—Orig. in Geneva, N.Y., by New York State Agr. Expt. Sta. Introd. in 1968. Golden Delicious x Jonathan. Cross made in 1943; original seedlings planted in 1946; first fruited in 1953; tested as NY 43013-1. Fruit: large; skin 50% to 80% red striping over yellow ground color, attractive; flesh firm, subacid, juicy; similar to Jonathan in flavor; dessert quality and processing quality outstanding; ripens with Golden Delicious; keeps well in storage up to 6 months. Fruit becomes greasy when overripe. Tree: resembles Golden Delicious in its strong wide crotches, but is more spreading like Jonathan; vigorous, but only of medium size; productivity comparable to McIntosh; crops produced mainly on spurs; triploid. Susceptible to fire blight, scab, and powdery mildew.
Jonagored.—Orig. at Halen, Belgium, by J. Morren. Introd. in 1985. U.S. plant patent 5937. Limb sport of Jonagold. Color tends to be brownish during the 3 weeks before harvest, but the brown coloring changes to an attractive solid red at harvest. Characteristics otherwise identical with Jonagold.
Jonagram.—Orig. in Mountain Grove, Mo., by Paul H. Shepard, Missouri State Fruit Expt. Sta. Introd. in 1956. Ingram x Jonathan; cross made in 1923; tested as Missouri A-1907. Fruit: medium to large, larger than Jonathan; roundish; stem thick, of medium length; skin thin, tough, smooth, pale bright yellow overlaid with red, faint stripe on some fruit, which have less color, dots small; flesh whiter than Jonathan but not as white as Ingram, firm, fine-grained juicy, crisp, tender, subacid; keeps well in storage; matures about 5 Sept. in southern Missouri, at Jonathan time. Tree: most outstanding characteristic is its late blooming habit, about 12 days after Jonathan and Delicious, and 6 days after Golden Delicious and Rome Beauty.
Jonalicious.—Orig. in Abilene, Texas, by Anna Morris Daniels. Introd. in 1960 by Stark Bros. Nurseries & Orchard Co., Louisiana, Mo. Plant patent 1777; 9 Dec. 1958; assigned to Stark Bros. Nurseries & Orchard Co. Chance seedling; discovered in 1933. Fruit: large, about 3 1/2 inches in diam.; round; skin thick, tough, smooth, dots conspicuous, ground color yellow, later blushed a bright, nearly solid red; flesh juicy, tinged yellow, texture firm, tender, crisp, subacid, aroma distinct, quality very good; keeps well for about 120 days in ordinary storage; ripens just after Jonathan, end of September in the midwestern U.S.; resembles Jonathan. Tree:

vigorous; upright; size medium; hardy; productive.

Jonamac.—Orig. in Geneva, N.Y., by Roger D. Way, New York State Agr. Expt. Sta. Introd. in 1972. McIntosh x Jonathan; cross made in 1944, selected in 1955, tested as New York 44428-5. Fruit: McIntosh type; size medium; round; skin attractive dark red blush, some stripe, 90% red, heavy waxy bloom; flesh nearly white, moderately firm, soft when fully ripe, eating quality good, superior to McIntosh; fair storage quality; ripens 8 days before McIntosh; less prone to fruit drop than McIntosh; dessert variety, not suitable for processing. Tree: size medium; upright-spreading; moderate vigor, productive; susceptible to apple scab; moderately tolerant to fire blight; good pollenizer for early and midseason varieties.

Jonared (*Jon-A-Red*).—Orig. in Peshastin, Wash., by William Uecker. Introd. in 1934. Plant patent 85; 16 Jan. 1934; assigned to Stark Bros. Nurseries & Orchards Co., Louisiana, Mo. Bud mutation of Jonathan; discovered in 1930. Fruit: size medium to large; skin colors earlier than parent, becoming solid red, smooth, thick, rough; flesh white tinged with yellow, firm, crisp, tender, juicy, aromatic; keeps well; ripens from mid-September to October in the midwestern U.S. Tree: substantially the same as Jonathan.

Jongrimes.—See Stark JonGrimes.

Jonica™ (*Schneica cv.*).—A red-fruited mutation of Jonagold, orig. in Germany. U.S. plant patent 7146, issued to Georg Schneider 13 Feb. 1990. Fruit: similar to Jonagold but with deeper overcolor and faded red hue; shape is less oblong that Jonagold; flavor medium sweet.

Jonnee.—Orig. in Caldwell, Idaho, by Stanley Robison and Ivan E. Wallace. Introd. in 1967. Plant patent 2934; 21 Oct. 1969; assigned to Hill Top Orchards & Nurseries, Hartford, Mich. Bud mutation of Blackjon; discovered in 1964. Fruit: resembles Blackjon but larger, and coloring 7 to 10 days earlier, with more intense color than Blackjon. Tree: identical to Blackjon.

Jono.—Orig. in Stillwater, Okla., by Herman A. Hinrichs, Agr. Expt. Sta., Oklahoma State Univ. Introd. in 1971. Summer Champion x Jonathan. Cross made in 1949, selected in 1960, tested as S16-9. Fruit: size medium; roundish, slightly conic; skin ground color yellow, 50% to 75% blushed with red, some striping, with darker red, waxy, smooth; flesh yellowish, moderately firm, juicy; quality similar to Jonathan; ripens 3 to 4 weeks before Jonathan. Tree: size medium; spreading; vigorous; hardiness equal to Jonathan or Summer Champion; tolerant to fire blight.

Jonwin.—Orig. in Ettersburg, Calif., by Albert F. Etter. Introd. in 1944. Plant patent 710; 17 Sept. 1946; assigned to George C. Roeding, Jr., California Nursery Co., Fremont, Calif. Jonathan x Baldwin; selected in 1944. Fruit: size and shape resemble Baldwin; stem length medium; skin yellow, striped red; flesh almost white, tender, juicy, crisp; core medium large, open; ripens in August and September, with Jonathan. Tree: vigorous; foliage larger and darker green than that of Jonathan.

Jored (*King Jonagold*).—Orig. in St. Truiden, Belgium, by Johann Nicholai. Discovered in 1985 as a whole-tree, red-fruited mutation of Jonagold with 80% to 90% red blush color. Plant patent 8851; assigned to TRECO®, Woodburn, Ore.

Jork 9 (*J9*).—See rootstocks for apple.

Joyce.—Orig. in Ottawa, Ontario, Canada, by Canada Dept. Agr., Res. Sta. Introd. about 1924. McIntosh o.p.; selected in 1912. Fruit: sizes well; attractive; equal to McIntosh in quality; same season as Wealthy. Tree: productive.

Jubilee.—Orig. in Summerland, British Columbia, Canada, by R.C. Palmer, Canada Dept. Agr., Res. Sta. Introd. in 1939. McIntosh x Grimes Golden; cross made in 1926; original tree planted in 1928; first fruited in 1934; selected in 1936 by A.J. Mann. Fruit: medium to large; oblong conic to oblate; skin yellow, overspread with deep red, similar to Baldwin; flesh cream-colored, firm, crisp, buttery, juicy, flavor mild, subacid, quality fair to good; ripens 3 weeks after McIntosh; stores well until February. Tree: vigorous; as hardy as McIntosh. Named in 1939 in honor of the British Columbia Fruit Growers' Association's Golden Jubilee Convention.

July Delicious.—Orig. in Baileyton, Cullman County, Ala., by R.L. Baker.

Introd. in 1948. Hackworth x Delicious. Fruit: small to medium; conic; skin red; flesh cream, subacid, texture medium; ripens early, about 60 days before Delicious; resembles Delicious. Tree: upright; vigorous.

Julyred.—Orig. in New Brunswick, N.J., by New Jersey Agr. Expt. Sta. Introd. in 1962. N.J. 8 (Petrel x Early McIntosh x [Melba x (Williams x Starr)]; cross made by G.W. Schneider; selected in 1955; introd. by L.F. Hough and Catherine H. Bailey of the above station; tested as N.J. 21. Fruit: large, averaging 2 3/4 inches in diam.; one-half to three-fourths of the skin washed and splashed red, attractive; firmer, has better shelf life and better shipping ability than most other early summer apple varieties; ripens about with Melba.

June Wealthy.—Orig. in Findlay, Ohio, by Earl C. Ohl. Introd. in 1947. Plant patent 765, 14 Oct. 1947; assigned to Stark Bros. Nurseries & Orchards Co., Louisiana, Mo. Wealthy o.p.; discovered about 1937. Fruit: size variable, up to 2 1/2 inches in diam., 2 in. long; oblate; skin light yellow, brightly splashed with deep red, thin, tough, glossy, smooth; dots many, conspicuous; flesh white with yellow-greenish tinge, firm, juicy, tender, crisp, subacid; aroma distinct; quality good; keeping quality medium; ripens early, first 2 weeks in July in Ohio, with Yellow Transparent; resembles Wealthy. Tree: size medium; vigorous, upright; hardy; productive and regular bearer.

Jupiter (*Malling Jupiter*).—Orig. in East Malling, Kent, England, by F.H. Alston in the breeding program at the East Malling Res. Sta. Introd. in 1981. Protected by Plant Breeders' Rights; owned by Plant Breeding International, Cambridge. Cox's Orange Pippin x Starking Delicious. Seed germinated in 1966. Tested as A 133/8. Fruit: large, 75 mm; attractive, well-colored, orange-red blush and stripes on a greenish-yellow ground (similar to Cox's Orange Pippin) with a smooth skin finish; some cracking around calyx; round to conic; flesh white, fine texture, juicy; flavor somewhat similar to Cox's Orange Pippin, but only fair quality; in 1.5C storage, susceptible to low-temperature breakdown, scald, and bitter pit, but remains firm until April in controlled atmosphere storage. Tree: vigorous with wide branch angles, trees 50% larger than Cox's Orange Pippin; high yielding; no preharvest drop; triploid, pollen not viable. Cox's Orange Pippin will not pollenize Jupiter; suitable pollenizers are Discovery, Spartan, Golden Delicious, Gala, Greensleeves, and Jester.

Jurale.—Orig. in Meriden, Conn. by Bernard Jurale. Introd. in 1974. Plant patent 3475; 12 Feb. 1974. Baldwin o.p., believed to be crossed with Red Astrachan. Fruit: medium-large; normal Baldwin shape; skin thin; not juicy; firm; tart and sweet; aromatic, spicy; juice pH 2.7 to 3; ripens about 1 Sept. in Connecticut; juice can be processed for sparkling beverage. Tree: size like Baldwin; vigorous; bears annually.

Kapai Jonathan.—Orig. in Frimley, Hastings, Hawkes Bay, New Zealand, by A. Tomlinson. Introd. in 1931. Bud mutation of Jonathan; selected in 1929. Fruit: resembles Jonathan, having improved color; skin bright dark red over entire fruit.

Karmijn (*Karmijn de Sonnaville, Karmine*).—Orig. in Wageningen, The Netherlands, by P. de Sonnaville. Introd. in 1971. Cox's Orange Pippin x Jonathan, crossed in 1949. U.S. plant patent pending; assigned to Carlton Plants, Dayton, Ore. Fruit: large; oblong-conical; color yellow-green with carmine red blush; russet can cause skin cracking; flesh color yellow; flesh texture firm, fine-grained; flavor is one of the best, with high sugar and acid; harvest with Delicious; storage life good through March, must be stored 4 weeks before good quality develops. Tree: moderately vigorous; horizontal branching; moderately precocious; low yields; bloom season mid to late; triploid, does not produce viable pollen.

Karmijn de Sonnaville.—See Karmijn.

Karmine.—See Karmijn.

Katja.—Orig. in Kristianstad, Sweden, by Balsgard Fruit Breeding Inst. Introd. in 1966. James Grieve x Worcester Pearmain. Cross made in 1947, selected in 1955, tested as BM 24353. Fruit: size medium; conical; skin nearly solid red, thick; flesh yellow-white, juicy; keeping and shipping quality good; ripens about 10 to 15 Sept., a week ahead of James Grieve; recommended for commercial orchards. Tree: size medium, upright, spreading; vigor moderate; hardy; productive and bears annually; seems to be tolerant to mildew and scab.

Keepsake.—Orig. in St. Paul, Minn., by C. Stushnoff, S. Munson, L.B. Hertz, W. Gray, and D.K. Wildung, Univ. of Minnesota. Introduced in 1979. MN447 (Malinda o.p.) x Northern Spy; cross made in 1936; selected in 1947; tested as MN 1593. Fruit: small to medium, 56 to 69 mm in diam.; red, some scarfskin; conical with angular sides; attractive; flesh light yellow, fine-grained, hard, very crisp, juicy; flavor very sweet, aromatic; quality good; harvest late October; storage life very long; shelf life very long. Tree: moderately vigorous, spreading; very winter hardy; productive; moderately resistant to apple scab, cedar apple rust, and fire blight.

Kemp Delicious.—See Topspur Delicious.

Kendall.—Orig. in Geneva, N.Y., by Richard Wellington, New York State Agr. Expt. Sta. Introd. in 1932. McIntosh x Zusoff; cross made in 1912. Fruit: large; round-oblate-conic; skin handsome, nearly solid dark red, bloom heavy; flesh whitish, crisp, tender, fine-grained, juicy, flavor sprightly; quality not as good as McIntosh; core small; quality very good; season about that of McIntosh, but keeps longer; hangs well on tree. Tree: large; vigorous; upright spreading; very productive; hardy; flower large, white, blooms in midseason; quite susceptible to fire blight and bitter pit.

Kent (*Malling Kent*).—Orig. in East Malling, Kent, England, by H.M. Tydeman, East Malling Res. Sta. Introd. in 1974. Protected by Plant Breeders' Rights; owned by Plant Breeding International, Cambridge. Cox's Orange Pippin x Jonathan. Tested from 1964 to 1974 as A 1379. Fruit: medium, most fruit 60 mm in diam. or larger; 75% skin surface dark coppery red on a green ground, occasionally extensive russet; shape slightly conic; texture firm; flavor very acid; late ripening; in 1.7C storage, keeps well until mid-April. Tree: medium size; upright; precocious cropping; little preharvest drop; blooms with Cox's Orange Pippin and Golden Delicious. This is not the same variety as the Kent described in the *Register of New Fruit & Nut Varieties* 2nd ed., 1972.

Kestrel.—Orig. by breeding program at Res. Sta, Agr. Canada, Kentville, Nova Scotia, Canada; cross made by C.J. Bishop in 1950, selected by R.P. Longley in 1961, introduced in 1975 by A.D. Crowe. New York Red Spy x Macoun. Tested as C13-30-88. Fruit: medium, 55 to 75 mm; shape round-conic, oblate; color almost completely washed or lightly striped with carmine to deep currant red on a pale greenish-yellow background; attractive; flesh creamy white, crisp, fine, juicy, exceptionally good texture; flavor subacid, mildly aromatic; eating quality very good to best; harvest time midseason, a few days after McIntosh; retains quality in storage; good shelf life. Tree: moderately vigorous, well-spurred, upright-spreading; annually productive. Has not had insect or disease problems in Nova Scotia, Canada, but not tested for actual resistances.

Keystone.—See Gourmet Golden.

Kidd's 21.—See Freyberg.

Kidd's D.5.—See Telstar.

Kidd's D.8.—See Gala.

Kidd's Orange (*Kidd's Orange Red*; *Delco* in New Zealand; *Kidd's Oranje Roode* in Holland).—Orig. in Greytown, Wairarapa, New Zealand, by J.H. Kidd. Introd. in 1924. Delicious x Cox's Orange Pippin. Fruit: medium to large; conical; skin greenish-yellow with red blush, thick; flesh cream, firm, crisp, juicy, sweet, aromatic, texture fine, quality good; stores well, but short shelf-life after removal from storage, susceptible to mealy breakdown; ripens before and coinciding with Jonathan; hangs well on tree and can be harvested over long period without fruit drop. Tree: vigorous, stronger than Delicious; very productive, regular bearer; long stem allowing easy spray penetration, reducing need for fruit thinning; susceptible to European canker; best suited to warm climates and would be expected to perform well in Italy, south of France, and warmer states of the U.S. Russeting increases in areas susceptible to spring frosts.

Kidd's Orange Red.—See Kidd's Orange.

Kidd's Oranje Roode.—See Kidd's Orange.

Killand.—Orig. in Mandan, N.D., by W.P. Baird, Northern Great Plains Field Sta. Introd. in 1957. McIntosh x Dolgo; selected in 1951; tested as Mandan 51-39. Fruit: size medium or below, averaging 4 1/2 fruit per lb; slightly oblate; stem medium long; skin bright red with darker red stripes, dots distinct, whitish; flesh firm, subacid, juicy, pleasant tasting, rates very good in cooking tests; matures after mid-September; keeps very well in storage. Tree: hardy; of moderate growth, spreading.

Kimball McIntosh.—Orig. in Littleton, Mass., by R. Kimball. Apparently not introd. Bud mutation of McIntosh. Fruit: large. A periclinal chimera 2-4-4-4, 2-4-4, or 2-4-2 determined by Haig Dermen, U.S. Dept. of Agriculture, Beltsville, Md.

King Jonagold.—See Jored.

King Jonathan.—See Kingjon.

King Red Jonathan.—See Kingjon.

King Luscious.—Orig. in Hendersonville, N.C., by Will Mills Dalton, Will Dalton & Son Nursery. Introd. in 1960 by William Edward Dalton and Bountiful Ridge Nurseries, Princess Anne, Md. Plant patent 1994; 6 Dec. 1960; assigned to Wm. E. Dalton and the above nursery. Chance seedling; discovered in 1935. Fruit: very large; skin highly colored, being deep red at maturity with a beautiful bloom, medium thick, tough, glossy; flesh white with yellowish tint, flavor excellent, firm, crisp, subacid; aroma wanting; ripens with Rome Beauty and Stayman Winesap. Tree: bears annually; semi-dwarf; apparently tolerant to apple scab; blooms 1 week after Rome Beauty. Tree: medium vigor; size medium; spreading; low; hardy; productive.

Kingjon (*King Jonathan*, *King Red Jonathan*).—Orig. in Wenatchee, Wash., by A.W. King. Introd. by various commercial firms since its discovery in 1933 as a limb sport on a 50-year-old tree of Jonathan. Fruit: resembles parent, but skin colors earlier and darker, at first red striped, then becoming solid red. Tree: standard type. A widely distributed and excellent color variant of Jonathan.

Kingscourt.—Orig. in Saskatoon, Saskatchewan, Canada, by C.F. Patterson, Univ. of Sask. Introd. in 1960 for home gardens. Columbia x Melba; selected in 1959; tested as A59-24. Fruit: up to 2 inches long and 2 inches in diam.; skin light green, slightly splashed with red; flesh firm, subacid, quality good; ripens in midseason to late. Tree: hardy.

Kinsei.—A selection from Mr. Majima Sato in Japan in 1972. Parents are Golden Delicious x Ralls Janet. It is not patented in the U.S. but it has patent #238 in Japan. Large, firm, creamy white flesh, yellow fruit resembling Golden Delicious but with more rough russet. Flavor sweet, slightly aromatic, and low acid. Good storage potential. Ripens after Golden Delicious. Tree: vigorous.

Kirby Delicious.—Orig. in Monroe, Mich., by C.I. Kirby, Ilgenfritz Nurseries. Parentage unknown; discovered about 1925. Introd about 1932. Fruit: Delicious type; skin solid red; keeps longer than Delicious; ripens 15 to 20 days before Delicious; hangs on tree better than Delicious, which it resembles. Tree: standard type.

Kirtland.—Orig. in Wooster, Ohio, by J.B. Keil, Ohio Agr. Expt. Sta. Introd. in the early 1930s. Ingram o.p.; selected about 1915. Fruit: medium to large; roundish-oblate to oblong; skin yellow mottled with bright red, attractive; flesh yellow, firm, juicy, sweetish to subacid; ripens in October to late winter. Tree: blooms late. Now obsolete.

Kitakami.—Orig. at Morioka, Japan, by Y. Yoshida, T. Haniuda, S. Tsuchiya, T. Sanada, and S. Sadamori, Morioka Branch, Fruit Tree Res. Sta. Introd. in 1982. Tohoku #2 (McIntosh x Worcester Pearmain) x Redgold. Japanese Reg. no. 366; 24 Feb. 1983. Tested as Morioka #15. Fruit: small, 150 to 200 g; oblate; skin solid bright red on yellow-green ground color, smooth, thick; flesh whitish-yellow, coarse, crisp, very juicy; flavor mild subacid, little aroma; good dessert quality. Ripens about 3 weeks before Gala. Tree: vigorous; upright; precocious; regularly productive. Diploid; blossoms after Jonathan and Fuji.

Kogetsu.—Orig. in Nanbu-machi, Aomori, Japan, by S. Taniuchi. Introd. in 1981. Golden Delicious x Jonathan. Japanese Reg. No. P.130. Fruit: medium-large, 300 to 350 g; round; skin 95% red; flesh creamy white, with good sugar/acid balance, excellent flavor, some drop. Harvest with Gala; stores well.

Kopps Delicious.—See Okanoma.

Korona.—Orig. at Kentville, Nova Scotia, Canada. Introd. in 1987. 34-91-13 (Mother x Red Rome Beauty) x Scotia (McIntosh o.p.). Fruit: medium to large; round-oblate, fairly regular; skin 80% to 95% attractive bright red over light greenish-yellow ground color; lenticels somewhat conspicuous; flesh light greenish cream, firm, tender, moderately juicy; aromatic subacid, very good. Matures a few days before McIntosh; not suitable for processing. Tree: small; spreading; precocious; annually productive.

Kosciusko.—Orig. in Kosciusko, Miss., by J.C.C. Price, Mississippi Agr. Expt. Sta., State College. Recommended for home gardens in 1938. Parentage unknown; discovered about 1922. Fruit: large; oblate; calyx open; skin dark red; ripens 15 July to 15 Aug. Tree: produces well in Mississippi. Primary value is for home gardens.

Kress McIntosh (*Stark McIntosh*).—Orig. in Hamilton, Mont., by Ben Kress. Introd. in 1934. Seedling of McIntosh; discovered about 1920. Fruit: large; skin bright red, over all blush; flesh white, juicy, crisp, subacid; ripens August through September in the midwestern U.S.; recommended for fresh market; pies and cider. Tree: extremely hardy; spreading; bears at an early age.

Krimskaja Zeleba.—See Zeleba.

Lady Williams.—A chance seedling found in Western Australia in 1949. Parents unknown but thought to be Granny Smith and Rockewood. Fruit: medium, round with pinkish red color. Flesh tart, white, firm, and crisp. Especially long storage life. Tree: medium, somewhat upright and precocious. Maturity late, three weeks after Granny Smith.

Lakeland.—Orig. in Excelsior, Minn., by the Univ. of Minnesota Fruit Breeding Farm. Introd. in 1950. Malinda o.p.; seed collected in 1907; selected in 1927; tested as Minn. 978. Fruit: size medium; oblate to roundish oblate; skin attractive bright, medium dark red, obscurely striped, medium tender; stem short, medium thick; flesh creamy yellow, sometimes slightly tinged red, texture fine, juicy, medium tender, mildly acid changing to subacid in cold storage; quality very good for sauce and pies, good for dessert; stores well to mid-December or later; no preharvest drop; Wealthy type; does not require thinning; nonclustering fruiting habit. Tree: bears annually; vigorous; hardy; susceptible to cedar rust, moderately resistant to fire blight and apple scab diseases.

Laking.—Orig. in Guelph, Ontario, Canada, by Charles Laking. Introd. about 1925. Parentage unknown; discovered in 1920. Fruit: size medium; skin yellowish-green, washed and streaked with red, appearance attractive; quality good; resembles King but keeps longer. Tree: more hardy than King. Apparently no longer propagated.

Lalla Delicious.—Orig. in Lalla, Tasmania, by Frank Walker. Introd. in 1928. Bud mutation of Delicious; discovered in 1914; now growing at the New York State Agr. Expt. Sta., Geneva. Fruit: a solid red striped Delicious; ripens in midseason. Tree: standard type; performs well at Geneva; appears to be more productive than Delicious, which it resembles.

Lambton.—Orig. in Saskatoon, Saskatchewan, Canada, by C.F. Patterson, Univ. Sask. Introd. in 1960 for home gardens. Columbia x Melba; selected in 1946; tested as A60-201. Fruit: up to 2 1/2 inches long and 1 3/4 inches in diam. under field conditions without irrigation in Saskatoon; slightly wedge-shaped; skin pale yellow with slight brownish blush; flesh white, mildly subacid, quality good when sound, but tends to break down prematurely; ripens in midseason. Tree: hardy.

Lamont.—See Starkspur Cortland.

Lantz Haralson.—See Haralred.

Large Transparent.—See Lodi.

Large Wealthy.—See Stevenson Wealthy.

Large Yellow Transparent.—See Perrine Yellow Transparent.

Law Red Delicious.—See Hawkes Bay Red Delicious.

Law Rome Beauty.—Orig. in Tonasket, Wash., by Walter A. Law. Introd. in 1958. Plant patent 1550; 8 Jan. 1957; assigned to Stark Bros. Nurseries & Orchards Co., Louisiana, Mo. Bud mutation of Rome Beauty; discovered in 1952 on a 35-year-old tree. Fruit: large, about 3 inches in diameter; shape somewhat longer and more conical than that of Rome Beauty;

skin with overall bright red color, which develops about 14 days before maturity, thick, tough, glossy, waxy; flesh white with greenish tint, firm, fine, crisp, subacid, no aroma; quality good; core large; dessert and culinary uses; keeps well in ordinary storage; ripens during third week in October; resembles Rome Beauty. Tree: large, vigorous, hardy, productive, bears regularly.

Lawfam.—Orig. in Ottawa, Ontario, Canada, by Canada Dept. Agr., Res. Sta. Introd. about 1935. Lawver x Fameuse; selected in 1921. Fruit: skin highly colored red; uniform; quality good; keeps longer than McIntosh. Tree: strong grower with good form and strong crotches; as hardy as McIntosh.

Lawspur.—Orig. in Unity Township, Columbiana County, Ohio, by D.E. Simmons. Plant patent 4096; 30 Aug. 1977; assigned to Hilltop Orchards & Nurseries, Hartford, Mich. Discovered in 1973 as a limb mutation of the Law Red Rome, tested under DS55. Fruit: typical Law Red Rome; subacid, quality good; medium grained texture; excellent for storage and shipping. Tree: upright spreading with strong spurring and strong dwarfing habit.

Laxton's Epicure.—Orig. in England by Laxton Nursery Co. Introd. in England in 1929. Introd. in the U.S. in 1937 as P.I. 123744. Cox's Orange Pippin x Wealthy. Fruit: 2 1/2 x 2 1/4 inches in diam.; round; regular; resembles Charles Ross; skin green, streaked with crimson; flesh tender, juicy, sweet; seeds large and long usually abundant. Tree: productive; flowers self-sterile; flowering earlier than Cox's Orange Pippin; harvest September.

Leafland.—Orig. in Carlos, Alberta, Canada, by Robert Erskine. Introd. in 1964. Parentage unknown, from mixed seed received from Morden Expt. Farm, Morden, Manitoba, Canada; selected in 1959. Fruit: resembles Heyer 12, but stores better; good cooking quality. Tree: very hardy.

Leelanau.—Orig. in Leelanau County, Mich., by C.W. Leofler. Introd. about 1928. Reported to be McIntosh x Wagener; selected in the early 1920s. Fruit: skin scarlet over a whitish ground color; matures relatively early, about 1 Sept. in northern Mich.; keeping quality poor. Tree: size medium; heavily spurred; resembles Wagener.

Legana.—Orig. in Legana, Northern Tasmania, Australia, by James Bulman. Introd. about 1950. Democrat x Delicious. Discovered in the late 1930s. Introd. in the U.S. as P.I. 203226 and P.I. 242012. Fruit: medium to large; round-conic; skin yellow ground color almost wholly covered with deep carmine, thick, tough, smooth; flesh yellow with greenish tinge, firm, crisp, juicy, aromatic, quality excellent; season May to June. Tree: not particularly subject to any of the pests that attack commercial varieties in Australia; as tree develops there is tendency toward alternate cropping under the short spur system of pruning.

Lemcke Rome.—See Nero Rome.

Lemon.—See Linonoe.

Levering Limbertwig.—Orig. in Ararat, Va., by S.R. Levering. Apparently not introd. Bud mutation of Limbertwig. Fruit: large. A periclinal cytochimera 2-4-4-4; breeds like a tetraploid; chimeral nature determined at the New York State Agr. Expt. Sta., Geneva; accessioned in 1957 by this station.

Liberty.—Orig. in Geneva. N.Y., by Robert Lamb, New York State Agr. Expt. Sta. Introd. in 1978. Macoun x PRI 54-12. Selected in 1961 or 1964. Fruit: a McIntosh type with good crisp flesh texture and quality. Tree: resistant to scab, cedar apple rust, fire blight, and powdery mildew, very productive.

Lina.—Orig. in Brookings, S.D., by N.E. Hansen, South Dakota Agr. Expt. Sta. Introd. in 1933. Malinda o.p. Fruit: averages 2 1/2 inches in diam.; skin yellow; somewhat conical; flesh juicy, subacid, flavor good; matures late. Now apparently obsolete.

Linda.—Orig. in Ottawa, Ontario, Canada, by Canada Dept. Agr., Res. Sta. Introd. about 1935. Langford Beauty o.p.; selected in 1914. Fruit: large; skin an attractive deep crimson; flavor sprightly; quality good in midwinter. Tree: vigorous; comes into bearing at an early age; not fully hardy in Ottawa and St. Lawrence River Valleys of Ontario and Quebec unless topworked on a hardy framework.

Lindel.—Orig. in Smithfield, Ontario, Canada, by S.J. Leuty and H.B. Heeney,

Canada Dept. of Agr., Expt. Sta. Introd. in 1971. Richared Delicious x Linda, cross made in 1939, selected in 1953 by the late D.S. Blair, tested as T-397. Fruit: 2 1/2 to 2 3/4 inches in diam.; round, conic; skin moderately thin, yellowish ground color, washed with medium red striped overcolor; flesh cream color, juicy, texture slightly coarse but firm, acid moderate; ripens late October, about 1 week later than Delicious, remaining in good condition under refrigeration until March; recommended as fresh fruit and for processing Tree: large, spreading; vigorous; slightly hardier than McIntosh; productive.

Linonoe (*Lemon*).—Orig. in Russia. Introd. in the U.S. in 1922 by N.E. Hansen, South Dakota Agr. Expt. Sta., Brookings, S.D. Fruit: skin greenish-yellow; large, up to 3 inches in diam.; quality poor; keeps well all winter. Tree: small; hardy; productive.

Little Giant.—Orig. near Milton-Freewater, Oregon, by J.D. Slusarenko. Plant patent 3974; 16 Nov. 1976. Discovered as a bud mutation growing in an orchard of Starkrimson Delicious. Fruit: same qualities as Starkrimson Delicious. Tree: dwarf; compact; spreading; up to 9 feet tall.

Lizakowsky.—See Saint Clair.

Lizzie.—See P.16 in apple rootstocks.

Lobo.—Orig. in Ottawa, Ontario, Canada, by Canada Dept. Agr., Res. Sta. Introd. about 1930. McIntosh o.p.; selected in 1906. Fruit: quality not quite as good as McIntosh but earlier in ripening; resembles McIntosh considerably in appearance. Tree: hardy; productive; bears at an early age.

Lodi (*Early Golden, Golden Lodi, Improved Early Transparent, Large Transparent*).—Orig. in Geneva, N.Y., by Richard Wellington, New York State Agr. Expt. Sta. Introd. in 1924. Montgomery x Yellow Transparent; cross made in 1912. Fruit: larger, keeps longer, does not grow mealy and soften at center as quickly as Yellow Transparent; skin yellow, thick; flesh firm, crisp, flavor rich, sprightly; ripens later than Yellow Transparent, ripens in July in the midwestern U.S.; culinary and fresh uses; resembles Yellow Transparent. Tree: large; tends to bear in alternate years, can be made annually productive through proper use of blossom thinning sprays.

Lonely.—See Hardi-Spur Delicious.

Loop Baldwin.—Orig. in North East, Pa., by H.L. Loop. Apparently not introd. Considered to be a bud mutation of Baldwin. Fruit: skin possibly with more red color than Baldwin. Accessioned in 1934 by the New York State Agr. Expt. Sta., Geneva.

Loop Rome (*Giant Rome*).—Orig. as a tree planted about 1922 in North East, Pa., by H.S. Loop. Apparently not introd. Considered to be a bud mutation of Rome. Fruit: large; asymmetrical; quality similar to parent. A periclinal cytochimera 2-2-2-4. Tree: good cropper; tends to revert to diploid. Chimeral nature determined at the New York State Agr. Expt. Sta., Geneva; accessioned in 1931 by this station.

Loop Russet Baldwin.—Orig. in North East, Pa., by H.L. Loop. Apparently not introd. Considered to be a bud mutation of Baldwin. Fruit: russeted; tends to revert to normal red fruit. Accessioned in 1934 by the New York State Agr. Expt. Sta., Geneva.

Loop Spy (*Giant Spy*).—Orig. in North East, Pa., by H.S. Loop. Apparently not introd. Considered to be a bud mutation of Northern Spy; discovered in 1925 in orchard of A.I. Loop planted about 1908. Fruit: large; flesh more tender, softer, and more subject to bitter pit than parent. A periclinal cytochimera 2-2-4-4. Tree: poor cropper. Chimeral nature determined at the New York State Agr. Expt. Sta., Geneva; accessioned in 1931 by this station.

Loop Striped Rome.—Orig. in North East, Pa., by H.L. Loop. Apparently not introd. Considered to be a bud mutation of Rome. Fruit: similar to Rome but possibly with less red skin color.

Loop Wealthy (*Giant Wealthy*).—Orig. in North East, Pa., by H.S. Loop. Apparently not introd.; included here because of its primary interest to fruit breeders. Considered to be a bud mutation of Wealthy; discovered in orchard of G.C. Smith & Sons planted about 1909. Fruit: large; similar to parent. Produces diploid gametes; a periclinal cytochimera 2-4-4-4. Tree: not as productive as parent. Chimeral nature determined at the New York State Agr. Expt. Sta., Geneva; accessioned in 1947 by this station.

Lovejoy Cortland.—Orig. in Springvale, Maine, by R.H. Lovejoy. Apparently not introd. Bud mutation of Cortland. Fruit: large. A periclinal cytochimera 2-4-4-4; breeds like a tetraploid; chimeral nature determined at the New York State Agr. Expt. Sta., Geneva; accessioned in 1957 by this station; now lost.

Luke.—Orig. in Rosthern, Saskatchewan, Canada, by J. Luke. Introd. in 1961. Parentage unknown; Luke obtained this as a selection from the Expt. Farm at Rosthern when it was closed in 1940. Fruit: large, to 3 1/2 inches in diam.; oblong to rounded; skin well colored like Delicious; flesh greenish-white, somewhat coarse and granular, firm, moderately juicy; good as sauce or baked, dessert quality only fair; resembles Delicious in shape and Northern Spy in color. Tree: tall; rounded to spreading; structurally very strong; hardiness exceptional; annually productive. Promising for some northern climates, yet the season at Saskatoon, Saskatchewan, Canada, is not long enough for fruit to mature properly. Tested at the Expt. Farm, Canada Dept. Agr., Morden, Manitoba, Canada.

Lustre Elstar.—See Elstar.

Lysgolden.—See Goldenir.

M.1, M.2, and other rootstocks of this series.—See rootstocks for apple.

MM 102, MM 103, and other rootstocks of this series.—See rootstocks for apple.

Maayan.—Orig. in Rehovot, Israel, by Chanan Oppenheimer. Introd. in 1967. F_2 selection (Calville St. Sauveur x Damascus) x Delicious. Fruit: medium size, 60 to 80 mm; round; skin yellow ground cover with over 50% covered with red, attractive; flesh juicy, firm, aroma good, texture good, flavor sweet, acid; of the Delicious group in taste; keeping quality good, keeps at room temperature at least 2 weeks, in cold storage at least 2 months; ships well; ripens from early June to late July. Tree: large; bears on spurs only; by bending branches, young trees can be induced to bear in third year; blooms end of February to beginning of March. This variety has been grown successfully in small orchards. Maayan is a Hebrew girl's name meaning spring.

MAC-9.—See Mark in apple rootstocks.

Macfree.—Orig. in Trenton, Ontario, by L.P.S. Spangelo, S.J. Leuty, H.B. Heeney, and L.L. Modderman, Smithfield Experimental Farm, Agr. Canada. Introd. in 1974. McIntosh x PRI 48-177 (from crosses at the Univ. of Illinois involving Rome Beauty, Jonathan, and *M. floribunda*); cross made in 1953, selected in 1963, tested as 0-532 and 53-08-02. Fruit: size equal to or slightly smaller than McIntosh; round to slightly conic, somewhat unequal, smooth; skin greenish yellow ground color, covered with 75% medium to dark red; flesh white, may be some green tinge, juicy, slightly coarse, tough; quality medium to good; ripens second week of October in Trenton; keeps 3 to 4 months. Tree: spreading; vigorous; hardy; resistant to *Venturia inaequalis.*

Macoun.—Orig. in Geneva, N.Y., by Richard Wellington. Introd. in 1923. McIntosh x Jersey Black; seed borne in 1909; fruit first described in 1918. Fruit: similar to McIntosh, but smaller; more ribbed and develops a black red color at maturity; flesh white, richly flavored, aromatic; high dessert quality; ripens 1 month after McIntosh. Tree: upright, may need extra branch thinning to develop a spreading top; however, upright habit minimized as tree comes into bearing.

MacSpur.—Orig. in Summerland, British Columbia, Canada, by Mervyn Greenslade. Introd. in 1970 by Hilltop Orchards & Nurseries, Hartford, Mich. Plant patent 2982; 28 July 1970, assigned to Hilltop Orchards & Nurseries. Whole-tree sport of Summerland McIntosh, discovered in 1964. Fruit: same as Summerland McIntosh and ripening date same. Tree: spur type; very compact growth; upright; vigor equal to McIntosh, productive, highly resistant to powdery mildew, extremely hardy. The British Columbia Fruit Growers' Assn. selected this variety as the official McIntosh strain for British Columbia.

Macy.—Orig. in Ames. Iowa, H.L. Lantz, Iowa Agr. Expt. Sta. Introd. in 1921. Seedling of (Northwestern Greening x Wealthy). Fruit: medium to large; roundish; skin pale green or yellow, striped and mottled with red, medium thick, medium tough, slightly waxy; flesh firm, resembling

Wealthy in texture and amount of juice, pleasantly subacid, flavor good to very good; tends to drop badly. Tree: vigorous; spreading; productive.

Maehara Starking.—Orig. in Kelowna, British Columbia, Canada, by T. Maehara. Introd. in 1955. Bud mutation of Starking Delicious; discovered in 1954 on a 16-year-old tree. Fruit: Delicious type; skin very dark red, appearing some 10 days before that of parent, some striping. Tree: standard type.

Maga (*Into*).—Orig. in Brookings, S.D., by N.E. Hansen, South Dakota Agr. Expt. Sta. Introd. in 1933. McIntosh x Virginia crab apple; selected about 1919. Fruit: averages 2 1/4 inches in diam.; oblate; skin striped red; flesh subacid, quality good, similar to McIntosh in flavor; ripens in August. Probably obsolete.

Magnolia Gold.—Orig. in Fort Valley, Ga., by Wm. J. Wilson, Magnolia Farms. Plant patent 3110. Introd. in 1970 by Bountiful Ridge Nursery, Princess Anne, Md. Golden Delicious o.p. Discovered in the 1960s. Fruit: size medium, round-elongate; skin yellow, smooth, russet-free; flesh mellow, good quality; recommended as a Golden Delicious type where russeting is a problem; ripens with Golden Delicious. Tree: large, upright to spreading; vigorous and productive; needs a pollenizer for best results; is a good pollenizer for other varieties.

Maigold (*Gold-N-Rose*).—Orig. in Wadenswil, Switzerland, by breeding program of the Swiss Federal Agr. Res. Sta. Franc-Roseau x Golden Delicious; cross made in 1944; introduced in 1964. Fruit: medium, 60 to 80 mm; skin bright red on exposed side on pale green ground color; no russet; does not shrivel in storage; flesh yellow at maturity, crisp, juicy, very firm; flavor mildly subacid, sweet, pleasant. Harvest late, with Winesap; exceptionally long storage life in air. Tree: vigorous; precocious; productive; annual cropping; blossom season late.

Majestic.—Orig. in Sebastopol, Calif., by Wm. E. Silva. Introd. in 1965. Rome x Cook seedling; selected in 1963; tested as Silva 6. Fruit: large; round; symmetrical; skin mottled red to striped; stem slender, long; flesh white, crisp, firm, aromatic; matures before Rome. Tree: regular bearer.

Malling Bountiful.—See Bountiful.

Malling Fiesta.—See Fiesta.

Malling Greensleeves.—See Greensleeves.

Malling Jester.—See Jester.

Malling Jupiter.—See Jupiter.

Malling Kent.—See Kent.

Malling Redsleeves.—See Redsleeves.

Malling Suntan.—See Suntan.

Manalta.—Orig. in Carlos, Alberta, Canada, by Robert Erskine. Introd. in 1963. Parentage unknown, from mixed seed received from Morden Expt. Farm, Morden, Manitoba, Canada. Fruit: to 2 3/4 inches in diam.; globose; skin yellow ground color half covered with red blush; flesh good cooking quality, fair for dessert after storage. Tree: very hardy

Manan.—Orig. in Morden, Manitoba, Canada, by Canada Dept. Agr., Res. Sta. Introd. in 1930. Antonovka o.p. Fruit: size medium, resembling Northern Spy; skin greenish-yellow, washed with deep red; flesh white, fine-grained, juicy, tart; quality fair; season October to January. Tree: upright vigorous; hardier than most standard apples; productive.

Manbee.—Orig. in Morden, Manitoba, Canada, by Canada Dept. Agr., Res. Sta. Introd. in 1938. Change seedling discovered about 1936. Fruit: averages about 2 3/8 inches in diam.; skin moderately bright red; flesh white, crisp, subacid, quality fair; ripens in late August; promising as an early, small apple for northern regions. Tree: vigorous, hardy, very productive.

Mance Winesap.—See Starkspur Winesap.

Manchu.—See rootstocks for apple.

Mandan.—Orig. in Mandan, N.D., by William A. Oitto, Northern Great Plains Res. Center. Introd. in 1965. Duchess x Starking Delicious; tested as Mandan 49-4; sibling of Garrison. Fruit: medium to large, average about 2 3/4 inches in diam.; round conic to oblong conic; skin orange-red to medium red with dark red splotches over greenish-yellow ground, grayish dots prominent; flesh yellowish, firm crisp, very juicy, mildly subacid; excellent for dessert purposes, good for culinary uses; ripens early to mid-September; keeps well in common storage through November. Tree: vigorous, spreading; annually productive; hardy; considerable resistance to fire blight. Primarily for the home garden.

Manitoba.—Orig. in Morden, Manitoba, Canada, by Canada Dept. Agr., Res. Sta. Introd. in 1931. Duchess of Oldenburg o.p.; selected in 1925. Fruit: medium to fairly large; roundish oblate; skin yellow to yellow amber, slight blush; flesh white, crisp, juicy, texture fine-grained, melting, flavor spicy, subacid; quality very good for cooking; season November through February. Tree: roundish; vigorous; only moderately hardy.

Manitoba Spy.—Orig. in Morden, Manitoba, Canada, by Canada Dept. Agr., Res. Sta. Introd. in 1931. Patten Greening o.p.; selected in 1927. Fruit: large; skin dull, dark red; flesh tender, juicy, somewhat coarse, flavor mild, sprightly, good for cooking; season November to February; resembles Northern Spy. Tree: upright spreading; hardy; vigorous; productive.

Manred.—Orig. in Morden, Manitoba, Canada, by Canada Dept. Agr., Res. Sta. Introd. in 1930. Anisim o.p. Fruit: small; skin yellow with red blush and gray bloom, thin; flesh white, fine-grained, subacid, quality fair to good; season October to November. Tree: spreading; strong grower; hardy; productive.

Mantet.—Orig. in Morden, Manitoba, Canada, by Canada Dept. Agr., Res. Sta. Introd. in 1929. Tetofsky o.p.; selected in 1928. Fruit: medium size; roundish; skin heavily washed and striped bright red; flesh fine-grained, tender, juicy, aromatic, sweet, very pleasant; fresh eating quality excellent; season early to late August. Tree: upright; hardy; productive.

Manton.—Orig. in Morden, Manitoba, Canada, by Canada Dept. Agr., Res. Sta. Introd. in 1929. Antonovka o.p. Fruit: medium size; skin greenish-amber washed with rich red, russet around stem cavity; flesh cream, crisp, very juicy, pleasant; season January to May. Tree: upright; vigorous; productive.

Margaret.—Orig. in Lander, Wyo., by J.M. Steinbrech, Wyoming Agr. Expt. Sta. Introd. in 1933. Fruit: large; oblong; skin light green, overlaid with red stripes; dessert and cooking quality good; ripens about 25 Sept.; resembles Roberts. Tree: vigorous.

Margaret Pratt.—Orig. in Owen Sound, Ontario, Canada, by C.A. Fleming. Introd. about 1938. Parentage unknown; discovered before 1923 in the garden of Margaret Pratt. Fruit: resembles Red Astrachan but larger; should be picked early because it breaks down if allowed to take on full color; used for cooking. Tree: appearance and performance similar to Red Astrachan; good bearer.

Mari.—See Early Cortland.

Mark (*MAC-9*).—See rootstocks for apple.

Marshall McIntosh.—Orig. in Fitchburg, Mass., by George and Alfred Marshall, Marshall Farm. Introd. in 1975. Bud mutation of McIntosh; discovered in 1967. Fruit: colors a full bright red earlier than other McIntosh strains; a blush strain that develops more and earlier red color than other strains. Fruit may be small and the flesh greener and vigor may be lower than the standard strain. It is very susceptible to low O_2 injury in CA storage.

Mart.—Orig. in Lander, Wyo., by J.M. Steinbrech, Wyoming Agr. Expt. Sta. Introd. in 1933. Fruit: large; oblate; skin green with blush on the sunny side, with a few dull crimson splashes; flesh juicy; ripens about 25 Sept.; keeps medium well. Tree: head clearly round, formed by multiple main branches originating low on main stem; upright spreading; moderately vigorous growth; hardy; close (dense); twigs slender; internodes medium spaced; buds blunt (obtuse); free; heavy pubescence; bark dark brown; lenticels large, oval. Leaf: oval, average 6.9 x 10 cm; edges unevenly serrate to almost medium toothed; top smooth; bottom lightly tomentose all over; much lighter colored; veins prominent on lower side; petiole average 3.6 cm; strong; completely tomentose. Fruit: requires thinning.

Martin.—Orig. in Gorum, Natchitoches Parish, La., by Charles A. Martin. Introd. in 1943. Plant patent 564; 5 Jan. 1943. Parentage unknown; discovered in 1936. Fruit: size medium to large, average 2 3/4 inches in diam.; globose; skin yellow with red splashing, glossy, thick, tough, waxy, dots large; flesh yellowish, firm, tender, crisp, subacid, mild; aroma pronounced; quality fair to good; ripens early, about 15 Aug. Tree: grows successfully in Louisiana; apparently fire blight resistant.

Maruba-kaido N-1.—See rootstocks for apple.

Maud.—Orig. in Ames, Iowa, by S.A. Beach, Iowa Agr. Expt. Sta. Introd. in 1921. McIntosh x Longfield. Fruit: small; skin solid red; tenaciously clinging to tree even after full maturity; quality fair to poor; ripening in August. Tree: original tree appears vigorous; healthy; productive.

Mavis.—Origin unknown. Apparently not introd.; used in breeding program at New Jersey Agr. Expt. Sta., New Brunswick. Parentage unknown; acquired by M.A. Blake from a commercial nursery about 1926. Fruit: medium; oblate; skin one-quarter to one-half blushed, rough; quality fair; keeps firm in storage until March; ripens 2 weeks before Stayman Winesap.

May Sweet (*Yellow Rose*).—Orig. in Hammond, N.Y., by Leslie Richardson. Introd. in 1960 by Fred L. Ashworth, nurseryman, Heuvelton, N.Y. Parentage unknown; discovered in 1958. Fruit: large; appears resistant to scab; keeps well in cold storage; resembles Tolman Sweet. Tree: hardy.

McClintock Grimes (*Ballou Grimes, Unprofitable Grimes*).—Orig. in Lafayette, Ind., by J.A. McClintock, Purdue Univ. Not introd. Considered to be a bud mutation of Grimes. Fruit: large. A periclinal cytochimera 2-2-4-4. Chimeral nature determined at the New York State Agr. Expt. Sta., Geneva; accessioned in 1948.

McCloy.—Orig. in Kinistino, Saskatchewan, Canada, by John McCloy. Introd. in 1948 by Percy H. Wright, Saskatoon, Saskatchewan, Canada. Open-pollinated seedling of a Russian apple discovered about 1945. Fruit: skin tender; dessert quality good; ripens early; shape of Delicious; poor keeper. Tree: extremely hardy, to –50F; yields heavy crops.

McIntosh-Pound.—Orig. in University Park, Pa., by C.E. Myers, Pennsylvania State Univ. Introd. in 1955 by the Converse Co. Nursery, Amherst, N.H. Pound x McIntosh; from crosses made in 1926 by students in a course (Advanced Plant Breeding) under the direction of Myers; seedlings from this cross were distributed. Fruit: flattened, similar to Winesap; flesh very white, firm, subacid, juicy, flavor of Jonathan; ripens in mid-September. Tree: size medium; tends to be spreading; foliage heavy.

McLemore.—Orig. in Muskogee, Okla., by J.V. McLemore. Introd. in 1968. Seed of Delicious planted in 1948; propagated in 1958 by Oklahoma State Univ.; released and named in 1969 by this university. Fruit: Delicious type; medium size; skin with a red blush, sometimes slightly striped; flesh yellowish, crisp, good flavor; dessert quality good, makes excellent pies, and shows good potential for commercial processing; has stored well in a refrigerator for 6 months; a summer apple. Tree: medium spreading; vigorous; hardy; productive.

Mclicious.—Orig. in Plattekill, N.Y., by Thomas A. Zalesak, of Wallkill, N.Y. Introd. in 1963. Early McIntosh x Golden Delicious; selected in 1957. Fruit: large, 2 1/2 to 3 inches in diam.; shape of Golden Delicious; skin golden yellow overlaid with red, no russet; flesh pleasantly tart yet sweet, white, crisp, fine-grained, very aromatic; ripens very early, during the second week of August in Ulster County, N.Y.; resembles Delicious. Tree: upright, spreading; very vigorous.

McQuarrie.—Orig. in Riverside, Wash., by E. Barclay Brauns, Wanatchee, Wash. Introd. in 1967. Plant patent 2750; 11 July 1967. Parentage unknown, chance seedling. Fruit: size medium; round-conic, irregular in shape; skin light yellow ground color blushed with bright red, dots numerous, tough; flesh tinged with yellow, firm, crisp, juicy, subacid; quality good; ripens 2 weeks before Jonathan. Tree: large; vigorous; spreading, open.

McShay.—Orig. in West Lafayette, Ind., by a cooperative breeding program consisting of Agr. Expt. Stations of Oregon (S.A. Mehlenbacher and M.M. Thompson), Purdue Univ. (E.B. Williams, J. Janick, and F.H. Emerson), Univ. of Illinois (S.S. Korban and D.F. Dayton) and Rutgers Univ. (L.F. Hough and C. Bailey). Plant patent 4724; 26 May 1981; assigned to Carlton Plants, Dayton, Ore. McIntosh x PRI 612-4. Ancestry includes Starking (giant limb) and Golden Delicious; this is a fifth-generation derivative of C.S. Crandall's Rome Beauty x *M. floribunda* 821 cross. Selected in 1970; tested as PRI 1773-8. Named in honor of late J.R. Shay, co-founder of the PRI program. Fruit: medium, 65 mm in diam.; round-conic; skin currant-red on uranium-green ground color, smooth, thin, no russet; flesh chartreuse-green, fine-grained, juicy; flavor mildly subacid; quality very good. Ripens 2 weeks before Jonathan; stores 2 to 3 months at 1C in air. Tree: vigorous; upright; annual cropping; field immune to apple scab (V_f gene); leaves, but not fruit, susceptible to powdery mildew. Flowers after McIntosh, before Rome Beauty; pollen viable.

Mead Gravenstein.—Orig. in Lunenburg, Mass., by R.J. Fiske on the Mead farm. Introd. in 1939. Bud mutation of Gravenstein; discovered in 1927. Fruit: skin solid red with very little striping; resembles Gravenstein. Appears to be superior to other sports of Gravenstein as tested at the New York State Agr. Expt. Sta., Geneva.

Medina—Orig. in Geneva, N.Y., by Richard Wellington, New York State Agr. Expt. Sta. Introd. in 1922. Deacon Jones x Delicious; cross made in 1911. Fruit: larger than Delicious; skin golden-yellow ground color, blushed with attractive red; flesh rather coarse; prolongs season of Delicious, which it resembles.

Melba.—Orig. in Ottawa, Ontario, Canada, by Canada Dept. Agr., Res. Sta. Introd. about 1924. McIntosh o.p.; selected in 1909. Fruit: attractive; quality high; ripens during the summer. Tree: hardy; bears at a young age; productive. Awarded the Silver Wilder Medal by the American Pomological Society in 1927.

Melrose.—Orig. in Wooster, Ohio, by Freeman S. Howlett, Ohio Agr. Expt. Sta. Introd. in 1944. Jonathan x Delicious; selected in 1937. Fruit: medium to large; skin attractive; quality better than Rome Beauty; dessert and culinary quality excellent; does not develop Jonathan spot; harvest season nearly 3 weeks later than Jonathan, and 11 days after Delicious, preceding Rome Beauty and Gallia Beauty by a few days; resembles Jonathan in color and shape, but less tart; late keeping, maintaining excellent flavor into May in storage; a good late fall variety to supplement Stayman Winesap, Rome Beauty, and Gallia Beauty in regions of long growing season, and in areas where the parents are grown. Tree: very productive, being surpassed only by Stayman Winesap; bears at an early age.

Merchant.—Orig. in Tift County, Ga., by Major C. Collins, Tifton, Ga. Introd. in 1977. Unknown seedling; discovered in 1960s. Fruit: 2 1/2 to 2 1/4 inches in diam.; round to oblate, symmetric, smooth in shape; skin ground color pale yellow-green, half covered by reddish-brown blush; flesh white, fine-textured, crisp; flavor nonacid, aromatic, semisweet. Tree: moderately vigorous; is in midbloom when Red Delicious is just beginning to bloom. Recommended as a dooryard variety.

Merton Beauty.—Orig. in Bayfordbury, Hertford, Herts., England, by the John Innes Institute. Introd. in 1958. Ellison's Orange x Cox's Orange Pippin; selected in 1946. Fruit: size medium; skin bright crimson; flavor aromatic; season Sept. and Oct.; a very attractive, high quality fruit. Tree: moderately vigorous; spreading; spurs freely.

Merton Charm.—Orig. in Merton, England, by M.B. Crane and A.G. Brown, John Innes Institute, Norwich, England. Introd. in 1961. Rogers McIntosh x Cox's Orange Pippin. Cross made in 1933, tested as 2326. Fruit: size medium; round , slightly flattened; skin greenish-yellow, 50% covered with red blush, smooth, no russet, very slightly greasy; flesh tender but crisp, juicy, sweet; ripens Sept. and Oct. Tree: size medium, spreading; moderately productive, regular bearer.

Merton Delight.—Orig. in Bayfordbury, Hertford, Herts., England, by the John Innes Institute. Introd. in 1953, into the U.S. as P.I. 205461. Cox's Orange Pippin x Golden Russet. Fruit: size medium to fairly large, averaging 2 1/4 to 2 3/4 inches in diameter; roundish oblate; skin 80% striped pinkish-red over yellow; flesh cream-colored, tender, soft, crisp, moderately juicy, sweet to subacid, spicy, aromatic, dessert quality good; ripens first week in Aug. in Maryland, in late Aug. in New York; resembles Cox's Orange Pippin. Tree: moderately vigorous; spreading; spurs freely.

Merton Joy.—Orig. in Bayfordbury, Hertford, England, by A.G. Brown, John Innes Institute, Norwich, England. Introd. in 1965. Cox's Orange Pippin x Sdlg 855 (Strumer Pippin x Cox's Orange Pippin). Cross made in 1940, tested as 4140. Fruit: medium to large; uniform; skin golden yellow with two-thirds blushed with orange-red, flecked with darker red, russet in stalk cavity only, attractive; flesh very juicy, sweet, aromatic, texture excellent; ripens late October in Ohio. Tree: large, moderately spreading; vigorous; productivity heavy, annual bearer; healthy foliage, almost free from mildew.

Merton Knave.—Orig. in Bayfordbury, Hertford, England, by A.G. Brown, John Innes Inst., Norwich, England. Introd. in 1973. Plant variety rights, Grant No. 593; 15 Dec. 1971; assigned to John Innes Inst. and NASO. Early Crimson x Laxton's Epicure; cross made in 1948; tested as 4663. Fruit: 60 x 70 mm; round to slightly conical; skin pale greenish-yellow, two-thirds overlaid with bright red blush; flesh greenish-white, firm, crisp; flavor subacid, rich; ripens between Beauty of Bath and Worcester Pearmain. Tree: bears annually; productivity good.

Merton Pippin.—Orig. in Merton, England, by the John Innes Institute. Introd. there in 1948; introd. in the U.S. in 1953 as P.I. 205462. Cox's Orange Pippin x Sturmer Pippin. Fruit: large, up to 2 3/4 inches in diam.; round-oblate; skin with greenish ground color, 90% light red stripe, slightly dull but fairly attractive; flesh whitish-cream, slightly acid, quality fair; resembles Cox's Orange Pippin. Tree: moderately strong grower; upright.

Merton Prolific.—Orig. in Merton, England, by the John Innes Institute. Introd. in 1948. Northern Greening x Cox's Orange Pippin; selected in 1935. Fruit: skin greenish-yellow with crimson blush and stripes; season December and January. Tree: tends to crop too heavily.

Merton Russet.—Orig. in Merton, England, by the John Innes Institute. Introd. in 1948. Sturmer Pippin x Cox's Orange Pippin; selected in 1935. Fruit: skin yellow covered with golden russet; flesh firm, juicy, aromatic; season January to March; tends to be small for commercial use. Tree: growth moderate to weak; spurs very freely; susceptible to mildew.

Merton Worcester.—Orig. in Merton, England, by the John Innes Institute. Introd. in 1948. Cox's Orange Pippin x Worcester Pearmain; selected in 1935. Fruit: size medium; skin greenish-yellow almost completely covered with bright crimson blush; flavor good, aromatic; season September and October. Tree: produces crops regularly on both young and old wood.

Metzgar.—Orig. in Seattle, Wash., by Ludwig Metzgar, Seattle, and Cecil C. Clark, Wapato, Wash. Introd. in 1948. Plant patent 972; 15 Aug. 1950; assigned to May Nursery Co., Yakima, Wash. Delicious o.p.; selected about 1938. Fruit: large, 3 3/8 inches long, 6 to 8 oz; oblong; skin thin, tough, glossy, up to 50% of surface red over yellow, attractive; flesh creamy white, firm, crisp, juicy, somewhat tart; aroma distinct; quality good; good shipper and canner; ripens in early August; resembles Delicious. Tree: spreading, very susceptible to fire blight, needs little thinning.

Miami.—Orig. in New Carlisle, Ohio, by H.N. Scarff. Introd. in 1935 by Scarf Nurseries, New Carlisle. Bud mutation of Stark; discovered in 1930. Fruit: skin solid bronze red; flesh fine-textured; high quality for eating and culinary use; late keeping; resembles Stark. Tree: medium; spreading; moderately vigorous; easily pruned.

Michaelmas Red.—Orig. in East Malling, Kent, England, by Henry M. Tydeman, East Malling Res. Sta. Introd. in England in 1945 and in Canada and the U.S. in 1946. McIntosh Red x Worcester Pearmain; selected about 1940. Fruit: skin bright red; flesh juicy, pleasantly flavored; ripens in England in late September, after Worcester Pearmain; resembles McIntosh Red. Tree: moderately vigorous; upright; spreading; spurs very freely.

Michal.—Orig. in Rehovot, Israel, by Chanan Oppenheimer. Introd. in 1967. F_2 selection (Calville St. Sauveur x Damascus) x Delicious. Fruit: medium size, 60 to 70 mm; round with deep cavity; skin yellow ground cover one-quarter covered light red; flesh juicy, firm, subacid, aroma and texture good, flavor of Jonathan group; keeping quality good; keeps at room temperature at least 2 weeks, becoming sweeter; in cold storage at least 2 months; ships well; ripens middle of June to end of July. Tree: low, widespread, making harvest easy; bears on spurs and on 1-year-old branches; very productive; bears in third year; blooms end of February until beginning of March. Michal is a Hebrew girl's name meaning brook.

Michinoku.—Orig. in 1981 in Hirosaki, Aomori, Japan, by K. Maeda. Kitakami x Tsugaru. Fruit: large, 270 to 300 g; skin solid red with prominent lenticels, often much russeted in cavity, very attractive; flesh coarse, firm, crisp, juicy; good flavor, sweet. Harvest early, with Gravenstein, 4 weeks before Delicious; fruit tend to drop before harvest. Susceptible to water core.

Miller Giant.—Orig. in Union Ridge, Cabell County, W.Va., on the farm of a Mr. Miller by Harry Kaiser of Athallia, Ohio. Introd. about 1922. Considered to be a seedling of Rome Beauty; seed planted about 1900; selected in 1914. Fruit: large, averaging about 3 inches in diam.; quality similar to Rome Beauty; ripens about 10 days before Rome Beauty, and keeps about as well as that variety. Tree: very vigorous; susceptible to scab.

Miller Red.—See Beacon and Ozark Ruby.

Miller Ruby Rome.—See Ruby Rome Beauty.

Miller Spur.—See Miller SturdySpur Delicious.

Miller Spur Delicious.—See Miller SturdySpur Delicious.

Miller SturdySpur Delicious (*Miller Spur, Miller Spur Delicious*).—Orig. in Paw Paw, W.Va., in Martins Mountain Orchard, Consolidated Orchard Co., by Gilbert M. Miller. Introd. in 1963 by Hill Top Orchards and Nurseries, Hartford, Mich. Plant patent 2433; 11 Aug. 1964; by Gilbert Miller, Spring Gap, Md., assignor to Consolidated Orchards, Co., Paw Paw. Origin uncertain, developing as a tree mutation in an orchard of Starking Delicious; discovered in 1957 as a whole tree; mutated tree planted in 1948. Fruit: large; skin a solid dark red, appearing very early, no stripes; stores very well; an excellent Delicious type. Tree: spur-type growth; sturdy, strong, semi-dwarf; productive; hardy.

Mills Red Rome.—Orig. in Sodus, N.Y., by Charles H. Mills. Introd. in 1952. Bud mutation of Rome Beauty; discovered in 1935. Fruit: identical with Rome Beauty, except more red color; one of the most attractively colored of the Rome color sports; indistinct striping; less red pigment than Law, Barkley, and Frimley. Tree: identical with Rome Beauty.

Milton.—Orig. in Geneva, N.Y., by Richard Wellington, New York State Agr. Expt. Sta. Introd. in 1923. Yellow Transparent x McIntosh; cross made in 1909. Fruit: skin pinkish-red, attractive; characteristic McIntosh flavor and aroma; sometimes has unattractive bulge on one side; season after Early McIntosh and preceding McIntosh by a month or 6 weeks; resembles McIntosh. Tree: vigorous; hardy; annual bearer; tolerant to cedar apple rust.

Minjon.—Orig. in Excelsior, Minn., by the Univ. of Minn. Fruit Breeding Farm. Introd. in 1942. Parentage unknown, but probably Wealthy x Jonathan; seed planted in 1912 or 1913; selected in 1923; tested as Minn. 700. Fruit: skin solid, dark red; flesh stained red as in Wealthy, quality good; ripens just after Wealthy; resembles Jonathan in size, shape, and skin color; for home and commercial use. Tree: very hardy; vigorous; productive and requiring thinning.

Minnehaha.—Orig. in Excelsior, Minn., by the Univ. of Minn. Fruit Breeding Farm, from seed collected from the Seth Kenney orchard of Morristown, Minn., in 1908. Introd. in 1920. Malinda o.p.; selected in 1914; tested as Minn. 300. Fruit: size medium; oblate, sometimes slightly irregular; deep red striped and splashed to completely cover a yellow ground; flesh firm, fine-grained, crisp, tender, white or slightly tinged with yellow, medium juicy, flavor rich, aromatic, subacid, quality good; season November to February. Tree: hardy, vigorous and productive; well-shaped, rounded, moderately dense head.

Minnesota Red Duchess.—See Daniels Red Duchess.

Minnetonka Beauty.—Orig. in Excelsior, Minn., by I.E. Soderlund. Introd. in 1941. Plant patent 474; 10 June 1941. Northwestern Greening o.p.; selected in 1928. Fruit: averages 3 inches in diam., 2 3/8 in. long; roundish conic, stem medium to long; no bunching on branches; calyx small, closed; skin red over yellow, thin, tough, russet dots large; flesh yellowish, firm, tender, juicy, mildly subacid, quality good; ripens late, with Haralson and Northwestern Greening; keeps well in common storage to May; for dessert and culinary purposes. Tree: vigorous; size medium; spreading; dense; rounded; rapid growing; hardy; productive; regular bearer.

Moira.—Orig. by L.P.S. Spangelo, S.J. Leuty, and H.B. Heeney at Agr. Canada Res. Sta., Ottawa, and Smithfield Experimental Farm, Trenton, Ontario, Canada. Introd. in 1978. McIntosh x Dg 22-81 [Jonathan x (Rome Beauty x *Malus floribunda* 821 sib)]. Tested as Ottawa 548. Fruit: smaller than McIntosh, up to 76 mm in diam.; round conic; skin color medium lively red on light green to yellow ground color; prominent white lenticels; flesh creamy white, slightly tinged with green; crisp; moderately acid, processed fruit was rated fair for juice and poor for sauce and slices; harvest with Delicious; storage life 23 months at 2C, shorter than McIntosh. Tree: moderately vigorous, spreading; high yield efficiency; blooms two to three days after McIntosh; resistant to apple scab and cedar apple rust; susceptible to fire blight; fruit very susceptible to quince rust.

Mollie's Delicious.—Orig. in New Brunswick, N.J., by New Jersey Agr. Expt. Sta. Introd. in 1966. (Golden Delicious x Edgewood) x (Red Gravenstein x Close). Cross made in 1948 by G.W. Schneider; selected in 1956; tested as NJ 28; introd. by L.F. Hough and Catherine H. Bailey, of the above station. Fruit: very large; conic; skin light yellow ground color with one-half to two-thirds blushed with red, very attractive; flesh very good quality; good shelf life; ripens 3 weeks before Golden Delicious. Tree: vigorous; productive.

Monark.—Orig. in Sebastopol, Calif., by Wm. E. Silva. Introd. in 1965. Red June x King; selected in 1964; tested as Silva 2. Fruit: medium to small; round-ovate; skin red; flesh white with red streaks, firm, fine-grained; ripens in early season, mid-August; for local markets. Tree: small; bears very heavily, requires thinning, comes into production early.

Monona.—Orig. in Ames, Iowa, by the Iowa Agr. Expt. Sta. Introd. about 1922. Wolf River x Harrington; tested as Iowa 473. Fruit: large; roundish; skin nearly solid red, smooth, dots large; flesh tender, smooth, juicy, mild, pleasantly subacid; season October to January; hangs well until ripe; sometimes needs thinning for best development. Tree: vigorous.

Monroe.—Orig. in Geneva, N.Y., by U.P. Hedrick, New York State Agr. Expt. Sta. Introd. in 1949. Jonathan x Rome Beauty; seed produced in 1910; tested as N.Y. 1546. Fruit: medium to large; roundish-conic; skin nearly solid red, appearance excellent; flesh yellowish, firm, crisp, juicy, subacid; quality good; primarily for processing, excellent for cooking, also for the fresh-fruit market; not subject to drop; resembles Jonathan. Tree: size medium; vigorous; upright, spreading; a reliable annual bearer; very productive; subject to powdery mildew, which is readily controlled by including sulfur in the spray schedule up to the time of terminal growth cessation. Named for Monroe County's, N.Y., extensive plantings in commercial orchards.

Morden Russet.—Orig. in Morden, Manitoba, Canada, by Canada Dept. Agr., Res. Sta. Introd. in 1935. Anisim o.p. Fruit: small; round; skin golden brown, russeted; flesh amber yellow, fine-grained, subacid, quality fair to good; resembles Golden Russet; season October to January. Tree: upright; moderately vigorous; productive.

Morgan Spur Delicious.—Orig. in Twisp, Wash., by Harry Morgan (deceased). Introd. in 1969 by Columbia & Okanogan Nursery Co., Wenatchee, Wash. Bud mutation of Shotwell Delicious. Discovered in 1957, tested as 411 by Columbia & Okanogan Nursery Co. Fruit: medium to large; shape similar to Delicious; skin dark red with stripes, similar to Shotwell Delicious; flesh characteristics same as Delicious; ripens 10 days later than standard Delicious. Tree: spur-type; three-fourths size of standard tree; good limb angles for a spur; vigor good; hardiness equal to other

Delicious varieties; productivity good with more fruit spurs per foot of wood. Recommended for planting only in areas with good coloring weather.

Moris.—Orig. in Morden, Manitoba, Canada, by Canada Dept. Agr., Res. Sta. Introd. in 1930. Anis o.p. Fruit: small; skin yellow, heavily blushed, many red stripes, greasy, thin but tough; flesh amber white; very juicy, texture smooth, pleasantly tart; fair to poor quality; season midwinter; used as dessert or sauce; resembles Blushed Calville. Tree: spreading; moderately vigorous; hardy; productive.

Morspur.—Orig. in Zillah, Wash., by Charles F. Morrison, Morrison Nursery. Introd. in 1963. Bud mutation of Starking Delicious; discovered in 1959. Fruit: Delicious type; skin bright red, coloring early. Tree: spur type.

Mortof.—Orig. in Morden, Manitoba, Canada, by Canada Dept. Agr., Res Sta. Introd. in 1929. Tetrofsky o.p. Fruit: size medium; skin greenish-yellow, washed with red; flesh white, fine-grained, melting, juicy; flavor pleasant, subacid; season September to December. Tree: vigorous; hardier than Mantet; productive.

Mosebar Black Winesap.—See Mosebar Winesap.

Mosebar Red Winesap.—See Mosebar Winesap.

Mosebar Winesap (*Crimson Winesap, Dexter Winesap, Gensinger Winesap, Mosebar Black Winesap, Mosebar Red Winesap, Ruble Winesap, Welch Winesap*).—Orig. in Yakima, Wash., by Howard S. Mosebar. Introd. in 1924. Bud mutation of Winesap; discovered in 1921 or 1922. Fruit: skin extremely dark red, in some districts almost black; all fruit on a tree colors about the same time. Fruit and tree otherwise identical to Winesap.

Mount.—Orig. in Morden, Manitoba, Canada, by Canada Dept. Agr., Res. Sta. Introd. in 1937. Antonovka o.p. Fruit: large; round oblate; skin pale yellow, washed scarlet with red stripes, rough; flesh white, crisp, juicy, tart; quality fair, good for cooking; season November through January. Tree: upright, spreading; vigorous; productive, very hardy.

Murray.—Orig. by L.P.S. Spangelo, S.J. Leuty, and H.B. Heeney at Agr. Canada, Res. Sta., Ottawa, and Smithfield Experimental Farm, Trenton, Ontario, Canada. Introd. in 1980. McIntosh x 52-05-26 [Platt Melba x R16T19 (Wolf River x *Malus atrosanguinea* 804)]. Tested as Ottawa 628. Fruit: medium; red blush stripe on green to yellow ground color; flesh white, fine texture, juicy, moderately acid; quality medium good; processed fruit was rated poor for juice, sauce, and slices; harvest 7 to 10 days before McIntosh; storage life short, less than 1 month at 2C. Tree: moderately vigorous, spreading; medium yields; blooms early with McIntosh; moderately resistant to apple scab, but occasional scab spots have been observed; resistant to cedar apple rust; very susceptible to quince rust.

Mutsu.—Orig. in Kuroishi, Aomori Prefecture, Japan, by the Aomori Apple Expt. Sta. Introd. in the U.S. in 1948; P.I. 195834 and P.I. 223602; the following information pertains to P.I. 195834. Golden Delicious x Indo; cross made in 1930; fruited for the first time in 1937; named by and published by the Japanese Hort. Assn. in 1948. Patented in Japan 14 Sept. 1949. Fruit: larger than either parent, up to 600 g; stem thick; skin smooth, dull, with scarfskin, golden yellow, blushed with orange on exposed cheek; dots numerous; core small, flesh yellowish-white, crisp but rather coarse compared with Golden Delicious, mildly subacid; very good for dessert; season after Golden Delicious; free from russeting; free from shriveling in storage. Tree: larger than either parent; very vigorous; bears annually, a triploid. Susceptible to blister spot.

Muz (*Muzalma*).—Introd. from Chinese Turkestan by U.S. Dept. of Agriculture as P.I. 30326, distributed for trial in 1941. Fruit: large; oblate to roundish; skin greenish-yellow; flesh crisp, firm, subacid, tart when green, with most of acidity disappearing when fully ripe; juicy, fair to good quality. Tree: semi-dwarf; able to withstand drought, alkali, and neglect; promising for hot desert regions under irrigation; sometimes used as dwarfing interstock.

Muzalma.—See Muz.

Mystery.—Origin unknown. Tree was sent by Percy H. Wright, of Saskatoon,

Saskatchewan, Canada, to A.J. Porter, Honeywood Nursery, Parkside, Saskatchewan, in 1949; parentage and source unknown. Introd. in 1964 by Honeywood Nursery. Fruit: up to 2 1/2 inches in diam.; skin green overlaid with red when ripe; eating quality fresh only fair, cooking quality excellent; ripens third week in September; keeps well to December in ordinary storage. Tree: very hardy, more so than Heyer 12 or Rescue.

Naomi.—Orig. in Rehovot, Israel, by Chanan Oppenheimer, Agr. Res. Inst. Introd. in 1955. F_2 (Calville St. Sauveur x an unnamed local seedling). Fruit: roundish; skin with less color but fruit with better flavor than Vered; ripens in June. Tree: resistant to delayed foliation; best adapted to dwarf rootstock M.9; on M.9 good shape and very prolific; hardy. Now obsolete.

Nebuta.—Orig. in 1981 in Hirosaki, Aomori, Japan, by K. Maeda. Kitakami x Tsugaru. Fruit: medium-large, 250 to 300 g; round to round conic; skin solid dark red; flesh pale yellow, medium firm, somewhat acid. Harvest early, with Gravenstein, 5 weeks before Delicious; tends to drop before harvest. Susceptible to water core.

Neff Delicious.—See Hi-Early.

Neipling Early Red Stayman.—See Neipling Red Stayman.

Neipling Red Stayman *(Early Red Stayman, Neipling Early Red Stayman)*.—Orig. in Waterford, N.J., by F.S. Neipling. Introd. in 1942 by Bountiful Ridge Nurseries, Princess Anne, Md. Bud mutation of Blaxtayman; discovered in 1934. Fruit: skin colors 10 to 14 days earlier than parent and Stayman Winesap, bright, attractive; resembles Scarlet Staymared and Improved Blaxtayman 201; crop can be harvested at one picking. Tree: growth and habits same as Stayman; generally open-spreading and symmetrical; strong scaffold branches bearing fruit mostly toward the outer one-half of tree with some fruit on the center part of tree.

Nelson.—Orig. in Lander, Wyo., by J.M. Steinbrech, Wyoming Agr. Expt. Sta. Introd. in 1933. Fruit: large; oblong; skin yellowish with bright, deep red stripes; appearance very attractive; flesh very tender, quality very good; matures 25 Sept.; does not keep well. Tree: good bearer.

Nepean.—Orig. in Ottawa, Ontario, Canada, by Canada Dept. Agr., Res. Sta. Introd. for trial about 1923; not planted commercially. Salome o.p.; discovered in 1910. Fruit: medium to large; oblong; cavity deep; stem stout; basin open, slightly wrinkled; skin thick, tough, greenish-yellow washed with bright pink-red, dots indistinct; flesh yellow, tender, moderately juicy; flavor pleasant; core large; season winter.

Nero Red Rome.—See Nero Rome.

Nero Rome (*Nero Red Rome, Flaming Rome, Lemcke Rome, Tieton Rome*).—Orig. in Tieton, Wash., by Robert S. Lemcke. Introd. in 1954. Plant patent 1414; 30 Aug. 1955; in name of Robert S. Lemcke, Tieton, Wash., and Jacob A. Snyder, Wenatchee, Wash.; assigned to Columbia & Okanogan Nursery Co., Wenatchee, Wash. Apparently a bud mutation, discovered in an orchard of Rome Beauty in 1951; selected in 1951. Fruit: resembles Rome Beauty; medium to large; averaging 3 to 3 1/2 inches in diam.; long and blocky; skin bright red with heavy red striping effect, coloring early; flesh firm, nearly white, crisp, subacid; quality good; keeps well. Tree: moderately vigorous; medium size.

Nertschinsk.—See rootstocks for apple.

Neubert.—Orig. in Tieton, Wash., by A. Neubert. Bud mutation of Winesap; discovered in 1930s. Fruit: skin with more red than parent, which it resembles. No longer propagated.

Neumann James Grieve (*Red James Grieve*).—Orig. in Zieverich, Bergheim, Germany, by Erich Neumann. Introd. in 1958 in Germany. Plant patent 1685; 18 Feb. 1958; assigned to Wilhelm Fey, Mechenheim, Germany. Bud mutation of James Grieve; discovered in 1953. Fruit: identical with parent except for color differences: solid, bright, red skin. Tree: vigorous; tall.

New Jonagold.—A red-fruited mutation of Jonagold discovered by M. Saito, Aomori Prefecture, Japan. Japanese Reg. No. P 63. Deep red stripe, but only slightly redder overall than Jonagold; harvest slightly earlier than Jonagold.

Newfane.—Orig. in Geneva, N.Y., by Richard Wellington, New York State Agr. Expt.

Sta. Introd. in 1927. Deacon Jones x Delicious; cross made in 1911. Fruit: large; oblong-conic; skin attractive red; flesh tender, medium juicy, mildly flavored with a pronounced Delicious aroma; dessert use; ripens with Delicious, which it resembles.

NewGold.—Orig. in New Brunswick, N.J., by L.F. Hough, Rutgers Univ. Introd. in 1987. Golden Delicious x Malling 2439 (Prince Rudolf x James Grieve); cross made in 1974. Tested as NJ89. Rights sold to Star Fruits, France. Fruit: medium to large; shape oblong conic; skin light green, similar to Golden Delicious; flesh texture fine and crisp; flesh color light straw; flavor excellent. Harvest after Golden Delicious; storage life good. Tree: very high vigor; precocity very good; very productive; very hardy; bloom mid to late.

Newjon.—See Nu-Jon.

Newtosh.—Orig. in Ottawa, Ontario, Canada, by Canada Dept. Agr., Res. Sta. Introd. in 1923. McIntosh x Yellow Newtown; selected in 1922. Fruit: resembles Yellow Newtown in flesh and flavor; resembles McIntosh in skin color; keeps longer than McIntosh in cold storage but susceptible to low-temperature breakdown. Tree: produces large percentage of extra fancy fruit; recommended for Ontario, Quebec, and New Brunswick, Canada.

Newtown Delicious.—Orig. in White Salmon, Wash., by Edwin P. Wray. Introd. about 1937. Plant patent 61; 18 Apr. 1933. Yellow Newtown x Delicious; selected about 1930. Fruit: stem very short; skin attractive light red on yellow ground; dots yellow, numerous prominent; surface ridged; stem and calyx cavity very deep, dark green; flavor good; keeps well and late, until March; superior for culinary use. Tree: growth vigorous; heavy bearer.

Niagara.—Orig. in Geneva, N.Y., by the New York State Agr. Expt. Sta. Introd. in 1962. Carlton x McIntosh; selected in 1950; tested as N.Y. 50-2. Fruit: McIntosh type; appearance and flavor similar to McIntosh; ripens 2 weeks before that variety; slight tendency to drop as it nears maturity, but not as serious as with McIntosh, with a stop-drop spray of benefit in most years. Tree: growth habit and productivity similar to McIntosh; diploid; full bloom about with McIntosh.

Nickell.—See Excells Red.

Nicobel™ Jonagold.—Mutation of Jonagold; only slightly better red color than Jonagold, with 50% red blush over yellow ground color, but less red than Jored (King Jonagold). Assigned to TRECO®, Woodburn, Ore.

Nittany.—Orig. in Biglerville, Pa., by Fruit Res. Lab., Pennsylvania State Univ. Introd. in 1977. York Imperial or one of its red sports o.p. Fruit similar to York Imperial; processing type; medium, 175 to 250 g; color dark pink or orange-red blush, narrow striping on a light clear yellow ground color; shape oblique; flesh texture firm; flesh color deep yellow; soluble solids slightly less than York Imperial; exposed flesh turns brown much more slowly than York Imperial; flesh retains shape during cooking; harvest first 3 weeks of October; longer storage life, 6 months without loss of quality; essentially free of corking. Tree: vigorous; productive; biennial tendency; susceptible to fire blight; bloom date late, similar to Imperial.

Noran.— Orig. from the breeding program at Agr. Canada, Res. Sta., Beaverlodge, Alberta. Introd. in 1976. Columbia (*Malus baccata* x Broad Green) x Redant (Antonovka o.p.). Selected in 1961. Fruit 60 x 55 mm, uniform sizes; background color green with a brick-red blush on exposed side; flesh cream-colored, crisp, slightly tart; good cooking; harvest late September; stores in air at 4C until mid-February. Tree: vigorous, upright-spreading; very hardy, precocious; moderate to good annual cropping.

Norcue.—Orig. from the breeding program at Agr. Canada, Res. Sta., Beaverlodge, Alberta. Introd. in 1976. Heyer 12 x Rescue. Selected in 1958. Fruit: medium-small, 55 mm in diam.; ground greenish-yellow, almost completely overlaid with dull red stripes; cream-colored with tinges of pink under the skin; moderately crisp and juicy with full, very sweet flavor; harvest before Heyer 12. Tree: vigorous, upright-spreading; very hardy; precocious; annual cropping.

Norda.—Orig. from the breeding program at Agr. Canada Res. Sta., Beaverlodge,

Alberta. Introd. in 1976. Rosilda (Prince McIntosh) x Trail (Northern Queen x Rideau). Selected in 1960. Fruit: large, variable sizes, up to 68 mm; background color green to yellow overlaid by red streaks; flesh cream-colored, crisp, moderately juicy, very good flavor. Harvest late September; stores at 4C until mid-January. Tree: vigorous, upright-spreading, moderately hardy; relatively precocious; moderately productive.

Noret.—Orig. from the breeding program at Agr. Canada Res. Sta., Beaverlodge, Alberta. Introd. in 1976. Rescue x Mantet. Selected in 1960. Fruit: medium, 55 x 50 mm; ground color green to yellow, overlaid almost completely with shiny dark red; flesh cream-colored, moderately firm, slightly tart, good flavor. Harvest early, 5 days after Dawn. Tree: small, upright-spreading; very winter hardy, hardier than Osman and Heyer 12; precocious; annual cropping.

Norhey.—Orig. from the breeding program at Agr. Canada, Res. Sta., Beaverlodge, Alberta. Introd. in 1976. Heyer 12 x Dr. Bill. Selected in 1960. Fruit: similar to Heyer 12; large, 65 x 60 mm; ground color green changing to yellow; flesh deep cream color, crisp, juicy, slightly tart; harvest early September; long storage. Tree: vigorous, upright, good branching habit; very winter hardy, hardier than Heyer 12; precocious; heavy annual cropping.

Norman Sweet.—Orig. in Mannsville, N.Y., by Mrs. Carl Norman. Introd. in 1962 by Fred L. Ashworth, nurseryman, Heuvelton, N.Y. Parentage unknown; discovered in 1960. Fruit: size medium; skin red; flesh white, sweet, flavor good; ripens in August; keeps well into winter.

Norson.—Orig. from the breeding program at Agr. Canada, Res. Sta, Beaverlodge, Alberta. Introd. in 1976. Haralson x Rescue. Selected in 1958. Fruit: medium-small, 50 mm in diam; green to yellow ground color, almost completely overlaid by a very attractive dark red; flesh cream-colored, crisp, juicy; pleasant flavor good quality, best dessert; harvest late September; excellent keeping ability. Tree: moderately vigorous, upright-spreading; very hardy; relatively precocious; annual, heavy crops.

Norspy.—Orig. in Ottawa, Ontario, Canada, by the Canada Dept. Agr., Res. Sta. Introd. in 1929. Northwestern Greening x Northern Spy. Fruit: medium to large; stem length medium; calyx open; skin yellow washed with carmine; dots few, distinct, white; flesh yellowish, moderately juicy, firm, flavor pleasant, subacid; core small; quality good; season late December to March or April. No longer of commercial value.

Northern Lights.—A very hardy midseason variety suitable for the north central plains. Orig. in Geneva, N.Y.; introd. in 1990 by R.D. Way, S.K. Brown, and K.G. Livermore, New York State Agr. Expt. Sta., Cornell Univ., and by A.A. Boe, North Dakota State Univ., Fargo. Not patented. Haralson x McIntosh; cross made in 1938; selected in 1947; tested as NY 17207. Named because tree is very hardy in North Dakota. Fruit: large, 65 to 75 mm in diam.; skin 60% to 100% bright, attractive red, with conspicuous white dots; color pattern striped; shape round-conic; flesh semi-firm, white; flavor subacid to slightly acid; eating quality fair. Harvest season same as McIntosh; storage life 3 months at –0.5C. Tree: very hardy, large, vigorous, productive but somewhat biennial in cropping. Pollen: effective pollenizer when tested on other varieties.

Northern Spy Double Red.—See Greene Spy.

Northern Star.—See Hokuto.

Nova Easygro.—Orig. in Kentville, Nova Scotia, Canada, by Agriculture Canada, Res. Sta. Introd. in 1971. Spartan x Prog. 565 [Ill. 5 (Fanny x Jefferis) x PRI 27-435 (Wealthy x R12740-7A)]. Cross made in 1956 by L.E. Aalders, pollen supplied by D.F. Dayton, Univ. of Illinois; tested as Pn6-16. Fruit: size medium; round, regular; skin pale greenish-yellow, striped or blushed medium red, dots inconspicuous, moderately thin, fairly smooth; flesh creamy white, moderately fine, firm, crisp, moderately juicy, flavor subacid, pleasant; ripens about with Cortland, which it most nearly resembles, but is slightly smaller and duller in color; keeps well; recommended for home gardens. Tree: spreading; vigor and productivity moderate; regular bearer; resistant to scab.

Novamac.—Orig. in Kentville, Nova Scotia, Canada, by A.D. Crowe in the breeding program of Agr. Canada, Res. Sta. Introd. in 1978. McIntosh x PRI 1018-3. Cross made by D.F. Dayton and E.B. Williams at Sturgeon Bay, Wis., in 1963. Fruit: medium-large, 70 mm in diameter; shape round to round conic, regular; 80% red stripe on greenish yellow ground; very attractive; flesh white, crisp, fine, tender, juicy; flavor subacid, similar to McIntosh; very good. Harvest early fall, with McIntosh. Tree: moderately vigorous, upright spreading, moderately productive; annual cropping; highly resistant to apple scab; resistant to cedar apple rust and fire blight; susceptible to mildew.

Novaspy.—Orig. by A.D. Crowe at Agr. Canada, Res. Station, Kentville, Nova Scotia. Introd. in 1986. Nova Easygro x 44411-1 (Red Spy x Golden Delicious), tested as S34-38-122. Fruit: medium; globose-conical, slightly ribbed; skin 30% to 85% striped to blushed dark red; flesh creamy yellow, fine-textured, very firm, crisp, moderately tender, juicy, moderately acid; flavor very good. Matures between Delicious and Northern Spy; long storage in air. Excellent processor; released as dual-purpose variety. Tree: vigor medium; upright-spreading; precocious; production moderate, annual; hardy in Nova Scotia. Resistant to apple scab.

Novole.—See rootstocks for apple.

Nu-Jon (*Newjon*).—Orig. in Entiat, Wash., by A.J. Marr. Introd. in 1949. Plant patent 1075; 19 Feb. 1952; assigned to Columbia & Okanogan Nursery Co., Wenatchee, Wash. Parentage unknown; discovered in 1945. Fruit: medium to large; roundish conic to globular; skin deep red, striped, glossy, smooth; flesh white to slightly yellowish, crisp, tender, juicy, subacid; quality good; keeps well; ripens 10 days before Jonathan; resembles Stayman Winesap. Tree: not subject to mildew; medium to large, vigorous.

Nugget.—Orig. in Princess Anne, Md., by Homer S. Kemp, Bountiful Ridge Nurseries. Introd. in 1966. Chance seedling of Golden Delicious, discovered in 1954. Fruit: resembles Golden Delicious, being more productive and bearing fruit in second year on seedling rootstock; ripens with Golden Delicious. Tree: semi-dwarf; spur-type; hardy; precocious.

Nured Jonathan.—Orig. in Wenatchee, Wash., by Robert E. Snyder, Columbia & Okanagan Nursery Co. Introd. in 1966; plant patent 2650; 5 July 1966; assigned to Columbia & Okanogan Nursery Co., Wenatchee, Wash.; bud mutation on Blackjon tree; discovered in 1953. Fruit: resembles Blackjon, but colors 10 days to 2 weeks earlier than parent, produces greater percentage of well-colored fruit. Tree: appears identical to parent variety.

Nured McIntosh.—See Summerland McIntosh.

Nured Rome (*SeeandO Red Rome 262*).—Orig. in Wenatchee, Wash., by J.A. Snyder, Columbia and Okanogan Nursery Co. Introd. in 1943. Bud mutation of Rome Beauty; selected in 1941. Skin solid bright red, darker than Rome Beauty, otherwise identical to standard Rome Beauty. Tree: resembles Rome Beauty.

Nured Royal Delicious (*Improved Royal Red Delicious*).—Orig. in Tonasket, Wash., by L. Hutchinson, Columbia and Okanogan Nursery Co., Wenatchee, Wash. Introd. in 1976 by this nursery. Bud mutation of Royal Red Delicious; discovered in 1966. Fruit: color more intense than Royal Red; colors as early as Topred Delicious. Tree: resembles Royal Red.

Nured® Spur Delicious.—Orig. by Harry Kemp, Othello, Wash. A spur mutation of Bisbee Delicious. Assigned to Columbia and Okanagan Nursery, Wenatchee, Wash.

Nured Stayman (*C & O Blaxtayman, Double Red Stayman, Red Stayman 201, Stayman 66*).—Orig. in Wenatchee, Wash., by J.A. Snyder, Columbia & Okanogan Nursery Co. Introd. in 1944. Bud mutation of Stayman Winesap; discovered in 1942. Fruit: skin solid bright red; resembles parent.

Nured® Winesap.—A red-fruited mutation of SeeandO Winesap. Fruit: earlier coloring and darker red than Winesap; striped; russet resistant.

Oberle.—Orig. in Virginia by George Oberle, Virginia Polytechnic Inst., Blacksburg. Introd. in 1987. Parentage unknown; tested as SC-2-11. Fruit: large, to

85 mm; skin 50% to 75% striped or washed with bright red; lenticels conspicuous at calyx end; occasionally russeted; round, occasionally ribbed; flesh creamy white, crisp, tender, somewhat coarse, juicy, good. Matures very early, two to four days before Close, requiring more than one picking. Tree: moderate vigor; upright spreading, somewhat spurry and compact; precocious; annually productive.

Ogden.—Orig. in Geneva, N.Y., by New York State Agr. Expt. Sta. Introd. in 1928. Zusoff x McIntosh; cross made in 1912; first full crop in 1924. Fruit: large; roundish oblate; skin dark red with heavy bloom, similar to McIntosh but with considerable scarfskin; flesh white, often stained with red, coarse, tender, juicy, aromatic, sweet; season a little earlier than McIntosh; good baking apple, resembles McIntosh. Tree: hardy and productive; requires thinning; alternate bearing.

Ohlson.—Orig. in Puyallup, Wash., by E.A. Ohlson. Introd. in 1951. Chance seedling; discovered about 1935. Fruit: very large, 4 inches or more in diam., light weight; oblong, slightly angular; calyx large, closed; skin tender, thin, smooth, pale yellow, mottled or striped bright carmine; dots small and scattered; flesh pale yellow, moderately coarse but crisp, somewhat juicy, subacid, quality good to very good; dessert and culinary purposes; ripens in fall, with Gravenstein in early September at Puyallup; keeps well in common storage until January; primarily for home orchards and roadside markets. Tree: very hardy; vigorous, open; produces good and regular crops; propagates well on M.9 and M.7 rootstocks.

Okanoma (*Kopps Delicious, Okanoma Delicious, Okanoma Red, Spurred Okanoma Red, Spurred Red Delicious*).—Orig. in Omak, Wash., by Nettie Kopp; known in the Okanogan–Omak area since about 1921. Introd. in 1947. Bud mutation in a Delicious planting. Fruit: Delicious type; skin a solid blush, staying a bright red rather than turning to a dark maroon as in many Delicious sports, but stripes may be present; resembles Richared Delicious. Tree: spur-type growth; semistandard; bears heavily.

Okanoma Delicious.—See Okanoma.

Okanoma Red.—See Okanoma.

Oregon Spur II (*Wells & Wade Spur*).—Orig. in Orondo, Wash., by Lester Green, Wells & Wade Fruit Co., Wenatchee, Wash. Introd. in 1982. Plant patent 4819; assigned to Van Well Nursery, Wenatchee, Wash. Bud mutation of Oregon Spur Delicious; discovered in 1973. Fruit: resembles Oregon Spur Delicious but colors 15 days ahead of parent and has more fruit spurs per lineal foot; striped color.

Oregon Spur Delicious.—Orig. in Milton-Freewater, Ore., by Wayne M. Trumbull. Introd. about 1966. Plant patent 2816; 4 June 1968; assigned to Van Well Nursery, Wenatchee, Wash. Bud mutation of Red King Delicious. Fruit: characteristics resemble parent, except in coloring; color 10 to 14 days earlier than parent and is deep red with prominent underlying stripe formation. Tree: spur-type growth similar to Wellspur Delicious.

Orenco.—Orig. in Orenco, Ore., by the Oregon Nursery Co. Introd. in 1920. Parentage unknown. Fruit: size medium; shape like Tompkins King; skin solid, beautiful, bright red; flesh white, juicy, crisp, sweet, mild; quality excellent; ripens 1 to 15 Sept. Tree: moderately vigorous; upright, spreading; healthy.

Orin.—Orig. before 1942 by T. Otsuki, apple grower in Fukushima Prefecture, Japan. Golden Delicious x Indo. Fruit: medium-large, 250 to 300 g; oblong to conic; yellow-green with prominent russeted lenticels, sometimes with red-orange blush; flesh firm, aromatic, sweet, juicy, excellent quality; harvest 1 week before Fuji; useful only in long-season areas, excellent shelf life. A triploid.

Oriole.—Orig. in Excelsior, Minn., by the Univ. Minn. Fruit Breeding Farm. Introd. in 1949. Parentage unknown; seed planted in 1914; selected in 1923; tested as Minn. 714. Fruit: large to very large; roundish, oblate; skin yellow-orange, striped and splashed red; similar to Duchess of Oldenburg in form and general appearance; flesh very soft, fine-grained, juicy, aromatic, subacid; quality excellent for cooking or dessert; a summer variety with short shelf life. Tree: medium tall; rounded; productive.

Orleans.—Orig. in Geneva, N.Y., by Richard Wellington, New York State Agr. Expt. Sta. Introd. in 1924. Deacon Jones x Delicious cross made in 1911. Fruit: more attractive and larger than Delicious; keeps in common storage 6 weeks longer than Delicious; similar to Medina but keeps longer; Delicious type. Tree: similar to Medina; thrives in some soils to which Medina is not adapted.

Ostem.—Orig. in Morden, Manitoba, Canada, by Canada Dept. Agr., Res. Sta. Introd. in 1936. Ostrakoff o.p. Fruit: large; ovate; skin greenish-yellow washed with red stripes; flesh greenish-white, juicy, spicy; quality good; season December to March. Tree: upright; vigorous; productive.

Ottawa.—Orig. in Ottawa, Ontario, Canada, by Canada Dept. Agr., Res. Sta. Introd. in 1921. Seedling of Swayzie. Fruit: size medium or above, oblong-conic, slightly angular; stem length medium; skin pale greenish-yellow, washed with rich crimson, rather thick, tough; dots numerous, white, distinct; flesh dull white to yellowish, juicy, subacid, spicy, quality very good; season midwinter to late winter.

Ottawa 3.—See rootstocks for apple.

Oxbo.—Orig. in Brookings, S.D., by N.E. Hansen, South Dakota Agr. Expt. Sta. Introd. in 1922. Roxbury Russet x Duchess of Oldenburg. Fruit: up to 2 1/2 inches in diam.; flesh white, juicy, subacid; ripens in late fall. Now obsolete.

Ozark Gold.—Orig. in Mountain Grove, Mo., by Mountain Grove Fruit Expt. Sta. Introd. in 1970. Golden Delicious x H1291 (Red Delicious x Conrad), cross made by Paul Sheppard. Fruit: skin waxy golden, very attractive; flesh flavor mild, subacid; flesh firmer than Golden Delicious; ripens late August, 2 weeks before Golden Delicious. Tree: vigorous; open spreading; highly productive.

Ozark Ruby (*Miller Red*).—Orig. in Green Forest, Ark.; first called Miller Red after the grower's name, Miller. Introd. in 1949 by the Ozark Nursery Co., Tahlequah, Okla., as Ozark Ruby. Parentage unknown; discovered in 1936. Fruit: large, resembles Idared and Jonathan, but larger and skin with more red; ripens 10 to 15 days before Jonathan with Summer Champion; color, flavor and quality superior to Summer Champion; flesh firm and juicy; highly susceptible to blotch and bitter rot. Tree: somewhat upright compared to most varieties; foliage medium green, rather roundish; terminal growth steady; terminal buds, large; bark of limbs and new growth dark red to light brown; rate of growth and life expectancy average.

P.1.—See rootstocks for apple.

P.2.—See rootstocks for apple.

P.16 (*Lizzy*).—See rootstocks for apple.

P.22.—See rootstocks for apple.

P.60.—See rootstocks for apple.

Pacific Gold.—Orig. in Myrtle Point, Ore., by J.E. Miller, Paradise, Calif. Introd. in 1948. Chance seedling; selected in 1944. Fruit: very large; flesh tender, crisp, quite sweet; dessert quality fine, good for culinary purposes; ripens in late fall or early winter. Tree: size medium; bears young; yields heavy crops annually.

Pacific Pride.—Orig. in Mount Vernon, Wash., by Ennals Ralph Gage. Introd. in 1959. Plant patent 1975; 4 Oct. 1960; assigned to Sherm Fox, Pacific Pride Gardens, Bow, Skagit County, Wash. Parentage unknown; chance seedling discovered in 1909. Fruit: averages about 3 3/4 inches long, up to 4 inches in diam.; conical, ribbed; skin thick, tough, smooth, oily, striped red over yellow; flesh slightly yellow, juicy, firm, crisp, tart, quality fair; ripens 4 to 6 weeks after Gravenstein, 10 to 24 Oct. at place of origin; keeps very well; useful principally for culinary purposes; resembles Gravenstein and King. Tree: size medium; very vigorous, spreading; hardy; productive; bears regularly.

Par-Fect Spur Criterion (*Fechtig*).—Orig. near Albany, Ore., by A.D. Fechtig. A semi-spur mutation of Criterion. Plant patent 6043; 1 Nov. 1988; assigned to TRECO®, Woodburn, Ore. Fruit: medium to large; variable in shape; green to yellow ground color with pink to rose blush; skin tough; texture firmer than Criterion, coarse; flavor good, aromatic; harvest date early to mid October, 1 week earlier than Criterion; storage life good; excellent shelf life. Tree: spur growth habit; tree 25% smaller than Criterion; precocious; strongly biennial; spurs break off easily when picking.

Park.—Orig. in Saskatoon, Saskatchewan, Canada, by C.F. Patterson, Univ. of Sask. Introd. in 1960 for home gardens. Osman x Wealthy; selected in 1945; tested as A58-302. Fruit: up to 2 1/2 inches long and 2 inches in diam.; under nonirrigated field conditions at Saskatoon; oblate; skin medium red, dots conspicuous; flesh mildly acid, quality good. Tree: hardy.

Parkdale Beauty.—Orig. in Parkdale, Ore., by Paul Klindt. Introd. in 1984. Plant patent applied for; to be assigned to Al Bacharach. Seedling of Golden Delicious. Fruit: large; elongated with pronounced points; skin waxy, shiny gloss, golden-colored, attractive blush, no russet; flesh soft; good shelf life. Tree: diploid and self-fruitful; pollenizer for Delicious and Granny Smith; nonspur, compact; annually productive; ripens just ahead of Golden Delicious.

Parrish Delicious.—Orig. in Topinabee, Mich., on the Chas. Parrish farm, by Roy E. Gibson, Greening Nursery Co., Monroe, Mich. Introd. in 1946. Bud mutation of Delicious; discovered in 1934 by Gibson. Fruit: skin solid, bright red; shape characteristic of Delicious; flesh fine-grained, crisp, juicy, quality good; does not tend to darken in storage; resembles Richared. Tree: vigorous; upright.

Patricia.—Orig. in Ottawa, Ontario, Canada, by Canada Dept. Agr., Res. Sta. Not introd. although tested rather widely including commercial orchards. Included for record purposes as variety cited at least twice in the literature; tetraploids have been produced from it, therefore useful in breeding studies. McIntosh o.p.; selected in 1920. Fruit: handsome; small unless thinned; high quality; about the same season as McIntosh. Tree: heavy bearer; hardier than McIntosh.

Patterson.—Orig. in Saskatoon, Saskatchewan, Canada, by C.F. Patterson, Univ. of Sask. Introd. in 1960 for home gardens. Columbia x Melba; selected in 1946; tested as Sask. A59-382. Fruit: up to 2 1/2 inches in diam.; at Saskatoon under field conditions without irrigation; slightly oblate; skin greenish-yellow, blushed; flesh very mildly subacid, flavor and quality excellent; keeps well; ripens in midseason. Tree: extremely hardy. Named in memory of Dr. Patterson.

Paulared.—Orig. in Sparta, Mich., by Lewis Arends. Introd. in 1967. Plant patent 2800; 12 Mar. 1968; assigned to Hill Top Orchards & Nurseries, Hartford, Mich. Chance seedling discovered in 1960. Fruit: average size 2 1/2 x 2 3/4 inches; roundish, oblate; skin ground color light yellow, covered almost entirely with bright red, beginning as a light stripe and developing into an attractive solid red; heavy bloom; flesh white, firm, crisp, juicy, fine-grained, subacid and sprightly; ripens about 3 weeks ahead of McIntosh; hangs well on tree, can be picked in one or two pickings. Tree: size medium; upright, forming very wide crotch angle on a sturdy central leader; vigorous; hardy; highly susceptible to powdery mildew; productive, bearing both laterally on spurs and on terminal buds; has much blind wood; very biennial.

Payette.—Orig. in Moscow, Idaho, by Leif Verner, Idaho Agr. Expt. Sta. Introd. in 1944. Ben Davis x Wagener; selected in 1936; tested as Idaho 7-11. Fruit: large; round to oblate, sides usually symmetrical but sometimes unequal; skin greenish-yellow to yellow with dark to bright red blush, lightly and inconspicuously streaked with carmine, smooth, fairly thick, inclined to become oily in storage; flesh creamy white, often somewhat greenish, fine-grained, tender very juicy, mildly subacid, aromatic; high dessert quality; somewhat susceptible to storage scald when picked before fully ripe; as grown in Moscow, Payette should be picked about a week or 10 days later than Jonathan. Tree: vigorous, wide spreading, strong, well-formed crotches; productive; while tree is young, the fruit should be thinned rather lightly to avoid development of excessive size.

Peace Garden.—Orig. in Mandan, N.D., by W.P. Baird, Northern Great Plains Field Sta. Introd. in 1957. Malinda x Duchess of Oldenburg; selected in 1948; tested as Mandan 48-50. Fruit: size medium, averaging about four per lb.; round to slightly oblate; stem short; skin waxy, bright glossy red with carmine stripes; dots small, whitish, obscure, flesh very firm, juicy, subacid; good eating quality especially if held in storage until winter; rates well in cooking tests; harvest latter part of September; if properly stored, its best season of use is winter and early spring; hangs well on tree. Tree: hardy; of moderate growth, slightly spreading.

Peace Valley.—See Starkspur Law Rome.

Penn Stayman.—Orig. in University Park, Pa., by Dept. of Hort., Pennsylvaina State Univ. Introd. in 1967. Assigned to Adams County Nursery & Fruit Farm, Aspers, Pa. Bud mutation of Stayman; discovered in 1955. Fruit: resembles Stayman, but skin is smooth and brighter colored, does not crack and does not develop scarfskin. Tree: growth habit same as Stayman.

Perrine Yellow Transparent (*Giant Transparent, Giant Yellow Transparent, Grandparent, Large Yellow Transparent*).—Orig. in Centralia, Ill., by D.B. Perrine. Introd. in 1961. Considered to be a bud mutation of Yellow Transparent; discovered about 1930. Fruit: larger than parent; flattened; quality somewhat better than parent; ripens up to 7 days before parent; otherwise, similar to parent and Wrixparent. A periclinal cytochimera 2-2-4-4 and 2-4-4-4; selection of latter produces diploid gametes. Tree: low, spreading, crotches wide; bears annually, productive; less subject to fire blight than parent. Chimeral nature determined at the New York State Agr. Expt. Sta., Geneva; accessioned in 1948 by this station.

Perrine York.—Orig. in Centralia, Ill., by D.B. Perrine. Not introd. Bud mutation of York Imperial. Fruit: large. A periclinal cytochimera 2-4-4; chimeral nature determined by Jules Janick, Purdue Univ. Agr. Expt. Sta., Lafayette, Ind., and by the New York State Agr. Expt. Sta., where it was accessioned in 1955. No commercial plantings. Used as parent in breeding for triploids.

Peru Red Spy.—See Schoharie Spy.

Pettingill.—Orig. in Seal Beach, Calif., by Clark Pettingill. Introd. in 1949. Plant patent 1352; 22 Feb. 1955; assigned to Armstrong Nurseries, Ontario, Calif. Chance seedling; discovered about 1935; first propagated in 1942. Fruit: large; apex broadly depressed; skin red overlying green, medium thick; flesh nearly white, firm, crisp, juicy, moderately acid; ripens early September to mid-October; ships and keeps well. Tree: large; upright; vigorous; very productive; regular bearer; short chilling requirement, being suitable to southern California and similar climates.

Phantasy.—See Fantazja.

Pink Lady.—Orig. from the breeding program of the Western Australian Dept. Agr. Introd. in 1989. Golden Delicious x Lady Williams. Selected in 1979 by John Cripps. Plant patent applied for. Fruit: medium to large; shape asymmetrical or oblong, with prominent ribbing; skin color attractive red, blush pattern, yellow ground; texture firm; flavor sweet, tart. Harvest 10 days after Granny Smith; storage life long. Tree: very vigorous; prolonged bloom period.

Pink Pearl.—Orig. in Ettersburg, Calif., by Albert F. Etter. Introd. in 1944. Plant patent 723; 4 Mar. 1947; assigned to George C. Roeding, Jr., Calif. Nursery Co., Fremont. Surprise o.p.; selected in 1944. Fruit: size of Northfield, varying from 2 to 2 3/4 inches in diam.; shape of Yellow Bellflower; skin transparent, which glows pink from flesh beneath, outstanding bouquet when skin is broken; flesh crisp, juicy, pink; core small; ripens in September.

Pioneer Mac™—Discovered in 1976 by Ernest Greiner, Marlboro, N.Y. Introd. in 1989. Plant patent 7002; 29 Aug. 1989; assigned to Adams County Nursery, Aspers, Pa. McIntosh o.p.; not a McIntosh sport. Fruit: appearance almost indistinguishable from McIntosh, perhaps better color than Rogers McIntosh; 90% red; eating qualities same as McIntosh. Harvest 10 days after Rogers McIntosh, with much less preharvest drop; excellent storage capacity. Tree: nonspur growth habit, less vigorous than Rogers McIntosh; very productive, similar to McIntosh, with less tendency to biennial bearing.

Piotet.—Orig. in Ottawa, Canada, by Canada Dept. Agr., Res. Sta. Introd. in early 1930s by Seager Wheeler, Rosthern, Saskatchewan, Canada. Pioneer x Tetofsky; cross made at the Central Expt. Farm. Fruit: small; dessert quality fair to good; matures early, during late August. Tree: hardy; yields regularly; bears on young trees.

Pittman Delicious.—Orig. in Okanogan, Wash., by John W. Pittman. Introd. in 1956 by H.R. Hamilton of Okanogan. Bud mutation of Delicious; discovered in 1948. Fruit: skin solid red, darker than Delicious. Tree: shows some tendency toward heavy spurring of lateral wood.

Pocomoke.—Orig. in Vineland, N.J., by Paul Sanguinetti. Introd. in 1968 by Bountiful Ridge Nurseries, Princess Anne, Md. Parentage unknown; discovered in 1966. Fruit: medium to large; Stayman Winesap shape; skin highly colored, very waxy, smooth finish; flesh white, juicy, high quality; ripens last of Delicious season. Tree: medium to large; growth habit like Stayman Winesap; vigorous; hardy; productive and annual bearer; drought resistant.

Poposia.—Orig. in Lander, Wyo., by J.M. Steinbrech, Wyoming Agr. Expt. Sta. Introd. in 1931. Fruit: skin streaked with green and red, tender; calyx open; flesh slightly subacid, flavor delicate; ripens first of October.

Porcupine.—Orig. in Brevo, Mich., by P.H. Brae. Introd. in 1922. Chance seedling; discovered at base of Porcupine Mountain. Fruit: slightly longer than Wealthy; skin dark red; quality good, flavor similar to Wealthy; ripens between Duchess of Oldenburg and Wealthy. Tree: hardy; blooms later than Northern Spy.

Portia.—See Caravel.

Potomac (*Browning Beauty*).—Orig. in Wallingford, Ky., by Frank Browning. Introd. in 1970 by Bountiful Ridge Nursery, Princess Anne, Md. Discovered in 1962 as an odd single tree in a planting. Fruit: medium to large; skin bright red, blushed with purple; flesh firm, tart, mellow. Tree: size medium, spreading; vigorous; productive.

Prairie Gold.—Orig. in Mandan, N.D., by W. P. Baird, U.S. Northern Great Plains Field Sta. Introd. in 1952. Robin x Duchess of Oldenburg; selected in 1935. Fruit: a large crab apple or a small standard apple; oblate; ridged; skin yellow; flesh firm, juicy, slightly tart to subacid, flavor pleasant; resembles Grimes Golden. Tree: semi-dwarf; spreading; very hardy.

Prairie Spy.—Orig. in Excelsior, Minn., by the Univ. of Minnesota Fruit Breeding Farm. Introd. in 1940. Parentage unknown; seed planted in 1914; selected in 1923; tested as Minn. 1007. Fruit: high dessert and culinary quality; flesh crisp and juicy; matures in October. Resembles Northern Spy. Tree: hardy; vigorous.

Priam.—Orig. in Lafayette, Ind., by a cooperative breeding program consisting of the Agr. Expt. Stations of Indiana (E.B. Williams, Jules Janick, F.H. Emerson), Illinois (D.F. Dayton, J.B.Mowry), New Jersey (Catherine Bailey, L.F. Hough), the U.S. Dept. of Agr., and Station d'Arboriculture Fruitiere, INRA, Beaucouze, Angers, France (L.M. Decourtye). Introd. in 1974, PRI 4-126 [Golden Delicious x 26829-2-2 (F_2)] x Jonathan; 26829 is the result of a sib cross of two selections of Rome Beauty x *Malus floribunda* 821. Cross made in 1951, first fruited in 1956, tested as 370-16. Fruit: 2 1/2 to 3 3/4 inches in diam; round-conical, very regular; skin green-yellow ground color, moderately tough, no russeting; flesh fine, crisp, juicy, medium sweet, acidity strong, aromatic, quality good; keeping quality at least 3 months at 2C; ripens about 1 week before Delicious, may be too acid for U.S. tastes before storage, optimum edible quality October to January; a dessert variety. Tree: vigor medium; spreading, open; productivity heavy; slight mildew susceptibility, immune to apple scab (*Venturia inaequalis*); blooms 1 week before Golden Delicious. Named for Priam, the last king of Troy, the father of Paris.

Prima.—Orig. in Urbana, Ill., by the Agr. Expt. Stations of Illinois, New Jersey, and Indiana and the U.S. Dept. of Agriculture. Responsible parties were D.F. Dayton and J.B Mowry of Illinois; L.F. Hough and Catherine Bailey of New Jersey; and E.B. Williams, Jules Janick, and F.H. Emerson of Indiana. Introd. in 1970. Ill. seedling 14-510 x N.J. seedling 123249. Cross made in 1957, selected in 1963; tested as Co-op 2 and 1225-100. Fruit: size medium; slightly oblate, slightly lopsided; skin 60% to 95% bright-red blush over yellow ground cover, thick, medium tough, smooth, waxy, attractive, dots small to medium, no russeting; flesh yellow, crisp, flavor mediocre, subacid, texture coarse, quality mediocre; tends to have bitter pit; ripens 3 weeks before Jonathan, 1 month before Delicious;

recommended for dessert use. Tree: size medium on standard rootstock, spreading; vigor good; hardy; productivity good; carries *Vf* gene for resistance to apple scab; tolerant to fire blight and apple blotch; adapted to central and southern midwest of the U.S. The prefix PRI in the name is an acronym from the three institutions involved: Purdue, Rutgers, and Illinois.

Prime Gold.—Orig. in Zillah, Wash., by Bud Hoekman. Introd. in 1965. Parentage unknown, probably a chance seedling of Golden Delicious. Plant patent 2806; 23 Apr. 1968. Discovered in 1965. Fruit: resembles Golden Delicious, shape and flavor being similar but not as good; firmer and keeps longer than Golden Delicious; does not russet. Tree: lenticels remain white; blooms somewhat later than Golden Delicious.

Primerouge.—See Akane.

Primicia.—Orig. in Caçador, Santa Catarina, Brazil, by F. Denardi, L.F. Hough, and A.P. Camilo, Santa Catarina Expt. Station. Introd. in 1988. Parentage: D1Rl01T117 x D1R103T245; crossed in 1975 at New Jersey Agr. Expt. Station; a sixth-generation derivative of C.S. Crandall's Rome Beauty x *M. floribunda* 821 sib cross; ancestry includes Yellow Newtown, Golden Delicious, Crandall, Cortland, Melba, Wealthy, Starr, and Rome Beauty. Selected in 1982; tested as Malus 29. Fruit: medium size, 150 g; round to round-conic, no russet, bright red over yellow ground; attractive; flesh light cream, moderately juicy, firm; flavor mildly acid; ripens 3 weeks before Gala® and 6 weeks before Golden Delicious; early ripening results in no bitter pit; stores well in cold storage. Tree: medium vigor; branches 90° angles with the trunk; open, spreading; resistant to apple scab; moderately resistant to powdery mildew. Lower winter chilling requirement than Gala®; does not require chemical treatment to break dormancy in Caçador. Precocious; heavy, annual cropping; overripe fruit drop from the tree. Flowering: several days before Gala®.

Prince.—Orig. in Tifton, Ga., by Major C. Collins. Introd. in 1977. Probably Golden Delicious o.p. Fruit: 2 3/4 to 2 1/2 inches in diam.; smooth, symmetric, round to conic; skin yellow-green when ready to be picked in mid-August, but will turn yellow on the tree, overspread up to 40% with a light blush of pink; flesh cream-colored, slightly coarse, juicy, quality good. Tree: resembles Golden Delicious; susceptible to *Glomerella* leaf blotch. Recommended as a dooryard variety.

Princesa.—Orig. in Caçador, Santa Catarina, Brazil, by F. Denardi, L.F. Hough, and A.P. Camilo, Santa Catarina Expt. Station. Introd. in 1988. NJ56 (NJ440249 x NJ39) x Anna; crossed in 1977; ancestry includes Adassia Red, Golden Delicious, Edgewood, Cox's Orange Pippin, Yellow Newtown, Red Rome, and Jonathan. Selected in Brazil in 1984. Fruit: average weight 160 g; shape round, crowns at calyx end; nearly 100% red over a yellow ground, no russet, attractive; flesh whitish-cream, firm, juicy; flavor sweet, subacid; harvest early, 3 to 4 weeks after Anna. Tree: moderately vigorous with strong, upright-spreading lateral branches; chilling requirement low, similar to Anna, about 400 h below 7C; flowering, 3 to 4 weeks after Anna; highly productive; precocious cropping; fruit hang well after fully ripened; susceptible to apple scab and powdery mildew.

Priscilla.—Orig. in Lafayette, Ind., by a cooperative breeding program consisting of the Agr. Expt. Stations of Indiana (E.B. Williams, Jules Janick, F.H. Emerson), Illinois (D.F. Dayton, J.B. Mowry), New Jersey (L.F. Hough, Catherine Bailey), and the U.S. Dept. of Agriculture. Introd. in 1972. Starking Delicious x seedling 610-2. Seedling 610-2 has a complex pedigree being derived from a sib cross of Rome Beauty x *Malus floribunda* 821 crossed successively to Golden Delicious and McIntosh. Cross made in 1961, first fruited in 1966, tested as Co-op 4 and 1659-1. Fruit: tends to be small; slightly oblate, regular; skin light yellow ground color, covered with 75% to 90% bright red blush, smooth, waxy with small white dots, medium thick, but not tough, no russeting; flesh crisp, flavor poor, texture medium grained, quality mediocre, mild subacid, aromatic; keeping quality 2 to 3 months at 34F; harvest 10 days after Prima, 1 week before Jonathan, 2 weeks before Delicious, little tendency to drop before maturity; a dessert variety. Tree: moderately spreading; vigorous; productive; flowers with or just after McIntosh, pollen fertile, should make good pollenizer for Prima; resistant to all presently known races of the scab organism *Venturia inaequalis* (Cke.) Wint.; tolerant to fire blight and powdery mildew; adapted to midwest and northern U.S. areas. Named in honor of Priscilla Hovde, wife of Frederick Boyd Hovde, 7th president of Purdue Univ.

Pristine™ (*Co-op 32*).—A disease-resistant early apple from the Purdue–Rutgers–Illinois cooperative breeding program. Parentage: Co-op 10 x Camuzat; tested as PRI 2946-1; introd. in 1994 by J. Janick, J.A. Crosby, P.C. Pecknold, Joseph Goffreda, and S.S. Korban. U.S. plant patent applied for; assigned to Purdue Res. Foundation, West Lafayettte, Ind. Fruit: most resembles Yellow Transparent; size medium, 65 to 70 mm; oblate round to round. Skin thin; pale green-yellow to cream-colored, maturing to deep yellow with moderate orange blush on exposed surface; finish smooth, glossy, with inconspicuous lenticels. Mildly acid, slightly spicy, moderately rich; retains excellent quality and texture for 4 to 6 weeks or more in refrigerated storage. Harvest just after Lodi, late July at West Lafayette; two pickings required. Tree: moderate vigor; round-spreading; semi-spur; limber wood that droops under crop load; moderately productive. Immune to apple scab (*Vf* gene); moderately resistant to fire blight; slightly resistant to cedar apple rust; resistant to powdery mildew.

Prolific.—Orig. in Saskatoon, Saskatchewan, Canada, by C.F. Patterson, Univ. of Sask. Introd. in 1960 for home gardens. Columbia x Wealthy; selected in 1944; tested as A58-307. Fruit: up to 2 inches long and 1 7/8 inches in diam. under nonirrigated field conditions at Saskatoon; conical; skin greenish with light blush and some bloom; flesh very mildly subacid, quality good; keeps fairly well; resembles Brightness. Tree: yields heavy crops; hardy.

Purdue Black-Jon.—Orig. in Lafayette, Ind., by Purdue Univ. Introd. in the 1940s. Parentage unknown; original scionwood from a Jonathan tree of unknown origin planted in Purdue Univ. orchards probably in the 1920s; used in conjunction in studies with Virginia crab apple stocks. Fruit: skin very dark red; flesh firm, dry; long storage life; apparently when grown on seedling rootstocks without the use or influence of Virginia crab, then it is not as well-colored as Jonared or Blackjon and does not have as good storage qualities. Not to be confused with Blackjon. No longer being propagated.

Puritan.—Orig. in Amherst, Mass., by F.C. Sears, Massachusetts Agr. Expt. Sta. Introd. in 1953. McIntosh x Red Astrachan; cross made about 1931. Fruit: size medium; skin attractive solid red; flesh white, a bit tart but of very good quality; season of Early McIntosh, ripening 3 weeks before McIntosh; McIntosh type. Tree: wide crotch angles; tendency to bear biennially; pollenizes McIntosh.

Quebec Belle (*Belle de Quebec*).—Orig. in Hemmingford, Quebec, Canada, by D.M. Laurie. Introd. in 1956 by W.H. Perron & Co., Montreal, Quebec. Northern Spy o.p.; discovered in 1948. Fruit: large; skin dark red; keeps very late; resembles Delicious in appearance and Northern Spy in quality. Tree: appears to be hardy in Quebec.

***Querina*.**—See Florina.

Quindell.—Orig. in Green Forest, Ark., by Ralph Banta. Introd. in 1965 by Green Forest Nursery. Parentage unknown; discovered in 1934. Fruit: large; skin red over yellow ground; flesh white, firm, flavor of Delicious with some tartness; dessert and culinary uses. Tree: rather compact; semi-dwarf; produces good annual crops; comes into bearing early.

Quinte.—Orig. in Ottawa, Ontario, Canada, by Canada Dept. Agr., Res. Sta. Introd. in 1964. Crimson Beauty x Red Melba; selected by Lloyd P.S. Spangelo from progeny grown at the Expt. Farm, Smithfield, Ontario; tested as Ottawa T-441. Fruit: medium, may require thinning in some districts to realize a good size; round to slightly conic; skin attractive yellow, heavily washed in red; flesh cream, very tender; very suitable for dessert purposes, equal to Melba and superior to Crimson Beauty and Close; outstanding in shipping quality; ripens 7 to 10 days before Melba and 3 to 4 days after Crimson Beauty; developed primarily for eastern Canada. Tree: annually bears good crops; as hardy as McIntosh.

Rafzubin.—See Rubinette.

Rahel.—Orig. in Rehovot, Israel, by Chanan Oppenheimer, Agr. Res. Sta., Div. Subtrop. Hort., National and Univ. Inst. of Agr. Introd. there in 1953. Bian x White Astrachan. Selected in 1951. Fruit: small; skin green; flesh firm, flavor good; ripens July. Tree: medium. Now obsolete.

Raikes McIntosh.—Orig. in Kelowna, British Columbia, Canada, by E. Raikes. Introd. in 1971. Mutation of Summerland McIntosh, discovered in 1968. Fruit: same as McIntosh, except perhaps more oblate; ripens same season as McIntosh, 20 Sept. Tree: compact growth; productive and precocious bearer.

Ranger.—Orig. in Ottawa, Ontario, Canada, by Canada Dept. Agr., Res. Sta. Introd. in 1964. Crimson Beauty x Melba; selected in 1944 by D.A. Blair; tested as Ottawa 342, early testing by Blair, tested later and released by Lloyd S.P. Spangelo. Fruit: larger than Caravel; roundish-conic; skin attractively red striped over a yellow ground color; flesh white, tender, flavor and general dessert qualities equal to Melba; ripens 3 to 4 days after Crimson Beauty and 7 to 10 days before Melba; developed primarily for eastern Canada. Tree: vigorous; hardy; bears biennially; branching angle narrow.

Raritan.—Orig. in New Brunswick, N.J., by New Jersey Agr. Expt. Sta. Introd. in 1966. Parentage: (Melba x Sonora) x [Melba x (Williams x Starr)]; cross made in 1949 by G.W. Schneider; selected in 1956; tested as NJ 25; introd. by L.F. Hough and Catherine H. Bailey. Fruit: large; skin yellow ground color with one-half to three-quarters washed and splashed with red; flesh good quality for summer apple; good shelf life; ripens about with Puritan. Tree: early bearing; may be biennial; natural spur type.

Razor Golden Delicious.—Orig. near Millington, Tenn., by H.L. Jones. Introd. in 1970. A russet-skinned mutation of Golden Delicious. The russet is uniquely light colored, uniform, and attractive. Resembles Golden Delicious in all other respects.

Real McCoy.—Orig. in Yakima, Wash., by J.D. McCoy. Introd. in 1971. Bud mutation of Starking Delicious, discovered in 1956. Assigned to May Nursery Co., Yakima. Fruit: similar to Starking Delicious, colors earlier and darker. Tree: standard; highly productive.

Red Atlas.—Orig. in Kakabeka Falls, Ontario, Canada, by William Sovereign. Introd. in 1950. Bud mutation of Atlas; discovered in 1945. Fruit: skin more heavily washed with deep red than average Atlas; otherwise, similar to Atlas in shape, flesh texture, and quality. Tree: very upright; vigorous.

Red Baldwin.—See Sawyer Baldwin.

Red Baron.—Orig. in Excelsior, Minn., by Hort. Res. Sta. Introd. in 1969. Golden Delicious x Red Van Buren. Cross made in 1926 by W.H. Alderman and selected by him in 1940; tested as Minn. 1500; released in 1969 by Cecil Stushnoff. Fruit: size medium; round truncate, regular; skin creamy yellow ground cover with 80% to 100% bright red blush, glossy, medium thick, tough, attractive; flesh crisp, juicy subacid; ripens 20 Sept., same season as Wealthy; good storage quality; for fresh fruit, pies, and sauce. Tree: size medium; medium open; vigor moderate; hardiness equal to Haralson; very productive; tolerant to fire blight.

Red Bouquet.—Orig. in Chelan County, Wash., by Herbert Diede, Wenatchee, Wash. Introd. in 1968. Plant patent 2956; 23 Dec. 1969; assigned to Stark Bros. Orchards & Nursery Co., Louisiana, Mo. Bud mutation of Starking Delicious; discovered in 1954. Fruit: skin nearly solid red stripe, more red than Starking Delicious, tough; flesh characteristics same as Starking Delicious. Tree: medium large, spreading; vigorous; hardiness same as Starking Delicious; productive; flower attractive purplish-pink; recommended for home or commercial planting.

Red Braeburn.—See Hillwell.

RedChief (*Campbell Spur*).—Orig. in Grant County, Columbia Basin, Wash., by Fred Campbell, Othello, Wash. Introd. in 1972. Plant patent 3578; 25 June 1974, assigned to Hilltop Orchards and Nurseries,

Hartford, Mich. Bud mutation of Starkrimson Delicious, discovered in 1967; tested by Hilltop Orchards and Nurseries as FC-32. Fruit: 3 inches in diam.; Delicious shape; skin bright cherry red, striped, attractive; flesh white to creamy white, quality excellent; ripens 12 to 22 Sept. Tree: semi-dwarf; upright to spreading; often low in vigor; hardy; productivity heavy.

Red Delight (*Bolding*).—Orig. in Sylacauga, Ala., by Oren T. Bolding, Sr. Introd. in 1960. Plant patent 1974; 4 Oct. 1960; assigned to the Commercial Nursery Co., Decherd, Tenn. Considered to be a seedling of Yellow Transparent; discovered in 1957; tested as Commercial 60, Commercial 61, and Commercial 1961. Fruit: large, up to 1 lb each; oblate; skin deep red with yellow ground color; flesh white, firm, juicy, subacid, flavor mild; keeps well in common storage; for dessert purposes; resembles Stayman Winesap and Mammoth Black Twig. Tree: vigorous; bears well farther south than some other apple varieties because it blooms late; productive, tends to bear annually.

Red Dougherty.—Orig. in Twyford, Hawkes Bay, New Zealand, by C.F. Bixley. Introd. in 1930. Bud mutation of Dougherty; selected in 1928. Fruit: identical to Dougherty except for brilliant scarlet coloration over whole fruit; very sweet, but lacking in character; texture is hard without being crisp; storage quality excellent. Tree: identical to Dougherty; more suited to warm climates.

Red Duchess.—See Red Van Buren.

Red Gala.—See Royal Gala.

Red Graham.—See Graham Spy.

Red Haralson.—See Haralred.

Red James Grieve.—See Neumann James Grieve.

Red Jonathan.—See Blackjon.

Red-Leaved Paradise (*Budagovsky 9, Bud.9, B9*).—See rootstocks for apple.

Red McIntosh.—See Rogers McIntosh.

Red of Eger (*Egri piros*).—Orig. in Eger, Hungary, by Kovacs Sandor, Univ. of Hort., Budapest, Hungary. Introd. in 1967. Parentage unknown, discovered in 1960, selected in 1964. Fruit: medium-large; roundish elongated; skin bright red, waxy, very attractive; flesh snow-white, firm, flavor excellent; ripens first part of August. Tree: medium; rounded; vigorous; hardy; productive; tolerant to mildew; self-fertile.

Red Queen Delicious.—Orig. in Paw Paw, W.Va., by Henry W. Miller, Consolidated Orchards. Introd. in 1957. Bud mutation of Starking Delicious; discovered in 1952. Fruit: identical with parent except for skin color, which colors early in July, showing a full bright red by the first of August. Tree: nonspur, standard type.

Red Saelet.—Orig. in Belgium by Edwin Saels, Herk-de-Stad. Introd. in 1976. Plant patent 3878; 4 May 1976. A natural mutation found in an orchard of low-stem Golden Delicious trees. Fruit: resembles Golden Delicious; flesh firm, bittersweet, crisp and juicy; fruit ripens about 8 days after Golden Delicious; due to firmness it is a better shipper, preserves well, and is especially suited for industrial production. Tree: less vigorous than Golden Delicious, but productivity is same.

Red Sauce.—Orig. in Geneva, N.Y., by Richard Wellington, New York State Agr. Expt. Sta. Introd. in 1926. Deacon Jones x Wealthy; cross made in 1910; first full crop in 1917. Fruit: large; roundish conic; skin nearly solid red; flesh coarse, briskly subacid, often red to coreline; makes a red sauce; season October. Tree: vigorous; upright; spreading; heavy annual bearer.

Red Scarlet.—Orig. in Sebastopol, Calif., by Wm. E. Silva. Introd. in 1965. Plant patent 2464; 15 Dec. 1964. Red Bellflower x Baldwin; selected in 1963. Fruit: up to 3 inches in diam.; oblong-ovoid, usually symmetrical; skin under color yellow, usually completely red, thin, smooth, not waxy; flesh whitish-yellow, juicy, firm, sweet, soluble solids 16.4%; prominent aroma; keeps well; for dessert and culinary uses; ripens in mid-August; resembles Bellflower. Tree: highly productive; vigorous, upright.

Red Sharon.—Orig. in Guthrie Center, Iowa, in the Howard Gillispie orchard by Harvey Lantz, Iowa Agr. Expt. Sta., Ames. Introd. in 1952. Bud mutation of Sharon; discovered about 1950. Fruit: skin near solid to solid red, attractive, smooth, tender; ripens after Redgold and before Jonadel. Tree: hardy; excellent foliage; in nursery has vigor equivalent to standard varieties in southwest Iowa.

Red Soviet.—Orig. in Russia. Introd. as scionwood into the U.S. from Russia in 1934 by N.E. Hansen, South Dakota Agr. Expt. Sta., Brookings. Parentage unknown. Tree: extremely hardy; vigorous. Probably obsolete.

Red Spy.—See Greene Spy.

Red Standard.—Orig. in Russia. Introd. in the U.S. in 1934 by N.E. Hansen, South Dakota Agr. Expt. Sta., Brookings; renamed by him because original name not known. Parentage unknown; selected in 1934 from this imported collection. Fruit: quality fair to poor. Tree: hardy.

Red Stark.—Orig. in Kibbie, Mich., by A.G. Spencer & Son. Introd. in 1927 by Greening Nursery Co., Monroe, Mich. Bud mutation of Starking Delicious. Fruit: skin solid red; otherwise, very similar to parent.

Red Statesman.—Orig. in Lenswood, South Australia, by H.K. Caldicott. Introd. in 1950. Bud mutation of Statesman; discovered about 1940. Fruit: skin with brilliant red; keeps late; general improvement over Statesman.

Red Stayman 201.—See Nured Stayman.

Red Striped Graham.—See Striped Graham Spy.

Red Summer Rambo.—Orig. in Chambersburg, Pa., by Ray Hess. Introd. in 1958 by the Adams County Nursery, Aspers, Pa. Summer Rambo o.p.; selected in 1957. Fruit: skin a bright, solid red; differs from Summer Rambo by having more color and coloring earlier; ripening between early summer and fall. Tree: like Summer Rambo.

Red Thorle.—Orig. in Remuera, Auckland, New Zealand, in the garden of E.T. Wilson, by D.W. McKenzie, Fruit Research Division, New Zealand. Introd. in 1950. Bud mutation of Thorle Pippin; formerly listed as Variety 39. Fruit: medium, 2 x 2 3/4 inches in diam., variable; flat smooth; transverse outline round, not crowned, regular; basin smooth, very shallow with open eye; calyx markedly tomentose, extending onto basin and giving appearance of being dipped in white flour; stem medium length; skin greenish-yellow striped with crimson, thick; flesh greenish-white, crisp, tender, juicy, sour and sprightly, faintly aromatic, slightly astringent, quality fair; storage good; ripening with Close, requiring several pickings. Tree: vigorous; spreading, pendulous; spur-bearing habit; long pliant fruiting arms; susceptible to apple scab; tolerant to mildew, woolly aphis; severe cracking at calyx end where spring frosts occur; regular and moderate producer, heavy for early variety.

Red Transparent.—See Close.

Red Van Buren (*Double-Red Duchess, Red Duchess, Van Buren*).—Orig. in New York by J.P. Van Buren. Introd. in 1937. Parentage unknown; discovered in 1914; evidence seems to indicate this variety did not arise as a bud mutation of Duchess. Fruit: similar to Duchess in flavor and quality, but larger and a solid, dark-red blush, attractive. Named after B.D. Van Buren, former President of the New York State Fruit Testing Assn. However, this variety is not related to the Van Buren variety described by W.H. Ragan in U.S. Dept. Agr. Bul. 56, 1905.

Red Warrior.—Orig. in western Maryland by Henry W. Miller, Sr. of Paw Paw, W.Va. Introd. in 1938 by the Bountiful Ridge Nurseries, Princess Anne, Md. Parentage unknown; discovered growing on Warrior Mountain about 1929. Fruit: large; skin color fine, deep, dark red appearing in August well before the better known varieties develop a dark color; excellent carrying qualities, being fully as firm as Rome Beauty; table quality poor, being a bit tough and dry with poor flavor, much better when cooked. Tree: vigorous; regular bearer; resembles McIntosh.

Red Wealthy.—See Case Wealthy.

Red Willow.—Orig. in Columbiana County, Ohio, by E.E. Leeper, Louisville, Ohio. Introd. in 1942. Bud mutation of Willow; discovered in 1939. Fruit: red skin color better than parent; other characteristics similar to Willow. No longer being propagated.

Red Winesap.—See Strand Winesap and Virginia Winesap.

Red York Imperial.—See Colora Red York and Yorking.

Red Yorking.—See Yorking.

Redant.—Orig. in Morden, Manitoba,

Canada, by Canada Dept. Agr., Res. Sta. Introd. in 1936. Antonovka o.p. Fruit: large; round; skin deep yellow, washed with red, some striping; flesh white, fine-grained, somewhat soft, mild, sweet, quality good; season October to January. Tree: upright, spreading; vigorous; productive.

Redbow.—Orig. in Bathurst, New South Wales, Australia, by F.T. Bowman, N.S.W. Dept. Agr. Introd. in 1960. Democrat x Jonathan; selected in 1956. Fruit: skin a brilliant red; highly resistant to black spot; does not store well; resembles Democrat. Tree: heavy cropper.

Redfree.—Orig. in Lafayette, Ind., by Purdue, Rutgers, and Illinois Univs. Raritan x PRI 1018-101. Plant patent 4322; assigned to Purdue Research Foundation. Introd in 1981. Variety's most valuable characteristics: immune to apple scab and apple rust; good resistance to mildew, fire blight, and European red mites. Very early maturity. Fruit is small, red stripes on greenish-yellow, attractive. Flavor sweet, insipid; texture crisp. Little russet; tree has much blind wood, light producer.

Red Jon.—See Blackjon.

Redsleeves (*Malling Redsleeves*).—Orig. from breeding program in East Malling Res. Sta., East Malling, Kent, England, by F.H. Alston. Introd. in 1986. Exeter Cross x TSR15T3 (a PRI scab-resistant selection). Tested as A232/2. Owned by Plant Breeding International, Cambridge. Fruit: medium size, 60 to 65 mm in diam.; attractive, predominantly bright red on yellow-green background; good skin finish; flavor pleasant, sweet; flesh texture crisp and juicy; best suited for home orchards; ripens early, in late August to early September; shelf life very good, at 18C it retains good texture for at least 1 month after harvest. Tree: precocious cropping; compact growth habit, requiring little pruning; thinning is necessary; on M.27 stocks, especially suited for small gardens.

Redsnow.—Orig. in Morpeth, Ontario, Canada, by H. Quackinbush, Fort William, Ontario. Introd. in 1945. Parentage unknown; discovered in 1935. Fruit: 2 x 2 1/4 inches in diam.; skin bright red to dark red; flesh white to pinkish, hard; quality good for sauce; ripens 10 Sept. Tree: upright; moderately hardy.

Redspur Delicious (*Show, Show Red*).—Orig. in Yakima, Wash., by Elon J. Gilbert of the Richey & Gilbert Co. Introd. in 1957. Plant patent 1822; 3 Feb. 1959; Loren R. Frazier and Philip J. Jenkins, Parker, Wash. Whole tree variation in a Starking Delicious orchard; discovered in 1954. Fruit: skin red, appearing very early; resembles a well-colored Richared. Tree: close internodal growth, resulting in a tree about two-thirds the size of a standard one, similar to Okanoma and Starkrimson Delicious.

Redsumbo.—Orig. in Cory, Caly County, Ind., by Lorne J. Doud, L.V. Doud & Son of Wabash, Ind. Introd. in 1945. Bud mutation of Summer Rambo; discovered in 1930. Fruit: skin lightly splashed with bright red; quality good, desirable for eating fresh and for cooking; ships well; smaller and firmer but resembles Summer Rambo; ripens slightly later than Summer Rambo. Tree: resembles parent.

Redwell.—Orig. in Excelsior, Minn., by the Univ. of Minnesota. Fruit Breeding Farm. Introd. in 1946. Scott's Winter o.p.; seed planted about 1911; selected in 1923; tested as Minn. 638. Fruit: size medium to above; roundish; skin very attractive, bright medium red over yellow, dots russet, tender, tends to bruise easily; flesh cream, mild; quality good for baking, dessert, sauce; matures about mid-October; keeps well until January; resembles Wealthy. Tree: size medium; strong framework; hardy; slightly susceptible to fire blight; medium susceptibility to cedar rust and scab; annual bearer; blooms in midseason.

Redwin Delicious.—Orig. in Peshastin, Wash., by J.L. Johnson. Introd. in 1928. Bud mutation of Delicious; discovered in 1925. Fruit: colors early, retaining dark red stripes at full maturity. Tree: standard.

Redwin Spy.—Orig. in Paris, Ontario, Canada, by A.W. Pate, Brantford, Ontario, Canada. Introd. in 1934 by the nursery of E.D. Smith Sons, Winona, Ontario. Generally considered to be a bud mutation of Northern Spy; discovered about 1917. Fruit: size above average; where exposed to sun, skin a bright red, but colors poorly in shade; unless well-colored, it has a dull, dark, unattractive appearance; skin color acquired about 10 days earlier than that of

Northern Spy; ships well; keeps well in storage; resembles Northern Spy. Tree: strong grower; similar to Northern Spy; limited commercial plantings, although one of the better red Spys, as Northern Spy is preferred by processors.

Regal.—See Regal Chelan Spur.

Regal Chelan Spur (*Regal, Regal Spur Red*).—Orig. in Brewster, Wash., by Cal Cooper. Introd. in 1970. Bud mutation of Wellspur Delicious; discovered in 1963. Fruit: medium, long Delicious shape; skin bright red with definite stripes; flesh white; a fresh-fruit variety; ripens about 145 days after blossom. Tree: 65% standard; upright spreading; vigor moderate; productivity excellent. Recommended for better coloring areas.

Regal Gala® (*Fulford Gala®, Gala Must*).—A red mutation of Gala®. Discovered near Hastings, New Zealand, by W. Fulford. Plant patent 3309; 20 Feb. 1973. Fruit: a bright reddish orange blush over a yellow ground; shape ellipsoidal. Matures about 5–7 days before Gala. Tends to become soft a little sooner than Gala, otherwise same as Gala.

Regal Spur Red.—See Regal Chelan Spur.

Regali® (*Delkistar*).—Orig. at Malicorne, France, in the breeding program of Georges Delbard Nursery. Kidd's 28 x Starkrimson Delicious. Fruit: medium to large size, 75 to 85 mm, sometimes more than 200 g; truncated cone; skin dark red, striped on yellow-green ground color, dots conspicuous, slight russet, stem short; very firm; very good eating quality; well suited for processing. Harvest 10 days after Gala; requires two or three pickings. Tree: vigorous; upright-spreading; productive. Blossoms with Golden Delicious; pollenized by Golden Delicious, Delicious, Granny Smith, Gala, and Golden Gem.

Regent.—Orig. in Excelsior, Minn., by E.T. Anderson, T.S. Weir, and Shirley T. Munson, Univ. of Minnesota Fruit Breeding Farm. Introd. in 1963. Daniels Red Duchess x Delicious; cross made in 1924; selected in 1937 by W.H. Alderman; tested as Minn. 1430. Fruit: size medium; oblong-conic, somewhat irregular; skin bright red stripes over light yellow, attractive, glossy, tough, medium thick; dots white or corky, medium small; flesh creamy white, crisp, very juicy, flavor suggestive of Delicious but more sprightly, quality high for fresh use, very good for pie and sauce; maintains fine dessert qualities with little change from harvest to late winter in storage, being superior to Haralson in appearance and storage quality, but some breakdown of flesh next to skin when stored at 35F; ripens late, in mid-October; resembles Delicious. Tree: vigorous; tends to bear when young; good yields usually produced annually; fruit hangs well, rarely drops before harvest; usually susceptible to fire blight and scab; very resistant to cedar apple rust; lacks extreme hardiness.

Reid.—Orig. in Silver Park, Saskatchewan, Canada, Mrs. F.E. MacDonald of Melfort, Saskatchewan. Introd. in the 1950s by Skinner's Nursery, Dropmore, Manitoba. Parentage unknown; seed brought from Ontario, Canada. Fruit: about 2 inches in diam. and 2 1/2 inches long; flavor good; keeps well in storage. Tree: considered hardy.

Rescue (*Scott 1*).—Orig. in Scott, Saskatchewan, Canada, by Canada Dept. Agr., Res. Sta. Introd. in 1933. Blushed Calville o.p. Fruit: 1 1/2 or more inches in diam.; round ovate; skin greenish-yellow, washed and striped red; flesh yellowish-white, firm; flavor sweet, subacid, pleasant; season late August; quality good. Tree: medium tall; rounded; healthy; very hardy; adapted to northern areas.

Reward.—Orig. in Indian Head, Saskatchewan, Canada, by Canada Dept. Agr., Res. Sta. Introd. about 1945. Moscow Pear o.p.; selected in 1929. Fruit: resembles Battleford somewhat but juicier and of better quality; about 2 1/2 inches in diam. Tree: moderately hardy.

Richardson.—Orig. in Hammond, N.Y., by Leslie Richardson. Introd. in 1959 by Fred L. Ashworth, Heuvelton, N.Y. Chance seedling; discovered in 1956. Fruit: medium; skin pale green or yellowish, blushed, like Pease; flesh yellowish, rich and mellow flavor; ripens in early fall; resembles Pease. Tree: vigorous; produces well and regularly.

Richared Delicious.—Orig. in Monitor, Wash., by J.L. Richardson. Introd. in 1926 by the Columbia & Okanogan Nursery Co., Wenatchee, Wash. Bud mutation of Delicious; discovered about 1915. Fruit: skin colors earlier and brighter red than Delicious, which it resembles. Tree: standard type.

Richelieu.—Orig. by R.L. Granger and G.L. Rousselle, Agr. Canada Research Station, St. Jean-sur-Richelieu, Quebec, Ottawa 521 x 11-51; ancestry includes Melba, McIntosh, Jonathan, Rome Beauty, and *Malus floribunda* 821; cross made in 1964 in Ottawa; selected in Trenton, Ontario; and evaluated at Frelighsburg, Quebec. Tested as Ottawa 635. Described in 1983 by G.L. Rousselle; named in 1990 by R.L. Granger. Fruit: medium-large, 65 to 70 mm in diam.; color 60% to 65% red on light green ground color; shape oblong-conical; flesh white, juicy, crisp, mild to subacid, high sugar and aroma; quality good. Introduced as midseason dessert apple, also suitable for culinary use. Harvest 1 week before McIntosh; storage life 3 months or more in air. Tree: vigor medium; spreading; precocious; annually productive; resistant to apple scab (*Vf*); moderately resistant to powdery mildew and to fire blight; very susceptible to cedar apple rust and quince rust.

Richland.—Orig. in Wyndmere, N.D., by R.L. Wodarz. Introd. in 1953. Adno o.p. Fruit: large, 3 to 3 1/2 inches in diam.; skin greenish-yellow with red over-stripes; rounded conic; flesh moderately juicy, firm, creamy white, texture medium, quality fair; ripens 10 Aug.; storage life very short. Tree: hardy; bears annually; young trees upright.

Rings Wealthy.—Orig. in Wolcott, N.Y., by Fred Rings. Introd. in 1959. Bud mutation of Wealthy; discovered in 1936. Fruit: skin solid dark red blush, being somewhat darker than parent; otherwise resembles Wealthy. Tree: identical to parent.

Roanoke.—Orig. in Blacksburg, Va., by George D. Oberle, Virginia Polytechnic Inst. Introd. in 1967. Red Rome x Schoharie. Cross made in 1949; first fruited and selected in 1957; tested as VPI 4. Fruit: rounded, symmetrical; skin near solid red, smooth, fine finish; flesh above average in firmness, texture medium; flavor mild, subacid with some suggestion of Northern Spy aroma; quality above average; keeps well in storage; resembles Red Rome; ripens about 20 Sept. in Blacksburg or 5 days before Delicious. Tree: round topped; vigor equal to Red Rome; heavy producer of fruit spurs and does not produce the blind wood or nodes characteristic of Red Rome; flowers late with Red Rome; good pollen.

Rob Roy.—See Early McIntosh.

Roberts (*John Roberts*).—Orig. in Lander, Wyo., by J.M. Steinbrech, Wyoming Agr. Expt. Sta. Introd. in 1932. Fruit: size above medium; oblong; skin light green with red stripes; flesh mild, quality good; very fine for eating fresh and cooking; season 10 Sept. Tree: upright; vigorous.

Robijn.—See Rotyp in apple rootstocks.

Robinson McIntosh.—Orig. in Oswego, N.Y., on the Anderson & Robinson Fruit Farm. Not introd. Considered to be a bud mutation of McIntosh. Fruit: large. A periclinal cytochimera 2-2-4-4. Chimeral nature determined at the New York State Agr. Expt. Sta., Geneva; accessioned in 1950 by this station. No commercial plantings; academic interest only.

Robusta 5.—See rootstocks for apple.

Rodney.—Orig. in Wyndmere, N.D., by R.L. Wodarz. Introd. in 1954. Okabena o.p. Fruit: averages 3 inches in diam.; rounded; skin red, little striping; flesh white, texture medium fine, juicy, fine quality for fresh use; ripens 10 Sept.; storage life averages about 2 months. Tree: size medium; spreading; hardy.

Rogers McIntosh (*Double Red McIntosh, Red McIntosh, Rogers Red McIntosh*).—Orig. in Dansville, N.Y., by Isaac C. Rogers. Introd. in 1932. Bud mutation of McIntosh. Fruit: McIntosh type; skin dark solid red; in all other respects, similar to its parent.

Rogers Red McIntosh.—See Rogers McIntosh.

Rome Beauty Double Red.—See Cowin Rome Beauty.

Rose Red Delicious.—Orig. in Tieton, Wash., by S.M. Coke. Introd. about 1972. Plant patent 3485; 19 Feb. 1974, assigned to H. Barry Rose, Alhambra, Calif. Parentage unknown, believed to be Starking Delicious. Fruit: typical Delicious shape; skin solid red, colors early, sometimes covered with netted russet; flesh crisp, sweet, flavor good; stores well; ripens late August to early September; good shipping qualities. Tree: semi-dwarf; spreading; vigorous; heavy producer; hardy; tolerant to lead-arsenic toxicity.

Rosebrook Gravenstein.—Orig. in Sebastopol, Calif., by a Mr. Rosebrook. Introd. about 1925. Considered to be a bud mutation of Gravenstein; discovered about 1920. Fruit: skin red, coloring early; hangs on tree better than parent, which it resembles.

Roseman.—Orig. in Roseisle, Manitoba, Canada, by J. Otto. Introd. in 1940. Parentage unknown. Fruit: medium to 2 3/4 inches in diam.; skin slightly blushed and striped bright red; flesh white, tender, flavor mild, sweet; fair to good as dessert, very good to excellent as sauce; a fall to early winter apple. Tree: medium tall; round; sturdy; hardy in southern Manitoba; annually productive. Now obsolete.

Rosthern 15.—Orig. in Rosthern, Saskatchewan, Canada, Canada Dept. Agr., Res. Sta. Introd. in 1935. Parentage unknown. Fruit: up to 2 1/2 inches in diam.; skin yellowish with red blush; flesh yellow-white, firm, quality good as dessert or sauce; season early September; season of storage short. Tree: moderately hardy; vigorous; yields well.

Rotyp (*Robijn*).—See rootstocks for apple.

Rouville.—Orig. by R.L. Granger and G.L. Rousselle, Agr. Canada Res. Sta., St. Jean-sur-Richelieu, Quebec. Introd. in 1983. 52-05-312 x 69-52; ancestry includes Red Melba, Melba, McIntosh, Wolf River, and *Malus atrosanguinea* 804. Cross made in 1962 in Ottawa, selected in 1972 in Trenton, Ontario, Canada, and evaluated at Frelighsburg, Quebec, Canada. Tested as Ottawa 627. Described by G.L. Rouselle in 1983; named by R.L. Granger in 1991. Fruit: very large, 70 to 80 mm in diam.; skin 75% red on pale green-yellow ground; shape oblate, symmetrical, somewhat ribbed; flesh white to creamy, juicy, slightly coarse; flavor subacid, high sugar and tannin; quality fair to good; dual purpose, including processing for juice; storage life 2 months. Tree: vigorous, semi-spreading; precocious; annually productive; cold hardy; resistant to apple scab in most years but foliage susceptible to race 5.

Royal Empire™ (*Teeple Red Empire*).—A limb, red-fruited mutation of Empire. Discovered in 1985 by John and Russell Teeple, Sodus, N.Y. Plant patent applied for; assigned to Cornell Research Foundation.

Royal Gala (*Red Gala, ten Hove Gala*).—Orig. in Matamata, New Zealand, by H.W. ten Hove. Introd. in 1974 by D.W. McKenzie, Fruit Res. Orchard, Havelock North, New Zealand. Bud mutation of Gala, discovered in 1971. Skin bright scarlet striping over golden yellow ground color, glossy, very attractive, much more red than parent. Otherwise identical to Gala.

Royal-Red Delicious.—Orig. in Wenatchee, Wash., by Walter D. Plough. Introd. in 1952. Plant patent 1278; 18 May 1954; assigned to Columbia & Okanogan Nursery Co., Wenatchee. Bud mutation of Richared Delicious; discovered in 1946. Fruit: resembles parent but skin with more overall red, which appears about 10 days earlier than parent. Tree: vigorous, large, spreading; hardy and productive.

Rubinette® (*Rafzubin* cv.)—A high-quality introduction from a Swiss breeding program. Parentage: Golden Delicious x Cox's Orange Pippin. U.S. plant patent 6588; assigned to North American Tree Co., Portland, Ore. Fruit: dull red stripe over golden ground, much russet; long, slender stem; shallow stem cavity; fruit size small. Flavor delicious, very sweet with slight tang, aromatic; storage life 3–4 months. Tree: habit very similar to Golden Delicious; moderately susceptible to mildew and moderately resistant to scab.

Rubinstar (*Rubinstar Jonagold*).—Orig. at Gaiberg, Germany, in 1980 as a bud mutation of Jonagold. U.S. plant patent applied for. Identical with Jonagold, except that it matures about 1 week earlier and at maturity exhibits a much more intense red blush coloration over 70% to 90% of the surface with little or no striping. Flavor same as Jonagold. Fruit less prone to sunburn than other red strains of Jonagold.

***Rubinstar Jonagold*.**—See Rubinstar.

Ruble Winesap.—See Mosebar Winesap.

Rubra Precoce.—Orig. in Ferrara, Italy, by Nino Breviglieri, Instituto di Coltivazioni Arboree, Firenze, Italy. Introd. there in 1960; in U.S. in 1962 as P.I. 282934. Alexander x Jonathan. Tested as cross 2752, made in 1948. Fruit: resembles Alexander; large to medium large; skin bright red; flesh creamy white, juicy, fine-grained, firm; flavor acid, sprightly, aromatic; mature mid-August, just after Gravenstein. Tree: vigorous; fruits heavily at an early age, even on 1-year-old branches.

Ruby.—Orig. in Wooster, Ohio, by the Ohio Agr. Expt. Sta. Introd. in 1953. Gallia Beauty x Starking Delicious. Fruit: large; high color; keeps well in storage; season late winter, about with Gallia Beauty; excellent for cooking, fair for dessert. Tree: blooms late; apparently productive at an early age; susceptible to scab.

Ruby Jon.—A limb mutation of Jonee producing fruit with superior red color. Discovered by Bill Jackson in Bowling Green, Ky., in 1986. Not patented; assigned to Summit Sales, Lawrence, Mich.

Ruby Red.—Orig. in Stephenville, Texas, by J.E. Fitzgerald, Fitzgerald Nursery. Introd. in the early 1930s. Parentage unknown. Fruit: similar to Delicious, but deeper red and smaller; a summer-type; comparable in quality to Yellow Transparent. Tree: tends to be a weak grower when young; propagated by double-working, using a Delicious or other compatible interstock.

Ruby Red Rome.—See Ruby Rome Beauty.

Ruby Rome.—See Ruby Rome Beauty.

Ruby Rome Beauty (*Black Rome Beauty, Cherry Red Rome Beauty, Miller Ruby Rome, Ruby Red Rome, Ruby Rome*).—Orig. in Milton-Freewater, Ore. Introd. in 1928 as Black Rome Beauty by the Milton Nursery Co., Milton-Freewater; limb sport of Rome; discovered about 1912. Fruit: appears to be identical to Rome Beauty except that the skin is a solid, dark red. Tree: similar to standard Rome Beauty.

Runkel.—A chance seedling of unknown parentage discovered in 1940 in Lincoln Park, Mich. U.S. plant patent 3308; 20 Feb. 1973. Fruit: skin fully red, thin, tender, smooth, waxen; flesh yellowish-cream-colored; flavor mildly subacid, rich, sweet; eating quality good; long storage life. Tree: vigorous, hardy, very productive, annual bearing.

Russian White.—Imported from Russia by N.E. Hansen, South Dakota Agr. Expt. Sta., Brookings. Introd. in 1924 into the U.S. May be an old Russian variety renamed. Fruit: flesh white, juicy, slightly subacid, cooks well; ripens in late August. Tree: productive; hardy.

Rutherford.—Orig. in Saskatoon, Saskatchewan, Canada, by C.F. Patterson, Univ. of Sask. Introd. in 1959. Elsa x Melba; selected in 1945; tested as A59-375. Fruit: up to 2 1/2 inches long and 2 1/4 inches in diam. under nonirrigated field conditions in Saskatoon; oblate; skin pale yellow; flesh white, juicy, subacid, quality good; ripens in midseason; resembles Heyer 12 but of better quality. Tree: comes into bearing early; produces heavy crops; hardy.

Ryan Red.—Orig. in Wapato, Wash., by James B. Ryan. Introd. in 1958. Bud mutation of Starking Delicious; discovered in 1956. Deep early coloring of skin, becoming a solid red and very intense at maturity; otherwise, very similar to parent.

Ryan-Spur Red™ (*Ryanred Spur Delicious*).—Orig. in Wapato, Wash., by Carlton Nursery Co., Dayton, Ore. Introd. in 1974. Whole tree mutation of Ryan Red, discovered in 1968, selected in 1969. Fruit: large; elongated; skin deep red, with slight stripe at maturity; flesh white, crisp; for fresh market; ripens same as Ryan Red. Tree: spur-type, 1/2 size on seedling root; upright, compact; vigorous; productivity excellent.

Ryanred Spur Delicious.—See Ryan-Spur Red.

Saint Clair (*Lizakowsky*).—Orig. in Lebanon, Saint Clair County, Ill., by John J. Lizakowsky. Introd. in 1947. Parentage unknown; raised from seed of an apple purchased from a train vendor in 1913 or 1914. Fruit: resembles Wealthy; large; quality good; matures in summer; skin nearly solid red when grown in southern U.S. Tree: resembles Wealthy; strong; productive; more vigorous under climatic conditions of the Tennessee and central Mississippi Valleys.

Sali™ Delicious.—A whole-tree mutation of Delicious discovered in Moxee, Wash., by Kasper Sali. U.S. plant patent 7237; 29 May 1990. Fruit: color solid blush; develops red color earlier than other strains of Delicious; develops sugar about 10 days earlier than other Delicious strains. Tree: standard nonspur growth habit.

Sandow.—Orig. in Ottawa, Ontario, Canada, by Canada Dept. Agr., Res. Sta. Introd. about 1935. Northern Spy o.p.; selected in 1912. Fruit: medium to large; round-oblate; skin yellow overspread with dull scarlet, dots white, few; flesh yellowish, juicy, tender, crisp, slightly subacid, aromatic; quality very high, core size medium; season midwinter to later. Tree: slightly more hardy than Northern Spy; susceptible to fire blight.

Sansa.—Orig. in Morioka, Iwate, Japan, by Y. Yoshida, T. Haniuda, S. Tsuchiya, T. Sanada, T. Masuda, and H. Bessho, Morioka Branch, Fruit Tree Res. Sta., in cooperation with Don McKenzie, DSIR Res. Sta., Havelock North, New Zealand. Gala x Akane; cross made in New Zealand in 1969; selected at Morioka in 1981; tested as Morioka #42. Introd. in 1986. U.S. plant patent 6519; 10 Jan. 1989. Fruit: medium, 200 to 250 g; conical; skin bright red on yellow-green ground, some russet around basin; flesh whitish yellow, crisp, fine texture, not firm but is crisp, juicy, slow browning; good sugar/acid balance; excellent dessert quality; quality still excellent after one month storage. Ripens early, 1 week before Gala. Tree: vigor moderate to moderately weak; upright-spreading; precocious; very productive. Some genetic chlorosis of leaves. Blooms with Gala. Resistant to *Alternaria* blotch and rust (*Gymnosporangium yamadae*).

Saska.—Orig. in Rosthern, Saskatchewan, Canada, by Seager Wheeler. Introd. in 1934. Chance seedling. Fruit: 1 1/2 inches in diam.; skin covered with bright red; cans well; season early September. Tree: upright; spreading; productive; fully hardy at Rosthern. Now obsolete.

Saugahatchee.—Orig. in Loachapoka, Ala., by J.D. Webster. Introd. in 1956. Assigned to Gray Stone Nursery, Loachapoka, Ala. Fruit: large; blocky shape, round oblate; skin 90% red, striped; flesh whitish, firmer than most early varieties, flavor subacid, quality fair to good; short storage life; Gravenstein type, ripens with Gravenstein. Tree: annual bearer.

Sawyer Baldwin (*Baldwin Double Red, Baldwin Scarlet, Black Baldwin, Double-Red Baldwin, Red Baldwin*).—Orig. in Salisbury, N.H., by E.N. Sawyer. Introd. in 1927. Bud mutation of Baldwin; discovered in 1924. Resembles Baldwin, except for deeper red skin color, too dark in some localities.

Sayaka.—Orig. in Matsumoto, Nagano, Japan, by K. Hatakoshi. Introd. in 1984. Jonathan x Sekai-ichi. Fruit: very large, 300 to 400 g; purplish red stripe over yellow ground with prominent lenticels, flesh yellow, medium firm with some water core, subacid; harvest 5 days before Gala. Tree: medium vigor; spreading; productive.

Scarlet.—Orig. in Matsumato, Nagano, Japan, by K. Hatakoshi. Introd.in 1984. Akane x Starking Delicious. Japanese Reg. no. P528. Fruit: 250 to 300 g; skin brilliant red stripe on yellowish-green ground, with no russet; flesh white, firm, juicy, low acid, sugar content 13.5°Brix; harvest early with Gravenstein; stores 1 month. Tree: medium to medium weak vigor; spreading; medium productivity.

Scarlet Gala.—Orig. near Cumberland, Ky., by John Creech. Plant patent 6172; 10 May 1988; assigned to C&O Nursery, Wenatchee, Wash. A very bright red-fruited mutation of Gala with striped color pattern; otherwise similar to Gala.

Scarlet Pimpernel.—See Stark Earliest.

Scarlet Spur (*Snipes*).—Orig. in Yakima, Wash., by Don Snipes, Snipes Orchard Co. Introd. in 1983. Plant patent 4839; assigned to Van Well Nursery, Wenatchee. Bud mutation of Oregon Spur Delicious; discovered in 1976. Fruit: colors 35 days ahead of parent Oregon Spur Delicious with blush-type coloring formation.

Scarlet Spur Delicious.—A very early coloring spur-type sport of Delicious, discovered by William G. Evans. U.S. plant patent 6839; 20 Apr. 1982. Tree: spur growth habit; heavy, annual cropping.

Scarlet Staymared.—Orig. in Wenatchee, Wash., by John H. Dickey. Introd. in 1936. Plant patent 57; 21 Mar. 1933; assigned to Stark Bros. Nurseries & Orchards Co.,

Louisiana, Mo. Bud mutation of Stayman Winesap; discovered in 1930. Skin solid red, with deeper red stripes, which appears about 30 days before Stayman Winesap color; otherwise, identical to Stayman Winesap.

Schneica.—See Jonica.

Schoharie Spy (*Harper Red Spy, Peru Red Spy*).—Orig. in Peru, Clinton County, N.Y., by W.B. Harper. Introd. between 1936 and 1940. Bud mutation of Northern Spy; discovered in 1936. Identical to Northern Spy, except for red-blushed skin, somewhat duller than Northern Spy.

Schulke Red Delicious.—Orig. in Brewster, Wash., by C.W. Schulke, Wenatchee, Wash. Introd. in 1971. Whole tree mutation of Starking Delicious, discovered in 1962. Identical with Starking Delicious except fruit colors 3 to 4 weeks earlier.

Scotia.—Orig. in Kentville, Nova Scotia, by Canada Dept. Agr., Res. Sta. Introd. in 1961. McIntosh Red o.p.; selected in 1948; tested as Nova Scotia K 33-125-70. Fruit: size above medium, 2 1/2 inches or more in diam.; quality excellent for dessert, good for cooking; skin red, attractive; ripens 1 week before Gravenstein; McIntosh type for season before McIntosh ripens. Tree: upright; moderately vigorous; bears annually.

Scott 1.—See Rescue.

Scottie.—Orig. in Baldur, Manitoba, Canada, by Wm. Scott. Introd. in 1960 by Ed Robinson, Gaybird Nursery, Wawanesa, Manitoba. Parentage unknown; discovered in 1957. Fruit: large; skin greenish-yellow with red cheek; flesh greenish, firm; good for pies or sauce; ripens 1 Sept. Tree: size medium; upright; hardy.

Secor.—Orig. in Ames, Iowa, by S.A. Beach, Iowa Agr. Expt. Sta. Introd. in 1922. Salome x Jonathan; cross made in 1906; selected in 1918. Fruit: skin red striped; juicy; sprightly flavored; matures 10 days later than Jonathan; hangs well; free from soft scald and Jonathan spot in storage; keeps well until April or May; best quality of any variety in its season. Tree: hardy; moderately productive; vigor about like Jonathan.

Sekai-ichi (*World #1*).—Orig. in Kuroishi, Aomori, Japan, in the breeding program of the Aomori Apple Res. Sta. Introd. in 1974. Delicious x Golden Delicious; tested as 6/7-1. Fruit: very large, 500 to 600 g, occasionally more than 1000 g unless it is cropping heavily, then only moderately large; color red striped, similar to Starking, on greenish-yellow ground; finish smooth; shape round conic, regular, less elongated than Delicious; flesh medium firm, fine-grained, aromatic; taste similar to Delicious. Harvest with Delicious; stores for 80 to 120 days. Tree: very vigorous; upright-spreading; not precocious; not very productive, tends to be biennial; blooms with Delicious; self-unfruitful. Susceptible to preharvest drop; susceptible to apple scab and to *Alternaria* leafspot.

Semla.—Orig. in Brookings, S.D., by N.E. Hansen, South Dakota Agr. Expt. Sta. Introd. in 1940. Wolf River o.p. Fruit: large; skin red striped over yellow; flesh pleasantly subacid. Variety apparently obsolete.

Senshu.—Orig. in Japan in the breeding program of the Akita Fruit Tree Expt. Sta. by S. Tanno, T. Taguchi, J. Tanba, H. Suzuki, and K. Kon. Toko x Fuji; selected in 1974. Introd. in 1980. Japanese Reg. no. P 42. Fruit: small to medium, round; skin bright striped red on greenish-yellow ground; flesh yellowish white, firm, juicy, excellent quality. Harvest about 10 days before Delicious; 20- to 30-day shelf life. Tree: medium vigorous; spreading; medium productivity; resistant to powdery mildew.

September Ruby.—Orig. at Morden, Manitoba, Agr. Canada Research Station. Cross made by C.R. Ure; selected at Beaverlodge, Alberta, Canada; introduced by Prairie Fruit Breeding Coop. Committee in 1986. Rescue x Haralson; tested as PF36. Fruit: medium; round-oblate; skin bright red on light-green ground color; flesh greenish, moderately acid. Fruit is suitable for dessert and for juicing. Harvest in early September in the Canadian prairies. Tree: vigor moderate; annually productive; moderately susceptible to fire blight; hardy to zone 2A.

Shamrock.—Orig. in Summerland, British Columbia, by W.D. Lane and R.A. MacDonald, Agr. Canada Res. Sta. Introd. in 1986. McIntosh 10C-843-1 (irradiated spur type) x Starkspur Golden Delicious;

cross made by K. Lapins in 1970; selected in 1978; tested as 8C-l-15. Fruit: medium, about equal to Golden Delicious; shape round, slightly oval; green with white lenticels, very similar to Granny Smith, attractive; sometimes a slightly brown-red color develops on the sun-exposed side; flesh juicy, firm, sweet, unlike Granny Smith, bruise resistant; eating quality fair; harvest 3 days before McIntosh. Storage life: at 0 to 3C keeps in good condition for 7 months. Tree: medium, smaller than McIntosh; compact, spurry growth habit; very precocious; moderately productive; bloom time early with McIntosh.

Sharon.—Orig. in Ames, Iowa, by S.A. Beach, Iowa Agr. Expt. Sta. Introd. in 1922. McIntosh x Longfield; cross made in 1906; selected in 1920. Fruit: skin red striped; high quality; keeps in storage through January; for local markets; resembles McIntosh; its principal fault is lack of red skin color. Tree: vigorous; productive; hardy; drops fruit too freely.

Sharp Red Delicious.—Orig. in Wenatchee, Wash., by Del Sharp, Van Well Nursery. Introd. in 1969. Bud mutation of Shotwell Delicious. Discovered in 1963. Bright red blush with a fine stripe; coloring earlier than parent. Tree: standard; vigor moderate.

Shawnee.—Orig. in Woodbridge, Conn., by S.L. Kaplan. Plant patent 4050; 10 May 1977. Macoun x Red Delicious; cross made in 1962. Fruit: oblate, 2 3/4 to 3 inches in diameter; uniform; medium thick skin, dull dark red blush over entire skin surface; flesh white, juicy, subacid, sweet, palatable before fruit reaches full maturity. Tree: medium, medium vigor; spreading; branches grow straight, many parallel to ground.

Shelred.—Orig. in Cashmere, Wash., by Frank Schell. Introd. in 1953 by the Columbia & Okanogan Nursery Co., Wenatchee, Wash. Richared Delicious x Golden Delicious; selected about 1948. Fruit: flavor exceptionally fine; shape and color unsatisfactory. Variety no longer propagated.

Shenandoah.—Orig. in Blacksburg, Va., by Virginia Polytechnic Inst. Introd. in 1967. Winesap x Opalescent. Cross made in 1942 by R.C. Moore; first fruited in 1948; selected in 1948 by George D. Oberle; tested as VPI2. Fruit: round conic; skin near solid red, waxy, good finish, moderately thick, tough; flesh very firm, flavor sprightly to tart; long storage life; resembles Winesap; ripens about 10 days before Delicious; selected for possible use in commercial processing for sauce and canned slices. Tree: vigor medium; heavy producer of spurs; flowers produce fertile pollen but in rather sparse quantity.

Shizuka.—Orig. in Kuroishi, Aomori, Japan in the breeding program of the Aomori Apple Res. Sta. Introd. by T. Murakami. Golden Delicious x Indo. Plant patent applied for. Fruit: large, skin yellow with pink blush; flesh fine-textured, sweeter than Mutsu with low acidity; harvest with Delicious; stores until April, longer than Mutsu. Tree: triploid, pollen not viable.

Shlomit.—Orig. in Rehovot, Israel, by Chanan Oppenheimer. Introd. in 1967. F_2 selection (Calville St. Sauveur x Damascus) x Delicious. Fruit: large, 6 to 8 cm; round to conic; skin ground color yellow, partially covered with red blush; flesh firm, melting, aroma good; flavor sweet, relatively low acid; flavor of the Golden Delicious group, but sweeter; keeps at room temperature at least 10 days, in cold storage at least 2 months; ripens beginning of June to middle of July. Tree: medium size; upright; early bearer, on spurs; very productive; blooms between end of February and the middle of March.

Shotwell Delicious.—Orig. in Wenatchee, Wash., by Henry Shotwell. Introd. in 1934. Plant patent 90; 3 Apr. 1934; assigned to Columbia & Okanogan Nursery Co., Wenatchee. Bud sport of Delicious; discovered in 1928. Fruit: large; conic; skin striped a more intense red, appearing earlier than Delicious.

Show.—See Redspur Delicious.

Show Red.—See Redspur Delicious.

SilverSpur Red Delicious.—Orig. in Zillah, Wash., by W.E. Silvers. Plant patent 4159; 29 Nov. 1977; assigned to Claud Callahan and Marjorie McCormick, Yakima, Wash. Discovered as a whole tree compact spur sport of Hi-Early. Fruit: bright cherry red striped early color; long fruit; matures 2 weeks before Hi-Early; juicy, extra firm;

ability to keep for long period of time, about 300 days in ordinary storage. Tree: semi-dwarf; medium vigor; spreading, productive.

Sinta.—Orig. in Summerland, British Columbia, Canada, by K.O. Lapins, Canada Dept. Agr. Res. Sta. Golden Delicious x Grimes Golden. Cross made in 1955, selected in 1965. Fruit: medium size; roundish oblong, skin yellow with light blush, smooth; flesh cream-colored, moderately firm, crisp, juicy, quality very good; recommended for local roadside market to precede Golden Delicious; ripens 3 weeks before Golden Delicious. Tree: medium size, spreading; moderate vigor; precocious bearer, moderately productive.

Sir Prize.—Orig. in Lafayette, Ind., by a cooperative breeding program consisting of the Agr. Expt. Sta. of Indiana (E.B. Williams, J. Janick, F.W. Emerson), Illinois (D.F. Dayton, J.B. Mowry), and New Jersey (L.F. Hough, C.H. Bailey). Introd. in 1975. Plant patent 3988; 14 Dec. 1976; assigned to Purdue Research Foundation, West Lafayette, Ind. Doud Golden Delicious (2-4-4) x PRI 14-152 (Golden Delicious x F_2 26829-2-2. F_2 26829-2-2 is the result of a cross between two selections of Rome Beauty x *Malus floribunda* 821. Cross made in 1955; first fruited in 1961; tested as Co-op 5. Fruit: large, 3 to 3 1/4 inches in diam.; ovate to conic, slightly ribbed; skin lemon yellow with a slight red blush occasionally, very smooth; no russeting, thin, waxy, does not shrivel in storage; flesh lemon yellow, fine, tender, juicy, flavor rich, subacid; quality excellent, a dessert cultivar; thin skin and bruise susceptibility make it unsuitable for commercial orchards under current handling and grading practices, recommended for home gardens; matures in Golden Delicious season. Tree: vigorous; productive; little tendency for biennial bearing; typical triploid characteristics; pollen viability is probably low; immune to apple scab, moderately resistant to powdery mildew and cedar apple rust; appears well adapted to the the midwestern U.S.

Skyline Supreme.—Orig. in Hendersonville, N.C., by Wm. E. Dalton. Introd. in 1960 by Bountiful Ridge Nurseries, Princess Anne, Md. Bud mutation discovered in 1956 as a whole tree in an orchard of Starking Delicious. Fruit: skin red-blushed, no striping, color appearing about 10 days before Richared under Appalachian conditions; shape typically Delicious; ripens with Delicious. Tree: standard type.

Skyspur.—Orig. in Woodside, Del., by S.H. Derby & Sons. Introd. in 1966 by Bountiful Ridge Nurseries, Princess Anne, Md. Bud mutation of Skyline Supreme; discovered in 1957. Fruit: large; Delicious shape; skin, flesh characteristics, and ripening date same as parent. Tree: semi-dwarf; open-spreading, precocious; needs pollenizer.

Smith Jonathan.—Orig. in Rock Island, Ill., by W.W. Smith. Apparently not introd. Bud mutation of Jonathan. Fruit: large. A periclinal cytochimera 2-2-4-4; chimeral nature determined at the New York State Agr. Expt. Sta., Geneva; accessioned in 1953 by this station.

Smoothee (*Gibson Golden Delicious*).—Orig. in Evans City, Pa., by Carl R. Gibson. Introd. in 1967. Plant patent 2803; 16 Apr. 1968; assigned to Hill Top Orchards & Nurseries, Hartford, Mich. Whole tree mutation of Golden Delicious; discovered in 1958. Fruit: resembles Golden Delicious but less lenticel rupture and with more red blush, more tolerant to russeting.

Snipes.—See Scarlet Spur.

Sonora.—Origin unknown. Apparently not introd.; used in breeding program at New Jersey Agr. Expt. Sta., New Brunswick. Parentage unknown; acquired by M.A. Blake from a commercial nursery about 1926. Fruit: skin blushed and striped one-half to three-quarters of surface, bruises easily; flesh medium firm, slightly tough, only moderately juicy, becomes mealy; ripens unevenly, about with Early McIntosh.

South Dakota Golden (*Golden*).—Orig. in Brookings, S.D., by N.E. Hansen, South Dakota Agr. Expt. Sta. Introd. in 1944. Grimes Golden x Oldenburg. Fruit: 2 3/4 inches in diam.; round, tapering; juicy, with fair quality; season late fall.

Spangelo.—Orig. in Morden, Manitoba, Canada, by Canada Dept. Agr., Res. Sta. Introd. in 1930. Crusoe o.p. Fruit: large; skin pale green, washed with bright red, thin but rough; flesh white, smooth, juicy; flavor mild; season early winter. Tree: vigorous; upright; hardy; productive.

Sparger.—Orig. in Mt. Airy, N.C., by John A. Young, Greensboro Nursery, Greensboro, N.C. Introd. about 1925. Parentage unknown. Fruit: size medium; round; skin red, dots white; flesh subacid. Tree: bears well annually. Apparently now obsolete.

Spartan.—Orig. in Summerland, British Columbia, Canada, by R.C. Palmer, Canada Dept. Agr., Res. Sta. Introd. in 1936. McIntosh x Newtown Pippin; cross made in 1926; original tree planted in 1928; first fruited in 1932; selected in 1936 by A.J. Mann of above station. Fruit: size above medium; skin highly colored, solid dark red; flesh firm, crisp, white, juicy, quality and texture very good, fully equal to Empire; matures about 2 weeks later than McIntosh, about Delicious time in Washington; keeps well in ordinary cold storage until February; picks easily; more resistant to preharvest drop than McIntosh; resembles McIntosh. Tree: regular and heavy cropping; if harvesting delayed, fruit develops musty flavor at end of storage season.

Spencer.—Orig. in Summerland, British Columbia, Canada, by Canada Dept. Agr., Res. Sta. Introd. in 1959. McIntosh x Golden Delicious; cross made by R.C. Palmer in 1926; selected by A.J. Mann in 1938; first trial plantings made in 1941; tested as Summerland S-5-4. Fruit: large; oblong; skin moderately colored, blush a bright rather light scarlet, attractive; flesh crisp, firm, juicy, white, subacid; core smaller than that of McIntosh; quality extremely high; shows a tendency for breakdown in cold storage in some areas, but in other areas, it keeps well until March; matures about 3 weeks later than McIntosh; requires moderate thinning. Tree: very hardy, being slightly hardier than Spartan or McIntosh; vigorous; upright, spreading.

Spigold.—Orig. in Geneva, N.Y., by the New York State Agr. Expt. Sta. Introd. in 1962. Red Spy x Golden Delicious; selected in 1953; tested as NYE6. Fruit: very large; skin 75% red striped; flesh very firm but bruises easily; fresh quality exceptionally high; released as a Northern Spy replacement; similar to this variety in flavor and flesh color in processing tests; texture adequate as canned or frozen slices; less bitter pit than Spy; ripens 1 week before Northern Spy. Tree: very vigorous; needs a pollenizer, not very productive. Triploid; pollen not viable. Extremely suitable for home gardens because of its outstanding dessert quality.

Spijon.—Orig. in Geneva, N.Y., by New York Agr. Expt. Sta. Introd. in 1968. Red Spy x Monroe. Cross made in 1944, selected in 1953, tested as NY 44408-5. Fruit: large, 3 to 3 1/4 inches in diam.; somewhat flat at both ends; skin attractive bright, sometimes dark crimson, 90% to 100% red, an even washed blush with no striping; flesh firm, light yellow, subacid to slightly acid, eating quality very good; stores up to 3 months; has shown no bitter pit; recommended for processing and fresh markets; ripens with Northern Spy. Tree: upright-spreading, with tendency to droop due to terminal bearing; less vigorous than Red Spy; productivity comparable to McIntosh, bears annually.

Splendour.—See Stark Splendor.

Spur Granny Smith.—See Granspur.

Spur Goldblush™ (*Griffspur*).—A semi-spur mutant of Stark Blushing Golden™, discovered as a limb sport by R.B. Griffith, Cobden, Ill. U.S. plant patent 7878; 2 June 1992. Fruit: identical to Stark Blushing Golden™—skin golden-red blush on creamy yellow ground; highly russet resistant; crisp; thin-skinned; very tart flavor at harvest; excellent eating quality after storage. Long storage life. Spur Goldblush™ differs from parent in having about 50% greater density of fruiting spurs, wider crotch angles, and greater percent budbreak. Moderately susceptible to scab and fire blight.

Spur Royal Red Delicious.—See Spured Royal Delicious.

Spur Winesap.—Orig. in Cashmere, Wash., by Robert E. Schell, Columbia & Okanogan Nursery, Wenatchee, Wash. Introd. in 1972. Bud mutation of Winesap, discovered in 1964. Fruit: same as Winesap. Tree: semi-dwarf; somewhat upright; recommended for home gardens.

Spur Winter Banana.—Orig. in Chelan, Wash., by Cecil Housden, Mgr., Chelan Apple Co. Introd. in 1971 by Columbia and Okanogan Nursery, Wenatchee, Wash.

Whole tree mutation of Winter Banana, discovered about 1962. Fruit: same as parent; ripens September to October. Tree: half normal size; upright growth; vigorous; productive; recommended as a pollenizer or for grafting into older trees for a pollenizer.

Spured Royal Delicious (*Spur Royal Red Delicious*).—Orig. in Naramata, British Columbia, Canada, by A.G. Staniforth, Columbia and Okanogan Nursery Co., Wenatchee, Wash. Introd. in 1976 by this nursery. Plant patent 3864; 13 Apr. 1976; assigned to this nursery. Bud mutation of Royal Delicious; discovered in 1967. Fruit: similar to its parent. Tree: spur type.

SpureeRome.—Orig. in Paw Paw, W.Va., by Henry Miller, Consolidated Orchards Co. Introd. in 1970 by Hilltop Orchards & Nurseries, Hartford, Mich. Plant patent 3198; 30 May 1972; assigned to Hilltop Orchards & Nurseries. Whole tree sport of Gallia Beauty, discovered in 1963. Tree: spur type; semi-upright; vigorous.

SpurMac.—Orig. in Casnovia, Mich., by Mr. and Mrs. Chas. Saur. Introd. in 1973 by Bountiful Ridge Nursery, Princess Anne, Md. Parentage unknown, discovered in 1960. Fruit: typical McIntosh, except more evenly colored and firmer, holds and stores better than McIntosh; ripens 1 week after McIntosh. Tree: semi-dwarf, spur type; vase-formed; vigorous, hardy and productive.

Spurred Okanoma Red.—See Okanoma.

Spurred Red Delicious.—See Okanoma.

Starcher.—Orig. in Lee County, Ala., by G.C. Starcher, Alabama Polytechnic Inst., Auburn. Introd. about 1922. Considered to be Horse o.p.; discovered in 1920. Fruit: shape flat; skin greenish-yellow at maturity; flesh tart until mellow, good for culinary usage; ships well; ripens in late summer. Tree: produces heavily; sturdy; quite disease resistant.

Stark Adina.—See Adina.

Stark Bounty.—Orig. in Batchtown, Ill., by Otto Klass. Introd. in 1975. Plant patent 3550; 14 May 1974; assigned to Stark Bros. Nurseries & Orchards Co., Louisiana, Mo. Parentage unknown, discovered in 1955. Fruit: large; skin solid red; flesh firm, crisp, juicy; flavor resembles Stayman Winesap; for processing or as fresh fruit; ripens with Winesap. Tree: medium large; upright spreading; vigorous; hardy; very productive.

Stark Double-Red Rome Beauty.—See Cowin Rome Beauty.

Stark EarliBlaze.—Orig. in New Lenox, Ill., by Worth B. Mooney. Introd. in 1957. Plant patent 1721; 24 June 1958; assigned to Stark Bros. Nurseries & Orchards Co., Louisiana, Mo. Parentage unknown; discovered in 1949. Fruit: large; skin highly red colored; flesh white with greenish tint, firm, crisp; flavor is sprightly, quality excellent; matures in Wealthy season, after Blaze; ships very well; makes excellent sauce; resembles Wealthy. Tree: medium large; medium vigor; upright; hardy; productive; regular bearer.

Stark Earliest (*Scarlet Pimpernel*).—Orig. in Orfino, Idaho, by Douglas Bonner. Introd. in 1944. Plant patent 642; 19 Sept. 1944; assigned to Stark Bros. Nurseries & Orchards Co., Louisiana, Mo. Parentage unknown; discovered about 1938. Fruit: size averages 2 1/4 inches in diam.; oblate; calyx open; skin yellow, cherry-red blush, attractive, thin, tough, smooth, waxy, dots russet; flesh yellow, firm, tender, sprightly subacid; ships well; ripens early, before Yellow Transparent and Close, about with June Wealthy, harvested over an extended period; a good culinary variety. Tree: large; vigorous, spreading; hardy; bears regularly.

Stark JonGrimes (*Jongrimes*, *Hoosier Seedling*).—Orig. in Bloomfield, Ind., by Roland S. Rogers. Introd. in the 1920s. Plant patent 794; 23 Mar. 1948; assigned to Stark Bros. Nurseries & Orchards Co., Louisiana, Mo. Parentage unknown. Fruit: size medium; round, oblate to oblong; skin yellow, streaked with red; flesh tart, dessert quality fair to medium, cooking quality good; keeps about as well as Duchess; ripens 2 to 3 weeks ahead of Jonathan; hangs in dense clusters unless thinned; storage season late fall; resembles Jonathan only in having some red on skin; bears no resemblance to Grimes. Tree: vigorous; upright; rapid growing; productive, regular bearer.

Stark Jumbo.—Orig. in Hood River, Ore., by D.A. Hanners. Introd. in 1971. Bud mutation of Eve's Delight (a bud mutation of Spokane Beauty); discovered in 1962. Fruit: very large, conic, skin bright red; flesh satiny white with greenish tint; flavor subacid; good culinary qualities; ripens in October in Missouri. Tree: medium large; spreading; open; productive; medium vigor.

Stark McIntosh.—See Kress McIntosh.

Stark Puregold (*Goldensheen #2, Stark Ultragold*).—Orig. in Louisiana, Mo. by Paul Stark, Jr. Introd. in 1980. Plant patent 4731; 2 June 1981; assigned to Stark Bros. Nurseries, Louisiana, Mo. Introd. in 1980. Bud mutation of Goldensheen (*Belgolden*). Fruit: resembles Goldensheen and similar to Golden Delicious, but skin smooth, resistant to russet. Tree: identical to Belgolden except that it is a standard size tree; Belgolden is a spur type.

Stark Splendor (*Splendour*).—Orig. in Napier, New Zealand, by Charles L. Roberts. Introd. in 1967. Plant patent 2460; 1 Dec. 1964; assigned to Stark Bros. Nurseries & Orchards Co., Louisiana, Mo. Parentage unknown; discovered in 1948. Fruit: large, averages about 3 inches in diam.; ovoid-conical, slightly ribbed; skin bright red, waxy, thin, tender, glossy; dots few, large, cream-colored, lenticular; flesh yellowish, tender, crisp, very juicy, fine-grained, aroma distinct; does not shrivel in cold storage; ripens in September and October; resembles Golden Delicious in size, shape, texture, and quality. Tree: size medium; vigor medium; spreading; hardy; very productive, bearing regularly; blooms with Golden Delicious. Nearly immune to fire blight.

Stark Summer Delicious.—Orig. in Morgantown, W.Va., by Elery R. Barker. Introd. in 1955. Plant patent 1422; 27 Sept. 1955; by mesne assignments to Stark Bros. Nurseries & Orchards Co., Louisiana, Mo. Delicious o.p.; discovered in 1940. Fruit: similar to Delicious in shape and flavor, but ripens 4 weeks earlier, about with Wealthy; high dessert quality; attractive appearance; dots more pronounced than those of Delicious. Tree: vigorous; large leaves.

Stark Summer Gold.—Orig. in Clarklake, Mich., by Floyd Avis. Introd. in 1962. Plant patent 2179; 2 Oct. 1962; assigned to Stark Bros. Nurseries & Orchards Co., Louisiana, Mo. Chance seedling discovered in 1948. Fruit: large, up to 3 1/2 inches in diam., 2 1/2 inches long; skin smooth, glossy, waxy; dots few but conspicuous, ground color yellow with a scarlet blush resembling that of Maiden Blush; flesh white with yellowish tint, firm, flavor mild, sprightly, no aroma, quality good; ripens during season of Maiden Blush and about 2 weeks before McIntosh; resembles McIntosh in size and shape. Tree: very productive; medium large; hardy; regular bearing.

Stark Summer Treat.—Orig. in Cream Ridge, N.J., by L. Fredric Hough and C.H. Bailey, Rutgers Univ., New Brunswick, N.J. Plant patent 4740; 9 June 1981; assigned to Stark Bros. Nurseries. Introd. in 1980. NJ 109055 x Mollie's Delicious. Selected in 1971; tested as NJ 49. Fruit: resembles Mollie's Delicious; large, 3 inches in diam., round conic to conic; skin covered with 75% red blush; flesh crisp, juicy; high quality.

Stark® SunCrisp™.—A high-quality, Golden Delicious type orig. from the apple breeding program at Rutgers Univ., New Brunswick, N.J. Golden Delicious x (Cortland x Cox's Orange Pippin). Selected in 1971; tested as NJ 55; Golden Delicious x (Cortland x Cox's Orange Pippin); introd. in 1992. Plant patent applied for. Fruit: size medium to large, 300 to 400 g; skin very attractive, 40% orange-red blush, sometimes pink, on pale yellow-green ground; shape round-conic, slightly irregular. Flesh: cream-colored, slow to oxidize; very firm, crisp, and juicy; flavor sweet, mildly subacid, aromatic; eating quality excellent. Harvest season 1 week after Delicious. Storage life medium, up to 6 months at –0.5C. Tree: tends to overcrop, thinning required; not disease resistant.

Stark Supreme Staymared.—Orig. in Hancock County, Ind., by Roy Tuttle, Greenfield, Ind. Plant patent 4060; 14 June 1977; assigned to Stark Bros. Orchards & Nurseries Co., Louisiana, Mo. Discovered in 1964 as a bud mutation of Scarlet Staymared. Fruit of deeper overall red color; absence of scarfskin, smoother and less tendency to crack, lenticels less prominent.

Stark Ultragold.—See Stark Puregold.

Starking.—See Starking Delicious.

Starking Delicious (*Double Red Delicious, Extra Red Delicious, Starking*).—Orig. in Monroeville, N.J., by Lewis Mood. Introd. in 1924. Bud mutation of Delicious; discovered in 1921. Similar to Delicious except for added skin color.

Starking FullRed Delicious—Orig. in Milton-Freewater, Ore., by Jack D. Slusarenko. Introd. in 1972. Plant patent 3567; 11 June 1974; assigned to Stark Bros. Nurseries & Orchards Co., Louisiana, Mo. Bud mutation of unknown strain of Delicious, discovered in 1961. Fruit: large; conical-angular; skin red, striped, colors 4 days before Super Starking Delicious; flesh white with yellowish tinge, firm, crisp, quality good; recommended as a market variety. Tree: medium large; spreading; vigorous; hardy; productive.

Starkrimson Delicious (*Bisbee Delicious*).—Orig. in Hood River, Ore., by Roy A. Bisbee. Introd. in 1956. Plant patent 1565; 12 Feb. 1957; assigned to Stark Bros. Nurseries & Orchards Co., Louisiana, Mo. Whole tree variation; discovered in 1953. Fruit: matures slightly later than Starking Delicious; skin a solid blush, gradually darkening as season progresses but still holding an attractive bright glossy sheen; size same as parent, but tends to be a little longer. Spur type tree.

Stark's Red Duchess.—See Daniels Red Duchess.

Starkspur Compact Mac (*Wijcik*).—Orig. in Kelowna, British Columbia, Canada by Anthony Wijcik. Plant patent 4382; 20 Feb. 1979, assigned to Stark Bros. Nurseries, Louisiana, Mo. Introd. in 1981. Bud mutation of standard McIntosh. Discovered and selected in 1963. Fruit: resembles standard McIntosh. Tree: very compact, erect, rigid; less than 25% size of standard McIntosh; leaves large, wide, green, close internodes; many spurs. Compact spur-type growth is simply inherited.

Starkspur Compact Red Delicious (*Cascade Compact Red Delicious*).—Orig. in Yakima, Wash., by John Garretson. Plant patent 4801; Jan. 1982, assigned to Carlton Nursery, Dayton, Ore. Licensed to Stark Bros. Nurseries, Louisiana, Mo. Introd. in 1981. Bud mutation of Starking Delicious. Discovered and selected about 1968. Fruit: identical to Starkrimson Red Delicious. Tree: compact, heavily spurred; 25% less than Starkrimson in size on same rootstock.

Starkspur Cortland (*LaMont*).—Orig. in Albion, N.Y., by George LaMont. Introd. in 1980. Assigned to Stark Bros. Nurseries, Louisiana, Mo. Bud mutation of Cortland, discovered and selected about 1970. Fruit: resembles Cortland, but with thicker fruit stem. Tree: spur type and compact.

Starkspur Dixiered Delicious (*Hared*).—Orig. in Long Creek, S.C., by Bruner Hare. Plant patent 5547; assigned to Stark Bros. Nurseries, Louisiana, Mo. Bud mutation of Oregon Spur. Discovered and selected in 1975. Fruit: resembles Starkspur Ultrared Delicious; very early coloring intense red, colors better than Ultrared, Oregon Spur, Redchief, and Topred, holds color well in summer heat. Tree: compact, spur type.

Starkspur Golden Delicious.—Orig. in Yakima, Wash., by Philip J. Jenkins. Introd. in 1961. Plant patent 2024; 14 Feb. 1961; assigned by mesne assignments to Stark Bros. Nurseries & Orchards Co., Louisiana, Mo. Bud mutation of Golden Delicious; discovered in 1959. Fruit: characteristics of Golden Delicious, being broader at base and longer than Golden Delicious, and having five large, relatively conspicuous points at the calyx end. Prone to russet. Tree: semi-dwarf; compact spurry type of growth; spurs, flower buds and shoots thicker than Golden Delicious; productive.

Starkspur Golden Russet (*Holmberg*).—Orig. in Gales Ferry, Conn., by Rick Holmberg. Introd. in 1981. Assigned for 10 years to Stark Bros. Nurseries, Louisiana, Mo. Bud mutation of Golden Russet; discovered and selected in early 1970s. Fruit: identical to Golden Russet. Tree: compact, spur type.

Starkspur Law Rome (*Peace Valley*).—Orig. in Rogers, Ohio, by Dan Simmons. Plant patent 4793; Nov. 1981, assigned to Stark Bros. Nurseries, Louisiana, Mo. Introd. in 1981. Bud mutation of Stark Red Rome Beauty; discovered and selected in 1976. Fruit: resembles Law strain Red Rome Beauty and Lawspur; intense red coloring of skin. Tree: spur-type, more compact than Lawspur and much more than Stark Red Rome Beauty.

Starkspur Lodi.—Orig. in Trinidad, Wash., by Thomas Sigurdson, Vantage, Wash. Introd. in 1964. Bud mutation of Lodi; discovered in 1960; patent applied for; assigned to Stark Bros. Nursery & Orchards Co., Louisiana, Mo. Fruit: large; skin bright golden yellow; flesh white, flavor rich; ripening early summer, 3 to 4 days after Yellow Transparent; keeps well. Tree: semi-dwarf; hardy; spur-type growth habit.

Starkspur McIntosh.—Orig. in Oyama, British Columbia, Canada, by Bernie Gatzke. Introd. in Canada in 1970. Assigned to Stark Bros. Nurseries & Orchards Co., Louisiana, Mo. Bud mutation of Summerland McIntosh, discovered in 1968. Fruit: same as McIntosh, ripening same time. Tree: semi-dwarf, spur-type; upright growth; moderately vigorous, hardy, productive.

Starkspur Prime Red—Orig. in Brewster, Wash., by Calvin L. Cooper. Introd. in 1975. Plant patent 3882; 11 May 1976; assigned to Stark Bros. Nurseries & Orchards Co., Louisiana, Mo. Bud mutation of Topred Delicious, induced by x-irradiation in 1964, discovered in 1971. Fruit: large, 3 to 3 1/4 inches; ovoid, truncate at base; skin bright red, striped; flesh white with greenish tinge, crisp, mild; a dessert variety; ripens 25 Sept. Tree: spur-type; medium small; upright, vase-shape, more compact than Starkrimson Delicious; vigor moderate; very productive.

Starkspur Red Delicious.—Orig. in Winthrop, Wash., by Walter A. Wood. Introd. in 1962. Plant patent 1930; 5 Apr. 1960; assigned to Stark Bros. Nurseries & Orchards Co., Louisiana, Mo. Bud mutation of Starking Delicious; discovered in 1953. Fruit: similar in size and shape to parent, but skin over-all striped red; color developing about 1 week earlier and more completely than Starking Delicious. Tree: smaller in size and more upright than parent, forming spurs earlier and more regularly than the parent.

Starkspur Supreme.—Orig. in Yakima, Wash., by Frank Paganelli. Introd. in 1972. Plant patent 3541; 7 May 1974; assigned to Stark Bros. Nurseries & Orchards Co., Louisiana, Mo. Bud mutation of Starking Delicious; discovered in 1967. Fruit: size, shape and flavor similar to Starkrimson Delicious; skin color bright cherry red, retained through harvest season unlike other varieties, thick, tough, glossy; flesh satiny white with greenish tint, firm, fine, crisp; harvested starting about 1 week after Delicious. Tree: size medium small, upright; spur type; hardy; very productive.

Starkspur Ultramac (*Dewar*).—Orig. in Oyama, British Columbia, Canada, by Mac F. Dewar. Plant patent 4383; 20 Feb. 1979; assigned to Stark Bros. Nurseries, Louisiana, Mo. Introd. in 1980. Whole tree bud mutation of Summerland Red McIntosh. Discovered in 1968. Fruit: resembles Starkspur McIntosh with excellent overall red color. Tree: spur type, more compact and spurry than Starkspur McIntosh, but less so than Macspur and Morspur McIntosh.

Starkspur® UltraRed Delicious (*Flanagan Delicious*).—Orig. in Thurmont, Md., by Russell N. Flanagan, Flanagan Orchards. Introd. in 1974. Plant patent 3557; 28 May 1974, assigned to Stark Bros. Nurseries & Orchards Co., Louisiana, Mo. Bud mutation of Starking Delicious, discovered in 1964. Fruit: large; conic; skin thick, tough, smooth, glossy, full red, colors early; flesh juicy, satiny white with yellowish tint, flavor mild; ripens 1–2 weeks after Delicious. Tree: spur type, semi-dwarf, upright; vigorous, hardy, productive; susceptible to scab and quince rust; tolerant to fire blight.

Starkspur UltraStripe Delicious (*Jenred*).—Orig. in Royal City, Wash., by Zona Jenkins. Plant patent 5472; May 1985, assigned to Stark Bros. Nurseries, Louisiana, Mo. Introd. in 1981. Bud mutation of Oregon Spur Delicious; discovered and selected in 1976. Fruit: resembles Redchief; intense red stripe coloring ahead of Redchief and Oregon Spur.

Starkspur Winesap (*Mance Winesap*).—Orig. in Naches, Wash., by Rudolph M. Mance. Introd. in 1963. Plant patent 2313; 19 Nov. 1963; assigned to Stark Bros. Nurseries & Orchards Co., Louisiana, Mo. Whole tree bud sport of Winesap. Similar to Stark Red Winesap, but distinguished by its smaller, semi-dwarf tree size, distinctive habit of bearing fruit beginning at an earlier age, greater productivity, and a distinctive spur-type growth and development.

Starkspur Winter Banana (*Frecon*).—Orig. in Boyertown, Pa., by Richard Frecon. Introd. in 1983. Plant patent 4901; Oct. 1982, assigned to Stark Bros. Nurseries and Orchard Co., Louisiana, Mo. Bud mutation of standard Winter Banana. Fruit: resembles Winter Banana. Tree: spur-type, not as compact as Spur Winter Banana, but free of skin russet like standard Winter Banana; excellent pollenizer.

State Fair.—Orig. in St. Paul, Minn., by C. Stushnoff, S. Munson, L.B. Hertz, W. Gray, and D.K. Wildung, Univ. of Minnesota. Introd. in 1978. Mantet x Oriole; cross made in 1949; selected in 1959; tested as MN 1639. Fruit: medium, 65 mm in diam.; skin 80% to 100% attractive bright red, glossy finish; round; flesh texture crisp, juicy, flavor moderately acid, sprightly, aromatic; ripens in early to midseason; very long storage, long shelf life under Minnesota conditions but not in Washington. Tree: moderately vigorous, round spreading; fully hardy, no cold injury following –40C; productive and precocious; no premature fruit drop; susceptible to apple scab, highly susceptible to fire blight.

Stauffer.—See Clear Gold.

Staybrite Stayman®.—A smooth-skinned mutation of Double Red Stayman. Assigned to Newark Nursery, Hartford, Mich. Similar to Stayman except mostly full red color, less scarfskin, and less russet. Tree: nonspur, vigorous, productive, triploid.

Stayman 66.—See Nured Stayman.

Staymared (*Dark-Red Staymared*).—Orig. in Covington, Va., by B.C. Moomaw. Introd. in 1929, by Stark Bros. Nurseries & Orchards Co., Louisiana, Mo. Bud mutation of Stayman Winesap; discovered in 1926. Fruit: skin solid, dark red, sometimes dull mahogany; resembles parent in all other characteristics.

Stephens.—Orig. in Mandan, N.D., by W.P. Baird, U.S. Northern Great Plains Field Sta. Introd. in 1952. Wealthy o.p.; selected in 1927. Fruit: size medium; roundish oblate; skin striped with red on greenish-yellow ground color; flesh subacid to sweet, flavor pleasant; resembles Wealthy. Tree: hardy.

Stevenson.—Orig. in Morden, Manitoba, Canada, by Canada Dept. Agr., Res. Sta. Introd. in 1931. Patten o.p.. Fruit: large; skin greenish-yellow, washed and striped with red; skin thick, tough; flesh cream, crisp, juicy, spicy; good for cooking; season winter. Tree: upright spreading, with strong crotches; vigorous.

Stevenson Wealthy (*Large Wealthy*).—Orig. in Cayuga, N.Y., by J.R. Stevenson. Apparently never introduced; included here because of its primary interest to fruit breeders. Bud mutation of Wealthy; discovered as a branch about 1934. Fruit: larger than Wealthy; skin red. A periclinal cytochimera 2-4-4-4; produces diploid gametes. Chimeral nature determined at the New York State Agr. Expt. Sta., Geneva; accessioned in 1938 by this station.

Stewart.—Orig. in Cottage Grove, Ore., by C.E. Stewart, Lane County fruit inspector. Introd. in 1935. Parentage includes Gravenstein, Delicious, and King; selected in the early 1930s. Fruit: very large; resembles Delicious in shape, color and sweetness but ripens a month earlier; skin waxy; similar to Gravenstein in flavor and cooking qualities. Tree: grows rapidly, precocious; bears annually thereafter.

Stirling.—Orig. in Summerland, British Columbia, Canada, by R.C. Palmer, Canada Dept. Agr., Res. Sta. Introd. in 1936. Yellow Newtown o.p., with the other parent almost certainly Rome Beauty; seed collected in 1924 by R.C. Palmer; original tree planted in 1927; first fruited in 1932; selected in 1933 by A.J. Mann; named in 1936. Fruit: above medium; skin attractively striped and blushed with bright red; flesh cream, moderately firm, crisp, juicy, somewhat coarse in texture, core small, very high quality, highly aromatic, matures late, with Yellow Newtown and Rome Beauty; keeps almost as long as Rome Beauty, which it resembles. Tree: compact in form but not spur-type; consistent annual bearer; hardy under conditions at Summerland, British Columbia. Not being planted and none in production in Summerland.

Stirling Delicious.—Orig. in Kelowna, British Columbia, Canada, by Peter Stirling. Introd. in 1962 by Carlton Nursery Co., Forest Grove, Ore. Discovered in 1957 as a bud mutation in an orchard of Starking Delicious; tested as Stirling Red Delicious No. 2. Fruit: skin dark red, faintly striped, coloring early. Tree: standard type.

Stontosh.—Orig. in Ottawa, Ontario, Canada, by Canada Dept. Agr., Res. Sta. Introd. in 1922. Stone x McIntosh. Fruit: size above medium to large; roundish, conic; cavity narrow; stem short to medium, moderately stout; calyx partly open; skin yellow, washed with deep red, predominantly deep red, resembles Stone, moderately thick, tough, dots numerous, conspicuous, pale yellow; bloom thin, pinkish; flesh dull white, tender, juicy, subacid, pleasant, spicy; quality good; season late December to March or April; no marked resemblance to McIntosh, except in core and seeds; resembles Stone somewhat in color of skin and about cavity.

Strand Winesap (*Red Winesap*).—Orig. in Cowiche, Wash., by C.G. Strand. Introd. in 1928 by the May Nursery Co., Yakima, Wash. Appeared as a complete tree in an orchard of Winesap; discovered in 1927. Fruit: skin full red, blush-type; long-type Winesap, which it resembles; appears to be a better fruit in eastern U.S. than in area of origin; mildew marks and storage shrivel tend to appear on fruit in the Pacific Northwest. Tree: resembles Winesap.

Striped Graham Spy (*Red Striped Graham*).—Orig. in Manistee, Mich., by Thomas Elwood Graham. Introd. in 1936. Plant patent 293; 4 Oct. 1938; assigned to Greening Nursery Co., Monroe, Mich. Bud mutation of Northern Spy; discovered in 1926. Skin brighter red than parent but striped red over a yellow ground color.

Sumac.—Orig. in Summerland, British Columbia, Canada, by Agr. Canada, Res. Sta. breeding program. Complex pedigree, including Vista Bella and Jerseymac. Selected in 1968 and introd. in 1987. Tested as BC 9F-22-42. Fruit: small; color attractive; better quality than Vista Bella; summer apple, harvest 3 days before Vista Bella; storage life very short. Tree: good growth habit; more productive than Vista Bella.

Summer Champion (*Holland*).—Orig. in Weatherford, Texas, by R.B. Alexander. Introd. about 1925. Parentage unknown; discovered in 1921. It appears that Summer Champion was originated in 1921 and Holland in 1939, the latter being the same as or a bud mutation of the former; there is not enough difference between the two to continue listing each name, and thus Holland, because it was named later than Summer Champion, is placed as a synonym of that variety. Fruit: large; skin bright red; ripens in mid-summer. Tree: bears at an early age; produces heavy crops.

Summer Granny.—See Goldsmith.

Summer Red.—See Coffman.

Summer Scarlet.—Orig. in Montour Falls, N.Y., by Frank Wheat. Introd. in 1963 by J.E. Miller Nurseries, Canandaigua, New York. Chance seedling, discovered in 1954. Fruit: medium to large; skin red overall when well grown; flesh firm, crisp, juicy; ripens in mid-August; storage quality good for an early ripening variety. Tree: vigorous; bears early; hardy.

Summerdel (*Applethorpe Summerdel*).—Orig. in Applethorp, Queensland, Australia, by L.B. Baxter and A.D. McWate, Granite Belt Horticultural Research Station. Introd. in 1989. Delicious x Earliblaze; cross made in 1969 by C.W. Winks; tested as 64-111. Fruit: medium, 65 to 75 mm; globose to flat-globose, often lobed; skin 75% or more medium to cherry red with darker red overstripes; smooth, covered with thin waxy bloom, lenticels not conspicuous; slight stem-end russet; round to slightly conical; flesh creamy white to yellowish coarse to medium, firm, juicy; flavor sweet subacid; aromatic. Ripens 2 to 3 weeks before Delicious; fruit hangs exceptionally well; shelf life about 14 days. Tree: vigor medium; mature size smaller and more compact than Delicious and Granny Smith; semi-spreading; precocious; production regular. Susceptible to scab and powdery mildew. Flowers just before and with Jonathan and Delicious.

SummerGlo.—Orig. in Calhoun County, Mich., by Frank H. McDermid, Battle Creek, Mich. Introd. in 1967. Plant patent 2702; 24 Jan. 1967; assigned to Stark Bros. Nurseries & Orchards Co., Louisiana, Mo. Chance seedling discovered in 1935. Fruit: medium large; oblong; skin smooth, thick, tough, 65% covered with bright red blush; flesh white, juicy, firm, texture fine, flavor excellent for season; McIntosh type; ripens 5 weeks before Jonathan; for fresh fruit or canning. Tree: large; upright, spreading; vigorous; hardy; productivity moderate.

Summerland.—Orig. in Summerland, British Columbia, Canada, by Canada Dept. of Agr. Introd. in 1969. McIntosh x Golden Delicious. Cross made in 1926, selected in 1939, tested as S-4-8. Fruit: medium to large; uniform, roundish, slightly conic; stem relatively long, straight and stiff, skin bright to dark solid-red blush, attractive; flesh white, very fine, moderately tender, juicy, subacid, flavor good; ripens one week after Golden in Summerland, keeps four months. Tree: spreading, with strong limbs and crotches; vigorous; hardy; biennial bearing, with very heavy set in alternate years.

Summerland McIntosh (*McIntosh (Summerland)*, *Nured McIntosh*).—Orig. in Summerland, British Columbia, Canada, by Canada Dept. Agr., Res Sta. Introd. in 1929. Bud mutation of McIntosh; selected in 1928. Fruit: skin solid red, not susceptible to striped sporting; otherwise similar to parent; appears to be indistinguishable from Rogers McIntosh. Tree: standard growth habit identical to McIntosh.

Summerland Red Rome.—Orig. in Apex Orchards, Kelowna, British Columbia, by Canada Dept. of Agr. Res. Sta., Summerland, British Columbia, Canada. Introd. in 1960. Bud mutation of Cowin Rome Beauty; discovered in 1958. Fruit: resembles Red Rome, characterized by its high degree of color stability; skin bright red blush, almost completely free of reverting to striping. Tree: same as Red Rome.

Summerred.—Orig. in Summerland, British Columbia, Canada, by K.O. Lapins, Canada Dept. Agr., Res. Sta. Introd. in 1964. Summerland (McIntosh x Golden Delicious) o.p.; selected in 1961; tested as Summerland 9E-11-40. Fruit: medium large; oblong; skin highly colored with a bright, solid blush; flesh cream, medium firm, crisp, juicy, flavor very good; ripens 3 to 4 weeks before McIntosh or a few days before Tydeman's Early Worcester; matures evenly on tree. Tree: vigorous; spreading; blooms relatively early; bears early and fairly regularly.

Sundale.—See Goldspur.

Sundale SturdySpur.—See Goldspur.

Sundance.—Orig. in Lander, Wyo., by J. M. Steinbrech, Wyoming Agr. Expt. Sta. Introd. in 1932. Parentage unknown. Fruit: size medium; oblate; skin yellow, speckled with red; slightly subacid, flavor good; ripens 15 Sept.; keeps medium well. Tree: exceptional bearer.

Sundowner.—Orig. from breeding program of the Stoneville Res. Sta., Western Australia. Golden Delicious x Lady Williams. Introd. in 1979. U.S. rights assigned to Carlton Nursery, Dayton, Ore. Fruit: medium, up to 70 mm; shape oblong; skin color crimson, with pronounced white lenticels; flesh white, very firm; flavor sweet, tart, mildly aromatic; eating quality very good; harvest date very late, four weeks after Granny Smith; storage life up to 5 months; shelf life long. Tree: very vigorous, very productive; low chilling, 500 h; early bloom.

Sungold.—Orig. in Bridgeton, N.J., by Alfred and Anthony Caggiano. Introd. in 1963; assigned to the New Jersey Apple Inst. Parentage unknown; selected in 1960. Fruit: little or no russeting; cold storage quality better than Golden Delicious, which it resembles. Finish tends to be splotchy.

Sunrise.—A high-quality, late-summer apple of excellent appearance from the breeding program of the Agr. Canada Res. Sta., Summerland, British Columbia. Introd. in 1991 by David Lane and Richard MacDonald. Parentage: 10C-10-19 (= McIntosh x Golden Delicious) x PCF 3-120 (a chance seedling). Tested as 8C-27-96. U.S. plant patent applied for. Fruit: skin 50% bright, pinkish-red stripes on pale-yellow ground; large, mostly over 200 g; shape somewhat irregular, ribbed. In Summerland, quite resistant to sunburn. Flesh firm, crisp, juicy; flavor mild, sweet, low acid, with little aroma. Controlled-atmosphere storage life 2 months. Tree: vigorous, spurry, precocious, productive. Relative tolerance to powdery mildew.

Suntan (*Malling Suntan*).—Orig. from breeding program at East Malling Res. Sta, East Malling, Kent, England, by F.H. Alston and R. Watkins. Introd. in 1974. Protected by Plant Breeders' Rights. Cox's Orange Pippin x Court Pendu Plat; tested as A 1585. Fruit: 65 mm in diam. and larger; color orange red blush and stripes on a greenish golden ground, russet, appearance similar to Cox's Orange Pippin; shape slightly flattened; high acidity, fully aromatic flavor; high quality; ripens late; at 3.3C, stores well until late March, but subject to low-temperature breakdown at 1.7C. Tree: vigorous, spreading; high yields every year; triploid, pollen not viable; late flowering, 6 days later than Cox's Orange Pippin; avoids spring frost; can be pollenized by Golden Delicious and Kent; trees on M.7 not precocious.

Super Chief™ Delicious.—A mutation of Redchief (Campbell) Delicious. Plant patent 6190; 31 May 1988; assigned to Van Well Nursery, Wenatchee, Wash. Fruit: stripe color pattern; color develops about 18 days before Campbell; shape long conic, typey; retains the qualities of Redchief Delicious. Tree: very compact spur habit; precocious.

Super Starking Delicious.—Orig. in Hood River, Ore., by Frank L. Rypczynski. Introd. in 1963. Plant patent 2440; 25 Aug. 1964; assigned to Stark Bros. Nurseries & Orchards Co., Louisiana, Mo. Bud mutation of Starking Delicious; discovered in 1955. Skin striped red early in season, later substantially solid bright red, appearing earlier than parent (some 30 days before maturity).

Superior.—Orig. near Blaney Park, Mich., by Joseph E. Lang. Introd. in 1930. Duchess of Olderburg x Wealthy; discovered in 1928. Fruit: fine quality; keeps well into December; resembles Duchess of Oldenburg. Tree: somewhat blight tolerant; inclined to overbear if not thinned; crotches strong.

Swanson Winesap (*Improved Red Winesap*).—Orig. in Peshastin, Wash., by Anton Swanson. Introd. in 1947 by Tualatin Valley Nurseries, Sherwood, Ore. Bud mutation of Winesap; discovered in 1935; tested as Swanson 1 and Wash. 36. Skin blushed red more than Winesap. Tree: standard type; identical to Winesap.

Sweet Delicious.—Orig. in Geneva, N.Y., by New York State Agr. Expt. Sta. Introd. in 1922. Deacon Jones x Delicious; cross made in 1911. Fruit: large; skin color attractive; sweet aromatic flavor of Delicious; home use for dessert and baking; season 10 days after Delicious; resembles Delicious in shape, but lipping of stem and dull grayish scarfskin over surface are like Deacon Jones. Tree: vigorous; dense; upright spreading; productive.

Sweet McIntosh.—Orig. in Geneva, N.Y., by Richard Wellington, New York State Agr. Expt. Sta. Introd. in 1922. Lawver x McIntosh; cross made in 1909. Fruit: externally resembles Lawver but sweet flavor suggests McIntosh; home use (baking and fresh). Tree: vigorous; upright, slightly spreading; heavy bearer.

Sweet Sixteen.—Orig. in St. Paul, Minn., by C. Stushnoff, S. Munson, L.B. Hertz, W. Gray, and D.K. Wildung, Univ. of Minnesota. Introd. in 1979. MN 477 (Malinda o.p.) x Northern Spy; cross made by Alderman in 1937; selected in 1950; tested as MN 1630. Fruit: medium to large, 69 to 89 mm in diam.; shape round-conic; rosy red, attractive smooth finish; quality good with unique, pleasing flavor combined with high sugar, moderate acid and crisp, fine-textured, juicy flesh; ripens with McIntosh; good storage, handling and culinary properties. Tree: moderately vigorous, upright; very winter hardy, as hardy as McIntosh; consistently productive; much less subject to preharvest drop and bruising than McIntosh; resistant to fire blight; not highly susceptible to apple scab or cedar apple rust.

Swiss Gourmet™ (*Arlet*).—Orig. in Wadenswil, Switzerland, by B. Krapf and C. Papperswill, Swiss Federal Res. Sta. Introd. in 1984. Golden Delicious x Idared. U.S.plant patent 6689; 1 Sept. 1989; assigned to TRECO®, Woodburn, Ore. Fruit: medium to large; 75% to 100% red, bright, striped, attractive, occasional russet; better color than Elstar; skin oily when over-mature; shape round to conic; flesh creamy white; texture fine and firm; taste sweet, tart, very good, less acid than Elstar; harvest 10 days before Jonathan; storage life longer than Elstar. Tree: medium vigor, precocious; yields better than Elstar. Bloom season early to midseason; diploid.

Tamar.—Orig. in Rehovot, Israel, by Chanan Oppenheimer, Agr. Res. Inst. Introd. in

1960. F_2 (Calville St. Sauveur x an unnamed local seedling). Fruit: large; roundish; skin mostly yellow; flesh firm, flavor good, quality very high; ripens in July. Tree: less resistance to delayed foliation than Vered and Naomi.

Tangier.—Orig. in Princess Anne, Md., by Bountiful Ridge Nurseries. Introd. in 1968. Delicious o.p.; discovered in 1964. Fruit: size medium; shape of Delicious; skin highly colored red, similar to Delicious; flesh creamy white, sweet, quality good; ripens 1 week after Lodi. Tree: size medium; vigorous; hardiness moderate; productivity very good.

Tangowine.—Orig. in Petitcodiac, New Brunswick, Canada, by Charles A. Stultz. Introd. in 1965. Chance seedling; discovered sometime before 1950. Fruit: size medium; 2 1/2 to 2 3/4 inches in diam.; round to slightly elongated; stem long; cavity medium, calyx closed, basin medium furrowed; skin dark red over yellow, medium bright, no russeting, bloom heavy, thickness medium; flesh texture smooth, medium tender, moderately juicy, subacid; quality good; keeps well in common storage; ripens with McIntosh; some tolerance to scab. Tree: large; vigorous, spreading, moderately open; hardy; medium productive, moderate annual bearer; many spurs.

Taylor.—See Flamespur Red Rome.

Telamon.—Orig. at East Malling Res. Sta., East Malling, Kent, England. Introd. by K.R. Tobutt in 1989. Wijcik McIntosh x Golden Delicious. U.S. plant patent 6224; 1988. Fruit: medium; conical, regular; skin crimson; flesh firm; mild flavored; long storage life. Harvested a week or two after Cox's Orange Pippin. Tree: vigor and growth habit resemble Wijcik McIntosh; almost no lateral extension shoots; precocious; annually productive.

Teeple Red Empire.—See Royal Red Empire.

Telstar (*Kidd's D.5*).—Orig. in Greytown, Wairarapa, New Zealand, by J.H. Kidd. Introd. in 1966. Kidd's Orange x Golden Delicious. Selected in 1939. Fruit: size medium; conical; skin greenish-yellow, heavily striped with red, thin, bruises easily; flesh yellow, fine, firm, crisp, tender, sweet; excellent quality; storage performance excellent if dehydration prevented; susceptible to russet and apple mosaic. Tree: vigorous, resembles Kidd's Orange (*Kidd's Orange Red*); heavy spur-bearing habit; tendency to mildew; regular and heavy producer; harvest midseason.

Ten Hove Gala.—See Royal Gala.

Theimer.—Orig. in Cedar Falls, Iowa, by Franz Theimer. Introd. about 1925. Parentage unknown; discovered about 1880. Fruit: medium; oblong; flavor sweet, subacid; ripens early, 60 days after bloom.

Thew Gold.—See Thewgold Delicious.

Thewgold Delicious (*Thew Gold*).—Orig. in La Grangeville, N.Y., by Gerald R. Thew. Introd. in 1966. Plant patent 2616; 1966, assigned to Columbia & Okanogan Nursery, Wenatchee, Wash. Found as a chance seedling in 1960. Fruit: 2 3/4 inches in diam.; skin bright yellow sometimes having a reddish blush, similar to Golden Delicious, thin, tough, waxy; flesh firm, more tart than Golden Delicious; small fruit core; matures after Golden Delicious; prolonged tendency of the fruit to hang on the tree; keeping quality excellent; fruit colors well while in storage, remains firm, without shriveling, resulting in superior handling, packing and shipping qualities; resistant to bruises. Tree: medium; vigorous; hardy; productive; regular bearer; very susceptible to fire blight; freedom from rust.

Thome Empire™.—Discovered in Thome Farms, Comstock Park, Mich. A whole-tree, red-fruited mutation of Empire; introduced 1989, tested and patented as TF 808; plant patent 8010; 1992. Fruit: fully colored 3 weeks before Empire; deep rich burgundy, slightly darker than Empire.

Thorberg.—Orig. in Mandan, N.D., by W.P. Baird, Northern Great Plains Field Sta. Introd. in 1957. Duchess of Oldenburg x Starking Delicious; selected in 1951; tested as Mandan 51-38; sibling of Garrison and Mandan. Fruit: medium to large, averaging about 3 1/2 per lb; slightly oblate; stem long; skin an attractive red, with carmine stripes and distinct tan dots; flesh firm, slightly tart, juicy, pleasant tasting, with some of the flavor of Delicious; very suitable for culinary purposes; matures 2 weeks before Delicious; best season of use is fall and early winter. Tree: hardy; growth moderate; medium spreading.

Tieton Rome.—See Nero Rome.

Tift Spur Rome.—Orig. near Brewster, Wash., by Calvin L. Cooper. Discovered as a limb sport of Law Rome Beauty in orchard of Paul G. Tift. Plant patent 7363; 1990; assigned to TRECO®. Differs from Law Rome Beauty in that most axillary buds become fruit spurs, tree structure is more compact, the bourse is unusually enlarged, and the fruit is almost totally red, coloring several days earlier.

Tohoku No. 3.—See Akane.

Toko.—Orig. in Kuroishi City, Aomori-Prefecture, Japan, by Aomori Apple Expt. Sta. Introd. there in 1963; in the U.S. as P.I. 287529 in 1963. Golden Delicious x Indo; selected in 1939; tested as Golden x Indo 24 go. Fruit: intermediate between parents in shape and color; flesh fine, tender; subject to core mold; storage period 2 months longer than Golden Delicious; good for processing.

Tolmo.—Orig. in Brookings, S.D., by N.E. Hansen, South Dakota Agr. Expt. Sta. Introd. in 1932. Tolman Sweet x Duchess of Oldenburg. Fruit: up to 2 3/4 inches in diam.; skin color similar to Duchess of Oldenburg; flesh quality good, subacid. Tree: vigorous; hardy.

Tom MacLean.—Orig. in Margaretville, Nova Scotia, Canada. Parentage unknown, possibly Red Astrachan o.p. Fruit: most nearly resembles Mantet; medium to medium-small; globose; skin color bright red striped on light greenish-yellow ground; finish smooth; flesh creamy white, medium firm, subacid, sprightly, very good flavor. Ripens very early, with Close. Tree: medium-vigorous; upright spreading; precocious; hardy.

Tonasket Red Delicious.—See Early Red Rich.

Topred Delicious.—Orig. in Tonasket, Wash., by Leonard R. Hutchinson, Columbia & Okanogan Nursery Co., Wenatchee, Wash. Introd. in 1959. Plant patent 1916; 1960; assigned to C&O. Bud mutation of Shotwell Delicious; discovered in 1954. Fruit: skin colors earlier than parent, being one of the earliest coloring of recent red sports; otherwise, very similar to Shotwell Delicious.

Topspur Delicious (*Kemp Delicious*).—Orig. near Royal Camp, Wash., by H.M. Kemp. A red-fruited limb mutation of Bisbee Delicious, discovered in 1973. Introd. in 1983. Plant patent 5334; 1989; assigned to C&O Nursery, Wenatchee, Wash.

Toshfor.—Orig. in Ottawa, Ontario, Canada, by Canada Dept. Agr., Res. Sta. Introd. in 1926. McIntosh x Forest. Fruit: size medium; roundish; skin yellow washed with crimson, moderately thick, moderately tough; dots fairly numerous, distinct, pale yellow; flesh yellowish with traces of red, subacid, juicy, pleasant, firm, crisp; season December to March.

Toshkee.—Orig. in Ottawa, Ontario, Canada, by Canada Dept. Agr., Res. Sta. Introd. in 1923. McIntosh x Milwaukee. Fruit: medium to large; oblate conic, slightly ribbed; cavity open, deep; stem stout; skin pale yellow, blushed, moderately thick, moderately tough, dots few and white, bloom bluish; flesh dull white or yellowish, tender, crisp, juicy, subacid, pleasant; quality above medium to good; keeps better than McIntosh; resembles McIntosh in color, flesh, flavor. Tree: grows slowly, dwarfish; bears annually.

Toshlaw.—Orig. in Ottawa, Ontario, Canada, by Canada Dept. Agr., Res. Sta. Introd. in 1925. McIntosh x Lawver. Fruit: medium to large; roundish, uniform; cavity narrow, very shallow, sometimes closed; stem length medium, stout; calyx closed or partly so; skin greenish-yellow, splashed dull crimson, moderately thick, moderately tough, dots obscure, bloom thin and pinkish; flesh dull white, firm, crisp, juicy, subacid, pleasant; quality medium to good; no marked resemblance to either parent.

Trajan.—Orig. from breeding program at East Malling Res. Sta., East Malling, Kent, England. Introd. by K.R. Tobutt in 1989. Golden Delicious x Wijcik McIntosh. U.S. plant patent 6226; July 1988. Fruit: medium to medium-small; skin bicolored solid bright red and green; oval-conical; flesh crisp, juicy, sweet, pleasant. Harvested in midseason, about with Cox's Orange Pippin. Can be kept in refrigerated storage for about 3 months. Tree: more vigorous than Wijcik McIntosh; produces some side shoots; strong biennial bearing tendency. Intended for home gardeners.

TRECO® Red Gala No. 42 (*Cooper* cv.).—Discovered by Cal Cooper, Brewster, Wash. U.S. plant patent 7396; 1990; assigned to TRECO, Woodburn, Ore. Tested as Spur Gala-Go-Red. Fruit: similar to Gala except skin color is bright overall red in a muted striped pattern; sweet, firm texture, long storage life.

Trent.—Orig. by L.P.S. Spangelo, S.J. Leuty, and H.B. Heeney at Agr. Canada Res. Sta., Ottawa and Smithfield Experimental Farm, Trenton, Ontario, Canada. Introd. in 1979. McIntosh x R18T40 [Jonathan x (Rome Beauty x *M. floribunda* 821 sib)]. Tested as Ottawa 531. Fruit: medium to large size; skin yellow; flesh moderately juicy, firm, cream-colored with greenish tinge, slightly coarse, moderately acid; processed fruit rated good for sauce and slices, fair for juice; susceptible to bitter pit; harvest with Northern Spy; storage life 6 months at 2C. Tree: vigorous, semi-spreading; medium yield efficiency; blooms 2 or 3 days after McIntosh; resistant to apple scab; susceptible to cedar apple rust; very susceptible to quince rust.

Trenton.—Orig. in Albury, Ontario, Canada, by P.C. Dempsey. Introd. in 1922. Northern Spy x Golden Russet. Fruit: size medium; roundish; skin yellow, washed with dull red, thick but tender, dots obscure; flesh yellowish, crisp, subacid, moderately juicy, coarse; season late September and early October

Tropical Beauty.—Orig. in Maidstone, South Africa, by Meredith B. Strapp. Seedling of unknown parentage planted about 1930; discovered in early 1930s by J.L. Alkins, Verulum, South Africa, and W.T. Lowen, Natal, South Africa; budwood distributed and promoted in 1953 by F.B. Harrington, Natal, South Africa; named and introduced in Australia in 1958 by Langbecker Nurseries; propagation rights outside of Australia and New Guinea assigned to Stark Bros. Orchard & Nursery Co., Louisiana, Mo.; brought into the U.S. as P.I. 281542 but has not been introduced commercially. Fruit: under Australian conditions: size medium; skin red; flesh crisp; eating quality very good, resembles Rome. Under U.S. conditions (Louisiana, Mo.): large; irregular in form; skin dull red; flesh not too crisp, flavor fair, quality poor. Susceptible to bitter rot. Tree: under Australian conditions: vigorous growth under tropical and subtropical conditions, short dormant period. Under U.S. conditions (Louisiana, Mo.): characteristics resemble Rome. Restricted to warmer climates.

Truax Greening.—Orig. in Hammond, N.Y., by F.E. Truax. Introd. in 1957 by Fred L. Ashworth, nurseryman, Heuvelton, N.Y. Parentage unknown; may be Tolman Sweet o.p.; discovered in 1955. Fruit: large; skin green, sometimes striped red; flesh subacid; keeps late in ordinary cold storage; resembles Coles Quince. Tree: very hardy; type of growth like that of Tolman Sweet.

Tru-Brite Rome Beauty.—Orig. in Moxee, Wash., in the Harry Boorsma Orchard by Everett Brandt, Carlton Nursery, Dayton, Ore. Introd. in 1983. Bud mutation of Law Rome Beauty; discovered in 1972. Fruit: resembles Law Rome Beauty, but colors earlier and more uniformly than parent.

Tsugaru.—Orig. in Kuroishi, Aomori Prefecture, Japan, by the breeding program at the Aomori Apple Res. Sta. Introd. in 1975; Japanese Reg. no. P 277. Golden Delicious o.p. Fruit: large 300 to 350 g; shape round; color with wide red stripes over yellow-green, not highly attractive; flesh firm, tender, subacid, mild, sweet flavor; juicy; harvest 3 weeks before Delicious; stores 2 months. Subject to preharvest drop.

Turley.—Orig. in Orleans, Ind., by Joe E. Burton, using funds of the Indiana Hort. Society. Introd. in 1922. Winesap o.p.; selected about 1910. Fruit: large; shape uniform; skin bright red; does not crack; flesh white, crisp, juicy; ripens 1 week before Stayman Winesap, which it resembles; good quality for cooking, but lacks the high dessert quality of Stayman Winesap. Tree: vigorous; upright; scaffold branches strong; annual bearer; productive; fruit spurs susceptible to winter injury; not considered a good pollenizer.

Turner Red.—See Turner Delicious.

Turner Delicious (Turner Red).—Orig. in Salmon Arm, British Columbia, Canada, by R. Turner. Introd. about 1925. Bud mutation of Delicious. Fruit: skin bright, solid red; in all other respects similar to the parent; appears to be indistinguishable from Richared. Tree: identical in growth habit to standard Delicious.

Tydeman Early.—See Tydeman's Early Worcester.

Tuscan.—Orig. from breeding program at East Malling Res. Sta. East Malling, Kent, England. Introd. by K.R. Tobutt in 1989. Wijcik McIntosh x Greensleeves. U.S. plant patent 6225; July 1988. Fruit: medium to medium-large; skin green with slight orange flush and traces of russet; shape rather irregular, oblong; flesh crisp, juicy, somewhat tart. Harvested early midseason, 1 week or more before Cox's Orange Pippin. Storage season about 1 month. Tree: more vigorous than Wijcik McIntosh; produces side shoots.

Tydeman's Early Worcester (*Early Worcester, Tydeman Early, Tydeman's Red*).—Orig. in Maidstone, Kent, England, by Henry M. Tydeman, East Malling Res. Sta. Introd. in England in 1945 and in Canada and the U.S. in 1946 as P.I. 241965. McIntosh x Worcester Pearmain; seedling planted in 1929; first fruited in 1936; selected about 1940. Fruit: intermediate between parents; large; skin solid bright red, attractive; flesh firm, quality very good, flavor pleasant for eating fresh; stores very well at 32F; ripens early, before Worcester Pearmain and 3 weeks before McIntosh; resembles McIntosh. Tree: bears regularly; appears to be susceptible to fire blight; tendency to produce long, spindly branches.

Tydeman's Late Orange.—Orig. in East Malling, Kent, England, by Henry M. Tydeman, East Malling Res. Sta. Introd. in England in 1945 and in Canada and the U.S. in 1946. Laxton's Superb x Cox's Orange Pippin; selected about 1940. Fruit: a very late maturing, eating apple of high quality; keeping, in England, until March or April; resembles Laxton's Superb; originally named Tydeman's Late Cox and was so carried in the East Malling Research Sta. Annual Rpt. for 1945 (vol. 33). Tree: fairly vigorous; upright; slightly spreading.

Tydeman's Red.—See Tydeman's Early Worcester.

Ulrich Delicious.—See Earlired Delicious.

Unity.—Orig. in Unity, Saskatchewan, Canada, by John Lloyd, John Lloyd's Fruit Nursery. Introd. in 1967. Parentage unknown. Fruit: 2 to 2 1/4 inches in diam.; skin partially covered with red blush; flesh white, flavor and quality good; keeps well; ripens after McIntosh. Tree: tall, upright; vigor moderate; has withstood -40F; productive; recommended for northern districts only.

Unprofitable Grimes.—See McClintock Grimes.

Vahldiks Cox Sämling No. III.—See Holstein.

Vallee Spur Delicious™.—Orig. near Prescott, Wash., by J.E. Valle. A mutation of Redchief Delicious (Campbell cultivar). Plant patent 6702; 1989; assigned to TRECO®, Woodbum, Ore., and to Cameron Nursery. Fruit: 100% red with fine stripes; color develops earlier than other Delicious strains. Tree: semi-dwarf, spur growth habit, heavy cropping.

Valmore.—Orig. in Visalia, Calif., by Valentine Moore. Introd. about 1934. Plant patent 238; 1937. Chance seedling; discovered about 1924. Fruit: large; conical; skin striped with red, medium thickness, waxy; flesh yellowish-white, texture fine, juicy, flavor subacid, aromatic; excellent for cooking; ripens early, with White Astrachan; resembles Stayman Winesap. Tree: vigorous, upright; productive; somewhat resistant to delayed foliation.

Van Buren.—See Red Van Buren.

Vance Delicious.—Orig. in Albemarle County, Va., by R.G. Vance, Waynesboro Nurseries, Waynesboro, Va. Introd. in 1935. Bud mutation of Delicious; discovered in 1930 on a 10-year-old tree. Fruit: skin solid red, coloring at least 2 weeks earlier than Delicious, which it resembles. Tree: standard type.

Vered.—Orig. in Rehovot, Israel, by Chanan Oppenheimer, Agr. Res. Sta., Div. Subtrop. Hort., National and Univ. Inst. of Agr. Introd. there in 1953; in the U.S. about 1956. F_2 (Calville St. Sauveur x unnamed local seedling); cross made in 1939; selected in 1947. Fruit: small, averaging about 100 g; conic; skin splashed with red; flesh sprightly; ripens early, end of May; soft and does not transport well; for local market. Tree: chilling requirement very slight, suitable for subtropical conditions; must be grown on strong rootstock; self-fruitful;

sets practically every flower and must be heavily thinned to obtain commercial size.

Vermont Spur.—Orig. in the Loomis Orchards, Panton, Vermont. Introd. in 1969 by Kraus Nursery, Carlisle, Ontario, Canada. Bud mutation of Starking Delicious, discovered about 1962. Fruit: very typey Delicious shape, skin cream color, blushed red; flesh cream white contains less chlorophyl than other spur Delicious mutants grown in same area. Tree: size medium; spur type; upright; vigor good; productivity good.

Victory.—Orig. in Excelsior, Minn., by Charles Haralson, Minnesota Agr. Expt. Sta. Introd. in 1943. Probably McIntosh o.p.; first fruited in 1918; tested as Minn. 396. Fruit: holds well on tree; flesh and skin not as tender as McIntosh, which it resembles; skin medium red; flesh white, fine-grained, juicy, aromatic; for dessert purposes. Tree: spreading, vigorous, productive. Susceptible to fire blight in Manitoba.

Viking (*D. B., Gurney Viking*).—Orig. in Yankton, S.D., by George W. Gurney, Gurney Seed & Nursery Co. Introd. in 1925. Selected from scions sent from Scandinavian countries. Fruit: large; quality good; skin mostly deep red although partly striped over a green ground color, very attractive; season fall. Tree: exceptionally hardy; bears early. No longer commercially available.

Virginia Winesap (*Red Winesap*).—Orig. in Troutville, Va., by Garland J. Hopkins. Introd. in 1924 by Stark Bros. Nurseries & Orchards Co., Louisiana, Mo. A superior selection of Winesap; discovered in 1922. Fruit: large; skin a very dark red. Tree: very vigorous; standard type.

Virginiagold.—Orig. in Blacksburg, Va., by R.C. Moore and G.D. Oberle, Virginia Agr. Expt. Sta., Virginia Polytechnical Inst. and State Univ. Introd. in 1972. Albemarle Pippin x Golden Delicious. Cross made in 1944, first fruited in 1952, selected in 1956, tested as VPI 8. Fruit: medium to large; round to round conic, sometimes angular to irregular; skin clear waxy yellow with bright pinkish blush and white dots, very attractive; flesh very firm, creamy yellow, texture medium, crisp, juicy; long storage life; for fresh market and commercial processing; ripens 2 or 3 weeks later than Delicious. Tree: size medium; round topped, sturdy; vigor moderate; very productive but will alternate; appears somewhat susceptible to late summer rots but very tolerant of conditions that cause severe lenticel and skin russeting on Golden Delicious under eastern conditions; pollen viable, blooms in Delicious season.

Vista Bella.—Orig. in New Brunswick, N.J., by L. Fredric Hough and Catherine H. Bailey. Introd. in 1974. 77349 (67634 x 122137) x Julyred. 67634 = Melba x Sonora, 122137 = (Williams x Starr) x USDA34. Cross made in 1956; selected in 1962; tested as NJ36. Fruit: medium large; round to slightly oblate; skin yellow-green ground color with up to 75% blushed with medium red, attractive; flesh creamy white, crisp; ripens with Melba, better shelf life than other varieties of same season; dessert variety. Tree: size standard; upright spreading; productivity good, but will alternate if allowed to overbear; widely adapted.

Voelker Solid Red Delicious.—Orig. in Tieton, Wash., by L.M. Voelker. Introd. in 1954. Bud mutation of Delicious; discovered in 1950. Fruit: skin solid red, even in shadiest part of tree; resembles Richared Delicious. Tree: similar to Delicious.

Volga.—Orig. in Brookings, S.D., by N.E. Hansen, South Dakota Agr. Expt. Sta. Introd. in 1933. Virginia x Anisim. Fruit: up to 2 3/8 inches in diam.; roundish conical; skin color fairly attractive, bright red with deeper red stripes; flesh fine-grained, juicy, subacid, light red outside core line, quality fair; season late fall.

Wahoya.—Orig. in Brookings, S.D., by N.E. Hansen, South Dakota Agr. Expt. Sta. Introd. in 1938. Nevis crab x Wolf River. Fruit: averages 2 3/8 inches in diam.; skin blushed red over green ground color; flesh mildly acid, good for culinary use; keeps well in cool storage.

Wakaga.—Orig. in Brookings, S.D., by N.E. Hansen, South Dakota Agr. Expt. Sta. Introd. in 1938. Nevis crab x Wolf River. Fruit: averages 2 1/2 inches in diam.; skin green, blushed red; flesh slightly acid, used for sauce; keeps well in cool storage. Tree: hardy.

Wakapala.—Orig. in Brookings, S.D., by N.E. Hansen, South Dakota Agr. Expt. Sta.

Introd. in 1928. Mercer x Tolman Sweet. Fruit: 2 1/8 inches in diam.; skin yellow, lightly striped with red; flesh white, subacid, with spicy sweet fragrance; season winter.

Waldorf.—Orig. in Arena, Wis., and propagated in Kenilworth, Ill., by Wm. S. Sawle of Kenilworth. Introd. in 1957. Plant patent 1587; 1957; by Wm. S. Sawle, Jr. Parentage unknown. Fruit: size medium; oblate; skin dark green with red blush; flesh juicy, crisp; harvest period 80–100 days after bloom. Tree: vigorous; compact and symmetrical; hardy; productive; retains fruit without dropping for longer periods than usual.

Waltana.—A mid-late introduction with intermediate chilling requirement. Orig. in California in early 1990s. Sold by L.E. Cooke Nursery, Visalia, Calif. Fruit: size medium to large; color thin red striping over greenish ground; shape variable. Flesh: texture crisp; hard; juicy; flavor slightly tart. Harvest season 4 weeks after Delicious; use dessert and cooking. Tree: vigorous, healthy, regularly productive; chilling requirement 900 h.

Warder.—Orig. in Wooster, Ohio, by Freeman S. Howlett, Ohio Agr. Expt. Sta. Introd. in 1937. Rome Beauty o.p. Fruit: attractive, red overcolor; dessert quality somewhat better than that of Rome Beauty; at Wooster, ripens in season of Grimes Golden to Jonathan (October); resembles Ensee. Propagation discontinued due to unsatisfactory fruit quality and color.

Washakie (*Chief Washakie*).—Orig. in Lander, Wyo., by J.M. Steinbrech, Wyoming Agr. Expt. Sta. Introd. in 1932. Fruit: large; oblate; skin nearly red, greenish on calyx end; appearance attractive; flesh fine-grained, subacid, flavor good, very firm; keeps well; harvest about 110 days after bloom.

Watopa.—Orig. in Brookings, S.D., by N.E. Hansen, South Dakota Agr. Expt. Sta. Introd. in 1939. Elk River x Jonathan. Fruit: about 3 inches in diam.; skin greenish-yellow striped with red; flavor mildly acid to neutral. Tree: very productive.

Watson Delicious.—See Hi-Red.

Watson Jonathan.—Orig. in Vernon, British Columbia, Canada, by J.K. Watson. Introd. about 1950. Bud mutation of Jonathan; discovered in 1940. Fruit: skin solidly blushed, bright, very attractive, being equivalent in color to Jonared; resembles Jonathan. Tree: standard growth habit identical to Jonathan.

Watts.—Orig. in Morden, Manitoba, Canada, by Canada Dept. Agr., Res. Sta. Introd. in 1931. Patten o.p. Fruit: large; skin greenish-yellow, blushed, some red striping; flesh juicy, tender, sweet, resembling Tolman; season winter. Tree: vigorous; moderately productive.

Waukon.—Orig. in Moorhead, Minn., by G.H. Sikes. Introd. in the early 1920s. Parentage unknown. Fruit: size medium; shape oblate, irregular; skin ground color yellow striped, splashed brilliant red; flesh firm, white, tender, melting, flavor mild, aromatic, nonacid; quality fair to good as dessert fruit, good to very good as sauce; Wealthy type. Tree: strong, sturdy; hardy; productive.

Wayne.—Orig. in Geneva, N.Y., by the New York State Agr. Expt. Sta. Introd. in 1962. Northwestern Greening x Red Spy; selected in 1951; tested as New York 44420-1; diploid. Fruit: large; dual-purpose, fresh and processing, being outstanding for canned and frozen slices and sauce, rating higher than Northern Spy; skin solidly blushed and washed light scarlet, no striping, appearance excellent; ripens directly after McIntosh, 2 to 3 weeks before Rhode Island Greening is ready for processing; not subject to bitter pit. Tree: upright, like Spy; blooms late; good pollenizer for late-blooming varieties; bears precociously, maturing commercial crops within 5 to 6 years. Very susceptible to fire blight.

Wayne Spur Delicious.—Orig. in Orchard Gap, Carroll County, Va., by S.R. Levering of Ararat, Va. Introd. in 1962. Considered to be a bud mutation of Starking Delicious; discovered in 1957. Fruit: skin with more blush and with less noticeable striping than Starking Delicious; firmer and coloring a few days before that variety, with longer storage life. Tree: Spur-type habit of growth.

Waziya.—Orig. in Brookings, S.D., by N.E. Hansen, South Dakota Agr. Expt. Sta. Introd. in 1938. Nevis crab x Northwestern Greening; sibling of Wetonka and Wecota. Fruit: small, averaging 2 1/4 inches in diam.; oblate; skin green; flesh sharply acid, not cooking satisfactorily; keeps very late in cold storage; fragrance of a wild crab apple.

Wealthy Double Red.—See Case Wealthy.

Webster.—Orig. in Geneva, N.Y., by Richard Wellington, New York State Agr. Expt. Sta. Introd. in 1938. N.Y. 26 (Ben Davis x Jonathan) x N.Y. 19 (Ben Davis x Jonathan); cross made in 1912; first full crop in 1921; triploid. Fruit: medium to large; oblong conic; skin nearly solid, bright red, attractive; quality better for culinary than for dessert purposes; resembles Wolf River in shape, but much larger.

Wedge.—From seed originally planted by Wyman Elliot, Minneapolis, Minn., and later selected in 1912 by the Univ. of Minnesota Fruit Breeding Farm. Introd. in 1922. Probably Ben Davis o.p.; tested as Minn. 207. Fruit: large; roundish oblate to roundish conic; color usually a good solid red, but sometimes striped and splashed on under-colored specimens; flesh fine-grained, firm, crisp, juicy, subacid, quality good. Tree: vigorous; large, upright, rounded; productive; hardy.

Welch Winesap.—See Mosebar Winesap.

Welday Jonathan.—Orig. in Smithfield, Ohio, by W.J. Welday. Apparently never introd. Considered to be a bud mutation of Jonathan; discovered in 1930. Fruit: large; quality good. A periclinal cytochimera 2-2-4-4. Tree: considered a good cropper. Chimeral nature determined at the New York State Agr. Expt. Sta., Geneva; accessioned in 1932 by this station.

Wellington.—Orig. in Geneva, N.Y., by the New York State Agr. Expt. Sta. Introd. in 1955. Cortland x Crimson Beauty; selected in 1937. Fruit: large; skin red-striped, attractive; quality fair to good; ripens evenly and very early, about 5 or 6 days after Crimson Beauty; fruit matures at one time, thereby requiring only one picking; for dessert, excellent for sauce; good commercial prospects for its season. Tree: annually productive, upright, spreading.

Wells & Wade Spur.—See Oregon Spur II.

Wellspur Delicious *(Azwell)*—Orig. in the Azwell orchard of Wells & Wade Fruit Co., Wenatchee, Wash., by Lester Green and Clifford Nelson. Introd. in 1958. Considered to be a bud mutation of Starking Delicious; discovered in 1952. Fruit: skin solid red type, coloring early; resembles Starkrimson Delicious. Tree: spur type; bears heavily; size substandard.

West Virginia Red York.—Orig. in Roanoke, W.Va., by John L. Hevener, Roanoke Nursery. Introd. in 1962. Plant patent 2288; 1963. Bud mutation of York Imperial; discovered in 1950. Skin solid red. Harvest about 1 or 2 weeks after York Imperial; resembles Yorking.

Western Giant.—Orig. in Paskenta, Calif., by Earl Davies. Introd. in 1948; assigned to United Nurseries, Red Bluff, Calif. Parentage unknown; discovered in 1947. Fruit: 4 1/2 to 5 inches in diam; 12 to 14 oz; skin thick with a deep blush flecked with red; flesh creamy white; ripens before Astrachan at the 4000-ft. level on the west side of the Sacramento Valley in Tehama County, around 5 to 12 July; keeps well in cold storage; resembles Alexander.

Whetstone.—Orig. in Mountain Grove, Mo., by Paul H. Shepard, Missouri State Fruit Expt. Sta. Introd. in 1935. Conard x Delicious; cross made in 1912; selected in 1930; tested as A-1289. Fruit: large to very large; long; conic, slightly knobbed; uniform in size and shape; skin well colored; quality fair; stem long; ripens about 10 days after Winesap; keeps well in storage until May; sizes well; does not drop when dry weather occurs in early fall. Tree: vigorous; leaves large; twigs thick; spreading; heavy production; fruiting habit similar to McIntosh.

Whitten.—Orig. at the Turner Sta. Fruit Farm by J.C. Whitten, Missouri Agr. Expt. Sta., Columbia, Mo. Introd. in 1926. Ingram x Delicious; selected in 1925. Fruit: skin yellow, blushed and striped with red; texture coarse, flavor fair; matures in late September; resembles Ingram, but larger. Tree: blooms late.

Wijcik.—See Starkspur Compact Mac.

Williams' Pride.—Orig. in West Lafayette, Ind., by a cooperative breeding program consisting of the Agr. Expt. Sta. of Indiana (J. Janick, F.H. Emerson, P.C. Pecknold, and J.A. Crosby) and Illinois (S.S. Korban and D.F. Dayton). Introd. in 1987. Plant patent 6268; 6 Sept. 1988; assigned to Purdue Research Foundation, West Lafayette, Ind. PRI 1018 x NJ50; cross made in 1973; a sixth-generation derivative of C.S. Crandall's Rome x *M. floribunda* 821 sib cross. Pedigree includes Mollie's Delicious, Julyred, Jonathan, Melba, Red Rome, Wealthy, and Starr. Selected in 1979 by E.B. Williams; named in his honor, tested as PRI 2845-1 and as Co-op 23. Fruit: medium-large, 74 mm in diam.; oblate to round, uniform shapes; skin ground color sap green, overcolor darkens to 90% ruby red, smooth, slightly waxy, no russeting; flesh light cream-colored, medium-grained, firm, crisp and juicy; flavor moderately subacid, slightly spicy, rich; quality excellent for the season; when stored at 1C quality and texture are retained 6 weeks; a summer dessert apple, ripens 7 to 8 weeks before Delicious; requires more than one picking. Tree: vigorous, spreading; moderately heavy cropping; annual bearer; fruit hang well to tree after ripe; extended bloom period from very early to late midseason; frost killing of blossoms less severe than Delicious, Jonathan, and Golden Delicious; immune to apple scab, apparently resistant to cedar apple rust, and moderately resistant to fire blight and powdery mildew; only slightly predisposed to bitter pit and water core.

Williams Rome.—Orig. in Yakima Valley, Wash., by Dan Williams. Introd. in 1931; propagated first by May Nursery Co., Yakima, Wash. Considered to be a bud mutation of Rome Beauty; discovered in 1931. Skin red with definite stripes, tends to color earlier than parent; Rome Beauty type. Tree: standard.

Willis Williams.—Orig. in Daleville, Va., in the Garland Orchards. Introd. in 1961 by Virginia Agr. Expt. Sta., Blacksburg. Bud mutation of Williams Red; discovered in 1954 by R.C. Moore, Virginia Polytechnic Inst., Blacksburg. Fruit: skin brighter, more intense, near solid red, with better finish, being of a beautiful appearance and better commercial appeal than parent; otherwise, all characteristics typical of parent. Tree: characteristics typical of parent variety.

Willow Twig Double Red.—Orig. in Hardin, Ill., by C.F. Braden. Introd. in 1929 by Stark Bros. Nurseries & Orchards Co., Louisiana, Mo. Bud mutation of Willow Twig; discovered in 1927. Fruit: resembles parent except for darker red skin. Tree: identical to Willow Twig.

Wilson Juicy.—Orig. in Hammond, N.Y., by John J. Wilson. Introd. in 1956 by Fred L. Ashworth, Heuvelton, N.Y. Parentage unknown; arose from a 100-year-old stump on which the scion variety died; may be a seedling of Pomme de Neige; discovered in 1955. Fruit: as large as McIntosh; skin yellow overlaid with a lavender-pink blush; flesh white, subacid, very juicy, flavor distinctive; keeps well in cold storage.

Winchester.—Orig. in Blacksburg, Va., by R.C. Moore and G.D. Oberle, Virginia Agr. Expt. Sta., Virginia Polytechnic Inst. & State Univ. Introd. in 1971. York Imperial x Lowry. Cross made in 1940, selected in 1959, tested as VPI 3. Fruit: medium to large; medium oblate to round oblate, truncate, slightly oblique, skin medium red wash, stripe and blush, flesh deep yellow, crisp, medium juicy, long storage life; for commercial processing; harvest 3 weeks later than Delicious. Tree: size medium, round, vigor moderate, very productive but will alternate; pollen viable, blooms late with York Imperial, 2 days before Rome Beauty. Recommended for areas in Virginia and elsewhere where York Imperial grows successfully.

Winesap Stark Double-Red.—Orig. in Louisiana, Mo., by Stark Bros. Nurseries & Orchards Co. Introd. in 1951. Bud mutation of Winesap, discovered in 1946. Fruit: a better all-red skin color than parent and larger; no red in vascular system or flesh; resembles Winesap. Tree: more vigorous than parent.

Winter Queen.—Orig. in Saskatoon, Saskatchewan, Canada, by C.F. Patterson, Univ. of Sask. Introd. in 1960 for home gardens. Osman x Wealthy; selected in 1946; tested as A58-312. Fruit: up to 2 1/2 inches in diam. under nonirrigated field conditions at Saskatoon; skin greenish, partially overlaid with dark, dull red; flesh white, subacid, quality very fair; keeps well, being a winter variety; ripens late. Tree: hardy.

Wisconsin Viking.—Orig. in Sturgeon Bay, Wis., by F.A. Gilbert, Peninsular Expt. Sta.

Introd. in 1969. Plant patent 3420; 1974. N.J. 148849 x PRI 187-4; cross made in 1956; selected in 1963; tested as Wis. 1033-5. Fruit: size above medium; ovate to oblong; skin washed red over 80% to 95%, smooth, waxy, glossy, bright; flesh white, juicy, fine-grained, quality very good, subacid, aromatic; a dessert variety; ripens early, 4 to 5 weeks ahead of McIntosh. Tree: size medium; upright; spreading; vigorous; hardy; may have tendency to biennial bearing; moderate tolerance to scab and fire blight.

World #1.—See Sekai-ichi.

Wright.—Orig. in Mountain Grove, Mo., by Paul H. Shepard, Missouri State Fruit Expt. Sta. Introd. in 1942. Ben Davis x Jonathan; selected in 1935; tested as A-827. Fruit: large; roundish; skin dark red over yellow, obscure darker red stripes, attractive; good flavor and keeping qualities; ripens with Jonathan; resembles Jonathan. Tree: thrifty grower on mediocre soil; somewhat tolerant to fire blight, not as susceptible to scab as Jonathan; produces large crops.

Wrixparent.—Orig. in Magnolia, Kent County, Del., by Wrixhem McIlvaine. Introd. in 1940. Plant patent 388; 1940; assigned to Bountiful Ridge Nurseries, Princess Anne, Md. Considered to be a bud mutation of Yellow Transparent; discovered in 1920. Fruit: larger than parent, averaging 2 1/2 inches in diam.; oblate, angular; calyx closed, persistent; skin thick, tough, smooth, waxy, light green; dots large, numerous; flesh white with greenish tinge, juicy, crisp, texture coarse, subacid, no aroma, quality good; ripens early, 10 days before Yellow Transparent; cooking quality good; fresh eating quality fair; resembles Yellow Transparent; tetraploid, a periclinal cytochimera 2-4-4-4. Tree: more spreading than parent. Chimeral nature determined at the New York State Agr. Expt. Sta., Geneva.

Yael.—Orig. in Rehovot, Israel, by Chanan Oppenheimer, Agr. Res. Sta., Div. Subtrop. Hort., National and Univ. Inst. of Agr. Introd. there in 1953. Calville St. Sauveur x unnamed local seedling; selected in 1951. Fruit: large; round; skin green; flavor fair; ripens July. Tree: medium; very prolific.

Yakhontowoye.—Orig. in Russia; imported into the U.S. by N.E. Hansen, South Dakota Agr. Expt. Sta., Brookings. Introd. here in 1934. Fruit: 2 inches in diam.; ripens in fall; resembles Almata. Tree: bark red; hardy; vigorous.

Yakima Newtown.—Orig. in Yakima, Wash., by Curtis C. Aller. Introd. in 1949. Plant patent 819; 11 Jan. 1949; assigned to May Nursery Co., Yakima, Wash. Bud mutation of Yellow Newtown. Fruit: more elongate, brighter green, more pronounced cavity and longer stem, more resistant to bitter pit and scald than parent. Tree: similar to parent but with long fruit spurs.

Yataka.—Orig. near Masuda-town, Akita, Japan, by Tadao Hitaragi. An early-ripening bud mutation of Fuji discovered in 1982. Registered in Japan in 1987. U.S. plant patent 7001; 1989; assigned to ProTree Nurseries, Brentwood, Calif. Matures 2 to 3 weeks earlier than Fuji; otherwise same as Fuji. An unstable chimera in form first introduced.

Yeager Sweet.—Orig. in Fargo, N.D., by A.F. Yeager, North Dakota Agr. Expt. Sta. Introd. about 1934. Patten Greening o.p. Fruit: ripens early; flesh sweet; skin and flesh yellow. Tree: hardy; productive in the northern tier of states.

Yellospur Delicious.—Orig. in Oroville, Wash., by Oscar Thornton and Ronald Thompson. Introd. in 1962. Bud mutation of Golden Delicious; discovered in 1959. Fruit: resembles Golden Delicious, more susceptible to russet. Tree: spur-type growth; size about three-fourths of standard type.

Yellow Beauty.—Orig. in Saskatoon, Saskatchewan, Canada, by C.F. Patterson, Univ. of Sask. Introd. in 1960 for home gardens. Columbia x Wealthy; selected in 1943; tested as A58-5. Fruit: up to 2 inches long and 1 3/4 inches in diam. under nonirrigated field conditions in Saskatoon; skin yellow; flesh mildly subacid, pleasing, quality good; ripens in midseason to late. Tree: yields good crops; hardy.

Yellow Hackworth (*Golden Hackworth*).—Orig. in or near Morgan County, Ala.; originator unknown. Introd. about 1938. Probably Hackworth selfed; discovered shortly after 1900. Fruit: flavor slightly milder than Hackworth; skin yellow with no red; otherwise identical to Hackworth. Tree: adapted (on its own roots) to poorer soils and hotter climates of the South than most other midsummer varieties.

Yellow Newtown.—See Newtown.

Yellow Rose.—See May Sweet.

Yellow Siberian.—See rootstocks for apple.

Yellow Sweet.—Imported from Russia by N.E. Hansen, South Dakota Agr. Expt. Sta., Brookings. Introd. in 1924 in the U.S.; may be an old Russian variety renamed. Fruit: flesh yellow, sweet, juicy, quality good, cooks well; ripens in early August in Brookings; averages 2 inches in diam.; resembles Yellow Transparent. Tree: hardy; very susceptible to fire blight.

Yoko.—Orig. in Gunma Prefecture, Japan. Introd. in 1981, Japanese Reg. no. 129. Golden Delicious o.p. Fruit: large, 300 g; shape round oval; color solid red with slight stripe, tendency to russet; flesh firm, white, sweet, slightly coarse; harvest late, 5 days after Mutsu; storage life excellent.

York-A-Red (*Double Red York Imperial*).—Orig. in Hedgesville, W.Va., by Paul L. Lingamfelter. Introd. in 1937. Plant patent 258; 20 July 1937; assigned to Stark Bros. Nurseries & Orchards Co., Louisiana, Mo. Bud mutation of York Imperial; discovered in 1931. Fruit: identical with York Imperial, except for all-over red skin; color appearing early in the season; fruit more uniform than parent. Tree: identical to York Imperial.

York Imperial Sport.—See Imperial Lady Clarendon.

Yorking (*Double Red York, Red York Imperial, Red Yorking, Yorking York*).—Orig. in Shippensburg, Pa., by the Allison Estate. Introd. in 1932. Plant patent 125; 28 May 1935. Bud mutation of York Imperial; discovered about 1925. Fruit: skin all red instead of partly red as on parent; otherwise, characteristics similar to parent. Tree: typical of York Imperial.

Yorking York.—See Yorking.

Young-Bearing Jonathan.—Orig. in Vera, Mo., by Lloyd C. Stark. Introd. in 1932 by Stark Bros. Nurseries & Orchards Co., Louisiana, Mo. Bud mutation of Jonathan; discovered in 1924. Fruit: resembles Jonathan in every respect. Tree: resembles Jonathan except that trees come into bearing at an earlier age.

YP.—See rootstocks for apple.

Zeleba (*Krimskaja Zeleba*).—Orig. in Crimea, Russia; imported into U.S. by N.E. Hansen, South Dakota Agr. Expt. Sta., Brookings, South Dakota. Introd. here in 1922. Fruit: large, up to 3 1/2 inches in diam.; round; skin red; flesh white, subacid; makes fine sauce; season December. Now obsolete.

not named.—Orig. in Bone Gap, Ill., by Lura B. Ellis. Not commercially introd. Plant patent 2608; 15 Mar. 1966; assigned to Stark Bros. Nurseries & Orchards Co., Louisiana, Mo. Parentage unknown. Fruit: 2 1/2 x 2 3/4 inches in diam.; globose, uniform; skin highly colored red, many small white dots, thin, tough, smooth, waxy; flesh white, tinged yellow near center, subacid, firm, quality good; resembles Jonalicious and ripens 2 weeks earlier. Tree: medium-large; vigorous; productive and regular bearer; flowers medium-large, white. Included here because variety has been patented.

not named.—Orig. in Los Angeles, Calif., by Marino La Pietra. Plant patent 801; 6 July 1948. Parentage unknown; discovered in 1935. Fruit: large; skin dull red overall; flesh crisp, juicy; seedless; ripens late November. Tree: bears well in the coastal regions of southern Calif. Included here because of patent.

not named.—Orig. in Cobden, Ill., by Corliss R. Maxam. Plant patent 3035; 2 Mar. 1971. Discovered in a planting of Starking Delicious. Fruit: deeper red color than Starking Delicious and uniform in color; matures all at same time; stems red. Included here because of patent.

not named.—Orig. in Brewster, Wash., by Calvin L. Cooper, Pateros, Wash. Introd. in 1968. Plant patent 2790; 23 Jan. 1968. Bud mutation of Tydeman's Early Worcester. Fruit: identical to parent and ripens same time. Tree: less vigorous than parent and bears earlier; branches do not have the undeveloped buds characteristic of parent tree, less distance between spurs and tendency to set fruit better. Variety is cited here because of patent.

Alnarp 2 (*A.2*).—Orig. in Alnarp, Sweden, by Carl G. Dahl and Emil Johansson, Alnarp Fruit Tree Sta. Introd. there in 1944; imported into U.S. in 1949 under P.I. 184382. Selected in 1920 from a group of mixed Doucin trees. Used only as a clonal rootstock: vigorous; very hardy; strongly anchored; propagated easily in stool beds; useful under a dwarfing interstem in regions with challenging winters.

Bemali.—Orig. at Balsgard, Sweden, by Viktor Trajkowski, Swedish Fruit Breeding Sta. Introd. in 1974. Manks Codlin x M.4; cross made in 1948; tested as 10342, C204, and BM342. Slightly less dwarfing, less precocious, and less productive than M.26. Anchorage better than M.9, not as good as M.26. Moderately resistant to fire blight. In the stoolbed, each mother plant generally produces four or five rooted shoots; few spines; few root primordia on liners. Moderate number of burr knots. More winter hardy than M.9.

Budagovsky 57-490 (*Bud.490*).—Orig. at Michurinsk, former Soviet Union, by V.I. Budagovsky, Michurinsk College of Agriculture. Red-Leaved Paradise (B9) x Bud. 13–14. A semi-vigorous clonal rootstock with winter hardiness similar to that of Common Antonovka. More precocious than MM.111, less so than MM.106. Some burr knots develop, usually less than on MM.111. Easily propagated by hardwood cuttings or in the stoolbed. Foliage bronze; wood, but not inner bark, red. No suckers. Susceptible to scab and powdery mildew. On an unbudded tree, most terminals carry fruit buds. Facultative parthenocarpic.

Budagovsky 57-491 (*Bud.491*).—Orig. at Michurinsk, the former Soviet Union, by V.I. Budagovsky, Michurinsk College of Agriculture. A very dwarfing clonal rootstock with greater winter hardiness than Common Antonovka. Wood is very brittle; trees require firm, permanent support. Gives tree size control similar to M.27. Induces very early, very heavy fruiting. Wood is pink. Few burr knots. Propagated in stoolbeds. No suckers. Very susceptible to fire blight, woolly aphids; not resistant to *Phytophthora.*

C6.—Orig. at Louisiana, Mo., by Harry Guengerich, Stark Bros. Nurseries. Introd. in 1974. Selected from population of open-pollinated seedlings of Clark Dwarf (M.8) for use as dwarfing interstem tolerant of common latent viruses. Plant patent 3057; 10 Aug. 1971; assigned to Stark Bros. Nurseries. Tree size control similar to that of M.9 when used as interstock or as rootstock. Less brittle than M.9; similar in susceptibility to fire blight and woolly aphids. Tolerant of common latent viruses. Induces early, heavy production. Very few suckers and burr knots. Compatible with wide range of cultivars.

Crab C.—Orig. in Maidstone, Kent, England, by the East Malling Res. Sta. Date of introd. unknown. From Todman, Borough Green, Kent, in batch of seedling stocks called "True French Crab," selected in 1914. Used only as a clonal rootstock: crab apple; summer shoots very vigorous; moderately susceptible to woolly aphids; dormant wood deep reddish on sun-exposed portions. Leaf: broad, obovate, dull, halves upfolded; teeth broad, irregular.

Geneva 11 (*G.11*).—A disease-resistant rootstock introd. in 1993 by J.N. Cummins and H.S. Aldwinckle, New York State Agr. Expt. Sta., Geneva. Malling 26 x Robusta 5; cross made in 1978; tested as 7826R5-011 and as CG.11. Plant patent applied for by Cornell Research Foundation; being propagated by licensed nurseries. Tolerant to *Phytophthora cactorum*; moderately resistant to fire blight, including virulent strain Ea 266; moderately susceptible to woolly apple aphids and powdery mildew. Trees on G.11 are well-anchored, precocious, and productive; few burr knots and suckers; moderate number of short spines on liners. Readily propagated in layerbeds and by cuttings. Spring budbreak late, like M.26.

Geneva 65 (*G.65*).—A very dwarfing, disease-resistant rootstock introd. in 1991 by J.N. Cummins and H.S. Aldwinckle, New York State Agr. Expt. Sta., Cornell Univ., Geneva. M.27 x Beauty Crab; cross made in 1974; selected in 1984. Tested as CG.65. Plant patent 8543; 18 Jan. 1994; assigned to Cornell Res. Foundation. More dwarfing than Malling 9 EMLA but less dwarfing than M.27. Resistant to crown rot, fire blight, apple scab, and powdery mildew; moderately susceptible to woolly apple aphids. Compatible with all varieties tested: Delicious, Golden Delicious, McIntosh, Northern Spy, Mutsu, Liberty, and Empire. Good anchorage; not brittle; training support required. Burr knots few. Moderately light suckering.

Jork 9 (*J9*).—Orig. at Jork, Germany, by K.-H. Tiemann and H.J. Darnmann, Fruit Tree Res. Sta. Introd. as a clonal rootstock in 1981. Selected from population of open-pollinated seedlings of M.9. Compared to M.26, J9 is slightly more dwarfing, induces earlier and heavier production, and is slightly more brittle. Suckers are rare. Extremely susceptible to fire blight and to woolly aphids; susceptible to scab and powdery mildew. Somewhat more winter hardy than M.9, less hardy than M.11. Root primordia are conspicuous on 2-year-old wood; many burr knots develop in the orchard; radial growth of burr knots is very rapid and girdling may occur. In the stoolbed, J9 is easily propagated, very productive of clean-shanked liners.

M.1 (*Malling I, Broad-leaved English Paradise*).—Orig. in Maidstone, Kent, England, by the East Malling Res. Sta. Date of introduction not known. Parentage unknown; selected in 1912; type plant from Barnham Nurseries, England. Used only as a rootstock: semi-dwarfing; useful on poor soils and with weaker varieties; incompatible with Wealthy and some other varieties; growth vigorous. Winter hardy but intolerant of dry soil conditions. Stem: summer shoot zig-zagged, greenish, slightly woolly; autumn wood purple-brown; lenticels numerous, small, inconspicuous. Leaf: large, broad, dark green, crinkled surface; stands almost horizontal and well away from stem; curved so apex points downward; stipules large. Roots easily from stool layering.

M.2 (*Malling II, Doucin, English Doucin, English Paradise*).—Orig. in Maidstone, Kent, England, by the East Malling Res. Sta. Date of introduction not known. Parentage unknown; selected in 1912; type plant from Bath's; common in England, France, Germany. Used only as a clonal rootstock: semi-dwarfing (75%); sometimes requires support in early years; does not root readily from stool layering. Summer shoot: exceptionally rigid, strong ones being spined, lenticels very conspicuous and large. Leaf: dark green, uniformly shaped, petiole flat; set close together on stem; stipules small; early defoliation.

M.3 (*Malling III, Dutch Doucin, Hollyleaf, Hollyleaf Paradise*).—Orig. in Maidstone, Kent, England, by the East Malling Res. Sta. Date of introduction not known. Parentage unknown; selected in 1912; type plant from Clibran's. Used only as a clonal rootstock: semi-dwarfing; very free rooting, often forming knots on older trunks near surface of the ground; suckering sometimes objectionable. Summer shoot: strong, but of whippy nature with drooping lateral branches; autumn wood dark purple-brown, with few inconspicuous lenticels, some short gray hair. Leaf: smaller than usual; long and gradually pointed; margin ragged and deeply serrate; tip curved; thin and inflexible, recalling character of holly leaf.

M.4 (*Malling IV, Dutch Doucin, Holstein Doucin*).—Orig. in Maidstone, Kent, England, by the East Malling Res. Sta. Date of introduction not known. Parentage unknown; selected in 1912; type plant from Little and Ballantyne. Used only as a clonal rootstock: only slightly dwarfing (75%); early cropping tendency; root system reported poor in the U.S.; tree tends to blow over in some districts; readily propagated. Resistant to collar rot. Summer shoot: short, light yellow-brown, with fine cracks like aged varnish, prominent bumps near buds; in autumn, light yellow wood distinctive; growth starts several weeks in advance of most other Malling stocks, except M8. Leaf: light green, surface rough; fleshy; edges wavy, slightly upturned; almost circular in outline.

M.5 (*Malling V, Doucin Amélioré, Improved Doucin, Red Paradise*).—Orig. in Maidstone, Kent, England, by the East Malling Res. Sta. Date of introduction not known. Parentage unknown; selected in 1912; type plant from W.R. Pierce, Canterbury, Kent. Used only as a clonal rootstock: semi-dwarfing; easily propagated from mound layers; some grafted trees tend to lean. Summer shoot: lenticels very few, large, scattered; buds and petioles reddish; in autumn, of purplish cast, with bright glabrous buds. Leaf: small; apex sharp and prominent; stipules small.

M.6 (*Malling VI, Nonsuch Paradise, Rivers' Nonsuch Paradise*).—Orig. in Maidstone, Kent, England, at the East Malling Res. Sta. Date of introduction not known. Parentage unknown; selected in 1912; type plant from Barnham Nurseries, England. Used only as a clonal rootstock: vigorous; roots and propagates readily from mound layering; prone to suckering in orchard. Summer shoot: yellow-green, with numerous but inconspicuous lenticels; of mealy appearance from pubescence. Leaf: light green, large, broad, rounded; petiole furrow deep; stipules very large; leaf fall in autumn very late.

M.7 (*Malling VII*).—Orig. in Maidstone, Kent, England, by the East Malling Res. Sta. Date of introduction not known. Parentage unknown; selected in 1912; one collection from Clibran's amalgamated with part of one other. Used only as a clonal rootstock: semi-dwarfing (55% to 65%); cropping early; roots and propagates readily from mound layering; grows well over wide range of soil temperatures. Tends to form root suckers. Summer shoot: slender; flexible; internodes long; light reddish purple, lenticels conspicuous. Dormant shoots: bud small, blackish brown, distinctive. Leaf: thin, papery; circular in outline; lower leaves lobed; petiole upright.

M.8 (*Malling VIII, Clark, Clark Dwarf, French Paradise, Red Paradise*).—Orig. in Maidstone, Kent, England, by the East Malling Res. Sta. Date of introduction not known. Parentage unknown; selected in 1912; type plant from Seabrook, Chelmsford, Essex, England. Used only as a rootstock: indistinguishable from Clark Dwarf; promising as a very dwarfing interstock (stem) in the U.S., in areas where low winter temperatures are a limiting factor. Summer shoot: vigor medium; whiplike; bright reddish, later becoming dark purple; glabrous and shiny, except for the young tips; lenticels numerous. Leaf: small, narrow, often concave, finely serrated; petiole very short. Dormant wood; recognized by narrow, glabrous, reddish buds.

M.9 (*Malling IX, Dieudonne, Jaune de Metz, Yellow Metz, Yellow Paradise of Metz*).—Orig. in Maidstone, Kent, England, by the East Malling Res. Sta. Date of introduction not known. Parentage unknown. Selected in 1914; type plant from Barbier, Orleans; common in France and Germany. Used only as a clonal rootstock: dwarfing, with restricted root system; some support must be provided for orchard trees; incompatible with some ornamental crab apple varieties; promotes earlier growth and fruit maturity, as well as precocious cropping; roots and propagates readily from stool layering. It is extensively used throughout the world. Summer shoot: very straight; reddish-yellow with distinct silvery glaze. Leaf: large; oblong; with rounded and scalloped teeth; shiny upper surface; upright.

M.9-T337 (*Malling 9-T337, NAKB 337*).—A Dutch selection of M.9. In the orchard, tree vigor is intermediate in the M.9 range. Leaf medium large; shoots fairly stocky; few side shoots in the stoolbed.

M.10 (*Malling X, Doucin U.1*).—Orig. in Maidstone, Kent, England, by the East Malling Res. Sta. Date of introduction not known. Parentage unknown. Selected in 1914; type plant from Späth, Berlin. Used only as a clonal rootstock. It is not in commercial production. Summer shoot vigorous, internodes 1.5 to 2.0 cm, tip gray-black, bark dark maroon, tip only slightly hairy; lenticels small, gray to yellowish; leaf ovate, 3.8 x 6.0 cm, dark green, the basal halves with offset insertion on petiole, blade infolded and twisted, margin with sharp serrate teeth in irregular order; petiole 1.6 cm long, round in cross-section, exposed portion dark red and hairy; stipules large, 1.4 x 0.3 cm. Winter shoot only slightly hairy, many lenticels, buds very lightly pubescent.

M.11 (*Malling XI, Black Doucin, Green Doucin, Grüner Doucin, Prachts Doucin, Wesseling's Paradise, Winter Sunshine*).—Orig. in Maidstone, Kent, England, by the East Malling Res. Sta. Date of introduction not known. Parentage unknown; selected in 1912; type plant from

Todman, Borough Green, Kent. Used only as a clonal rootstock. It is not in commercial production. Summer shoot very long, internodes 1.0 to 1.5 cm, rigidly upright, straight, with very pubescent tip, bark dark brown; lenticels numerous, grayish-yellow, small, round; leaf broadly ovate, green with reddish nerves, 4.2 x 6.3 cm, blade flat with constricted apical point, margin with broad teeth in two series; petiole short, 1.2 cm, red, slightly hairy, stipules about 1 cm long, slender. Winter shoot with many, small yellow-brown lenticels, buds reddish brown, only slightly pubescent.

M.12 (*Malling XII*).—Orig. in Maidstone, Kent, England, by the East Malling Res. Sta. Date of introduction not known. Parentage unknown; selected in 1912; type plant from Veitch, England. Used only as a clonal rootstock: produces a standard tree without dwarfing effects; very vigorous; difficulty in nursery propagation has been reported, owing to poor rooting. Summer shoot: stiff, erect, chocolate colored. Leaf: concave; surface broadly undulated, very shiny; petiole kinked at base.

M.13 (*Malling XIII, Black Doucin, Doucin U.2, Swarzer Doucin*).—Orig. in Maidstone, Kent, England, by the East Malling Res. Sta. Date of introduction not known. Parentage unknown. Selected in 1914; type plant from Späth, Berlin. Used only as a clonal rootstock: produces a semi-standard tree, with extensive and well-anchored root system; tolerant to wet soils. It is sometimes used as a rootstock in a 3-piece tree. Summer shoot: ragged. Dormant wood: black-brown; lenticels prominent, white. Leaf: ovate; distinct because of irregular lobed and serrated margins; erect.

M.14 (*Malling XIV, Doucin U.3*).—Orig. in Maidstone, Kent, England, by the East Malling Res. Sta. Date of introduction not known. Parentage unknown; selected in 1914; type plant from Späth, Berlin. Used only as a clonal rootstock. It is not in commercial production. Summer shoot medium long, internodes 2.0 cm, slender, upright, seldom and weakly side branching; bark very dark reddish-brown; tip very downy; lenticels small, round, yellow-gray, in moderate number; leaf short ovate, dark green with reddish nerves, 4.7 x 6.5 cm, heavily pubescent, margin irregularly lobed with occasional prominent obtuse teeth; stipules oblanceolate. Winter shoot uniformly dark brown; lenticels small, yellow-gray; buds with very slight pubescence, but very broad and thick; leaf scar small, flattened.

M.15 (*Malling XV, Doucin U. 2*).—Orig. in Maidstone, Kent, England, by the East Malling Res. Sta. Date of introduction not known. Parentage unknown; selected in 1914; type plant from Späth, Berlin. Used only as a clonal rootstock: very vigorous; summer shoot long, internodes 1.5 to 2.0 cm, rigid, upright; bark dark reddish-brown, pubescent even at some distance from apex; lateral branches frequent; lenticels quite visible, medium, abundant, light gray; leaf broadly ovate, 5.2 x 7.5 cm, dark green, but the nerves conspicuously lighter, many red dots on upper side, with the base circular in outline, apex prominent and acute, teeth in two series, flat, wide, scalloped; petiole 1.5 cm long, dark red, very hairy; stipules 1.4 cm long, prominent. Winter shoot uniformly brownish-black, lenticels light gray, medium; buds rounded, very pubescent.

M.16 (*Malling XVI, Doucin U. 3, Ketziner Ideal*).—Orig. in Maidstone, Kent, England, by the East Malling Res. Sta. Date of introduction not known. Parentage unknown; selected in 1914; type plant from Späth, Berlin. Used only as a clonal rootstock: very vigorous; root system sturdy, well anchored; dwarfing tendency little or none; tolerant to wet soil. Summer shoot: very erect; growing tip light green in contrast to dull, dark, olive green of mature tissue; dormant shoot blackish-green. Leaf: teeth poorly marked on lower leaves.

M.17 (*Malling XVII*).—Orig. in Maidstone, Kent, England, at the East Malling Res. Sta., from Prof. Sprenger of Wageningen, The Netherlands. Date of introduction not known. Parentage unknown; selected in 1924. Used only as a clonal rootstock: semi-dwarfing; summer shoot of medium length, internodes 1.0 to 1.5 cm, growing somewhat curved as Vs from stool, reddish-brown with lighter colored basal bark; apex only pubescent; very few side branches and these only on very vigorous shoots; leaf green, elliptical, 3.4 x 6.0 cm, slightly shiny, nerves much lighter in color, most of margin entire or shallowly scalloped, teeth very small, poorly defined; apex long, narrow and prominent on lower leaves, but hardly apparent on upper part of shoot; petiole thin, medium hairy, thickened, and dark red at base.

M.18 (*Malling XVIII*).—Orig. in Maidstone, Kent, England, at the East Malling Res. Sta., from Prof. Sprenger of Wageningen, The Netherlands. Date of introduction unknown. Parentage not known; selected in 1924. Used only as a clonal rootstock: summer shoot thick, sturdy, upright, internodes 1.0 to 1.5 cm, with frequent lateral branches, dark reddish-brown to greenish-brown below, pubescence very heavy, obscuring bark color; lenticels strikingly few, very small, round, gray; leaf green, short ovate to cordiform, infolded 3 x 5.5 cm, with reddish nerves and numerous dark red dots, medium covering of short hairs on under side, with short and broad apex, teeth small, finely pointed, curved; petiole 1.3 cm long, red on underside, pubescence medium. Winter shoot with heavy down, lenticels very few or not apparent.

M.25 (*Malling XXV, M. XXV, Im 238*).—Orig. in Maidstone, Kent, England, by the East Malling Res. Sta. Introd. in 1952. Northern Spy x M.2; selected in 1930. Used only as a clonal rootstock: shoot erect; stocky, internodes short; growth very vigorous; buds large, reddish. Leaf: small flat to convex, often crinkled, with apex turned downward; budding out early; fall coloration dull coppery to yellow, tips green. Bred in woolly aphid resistant series of Malling-Merton stocks but not highly resistant so omitted from MM series and added to Malling stocks.

M.26 (*Malling 26*).—Orig. in Maidstone, Kent, England, by H.M. Tydeman, East Malling Res. Sta. Introd. in 1959. M.16 x M.9; selected in 1929; one of a series known as the "Nine Crosses"; tested as Malling 3436. Used only as a clonal rootstock: semi-dwarfing; intermediate in vigor between the dwarfing M.9 and the semi-dwarfing M.7; better anchored than M.9 but not as good as M.7; very early bearing; resistant to *Phytophthora cactorum* Leb. & Cohn.

M.27 (*Malling 27*).—Orig. in East Malling, England, by H.M. Tydeman. Plant patent 3793; 7 Oct. 1975; assigned to National Seed Development Organization; M.13 x M.9; cross made in 1929; selected in 1934; tested as Malling 3431. Used only as a clonal rootstock: dwarfing; grafted varieties are significantly smaller than the same varieties on M.9, but produce a higher proportion of fruit relative to wood than the same varieties on M.9 rootstocks; the bark of M.27 is thicker than that of M.9; M.27 has fewer palisade cells per unit area of leaf in July than equivalent leaves of M.2, M.4, M.12 and M.26. M.27 has fewer stomata per unit area of the lower surface of the leaf than equivalent leaves of the rootstocks M.25, MM.106 and MM.111. Scion varieties come into fruit production more quickly on Malling 27 than on M.7 or M.16.

MM.111 (*Malling-Merton 111, Im 524*).—Orig. in Merton, England, by the John Innes Hort. Inst. and the East Malling Res. Sta., Maidstone, Kent, England. Introd. in 1952; introd. into the U.S. in 1954 as P.I. 212252. Northern Spy x Merton 793; selected in 1932. Used only as a clonal rootstock. It is a semistandard (90%) and used on light soils. Summer shoot erect, stout, with swollen nodes; buds ragged in outline, ashen gray; leaf blade up-cupped with raised and wavy margins, coarse serrations tipped with red; petiole base lumpy; fall coloration dull green to mottled yellow; resistant to woolly aphid; easily propagated; not precocious, but cropping is improved.

MM.102 (*Malling-Merton 102, Im 328*).—Orig. in Maidstone, Kent, England, by the East Malling Res. Sta. Date of introd. unknown; introd. into the U.S. in 1954 as P.I. 212248. Northern Spy x M.1; selected in 1931. Used only as a clonal rootstock: summer shoots have a zig-zagged pattern; bud bases sharp, buds slightly ragged; leaf with down-turned and twisted tip, serrations reflexed, tips reddish brown; leafing out late; autumn coloration dull green to bronzy yellow; resistant to woolly aphid.

MM.103 (*Malling-Merton 103, Im 665*).—Orig. in Maidstone, Kent, England, by the East Malling Res. Sta. Date of introd. unknown. Northern Spy x Ben Davis; selected in 1933. Used only as a clonal rootstock:

summer shoot slender, flexible, with prominent ridges; buds ashen gray; leaf flat or only convex toward the base, almost rectangular, petiole very long; autumn coloration deep wine-red; resistant to woolly aphid.

MM.104 (*Malling-Merton 104, Im 425*).— Orig. in Maidstone, Kent, England, by the East Malling Res. Sta. Introd. in 1952; introd. into the U.S. in 1954 as P.I.212249. M.2 x Northern Spy; selected in 1930 to 1931. Used only as a clonal rootstock. It is a semidwarf (80% to 90% of seedling). Summer shoot with bright red buds; lenticels numerous, conspicuous; leaf upfolded, markedly concave; petiole wine-red, outturned; autumn coloration pale green to dull yellow; excellent propagation from stool layering; resistant to woolly aphid; but very susceptible to collar rot. Use of this rootstock was discontinued because of collar rot.

MM.105 (*Malling-Merton 105, Im 151*).— Orig, in Merton, England, by the John Innes Hort. Inst. and the East Malling Res. Sta., Maidstone, Kent, England. Date of introd. unknown. M.2 x Northern Spy; selected in 1930 to 1931. Used only as a clonal rootstock: summer shoot with metallic sheen; bud red; bark with slightly raised lenticels; leaf blade flat, with long acute tips, base of leaf wedge shaped with unequal sides; leafing out and defoliation early; fall coloration bright coppery red; excellent propagation from stool layering; resistant to woolly aphid.

MM.106 (*Malling-Merton 106, Im 332*).— Orig. in Maidstone, Kent, England, by the East Malling Res. Sta. Introd. in 1952; introd. to the U.S. in 1954 as P.I. 212250. Northern Spy x M.1; selected during 1930 to 1932. Used only as a clonal rootstock: summer shoot pubescent, with swollen nodes; lenticels indistinct; buds ashen gray; leaf blade flat, horizontal, upper surface very glossy; petioles out-turned; fall coloration grayish-green to dull yellow; resistant to woolly aphid; easily propagated; very precocious and productive. It grows late in the season, making it more susceptible to winter injury.

MM.109 (*Malling-Merton 109, Im 74*)— Orig. in Merton, England, by the John Innes Hort. Inst. and the East Malling Res. Sta., Maidstone, Kent, England. Introd. in 1952; introd. into the U.S. in 1954 as P.I. 212251. M.2 x Northern Spy; selected in 1930 to 1931. Used only as a clonal rootstock. It is as vigorous as a seedling rootstock. Summer shoot with reddish buds; lenticels numerous, slightly raised; leaf blade concave, up-turned, with undulated surface, margin sharply serrate; stipules very small; leafing out and defoliation early; in autumn shoot tips green, leaves dull yellow; easily propagated from stool layering; resistant to woolly aphid; improved cropping. The use of this rootstock was discontinued because of excess vigor and poor anchorage.

MM.110 (*Malling-Merton 110, Im 319*).— Orig. in Maidstone, Kent, England, by the East Malling Res. Sta. Date introd. unknown. Northern Spy x M.1; selected in 1930 to 1932. Used only as a clonal rootstock: summer shoot stiff, spreading, ridged, internode long and straight, lenticels very numerous; leaf margins convex, serrations fine, acute, pointing downward; stipules large, serrate, widely flanged; resistant to woolly aphid; easily propagated.

MM.112 (*Malling-Merton 112, Im 605*).— Orig., in Maidsone, Kent England, by the East Malling Res. Sta. Date of introd. unknown. Northern Spy x Winter Majetin; selected in 1933. Used only as a clonal rootstock: summer shoot slightly zig-zagged, lightly pubescent; leaf blade flat or slightly convex, crinkled, tips irregularly reflexed; petioles short, stipules large and serrate; defoliation very late; fall coloration deep blue-green; resistant to woolly aphid; easily propagated.

MM.113 (*Malling-Merton 113, Im 304*).— Orig. in Maidstone, Kent, England, by the East Malling Res. Sta. Date introd. unknown. Northern Spy x M.12; selected in 1930 to 1931. Used only as a clonal rootstock: summer shoot slender, spreading, with swollen nodes, lenticels sunken, brownish-orange; leaves large, uncupped, serrations fine, rounded; petiole short, upturned; leafing out and defoliation late; fall coloration bluish-green with paler tips; propagates very poorly from stool layering; resistant to woolly aphid.

MM.114 (*Malling-Merton 114, lm 306*).— Orig. in Maidstone, Kent, England, by the East Malling Res. Sta. Date introd. unknown. Northern Spy x M.12; selected in 1930 to 1931. Used only as a clonal rootstock: summer shoot thick, internodes short, ridged; leaf turned downward and curved over; petiole kinked near base; fall coloration dull grayish-green; propagates very poorly by stool layering; resistant to woolly aphid.

MM.115 (*Malling-Merton 115, lm 671*).— Orig,. in Maidstone, Kent, England, by the East Malling Res. Sta. Date introd. unknown. Northern Spy x Ben Davis; selected in 1930 to 1931. Used only as a clonal rootstock: summer shoot stout, spreading, internodes long, swollen nodes prominent; buds ashen gray, ragged; leaf large, almost oblong, with reflexed tip and up-turned and undulated borders; fall coloration deep red; resistant to woolly aphid; easily propagated.

Manchu.—Orig. in Brookings, S.D., by N.E. Hansen, South Dakota Agr. Expt. Sta. Introd. about 1960. Grown from seed gathered by Hansen in 1924 in a mountain region about 50 miles east of Harbin, Manchuria; *Malus baccata* variety *mandschurica* Schneid., Manchurian crab apple. Seedlings of this clone are used for rootstock purposes, as they are extremely hardy and produce a semi-dwarf tree. Both clone and open-pollinated seedlings from this clone are known as Manchu.

Mark (*MAC-9*).—Orig. in East Lansing, Mich., by R.F. Carlson. Introd. in 1979. Plant patent 4678; 5 Nov. 1979; assigned to Michigan State Univ., sublicensed to TRECO®. M.9 o.p.; selected in 1959. Compared to M.26, Mark is more dwarfing, more precocious, more productive, better anchored, and somewhat less sensitive to fire blight. Mark has demonstrated more winter hardiness than any of the English rootstocks. Tolerant of *Phytophthora;* susceptible to fire blight and woolly apple aphids. Tolerant of heavy soils; irrigation required on light sands. Mark appears to be tolerant of common latent viruses. In the nursery, staking of budlings is recommended to prevent blowouts; scions defoliate early, permitting early digging. Moderate numbers of burr knots initially, but rapid radial extension may result in girdling. Soil line proliferation leads to tree decline. Readily propagable in stoolbeds and by micropropagation.

Maruba-kaido N-1.—A clonal selection of the weeping form of Maruba-kaido (*Malus prunifolia ringo*), the standard vigorous rootstock of Japan. Semi-vigorous, hardy, resistant to woolly aphids and *Phytophthora.* Greatest potential value appears to be as rootstock to support interstem system. Tolerant of drought and water-logging. Sensitive to common latent viruses.

Nertschinsk.—Orig. in Morden, Manitoba, Canada, by C.R. Ure. *Malus baccata* o.p.; selected in 1961. Valuable as a dwarfing interstem that is tolerant of extremely low midwinter temperatures. Very susceptible to fire blight. Very early budbreak. Virus sensitivities unknown.

Novole.—Orig. in Geneva, N.Y., by J.N. Cummins, H.S. Aldwinckle, and R. Byers. Selected from a population of *Malus prunifolia* o.p.; putative pollen parent is *M. sieboldii;* tested as NY613. Plant patent 5333; 13 Nov. 1984. Introd. in 1983 as a clonal trunk and rootstock nonpreferred by pine and meadow voles; both vole species consistently reject Novole tissue in free-choice tests. Not tested with related vole species. Novole is resistant to *Phytophthora cactorum* and apple scab, and moderately resistant to powdery mildew. Very sensitive to apple stem grooving virus. Propagated easily in the stoolbed, by hardwood cuttings, and by micropropagation. Burr knots are absent. Scion cultivars grafted directly on Novole produced very large, productive trees. When a 20-cm virus-free dwarfing interstem is high-worked at about 40 to 50 cm above soil level, a very productive dwarfed tree results. As a scion variety, Novole produces many flowers over 7 to 9 days, beginning 2 days after king blossom opening of Delicious and just before Golden Delicious; pollen is abundant and flowers are attractive to bees.

Ottawa 3.—Orig. in Ottawa, Ontario, Canada, by Lloyd Spangelo, S.O. Fejer, S.J. Leuty, and R.L. Granger, Agr. Canada Res. Sta. Introd. in 1974 as a hardy dwarfing rootstock. Not patented. M.9 x Robin Crab. Used either as a rootstock or interstock, slightly less dwarfing than M.9, more dwarfing than M.26. Nonbrittle. Liners are sturdy, completely free of spines. Unusually large leaves. Micropropagation is efficient; very poor rooting in stoolbed; can be propagated by root cuttings. Induces very early, very efficient fruit production. Produces no burr knots, no suckers. More winter hardy than M.26. Very resistant to *Phytophthora* spp.; susceptible to fire blight and to woolly aphids. Sensitive to apple stem grooving virus. A highly useful parent in rootstock breeding; transmits *Phytophthora* resistance, dwarfing capacity, freedom from spines, and distinctive large leaves; each five-carpelled fruit produces 25 to 35 viable seeds.

P.1.—Orig. in Skierniewice, Poland, by S. Zagaja and A. Czynczyk, Res. Inst. of Pomology. M.4 x Common Antonovka. A clonal root stock slightly more dwarfing than M.7. Slightly less winter hardy than Common Antonovka. Many small burr knots are produced just below graft union; subsequent coalescence may result in girdling. No suckers. Good anchorage. Very easily propagated in stoolbed. Moderately susceptible to fire blight, *Phytophthora*, and woolly aphids.

P.2.—Orig. in Skierniewice, Poland, by S. Zagaja and A. Czynczyk, Res. Inst. of Pomology. M.9 x Common Antonovka. A clonal rootstock about as dwarfing as M.26. Slightly less winter hardy than Common Antonovka. Burr knots are rare. Induces very early, heavy fruiting. Stoolbed propagation is satisfactory. Late budbreak. Moderately susceptible to fire blight; susceptible to woolly aphids; resistant to *Phytophthora*. May be susceptible to tomato ringspot virus.

P.16 (*Lizzy*).—A very dwarfing rootstock introd. by S. Zagaja and A. Czynczyk, Res. Inst. for Pomology, Skierniewice, Poland. M.9 x Common Antonovka. Induces tree vigor intermediate between M.27 and M.9. Shoots stocky with numerous short spines. Leaves large, dark green, dull. Budbreak very early. Low-temperature tolerance similar to M.9. Induces very early, heavy production. Burr knots and suckers rare. Moderately easily propagated in stoolbed. Susceptible to fire blight and woolly apple aphids.

P.22.—Orig. in Skierniewice, Poland, by S. Zagaja and A. Czynczyk, Res. Inst. of Pomology. M.9 x Common Antonovka. A clonal rootstock about as dwarfing as M.27. Slightly less winter hardy than Common Antonovka. Induces very early heavy fruiting. Burr knots rare. Late budbreak. No suckers. Moderately susceptible to fire blight; very susceptible to woolly aphids; resistant to *Phytophthora*.

P.60.—A very dwarfing, red-leaved, vegetatively propagated apple rootstock, adapted to northern areas. Introd. in 1991 by S. Zagaja, T. Jakobowski, and A. Przbyla, Res. Inst. of Pomology, Poland. Alnarp 2 x Red-Leaved Paradise (Budagovsky 9); selected from a population of 23 seedlings. Not patented. Under Polish conditions, P.60 is similar to M.9 in dwarfing influence. Compared to M.26, P.60 is more prolific in the stoolbed, better rooted, and produces fewer spines; in the orchard, P.60 produces twice as many burr knots and is similar in winter hardiness. Leaves, fruit flesh, inner bark, and wood are pinkish-red, similar to the pollen parent. P.60 is susceptible to fire blight and woolly apple aphids and moderately susceptible to *Phytophthora*, apple scab, and powdery mildew. Wood is brittle; trees in the orchard require permanent support.

Red-Leaved Paradise (*Budagovsky 9, Bud.9, B9*).—Orig. in Michurinsk, the former Soviet Union, by V.I. Budagovsky, Michurinsk College of Agr. Introd. in 1946 as a very dwarfing rootstock; very useful dwarfing interstock. M.8 x Red Standard. Not patented. Slightly more dwarfing than M.9. Not as brittle as M.8. Red wood. Trees on Red-Leaved Paradise bear early, heavily. Anchorage fair. Susceptible to blight, woolly aphids; very resistant to *Phytophthora*. Tolerant of common latent viruses. Fair stoolbed subject. Suckers and burr knots are rare. Slightly more winter hardy than M.26.

Robusta 5.—Orig. in Ottawa, Ontario, Canada, by Canada Dept. Agr., Res. Sta. Introd. in 1947. Seedling raised from seed obtained in 1927 from Russia through courtesy of the Arnold Arboretum. Used only as a rootstock: under Ottawa conditions, it has remained fully hardy, showing a high resistance to fire blight and crown rot; compatible with all commercial varieties grown in Canada; propagated by softwood cuttings by techniques developed by Div. Hort.; classed as a semi-vigorous stock; scion trees smaller than on French crab. It is vigorous and produces trees comparable in size to seedling rootstocks.

Rotyp (*Robijn*).—Orig. in Zuid-Beveland, The Netherlands, by W.J. Robijn, Goes, The Netherlands. Patented in The Netherlands as CN IX RG, but patent relinquished 5 June 1968. Introd. in The Netherlands about 1962, in Canada in 1965, and North America in 1966 by Traas Nursery, Langley, B.C., Canada. Parentage unknown; discovered in 1936 as a seedling. Used for rootstock purposes. Very little interest in variety in Holland because it is too vigorous. Vigor similar to M.26, but tree union is weak as with M.26 and M.9. Distinguished by the olive-green- to bronze-colored wood; small number of lenticels; large, oval, very shiny and cup-shaped leaves.

Yellow Siberian.—Orig. in Brookings, South Dakota, by N.E. Hansen, South Dakota Agr. Expt. Sta. Introd. in 1938. *Malus baccata;* parentage unknown. Fruit: crab; skin yellow. Tree: bears alternately; productive; hardy; commonly used as a seed source for rootstocks, promoting upright, vigorous growth of scion varieties.

YP.—Orig. in Piikkio, Finland, by J. Sakö, Agr. Res. Center, Dept. of Horticulture. Introd. in 1977. *Malus baccata* o.p.; tested as Mb4. Clean rootstock; fairly easily propagated in stoolbed; easily raised by soft wood cuttings and root cuttings. Vigorous, similar to Alnarp 2. Induces early fruiting and very good level of productivity in scion variety. Compatible with broad spectrum of scion varieties. Much more winter hardy than either Alnarp 2 or M.26; induces early dormancy in scion. Leaf fall early; budbreak early.

Apricot

Alfred.—Orig. in Geneva, N.Y., by Robert C. Lamb, New York State Expt. Sta., Geneva. Introd. in 1965. Open-pollinated seedling of selection from (Doty x Geneva); selected in 1949; tested as N.Y. 345. Fruit: small to medium size, 1 3/8 inches in diam.; roundish; skin and flesh orange; freestone; quality good. Tree: moderately vigorous; hardy.

Alves.—Orig. in Le Grand, Calif., by F.W. Anderson, Merced, Calif. Introd. in 1968. Plant patent 2789; 16 Jan. 1968; assigned to Reedley Nursery, Reedley, Calif. F_2 Perfection x Blenheim; tested as Anderson 38P245. Fruit: size medium; skin yellow ground color, highly blushed with red; flesh yellow, firm, freestone; resembles Mono; ripens about 2 weeks before Royal, 1 week before Pomo. Tree: regular and productive bearer; flowers white.

Amber.—Orig. in Chilliwack, British Columbia, Canada, by the British Columbia Nurseries Co. Introd. in 1930. Parentage unknown. Fruit: larger than Moorpark; somewhat similar in shape to that variety; skin with attractive pink blush, tender, thin; flesh amber when cooked, quality good, aromatic. Tree: yields well.

Anda.—Orig. in Brookings, S.D, by N.E. Hansen, South Dakota Agr. Expt. Sta. Introd. in 1936. Seedling of *Prunus mandshurica* seed brought from northern Manchuria in 1924. Fruit: freestone; of good size; firm even when cooked; ripens late. Tree: large; productive; hardy.

Autumn Royal.—Orig. in Hollister, Calif., by Andrew W. Kennon. Plant patent 2906; 15 July 1969; assigned to L.E. Cooke Co., Visalia, Calif. Introd. in 1970. Limb mutation of Blenheim; discovered in 1967. Fruit: size medium; apical end flattened, cavity very large; skin yellow with very slight orange blush, pubescence scant; flesh yellow, subacid, freestone; ripens late, about 15 Sept. retaining same quality as parent. Tree: large, vigorous, spreading, vase-formed, dense in growth; productive and regular bearer.

Avimag.—See Cadaman.

Big Davis.—Orig. in Clarendon, Texas, by A.L. Bruce, Bruce Nurseries of Alanreed, Texas. Introd. in 1924. Parentage unknown; disc. in 1920. Fruit: large; quality good; resembles Moorpark. Tree: prolific; hardy.

Blenril.—Orig. in Prosser, Wash., by the Washington Agr. Expt. Sta. and the U.S. Dept. of Agriculture. Introd. in 1957. Blenheim x Riland; selected in 1950. Fruit: size medium; shape similar to Riland; halves equal; flesh yellow-orange, firm, texture fine; fresh fruit quality excellent; freestone; about 60% of skin blushed an attractive Riland-type red over a bright yellow-orange ground color; matures about with Riland. Tree: vigorous; spreading to slightly upright; productive; blooms with Wenatchee; self-fruitful; pollinates Earliril.

Border Queen.—Orig. in Caldwell, Kan., by A.M. Mallory. Introd. but year unknown. Parentage unknown; discovered in the 1910s. Fruit: resembles Moorpark.

Bowers.—Orig. in Yakima County, Wash., by William I. Bowers. Introd. in 1943. Plant patent 630; 27 June 1944. Parentage unknown; discovered in 1930. Fruit: up to 2 inches in diam.; nearly oval; skin smooth, yellow, highly blushed; suture usually shallow; flesh texture fine, yellow to orange, flavor good, sweet; pit large, free; ripens early, about with Riland; resembles Riland. Tree: productive; vigorous.

Brooks.—Orig. in Homedale, Idaho, by Bert Brooks and Glenn M. Brooks, Lafayette Nursery Co., Lafayette, Ore. Introd. in 1937. Plant patent 498; 30 Dec. 1941. Parentage unknown; discovered in 1935 when seedling tree was 6 years. old. Fruit: very large; sugar content high; freestone; skin and flesh reddish; cans and ships well; ripens evenly.

Bryan (*Dr. Bryan*).—Orig. in Dublin, Texas. Introd. in the mid-1930s. Chance seedling. Discovered about 1932. Fruit: size medium; global; flesh very sweet, freestone; ripens 30 May; recommended for central Texas.

Tree: size medium; regular and productive bearer; susceptible to bacterial spot.

Cadaman (*Avimag*).—Not compatible.

Castlebrite.—Orig. in Fresno, Calif., by J.H. Weinberger, Agr. Res. Serv., U.S. Dept. of Agriculture, Fresno. Introd. in 1971. Open-pollinated seedling of B60-12 (Perfection x Castleton); tested as K111-6. Fruit: medium, 1 3/4 inch in diam.and length, uniform; orange color with slight red blush; acid until fully mature; freestone; ripens end of May, with exceptional color and firmness. Tree: vigorous; productive under San Joaquin Valley conditions. Flower: large, white, self-fertile. Recommended as early shipping variety.

Castleton.—Orig. in Fresno, Calif., by John H. Weinberger, U.S. Hort. Field Sta., U.S. Dept. of Agriculture Introd. in 1963. Perfection x Newcastle; cross made in 1955; first fruited in 1960; tested as F 69-44. Fruit: size medium, averaging 8 to 10 per lb; slightly ovate, rounded at apex, flattened at stem end, halves nearly equal, averaging 2 inches in length, 1 3/4 inches in cheek diam.; skin yellow-orange with a very slight blush; flesh yellow, flavor fair to good, texture medium, medium firm; ripens evenly, with Newcastle or about 3 weeks before Blenheim; above average resistance to pit burning; when canned, texture resembles Blenheim but more acid and less highly flavored; when dried, color attractive but more acid and less highly flavored than Blenheim; ships well to nearby markets (e.g., Los Angeles). Tree: vigorous; very productive; fruit sets on terminals as well as on spurs; flowers large, white, self-fruitful. Recommended for trial as an early shipping apricot for nearby markets, combining improved pit-burning resistance with earliness and firmness of flesh.

Chow.—Orig. in Brookings, South Dakota, by N.E. Hansen, South Dakota Agr. Expt. Sta. Introd. in 1936. Seedling from seed gathered in northern Manchuria; selected in 1936. Fruit: small; flesh dry, quality fair to poor. Tree: very hardy. Variety now obsolete.

Citation.—See rootstocks for apricot.

Coffing.—Orig. in Covington, Ind., by Coffing Bros. Orchard. Introd. in 1959 by C.M. Hobbs & Sons, Indianapolis, Ind. Parentage unknown. Fruit: size medium; round to slightly elongated; skin yellow ground color with red blush; flesh yellow, firm, quality good; ripens late August to early September. Tree: large; upright; vigor good; hardy; productive.

Curtis.—Orig. in Charlotte, Mich., by Frank J. Curtis. Introd. in 1967. Chance seedling; discovered in 1942; selected and tested by Stanley Johnston, Mich. Agr. Expt. Sta., South Haven. Fruit: size medium; round; skin rich golden with a bright blush, attractive; flesh medium firm, flavor excellent when fully ripe; ripens about 27 July at South Haven, inclined to drop when mature. Tree: tall, open, slow growing. Recommended only for home and local use.

Cutler (*Cutler Late*).—Orig. in Salt Lake City, Utah, on the farm of A. P. Cutler. Introd. in 1938 by Farres L. Smith, Nurseryman, of Centerville, Utah. Parentage unknown; discovered in 1931. Fruit: medium; skin thick, adherent, no blush; flesh greenish-yellow to orange at center, ripening unevenly, juicy, subacid, pleasantly flavored, quality fair; pit large, freestone; ripens very late, 10 to 14 days after Chinese; resembles Moorpark. Tree: more susceptible to winter injury than Moorpark; moderately vigorous; upright spreading; no longer being propagated.

Cutler Late.—See Cutler.

Dallas.—Orig. in Dallastown, Pa., by a Mr. Kauffman. Introd. in 1922 by Enterprise Nurseries, Wrightsville, Pa. Parentage unknown; original material secured about 1912 from Mr. Kauffman by George E. Stein of Enterprise Nurseries. Fruit: large; skin yellow, occasionally blushed red; quality fair to good; reportedly ripens during week of 10 to 17 July in York County, Pa., area; resembles Moorpark. Tree: bears well; long-lived.

Deatrick.—Orig. in Franklin County, Pa., by Adams County Nursery & Fruit Farms, Aspers, Pa. Introd. in 1955. Parentage unknown; selected in 1953. Fruit: averages 1 3/4 inches in diam. with a heavy crop; round; skin golden, slightly blushed. Tree: hardy, setting crops under eastern climatic conditions; vigorous; tends to be bushy.

Doty.—Orig. on the farm of Floyd Doty, Oaks Corners, N.Y. Introd. in 1944. Parentage unknown; tree planted before Mr. Doty purchased the farm. Fruit: size medium; skin attractive, light golden yellow with few reddish markings; flesh sweet, nearly free from fiber; home and local market. Tree: large; hardy.

Dr. Bryan.—See Bryan.

Earle Orange (*Erle Orange, Stark Earli-Orange*).—Orig. in Grandview, Wash., by William L. Roberts. Introd. in 1945. Plant patent 674; 2 Apr. 1946; assigned to Stark Bros. Nurseries & Orchards Co., Louisiana, Mo. Parentage unknown; discovered in 1920. Fruit: size medium, averages 1 3/4 inches in diam.; skin with crimson blush, medium tough; flesh deep orange, firm, juicy, mildly subacid, ripens fairly evenly; pit size medium large, free; ripens 10 days before Riland at Grandview. Tree: upright; blooms late.

Earligold.—Orig. in Ontario, Calif., by W.H. Kemple. Introd. in 1938. Parentage unknown; selected in 1928. Fruit: resembles Newcastle; usually matures about 2 weeks before Newcastle in southern Calif. Tree: resistant to delayed foliation.

Earliril.—Orig. in Prosser, Wash., by the Washington Agr. Expt. Sta. and the U.S. Dept. of Agriculture; Introd. in 1957. Open-pollinated seedling of Riland; selected in 1950. Fruit: size medium, with equal halves; flesh orange-yellow, firm, texture fine, flavor good; soluble solids and acid content comparable to Blenheim; freestone; about 60% of skin covered with an attractive red blush over a yellow ground color; matures about 1 week before Riland. Tree: vigorous; upright-spreading; productive; hardy to –19F; self-unfruitful, not cross-pollinated by Riland.

Early Bee.—Orig. in McFarland, Calif., by T.A. Sand, Del Rancho Fortuna Nursery. Introd. in 1949. Parentage unknown; discovered in 1947. Fruit: ripens 10 days before Royal or Blenheim, which it resembles. Tree: resembles Tilton.

Early Blenheim.—Orig. in Summerland, British Columbia, Canada by K.O. Lapins, Canada Dept. Agr., Res. Sta. Introd. in 1970. Mutation of Blenheim produced by treating scions with thermal neutrons, discovered in 1956. Fruit: large, roundish; skin yellow with slightly green shoulders; flesh yellow, fairly soft, juicy, flavor good, ripens first week of August or 1 week before Blenheim. Tree: large, spreading; vigorous; moderately hardy; moderately productive, bears annually; flowers self-sterile, requires cross-pollinization; little or no fruit thinning required.

Early Divinity.—Orig. in Adelaide, South Australia, by H. Bain, Dept. of Agr. Introd. in 1962. Chance seedling; discovered about 1960. Fruit: medium to large; skin, flesh and flavor resemble Moorpark, but unlike Moorpark, ripens evenly, 1 to 2 weeks before Newcastle Early. Tree: large, framework thin; productivity light; only early variety in South Australia with reasonable flavor and color.

Early Gold.—Orig. in Trinidad, Wash., by E.E. Williams. Introd. in 1963. Plant patent 2256; 30 Apr. 1963; assigned to Pacific Coast Nursery, Portland, Ore. Open-pollinated seedling of Blenheim, F_2. Fruit: 2 to 2 3/8 inches in diam.; slightly ovate; skin light reddish blush over 25% of surface; flesh orange, firm and juicy; matures about 2 weeks before Blenheim; ripens irregularly, often a serious defect. Tree: vigorous; large; upright; hardy; productive, regular bearing; similar to parent but leaves are larger; flower pinkish.

Early Royal (*Resistant Royal*).—Orig. in Cabazon, Calif., by B.A. Van Duine. Not introd.; included because it has been used as breeding stock and is one of the parents of Nugget. Bud mutation of Royal; discovered in 1934. Fruit: ripens 2 to 3 weeks earlier than Royal. Tree: very resistant to prolonged dormancy due to insufficient winter chilling; in southern Calif. it usually blooms and leafs out 2 to 3 weeks earlier than Royal; bears fruit only at tips of branches, making this variety of little value except for breeding purposes.

Erle Orange.—See Earle Orange.

Ernie Fehr.—Orig. in Lewiston, Idaho, by Emma Marie Fehr. Introd. about 1938. Plant patent 503; 24 Feb. 1942. Parentage unknown. Fruit: large, 2 to 2 1/4 inches in diam.; flesh firm, color deep orange, medium juicy, subacid, flavor mild; cans well; ripens early, 7 to 10 days before Moorpark, from pit outward. Tree: quite productive; vigorous.

Farmingdale.—Orig. in Geneva, by Robert C. Lamb, New York State Agr. Expt. Sta. Introd. in 1965. Open-pollinated seedling of selection from (Doty x Geneva); selected in 1949; tested as N.Y. 346. Fruit: medium, 1 1/2 inches in diam.; roundish; skin and flesh orange; freestone; quality good. Tree: very vigorous; hardy.

Ferciana.—See Ishatara.

Flameburst™ (*Suapritwo*).—Orig. in Fresno, Calif., by John Weinberger, Superior Farming Co. Introd. in 1987. Plant patent pending; assigned to Superior Farming Co. Annabelle x open-pollinated seedling of Lulubelle. Fruit: medium to large; round to slightly oblong; skin bright red blush on orange background; flesh bright orange, medium firm, juicy; excellent eating quality, superior to Castlebrite; ripens 7 to 10 days after Castlebrite. Pollen-sterile. Tree: vigorous; productive. Suitable for fresh-market shipping.

Flaming Gold.—Orig. in Modesto, Calif., by C. Floyd Zaiger. Introd. in 1967. Plant patent 2822; 30 July 1968; assigned to Dave Wilson Nursery, Modesto. Seedling of Perfection; discovered in 1962. Fruit: resembles Perfection; large; skin highly colored, thin, no tendency to crack, colors 2–4 weeks before ripening; flesh very firm, quality good; for canning, drying, shipping and local market; ripens 20 to 25 May or 15 to 25 days earlier than Perfection. Tree: large; productivity heavy to medium.

Flavor Giant.—Orig. in Modesto, Calif., by Floyd, Gary, and Grant Zaiger and Leith Gardner. Introd.in 1984. (Seedling of Flaming Gold) x King Cot; tested as 48-G-472. Plant patent 5308; 30 Oct. 1984; assigned to Zaiger's Genetics. Fruit: large; globose; skin yellow; flesh dark yellow, firm; mild flavor. Tree: vigorous; upright; productive.

Franciscan.—Orig. in Fresno, Calif., by Marshall E. Francisco. Introd. in 1948. Plant patent 808; 12 Oct. 1948. Parentage unknown. Fruit: similar to Royal but shaped like Tilton; flesh firm; good shipping, drying and canning quality; ripens from the outside skin inwardly to pit, in Blenheim time, late June or first of July; resembles Tilton. Tree: fruit buds large; vigorous; tolerant to heat and cold; heavy producer.

Frascati.—See Geneva.

Geneva (*Frascati*).—Orig. in Geneva, N.Y., by Richard Wellington, New York State Agr. Expt. Sta. Introd. in 1934. Seed brought into the U.S. by the U.S. Dept. of Agriculture from Frascati, Italy; P.I. 34265; seedling grown and fruited at Geneva. Fruit: large; oval; flesh bright orange, freestone, firm, sweet, quality high. Tree: appears to be hardier than most commercial apricot varieties.

GF31.—See rootstocks for apricot.

GF677.—See rootstocks for apricot.

Giant Kohl.—Orig. in Clarks, Neb., by W.G. Kohl. Introd. in 1962 by Gurney Seed & Nursery Co., Yankton. Open-pollinated seedling of Manchurian apricot (*Prunus mandschurica*). Fruit: large; quality fine. Tree: unable to get satisfactory pollination; no longer being propagated.

Gold Dollar.—Orig. in Trinidad, Wash., by Edgar E. Williams. Introd. in 1963. Plant patent 2254; 30 Apr. 1963; assigned to Pacific Coast Nursery, Portland, Ore. Bud mutation of Moorpark; discovered in 1945. Fruit: similar to Moorpark except skin being a mottled blush over one-half of surface and ripens about 3 weeks earlier. Tree: vigorous; large, spreading; hardy; productive; regular bearer.

Gold Kist.—Orig. in Modesto, Calif., by C. Floyd Zaiger. Introd. in 1967. Plant patent 2826; 6 Aug. 1968; assigned to Dave Wilson Nursery, Modesto. Seedling of Perfection; discovered in 1962. Fruit: medium to large; skin high red color on side exposed to sun, thin, slight tendency to crack, colors 2 to 4 weeks before ripening; flesh orange to light orange, flavor mild, quality good, firmer than Perfection; for canning, freezing, drying and fresh for shipping and local market; ripens 3 to 4 weeks before Perfection. Tree: large to medium; upright; vigorous; very productive and regular bearer.

Goldbar.—Orig. in Prosser, Wash., by T.K. Toyama, Irrigated Agr. Res. and Ext. Center, Washington State Univ. Introd. in 1989. Plant patent pending; assigned to Washington State Univ. Goldrich x Blenril. Cross made in 1972; selected in 1976; tested as PA 7201-1. Fruit very large; sides slightly compressed; skin light orange with reddish-purple blush covering 20% of surface; flesh orange, moderately juicy, firm; quality fair; ripens 2 weeks before Wenatchee in Prosser; for fresh market. Tree: vigorous. Flowers require cross pollination; Goldrich, Perfection, and Goldstrike are effective pollinizers.

Goldcot.—Orig. in South Haven, Mich., by South Haven Expt. Sta., Michigan Agr. Expt. Sta., South Haven. Introd. in 1967. Open-pollinated seedling of Perfection. Planted in 1948; selected in 1952; tested as SHA 6. Fruit: size medium; nearly round; skin moderately thick, tough, golden colored; flesh medium-orange, medium firm; shipping quality good; ripens about 20 July in South Haven; for fresh use, home processing, and baby food processing, skin may darken too much for commercial canning of halves or whole fruit; adequate thinning must be done to obtain fruit of satisfactory size and to prevent biennial bearing. Tree: exceptionally strong; wide crotch angles; flowers partially self-fruitful; above average ability to crop satisfactorily in a cold, humid climate, not recommended for areas where well-known standard varieties can be grown.

Golden Amber.—Orig. in Visalia, Calif., by Willard Clayton Hill. Introd. in 1970. Plant patent 3067; 5 Oct. 1971, to be assigned to L.E. Cook Co., Visalia. Tilton x Royal. Cross made in 1937, selected in 1948. Fruit: large; uniform, symmetrical; skin yellow, tenacious to flesh; flesh yellow, firm, subacid, quality excellent; for fresh fruit, canning and drying; ripens 1 July through 25 July, therefore recommended as a home garden variety. Tree: large; upright; vigorous; hardy; very productive; tolerant to pit burn; flowers self-fertile.

Golden Delicious.—Orig. in Trinidad, Wash., by Edgar E. Williams. Introd. in 1963. Plant patent 2255; 30 Apr. 1963; assigned to Pacific Coast Nursery, Portland, Ore. Open-pollinated seedling of Blenheim; discovered in 1942. Fruit: similar to Blenheim except skin blushed over about 40% of surface and ripens about 10 to 14 days earlier. Tree: vigorous; large; upright; hardy; productive; regular bearer.

Golden Gem.—Orig. in Webster, New York, on property owned by Mrs. Ward Metcalf. Ed Miller, J.E. Miller Nurseries, Canadaigua, N.Y., was asked to propagate trees. Introd. in 1963 by this nursery. Chance seedling, planted about 1920; selected in 1961. Fruit: large, 2 to 2 1/2 inches in diam.; skin golden yellow with a bright orange blush, attractive; flesh sweet, sprightly; resembles Moorpark. Tree: extremely hardy; very productive; bears regularly.

Golden Giant.—Orig. in Grand Junction, Colo., by David A. Moore. Introd. in 1948. Plant patent 1053; 18 Dec. 1951; assigned to Inter-State Nurseries, Hamburg, Iowa. Parentage unknown; discovered about 1907. Fruit: large; oblong, compressed laterally; suture distinct; skin yellow, attractive, medium thick; flesh firm, melting, flavor mild, quality fine, freestone; resembles Moorpark. Tree: very productive; bears annually.

Golden Nugget.—Origin unknown; appeared in a bundle of trees of Moorpark shipped to Ackerman Nurseries, Bridgman, Mich. Introd about 1930. Fruit: resembles Moorpark. Tree: not hardy. Variety now obsolete.

Goldrich.—Orig. in Prosser, Wash., by Washington State Agr. Ext. Serv. and U.S. Dept. of Agriculture. Introd. in 1971. Sun Glo x Perfection. Cross made in 1954 and selected in 1958 by H.W. Fogle; tested as Prosser 5-61 by T.K. Toyama, Prosser (U.S. Dept. of Agriculture). Fruit: larger than Sun Glo; oval, halves nearly equal; skin bright orange ground color that must be allowed to develop before harvest to assure quality and low acidity, waxy, almost no blush; flesh deep orange, firm, texture fine, acid until fully mature; ripens 6 to 8 days before Wenatchee. Tree: vigorous; productive; hardier than Wenatchee; in Maryland more resistant to bacterial spot than Wenatchee and Blenril; flowers are only partially self-fertile and should be cross-pollinated, avoiding late-flowering varieties, inter-fertile with Wenatchee, Earliril, Blenril, Riland, and Rival; tested at Beltsville, Md.; Byron, Ga.; and Fresno, Calif. Recommended as a replacement for Wenatchee because of earlier and more uniform ripening, more attractive and firmer fruit, and hardiness of tree.

Goldstrike.—Orig. in Prosser, Wash., by T.K. Toyama, Irrigated Agr. Res. and Ext.

Center, Washington State Univ. Introd. in 1989. Plant patent pending; assigned to Washington State Univ. Goldrich x (Earliril x Blenril). Cross made in 1972; selected in 1975; tested as PA 7221-1. Fruit:very large; round to oval; skin 20% reddish purple blush on light orange background; flesh orange, firm, moderately juicy; quality fair; ripens 2 weeks before Wenatchee at Prosser; for fresh market. Tree: vigorous. Flowers require cross-pollination; Goldrich, Perfection, Rival, and Goldbar are effective pollenizers.

Haffer.—Orig. in Chambersburg, Pa., by J.U. Ruef, Pennsylvania Agr. Expt. Sta. Introd. in 1943 by Worley's Nurseries, York Springs, Pa. Parentage unknown; discovered by J.U. Ruef. Tree: bears annually in area of origination; withstands adverse weather conditions, which is the variety's only asset. For home gardens.

Haggith.—See rootstocks for apricot.

Hansen 536.—See under almonds. Incompatible.

Hansen 2168.—See under almonds. Incompatible.

Harbin.—Orig. in Brookings, S.D., by N.E. Hansen, South Dakota Agr. Expt. Sta. Introd. in 1944. Seedling from seed gathered in northern Manchuria; selected in 1944. Fruit: quality fair to poor. Tree: very hardy.

Harcot.—Orig. in Harrow, Ontario, Canada, by Richard E.C. Layne, Agr. Canada, Res. Sta. Introd. in 1977. [(Geneva x Naramata) x Morden 604] x NJA1 (Phelps x Perfection); cross made by L.F. Houch and C.H.Baile 1963; selected by Layne at Harrow in 1968; tested as H6301007 and HW401. Fruit: medium to large, 4.4 to 5.1 cm, roundish; skin orange with slight red blush on sun exposed side; flesh orange, firm, sweet, juicy, excellent flavor; usually freestone; ripens 16 July, 5 days before Goldcot; kernals sweet. Tree: large; upright to spreading; vigorous; cold hardy; tolerant to bacterial spot, brown rot and perennial canker (*Leucostoma* spp.). Recommended for fresh market in Ontario.

Hardy Iowa (*Iowa*).—Orig. in Glenwood, Iowa, by the Henry Field Seed Nursery Co., Shenandoah, Iowa. Introd. in 1950. Chance seedling discovered in 1948. Fruit: small; very sweet; skin light yellow, thin; ripens in late midseason, later than Superb, which it resembles. Tree: very hardy; prolific annual bearer; blooms later than most varieties, thus escaping many late frosts. Good for home gardens.

Hardy.—Orig. in Fayetteville, Ark., by Mr. Gibson, Crider Bros. Nursery. Introd. about 1940. Parentage unknown; discovered in 1939 Fruit: resembles Moorpark. Tree: very hardy; bears in Ark. where other varieties often fail to produce. Variety now obsolete.

Harglow.—Orig. in Harrow, Ontario, Canada, by Richard E.C. Layne, Agr/ Canada Res. Sta. Harrow Res. Sta., Agr. Canada. Introd. in 1982. V51092 [(Reliable x o.p.) x o.p.] x Sun Glo. Selected in 1975; tested as HW 425. Fruit: resembles Harlayne; size medium; orange skin, with 20% blush, firm orange flesh; freestone; good fresh and processed quality; skin resistant to cracking; ripens 5 days after Veecot, 26 July in Harrow. Tree: hardy, productive, late blooming; good resistance to bacterial spot, brown rot, and perennial canker (*Leucostoma* spp.). Recommended for fresh market and home processing in Ontario

Hargrand.—Orig. in Harrow, Ontario, Canada, by Richard E.C. Layne, Agr. Canada Res. Sta. Introd. in 1980. V51092 [(Reliable x o.p.) x o.p.] x NJA1 (Phelps x Perfection). Cross made by C.H. Bailey at Agr. Canada Res. Sta. Summerland in 1966; selected by Layne at Harrow in 1972; tested as HW 410. Fruit: large to very large, 5.4 to 6.4 cm, oblate to round; skin firm, dull orange color, 10% to 20% blush; moderately resistant to cracking; flesh very firm, orange, fine texture and good flavor, quality good; freestone; ripens 10 days after Harcot, and 5 days after Veecot, 26 July in Harrow. Tree: size medium, open-spreading; vigor medium; hardiness good, comparable to Goldcot; vigor medium; productivity medium. Tolerant to perennial canker (*Leucostoma* spp.), brown rot, bacterial spot and skin cracking. Recommended for fresh market and home processing in Ontario.

Harlayne.—Orig. in Harrow, Ontario, Canada, by Richard E.C. Layne, Agr. Canada Res. Sta. Introd. in 1980. V51092

[(Reliable x o.p.) x o.p.] x Sun Glo. Cross made in 1965; selected in 1970; tested as HW 407. Named in honor of R.E.C. Layne, apricot breeder, Harrow Res. Sta. Fruit size medium; 5 cm x 4 cm, oblong; skin orange, with 20% red blush, moderately resistant to skin cracking; flesh orange, very firm, good texture and flavor; freestone; ripens 28 July in Harrow, 4 days after Harogem and 7 days after Veecot. Tree: cold hardy, large, spreading, vigorous, productive, consistently productive; tolerant to perennial canker (*Leucostoma* spp.), bacterial spot, and brown rot. Recommended for fresh market and home processing in Ontario.

Harogem—Orig. in Harrow, Ontario, Canada, by Richard E.C. Layne, Agr. Canada, Res. Sta. Introd. in 1979. Rouge du Roussillon x NJA 2 (Morden 604 open-pollinated). Cross made by L.F. Hough and C.H. Bailey, Rutgers Univ., in 1963; selected by Layne at Harrow in 1969; tested as HW405. Fruit: medium size; ovate, compressed sides; skin 60% glossy red blush on bright orange background, very attractive; flesh orange, very firm, moderately juicy, smooth texture; good flavor; freestone; ripens 24 July in Harrow, and with Veecot. Tree: medium vigor; spreading; consistently productive; medium hardy like Veecot. Resistant to brown rot, perennial canker (*Leucostoma spp.),* skin cracking, and preharvest drop; moderately susceptible to bacterial spot. Recommended for fresh market in Ontario.

Harriet.—Orig. in Saint Paul, Minn., by Carl Weschcke. Introd. in 1942. Plant patent 476; 24 June 1941. Parentage unknown; selected about 1933. Fruit: size variable; shape irregular, with pronounced apical point; skin deep apricot with flecks of red; flesh firm, flavor good; pit very thick; ripens first of August. Tree: vigorous; hardy.

Harval.—Orig. in Harrow, Ontario, Canada, by Richard E.C. Layne, Agr. Canada Res. Sta. Introd. in 1989. Veecot x HW435 [Rouge du Roussillon x NJA 2 (Morden 604 open-pollinated)]. Cross made in 1969; selected in 1974; tested as HW437. Fruit: medium to large; skin 30% red blush on orange background, attractive; flesh orange, very firm; good quality; freestone; ripens 2 Aug. in Harrow, 5 days after Harlayne. Resistant to bacterial spot, perennial canker (*Leucostoma* spp.), skin cracking, and preharvest drop; fruit and spurs moderately susceptible to brown rot. Recommended for fresh market and home processing in Ontario.

Henderson.—Orig. in Geneva, N.Y., by George W. Henderson. Introd. in 1935. Parentage unknown. Fruit: large; roundish; flesh yellow, slightly fibrous, sweet, quality good; semi-clingstone. Tree: strong; vigorous.

Homedale.—See Stark Double Delight.

Hulan.—Orig. in Brookings, S.D., by N.E. Hansen, South Dakota Agr. Expt. Sta. Introd. in 1936. Seedling of seed gathered in northern Manchuria; selected in 1936. Fruit: quality very poor. Tree: very hardy.

Improved Cluster.—See Little Sam.

Improved Superb.—See Scheighing.

Iowa.—See Hardy Iowa.

Ishtara® (*Ferciana*).—See rootstocks for apricot.

Jannes.—Orig. in Seattle, Wash., by Jannes Wynn. Introd. in 1958. Open-pollinated seedling of Riland, discovered in 1940, selected in 1941. Fruit: large; round; skin blushed, very tough; flesh yellow to orange; a fresh or Canning variety; ripens 1 to 15 Aug. Tree: about 20 ft. high; spreading; very productive.

Jerseycot.—Orig. in Cream Ridge, N.J., by S.A. Mehlenbacher, L.F. Hough, and C.H. Bailey, New Jersey Agr. Expt. Sta. Introd. in 1986. RR18-64 x CR3-135; pedigree includes Scout, McClure, Perfection, and Phelps. Cross made in 1970; selected in 1977; tested as NJA44. Fruit: medium; round; skin pale orange with no blush; flesh bright orange, sweet, juicy, soft, good quality, with distinctive cantaloupe aroma; usually freestone; ripens early, 2 weeks before Goldcot at Cream Ridge; for U-pick and roadside markets. Pit: medium; tan surface grainy; kernel bitter. Tree: upright to spreading; unusually consistent copper; cold hardy in New Jersey. Blossom time early, 1 day before Veecot. Resistant to bacterial spot; moderately resistant to brown rot and skin cracking; *Leucostoma* cankers have not been observed.

Kaleden.—Orig. in Kaleden, British Columbia, Canada, by a Mr. Harrison. Introd. in

1922. Parentage unknown; bud mutation of a large-fruited, deep orange colored apricot resembling Chinese; discovered in 1921. Fruit: flesh texture unsuitable for canning, breaks at stem end; subject to marking from handling. Tree: annual bearer although sometimes produces a shy crop. Variety now obsolete.

Kidd's Early.—See Story's Early.

King.—Origin unknown. Introd. in 1964 by Stribling's Nurseries, Merced, Calif. Assumed to be a seedling of Perfection. Fruit: resembles Perfection, being larger, having higher color and maturing 5 days earlier. Tree: self-sterile and difficult to pollenize.

Lalin.—Orig. in Brookings, S.D., by N.E. Hansen, South Dakota. Agr. Expt. Sta. Introd. in 1936. Seedling from seed gathered in northern Manchuria; sister of Anda. Fruit: quality fair to poor. Tree: very hardy.

Lamale.—Orig. in Monroe, Mich., by a Mr. Lamale, Erie, Mich. Introd. about 1930. Considered to be a chance seedling; discovered about 1925. Fruit: skin with a red blush; flesh a deep yellow, flavor excellent; ripens 10 to 18 July. Tree: notable for a fairly consistent ability to set fruit in locality of origin, probably due to a tendency to come into bloom late.

Leslie.—Orig. in Morden, Manitoba, Canada, by the Canada Dept. Agr., Res. Sta.; selected by A. J. Mann, Res. Sta., Summerland, British Columbia Introd. in 1943. (Scout x McClure seedling); cross made in 1936 at Morden; original tree planted in 1938 and selected in 1942 at Summerland; named in honor of W.R. Leslie, Superintendent, Res. Sta., Morden. Fruit: color pale yellow; inclined to be somewhat soft in handling; resembled Tilton. Variety proved to be of poor quality and unproductive; now obsolete.

Little Sam (*Improved Cluster*).—Orig. in De Leon, Texas, by Stanley Carruth. Introd. in 1933. Open-pollinated seedling of Cluster; discovered in 1931. Fruit: quality poor; resembles Cluster. Tree: very productive.

Malan Royal.—Orig. in Wellington, South Africa, on J.F. Malan's farm, Oakdene, by P.A. Roux, Bovlee, Wellington, South Africa. Introd. in 1930s. Bud mutation of Royal. Fruit: resembles Royal; size medium; flesh orange yellow; ripens 10 days before Royal. Tree: spreading.

Manchu.—Orig. in Brookings, S.D., by N.E. Hansen, South Dakota. Agr. Expt. Sta. Introd. in 1936. Seedling of *Prunus mandshurica* from seed brought from northern Manchuria in 1924. Fruit: large; skin yellow; cooking quality good. Tree: large; heavy bearing; hardy.

Mandarin.—Orig. in Brookings, S.D., by N.E. Hansen, South Dakota Agr. Expt. Sta. Introd. in 1936. seedling from seed gathered in northern Manchuria; selected in 1936. Fruit: quality fair to poor. Tree: very hardy.

Mantoy.—Orig. in Mandan, N.D., by W.P. Baird, Northern Great Plains Field Sta. Introd. in 1957. Derived from seed (P.I. 65075) obtained in Manchuria by U.S. Dept. of Agriculture Plant Introduction Section and sent to Mandan in 1925; selected in 1943; tested as Mandan 43-1. Fruit: flesh golden yellow, medium firm, sweet, juicy, freestone, quality good both for fresh and culinary uses; usually matures 1 Aug.; size large for a Manchurian apricot, averaging about 22 per lb. Tree: hardy, vigorous; medium spreading; main fault, as with all other apricot varieties tested at Mandan, is that fruit buds are subject to winter killing and late spring frost may injure flowers; hence, regular annual crops cannot be expected.

Marianna GF 8.1.—See rootstocks for apricot.

Max-Gold.—Orig. in Zillah, Wash., by Arthur D. MacKelvie. Introd. in 1957. Wenatchee Moorpark x Tilton; selected in 1942. Fruit matures early; fresh yellow, firm; resembles Tilton. Tree: hardy.

Modesto.—Orig. in Le Grand, Calif., by F.W. Anderson, Merced, Calif. Introd. in 1964. Plant patent 2543; 6 July 1965; assigned to Irvin M. Burchell, Burchell Nursery, Modesto, Calif. F_2 (open-pollinated seedling of Perfection); selected in 1959; tested as Anderson 28H106. Fruit: medium to large, averaging 2 inches in diam.; globose, compressed laterally; suture distinct but shallow; skin thickness medium, tenacious to flesh; flesh firmer than Blenheim, meaty, flavor delicate, subacid, moderately juicy, fibers few; eating quality very good; freestone; ripens evenly, 10 days before Blenheim; ships well. Tree: moderately vigorous; medium, spreading; productive; regular bearer; leaf glands globose; flower medium, pinkish before open, almost white at full bloom.

Mohr.—Orig. in New Jersey by P.P. Mohr of Fogelsville, Pa. Introd. about 1935. Parentage unknown; Mohr secured scionwood from A.C. Hittle who secured the wood somewhere in New Jersey about 1910. Fruit: large. Tree: buds not considered as hardy as those of peach.

Mono.—Orig. in Le Grand, Cailf. by F.W. Anderson, Merced, Calif. Introd. in 1965 by Reedley Nursery, Reedley Calif. Plant patent 2515; 18 May 1965; assigned to Kim Bros., Reedley Nursery. Seedling of Perfection open-pollinated; tested as Anderson 38P1. Fruit: medium, averages 1 7/8 inches in suture diam.; globose to slightly oblong, symmetrical; flesh firm, meaty, acid to subacid, moderately juicy, good quality; stone small to medium, free; shipping and keeping qualities very good; suitable for dessert, long-distance shipping, drying, freezing, canning; ripens 1 week before Newcastle. Tree: size medium; moderately vigorous; very productive, bears regularly.

Moongold.—Orig. in Excelsior, Minn., by A.N. Wilcox, T.S. Weir, and Shirley Trantanella, Minnesota Agr. Expt. Sta. Fruit Breeding Farm. Introd. in 1961. Superb x *Prunus sibirica* L. 'Manchu'; cross made in 1940; sibling of Sungold; tested as Minn. 15. Fruit: oblate; medium, averages 1 3/4 inches in diameter; skin orange, rather tough, medium thick, medium juicy, orange-yellow, sweet, slightly subacid, very pleasant, quality fair, fine for dessert and processing; freestone; ripens in late July, before Sungold; hangs well on tree until ripe. Tree: size medium; rather spreading; vigorous and cold hardy; intended for the Minnesota climate; self-unfruitful, but pollinated by Sungold. For home gardens.

Morden 601.—Orig. in Morden, Manitoba, Canada, by Canada Dept. Agr., Res. Sta. Introd. in 1940. F_2 selection of (a Siberian apricot x Moorpark); selected in 1940. Fruit: flesh deep yellowish-amber, meaty, tender, with good flavor; canning quality good; more rounded in outline than Scout, which it resembles; over 1 1/2 inches in diam. Tree: hardy; productive; spreading.

Morden 604.—Orig. in Morden, Manitoba, Canada, by Canada Dept. Agr. Res. Sta. Introd. in 1946. Scout x McClure; selected in 1945. Fruit: large; skin golden yellow; flesh thick, deep orange, moderately firm, medium texture, flavor sweet, pleasing; medium quality as dessert, preserves or jams; season mid-August. Tree: upright, spreading; rounded; vigorous; hardy.

Mr.S.2/5.—See rootstocks for apricot.

Myran®.—See rootstocks for apricot.

Nemared.—See rootstocks for apricot.

Ninguta.—Orig. in Brookings, S.D., by N.E. Hansen, South Dakota Agr. Expt. Sta. Introd. in 1937. Seedling of *Prunus sibirica* from seed brought from northern Manchuria in 1924. Fruit: large; skin yellow with red blush; flesh very mild flavor; season late. Tree: heavy bearer; hardy.

not named.—Orig. in Hanford, Calif., by Judson L. Crandall. Plant patent 1551; 15 Jan. 1957. Second-generation seedling of Tilton. Fruit: matures about 2 weeks before Tilton; otherwise, very similar to Tilton. Tree: blooms about 2 weeks before Tilton. Appears identical to Tilton; included here because it has been patented.

Nugget.—Orig. in Ontario, Calif., by Herbert C. Swim, Armstrong Nurseries. Introd. in 1956. Plant patent 1651; 8 Oct. 1957; assigned to Armstrong Nurseries. Early Royal x Casaba; selected in 1949. Fruit: large; skin well-colored with a red blush; flesh yellow, flavor good, freestone; resembles Royal but much larger, more highly colored, and ripens earlier; maturation period from peak bloom to tree ripe is about 10 days less than that of Royal. Tree: regular and heavy producer; vigorous; large.

Patterson.—Orig. in Le Grand, Calif., by F.W. Anderson, Merced, Calif. Introd. in 1968. Plant patent 2877; 1 Apr. 1969; assigned to Burchell Nursery, Modesto, Calif. F_2 seedling of Perfection x unknown; tested as Anderson 31H155. Fruit: resembles Modesto; compared to Modesto, the flesh is firmer whereby the fruit holds its shape and texture for a longer period after canning; ripens 2 to 3 days later than Modesto. Tree: resembles Modesto, blooming 2 to 3 days later.

Perfection.—Orig. in Waterville, Wash., by John and Bertha Goldbeck. Introd. in 1937.

Parentage unknown; seed planted in 1911. Fruit: large; oval, blocky; skin pebbled, light yellow-orange, no blush; flesh bright orange, firm, quality fair; pit air-free; ripens about 1 week after Wenatchee. Tree: vigorous, hardy; blooms early; self-unfruitful.

Phelps.—Orig. in the Yakima Valley, Wash., by B.F. Phelps. Introd. about 1938. Parentage unknown; discovered about 1934. Fruit: very large; skin blushed; ripens very early and evenly; excellent shipper; flesh yellow, quality good.

Piet Cillié.—Orig. in Stellenbosch, Republic of South Africa, by P.A.L. Steyn, Fruit and Fruit Tech. Res. Inst., Dept. of Agr. Tech. Serv. Introd. in 1975. Royal x Bulida, cross made in 1952, selected in 1961, tested as SN35/9. Fruit: larger than Royal; round, flattened at the apex; flesh light yellow, firm, flavor good; keeps well; canning quality good when lye-peeled; ripens 12 to 14 days before Royal. Tree: slightly upright; vigorous; very productive; apricot appears to be the best rootstock.

Pinkerton.—Orig. in Le Grand, Calif., by F.W. Anderson, Merced, Calif. Introd. in 1967. Plant patent 2880; 22 Apr. 1969; assigned to Reedley Nursery., Reedley, Calif. F_2 seedling of Perfection; tested as Anderson 43P215. Fruit: size medium; oblong, unsymmetrical; skin medium thick, yellow, blushed and slightly mottled with red on sunny side; flesh yellow and very firm; excellent shipping and processing qualities; freestone; ripens 10 days before Pomo. Tree: very productive and regular bearer.

Pomo.—Orig. in Le Grand, Calif., by F.W. Anderson, Merced, Calif. Introd. in 1962. Plant patent 2192; 20 Nov. 1962; assigned to Kim Bros., Reedley Nursery, Reedley, Calif. F_2 seedling of Perfection; selected in 1960; tested as Anderson 7L8A. Fruit: size medium, 6 to 11 fruit per lb, averaging 2 1/8 inches in diam., larger than Royal, smaller than Perfection, similar to Tilton; flesh meaty, firm, ripens evenly; flavor delicate, subacid, aroma distinct, eating quality good; freestone; shipping quality good; keeping quality medium; intended for fresh-fruit market; ripens early, last week of May and first week in June, a week before Perfection. Tree: very large; productive; bears regularly.

Puget Gold.—Orig. in Anacortes, Wash., by R.A. Norton, J. King, and G.A. Moulton, Northwestern Washington Res. and Ext. Center, Mt. Vernon. Introd. in 1987. Chance seedling of unknown parentage, discovered by Jean Copeland. Fruit: medium; roundish ovate; skin light red to red-orange blush on clear yellow background; flesh deep yellow-orange, firm, sweet; freestone; ripens about 8 Aug. at Mt. Vernon, Wash., 1 week later than Rival; for home gardens. Tree: moderate to low vigor; spreading; prolific and consistent cropper; partially self-fruitful. Blooms prolifically, 3 days before Skaha. Tolerant to brown rot, *Coryneum* blight, and blossom blight caused by *Pseudomonas syringae*. Particularly adapted to a cool, maritime climate.

Rayburn.—Orig. in Hayworth, Ill., by William Rayburn. Introd. about 1922. Parentage unknown, but apparently a seedling of a Russian variety. Fruit: large; quality good. Tree: hardy. Now obsolete.

Redsweet.—Orig. in Fresno, Calif., by Haig Culbashian. Introd. in 1961. Plant patent 1732; 15 July 1958; assigned to Armstrong Nurseries, Ontario, Calif. Bud mutation of Blenheim; discovered in 1954. Fruit: size average; skin almost completely covered with a deep red blush; ripens in Blenheim season, the last of June, in Fresno; freestone; shipping quality good; resembles Blenheim. Tree: very heavy producer; flower pink, rather small.

Reeves.—Orig. in Burbank, Calif., by William H. Reeves. Introd. in 1947. Plant patent 693; 4 June 1946; assigned to Armstrong Nurseries, Ontario, Calif. Parentage unknown; discovered about 1939. Fruit: ripens about 2 weeks before Newcastle; superior in flavor, skin color, and size to Newcastle; poor shipper. Tree: chilling requirement similar to that of Newcastle.

Reliable.—Orig. in Summerland, British Columbia, Canada, by A.J. Mann, Canada Dept. of Agriculture, Res. Sta. Introd. in 1945. Wenatchee Moorpark x Hewetson seedling; cross made in 1937; selected in 1942. Fruit: skin orange with a blush; quality fair; texture dry and firm, maintaining its firmness on the tree for about 10 days after attaining full color; ships well; not suitable for canning. Tree: upright, spreading; vigorous; hardy; bears regularly and heavily.

Resistant Royal.—See Early Royal.

Riland.—Orig. in Rock Island, Wash., by Harry Yount. Introd. in 1932. Plant patent 74; 26 Sept. 1933; Harry Yount, Douglas County, Wash., and Andrew T. Gossman, Wenatchee, Wash.; assigned to Gossman. Parentage unknown; seed planted in 1923 by Yount. Fruit: large; flat, oval; skin light yellowish with a bright, deep red blush over one-half or more; flesh flavor rich, plum-like, somewhat acid, texture slightly coarse, firm, melting; said to keep well; freestone; ripens early; resembles Yakimene. Name coined from the words Rock Island. Tree: young trees have vigorous growth and upright form.

Rival.—Orig. in Prosser, Wash., by Washington State Agr. Ext. Serv. and U.S. Dept. of Agriculture. Introd. in 1971. W1-24 (Tilton x Riland) x Perfection. Cross made in 1953 and selected in 1957 by H.W. Fogle; tested as Prosser 4-12 by T.K. Toyama, Prosser (U.S. Dept. of Agriculture). Fruit: large; oval and symmetrical; skin light orange, develops a pronounced blush on exposed cheeks; flesh deep orange, firm, texture fine, flavor mild, acidity low; ripens 3 to 5 days before Wenatchee. Tree: vigorous; hardier than Wenatchee; productive; flowers are self-incompatible, requiring cross-pollination; late-flowering varieties should not be used as pollinizers, inter-fertile with Wenatchee, Perfection, Sun Glo, Earliril, and Goldrich. Recommended as a replacement for Wenatchee (Wenatchee Moorpark) because of earlier ripening, more attractive and firmer fruit, and tree hardiness.

Robust.—Orig. in Morden, Manitoba, Canada, by A. Spangelo, Morden Nurseries. Introd. in 1941. Seed brought from Manchuria; seedling selected in 1940. Fruit: size medium; skin yellow; flesh yellow, meaty, smooth texture; flavor mild, sweet, pleasant; very fine as dessert or jam; season late July and early August. Tree: tall, rounded; vigorous; hardy; strong crotches.

Sansin.—Orig. in Brookings, S.D., by N.E. Hansen, S. Dakota. Agr. Expt. Sta. Introd. in 1936. seedling from seed gathered in northern Manchuria; selected in 1936. Fruit: small; quality fair. Tree: very hardy.

Scheighing (*Improved Superb*).—Orig. in Guthrie, Okla., by David Scheighing, Jr. Introd. in early 1940s. Considered to be an open-pollinated seedling of Superb. Fruit: larger and juicer than Superb. Tree: bears annual crops; blooms late. No longer being propagated.

Scout.—Orig. in Morden, Manitoba, Canada, by the Canada Dept. of Agriculture, Res. Sta. Introd. in 1937. Seed of *Prunus armeniaca* sent by the Expt. Sta. of the Eastern Siberian Railway, Echo, Manchuria; P.I. 132840; planted in 1930. Fruit: medium to large; flat, oblong-cordate, pointed; skin bronzy golden; flesh deep yellow, fine, tender; freestone; quality fair to good as dessert, good canned, good as jam; season late July and early August. Tree: tall, upright, rounded; vigorous; hardy; productive.

Shamrock.—Orig. in Shamrock, Texas, by J.W. Blake, Jr., Bruce Nurseries, Alanreed, Texas. Introd. in 1960. Parentage unknown; selected in 1958. Fruit: size medium; flesh yellow, quality good; resembles Early Golden. Tree: blooms late; bears prolific crops regularly; hardy.

Sing.—Orig. in Brookings, S.D., by N.E. Hansen, South Dakota. Agr. Expt. Sta. Introd. in 1936. Seedling of *Prunus mandshurica* from seed brought from northern Manchuria in 1924. Fruit: large; cooks to a rich yellow sauce. Tree: large; productive; hardy.

Sino.—Orig. in Brookings, S.D., by N.E. Hansen, South Dakota. Agr. Expt. Sta. Introd. in 1936. Seedling from seed gathered in northern Manchuria; selected in 1936. Fruit: quality fair to poor. Tree: very hardy. Variety now obsolete.

Skaha.—Orig. in Summerland, British Columbia, by K.O. Lapins, Agr. Canada, Res. Sta. Introd. in 1973. Open-pollinated seedling of Perfection, selected in 1956, tested as 4E-28-14. Fruit: large, round; skin orange with slight blush, attractive; flesh moderately firm; flavor fair to good; ripens early, 5 days before Wenatchee (last week of July at Summerland). Tree: large, wide, spreading; very vigorous and productive; as hardy as Wenatchee.

Southerner (*Suaprione*).—Orig. in Fresno, Calif., by John Garabedian. Introd. in 1971. Plant patent 3345; 22 May 1973; assigned to Superior Farming Co.; trademarked Southerner. Blenheim-Royal open-pollinated. Fruit: large; flesh yellow-orange, firm; ripens 30 days before Blenheim-Royal, which it resembles. Well-adapted to shipping.

Southwick.—Orig. in Sherwood, Ore., by Tualatin Valley Nursery. Introd. in 1945. Parentage unknown; original tree from seed planted in 1852. Fruit: 1 to 1 1/4 inches in diam.; suture deep; skin bright orange, no red blush; ripens unevenly; flavor good; resembles Wenatchee Moorpark. No longer being propagated.

Starglow.—Orig. in Lansing, Mich. Introd. in the 1940s by Stahelin Nursery Co., Bridgan, Mich. Parentage unknown. Fruit: quality good. Tree: very hardy. Variety did not produce well in South Haven, Mich.

Stark Double Delight (*Homedale*).—Orig. in Homedale, Idaho, by Garfield Shults. Introd. in 1984. Wilson x Earle Orange; selected in 1968. Fruit: resembles Wilson; size medium; flavor better than Wilson; freestone. Tree: very productive, consistent bearer; matures early July.

Stark Earli-Orange.—See Earle Orange.

Stark Giant Tilton.—Orig. in Tieton, Wash., by R.C. Freimann. Plant patent 3927; 29 June 1976; assigned to Stark Bros. Orchards and Nurseries Co., Louisiana, Mo. Originated as a bud mutation of Tilton apricot. Fruit: identical to Tilton, except fruit is larger. Tree: large, vigorous; spreading; productive.

Steindorf (*Steindorf Blenheim*).—Orig. in Gilroy, Santa Clara Valley, Calif., by Herbert W. Steindorf. Introd. in 1967. Plant patent 2742; 16 May 1967. Bud mutation of Blenheim. Discovered in 1948; tested as apricot No. 5. Fruit: same as Blenheim but ripening 30 days later. Tree: same as Blenheim but more vigorous; blooms 30 days later than Blenheim.

Steindorg Blenheim.—See Steindorf.

Story's Early (*Kidd 's Early*).—Orig. in Renmark, South Australia, by F.W. Story, Story's Nurseries. Introd. about 1945. Bud mutation of Moorpark; discovered in 1936. Fruit: skin tough; flesh firmer and ships better than Moorpark; suitable for fresh market, canning and drying; ripens 14 days before Moorpark, and more uniformly than that variety; colors to market stage 7 to 10 days before softening. Tree: stronger growth habit than Moorpark; tendency to long leader growths, which do not branch on twig in first season; leaders tend to become willowy if growth excessive; susceptible to shot hole disease.

Suaprione.—See Southerner.

Suapritwo.—See Flameburst.

Sun Glo.—Orig. in Entiat, Chelan County, Wash., by Otto H. Heider. Introd. in 1946. Plant patent 751; 12 Aug. 1947; assigned to Columbia & Okanogan Nursery Co., Wenatchee, Wash. Parentage unknown; discovered about 1942. Fruit: somewhat elongated and blocky; large, 2 inches in diam.; skin solid orange, smooth, glossy; flesh firm, juicy, sweet; pit small, free; ripens uniformly and about 7 days before Wenatchee Moorpark. Tree: apparently identical to Wenatchee Moorpark.

Sundrop.—Orig. in Summerland, British Columbia, Canada, by W.D. Lane, Agr. Canada, Res. Sta. Introd. in 1975. Open-pollinated seedling of Perfection, cross made in 1946, selected in 1956 and tested as 4E-55-9 by K.O. Lapins. Fruit: size medium, round, skin bright orange; flesh firm; for fresh fruit market; ripens 7 days before Blenheim. Tree: vigorous; moderately hardy; very productive. Promising in New Zealand and recommended for planting there as a commercial variety.

Sungold.—Orig. in Excelsior, Minn., by A.N. Wilcox, T.S. Weir, and Shirley Trantanella, Minnesota Agr. Expt. Sta. Fruit Breeding Farm. Introd. in 1961. Superb x *Prunus sibirica* L. 'Manchu'; cross made in 1940; sibling of Moongold; tested as Minn. 36. Fruit: round to slightly elongated, averaging 1 1/4 inches in diam.; skin medium thin, tender, gold with orange blush; flesh clear orange, tender, medium juicy, flavor very mild and sweet, quality very good; fine for dessert and processing; freestone; ripens in early August; hangs well on tree until ripe. Tree: size medium; vigorous, rather upright; exceptionally hardy; very productive; self-unfruitful, but pollinated by Moongold; intended for the Minnesota climate. For home gardens.

Sunshine.—Orig. in Brookings, S.D., by the South Dakota. Agr. Expt. Sta. Introd. in 1950. Open-pollinated seedling of Manchurian apricot; selected in 1940. Fruit: large; quality good; resembles Manchu. Tree: vigorous; hardy; moderately productive.

SweetHeart.—Orig. near Le Grand, Merced County, Colo., by F.W. Anderson. Plant patent 4080; 19 July 1977; assigned to Stark Bros. Orchards & Nurseries Co., Louisiana, Mo. Seed parent unknown, from Iran; open-pollinated seedling of parent variety. Fruit: small; medium thick skin, abundant juice with high sugar content; good eating quality, firm and crisp; freestone; good shipping and keeping quality; kernel sweet. Tree: large; vigorous; spreading; productive.

Thompson Early.—Orig. in Fremont, Saunders County, Neb. Introd. about 1936. Apparently a chance seedling from seed brought from Denmark, discovered on the farm of Carl Thompson. Fruit: size medium; round; flesh yellow, quality good; ripens in late July at place of origin. Tree: heavy producer. Variety now obsolete.

Tola.—Orig. in Brookings, S.D., by N.E. Hansen, South Dakota. Agr. Expt. Sta. Introd. in 1936. Seedling of *Prunus mandshurica* from seed brought from northern Manchuria in 1924. Fruit: large; heavy bearer; hardy.

Tomcot.—Orig. in Prosser, Wash., by T.K. Toyama, Irrigated Agr. Res. and Ext. Center, Washington State Univ. Introd. in 1989. Plant patent pending; assigned to Washington State Univ. Rival x P63-265 (Goldrich open-pollinated). Cross made in 1970; selected in 1974; tested as PA 7005-8. Fruit large; oval to round; skin has trace of blush on light orange background, attractive; flesh light orange, firm, moderately juicy; quality very good; ripens very early, 18 days before Wenatchee in Prosser; for fresh market. Tree: vigorous; blooms early, 2 or 3 days before Wenatchee; flowers partially self-fertile.

Tracy.—Orig. in Le Grand, Calif., by Frederic W. Anderson, Merced, Calif. Introd. in 1971. Plant patent 3062; 10 Aug. 1971; assigned to Burchell Nursery, Modesto, Calif. Open-pollinated seedling of Perfection F_2. Tested as Anderson 39P115. Fruit: large to medium; freestone; compared to Blenheim, larger, skin tougher, flesh firmer and less pit burn; ripens about 6 days later than Modesto; more adaptable to mechanical harvesting, processing and packing then Blenheim; keeping quality good with long shelf life. Tree: large to medium; spreading, open-growth; vigorous; regular and productive bearer.

Tri-Gems.—Orig. in Modesto, Calif., by Floyd, Gary, and Grant Zaiger and Leith Gardner, Modesto, Calif. Introd. in 1989. Open-pollinated seedling of Flaming Gold. Plant patent 6755; 25 Apr. 1989; assigned to Zaiger's Genetics. Fruit: medium; globose; skin yellow-orange; flesh butter-yellow, very firm; flavor subacid, mild. Tree: vigorous; semi-spreading; productive.

VALnur.—Orig. in Wenatchee, Wash., by Peter Van Well, Van Well Nursery. Introd. in 1958. Riland x Wenatchee; selected in 1942. Fruit: medium-large; round, sides unequal; skin greenish-yellow, slightly blushed, dull; flesh light orange, soft, quality mediocre; freestone; ripens about 1 week before Wenatchee. Tree: vigorous; spreading; productive.

Veecot.—Orig. in Vineland Station, Ontario, Canada, by O.A. Bradt, Ontario Dept. Agr., Res. Inst. Introd. in 1964. Open-pollinated seedling of (Reliable open-pollinated); selected in 1958; tested as Ont. 51091. Fruit: medium; skin bright orange, attractive, coloring well before maturity; flesh dark orange, rather dry, high quality and attractive appearance when canned; only fair quality fresh; ripens 1 week after Viceroy, or first week in August; hangs well on tree until fully mature; very resistant to cracking and splitting; resembles Reliable. Tree: vigorous, spreading; productive annually; relatively resistant to bacterial spot on both leaves and fruit, moderately susceptible to brown rot. Recommended for fresh market and home processing in Ontario.

Velvaglo.—Orig. in Vineland Station, Ontario, Canada, by O.A. Bradt and E.T. Andersen, Hort. Res. Inst. of Ontario. Introd. in 1978. Veecot x Farmingdale. Cross made in 1960; selected in 1968; tested as V60091. Fruit medium; skin light red blush on bright orange background; flesh orange; flavor fair to good; freestone; ripens uniformly with Goldcot in last week of July at Vineland Station. Tree: vigorous; spreading; productive. Moderately resistant to bacterial spot, brown rot and *Leucostoma* canker; resistant to preharvest drop and to skin cracking.

Viceroy.—Orig. in Vineland Station, Ontario, Canada, by O.A. Bradt, Ontario Dept. Agr., Res. Inst. Introd. in 1964. Geneva x Naramata; cross made in 1949; tested as Ont. 49056. Fruit: medium size; skin yellow, blushed; flesh soft, juicy, flavor good; good dessert quality, soft when canned; suceptible to skin cracking in wet seasons; ripens during last week of July, 4 or 5 days before Perfection; resembles Naramata. Tree: large; vigorous; productive annually; medium hardy. Recommended only for home gardens in Ontario.

Victoria (*Wilson Delicious*).—Orig. in Hannibal, Mo., by H.J. Jenner, Sims Fruit and Nursery Co. Introd. in 1948. Parentage unknown; selected in 1943. Fruit: large size is its most valuable characteristic; skin red. Tree: blooms late.

Vivagold.—Orig. in Vineland Station, Ontario, Canada, by O.A. Bradt and E.T. Andersen, Hort. Res. Inst. of Ontario. Introd. in 1978. Veecot x V49024 (Geneva x Gibb). Cross made in 1960; selected in 1965; tested as V60102. Fruit: medium; skin bright orange, attractive; flesh orange. somewhat dry; flavor good; freestone; ripens 1 week after Veecot at Vineland Station; for fresh market and home canning. Tree: moderately vigorous; spreading; moderately productive. Resistant to skin cracking and preharvest drop; moderately resistant to brown rot and bacterial spot. Recommended for home processing in Ontario.

Westcot.—Orig. in Morden, Manitoba, Canada, by L. Kerr, Agr. Canada Res. Sta. Introd. in 1982. (Scout open-pollinated) x (McLure open-pollinated). Fruit: medium; ovate; skin light reddish blush on yellow orange background; flesh orange, smooth, juicy, sweet; flavor mild; freestone. Tree: vigor moderate; tree shape irregular; cold hardy; more bud-hardy than Scout or Morden 604; medium productivity (but best in very cold locations). At Morden, flowers 25 Apr. to 15 May; fruit ripens 7 to 15 July.

Westley.—Orig. in Le Grand, Calif., by F.W. Anderson, Merced. Introd. in 1973. Plant patent 3626; 24 Sept. 1974; assigned to Burchell Nursery, Modesto, Calif. F_2 of Perfection x Tilton, cross made in 1965, selected in 1972, tested as 42P620. Fruit: medium to large; flesh orange, medium firm, sweet, flavor good; ripens a week after Tilton; for canning, freezing, drying and fresh. Tree: larger and more productive than Tilton; vigor moderate.

Wilson Delicious.—See Victoria.

Worley (*Worley's Peach*).—Orig. in York Springs, Pa., by Chester B. Worley, Worley's Nurseries. Introd. in 1939. Parentage unknown; discovered in 1930. Fruit: large; skin bright orange, attractive; flesh thick, juicy and slightly soft, freestone, flavor slightly sweet and mild; early ripening. Tree: hardy; vigorous grower; blooms in midseason; productive.

Worley's Peach.—See Worley.

Zun.—Orig. in Brookings, S.D., by N.E. Hansen, South Dakota. Agr. Expt. Sta. Introd. in 1936; seedling from seed gathered in northern Manchuria; selected in 1936. Fruit: quality fair to poor. Tree: very hardy.

Rootstocks for Apricot

Avimag.—See Cadaman.

Cadaman (*Avimag*).—Not compatible.

Citation.—See under peaches. Apricot trees dwarfed by 20% to 30%.

GF 31.—Orig. in Pont-de-la-Maye, near Bordeaux, France, by R. Bernhard, INRA Fruit Res. Sta. Selected from a *Prunus cerasifera* x *P. japonica* cross. Introd. as a vigorous, productive stock for apricots. Compatible with very broad range of apricot varieties. Tolerant of high soil moisture and high calcium. Very susceptible to verticillium.

GF677.—See under peaches. Compatible under Bergeron, Luizet, Paviot, and Polonais, but incompatible with most varieties.

Haggith.—Orig. in Ruthven, Ont., Canada, on the farm of Murray Haggith, by R.E.C. Layne and T.B. Harrison, Agricultural Canada, Res. Sta., Harrow, Ont. Introd. in 1974. Chance seedling, selected in 1950. Fruit: small, less than 1 1/2 inches; ovate; skin yellow, thick; flesh orange, soft; drops readily when ripe; ripens 21 July at Harrow, with Goldcot; recommended for use as seedling rootstock for apricot. Tree: large; spreading; vigorous; very hardy; very productive, bears regularly; self-sterile; tolerant to perennial canker (*Leucostoma* spp.) and bacterial spot, susceptible to brown rot. Recommended rootstock for apricot in Ontario.

Hansen 2168.—See under almonds. Incompatible.

Hansen 536.—See under almonds. Incompatible.

Ishtara® (*Ferciana*).—See under plums. Semi-dwarfing. Induces early, heavy fruiting of Bergeron; compatible with most varieties, including those of the Roussillon group. Anchorage not as good as that of Myran.

Marianna GF 8.1.—See under plums. Not widely tested under apricots; known to be compatible under some, but not all, varieties.

Mr.S.2/5.—See under peaches. Compatible with apricot.

Myran®.—See under peaches. Suitable as a vigorous rootstock for certain apricot varieties.

Nemared.—See under peaches. Compatible with apricot.

ATEMOYA

African Pride.—See Kaller.

Kaller (*African Pride*).—Orig. in Israel. Introd. in 1960 by Langbecker Nurseries, Bundaberg, Queensland, Australia, as African Pride. Imported into the U.S. in 1963; first fruiting in Florida in 1965. Known in Australia as a custard apple, as grown in Florida atemoya; it has all the appearances of a hybrid between *Annona squamosa* x *Annona cherimola*. Fruit: large, up to 20 oz; skin dull green, smooth to medium rough; seeds few; quality good. Tree: smaller than other varieties; bears biannually in Florida, the first crop in September and October, the second in March and April; sheds leaves in March and April for a short dormant period, then leafs out and flowers in May; bloom for second crop comes in September. In Australia is productive, low growing, and heavily branched. In Florida recommended as a garden variety; production there is not high.

AVOCADO

Acme.—See Booth 2.

Adi.—Orig. at Akko Expt. Station, Western Galilee, Israel, by E. Lahav, D. Zamet, and S. Gazit. Seedling of Horshim caged with Hass and resembles both cultivars. Introd. in 1992, and was to be patented. Fruit: obovate (pear-shaped) to slightly necked, of small to medium size, 6 to 10.5 oz, usually 8 oz; green when mature; skin pebbled, of medium thickness, easily separated from flesh, slightly stonier than that of Hass; flesh pale yellow with a green rim, buttery texture, and excellent flavor, slightly less nutlike than Hass; seed round to slightly elongated, small (<10% of fruit weight) and tight in cavity; season begins and ends earlier than Hass (which is early October to March). Tree: small to medium, sturdily branched, with short and compact flowering racemes; similar to Hass in cold sensitivity. Yields appear good, with a tendency toward alternate bearing. Flowering group A; distributed for evaluation throughout Israel but not recommended for hot summer areas, where fruit size might be reduced.

Ajax.—Orig. in Homestead, Fla., by Will Booth. Introd. in 1934 when propagated by Coral Reef Nursery, Homestead. Open-pollinated seedling of (Guatemalan x West Indian types); discovered about 1927. Fruit: medium large, 16 to 30 oz; round to obovoid; skin green, thick, glossy, slightly pebbled; oil content 9% to 13%; flavor good, rather dry; seed medium large; resembles Collinson. Tree: vigorous and productive. Flowering group B. Minor commercial cultivar in Florida.

Alexandria.—Orig. in Duarte, Calif., by W.A. Spinks. Introd. in 1925. Chance seedling; Guatemalan type; first fruited in 1921. Fruit: large, 16 to 24 oz; oval; skin green, glossy, slightly rough; seed small; season July. Now obsolete.

Alice.—Orig. in Carlsbad, Calif., by Sam Thompson. Introd. in 1952. Open-pollinated seedling of Guatemalan type; selected in 1944. Fruit: averages 8 oz; pyriform; skin smooth, medium thick, medium green; flavor good, oil content 18%, keeping quality poor; seed 2 oz. Tree: tall; spreading; vigorous; cold resistance similar to Fuerte; productive. Obsolete.

Alice Mirandi.—See Carr.

Alma.—Orig. in Sebring, Fla., by Heinrich Rohde. Introd. in 1944. Parentage unknown; discovered in 1922; Guatemalan x West Indian type. Fruit: 12 to 20 oz; rounded; skin purple, thick, rough; oil content 20%; flavor and quality good; seed large; season 15 Jan. to 15 Mar. Formerly recommended for home planting as a late winter variety because of resistance to cold. No longer propagated.

Amanda.—Orig. in Vista, Calif., by Paul L. Kovach. Introd. in 1956. Chance seedling of hybrid origin; seed planted in 1947. Fruit: 8 oz; pyriform; skin green, thick, does not peel from flesh; flavor very good, oil content 22%; ripens slowly. Tree: upright; bears well. Obsolete.

Ambassador.—Orig. in Whittier, Calif., by A.R. Rideout. Introd. in 1922. Lyon x Puebla; Guatemalan type. Fruit: 14 to 16 oz; oblong to pyriform; skin rough, purple; seed large; season July to September. Now obsolete.

Ames.—Orig. in Pala, Calif., by J.H. Ames. Introd. in 1956. Open-pollinated seedling of Topa Topa; planted in 1952. Fruit: 7 oz; pyriform; skin green, smooth, thin, peels readily from flesh; flavor fair, oil content 9%; season mid-September to mid-October. Tree: growth similar to Fuerte. Now obsolete.

Amigo.—Orig. in Whittier, Calif., by C.H. Hamburg. Introd. in 1922. Chance seedling discovered in 1917; Guatemalan type. Fruit: 12 to 14 oz; round; skin purple, rough; seed large; oil content 14.7%; season April to May. Now obsolete.

Arctic.—Orig. in Tustin, Calif., by B.H. Sharpless. Introd. in 1925. Open-pollinated seedling of Sharpless; first fruited in 1923. Fruit: 16 to 18 oz; pyriform; skin dark purple, rough; flavor good; seed size medium; matures in late fall, after Sharpless. Now obsolete.

Arturo.—Orig. in Fallbrook, Calif., by A.R. Chenoweth. Introd. in 1946. Plant patent 667; 15 Jan. 1946. Mexican seedling; discovered in 1933. Fruit: size medium, 6 to 10 oz; long oval; skin dull green; soft leathery, surface slightly rough, no corkiness in early part of season but develops later on, excellent peeling quality; flesh light yellow, fiber-free, soft, melting, flavor mild, quality very good, oil content 11.5% as early as mid-September, progressing to 20%; seed 1 1/2 to 2 1/2 oz, tight in cavity; ripens from September into December. Tree: symmetrical; tends to develop a strong central leader with long drooping lateral branches; more cold resistant than Fuerte; blooms in February. Planted in home gardens in southern California.

Arue.—Orig. in the Society Islands. Introd. in 1932 into the U.S. as P.I. 99805 by U.S. Dept. of Agriculture Sta., Miami, Fla. Parentage unknown. Fruit: large, averaging 20 to 30 oz; quality fair; skin rough for a summer variety; seed large; matures in May and June, much earlier than other Florida avocado varieties regularly, but with low yields; flowering group A; minor commercial cultivar.

Atkins.—Orig. in Carpinteria, Calif., by J.D. Atkinson. Introd. in 1950. A Mexican x Guatemalan hybrid type; tree planted before 1937. Fruit: 16 oz; roundish pyriform; skin green, smooth, thin; flavor good; seed large. Now obsolete.

Atmore.—Orig. in Sespe, Calif., by F.E. Atmore. Introd. in 1934. Chance seedling discovered about 1927; Guatemalan type. Fruit: 13 oz; obovoid; skin black, thin; flavor fair, oil content 11%, considerable fiber; season March to July. Now obsolete.

Aurel.—Orig. in Vista, Calif., by Paul L. Kovach. Introd. in 1949. Hybrid parentage unknown; seed planted in 1945. Fruit: averages 11 oz; pyriform; skin dark green, thin, smooth peels easily; flavor fair, oil content 16%; seed averages 1 1/4 oz, tight in cavity; season April and May. Tree: upright, slender.Now obsolete.

Avila.—Orig. in Quebradillas, Puerto Rico. Chance seedling of West Indian type. Fruit: very large, 30 oz; pyriform; skin light green; seed medium to large; season October to December.

Avon.—Orig. in Avon Park, Fla., by W.F. Ward. Introd. in 1940. Hybrid of Guatemalan x West Indian types; seedling discovered in 1927. Fruit: 12 to 20 oz; obovate with obliquely flattened base; skin medium green, thin, pliable, glossy, attractive; flesh light yellow, oil content 9%; seed very large, 22% of fruit weight; season December and January. Tree: branches able to stand a great deal of bending without breakage; consistently heavy bearer; not cold resistant. Flowering group A. No longer propagated.

Bacon.—Orig. in Buena Park, Calif., by James E. Bacon. Introd. in 1951. Chance seedling, Mexican–Guatemalan hybrid type; planted in 1928. Fruit: 7 to 12 oz; ovoid; skin green, thin, smooth; flavor good; flesh an unusually pale yellow-green; oil content high; flavor good; keeping quality good; matures November to January in Orange County and December to March in Ventura County. Tree: upright, slender; consistent, heavy producer; frost tolerance excellent. Flowering type B, valuable pollinizer of other cultivars; minor commercial cultivar in southern California.

Baird.—Orig. in Carlsbad, Calif., by Robt. W. Baird. Introd. in 1933. Chance seedling of a Guatemalan type; selected in 1928. Fruit: 10 to 14 oz; pyriform; skin green, pebbled, leathery; flavor fair to good, oil content 15.3%; seed large, 2 1/4 oz; season April to June. Now obsolete.

Bannon.—Orig. in Altadena, Calif., by Thomas C. Bannon. Introd. in 1950. Open-pollinated seedling of Rochwood; discovered in 1946. Fruit: small, 4 oz; ovoid; skin black, thin, smooth; flavor good, oil content 14.8%; seed large, 2 3/4 oz; season October to December. Tree: hardy; vigorous; tall, slender.

Batangas.—Orig. in Batangas, Philippines, by Mrs. Aurlie Torres, St. Thomas. Introd. there about 1935; imported into the U.S. about 1947. Chance seedling, West Indian type; seed planted in 1928. Fruit: 15 oz; shape bottle necked; skin light green, turning to a bright maroon with greenish-yellow stripes, glossy, smooth; quality fine; seed 1 1/4 oz, medium loose in cavity; season July to September. No longer propagated in U.S.

Benedict.—Orig. in Bangor, Butte County, Calif., by "Duke" Hornung. Introd. in the late 1920s. Chance seedling; originated in same planting as did Duke; Mexican type; seed imported in 1912, apparently from Mexico, by a Mr. Benedict for Sunnyslope Avocado Nursery, Bangor; following abandonment of this nursery, Hornung purchased the area, noted the seedlings, selected several; from one of these came Benedict, being selected in 1924 by Hornung; evaluated in 1924 and later by J. Eliot Coit, Vista, Calif. Fruit: size medium, 6 to 8 oz; pyriform to obovoid; skin black, smooth, glossy; flavor good, oil content 15%; seed small; ripens during September and October. Tree: productive; quite hardy; resistant to wind damage; adapted to hot southern Calif. interior valleys. Now obsolete.

Benik.—Orig. in the finca La Polvora, Antigua, Guatemala. Imported into the U.S. in 1917 by Wilson Popenoe, U.S. Dept. of Agriculture, as P.I. 44626. Introd. in 1924 in the U.S. Fruit: very large, about 20 oz; broadly obovoid to pyriform; skin purplish-maroon, thin, pebbled, somewhat pliable, very attractive; flesh yellow, free from discoloration, flavor rich and pleasant, quality very good; seed size medium, 3 oz, tight in cavity; ripens in midseason. Tree: vigorous; branchlets rather short; bears well. Flowering group A. Not grown commercially.

Bernard.—Orig. in Brentwood Heights, Calif., by H.C. Oakley. Introd. in 1934. Chance seedling discovered in 1927; Guatemalan type. Fruit: 15 oz; pyriform; skin dark purple, thick; flavor very good, oil content 23.1%; seed 2 1/2 oz; season June to August. Now obsolete.

Bernhardt.—Orig. in Santa Paula, Calif., by Edwin C. Bernhardt, Foothill Avocado Nursery. Introd. in 1964. Plant patent 2403; 12 May 1964. Fuerte x seedling of Irving; cross made in 1954; Mexican type. Fruit: medium to small, 7 to 9 oz; pyriform, 3 1/2 to 5 in. long, 1 3/4 to 2 inches in diam.; skin green, slightly pebbled, leathery, tough, peels easily; flesh light yellow, firm, consistency butter-like, slightly milder than Fuerte, flavor very good; seed small, not adhering to flesh; stores well; ripens in fall (mid-October to end of November), resembles Fuerte. Tree: upright; blooms in February and March. Now obsolete.

Biscayne.—Orig. in Miami, Fla., by the U.S. Plant Introd. Sta. Introd. in 1951. Open-pollinated seedling of a West Indian type tree; selected about 1945. Fruit: large; skin greenish-yellow, apparently too much so for commercial purposes; ripens in late August. Tree: weak. Not propagated commercially.

Bitte.—Orig. in Redland, Fla., by Seymour Dane. Introd. in 1930. Chance seedling discovered before 1912; West Indian type. Fruit: size medium; obovate; skin green, smooth, leathery; seed size medium, tight in cavity; season October and November. Tree: bears heavily. Flowering group B. Now obsolete.

Black Prince.—Orig. in Homestead, Fla., by Issac Lasso. Introd. in 1950. Parentage unknown. Fruit: 16 to 32 oz, skin medium thick, leathery, granular; flesh bright yellow; seed small, tight in cavity; season August–September. Major commercial cultivar in Florida.

Blair.—Orig. in Naranja, Fla., by J.R. Hickson. Introd. in 1939. Open-pollinated seedling of Taylor; Guatemalan type. Fruit: medium; obovate; skin green; not greatly subject to scab; resembles Lula; season December and January. Tree: bears regularly. Minor commercial cultivar in Florida.

Boley.—Orig. in Highlands, Calif., by J.D. Boley. Introd. in 1945. Chance seedling, Mexican–Guatemalan hybrid type; discovered about 1930. Fruit: 9 to 12 oz; pyriform; skin green, smooth, leathery; flavor rich. A vigorous seedling, which grew well and produced fruit; trees subsequently budded or grafted have failed to produce well; limbs characteristically appear to be fasciated on many trees. Now obsolete.

Bondoso.—Orig. in Vista, Calif., by Paul L. Kovach. Introd. in 1952. Chance seedling; selected in 1950. Fruit: averages 8 to 12 oz; pyriform; skin green, smooth; flavor good; oil content medium; ripens the last of October; seed 1/2 oz. Tree: size medium; upright; bears well. Now obsolete.

Bonita.—Orig. in Homestead, Fla., by C.

Santini. Introd. in 1936. Parentage unknown; selected in 1930. Fruit: medium to large; obovate; skin green; not considered promising for northern U.S.'s markets, but sells well locally; season December and January. Tree: bears heavily. Flower group B. Now obsolete.

Bonita.—Orig. in Santa Barbara, Calif., by Horace F. Pierce at the Vista del Mundo Ranch. Introd. in 1939. Chance seedling, Guatemalan type. Fruit: 13 oz; skin green, peels easily from flesh; flavor good, oil content 18.5%; season 1 Sept.; seed 2 1/2 oz. Now obsolete.

Booth 1.—Orig. in Homestead, Fla., by Will Booth, as a seedling of an unknown Guatemalan parent in a mixed planting of Guatemalan and West Indian avocados; propagated commercially in 1935. Fruit: rounded obovate, medium large, 16–24 oz; skin nearly smooth, glossy dark green, medium-thick and brittle; flesh light yellow, of fair quality, oil content 8% to 12%; seed large and loose in cavity; season late, December to January. Tree moderately vigorous and a heavy bearer. Flowering group A. Minor commercial cultivar in Florida.

Booth 2 (*Acme*).—Orig. in Homestead, Fla., by William Booth. Introd. about 1940. Chance seedling of a Guatemalan type arising in a mixed planting of Guatemalan and West Indian types; seed planted in 1920; first fruited in 1927. Fruit: averages 1 lb; obovate to round; skin green; season February and March. Flowering group B.

Booth 3.—Orig. in Homestead, Fla., by William Booth. Introd. in 1940. Open-pollinated seedling of a Guatemalan type (probably crossed to a West Indian type); selected in 1927. Fruit: 12 to 14 oz; rounded obovoid; skin dark green, glossy, nearly smooth, quite thick; oil content 10% to 12%; quality good; seed size medium; season December and January. Tree: resembles Booth 7; bears heavily; flowering group b.

Booth 5.—Orig. in Homestead, Fla., by William Booth. Introd. about 1940. Chance seedling of a Guatemalan type arising in a mixed planting of Guatemalan and West Indian types; seed planted in 1920; first fruited in 1927. Fruit: averages 16 oz; obovate; skin green; season December. Flowering group B. Major commercial cultivar in Florida.

Booth 7.—Orig. in Homestead, Fla., by William Booth. Introd. in 1935. Open-pollinated seedling of a Guatemalan type (probably crossed with a West Indian type); selected in 1927. Fruit: 10 to 20 oz; round-obovate; skin green, glossy, thick, slightly pebbled; oil content 10% to 14%; seed size medium; season November to January. Tree: bears prolifically; flowering group B. Major commercial cultivar in Florida.

Booth 8.—Orig. in Homestead, Fla., by William Booth. Introd. in 1935. Open-pollinated seedling of a Guatemalan type (probably crossed with a West Indian type); selected in 1927. Fruit: medium, 14 to 18 oz; oblong ovate, similar to Hickson; skin dull green, rough, thick; oil content 6% to 8%; seed medium large; marketing season November and December. Tree: very prolific bearer; flowering group B. Major commercial cultivar in Florida.

Booth 10.—Orig. in Homestead, Fla., by William Booth. Introd. about 1935. Parentage unknown. Fruit: medium large, 16 to 24 oz; skin green, slightly rugose, glossy, thick, very granular, separates readily from flesh, lenticels numerous and light green; flesh yellow, green next to skin; seed tight in cavity; season October and November. Flowering group B.

Brogdon.—Orig. in Winter Haven, Fla., by Tom W. Brogdon. Introd. in 1951. Parentage unknown; discovered in late 1930s; tested by Univ. of Florida, Sub-Tropical Expt. Sta., as PI 4761; Mexican race. Fruit: 7 to 20 oz; somewhat pear shaped; skin purple, very thin; susceptible to handling injury; flesh yellow, buttery; ripens late July to August. Tree: small to medium; vigor moderate; productivity low; very hardy; susceptible to anthracnose; recommended for central Florida.

Byars.—Orig. in Homestead, Fla., by E.C. Byars, Sr. Introd. in 1948. Open-pollinated seedling of Collinson; selected in 1932. Fruit: 16 to 40 oz; elliptical; skin slightly rugose, olive green, stippled with light green lenticels; flesh yellow with a wide green area next to skin; seed size medium, fairly tight in cavity; season December through February.; resembles Collinson. Flowering group B. Obsolete.

Caliente.—Orig. in Mexico. Introd. about

1948 into the U.S. by M. Garcia, Duarte, Calif. Chance seedling of Mexican type. Fruit: 4 to 9 oz; long pyriform, necked; skin deep mahogany red, smooth, glossy; flavor good, oil content, 18%; seed size medium, often loose in cavity; marketability poor; season October to December. Now obsolete.

Calmayo.—Orig. in Carlsbad, Calif., by John R. Howe at the Pyle Ranch. Introd. in 1943. Chance seedling of Mexican type. Fruit: 8 oz; skin dark green; quality good, oil content 13.7%; seed 2 oz; season October to January. No longer propagated.

Camel.—Orig. in Whittier, Calif., by A.R. Rideout. Introd. about 1925. Chance seedling of Guatemalan type; selected in 1922. Fruit: 12 to 16 oz; elongated oval; skin green, rough and pebbly, thick; flavor very good, tendency to darken next to seed; seed large, 3 1/2 oz, often loose in cavity; season May to July. Now obsolete.

Carlos.—Orig. in Villa Park, Calif., by Carl S. Crawford. Introd. in 1950. Chance seedling discovered in 1944. Fruit: averages 8 3/4 oz; pyriform; skin olive green; flavor quite good, oil content averages. 19.2%; keeps well; seed averages 1 oz; ripens during June into September. Now obsolete.

Carolyn.—Orig. in Camarillo, Calif., by C.J. Dailey. Introd. in 1941. Open-pollinated. seedling of Linda planted in 1934. Fruit: 14 oz; oval; skin green, thick, pebbled; flavor good; season September and October. Tree: consistent bearer; hardy.

Carr (*Alice Mirandi*).—Orig. in Rivera, Calif., by John Carr. Introd. about 1930. Chance seedling of Mexican type; discovered about 1925. Fruit: 4 to 7 oz; oval to pyriform; skin dark green, smooth, thin; flavor fair; seed size medium, sometimes used for production of seedlings for rootstock purposes; season October Tree: upright; resistant to frost. Now obsolete.

Carroll.—Orig. in Carlsbad, Calif., by G. Maltby. Introd. in 1932. Open-pollinated seedling of Anaheim; selected in 1926. Fruit: large, 21 oz; pyriform; skin green, pebbled; flavor very good, oil content 21.7%; seed 4 oz; season March and April. Now obsolete.

Case.—Orig. in Wailuku, Maui, Hawaii, by D.H. Case. Introd. in 1922. Open-pollinated seedling of Dalna. Fruit: Guatemalan; 15 to 20 oz; long, pyriform, slightly necked; skin dark green, glossy, nearly smooth, attractive; flesh mild, fiberless; seed medium; good quality and shelf life; ripens January to March. Tree: dwarf. Recommended for low and medium elevations up to 2000 ft.

Catalina.—Orig. in Cuba. West Indian type. Fruit: large, 18 to 26 oz; skin light green, smooth, glossy, thick; seed small, loose in cavity. Flowering group A; minor commercial cultivar in Florida.

Cecil.—Orig. in National City, Calif., by Frank Dale. Introd. in 1934. Chance seedling of (Guatemalan x West Indian hybrid type); discovered in 1929. Fruit: 7 1/2 oz; pyriform; skin green; flavor good, oil content 23.9%; seed size medium. Now obsolete.

Chapala.—Orig. in Vista, Calif., by Paul L. Kovach. Introd. in 1944. Chance seedling of Guatemalan type. Fruit: 12 oz; pyriform; skin green, smooth, medium thick, peels readily; flavor good, oil content 16.2%; seed 2 1/4 oz; season June to August. Now obsolete.

Choquette.—Orig. in Miami, Fla., by R.D. Choquette. Introd. in 1939. Guatemalan x West Indian types; selected in 1934 from seed sown in 1929. Fruit: 30 to 40 oz; oval; skin nearly smooth, glossy, dark green, leathery; flesh thick, yellow; seed size medium, tight in cavity; oil content 13%, quality very good; season January to March. Tree: alternate bearing habit; resistant to common avocado diseases. Of major commercial importance in Florida. Flowering group A.

Clifton (*Monica*).—Orig. in Glendora, Calif., by C.B. Shoemaker. Introd. about 1949. Chance seedling of Mexican type; discovered in 1933. Fruit: small, averaging 7 oz; oval; skin turns purplish-black at maturity; seed 2 oz; season September to December. Tree: hardy; production fair; grown commercially in a small way in the inland areas of California during the 1940s and 1950s. No longer propagated.

Coit.—Orig. in Vista, Calif., by J. Eliot Coit. Introd. in 1936. Guatemalan x Mexican types; registered with the Calif. Avocado Soc. in 1939. Fruit: skin light green, thin, smooth; pyriform; 10 to 16 oz; seed size medium; flavor good; oil content 15%; season May to 1 Aug. (in southern Calif.); resembles Fuerte. Tree: upright. No longer propagated.

Collinred Seedling B.—Orig. in Coconut Grove, Fla., by the Plant Introduction Garden, U.S. Dept. of Agriculture. Introd. in 1934; P.I. 106941. Seedling of Collinred (P.I. 62376); hybrid of Guatemalan x West Indian types; from seed planted in 1925 and selected in 1930. Fruit: 16 to 30 oz; obovate, slightly flattened at the distal end; skin dull light green, often with deep maroon blush, thin, tough; flesh light yellow, of very good quality; oil content 6% to 8%; seed medium large, loose in cavity with loose seed coats; season October; flowering group A. Does not ship well. No longer propagated.

Collinred.—Orig. in Miami, Fla., by the Plant Introduction Garden, U.S. Dept. of Agriculture. Introd. in 1929. Seedling of Collins (P.I. 19080); planted in 1916; P.I. 62376. Fruit: size medium; roundish oblate; skin surface roughened by deeply set dots; skin thin, dark purplish-brown, separating readily from flesh; flesh deep yellow, green near skin, smooth, no fiber, flavor rich and pleasant; stem short, thick. Tree: tall, spreading; flowering group A. No longer propagated.

Collinson.—Orig. in Miami, Fla., by the Plant Introduction Garden, U.S. Dept. of Agriculture. Introd. in 1922. Open-pollinated seedling of Collins, apparently crossed with a West Indian type; seed planted in 1915, first crop in 1920. Fruit: 18 to 30 oz; short obovoid to elliptical; skin dark green, glossy, smooth, leathery; oil content 10% to 16%; flavor good; seed medium, tight in cavity; season November and December. Tree: Class A flowering type, but pollen not produced; rapid, vigorous grower; leaves and fruit resistant to scab; fairly productive. Not recommended for commercial plantings in Florida. No longer propagated, but still carried as a minor commercial cultivar by the Florida Avocado Administrative Committee.

Corona.—Orig. in Goleta, Calif., by J.E. Coit of the Thomas B. Bishop Co., Corona del Mar Rancho. Introd. in 1950. Parentage unknown; discovered in 1944. Fruit: skin green, quite smooth, peels easily; oval; weight averages 7 oz; seed weight 3/4 oz; flavor good, oil content 23.8%; keeping quality good; season June and July. Tree: moderately spreading. Planted in dooryards in California.

Courtright.—Orig. in Lakeland, Fla., by R.M. Courtright. Introd. in 1942. Parentage unknown; Mexican type; discovered in 1941. Fruit: 8 to 12 oz; pyriform; skin purple, thin; flesh yellow, quality fair, oil content 15%; seed fairly large, loose in cavity; season September and October. Flowering group A.

Covacado.—See Mayo.

Creelman.—Orig. in Camarillo, Calif. Introd. in 1961. Open-pollinated seedling of (Fuerte x Hass); seed selected and collected in 1951 by Royce S. Bringhurst, Calif. Agr. Expt. Sta., then at Los Angeles; in 1952 and 1953 some seedlings sent to Calvin C. Delphey, Farm Advisor, Ventura, Calif.; planted in the A. Creelman orchard in Camarillo; selected as promising in 1957 by Delphey; tested as UCLA 51007. Fruit: pear-shaped; skin green, very attractive; ripens in Ventura County from 15 Mar. to 15 July; 10 to 12 oz, with seed weighing about 16% of fruit weight; flesh yellowish, very little fiber, quality good; resembles Fuerte in attractiveness, color, and shape. Tree: moderately spreading; flowering group B.

Danish West Indies Bank.—See Frederiksted Bank.

Darwin.—Orig. in Santa Ana, Calif., by Darwin Speck. Introd. in 1946. seedling of Mexican type; discovered on the Santa Ana property of R.T. Brown (father-in-law of Speck); tested by Armstrong Nurseries, Ont., Calif.; tests and further propagation discontinued in 1947. Fruit: averages 6 to 8 oz; ovoid to somewhat pyriform; skin purplish-black; flavor fair, strong, somewhat anise-like, oil content 20.1% ripens in October in Santa Ana; resembles Blackbird but larger. Tree: hardy, productive. While size and general appearance were good quality was not good enough for further propagation.

Dawn.—Orig. in Homestead, Fla., by R.O. Shore of J.R. Brooks Son. Introd. in 1959. Open-pollinated seedling of Peterson. Fruit: small, 12 oz; ovate; skin medium green, smooth, medium thick; flavor bland; ripens during July and August; seed small. Not of commercial importance.

Decem.—Orig. in Carlsbad, Calif., by person or persons unknown. Introd. in 1947 when it was first planted. Chance seedling, possibly of Fuerte; Fruit: 6 to 8 oz; pyriform; skin green, smooth, leathery; seed size medium; ripens in early fall with Fuerte-resembles Fuerte in appearance and size. Tree: moderately consistent bearer in some areas such as Ventura County, Calif.; leaves with a characteristic yellow-green blotch, which readily identifies all known Decem trees, but it does not appear to be detrimental or serious to tree growth or production of fruit. Now obsolete.

Dewey.—Orig. in Fullerton, Calif. Introd. about 1945. Chance seedling of (Mexican x Guatemalan types); seed planted in 1933; tested by H.L. Gillespie, San Gabriel, Calif. in the 1950s. Fruit: 3 to 5 oz; pyriform; skin thin, medium green; flavor good to very good oil content 24.5% the end of October; keeping quality poor; seed 3/4 oz; season late August through November Tree: hardy; good yields. Now obsolete.

Diamond.—Orig. in Fallbrook, Calif., by Joe Diamond. Introd. in 1955. Open-pollinated seedling of Fuerte; seed planted in 1942. Fruit: medium, averaging 8 oz; pyriform; skin purplish-green, medium smooth, peels well from flesh; flavor very good; oil content 20%. Tree: upright; productive; fairly hardy.

Dolores.—Orig. in LaHabra Heights, Calif., by J. Scott Thompson. Introd. in 1932. Chance seedling of Guatemalan type; seed imported from Guatemala; selected in 1924. Fruit: averages 10 oz; pyriform; skin green, thin, smooth; flavor fair; oil content 22.5%; keeping quality good; seed large; ripens April into June. Now obsolete.

Dr. DuPuis #2.—Orig. in Miami, Fla., by John G. DuPuis, Jr. Introd. in 1963. Plant patent 2528; 8 June 1965. Bud mutation discovered on a West Indian seedling Fruit: medium, 10 to 16 oz; averages about 12 oz; oblong to pyriform, up to 5 inches long, up to 3 1/2 inches in diam.; skin green, smooth, light yellow dots; flesh light yellow, firm, free of fiber, flavor rich and nutty; seed small, nearly round, tight in cavity; ripens early mid-June to end of July; ships very well. Tree: moderately upright, symmetrical; sets and holds fruit under adverse weather conditions; blooms in February. Flowering group A; major commercial cultivar in Florida.

Dr. John G. DuPuis.—Orig. in Miami, Fla., by John G. DuPuis. Introd. in 1955. Plant patent 1369; 29 Mar. 1955. Parentage unknown, of West Indian origin; discovered in 1941; tested as DuPuis 1. Fruit: large, 16 to 32 oz; up to 6 inches long and 4 1/2 inches in diam.; plumply pyriform; skin green, smooth, dots scattered and cream colored; stem heavy, strong, relatively short, up to 1 1/2 inches long; flesh yellowish, firm, free of fiber, flavor delicate and nutty; seed medium, adhering tightly in cavity; season July and August; ships well; no excessive drop. Tree: productive annually; resistant to scab. Obsolete.

Duke.—Orig. in Bangor, Butte County, Calif., by "Duke" Hornung. Introd. in the late 1920s. Chance seedling; Mexican type; seed imported in 1912, apparently from Mexico, by a Mr. Benedict for Sunnyslope Avocado Nursery, Bangor; following abandonment of this nursery, Hornung purchased the area, noted the seedling, and selected several; from one of these came Duke; selected in 1924 by Hornung; evaluated in 1924 and later by J. Eliot Coit, Vista, Calif. Fruit: size medium, 8 to 12 oz; pyriform; skin yellowish-green, glossy, nearly smooth, very thin; quality fair to good; poor shipper because of thin skin; seed large, often loose in cavity; primarily for local markets; ripens early, September to November. Tree: vigorous; productive; very hardy; wind resistant; blooms early; Duke seedlings and selected rooted clones of these seedlings show partial resistance to phytophthora root rot.

Dunedin (*Skinner*).—Orig. in Dunedin, Fla., by L.B. Skinner. Introd. in 1925. Open-pollinated seedling of Winslow; selected in 1920. Fruit: 12 to 20 oz; rounded obovoid; skin light green, dull, smooth, fairly thick; flavor very good; oil content 12% to 16%; seed fairly large; season 15 Dec. through January. Flowering group A.

DuPuis 3 (*Katherine*).—Orig. in Miami, Fla., by John G. DuPuis, Jr. Introd. in 1963. Plant patent 2529; 8 June 1965. Bud mutation discovered on a West Indian seedling. Fruit: 12 to 20 oz, averages. about 15 oz, up to 6 inches long and 4 inches in diam.; broadly oval to pyriform, nearly round; skin green, slightly rough, thick, dots conspicuous; flesh deep yellow, firm, free of fiber, delicate nutty flavor; seed medium to large, adhering tightly to flesh in cavity; ripens early, from mid-June through July; ships very well. Tree: semi-upright; blooms in February; sets and holds fruit in adverse weather conditions.

Edmonds.—Orig. near Homestead, Fla., by Tom Edmonds. Introd. in 1935. Considered to be (Guatemalan x West Indian types) discovered in 1920. Fruit: 24 to 36 oz; oval; skin green, smooth; seed rather small; quality good; season September and October. Flowering group B. No longer propagated.

Edranol.—Orig. in Vista, Calif., by E.R. Mullen. Introd. in 1932. Open-pollinated seedling of Lyon; seed obtained from A.R. Rideout; selected in 1930. Fruit: 10 to 12 oz; skin green, thick, russets badly in interior areas; pyriform; quality high; flavor excellent, oil content high; seed small; season at Santa Barbara, Calif., from March to November; season at Vista from February to July. Tree: tall; slender; vigorous; bears heavily; adapted to coastal areas in California where it is a minor commercial cultivar.

Elsie.—Orig. in Whittier, Calif., by Charles H. Hamburg. Introd. in 1935. Plant patent 816; 28 Dec. 1948; assigned to Cecil Knowlton. Chance seedling from a Guatemalan type; selected in 1930. Fruit: 8 to 14 oz; pyriform; skin green, relatively thin; flesh firm flavor fine; oil content high; ripens during June and August in Whittier; seed size medium, tight in its cavity; resembles Fuerte. Tree: upright, central leader; vigorous. Planted in dooryards in California.

Emerald.—Orig. in San Diego, Calif., by Laura McFadden. Introd. in 1951. Chance seedling of Mexican type; discovered in 1928. Fruit: medium, average weight 7.5 oz; with seed weight of about 2 oz; pear-shaped; skin bright green, attractive; oil content medium quality good; season mid-December to April; tends to drop badly near maturity. Tree: tall, rather upright; moderately hardy to frost at Riverside, Calif.; leaves resemble Puebla. No longer propagated.

Encanada.—Orig. in LaHabra, Calif., by H.B. Griswold. Introd. in 1941. Chance seedling of (Mexican x Guatemalan types). Fruit: averages 8 1/2 oz; pyriform; skin green, medium thick with greenish yellow surface dots; flavor fair to good; seed 1 oz, season July to September. No longer propagated.

Erey.—Orig. in Baldwin Park, Calif., by O. Ruf. Introd. about 1922. Chance seedling of Guatemalan type; selected in 1915. Fruit: 15 to 18 oz, pyriform with some crook necked; skin green, smooth, leathery, flavor good; seed small; season April to August Tree: small, upright, weak grower; very productive.

Ettinger.—Orig. in Kefar Malal, Israel. Introd. there in 1947 after observation by members of the Committee for Subtrop. Fruits; introd. in the U.S. in 1954; P.I. 218196. seedling of Fuerte; selected in 1940. Fruit: Fuerte type; medium, 6 to 12 oz, pyriform; skin bright green, surface slightly rough, some tendency to corkiness, tendency to end-spot which appears early, peeling quality excellent; flesh clear light yellow, free of fiber, soft, melting, ripens uniformly, quality excellent; seed large, 1 1/2 to 2 1/2 oz, loose in cavity, seedcoats adhere to flesh; ripens early, about 1 month before Fuerte in Israel; ripens slightly before and with Fuerte in Riverside, Calif.; important commercial variety in Israel and used there for domestic and export markets; not propagated in California. Tree: very prolific; vigorous; produces a strong central leader, drooping lateral branches.

Frederiksted Bank (*Danish West Indies Bank*).—Orig. in St. Croix, U.S. Virgin Islands by Ira Ross and Winifred de Castro, discovered by U.S. Dept. of Agriculture Res. Serv., U.S. Virgin Islands. Introd. in 1956. Parentage unknown, probably West Indian variety Fruit: 4 1/2 to 5 inches in diam.; average weight 1 1/2 lb; round; no neck, skin develops reddish or purplish cast when ready to harvest; flesh yellow, attractive thick, nonwatery; flavor nutty and excellent; quality excellent; ripens 1 Oct. to 15 Nov. Tree: orig. tree planted in backyard of the Danish West Indies Bank, also known as the Frederiksted Bank, from seed of avocado bought from a sloop coming from the British Virgin Islands; tree badly damaged by hurricane Betsy in 1956 and cut down in 1957 to make room for construction; budwood obtained by above station; pollination type unknown.

Frowe.—Orig. in Honolulu, Hawaii, by Chester E. Frowe. Introd. in 1957. Open-pollinated seedling of West Indian type; selected in 1952. Fruit: elliptical; 18 to 30 oz; skin smooth, dark purple; seed size medium loose in cavity; flesh buttery, with distinctive nutty flavor, quality excellent; ripens September to December. Tree: prolific, low headed and spreading.

Fuchs (*Fuchsia, Fuchs Early*).—Orig. in Homestead, Fla., by C.T. Fuchs, Sr. Introd. in 1926. Parentage unknown, of West Indian type; first fruited in 1916. Fruit: 10 to 18 oz; pyriform to oblong; skin light green, smooth, appearance attractive; flesh greenish-yellow, oil content 4% to 6%, quality good; seed medium size, loose in cavity; very early in maturity, June to August; shipping quality very poor. Tree: grows vigorously; bears at young age and regularly; flowering group A; minor commercial cultivar in Florida.

Fuchsia.—See Fuchs.

Fuchs Early.—See Fuchs.

Fuchs-20.—Orig. at Institute of Horticulture, Agricultural Research Organization, The Volcani Center, Bet Dagan, Israel, by A. Kadman and A. Ben Ya'acov. Introd. in 1981. Seedling of the Florida Fuchs (West Indian) cv. that survived irrigation with highly saline water (1330 ppm NaCl) and could be propagated from cuttings. Fruit: ellipsoid, of medium to large size (13 to 16 oz); flesh very good in flavor, with 12% to 13% oil content; skin smooth, pale green covered with numerous yellowish lenticels during ripening season (October). Tree: large and vigorous, to 33 ft height, with large, dark-green leaves; free of sunblotch viroid when selected. Yields are relatively low. This cultivar's primary use is as an asexually propagated rootstock for commercial cultivars planted in soils of high salinity. Cuttings rooted in an experiment at rates higher than 70%, and showed high uniformity and tolerance to salinity. Trees of various cultivars grafted on Fuchs-20 rootstocks have shown excellent performance in a commercial orchard irrigated with water containing 300–400 ppm of chlorine.

Gae Fuerte.—Orig. in Yorba Linda, Calif., by George Kellogg. Introd. in 1954. Bud mutation of Fuerte; discovered about 1935. Fruit: averages 7.6 oz; seed averages 1.7 oz; extends the Fuerte season, sometimes maturing as late as August; resembles Fuerte. Tree: productive. No longer propagated.

Gano.—Orig. in Whittier, Calif., by Jennie C. Gano. Introd. in 1935. Plant patent 100; 7 Aug. 1934. Open-pollinated seedling of Colorado selected in 1918. Fruit: pyriform, rather long, 5 in.; skin thick, tough, dull green; size averages. 1 lb.; flesh light yellow, firm, little fiber, oil content about 20%, flavor excellent; season July and August. Tree: vigorous, upright; stands frost slightly better than most Guatemalan types, but less resistant than Fuerte; bears well; leaves broader at base than Fuerte, with pronounced veining. Now obsolete.

Gem.—Orig. in Fullerton, Calif., by Lawrence W. Sherwood, Sherwood Specialty Nursery. Introd. in 1947. Plant patent 773; 30 Dec. 1947. Chance seedling; discovered among a number of trees grown from seeds of a Mexican seedling of unknown parentage planted about 1907. Fruit: 8 to 12 oz, averaging 4 1/2 inches long and 2 1/2 inches in diam.; pyriform; skin bright purple, attractive, smooth, glossy; seed small; flesh pale yellow, flavor mild, buttery, oil content high, quality very good; ripens 15 Sept. in southern Calif. Tree: upright spreading; bears well; blooms 1 1/2 to 2 months before Fuerte. Now obsolete.

Gil.—Orig. at Akko Expt. Station, Western Galilee, Israel, by E. Lahav, D. Zamet, and S. Gazit. Seedling of Tova by unknown pollen parent. Introd. in 1995, to be patented. Fruit: pear-shaped with a short neck, medium to relatively large, 9 to 14 oz, flesh pale yellow with green margins near the skin, buttery firm to slightly doughy at end of harvest season; flavor good, somewhat nutlike but with occasional slight bitterness; skin black, pebbled, similar or slightly thicker than that of Hass, easily separated from flesh; seed relatively large (16% to 18% of fruit weight), tight in cavity; season about 6 months, from end of December to June. Tree: upright and medium-small (smaller than Hass), without anise scent in leaves, similar to Hass in cold resistance. Flowering group A; planted throughout the avocado growing region in Israel; recommended for high summer temperature areas where the relatively large fruit size may be reduced.

Gossman.—Orig. in Homestead, Fla., by Otis L. Gossman. Introd. in 1958. Parentage unknown; discovered before 1958. Fruit: flavor and texture good; ripens late; somewhat resembles Lula; flowering group B; minor commercial cultivar in Florida.

Gottfried.—Orig. in Miami, Fla., by the U.S. Plant Introduction Station. Introd. in 1920. Open-pollinated seedling of a Mexican seedling growing at the residence of Edward Gottfried, Key Largo; first fruited in 1914; selected in 1918; P.I. 46337. Fruit: pyriform; 12 to 20 oz; skin smooth, purplish-maroon when mature; ripens August to October. Tree: vigorous; hardy; does not bear well on the lower Florida east coast, better on its west coast; flowering group A; for home gardens in marginal (cold winter) cultural areas.

Graham.—Orig. in LaHabra, Calif., by Melville M. Graham. Introd. in 1945. Plant patent 662; 27 Nov. 1945; assigned to Armstrong Nurseries, Ont., Calif. Open-pollinated seedling of Lyon; selected in 1944; Guatemalan type. Fruit: medium to large, averaging 16 to 18 oz, with a range from 11 to 29 oz; pyriform; skin dark green, oil content high, averaging 25% to 27%, flavor excellent, good keeping quality, flesh resistant to discoloration when exposed to air; matures April to June. Tree: vigor medium; habit moderately upright; alternate bearing; fairly tender to frost; more easily propagated than Lyon. Nursery trees sold 1945 to 1951, mostly for home plantings; discontinued in 1952 because of lack of demand.

Greengold.—Orig. in Haleakala Branch Station, Hawaii by R.A. Hamilton, Univ. of Hawaii. Introd in 1982. Sharwill x open-pollinated seedling. Cross made in 1962; selected in 1970; tested as HAL R27T8. Fruit: size 12 oz; oval; skin green; seed 16% of fruit weight; flesh quality good; ripens February to May, good handling quality. Tree: medium, upright; very productive, tendency to alternate bearing; vigor medium; tolerant to ethylene dibromide fumigation, anthracnose, thrips, and rust mite damage.

Gulfstream.—Orig. in West Palm Beach, Fla., by G.W. Smith. Introd. in 1944. Parentage unknown but assumed to be a hybrid of the West Indian race with a variety of the Guatemalan race; seed planted in 1934; first fruited in 1940. Fruit: size medium, averaging 16 to 18 oz; shape pyriform; skin smooth, bright light green, glossy; flesh bright yellow, oil content 11%, quality good; season mostly October with a range to December; seed large. Tree: flower type B; tends to alternate bearing. No longer propagated.

Gwen.—Orig. in Univ. of California's avocado breeding program by B.O. Bergh; seedling of Hass. Introd. in 1982. Plant patent 5298; 23 Oct. 1984. Fruit: size medium, 10 oz; flavor excellent; skin green, moderately thick and rough, easily separated from flesh; seed small. Tree: heavy bearer (approx. 20,000 lbs/acre). Flowering group A.

Hall.—Orig. in Miami, Fla., by Willis Hall. Introd. in 1938. Parentage unknown (probably a West Indian x Guatemalan hybrid); selected in 1937. Fruit: large; pyriform; skin bright green, handsome; season November to March, resembles Monroe in being excellent for the limited fancy fruit market. Tree: alternate but heavy bearer; flowering group B; major commercial cultivar in Florida.

Harms.—Orig. in Carlsbad, Calif., by Carl F. Harms. Introd. about 1952. Parentage unknown, but of Guatemalan race; tree planted about 1940; selected about 1946. Fruit: pyriform; 7 to 9 oz; skin medium pebbled, green; flavor good, with high oil content; season July to October; keeping quality very good; seed size medium. Tree: blooms late; appears resistant to thrips; claimed to have cold resistance; still planted in California dooryards.

Harris.—Orig. in Homestead, Fla., by W. K.

Walton. Introd. in 1940. Open-pollinated seedling of Wagner; selected in 1935. Fruit: size medium; obovate; skin dull, dark green; pebbled, resembles Booth 7; flesh light yellow, with some fiber, quality good, oil content 10% to 16%; season mid-December to mid-February; seed medium to large, tight in cavity; resembles Wagner. Tree: heavy bearer; flowering group A. No longer propagated.

Hartman.—Orig. in Vista, Calif., by Paul L. Kovach. Introd. in 1947. Parentage unknown; selected in 1936. Fruit: oblate; skin green, smooth, peels easily from flesh; flavor good; oil content about 22%; matures during late July and early August; resembles a large Fuerte; good market acceptance. Tree: vigorous; upright, stately; symmetrical; budded trees mature fruit earlier. No longer propagated.

Hass.—Orig. in LaHabra, Heights, Calif., by Rudolph G. Hass. Introd. in 1936. Plant patent 139; 27 Aug. 1935. Parentage unknown (but of Guatemalan type); selected in 1934. Fruit: size medium, 10 oz; flesh creamy, no fiber; flavor excellent; oil content 23.7%; skin leathery, rough, purple when ripe, thick; seed small, tight in cavity; keeping qualities excellent; good shipper; ripens in summer. Tree: heavy bearer; starts bearing second year; thrifty grower; propagated readily; flowering group A. The major commercial cultivar in California, with 88% of avocado acreage in 1996.

Hazzard.—Orig. in Vista, Calif., by A.G. Hazzard. Introd. in 1932. Probably an open-pollinated seedling of Lyon; discovered in 1928. Fruit: pyriform, decidedly oblique; 12–15 oz; skin thin, green, rather rough, peels freely; seed small to large; flavor good to excellent; oil content high; season April to July; suitable only for home gardens, where it is still grown, as the crooked neck makes it unsuited for commercial packing.

Hellen.—Orig. in Santa Monica, Calif., by Mrs. S.L. Duey. Introd. in 1941. Parentage unknown. Fruit: 9–14 oz; pyriform to round; skin dark green; seed tight; oil content 21%; flavor good; season June to September. Tree: upright; round. No longer propagated.

Henry Select.—See Henry's Select.

Henry's Select *(Henry Select)*.—Orig. in Escondido, Calif., by Jesse L. Jones and Charles C. Henry; introd. in 1937. Plant patent 234; 9 Feb. 1937; taken out by Henry, assignor of one-half to Jones. Parentage unknown but may be a Mexican seedling; discovered in 1931. Fruit: broadly ovate; 12–14 oz; skin deep maroon-purple, thin, smooth, glossy; flesh smooth, no fiber; nutty flavor; oil content 18%; seed small, conical, loose in cavity; season September to November at Escondido. Tree: usually does not bear well nor consistently, but produces satisfactorily at Escondido. Obsolete.

Herman.—Orig. in South Miami, Fla., by Fred Herman. Introd. in 1940. Parentage unknown; selected in 1937. Fruit: quality good; season November through January; resembles Lula, but skin smoother. Tree: precocious and heavy bearer; flwering group A; minor commercial cultivar in Florida.

Hickson.—Orig. in Naranja, Fla., by J.R. Hickson. Introd. in 1936. Parentage unknown; selected in 1934. Fruit: quality excellent; ships well; season November to January; similar to Booth 8 in shape, but brighter green skin. Tree: fair bearing ability; class B for pollination purposes.

Iriet.—Orig. and tested at the Akko Expt. Sta. in western Galilee by E. Lahav, D. Zamet, and S. Gazit. Hass open-pollinated. Plant patent applied for. Fruit: similar to Hass but larger, 300 to 500 g; peel thinner than Hass, thicker than Fuerte; pear-shaped, uniform; skin black, glossy, slightly pebbly; seed round to slightly elongated, 10% to 15% of fruit weight. Flesh green-yellow; buttery texture; excellent nut-like flavor. Storable for 14 days at 5C, 95% relative humidity; softens at room temperature in about 9 days; shelf life 3 to 5 days. Tends to be biennial; season begins in off years in December, in on years in March. Tree: canopy small to medium, slightly weeping form; precocious, fruiting second year after planting. Flowering group B.

Irving.—Orig. in Carlsbad, Calif., by Irving Cummings. Introd. in 1939. Chance seedling; discovered in 1920. Fruit: 8–10 oz; pyriform; seed small; skin green, smooth, glossy; flavor excellent; oil content high;

season December to February; resembles Fuerte. Tree: weak grower, tends to droop (weeping); bears erratically; appears equally satisfactory for both inland and coastal districts in California. Not planted since 1960.

Itzamna.—Orig. in Santa Maria de Jesus, Guatemala; imported into U.S. in 1916 by F.W. Popenoe, U.S. Dept. of Agriculture. Introd. in the U.S. in 1932; P.I. 43486, P.I. 55736. Parentage unknown; of Guatemalan type. Fruit: 12–18 oz; pyriform to round; skin light green, thick; seed small; flesh quality good in California, poor to fair in Florida; oil content 15%; season September to December in southern California, 15 Feb. to 15 Apr. in Florida; storage quality sometimes poor. Tree: alternate bearing; fairly susceptible to anthracnose; flowering group B.

Jalna.—Orig. in Vista, Calif., by J. Eliot Coit. Introd. in 1936 by Armstrong Nurseries, Ontario, Calif. Parentage unknown (but of Mexican type); selected in 1933. Fruit: pyriform; skin green, thin, end-spots; quality fair; matures in early fall. Tree: heavy bearer, but tends to alternate; good frost tolerance; minor commercial cultivar in California.

Kahaluu.—Orig. in Kahaluu, N. Kona, Hawaii. Introd. in 1955. Guatemalan type x West Indian type; selected in 1940. Fruit: obovate to pyriform; 12–18 oz; skin smooth, green; seed small, tight in cavity; oil content high; quality excellent; ripens November to January; resembles Fuerte. Tree: large; upright; vigorous; annual bearer, but only moderately productive.

Katherine.—See DuPuis 3.

Kilgore Special.—Orig. in Clearwater, Fla., by Barnard Kilgore. Introd. in 1936. Open-pollinated seedling of Fuerte; selected in 1930 from seed sown in 1927. Fruit: 12–18 oz; obovate; skin green, smooth, leathery; flesh pale yellow; quality only fair; seed size medium; season December and January. Tree: bears heavily; blooms very early; flowering group B; appears more resistant to anthracnose than Fuerte; about as hardy as Fuerte. No longer propagated.

Kimball.—Orig. in Monrovia, Calif., by Ingegerd Isakson, Arcadia, Calif. Introd. in 1953. Plant patent 1438; 20 Dec. 1955; assigned by direct and mesne assignments of 51% to Arthur S. Kimball, Monrovia, and 49% to Weldon C. Wilkins, El Monte, Calif. Parentage unknown; Mexican type; seed secured from Gorden Maddock of Duarte, Calif., planted in early 1930s. Fruit: small, averaging about 6 oz; skin medium green, thin, smooth; flavor very good; seed size medium; season September and October; resembles Duke. Tree: a little more hardy than Mexicola; bears profusely. Now obsolete.

Leucadia.—Orig. in Encinitas, Calif., by J. Eliot Coit. Introd. in 1932. Parentage unknown (but of Mexican type); selected in 1929. Fruit: size good; skin handsome, purple, thin, smooth; resembles Puebla. Tree: moderately spreading. No longer propagated.

Lindgren.—Orig. in Goulds, Fla., by Mrs. Augusta Lindgren. Introd. in 1939. Hybrid of Guatemalan x West Indian types; selected in 1932 from seed sown in 1928. Fruit: size medium, 14–20 oz; round-oval; skin dark, bright green, nearly smooth, thick, fragile; flesh yellow; seed small; season mid-December to mid-February. Tree: bears regularly; flowering group A.

Lisa.—Orig. in Goulds, Fla., by H.E. Kendall. Introd. in 1974. Plant patent 3737; 1 July 1975. Chance seedling of West Indian and Mexican types; discovered in 1969. Fruit: small, 9 to 14 oz, size increases as harvest season progresses; pyriform to oval; skin dark green, thin, shiny, slightly pebbled, free of corky lenticel; flesh greenish-yellow, smooth, nearly free of fiber; flavor excellent, nutty, strong, sometimes slightly sweet, resembles that of Mexican hybrids; seed size small to medium, tight in cavity; ripens late August through October, softens after 9 days at 70F. Tree: about 18 ft; vigorous; 24 fruit per shoot; blooms in March; flower type B; easily propagated.

Lula.—Orig. in Miami, Fla., by George B. Cellon, nurseryman. Introd. in 1921. Open-pollinated seedling of Taft; selected in 1919. Fruit: pyriform; skin almost smooth, dark, glossy green; quality good; large seed, tight in cavity; fair market appeal; very susceptible to avocado scab. Tree: vigorous and upright, prolific and regular bearer; class "A" for pollination purposes.

MacArthur.—Orig. in Monrovia, Calif., by Thomas H. Shedden. Introd. in 1922. Parentage unknown. Fruit: large, 14–18 oz; pyriform; skin green; flavor fair to good; oil content medium; season August to October. Tree: vigorous; spreading; high N-requiring variety; bears well in coastal districts of California and is a minor commercial variety in Ventura and Santa Barbara counties.

Macpherson.—Orig. in Encinitas, Calif., by James H. Macpherson. Introd. in 1942. Plant patent 433; 26 Nov. 1940. Parentage unknown; selected in 1936. Fruit: 10–12 oz; 4 1/2–5 inches long, asymmetrically pyriform; appearance and flavor good; skin green, smooth; flesh deep, golden yellow; oil content up to 23%; seed small, tight in cavity; season December to April. Tree: large; vigorous grower; hardy. Now obsolete.

Major.—Orig. in Homestead, Fla., by Herbert Katz. Introd. in 1942. Guatemalan type x West Indian type; selected about 1942. Fruit: 30–40 oz; elliptical; skin bright green, pebbled, rough, attractive; flesh yellow; oil content medium; quality good; seed small, fairly tight in cavity; flowerin group B. No longer propagated.

Marfield.—Orig. in Homestead, Fla., by H.A. Marsh and Austin Fields. Introd. in 1943. Parentage unknown; West Indian type; selected in 1938. Fruit: 12–20 oz; pyriform with small neck; skin smooth, leathery, bright green; flesh yellow; seed size medium, tight in cavity; season late June and July. Tree: bears well. Flowering group A. No longer propagated.

Marguerite.—Orig. in Homestead, Fla., by Herbert Katz. Introd. in 1948. Hybrid of Guatemalan x West Indian types; selected in 1942. Fruit: 20–30 oz; oval; skin nearly smooth, thin, bright green; flesh yellow; quality fair to good; oil content 8%; seed size medium, tight in cavity; season October; resembles Monroe. Tree: fairly heavy bearer. Flowering group B.

Mary Martin.—Orig. in San Diego, Calif., by John Martin Reinecke. Introd. in 1943. Plant patent 576; 20 Apr. 1943; assigned to Armstrong Nurseries, Ontario, Calif. Open-pollinated seedling of Linscott; selected about 1938. Fruit: medium large, 9–15 oz; broadly obovoid, no neck; skin smooth, medium thick, tough, green; flesh texture fine; flavor nutty, rich, aroma distinct; quality very good; oil content up to 23%; seed medium-large, 1 1/2–2 1/2 oz; ships well; season June through August; resembles Nabal. Tree: upright; round. No longer propagated.

Mayapan.—Orig. in Purula, Dept. Baja Vera Paz, Guatemala. Imported in 1917 into the U.S. by Wilson Popenoe, U.S. Dept. of Agriculture, as P.I. 44680. Introd. in 1924 in the U.S. Fruit: size medium, 13–16 oz; round to roundish obovoid; skin greenish-purple to dull purple, very thick, separates readily from flesh; flesh cream-yellow; flavor pleasant; quality very good; seed size medium, 1 1/3–2 oz, tight in cavity; midseason in ripening. Tree: very productive.

Mayo *(Covacado)*.—Orig. in Covina, Calif. Introd. in 1947. Parentage unknown; Mexican type; discovered in the early 1940s; propagated by Harry Mayo and promoted by Stuart Coulter. Fruit: 4–8 oz; oval; skin light green with a tendency for brown blemishes, very thin, peels easily from flesh; season early October through November; rather low oil content by California standards, but seems to taste good at 7% and will obtain at least 10% oil content. Tree: quite resistant to frost; upright grower; production extremely heavy and consistent; not grown commercially.

Mesa.—Orig. in Carpinteria, Calif., by Stanley L. Shepard. Introd. in 1936. Parentage unknown; discovered in 1936. Fruit: oval; skin green with a slight copper color; quality good; matures in early summer. Tree: vigorous; upright-spreading; produces good crops annually; not too susceptible to frost. Obsolete.

Middleton.—Orig. in Pomona, Calif., by E.E. Middleton. Introd. in 1941. Parentage unknown; Mexican type; selected about 1932. Fruit: 8 oz; pyriform; skin dark purple; oil content medium; flavor excellent; high quality; season just before Fuerte. Tree: tall, slender; some frost resistance. No longer propagated.

Miguel.—Orig. in South Dade County, Fla., by H.E. Kendall. Introd. in 1974. Plant patent 3734; 24 June 1975. Chance seedling, appears to be West Indian parentage; discovered in 1970. Fruit: 20 to 48 oz; elliptical, cleft at stylar end; skin dark green, smooth, shiny, slightly leathery, peels easily, free of cork lenticels; flesh yellow, smooth, buttery, free of fiber; flavor excellent, nutty; seed medium, tight in cavity; ripens late August through September; softens after 6 to 10 days at 70F. Tree: fairly spreading; vigorous; blooms late January to March; flower type B; one to two fruit per stem; major commercial cultivar in Florida.

Mitchell.—Orig. on the Mitchell place between Cutler and South Miami, Fla. Introd. in 1930. Parentage unknown; West Indian type. Fruit: 2–3 lb; short obovate; skin very smooth, dark green, pliable; flesh yellow; seed size medium; season August and September. Tree: mentioned as bearing well; flowering class A.. No longer propagated.

Monica.—See Clifton.

Monroe.—Orig. in Homestead, Fla., by J.J.L. Phillips. Introd. in 1937. Plant patent 261; 24 Aug. 1937; taken out by J.R. Byrum, assigned to Redland Avocado Grove, Homestead. Parentage unknown; probably a West Indian x Guatemalan hybrid; selected in 1935. Fruit: large, 26–36 oz; skin dark green, glossy, attractive; flesh yellow, slightly green near skin, fiberless; oil content 10% to 30%; flavor nutty; season December and January. Tree: heavy bearer; flowering group B.

Mundo.—Orig. at the Vista del Mundo ranch in Santa Barbara, Calif., by H.F. Pierce. Introd. in 1938. Open-pollinated seedling of Lyon; selected about 1932. Fruit: skin light green; oval to round; averages 17.5 oz; seed averages 3 oz, tight to flesh; flavor good; oil content averages 20.8%; ripens during August. Tree: heavy and regular bearer. No longer propagated commercially.

Murray Red.—Orig. in Stuart, Fla., on property of James W. Mills. Introd. in 1969. Plant patent 2948; 9 Dec. 1969, by Howard E. McCormick, Stuart, Fla. Parentage unknown. Seedling discovered in 1967 by Mr. McCormick, Rio Nursery, Stuart. Fruit: size medium, 12 to 20 oz, 4 to 7 inches long, pyriform; skin brightly covered with red over yellow ground color, flecked appearance, nearly smooth; seed medium tight in cavity; flesh yellowish-green, flavor pleasant, mild, nut-like, faintly sweet, not watery or greasy; ripens latter part of September through October. Tree: upright; hardiness moderate.

Nabal.—Orig. in Guatemala, imported into the U.S. by F.W. Popenoe, U.S. Dept. of Agriculture. Introd. in California in 1927, in Florida in 1937. Parentage unknown; Guatemalan type. Fruit: medium, 16–24 oz; almost round; skin smooth, dark green; flesh yellow; good quality; oil content 10% to 15%; seed medium-small, tight in cavity; season January and February. Tree: strongly alternate bearer. Flowering group B.

Nadir.—Orig. in Homestead, Fla., in the Frank Upchurch grove. Introd. in 1950; named by W.H. Krome, Homestead. Selected in 1942 by Krome; first propagated in 1943; a West Indian type, possibly hybridized with a Guatemalan type. Fruit: 10–20 oz; pyriform; skin green, smooth, thin; quality fair; seed small, tight in cavity; midseason, July and August, slightly ahead of Waldin; flowering group A; major commercial cultivar in Florida.

Nena.—Orig. in Atlixco, Puebla, Mexico, by Wilson Popenoe, Escuela Agricola Panamericana, Tegucigalpa, Honduras. Introd. in Honduras in 1964. Parentage unknown; Mexican type; selected in 1948. Fruit: large for this type, about 3 1/4 inches long and up to 2 1/2 inches in diameter, 6–8 oz; skin dark purple, glossy, smooth, very attractive; lenticels few and yellowish; flesh cream changing to greenish near skin, very slight fiber markings; buttery, nutty flavor; oil content 19% to 25%; quality good; seed small fc• this type, tight in cavity, both seedcoats adhering closely; season October and November. Tree: erect, broad; branches stiff; productivity appears to be quite satisfactory.

Nichols.—Orig. in Los Angeles, Calif., by Royce S. Bringhurst, California Agr. Expt. Sta. Introd. in 1962. Open-pollinated seedling of (Fuerte x Mexicola); planted in 1953 in California Agr. Ext. Serv. test plot at the Ernest R. Nichols orchard by Nichols and Calvin C. Delphey, farm advisor, Ventura County. Fruit: pear-shaped; 12–15 oz; skin green; flesh color creamy; quality excellent to good, very little fiber; oil content 13% in January, over 20% in April; resembles Fuerte. Tree: produces heavy crops. Not propagated commercially.

Norman.—Orig. in the Hyman and Katzin Grove, Homestead, Fla., by the Florida East Coast Fertilizer Co. of Homestead. Introd. in 1957. Parentage unknown; Guatemalan type. Fruit: 16–19 oz; skin bright green; quality good; season January. Tree: tall but more spreading than Lula or Taylor; yields averaged 5 bushels per year for 1953–58. Flowering group A; not propagated commercially.

Nowels.—Orig. in Huntington Park, Calif., by M.E. Nowels. Introd. in 1940. Chance seedling. Fruit: 5–10 oz; small when heavy crops are produced; skin green, russets; flavor good; oil content high; season, October to December; less frost tolerance than Fuerte. Tree: open-spreading; tends to bear in alternate years. Not propagated commercially.

Olivia.—Orig. in Leucadia, Calif., by H.E. Kemp and Olivia R. Kemp. Introd. in 1952. Parentage unknown; Guatemalan type; seed planted about 1932; selected in 1949. Fruit: weight about 8 1/2 oz; skin green, medium pebbled, peels easily; flavor good; oil content 24.5%; seed averages 2 oz; season August and September. Tree: upright; quite hardy. Not propagated commercially.

Peterson.—Orig. in Homestead, Fla., on the Murray Place by Peter Peterson of Modello, Fla. Introd. about 1930. West Indian type; discovered before 1912; first fruited in 1920 at Modello. Fruit: 14–20 oz; obovoid; skin green, smooth, glossy; flavor good; oil content 4% to 7%; season September through November. Flowering group A; minor commercial cultivar in Florida.

Pinkerton.—Orig. in Saticoy, Ventura County, Calif., by J. Pinkerton, Ventura. Introd. in 1974. Plant patent 3712; 29 Apr. 1975. Seedling of Hass x Rincon; discovered in 1960. Fruit: medium-large; pyriform; skin green, leathery, peels easily; flesh smooth, buttery, fiberless; oil content up to 25%; seed small; excellent commercial variety; possible Fuerte replacement; ripens November through January; hold until late June on tree. Tree: large; spreading; very vigorous; heavy bearer.

Pollock.—Orig. in Miami, Fla., by H.S. Pollock before 1896. Introd. in 1901. West Indian type. Fruit oblong to pyriform, very large 30–50 oz, occasionally to 5 lbs; skin smooth, light glossy green; flesh a rich yellow blending with green near the skin, of excellent quality; oil content 3% to 5%; seed size large, loose in cavity; season early, July to September. Tree: vigorous but a shy bearer. Flowering type B; major commercial cultivar in Florida.

Ponteprino.—Orig. in Anaheim, Calif., by Meyer E. Ponteprino. Introd. in 1952. Plant patent 1080; 25 Mar. 1952. Parentage unknown, Guatemalan type; discovered in 1937. Fruit: 8–16 oz; pyriform; skin dark green, smooth, medium thickness, peels readily; flesh rich yellow; flavor very good; oil content high; seed small, 1 1/2 oz; harvest season May to October; keeps well. Tree: upright; yield good. No longer propagated.

Reed.—Orig. in Carlsbad, Calif., by James S. Reed. Introd. in 1960. Plant patent 1967; 16 Aug. 1960. Chance seedling; hybrid type; planted in 1948. Fruit: small, averaging about 3 inches long and 8–10 oz; ripens in April; flesh smooth, firm, buttery, with oil content averaging 18.9% to 20%; may remain on tree for a relatively long time after reaching maturity; skin medium smooth, medium green when mature, thick, peels easily; ships well; keeps well; resembles a small Nabal. Tree: slender, withstands frost to 28F; resistant to salt burn and sun-blotch. Still of minor commercial importance in California.

Regina.—Orig. in Los Angeles, Calif., by V.L. Gerardi. Introd. in 1948. Chance seedling; discovered in 1944. Fruit: small, averaging 5 oz; pyriform; skin green, tendency to develop corky spots; flavor good; oil content high; seed small; season February and March; resembles Lodge. Tree: alternate bearer; adapted only to coastal regions. Not propagated commercially.

Rincon.—Orig. in Carlsbad, Calif., by Sam Thompson. Introd. in 1948. Chance seedling; discovered before 1945. Fruit: pyriform; 6–10 oz; seed 1–1 1/2 oz; skin thickness medium, smooth, leathery, peels well;

flesh fine; oil content medium; flavor rich; season January to April; resembles Fuerte. Tree: medium to small; weak grower; consistent and heavy bearer; minor commercial variety in coastal areas.

Routh.—Orig. in San Gabriel, Calif., by Herbert L. Gillespie. Introd. in 1953. Plant patent 1208; 4 Aug. 1953. Parentage unknown; Guatemalan type; seed planted in 1942; first fruited in 1948. Fruit: matures over a long period, making for market flexibility; skin bright green, medium smooth surface; pyriform in coastal areas to slightly necked inland; 5.5–8.5 oz. Tree: compact; flowering and fruit set prolonged through several months. Not propagated commercially.

Roy.—Orig. in Kainaliu, Hawaii, on the property of T.C. White by W.T. Pope, Hawaii Agr. Expt. Sta., Honolulu. Introd. in 1926. Macdonald x an unidentified West Indian variety. Fruit: Guatemalan type; flavor nutty, good; oil content 27%; average weight 30 oz; skin deep purple; matures in late fall. No longer propagated.

Ruehle.—Orig. in Homestead, Fla., at the Florida Subtropical Expt. Sta. and described by John Popenoe of this station. Introd. in 1963. Open-pollinated seedling of Waldin; seed planted in 1933; first propagated in 1946; West Indian type; tested as Fla. 27-1. Fruit: averages about 15 oz, ranging from 10–20 oz; pear-shaped; skin bright green, fairly thin; flesh yellow with a thin greenish layer next to skin; flavor good; oil content low, from 2% to 5%; seed relatively small, averaging 12% of fruit by weight; ripens with Pollock, about the last week of July, with acceptable fruit on the tree through September; fairly resistant to avocado scab, *Cercospora* spot, and anthracnose. Tree: production very heavy; growth habit more erect than Waldin, more spreading with a better shape than Lula; blooms early, flower type "A"; in home gardens, satisfactory crops produced without cross-pollination; lacks hardiness. Named in memory of George D. Ruehle (1898–1962), director of the station for 18 years. Not propagated commercially.

Russell.—Orig. in Islamorado, Fla., by John Russell. Introd. in 1935. Seedling of West Indian type from seed of fruit brought from Cuba about 1880. Fruit: large, 24–36 oz; often 12 inches long; clavate; skin green, smooth, leathery; flesh yellow; seed size medium; cavity low in the broad end of fruit, often a solid neck of flesh 5–6 inches long; season August and September. Tree: vigorous; moderately productive. Flowering group A. Not propagated commercially.

Rusterholz.—Orig. in Whittier, Calif., by Katherine Lillian Rusterholz. Introd. in 1950. Plant patent 969; 25 July 1950. Parentage unknown; probably a seedling of Lyon; Guatemalan type discovered in 1925. Fruit: medium to large, 4–5 inches long, 8–16 oz; oval to pyriform; skin purple, thin, pliant; seed size medium, tight to flesh; flesh smooth, creamy, light green, no fiber; season late November to January; resembles Lyon. Tree: upright; some frost resistance. Not propagated commercially.

Ryan.—Orig. in Whittier, Calif., by A.R. Rideout. Introd. in 1936. Hybrid between Guatemalan and Mexican types; possibly an open-pollinated seedling of Amigo. Fruit: pyriform; 10–14 oz; skin green, slightly rough, leathery; seed large; quality fair to good; oil content high; season May to September. Tree: vigorous; propagates readily; suitable for planting in inland areas of California. No longer propagated commercially.

Schmidt.—Orig. in Mexico and imported as budwood in 1911 by Carl Schmidt for the West India Gardens, Altadena, Calif. Introd. in Florida in 1922. Parentage unknown; Guatemalan type. Fruit: medium to large, 16–26 oz; pyriform; skin rough, pebbled, dark green; flesh light yellow; good flavor; oil content medium; seed size medium, adheres tightly; season 1 Feb. to 1 Apr.; does not store well; easily injured by frost. Tree: weak grower; poor bearer. No longer propagated.

Scott.—Orig. in Santa Ana, Calif., by Charles C. Scott. Introd. in 1959. Plant patent 1815; 3 Mar. 1959. Apparently a seedling of the Mexican avocado (*Persea drymifolia*); discovered in 1952. A rootstock clone: vigorous; tolerant to sodium bicarbonate and sodium chloride; apparently resistant to lime-induced chlorosis; cuttings root easily. Not propagated commercially.

Sharwil.—Orig. in Redland Bay, Queensland, Australia, by Frank V. Sharpe of Clayfield, Brisbane. Introd. there in 1954; now grown in three states in Australia, in Spain, and in the U.S. in Florida and California. Chance seedling; selected in 1951. Fruit: 7–13 oz; somewhat more ovoid than Fuerte; skin green, thickness medium; seed small; market quality excellent; season August to November in Australia (Fuerte's season there April to July); resembles Fuerte. Tree: hardy; growth dwarf; produces crops regularly; flowering group A.

Shepard.—Orig. in Carpinteria, Calif., by Stanley L. Shepard. Introd. in 1958. Open-pollinated seedling of Edranol; Guatemalan type; selected in 1952. Fruit: small to medium, 7 1/2 oz; pyriform; skin green, thick, peels easily; appearance very good; quality high; flavor excellent; oil content medium; keeping quality good; apparently frost resistant; resembles Edranol. Tree: spreading; heavy producer. Not propagated commercially.

Silliman.—Orig. in LaHabra, Calif., on the Silliman place. Introd. about 1945. Chance seedling planted about 1930; Mexican Guatemalan hybrid type; selected about 1938; propagated by Walter Beck, LaHabra. Fruit: 4–8 oz; pyriform; skin green, leathery, corky; flavor good; seed size medium; season March to June. Tree: large; spreading; upright; central-leader type; not a consistent producer; more susceptible to frost than Fuerte. Not propagated commercially.

Simmonds.—Orig. in Washington, D.C., by U.S. Dept. of Agriculture, probably from a seed of Pollock sent from Miami in 1908. The seedling was sent back to Miami where it grew to maturity and was distributed, without a name, in 1916–17. It was propagated commercially in 1921 and was named in 1923. West Indian type. Fruit: oblong-oval to pyriform, large, 18–24 oz; skin smooth, light to yellowish green, flesh rich yellow, green near the skin, of very good quality; oil content 3% to 6%; midseason, ripening 15 July to 15 Sept. Tree: compact, spreading growth less vigorous than Pollock, and more regular in bearing. Flowering type A; major commercial cultivar in Florida.

Simpson.—Orig. in Richmond, Fla., by Robert Simpson. Introd. in 1933. Parentage unknown (considered to be a West Indian x Guatemalan hybrid); selected in 1925. Fruit: large; obovate; skin light green, smooth; season November to December. Tree: flowering group B; good crops produced consistently; minor commercial cultivar in Florida.

Skinner.—See Dunedin.

Stewart.—Orig. in Mentone, San Bernardino County, Calif., on the Stewart Ranch. Introd. in 1956; first propagated for trial at the California Citrus Res. Center, Riverside, in 1952. Chance seedling possibly of Mexicola; Mexican type. Fruit: medium, 6–10 oz; pyriform; skin green, but at maturity a rich dark purple, slightly rough, raised lenticels sparse near base but very numerous at apex, corkiness none to very slight, texture soft-leathery, excellent peeling quality; flesh clear, bright, light yellow shading to green toward skin; firm, but melting; excellent quality; ripens from early October to mid-December; oil content about 8% by first of October to about 16% in mid-November. Tree: spreading; strong; vigorous; bears good crops. Commercial cultivar in South Africa.

Susan.—Orig. in Baldwin Park, Calif., by J.M. Allred, Covina, Calif. Introd. in 1953. Plant patent 1474; 1 May 1956. Parentage unknown; discovered in 1944; selected in 1949; Mexican type. Fruit: 4–8 oz, average of 5 1/2 oz; ovoid, 3 1/4 inches long, 2 1/2 inches in diam.; skin thin, smooth, green; flesh yellow; flavor nutty and rated good plus; consistency buttery; oil content up to 17%; season September to mid-November; resembles Fuerte. Tree: frost resistant; produces well. Limited commercial production in California.

Taylor.—Orig. in Miami, Fla., by U.S. Dept. of Agriculture. Open-pollinated seedling of Royal; Guatemalan type. Fruit obovate to pyriform, small to medium size, 12–18 oz; skin rough, dark green, glossy; seed size medium, tight in cavity; flesh light yellow, of very good flavor, oil content 13% to 17%; late, ripening December to February. Tree: upright, vigorous, cold-hardy. Flowering type A; minor commercial cultivar in Florida.

Thille.—Orig. in Santa Paula, Calif., by J.

Nicholas Thille. Introd. in 1954. Open-pollinated seedling of Hass; seed planted in 1946; tested as Hass 4 and Thille 4. Fruit: size medium; oblate; skin green, thick, peels reasonably well, appearance good, tends to russet late in season in inland areas; oil content high; flavor good; season July to November. Tree: tall; upright; vigorous; good yields. Not propagated commercially.

Tomko.—Orig. in Carlsbad, Calif., by Sam Thompson. Introd. in 1944. Plant patent 628; 6 June 1944; assigned to Paul L. Kovach, Vista, Calif. Parentage unknown, but considered to be a seedling of (Cantel x Fuerte). Fruit: 11 oz; pyriform, neck prominent, stem offset; ripens after Fuerte; skin green, glossy, leathery, smooth, peels easily; flesh buttery; oil content 24%; seed tight in cavity; ripens from mid-May to mid-July or later, after Fuerte. Tree: upright spreading. Obsolete.

Tonnage.—Orig. in Homestead, Fla., by Dan Roberts. Introd. in 1930. Open-pollinated seedling of Taylor; Guatemalan type; selected in 1921. Fruit: 16–24 oz; pyriform with stem placed obliquely, often necked; skin dark green, pebbled, rather glossy, thick; flesh pale yellow; oil content 8% to 10%; seed size medium, fairly tight; season 15 Oct. to 1 Dec. Tree: very productive; flowering group B; major commercial cultivar in Florida.

Trapp.—Orig. in Coconut Grove, Fla., by H.A. Trapp. Introd. in 1901. West Indian type. Fruit: roundish-ovate or slightly pyriform, medium to large size, 12–24 oz; skin smooth, light to yellowish green; seed size large, loose in cavity; flesh golden yellow to pale green near the skin, of excellent flavor; oil content 3% to 6%; late midseason, ripening September to November. Tree: comparatively weak grower and sometimes carries few fruit to maturity, but usually a consistent bearer of good crops. Flowering type B; minor commercial cultivar in Florida.

Vaca.—Orig. in Homestead, Fla., by W.H. Krome. Introd. about 1945. Chance seedling discovered in 1932. Fruit: 16–24 oz; obovoid to pyriform; skin light green, glossy, smooth, medium thick, slightly granular, separates readily from flesh, lenticels slightly raised; flesh yellow, of very good quality; seed loose in cavity; season October to February.

Waldin.—Orig. in Homestead, Fla., by B.A. Waldin. Introd. in 1917. West Indian type. Fruit: oblong to oval and characteristically flattened on one side at the blossom end; of medium to large size, 14–28 oz; skin smooth, pale green to greenish yellow; seed size medium to large, tight in cavity; flesh pale to greenish yellow, of good flavor, oil content 6% to 10%; midseason, August to September. Tree: precocious with a tendency to overbear; hardier than most West Indians. Flowering type A; major commercial cultivar in Florida.

Winter Mexican.—Orig. in Palm Beach, Fla. Introd. in 1936. Hybrid of Guatemalan x Mexican types; discovered in 1922. Fruit: 12–18 oz; oblong to pyriform; skin thick, leathery, dark green; seed size medium, tight in cavity; season December and January. Tree: very vigorous, bearing heavily and regularly; hardy; resistant to scab; susceptible to anthracnose; flowering group B. Obsolete.

Wright.—Orig. in Whittier, Calif., by Carter Barrett. Introd. in 1949. Chance seedling of Mexican type; planted about 1930; selected about 1945. Fruit: 5–8 oz; pyriform; skin green, very thin; flavor good; oil content high; season early, October and November. Tree: large; spreading; foliage sparse; frost tolerance excellent; production generally light. Not propagated commercially.

Wurtz.—Orig. in Encinitas, Calif., by Roy Wurtz. Introd. in 1948. Chance seedling planted about 1935; selected about 1940. Fruit: Mexican–Guatemalan hybrid type; quality good; skin green, medium thick; long oval; size 7–10 oz. Tree: production fair to good; distinctive weeping growth habit, with dense cover of leaves and small limbs.

Yon.—Orig. in Orlando, Fla., by J. Hugh Yon. Introd. in 1937. Hybrid of Guatemalan x West Indian types; selected in 1932 from seed planted in 1926. Fruit: 2–4 lb; oval-oblong; skin dark bronzed-green, almost smooth, thick, leathery; seed medium-small; flesh very thick; season December to February in Florida. Tree: hardy; poor bearer. Flowering group B. Obsolete.

Zutano.—Orig. in Fallbrook, Calif., by W.L.

Ruitt. Introd. in 1941. Parentage unknown; probably a hybrid between Mexican and Guatemalan types; selected in 1926. Fruit: 8–12 oz; pyriform; skin very thin, light green, russets in inland areas; oil content medium; flavor fair to good; season October to December in San Gabriel Valley of Los Angeles County, Calif.; season January and February in Ventura County, Calif. Tree: consistent producer; more hardy than Fuerte; vigorous; upright grower. Still grown in marginal areas in California.

BANANA

There are more than 1000 clone names listed for bananas and plantains (see R.H. Stover and N.W. Simmonds. 1987. *Bananas*. 3rd ed. Longman Scientific and Technical, Harlow, Essex, England). All are natural clones and many of the names refer to the same clone in different places. The Williams Hybrid listed below is but one of more than 50 names applied to the several closely related cultivars within the Cavendish Group. The currently most widely grown export banana is the Grand Nain variety, which also belongs to the Cavendish Group, but has a plant height about 2 ft shorter than Williams Hybrid.

Goldfinger.—Orig. as a tetraploid selection in the banana program of the Fundación Hondureña de Investigación Agrícola (FHIA) in La Lima, Honduras, from a cross between the Dwarf Prata triploid and the bred SH-3142 diploid. Patent 8983. Bunch: large, heavy, cylindrical, with 8–12 hands. Fruit: thicker and more angular than the standard Cavendish export banana. Plant: vigorous growth habit; strong psuedostems 8–13 ft high; resistant to all races of Panama disease, to black Sigatoka leaf spot, to the burrowing nematode, and to crown rot. This is the only bred banana that is being planted commercially.

Williams Hybrid.— Orig. in Coffs Harbor, New South Wales, Australia. Introd. into Hawaii in 1954 by J.H. Beaumont, Hawaii Agr. Expt. Sta. Bud mutation of Chinese (Cavendish). Bunch: large, heavy, cylindrical, with 9–14 hands; flower bracts shed more readily than Chinese; flowering axis below bunch tends to be curved rather than straight. Fruit: larger and more attractive than parent; quality good; ripens about with Chinese, which it resembles. Plant: growth habit relatively strong; pseudostems 7–12 ft high, with considerable red pigmentation; leaves longer and wider than Chinese; resistant to Panama disease.

Barbados Cherry

B-17.—Orig. in Rio Piedras, Puerto Rico, by the Puerto Rico Agr. Expt. Sta. Introd. in 1956. Chance seedling; selected in 1954. Fruit: acid, weight 9–12 g, large; high ascorbic acid content (1325–2250 mg/100 g juice). Tree: yield high, 8.1 to 12.1 t/ha/year, from 4-year-old trees.

Beaumont.—Orig. in Honolulu, Hawaii, by Henry Y. Nakasone, Univ. of Hawaii, Honolulu. Introd. in 1961. Parentage unknown; selected in 1958; tested as Hawaii 3B-T21. Fruit: acid type; orange-red when ripe; average weight 4.9 g, 11% soluble solids, pH 3.3, and high ascorbic acid content (2820 mg/100 g fruit). Tree: compact; low; branches spreading; foliage dense.

Florida Sweet *(Sweet)*.—Orig. in Homestead, Fla., by the Florida Subtropical Expt. Sta. Introd. in 1952. Chance seedling; selected in 1946. Fruit: large, 1 1/4 inches in diam., average weight 4 g, skin thick; flesh very juicy, flavor semi-sweet; vitamin C content high (1500–2000 mg/100 g juice). Tree: erect; open-type growth; outstanding yields (33.6 t/ha/year). Relatively good tolerance to cool temperatures and has moderate degree of tolerance to Cercospora leaf spot.

Haley.—Orig. in Honolulu, Hawaii, by Henry Y. Nakasone, Univ. of Hawaii. Introd. in 1966. Parentage unknown; discovered in 1958; tested as R3A-T8. Fruit: purplish-red when ripe; average weight 4.5 g; 8.2% soluble solids, pH 3.2, vitamin C 2663 mg/100 g juice, acid variety. Tree: upright; branches long.

Hawaiian Queen.—Orig. in Honolulu, Hawaii, by Henry Y. Nakasone, Univ. of Hawaii, Honolulu. Introd. in 1963. Open-pollinated seedling of B-17; selected in 1958; tested as Hawaii 8E-T32. Fruit: sweet type, average weight 6.1 g, 9.6% soluble solids, pH 3.5, vitamin C 1577 mg/100 g juice.Tree: very productive; upright; spreading; resembles Manoa Sweet.

Manoa Sweet.—Orig. in Honolulu, Hawaii, by Henry Y. Nakasone, Univ. of Hawaii, Honolulu. Introd. in 1963. Open-pollinated seedling of B-17; selected in 1958; tested as Hawaii R4-T43. Fruit: sweet type; orange-red, average weight 6.2 g, 10% soluble solids, pH 3.6, vitamin C 1537 mg/100 g juice. Tree: upright; spreading; very productive.

Maunawili.—Orig. in Maunawili, Oahu, Hawaii, by Henry Y. Nakasone, Univ. of Hawaii, Honolulu. Introd. in 1960. Parentage unknown; selected in 1957; tested as Hawaii 8. Fruit: acid type; small; red; average weight 4.3 g, 8.9% soluble solids, pH 3.1, and vitamin C 2010 mg/100 g juice. Relatively tolerant to Cercospora leaf spot. Tree: upright; single trunk; leaf small; narrow.

Red Jumbo.—Orig. in Honolulu, Hawaii, by Henry Y. Nakasone, Univ. of Hawaii. Introd. in 1961. Parentage unknown; tested as Hawaii 3B-Tl. Fruit: acid type; large, 8–12 g; purplish-red; 8.7% soluble solids, pH 3.1, and vitamin C 2183 mg/100 g juice. Tree: compact; well-branched; foliage dense; low growing.

Rehnborg.—Orig. in Honolulu, Hawaii, by Henry Y. Nakasone. Introd. in 1966. Parentage unknown; discovered in 1958; tested as R22-T40. Fruit: dark red when ripe; acid variety; average weight 6.2 g, 8.3% soluble solids, pH 3.3, and vitamin C 1717 mg/100 g juice. Tree: compact; dense; spreading; productive.

Sweet.—See Florida Sweet.

Tropical Ruby.—Orig. in Honolulu, Hawaii, by Henry Y. Nakasone, Univ. of Hawaii. Introd. in 1963. Seedling of B-17 (Puerto Rico); discovered in 1958. Fruit: large; resembles Manoa Sweet; good yield; sweet variety; average weight 7.2 g, 9.0 soluble solids, pH 3.5, and vitamin C 1437 mg/100 g juice. Plant: suitable for home gardens.

BEACH PLUM

Autumn.—Orig. in Chappaquiddick Island, Mass., by Russell B. Stearns, Edgartown, Mass. Introd. in 1957. Chance seedling; selected in 1950. Fruit: medium; skin purple; pit size small; processes very well. Tree: bears regularly; fine ornamental qualities when in bloom; foliage turns red in autumn.

Cotuit.—Orig. in Tern Island, Cotuit, Mass., by Wilfred Wheeler of Hatchville, Mass. Introd. during the 1940s. Chance seedling; selected about 1939. Fruit: medium, 3/4 to 1 inch in diameter; skin amber, very attractive; flesh sweet, flavor good when ripe; ripens early, often in July. Tree: very productive.

Eastham. —Orig. in Eastham, Mass., by J.M. Batchelor, U.S. Soil Conservation Serv., Hill Culture Div., Washington, D.C. Introd. in the 1940s. Seedling of *Prunus maritima*; discovered in 1938. Fruit: large; flavor good. Tree: production heavy.

Hancock.—Orig. in Fort Hancock, N.J., by J.M. Batchelor, U.S. Soil Conservation Serv., Hill Culture Div., Washington, D.C. Introd. in the 1940s. Seedling of *Prunus maritima;* discovered about 1938. Fruit: skin blue; flesh golden; sweet with little acidity, juicy; flavor excellent for eating fresh; pit small; ripens early.

Makonikey.—Orig. in West Tisbury, Mass., by Russell B. Stearns of Edgartown, Mass. Introd. in 1960. Open-pollinated seedling; discovered in 1949. Fruit: resembles Stearns. Plant: free-growing.

Northneck.—Orig. in Chappaquiddick Island, Mass., by Russell B. Stearns, Edgartown, Mass. Introd. in 1957. Chance seedling; selected in 1947. Fruit: large; quality excellent; pit large. Tree: not a regular bearer; outstanding as a roadside ornamental for bank planting because of spreading by root offshoots.

Safford.—Orig. in Plum Island, Newburyport, Mass., by J.M. Batchelor, U.S. Soil Conservation Serv., Hill Culture Div., Washington, D.C. Introd. in the 1940s. Seedling of *Prunus maritima;* discovered about 1938. Fruit: skin blue. Tree: a very poor grower; hard to propagate.

Stearns.—Orig. in Chappaquiddick Island, Mass., by Russell B. Stearns of Edgartown, Mass. Introd. in 1957. Chance seedling; selected in 1947. Fruit: processes well. Tree: fine ornamental qualities.

Wheeler 6.—Orig. in Truro, Mass., by Wilfred Wheeler of Hatchville, Mass. Introd. in the 1940s. Chance seedling; discovered about 1937. Fruit: medium, averages 1 1/8 inches in diam.; skin light blue, very attractive; flesh sweet, flavor good when ripe. Tree: productive.

Black Walnut

The black walnut, *Juglans nigra,* is among the few temperate zone fruit crops native to the North American continent. A high-energy food prized by man and squirrel alike, black walnuts have been collected from the wild for centuries. Today, America's black walnut industry still relies on the annual collection of 20 million pounds of nuts by the hands of thousands of rural families across the midwest.

Cultivar development began in 1880, when Pennsylvania nurseryman Joseph W. Thomas propagated the first black walnut—a native selection he named 'Thomas'. Since that time, black walnut enthusiasts have searched the nation for native trees that merit propagation. Hundreds of selections have been named but few have been propagated widely or, for that matter, deserve propagation. Previous lists of black walnut cultivars included many selections that were entered into early nut contests sponsored by the Northern Nut Growers Association. Many of these were not propagated widely and never achieved cultivar status. In preparing this cultivar list, only selections that have been propagated widely over an extended time period, both in the past and the present, have been included.

Cultivar development in black walnut is still in its infancy. The majority of cultivars listed here are selections made from natural tree stands. A smaller number of cultivars originated as open-pollinated seedlings of native selections. There have been no systematic attempts at breeding new black walnut cultivars. Outside of reports of nut weight and percent kernel, cultivar data for many black walnuts is scarce. Reliable reports of tree phenology and tree performance have been incorporated in the cultivar descriptions.

Adams.—Native selection from Scotts, Mich., made by Henry Adams in 1920. Nuts average 11 g with 26% kernel. Cracks well. Won first prize in the 1920 Contest of Northern Nut Growers Assn. Obsolete.

Alley.—Native selection made by Amy A. Alley of La Grangeville, N.Y., in 1918. Small nuts average 11 g and produce 25% kernel. Good cracker. Obsolete.

Arkansas.—See Cooper.

Asbury.—Native selection by C.D. Asbury of Augusta, Ky. First displayed in 1926. Large nut weighing 20 g with 24% kernel. Good cracker. Obsolete.

Beck.—Native selection by Daniel Beck, Hamilton, Mich., in 1928. Small nut weighing 12 g with 33% kernel. Thin shell, excellent cracker. Nuts ripen early. Protogynous flowering habit. Bears on lateral branches. Tree is vigorous and disease resistant. Regular nut production. Recommended in Illinois, Indiana, and Michigan.

Benton.—Native selection found in Wassaic, N.Y., in 1936 by William A. Benton, Benton & Smith Nut Tree Nursery (now defunct), Millerton, N.Y. Medium-sized nuts weigh 13 g and produce 29% kernel. Medium shell thickness, cracks well. Late ripening. Obsolete.

Boellner.—See Kwik-Krop.

Bicentennial.—Orig. as controlled cross (Weschcke x Rohwer) made in 1945 by F.L. Ashworth, Heuvelton. Introd. in 1976. Large nut weighing 18 g. Ripens 15 Oct. in New York, about 1 week after Weschcke. Tree vigorous with medium nut productivity. Introduced to help widen the range of black walnuts in Canada.

Booth.—Native selection from Botetourt Co., Va., discoved in 1931 by H.F. Stoke on the farm of R.A. Booth, Bonsack, Va. Parent tree stood near Virginia Highway 604 between Bonsack and Cloverdale. Nut weighs 15 g and produces 29% kernel. Uniformly well-filled nut with thin shell. Tree retains foliage in autumn longer than most other eastern black walnut cultivars. Obsolete.

Bowser.—Seedling selection orig. in Xenia, Ohio, by John Mitchell Davidson and

Homer G. Bowser. Tree planted by Davidson in 1949, selected by Davidson in 1956, and propagated and evaluated further by Bowser. Small ovate nut averages 15 g with 31% kernel. Thin shell with good shelling quality. Inconsistent kernel quality. Protogynous flowering. Early ripening.

Brown Nugget.—See Mintle.

Burns.—Seedling selection from Ont., Canada. First reported in 1973. Small nut averaging 12 g with 36% kernel. Excellent cracking quality with thin shell. Kernels extracted in halves.

Clermont.—Native seedling found in 1912 by Albert M. Kirtnes in Clermont County, Ohio. Introd. in 1959 by Sterling Rouse, Pleasant Acres Nut Nursery, Walton, Ky. Medium-sized nuts are transversely oblong weighing 14 g and producing 30% kernel. Kernel pieces difficult to remove from shell.

Cochrane.—Seedling selection made by Ferdinand Huber of Cochrane, Wis., in 1940. Open-pollinated seedling of Huber. Nut averages 14 g with 32% kernel. Cracks well with high percentage of quarters. Susceptible to anthracnose.

Cooper (*Arkansas*).—Native seedling found in 1920 by M.L. Cooper in Summers, Ark. Nut weighs 17 g with 27% kernel. Outer hull very thick, making the nut hard to clean. Once recommended in Arkansas and Oklahoma in the 1940s, now obsolete.

Cranz.—Native selection made by Eugene F. Cranz, Ira, Ohio. Small, elliptic nuts weigh 13 g with 31% kernel. Protogynous flowering. Bears on lateral branches. Midseason nut maturity.

Creitz.—Native selection made in 1926 by W.A. Creitz of Cambridge City, Ind. Nut average 16 g with 29% kernel. Thick husk. Good cracking quality. Resembles Ohio.

Davidson.—Native selection found in 1946 by W.J. Davidson of Xenia, Ohio. Small, very widely obovate nuts average 13 g and 35% kernel. Protogynous flowering habit. Bears on lateral branches. Nuts ripen early.

Drake.—Native selection found in 1926 by J.B. Drake of Ray, Ohio. Large, transversely-oblong nut weighs 19 g and produces 30% kernel. Nuts ripen early.

Edras.—Native selection found in 1926 by Gerald Adams of Moorhead, Iowa. Medium-sized nuts average 15 g and produce 31% kernel. Good cracking quality.

Eldora.—Native selection found in Eldora, Iowa. Large nut weighs 19 g and produces 30% kernel. Protandrous flowering. Early ripening.

Elmer Myers.—See Myers.

El-Tom.—Resulted from a controlled cross of Myers x Thomas. Nuts average 16 g with 29% kernel. Recommended in Michigan.

Emma K.—Native selection found by J.C. McDaniel, Univ. of Illinois in the 1960s. Elliptic nut averaging 17 g with 34% kernel. Excellent shelling characteristics. Bears on lateral branches. Midseason ripening. Recommended in Illinois, Kansas, Michigan, and Missouri.

Farrington.—Native selection found in 1954 in Fayette County, Ky., by O.M. Farrington. Very large nut weighing 23 g, producing only 22% kernel. Thick shell but cracks well. Protogynous flowering. Midseason ripening.

Football.—Seedling selection made during the 1970s by Leon Falconer of Sarcoxie, Mo. Large, elliptic nut weighs 19 g and produces 30% kernel. Susceptible to anthracnose. Bears on lateral branches. Midseason ripening.

Grundy.—Native selection discovered in 1921 by John W. Rohwer on Grundy Center, Iowa. Large nut weighs 19 g producing 26% kernel. Protandrous flowering. Midseason ripening.

Hare.—Native selection discovered in 1926 by F.M. Hare of Rushville, Ill. Large, widely ovate nut averages 20 g yielding 27% kernel. Good cracking quality. Protogynous flowering. Late ripening. Recommended in Nebraska.

Hay.—Orig. as a cross of Thomas x Myers made by Leander Hay of Gilliam, Mo. Nuts weigh 20 g and yield 31% kernel. Protogynous flowering. Midseason ripening.

Horton.—Native selection found in 1932 by E.F. Huen on the Horton farm near Whitten, Iowa. Nuts weigh 18 g with 29% kernel. Protogynous flowering. Obsolete.

Huber.—Seedling selection made in 1929 by Ferdinand Huber of Cochrane, Wis. Seed planted in 1900. Nuts weigh 17 g with 31% kernel. Obsolete.

Impit.—Seedling selection found by J.U. Gellatly on the Impit farm in Penticton, B.C., Canada, in 1935. Seed collected in 1927 from a native seedling growing in Ontario, Canada. Ovate nut weighs 17 g and yields 25%. Good shelling characteristics. Husk thick. Obsolete.

Korn.—Native selection discovered in 1920 by G.J. Korn on the farm of J.D. Boone in Berrien Springs, Mich. Named by N.F. Drake, Michigan Agr. Expt. Sta., East Lansing. Widely elliptic nut weighs 16 g and produces 26% kernel. Fair cracking quality. Early ripening.

Krause.—See Krouse.

Krouse (*Krause*).—Native selection found in 1940 on the farm of Russell Krouse, Toddville, Iowa, by Joe Crumley, Univ. of Iowa, Iowa City, Iowa. Medium-sized nut averages 19 g with 27% kernel. Protandrous flowering habit. Early ripening. Recommended in Nebraska.

Kwik-Krop (*Boellner*).—Native selection found in 1928 on the Boellner family homestead near Eldorado, Kan. Introd. by Louis B. Boellner of Roswell, N.M. Plant patent 1063; approved 1 Jan. 1952; assigned to Stark Bros. Nurseries & Orchards Co., Louisiana, Mo. Medium-sized, elliptic nut weighs 17 g and yields 31% kernel. Shell pointed on both ends. Pronounced depression in shell along the suture near the base of nut. Easy to crack. Bears on lateral branches. Early ripening.

Lamb.—Native selection discovered in 1926 by George N. Lamb in Ada, Mich. Trees selected for its ability to produce curly-grained wood prized for veneer. However, most grafted trees of this cultivar have not developed the curly grain characteristic.

Mintle (*Brown Nugget*).—Native selection found in 1930 by J.R. Mintle in Glenwood, Iowa. Small nut weighs 14 g and produces 31% kernel. Thin, smooth shell. Excellent cracking quality. Strongly alternate bearing. Recommended in Nebraska.

Myers (*Elmer Myers*).—Native selection made in 1926 by Elmer R. Myers in Bellefontaine, Ohio. Medium-sized nuts weigh 15 g and produce 33% kernel. Very thin shell. Protogynous flowering. Early ripening. Recommended in New York.

Norris.—Native selection found in 1940 by Spencer B. Chase, Div. of Forestry Relations, Tennessee Valley Authority. Tree found on the farm of R.B. Cox, Fountain City, Tenn. Nuts weigh 19 g and produce 28% kernel. Obsolete.

Ogden.—Native selection discovered in 1926 by Mrs. Joe Ogden in Bedford, Ky. Large, transversely oblong nuts average 20 g per nut producing 26% kernel. Good cracking quality. Protogynous flowering. Late ripening. Recommended in Oklahoma.

Ohio.—Native selection found in 1915 by Charles Arbogast in McCutcheonville, Ohio. Medium-sized nuts weigh 16 g, yielding 27% kernel. Thick husk. Protogynous flowering. Midseason ripening. Good resistance to anthracnose. Recommended in Nebraska.

Patuxent.—Native selection found in 1931 by Herbert Miller of Gaithersburg, Pa. Large nut weighs 21 g, producing 29% kernel. Obsolete.

Peanut.—Native selection discovered in 1891 by H.C. Cline in Lyndon, Ohio. Selected for tendency to produce single-lobed nuts. Nuts average 17 g and produce 26% kernel. Protogynous flowering. Early ripening. Obsolete.

Pinecrest.—Native selection found in 1931 by Herbert Miller in Richfield, Pa. Large nut averages 19 g and produces 27% kernel. Good cracking quality. Midseason ripening. Obsolete.

Pritchett.—Native selection found in 1932 in East River, W.Va. Introduced by J.E. Pritchett, Bluefield, W.Va., in 1966. Large nut weighing 22 g and yielding 29% kernel. Midseason ripening. Good resistance to anthracnose.

Rohwer.—Native selection discovered in 1926 by John W. Rohwer in Grundy Center, Iowa. Very widely obovate nut averages 21 g yielding 25% kernel. Protogynous flowering. Late ripening. Recommended in Nebraska and Pennsylvania.

Sauber.—Seedling selection made by Paul J. Sauber of Barberton, Ohio. Introduced in 1978 by Wilmoth Nursery, Elizabethtown, Ky. Small, widely elliptic nut averages 15 g and yields 33% kernel. Excellent shelling qualities. Outstanding kernel quality.

Schreiber.—Native selection found in 1940 by R.H. Schreiber in New Albany, Ind.

Large nut weighs 21 g and produces 30% kernel. Obsolete.

Snyder.—Native selection found in 1934 by C.H. Synder on his farm in Newfield, N.Y. Large, very widely obovate nut weighs 21 g yielding 26% kernel. Adapted to the northern edge of the black walnut's natural range. Recommended in New York.

Sol.—Native selection found in 1932 by Fred Bolten of Linton, Ind. Introd. in 1951. Large, very widely obovate nut weighs 21 g but produces only 24% kernel. Fills and cracks out better than Thomas in Indiana. Moderate resistance to anthracnose.

Sparks 127.—Seedling selection made by Archie Sparks of Beaver, Iowa. Small, widely elliptical nut weighs 14 g and produces 32% kernel. Protogynous flowering. Bears on lateral branches. Early ripening. Prone to overproduction and alternate bearing.

Sparks 147.—Seedling selection made by Archie Sparks of Beaver, Iowa. Medium-sized, elliptical nut weighs 17 g and yields 38% kernel. Nut strongly pointed on both ends but basal point distinctively long. Thinnest shell of any walnut cultivar. Bears on lateral branches. Midseason ripening.

Sparrow.—Native selection made in 1935 by Harry C. Sparrow of Lomax, Ill. Medium-sized, widely depressed obovate nut weighs 17 g and produces 29% kernel. Husk very thin and easily removed. Excellent shelling quality. Protandrous flowering. Early ripening. Good resistance to anthracnose. Recommended in Kansas, Indiana, Illinois, Michigan, Missouri, Nebraska, and New York.

Stabler.—Native selection made in 1911 by Henry Stabler of Sandy Springs, Md. Medium-sized, widely ovate nut weighs 16 g and yields 31% kernel. Some nuts develop with a single lobe. Protogynous flowering. Midseason ripening. Susceptible to husk fly maggot.

Stambaugh.—Native selection discovered in 1926 near Browning, Ill., by H.V. Stambaugh of Galesburg, Ill. Medium-sized, transversely oblong nut weighs 19 g and produces 31% kernel. Protogynous flowering. Susceptible to anthracnose. Recommended in New York.

Surprise.—Seedling selection made during the 1970s by Leon Falconer of Sarcoxie, Mo. Large, depressed-obovate nut weighs 19 g and yields 34% kernel. Bears on lateral branches. Midseason ripening.

Tasterite.—Native selection discovered in 1924 by Samuel H. Graham of Ithaca, N.Y. Small nut weighs 13 g and produces 28% kernel. Cracking quality very good. Adapted to the northern edge of the black walnut's natural range. Obsolete.

Ten Eyck.—Native selection discovered in 1915 by E.M. Ten Eyck of Rahway, N.J. Medium-sized nut weighs 17 g and produces 30% kernel. Late leafing. Obsolete.

Thomas.—Native selection discovered by Joseph W. Thomas of King of Prussia, Pa. First propagated in 1880. Most widely propagated black walnut cultivar to this day. Medium-sized, oblate nut weighs 19 g and yields 24% kernel. Precocious and prolific. As tree matures, kernels do not fill as well. Susceptible to anthracnose. Tree grows vigorously but is subject to wind damage. Thomas is the standard to which all other cultivars are compared. Recommended in Nebraska, Michigan, and New York.

Throp.—Native selection discovered in 1926 by Ralph Throp of Greensburg, Ind. Small, elliptical nut weighs 11 g and yields only 22% kernel. Selected primarily for the propensity to produce single lobed nuts. Protandrous flowering. Late ripening. Propagated primarily as a curiosity.

Todd.—Seedling selection introduced in 1934 by L.H. MacDaniels, New York State Agr. Expt. Sta., Ithaca. Tree discovered about 1925 on the farm of Albert Todd, Wakeman, Ohio. Large nut weighs 25 g and yields 24% kernel. Protogynous flowering. Midseason ripening.

Vandersloot.—Native selection found in 1926 by C.E. Vandersloot of Muddy Creeks Forks, Pa. Large, widely depressed obovate nut weighs 20 g and yields 24% kernel. Protogynous flowering. Midseason ripening.

Victoria.—Native selection found in 1944 by W.G. Tatum of Lebanon, Ky. Medium-sized, oblate nut weighs 15 g and yields 26% kernel. Late ripening. Vigorous tree growth. Fair resistance to anthracnose.

Wanda.—Seedling selection made in 1932 by Gerald Adams of Moorhead, Iowa. Originated as a open-pollinated seedling of Edras. Medium-sized nut weighs 15 g and produces 31% kernel. Obsolete.

Ward.—Native selection found in 1936 by Willard B. Ward of West Lafayette, Ind. Named and introd. in 1955 by J. Ford Wilkinson, Indiana Nut Nursery, Rockport, Ind. Large nut weighs 25 g and yields 26% kernel.

Weschcke.—Seedling selection made by Carl Weschcke of River Falls, Wis. Seed source was a native tree growing in St. Paul, Minn. Plant patent 1229; 17 Nov. 1953. Elliptical nut, pointed at both ends. Thin shell, cracks easily. Selected to mature in the northern most range of black walnut. Obsolete.

Wiard.—Native selection made in 1926 by Everett W. Wiard of Ypsilanti, Mich. Small, very widely obovate nut weighs 12 g, producing 25% kernel. Thick husk. Very slow coming into bearing, shy bearer. Susceptible to anthracnose.

BLACKBERRY

Acme Thornless Young.—Orig. in Chino, Calif., by Elmer L. Pollard and Jubal E. Sherrill. Introd. in 1930. Plant patent 4; 20 Oct. 1931. Bud mutation of Young; discovered in 1928. Fruit: characteristics of Young. Bush: thornless; has all the other characteristics of parent, but is more susceptible to disease; dewberry type.

Adrienne.—An early ripening, genetically spineless trailing blackberry with excellent fruit quality. Orig. in Kent, England, by D.L. Jennings, Medway Fruits. Silvan x unnamed seedling; introd. in 1995. Worldwide marketing rights held by NSA Plants, West Malling, Kent. Fruit: size between 6.0 and 6.5 g, similar to Silvan; firm, long, conic; attractive, bright, regular appearance; ripens early July in southern England, 1 week after Silvan and 1 week before Waldo; excellent flavor; better adapted to distant markets than Silvan. Plant: high yield potential; first-year canes stout, spine-free, vigorous, and spreading; second-year canes short, stout, good fruit exposure.

Alabama Sweet.—Orig. in Waverly, Ala., by Paul Hoffman. Introd. about 1924. Parentage unknown. Fruit: large; very sweet. Bush: dewberry type.

Albino White.—Orig. in Texas. Introd. about 1930 by Henderson's Nursery, Athens, Texas. Fruit: white; quality good. Bush: dewberry type; hardy.

Arapaho.—An early ripening, genetically spineless, erect blackberry with excellent fruit quality. Orig. in Fayetteville, Ark., by J.N. Moore and J.R. Clark, Dept. of Horticulture, Univ. of Arkansas. Ark. 631 x Ark. 883. (Both parents are heterozygous for recessive spineless genes derived from Merton Thornless via Thornfree). Cross made in 1982; selected in 1985; tested as Ark. 1536; introd. in 1993. Plant patent 8510; 21 Dec. 1993; assigned to Univ. of Arkansas. Fruit: medium size; bright glossy black; firm; small seed size; short conic shape; earlier ripening than any other spineless blackberry cultivars; concentrated season of 4 weeks; high soluble solids concentration; stores well; excellent flavor fresh and processed. Plant: productivity similar to Navaho; moderately vigorous; very erect spineless canes; produces more primocanes from roots than Navaho; good cold hardiness to –24C. No symptoms of either orange rust or rosette have been observed despite high levels of inoculum of the respective causal organisms.

Aulo.—Orig. in Morgan City, La., by B.M. Young. Introd. about 1926. Considered to be Austin x Logan; direction of cross unknown. Fruit: color and flavor similar to Logan. Bush: vigorous. Now obsolete.

Aurora.—Orig. in Corvallis, Ore., by George F. Waldo, U.S. Dept. of Agriculture and Oregon Agr. Expt. Sta. Introd. in 1961. U.S. – Ore. 616 x U.S.–Ore. 73; cross made in 1946; selected in 1949; tested as U.S.–Ore. 966. Fruit: large; firm; flavor excellent; ripens very early, with a season averaging from 21 June to 15 July; desirable market variety, processes very well; resembles Marion. Bush: canes medium vigorous, very pliable, making training easy; majority of fruit borne on basal 4–5 ft of canes, thus close planting is necessary to secure high yields since individual plant yields average about 50% of those for Boysen or Marion; hardy under western Oregon conditions.

Aus-Lu.—Orig. in Salisbury, Md. Introd. in 1923 by E.W. Townsend & Sons, Salisbury. Considered to be Mayes x Lucretia. Fruit: large; mild, sweet; ripens early. Bush: productive; dewberry type.

Austin Thornless *(Thornless Austin)*.—Orig. in Tecumseh, Okla., by J. Parker. Introd. in 1924. Reported to be an open-pollinated seedling of Mayes; selected about 1918. Fruit: resembles Mayes. Bush: dewberry type; shy bearer; thornlessness dominant; an octoploid (2n = 56). Received such extensive damage from rabbits feeding on variety during the winter that variety is no longer propagated.

Bailey.—Orig. in Geneva, N.Y., by George L. Slate, New York State Agr. Expt. Sta. Introd. in 1950. Parentage unknown, as the label was lost; selected about 1929. Fruit: large; medium firm; quality good; subacid. Bush: reliable productivity.

Bauer Thornless Logan.—Orig. in San Gabriel, Calif., by Beulah E. Bauer and Gordon R. Bauer. Introd. in 1934. Plant patent 82; 14 Nov. 1933. Bud mutation of Logan; discovered in 1929. Fruit: similar to parent, except season earlier and longer. Bush: canes and leaves thornless; canes strong.

Big-Ness.—Orig. in College Station, Texas, by S.H. Yarnell, Texas Agr. Expt. Sta. Introd. in 1946. Selected in 1940 from F_3 of (*Rubus rubrisetus* x Nessberry); tested as Tex. R40-51. Fruit: large; season early; resembles Mammoth. Bush: vigorous; resistant to little-fruit.

Black Satin.—Orig. in Carbondale, Ill., by U.S. Dept. of Agriculture and Southern Illinois Univ. Introd. in 1974. SIUS 47 (US 1482 x Darrow) x Thornfree; cross made in 1964; selected in 1967 by J.W. Hull; tested as SIUS 64-21-11. Fruit: large; skin black, loses glossiness at maturity; less astringent than Thornfree; ripens 7–10 days later than Dirksen Thornless, 2 weeks earlier than Thornfree. Bush: thornless; semi-erect; hardy; more vigorous and productive than Dirksen Thornless, otherwise very similar.

Bowen.—Orig. in Burlingame, Calif., by John C. Bowen. Introd. in 1944. Plant patent 635; 1 Aug. 1944. From a wild thornless blackberry; discovered in 1940. Fruit: large; long; black; matures early; seeds very small; resembles Cory Thornless. Bush: thornless; does not sucker; quite productive. May be similar to, or identical with, Cory Thornless.

Boysen *(Boysenberry)*.—Orig. in Napa, Calif., by Rudolph Boysen. Introd. in 1935 by Walter Knott, Knott's Berry Farm, Buena Park, Calif. Parentage unknown but is of type as might be expected from a Logan x Lucretia cross; some consider Himalaya as the female parent; seedling grown about 1920; bushes transplanted to Anaheim, Calif., in the late 1920s. Fruit: black; flavor very fine but not as sweet as Young; size extremely large, 1 1/4 inches or more long; rather soft; aroma distinct; covered with a dusty bloom; bears through August in some areas, ripens 1 week before Young. Bush: trailing (dewberry type); more vigorous than Young; high yields. In California, Boysen and Nectar are different varieties.

Boysenberry.—See Boysen.

Brainerd.—Orig. in Atlanta, Ga., by George M. Darrow, U.S. Dept. of Agriculture. Introd. in 1932. Himalaya x eastern erect-growing variety (possibly Georgia Mammoth); selected in 1920. Fruit: large; quality high; good for canning, preserving, and the frozen-pack trade. Bush: extremely productive; vigorous; very hardy.

Brazos.—Orig. in College Station, Texas, by J. Benton Storey, Texas Agr. Expt. Sta. Introd. in 1959. Second-generation selection resulting from Lawton x Nessberry; selected in 1950; 4n = 28 chromosomes; Nessberry is an amphidiploid second-generation selection from a cross between a wild dewberry and Brilliant red raspberry (*Rubus rubrisetus* x *R. strigosus*). Fruit: large, with size maintained longer during harvest than Lawton; matures early, earlier than these two varieties; harvest season as long as that of Lawton but longer than that of Humble; attractive; for fresh market; outyields Humble and Lawton. Bush: displays greater vigor and less evidence of disease than Humble and Lawton; canes erect.

Brewer.—Orig. in Rahway, N.J., by C.H. Brewer. Introd. in the early 1920s. Chance seedling; discovered in 1909. Fruit: medium to large; firm; quality fair; ripens over a long period. Bush: vigorous; productive; free from rust.

Brison.—Orig. in College Station, Texas, by H.H. Brown and E.T. Graham, Texas A&M Univ. Introd. in 1977. F_2 progeny of Brainerd x Brazos, backcrossed to Brazos in 1966. Selected in 1968; evaluated by T.E. Boswell, J.W. Worthington, and J.A. Lipe. Berry: firmer and better quality, with smaller seed size than Brazos; ripens same season as Brazos (mid-May); intended as companion or replacement of Brazos. Plant: similar to Brazos, except flower petals lavender instead of white; canes moderately upright, slightly more than Brazos; productivity equal to or better than Brazos. Recommended for south-central Texas; not recommended for southeastern Texas because of fungal disease or for northwestern Texas because of low winter temperatures. Should be treated with pesticides for insect, disease, and weed control.

Cameron.—Orig. in Raleigh, N.C., by Carlos F. Williams, North Carolina Agr. Expt. Sta. Introd. in 1938. Young x Lucretia; cross made in 1930. Fruit: shiny black; larger and sweeter than Lucretia, which it resembles. Bush: productive; vigorous; disease resistant; dewberry type. No longer propagated in North Carolina.

Carolina.—Orig. in Raleigh, N.C., by Carlos F. Williams, North Carolina Agr. Expt. Sta. Introd. in 1955. Austin Thornless x Lucretia; cross made in 1938; selected in 1942. Fruit: quality good both fresh and processed; resembles Lucretia. Bush: dewberry type; thorny; more vigorous, more productive, and more resistant to septoria leaf spot than Lucretia.

Cascade.—Orig. in Corvallis, Ore., by George M. Darrow and George F. Waldo, U.S. Dept. of Agriculture and Oregon Agr. Expt. Sta. Introd. in 1940. Zielinski x Logan; selected in 1935. Fruit: bright, deep red; excellent flavor, high dessert quality; excellent for canning or in frozen pack; flavor characteristic of the native blackberry (*Rubus macropetalus*). Bush: vigorous; productive; hardy only in Pacific Coast area; canes trailing.

Chehalem.—Orig. in Corvallis, Ore., by George F. Waldo, U.S. Dept. of Agriculture and Oregon Agr. Expt. Sta. Introd. in 1948. Santiam x Himalaya; cross made in 1936; seedling planted in 1937; selected in 1939; tested as Oregon 731. Fruit: smaller than Logan; skin bright, shiny black; flavor excellent; excellent for frozen products; "seeds" small. Bush: adapted to northern Pacific Coast; vigorous and productive in moist, rich soils.

Cherokee.—Orig. in Fayetteville, Ark., by J.N. Moore, Elvin Brown, and W.A. Sistrunk, Arkansas Agr. Expt. Sta. Introd. in 1974. Darrow x Brazos; cross made in 1965 by Moore; selected in 1968; tested as Ark. 531. Fruit: large; skin black, attractive; flesh firm, quality high (including canned and frozen); ripens early, before Raven. Bush: vigorous; erect, even in young plantings; productive; fruit borne high, well-adapted to mechanical harvest; suckers freely from root cuttings; prolific plant maker; same reaction to major diseases as Raven. Adapted throughout Arkansas.

Chester Thornless.—Orig. in Carbondale, Ill., by J.W. Hull, U.S. Dept. of Agriculture and Southern Illinois Univ. Introd. in 1985. SIUS 47 x Thornfree; cross made in 1968; tested as SIUS 68-6-17. Fruit: large, high quality, very firm; does not soften or lose color in hot weather; ripens late, with Thornfree. Plant: vigorous; semi-trailing thornless canes; very productive; cold hardy. Resistant to cane blight. Propagates readily by tip layering, softwood cuttings, or micropropagation.

Cheyenne.—Orig. in Fayetteville, Ark., by J.N. Moore, E. Brown, and W.A. Sistrunk, Arkansas Agr. Expt. Sta. Darrow x Brazos; cross made 1965; selected in 1968; tested as Ark. 544. Fruit: very large, maintains size throughout the season better than most cultivars; firm; attractive; slightly better flavored than Comanche; harvest period averages 25 days, beginning 1 June in southern Arkansas; fruit are borne high on cane and accessible to machine harvesting. Bush: canes are vigorous, erect, and moderately thorny; moderately resistant to anthracnose, resistant to orange rust, susceptible to rosette.

Choctaw.—Orig. in Fayetteville, Ark., by J.N. Moore and J.R. Clark, Dept. of Horticulture, Univ. of Arkansas. Introd. in 1988. (Darrow x Brazos) x Rosborough; cross made in 1975; selected in 1978; tested as Ark. 876. U.S. plant patent 6678; 21 Mar. 1989; assigned to Univ. of Arkansas. Fruit: medium-large (5.7 g), short conic, glossy black, similar in firmness to Shawnee; good flavor; small seed size. Fruit ripens very early, 2 weeks before Shawnee. Plant: very erect, very prolific; thorned; produces canes from root cuttings; very productive in warm winter locations. Immune to orange rust, moderately resistant to anthracnose, moderately susceptible to powdery mildew, susceptible to rosette; hardy to –20C.

Comanche.—Orig. in Fayetteville, Ark., by J.N. Moore, Elvin Brown, and W.A. Sistrunk, Arkansas Agr. Expt. Sta. Introd. in 1974. Darrow x Brazos; cross made in 1965 by Moore; selected in 1968; tested as Ark. 527.

Fruit: very large; skin black, attractive; flesh firm, quality good; ripens 4 days before Cherokee. Bush: erect; vigorous; productive; suckers freely from roots; adapted to mechanical harvest or home gardens; same reaction to major diseases as Raven. Adapted throughout Arkansas.

Darrow.—Orig. in Geneva, N.Y., by George L. Slate, New York State Agr. Expt. Sta. Introd. in 1958. N. Y. 15826 (Eldorado x Brewer) x Hedrick (Eldorado x Brewer); cross made in 1940; selected in 1946; began fruiting as a selection in 1950. Fruit: large, 1 inch long, 3/4 inch wide; long conic, irregular; skin black, glossy; flesh firm, mildly subacid; quality good; begins ripening early or about with Eldorado, continuing over a long period; secondary fruiting laterals produce a few berries in late August and early September. Bush: vigorous; reliable and heavy producer; apparently more hardy than other selections and varieties grown at the station; resistance to orange rust not known. Named in honor of George M. Darrow, U.S. Dept. of Agriculture.

Delicious.—Orig. in Texas. Introd. in 1922 by E.W. Townsend & Sons, Salisbury, Md. Parentage unknown; received by E.W. Townsend & Sons from southern U.S. Fruit: large; flesh firm; delicious flavor; ripens later than Mayes. Bush: very hardy; dewberry type.

Delsweet.—Orig. in Stayton, Ore. Introd. in the 1930s. Seedling of Evergreen (Oregon Evergreen); selected before 1930. Fruit: considered to be more productive than Evergreen. Now obsolete.

Dew-Black *(Early Wonder, Texas Wonder)*.—Orig. in Stephenville, Texas, by Ross R. Wolfe, Wolfe Nursery. Introd. in 1926. Parentage unknown; selected about 1922. Fruit: rather large; ripens early; resembles Lawton. Bush: erect; vigorous; not too hardy in central and northern Oklahoma; rather susceptible to red-necked cane borer.

Dirksen Thornless.—Orig. in Carbondale, Ill., by U.S. Dept. of Agriculture and Southern Illinois Univ. Introd. in 1974. SIUS 47 (US 1482 x Darrow) x Thornfree; cross made in 1964; selected in 1967 by J.W. Hull; tested as SIUS 64-21-8. Fruit: clusters contain 20 or more large blackberries, which become slightly dull in appearance at maturity; flavor markedly less astringent than Thornfree; ripens first 2 weeks of July, 3 weeks before Thornfree. Bush: thornless; semi-erect; hardier, more vigorous, and more productive than Thornfree; vigorous primocanes of mature plants may grow 12–15 ft in length and 1 1/2 inches in diam.; does not sucker readily; tolerant to septoria leaf spot and anthracnose, somewhat tolerant to mildew.

Douglass.—A high-yielding, genetically spineless trailing blackberry with excellent fruit quality. Orig. in Hillsboro, Ore., by B.S. Douglass. Lawrence x Sander (both derived from Boysenberry and several ORUS blackberry selections; Sander also derived from Austin Thornless, Chehalem, Marion, and Lincoln). Cross made in 1984; selected in 1986; introd. in 1993. U.S. plant patent 8423; Oct. 1993. Fruit: medium to large size; medium-large drupelets with small seeds; glossy black; medium firm, similar to Marion; skin medium tender; bluntly conic to bluntly pointed cylindrical; midseason ripening; easy to harvest; adapted to machine harvest; excellent flavor typical of *Rubus ursinus* derivatives; fresh and processing. Plant: vigorous trailing canes require trellis; small incipient spines tend to disappear as canes grow. Moderately resistant to leaf and cane spot.

Earli-Ness.—Orig. in College Station, Texas, by S.H. Yarnell, Texas Agr. Expt. Sta. Introd. in 1946. Plant selected in 1940 from F_3 of *Rubus rubrisetus* x Nessberry; tested as Tex. R40-4. Fruit: ripens early. Bush: vigorous; productive; resistant to little-fruit; resembles Early Wonder.

Early June.—Orig. in Experiment, Ga., by B.O. Fry, Georgia Agr. Expt. Sta. Introd. in 1959. USDA 266 x North Carolina 36; selected in 1951. Fruit: large; round; flesh jet-black, medium firm; flavor excellent, soluble solids high, acid enough for good jam, jelly, pies; ripens early, during 1–15 June at place of origin; clusters large. Bush: very vigorous; trailing; semi-thornless; fruiting branches long, making for easy picking; appears to be partially resistant to anthracnose and leaf spot. Tested successfully in Georgia, Mississippi, and Alabama;

since Early June is tender to cold, it is recommended only for the southern half of the blackberry belt. Recommended for home gardens because of early maturity and fine fruit quality.

Early Wonder.—See Dew-Black.

Ebano.—Orig. in Pelotas, Brazil, by Maria do Carmo M. Bassols and James N. Moore, UEPAE de Cascata and Univ. of Arkansas. Introd. in 1981. F_2 of Comanche x (Thornfree x Brazos); cross made in 1972; selected in 1975; tested as Black44. Fruit: glossy black, medium-large (5.0–6.5 g), firm; seeds smaller than Thornfree; good fresh and processed quality; ripens very late, 40 days after Brazos. Excellent quality for production of jams, jellies, yogurt, and ice cream. Plant: canes semi-erect, genetically thornless; vigorous; productive; adapted to mild, winter climates (<400 hours chilling).

Ebony King.—Place of origin and originator unknown. Introd. about 1940 by Krieger's Wholesale Nursery, Bridgman, Mich. Parentage unknown. Fruit: large as Eldorado, which it resembles; skin black, glossy; flavor sweet, tangy, good; ripens early. Bush: upright; hardy; resistant to orange rust.

Edward Langley.—Orig. in Bedford, England, by E.M. Langley. Introd. in 1919 by Laxton Bros. and in the U.S. about 1925. A selection from *Rubus calvatus*; year of selection not known. Fruit: size medium; shining black when mature; flavor very good; ripens early. Bush: produces very heavy crops.

Fertodi botermo.—A large-fruited trailing hybrid berry. Orig. at Fertod Res. Sta., Hungary, by L. Kollanyi and M. German. F_3 seedling of Logan x (*Rubus caesius* x Lloyd George); selected in 1968; tested as F. 9086/21; introd. in 1980. Fruit: large; dark red; long shape; early ripening; acid flavor. Plant: very productive; moderately vigorous; trailing habit.

Fertodi Hungaria.—A large-fruited hybrid berry with a mild, sweet flavor suitable for fresh market. Orig. at Fertod Res. Sta., Hungary, by A. Porpaczy and L. Kollanyi. F_2 seedling of Logan x (*Rubus caesius* x Lloyd George); selected in 1963; tested as F. 9001/188; introd. in 1965. Fruit: light red; early ripening. Plant: very tall.

Flint.—Orig. in Experiment, Ga., by B.O. Fry, Georgia Agr. Expt. Sta. Introd. in 1957. Brainerd x Eldorado; cross made in 1948; tested as Ga. 9-1. Fruit: black; borne on numerous large clusters; firm, does not lose juice, change color, or mold as quickly after picking as does the fruit of most other varieties; matures during midseason; quality high. Bush: vigorous; thorny; very resistant to disease; blooms over a long period.

Flordagrand.—Orig. in Gainesville, Fla., by R.H. Sharpe and J.S. Shoemaker, Florida Agr. Expt. Sta. Introd. in 1958. F_2 (Regal Ness x *Rubus trivialis* selection); selected in 1955; tested as Fla. 77. Fruit: large; oblong; drupelets fairly large; skin shiny black; flesh somewhat tart, softer than strawberries; juice very red; aroma delightful; very acceptable for pies, preserves, jellies; ripens very early, several days before Advance and Regal Ness, or about 1 month before local wild berries. Bush: canes trailing, long, purplish-red, thorny; vigorous; evergreen; productivity high; adapted to hot, humid summers; winter-chilling requirement very short; free from double blossom disease; chromosome number, $2x = 14$; flower size medium, petals white, tinged pink; self-sterile; recommended for central Florida home gardens and local markets.

Gem.—Orig. in Experiment, Ga., by B.O. Fry, Georgia Expt. Sta. Named and introd. in 1967. Flint x Early June. Selected in 1959. Fruit: very large; round; skin black; flesh firm, flavor good; ripens about 15 June at Experiment, Ga. Bush: primocanes thorny; trailing; vigorous; hardy; nonsuckering; fruiting laterals bearing 10–20 berries; propagation by tip layering and tip cuttings under mist. For home gardens and local market.

Georgia Thornless.—Orig. in Experiment, Ga., by B.O. Fry, Georgia Expt. Sta. Named and introd. in 1967. F_1 seedling (USDA 1445 x Early June), selfed. Selected in 1958. Fruit: large; blunt conic; skin black; flesh firm, flavor good; ripens 10 June at Experiment, Ga. Bush: thornless; semi-upright; primocanes large, long; nonsuckering, fruiting laterals bearing 10–15 berries; susceptible to cold injury with temperature below 10F; propagation by tip layering or tip cuttings under mist. For home use.

Guilford.—See Guilford Improved.

Guilford Improved *(Guilford).*—Orig. in Guilford County, N.C., by John A. Young, Sr., Greensboro Nurseries, Greensboro, N.C. Introd. about 1922. Chance seedling; discovered about 1912. Fruit: large; sweet; juicy. Now apparently obsolete.

Hedrick.—Orig. in Geneva, N.Y., by George L. Slate, New York State Agr. Expt. Sta. Introd. in 1950. Eldorado x Brewer; cross made in 1929. Fruit: large; medium firm; pleasantly tart; resembles Eldorado but shape is more irregular. Bush: reliably productive.

Hull Thornless.—Orig. in Carbondale, Ill., by John W. (Jack) Hull, U.S. Dept. of Agriculture, Illinois Agr. Expt. Sta., Urbana, and Ohio Agr. Res. and Development Center, Wooster. Introd. in 1981 and named in honor of Hull, who spent his entire professional career breeding and studying blackberries and raspberries. SIUS 47 (US 1487 x Darrow) x Thornfree; cross made in 1968; selected in 1972; evaluated by Gene Galletta, A.D. Draper, R.G. Hill, R.C. Blake, and R.M. Skirvin; tested as SIUS 68-6-6. Berry: large like Black Satin, oblong; skin jet black, colors well in heat; flesh firmer and sweeter than Black Satin; keeping quality good; ripens few days after Black Satin. Bush: similar to Black Satin but hardier.

Humble.—Orig. in Tyler, Texas, by E.C. Humble. Introd. about 1942. Parentage unknown; discovered about 1938. Fruit: size medium; flesh soft, flavor sweet, quality good; ripens 23 May–17 June. Bush: 3–4 ft; erect; vigor moderate; productivity medium to good; upright for mechanical harvest; resistant to rosette, susceptible to orange rust. Recommended for eastern Texas.

Illini Hardy.—Orig. in Urbana, Ill., by Robert M. Skirvin, Univ. of Illinois. Introd. in 1990. NY95 x Chester; tested as Ill. 4-1. Plant patent 8333. Fruit: medium size; elliptical; shiny black; good flavor; resembles Darrow. Plant: very vigorous; erect thorny canes; late-ripening; very cold hardy (to –29C).

Jerseyblack.—Orig. in New Brunswick, N.J., by J. Harold Clarke, New Jersey Agr. Expt. Sta. Introd. in 1953. Evergreen x Eldorado; selected in 1942; tested as N.J. 16. Fruit: resembles Eldorado. Bush: vigorous; apparently rust resistant; produces very few nubbin fruit; productive, with the season beginning 15 July and peaking 25 July.

John Innes.—Orig. in Merton Park, London, England, by M.B. Crane, John Innes Inst. Introd. in England in 1933 and in the U.S. in 1944. *Rubus ulmifolius inermis* x *R. thyrsiger*; selected in 1924. Fruit: large; very firm; sweet; ripens late; resembles Oregon Evergreen; valuable for home use as a late-ripening variety. Bush: produces heavy crops.

Kaiteri.—A hybrid berry with high acid content, suited for processing. Orig. in Nelson, New Zealand, by H. Hall, Riwaka Res. Center. SCRI 78154E3 x SCRI 79102E8; cross made in 1982 by D.L. Jennings, Scottish Crop Res. Inst.; selected in New Zealand in 1985; tested as 8242B68; introd. in 1991. Fruit: medium size; wine red color; very soft, especially when overripe; strong flavor; ripens as much as 6 days before Boysenberry. Plant: moderate numbers of medium-strength spiny canes.

Kayberry.—Orig. in Chehalis, Wash., by Maggie Phillips Kelly. Introd. in 1948. Native blackberry (*Rubus macropetalus*) x Logan; discovered in 1940. Fruit: small seeds; soft core; good shipper and freezer; excellent for pies, jam, and jelly; resembles California Mammoth. Bush: heavy producer.

Kiowa.—Orig. in Clarksville, Ark., by J.N. Moore and J.R. Clark, Dept. of Horticulture, Univ. of Arkansas. Introd. in 1995. [(Thornfree x Brazos) x Comanche] x [(Ark. 557 x Wells Beauty) x Rosborough]; cross made in 1980; selected in 1983; tested as Ark. 1380. U.S. plant patent applied for. Fruit: very large, averaging >10 g; glossy black; blocky oblong; flesh more firm than Shawnee; quality good; ripens 3 days after Shawnee and has a long fruiting period. Plant: erect; medium vigor; productive; suckers freely from roots; appears immune to orange rust, reaction to other diseases unknown. Adapted throughout Arkansas and recommended for trial in areas where Shawnee is adapted.

Kosmo.—See Kosmos.

Kosmos *(Kosmo).*—Orig. in Oregon City,

Ore., by Percy W. Meredith. Introd. in 1932. Plant patent 39; 25 Oct. 1932. Parentage unknown; discovered in a field of Logan plants. Fruit: large; jet black; very soft; resembles Mammoth. Bush: very susceptible to leaf spot; trailing. Apparently now obsolete.

Kotata.—Orig. in Corvallis, Ore., by George Waldo, U.S. Dept. of Agriculture and Oregon State Univ. Introd. in 1984. (Pacific x Boysen) x (Jenner-1 x Eldorado); selected in 1950; tested as OR-US 1050. Fruit: large, equal to Marion in size; glossy black, attractive; easily detached; firmer than Marion or Boysen; good shelf life; good flavor. Superior for fresh shipping. Plant: vigorous; thorny trailing canes; yields equal Marion in most years; as resistant to cane and leaf spot as Marion and more winter-hardy than Marion or Boysen.

Lincoln *(Lincoln Mountain, Lincoln Wild)*.—Orig. in Siletz, Lincoln County, Ore., by Norwood J. Nute. Introd. in 1957. Plant patent 1936; 19 Apr. 1960. Seedling of a native Oregon trailing blackberry (*Rubus macropetalus* Dougl.); discovered in 1951. Fruit: large, only when grown under the cool, moist conditions of the Oregon coast; reddish; quite soft; flavor good, acid content low; apparently does not process well; resembles Cascade. Bush: trailing; production only fair.

Lincoln Logan.—Orig. in Lincoln, New Zealand, by H.K. Hall, M.H. Quazi, and R.M. Skirvin, Crop Res. Division, DSIR, Christchurch. Introd. in 1986. A tissue culture-derived genetically thornless (non-chimeral) sport of Thornless Loganberry. Fruit: color, flavor, and texture indistinguishable from Thornless Loganberry. Plant: morphologically similar to Thornless Loganberry; primocanes vigorous, green, prickle (thorn)-free; produces only thornless shoots from roots. Seedling populations include both thorned and thornless offspring.

Lincoln Mountain.—See Lincoln.

Lincoln Wild.—See Lincoln.

Loch Ness.—Orig. in Invergowrie, Scotland, by D.L. Jennings, Scottish Crop Res. Inst. Parentage complex. Introd. in 1988. Fruit: large; glossy black; good quality; ripens with or slightly earlier than Hull Thornless. Plant: semi-erect; spine free; less vigorous than Chester, Hull, and Black Satin; cold hardiness as good as or better than Chester and Hull.

Lowden.—Orig. in Hamilton, Ontario, Canada, by Edward Lowden. Introd. in 1939. Possibly a mutation of Snyder. Discovered about 1926. Fruit: core small; color uniformly. Bush: hardy; productive; resistant to orange rust.

Malling Sunberry.—Orig. in East Malling, Kent, England, by E. Keep, J.H. Parker, and V.H. Knight, Inst. of Horticulture Res. (East Malling Res. Sta.). Introd. in 1981. *Rubus ursinus* x selfed tetraploid sport of Malling Jewel; cross made in 1967; tested as EM 2143/9. Fruit: medium-large, similar to Loganberry; very dark red, glossy; blunt conical; ripens early July until mid-August; excellent fresh flavor; canning quality fairly good; flavor and color after thawing good, but drip loss high. Moderately easy to pick, although sometimes concealed by large leaves on the lateral; plug (receptacle) remains with fruit when picked. Plant: canes very vigorous, with short leafy laterals; productive.

Marion.—Orig. in Corvallis, Ore., by George F. Waldo, U.S. Dept. of Agriculture and the Oregon Agr. Expt. Sta. Introd. in 1956. Chehalem x Olallie; cross made in 1945; selected in 1948; tested as US–Oregon 928. Fruit: medium to large; round in cross diameter, somewhat longer than wide; individual drupelet and seed size medium; skin bright black, not as bright as Chehalem or Evergreen, but more attractive than Boysen; flesh medium firm; quality high with excellent flavor, generally superior to Boysen or Evergreen; excellent for local markets, canning, freezing, pies, ice cream flavoring, jams, jellies; harvest season more favorable than Evergreen, ripening in Boysen season and before Evergreen in Oregon; resembles Himalaya or Chehalem. Bush: trailing; vigorous grower; generally only a few long canes, up to 20 ft long; spines large and numerous; fruiting laterals long and strong, with many fruits; very productive, more so than Boysen, higher yields for a longer period than Boysen; adapted to western Oregon and western Washington. Named after Marion County, in which it was tested extensively.

May Hardiest.—Orig. in Shenandoah, Iowa, by the May Seed & Nursery Co. Introd. in 1924. Chance seedling. Fruit: large; skin black. Bush: very hardy.

Merton Early.—Orig. in Merton Park, London, England, by the John Innes Inst. Introd. in 1936. Selection from *Rubus nitidioides*. Fruit: large; skin black; flavor very good; ripens early. Bush: moderately vigorous; produces heavy crops; growth dies after fruiting; valuable home variety producing good-quality dessert fruit.

Merton Thornless.—Orig. in Merton Park, London, England, by M.B. Crane, John Innes Inst. Introd. in England in 1936, introd. in the U.S. in 1944. Plant patent 571; 9 Mar. 1943; assigned to Armstrong Nurseries, Ontario, Calif. John Innes (*Rubus ulmifolius inermis* x *R. thyrsiger*) x a thornless selfed John Innes seedling. Fruit: large; round; drupelets large; long midseason ripening. Bush: thornless; vigorous grower; crop production in U.S. has not been very high; tetraploid. Useful in breeding thornless plants. No longer propagated for its fruit.

Midnite.—Orig. in Clermont County, Ohio, by B.E. Spurgeon. Introd. in 1957 by Scarff's Nursery, New Carlisle, Ohio. Chance seedling; discovered about 1945. Fruit: large; skin black; clusters well-formed; core small; juicy; quality high; resembles Eldorado. Bush: upright grower; production heavy.

Monroe.—Orig. in Austin, Texas. Introd. in 1920 by the Austin Nursery Co., Austin. Parentage unknown; plant received by this nursery from Florida about 1900. Fruit: large; long. Bush: southern dewberry type; vigorous on sandy soils.

Murrindindi.—A spineless, trailing blackberry with good fruit quality. Orig. in Toolangi, Australia, by G. McGregor, Inst. for Horticultural Development, Victoria. Silvan x a thornless blackberry hybrid of complex origin; cross made by D.L Jennings, Scottish Crop Res. Inst.; selected at Toolangi; tested as A85-15-A3; introd. in 1994. Fruit: similar to Silvan but slightly smaller with smaller drupelets; ripens after Silvan and before Smoothstem. Plant: canes shorter than Silvan with shorter internodes; spineless.

Navaho.—Orig. in Fayetteville, Ark., by J.N. Moore and J.R. Clark, Dept. of Horticulture, Univ. of Arkansas. Introd. in 1988. (Thornfree x Brazos) x (Ark. 550 x Cherokee); cross made in 1977; selected in 1980; tested as Ark. 1172. U.S. plant patent 6679; 21 Mar. 1989; assigned to Univ. of Arkansas. Fruit: medium (5.0 g); short conic; glossy black; very firm; stores well; high soluble solids, excellent flavor fresh and processed; ripens late, 7 days after Shawnee. Plant: genetically thornless, erect canes; suckers poorly from roots; not as productive as Shawnee. Plants and fruit moderately resistant to anthracnose; susceptible to orange rust; moderately resistant to rosette. Plants have shown good cold hardiness to –25C.

Nectar *(Nectarberry)*.—Orig. in El Monte, Calif., by Howard G. Benedict. Introd. in 1937. Said to have been a seedling of Young but thought by many to be a chimera of Boysen, which it resembles; discovered before 1936. Fruit: diameter greater than Boysen; many claim the berry is sweeter than Boysen; about 9 drupelets around the core at the calyx end instead of 10 or usually 11 as in Boysen; drupelets larger than those of Boysen; drupelets not acute; base of style usually set in a depression. Bush: trailing (dewberry type); production and habit almost if not identical with Boysen.

***Nectarberry*.**—See Nectar.

Nessberry.—Orig. in College Station, Texas, by H. Ness, Texas Agr. Expt. Sta. Introd. in 1921. Original cross made in 1912. F_3 of (*Rubus rubrisetus* x *R. strigosus* Brilliant), nine similar clones were included as the variety in 1921. Fruit: larger than Logan; size variable; hemispherical; skin deep red to blood-red, turning brown when overripe, shiny; drupelets large; keeping quality good; flavor of raspberry; high acidity; difficult to pick from receptacle, retains core like blackberry, and calyx adheres to fruit; borne on elongated pedicels like blackberry parent. Bush: very vigorous; growth habit like raspberry; canes up to 15 ft in length, with weak, short prickles; leaf of three to five ovate leaflets; productive and fully self-fertile.

Oklawaha.—Orig. in Gainesville, Fla., by R.H. Sharpe, J.S. Shoemaker, and P.J. Westgate, Florida Agr. Expt. Sta. Introd. in

1964; distributed by Florida Foundation Seed Producers, P.O. Box 14006, University Station, Gainesville, FL 32603. F_2 (Regal Ness x *Rubus trivialis*) open-pollinated; cross made in 1953 by Sharpe; sibling of Flordagrand; tested as Fla. 24. Fruit: about 1 1/8 inch x 1/2 inch in diam.; somewhat blunt; skin black, somewhat glossy, turning reddish if exposed to sun after picking; juice deep red; aroma delicate and delightful; drupelets slightly smaller than Flordagrand; ripening starts about 2–5 days before Flordagrand, which it resembles; harvest dates vary: 28 Apr.–16 May in late, cool season, 9–24 Apr. in early, warm, dry season. Bush: vigorous; canes trailing; semi-evergreen to evergreen; yield of 8 pints or more per plant in alternate rows of Oklawaha and Flordagrand, both an effective pollinator for each other; largely self-fruitful. Named for a river in Florida.

Olallie.—Orig. in Corvallis, Ore., by George F. Waldo, U.S. Dept. of Agriculture and the Oregon Agr. Expt. Sta. Introd. in 1950. Black Logan x Young; selected in 1938; tested as Oregon 609. Fruit: large; slightly longer and more slender than Boysen; skin shiny black, attractive; flesh firm; flavor good, but peak quality reached only at full maturity; resembles Mammoth. Excellent for processing. Bush: very productive; canes vigorous, thorny, trailing; adaptable to California, western Oregon, and Gulf Coast conditions; winter-chilling requirements shorter than for Boysen, thus recommended for southern California; grown extensively near Watsonville, Calif.; resistant to verticillium wilt and powdery mildew; very susceptible to systemic orange rust when grown in a humid climate.

Olympic.—Orig. in Vashon, Wash., by Peter Erickson. Introd. in 1930. Plant patent 247; 20 Apr. 1937; by Hallack F. Greider, Vashon. Young x Plum Farmer (?); selected about 1929. Fruit: claimed to be larger, darker, and sweeter than Young, but apparently similar or identical to Young. Bush: claimed more productive and hardier than Young.

Pacific.—Orig. in Corvallis, Ore., by George M. Darrow and George F. Waldo, U.S. Dept. of Agriculture and Oregon Agr. Expt. Sta. Introd. in 1940. Zielinski x Logan; selected in 1935. Fruit: small; deep red; excellent for canning and frozen pack; firmer than Cascade, with more acid; flavor characteristic of the native blackberry (*Rubus macropetalus*). Bush: vigorous; productive; trailing; adapted to western Oregon and Washington. Similar, if not identical, to Cascade.

Perron's Black.—Orig. in Quebec, Canada, by W.H. Perron & Co. Introd. in 1987. Selected from wild *Rubus canadensis* on the Appalachian plateau of southern Quebec. Fruit: black; slightly conical, 30 x 20 mm; high sugar content, aromatic. Plant: thornless, trailing canes; vigorous; no suckering; cold hardy.

Pink Crystal.—Orig. in Spencer County, Ind., by Albert B. Cook, Grandview, Ind. Introd. in 1976 by Herschel L. Boll, Champaign, Ill. Parentage unknown; discovered about 1914 in woods. Berry: medium to large, oblong; skin pink, translucent; flesh slightly soft; apparently must be pollinated by a black blackberry; ripens about with Darrow. Bush: low to medium in productivity, occasionally producing very large clusters of fruit; vigorous, grows up to 7 ft in height; canes erect; produces medium number of suckers, slightly more than Darrow.

Ranger.—Orig. in College Park, Md., by I.C. Haut, Agr. Expt. Sta., Univ. of Maryland. Introd. in 1964. Eldorado x Dew-Black; cross made in 1953; selected in 1958; tested as Md. 28C. Fruit: large; oblong; skin black; flesh moderately soft, quality excellent; ripens early, 2 days after Raven. Bush: vigorous, partially trailing first year; vigor good; productivity excellent.

Ranui.—A hybrid berry producing high-quality fruit for fresh-market use. Orig. in Nelson, New Zealand, by H. Hall, Riwaka Res. Center. Aurora x Marion; cross made in 1983; selected in 1985; tested as 833158; introd. in 1990. Fruit: medium size; medium to dark red, firm; small drupelets; longer and thinner shape than Boysenberry; ripens 2–3 weeks before Boysenberry; when harvested before complete ripeness, it retains pleasant, mild flavor. Plant: many medium-strength canes; good budbreak; even lateral development.

Raven.—Orig. in College Park, Md., by I.C. Haut, Maryland Agr. Expt. Sta. Introd. in 1962. Dew-Black x Eldorado; selected in 1953. Fruit: large; attractive; very high quality, including canned and frozen packs; flavor rich; ripens early. Bush: erect; vigorous; very productive; hardy from Maryland southward and westward.

Redlich.—Orig. in Vittoria, Norfolk County, Ont., Canada, on the farm of Felix Redlich by Bion T. Forse of Simcoe, Ont. Introd. in 1950, both in Canada and the U.S.; P.I. 188872. Selection from a native wild species; discovered in 1946. Fruit: extremely large; very fine flavor. Bush: large; vigorous; produces very good crops.

Regal.—Orig. in Sebastopol, Calif., by William E. Silva. Introd. in 1965. Plant patent 2469; 2 Feb. 1965. Mammoth x Crandall; selected in 1962. Fruit: about 1 3/8 to 1 3/4 inches long, about 5/8 inch in diam.; generally cylindrical; 85–90 drupelets/fruit; firm; subacid, up to 12.3% total solids; flavor very good; prominent aroma; mature berry holds well on bush for 8 days or more; season about 20 June to end of July in Sebastopol area; ships well; resembles Mammoth. Bush: yields very well; very vigorous.

Regal-Ness.—Orig. in College Station, Texas, by S.H. Yarnell, Texas Agr. Expt. Sta. Introd. in 1946. Selected in 1940 from F_3 of *Rubus rubrisetus* x Nessberry; tested as Tex. R40-202. Fruit: matures early; resembles Earli-Ness. Bush: vigorous; resistant to little-fruit.

Riwaka Tahi.—A firm-fruited hybrid berry with Boysenberry-like flavor. Orig. in Nelson, New Zealand, by H. Hall, Riwaka Res. Center. Aurora x Boysenberry; cross made in 1983; selected in 1985; tested as 832F74; introd. in 1990. Fruit: medium size; wine-red color; firm; many tightly packed drupelets; moderate force required to separate from calyx; ripens up to 2 weeks before Boysenberry; fresh market. Plant: lower yield than Boysenberry; fruit numbers per lateral similar to Boysenberry; medium-strength, spiny canes; good budbreak, but may show uneven lateral development.

Rosborough.—Orig. in College Station, Texas, by H.H. Bowen and E.T. Graham, Texas A&M Univ. Introd. in 1977. F_2 progeny of Brainerd x Brazos, backcrossed to Brazos in 1966. Selected in 1968; evaluated by T.E. Boswell, J.W. Worthington, and J.A. Lipe. Berry: firmer and better quality, with smaller seed size than Brazos; ripens same season as Brazos (mid-May); intended as companion or replacement for Brazos. Plant: similar to Brazos, except flower petals lavender instead of white; canes moderately upright, slightly more than Brazos; productivity equal to or better than Brazos. Has done well at College Station. Not recommended for southeastern Texas because of fungal disease or for northwestern Texas because of low winter temperatures. Should be treated with pesticides for insect, disease, and weed control.

Rusk's King.—Orig. in Grants Pass, Ore., by W.H. Rusk. Introd. in 1942. Parentage unknown; discovered in 1939. Fruit: large; dark red; juicy, no core; flavor pleasing, not too tart; quality good for use as fresh fruit, in jams, and for freezing; seeds few. Bush: productive; vigorous; hardy. Similar, if not identical, to Boysen.

Santa Rosa.—Orig. in Santa Rosa, Calif., by Luther Burbank. First announced in print in 1920. Parentage unknown. Fruit: medium-large; skin black; flesh firm; very sweet; quality excellent; ripens late, September to November. Bush: very vigorous; very productive; canes thornless, long; begins to bear 1 year later than other blackberry varieties.

Sebastopol.—Orig. in Santa Rosa, Calif., by Luther Burbank. First announced in print in 1920 by L. Burbank. Parentage unknown. Fruit: similar to Santa Rosa but sweeter, later, and a little larger.

Shawnee.—Orig. in Fayetteville, Ark., by J.N. Moore, W.A. Sistrunk, and J.B. Buckley, Dept. of Horticulture, Univ. of Arkansas. Introd. in 1985. Cherokee x (Thornfree x Brazos); cross made in 1971; selected in 1974; tested as Ark.730. U.S. plant patent 5686; 11 Mar. 1986; assigned to Univ. of Arkansas. Fruit: very large (7.7 g), maintains size through season; shiny black; medium firm; good flavor. Plant: late-ripening; long harvest period; very productive; thorned canes; very erect cane growth; vigorous; prolific; produces canes readily from cuttings. Immune to orange rust, moderately resistant to anthracnose, susceptible to rosette; hardy to –23C.

Siberian.—Orig. from seeds brought from Siberia by N.E. Hansen, South Dakota Agr. Expt. Sta., Brookings, S.D. Introd. in 1930. Seeds obtained at the Irtysh River steamer landing in Katschiskaya in 1908; P.I. 24377. Fruit: small; skin black; quality poor; only a few drupelets per fruit. Bush: dewberry type; lacks hardiness. Variety now obsolete.

Silvan.—Orig. in Victoria, Australia, by G.R. McGregor and K.H. Kroon, Res. Inst., Dept. of Agriculture, Victoria. Introd. in 1984. U.S.-Ore 742 x Marion; seed obtained from G.F. Waldo in 1952; selected in 1964. Fruit: shiny purple-black; 6–8 g; softens after harvest; excellent flavor, sweeter and less acid than Boysen, Marion, and Young; ripens 2 weeks earlier than Marion. Plant: primocanes trailing; densely armed; vigorous; high-yielding. Tolerant to anthracnose and crown gall and stress-tolerant to wind, drought, and heavy soils; cold hardiness similar to Marion.

Smoothstem.—Orig. in Beltsville, Md., by D.H. Scott and D.P. Ink, Crops Res. Division, U.S. Dept. of Agriculture. Introd. in 1966. US 1482 (Merton Thornless x US 1411) open-pollinated; seed collected in 1955; selected in 1958. Fruit: large; blunt conic; skin jet black; firm; good flavor; matures about 1 Aug. or 1 month later than Eldorado. Bush: canes thornless; semi-upright; 8–10 ft long, up to 1 1/4 inches diam. at base; fruiting laterals bearing 35–40 fruit; nonsuckering; winter hardy at Beltsville and southward; propagation by tip rooting or cuttings; productive. For local market and home garden.

Taranaki.—A firm-fruited hybrid berry with typical blackberry-like flavor for fresh market. Orig. in Nelson, New Zealand, by H. Hall, Riwaka Res. Center. Aurora x Cherokee; cross made in 1982; selected in 1985; tested as 821C93; introd. in 1990. Fruit: medium size; turns dark black about 1 week before fully ripe, difficult to distinguish when fully ripe; very firm; many tightly packed drupelets; ripens up to 2 weeks before Boysenberry. Plant: lower yield than Boysenberry; moderate numbers of spiny canes of medium strength; foliage often shows genetic chlorosis.

Tasman.—A spineless Boysenberry mutation. Orig. in Nelson, New Zealand, by H. Hall, Riwaka Res. Center, in the Boysenberry Clonal Reselection Program; introd. in 1994. Fruit: size, shape, and color intermediate between Boysenberry and Youngberry; extends Boysenberry harvest season. Plant: high-yielding; initial slow establishment.

Tayberry.—Orig. in Scotland, by D.L. Jennings, Scottish Crop Res. Inst., Dundee, Scotland. Introd. in 1980. Plant patent 4424; 12 June 1979; assigned to NSDO, Newton House, Newton, Cambridge, U.K. Aurora (blackberry) x SCRI 626/67 (raspberry); tested as 69102/51. Berry: large, long conical; good flavor and color. Plant: large; semi-spreading; vigorous; hardy; highly productive; susceptible to leaf and bud mite.

Texas Wonder.—See Dew-Black.

Thornfree.—Orig. in Beltsville, Md., by D.H. Scott and D.P. Ink, Crops Res. Division, U.S. Dept. of Agriculture. Introd. in 1966. US 1410 (Brainerd x Merton Thornless) x US 1414 (Merton Thornless x Eldorado). Cross made in 1956; selected in 1959. Fruit: medium-large; blunt conic; skin glossy black; flavor good, tart; adheres firmly; ripens 3 weeks later than Eldorado. Bush: canes thornless; semi-upright; 7–8 ft long; fruiting laterals with 20–30 berries; very productive in fertile soils; winter hardy at Corvallis, Ore., and Beltsville; propagation by tip rooting or cuttings. For local market and home garden.

Thornless Austin.—See Austin Thornless.

Thornless Boysen.—Orig. in El Monte, Calif., by D.L. Duffin. Introd. in 1938. Bud mutation of Boysen; discovered in 1936. May be identical with Acme Thornless Young. Now obsolete.

Tom Doyle.—Orig. in Washington, Ind., by T. Doyle. Introd. in 1976. Plant patent 4094; 23 Aug. 1977. Of unknown parentage, discovered as a chance seedling in 1975. Fruit: round; similar in taste to native blackberry; ripens over an extended period. Bush: large; vigorous; upright-arching shrub; unusually long primocanes; thorns and prickles totally absent; appears to be resistant to rust, leaf spot, and cane blight.

Tummelberry.—Orig. in Invergowrie, Scotland, by D.L. Jennings, Scottish Crop Res. Inst. Introd. in 1983. Tayberry x SHRI 69102/18 (a sibling of Tayberry); tested as SHRI 73153C4. Fruit: medium-large; short conical; skin more red-purple than Tayberry; flavor slightly acid without aromatic characteristics of Tayberry; ripening over a long period starting in July, 1 week later than Tayberry, and finishing 1 week later than Tayberry; plug (receptacle) remains with fruit when picked. Plant: vigorous; sturdy canes produced in moderate to large numbers, spreading to semi-erect, but slightly more erect than Tayberry; yielded slightly more than Tayberry in season that followed a severe winter and slightly less in a season that followed a mild winter; fruiting laterals about 30 cm long and display fruit well. Primocanes dark green with anthocyanin pigments less intense than in Tayberry; dense, moderately sized prickles. Susceptible to raspberry leaf and bud mite.

Waldo.—Orig. in Corvallis, Ore., by F.J. Lawrence, U.S. Dept. of Agriculture and Oregon State Univ. Introd. in 1989. ORUS 1122 x ORUS 1367; selected in 1976; tested as ORUS 2031. Fruit: large, equal in size to Marion; firmer than Marion; glossy black; good shelf life; good quality fresh and processed. Plant: thornless; growth habit similar to Evergreen; moderately vigorous; yields high (9–12 $t \cdot ha^{-1}$ in Oregon); more resistant to cane and leaf spot than Marion or Boysen; slightly less cold hardy than Marion.

White.—See White Pearl.

White Knight.—See White Pearl.

White Pearl *(White, White Knight)*.—Orig. in Berrien City, Mich., by Herschel L. Boll, Champaign, Ill. Introd. in 1974. Parentage unknown. Discovered in 1954 along a fence row. Berry: medium to large, oblong, white, translucent; flesh firm; apparently must be pollinated by a black blackberry; ripens about with Darrow. Bush: low to medium productivity; vigorous, growing up to 8 ft in height; canes erect; produces medium number of suckers, slightly more than Darrow.

Williams.—Orig. in Willard, N.C., by C.F. Williams, U.S. Dept. of Agriculture, Crops Res. Div. and the North Carolina Agr. Expt. Sta. and introduced in 1962 jointly with J.P. Overcash, Mississippi Agr. Expt. Sta., State College, Miss. Himalaya x Taylor; cross made in 1935; seedlings grown at Willard and selected in 1939; tested in North Carolina and Mississippi as NC 436. Fruit: size medium, maintaining good size during the picking season; skin glossy black; flesh medium firm; flavor acid, dessert quality good; ripens in late June in North Carolina and Mississippi, after Boysen, Carolina, and Lucretia but before Flint. Bush: semi-erect; very vigorous; long lateral canes thorny, similar to Himalaya; new primary canes often 8–10 ft long by mid-June; resistant to most cane and leaf diseases; very productive when trained on wire trellis or stakes at Raleigh, N.C., or grown in hedgerows at State College, Miss.; yields at Raleigh averaged 5.3 quarts/plant compared to 3.1 quarts for Brainerd; propagated by tip-layering and by softwood cuttings rooted under mist. Cultivar released as a home garden and local-market berry.

Womack.—Orig. in College Station, Texas, by H.H. Bowen and E.T. Graham, Texas A&M Univ. Introd. in 1977. F_2 progeny of Brainerd x Brazos, backcrossed to Brazos in 1966. Selected in 1968; evaluated by T.E. Boswell, J.W. Worthington, and J.A. Lipe. Berry: firmer and better quality, with smaller seed size than Brazos; ripens same season as Brazos (mid-May); intended as companion or replacement for Brazos. Plant: similar to Brazos, except flower petals lavender instead of white; canes moderately upright, slightly more than Brazos; productivity equal to or better than Brazos. Recommended for the West Cross Timbers region. Not recommended for southeastern Texas because of fungal disease or for northwestern Texas because of low winter temperatures. Should be treated with pesticides for insect, disease, and weed control.

Young *(Youngberry)*.—Orig. in Morgan City, La., by B.M. Young. Introd. in 1926. Mayes x Phenomenal; selected about 1905. Fruit: very large; excellent for frozen pack and jam; flavor very sweet; wine-colored, appearance outstanding. Bush: trailing;

high yielder; early production; vigorous; adapted to East Texas and southern Arkansas; anthracnose is serious in lower south; susceptible to rosette in many southern areas; sunscald or winter injury is occasional when trellised.

Youngberry.—See Young.

Zielinski.—Orig. in Salem, Ore., by B.C. Zielinski. Introd. in 1930. Open-pollinated seedling of a native wild blackberry (*Rubus macropetalus*); discovered in 1926. Fruit: small; flesh medium firm; flavor very agreeable; appearance fine. Used only for breeding purposes. Variety is pistillate.

BLUEBERRY

Aliceblue.—Orig. in Gainesville, Fla., by W.B. Sherman and R.H. Sharpe, Fruit Crops Dept., Univ. of Florida. Selected in 1970 from an open-pollinated seed lot of Beckyblue; tested as Fla. A. Fruit: medium-small, round, firm; medium-blue; flavorful with medium-dry small scars; ripens late May–early June; recommended for shipping or local market use. Plant: vigorous; tall, spreading, similar to Tifblue; not self-fertile and should be interplanted with other rabbiteyes for cross-pollination; some resistance to powdery mildew. Aliceblue has a lower chilling requirement than Bluegem.

Angola.—Orig. in Raleigh, N.C., by E.B. Morrow, North Carolina Agr. Expt. Sta., and George M. Darrow, U.S. Dept. of Agriculture. Introd. in 1952. Weymouth x F-6 (Stanley x Crabbe 4); (Crabbe 4, a native selection from eastern North Carolina); cross made in 1934; selected in 1940; tested as N.C. 246. Fruit: cluster loose; berry size medium; globose; skin dark blue; flesh somewhat soft; flavor and aroma good; dessert quality fair; scar medium; equal in size and color to Weymouth; ripens earlier than Weymouth in eastern North Carolina. Plant: vigorous; open; productive; highly resistant to canker; leaf large. Since 1953 not recommended commercially because of dark color, lack of fruit firmness, and finicky reaction to replant soil situations.

Ashworth.—Orig. in Lisbon, N.Y., by Fred L. Ashworth, St. Lawrence Nurseries, Heuvelton, N.Y. Introd. for its good breeding characteristics; not commercial. Discovered in 1947 as a wild highbush blueberry plant (*Vaccinium corymbosum*); named by U.S. Dept. of Agriculture (USDA), Beltsville, Md. Fruit: small, its chief disadvantage; ripens early. Bush: a pure tetraploid; transmits hardiness, reported to have borne fruit after temperatures as low as –50F; resistant to frost injury during bloom time; used extensively for breeding purposes by USDA, Beltsville.

Atlantic.—Orig. in Whitesbog, N.J., by F.V. Coville, U.S. Dept. of Agriculture. Introd. in 1941. Jersey x Pioneer; cross made in 1925. Fruit: cluster loose; berry large; oblate; skin blue, attractive; flesh firm; aroma slight; flavor good; dessert quality medium; resistant to cracking; ripens late, just before Jersey. Bush: vigorous; open spreading; very productive; leaf large; subject to bacterial dieback in Oregon and Washington.

Augusta.—Orig. in Orono, Maine, by Agr. Expt. Sta., Univ. of Maine. Introd. in 1975. Open-pollinated seedling selected by Leslie Whitten from clone 3302, which was selected at Frankfort, Maine, in 1933; tested as ME2. Fruit: large, oval; attractive; good eating quality fresh or frozen; berries borne well off the ground, mature uniformly for a single harvest, thus adaptable to mechanical harvest; ripens midseason. Plant: shoots green, 18 cm high; foliage thick; productive; roots readily; self-incompatible. Susceptible to witches' broom caused by *Pucciniastrum goeppertianum* (Kuhn.) Kleb.; tolerant to redleaf, blight, and *Botrytis cinerea* Pers. The first clone of lowbush blueberry *Vaccinium angustifolium* Ait. named and released for commercial production through Res. Sta., Agriculture Canada, Kentville, Nova Scotia, blueberry breeding program.

Austin.—Rabbiteye. Released by the U.S. Dept. of Agriculture in 1996. Named by Arlen Draper in honor of Max E. Austin, retired blueberry breeder from the Coastal Plains Expt. Sta., Univ. of Georgia, Tifton. T110 (Woodard x Gardenblue) x Brightwell; tested as T339. Fruit: large; light blue; scar, firmness, and flavor good. At Poplarville, Miss., and Tifton, Ga., it flowers and ripens about the same time as Climax. Bush: upright; yield high. Fruits and leafs well as far south as Gainesville, Fla.

Avonblue.—Orig. in Gainesville, Fla., by W.B. Sherman and R.H. Sharpe, Fruit Crops Dept., Univ. of Florida. Introd. in 1977. [E66 Berkeley x F-72 (Wareham x Pioneer)] x FLA. 1-3 (Michigan–USDA Highbush) x Sel. [Earliblue x Sel. (*V. ashei* Callaway or Coastal x *V. darrowi*)]; cross made in 1962; selected in 1964; tested as Fla. 4-65. Fruit: oblate; light blue; firm, dry stem scar; 65 berries/cup (0.24 liter); ripens after Sharpblue in mid-May; adaptable to mechanical harvest; for shipping and local market; high quality. Plant: spreading; slightly smaller than Flordablue; fruit set extremely heavy, pruning necessary; blossoms in late February to March. Yield medium to high; resistant to stem canker (*Botryosphaeria corticis* Demaree & Wilcox) and to bud mites.

Baldwin.—Orig. in Tifton, Ga., by M.E. Austin, Univ. of Georgia, and A.D. Draper, U.S. Dept. of Agriculture. Introd. in 1985. Tifblue x GA 6-40 (Myers x Black Giant); tested as T-117. Fruit: medium; skin dark blue; scar small; medium firm; flavor good. Very productive, late-ripening rabbiteye blueberry (*Vaccinium ashei*). Chilling requirement 450–500 hours. Intended primarily for pick-your-own because of long ripening period.

Beckyblue.—Orig. in Gainesville, Fla., by W.B. Sherman and R.H. Sharpe, Fruit Crops Dept., Univ. of Florida. *Vaccinium ashei* Reade (6x) x E96 (an unnamed tetraploid highbush); cross made in 1965; selected in 1968; tested as Fla. X. Fruit: round, medium; firm; medium-blue; flavorful with dry, small scars; recommended for shipping or local market use; blooms in late February to early March, 1 week before Bluegem; ripens late May to early June. Plant: vigorous; tall; spreading; similar in size to Tifblue; not self-fertile and should be interplanted with other rabbiteyes for cross-pollination; some resistance to powdery mildew.

Berkeley.—Orig. in Weymouth, N.J., by F.V. Coville and George M. Darrow, U.S. Dept. of Agriculture, and Franklin A. Gilbert, New Jersey Agr. Expt. Sta. Introd. in 1949. Stanley x GS-149 (Jersey x Pioneer); cross made in 1932; selected in 1938. Fruit: cluster loose; berry very large; oblate; skin very light blue, very attractive; flesh firm, aroma slight, slightly subacid; dessert quality medium; not subject to cracking; stores well; scar very good; ripens late midseason, about 1 week after Stanley and 1 week before Jersey. Bush: productive; easy to propagate.

Black Giant.—Orig. in Crestview, Fla., by M.A. Sapp. Introd. in the 1920s. Selected in the wild and probably set in an original planting in 1893; *Vaccinium ashei*, rabbiteye blueberry; used in breeding. Fruit: one of the largest of the wild selections; skin black; flesh soft; flavor good. Bush: extremely vigorous; about 15 ft high x 15 ft wide. No longer propagated.

Bladen.—Southern highbush. Selected at Castle Hayne, N.C., a test location for North Carolina State Univ. NC1171 x NC SF-12-L; tested as NC2055; released in 1994. Berry: size medium (about the same as Croatan); dark blue; scar, firmness, and flavor good. Flowers a few days before Croatan in eastern North Carolina, and ripens about the same time as Reveille and O'Neal and somewhat before Croatan. Bush: upright; vigorous; good productivity; not completely self-fruitful and should be interplanted with another clone, such as Reveille; resistant to cane canker; field tolerance to stem blight.

Blomidon.—Orig. in Res. Sta., Agriculture Canada, Kentville, Nova Scotia. Introd. in 1982. The first named lowbush blueberry from controlled breeding. 451 (a selection obtained from a lowbush blueberry field at Kirkhill, Nova Scotia) x Augusta; selected in 1970; tested as 70-36. Fruit: large, uniform, ovate; flavor fair; ripens uniformly with 95% mature by 10 Aug. Plant: vigorous; upright to spreading; productive; easily propagated; named after Cape Blomidon, a landmark in the area where variety was developed.

Bluebelle.—Orig. in Tifton, Ga., by Georgia Coastal Plain Expt. Sta. and U.S. Dept. of Agriculture. Introd. in 1974. Callaway x Ethel; cross made in 1946 by C.M. Darrow; selected in 1951 by W.T. Brightwell; tested as T-14. Fruit: large; round; skin light blue; excellent flavor; ripens over a relatively long period beginning in midseason. Bush: upright; moderately vigorous; productive; self-unfruitful.

Bluechip.—Orig. at Garland, N.C., by J.R. Ballington, A.D. Draper, G.J. Galletta, and C.M. Mainland, North Carolina Agr. Res. Serv., Raleigh, and U.S. Dept. of Agriculture, Beltsville, Md. Introd. in 1979. Croatan x US 11-93; cross made at Beltsville by G.M. Darrow; selected in 1957 by G.W. Schneider, F.E. Correll, and G.M. Darrow; tested as NC 57-31. Berry: superior in size, color, firmness, and flavor to standard cultivars; ripening follows Harrison on mature plants, younger plants mature just ahead of Harrison. Bush: self-fruitful; productive; upright; easily trained; cane-canker resistant; susceptible to stem blight.

Bluecrop.—Orig. in Weymouth, N.J., by F.V. Coville and O.M. Freeman, U.S. Dept. of Agriculture (USDA) and the New Jersey Agr. Expt. Sta. Introd. in 1952. GM-37 (Jersey x Pioneer) x CU-5 (Stanley x June); cross made in 1934 by Coville and Freeman; seedling raised in Weymouth, N.J.; selected in 1941 by J.H. Clarke (then of the New Jersey Agr. Expt. Sta.) and George M. Darrow, USDA; tested as N.J. 17-19. Fruit: cluster large, medium loose; berries roundish-oblate; skin very light blue; flesh very firm; subacid, flavor good; resistant to cracking; dessert quality above medium; moderately aromatic; scar small; ripens midseason, about 4 days before Berkeley; stem sometimes clings to berry. Bush: upright; spreading; vigorous; leaf size medium to below medium; very consistent producer; hardy; drought resistant. For many years, the world's most important blueberry cultivar.

Bluegem.—Orig. in Gainesville, Fla., by R.H. Sharpe and W.B. Sherman, Univ. of Florida, Gainesville. Introd. in 1970. Open-pollinated seedling of Tifton 31 (unnamed selection); selected in 1954; tested as Fla. 6-164. Fruit: rabbiteye type; size medium, averaging 110 berries/cup; skin light blue, heavily waxed; picking scar very small and dry; flesh firm; recommended as a home-garden variety and as a commercial variety; ripens mid-June, fruit tends to ripen at one time, a good characteristic for mechanical harvesting. Bush: size medium, moderately spreading; productive; vigor between Woodard and Tifblue; should be cross-pollinated, recommended that Woodard be interplanted in alternate rows.

Bluegold.—Orig. at Beltsville, Md., by A.D. Draper. Introd. in 1988. Bluehaven x ME-US-5 (Ashworth x Bluecrop), *Vaccinium corymbosum*. Selected in 1972 in Weymouth, N.J., by A.D. Draper, D.H. Scott, and G. Jelenkovic from cross made in 1968; evaluated by N. Vorsa and G.J. Galletta as G-355. Fruit: light blue color; scar good; firm; flavor good; late-ripening, with Jersey. Bush: vigorous; low-growing; highly productive.

Bluehaven.—Orig. in South Haven, Mich., by Stanley Johnston and J.E. Moulton, Michigan State Univ. Introd. in 1967. (Berkeley x 19-H). Cross made in 1948; selected in 1952. Fruit: large; round; skin light blue; flesh firm; flavor excellent; picking scar very small and dry; ripens about 15 July at South Haven; holds quality well on bush; can be picked over 4- to 6-week period. Bush: hardy; productive; recommended for any area where the highbush blueberry thrives.

Bluejay.—Northern highbush. Released in 1978 by Michigan State Univ. Berkeley x Michigan 241 (Pioneer x Taylor); tested as HBS400. Fruit: size medium; light blue; scar and firmness good; ripens 5–7 days before Bluecrop; stores well on the bush. Plant: vigorous; upright; resistant to mummyberry in New Jersey.

Blueray.—Orig. in Weymouth, N.J., by F.V. Coville and O.M. Freeman, U.S. Dept. of Agriculture, and J.H. Clarke, New Jersey Agr. Expt. Sta. Introd. in 1955. GM 37 (Jersey x Pioneer) x CU 5 (Stanley x June); cross made in New Jersey in 1934; seedling raised in Washington, D.C.; selected in New Jersey in 1941 by Clarke and George M. Darrow; tested as US 18-96. Fruit: cluster small, tight; berry very large; skin medium light blue; flesh firm, aromatic, flavor fine, noncracking; scar medium; ripens midseason with Ivanhoe and Rancocas; resembles Dixi. Bush: hardy; very productive, more consistent than Ivanhoe in New Jersey; very easily propagated by hardwood cuttings; promising from New Jersey northward.

Blue Ridge.—Orig. at White Lake, N.C., by G.J. Galletta and A.D. Draper. Introd. in 1987. Southern highbush, mainly *Vaccinium corymbosum*. Patriot x US 74 (*V. darrowi* cv. Fla. 4B x Bluecrop). Selected in 1975; tested as NC 1440 by J.R. Ballington, C.M. Mainland, S.D. Duke, and A.D. Draper. Fruit:

medium-large; excellent color, firmness, and flavor; picking scar fair to poor; ripens early midseason to midseason. Plant: vigorous; erect; widely adapted. Chilling requirement 500–600 hours. Susceptible to cane canker and mummy berry; field tolerant to stem blight.

Bluetta.—Orig. in New Brunswick, N.J., by New Jersey Agr. Expt. Sta. and U.S. Dept. of Agriculture, Beltsville, Md. Introd. in 1968. No. 3 (North Sedgwick lowbush x Coville) x Earliblue; cross made in 1947 by G.M. Darrow; selected in 1952 by G.M. Darrow, D.H. Scott, and L.F. Hough, of the above stations. Fruit: size medium; skin light blue; flesh firm; broad stem scars; more flavor than Weymouth; ripens early, with Weymouth. Bush: short, compact-spreading; vigor medium; consistently productive; more resistant to spring frosts than Weymouth.

Bonita.—Orig. in Gainesville, Fla., by P.M. Lyrene and W.B. Sherman. Introd. in 1985. Beckyblue open-pollinated; *Vaccinium ashei*. Fruit: medium-large; light blue; excellent scar; firm; flavor tart until fully ripe. Early ripening, mechanically harvestable rabbiteye blueberry. Self-incompatible. Moderately susceptible to phytophthora root rot. Chilling requirement same as for Climax.

Bounty.—Orig. at Cypress Creek, N.C., by G.J. Galletta and A.D. Draper. Tested as NC 1074; *Vaccinium corymbosum*. Introd. in 1987. Murphy x G-125. Selected in 1970; tested by J.R. Ballington, C.M. Mainland, S.D. Duke, A.D. Draper, and G.J. Galletta. Fruit: very large; color, stem scar, and firmness above average; flavor good; ripens early midseason to midseason. Bush: consistently productive; field tolerance to cane canker and stem blight.

Brightwell.—Orig. in Alapaha, Ga., by Georgia Agr. Expt. Sta., Tifton, and U.S. Dept. of Agriculture. Introd. in 1981. Named in honor of W.T. Brightwell, who bred rabbiteye blueberries for more than 30 years at Georgia Coastal Plain Expt. Sta. Tifblue x Menditoo; cross made by W.T. Brightwell and J.N. Moore in 1963; tested as T-99. Berry: medium; skin blue, small dry scars; flavor good; ripens after Climax, about same season as Woodard. Plant: productive; vigorous; upright; new cane production adequate; self-unfruitful.

Brigitta.—Northern highbush. Selected at the Horticultural Res. Inst., Victorian Dept. of Agriculture, Knoxfield, Victoria, Australia. Tested as Knoxfield 11A and released in 1977. Orig. in Australia from seed obtained by open pollination of Lateblue at Michigan State Univ. Berry: large; medium blue; very firm; flavor good; scar small and dry; ripens late, between Bluecrop and Lateblue in Victoria; ships and stores well. Bush: upright; vigorous; highly productive; easy to propagate by softwood and hardwood cuttings. As of 1966, Brigitta was the most important blueberry variety developed in Australia. Has been introduced into the U.S. and Chile.

Briteblue.—Orig. in Tifton, Ga., by Georgia Coastal Plain Expt. Sta. and Crops Res. Div., U.S. Dept. of Agriculture. Introd. in 1969. Ethel x Calloway; cross made in 1950 by W.T. Brightwell; selected in 1955 by Brightwell and G.M. Darrow; tested as T-71. Fruit: large; skin light blue; flavor fair; very firm, handling well for distant marketing; ripens later than Tifblue. Bush: open-headed and spreading; produces only enough new canes from the crown for renewal of the top. Adapted to the Gulf Coast and southeastern U.S.

Brunswick.—Orig. in Res. Sta., Agriculture Canada, Kentville, Nova Scotia. Introd. in 1977. Selected in 1965 from a native stand of lowbush blueberries in a commercial field owned and operated by Shepoddy Blueberry & Lumber Co., Fairfield, N.B.; tested as 734. Fruit: large; globose; attractive; skin medium blue; clusters compact with an average of 20 berries; matures uniformly; ripens midseason. Plant: vigorous; shoots up to 30 cm in height; productive; roots easily; crosses readily with Augusta; tolerant of common foliage diseases. Brunswick is the second lowbush blueberry named and released from Kentville Sta. blueberry program.

Burlington.—Orig. in Whitesbog, N.J., by F.V. Coville, U.S. Dept. of Agriculture. Introd. in 1941. Rubel x Pioneer; cross made in 1916. Fruit: cluster medium tight; berry small; skin blue; flesh firm; aroma slight; dessert quality very good; scar very good;

resistant to cracking; stores and ships well; matures late, lasting 1 week after Rubel. Bush: vigorous; upright spreading; hardy; cold storage quality good; healthy.

Cabot.—Orig. in Whitesbog, N.J., by F.V. Coville, U.S. Dept. of Agriculture. Introd. in 1920. Brooks x Chatsworth; cross made in 1913. Fruit: small; scar only fair, tears; flavor and texture poor; ripens early, season long. Bush: not vigorous; low-spreading. Being discarded.

Callaway.—Orig. in Tifton, Ga., by W.T. Brightwell, Georgia Coastal Plain Expt. Sta. and the U.S. Dept. of Agriculture. Introd. in 1949. Myers x Black Giant; selected in 1943. Fruit: large; skin medium-dark blue; flesh soft; flavor good; scar medium-large; moist. Bush: slow upright growth, few new stems produced; productive; rabbiteye type; being replaced by new varieties.

Cape Fear.—Orig. at Castle Hayne, N.C., by G.J. Galletta and A.D. Draper. Southern highbush, mainly *Vaccinium corymbosum*. US75 (*V. darrowi* Fla 4B x Bluecrop) x Patriot. Selected in 1975; tested as NC 1525 by J.R. Ballington, C.M. Mainland, S.D. Duke, and A.D. Draper. Fruit: very large; good color, scar, and firmness. Bush: precocious; productive; vigorous; semi-upright; roots readily from cuttings. Susceptible to cane canker; field-resistant to stem blight. Chilling requirement 500–600 hours.

Catawba.—Orig. in Whitesbog, N.J., by F.V. Coville, U.S. Dept. of Agriculture. Introd. in 1932. F_2 of (Brooks x Russell); cross made in 1913. Fruit: Catawba-grape color. Bush: semi-dwarf. No longer propagated.

Centurion.—Orig. in Beltsville, Md., by G.M. Darrow, Agr. Res. Serv. W-4 (native Georgia selection) x Callaway; selected in 1961; tested as NC 854. Fruit: using Tifblue as the standard, Centurion is later by 1–2 weeks, better-flavored, darker in fruit color, slightly less firm, and yields as well; Centurion adds one full week or more to the rabbiteye ripening season. Plant: vigorous; productive; should be interplanted with a suitable pollinator variety.

Chandler.—Northern highbush. Selected in Weymouth, N.J.; tested as G-139; released by the U.S. Dept. of Agriculture in 1994. Darrow x M-23. Berry: large; light blue; small, dry scar; firmness and flavor good; ripens late midseason over a long period, about 2 weeks earlier than Elliott in Oregon. Plant: vigorous; upright; well-branched. Consistently produces high yields in high-chill areas where winters are not severe.

Chaucer.—Orig. in Gainesville, Fla., by P.M. Lyrene and W.B. Sherman. Introd. in 1985. *Vaccinium ashei*. Berry: light blue; medium firm; flavor good; poor scar limits shipping potential. Early ripening. Self-incompatible. No major disease problems other than those affecting all rabbiteyes. Low chilling.

Chignecto.—Orig. in Res. Sta., Agriculture Canada, Kentville, Nova Scotia. Introd. in 1977. Selected in 1964 from a native stand of lowbush blueberries in a commercial field owned and operated by Dickinson Brothers, West Brook, N.S.; tested as 508. Fruit: somewhat oblong; weight less than Augusta and Brunswick, but more berries per cluster; skin powdery blue; scar very small; matures unevenly. Plant: more vigorous than Augusta and Brunswick; shoots up to 80 cm under greenhouse conditions; easily propagated; less productive than Augusta and Brunswick; tested as both a male and female in crosses with Augusta and Brunswick; as a male was compatible with Augusta; as a female was compatible with Augusta and Brunswick; tolerant to foliage diseases. The third lowbush blueberry named and released from the Kentville Sta. blueberry program.

Chippewa.—Based on highbush x lowbush crosses. Released in 1996 by the Univ. of Minnesota; selected at Elk River, Minn.; tested as MN393. B18A (G65 x Ashworth) x US3 (Dixi x Michigan Lowbush No. 1). Berry: large (about the same as Northblue); firm; sweet; light blue. Plant: more upright than Northblue; productive; ripens midseason. It is somewhat self-fruitful, but benefits from cross-pollination.

Choice.—Orig. in Gainesville, Fla., by P.M. Lyrene and W.B. Sherman. Introd. in 1985. *Vaccinium ashei*. Fruit: size medium; light blue; excellent scar; firm; very good flavor; late-ripening; machine-harvestable. Self-incompatible. Somewhat susceptible to phytophthora root rot. Chilling requirement similar to Tifblue.

Clara.—Orig. 20 miles west of Valdosta, Ga., by J.T. Bush. Introd. in the early 1930s. Parentage unknown; discovered in 1929. Fruit: skin medium blue; flavor good. Bush: rabbiteye type; vigorous, to 10 ft high.

Climax.—Orig. in Tifton, Ga., by Georgia Coastal Plain Expt. Sta. and U.S. Dept. of Agriculture. Introd. in 1974. Callaway x Ethel; selected by W.T. Brightwell; tested as T-90. Fruit: size medium; skin medium dark blue; flesh good flavor; ripens early over short period, about 80% may be harvested at one time; recommended for mechanical harvest. Bush: upright, spreading, with stems originating in a relatively small area; produces only enough new stems to renew plant; self-unfruitful.

Coastal.—Orig. in Tifton, Ga., by W.T. Brightwell, Georgia Coastal Plain Expt. Sta. and the U.S. Dept. of Agriculture. Introd. in 1949. Myers x Black Giant; selected in 1943. Fruit: medium-large; skin medium-dark blue, thin, splits easily in adverse weather; flavor good; scar medium-small. Bush: vigorous; medium upright, producing several stems for plant renewal; height up to 6 ft; productive; rabbiteye type; being replaced by newer varieties.

Collins.—Orig. in New Brunswick, N.J., by the U.S. Dept. of Agriculture and the New Jersey Agr. Expt. Sta. Introd. in 1959. Stanley x Weymouth; cross made in 1936 by F.V. Coville; selected by J.H. Clarke, E.M. Meader, and George M. Darrow; tested as USDA 18-116; a sister selection of Earliblue. Fruit: cluster size medium, rather compact; attractive; berries as large as Earliblue; oblate; flesh very firm, light blue; highly flavored, sweet to mildly subacid; ripens early, midway between Earliblue and Bluecrop, filling the gap in the ripening season of large-fruited varieties; does not drop; resistant to cracking. Bush: erect; well-shaped; vigorous; not consistently productive; appears to be as hardy as Berkeley and Pemberton under normal conditions. Recommended for trial as a second early, large-fruited variety for northeastern U.S. Named in honor of Lester Collins (1880–1957), amateur horticulturist and fruit grower, president of the New Jersey State Board of Agr., the Blueberry Cooperative Assn., the American Cranberry Growers Assn., and the New Jersey Peach Council.

Concord.—Orig. in Whitesbog, N.J., by F.V. Coville, U.S. Dept. of Agriculture. Introd. in 1928. Brooks x Rubel; cross made in 1916. Fruit: small; scar poor; midseason. Bush: productive; expensive to prune.

Cooper.—Orig. at Poplarville, Miss., by J.M. Spiers and A.D. Draper. Southern highbush, mainly *Vaccinium corymbosum*. Introd. in 1987. G-180 (*V. corymbosum*) x US 75 (*V. darrowi* Fla. 4B x Bluecrop). Selected in 1977; tested as MS-78. Fruit: good color, firmness, scar, and flavor; early ripening. Low chilling. Moderately productive.

Coville.—Orig. in Weymouth, N.J., by F.V. Coville and George M. Darrow, U.S. Dept. of Agriculture and Franklin A. Gilbert, New Jersey Agr. Expt. Sta. Introd. in 1949. GM-37 (Jersey x Pioneer) x Stanley; cross made in 1930; selected in 1936. Fruit: cluster loose; berry very large; flesh firm; flavor tart until ripe, aromatic; dessert quality good; scar good; resistant to cracking; does not drop; ripens late, 30 days after Earliblue. Bush: vigorous; productive; easy to propagate.

Croatan.—Orig. in Raleigh, N.C., by the North Carolina Agr. Expt. Sta. and the U.S. Dept. of Agriculture. Introd. in 1954. Weymouth x F-6 (Stanley x Crabbe 4), being of the same parentage as Angola, Wolcott, and Murphy; cross made in 1934 by F.V. Coville; tested as N. C. 257. Fruit: ripens later than Angola and Wolcott but before Murphy; clusters long, fairly loose, attractive, slightly larger than those of Angola, Wolcott, Weymouth, and Murphy; berry medium to large, globose; scar excellent; flesh medium firm, aromatic, sweet to subacid depending on degree of ripeness; good dessert quality; may produce 1 1/2 to 2 times the yield of Wolcott and Murphy; tends to ripen quickly in some seasons and must be picked on a shorter schedule for best results. Bush: resistant but not immune to canker; more resistant to bud mite than Wolcott, Murphy, and Weymouth; vigorous; precocious; productive; dense leaf cover. Since 1960 considered a standard variety of North Carolina. For many years, it has accounted for more than half the blueberry acreage in North Carolina.

Darrow.—Orig. in Beltsville, Md. Introd. in 1965. F-72 x Bluecrop; cross made in 1949 by G.M. Darrow, U.S. Dept. of Agriculture (USDA), Beltsville; selected by him in 1955; evaluated and named by D.H. Scott and J.N. Moore, USDA, Beltsville, and by H.H. Bowen and L.F. Hough, New Jersey Agr. Expt. Sta., New Brunswick; tested as G-71. Fruit: cluster medium, attractive; berry as large as, to larger than, Coville; skin light blue; flesh firm, aromatic, highly flavored, tart to mildly tart depending on maturity; neither drops nor cracks during wet weather; scar medium; ripens about with Coville, which it resembles. Bush: erect; vigorous; consistently productive, more so than Coville.

Delite.—Orig. in Tifton, Ga., by Georgia Coastal Plain Expt. Sta. and Crops Res. Div., U.S. Dept. of Agriculture. Introd. in 1969. T-14 x T-15; cross made in 1954 by W.T. Brightwell; selected in 1959 by Brightwell and D.H. Scott; tested as T-94. Fruit: large; round; skin light blue; flavor excellent, are not tart before fully ripe as are most varieties of rabbiteye blueberries; ripens a few days later than Briteblue. Bush: upright with numerous canes from the crown. Adapted to the Gulf Coast and southeastern U.S.

Denise.—Northern highbush. Selected at the Horticultural Res. Inst., Victorian Dept. of Agriculture, Knoxfield, Australia; tested as Knoxfield 47; released in 1977. Orig. in Australia from seed obtained from Michigan State Univ. Berry: very large; medium dark blue; remains poorly pigmented around the pedicel until fully mature; scar medium; firmness good; flavor excellent, but develops off-flavor when over-mature; ripens with or slightly before Bluecrop. Bush: vigorous; highly productive but sprawling; requires pruning to obtain upright growth; easy to propagate from softwood and hardwood cuttings.

Dixi.—Orig. in Weymouth, N.J., by F.V. Coville, U.S. Dept. of Agriculture. Introd. in 1936. (Jersey x Pioneer) x Stanley; cross made in 1930. Fruit: cluster medium-tight; berry large; round-oblate; skin blue; flesh firm, aromatic; dessert quality good; flavor pronounced; susceptible to cracking; scar large, poor; season late, after Jersey. Bush: productive; open-spreading; vigorous.

Duke.—Orig. in Weymouth, N.J., by A. Draper, G. Galletta, G. Jelenkovic, and N. Vorsa. *Vaccinium corymbosum*. Introd. in 1987. G-100 (Ivanhoe x Earliblue) x 192-8 [(Berkeley x Earliblue) x (Coville x Atlantic)]. Selected in 1972; tested as G-354. Fruit: medium to large; scars small, dry; good color; firm; flavor mild; good shipping quality; early ripening. Plant: vigorous; upright; self-fruitful; consistently productive; canes numerous, stocky, moderately branched; high chilling requirement. Flowers late, after Weymouth, with Bluecrop.

Earliblue.—Orig. in Weymouth, N.J., by F.V. Coville and O.M. Freeman, U.S. Dept. of Agriculture (USDA) and the New Jersey Agr. Expt. Sta. Introd. in 1952. Stanley x Weymouth; cross made in 1936 by Coville and Freeman; seedling raised in Weymouth, and selection made in 1943 by J.H. Clarke (then of the New Jersey station), E.M. Meader (New Hampshire Agr. Expt. Sta.), and George M. Darrow (USDA); tested as US 15-121. Fruit: cluster size medium, loose; berries large, oblate; skin light blue; flesh very firm, subacid, flavor good, resistant to cracking, moderately aromatic, dessert quality good; scar good; season very early with or before Weymouth. Bush: hardy; upright; vigorous; well-shaped; productive; a fine early variety for Maryland northward.

Elizabeth.—Orig. in Whitesbog, N.J., by Elizabeth White. Introd. in 1966. Parentage unknown. Discovered about 1950; tested as 3850-A. Fruit: cluster loose; berries very large; skin color fair; flavor excellent; good shipper; ripens over very long season, from midseason to very late. Bush: upright; spreading.

Elliott.—Orig. in Beltsville, Md., by U.S. Dept. of Agriculture. Introd. in 1973. Burlington x US 1 [Dixi x (Jersey x Pioneer)]; cross made in 1947 by George M. Darrow; selected by Arthur Elliott of Otter Lake, Mich., in 1948 for testing as E-70. Cluster: loose. Berry: size medium; skin light blue; flesh firm, flavor good, mild; ripens late, 7–10 days after Lateblue (extending season); can be machine-harvested in one or two pickings. Bush: vigorous; upright; productive; hardy; shows moderate resistance to *Monilina vaccinii-corymbosi*.

Ethel *(Satilla)*.—Orig. in Brunswick, Ga., by W.M. Walker. Introd. in 1944. Parentage unknown. Fruit: large; skin light blue; flavor fair when fully ripe; scar very small, dry. Bush: very compact growth; low, spreading type of growth up to 5 ft; very productive; rabbiteye type. Not grown commercially.

Flordablue.—Orig. in Gainesville, Fla., by R.H. Sharpe, Univ. of Florida. Introd. in 1975. A tetraploid derived from a series of crosses between *V. ashei*, *V. darrowi*, and *V. corymbosum*; cross made in 1965; selected in 1967; tested as Fla. 67-2. Cluster: medium tight. Fruit: size medium; round; skin light blue, scar size medium; flesh firm; ripens very early. Bush: spreading; vigorous; very productive; resistant to canker; susceptible to phytophthora root rot. Very low chilling requirement.

Garden Blue.—Orig. in Raleigh, N.C., by E.B. Morrow, North Carolina Agr. Expt. Sta. and George M. Darrow, U.S. Dept. of Agriculture. Introd. in 1958. Myers x Clara; rabbiteye type; selected in 1944; tested as N.C. 468. Fruit: medium, averaging 130–135/half-pint cup; round; flesh firm, blue; flavor good; scar small, dry; in eastern North Carolina, ripening begins in mid-June and ends by mid-July. Bush: consistently high-yielding; vigorous; upright; suckers readily; fruit well distributed on long shoots, making picking easy. Garden Blue appears adapted to home gardens because of its vigorous growth and good fruit quality.

Gem.—Orig. in Gate, Wash., by H.E. Drew. Introd. in 1952. Plant patent 1181; 21 Apr. 1953; assigned to Alberta M. Drew. Rancocas x June; cross made in 1943; selected in 1949. Fruit: cluster size medium, fairly open; berry size medium; oblate; skin dark blue; flesh firm, crisp, bluish, aroma slight, dessert quality medium; ripens early midseason, slightly before Rancocas; principal distinguishing characteristic is that all berries ripen virtually at the same time, within 2 or 3 days, but may remain in good condition on bush for 2 weeks or more; processes well; resembles Rancocas. Bush: erect; vigorous; produces well; apparently quite hardy.

Georgiagem.—Orig. in Tifton, Ga., by M.E. Austin and A.D. Draper. Southern highbush, mainly *Vaccinium corymbosum*. Introd. in 1987. G-132 x US-75; about 25% *V. darrowi*; pedigree includes Ashworth, Earliblue, and Bluecrop. Cross made in 1972; selected in 1977; tested as TH-285. Very good fruit color and quality; small stem scar; firm; pleasant flavor; early ripening. Low chilling, requiring at least 350 hours below 7C; tolerant of climate of southeastern U.S.; diseases have not been a problem. Easy rooting. Bush: moderately vigorous; moderately productive; semi-upright.

Greenfield.—Orig. in Whitesbog, N.J., by F.V. Coville, U.S. Dept. of Agriculture. Introd. in 1926. F_2 of Brooks (lowbush) x Russell (highbush); parents from Greenfield, N.H.; cross made in 1913. Fruit: small; skin dark blue; flesh sweet; ripens very early. Bush: medium, 2–3 ft high; production medium.

Gulfcoast.—Orig. at Poplarville, Miss., by J.M. Spiers and A.D. Draper. Southern highbush, mainly *Vaccinium corymbosum*. Introd. in 1987. G-180 (*V. corymbosum*) x US75 (*V. darrowi* x Bluecrop). Tested as MS-90. Fruit: good color, firmness, and flavor; scar medium; pedicel tends to adhere to harvested berry. Early ripening. Low chilling.

Harrison.—Orig. in Raleigh, N.C., by North Carolina Agr. Expt. Sta. and U.S. Dept. of Agriculture. Introd. in 1974. Croatan x US 11-93; cross made in early 1950s at Beltsville, Md.; selected in 1961 by G.J. Galletta and R.J. Knight near Garland, N.C.; tested as NC 61-3. Fruit: larger and firmer than Morrow and Wolcott; improved color and flavor over other standard varieties; flesh keeps well and has high quality; ripens early midseason between Croatan and Murphy. Bush: highbush; semi-upright; vigorous; productive; outyields Morrow and Wolcott 60% to 300%; self-fruitful; resistant to cane canker and tolerant to bud mite; susceptible to anthracnose. Recommended for areas subject to cane canker caused by *Botryosphaeria corticis*. Named for Gale Harrison in recognition of his cooperation in blueberry breeding.

Herbert.—Orig. in Weymouth, N.J., by F.V.

Coville and O.M. Freeman, U.S. Dept. of Agriculture (USDA) and the New Jersey Agr. Expt. Sta. Introd. in 1952. Stanley x G5-149 (Jersey x Pioneer); cross made in 1932; seedling raised in Washington, D.C., and selection made in 1938 in Weymouth by J.H. Clarke (then of the New Jersey station), George M. Darrow (USDA), and E.B. Morrow (North Carolina Agr. Expt. Sta.); tested as V-25. Fruit: clusters loose; ripens late, about the same time as Jersey, Rubel, and Dixi, earlier than Coville and later than Berkeley; berry very large, about equal to the best varieties; scar fair; hangs well on bush. Bush: blooms late; consistently productive.

Homebell.—Orig. in Beltsville, Md., and Tifton, Ga., by W.T. Brightwell, Georgia Coastal Plain Sta., and George M. Darrow, U.S. Dept. of Agriculture. Introd. in 1955. Myers x Black Giant; cross made in 1940 at Beltsville; selection made at Tifton. Fruit: medium large; skin medium-dark blue; flesh firm; flavor mild. Bush: rabbiteye type; very vigorous; productive; easy to propagate.

Ivanhoe.—Orig. in New Jersey by George M. Darrow, U.S. Dept. of Agriculture, and E.B. Morrow, North Carolina Agr. Expt. Sta. Introd. in 1952. Z-13 (Rancocas x Carter) x Stanley; cross made in 1933. Fruit: cluster medium loose; berry large; round-oblate; skin light blue, attractive; slightly later than Stanley; intermediate in color between Weymouth and Stanley; flesh firm, highly aromatic; flavor good to excellent; scar good. Bush: vigorous; productive; erect. No longer recommended in North Carolina due to susceptibility to canker.

Jersey.—Orig. in Whitesbog, N.J., by F.V. Coville, U.S. Dept. of Agriculture. Introd. in 1928. Rubel x Grover; cross made in 1916. Fruit: cluster very loose, long; berry size medium, round-oblate; skin blue; flesh firm, aroma lacking; dessert quality fair; keeps well; scar good; ripens late, with Rubel. Bush: vigorous; erect; hardy; productive; very widely grown.

Jubilee.—Southern highbush. Selected at the U.S. Dept. of Agriculture (USDA) Small Fruit Sta., Poplarville, Miss.; released in 1994; tested as MS351. Sharpblue x MS160 (G132 x US75). Berry: size medium; scar small; color, firmness, and flavor good. Flowers later than the earliest rabbiteye varieties at Poplarville, but ripens 2 weeks earlier. Adapted to USDA plant hardiness zone 8. Plant: vigorous; upright; very productive; should be interplanted with other southern highbush cultivars.

June.—Orig. in Whitesbog, N.J., by F.V. Coville, U.S. Dept. of Agriculture. Introd. in 1930. (Brooks x Russell) x Rubel; cross made in 1919. Fruit: size medium; ripens early midseason, after Collins and Earliblue, season short. Bush: moderate grower.

Katharine.—Orig. in Whitesbog, N.J., by F.V. Coville, U.S. Dept. of Agriculture. Introd. in 1920. Brooks x Sooy; cross made in 1913. Fruit: small; skin light blue; flavor pronounced; hard to pick, tears badly; ripens later than Pioneer.

Kenafter.—Orig. in Ottawa, Ontario, Canada, by Canada Dept. of Agriculture, Res. Sta. Selected and introd. by Canada Dept. of Agriculture, Res. Sta., Kentville, Nova Scotia. Parentage unknown; selected in 1944. Fruit: small; season late, seldom ripens fully at Kentville. Bush: productive; vigorous. Now obsolete.

Kengrape.—Orig. in Ottawa, Ontario, Canada, by Canada Dept. of Agriculture, Res. Sta. Selected and introd. by Canada Dept. of Agriculture, Res. Sta., Kentville, Nova Scotia. Parentage unknown; selected in 1935. Fruit: large; lacks flavor, similar to Berkeley; scar poor. Bush: low, wide, enabling it to remain below snow cover more easily than taller varieties. Promising as a parent in breeding program. Now obsolete.

Kenlate.—Orig. in Ottawa, Ontario, Canada, by Canada Dept. of Agriculture, Res. Sta. Selected and introd. by Canada Dept. of Agriculture, Res. Sta., Kentville, Nova Scotia. Parentage unknown; selected in 1940. Fruit: size medium; flavor good; clusters too tight for easy picking; last season. Bush: lacks vigor at Kentville. Now obsolete.

Keweenaw.—Orig. in South Haven, Mich., by Stanley Johnston, South Haven Expt. Sta. Introd. in 1952. Open-pollinated seedling of Katharine; seed gathered in 1927; selected in 1934; tested as Michigan HBS 46. Fruit: large; nearly round; skin light blue; picking scar rather large; flesh firm; flavor excellent; resembles Katharine. Bush: vigorous; upright; hardy; productive. Named for the Keweenaw Peninsula, Mich., where this variety made its good record when being tested. No longer recommended in Michigan.

Lateblue.—Orig. in Beltsville, Md., by U.S. Dept. of Agriculture and New Jersey Agr. Expt. Sta. Introd. in 1967. Herbert x Coville. Cross made in 1950 by G.M. Darrow; selected in 1956 by Darrow and D.H. Scott. Fruit: cluster size medium; skin light blue; small stem scars; flesh firm, highly flavored; smaller than and ripens 7 days after Coville; simultaneous ripening of fruit in short period. Bush: erect; vigorous; consistently productive.

Legacy.—From the cross Elizabeth x US75 (*Vaccinium darrowi* Fla.4B x Bluecrop) made by Arlen Draper at Beltsville, Md. Selected at Weymouth, N.J.; tested as G-290; released in 1993 by the U.S. Dept. of Agriculture and the New Jersey Agricultural Expt. Sta. Berry: size medium; scar and flavor superior; firmness good; ripens late midseason, after Bluecrop. Flowers somewhat earlier than optimum for Michigan and New Jersey. Plant: vigorous; upright; yield high. Best adapted to the warmer parts of the northern highbush production area.

Little Giant.—Hexaploid. *Vaccinium constablaei* x *V. ashei* clone T65. T65 is Walker x (Myers x Black Giant). Tested as US57; released by the U.S. Dept. of Agriculture in 1995. Berry: small; dark blue; scar very good; flavor good; ripening concentrated, typically about 2 weeks later than Bluecrop in northern production areas. Probably best suited for northern highbush production areas. Plant: medium height, about 2 m tall; medium productivity; should be interplanted with highbush cultivars for pollination. Well suited for freezing and processing, which is the use for which it was principally intended.

Magnolia.—Southern highbush. Released in 1994 from the U.S. Dept. of Agriculture Small Fruit Expt. Sta., Poplarville, Miss.; tested as MS162. FL78-15 (Harrison x Avonblue) x FL72-5. Berry: size medium; scar small; color, firmness, and flavor good. Flowers later than the earliest rabbiteye varieties at Poplarville, but ripens 2 weeks earlier. Plant: vigorous; productive; spreading; medium height; should be planted with other southern highbush cultivars for pollination. Adapted to U.S. Dept. of Agriculture plant hardiness zone 8.

Menditoo.—Orig. in Raleigh, N.C., by E.B. Morrow, North Carolina Agr. Expt. Sta., and George M. Darrow, U.S. Dept. of Agriculture. Introd. in 1958. Myers x Black Giant; rabbiteye type; selected in 1944; tested as N.C. 478. Fruit: large, averaging 95–100/half-pint cup; round; flesh medium-firm, dark blue to black; mildly and pleasantly flavored; free of grit cells; scar small, dry; ripens about 1 week later than Garden Blue, over a 6- to 8-week period or longer; clusters loose, medium. Bush: vigor medium; spreading. Menditoo is adapted for local market and home-garden use because of its large fruit size and high quality.

Misty.—Southern highbush. Tested as FL72-1; released in 1990 from the Univ. of Florida, Gainesville. FL67-1 x Avonblue. Fruit: large; light blue; scar, firmness, and flavor good. Plant: upright; vigorous; should be interplanted with other southern highbush for pollination; tends to produce excessive flower buds and usually requires winter pruning to reduce flowering potential. Partially to completely evergreen in central Florida. If it defoliates in winter, it may flower long before producing new leaves, in which case it is highly susceptible to stem blight. Chilling requirement about 300 hours.

Morrow.—Orig. in Ivanhoe, N.C., by the North Carolina Agr. Expt. Sta. and the Crops Res. Div., U.S. Dept. of Agriculture. Introd. in 1964. Angola x Adams; cross made in 1945; selected and tested during 1948–52 by E.B. Morrow, G.M. Darrow, and D.H. Scott; tested in North Carolina by G.J. Galletta and J.M. O'Neal during 1960–64; tested as N.C. 678. Fruit: averaged uniformly larger than Angola, Croatan, Murphy, and Wolcott; skin color superior, much lighter blue than Angola; scar large, moist, may tear under some circumstances; flesh mild, pleasant, as firm as Wolcott; flavor good; ripens 5–7 days before Croatan and Wolcott, and about 2 or 3 days before Angola in eastern North Carolina. Bush: medium; semi-upright; broad; canes thick, with similar tolerance to stem canker as Wolcott; fruit clusters usually borne upright at periphery of bush; leaf subject to septoria leaf spot; slow-growing when of fruiting age; propagated readily from hardwood or softwood cuttings. Released for use in North Carolina and southeastern U.S. as a canker-resistant, very early commercial variety with improved size and color as a replacement for Angola.

Murphy.—Orig. in Atkinson, N.C., by the North Carolina Agr. Expt. Sta. and F.V. Coville, U.S. Dept. of Agriculture. Introd. in 1950. Weymouth x F-6 (Stanley x Crabbe 4); cross made in 1934; planted in 1935; selected in 1940; tested as N. C. 262. Fruit: excellent cluster type, loose; berry medium, round to round-oblate; flesh firm, slightly aromatic; flavor good; dessert quality fair; scar fair; ripens early, about with June. Bush: low, spreading; vigorous; hardy; high resistance to canker; as productive as Weymouth.

Myers.—Orig. south of Valdosta, Ga., by J.T. Bush. Introd. in the 1930s. Parentage unknown; discovered about 1930. Fruit: size medium; skin dark blue; flavor good. Bush: rabbiteye type; productive.

Nelson.—Orig. at Weymouth, N.J., in 1965 by A.D. Draper and D.H. Scott. *Vaccinium corymbosum.* Introd. in 1988. Bluecrop x G-107 (F-72 x Berkeley). Tested as G-351 by N. Vorsa, G. Jelenkovic, and G.J. Galletta. Fruit: light blue; firm; high quality; late-ripening. Bush: vigorous; productive.

Northblue.—Orig. in Becker, Minn., by C. Stushnoff, Univ. of Minnesota, St. Paul. Introd. in 1983. B10 x US 3; selected in 1967; tested as Minn. 360. Berry: large; flesh flavor good, pleasing sugar : acid ratio, firm; holds well under refrigeration; processed quality good; skin dark blue, attractive. Plant: low-statured, 20–30 inches; productive; vigorous; foliage dark glossy green.

Northcountry.—Orig. in Becker, Minn., by J. Luby, D. Wildung, C. Stushnoff, S. Munson, P. Read, and E. Hoover. Introd. in 1986. B6 (G65 x Ashworth) x R2P4 (open-pollinated seed from *Vaccinium corymbosum* x *V. angustifolium* hybrids). Tested as MN 350. Fruit: average weight 0.8 g; soft; scar small; flavor similar to *V. angustifolium.* Cold hardy and productive. Intermediate between highbush and lowbush in many characters.

Northland.—Orig. in South Haven, Mich., by Stanley Johnston and J.E. Moulton, Michigan State Univ. Introd. in 1967. (Berkeley x 19-H). Cross made in 1948; selected in 1952. Fruit: size medium; round; skin medium blue; flesh firm; flavor good; picking scar small and dry; ripens 10 July at South Haven. Bush: 4 ft high at maturity; moderately spreading; very productive; hardy; can be harvested with machines, provided lower branches are removed each year; branches pliable.

Northsky.—Orig. in Becker, Minn., by C. Stushnoff, Univ. of Minn., St. Paul. Introd. in 1983. B6 x R2P4; cross made in 1968; selected in 1973; tested as Minn. 332. Berry: medium; flesh flavor good; skin sky-blue with heavy dusty bloom, attractive. Plant: very low-statured, 10–20 inches, allowing it to be covered with snow for winter protection.

Nui.—Northern highbush. E118 (Ashworth x Earliblue) x Bluecrop. Selected at Moanatuatua Research Farm, New Zealand; tested as E118-12; released in 1989. Fruit: very large; color and flavor good; ripens with Earliblue at Moanatuatua. Yield medium, but higher than Earliblue.

Olympia.—Orig. in Olympia, Wash., by Joseph Eberhardt, Eberhardt Blueberry Nurseries. Introd. in 1933. Fruit: large; flavor very good; resembles Harding. Bush: very tall; very productive.

O'Neal.—Orig. in Castle Hayne, N.C., by G.J. Galletta and A.D. Draper. Introd. in 1987. Wolcott x Fla4-15; southern highbush, mainly *Vaccinium corymbosum*; contains some genes from *V. angustifolium*, *V. ashei*, and *V. darrowi*. Selected in 1970; tested as NC 1688 by J.R. Ballington, C.M. Mainland, S.D. Duke, and A.D. Draper. Fruit: large, with good color, scar, firmness, and flavor; early ripening. Bush: productive; vigorous; semi-upright; widely adapted. Low chilling, about 400 hours. Resistant to stem canker race 1.

Ozarkblue.—Patented southern highbush variety released in 1996 by the Univ. of Arkansas, Fayetteville. Selected from the cross G-144 x FL64-76; tested as A-109. Believed to be about 81% *Vaccinium corymbosum*, 13% *V. darrowi*, and 6% *V. ashei*. Fruit: large; light blue; scar, firmness, and flavor good; ripens 10 days later than Bluecrop and 10 days earlier than Tifblue at Clarksville, Ark. Spring budbreak later than most rabbiteye and southern highbush cultivars at Clarksville. Plant: vigorous; semi-upright; consistent high yields.

Pacific.—Orig. in Olympia, Wash., by Joseph Eberhardt, Eberhardt Blueberry Nurseries. Introd. in 1932. Pioneer x Grover; selected in 1932. Fruit: large; matures early; resembles Pioneer. No longer propagated.

Patriot.—Orig. by Agr. Res. Serv., U.S. Dept. of Agriculture and Maine Agr. Expt. Sta., Orono. US 3 (Dixi x Michigan Lowbush 1) x Earliblue; cross made in Beltsville, Md.; seedlings planted in Jonesboro, Maine, in 1954; selected in Maine in 1957; tested as MeUS 32. Fruit: berries large, slightly flat, averaging 49/cup at midharvest and 70/cup in late harvest; scar is small, dry, recessed; color good; flavor very good; ripens with Collins or between Earliblue and Bluecrop. Plant: upright; vigorous; relatively open highbush; plant survival has been superior to present highbush cultivars in Maine and consistently more productive; resistant to *Phytophthora cinnamomi*. Recommended for home garden, market use, particularly for the northeastern United States. Named in recognition of the U.S. bicentennial.

Pearl River.—Pentaploid northern highbush x rabbiteye hybrid from the cross G-136 (highbush) x Beckyblue (rabbiteye). Tested as MS149; released in 1994 by the U.S. Dept. of Agriculture Small Fruit Sta., Poplarville, Miss. Berry: size medium; dark blue; scar, firmness, and flavor good. Flowers late at Poplarville, but ripens 1 week earlier than the earliest rabbiteye cultivars (Climax and Premier). Plant: vigorous, upright; productive; must be planted with southern highbush cultivars for pollination.

Pemberton.—Orig. in Whitesbog, N.J., by F.V. Coville, U.S. Dept. of Agriculture. Introd. in 1941. Katharine x Rubel; cross made in 1921. Fruit: cluster very loose; berry medium to large; roundish; skin darker blue than Atlantic and Jersey; flesh firm; aroma slight; dessert quality fair to good; scar poor; season late, just before Jersey and Rubel; difficult to pick; cracks some in wet weather. Bush: very productive; extremely vigorous; erect; leaf large.

Pioneer.—Orig. in Whitesbog, N.J., by F.V. Coville, U.S. Dept. of Agriculture. Introd. in 1920. Brooks x Sooy; cross made in 1912. Fruit: small; skin dark blue; flesh firm; flavor pronounced; scar medium to poor; ripens midseason. Bush: medium vigor; spreading.

Polaris.—Northern highbush. Selected at Elk River, Minn., from the cross B15 (G-65 x Ashworth) x Bluetta; tested as MN408; released in 1996 by the Univ. of Minnesota. Fruit: size medium; light blue; scar moderately small; very firm; excellent flavor; ripens early. Flowers at about the same time as Bluetta and about 5 days earlier than Bluecrop. Plant: upright; slightly smaller than Bluetta, but taller and less spreading than Northblue; reaches 1.3 m in height and spread; highly self-unfruitful and must be planted with other varieties for cross-pollination; yield equal to Northblue.

Powderblue.—Orig. in Beltsville, Md., by G.M. Darrow, Agr. Res. Serv. Tifblue x Menditoo; selected in 1963; tested as NC 901. Fruit: using Tifblue as the standard, Powderblue is in the same season and has better fruit color, more foliage disease resistance, similar yields and mechanical harvesting characteristics. Powderblue is similar enough to Tifblue in appearance, quality, and season that the two can be harvested together and used as pollinators for each other. Plant: vigorous; productive; should be interplanted with a suitable pollinizer variety.

Premier.—Orig. in Beltsville, Md., by G.M. Darrow, Agr. Res. Serv. Tifblue x Homebell; selected in 1965; tested as NC 898. Fruit: using Tifblue as the standard, Premier is earlier by 2 to 3 weeks, larger-fruited, better-flavored, and similar in other characteristics. Plant: vigorous; productive; should be interplanted with a suitable pollinizer variety; highly susceptible to blueberry gall midge in Florida.

Puru.—Northern highbush. E118 (Ashworth x Earliblue) x Bluecrop. Selected at Moanatuatua Research Farm, New Zealand; tested as E118-4; released in 1989. Fruit: very large; color and flavor good; ripens with Earliblue. Yield medium, but higher than Earliblue.

Rancocas.—Orig. in Whitesbog, N.J., by F.V. Coville, U.S. Dept. of Agriculture. Introd. in 1926. (Brooks x Russell) x Rubel; cross made in 1915; grown near Rancocas Creek, N.H. Fruit: cluster tight; berry small; oblate; flesh firm, crisp; aroma slight; flavor mildly subacid; dessert quality fair to good; ships well; early midseason, after Collins; cracks badly in wet weather. Bush: productive; erect; fair vigor; resistant to stunt virus disease and canker; leaf susceptible to June spot.

Redskin.—Orig. in Whitesbog, N.J., by F.V. Coville, U.S. Dept. of Agriculture. Introd. in 1932. F_2 of (Brooks x Russell); cross made in 1913. Fruit: small; skin albino, red on exposed side; ripens early. Bush: semi-dwarf.

Reka.—Northern highbush. E118 (Ashworth x Earliblue) x Bluecrop. Selected at Moanatuatua Research Farm, New Zealand; tested as E118-4; released in 1989. Fruit: size medium; ripens with Earliblue. Yield very high.

Reveille.—Orig. at Castle Hayne, N.C., by J.R. Ballington. Introd. in 1990. Southern highbush, mainly *Vaccinium corymbosum*. NC1171 {G111 (Bluecrop x Berkeley) x Fla. 61-7 [o.p. seedling of US 39 x (*V. corymbosum* x *V. darrowi*)]} x NC SF-12-L [Ivanhoe x NC 297 (Weymouth x Crabbe-4)]. Selected in 1979; tested as NC 2054 by J.R. Ballington, S.D. Duke, and C.M. Mainland. Fruit: small to medium; light blue; good picking scar; good quality; excellent firmness; slightly earlier ripening than O'Neal or Wolcott. Bush: excellent vigor and precocity; good productivity; narrowly erect habit; broad soil adaptation; resistant to stem canker; some field tolerance to stem blight. Chilling requirement 600–800 hours. Adapted to mechanical harvesting for fresh and processing markets.

Satilla.—See Ethel.

Scammell.—Orig. in Whitesbog, N.J., by F.V. Coville, U.S. Dept. of Agriculture. Introd. in 1931. (Brooks x Chatsworth) x Rubel; cross made in 1915. Fruit: cluster tight, long; berry size medium; ripens midseason. Bush: productive, especially in North Carolina; resistant to canker.

Sharpblue.—Orig. in Gainesville, Fla., by R.H. Sharpe, Univ. of Florida. Introd. in 1975. A tetraploid derived from a series of crosses between *V. ashei*, *V. darrowi*, and *V. corymbosum*; cross made in 1964; selected in 1966; tested as Fla. 66-11. Cluster: tight. Fruit: size medium; round-oblate; skin dark blue, scar size medium; flesh medium firm; ripens very early. Bush: spreading; vigorous; very productive; resistant to canker; very low chilling requirement.

Sierra.—Orig. at Beltsville, Md., in 1980 by A.D. Draper and G. Jelenkovic. Introd. in 1988. US 169 x G-156; US 169 originated by self-pollination of a pentaploid hybrid obtained by crossing diploid *Vaccinium darrowi* clone Fla.4B x US56, a hexaploid *V. ashei–V. constablaei* hybrid; G-156 is a highbush selection (*V. corymbosum*). Cross made in 1976; selected at Weymouth, N.J., in 1980; evaluated as G-640 by N. Vorsa and G. Galletta. Fruit: medium; medium blue; firm; small scar; very good quality; midseason. Bush: vigorous; productive; upright. High chilling.

Southland.—Orig. in Tifton, Ga., by Georgia Coastal Plain Expt. Sta. and Crops Res. Div., U.S. Dept. of Agriculture. Introd. in 1969. Garden Blue x Ethel; cross made in 1946 by G.M. Darrow; selected in 1951 by W.T. Brightwell and Darrow; tested as T-20. Fruit: medium large; skin light blue; flesh firm; flavor good; ripens late midseason, with Tifblue. Adapted to the Gulf Coast and southeastern U.S.

Southmoon.—Patented southern highbush from the Univ. of Florida, Gainesville. A seedling originated by pollination of FL80-46 with a mixture of pollens from four southern highbush selections. Tested as FL85-15; released in 1995. Berry: size large; color medium blue; scar, firmness, and flavor good. Flowers and ripens about 10 days later than Sharpblue. Yield high. Chilling requirement about 300 hours.

Spartan.—Orig. in Otter Lake, Mich., by G.M. Darrow and Arthur Elliott, U.S. Dept. of Agriculture, Beltsville, Md. Introd. in 1977. Earliblue x US 11-93; selected in 1956;

tested as E-13E. Berry: large; flesh firm, flavor good; early ripening. Bush: medium; upright; moderately productive; hardy; picks well mechanically.

Stanley.—Orig. in Whitesbog, N.J., by F.V. Coville, U.S. Dept. of Agriculture. Introd. in 1930. Katharine x Rubel; cross made in 1921. Fruit: cluster size medium, loose; berry size medium to small in latter part of season; oblate; flesh highly aromatic, firm; dessert quality excellent; flavor very pronounced; early midseason. Bush: vigorous; erect; main branches few; easy to prune; leaf large.

Star.—Patented southern highbush from the Univ. of Florida, Gainesville. Selected from the cross FL80-31 x O'Neal; tested as FL87-139-S. Fruit: large; dark blue; scar, firmness, and flavor good. Does not produce large numbers of flower buds. Flowers with Sharpblue, but entire crop matures during the first half of the Sharpblue season. Plant: moderately upright; medium vigor; leafs well in spring; yield medium. Chilling requirement about 400 hours.

Sunrise.—Orig. in Beltsville, Md., by A.D. Draper. Introd. in 1988. Mainly *Vaccinium corymbosum,* but with some *V. angustifolium.* From a G180 x ME-US 6620 cross made in 1974. Selected in 1978; tested as G482. Fruit: medium blue; scar good; firm; high quality; early ripening, with Bluetta. Bush: moderately vigorous; taller than Bluetta; medium yields.

Suwannee.—Orig. along the Suwannee River in Georgia near the Georgia–Florida boundary line by J.T. Bush. Introd. in the 1930s. Parentage unknown; discovered in 1931; rabbiteye type. Fruit: smaller and less flavorful than Clara.

Taylor.—Orig. in Grass Lake, Mich., on the farm of Mrs. Fred E. Taylor. Not introduced commercially but included because of value in breeding for earliness of fruit ripening. Wild seedling; discovered in 1931. Fruit: not large enough for a commercial variety; skin bright blue; picking scar small. Selection now obsolete.

Tifblue.—Orig. in Beltsville, Md., and Tifton, Ga., by W.T. Brightwell, Georgia Coastal Plain Sta. and George M. Darrow, U.S. Dept. of Agriculture. Introd. in 1955. Ethel x Clara; cross made in 1945 at Beltsville; selection made in Tifton; tested as USDA T-9. Fruit: skin a very light blue; flesh firm; flavor very good; scar very small and dry. Bush: rabbiteye type; vigorous. For many years, the most widely planted rabbiteye blueberry variety in the world.

Tophat.—Orig. in East Lansing, Michigan State Univ., by J.E. Moulton, Stanley Johnston, and R.L. Andersen. Selected in 1960 at South Haven Expt. Sta.; introd. in 1977, Mich. 19-H (Mich. HBS 120 x Mich. wild LBS 1) x Mich. 36-H (Mich. 19-H x Berkeley). Mich. HBS 120 is from Pioneer, open-pollinated. Native *Vaccinium angustifolium* Aiton, enters into parentage in pedigree given as LBS, lowbush selection. Fruit: berry medium-large; bright blue; firm; small scar and good flavor; very attractive to birds. Plant: of dwarf habit; spherical, about 30-cm spread; resembles lowbush species; small leaves turning red in fall; compact growth habit. Yields about 1/2 pint/bush. Requires moist and acid soil, pH 4.5–5.2, high organic matter. Recommended especially as an ornamental.

Toro.—Orig. at Weymouth, N.J., by A.D. Draper, D.H. Scott, and G. Jelenkovic. Introd. in 1987. *Vaccinium corymbosum.* Earliblue x Ivanhoe; cross made by Draper at Beltsville, Md., in 1969; selected in 1972 in New Jersey; evaluated by N. Vorsa and G.J. Galletta as G-360. Fruit: scar, color, firmness, and flavor good. Concentrated ripening in Bluecrop season. High-chill, upright, vigorous plant with consistent high yields.

Triumph.—Orig. in Olympia, Wash., by Joseph Eberhardt, Eberhardt Blueberry Nurseries. Introd. for trial purposes only in the 1930s. Grover x Harding; cross made in 1926; fruited for the first time in 1929. Fruit: large; quality good; season late. Variety no longer propagated.

True-Blue.—Orig. along the Savannah River near Savannah, Ga., by George D. Lowe of Baxley, Ga. Introd. about 1943. A seedling highbush (*V. australe*); discovered about 1938. Fruit: small; skin light blue; flavor good; scar large, moist. Bush: very slow growth; weak plants; very productive.

Walker.—Orig. in Brunswick, Ga., along the Satilla River, by W.M. Walker. Introd. in 1944. Parentage unknown. Fruit: small;

skin medium light blue; flesh tart until fully ripe; scar small, dry. Bush: upright; weak; slow growth for rabbiteye type, grows to 6 ft; disease resistant.

Wareham.—Orig. in East Wareham, Mass., by F.V. Coville, U.S. Dept. of Agriculture. Introd. in 1936. Rubel x Harding; cross made in 1915. Fruit: large; round; skin dark; little bloom; ordinary in appearance; cracks badly after rains; flavor good to excellent; does not ship well; ripens late, with Rubel or later. Bush: tall; upright-spreading; vigorous; hardy; productive; easily propagated and pruned.

Washington.—Orig. in Olympia, Wash., by Joseph Eberhardt, Eberhardt Blueberry Nurseries. Introd. in 1928. Pioneer x Rubel; selected in 1928. Fruit: large; flesh very firm; keeping quality good; resembles Rubel. No longer propagated.

Weymouth.—Orig. in Weymouth, N.J., by F.V. Coville, U.S. Dept. of Agriculture. Introd. in 1936. June x Cabot; cross made in 1928. Fruit: cluster loose; berry large; round-oblate; skin dark blue; aroma lacking; dessert quality usually poor; scar medium; tends to drop; season very early. Bush: very productive; vigor low; erect; spreading.

Windy.—Patented rabbiteye variety released in 1992 from the Univ. of Florida, Gainesville. Bluebelle x Florida-M. Tested as FL83-97. Fruit: medium to large; dark blue; scar, firmness, and flavor good; ripens 10 to 14 days earlier than Climax in Gainesville. Flowers very early in season. Bush: vigorous but spreading; yield medium, but often reduced in Gainesville due to spring freezes and extreme susceptibility to blueberry gall midge. Chilling requirement 300 hours.

Wolcott.—Orig. in Atkinson, N.C., by the North Carolina Agr. Expt. Sta. and F.V. Coville, U.S. Dept. of Agriculture. Introd. in 1950. Weymouth x F-6 (Stanley x Crabbe 4); cross made in 1934; planted in 1935; selected in 1940; tested as N.C. 255. Fruit: cluster loose; berry medium, generally round; flesh firm to slightly soft, but firmer than Weymouth; good flavor, dessert quality medium, aromatic; season early as Weymouth, short; scar small. Bush: vigorous; semi-upright; large; though highly resistant to canker when introduced, showed symptoms in 1953, and more susceptibility in 1966; as productive as Weymouth; leaf large. In 1966 comprised 80% of commercial acreage in North Carolina.

Woodard.—Orig. in Tifton, Ga., by W.T. Brightwell, Georgia Coastal Plain Expt. Sta. and the U.S. Dept. of Agriculture. Introd. in 1960. Ethel x Callaway; cross made in 1945; tested as Ga. T-1. Fruit: large, averaging 91 berries/half-pint cup; skin light blue, appearance excellent; flavor mild, slightly more acid than Homebell or Tifblue; earlier ripening than Tifblue, but later than Climax. Not suitable for shipping due to soft berry and wet scar. Bush: more spreading and generally productive at a younger age than Homebell or Tifblue; cross-pollination necessary for fruiting. Named in honor of Otis Woodard, in charge of fruit work at the Georgia Coastal Plain Expt. Sta. for many years before his retirement in 1960.

Butternut

Butternut cultivars are generally not available from commercial nurseries. The following cultivar list contains about one-third of the extant butternut cultivars and includes those considered most likely available as scion material.

Ayers.—Discovered by Corwin Davis about 1965 near Bellevue, Mich. Top scorer in evaluations, near 100% crackout in first crack.

Bear Creek.—Discovered in 1980 in Boone County, Iowa, by Brian Sparks, Beaver, Iowa. Top scorer in evaluations, near 100% crackout in first crack.

Chamberlin.—Discovered by Fred L. Ashworth in 1954 near Madrid, N.Y. Introd. in 1967 by Ashworth.

Craxezy.—Discovered by H.P. Burgart near Union City, Mich., in 1933. Considered one of the best crackers.

Kinneyglen.—Discovered in Ithaca, N.Y., by S.H. Graham. Introd. in 1942. Cracks easily.

Love.—Discovered in Howell, Mich., in 1930 by Frank Love. One of the better crackers, 19% kernel.

Sherwood.—Discovered in Des Moines County, Iowa, by S.W. and D.C. Snyder and introd. by them in 1924.

Stark Bountiful.—Discovered in Bowling Green, Mo., by William K. Erickson. Introd. in 1982. Plant patent 5116; assigned to Stark Bros. Nurseries, Louisiana, Mo.

Carambola

Golden Star.—Orig. in Homestead, Fla., by Subtropical Expt. Sta., Univ. of Florida. Introd. in 1965. Parentage unknown; seed received from Hawaii in 1935; selected in 1940. Fruit: 4–8 oz; 4–5 inches long, with four to six prominent longitudinal ribs; somewhat oval; skin bright golden-yellow, very waxy; flesh juicy, crisp, mildly subacid to sweet flavor, contains no fibers; high in carbohydrates and vitamins A and C; good for fruit and vegetable salads. Tree: small to 20 feet high; easily trained; unpruned is rounded, open crown; wind resistant; exhibits both flowers and fruit throughout the year. Trees at Homestead have borne 100–300 lb of fruit annually; injured at 27 to 29F.

Kary.—Introd. in 1992 by Richard A. Hamilton and Philip J. Ito, Univ. of Hawaii, HITAHR, Honolulu, Hawaii. A sweet, firm-fleshed starfruit with good shelf life. Selected in 1980 at the Poamoho experimental farm as DR11T1 from open-pollinated seedlings of Sri Kembangan. Trees are vigorous, have a round, compact canopy, and are productive. Thinning of fruit clusters is recommended for better fruit size. Fruit mature about 60 days after flowering; they are 10–15 cm in length, 7–8 cm across, with five distinct wings. Skin is deep yellow and flesh is bright orange-yellow when fully ripe. Soluble solids 17%.

CARISSA

Alles.—See Chesley.

Chesley *(Alles)*.—Orig. in Carlsbad, Calif., by Chesley Alles. Introd. in 1929. Parentage unknown; selected in 1928. Fruit: size medium; oval; flavor and texture fair.

Frank.—Orig. in Riverside, Calif., by J. Eliot Coit, Vista, Calif. Introd. about 1940. Chance seedling; discovered in 1937 on the Ben Frank ranch. Fruit: large; good quality. Plant: nearly self-unfruitful. Now obsolete.

Serena.—Orig. in Santa Barbara, Calif., by J. Eliot Coit. Introd. in 1932. Parentage unknown; selected in 1929. Fruit: oval and pointed; medium to large; ripe 8 months of the year; stem long, making it possible to pick with buttons on; keeps well.

Torrey.—Orig. in La Jolla, Calif., by Thomas W. Whitaker, Torrey Pines Hort. Field Sta., U.S. Dept. of Agriculture. Introd. in 1948. Spontaneous tetraploid of *Carissa grandiflora*; discovered in 1943. Fruit: medium; round to slightly oval; skin medium thick; flavor subacid. Plant: larger flowers and thicker growth habit than usual types; regular and heavy producer.

• *Revisions for this commodity were not received before the book went to press; therefore, the above variety descriptions are virtually identical to those published previously.*

CAROB

Amele.—Orig. in ancient times; long the principal variety grown near Bari, Italy, on the Adriatic Coast. Introd. into the U.S. in 1951 as P.I. 194137 by J. Eliot Coit, Vista, Calif., from budwood furnished by V. Carrante of the experiment station near Bari. Fruit: sugar content high, 53.8%, near Indio, Calif.; flavor good; harvested during October. Tree: upright; vigorous; blooms heavily, but does not set a heavy crop; female flower.

Bolser *(Horne, Nichols)*.—Orig. in Rialto, Calif., by Jerry Bolser of the San Bernardino County Agriculture Commission. Introd. in 1920. Discovered about 1920. Fruit: short, 5–6 inches long, rather plump; sugar content averages 42.3%; matures in October. Tree: bears well; self-fruitful; flowers perfect.

Clifford.—Orig. in Corona, Calif., by J. Eliot Coit. Introd. in 1959. Chance seedling. Fruit: pods slightly curved, light brown; length 13 cm, width 20 mm; sugar content 52.9%; flavor fair and neutral; ripens early October; worm-free. Tree: upright; compact; very precocious; heavy and fairly regular bearer; flowers reddish unopened, yellow opened; a very promising variety at Vista.

Conejo.—Orig. in Vista, Calif., by L.S. Epker. Introd. in 1958. Seedling selected in 1953. Fruit: size medium; matures 15 Sept. to 15 Oct.; sugar content averages 55.5%; flavor excellent. Tree: precocious, good bearer; self-fruitful (hermaphroditic).

Excelsior.—Orig. in California, by Armstrong Nurseries, Ontario, Calif. Introd. about 1930. Seedling selected in 1925. Fruit: small. Tree: bears tremendously heavy crops; pistillate. No longer propagated.

Gabriel.—Orig. in Rancho Santa Fe, Calif. Introd. in 1958. Seedling selected in 1953. Fruit: very large; handsome; sugar content averages low, 43.5%; harvested during October. Tree: very vigorous; precocious; bears heavily.

Grantham.—Orig. in Bari, Province of Puglia, Italy. Introd. into the U.S. in 1930 by Carl Grantham, Rialto, Calif. Discovered in ancient times. Fruit: none produced. Tree: male; blooms very heavily, but too late to pollinate some early blooming, pistillate varieties. An old variety commonly grown on the Adriatic Coast under various names.

Holmes.—Orig. in Vista, Calif., by J. Eliot Coit. Introd. in 1951. Unknown seedling. Discovered in 1951. Tree: male; low-growing, drooping; roundish-compact; vigorous; blooms from 15 Aug.–1 Feb., making it a potent pollinizer for all females; preferred by bees for both pollen and honey. Named after the superintendent and partner of the Cornell–Holmes extensive carob plantings near Riverside, Calif.

Horne.—See Bolser.

Molino.—Orig. in Rancho Santa Fe, Calif. Introd. in 1958. Seedling discovered in 1953. Fruit: matures early; worm-free; low sugar content, 39.4%, but has a good flavor. Tree: very precocious; yields heavily; hermaphroditic.

Nichols.—See Bolser.

Santa Fe.—Orig. in Santa Fe Springs, Los Angeles County, Calif., by Lawrence Holmes, California Carob Plantations. Introd. in 1922. Selection made from a wild tree in 1920; original tree still alive in 1946. Fruit: large, up to 7 inches long; slightly curved, thin; good sugar content, 47.6%; flavor very good; harvested during September and first 10 days in October; immune to worms. Tree: spreading; precocious and alternate bearer; flower perfect. An excellent self-fertile variety on coastal foothills.

Sfax.—Orig. in Tunis, Africa. Introd. in 1959 by J. Eliot Coit, Vista, Calif. Parentage unknown; secured from J.B. Blessis, Manzel bou Zelfa, Tunis; imported by U.S. Dept. of Agriculture as P.I. 187063; received 23 Feb. 1950. Fruit: very large; thick; sugar content about 51%; flavor good; ripens very early, 10–30 Sept. at Vista; milling quality good. Tree: yields heavily in alternate years; vigorous; spreading; twigs thick;

leaves large; claimed to be hermaphroditic, but at Vista, all anthers rudimentary and aborted, so considered to be a pistillate variety. In southern California, best suited to foothills some distance from the coast.

Sykea.—Orig. in Laconia, Greece. Introd. into the U.S. in 1950, at Vista, Calif. Fruit: appears to be immune to worm infestations. Tree: pistillate; precocious in bearing.

Tantillo.—Orig. in Sicily. Introd. in 1956 by J. Eliot Coit, Vista, Calif. Parentage unknown; imported by U.S. Dept. of Agriculture as P.I. 233580, 26 May 1956, from Italian Expt. Sta., Acireale, Sicily. Fruit: pods mostly straight; dark brown; length 13–15 cm, width 20 mm; flavor fair; ripens 15 Sept.–15 Oct.; worm-free. Tree: habit varies; very precocious; flowers red unopened; hermaphrodite. Very promising variety.

Tylliria.—Orig. in Cyprus in ancient times. Introd. into the U.S. about 1940 as P.I. 189008. Fruit: sugar content averages 47.4%, protein 5.81%; flavor good; harvest season late, in October. Tree: not precocious, with no commercial yields before 10 years from budding; pistillate. Good commercial variety; in California best adapted to foothills near the coast.

White.—Orig. in White Park, Riverside, Calif. Introd. in 1926. Seedling selected in 1926. Fruit: good. Tree: alternate bearer; pistillate.

• *Revisions for this commodity were not received before the book went to press; therefore, the above variety descriptions are virtually identical to those published previously.*

CHERIMOYA

Bays.—Orig. on the James H. Bays ranch, Ventura, Calif. Introd. in 1920 by William Weber. Parentage unknown. Fruit: round; large, 18–24 oz; skin light yellowish-green, covered with conical protuberances; pulp white, juicy, melting; subacid; quality good; season February to April. Tree: bushy; spreading; good bearer. Probably California's leading commercial cultivar (1996).

Booth.—Orig. in Hollywood, Calif., by John S. Armstrong on the A.F. Booth ranch. Introd. in 1921 by Armstrong Nurseries, Ontario, Calif. Parentage unknown. Fruit: conical; skin with slight to moderately developed protuberances; quality good; handles well; ripens late. Tree: bears well; major commercial cultivar, widely planted.

Carter.—Orig. in Encinitas, Calif., in the N.E. Carter orchard by James H. Macpherson, Williams and Macpherson Subtropical Nursery, Encinitas. Introd. in 1940. Parentage unknown; tree planted in 1931 by a Mr. Summerville; selected in 1938. Fruit: long, conical; skin smooth; ripens in March in Los Angeles County, in June in San Diego County; keeps well; a good shipper; resembles Ryerson. Tree: heavy bearer; sets fairly well without pollination in some areas; hardy.

Chaffey.—Orig. in West Los Angeles, Calif., by Mrs. A.M. Chaffey. Introd. in 1945. Seedling from a seed lot bearing P.I. 44841 collected in Oran, Salta, Argentina, by W.S. Damon; seedling received 9 June 1917 from the U.S. Dept. of Agriculture. Fruit: 3/4–1 lb; short, conical, 4 inches long; flesh light cream; quality excellent; does not bruise easily; seed small, 30–60 per fruit; season December to March along the coast of California. Tree: bears regularly.

Loma.—Orig. in Point Loma, San Diego, Calif., by the Point Loma Homestead. Introd. about 1945. Parentage unknown; discovered about 1926. Fruit: matures earlier than any other variety. For home gardens, not widely planted

McPherson.—Orig. in Orange, Calif., by Bill Weber. Introd. in 1933. Parentage unknown; original tree growing in the McPherson Brothers' orchard, after whom the variety was named. Fruit: size moderate; skin smooth but with some development of tubercles; quality good; midseason to late in maturity. Tree: vigorous; quite productive even without hand-pollination.

Ott.—Seed orig. in Mexico City, Mexico, and fruited at Whittier, Calif., by W.H. Ott. Introd. in 1946. Plant patent 656; 27 Mar. 1945; assigned to Armstrong Nurseries, Ontario, Calif. Open-pollinated seedling of unknown parentage. Fruit: averages 3 inches in diam., 7–12 oz, smaller than Booth or White; heart-shaped; skin green with brownish spots, smooth, tough, medium thick, peels easily; flesh yellow, very fine-textured; flavor very pronounced and distinct; high sugar content (up to 26.5%); eating quality excellent; seeds 24–36 per fruit; ships very well; season January through April. Tree: size medium; medium vigor; upright-spreading; fairly self-fruitful.

Ryerson.—Orig. in Chula Vista, Calif., by William H. Sallmon, who named it for Knowles A. Ryerson, dean emeritus, College of Agriculture, Univ. of California, Berkeley. Introd. in 1935. Parentage unknown; selected in 1928 by J. Eliot Coit of Vista, Calif. Fruit: shape regular; skin smooth, tough; ships well.

Sallmon.—Orig. in Chula Vista, Calif., by J. Eliot Coit. Introd. in 1931. Seeds, from which this variety arose, came from the Willism H. Sallmon ranch; parentage unknown. Fruit: large; skin smooth, tough; quality poor. No longer planted.

Whaley.—Orig. in Hollywood, Calif., by John S. Armstrong. Introd. in 1927 by Armstrong Nurseries, Ontario, Calif. Parentage unknown. Fruit: medium to large, 1 1/2–2 lb; short, conical; skin light green, surface tuberculate; flavor and quality good; membrane tends to develop around seed; ripens in midseason and later. Tree: somewhat smaller and less vigorous than some other varieties.

White ***(Dr. White)***.—Orig. on the White ranch, Lemon Grove, Calif., by James H. Macpherson, Macpherson and Williams Subtropical Nursery, Encinitas, Calif. Introd. in 1930. Parentage unknown; selected in 1928 from a tree planted in 1912. Fruit: large; skin semi-smooth; flavor fine, even after fruit is quite soft. Tree: bears well in San Diego County, Calif.; a major commercial cultivar.

CHERRY

***Prunus avium*, sweet; *Prunus cerasus*, sour**

Angela.—Orig. in Farmington, Utah, by U.S. Dept. of Agriculture and Utah Agr. Expt. Sta. Introd. in 1974. Open-pollinated seedling of Napa Longstem Bing; selected in 1969 by Bryce N. Wadley; tested as LSB-16. Fruit: sweet; large; Bing-shaped; skin black, shiny; flesh dark red, darker than Bing or Lambert, almost as firm as Bing, flavor good; stem medium-long; less cracking than either Bing or Lambert; no doubles observed in central Utah; ripens with Lambert; for fresh market or canning. Tree: large; more upright than spreading; vigorous; very productive, good crops every year in Utah; hardier than Bing or Lambert; tolerant to x-disease and necrotic rusty mottle; blooms 3–5 days after Lambert, escapes early spring frosts; pollen compatibility studies inconclusive, appears to be compatible with Bing and Lambert.

Attika® (*Kordia*).—Orig. in Holovousy, Czech Republic, by Research and Breeding Inst. of Pomology as a chance seedling. Introd. in 1982. Introd. to the U.S. in 1987 by VanWell Nursery, Wenatchee, Wash. Fruit: sweet; ripens about 10 days after Bing, heart-shaped; firm; quality the best; about 8 g; large; long stem; dark mahogany; flesh dark red-black; brix usually 17.5% or more. Tree: self-incompatible; pollinated by Van, but flowers open late so Hedelfingen, Lambert, Sam, and Sylvia pollinize better; vigorous and very spreading; fruit often born singly.

August Supreme.—Orig. in Wayne County, Ohio, by Menno Gerber. Introd. in 1936. Plant patent 164; 21 Jan. 1936; assigned to W.N. Scarff's Sons' Nursery, New Carlisle, Ohio. Parentage unknown; discovered in 1932. Fruit: sweet; ripens very late, 1–15 Aug. in region of origin; flesh firm; juicy, but juice does not exude when stem is removed. Tree: large; vigorous; productive.

Bada.—Orig. in Davis, Calif., by the California Agr. Expt. Sta. Introd. in 1964. Seedling of unknown parentage x Ord (Orb); cross made in 1940 by Guy L. Philp; selected in 1950 by R.M. Brooks; tested as UCD 19A-4 by Brooks and W.H. Griggs. Fruit: sweet; size about that of Royal Ann, averaging 1 inch in diam.; cordate, smooth, symmetrical; skin glossy, cream-colored with medium to high blush; flesh cream-colored, firmness similar to Royal Ann, meaty, texture fine, moderately juicy; flavor mild, less tart than Royal Ann; stone medium-large, free to slightly cling to flesh; resistant to cracking and doubling in many years of testing; stem medium-long; ripens an average of 4 days before Royal Ann; satisfactory for shipping, canning, brining; a good companion variety for Berryessa. Tree: early bearing; moderately vigorous; upright-spreading; productive to very productive; blooms relatively late, between Lambert and Jubilee, but overlaps Bing, Royal Ann, Van, Mona, Berryessa, and Larian for good cross-pollination.

Balaton™.—Orig. in Ujfeherto, Hungary, as a local variety identified by Ferenc Petho, Tibor Szabo, and Bertalan Pusztai. Released and trademarked by Michigan State Univ. in 1996. Fruit: sour; red skin and flesh; larger, sweeter, and firmer than Montmorency; good for processing; fruit ripens slightly later than Montmorency. Tree: self-fertile; trunk may be slightly more susceptible to mid-winter cold temperature injury than Montmorency; flowers open 1 or 2 days later than Montmorency in some orchard environments.

Bangandy.—Orig. near Le Grand, Calif., by F.W. Anderson, Merced, Calif. Introd. in 1972. Plant patent 3157; 9 May 1972; assigned to Fowler Nurseries, Newcastle, Calif. Open-pollinated seedling of Bing; discovered in 1965; tested as Anderson 44P255. Fruit: large, globose, resembles Bing; skin deep red; flesh firm, fine, crisp, meaty, deep red; flavor subacid, good; ripens with Bing; few doubles. Tree: size medium; upright to spreading; vigorous; very productive; intercompatible with Bing, Lambert, and Napoleon.

Bell Montmorency.—Orig. in the J.C. Bell orchard, Kibbie, Mich., by Roy Gibson, Greening Nursery Co., Monroe, Mich. Introd. in 1940. Orig. as a whole-tree mutation. Fruit: sour; very large; quality high. Tree: blooms later than Montmorency, thus tending to escape late frosts; produces very well.

Berryessa.—Orig. in Davis, Calif., by the California Agr. Expt. Sta. Introd. in 1964. Seedling of unknown parentage x Bush Tartarian; cross made in 1940 by Guy L. Philp; selected in 1949 by R.M. Brooks; tested as UCD 23A-3a by Brooks and W.H. Griggs. Fruit: sweet; slightly larger than Royal Ann; broad-cordate, symmetrical; skin glossy, cream-colored with medium to high blush, more resistant to bruising than Royal Ann; flesh cream-colored, moderately firm to firm, meaty, fine-textured, moderately juicy, flavor sweet and mildly tart; stone medium-large, slightly cling, but satisfactory for pitting; subject to some cracking when exposed to late rains; produces a relatively low percentage of doubles when doubling is a serious problem with Royal Ann and Bing; however, in most years, Berryessa has been free of cracking and doubling; stem medium to long, relatively thick; ripens an average of 6 days before Royal Ann; suitable for the same trade that uses Royal Ann. Tree: early bearing; moderately vigorous; upright-spreading; productive; satisfactorily cross-pollinated by Royal Ann, Van, Moreau, Starking Hardy Giant, Bada, and Jubilee.

Big K.—Orig. in Portland, Ore., by Julius Killick. Introd. in 1928. Parentage unknown. Fruit: sweet; skin dark red; flesh dark red; not better than Bing. Used in breeding studies; now obsolete.

Bingandy.—See Bing d'Andy.

Bing d'Andy *(Bingandy)*.—Orig. in Le Grand, Calif., by F.W. Anderson, Merced, Calif. Introd. in 1963. Plant patent 2198; 4 Dec. 1962; assigned to Fowler Nurseries, Newcastle, Calif. Open-pollinated seedling of Bing; selected in 1959; tested as Anderson 1C24. Fruit: sweet; large, averaging 1 inch in cheek diam.; shape uniform, symmetrical, globose to slightly oblate, suture shallow; skin maroon, medium thick, texture tough to medium; flesh dark red, firm, crisp, meaty, subacid; eating quality good to best; pit semi-freestone, large, averaging 7/16 inch long; shipping quality good; ripens about with Bing, eating ripe 10 June 1961, season 3–12 June; very similar to Bing in size, shape, quality, and handling, but with much less fruit cracking and doubling under California's interior valley conditions. Tree: generally larger and more productive than Bing; interfertile with Bing, Lambert, Royal Ann, and Van.

Black Flame.—Orig. in Flora, Ill., by J.E. Markham, Markham Nursery. Introd. in 1940. Open-pollinated seedling of Montmorency; selected in 1937. Fruit: sour; large; dark red; ripens middle of June. Tree: hardy.

Black Giant.—See Burbank Black Giant.

Black Russian.—Orig. in Clyde, Ohio, by a Mr. Burkholder. Introd. about 1944. Parentage unknown, found on Mr. Burkholder's farm. Fruit: sweet; size medium; heart-shaped; skin nearly black, cracks less than Lambert; flesh very dark red, firm, sweet; ripens just before Schmidt. Tree: vigorous; hardy tree and blossom buds; productive.

Black Sour.—Orig. in Rahway, N.J., by a Mrs. Dolechek who imported seeds from Rumania. Introd. in 1949 by Ackerman Nurseries, Bridgman, Mich. Parentage unknown; selected in 1935. Fruit: sour; skin dark red; resembles May Duke but darker red.

Brooks.—Orig. at Wolfskill Ranch, Univ. of California, Davis, by P.E. Hansche, W. Beres, J. Doyle, and W.C. Micke. Introd. in 1988. Rainier x Early Burlat. Plant patent 6676; 14 Mar. 1989; assigned to Regents of Univ. of California. Fruit: sweet; large; skin dark burgundy-red, occasionally some streaking and mottling over basal shoulders; flesh color variable, with rays and streaks of red and pink; firm; numerous fibers; nearly freestone without air space; susceptible to rain cracking. Flavor exceptional, sweet, well-balanced, rich; not aromatic; overall quality outstanding. Ripens midway between Early Burlat and Bing; fruit hangs well on tree. Tree slightly smaller than most sweet cherries; upright to upright spreading; hardy in San Joaquin Valley; very productive; annual bearing. Bloom date between Early Burlat and Bing.

***Brummers Red Sweet*.**—See Woehrman Red Giant.

Burbank Black Giant *(Black Giant, Burbank Giant)*.—Orig. in Sebastopol, Calif., by Luther Burbank. Introd. about 1932 by Stark Bros. Nurseries & Orchards Co., Louisiana, Mo. Parentage unknown. Fruit: sweet; large; skin black; flesh firm, blood red; pit small; ships well; resembles Bing but smaller and not as firm. Tree: fairly hardy. No longer propagated. Not the same as Giant, also a Luther Burbank selection.

***Burbank Giant*.**—See Burbank Black Giant.

***Bush Bing*.**—See Bush Tartarian.

Bush Tartarian *(Bush Bing, Early Bing, Giant)*.—Orig. in San Joaquin County, Calif. Introd. in 1922 by the Louis Vistica Nursery, Stockton, Calif.; apparently named by Guy L. Philp, California Agr. Expt. Sta., Davis. Parentage unknown. Fruit: sweet; skin black; flesh dark red to dark purple, meaty, firmer than that of Black Tartarian if picked early; shape similar to Bing; sizes well when trees are young, but becomes small as trees mature, apparently from overbearing; ripens between Black Tartarian and Bing seasons; picking season very short, usually only 2–3 days before becoming too soft to ship well. Tree: bushy; spreading; crinkle-free; pollinator for Bing, Lambert, Napoleon (Royal Ann), Chapman, Burbank, Sanguinetti.

***Carnefix Compact Montmorency*.**—See Starkspur Montmorency.

Carnival.—Orig. in Mission, British Columbia, Canada, by H.M. Eddie & Sons, Vancouver, B.C. Introd. in 1935. Chance seedling. Fruit: sweet; size medium; uniform; roundish-conic; flesh soft and tender, too soft for successful shipping; quality good; shows some tendency to rain cracking and dropping; ripens about 2 weeks before Bing; resembles Lambert. Tree: large; vigorous; upright; sets a medium to heavy crop.

Cashmere.—Orig. in Prosser, Wash., by Thomas Toyama, Washington Agr. Expt. Sta. and U.S. Dept. of Agriculture. Introd. in 1994. Plant patent pending; assigned to Central Washington Nursery Improvement Institute. Stella x Early Burlat; selected in 1976; tested as PC 7144-3. Fruit: sweet; medium-large; round to oblong; skin dark mahogany; flesh firm, dark red; juicy with excellent flavor if tree is not overcropped; ripens 8–10 days before Bing; medium slender stem; semi-free, small stone. Tree: self-fertile; flowers open later than Bing; vigorous; spreading, with horizontal branch habit; productive; outstanding characteristic is excellent cold hardiness; tendency to retain aborted fruit through harvest; more cracking-susceptible than Bing.

Chelan.—Orig. in Prosser, Wash., by Thomas Toyama, Washington Agr. Expt. Sta. and U.S. Dept. of Agriculture. Introd. in 1991. Plant patent 8545; plant patent date 1992, assigned to Central Washington Nursery Improvement Institute. Stella x Beaulieu; selected in 1978; tested as PC 7146-23. Fruit: sweet; medium-large, but smaller when overcropped; round to broadly cordate; skin glossy, dark mahogany; flesh exceptionally firm, dark red; juicy with good flavor; ripens 10–12 days before Bing; medium long, slender stem; semi-free, small stone. Tree: self-sterile; flowers open earlier than Bing; pollen-compatible with Bing and Rainier; vigorous; upright-spreading; very productive; less cracking-susceptible than Bing.

Chinook.—Orig. in Prosser, Wash., by Harold W. Fogle, Washington Agr. Expt. Sta. and U.S. Dept. of Agriculture. Introd. in 1960. Bing x Gil Peck; selected in 1954; tested as Wash. P. 1-246. Fruit: sweet; large; Bing-type; heart-shaped to round; stylar scar relatively large; skin mahogany, glossy at maturity; flesh medium to dark red, uniformly colored, firm; pit size medium, relatively free; soluble solids equal to or slightly higher, titratable acidity lower, than Bing; ripens 4–10 days before Bing, which it resembles; apparently about as susceptible to rain-cracking as Bing. Tree: vigorous; upright-spreading; productive; flowering 1 or 2 days and foliation 3 or 4 days before Bing; has not appeared to be more susceptible than Bing to frost damage; leaf petioles usually with a few fine hairs; at least in young trees, more hardy than Bing but less so than Van; self-unfruitful; has been hand-pollinated satisfactorily by Bing, Van, and Sam; effective pollinator for Bing.

Colt.—See rootstocks for cherry.

Compact Stella.—Orig. in Summerland, British Columbia, by K.O. Lapins, Agriculture Canada, Res. Sta. Introd. in 1973. Variety is result of irradiation of Stella with X-rays (4 kR). Selected in 1966; tested as Stella 35B-11. Fruit: sweet; medium-large; heart-shaped; skin black; flesh medium firm; flavor fair; ripens 6 days before Lambert (11 July at Summerland). Tree: semi-dwarf (1/2 standard size); spreading; vigor and hardiness moderate; very productive; self-fertile.

Coronation.—Orig. in Morden, Manitoba, Canada, by Canada Dept. of Agriculture, Res. Sta. Introd. in 1937. Open-pollinated seedling of Shubianka; selected in 1933; tested as Morden 500. Fruit: sour; large, up to 15/16 inch in diam. on young bushes; round obovate; skin dark red; flesh yellowish, firm, acid; juice pink; plentiful; quality fair to good for canning; season late July and early August. Bush: upright; round-headed; vigorous; moderately productive.

Corum.—Orig. in Eugene, Ore., by Gordon Corum. Introd. in 1961. Chance seedling; discovered in 1950. Fruit: sweet; skin yellow with attractive blush, moderately thick; flesh white with a yellow tinge, tender, meaty, crisp, flavor mild; slightly clingstone, pit small; ripens about 6 days before Napoleon (Royal Ann); processes very well; moderately resistant to cracking, equaling Napoleon and Lambert; stem slender; 1.5 to 2 .5 inches long, adheres well to fruit; resembles Napoleon. Tree: fairly vigorous; habit resembles Lambert; as hardy as Napoleon; preliminary tests indicate it as a possible pollinator for Bing, Lambert, Napoleon, Sam, Sue, Van, and, in some districts, Republican; blooms slightly before Napoleon and with Republican.

Craighead Early.—Orig. in Brigham City, Utah, by William Craighead. Introd. about 1925. Parentage unknown. Fruit: sweet; size above medium; round, slightly compressed, halves unequal; cavity broad, shallow; skin dark red to black, smooth, waxy, glossy, thick, moderately tough; flesh deep red, fairly firm, meaty, soft about pit, subacid, juicy; quality fair to good; flavor mild; pit round-oval, semi-freestone; ripens early, 1 week before Black Tartarian, and with Early Purple. Tree: medium; very unproductive, although grown in orchard surrounded by trees of Orb, Bing, Lambert, Major Francis, and Napoleon; leaves long, lanceolate, curled, resembling Early Purple. Obsolete because of unproductiveness.

Dayton Montmorency.—Orig. in Dayton, Ohio, by Clyde Mumma, Mumma's Fruit Farms. Introd. in 1941. Selection of Montmorency; discovered in 1936. Fruit: sour, large. Tree: bears a heavy crop every year; a more consistent and heavier bearer than other Montmorency trees.

Dropmore.—See rootstocks for cherry.

Dwarfrich.—Orig. in Cheyenne, Wyo., by A.C. Hildreth, Cheyenne Hort. Field Sta., U.S. Dept. of Agriculture. Introd. in 1952. Open-pollinated seedling of Vladimir; selected about 1948. Fruit: similar in shape, color, flavor, and flesh texture to Early Richmond, but may average slightly smaller. Bush: a true dwarf cherry; original bush, after 14 seasons in its permanent location, is less than 4 feet high; suckers badly on own roots, but can be budded readily on Mahaleb understock; hardy at Cheyenne and in western Iowa; blooms late; leaves smaller and glossier than those of standard sour cherry and seem to be slightly more resistant to leaf spot disease; suitable for small properties where it will serve both for fruit production and as an ornamental; once recommended for the northern and central Great Plains where standard cherry varieties are not sufficiently hardy and in regions adapted to standard sour cherry varieties wherever a dwarf sour cherry is desired. Flower: white, numerous; requires a pollinator; fruitful with Montmorency, English Morello, Vladimir and its seedlings, and perhaps also with Meteor and North Star. Vladimir has been used as an interstem.

Dwarf Siberian.—See rootstocks for cherry.

Early Bing.—See Bush Tartarian.

Early Burlat.—U.S. Dept. of Agriculture released for propagation in 1961. First fruited in this country at the U.S. Plant Introduction Sta., Chico, Calif., in 1948. Early Burlat is the commercial synonym for a variety introduced from Morocco in 1936 as P.I.

125106 Bigarreau Hatif de Burlat, from France in 1938 as P.I. 127378 Hatif Burlat, and in 1949 as P.I. 162449 Bigarreau Precoce de Burlat. Fruit: sweet; large; skin red to dark purplish-red; flesh medium firm, more so than Black Tartarian and Chapman, but not as firm as Bing; semi-freestone; stem length medium, thick; eminently suitable for shipping long distances; ripens 4–7 days before Black Tartarian and 2 days after Moreau in Stockton, Calif., cherry district, but too late in Oregon to compete in the early market; excellent market acceptance; doubling generally lower than Bing, but the majority of doubles drop before harvest; susceptible to cracking after rain. Tree: moderately vigorous, but less than Bing; pollinator for Bing, Napoleon, Mona, and Saint Charmez.

Early Montmorency.—See Richmorency.

Early Ruby.—Orig. in Visalia, Calif., by R.S. Moore. Introd. in 1977. Plant patent 4150; 15 Nov. 1977; assigned to Sequoia Nursery, Visalia, Calif. Discovered as a chance seedling of unknown parentage. Fruit: sweet; medium; oval to ovoid; similar to Black Tartarian, but dark red at maturity; flesh speckled white; matures 5 days after Black Tartarian; holds well on tree. Tree: medium; upright; blossoms later than Royal Ann; exceptional because of regular fruitfulness in hot, dry climate of San Joaquin Valley.

Early Seneca.—See Seneca.

Ebony.—Orig. in Moscow, Idaho, by Leif Verner, Idaho Agr. Expt. Sta. Introd. in 1946. Lambert x (probably) Republican; selected in 1940. Fruit: sweet; medium to large, averaging somewhat less than 1 inch diam.; broadly cordate; skin nearly black, moderately thick, separates from pulp; dots small to medium; stylar scar small; flesh deep mahogany to nearly black, firm, tender, juicy; juice deep purple; flavor very sweet; quality excellent; stone nearly free, medium; dorsal suture prominent, raised; ventral suture very wide, strongly ridged; ripens in early midseason, about with, or slightly before, Bing; slightly more resistant to cracking than Bing or Lambert; resembles Republican. Tree: bears heavily; upright-spreading; moderately vigorous; leaf medium, cupped or folded, oval, margin doubly serrate; buds small, pointed.

Fernwood Montmorency.—Orig. in Bravo, Mich., by Carne Tucker. Introd. in 1956 by the Greening Nursery Co., Monroe, Mich. Considered to be a bud mutation of Montmorency; discovered in 1940. Fruit: sour; about 10% larger than Montmorency; uniform in ripening time, which is the end of June. Tree: vigorous; upright; bears heavily and at an early age.

Giant.—See Bush Tartarian.

Gilbert Montmorency.—Orig. in Door County, Wis., by J. Vernon Gilbert. Introd. about 1937 by Willis Nursery Co., Ottawa, Kan. Bud mutation of Montmorency. Fruit: sour; larger than the average fruit of Montmorency. Tree: consistent bearer.

Gil Peck.—Orig. in Geneva, N.Y., by Richard Wellington, New York State Agr. Expt. Sta. Introd. in 1936. Napoleon x Giant; cross made in 1925; first full crop in 1933. Fruit: sweet; large; skin dark purplish-black; flesh firm, juicy, richly flavored; may occasionally crack, less than Napoleon; ripens in late midseason, between Noble and Gean d'Hedelfingen; resembles Giant in shape and quality. Named in memory of Gilbert W. Peck, Cornell Univ., Ithaca, N.Y.

Glacier.—Orig. in Prosser, Wash., by Thomas Toyama, Washington Agr. Expt. Sta. and U.S. Dept. of Agriculture. Introd. in 1990. Plant patent 8051; plant patent date 1991; assigned to Central Washington Nursery Improvement Institute. Stella x Early Burlat; selected in 1976; tested as PC 7144-7. Fruit; sweet, very large; round to heart-shaped; skin glossy, dark mahogany; flesh soft, dark red; juicy with good flavor; ripens 3–5 days before Bing; long slender stem; semi-free, small stone. Tree: self-fertile; flowers open slightly later than Bing; vigorous; upright-spreading; moderately productive.

Greening.—Orig. in New York. Introd. in 1930. Discovered along a fence row by a representative of the Greening Nursery Co., Monroe, Mich. Fruit: sweet; apparently more resistant to cracking than some other varieties; resembles Windsor.

Hansen.—See Kansas Sweet.

Hansen Sweet.—See Kansas Sweet.

Hartland™.—Introd. in 1992 by R.L. Andersen, S.K. Brown, R.D. Way, K.G.

Livermore, and D.E. Terry, New York State Agr. Expt. Sta., Geneva. Windsor open-pollinated; selected in 1958; tested as NY 3308. Plant patent pending; assigned to Cornell Res. Foundation. Fruit: sweet; round; size medium, about 20–30 mm in diam.; pit round-conic; skin purple; rain-induced cracking tolerance better than Bing. Soluble solids usually 14% to 15%; firmness medium, similar to Sam; flavor good; fruit removal force from peduncle about 490 g at maturity. Tree: moderately vigorous; habit open, very spreading with many lateral branches along apical portions of new growth; slightly drooping form in mature trees. Very precocious and productive. Not self-fertile; pollen compatibility Group VI, with Gold; midseason bloom.

Honey Heart.—Orig. in Santa Rosa, Calif., by Luther Burbank. Introd. in 1934. Plant patent 41; 8 Nov. 1932; Elisabeth Waters Burbank, executrix of the estate of Luther Burbank; assigned to Stark Bros. Nurseries & Orchards Co., Louisiana, Mo. Parentage unknown. Fruit: sweet; up to 1 inch in diam.; round, blunt cordate, symmetrical; skin yellow, blushed, thin, tough; flesh creamy white, firm, moderately tender, juicy, mildly sweet; juice colorless; slightly clingstone; ripens end of May; resembles Napoleon (Royal Ann). Tree: moderately vigorous; open-spreading; branches strong.

Hudson.—Orig. in Geneva, N.Y., by New York State Agr. Expt. Sta. Introd. in 1964. Oswego x Giant; selected in 1935; tested as NY 591. Fruit: sweet; size medium; skin very dark red; flesh very firm; flavor sweet, good; little cracking, even under extreme conditions; ripens very late, after Lambert, extending the cherry harvest about 10 days; best in its season. Tree: very large; open-spreading; vigorous; moderately productive; hardy.

Index.—Orig. in Prosser, Wash., by Thomas Toyama, Washington Agr. Expt. Sta. and U.S. Dept. of Agriculture. Introd. in 1994. Plant patent pending; assigned to Central Washington Nursery Improvement Institute. Stella x unknown; selected in 1978; tested as PC 7222-1. Fruit: sweet; large; broadly cordate with rounded apex end; skin glossy, dark mahogany; flesh is firm, crisp, dark red; juicy with sweet, low acid flavor; excellent quality; gets sugar early while fruit is still red and flesh is still pink; ripens 3–6 days before Bing; medium long, slender stem; semi-free, small stone. Tree: self-fertile; productive; flowers open earlier than Bing; pollen compatible with Bing and Chelan; vigorous; upright-spreading; very productive.

Jordan.—Orig. in East Jordan, Mich., in the Eveline Orchards. Introd. about 1937. Considered to be a bud mutation of Montmorency; discovered about 1930. Fruit: sour; ripens 1 week before Montmorency. Now obsolete.

Jubilee.—Orig. in Davis, Calif., by California Agr. Expt. Sta. Introd. in 1964. Lambert x Napa Long Stem Bing; cross made in 1940 by Guy L. Philp; selected in 1948 by Brooks; tested as UCD 26-15 by R.M. Brooks and W.H. Griggs. Fruit: sweet; slightly larger than Bing; cordate, symmetrical; skin glossy, red to dark red, tender to moderately tender, like Bing in texture; flesh red to dark red, fine-textured, moderately firm to firm, moderately juicy; juice light red to red; flavor ranges from very sweet and mild to sweet and mildly tart; stone free to slightly cling, medium large, ridged; susceptibility to cracking similar to Bing; produced practically no doubles even in years when doubling was a serious problem with Bing and Royal Ann; stem short to medium; ripens an average of 6 days before Bing; may be harvested over a long period as it has excellent flavor when light red to fully ripe; excellent for shipping, fresh consumption, canning; satisfactory for brining; supplements Bing. Tree: similar to Bing in growth and appearance; productive; vigorous; upright-spreading; a late average bloom period; effectively cross-pollinated by Bing, Royal Ann, Larian, Lambert, Berryessa, Bada, Early Burlat, Moreau, Starking Hardy Giant.

JuneRed Mont.—Orig. in Acme, Mich., by Raymond Burkholder. Introd. in 1982 by Bountiful Ridge Nurseries, Princess Anne, Md. Whole-tree mutation of Montmorency. Fruit: sour; similar to Montmorency, except matures earlier and has red juice. Tree: self-fruitful; more consistent bearer than Montmorency.

Kansas Sweet *(Hansen, Hansen Sweet)*.—Orig. in Wichita, Kan., by H.J. Hansen, Hansen Nursery. Introd. about 1935 by the Willis Nursery Co., Ottawa, Kan. Parentage unknown; discovered between 1920 and 1930; *Prunus cerasus*. Fruit: sour; skin red; flesh firm; somewhat sweet; resembles English Morello. Tree: fruits satisfactorily in the Great Plains area; hardy.

Kelleriis 16 *(Morellenfeuer, in Germany)*.—Orig. in Kvistgaard, Denmark, by Svend Poulsen, D.T. Poulsen's Nurseries. Introd. in 1956. German patent 1,039,292; assigned to Horstmann & Co., Germany. Open-pollinated seedling of (*Prunus cerasus* 'Ostheimer' x *Prunus avium* 'Früheste der Mark'); perhaps the male parent was Reine Hortense; selected in 1945. Fruit: sour-type large, hybrid, low acidity; may be eaten out-of-hand when fully ripe. Tree: very high yielding; self-fruitful.

***Kordia*.**—See Attika®.

Krassa Severa *(Northern Beauty)*.—Imported from Mitchurin Hort. Inst., Koslov, through Inst. of Plant Industry, Leningrad, Russia, by U.S. Dept. of Agriculture (USDA); received in U.S. 18 Apr. 1934; P.I. 105207; propagated at USDA Plant Introduction Sta., Chico, Calif. Origin unknown; translated varietal name: Northern Beauty, Beauty of the North; said to be *Prunus avium* x *P. cerasus*. Fruit: large; roundish-oblate; skin dark red to dark purple; flesh dark red; flavor acid, good; juice light red-purple; semi-clingstone, pit size medium; fine for home canning; ripens very late; resembles a sour cherry, being more like *P. cerasus* than the hybrid Duke cherry, resembling English Morello; a good sour cherry. Tree: upright, being very unlike the standard weeping English Morello tree; fruiting system very good, with much better differentiation of leaf buds than in English Morello; leaf spot disease moderate; slight to moderate susceptibility to brown rot fungus.

Kristin.—Orig. at Geneva, N.Y., by New York State Agr. Expt. Sta., Cornell Univ. Introd. in 1982. Emperor Francis x Gil Peck; cross made in 1938; selected in 1949; tested as NY 1599 at the Geneva station; at Bozeman, Mont., by Montana State Univ.; and at Ullensvang Res. Sta., Lofthus, Norway. Fruit: large; flesh quality high, aromatic, firm, flavor sweet, red; skin dark red or purplish-black, glossy, attractive; similar to Schmidt; ripens midseason at Geneva 2 days after Emperor Francis, in Montana before Lambert, and in Norway with Van and a few days before Sam; will crack if rainy weather occurs just before harvest, but cracked less in Norway than did Van. Tree: large; vigorous; hardier than Schmidt, Bing, and Chinook; productive; needs a cross-pollinator. Recommended as a possible replacement for Schmidt. Named by Ystaas of the Norway Res. Sta.

***Lambert Pollenizer*.**—See Peerless.

Lambush.—Orig. in Davis, Calif., by Guy L. Philp, California Agr. Expt. Sta. Introd. in 1945. Lambert x Bush Tartarian; cross made in 1936; selected in 1942; named in 1943; tested as Calif. 46. Fruit: sweet; large, averaging 1 inch in diam.; cordate, somewhat pointed, Lambert type; stem medium to short; skin red to dark red, tough, very attractive; flesh red to dark red, soft, juicy; juice light red; flavor mild, good, mildly tart; pit medium, free to slightly cling; cracks badly following rain; some doubles produced in some seasons; ripens early, several days after Chapman, around 15–20 May at Davis; resembles Lambert. Tree: productive. No longer recommended.

Lamida.—Orig. in Moscow, Idaho, by Leif Verner, Idaho Agr. Expt. Sta. Introd. in 1946. Open-pollinated seedling of Lambert; selected in 1940. Fruit: sweet; medium to large, averaging less than 1 inch in diam., slightly larger than parent; cordate, apex more pointed than Lambert; suture distinct; skin blackish-red, moderately thick, tough, separates from flesh, good finish; stylar scar small, oval depressed slightly; flesh deep red, tender, very juicy, fine-textured, moderately firm; flavor sweet, pleasing, mild; quality very good to excellent; stone semi-free, ovate, medium to large; more resistant to cracking than Lambert and comparable to Deacon; ripens with, or slightly after, Lambert; resembles Lambert. Tree: upright; moderately vigorous; leaves medium to large, petiole fairly long; glands 1–3, small; bud medium; pollinated by Black Tartarian, Republican, Van; does not set excessively heavy crops.

Lapins.—Orig. in Summerland, B.C., by K.O. Lapins and W.D. Lane, Res. Sta. Agriculture Canada. Introd. in 1983. Van x Stella. Selected in 1971. Fruit: sweet; resembles Lambert, but larger, firmer, and ripens 2 days later. Tree: upright; self-fertile; more productive than most commercial cultivars.

Larian.—Orig. in Davis, Calif., by the California Agr. Expt. Sta. Introd. in 1964. Lambert x UCD 50 (Bing x Bush Tartarian); cross made in 1946 by Guy L. Philp; selected in 1956 by R.M. Brooks; tested as UCD 53-20 by Brooks and W. H. Griggs. Fruit: sweet; large, averaging 1.02 inches in diam., whereas Black Tartarian averaged 0.89 inches in diam., Bing averaged 0.98 inches in diam.; round-cordate; skin dark red, tender; flesh and juice red to dark red, moderately firm to firm; sweet, mild to tart; stone slightly clinging; very resistant to cracking; produces no doubles in most years and very few in years when doubling is a serious problem with Bing and Royal Ann; ripens an average of 6 days after Black Tartarian and 12 days before Bing; excellent for fresh fruit consumption and shipping; limited canning and brining tests indicate it is apparently satisfactory. Tree: moderately vigorous; upright-spreading; productive; effectively cross-pollinated by Black Tartarian, Van, Mona, Berryessa, Bing, Bada, and Jubilee.

Late Lambert.—See Stark Royal Purple.

Macmar.—Orig. in Corvallis, Ore., by the Oregon Agr. Expt. Sta. Introd. in 1961. Parentage unknown, but apparently a chance seedling; discovered before 1920. Fruit: sweet; size medium; round-cordate, similar to Bing; skin yellow, about one-half blushed, tough, very firm, strongly adherent to fruit; fresh shipments satisfactory; not equal to Napoleon when canned; brining quality not determined. Tree: large; vigorous; appears to be as hardy as Bing and Napoleon.

Mac Mont.—See McClain Montmorency.

Mahaleb 900 *(Mh. 900)*.—See rootstocks for cherry.

Maynard.—Orig. in Amarillo, Texas, by John N. Maynard. Introd. in 1959 by Inter-State Nurseries, Hamburg, Iowa, which named it. Parentage unknown; discovered in 1929. Fruit: Duke type; sweet; red; large; flavor good. Tree: bears annually; large.

McClain Montmorency *(Mac Mont, Montmammoth, Montmore)*.—Orig. in Knoxville, Tenn., by Lee McClain, McClain's Washington Heights Nurseries. Introd. in 1925. Bud mutation of Montmorency. Fruit: sour; larger than Montmorency, which it resembles; color darker than that variety. Tree: somewhat dwarf; spreading.

Merchant.—Orig. in Hertford, England, by P. Matthews, John Innes Inst., Norwich, U.K. Introd. in 1976. Open-pollinated seedling of Merton Glory, Selected in 1970; tested as J 11376. Fruit: sweet; large; skin black, resembles a black Merton Glory; ripens early midseason. Tree: resistant to bacterial canker; very responsive to hormone spray; universal donor pollination group.

Merla.—Orig. in Hertford, England, by P. Matthews, John Innes Inst., Norwich, U.K. Introd. in 1976. Open-pollinated seedling of Merton Late. Selected in 1970; tested as J 11721. Fruit: sweet; late midseason; white cherry. Tree: precocious; productive; resistant to bacterial canker; universal donor pollination group.

Mermat.—Orig. in Hertford, England, by P. Matthews, John Innes Inst., Norwich, U.K. Introd. in 1976. Open-pollinated seedling of Merton Glory. Selected in 1970; tested as J 11207. Fruit: sweet; large; skin black; resembles Early Rivers. Tree: resistant to bacterial canker; early ripening, incompatibility group II.

Merpet.—Orig. in Hertford, England, by P. Matthews, John Innes Inst., Norwich, U.K. Introd. in 1976. Open-pollinated seedling of Merton Glory. Selected in 1970; tested as J 11253. Fruit: skin black, resembles a black Merton Glory; ripens early mid-season. Tree: resistant to bacterial canker; universal donor pollination group.

Merton Bigarreau.—Orig. in Merton Park, London, England, by the John Innes Inst. Introd. in 1947 in England and in the U.S. in the 1950s. Knight x Napoleon; selected in 1930. Fruit: sweet; large; round; skin deep mahogany-red; flesh dark red, tender, meaty; flavor extremely rich; stem short; ripens in midseason. Tree: bears heavily and regularly; vigorous; very spreading;

self-unfruitful, pollinated by Black Tartarian, Republican, Van.

Merton Bounty.—Orig. in Merton Park, London, England, by M.B. Crane, John Innes Inst. Introd. in 1947 in England, in the U.S. in the 1950s; P.I. 212493. Elton Heart x Bigarreau de Schrecken; selected in 1930; tested as Merton 490. Fruit: sweet; large; heart-shaped; deep purple; flavor rich. Tree: bears heavily and consistently; vigorous; upright.

Merton Crane.—Orig. in Merton, England, by M.B. Crane, John Innes Inst., Norwich, England. Introd. in 1961. Bigarreau Napoleon x Bigarreau de Schrecken. Cross made in 1925; tested as 519. Fruit: sweet; large; round; skin black; flesh deep crimson, firm, juicy; ripens mid-July. Tree: large; upright, stiff, and close-spurring; vigorous. Named in honor of M.B. Crane, head of John Innes Pomology Dept. until his retirement in 1953.

Merton Favourite.—Orig. in Merton Park, London, England, by the John Innes Inst. Introd. in 1947 in England and in the 1950s in the U.S. Knight x Bigarreau de Schrecken; selected in 1930. Fruit: sweet; large; oval; skin and flesh deep purplish-crimson to black; flavor excellent; cracks sometimes; ripens with or before Napoleon (Royal Ann). Tree: moderately vigorous; much branched; pollinated by Black Tartarian, Republican, and Van; bears well.

Merton Glory.—Orig. in Merton Park, London, England, by the John Innes Inst. Introd. in 1947 in England and in Canada in the early 1950s. Ursula Rivers x Noble; selected in 1945. Fruit: sweet; very large; heart-shaped; skin cream overspread with crimson; slightly soft; flavor good; useful for processing. Tree: strong; spreading.

Merton Heart.—Orig. in Merton Park, London, England, by the John Innes Inst. Introd. in 1947 in England and in the 1950s in the U.S. Bigarreau de Schrecken x Elton Heart; selected in 1930; tested as Merton 404. Fruit: sweet; large; heart-shaped; skin deep purplish-crimson; ripens early, after Early Rivers and about 2 weeks before Napoleon; flesh dark red, juicy, meaty; flavor extremely rich. Tree: vigorous; very upright; markedly resistant to bacterial canker and blossom wilt; pollinated by Black Tartarian, Republican, and Van.

Merton Late.—Orig. in Merton, England, by M.B. Crane, John Innes Inst., Norwich, England. Introd. in 1961. Hildesheim (Belle Agathe) x Bigarreau Napoleon (Royal Ann). Cross made in 1926; tested as 659. Fruit: sweet white; size medium; round; skin orange-yellow overspread with deep crimson, being very dark for a white variety; flesh orange-yellow, very firm; ripens very late, early September. Tree: large; spreading; vigorous; susceptible to bacterial canker; produces large amounts of lateral wood.

Merton Marvel.—Orig. in Merton, England, by M.B. Crane, John Innes Inst., Norwich, England. Introd. in 1961. Nobel x White Bigarreau. Cross made in 1930; tested as 868. Fruit: sweet; large; blunt, heart-shaped; skin black; flesh deep crimson, juicy, firm; ripens late midseason. Tree: medium to large; compact; considerably tolerant to bacterial canker; produces too much thin lateral wood.

Merton Premier.—Orig. in Merton Park, London, England, by M.B. Crane, John Innes Inst. Introd. in 1947 in England, in the early 1950s in Canada, and in 1954 in the U.S. as P.I. 212490. Emperor Francis x Bedford Prolific; selected in 1930; tested as Merton 418. Fruit: sweet; large; round; skin deep crimson-purple; flavor very good. Tree: moderately vigorous; upright.

Merton Reward.—Orig. in Merton Park, London, England, by the John Innes Inst. Introd. in 1958. Emperor Francis x Bedford Prolific; selected in 1930. Fruit: sweet; very large; blunt, heart-shaped; skin deep mahogany-red; flesh firm; ripens late. Tree: vigorous; upright; good resistance to bacterial canker, but shows tendency to blossom wilt.

Mesabi.—Orig. in Duluth, Minn., by Chris Knutson, Farmer Seed & Nursery Co., Faribault, Minn. Introd. in 1964. Plant patent pending. Bing x a chance seedling, considered to be sour cherry (*Prunus cerasus*); selected in 1954. Fruit: hybrid; skin quite red; flesh lighter red than skin; sugar content midway between sweet and sour cherries; pit smaller than Meteor; for sauce and pie; ripens in mid-July; resembles Meteor. Tree: heavy yields; very hardy;

spreading; 10–14 feet tall if unpruned; tends to bear when small, about 40 inches tall; grows slowly.

Meteor.—Orig. in Excelsior, Minn., at the Minnesota Fruit Breeding Farm. Introd. in 1952. Montmorency x a selection of a Russian type tracing back through two or more generations to the varieties Vladimir and Shubianka; cross made in 1935; first fruited in 1944; tested as Minn. 66. Fruit: sour; large; roundish-oval, flattened lengthwise; apex slightly depressed; stem medium in length and thickness; skin clear, light, bright red, very attractive, thin, tender; flesh light, bright yellow, medium firm, medium juicy; flavor a pleasant mild acid; quality very good; stone small, long oval, very free; very suitable for pie; ripens about midseason, about 7–10 days after Northstar. Tree: very hardy; a strong, vigorous grower; upright; moderately spreading; heavily spurred leaves very large, apparently highly resistant to leaf spot; blooms after Montmorency and Northstar; bears regularly and annually.

Meyer.—See Starking Hardy Giant.

Mh. 900.—See Mahaleb 900.

Mona.—Orig. in Davis, Calif., by the California Agr. Expt. Sta. Introd. in 1964. LaCima x Chapman; cross made in 1940 by Guy L. Philp; selected in 1948 by R.M. Brooks; tested as UCD 30-9a by Brooks and W.H. Griggs. Fruit: sweet; large, cordate, symmetrical; skin glossy red to dark red, tender; flesh red to dark red, tender to moderately firm, juice red; flavor sweet, mild, flavor superior to Black Tartarian, less tart than that variety; stone smooth, relatively large, free to slightly cling; subject to some cracking in late rains; some doubles in interior valleys in years when doubling a serious problem with Bing and Royal Ann; stem usually long, slender, light green; ripens about 1 day after Black Tartarian and 17 days before Bing; fresh-fruit market apparently is its best outlet, but also it has been rated excellent for canning; a little soft for a good brining cherry, although superior to Black Tartarian. Tree: vigorous; very productive; upright spreading; satisfactorily cross-pollinated by Bing, Black Tartarian, Berryessa, Van, Royal Ann, Early Burlat, Republican, Starking Hardy Giant, Moreau, and Jubilee.

Montearly.—Orig. in East Jordan, Mich., by Levi R. Taft. Introd. in 1932. Plant patent 30; 4 Oct. 1932; assigned to Stark Bros. Nurseries & Orchards Co., Louisiana, Mo. Parentage unknown; selected in 1928. Fruit: sour; one-third larger than Early Richmond; roundish-oblate; skin glossy, deep red at maturity, thick, tough; flesh dark crimson, tender, melting; juice abundant, red; stone size medium, free; ripens 10–14 days earlier than Montmorency, which it resembles; stem 1 to 1 1/8 inches long, thick. Tree: similar to Montmorency, being more stocky, with heavier branches and thicker leaves having larger ribs and veins and slightly darker color.

Montlate.—Orig. in East Jordan, Mich., by Levi R. Taft. Introd. in 1932. Plant patent 29; 4 Oct. 1932; Levi R. Taft; assigned to Stark Bros. Nurseries & Orchards Co., Louisiana, Mo. Bud mutation of Montmorency; selected in 1927. Fruit: sour; identical with Montmorency, except that it ripens 10–14 days later than parent. Tree: identical to Montmorency.

Montmammoth.—See McClain Montmorency.

Montmore.—See McClain Montmorency.

Moreau.—U.S. Dept. of Agriculture released for propagation in 1961 this variety new to the U.S. First fruited in this country at the U.S. Plant Introduction Sta., Chico, Calif., in 1949. Moreau is the commercial synonym for a variety introduced from Germany in 1937 as P.I. 125697, Bigarreau de St. Charmez, sometimes known also as Souvenir des Charmes and Bigarreau Sandrin. Fruit: sweet; outstanding characteristics are large size, early maturity, and suitability for shipping, being very similar to Early Burlat in all respects, except that Moreau ripens 2 days before Early Burlat and has much less doubling; susceptible to cracking after rain; preliminary tests indicate Moreau will pollinate Bing and vice versa.

Morellenfeuer.—See Kelleriis 16.

Musselman Montmorency.—Orig. in Gettysburg, Pa., at the Blue Ribbon Orchards, by the C.H. Musselman Co., Biglerville, Pa. Introd. in 1934. Bud mutation of Montmorency; discovered in 1934. Fruit: sour; large; resembles Montmorency.

Tree: heavy producer, bearing along entire branch.

Noble ***(Tradescant Heart, St. Margaret).***—An Old English variety of unknown origin. Introd. in the U.S. in 1943 by the New York State Agr. Expt. Sta. Parentage unknown. Fruit: sweet; large; heart-shaped; flesh dark purplish-red, firm; cracks less than most late varieties; season between Schmidt and Hedelfingen.

Northern Beauty.—See Krassa Severa.

NorthStar.—Orig. in Excelsior, Minn., by the Univ. of Minnesota Fruit Breeding Farm. Introd. in 1950. English Morello x Serbian Pie 1 (a selection from seed of sour cherry obtained in Serbia in 1918 by C.P. Bull of Univ. of Minnesota); cross made in 1933; selected in 1942; tested as Minn. 58. Fruit: sour, Morello type; about 0.75 inches in diam.; roundish heart-shaped, apex slightly depressed; stem slender, about 1 inch long; skin bright red changing to a dark, glistening mahogany-red at maturity, medium tender; flesh dark red at maturity, juicy, meaty, tender; flavor pleasantly acid; quality good; stone small, easily removed; ripens very early, beginning 5–10 July at place of origin, remaining on tree in good condition for about 2 weeks. Tree: small; productive; self-fruitful; resistant to leaf spot and apparently to brown rot diseases; very hardy in wood and fruit bud; very ornamental; suitable for the home garden with good commercial possibilities.

Oliver Long Stem Bing.—Orig. in Santa Cruz, Calif., by Mr. and Mrs. W.H. Oliver. Introd. in 1921. Parentage unknown; tree planted in 1914; selected in 1919. Fruit: sweet; flesh dark red to black; stem very long; matures 2 weeks later than Bing. Tree: bears moderately and consistently; blooms after Napoleon; pollinates Bing and Napoleon and is pollinated by them.

Olympus.—Orig. in Prosser, Wash., by Thomas Toyama, Washington Agr. Expt. Sta. and U.S. Dept. of Agriculture. Introd. in 1990. Plant patent 8033; plant patent date 1991; assigned to Central Washington Nursery Improvement Inst. Lambert x Van; selected in 1977; tested as PC 6659-2. Fruit: sweet, large; heart-shaped with round apex end and pronounced suture ridge; skin glossy, dark mahogany; flesh firm, dark red-black; juicy with low acid flavor; ripens 5–7 days after Bing; medium long stem; semi-free, medium stone. Tree: self-sterile; flowers open later than Bing but not as late as Star; pollen-compatible with Sam and Lambert; vigorous; spreading; very productive.

Parkhill.—Orig. in East Wenatchee, Wash., by John D. Parkhill. Introd. about 1930. Probably an open-pollinated seedling of Lambert; discovered in 1925. Fruit: sweet; flesh very firm; slightly smaller than Bing; ripens with Black Tartarian, which it resembles. Tree: vigorous; spreading; leggy.

Paul Rose.—Orig. in Elberta, Mich., by Paul Rose, Paul Rose Orchard Co. Introd. in 1933. Bud mutation of Schmidt; discovered in 1924. Fruit: sweet; large; skin pink, with a dark red line on the ventral suture; flesh white, juicy, tender; flavor fine; ripens in July in Michigan. Now obsolete.

Peerless ***(Lambert Pollenizer).***—Origin unknown, but thought to be somewhere on the West Coast. Introd. in 1937 in Montana. Parentage unknown. Fruit: sweet; slightly smaller than Lambert, which it resembles; skin dark red; flesh firm; satisfactory for canning; same harvest season as, or sometimes several days later than, Lambert. Tree: large, resembling Lambert; leaf petioles smooth, only occasionally one or two hairs present; small leaves sometimes present at base of pedicels; sepals sometime remain on fruit until harvest; pollinates Lambert and that variety pollinates Peerless; very susceptible to Lambert mottle virus; as hardy as Lambert.

Pezzi Tartarian.—See Sciutti Tartarian.

Rainbow Stripe.—Orig. in Yakima, Wash., by E. Remy & Son. Introd. in 1930. Bud mutation of Lambert; discovered in 1925. Fruit: sweet; skin white with narrow blood-red stripe along suture from stem to tip, extending through the flesh; otherwise resembles Lambert. Tree: vigorous; spreading; resembles Lambert.

Rainier.—Orig. in Prosser, Wash., by Harold W. Fogel, Washington Agr. Expt. Sta. and U.S. Dept. of Agriculture. Introd. in 1960. Bing x Van; selected in 1954; tested as Wash. P 1-680. Fruit: sweet; very large; slightly obovate, similar to Bing; skin yellow, highly blushed, attractive; flesh firm;

juice colorless; soluble solids equal to or slightly higher than Bing; quality high; stone medium to small, relatively free; ripens 3–7 days before Napoleon (Royal Ann) and Bing; stem thick. Tree: vigorous; spreading to upright-spreading; very productive; comes into bearing early; outstanding characteristic is its apparent extreme hardiness, equal to its parent Van; blooms and leafs out with Bing; leaf petioles hairless; self-unfruitful, but satisfactorily pollinated by Bing, Van, and Sam; Rainier is a satisfactory pollinator for Bing and Chinook.

Ransom.—Orig. in Farrer Place, Sydney, New South Wales, Australia, by F.T. Bowman, New South Wales Dept. of Agriculture. Introd. in 1955. Open-pollinated seedling of St. Margaret; selected during the 1940s. Fruit: sweet; large; skin reddish-black; not quite as firm or as highly flavored as St. Margaret, but more resistant to splitting; ripens several days before St. Margaret, which it resembles. Tree: produces heavy crops.

Regina.—Orig. in Farrer Place, Sydney, New South Wales, Australia, by F.T. Bowman, New South Wales Dept. of Agriculture. Introd. in 1955. St. Margaret x Black Eagle; selected during the 1940s. Fruit: sweet; large; skin dark red; flesh firm; flavor very good; resistant to splitting; ripens early, at the end of Early Lyons season; short stem; resembles Early Lyons. Tree: prolific bearer.

Reidler.—Orig. in Klingerstown, Pa., by Paul G. Reidler, Ashland, Pa. Introd. in 1963. Propagated by Nut Tree Nurseries, Downingtown, Pa. Chance seedling; discovered about 1941. Fruit: sweet; moderately large; fairly firm; excellent for pie and jelly; ripens 15–25 June at place of origin; a general-purpose sweet variety; black Oxheart type.

Richmorency *(Early Montmorency)*.—Orig. in Saint Joseph, Mich., by Lewis Benjamin Reber. Introd. in 1938. Plant patent 316; 28 Feb. 1939; assigned to Greening Nursery Co., Monroe, Mich. Bud mutation of Montmorency; discovered in 1929. Fruit: sour; size medium; high quality; characterized particularly by its early ripening, 7–10 days before Montmorency. Tree: produces well; identical to Montmorency.

Rival.—Orig. in Farrer Place, Sydney, New South Wales, Australia, by F.T. Bowman, New South Wales Dept. of Agriculture. Introd. in 1955. Open-pollinated seedling of St. Margaret; selected during the 1940s. Fruit: sweet; large; skin red-black; flesh firm, vinous-flavored; quality good; only slightly susceptible to splitting; ripens early, during the mid-pick of Early Lyons, which it resembles; stem short.

Rons.—Orig. in Young, New South Wales, Australia, by S.A. Thornell, New South Wales Dept. of Agriculture. Introd. there about 1928. Eagle x Bigarreau Noir de Guben; selected about 1928. Fruit: sweet; large; skin dark red; good resistance to splitting; quality excellent; ripens 7–10 days before St. Margaret; resembles Bigarreau Noir de Guben. Tree: very heavy and reliable producer.

Royalton™.—Introd. in 1991 by R.L. Andersen, S.K. Brown, R.D. Way, K.G. Livermore, and D.E. Terry, New York State Agr. Expt. Sta., Geneva. NY 1725 open-pollinated; selected in 1975; tested as NY 11390. Plant patent pending; assigned to Cornell Res. Foundation. Fruit: sweet; round-oblate; large, about 30 x 28 mm; round pit; purple skin. Fruit more tolerant of rain-induced cracking than Bing, but less tolerant than Sam; firmness similar to Ulster and Kristen. Fruit removal force from peduncle about 550 g at maturity. Tree: very vigorous; very upright; slow to begin bearing. Not self-fertile; pollen compatibility Group VIII with Schmidt; early mid-season bloom.

Salmo.—Orig. in Summerland, British Columbia, Canada, by K.O Lapins, Canada Dept. of Agriculture, Res. Sta. Introd. in 1970. Lambert x Van. Cross made in 1948; selected in 1956; tested as SJ-9-14. Fruit: sweet; Van type; large, round with short stem; skin black, moderately susceptible to cracking in rain; flesh moderately firm, flavor fair to good; juice dark; for fresh market; ripens first week of July or 1 week before Van or Bing. Tree: medium size, spreading; moderately vigorous; hardy; precocious-bearing, productivity moderate to heavy; tolerant to little cherry virus.

Sam.—Orig. in Summerland, British Columbia, Canada, by A.J. Mann, Canada Dept. of Agriculture, Res. Sta. Introd. in

1953. Open-pollinated seedling (V-160140) of Windsor open-pollinated; seed collected in 1938; original tree planted in 1940; selected in 1946; tested as Summerland S-6-10. Fruit: sweet; medium to large; flesh black, firm; fair to good quality; ripens about 1 week earlier than Bing; highly resistant to cracking; resembles Deacon in shape. Tree: vigorous; hardy; self-unfruitful but interfruitful with Bing, Lambert, and Van. Widely planted. The name Sam was formed from the initials of Summerland and of the originator.

Sanguinetti.—Orig. in Stockton, Calif., by Dave V. Sanguinetti. Not introd. commercially; included here because it has been used, particularly by the late Guy L. Philp, in the sweet cherry breeding program conducted by the Dept. of Pomology, California Agr. Expt. Sta., Davis. Bud mutation of Black Tartarian; discovered in 1931 as one-half of a tree of Black Tartarian on mazzard rootstock; leaves of the rootstock suckers differ from Sanguinetti, so it is assumed it arose as a bud mutation. Fruit: normally somewhat larger but with same maturity as Black Tartarian; stem length medium, heavy, thick; a little more pointed than parent, although not as much so as Lambert; must be picked within 2 or 3 days of maturity as otherwise the fruit pulls away easily from the stem; ships well to eastern markets. Tree: shy bearer; pollinator for Bing; blooms at same time as Black Tartarian.

Sciutti Tartarian *(Pezzi Tartarian)*.—Orig. in Stockton, Calif., by Antonio Sciutti. Introd. in 1941. Parentage unknown; discovered in 1924. Fruit: sweet; resembles Black Tartarian, but ripens 4–6 days earlier; larger and more uniform than that variety.

Seneca *(Early Seneca)*.—Orig. in Geneva, N.Y., by Richard Wellington, New York State Agr. Expt. Sta. Introd. in 1924. Early Purple x unknown; cross made in 1911; first full crop in 1922. Fruit: sweet; early, ripening 2 weeks earlier than Black Tartarian; large, round-cordate; skin purple-black; flesh soft, juicy, melting; rich flavor; pit free; skin does not crack. Tree: vigorous; productive; resembles Black Tartarian.

Sodus.—Orig. in Geneva, N.Y., by Richard Wellington, New York State Agr. Expt. Sta. Introd. in 1938. Napoleon x Giant; cross made in 1925; first full crop in 1935. Fruit: sweet; large; skin light-colored; resistant to cracking; firm-fleshed; ripens between Emperor Francis and Napoleon; resembles Victor. Tree: vigorous; spreading; hardiness medium.

Somerset™.—Introd. in 1993 by R.L. Andersen, S.K. Brown, R.D. Way, K.G. Livermore, and D.E. Terry, New York State Agr. Expt. Sta., Geneva. Van x Vic; cross made in 1960; tested as NY 6476. Plant patent pending; assigned to Cornell Res. Foundation. Fruit: sweet; cordate, very symmetrical; about 30 x 27 mm; skin purple with high sheen; pit round-conic with slightly protruding tip. Rain-induced cracking is lower than in Lapins, more than Early Rivers. Strong cherry flavor with good sugar : acid balance; fruit firm with excellent shelf life. Tree: moderately low vigor; spreading, with many lateral branches; very precocious and productive. Not self-fertile; pollen compatibility Group III, with Bing; blooms early, 1 day before Bing. At Geneva, good orchard tolerances to brown rot and bacterial canker.

Spalding.—Orig. in Moscow, Idaho, by Leif Verner, Idaho Agr. Expt. Sta. Introd. in 1946. Bing x (probably) Deacon; selected in 1942. Fruit: sweet; very large, equaling Bing, averaging 1 inch or more diam.; broadly cordate, somewhat more pointed than Bing; skin mahogany-red to nearly black, thick, tough, adherent to flesh, high luster; stylar scar small; flesh dark red streaked with black, coarse, very firm; somewhat tart until fully tree-ripe, then becomes sweet and rich; quality very good; clingstone, very large, plump, ventral suture very broad and ridged; slightly less susceptible to cracking than either Bing or Lambert; ripens late, 7–10 days after Bing. Tree: vigorous; spreading; leaf large, thick, flat, obovate or oval. Named for the Rev. Henry Harmon Spalding, early missionary of the Pacific Northwest, who introduced fruit-growing in northern Idaho.

Sparkle.—Orig. in Summerland, British Columbia, Canada, by A.J. Mann, Canada Dept. of Agriculture, Res. Sta. Introd. in 1944. Open-pollinated seedling of Empress

Eugenie; cross made in 1936; selected in 1942. Fruit: sweet; medium to small; flesh white, firm; flavor sweet, sprightly; appearance attractive, as skin has a bright luster, which suggested the varietal name; fairly resistant to cracking; rounder and less conical than Napoleon, which it resembles; matures 1 week earlier than Napoleon. Tree: vigorous; upright grower; a pollinator for Bing, Lambert, Napoleon, and Van; pollinated by the above and Deacon. Not recommended for commercial planting; for home gardens.

Star.—Orig. in Summerland, British Columbia, Canada, by A.J. Mann, Canada Dept. of Agriculture, Res. Sta. Introd. in 1949. Open-pollinated seedling of Deacon; selected in 1944. Fruit: sweet; dark red; resembling Deacon in shape; good size and quality; moderately firm; stem very long, thick, stiff; susceptible to splitting; ripens 10 days before Bing. Tree: not hardy; moderately heavy bearer; self-unfruitful and cross-sterile with Bing, Lambert, Royal Anne, but interfertile with Van and Sam.

Starking Hardy Giant *(Meyer)*.—Orig. in Cedarburg, Wis., by Ottilie R. Meyer. Introd. in 1949. Plant patent 764; 7 Oct. 1947; assigned to Stark Bros. Nurseries & Orchards Co., Louisiana, Mo. Originally named Meyer; renamed in 1949 by Stark Bros. Parentage unknown; discovered in 1925. Fruit: sweet; large; skin attractive, dark red; flesh firm, high quality; resembles Schmidt. Tree: large; vigorous; upright-spreading; extreme hardiness; very productive; bears regularly.

Starkrimson *(Zaicrimson)*.—Orig. in Modesto, Calif., by Floyd Zaiger. Plant patent 4519; 8 Apr. 1980; licensed to Stark Bros. Nurseries, Louisiana, Mo., for Canada, Mexico, and U.S. Introd. in 1980. Garden Bing x Stella. Fruit: resembles Stella; shape cordate; skin black-red; flesh red, firm, flavor sweet, slightly subacid at full maturity; high in sugar and retained up to 3 weeks on trees; quality excellent. Tree: dwarf; compact, spreading; vigorous; productive; flowers self-fertile; blooms with Bing at Modesto; chilling requirement 1000 hours.

Stark Royal Purple *(Late Lambert)*.—Orig. in Charlevoix, Mich., by Claude Fairman. Plant patent 627; 9 May 1944; assigned to Stark Bros. Nurseries & Orchards Co., Louisiana, Mo. Originally named Late Lambert; renamed in 1948 by Stark Bros. Bud mutation of Lambert; discovered in 1933. Fruit: sweet; ripens 10–14 days after Lambert, which it resembles. Propagation has been discontinued because no virus-free budwood has been secured.

Starkspur Montmorency *(Carnefix Compact Montmorency)*.—Orig. in Fruitland, Idaho, by Warren Carnefix. Introd. in 1973. Plant patent 3304; 13 Feb. 1973; assigned to Stark Bros. Nurseries & Orchards Co., Louisiana, Mo. Bud mutation of Montmorency. Fruit: sour; same as Montmorency; 3/4–7/8 inch in diam.; skin red, no cracking; flesh white; blooms 15–25 Apr.; ripens early July in central Missouri.Tree: small; hardiness same as Montmorency; productivity heavy; susceptible to mildew.

Stella.—Orig. in Summerland, British Columbia, Canada, by K.O. Lapins, Canada Dept. of Agriculture, Res. Sta. Introd. in 1968. Lambert x John Innes seedling 2420. Cross made in 1956; selected in 1963; tested as 2C-27-19. Fruit: sweet; large; heart-shaped, Lambert type; skin black; fairly susceptible to cracking in rain; flesh black, medium firm, relatively coarse; ripens first week in July at Summerland. Tree: upright; spreading; vigorous; productive; moderately tender in tree and fruit buds; self-fertile, also serves as a pollinizer for other varieties; recommended for areas with unfavorable conditions for cross-pollination, for home orchards, commercial orchards, and as a parent in breeding for self-fertility.

St. Margaret.—See Noble.

Sue.—Orig. in Summerland, British Columbia, Canada, by A.J. Mann, Canada Dept. of Agriculture, Res. Sta. Introd. in 1954. Bing x Schmidt; cross made in 1936; selected in 1946. Fruit: sweet; medium; roundish conic; resembles Napoleon in general appearance; skin with yellow ground color and an attractive bright red blush; juice colorless; nearly as firm as Napoleon; quality high; highly resistant to cracking, similar in this respect to Sam; canning characteristics good, comparable to Napoleon; slightly earlier than Napoleon, although heavy setting results in variable maturity. Tree: vigorous; upright; moderately compact; foliage rather dense; medium in hardiness; bears very heavily. Pollination: self-unfruitful; successfully pollinated by Bing, Lambert, Napoleon, Star, and Van; successful pollinator for Bing, Lambert, Sam, Napoleon, and Van.

Summit.—Orig. in Summerland, British Columbia, by K.O. Lapins, Agriculture Canada, Res. Sta. Introd. in 1973. Van x Sam; cross made in 1957; selected in 1964; tested as 4C-18-21. Fruit: very large; roundish heart-shaped; skin black-dark red; flesh fairly light color, medium soft; flavor sweet, mild, good; ripens 4 days before Bing (8 July at Summerland). Tree: large; spreading; vigorous; as hardy as Bing; productivity moderate.

Sunburst.—Orig. in Summerland, B.C., by K.O. Lapins and W.D. Lane, Res. Sta., Agriculture Canada. Introd. in 1983. Van x Stella. Selected in 1971. Fruit: resembles Stella; very large; resists rain splitting; not as firm as Bing or Van; matures with Van. Tree: self-fertile; very productive; good growth habit.

Surefire™.—A late-blooming tart cherry, introd. in 1993 by R.L. Andersen, S.K. Brown, R.D. Way, K.G. Livermore, and D.E. Terry, New York State Agr. Expt. Sta., Geneva. Borchert Black Sour x NY 6935 (Richmorency x Schattenmorelle); tested as NY 12716. Plant patent pending; assigned to Cornell Res. Foundation. Fruit: sour; symmetrical, slightly cordate; about 24 mm in diam. x 20 mm long; skin and flesh bright red; pit oblong-conic, small. Strong cherry flavor; firmness similar to Montmorency; slightly more acid than Montmorency. Fruit ripens about 65 days after full bloom. No rain-induced cracking. Tree: medium vigor; semi-upright, with relatively few lateral branches. Cropping on spurs and previous season's growth; self-fruitful; flowers 5–7 days after Montmorency.

Sweet Ann.—Orig. in Logan, Utah, by Utah Agr. Expt. Sta. and U.S. Dept. of Agriculture (USDA). Introd. in 1981. Orig. from open-pollinated Napa Long Stem Bing seeds collected in 1964. Selected and evaluated by Bryce Wadley, formerly with USDA; later studies conducted by Sherman V. Thomson, Utah State Univ. Tested as LSB-7. Fruit: sweet; medium size, yellow, one-half covered by pleasing blush; flesh sweet, excellent flavor, firm; pit small; sweeter, firmer, brighter red than Royal Ann; ripens with Bing; tolerant of western X-disease. Tree: shape similar to Bing; regularly sets heavy crop, many fruit to a cluster; serves as good pollinizer for Bing and Lambert; self-sterile; somewhat resistant to spring frosts; winter hardy; stems medium length; excellent for canning.

Sweet September.—Orig. in Orrville, Wayne County, Ohio, by Menno Gerber. Introd. in 1936. Plant patent 94; 17 Apr. 1934; assigned to Scarff's & Sons Nursery, New Carlisle, Ohio. Parentage unknown; discovered in 1930; original tree about 45 years old in 1932. Fruit: sweet; up to 3/4 inches in diam.; skin light to medium red, thin; flesh yellowish to light red, firm; clingstone; ripens late, last of August; keeps in good condition on tree for a long time; stem reddish. Tree: very large; upright-spreading; branches smooth, dull reddish-brown with many lenticels; branchlets thick, long, dark reddish-brown; productive.

Taylor.—Orig. in Wenatchee, Wash., by Robert Taylor. Introd. in 1970 by Van Well Nursery, Wenatchee. Chance seedling; discovered about 1950. Fruit: same as Bing, but more tolerant to cracking and ripens 4–5 days later. Tree: resembles Bing; flower is not self-fertile; used as a pollinizer for Bing.

Tradescant Heart.—See Noble.

Tulare.—Orig. in LeGrand, Calif., by Norman G. and Lowell G. Bradford. Plant patent 6407; 22 Nov. 1988. Introd. in 1988. Open-pollinated seedling of an open-pollinated seedling of Bing. Fruit: sweet; skin red. In central California, ripens in second week of May, 5 days before Bing.

Ulster.—Orig. in Geneva, N.Y., by the New York State Agr. Expt. Sta. Introd. in 1964. Schmidt x Lambert; selected in 1947; tested as N.Y. 1519. Fruit: sweet; large; skin very dark red; flesh firm, juicy, sweet, crisp; quality good; appears less susceptible to cracking than most sweet cherry varieties; designed for roadside markets; ripens in midseason, a few days before Schmidt, which it resembles. Tree: more productive than Schmidt; performs well in western New York and at the Pennsylvania Agr. Expt. Sta. in University Park.

Unark.—Orig. in Fayetteville, Ark., by Harry R. Rosen, Dept. of Plant Pathology, Univ. of Arkansas. Introd. in 1930. Parentage unknown;discovered in 1918. Fruit: sweet; skin yellow; resembles Saylor or Yellow Spanish. Tree: partially resistant to brown rot; hardiness above average; blooms late; apparently self-fertile and cross-compatible with Schmidt.

Utah Giant.—Orig. in Logan, Utah, by U.S. Dept. of Agriculture (USDA) and Utah Agr. Expt. Sta. Introd. in 1981. Open-pollinated seedling of Napa Long Stem Bing selected by Bryce Wadley, formerly with USDA; tested as LSB-88 by Sherman V. Thompson, Utah State Univ. Fruit: sweet; larger and firmer than Bing or Lambert; shape and stem length resemble Van; skin mahogany-colored; flesh dark red, sweet, firm, flavor outstanding; pit medium size, partially freestone; sets heavy in normal years, but susceptible to early frosts, often sets in large clusters; fruit doubling has not been observed; fruit splitting similar to Bing. Tree: only slightly damaged by severe winters of 1972 and 1978; blooms with Bing, 5–7 days before Angela or Star, self-sterile; partially resistant to western-X disease.

Valera.—Orig. in Vineland Station, Ontario, Canada, by G.H. Dickson, G.W. Eaton, C.D. Whitty, and G. Tehrani, Ontario Dept. of Agriculture, Hort. Res. Inst. Introd. in 1967. Hedelfingen x Windsor; selected in 1941; tested as V350427. Fruit: sweet; size medium; skin black, smooth; flesh dark, firm, sweet; quality good; ripens 13 July in Ontario, 1 day after Venus; recommended as a fresh variety; resembles Venus, but darker color and richer flavor. Tree: vigorous; bears early and consistently.

Van.—Orig. in Summerland, British Columbia, Canada, by A.J. Mann, Canada Dept. of Agriculture, Res. Sta. Introd. in 1944. Open-pollinated seedling of Empress Eugenie; cross made in 1936; selected in 1942. Fruit: Bing type; sweet; skin black, with bright luster; somewhat resistant to cracking, being less susceptible than Lambert; as large as Bing with comparable crop, but often smaller due to overbearing; slightly firmer and quite as good in quality as Bing; same season as Bing; stem short. Tree: vigorous; upright; heavy, annual bearer; hardier than Bing; a pollinator for Bing and Lambert; pollinated by Bing, Deacon, Lambert, and Napoleon. Widely grown and a parent of Ranier. Named in honor of J.R. Van Haarlem, Hort. Expt. Sta., Vineland, Ontario.

Vega.—Orig. in Vineland Station, Ontario, Canada, by G.H. Dickson, G.W. Eaton, C.D. Whitty, and G. Tehrani, Ontario Dept. of Agriculture, Hort. Res. Inst. Introd. in 1967. Bing x Victor; selected in 1937; tested as V31034; *Prunus avium*. Fruit: large; skin white, with red blush, smooth, very attractive; flesh white, slightly tart, firm; pit small, easily removed; ripens 11 July in Ontario, 5–8 days before Victor and Napoleon; recommended for brining and maraschino, too tart for dessert purposes until very ripe. Tree: upright; vigorous; hardy; very productive; brown rot has occasionally been a problem; can be pollinated by any sweet cherry recommended for Ontario. Introduced because variety is larger, firmer, and earlier than most white varieties.

Velvet.—Orig. in Vineland, Ontario, Canada, by Ontario Dept. of Agriculture, Hort. Res. Inst. Introd. in 1937. Open-pollinated seedling of Windsor; selected in 1925. Fruit: sweet; black; ripens late; resembles Schmidt. Tree: productivity low.

Venus.—Orig. in Vineland Station, Ontario, Canada, by Ontario Dept. of Agriculture, Hort. Res. Inst. Introd. in 1959. Hedelfingen x Windsor; selected in 1950; tested as Ontario 35042. Fruit: sweet; large; skin dark purplish-red, shiny; flesh red to black; quality high; not as firm as Vista nor Lambert and therefore less subject to cracking; ripens about 5 days after Vista and Black Tartarian. Tree: in heavy-crop years tendency to overset with reduction in fruit size, but under normal conditions for pollination, this does not occur often.

Vernon.—Orig. in Vineland, Ontario, Canada, at the Ontario Dept. of Agriculture, Hort. Res. Inst. Introd. in 1937. Open-pollinated seedling of Windsor; selected in 1925. Fruit: sweet; large; skin tough, reddish-black; flesh firm, meaty, texture fine;

quality good; brines very well; like Bing in susceptibility to cracking; ripens in midseason, with Bing; resembles Windsor. Tree: yields heavy crops.

Vic.—Orig. in Vineland Station, Ontario, Canada, by Ontario Dept. of Agriculture, Hort. Res. Inst. Introd. in 1959. Bing x Schmidt; selected in 1937; tested as Ontario 27026. Fruit: sweet; quality good; ripens with Windsor but of better size and darker color; satisfactory for freezing and maraschino purposes; superior to Windsor in canning tests. Tree: large; consistently heavy bearer. Released as a replacement for Windsor.

Victor.—Orig. in Vineland Station, Ontario, Canada, by Ontario Dept. of Agriculture, Hort. Res. Inst. Introd. in 1925. Open-pollinated seedling of Windsor; selected in 1923 from 1916 crosses. Fruit: sweet; skin white, heavily blushed, glossy, attractive; flesh creamy white, firm, meaty, subacid; processes well; ripens about 7–10 days before Napoleon, which it resembles. Tree: vigorous; productive; hardy, similar to Napoleon.

Viscount.—Orig. in Vineland Station, Ontario, by G.H. Dickson and G. Tehrani, Ontario Ministry of Agr. & Food, Hort. Res. Inst. of Ontario. Introd. in 1983. V 35024 (Hedelfingen x Bing) x V 35029 (Hedelfingen x Bing). Selected in 1960. Fruit: sweet; medium large; kidney-shaped; skin shiny black, attractive; resembles Bing; flesh red, firm, quality good, juice dark; ripens 16 July at Vineland, same as Bing. Tree: vigorous; spreading; productive; performed well on Mazzard and poorly on Mahaleb roots; pollination incompatible with Hudson and Rainier.

Vista.—Orig. in Vineland Station, Ontario, Canada, by Ontario Dept. of Agriculture, Hort. Res. Inst. Introd. in 1959. Hedelfingen x Victor; selected in 1947; tested as Ontario 35031. Fruit: sweet; large; skin dark with a very attractive finish; flesh firm; matures about with Black Tartarian; some cracking has occurred, but other factors outweigh it; released as a Black Tartarian replacement. Tree: vigorous; productive.

Viva.—Orig. in Vineland Station, Ontario, Canada, by G. Tehrani and G.H. Dickson, Ontario Ministry of Agr. and Food, Hort. Res. Inst. Introd. in 1972. Parentage unknown; selected in 1940; tested as V 35033; originally distributed as a selection from Hedelfingen x Victor; recent compatibility studies show Hedelfingen is not the seed parent; Viva belongs to incompatibility Group IV (S_2S_3). Fruit: sweet; size medium; skin dark red, tolerant to cracking; long stem; flesh semi-firm, sweet, quality good; ripens 9 July with Black Tartarian, 1 day after Vista. Tree: large, spreading; better shaped, better crop distribution, less susceptible to brown rot than Vista.

Vogue.—Orig. in Vineland Station, Ontario, by G. Tehrani and G.H. Dickson, Ontario Ministry of Agr. & Food, Hort. Res. Inst. Introd. in 1974. Parentage unknown; selected in 1940; tested as V35038; originally distributed as a selection from Hedelfingen x Victor; recent compatibility studies show Hedelfingen is not the seed parent of this cultivar; Vogue belongs to incompatibility Group IV (S_2S_3). Fruit: sweet; large; skin dark red, shiny, resistant to cracking; small pit; flesh firm, quality good; ripens 19 July, 1–2 days later than Bing and before Windsor. Tree: large; spreading; very productive, in heavy crop years tends to set in bunches and requires careful spraying for brown rot control.

Woehrman Red Giant *(Brummers Red Sweet)*.—Orig. in Hannibal, Mo., by John Brummer, Sims Fruit & Nursery Farm. Introd. in 1939. Parentage unknown. Selected in 1932. Fruit: sweet; large; skin bright red; ripens with Black Tartarian; resembles Bing. Tree: blooms late.

York Imperial.—Orig. in York, Pa., by H.B. Faber and J.A.C. Ziegler, Jr. Introd. in 1940. Plant patent 421; 17 Sept. 1940; assigned to White Rose Seed & Nursery Co., York, Pa. Napoleon x Schmidt *Prunus avium*. Fruit and tree: apparently similar to or identical with Napoleon.

Zaicrimson.—See Starkrimson.

not named.—Orig. in Walla Walla, Wash., by Percy A. Clarke. Plant patent 2572; 16 Nov. 1965; assigned to Stark Bros. Orchards & Nurseries Co., Louisiana, Mo. Bud mutation of Bing; discovered in 1927. Fruit and tree characteristics apparently same as Bing, essentially differentiated by its white, large, double flowers.

Rootstocks for Cherry

Camil ***(GM.79).***—Orig. at Kalmthout, Belgium, by P. Monin and R. Trefois, Fruit and Nut Res. Sta., Gembloux. Introd. in 1986. A *Prunus canescens* selection of unknown provenance. Original selection in 1964; tested as 3-8-K. Plant patent 5802; 18 Nov. 1986; assigned to TRECO, Woodburn, Ore. A semi-dwarfing rootstock for sweet cherries and tart cherries; tree size is reduced by about one-third compared to F12/1. Induces fruiting of sweet cherries earlier than does mazzard; Montmorency fruiting begins 1 year later than trees on mahaleb. Very susceptible to *Phytophthora cambivora, P. dreschleri,* and *P. megasperma.* Resistant to cherry leafspot. Readily propagated by greenwood cuttings and in micropropagation. Ungrafted tree is moderately vigorous, drooping habit.

Charger.—Orig. at East Malling, Kent, U.K., by H.M. Tydeman, East Malling Res. Sta. Selected from a population of seedlings of F1/3 selfed (*Prunus avium*); tested as Fd 1/57/4/13. Introd. in 1982. Productive in layer bed. Resistant to bacterial canker. Trees on Charger are standard size.

Cob.—Orig. at East Malling, Kent, U.K., by H.M. Tydeman, East Malling Res. Sta. *Prunus avium* x *P. pseudocerasus* Tested as Fb2/58/17. An easily propagated, vigorous clonal rootstock, used for *P. serrulata* and other ornamental cherries. Not recommended for sweet cherries.

Colt.—Orig. in Kent, England, by H.M. Tydeman, deceased, late of East Malling, England. Plant patent 4059; 14 June 1977; assigned to National Seed Development Organization, Newton, England. *Prunus avium* F 299/2 x *Prunus pseudocerasus* Lind; cross made in 1958; selected in 1966–67. Vigorous, sturdy, erect shoots with large, bright-green leaves; stems greenish. When used as understock, induces some dwarfing and higher fruit bud production; graft compatible with most cherry varieties. Relatively winter-tender. Preformed root initials make propagation by cuttings very effective.

Damil ***(GM.61/1).***—Orig. at Gembloux, Belgium, by P. Monin and R. Trefois, Fruit and Nut Res. Sta. Selected from open-pollinated population of *Prunus dawykensis*; tested as 4-1-K. Introd. in 1986. Plant patent 5803; 18 Nov. 1986; assigned to TRECO, Woodburn, Ore. Semi-dwarfing rootstock for sweet cherries and tart cherries, compatible with wide range of cultivars. Reduces tree size by about one-half. Induces fruiting of some sweet cherry varieties earlier than does mazzard; for Montmorency, not as precocious as mahaleb. Compared to mahaleb, more tolerant than mazzard to *Phytophthora cambivora* and *P. megasperma*; relatively tolerant of waterlogging. Resistant to cherry leafspot. Susceptible to the mycoplasma causing Western X-disease. Readily propagated by greenwood cuttings and in micropropagation. Nonsuckering; well-anchored. Ungrafted tree is of moderate vigor, semi-upright.

Dropmore.—Orig. in Dropmore, Manitoba, Canada, by F.L. Skinner, Skinner's Nursery. Introd. in 1952. F_2 [Kozlov Morello (*P. cerasus*) x *P. pennsylvanica*], the F_1 hybrid sometimes called *P. dropmoreana.* Fruit: 0.5 to 0.75 inch in diam.; round; skin deep maroon to black; flesh greenish-yellow; flavor bitter, nonedible. Bush: 10–12 feet tall; extremely hardy; normally propagated from seed; useful as rootstock for dwarfing cherry and sour cherry. Propagated by cuttings.

Dwarf Siberian.—Orig. in McMinnville, Ore., by Homer R. Brixey, Brixey Experimental Nursery. Introd. in 1964. Plant patent 2314; 26 Nov. 1963. Possibly a hybrid of *Prunus subhirtella* and *P. serrulata*; seeds secured in 1948 in Shantung Peninsula, near Tsingtao, China, by Glenn Brixey; seeds grown and tested in Oregon. For rootstock purposes only: a dwarfing understock; found to be compatible with Bing, Lambert, Republican, Royal Ann (Napoleon), and Van sweet cherry, with Montmorency sour cherry, with Late Duke and Royal Duke, with Kwanzan and Shirofugen flowering cherry, with mazzard and mahaleb rootstocks; hardy as mazzard; propagated by stooling, rooted cuttings, and double working.

Gisela®1.—Orig. at Giessen, Germany, by W. Gruppe and H. Schmidt, Justus Liebig Univ., Giessen. Selected from a population of hybrid seedlings of *Prunus fruticosa* 'Klon

64′ x *Prunus avium*. Tested as GI® 172-9. Propagation rights assigned to Inter-plant Patent Marketing, Inc., Niagara-on-the-lake, Ont., Canada. A triploid; dwarfing rootstock for sweet and sour cherries; finished, grafted tree size is about 20% compared to F12/1 Mazzard. Induces exceptionally early fruiting of scions of sweet cherries compared to mazzard. Anchorage is poor so tree must be supported; small tree amenable to pot culture; exceptionally high yield efficiency, but heavy cropping can reduce fruit size; requires close management to retain good vigor and fruit size. Sensitive to pollen-born viruses like Prunus necrotic ringspot virus and Prune dwarf virus and possibly others. Readily propagated by greenwood cuttings and micropropagation. Ungrafted tree is somewhat similar to *Prunus fruticosa*.

Gisela®4.—Orig. at Giessen, Germany, by W. Gruppe and H. Schmidt, Justus Liebig Univ., Giessen. Selected from a population of hybrid seedlings of *Prunus avium* x *Prunus fruticosa*. Orig. tested as GI® 473-10. Propagation rights assigned to Interplant Patent Marketing, Inc., Niagara-on-the-lake, Ont., Canada. A triploid; semi-dwarfing rootstock for sweet and sour cherries; finished, grafted tree size is about 35% compared to F12/1 Mazzard. Induces exceptionally heavy, early fruiting of scions of sweet cherries, earlier than mazzard. Sensitive to pollen-born viruses unknown. Propagated by greenwood cuttings and one of the easiest stocks to propagate by microtechniques. Grafted trees are nonsuckering and defoliate early in the autumn. Ungrafted tree is similar to *Prunus fruticosa*.

Gisela®5.—Orig. at Giessen, Germany, by W. Gruppe and H. Schmidt, Justus Liebig Univ., Giessen. Selected from a population of hybrid seedlings of *Prunus cerasus* 'Shattenmorelle' x *Prunus canescens*. Tested as GI® 148-2. Plant patent 9622; 13 Aug. 1996; assigned to Inter-plant Patent Marketing, Inc., Niagara-on-the -lake, Ont., Canada. A triploid; semi-dwarfing rootstock for sweet and sour cherries; finished, grafted tree size is about 50% compared to F12/1 Mazzard. Induces fruiting of scions of sweet cherries earlier than mazzard; wide-angled branching and greater lateral branching are induced in all scion cultivars; anchorage less than Gisela®6 and Gisela®7. Propagated by greenwood cuttings and in micropropagation. Nursery liner management needs intense care. Grafted trees are nonsuckering and defoliate early in the autumn. Ungrafted tree is similar to *Prunus canescens*.

Gisela®6.—Orig. at Giessen, Germany, by W. Gruppe and H. Schmidt, Justus Liebig Univ., Giessen. Selected from a population of hybrid seedlings of *Prunus cerasus* 'Shattenmorelle' x *Prunus canescens*. Tested as GI® 148–1. Plant patent 8954; 25 Oct. 1994; assigned to Inter-plant Patent Marketing, Inc., Niagara-on-the-lake, Ont., Canada. A triploid; semi-dwarfing rootstock for sweet and sour cherries; finished, grafted tree size ranges from about 60% to 95% compared to F12/1 Mazzard, depending on scion cultivar and orchard environment. Induces very early fruiting of scions of sweet cherries, much earlier than mazzard; wide-angled branching and greater lateral branching are induced in all scion cultivars. Readily propagated by greenwood cuttings and in micropropagation. Grafted trees are nonsuckering and defoliate early in the autumn. Ungrafted tree is similar to *Prunus canescens*.

Gisela®7.—Orig. at Giessen, Germany, by W. Gruppe and H. Schmidt, Justus Liebig Univ., Giessen. Selected from a population of hybrid seedlings of *Prunus cerasus* 'Shattenmorelle' x *Prunus canescens*. Tested as GI® 148–8. Plant patent 8852; 2 Aug. 1994; assigned to Inter-plant Patent Marketing, Inc., Niagara-on-the-lake, Ont., Canada. A semi-dwarfing rootstock for sweet and sour cherries; finished, grafted tree size is about 50% compared to F12/1 Mazzard. Induces very early fruiting of scions of sweet cherries much earlier than mazzard; wide-angled branching and greater lateral branching are induced in all scion cultivars. Somewhat more sensitive to Prunus necrotic ringspot virus than Gisela®5 and Gisela®6. Readily propagated by greenwood cuttings and in micropropagation. Grafted trees sucker moderately and defoliate early in the autumn. Ungrafted tree is similar to *Prunus canescens*.

Gisela®8.—Orig. at Giessen, Germany, by W. Gruppe and H. Schmidt, Justus Liebig Univ., Giessen. Selected from a population of hybrid seedlings of *Prunus cerasus* 'Shattenmorelle' x *Prunus canescens*. Tested as GI® 148–9. Plant patent 9623; 13 Aug. 1996; assigned to Inter-plant Patent Marketing, Inc., Niagara-on-the-lake, Ont., Canada. A semi-dwarfing rootstock for sweet and sour cherries; grafted tree size is about 40% compared to F12/1 Mazzard. Induces very early fruiting of scions of sweet cherries, much earlier than mazzard; wide-angled branching and greater lateral branching are induced in all scion cultivars. Somewhat more sensitive to Prunus necrotic ringspot virus than Gisela®5 and Gisela®6. Anchorage is not as strong as Gisela®6 or Gisela ®7. Readily propagated by greenwood cuttings and in micropropagation. Grafted trees sucker moderately and defoliate early in the autumn. Ungrafted tree is similar to *Prunus canescens.*

Gisela®10.—Orig. at Giessen, Germany, by W. Gruppe and H. Schmidt, Justus Liebig Univ., Giessen. Selected from a population of hybrid seedlings of *Prunus fruticosa* 'Klon 64' x *Prunus cerasus* 'Schattenmorelle'. Tested as GI® 173–9. Propagation rights assigned to Inter-plant Patent Marketing, Inc., Niagara-on-the-lake, Ont., Canada. A triploid; dwarfing rootstock for sweet and sour cherries; grafted tree size is about 60% compared to F12/1 Mazzard. Induces exceptionally early fruiting of scions of sweet cherries compared to mazzard. Montmorency fruiting is better on a per tree efficiency basis than on common mahaleb and fruiting begins sooner than trees on mahaleb. Sensitive to pollen-born viruses like Prunus necrotic ringspot virus, Prune dwarf virus, and possibly others. Readily propagated by greenwood cuttings and in micropropagation. Grafted trees are quite adaptable to different soil types; sucker profusely; winter injury to roots and ground level crown area severe in some climates.

Gisela®11.—Orig. at Giessen, Germany, by W. Gruppe and H. Schmidt, Justus Liebig Univ., Giessen. Selected from a population of hybrid seedlings of *Prunus canescens* x *Prunus cerasus* 'Leitzkauer'. Tested as GI® 195–1. Plant patent 9630; 20 Aug. 1996; assigned to Inter-plant Patent Marketing, Inc., Niagara-on-the-lake, Ont., Canada. A triploid; semi-dwarfing rootstock for sweet and sour cherries; grafted tree size is about 60% to 70% compared to F12/1 Mazzard. Induces fruiting of scions of sweet cherries earlier than mazzard; Montmorency fruiting begins earlier than trees on mahaleb. Sensitive to Prunus necrotic ringspot virus and Prune dwarf virus. Susceptible to crown rot (*Phytophthora cactorum*) but less so than common mahaleb stocks used in the U.S. Readily propagated by greenwood cuttings and in micropropagation. Grafted trees are nonsuckering; defoliate early in the autumn; ungrafted tree is similar to *Prunus canescens.*

Gisela®12.—Orig. at Giessen, Germany, by W. Gruppe and H. Schmidt, Justus Liebig Univ., Giessen. Selected from a population of hybrid seedlings of *Prunus canescens* x *Prunus cerasus* 'Leitzkauer'. Tested as GI® 195–2. Plant patent 9631; 20 Aug. 1996; assigned to Inter-plant Patent Marketing, Inc., Niagara-on-the-lake, Ont., Canada. A triploid; semi-dwarfing rootstock for sweet and sour cherries; grafted tree size is about 60% compared to F12/1 Mazzard. Induces fruiting of scions of sweet cherries earlier than mazzard; Montmorency fruiting begins earlier than trees on mahaleb. Readily propagated by greenwood cuttings and in micropropagation. Grafted trees are nonsuckering; defoliate early in the autumn; ungrafted tree is similar to *Prunus canescens.*

GM.9.—See Inmil.

GM.61/1.—See Damil.

GM.79.—See Camil.

Inmil *(GM.9).*—Orig. at Kalmthout, Belgium, by P. Monin and R. Trefois, Fruit and Nut Res. Sta. A *Prunus incisa* x *P. serrulata* clone of unknown provenance; tested as 9-14-K. Introd. in 1986. Plant patent 5801; 18 Nov. 1986; assigned to TRECO, Woodburn, Ore. Dwarfing rootstock for sweet and tart cherries, compatible with wide range of cultivars. Reduces tree size by 75% compared to F12/1. For some sweet cherry varieties, induces fruiting earlier than does mazzard; under Montmorency, as precocious as mahaleb.

Susceptible to *Phytophthora cambivora* and *P. megasperma*. Ungrafted tree is dwarfish, semi-upright.

MxM 2 (*Brooks-2*).—Orig. at Forest Grove, Ore., by Lyle Brooks. Selected from a population of mahaleb open-pollinated seed; vegetative attributes indicate pollen parent was mazzard, hence the "Mahaleb x Mazzard" designation and its "MxM" diminutive. Tested as MxM 2. Plant patent 8194; 6 Apr. 1993; assigned to Inter-plant Patent Marketing, Inc., Niagara-on-the-lake, Ont., Canada. A vigorous rootstock that is somewhat stronger growing than mazzard and has high compatibility for both sour and sweet cherries; earlier bearing than mazzard in some orchard environments; high production capacity. Propagated by hardwood cuttings. Resistant to *Phytophthora cambivora* and *P. megasperma* and somewhat tolerant to bacterial canker (*Pseudomonas syringae*).

MxM 39.—Orig. at Forest Grove, Ore., by Lyle Brooks. Selected from a population of mahaleb open-pollinated seed; vegetative attributes indicate pollen parent was mazzard, hence the "Mahaleb x Mazzard" designation and its "MxM" diminutive. Tested as MxM 39; assigned to Inter-plant Patent Marketing, Inc., Niagara-on-the-lake, Ont., Canada. A semi-dwarfing rootstock for sour (about 10% to 15% less than common mahaleb seedling rootstocks used in the U.S.) and for some sweet cherry cultivars; highly precocious and productive; excellent resistance to *Phytophthora cambivora* and *P. megasperma* and somewhat tolerant to bacterial canker (*Pseudomonas syringae*).

MxM 60 (*Brooks-60*).—Orig. at Forest Grove, Ore., by Lyle Brooks. Selected from a population of mahaleb open-pollinated seed; vegetative attributes indicate pollen parent was mazzard, hence the "Mahaleb x Mazzard" designation and its "MxM" diminutive. Tested as MxM 60. Plant patent 8132; 9 Feb. 1993; assigned to Inter-plant Patent Marketing, Inc., Niagara-on-the-lake, Ont., Canada. A slightly dwarfing rootstock, about 90% of mazzard; more precocious and better yielding than mazzard. Adaptable to diverse soils with good resistance to *Phytophthora cambivora* and *P. megasperma* and somewhat tolerant to bacterial canker (*Pseudomonas syringae*).

Mahaleb 900 (*Mh. 900*).—Orig. in Prosser, Wash., by Earle C. Blodgett, Irrigation Expt. Sta., Washington Agr. Expt. Sta. Introd. in 1946. The variety represents a mixture of clones selected for freedom of virus diseases, good vigor, and satisfactory seed sources. The original stocks were obtained as seed or seedlings from E.L. Reeves and the U.S. Dept. of Agriculture Soil Conservation Service in Washington. Fruit and tree follow description of the species *Prunus mahaleb*. Seed from these foundation trees and their progeny is widely disseminated for cherry rootstocks.

Cherry Plum

Alace.—Orig. in Brooks, Alberta, Canada, by Canada Dept. of Agriculture, Res. Sta. Introd. in 1942. Open-pollinated seedling of Opata; tested as Brooks 24. Fruit: slightly smaller than Opata; skin dark purplish; flesh greenish; resembles known parent; ripens midseason. Plant: bushy, low, spreading; vigorous; hardy; grows well in the Peach River area.

Beta.—Orig. in Saskatoon, Saskatchewan, Canada, by C.F. Patterson, Univ. of Sask. Introd. ln 1960 for home gardens. Open-pollinated seedling of Opata; selected in 1948; tested as C59-352. Fruit: slightly oval; skin thin, reddish-blue; flesh greenish, moderately firm, very mildly acid; quality good; freestone; resembles Opata. Tree: hardy.

Clark Hill Redleaf.—Collected by J.M. Thompson. and W. Knight in 1973 or 1974 near Augusta, Ga., at an old homesite on the Georgia side of the Savannah River about 1 km from Clarks Hill Dam. Introd. in 1989. Probably a seedling of an ornamental plum; resembles a redleaf *Prunus cerasifera*. Fruit: 20–30 mm, round; skin red with yellow to yellow-red flesh; quality insipid. Tree: vigorous; tolerant to high levels of calcium; tolerant to bacterial spot and bacterial canker; moderate resistance to plum leaf scald. Bloom is later than most Japanese plums; flowers light pink; leaves deep red, fading to purple-green during summer.

Convoy.—Orig. in Valley River, Manitoba, Canada, by W.J. Boughen, Boughen Nurseries. Introd. in 1943. Seedling of a sandcherry hybrid; selected in 1941. Fruit: somewhat larger than Compass; skin scarlet red; flesh yellowish; little harsh for dessert, good for canning; season end of August. Tree: vigorous; very upright, narrow; hardy; productive.

Cooper.—Orig. in Fargo, N.D., by A.F. Yeager, North Dakota Agr. Expt. Sta. Introd. in 1935. Open-pollinated seedling of F_1 (*Prunus besseyi* 'Compass' x *P. hortulana* var. *mineri*); selected about 1930. Fruit: larger, skin darker than Compass; round; flesh green; flavor mild, good; makes a better canned product than some other cherry plums, being similar to a white-fleshed sweet cherry, such as Napoleon. Tree: not too productive.

Deep Purple.—Orig. in Excelsior, Minn., by E.T. Anderson, Univ. of Minnesota Fruit Breeding Farm. Introd. in 1965. *Prunus besseyi* 'Sioux' x *P. salicina* 'Elephant Heart'; cross made in 1942; first fruited in 1947; tested as Minn. 440. Fruit: large, 1 1/2 inches long and 1 1/4 inches in diam.; oval conic, apical point pronounced; suture a faint line; skin deep purple to black, medium thick, tender; flesh dark rich purple, firm, meaty, medium juicy; pit small, flat, nearly free; flavor sweet, pleasant; ripens in mid-August; remains mature on tree in good condition up to 2 weeks; useful mostly for processing, compares favorably with Sapa for jam; more productive than Sapa. Tree: can be trained to produce a tree, but develops naturally into a wide-spreading bush; more vigorous than Sapa; very precocious; requires cross-pollination, such as Compass, as well as sandcherry seedlings; very hardy.

Delta.—Orig. in Saskatoon, Saskatchewan, Canada, by C.F. Patterson, Univ. of Sask. Introd. in 1960 for home gardens. Derived from Sapa and Oka; selected in 1950; tested as C59-354. Fruit: spherical; skin bluish-black; flesh dark red; very mildly acid; quality good.

Dura.—Orig. in Morden, Manitoba, Canada, by Canada Dept. of Agriculture, Res. Sta. Introd. in 1940. Open-pollinated seedling of Sapa; selected in 1936; tested as Morden 116. Fruit: oblong; skin dark green, mottled purplish; flesh maroon-purple, crisp, tender, meaty; sweeter and less clingstone than Sapa; quality very good as dessert or canned; season late August to October. Bush: low, spreading; very hardy, productive.

Eileen.—Orig. in Dropmore, Manitoba, Canada, by F.L. Skinner, Skinner's Nursery. Introd. in 1939. *Prunus besseyi* x *P. tomentosa*. Fruit: small; used for jelly. Primarily an ornamental variety.

Epsilon.—Orig. in Saskatoon, Saskatchewan, Canada, by C.F. Patterson, Univ. of Sask. Introd. in 1959 for home gardens. Derived from Sapa and Oka; selected in 1950; tested as C59-351. Fruit: spherical; skin bluish-black; flesh dark red, firm, sweetish; good quality.

Gamma.—Orig. in Saskatoon, Saskatchewan, Canada, by C.F. Patterson, Univ. of Sask. Introd. in 1960 for home gardens. Derived from Sapa and Oka; selected in 1950; tested as C59-353. Fruit: slightly oblate; skin slmost black; flesh very dark red, sweet; quality good; resembles Sapa.

Genevieve.—Orig. in Hampton, Iowa, by Earl Ferris Nursery Co. Introd. in 1926. Chance seedling. Fruit: large; nearly round; skin red, attractive; flesh flavor good. Tree: hardy; productive; blooms in late spring.

Giant.—Orig. in Milford, Conn., by De Ross and Duane Kellogg. Introd. in 1946. Plant patent 793; 23 Mar. 1948; by De Ross Kellogg; assigned to Stark Bros. Nurseries & Orchards Co., Louisiana, Mo. Chance seedling; discovered in 1940. Fruit: 1 1/2 inches in diam.; skin solid red, medium thick, medium tough; flesh yellow, red next to skin, firm, subacid; quality good; clingstone; suitable for market, culinary, and canning purposes; ships well; ripens 3 days before Compass, which it resembles. Tree: hardy; vigorous; produces regularly.

Heaver.—Orig. in Baljennie, Saskatchewan, Canada, by K.N. Heaver. Introd. in 1944. Open-pollinated seedling of Opata. Fruit: large, 1 1/2 x 1 3/8 inches in diam.; roundish; skin purplish-black; flesh greenish, firmer than Opata, juicy; sweet; quality good as dessert, very good when canned; season mid-August. Bush: vigorous; low, spreading; hardy; productive.

Hiawatha.—Orig. in Mandan, N.D., by W.P. Baird, Northern Great Plains Field Sta. Introd. in 1957. Open-pollinated seedling of Sapa; selected in 1942; tested as Mandan 42-1. Fruit: fairly large, averaging about 35 per lb; round; skin thin, mottled purple; flesh medium firm, juicy, fairly sweet; semi-freestone; usually ripens during mid-August; rated first in its class in taste tests, being very good for culinary uses, especially jam. Bush: fairly vigorous, spreading; like Sapa, it suffers severe winter injury in some seasons, but makes a good recovery. A superior variety to Sapa because of its better quality and larger size.

Honey Dew.—Orig. in Brookings, S.D., by South Dakota Agr. Expt. Sta. Introd. in 1950. *Prunus besseyi* x (*P. americana* x *P. salicina*) Shiro; discovered in 1930; tested as S. Dak. E 155. Fruit: ripens early; high quality; resembles Opata. Tree: small; bush-like; hardy.

Kappa.—Orig. in Saskatoon, Saskatchewan, Canada, by C.F. Patterson, Univ. of Saskatchewan. Introd. in 1960 for home gardens. Derived from Sapa and Oka; selected in 1950; tested as C59-356. Fruit: spherical; large; skin bluish-black; flesh dark red; sweet; excellent quality.

Little Eagles.—Orig. in Malaga, Wash., by David Foraker, Wenatchee. Introd. in 1962. Parentage unknown. Fruit: large; round; skin yellow; flesh yellow; flavor mild, but slightly sour at pit; stem 1 inch long and curved; ripens in August; for jams and nectar; flavor and color similar to ground cherry. Tree: large; upright; bushy; hardy; tolerant to insects and disease prevalent to area.

Mana.—Orig. in Brookings, S.D., by N.E. Hansen, South Dakota Agr. Expt. Sta. Introd. in 1933. *Prunus besseyi* x *P. salicina* 'Burbank'. Fruit: about 1 inch in diam.; round; flesh greenish-yellow; quality good; makes a well-flavored, red sauce. Now obsolete.

Manor.—Orig. in Morden, Manitoba, Canada, by Canada Dept. of Agriculture, Res. Sta. Introd. in 1945. Open-pollinated seedling of Sapa; selected in 1937; tested as Morden 119. Fruit: roundish; skin purplish; flesh dark red to purplish, meaty, tender, juicy, sweet; good as dessert, fair for canning; season mid-August. Bush: hardier and more upright than Sapa; particularly adapted to northern prairies.

Manorette.—Orig. in Moose Range, Saskatchewan, Canada, by Percy M. Wright, Carrot Valley Nursery, Sutherland, Sask. Introd. in 1964. Open-pollinated seedling of Manor; selected in 1952. Fruit: high quality; flavor mild, not astringent. Bush: high yield; moderately resistant to mildew.

Mansan.—Orig. in Morden, Manitoba, Canada, by Canada Dept. of Agriculture, Res. Sta. Introd. in 1935. *Prunus besseyi* x *P. hortulana*. Fruit: small; oval; skin dark carmine with heavy bloom; flesh greenish-yellow, firm; mildly subacid, pleasant; quality good for canning; season early September. Tree: resembles Compass but hardier; blooms late; to be productive requires a late-blooming variety for cross-pollination.

Mistawasis.—Orig. in Parkside, Saskatchewan, Canada, by A.J. Porter, Honeywood Nursery. Introd. in 1972. *Prunus salicina* x *Prunus besseyi*. Discovered in 1965. Fruit: to 1 1/4 inches in diam., slightly oblate; skin purple, thin; flesh red, moderately firm, flavor pleasant; ripens early September. Bush: 8–12 ft; upright-spreading; vigorous; uninjured at –50F; productivity moderate. Recommended for prairie regions of U.S. and Canada.

Mordena.—Orig. in Morden, Manitoba, Canada, by Canada Dept. of Agriculture, Res. Sta. Introd. in 1930. Open-pollinated seedling of Compass. Fruit: small; skin red to dark purplish-red; flesh yellow, juicy, sweet, sprightly; quality good for canning, excellent for jelly; season early September. Tree: spreading; vigorous; hardy; productive; late blooming.

Nicollet.—Orig. in Excelsior, Minn., by the Univ. of Minnesota Fruit Breeding Farm. Named and introd. in 1925. (*Prunus avium* x *P. pennsylvanica*) x *P. besseyi*; seed planted in 1912; selected in 1916 or 1917; tested as Minn. 144. Fruit: small; round to round oval; stem short, slender; skin red, thin, medium tough; flesh greenish-yellow, medium firm, tender, juicy; sour-cherry flavor; quality fair; pit small, typically cherry-like, freestone; ripens in midseason, last half of August at St. Paul. Tree: dwarf, bush-like, 8 ft tall; vigorous; productive; hardy; finely branched; leaves small.

Oka.—Orig. in Brookings, S.D., by N.E. Hansen, South Dakota Agr. Expt. Sta. Introd. in 1924. Open-pollinated seedling of Champa. Fruit: about 1 inch in diam.; skin tough, thick, keep purplish-red; flesh dark red, juicy, firm; quality good; season from mid-August to early September; fruit remains in edible condition on tree for 3 weeks after ripening; resembles Compass. Tree: hardy; dwarf; bears well.

Omega.—Orig. in Saskatoon, Saskatchewan, Canada, by C.F. Patterson, Univ. of Sask. Introd. in 1960 for home gardens. Derived from Sapa and Oka; selected in 1950; tested as C59-358. Fruit: spherical; skin bluish-black, very thin; flesh dark red, sweet; quality good; ripens early.

Prolific.—Orig. in Rosthern, Saskatchewan, Canada, by Seager Wheeler. Introd. in 1935. Open-pollinated seedling of Opata; discovered about 1930. Fruit: 1 to 1 1/8 inches in diam.; obovate, irregular; skin dark purple with light gray bloom; flesh greenish, firm, fine, juicy, mildly subacid; quality good; season mid-August; resembles Sapa. Bush: very hardy; produces good crops; less vigorous than Opata; for home gardens.

Queensland Earlisweet.—Orig. in Applethorpe, Queensland, Australia, by C.W. Winks, A.M. Hibberd, and B.L. Topp, Granite Belt Horticulture Res. Sta. Introd. in 1988. Early Jewel x Early Gem (both parents have *Prunus cerasifera* and *P. salicina* in pedigree); selected in 1981; tested as GB 3-91. Fruit: small, 32 mm; round; skin color bright red, becoming dark red at full maturity; flesh yellow, melting, sweet, juicy; flavor balanced with slightly acid skin; pit small; clingstone. Ripens 2 weeks before Wilson and Methley, 5 weeks before Santa Rosa. Tree: very vigorous; shape intermediate between upright and spreading; moderately productive. Flowering is early, 10 days before Wilson; pollen sterile; Mariposa and Donsworth are suitable pollinizers. Good field resistance to bacterial spot (*Xanthamonas campestris* pv. *pruni*).

Red Cortland.—Orig. in Blaney Park, Mich., by Joseph E. Lang. Introd. in 1934. Sapa x Superior; selected in 1933. Fruit: larger than Sapa; skin purplish-black; flesh reddish-purple; quality good. Tree: strong, rapid grower; productive; only moderately hardy.

Ruby.—Orig. in Rosthern, Saskatchewan, Canada, by Seager Wheeler. Introd. in 1935. Open-pollinated seedling of Compass; discovered about 1930. Fruit: skin red; of the Compass type, but rates lower in quality. Bush: very hardy; produces good crops; not vigorous; for home gardens.

Sacagawea.—Orig. in Mandan, N.D., by W.P. Baird, Northern Great Plains Field Sta. Introd. in 1957. Open-pollinated seedling of Sapa; selected in 1942; tested as Mandan 42-8. Fruit: averages about 40 per lb, being similar to Sapa; round; skin mottled purple; flesh dark red resembling known parent, medium firm, juicy, fairly sweet, partly freestone; usually matures about mid-August; rates high in cooking tests, makes excellent jam. Bush: Sometimes considered to be the hardiest variety of its class; very spreading.

St. Anthony.—Orig. in Excelsior, Minn., by the Univ. of Minnesota Fruit Breeding Farm. Named and introd. in 1923. *Prunus besseyi* x *P. salicina* Satsuma; selected in 1915; tested as Minn. 145. Fruit: large; round, slightly flattened at base; cavity large, flaring, deep; stem slender; skin dull purplish-red, large russeted spots, bloom heavy, medium thick, only slightly tough; flesh dark rich red, fine-grained, juicy, tender, medium firm, subacid; quality fair to good, especially fine for culinary use; stone size medium, semifree; ripens in mid-season; resembles Sapa. Tree: small, upright-spreading; vigorous; early and prolific bearing habit; hardy.

Sanoba.—Orig. in Brookings, S.D., by N.E. Hansen, South Dakota Agr. Expt. Sta. Introd. in 1929. Sapa x *Prunus besseyi* Bailey. Fruit: small, less than 1 inch in diam.; flesh red; quality good; pit round, small. Plant: upright; productive. Now obsolete. The name is coined from Sapa and Manitoba.

Sapalta *(Brooks Sapa, Sandhill Sapa, Sapa Seedling)*.—Orig. in Brooks, Alberta, Canada, by the Canadian Pacific Railway. Introd. in 1941. One of 1000 seedlings from the Northwest Nursery Co., Valley City, N.D.; probably a seedling of Sapa; name is a contraction of Alberta Sapa; selected in 1934. Fruit: similar to Sapa but with tendency toward freestone condition; smaller seed with slight ridge on each side. Bush: difficult to distinguish from Sapa.

Sigma.—Orig. in Saskatoon, Saskatchewan, Canada, by C.F. Patterson, Univ. of Saskatchewan. Introd. in 1960 for home gardens. Derived from Sapa and Oka; selected in 1950; tested as C59-357. Fruit: spherical; skin dark bluish; flesh dark red, sweetish, of very fair quality. Plant: more upright than some other varieties.

Skinner Favourite.—Orig. in Dropmore, Manitoba, Canada, by F.L. Skinner, Skinner's Nursery. Introd. in 1956. A yellow-fruited sandcherry (*Prunus besseyi* Bailey) x *P. salicina* Lindl. Fruit: size medium; oblong; skin with a purple-red cheek; flesh greenish-yellow; quality good for dessert use, excellent for jam. Tree: upright, averages 5–6 ft high; bushy.

Wessex.—Orig. in Moose Range, Saskatchewan, Canada, by Percy H. Wright, Carrot Valley Nursery, Sutherland, Sask. Introd. in 1964. Open-pollinated seedling of (*Prunus besseyi* x Mordel plum); selected in 1950. Fruit: quality good; dries like a prune. Bush: extremely hardy; highly productive; blooms late, escaping spring frosts.

Zeta.—Orig. in Saskatoon, Saskatchewan, Canada, by C.F. Patterson, Univ. of Sask. Introd. in 1960 for home gardens. Derived from Sapa and Oka; selected in 1950; tested as C59-355. Fruit: spherical; skin bluish; flesh dark red, very mildly acid; quality good.

Zumbra.—Orig. in Excelsior, Minn., by the Univ. of Minnesota Fruit Breeding Farm. Named and introd. in 1920. (*Prunus avium* x *P. pennsylvanica*) x *P. besseyi*; selected in 1916. Fruit: up to 1 1/8 inches in diam.; round to slightly ovate; cavity medium to large, flaring; skin medium thick, medium tough, nonastringent, dark reddish-black, attractive; flesh greenish-yellow, firm, crisp, medium juicy, mildly subacid, pleasantly flavored; quality good, excellent for cooking; stone small, semi-free to free; ripens second half of August. Tree: very vigorous; dwarf, bushy; hardy in northern Minnesota and southern Manitoba, Canada; blooms late, needs cross-pollination (e.g., Compass, Nicollet); a worthwhile ornamental.

Chestnut

Chestnut Species

Section *Castanea* (three nuts per bur)	
Castanea dentata (Marshall) Borkhausen	American chestnut
Castanea sativa Miller	European chestnut
Castanea mollissima Blume	Chinese chestnut
Castanea crenata Siebold and Zuccarini	Japanese chestnut
Castanea seguinii Dode	Chinese dwarf chestnut
Section *Balanocastanon* (one nut per bur)	
Castanea pumila (Linnaeus) Miller	
variety *pumila*	Allegheny Chinquapin
variety *ozarkensis*	Ozark Chinquapin
Section *Hypocastanon* (one nut per bur)	
Castanea henryi (Skan) Rehder & Wilson	Chinese Timber Chinquapin, Henry Chinquapin

Abundance.—Orig. in Eagle Creek, Ore., by Carroll D. Bush. Introd. in 1941. Parentage: scions were cut from the most vigorous seedlings grown from imported seed of *Castanea mollissima*. Nut: appearance attractive; cleans well; sweet; larger and more prolific than Honan, which it resembles.

Alaling.—Orig. in Auburn, Ala., by the Alabama Agr. Expt. Sta. Introd. in 1952. Open-pollinated seedling of *Castanea mollissima*; selected in 1943. Nut: medium; quality good; mahogany; matures late; keeps well in storage; drops freely from bur. Tree: blight resistant; vigorous.

Alamoore.—Orig. in Auburn, Ala., by the Alabama Agr. Expt. Sta. Introd. in 1952. *Castanea arenata* x *C. pumila*; selected in 1943. Nut: medium; quality excellent, flavor sweet; brown; keeps well in storage. Tree: blight resistant; very prolific; bears early.

Armstrong.—The original Chinese tree grew at the Univ. of Kentucky, and this cultivar is named after the university's plant breeder. Tree: consistent producer of good-quality nuts that ripen the first week of October in the southern U.S.

AU-Cropper.—Orig. at the Auburn Univ.–U.S. Dept. of Agriculture Hillculture Farm, Auburn, Ala., by J.D. Norton. Evaluated as 35A-4-4. Nut: 11 g, dark chocolate-brown, two or three nuts per bur. Midseason maturity; bur opens well. Tree: Chinese type, high yielding; resistant to chestnut blight.

AU-Homestead.—Orig. at the Auburn Univ.–U.S. Dept. of Agriculture Hillculture Farm, Auburn, Ala., by J.D. Norton. Evaluated as 54-26. Nut: 11.3 g, very dark chocolate-brown, two or three nuts per bur. Late uniform maturity; bur opens fairly well. Tree: Chinese type, high yielding; resistant to chestnut blight.

AU-Leader.—Orig. at the Auburn Univ.–U.S. Dept. of Agriculture Hillculture Farm, Auburn, Ala., by J.D. Norton. Evaluated as 54-13. Nut: 13 g, dark chocolate-brown, two or three nuts per bur. Midseason maturity; bur opens well. Tree: Chinese type, high yielding; resistant to chestnut blight.

AU-17.—Germplasm release orig. at the Auburn Univ.–U.S. Dept. of Agriculture Hillculture Farm, Auburn, Ala., by J.D. Norton. Nut: 13 g, medium brown, two or three nuts per bur. Midseason maturity; bur opens well. Tree: Chinese type, high yielding; resistant to chestnut blight and chestnut gall wasp.

AU-54-60.—Germplasm release orig. at the Auburn Univ.–U.S. Dept. of Agriculture Hillculture Farm, Auburn, Ala., by J.D.

Norton. Nut: 13.4 g, dark chocolate-brown, two or three nuts per bur. Midseason maturity; bur opens well. Tree: Chinese type, high yielding; resistant to chestnut blight and chestnut gall wasp.

Bartlett.—Orig. in Stamford, Conn., by F.A. Bartlett, Bartlett Tree Res. Laboratories. Introd. in 1934 when it was named by C.A. Reed of the U.S. Dept. of Agriculture (USDA). Selected from PI 36666 sent to Bartlett in 1916 from the Plant Introduction Garden, USDA, Chico, Calif.; 250 lb of seed was purchased by plant explorer Frank Meyer as Lee Tze from the Pang Shan region northeast of Peking; *Castanea mollissima*. Nut: resembles the American chestnut in size and sweetness. The original is still alive, but it is very difficult to propagate.

Big Boy.—Orig. in Eagle Creek, Ore., by Carroll D. Bush. Introd. in 1931. Chance seedling that appeared in his nursery. Nut: large; matures about midseason; pellicle often folds deeply into nut; nut falls before the burs and leaves; kernel bitter when fresh. Tree: rather a shy bearer.

Blackbeauty.—Orig. in Auburn, Ala., by the Alabama Agr. Expt. Sta. Introd. in 1952. Open-pollinated seedling of *Castanea mollissima*; selected in 1943. Nut: size above average; black; quality excellent; drops freely from bur; keeps well in storage. Tree: blight resistant; bears heavily.

Britoka.—Orig. in Westbank, British Columbia, Canada, by J.U. Gellatly, Gellatly Nut Nursery. Introd. in 1951. Open-pollinated seedling of Chinese chestnut (*Castanea mollissima*); selected in 1940. Nut: large, averaging 1 1/4 inches in diam.; dark brown; free of indentures; peels easily; husk thin, self-opening. Tree: timber-type tree; produces only fair crops.

Campbell #1.—One of a number of seedling selections of Layeroka dating from the 1980s that were made by Douglas Campbell. A good producer, but has little resistance to chestnut blight disease (caused by the fungus *Cryphonectria parasitica*).

Carolina.—A Chinese–American complex hybrid produced by R.T. Dunstan about 1950. Patent 7041. Nut: dark brown, 24–28 per lb.

Carpenter.—A Chinese–American complex hybrid produced by R.T. Dunstan about 1950. Nut: red-brown, 20–25 per lb. Tree: extremely vigorous and produces well in the southern U.S.

Carr *(Carrissima)*.—Orig. in Magnolia, N.C. Introd. in 1935. Grown from seed (*Castanea mollissima*) of PI 39721 procured from Tietsin, China, in 1914 and a small tree sent to R.D. Carr by the U.S. Dept. of Agriculture in 1915; selected in 1930; first grafted in 1932 by H.F. Stoke of Roanoke, Va. Nut: size good; too gray to be as attractive as some others; very sweet; quality excellent.

Carrissima.—See Carr.

Castiva.—Orig. in Nevada City, Calif., by C.E. Parsons, Felix Gillet Nursery. Introd. in 1930 by this nursery. Selection of *Castanea sativa* made in 1929. Nut: about the size and shape of Large American Sweet; quality fine; pellicle peels free from kernel; falls free from bur. Tree: bears heavily.

Ching Chow.—Orig. in Swarthmore, Pa., by J. Russell Smith. Sam Hemming selection of P.I. 70314; original seed brought from Nanking, China. Introd. about 1950. Seedlings of the Chinese chestnut (*Castanea mollissima*). Tree: most are blight resistant.

Chioka.—Orig. in Westbank, British Columbia, Canada, by J.U. Gellatly, Gellatly Nut Nursery. Introd. in 1932. Chance seedling; discovered in 1929. Nut: large. Tree: timber type; not as productive as other varieties. Now obsolete.

Colby.—Orig. in Urbana, Ill., by Arthur S. Colby, Illinois Agr. Expt. Sta. Not introduced commercially, but was named prematurely and name appears in the literature. Fuller x *Castanea mollissima*; cross made in 1931; tested as Ill. 31-4. Nut: large; flavor good; resembles Fuller but larger when well grown. Tree: susceptible to blight like American ancestors; usually proved incompatible in grafts, so never commercially propagated.

Colossal.—Probably the most commonly planted cultivar in the U.S., mostly in the Northwest. Introd. in 1925 by C.E. Parsons of the Felix Gillet Nursery of Nevada City, Calif. Nut: large (11–15 per lb); very good flavor; pellicle peels easily. Harvest in the northwestern U.S. is from early to mid-September. Tree: large, spreading, and globose at maturity; appears to be a Japanese x European hybrid; pollen-sterile. Mature trees (12 years old) will produce up to 150 lb of nuts per tree. Some resistance to chestnut blight disease.

Connecticut Yankee.—See Yankee.

Crane.—Orig. as a seedling of Hung Pan Lee, PI 113665 from the Chian Tzu Hsiao district, Chekiang Province, China (Peter Liu, collector). Propagation began in 1947; named in 1963 by J.W. McKay in honor of Harley L. Crane. Nut: dark cherry-red, nearly hairless, average 32 per lb; flavor and keeping quality excellent. Best adapted to the middle and southern U.S. chestnut-growing areas, where it flowers in midseason.

Douglas #1A.—Seedling selection of Chinese. Selected by Earl Douglas from open-pollinated seedlings from Manchurian trees planted near American. Nut: medium (about 40 per lb); ripens midseason. Very hardy in the northeastern U.S.

Douglas #2.—As above.

Douglas Manchurian.—Orig. as a seedling from nuts picked up by Earl Douglas in the Poconos (eastern Pennsylvania) sometime in the 1950s. Appears to be pure Chinese.

Eaton.—Open-pollinated seedling given to Frederick S. Eaton of Wallingford, Conn., by Arthur H. Graves. Probably a seedling of a Chinese x (Japanese x American) tree. Named for Eaton in 1970 by R.A. Jaynes. Nut: large (30–40 per lb); texture, flavor, and sweetness are among the best; matures 3–7 days earlier than most Chinese chestnuts.

Essate-Jap.—Orig. in New Haven, Conn., by Arthur H. Graves, Connecticut Agr. Expt. Sta. Introd. in 1960 [(*Castanea crenata* x *C. pumila*?) x *C. crenata*] x *C. crenata*; selected in 1959. Nut: ripens early, 5–20 Sept.; resembles the Japanese chestnut. Tree: erect; fast-growing; leaves rather slender, long pointed, glandular beneath; reddish color prominent, i.e., young leaves bronzy, twigs dark red, and buds with carmine hue; blooms early, 1–15 June; prolific; blight free; hardy in Rehder's zone 5. Tested in the U.S., Italy, and The Netherlands.

Ford's Sweet.—Seedling named by J. Ford Wilkinson of Rockport, Ind. Wilkinson received 3 lb of FP #462 Chinese chestnuts in 1933, and reported in 1954 that he was going to name one of the trees from this seed lot as a cultivar because of its superior nuts. FP #462 was purchased in a San Francisco market, and the nuts were reputed to be from "the region back of Tientsin."

Grimo 142 Q.—Open-pollinated seedling of Layeroka, selected by E. Grimo. Tree: pollen-sterile; moderately resistant to chestnut blight disease.

Hamden.—Orig. in Hamden, Conn., by Arthur H. Graves, Connecticut Agr. Expt. Sta. Introd. in 1962. A complex hybrid of *Castanea dentata*, *C. crenata*, and *C. mollissima*; selected in 1961. Nut: Chinese type; rather large, 2.5 x 3.0 cm, squarish; stylar end with white woolly pubescence; inner skin separating from embryo; good consistency; edible, sweet; matures late. Tree: an attractive ornamental, with thick, glossy leaves; late-flowering.

Heritage.—Orig. by R.T. Dunstan. Plant patent 6674; 14 Mar. 1989. Nut: small (10 g); American type, elongated; chocolate-brown; very sweet. Burs split on tree; harvest late, short. Tree: very vigorous; very upright, straight-boled, timber-type form; cropping sparse; leaves light green, American-type with acute base and very dentate margins.

Hobson.—Orig. from Chinese seed procured by the U.S. Dept. of Agriculture, northwest of Peking, China; sent to J. Hobson as a small seedling in 1917. First grafted in 1932 by H.F. Stoke, Roanoke, Va., although named in 1930. Introd. in 1935. Parentage unknown. Fruit: smaller and more attractive than Carr, but no sweeter; ripens 2 weeks later. Tree: considered to be more prolific than Carr.

Honan.—Orig. in Eagle Creek, Ore., by Carroll D. Bush. Introd. about 1935. A vigorous seedling selected from a planting of imported seed of *Castanea mollissima*. Nut: resembles Abundance, but not as large. Tree: resembles Abundance, but not as productive.

Jersey Gem.—Orig. in Montville, N.J., by Henry Hartman, Jersey Chestnut Farm, Wayne, N.J. Introd. in 1976. Orrin x Nanking; cross made in 1971; tested as 23-6. Nut: 30/lb, spherical; shell glossy, red-mahogany color, good seal; kernel sweet, crunchy, yellow meats with good flavor; slip skin; harvested first week in October, all nuts drop within a few days. Tree: trimmed to 12 ft, round; vigor moderate; very productive.

Kelsey.—Orig. in Clinton, Conn., by Josiah J. Kelsey, through Arthur H. Graves, Connecticut Agr. Expt. Sta. Introd. in 1960. Seedling of *Castanea mollissima;* scionwood of a seedling secured from Luther Burbank, Santa Rosa, Calif., by Kelsey, and upon reaching bearing age, it was crossed with another seedling. Nut: Chinese type; small; very dark brown; squarish; inner skin readily separable; meat sweet, yellow, crunchy. Tree: light-colored, scaly bark as in Chinese chestnut, twigs silvery; outer bark on larger branches curling back a little in lenticular patches, disclosing characteristic rather smooth, buff inner bark; blight-free; hardy in Rehder's zone 5. Tested in the U.S., Italy, and The Netherlands.

Kuling.—Chinese seedling 7919 grown in a U.S. Dept. of Agriculture orchard in Philema, Ga., from 1933 nuts of K'uei Lee (PI 108,552) from Hang Chow, Cheking Province, China. Nut: 35–40 per lb; drops free from the bur in midseason; keeping quality very good. Tree: medium-large, moderately upright, vigorous in the southern U.S.

Large American Sweet.—Orig. in Nevada City, Calif., by C.E. Parsons, Felix Gillet Nursery. Introd. in 1923. Parentage unknown; discovered in 1920. Nut: size uniform; meat smooth, sweet. Tree: heavy bearer.

Layeroka.—Skioka o.p. Has been called Chinese, but has characteristics of a Japanese x European hybrid. Susceptible to chestnut blight disease.

Linden.—Possibly a Japanese x European hybrid. Selected for pollinating Silverleaf in the northwestern U.S. Nut: medium; good flavor; harvested in early to mid-October.

Manoka.—Orig. in Westbank, British Columbia, Canada, by J.U. Gellatly, Gellatly Nut Nursery. Introd. in 1953. Open-pollinated seedling of the Chinese chestnut (*Castanea mollissima*) from seed imported from China; selected in 1940. Nut: large, averaging 1 3/8 inches in diam.; shell dark brown; kernel yellow, well-flavored, quality good; husk small, thin. Tree: hardy; timber-type tree.

Mayseptjan.—Orig. in Nevada County, Calif., by C.E. Parsons, Felix Gillet Nursery. Introd. in 1932. Parentage unknown; from a seedling brought from the Reihl orchard, Alton, Ill.; selected in 1930. Nut: large; quality good; matures in early fall. Tree: bears well.

Meiling.—Orig. as seedling 7880 from K'uei Lee (PI 108,553) from Hsin Teng, Chekiang Province, China (Peter Liu, collector) in 1935. Released by U.S. Dept. of Agriculture in 1949. Nut: 35–40 per lb; light tan, closely resembles Nanking; falls free from the bur in midseason; keeps well. Tree: upright; bears well in the southern U.S., but grows poorly in New England.

Milford.—Orig. in Milford, Del., by Frank B. Austin. Introd. in 1934. Selected seedling of *Castanea mollissima;* seed received from the U.S. Dept. of Agriculture in 1926. Nut: dark brown; slightly larger than native species; bur small, with rather short spines. Tree: prolific bearer. Now obsolete, as hurricane Hazel in 1954 destroyed all budded trees of this variety.

Mossbarger.—A hardy Chinese from Glendale, Ky. Tree: vigorous; bears early in the southern U.S.

Murden.—Orig. in Pleasant Ridge, Va., by Erwin Murden, possibly from seedlings on the neighboring Bonney property. Introd. about 1935. Selection of *Castanea crenata* made in 1933. Nut: very large, up to 2 1/4 inches in diam. Named and propagated by H.F. Stoke, Roanoke, Va. Not available from commercial sources.

Myoka.—Orig. in Westbank, British Columbia, Canada, by J.U. Gellatly, Gellatly Nut Nursery. Introd. in 1950. Open-pollinated seedling of Chinese chestnut (*Castanea mollissima*) from seeds imported from China; selected in 1940. Nut: large, averaging 1 3/8 inches in diam.; flavor good; peels well; resembles Manoka. Tree: very fine, large, timber tree.

Nanking.—Orig. as seedling 7930 from PI 108,552, K'uei Lee, from Hang Chow, Chekiang Province, China (Peter Liu, collector) in 1935. Nut: usually 30–35 per lb, dark tan, shape is uniform; ripening season medium-late. Very precocious. Widely planted in the southern U.S.

Nevada.—Selected as a pollinizer of Colossal because of its profuse catkin bloom at the time when female flowers of Colossal are fertile. Nut: medium, dark-brown shells, very sweet; harvest is in early October in the northwestern U.S. Tree: vigorous, upright.

Ok Kwang.—Japanese cultivar from Korea. Reputed to be resistant to chestnut blight disease and to the Oriental chestnut gall wasp (*Dryocosmus kuriphilus*).

Orrin.—A seedling grown on the farm of Orrin S. Good at Lockhaven, Pa. Seed source was Leeland Farms Nursery in Leesburg, Ga.; orig. may be the same as Nanking, Meiling, and Kuling. Nut: average 32 per lb; dark mahogany, nearly black; slightly pubescent tip; light-colored seed scar contrasts with the dark color, giving the nut a distinctive appearance; superior keeping quality; earlier maturing than Nanking. Tree: precocious; blight resistant.

Penoka.—Orig. in Westbank, British Columbia, Canada, by J.U. Gellatly, Gellatly Nut Nursery. Introd. in 1950. Open-pollinated seedling of the Chinese chestnut (*Castanea mollissima*) from seed imported from China; selected in 1940. Nut: large; flavor very fine. Tree: hardy.

Reliable.—Orig. in Roanoke, Va., by Harvey F. Stoke. Introd. in 1936. Seedling of *Castanea mollissima* selected in 1932. Nut: small. Tree: blooms late; bears regularly. Variety now obsolete.

Revival.—A Chinese–American complex hybrid; produced by R.T. Dunstan about 1950. Patent 5537; 20 Aug. 1985. Nut: 24–32 per lb, red-brown. Trees are reliable producers in the southern U.S.

Rosa.—Orig. in Linden, Calif., by a Mr. Rosa. Introd. about 1930. Parentage unknown; discovered during the period 1925–30. Nut: European; elongated, somewhat necked. Tree: production heavy.

Silverleaf.—Possibly a European or European x Japanese hybrid. Nut: smaller than Colossal; dark brown; peels easily; excellent, sweet flavor. At nut fall (early October in the northwestern U.S.) the underside of the leaf turns silver. Susceptible to chestnut blight disease.

Simpson.—Probably Chinese; orig. from seed picked up in Erie, Pa., by William Simpson. Tree: blight resistant, but not very winter hardy.

Skioka.—Grown by J.U. Gellatly in Westbank, British Columbia, Canada; introd. in 1950. Gellatly called tree Chinese, but has many Japanese characteristics. Nut: average 35 per lb; shell dark brown; kernel white, free from indentures. Tree: blooms late; hardy, but susceptible to chestnut blight.

Sleeping Giant.—Orig. in New Haven, Conn., in 1937 as a hybrid of a hardy, timber-type Chinese tree (PI 70315) with a Japanese x American hybrid. Cross made by Arthur H. Graves using a Japanese in an old planting on Long Island, N.Y., that had good resistance to chestnut blight. Graves named it in 1958 after a state park in Hamden, Conn.; released by the Connecticut Agr. Expt. Sta. in 1960. Consistent producer of large, handsome nuts (about 40 per lb) in burs that are often slightly conical. Tree: somewhat spreading; blight free. Hardy in zone 5 of Rehder, and tested in the U.S., Italy, and The Netherlands.

Stoke.—Orig. in Roanoke, Va., by H.F. Stoke from mixed seed sent him for stock purposes by the Div. of Forest Pathology, U.S. Plant Industry Sta., U.S. Dept. of Agriculture, Beltsville, Md. The seed was produced at the Plant Propagating Garden (Bell Station), Glenn Dale, Md. Introd. in 1936 by Stoke. Parentage largely or altogether Japanese, although possibly part Chinese. Nut: ripens at Roanoke very early, usually beginning during the third week in August. Tree: extremely precocious; prolific; little else to commend it. Probably obsolete.

Toumey.—A 1937 hybrid of Japanese chestnut MJ (PI 86873) crossed with the same Japanese x American hybrid used as a pollen parent of Sleeping Giant. Graves named it in 1961 after a forester at Yale Univ., New Haven, Conn.; released by the Connecticut Agr. Expt. Sta. in 1962. Nut: large (about 50 per lb); between bay and chestnut color, with darker lengthwise ridges; tawny hairs in triangular wings 1.0 to 1.5 cm toward the hilum; flavor fair.

Van Fleet.—Orig. in Godfrey, Ill., by E.A.

Riehl. Introd. in 1934. (*Castanea dentata* x *C. sativa*) Rochester x *C. dentata*; not S-8 Van Fleet hybrid; selected in 1926. Nut: larger than the native American chestnut (*C. dentata*); kernel very sweet, fine quality. Tree: productive. Named in memory of Walter Van Fleet (1857–1922), U.S. Dept. of Agriculture.

Vibbert.—Orig. in Hartford, Conn., by W.C. Deming. Introd. in 1935. Selected in 1931 from trees growing near Hartford; *Castanea crenata*. Nut: larger and sweeter than others of this species. Propagation discontinued in 1939.

Willamette.—A Chinese–American complex hybrid produced by R.T. Dunstan about 1950. Patent 7195. Nut: very large (18–22 per lb), dark brown. Reliable producer in the southern U.S.

Willoka.—Orig. in Westbank, British Columbia, Canada, by J.U. Gellatly, Gellatly Nut Nursery. Introd. in 1950. Open-pollinated seedling of the Chinese chestnut (*Castanea mollissima*); selected in 1940. Nut: large, up to 1 1/2 inches in diam.; kernel yellow, pleasant tasting; peels well. Tree: produces good crops.

Yankee *(Connecticut Yankee)*.—Orig. in Riverside, Conn., on land now owned by E.E. Hunt; first propagated by J. Russell Smith, Swarthmore, Pa., in the Sunny Ridge Nurseries of Northern Virginia in 1935. Introd. in 1935. *Castanea mollissima*; PI 36666 (as Bartlett) from the U.S. Dept. of Agriculture. Nut: choice but small. Tree: hardy in eastern U.S.

Zimmerman.—Orig. in Linglestown, Pa., by G.A. Zimmerman. Introd. in 1935. Grown as a nursery seedling resulting from nuts (*Castanea mollissima*) imported from Nanking, China; selected in 1930; first propagated by H.F. Stoke, Roanoke, Va., in 1933; first cataloged by J. Russell Smith, Swarthmore, Pa., in 1938. Nut: appearance attractive; cleaning quality good; sweet; trifle small; may be supplanted by other varieties having larger fruits.

Chestnut Seedlings

Dunstan seedlings.—Chinese–American complex hybrids produced by R.T. Dunstan about 1950. Nut: large (20–35 per lb). Tree: fairly cold hardy, reliable producers in the southern U.S. (not as far north as New England).

Sweet Hart.—Although technically not a cultivar, the open-pollinated seedlings sold under this name are widely distributed. The original Chinese trees were PI 36666, a Tiger Paw-type imported by Frank Meyer (the same importation that yielded Bartlett and Connecticut Yankee). The original trees were planted in 1917 by J. Henry Lienhard in Beach City, Ohio, and they fruited in 1921. Lienhard's son-in-law, J. Lester Hawk, then became the owner, and started Hawk & Son Nursery. He reported that they had 2511 seedling Chinese trees in 1943. Some of these were purchased by Mr. and Mrs. Bert Harter in 1946. In 1961, Oliver Diller noted that the nuts produced by these trees were excellent. From these trees, Boyd's Nursery produced and named the "Hart Chestnut," and subsequently the Sweet Hart.

Sweetheart.—See Sweet Hart.

Chinese Bush Cherry

Drilea.—Orig. in Morden, Manitoba, Canada, by Canada Dept. of Agriculture, Res. Sta. Introd. in 1938. Open-pollinated seedling of a white-fruited plant (*Prunus tomentosa*), secured at U.S. Dept. of Agriculture Northern Great Plains Field Sta., Mandan, N.D. Fruit: up to 5/8 inch in diam.; round; skin bright scarlet, thin, tender; flesh firm to melting, tender, sweet, pleasant, quality very good; pit small; season last half of July; good as dessert, canned, jelly, jam. Bush: upright, large, vigorous; dense; spreading; annual bearer, very productive; tolerant to dry conditions.

Orient.—Orig. in Excelsior, Minn., by Univ. of Minnesota Fruit Breeding Farm. Introd. in 1949. Self-pollinated seedling of *Prunus tomentosa* originally obtained from O.M. Jenson of Albert Lea, Minn., in 1925; selected in 1940; tested as Minn. 63. Fruit: about 1/2 inch in diam.; skin bright medium red, attractive; flesh meaty, juicy, pleasantly subacid; stone small, long oval, free; quality good; for dessert and jelly; ripens in mid-July. Tree: bushy; vigorous; productive; self-fruitful; bears second year after planting; ornamental.

CITRANGE

Carrizo.—Orig. in Riverside, Calif., by E.M. Savage, U.S. Dept. of Agriculture. Named in 1938. *Citrus sinensis* Washington x *Poncirus trifoliata*; sibling of Troyer. Used only as a rootstock: nucellar seedlings used; vigorous; shows some tolerance to the burrowing nematode (*Radolpholus similis*).

Troyer *(Citravel)*.—Orig. in Riverside, Calif., by E.M. Savage, U.S. Dept. of Agriculture. Named in 1934. *Citrus sinensis* Washington x *Poncirus trifoliata*: cross made in 1909; sibling of Carrizo. Fruit: many seeds. Used only as a rootstock: nucellar seedlings used; vigorous; apparently tolerant to quick decline; tolerant to citrus nematode (*Tylenchulus semipenetrans*); primary rootstock for oranges in California since 1950; satisfactory for Lisbon, but not for Eureka, lemons.

Yuma.—Orig. in Yuma Mesa Branch Sta. of the Univ. of Arizona. Introd. in 1960. Possibly seedling of Savage citrange; selected by R.D. Rodney and R.M. Allen. Fruit: small, 1.5 to 2.5 inches in diam.; oblate; skin yellow, pubescent, without furrows, thin; pulp yellow; seeds 15–25 fully developed seeds per fruit; ripens in late October. Tree: upright, to 30 ft tall; very productive; leaf trifoliate, with center leaflet double the size of the laterals. Unbudded seedlings tested by J.B. Carpenter show tolerance to *Phytophthora* fungi at U.S. Date and Citrus Field Sta., Indio, Calif. An experimental rootstock in the Arizona–California desert regions. Produces mostly nucellar seedlings.

CITRANGEDIN

Glen.—Orig. in Eustis, Fla., by U.S. Dept. of Agriculture. Introd. about 1932. Calamondin (*Citrus reticulata* x *Fortunella* sp.) x Willits citrange (*C. sinensis* x *Poncirus trifoliata*). Fruit: about 1 inch in diam.; calyx persistent; rind reddish-orange, thin, smooth, firm; pulp orange-yellow, juicy, translucent, very sharply acid; seeds 3–5, small. Tree: vigorous; upright; very ornamental when it is in crop; more tolerant to cold than many commercial citrus varieties.

CITRANGEQUAT

Sinton.—Orig. in Eustis, Fla., by Walter T. Swingle, U.S. Dept. of Agriculture. Introd. in 1923. Nagami kumquat x Rusk citrange; selected in 1909; tested by J.W. Benson in Sinton, Texas, as C.P.B. 48050. Fruit: small, averaging 1 1/4 inches in diam.; globose to somewhat elongated; rind light yellow to scarlet, very thin, sweet like a kumquat; pulp acid; few oil glands; nearly seedless; ripens October to December; suggested use as an ade or juice fruit.

Telfair.—Orig. in Eustis, Fla., by Walter T. Swingle, U.S. Dept. of Agriculture. Introd. in 1923. Nagami kumquat x Willits citrange; selected in 1909; first fruited at McRae, Telfair County, Ga.; tested as C.P.B. 48005. Fruit: round; small, averaging 1 1/4 inches in diam.; rind thin, somewhat bitter; oil glands small, prominent; segments 4 to 6; pulp yellow, very acid even when mature; seeds few. Tree: vigorous; upright.

Thomasville.—Orig. in Eustis, Fla., by Walter T. Swingle, U.S. Dept. of Agriculture. Introd. in 1923. Nagami kumquat x Willits citrange; selected in 1909; tested as C.P.B. 48010; named after Thomasville, Ga., where first tested. Fruit: globose to oval; small, averaging 1 1/2 inches in diam. and 2 inches long; rind thin, yellow; segments 7 or 8; pulp juicy, pleasantly acid when immature, sweetish at maturity, light green to amber; seeds 0 to 12, averaging 6, small; season October to December, but juicy and suitable for ade from July to October. Tree: vigorous; upright; hardy; disease resistant; useful as a rootstock or as an ornamental.

CITRUMELO

Sacaton.—Orig. in Sacaton, Ariz., by Walter. T. Swingle, U.S. Dept. of Agriculture. Introd. in early 1930s. *Citrus paradisi* x *Poncirus trifoliata*; planted in 1919; tested as 4554A. Fruit: up to 3 inches in diam.; skin yellow; flesh seedy, inedible; 50% polyembryonic. Tree: trifoliata leaf character; good tolerance to root rot caused by *Phytophthora parasitica*; excellent tree vigor on own root. As rootstock for lemon indicates very good tree growth and yield in heavy loam soil; with Washington Navel, orange scion on calcareous sandy loam soil (ph 8) growth similar to sour orange. Good though not outstanding rootstock at Indio, Calif. Budded trees develop iron chlorosis. Original bud source is lost and variety is propagated from seedlings.

Swingle.—Orig. in Eustis, Fla., by Walter T. Swingle. Introduced in 1974. *Citrus paradisi* x *Poncirus trifoliata*; cross made in 1907; tested as CPB 4475. Used only as a rootstock: seedlings vigorous; budded trees grow slower than trees on sour orange, especially during winter months; tolerant of tristeza, exocortis, and xyloporosis viruses, and footrot (*Phytophthora parasitica*). Salt tolerant and cold resistant. Asexually reproduced by seed, which is variably polyembryonic, depending on seed source and season. Very satisfactory for grapefruit; only moderate sized trees and yields with oranges. Most widely planted rootstock in Florida since the late 1980s.

Cranberry

Beaver.—Orig. in Beaver River, Nova Scotia, Canada, by E.L. Eaton, Canada Dept. of Agriculture, Res. Sta., Kentville, Nova Scotia. Introd. in 1956. Seedling of the native cranberry; selected in 1940. Fruit: consistently matures 7–10 days earlier than Early Black, with larger and more attractive fruit than that variety; keeps very well in cold storage. Vine: yields moderately well; susceptible to false blossom.

Beckwith.—Orig. in Whitesbog, N.J., by H.F. Bain and H.F. Bergman, U.S. Dept. of Agriculture. Introd. in 1950. McFarlin x Early Black; selected 1938–40. Fruit: large; cup-count 50–60; shape oval-oblong; skin deep red; keeping quality fair to good; late-maturing. Vine: productive, but susceptible to false blossom in New Jersey. Named after Charles S. Beckwith, formerly of Cranberry and Blueberry Lab., New Jersey.

Bergman.—Orig. in Whitesbog, N.J., by the U.S. Dept. of Agriculture (USDA). Introd. in 1961. Early Black x Searles; cross made in 1930 by H.F. Bain, U.S. Dept. of Agriculture; selected by F.B. Chandler and I.E. Demoranville, Cranberry Sta., Massachusetts Agr. Expt. Sta., East Wareham; tested as Mass. 8 in East Wareham. Fruit: size medium with cup count 65–80; short, pear-shaped; skin red; keeping quality good to excellent, storage shrinkage small, suitable for long storage; yield greater than Early Black or Howes; ripens in midseason; fruit rot less than Early Black or Howes. Bush: uprights medium length, holding berries up from ground; runners few; resistant to feeding by the leafhopper that spreads false blossom virus. Named in honor of H.F. Bergman, pathologist, retired from the USDA.

Crowley.—Orig. in Long Beach, Wash., by D.J. Crowley, Washington State Univ., Pullman. Introd. in 1961. McFarlin x Prolific; cross made in 1943; selected in 1954; tested as WSU seedling 72. Fruit: medium to large; round-oblate; skin medium to dark red with slight bloom, high pigment content; flesh white, low in astringency; recommended for processing and juice; ripens 2 weeks before McFarlin. Bush: moderately to highly vigorous; hardiness same as McFarlin; productive; adapted to Washington and Oregon.

Franklin.—Orig. in Whitesbog, N.J., by the U.S. Dept. of Agriculture. Introd. in 1961. Early Black x Howes; cross made in 1930 by H.F. Bain, U.S. Dept. of Agriculture; selected by F.B. Chandler and I.E. Demoranville, Cranberry Sta., Massachusetts Agr. Expt. Sta., East Wareham; tested as Mass. 31 in East Wareham. Fruit: size medium with cup count 57–90; nearly round; skin red to very dark red; keeping quality good to excellent; yield greater than Early Black or Howes; fruit rot no more than Early Black or Howes; ripens early. Bush: uprights medium length; runners few and short; resistant to feeding by the leafhopper that spreads false blossom virus. Named in honor of H.J. Franklin (1883–1958), former head, East Wareham Cranberry Sta.

Pilgrim.—Orig. in Whitesbog, N.J., by the U.S. Dept. of Agriculture. Introd. in 1961. Prolific x McFarlin; cross made in 1930 by H.F. Bain, U.S. Dept. of Agriculture; selected by F.B. Chandler and I.E. Demoranville, Cranberry Sta., Massachusetts Agr. Expt. Sta., East Wareham; tested as Mass. 17 in East Wareham. Fruit: large with cup count 43–66; long oval; skin purplish-red with yellow undercolor, bloom present; keeping quality good; fruit rot no more than Early Black or Howes; ripens late. Bush: prolific bearer, greater than Early Black or Howes, particularly on peat bottom land; medium number of runners; uprights medium to long; resistant to feeding by the leafhopper that spreads false blossom virus.

Stankavich.—Orig. in Bandon, Ore., by Joseph F. Stankiewicz. Introd. in 1926. Oregon Native x an Eastern variety; selected between 1914–17. Fruit: large, averaging 3/4 to 5/8 inch in diam.; uniform, global; skin color high; flesh with low acid content, good sugar content; ripens early. Bush: produces well; resembles Michigan Bennett.

Stevens.—Orig. in Whitesbog, N.J., by H.J. Bain, U.S. Dept. of Agriculture (USDA). Introd. in 1950. McFarlin x Potter; selected 1938–40. Fruit: large; cup-count 50–60; shape round-oval; skin color deep red; flesh firm, keeping quality good; medium-late ripening. Vine: vigorous; very productive. Being planted to some extent in Wisconsin; promising in New Jersey. Named after Neil Stevens, former USDA plant pathologist.

Wilcox.—Orig. in Whitesbog, N.J., by H.J. Bain, U.S. Dept. of Agriculture (USDA). Introd. in 1950. Howes x Searles; selected 1938–40. Fruit: size medium; shape oval; skin deep red; keeping quality fair; ripens mid to late season. Vine: very productive; vigorous; appears to be more resistant to false blossom disease than Beckwith and Stevens. Named after Raymond B. Wilcox, former USDA plant pathologist.

CURRANT

Atta.—Orig. in Brookings, S.D., by N.E. Hansen, South Dakota Agr. Expt. Sta. Introd. in 1925. Seedling of wild black currant. Fruit: large, averaging 1.3 g; round; skin black. Plant: large; bears well. Now obsolete.

Cascade.—Orig. in Excelsior, Minn., by W.H. Alderman, A.N. Wilcox, T.S. Weir, and F.E. Haralson, Univ. of Minnesota Fruit Breeding Farm. Introd. in 1942. Open-pollinated seedling of Diploma; selected in 1926; tested as Minn. 70. Fruit: larger than Red Lake; matures 1 week earlier than Red Lake; clusters not as long as those of Red Lake. Plant: productive; erect; vigor medium; moderately susceptible to leaf spot.

Consort.—Orig. in Ottawa, Ontario, Canada, by Canada Dept. of Agriculture, Res. Sta. Introd. in 1952. Kerry x *Ribes ussuriense*; selected in 1944; tested as Ottawa 396. Fruit: skin black; flavor strong; ripens late. Plant: immune from rust (*Cronartium ribicola*); self-fruitful, not requiring insect pollination; productive.

Coronet.—Orig. in Ottawa, Ontario, Canada, by Canada Dept. of Agriculture, Res. Sta. Introd. in 1948. *Ribes ussuriense* Jancz. x Kerry; selected in 1943; tested as Ottawa 393. Fruit: skin black; resembles standard black currant varieties. Plant: immune from rust (*Cronartium ribicola* A. Fisch.).

Cotswold Cross.—Orig. in Bristol, England, by G.T. Spinks, Univ. of Bristol, Res. Sta., Long Ashton. Introd. in 1933. Baldwin x Victoria; selected in 1929; sibling of Malvern Cross. Fruit: skin black; resembles Baldwin. Plant: main crop high yielding.

Crusader.—Orig. in Ottawa, Ontario, Canada, by Canada Dept. of Agriculture, Res. Sta. Introd. in 1948. *Ribes ussuriense* Jancz. x Kerry; selected in 1942; tested as Ottawa 381. Fruit: skin black; resembles standard black currant varieties. Plant: most valuable characteristic is its immunity from rust (*Cronartium ribicola* A. Fisch.).

Daniels September.—Orig. in England. Introd. in 1923. Possibly a bud mutation of Baldwin; discovered about 1915; bush purchased by Daniels Bros. (nurserymen), Norwich, England. Fruit: clusters long; berries large; skin black; ripens very late. Plant: indistinguishable from Baldwin except for later flowering and ripening; shows a marked tendency to revert to the ripening period of Baldwin.

Fleming.—Orig. in Pembina County, N.D., by James Fleming, Hamilton, N.D., and A.F. Yeager, North Dakota Agr. Expt. Sta., Fargo. Introd. in 1924. J. Fleming sent some of the better-yielding bushes of *Ribes americanum* growing along the Red River to Yeager, who selected in 1920 the one named Fleming. Fruit: large; skin black; quality better for jam and jelly than other black currants then being grown. Plant: canes stiffer, more erect than other varieties. No longer propagated by above station.

Holländische Rote.—See Viking.

Laleham Beauty.—Orig. in the garden of P.H. Salter, Staines, Middlesex, England. Introd. there in 1948. Considered to be a chance seedling of Boskoop Giant; discovered in 1940; Daniels September may have been the pollen parent. Fruit: black. Plant: produces an early crop ripening with Baldwin and a late one 2 weeks later.

Malvern Cross.—Orig. in Bristol, England, by G.T. Spinks, Univ. of Bristol, Res. Sta., Long Ashton. Introd. in 1946. Baldwin x Victoria; selected about 1939; sibling of Cotswold Cross. Fruit: skin black; hangs well; resembles Baldwin. Plant: high yield.

Mato.—Orig. in Brookings, S.D., by N.E. Hansen, South Dakota Agr. Expt. Sta. Introd. in 1925. Seedling of a wild black currant, *Ribes floridum*. Fruit: large, up to 9/16 inch in diam.; skin black. Plant: large; production good. Now obsolete.

Mendip Cross.—Orig. in Bristol, England, by G.T. Spinks, Univ. of Bristol, Res. Sta., Long Ashton. Introd. in 1933. Baldwin x Boskoop; selected in 1929. Fruit: skin black; ripens very early; characteristics intermediate between the parents. Plant: high yielding.

Norwegian Red Dutch.—See Viking.

Pawnee Golden.—Orig. in Brookings, S.D., by N.E. Hansen, South Dakota Agr. Expt. Sta. Introd. in 1938. Fourth-generation seedling of Crandall grown among native South Dakotan seedlings of Golden currant. Fruit: very large; quality good. Now obsolete.

Red Lake.—Orig. in Excelsior, Minn., by W.H. Alderman, Univ. of Minnesota Fruit Breeding Farm. Introd. in 1933. Parentage unknown; tested as Minn. 24. Fruit: clusters long, well-filled; berries large, borne on long stems; excellent for jelly; ripens late; equal to those of Perfection. Plant: upright; vigorous; superior to Perfection.

Rød Hollandsk Druerips.—See Viking.

Stephens *(Stephens 9).*—Orig. in Orilla, Ontario, Canada, by C.L. Stephens. Introd. in 1933. Parentage unknown. Fruit: very large clusters medium long; skin attractive red; subacid, quality excellent; comparable to Red Lake, superior to most varieties. Plant: vigorous; somewhat spreading; very hardy; productive.

Stephens 9.—See Stephens.

Tonah.—Orig. in Brookings, S.D., by N.E. Hansen, South Dakota Agr. Expt. Sta. Introd. in 1925. Seedling of a wild black currant. Fruit: average weight of berry, 1.2 g; skin black. Plant: large; yields heavy. Now obsolete.

Tor Cross.—Orig. in Bristol, England, by G.T. Spinks, Long Ashton Res. Sta. Introd. in 1962. Baldwin x "Rogue" Boskoop; selected in 1930. Fruit: skin black; ripens early, being more resistant to dropping and splitting than other early varieties; resembles Mendip Cross. Bush: high yielding.

Viking *(Holländische Rote, Norwegian Red Dutch, Rød Hollandsk Druerips).*—Orig. in Europe where it has been known for many years. Introd. in the U.S. from Norway by Glenn Gardner Hahn, U.S. Dept. of Agriculture. *Ribes petraeum* Wulf. x *R. rubrum* L. (direction of cross not known). Fruit: red; matures late. Plant: immune from white pine blister rust (*Cronartium ribicola*); productive; only red currant grown in white pine territory [according to W.H. Alderman, Univ. of Minnesota, St. Paul (1945)].

Wakapa *(Wakapa Golden).*—Orig. in Cottonwood, S.D., by N.E. Hansen, South Dakota Agr. Expt. Sta., Brookings. Introd. in 1938. Selection from a wild plant. Fruit: yellow; averages 5/8 inch in diam.; pleasantly subacid. Plant: very productive. Now obsolete.

Wakapa Golden.—See Wakapa.

Wanka.—Orig. in Brookings, S.D., by N.E. Hansen, South Dakota Agr. Expt. Sta. Introd. in 1925. Seedling of a wild black currant. Fruit: medium; skin red. Plant: large; good crops. Now obsolete.

Wapago Golden.—Orig. in Cottonwood, S.D., by N.E. Hansen, South Dakota Agr. Expt. Sta., Brookings, S.D. Introd. in 1939. Chance seedling of a native golden currant (*Ribes odoratum* Wendl.). Fruit: up to 3/4 inch in diam.; skin golden; flesh pleasantly acid, fragrance spicy. Plant: drought resistant. Now obsolete.

Wato *(Wato Golden).*—Orig. in Brookings, S.D., by N.E. Hansen, South Dakota Agr. Expt. Sta. Introd. in 1938. Seedling of a native golden currant (*Ribes odoratum* Wendl.). Fruit: skin translucent, shiny; flesh light orange-yellow; flavor pleasant, mildly acid. Now obsolete.

Wato Golden.—See Wato.

Willoughby.—Orig. in Parkside, Saskatchewan, Canada, by Walter Wil-loughby, Sr. Introd. in 1953. Parentage unknown; discovered about 1940. Fruit: skin black; quality good. Plant: resistant to mildew; self-fruitful.

Woga Golden.—Orig. ln Brookings, S.D., by N.E. Hansen, South Dakota Agr. Expt. Sta. Introd. in 1939. Chance seedling of a native golden currant (*Ribes odoratum* Wendl.) from seed gathered at Cottonwood, S.D. Fruit: up to 5/8 inch in diam.; skin golden; flesh acid. Now obsolete.

DATE

Abada.—See Abbada.

Abbada *(Abada, Black Abbada).*—Orig. in Brawley, Calif., in 1936 by D.G. Sniff, Indio, Calif. Chance seedling found in a riverbed in Brawley. Name coined from first names of wife and originator. Fruit: semi-dry type; medium, 38–52 x 20–22 mm, oblong to oblong-ovate; skin medium thick, deep red when immature, black with heavy frostlike bloom at maturity; flesh soft; flavor sweet and somewhat cloying; ripens early. Tree: vigor medium; productivity medium. Introduced because of attractive fruit, but quality is mediocre.

Amber Queen.—Orig. as Deglet Noor o.p. seedling on property of J.H. Northrop near Indio, Calif., between 1911 and 1917; further developed by B.H. Hayes after 1924. Fruit: semi-dry type; large, 41–47 x 21–24 mm, oblong to oval; skin medium thick, yellow when immature, amber at maturity; flesh soft; flavor good; ripens early. Tree: medium vigor; heavy trunk.

Amir Al-Haj.—See Amir Hajj.

Amir Haj.—See Amir Hajj.

Amir Hajj *(Amir Al-Haj, Amir Haj, Mirhage).*—Orig. in Iraq; introd. in the U.S. in 1929 by R.W. Nixon, U.S. Dept. of Agriculture. Fruit: soft type; below medium, 32–36 x 20–22 mm, oblong to oval; skin tender, yellow when immature, amber to red-brown at maturity; flesh soft; flavor delicate; quality good; ripens at midseason. Tree: medium vigor; heavy trunk; moderate productivity. Introduced for highly regarded quality of fruit; especially desirable in khalal stage.

'Amran.—See Sayer.

Ascherasi.—See Ashrasi.

Asharasi.—See Ashrasi.

Ashrasi *(Ascherasi, Asharasi).*—Orig. in Iraq; introd. in the U.S. by D.G. Fairchild, U.S. Dept. of Agriculture, in 1902. Fruit: semi-dry type; medium, 35–42 x 25–32 mm, ovate to wedge-shaped; skin thick, yellow with reddish-brown stippling around perianth when immature, dull amber with light brown areas near base at maturity; flesh firm and smooth; flavor rich and nutty; ripens at midseason. Tree: vigorous; heavy trunk. Yields usually somewhat low due to poor set. Introduced for fruit quality.

Badrahi.—See Badrayah.

Badraihi.—See Badrayah.

Badrayah *(Badrahi, Badraihi, Badurahi, Bedraihe, Bedraya).*—Orig. in Iraq; introd. to the U.S. in 1929 by R.W. Nixon, U.S. Dept. of Agriculture. Fruit: dry type; large, oblong with rounded base and apex; skin thick and tough, yellow when immature, near amber at maturity; flesh thick; flavor sweet and delicate; ripens at midseason. Tree: medium vigor; compact crown. Introduced for fruit quality. Not well characterized in the U.S.

Badurahi.—See Badrayah.

Barhee *(Barhi, Berhi, Birhi).*—Orig. in Iraq; introd. in the U.S. in 1913 by P.B. Popenoe (private importation for West India Gardens). Fruit: soft type; small, 32–37 x 23–30 mm, short ovate; skin medium thick, yellow when immature, amber to red-brown at maturity; flesh soft; flavor delicate, rich; quality very good; ripens late in season. Tree: very vigorous; very productive, yields 115–160 kg / tree.

Barhi.—See Barhee.

Bartamuda.—See Bentamoda.

Bartomoda.—See Bentamoda.

Bedraihe.—See Badrayah.

Bedraya.—See Badrayah.

Bentamoda *(Bartamuda, Bartomoda, Betamoda).*—Orig. in Sudan, imported to the U.S. in 1914 by S.C. Mason, U.S. Dept. of Agriculture. Fruit: semi-dry type (?); medium; oblong-elliptical; skin medium thick, dull red when immature, amber at maturity, dark brown when cured; flesh firm; flavor good but not outstanding; ripens late. Tree: low vigor; yields low. Introduced for fruit quality as dry date; in the U.S., appears to be semi-dry type and quality has not attained purported level.

Berhi.—See Barhee.

Betamoda.—See Bentamoda.

Birhi.—See Barhee.

Black Abbada.—See Abbada.

Black Sphinx.—See Sphinx.

Blond Beauty.—Orig. as Deglet Noor o.p. seedling on property of J.H. Northrop near Indio, Calif., between 1911 and 1917; further developed by E.F. Shields. Fruit: semi-dry type; large, 45–50 x 20–24 mm, oblong-elliptical; skin tender, yellow when immature, amber to reddish-brown at maturity; flesh soft and melting; flavor pleasing, "taffy-like"; ripens at midseason. Tree: medium vigor; prolific offshoot producer. Somewhat resistant to checking.

Brunette Beauty.—Orig. as Deglet Noor o.p. seedling on property of J.H. Northrop near Indio, Calif., between 1911 and 1917; further developed by E.F. Shields. Fruit: semi-dry type; medium, 42–27 x 20–23 mm, oblong-elliptical; skin medium tough, red when immature, nearly black when mature; flesh soft; flavor good; ripens late. Tree: medium vigor; thin trunk.

Dairee *(Dairi, Dayri)*.—Orig. in Mesopotamia; imported into the U.S. in 1913 by P.B. Popenoe. Introd. in the U.S. about 1934 by R.W. Nixon, U.S. Dept. of Agriculture, Indio, Calif. Fruit: semi-dry type; large, 36–45 x 19–23 mm, oblong to oblong-elliptical; skin tough, red when immature, brown to black when soft and ripe; flesh soft, amber, texture good, semi-translucent; flavor good, with a "heavy" sweetness; ripens in midseason, ripens well in seasons of high humidity and extreme heat. Palm: production good, 115–140 kg (on 18-year-old palms); appears very promising for the Southwest.

Dairi.—See Dairee.

Daqlah Baydahi.—See Deglet Beida.

Dayri.—See Dairee.

Degla beida.—See Deglet Beida.

Deglet Beida *(Daqlah Baydahi, Degla beida)*.—Orig. in Algeria; introd. to the U.S. in 1904 by W.T. Swingle, U.S. Dept. of Agriculture. Fruit: dry type; medium, 37–45 x 20–23 mm, oblong or narrowly oblong; skin moderately thick and relatively smooth, yellow when immature, very light pale brown or buff at maturity and when cured; flesh firm; flavor good but quality not as good as Thoory; ripens early. Tree: medium vigor; medium-heavy trunk. Quite salt-tolerant.

Deglet Noor *(Deglet Nour, Deglet Nur)*.—Orig. in Algeria; introd. to the U.S. in 1900 by W.T. Swingle, U.S. Dept. of Agriculture. Fruit: semi-dry type; large, 40–50 x 20–25 mm, oblong-ovate; skin medium thick, light red when immature, amber to medium-brown at maturity and when cured; flesh thick, firm, soft; flavor excellent, peculiarly distinctive; quality very good under favorable conditions; ripens late. Tree: vigorous; trunk slender to medium heavy; leaves olive-green; yields good (90–140 kg/tree) under favorable conditions. Checks more easily than most other commercial varieties. Relatively few offshoots produced. Chief commercial variety of date in the U.S.

Deglet Nour.—See Deglet Noor.

Deglet Nur.—See Deglet Noor.

Desert Dew.—Orig. in Yuma, Ariz., by C.D. McGinnis; introd. in 1946. Discovered in 1909 among some offshoots taken from Bard Valley, Calif., to the McGinnis date grove in Yuma. Fruit: semi-dry type; resembles Deglet Noor, being slightly larger and having better flavor before processing; skin thin, tender, red when immature, dull brown when ripe, purplish-brown when cured; flesh soft and melting; flavor rich and sweet but cloying; ripens early. Tree: medium vigor; resembles Deglet Noor. Resistant to cracking and checking.

El Toby.—Orig. from five very old palms discovered in 1923 near Yuma, Ariz., by M. Winsor. Fruit: semi-dry type; small, 30–40 x 17–19 mm, ovate or oblong-oval; skin medium thick, rather tough, bright red when immature, almost black at maturity; flesh soft; flavor rather strong; ripens midseason. Tree: low vigor; trunk medium to heavy; slow to come into bearing. Somewhat resistant to checking.

Empress.—Orig. in Cathedral City, Calif., by E.K. Daval. Open-pollinated seedling of Thoory; seed first planted in 1916. Fruit: dry type; large, 46–54 x 21–24 mm, oblong; skin medium thick, yellow when immature, amber to medium brown at maturity; flesh medium soft; flavor rich; quality very good; ripens midseason. Tree: medium vigor; general good productivity (135–180 kg/tree), occasionally reduced by poor set. Introduced for fruit quality and appearance.

Ghars.—See Rhars.

Halawi.—See Halawy.

Halawy *(Halawi, Hallawi, Hellawi).*—Orig. in Iraq; introd. in the U.S. in 1902 by D.G. Fairchild, U.S. Dept. of Agriculture; commercial plantings date from 1913 importations by P.B. Popenoe. Fruit: semi-dry type; below medium, 35–45 x 17–20 mm, oblong; skin thin, yellow when immature, translucent golden-brown at maturity; flesh soft; very rich flavor, sweet; quality very good; ripens early in season. Tree: medium vigor; productive (70–90 kg / tree). Introduced for fruit quality.

Hallawi.—See Halawy.

Harra.—See Horra.

Hayani.—See Hayany.

Hayany *(Hayani).*—Orig. in Egypt; introd. in the U.S. by D.G. Fairchild, U.S. Dept. of Agriculture, in 1901. Fruit: soft type; large, 45–55 x 22–28 mm, oblong-elliptical; skin medium thick, deep red when immature, almost black at maturity; flesh soft, watery; flavor mild, lacks distinct quality; ripens early. Tree: medium vigor; trunk slender; moderately arched leaves; good yields (110–140 kg / tree). Prolific offshoot production.

Haziz.—Orig. as Deglet Noor o.p. seedling on property of J.H. Northrop near Indio, Calif., between 1911 and 1917; further developed by B.H. Hayes, for whom it was named, after 1924. Fruit: semi-dry type; medium, 44–48 x 22–25 mm, oblong-ovate; skin medium thick, yellow when immature, amber to reddish-brown at maturity; flesh soft and melting; flavor mild and pleasing; ripens midseason. Tree: medium vigor; trunk medium heavy. Flowers early in season, sometimes leading to poor set (due to weather) and low yields.

Hellali.—See Hilali.

Hellawi.—See Halawy.

Herra.—See Horra.

Hilali *(Hellali).*—Orig. from Arabia; introd. to the U.S. in 1902 by D.G. Fairchild, U.S. Dept. of Agriculture. Fruit: soft type; small, 30–36 x 25–27 mm, broadly oval or somewhat ovate; skin thin, yellow when immature, amber at maturity, brown when cured; flesh soft and smooth; flavor mild, delicate; ripens very late. Tree: medium vigor; trunk medium heavy; short leaves. Outstanding characteristic is very late ripening period; fruit sometimes carries into January.

Honey.—Orig. as Deglet Noor o.p. seedling on property of E.K. Daval near Cathedral City, Calif., in 1916 or 1917. Fruit: semi-dry type; medium, 40–47 x 21–23 mm, oblong-oval; skin medium thick, yellow when immature, reddish-brown at maturity; flesh soft and melting; flavor mild and pleasing; ripens over long period, starting before Empress and continuing later. Tree: medium vigor; trunk medium heavy. Yields comparable to Deglet Noor.

Horra *(Harra, Herra, Hourra, Hurra, Hurrah).*—Orig. from Tunisia; imported to the U.S. by T.H. Kearney, U.S. Dept. of Agriculture, in 1905. Fruit: dry type; medium, 37–45 x 19–23 mm, oblong-ovate; skin thick, yellow when immature, dark purplish drab at maturity and when cured; flesh thick, firm; flavor marred by tannin; ripens midseason. Tree: medium vigor; trunk medium heavy. Quality in the U.S. has not approached reputed quality in Tunisia and Algeria.

Hourra.—See Horra.

Hurra.—See Horra.

Hurrah.—See Horra.

Ista'amran.—See Sayer.

Iteema *(Itima, Itime, Yatimeh, Ytima).*—Orig. from Algeria; introd. in the U.S. by W.T. Swingle, U.S. Dept. of Agriculture, in 1900. Fruit: semi-dry type; medium large, 38–50 x 20–26 mm, oblong-ovate; skin medium thick and a little tough, yellow when immature, light to reddish-brown at maturity; flesh soft and melting; flavor very sweet, rather cloying; ripens midseason. Tree: medium vigor; trunk medium heavy; yields 90–115 kg / tree. Quality good, but susceptible to damage from rain or high humidity.

Itima.—See Iteema.

Itime.—See Iteema.

Khadhrawi.—See Khadrawy.

Khadrawi.—See Khadrawy.

Khadrawy *(Khadhrawi, Khadrawi, Khudrawee).*—Orig. in Iraq; introd. in the U.S. in 1902 by D.G. Fairchild, U.S. Dept. of Agriculture. Fruit: soft type; below medium, 33–40 x 20–24 mm, oblong-ovate; skin medium thick, yellow when immature, reddish-brown at maturity; flesh soft; flavor rich, pleasant; quality very good; ripens early in season. Tree: below-average vigor; slow vertical growth; medium productivity. Introduced for fruit quality.

Khalas.—See Khalasa.

Khalasa *(Khalas, Khalaseh, Khalasi, Khlas, Khulas).*—Orig. from Arabia; introd. in the U.S. by P.B. Popenoe in 1913 (private importation for West India Gardens). Fruit: semi-dry type; under medium, 30–40 x 19–23 mm, oblong-oval; skin thin, yellow when immature, amber to reddish-brown at maturity; flesh very tender, melting; flavor rich and delicate; ripens midseason. Tree: moderately low vigor; trunk medium heavy. Imported for excellent quality and flavor, but yields low (55–70 kg / tree). Prolific offshoot production.

Khalaseh.—See Khalasa.

Khalasi.—See Khalasa.

Khasab.—See Khisab.

Khastawi.—See Kustawy.

Khir.—Orig. from Turkey; introd. in the U.S. in 1904 via the British Assistant Political Agency; identity uncertain. Fruit: soft type; medium, 35–42 x 18–21 mm, oblong; skin medium thick, yellow when immature, amber at maturity, brown when cured; flesh soft, becoming caramel-like; flavor mildly sweet, pleasing; ripens early. Tree: medium vigor; trunk medium heavy. Flowers early, sometimes leading to poor set from cold damage or lack of pollen.

Khisab *(Khasab).*—Orig. in Iraq; introd. to the U.S. in 1929 by R.W. Nixon, U.S. Dept. of Agriculture. Fruit: soft type; medium, round-oval; skin tough, deep red when immature, almost black at maturity; flavor and quality mediocre; ripens very late. Tree: medium vigor. Not well characterized in the U.S.

Khlas.—See Khalasa.

Khudrawee.—See Khadrawy.

Khulas.—See Khalasa.

Khustawi.—See Kustawy.

Kush Zebda *(Qush Zabad).*—Orig. in Arabia; introd. in the U.S. in 1913 by P.B. Popenoe (commercial importation for West India Gardens). Fruit: soft type; small, 28–35 x 19–22 mm, ovate; skin thin, tender, yellow when immature, chestnut-brown at maturity; flesh soft, melting; flavor very rich, distinctive; quality excellent; ripens at midseason. Tree: medium vigor; moderate productivity. Introduced for fruit quality.

Kustawi.—See Kustawy.

Kustawy *(Khastawi, Khustawi, Kustawi).*—Orig. in Iraq; introd. in the U.S. by D.G. Fairchild, U.S. Dept. of Agriculture, in 1902. Fruit: soft type; small, 28–38 x 19–23 mm, oblong-ovate or oblong-oval; skin medium thick, yellow when immature, light to reddish-brown at maturity; flesh soft, becoming caramel-like; flavor honey-like and pleasing; ripens midseason. Tree: vigorous; trunk medium heavy; full crown; yields 65–95 kg / tree. Relatively tolerant of high humidity. Similar to Khadrawy but lower in quality.

Lindy.—Orig. as Medjool o.p. seedling on property of F. Heiny, Brawley, Calif., in 1911. Fruit: soft type; large, 35–50 x 26–32 mm, oblong; skin medium thick, pink when immature, amber to dull reddish-brown with purple tint at maturity; flesh soft but firm; flavor very similar to Medjool; ripens early. Tree: vigorous; trunk medium heavy; similar to Medjool in most respects. Good-quality date, but extremely short fruit stalks are too short for proper handling.

Majhul.—See Medjhool.

Maktoom *(Maktum).*—Orig. from Iraq; introd. in the U.S. by D.G. Fairchild, U.S. Dept. of Agriculture, in 1902; commercial planting largely traceable to private importation (for West India Gardens) by P.B. Popenoe in 1913. Fruit: soft type; medium, 30–40 x 22–28 mm, oblong; skin medium thick, yellow when immature, amber to reddish-brown with purple cast at maturity; flesh soft, melting, slightly mealy, becoming caramel-like; flavor mild and delicately sweet; ripens late. Tree: medium vigor; trunk medium heavy or slightly more; foliage lighter and more whitish than most other varieties; yields 80–100 kg / tree. Introduced for fruit quality. Only a small amount of checking. Best adapted to use as fresh date.

Maktum.—See Maktoom.

Manakhir.—See Menakher.

Medjehuel.—See Medjhool.

Medjhool *(Majhul, Medjehuel, Medjool, Tafilat, Tafilatet, Tafilelt).*—Orig. in Bou

Denib, French Morocco, North Africa. Introd. by Walter T. Swingle, U.S. Dept. of Agriculture, in 1927. Fruit: soft type; large, 38–48 x 26–32 mm, oblong-oval to ovate; skin medium, orange-red when immature, amber at maturity, reddish-brown when cured; flesh soft; flavor mildly rich, pleasing; quality good; cures well; ripens early. Tree: high tolerance to rain and high humidity; prolific offshoot production. The second most important commercial variety in the U.S.

Medjool.—See Medjhool.

Menakher *(Manakhir, Monakhir).*—Orig. from Tunisia; introd. to the U.S. in 1905 by T.H. Kearney, U.S. Dept. of Agriculture. Fruit: semi-dry type; large, 45–52 x 22–24 mm, oblong; skin medium thick to thick, red with yellowish outcroppings when immature, dull amber at maturity, reddish-brown when cured; flesh firm; flavor pleasing, resin-like; ripens late. Tree: medium vigor; trunk heavy; long, heavy deep-green leaves. Imported for fruit quality. Unsuited for humid areas.

Mirhage.—See Amir Hajj.

Monakhir.—See Menakher.

Oga de Bedrichen.—See Saidy.

Peggy Ann.—Orig. as Medjool (?) o.p. seedling on property of F. Heiny, Brawley, Calif., in 1911; further developed by L.W. Eddins of Calipatria, Calif., and named for his daughter. Fruit: soft type; large, 46–50 x 25–28 mm, oblong; skin thin, greenish-yellow when immature, amber to reddish-brown at maturity; flesh soft; flavor mildly rich and pleasing; ripens midseason. Tree: resembles Medjool.

Qush Zabad.—See Kush Zebda.

Rashedi.—See Samany.

Rhars *(Ghars).*—Orig. in Algeria; imported to the U.S. in 1900 from W.T. Swingle, U.S. Dept. of Agriculture. Fruit: soft type; large, 45–55 x 20–24 mm, oblong-ovate; skin medium thick and medium tough, yellow when immature, amber to reddish-brown at maturity; flesh soft, melting; flavor rich and sweet but cloying; ripens very early. Tree: medium vigor; trunk heavy; strongly recurved leaves. Good quality but very susceptible to checking from rain or high humidity; yields 90–115 kg / tree when losses from checking do not occur. Vigorous offshoot producer.

Sa'amran.—See Sayer.

Sai.—See Sayer.

Saiar.—See Sayer.

Saidi.—See Saidy.

Saidy *(Oga de Bedrichen, Saidi, Sayd, Sayeh, Seewah, Sewi, Siwah, Siwi, Wahi).*—Orig. in Egypt; introd. to the U.S. in 1901 by S.C. Mason, U.S. Dept. of Agriculture as "Oga de Bedrichen"; commercial (private) importations made by Mason in 1920 and 1922 for Gillette–Rosenberger Date Gardens and Calizona Date Nurseries. Fruit: semi-dry type; medium, 35–48 x 22–28 mm, oblong-oval or oblong-ovate; skin medium thick and a little tough, yellow when immature, light to dull reddish-brown at maturity; flesh soft but rather firm; flavor heavily sweet and rich, a little cloying; ripens late. Tree: medium vigor; trunk heavy; leaves jade-green; yields 90–115 kg / tree. Said to be susceptible to saline conditions and checking.

Samani.—See Samany.

Samany *(Rashedi, Samani, Samiani).*—Orig. in Egypt; introd. into the U.S. by S.C. Mason, U.S. Dept. of Agriculture, in 1925. Fruit: soft type; very large, 50–60 x 25–35 mm, oblong-ovate; skin thin and tender, yellow when immature, dull amber to brown at maturity; flesh soft but rather coarse; flavor mildly sweet early in season, becoming rather insipid later; ripens midseason. Tree: vigorous; trunk medium heavy; leaves long and recurved. Quality good early in season (khalal, early rutab) but disappointing later in season, when most fruit in the U.S. is consumed. Sours easily.

Sambran.—See Sayer.

Samiani.—See Samany.

Sayd.—See Saidy.

Sayeh.—See Saidy.

Sayer *('Amran, Ista'amran, Sa'amran, Sai, Saiar, Sambran, Sayir, Sta'amran, Usta Imran).*—Orig. from Iraq; introd. in the U.S. by D.G. Fairchild, U.S. Dept. of Agriculture, in 1902. Fruit: soft type; medium, 40–48 x 19–22 mm, oblong to oblong-oval; skin rather thick, yellow with red near base when immature, amber to reddish-brown at maturity; flesh soft; flavor good but lacking in delicacy; ripens midseason. Tree: medium vigor; trunk medium heavy;

leaves jade-green; crown appears flattened above; yields 80–90 kg/tree. Fruit stalks susceptible to physiological breakdown.

Sayir.—See Sayer.

Seewah.—See Saidy.

Sewi.—See Saidy.

Siwah.—See Saidy.

Siwi.—See Saidy.

Sphinx *(Black Sphinx).*—Orig. in Phoenix, Ariz. Discovered and introd. about 1925 by Frank Brophy, Phoenix Date Co., Phoenix, Ariz. Open-pollinated seedling. Fruit: semi-dry type; large, 41–48 x 26–28 mm, oblong-conic; skin medium thick, medium red when immature, ripening and curing to brown-black, thin and delicate, adheres to flesh, checks easily; flesh amber, soft, slightly syrupy, smooth, melting, no fiber; flavor sweet, mild, and pleasing; quality very good; seed small to medium; ripens midseason to late, late September to November; resembles Hayany but has a higher sugar content and is not so perishable after it is cured and packed; injured about the same extent as Hayany during unfavorable ripening seasons. Tree: very productive (produces 115–140 kg/tree); trunk thicker than average; slow in growth, about the rate of Khadrawy.

Sta'amran.—See Sayer.

Sultany.—Orig. from Deglet Noor o.p. seedling on property of F. Heiny in Brawley, Calif., in the 1910s or 1920s. Fruit: soft type; below medium, 37–42 x 21–25 mm, oblong-oval; skin tender, light yellow when immature, amber to reddish-brown at maturity; flesh soft; flavor very sweet, somewhat cloying; ripens late. Tree: medium vigor; trunk medium heavy; resembles Deglet Noor, but less olive-green in foliage. Susceptible to checking.

Tabarzal.—Orig. uncertain; probably o.p. seedling found near Mecca, Calif., between 1900 and 1920; developed as variety by M.R. Sheets and J. Wise. Name selected due to alleged resemblance to Egyptian variety with that name. Fruit: soft type; large, 46–54 x 27–29 mm, oblong-oval to oblong-ovate; skin thin and tender, yellow when immature, amber to reddish-brown at maturity; flesh soft; flavor mild and pleasing; ripens early (a little later than Khadrawy). Yields similar to Deglet Noor.

Tadala *(Tadall, Tedalla, Teddala).*—Orig. in Algeria; imported to U.S. by W.T. Swingle, U.S. Dept. of Agriculture, in 1900. Fruit: soft type; large, 46–52 x 18–21 mm, narrowly oblong; skin medium tough, yellow when immature, amber at maturity and when cured; flesh soft; flavor pleasing, but indistinct; ripens early. Tree: medium vigor; trunk rather heavy. Fair quality, but not equal to Rhars. Somewhat resistant to damage from rain and humidity.

Tadall.—See Tadala.

Tafilat.—See Medjhool.

Tafilatet.—See Medjhool.

Tafilelt.—See Medjhool.

Tazerzait.—See Tazizoot.

Tazezait.—See Tazizoot.

Tazizaoot.—See Tazizoot.

Tazizaout.—See Tazizoot.

Tazizaut.—See Tazizoot.

Tazizoot *(Tazerzait, Tazezait, Tazizaoot, Tazizaout, Tazizaut).*—Orig. in Algeria; introd. to the U.S. in 1904 by T.H. Kearney (?), U.S. Dept. of Agriculture. Fruit: semi-dry type; large, 40–50 x 20–25 mm, oblong-elliptical; skin medium thick, yellow when immature, amber to reddish-brown at maturity; flesh soft and melting; flavor rather "heavy" and syrupy; ripens early. Tree: medium vigor; trunk slender to medium heavy; leaves deep green; yields 90–115 kg/year. Susceptible to rain damage. Poor production of offshoots.

Tedalla.—See Tadala.

Teddala.—See Tadala.

Thauri.—See Thoory.

Third Row.—See "T-R".

Thoory *(Thauri, Thuri, Tsuri).*—Orig. in Algeria; introd. in the U.S. in 1900 by W.T. Swingle, U.S. Dept. of Agriculture. Fruit: dry type; above medium, 37–45 x 20–23 mm, oblong; skin medium thick, yellow when immature, light grayish-brown at maturity; flesh firm, brittle, flavor pleasant, nut-like; quality very good for dry-type date; ripens late in season. Tree: vigorous; productive. Introduced for fruit quality.

Thuri.—See Thoory.

Triumph.—See "T-R".

"T-R" *(Third Row, Triumph).*—Orig. as Thoory o.p. seedling on property of E.K. Daval near Cathedral City, Calif., in 1916 or 1917; named after location in third row of planting. Fruit: dry type; large, 43–52 x 19–22 mm, oblong; skin medium thick, pale yellow with greenish tinge when immature, amber to reddish-brown at maturity; flesh soft and melting; flavor pleasing; ripens early. Tree: vigorous; resembles Thoory, but leaves have more curvature and crown is more open.

Tsuri.—See Thoory.
Usta Imran.—See Sayer.
Wahi.—See Saidy.
Yatimeh.—See Iteema.
Ytima.—See Iteema.
Zadie.—See Zahidi.
Zahdi.—See Zahidi.
Zaheedy.—See Zahidi.

Zahidi *(Zadie, Zahdi, Zaheedy, Zaydi, Zehedi).*—Orig. in Iraq; introd. to the U.S. in 1902 by D.G. Fairchild, U.S. Dept. of Agriculture. Fruit: semi-dry type; small, 34–40 x 23–25 mm, obovate; skin rather thick and tough, yellow when immature, light to reddish-brown with faded yellow near base at maturity; flesh firm, of smooth consistency; flavor not outstanding, lacking in delicacy; ripens midseason. Tree: vigorous; trunk medium heavy; leaves jade-green; crown compact; yields high (90–140 kg/tree). Large numbers of offshoots produced.

Zaydi.—See Zahidi.
Zehedi.—See Zahidi.

Minor Varieties Not Described

Imported

**Adebet et Teen
**Ahmar Msab
*Allona
Amhat
Ammary *(Amari, Ammaree)*
Amri *(Amiri, Amary)*
Aooshet *(Ausheh, **Aujeh)*
Apdandon *(Ap-e-dandan, **Hurshud, **Hurshut)*
Areshty *(Areshtee, Arechti, Arishti, Archeti, Arashti, Arichti, Rishti)*
*Ashag *(**Russel's No. 28)*
**Azerza
*Azmashi
Bahrab
**Banawish
*, **Banquet Maktoom
Baydh Hammam *(Bayd Hamam)*
Bayjoo *(Badjou)*
*Beach's 8-4
**Bent el Fgee
**Bent el Maroo
Bent Keballa *(Bent Kebela, Bint Khabala, Bint Qabbaleh, Bint Qabilah, Bint Qibleh, Bint Qubaleh)*
Besser Haloo *(Bisra Haloua, Busr Hulu)*
*Blue Thoory
*Boo Affar
*Boo Fagoos *(*Algerian Iteema)*
Boo Halas
Braim *(Buraym, Brim, Brem, Brehm, *Berhi)*
*Deglet Barca
**Dishtari
Doonga *(Denanga)*
Dubayni *(Dubaini, Deboenii, Debowenii, *Burni)*
Frahee *(Freyeh)*
Fursi *(Firse, Farisi)*
**Gaggar
*Gale's Palm
Gantar *(Guntar, Qintar)*
**Gasb Haloo
**Gasby
Gondeila *(Gondaila, Gondela, Gondila, Gundila)*
*Goondy *(*Goondee, *Goondie)*
*Gush
*, **Halooa *(**Halloua, **Heloua)*
**Halooa Bayda
Hamra Bischry *(Hamra Bechry, Hamra Bishri)*
Hamraya *(Hamraia, **Hamrawi, **Hamrayah, **Hamra)*
**Hamraya-2
Hasan Efendi *(Hasan Effendi, **Hussein Effendi)*
Jafari *(Jaafari)*
Jauzi *(Jauzi Ahmar, Juzi, Jozi, Jozee)*
**Kaiby
Kalara *(Kularu)*
**Karooy
Kenta *(Kanta, **Kentah)*
**Khadrawy of Baghdad
**Khalt Kebeer
Khatuni *(Khatooni)*
Koroch *(Kuroch, **Kuruch, **Korroo, **Begum Jurghi, **Rogani)*
Kseba *(Kasbeh, **Kesba, **Kesseba, **Kessebi, **Ksebba)*
Kush Batash *(Qush Batash)*
Kush Sheham *(Qush Sheham, **Qush Shahm)*

*Lagoo *(*Lagou, *Laqu)*
*Leonard's Unknown
*Lookzy *(**Lookzee, **Rhazee)*
Makelet el Leef *(Makilat al-Lif)*
**Medina
Mesh Degla
*Middleton's Unknown
Mokh Begry *(Moukh Begri)*
*Mozaty
**M'sillia
*Nagal *(*Naghl, *Naghal)*
Najl al Pasha *(Nagl el Basha, **Nazl el Basha)*
Nakhelet Feraoon *(Nakhlat Fir'aun)*
Nakleh el Pasha *(Neklet el Pahsa, **Nakhlet al Pahsa)*
Nakleh Zianeh *(Nakleh Aian)*
*Near Khalasa
**Nesheen
Okht Fteemy *(Oukht Ftimi)*
**Oogbales
Remta *(Remtsa)*
**Retbet Abdala
Retbet Hafsia *(Rutbat Hafsah)*
Retbet Regaya *(Retbet Regaia)*
Roghm Gazal *(Ruhm al Ghazel)*
*Russel's No. 17 *(*Hasawi)*
*Russel's No. 23 *(*Asabi al-Arus)*
*Russel's No. 32 *(*Swaidan)*
*Russel's No. 48 *(*Nabaity)*
*Russel's No. 50 *(*Khasawi al-Baghl)*
Safraia *(Safrayah)*
**Sba Aroosa
Sebaa Loosif *(Sayba Loosif)*
*Seewah
**Shitwi Asfar
Shukkar *(Shukar)*
**Sokria
*Sukkar Nabat *(*Shuccar Nabat)*
*Sukkari *(*Sukeri)*
Tafazween *(Tafazwin, Tafazaouine, Tefzaouine)*
**Tafazweent
**Takadet
Takermest *(Takarmist, Takermoust)*
**Tamzoohart
*Tantaboosht *(*Tantaboucht, *Tentaboosh)*
**Taoorkhet *(*Toorekhet)*
**Taremoont
**Taty
**Taurarket
Tenaseen *(Tenacine, Tennessin, Tanasin, Tanessin, Tenassine, Tinissin)*
Timjooert *(Timjouert, **Timjuhart, **Timdjehouert, **Tamedjohert, **Tinjuhart)*
**Toojat
*Towadant *(*Taoudant)*
Tozer Zaid Khala *(Tozeur Zid Kahala)*
*Tronja *(*Troundja, *Turunja)*
Zagloul *(Zaglul)*
**Zerza
**Zoozia
**Zrai
*16-23

American Origin

Amarillo
Andrate
Black Lucifer
Black Medjool
Black Royal
Deglar
Ford Date
Frau Damke
Heiny
Heinyana
Humbarger
Laguna
Long Yellow
Maktoom No. 1
Malakai
McGill's No. 1
Metzler's Black
*Plum Date
Precisoso
Rubidor
Shield's 6-7
Smoky
Sweetheart
Wieman's Date
Wonderful Black

**Name known to be incorrect.*
***Identity is uncertain.*

ELDERBERRY

Adams.—Orig. in Union Springs, N.Y., by William W. Adams. Introd. in 1926. Selected from wild bushes of *Sambucus canadensis* L. Fruit: clusters and berries exceptionally large; berries somewhat larger than Adams 2. Plant: only partially self-fruitful; cross-pollination required; bark of mature stems is greenish-gray in autumn; strong; vigorous; productive.

Adams 2.—Orig. in Union Springs, N.Y., by William W. Adams. Introd. in 1926. Selected from wild bushes of *Sambucus canadensis* L. Fruit: cluster exceptionally large; berries somewhat smaller than Adams 1. Plant: only partially self-fruitful; cross-pollination required; during autumn, bark of mature stems is reddish; strong; vigorous; somewhat more productive than Adams 1.

Ezyoff.—Orig. in Ithaca, N.Y., by Samuel H. Graham. Introd. in 1938. Parentage unknown; selected in 1934. Fruit: cluster size medium; berries larger than average; berries easily separated from cluster bush. Plant: vigor medium; productivity medium.

Kent.—Orig. in Kentville, Nova Scotia, by E.L. Eaton, Canada Dept. of Agriculture, Res. Sta. Introd. in 1957. Open-pollinated seedling of Adams; crossed in 1944; selected in 1947; tested as Nova Scotia 44-1. Fruit: size and quality equal to Adams but ripens 7–10 days earlier. Plant: vigorous; productive.

Nova.—Orig. in Kentville, Nova Scotia, Canada, by E.L. Eaton, Canada Dept. of Agriculture, Res. Sta. Introd. in 1959. Open-pollinated seedling of Adams 2; selected in 1946; tested as Nova Scotia 43-50. Fruit: large; matures early and uniformly in the cluster; sweeter than Victoria and Kent, which it resembles. Plant: suckers easily; readily propagated by dormant cuttings of 1-year-old canes.

Scotia.—Orig. in Kentville, Nova Scotia, Canada, by E.L. Eaton, Canada Dept. of Agriculture, Res. Sta. Introd. in 1959. Open-pollinated seedling of Adams 2; selected in 1948; tested as Nova Scotia 43-51. Fruit: large; ripens early and uniformly in the cluster; sweeter than Kent and Victoria, which it resembles. Plant: suckers easily; easily propagated by dormant cuttings of 1-year-old canes.

Victoria.—Orig. in Kentville, Nova Scotia, Canada, by E.L. Eaton, Canada Dept. of Agriculture, Res. Sta. Introd. in 1957. Open-pollinated seedling of Adams 2; selected in 1946; tested as Nova Scotia 43-43. Fruit: size and quality equal to Adams, but ripens 3–6 days earlier. Plant: moderately vigorous; productive; readily propagated by dormant cuttings.

York.—Orig. in Geneva, N.Y., by the New York State Agr. Expt. Sta. Named and introd. in 1964. Adams 2 x Ezyoff; selected in 1958; tested as N.Y. 12. Fruit: cluster heavy; berry large; during a 4-year period, averaged only 9.9% soluble solids, about 3% less than Johns, Scotia, and Victoria; ripens after Adams 1 and Adams 2. Plant: very large; more productive than Adams 1 and Adams 2.

FIG

Beall.—Orig. in Santa Clara Valley, Calif., by W.A. Beall. Introd. in 1924. Parentage unknown; seedling transplanted to Fresno, Calif., and first fruited there about 1922. Fruit: medium to large; skin purplish-black with amber pulp; excellent in Imperial Valley, San Diego County, and Fresno. Tree: produces two crops for fresh-fruit purposes; moderately vigorous; common type.

Conadria.—Orig. in Riverside, Calif., by Ira J. Condit, California Agr. Expt. Sta. Introd. in 1955. Adriatic x Calif. 72-80 (caprifig); selected in 1947; tested as Calif. 143-5. Fruit: flesh very firm; smaller eye, lighter dry color, and higher sugar content than Adriatic, which it resembles, also more resistant to fruit spoilage. Tree: vigorous; precocious; produces two crops; breba crop good, second crop very good. Widely planted in San Joaquin Valley for production of dried figs and in dooryards for fresh fruit. Leaves mostly five-lobed, practically free from leaf mosaic.

Desert King.—See King.

Hunt.—Orig. in Eatonton, Ga., by B.W. Hunt. Introd. about 1932. Ischia Green (Verte) x a caprifig in California; cross made before 1929. Fruit: breba crop: small to medium; elongated, with stalk up to 1/2 inch long; skin green; pulp red, somewhat dry, mealy, quality poor; crop small. Second-crop fruit: small to below medium; elongated, pyriform, neck short; stalk up to 3/4 inch long; skin bronze, dull; flesh amber with some red; flavor rich, sweet; quality fair to good; resembles Celeste. In Georgia, fruit parthenocarpic; in California, fruit incompletely parthenocarpic. Tree: vigorous; branches upright; hardy. Variety found only in collections.

Kearney.—Orig. in Kearney Park, Fresno, Calif., by G.P. Rixford, U.S. Dept. of Agriculture. Introd. in 1925. Parentage unknown; selected in 1925; tested as USDA Rixford 2830. Fruit: medium to large; skin green; pulp violet-purple. Tree: vigorous; spreading; abundant cropper; caprifig; not widely propagated commercially.

King *(Desert King).*—Orig. in Madera, Calif., by Sisto Pedrini, Western Evergreen Co., San Francisco. Introd. in 1940 by King Fig Plantation, San Francisco. Parentage unknown; discovered in 1930. Fruit: large; pyriform; skin dark green, thin, smooth; flesh pink, sweet, quality excellent; matures in cool coastal climates as far north as British Columbia, Canada; resembles Genoa; of the White San Pedro group. Tree: moderately vigorous; breba crop prolific; severe dropping of second crop in inland districts.

Tena—Orig. in Riverside, Calif., by W.B. Storey, Univ. of California, Riverside. Informally released in 1971. F_1 of Calimyrna, 158-46 x 271-1; pedigree 279-57. Fruit: skin greenish; white meat with amber pulp; oblate shape and medium-tight eye; average weight 52 g, range 46–60 g; average 19 figs/kg. An attractive midseason fruit with no splitting.

GOOSEBERRY

Abundance.—Orig. in Fargo, N.D., by A.F. Yeager, North Dakota Agr. Expt. Sta. Introd. in 1932. *Ribes missouriensis* x Oregon Champion; tested as N. Dak. 9. Fruit: size medium; round; skin purplish when ripe, thick, tough; borne on long stems. Plant: a strong grower; hardy; extremely productive.

Captivator.—Orig. in Ottawa, Ontario, Canada, by Canada Dept. of Agriculture, Res. Sta. Introd. in 1952. Spinefree x Clark; selected in 1935; tested as Ottawa 272. Fruit: large, equal in size to best American varieties. Plant: thornless.

Clark.—Orig. in Burlington, Ontario, Canada, by M.C. Smith. Introd. in 1922. Chance seedling; apparently a natural hybrid of European and American species. Fruit: very large; skin thick, rather tough, greenish-yellow, turning red when mature; flesh quality fair to good; free of mildew; easy to pick; ripens in late midseason. Plant: moderately vigorous; hardy; appears not to be as productive as Mable; propagated by layering. Clark and Clarke appear to be the same variety.

Colossal.—Orig. in Mankato, Minn., by Frank M. Schwab. Introd. in 1974 by Herschel L. Boll, Champaign, Ill. Open-pollinated seedling of unknown European-type gooseberry; discovered about 1940. Fruit: large, up to 1 1/2 inches in diam., round to round oblong, green; ripens about 1 July in Michigan. Plant: highly productive; grows to 4 ft in height and diam.; vigorous, hardy; has many spines; difficult to propagate.

Como.—Orig. in Excelsior, Minn., by the Univ. of Minnesota Fruit Breeding Farm. Introd. in 1922. Pearl x Columbus; cross made in 1908; tested as Minn. 43. Fruit: large; round to slightly oblong; skin greenish-yellow; flesh green, medium juicy, sweet, quality good; ripens midseason. Plant: vigorous; very productive; resistant to sunscald, leaf spot, and American powdery mildew.

Davidson *(Ross)*.—Orig. probably in Saskatchewan, Canada, by Mrs. A. Davidson, Indian Head, Sask. Introd. in 1936. Parentage unknown; discovered in 1932. Fruit: skin smooth; larger than Pixwell, up to 1 1/2 inches long, which it resembles. Plant: not as heavy in yield as Pixwell; susceptible to injury. No longer propagated.

Fredonia.—Orig. in Geneva, N.Y., by Richard Wellington, New York State Agr. Expt. Sta. Introd. in 1927. Open-pollinated seedling of an unknown English-type gooseberry; originated in 1910. Fruit: very large; roundish oval; skin dark red of the English type, attractive, thick, tough; flesh tender, mild, juicy, subacid to sweet; quality very good; keeps and ships well; ripens late. Plant: healthy; vigorous, upright-spreading; productive; apparently resistant to American powdery mildew.

Glenndale.—Orig. in Little Silver, N.J., by Walter Van Fleet (1857–1922), U.S. Dept. of Agriculture. Introd. in 1932. [(*Ribes missouriense* Nutt. x Red Warrington) x Triumph] x Keepsake; cross made about 1905; selected in 1913; tested near Washington, D.C. Fruit: as large as Downing; roundish-elliptic; skin smooth, dark reddish-purple, thin; seeds small; quality good; excellent for jam. Plant: vigorous; erect, up to 6–8 ft tall; holds foliage until late fall; resistant to leaf spot and American powdery mildew; best suited for southern limit of gooseberry growing in U.S.; very productive; propagated easily from cuttings. Named for Glenn Dale, Md.

Green King.—Introd. in British Columbia, Canada, about 1937 from England. Fruit: large; skin green; quality fine. Plant: prolific bearer.

Pankiw.—Orig. in Dufrost, Manitoba, Canada, by Nicholas Pankiw. Introd. in 1953. Seedling selected in 1945 from seed obtained from western Ukraine in 1938. Fruit: size medium; quality good. Plant: bears well and annually; grows well on heavy soil.

Pembina Pride *(Friesen's Mammoth,*

Thoreson).—Orig. in Morden, Manitoba, Canada, by J.A. Friesen. Introd. about 1935. Seedling of a gooseberry brought from central Europe. Fruit: medium to large; skin tough, green when mature; quality fair to good; resembles Oregon Champion. Plant: hardy; productive.

Perkins.—Orig. in St. Paul, Minn., by Alfred Perkins. Introd. about 1934. Parentage unknown; plants brought from Vancouver, British Columbia, about 1924. Fruit: clusters dense; berry large, up to 1 inch in diam.; long-oval; skin golden-yellow, almost translucent; flesh sweet, quality good. Plant: hardy. No longer propagated.

Perry.—Orig. in Fargo, N.D., by A.F. Yeager, North Dakota Agr. Expt. Sta. Introd. in 1932. Red River wild gooseberry (*Ribes missouriense* Nutt) x Oregon Champion; sibling of Pixwell. Fruit: quality very good; resembles Pixwell. Plant: extremely vigorous; upright; hardy; yields very well.

Pixwell—Orig. in Fargo, N.D., by A.F. Yeager, North Dakota Agr. Expt. Sta. Introd. in 1932. Oregon Champion x *Ribes missouriense* Nutt; cross of 1920; sibling of Perry. Fruit: medium; skin pale green, becoming pink; when ripe, thin; in clusters and borne on long pedicels; few thorns, easily picked; very good for all culinary purposes. Plant: hardy; very productive.

Ross.—See Davidson.

Spinefree.—Orig. in Ottawa, Ontario, Canada, by Canada Dept. of Agriculture, Res. Sta. Introd. about 1935. (F_2 of *Ribes oxyacanthoides* x Victoria) x Mabel; selected in 1925. Fruit: of reasonable commercial size; skin thick and bright red when ripe. Plant: thornless; upright. Spinefree is the female parent of Captivator.

Welcome.—Orig. in Excelsior, Minn., by the Minnesota Agr. Expt. Sta. Fruit Breeding Farm. Introd. in 1957. From open-pollinated seed of Poorman, collected in 1938; selected in 1944; tested as Minn. 206. Fruit: medium-large; skin light, dull red, glabrous; flesh pink; flavor tart; quality good; seeds few, small; ripens before Pixwell and Como. Plant: spines sparse, short, weak, missing on older wood; vigorous; productive; relatively free of disease, particularly anthracnose.

GRAPE

Adams *(Champagne)*.—Orig. in Hammonton, N.J., by Alvin Adams. Introd. in 1920 by Hugo Kind, nurseryman. Chance seedling of fox grape (*Vitis labrusca*); discovered in 1888. Berry: large; skin amber; flesh juicy, sweet, very foxy flavor, quality poor. Vine: vigorous; productive; similar to the native *V. labrusca* in being very disease and cold resistant.

African Queen.—Orig. in Brooks, Ga., by B.O. Fry and W. Ison. Introd. in 1988. Dixieland x Sugargate. Plant patent 7295; 14 Aug. 1990. Female muscadine. Berry: large, up to 28 mm; skin black, medium thick; dry scar; tangy flavor; 22% solids; ripens midseason. Cluster: four to 10 berries. Vine: vigorous; productive.

Aki Queen.—Large-fruited, seeded table grape. Orig. in Akitsu, Hiroshima, Japan, by Hiroyasu Yamane. Kyohou selfed. Introd. in 1991. Berry: very large (13 g); obovate; bright red; treatment with gibberellic acid can improve fruit set and induce seedlessness; foxy flavor. Cluster: winged-conical; loose. Vine: vigorous; downy and powdery mildew resistance higher than *V. vinifera* cultivars.

Aki Seedless.—Seedless black table grape. Orig. in Akitsu, Hiroshima, Japan, by Hiroyasu Yamane. Muscat Bailey A x Himrod; selected in 1972; tested as Akitsu-1; introd. in 1986. Berry: medium (3 g); short-elliptic; slipskin; no aroma. Cluster: large; winged-conical; well-filled. Vine: fairly vigorous; productive; moderately resistant to ripe rot and downy mildew; cold hardiness is low.

Alachua.—Orig. in Leesburg, Fla., by J.A. Mortensen and J.W. Harris, Central Florida Res. and Education Center. Introd. in 1990. Fry x Southland; selected in 1982; tested as Fla. CA9-48. A mechanically harvestable, black muscadine with uniform ripening and dry stem scar. Berry: size of Cowart; oval; ripens 26 Aug. at Leesburg. Vine: productive; moderately vigorous; disease resistant; may not need regular sprays, except in areas of heavy disease pressure; flower self-fertile; postharvest susceptibility to angular leaf spot.

Albemarle.—Orig. in Raleigh, N.C., by the U.S. Dept. of Agriculture and the North Carolina Agr. Expt. Sta. Introd. in 1962. Topsail x Burgaw; selected in 1951; tested as NC 9-305. Muscadine type. Berry: large; skin blue-black, smooth; flesh sweet; flavor very good; soluble solids average 19%; ripens about midseason, with Burgaw; superior to Burgaw, Duplin, and Tarheel in size, flavor, sugar content. Vine: vigorous; productive; resistant to leaf diseases, being as good as Burgaw, Duplin, and Tarheel in these respects; flower self-fertile. Suggested as a black pollinator; for home gardens where a single vine is desired.

Alden.—Orig. in Geneva, N.Y., by Richard Wellington, New York State Agr. Expt. Sta. Introd. in 1952. Ontario x Grosse Guillaume; cross made in 1925; selected in 1932; tested as N.Y. 13035. Cluster: large; tapering; loose; conical; shouldered. Berry: large; oval; reddish-black; non-slipskin; flesh juicy, meaty, tender; pleasing aroma; quality very good; season just after Concord. Vine: vigorous; productive, tends to overbear; adapted to wide range of soils; some winter injury. Recommended for home and local market use.

Almission.—Orig. in Fresno, Calif., by L.O. Bonnet. Introd. in 1941. Mission x Alicante Bouschet; selected about 1936. Cluster: loose; large. Berry: black; juice red; used for wine. Vine: vigorous.

Alwood.—Orig. in Blacksburg, Va., by Virginia Agr. Expt. Sta. Introd. in 1967. Fredonia x Athens. Cross made in 1949 by R.C. Moore; selected in 1955 by Moore for propagation and further testing, and in 1960 for naming and release; tested as V.P.I. 5-2. Cluster: medium; compact; cylindrical to tapering; often shouldered. Berry: medium; skin blue-black, heavy waxy bloom; slipskin medium thick and medium tough, not subject to cracking; flesh sweet; moderately foxy, compares in flavor to Worden or Concord; quality good; ripens a few days after Fredonia or about 2 1/2 weeks before Concord; ripens uniformly. Vine: vigor medium; cane internodes short, giving a compact or bushy type of growth; productive; hardy; tolerance to black rot, anthracnose, downy mildew, and powdery mildew. Named after William B. Alwood, first horticulturist of the Virginia Agr. Expt. Sta.

American Rose.—Large, very firm, seeded table grape. Orig. in Fowler, Calif., by John Paboojian, Jr. Plant patent 6829; 6 June 1989. Chance seedling of unknown parentage. Berry: variable color, depending on season and culture; but under favorable conditions, has a dominant color of reddish-brown with a small, greenish-white halo encircling the pedicel; large (9.2 g); neutral flavor; three seeds/berry. Cluster: medium; loose but well-filled; conical. Vine: vigorous; productive; one or two clusters/shoot. Released as an extremely firm, large, fresh-market grape.

Athens.—Orig. in Geneva, N.Y., by New York State Agr. Expt. Sta. Introd. in 1938. Hubbard x Portland; cross made in 1925; selected in 1932. Cluster: larger than Concord, to 300 g; conical with large shoulders; loose. Berry: medium; ellipsoidal; skin reddish-black, cracks easily; fair adherence; flavor foxy; ripens 3 weeks earlier than Concord; does not hold or ship well. Vine: vigorous; bears heavy crops, often only in alternate years; *V. labrusca* type.

Atkan.—Orig. in Brookings, S.D., by N.E. Hansen, South Dakota Agr. Expt. Sta. Introd. in 1925. Lady x North Dakota wild. Cluster: long. Berry: medium; skin white with pink tinge.

Aurelia.—Orig. in Greensboro, N.C., by Robert T. Dunstan. Introd. in 1963 by Southeast Nurseries, Raleigh, N.C. Chaouch x Seyve-Villard 12-375; selected in 1950; tested as D56. Cluster: long; tapered; shouldered; stem very long. Berry: large; oval; skin greenish-golden; flesh firm; sweet; high dessert-quality table variety; resembles Chaouch. Vine: extremely vigorous; must be pruned or cluster-thinned to prevent over-bearing; perhaps more robust than Seyve-Villard 12-375, which it resembles; appears to be resistant to mildew and phylloxera; somewhat susceptible to black rot; grows well from North Carolina to Oklahoma.

Aurore *(Seibel 5279)*.—Orig. in Aubenas, Ardeche, France, by Albert Seibel. Introd. into Canada in 1946, into U.S. in 1940s. Seibel 788 x Seibel 29. Cluster: long, to 10 inches; loose. Berry: spherical to flattened, to 1/2 inch in diam.; skin golden to light pink; pulp soft and juicy; foxy flavor; makes a light, delicate wine of fair quality but of foxy flavor; matures very early. Vine: buds-out early; vigorous and productive if pruned long; hardy; subject to black rot; appears most promising in the cool, humid areas of northeastern U.S.

Autumn Royal.—Black, late-ripening table grape. Orig. in Fresno, Calif., by David W. Ramming and Ronald Tarailo, U.S. Dept. of Agriculture Horticultural Crops Res. Lab. Not patented. Autumn Black x C74-1; selected in 1984; tested as A97-68; introd. in 1996. Berry: large (6.5 g); ovoid to ellipsoidal; skin black to purple-black with a light, waxy bloom, medium thick, medium tough, adheres to flesh; flesh translucent green-yellow and very firm; flavor sweet, neutral, good; quality best if harvested at 18% soluble solids or higher; two to four aborted seeds/berry that are medium to small and slightly detectable in some years; ripens after Ribier, about the first week of October in Fresno. Cluster: large (0.5–1.0 kg); conical with shoulders; well-filled and compact; pedicel medium in length with medium to poor attachment to the rachis. Vine: medium to slightly weak in vigor; shoots and canes brittle and easily broken by hand; production very good; should be spur-pruned.

Baco *(Baco 1)*.—Orig. in Bélus, Landes, France, by Francois Baco. *Vitis vinifera* Folle blanche x *V. riparia*; cross made in 1902; tested as Baco 24-23. Introd. in U.S. in 1951. Cluster: very small; cylindrical; compact. Berry: small; round; skin jet black; flesh soft; ripens early. Vine: rampant, canes long; leaves resemble *V. riparia*; very good resistance to downy mildew, somewhat susceptible to powdery mildew; buds-out and blossoms very early; must be pruned long to be productive; growth cycle short; best adapted to regions of short growing season; useful for arbor-type plantings.

Wine: highly colored; heavy; bitter; herbaceous character, requiring long aging to improve.

Baco 1.—See Baco.

Ballhill.—Orig. in Forestville, N.Y., by Frank J. Grisanti. Introd. in 1964. Plant patent 2405; 26 May 1964. Bud mutation of Fredonia. Cluster: larger than Fredonia and Concord. Berry: skin deep blue; flavor and eating quality excellent; keeping and shipping quality exceptional; ripens 2 weeks before Concord. Vine: extremely hardy; blooms 10 days after Concord; wood matures in early fall.

Bath.—Orig. in Geneva, N.Y., by Richard Wellington, New York State Agr. Expt. Sta. Introd. in 1952. Fredonia x N.Y. 10805 (Chasselas Rose x Mills); cross made in 1930; selected in 1937; tested as N.Y. 18149. Cluster: medium; cylindrical to tapering; compact. Berry: medium; round-oval; skin black; flesh tender, juicy; sweet; quality fair; ripens end of September in New York; lacks foxy flavor of Fredonia, which it resembles. Vine: vigorous; very productive, tends to overbear; fairly hardy.

Beaufort.—Orig. in Cincinnati, Ohio, by Byron T. Johnson. Introd. in 1982. Open-pollinated seedling of *Vitis rupestris* Ft. Worth #3, Richter (PI 232091); pollen parent apparently French hybrid. Tested as selection 64.RFW-1. Berry: small; skin dark blue-black; round; flavor light, resembling black currants; ripens 10 Aug. in Cincinnati. Cluster: small; conical; five/shoot. Vine: vigorous; productive; winter hardy; tolerant of insects and diseases. Produces a deeply colored red wine.

Beaumont.—Orig. in Cincinnati, Ohio, by Byron T. Johnson. Introd. in 1982. Open-pollinated seedling of Headlight; tested as selection 62.A-1. Berry: small; skin lavender-pink; round; ripens midseason. Cluster: well-filled; prominently shouldered; resembles Delaware; two or three/shoot. Vine: vigorous; moderately productive; fruit borne on first four nodes only; tolerant of common insects and diseases; winter hardy. Produces a neutral white wine.

Beauty Seedless.—Orig. in Davis, Calif., by H.P. Olmo, California Agr. Expt. Sta. Introd. in 1954. Scolokertek kiralynoje x Black Kishmish; cross made in 1941; first fruited in 1946; tested as Calif. K 5-69. Cluster: large; conical; heavily shouldered; often too compact. Berry: size of Sultanina; ovoid; skin bluish-black, waxy, good adherence; flesh tender, firm; flavor spicy; quality good at full maturity; seedless; ripens very early, with Perlette; subject to sun scald if planted on light-textured sandy soils. Vine: very vigorous, with thick trunk; foliage distinctive blue-green; very productive as spur-pruned cordon. The first black seedless cultivar produced by breeding.

Beaver.—Orig. in Mountain Grove, Mo., by Paul H. Shepard, Missouri State Fruit Expt. Sta. Introd. in 1947. Open-pollinated seedling of Triumph; selected in 1945; tested as G-388. Cluster: medium. Berry: medium; skin black; flesh firm, tender, juicy, separates easily from seeds; highly flavored, quality high; ripens 1 week before Moore's Early; hangs well without cracking or shattering. Vine: productive; vigorous; flower self-fertile.

Belair.—Orig. in Roodeplat, Republic of South Africa, by E.P. Evans and P.J.L. Ellis, Viticultural and Oenological Res. Inst. Barlinka x Black Spanish; tested as 74–1407; introd. in 1986. Berry: large (8 g); oblate; blue-black; firm; not juicy; neutral flavor. Cluster: large; oblong; well-filled and compact. Vine: vigorous; basal buds fruitful; productive. Released as a table grape with good storage potential that ripens in mid- to late-season.

Bellevue.—Seeded white table grape. Orig. in Stellenbosch, Republic of South Africa, by P.J.L. Ellis and E.P. Evans, Viticultural and Oenological Res. Inst. Olivette Blanche open-pollinated; tested as C2-17; introd. in 1982. Berry: large (9 g); round-oval. Cluster: large; more round than oblong; uniform. Vine: productive; vigorous. Developed as a replacement for Waltham Cross.

Berenda Red.—Orig. in Madera, Calif., by John M. Garabedian, Fresno, Calif. Introd. in 1971. Plant patent 3124; 18 Apr. 1972; assigned to Superior Farming Co., Bakersfield, Calif. Cardinal x unnamed seedless selection; cross made in 1963; selected in 1967. Berry: large-medium; ellipsoidal; skin port-wine red, medium thick, tough; flesh greenish-opaque, firm, seeded; good dessert and keeping qualities; ripens 7 days before Cardinal. Vine: medium-large; vigorous; productive.

Bianco One.—See Queen Arlene™.

Biscayne *(Fennell 113)*.—Orig. in Princeton, Fla., by J.L. Fennell, Lady Lake, Fla. Introd. for trial about 1948. *Vitis rufotomentosa* 9 x Feher Szagos; selected in 1942. Berry: quality similar to cultivars of *V. vinifera*. Cultivar lost by the originator.

Black Beauty.—Orig. in Newton, Ill., by L.A. Richards. Introd. in 1949. Parentage unknown; selected in 1940. *Vitis labrusca* type. Berry: large, up to 1 1/4 inch in diam.; round; skin black; good for jelly, wine, and dessert; ripens October and early November. Vine: self-pollinating; heavy producer; regular bearer.

Black Beauty.—Orig. in Brooks, Ga., by B.O. Fry and W.G. Ison. Introd. in 1988. Fry x 12-12-1 muscadine. Plant patent 7592; 16 July 1991. Female muscadine. Berry: large (31 mm); skin black, edible; dry scar; good flavor; 24% solids; ripens mid- to late season. Cluster: large. Vine: very vigorous; medium productivity; disease resistant.

Black Emerald.—Black, seedless grape for fresh market. Orig. in Fresno, Calif., by David W. Ramming and Ronald Tarailo. Not patented. A69-190 x C84-116; selected in 1987; tested as A12-137; introd. in 1994. Berry: round to slightly oval; medium (2.5 g); responds to gibberellic acid at a rate lower than that required for Thompson Seedless; flavor sweet and neutral, best when harvested with at least 18% solids; two to four aborted seeds that are very small and undetectable; ripens early, between Perlette and Flame Seedless. Cluster: medium in size and length; conical with a shoulder; well-filled to slightly compact. Vine: medium vigor; production good to high when cane-pruned. Recommended as an early ripening, fresh-market table grape.

Black Fry.—Orig. in Brooks, Ga., by B.O. Fry and W.G. Ison. Introd. in 1986. Fry x Cowart. Plant patent 5824; 9 Dec. 1986. Female muscadine. Berry: large (30 mm); skin black; 21% solids; good flavor; uniform ripening, early to midseason. Cluster: large; can be clipped in clusters. Vine: disease resistant; very productive.

Blackrose.—Orig. in Fresno, Calif., by Elmer Snyder and Frank N. Harmon, U.S. Hort. Field Sta. Introd. in 1951. (Damas Rose x Black Monukka) x Ribier (Alphonse Lavallée); cross made in 1941; first fruited in 1944; selected in 1944. Cluster: large; conical; well-filled but not usually compact unless girdled. Berry: large; high quality; skin jet black with a light grayish bloom, very attractive appearance; because of skin tenderness, better adapted to local markets than to distant ones; ripens in early midseason; a table grape for *V. vinifera* table-grape areas. Vine: vigorous; very productive.

Blanc Du Bois.—Orig. in Leesburg, Fla., by J.A. Mortensen, Central Florida Res. and Education Center. Introd. in 1987. Fla. D6-148 (self of Fla. A4-23) x Cardinal; selected in 1974; tested as Fla. H18-37. White wine cultivar for southern U.S. Berry: size of Suwannee; skin light green; flavor muscat-like; ripens 3 July at Leesburg. Vine: productive and vigorous; flower self-fertile; resistant to Pierce's disease; susceptible to anthracnose, black rot, and ripe rot. Does not require grafting, except in limey soil.

Bluebell.—Orig. in Excelsior, Minn., by Univ. of Minnesota Fruit Breeding Farm. Introd. in 1944. Beta x an unknown variety of eastern U.S.; selected in 1923; tested as Minn. 158. Berry: resembles Concord in size and color; high quality for dessert, juice, jelly; matures about mid-September. Vine: productive; vigorous; hardy.

Blue Boy *(Cook, Davenport Blue)*.—Formerly introduced as Cook. Stark Bros. Nurseries & Orchards Co., Louisiana, Mo., purchased exclusive propagation rights and have renamed this cultivar.

Blue Eye.—Orig. in Mountain Grove, Mo., by Paul H. Shepard, Missouri State Fruit Expt. Sta. Introd. in 1947. Ellen Scott x American; selected in 1947; tested as G-794. Cluster: compact; larger than Concord. Berry: large; round; skin black, tough; flesh sweet, firm; handles and keeps well; ripens about 10 days after Concord; berry larger than Concord; produces an attractive red juice. Vine: vigorous; flower self-fertile.

Blue Jay.—Orig. in Excelsior, Minn., by A.N. Wilcox, W.H. Alderman, and F.E. Haralson, Univ. of Minnesota Fruit Breeding Farm. Introd. in 1944. Parentage unknown; selected in 1923; tested as Minn. 69. Berry: nearly size of Concord; skin dark blue, bloom blue; good for juice and jelly; matures in mid-September. Vine: hardy; productive; flower pistillate, requiring cross-pollination.

Blue Lake.—Orig. in Leesburg, Fla., by L.H. Stover, Watermelon and Grape Investigations Lab., Florida Agr. Expt. Sta. Introd. in 1960. Florida 43-47 (open-pollinated selection of *Vitis smalliana* x Caco); cross made in 1950. Cluster: loose. Berry: skin uniformly blue with a light surface bloom; flesh spicy, slightly tart; ripens from late June to late July at Leesburg; adapted for fresh juice, pasteurized juice, jellies, preserves; when fully mature, distinctive from European dessert types; limited value for fresh market; excellent as a home-garden cultivar. Vine: resistant to Pierce's disease; flower self-fertile; in some seasons, fruit set may be improved by cross-pollination, e.g., by Lake Emerald.

Bluestar.—Orig. in Beltsville, Md., by D.H. Scott, U.S. Dept. of Agriculture and Maryland Agr. Expt. Sta. Introd. in 1967. Fredonia x Niagara. Cross made in 1946 by N.H. Loomis; selected in 1950 by D.H. Scott; tested as US-4606-48. Cluster: medium-large; tight; attractive. Berry: medium; skin blue; flavor mildly foxy, pleasant; soluble solids content varying between 16% and 21%; ripens between Fredonia and Concord, ripening more evenly; for roadside-stand sales and home gardens. Vine: tends to over-produce unless properly pruned; adapted to central area of the U.S.; more consistently productive than Fredonia.

Blush Seedless.—Orig. in Davis, Calif., by H.P. Olmo, California Agr. Expt. Sta. Plant patent 4856; 8 June 1982; assigned to Regents of Univ. of California. Introd. in 1981. Emperor x Z 4-87; cross made in 1966; first fruited and selected in 1970; tested as 31-37F. Cluster: large; uniformly long conical; borne at fourth or fifth node; well-filled to compact. Berry: ovoid; large (3.8 g); skin tender, bright red develops as blush; flesh meaty; flavor neutral; seed traces green and noticeable, but not flinty. Vine: foliage much resembling Emperor, but more fruitful and ripening several weeks before; good shipping and storage cultivar for late market.

Bokay.—Orig. in Mountain Grove, Mo., by Paul H. Shepard, Missouri State Fruit Expt. Sta. Introd. in 1947. Captain x Terret Monstre; selected in 1947; tested as G-994. Cluster: large; compact. Berry: large; skin yellow; flesh meaty, tender; pleasing flavor; ripens with Concord; keeps well on the vine or in storage for 2 months without breakdown; resembles Malaga. Vine: vigorous; productive; hardy; flower self-fertile.

Bonnet Seedless.—Orig. in Fresno, Calif., by Leon O. Bonnet. Introd. in 1934. Plant patent 88; 13 Mar. 1934. Muscat of Alexandria x Thompson Seedless; selected in 1932. Cluster: medium; globular; loosely branched; loose to poorly filled; peduncle reddish-green. Berry: ellipsoidal (18 x 23 mm); dull green; partially seedless; flavor neutral; pulp soft. Vine: vigorous; semi-erect; poor productivity; leaf deeply lobed and petiolar sinus open wide. Appears inferior to Thompson Seedless in California.

Bonnet Seedless Muscat.—Orig. in Fresno, Calif., by L.O. Bonnet. Introd. in 1941. Muscat of Alexandria x Sultana. Cluster: medium. Berry: small; variable in set; seeds partly hardened. Vine: produces irregular crops; weak grower.

Bonnoir.—Black, late-season, seeded table grape. Orig. in Stellenbosch, Republic of South Africa, by E.P. Evans and P.J.L. Ellis, Viticultural and Oenological Res. Inst. Barlinka x Alphonse Lavallée; tested as 11-2-63; introd. in 1986. Berry: large (8 g); short-oval; firm; not juicy; neutral flavor. Cluster: medium-large; oblong with well-developed shoulders; well-filled to slightly compact. Vine: vigorous; productive. Developed as a late-season, black, low-cracking variety.

Bountiful.—Orig. in Meridian, Miss., by N.H. Loomis, U.S. Dept. of Agriculture. Introd. in 1967. Creek x seedling of Topsail; tested as Meridian 45-16D. Muscadine type. Berry: slightly larger than Creek or Thomas, about 150 berries per lb; skin purple, glossy; rich vinous flavor; quality high; sugar content 10% to 20%; ripens 5–8 days later than Southland, 15–18 Sept. at place of origin; tends to shatter when fully ripe; excellent as fresh fruit and good for culinary use. Vine: very vigorous; productive; has produced very well in the central and southern portions of the Gulf Coast states and Georgia; suggested for trial commercially, for home use, and as pollinizer for cultivars having imperfect flowers.

Bounty.—Orig. in Stillwater, Okla., by Herman A. Hinrichs, Agr. Expt. Sta., Oklahoma State Univ. Introd. in 1975. Ellen Scott x Bailey; cross made in 1949; selected in 1954; tested as Okla. 220. Cluster: medium to large with two large shoulders. Berry: medium; skin blue-black with heavy bloom; juice red, 18% sugar, 1.1% tartaric acid; ripens evenly, about 20 Aug. at Stillwater. Vine: vigorous; dense foliage; self-fertile.

Brandis.—Orig. in Terryville, Conn., by R. Grem. Introd. in 1985. Origin: roadside chance seedling; appears to have *Vitis aestivalis, V. riparia,* and *V. labrusca* ancestry. Berry: medium; skin blue; flesh pulpy until very ripe; pleasant flavor. Cluster: medium cylindrical; loose to moderately compact; small shoulder. Vine: vigorous; productive compared to other wild vines; tolerant of marginal soils; slightly less hardy than Concord; female; resistant to black rot; moderately resistant to powdery and downy mildew. Produces pleasant, fruity wine.

Brocton.—Orig. in Geneva, N.Y., by S.A. Beach, New York State Agr. Expt. Sta. Introd. in 1919. Brighton x N.Y. 125 (Winchell x Diamond); cross made ln 1899. Berry: white; quality good; flavor less foxy than Niagara, which it resembles.

Bronx Seedless.—Orig. in Geneva, N.Y., by A.B. Stout, New York State Agr. Expt. Sta., in cooperation with the New York Botanical Garden. Introd. in 1937. N. Y. 8536 (Goff x Iona) x Sultanina; cross made in 1925; selected in 1931. Cluster: large; long conical; loose. Berry: small; oval; red; soft rudimentary seeds; flesh juicy, mild, soft; quality good; cracks easily during wet weather; ripens 1 week before Concord. Vine: fairly hardy; productive; vigorous; susceptible to anthracnose, downy mildew, and cluster spoilage. Recommended where cracking is not prevalent.

Brownie.—Orig. in Experiment, Ga., by J.G. Woodroof, Georgia Agr. Expt. Sta. Introd. in 1933. Open-pollinated seedling of San Monta; selected in 1930. Muscadine type. Berry: very high sugar content, one of highest muscadine cultivars; quality excellent.

Bruce's Sport.—Orig. in Merbein, Victoria, Australia, by A.J.J. Bruce. Introd. in 1934. Bud mutation of Sultanina (Thompson Seedless). Cluster: about the same size or smaller than Sultanina. Berry: smaller than Sultanina; rounded; skin paler than Sultanina; flesh white, less tendency to darken during drying than Sultanina; flavor slightly "flat" and less rich than Sultanina; seedless; recommended as a raisin variety. Vine: less vigorous, productive, and hardy than Sultanina; susceptible to downy mildew, oidium (powdery mildew), and anthracnose; first season's leaves same as Sultanina, but after flowering the leaves become variegated and develop irregular, whitish translucent areas.

Bryant.—Orig. in Mountain Grove, Mo., by Paul H. Shepard, Missouri State Fruit Expt. Sta. Introd. in 1947. Muench x *Vitis vinifera* Terret Monstre; selected in 1947; tested as G-2589. Cluster: large; loose. Berry: size of Concord; skin black; flesh firm, juicy, tender; flavor sweet, pleasing; ripens late, 4 weeks after Concord. Vine: very vigorous; productive; flower self-fertile.

Buffalo.—Orig. in Geneva, N.Y., by Richard Wellington, New York State Agr. Expt. Sta. Introd. in 1938. Herbert x Watkins; cross made in 1921; selected in 1928; tested as N.Y. 10830. Cluster: medium-large; usually well-filled to compact; shouldered; conical. Berry: medium, 18–19 mm long, 16–18 mm wide; skin reddish-black, tough, medium thick, slipskin; bloom heavy; flesh greenish, translucent, medium juicy, tender; flavor sweet, not aromatic, tart with pleasing flavor, spicy; quality excellent; ripens about 1 week before Concord; seed size medium; resembles Herbert. Vine: vigorous; productive; drooping; hardy. Considered very fine for home garden and local market.

Burgaw.—Orig. in Willard, N.C., by Charles Dearing, U.S. Dept. of Agriculture and North Carolina Agr. Expt. Sta. Introd. in 1946. Thomas x V19 R7 B2 (Scuppernong x New Smyrna); cross made in 1912; tested as V4 R27 B4. Muscadine type. Cluster:

loose; open; consisting of one to six berries. Berry: medium; round; skin reddish-black, shiny, medium thick, tough; flesh sprightly, fairly sweet, tenacious to seeds; seed fairly small; resembles Thomas. Vine: very vigorous; lateral growth rather short; foliage dense, disease-resistant, broadly cordate; growing tips reddish-green; flower self-fertile, can be used as a pollinator, self- and cross-fruitful; very productive annually; resembles Thomas. Named for Burgaw, N.C., location of an experimental vineyard.

Calmeria.—Orig. in Fresno, Calif., by Elmer Snyder and Frank N. Harmon, U.S. Dept. of Agriculture. Introd. in 1950. Open-pollinated seedling of Ohanez (Almeria); selected in 1941. Cluster: large; well-filled. Berry: large; ovoid elongated; skin greenish-white, tough; flesh white, meaty; medium sugar content and low acidity; resembles Ohanez; less susceptible to fruit spotting than Ohanez, but more subject to sunburn. Vine: vigor medium; very productive; flower with upright stamens and self-fertile. A very late-maturing cultivar well adapted for cold storage, similar to Ohanez.

Calzin.—Orig. in Davis, Calif., by H.P. Olmo, California Agr. Expt. Sta. Introd. in 1958. Zinfandel x Refosco; cross made in 1937; tested as Calif. G22-2 and Calif. Oakville 4. Cluster: large; long conical; heavily shouldered; compact. Berry: medium; spherical; skin bluish-black, heavy bloom; resistant to spoilage; pulp soft, very juicy; flavor of Zinfandel, very high in tannin. Vine: similar to Zinfandel but more tolerant to red spider mite; productivity above average. For red wine production of Zinfandel type.

Canada Muscat. Orig. in Geneva, N.Y., by Richard Wellington, New York State Agr. Expt. Sta. Introd. in 1961. Muscat Hamburg x Hubbard; cross made in 1928; seed planted in 1930; first described in 1933; propagated for wider testing in 1937 as N.Y. 17806. Cluster: medium; compact. Berry: greenish-yellow; Muscat flavor. Vine: moderate vigor and productivity; may suffer winter injury below –15F; fairly susceptible to downy mildew. Recommended for making sweet, muscatel wine. Grown commercially on limited scale in Ontario grape district, Canada.

Canadice.—Orig. in Geneva, N.Y., by R.M. Pool, J.P. Watson, K.H. Kimball, and J. Einset, Agr. Expt. Sta. Bath x Himrod; cross made in 1954; selected in 1962; tested as NY 45625. Cluster: compact; large. Berry: medium; ripens early, usually fully mature at Geneva by 1 Sept.; flavor is *V. labrusca* type, comparable to Delaware; slip-skin, tender and edible; seedless; stores well; tolerant of postharvest fumigation with SO_2. Vine: strong growing; resistant to phylloxera; not immune to downy mildew, powdery mildew, and black rot, but resistance is high enough so that a spray program that controls these diseases on Concord will also control them on Canadice; vines are winter hardy, enduring temperatures to –29C; a seedless cultivar adapted to northeastern North America.

Canner.—Orig. in Davis, Calif., by H.P. Olmo, California Agr. Expt. Sta. Introd. in 1958. Hunisa x Sultanina; cross made in 1931; tested as Calif. A4-72. Cluster: large; long conical; prominent shoulders; compact. Berry: larger than Sultanina; ovoid elongated; seedless; skin dark green, tender, resistant to cracking; ripens 2 weeks later than Sultanina; low in sugar and acid; pulp greenish-white, very crisp; detaches very easily and cleanly from capstem, thus useful for canning and fruit cocktail; processes very well. Vine: very vigorous, more so than Sultanina; very productive; quite susceptible to powdery mildew late in the season; young shoots maroon; leaf dark green, five-lobed, almost glabrous, petioles wine-red.

Cape Fear.—Orig. in Willard, N.C., by Charles Dearing, U.S. Dept. of Agriculture and North Carolina Agr. Expt. Sta. Introd. in 1946. Burgaw x V20 R36 B4 [Vl9 R7 B2 (Scuppernong x male) x Kilgore]; cross made in 1915 or 1916; tested as 55B4. Muscadine type. Cluster: six to eight medium to large berries. Berry: slightly oval; good adherence; skin greenish-yellow, thin to medium thick; pulp smooth, soft; fairly acid; moderately sweet; very late ripening; resembles Scuppernong. Vine: average vigor; growth open; upright; fair disease resistance; leaves crinkly; growing tips yellow-green; flower pistillate. Named after locality in southeastern North Carolina.

Cardinal.—Orig. in Fresno, Calif., by Elmer Snyder and Frank N. Harmon, U.S. Dept. of Agriculture Hort. Field Sta. Introd. in 1946. Flame Tokay x Ribier (Alphonse Lavallée); selected in 1943. Cluster: very numerous, three/cane; medium; short conical; very irregular in set; well-filled to ragged; peduncle thick, bright green; often with few, interspersed, greenish shot berries. Berry: matures very early; large; spherical to oblate; often lobed or creased; skin dark red to reddish-black, thin; flesh firm, crisp, greenish; seeds few, greenish, often hollow and inviable. Vine: very vigorous to rampant; buds-out medium; thick trunk; requires heavy flower cluster removal; very productive; short spur pruning on cordon.

Carlos.—Orig. in Raleigh, N.C., by W.B. Nesbitt, V.H. Underwood, and D.E. Carroll, North Carolina Agr. Expt. Sta. Introd. in 1970. Howard x NC 11-173 (Topsail x Tarheel). Cross made in 1951 by Carlos F. Williams; selected in 1954 by Williams (U.S. Dept. of Agriculture–North Carolina State Univ.). Tested as NC 57-56. Muscadine type. Cluster: intermediate. Berry: 1/2 inch in diam.; round; skin bronze, attractive; dry stem scar; flavor similar to Scuppernong; ripens midseason with Scuppernong; for wine or fresh fruit. Vine: vigorous; very productive; hardy; perfect-flowered, self-fertile; intermediate resistance to bitter rot, angular leaf spot, and powdery mildew; susceptible to black rot and rust; suitable for mechanical harvesting.

Carmine.—Orig. in Davis, Calif., by H.P. Olmo, Agr. Expt. Sta., Univ. of California. Introd. in 1975. Plant patent 3929; 29 June 1976; assigned to the Regents of the Univ. of California, Berkeley. F2-7 (Carignane x Cabernet Sauvignon) x Merlot; cross made in 1946; first fruited in 1951; tested as M51-50 and 802. Cluster: 11.6 x 14.5 cm; average 267 g; short conical; loose to very compact; usually well-filled; shouldered in one-half of clusters. Berry: average 1.9 g; uniform in size from cluster base to apex; spherical; skin black with bluish bloom, similar to Cabernet Sauvignon in skin and pulp texture and color; two to three seeds/berry; ripens 1–3 weeks before Cabernet Sauvignon; holds well on vine; no shriveling or spoilage; makes outstanding dry red wine, accentuates Cabernet flavor and aroma, improves with aging, exceptionally long life. Vine: upright; canes short, few, like Carignane; leaves dark green, glabrous, thin; easily trained; short pruned; more productive than Cabernet Sauvignon; widely adapted, from cooler coastal regions of California to areas as warm as Fresno, Calif.

Carnelian.—Orig. in Davis, Calif., by H.P. Olmo, Agr. Expt. Sta. Introd. in 1973. Plant patent 3625; 24 Sept. 1974; assigned to the Regents of the Univ. of California. F2-7 (Carignane x Cabernet Sauvignon) x Grenache; cross made in 1949; selected in 1954. Cluster: compact; average 455 g; short conical; second cluster on cane almost as large as primary. Berry: about the size of Carignane; black with bluish bloom; seeds very small; sugar content moderate, but acidity remaining high at maturity. Vine: vigorous, resembles Grenache; foliage light yellow-green, glabrous, glossy. Recommended for the production of light-bodied red table wines or rose in the warm Central Valley of California, superior to the standard Carignane and Grenache. Wine matures early, but also improves with aging.

Carolina Blackrose.—Orig. in Greensboro, N.C., by Robert T. Dunstan. Introd. in 1964 by Southeast Nurseries, Raleigh, N.C. Blackrose x Aurelia; selected in 1957; tested as D200. Cluster: large; heavily shouldered. Berry: large; oval; skin black; flesh firm; dessert quality excellent; resembles Blackrose. Vine: vigorous; canes large, robust, somewhat upright; appears resistant to mildew and phylloxera, somewhat susceptible to black rot; prune close to prevent over-bearing.

Cascade *(Seibel 13053)*.—Orig. in Aubenas, Ardeche, France, by Albert Seibel. Seibel 7042 x Seibel 5409. Cluster: small. Berry: black; ripens very early, with Marechal Foch. Vine: vigorous; very productive; hardy; susceptible to downy mildew; adapted to moist areas of short growing season where *V. vinifera* cultivars are not successful. Wine low in color but considered of passable quality when blended with better cultivars.

Cayuga White.—Orig. in Geneva, N.Y., by John Einset and W.B. Robinson, Agr. Expt. Sta. Introd. in 1972. Seyve-Villard 5-276 x Schuyler; cross made in 1945; selected in 1952; tested as New York 33403 or G.W. 3. Cluster: medium-large; medium compact; long and slightly tapering. Berry: medium-large; roundish to ellipsoidal; skin white, resistant to cracking; flesh meaty, somewhat astringent; ripens 24–29 Sept. (Concord season); soluble solids from 17.0% to 19.0%; recommended for a quality dry white wine resembling White Riesling. Vine: vigorous; highly productive; flower perfect with upright stamens.

Centennial Seedless.—Orig. in Davis, Calif., by H.P. Olmo and Albert T. Koyama, California Agr. Expt. Sta. Plant patent 4784; 3 Nov. 1981. Introd. in 1980. Named to commemorate the centennial of the Dept. of Viticulture and Enology. Gold x Q 25-6 (F_2 sel. Emperor x Pirovano 75). Cross made in 1966; first fruited in 1970; selected and tested as seedling 31-123F. Cluster: medium to large; conical; well-filled to loose; self-thinning at fruit set. Berry: ovoid elongated; yellow-green; mostly parthenocarpic, with less seed development than Thompson Seedless; very large for a seedless grape, 1.6 x 2.5 cm; ripens early August at Fresno, Calif.; adherence good; produces large raisins of excellent quality with mild Muscat flavor; very responsive to application of gibberellic acid in increasing berry size. Vine: vigorous; very large leaves; fruitful with spur (two to three buds) pruning on bilateral cordon.

Centurion.—Orig. in Davis, Calif., by H.P. Olmo, Agr. Expt. Sta., Univ. of California. Introd. in 1975. Plant patent 3870; 20 Apr. 1976; assigned to the Regents of the Univ. of California, Berkeley. F2-7 (Carignane x Cabernet Sauvignon) x Grenache; cross made in 1949; first fruited in 1954; tested as B5. Cluster: long conical; occasional small wing; loose to well-filled; 281 g; two to three / cane. Berry: size and shape similar to Carignane; pronounced but complex flavor; ripens last half of September; recommended for dry, dark-red table wine and also port wine, and wine improves on aging; sugar about 24%; acidity 0.76% to 0.94% as tartaric; average pH 3.3. Vine: very productive, 10–12 tons / acre; vigor moderate; short pruning recommended. For areas such as the warm San Joaquin Valley of California.

Century I.—Orig. in Blacksburg, Va., by George D. Oberle, Virginia Polytechnic Inst. and State Univ. Introd. in 1973. Seyve-Villard 20-347 x Dunstan 3 (Chasselas Violet x Golden Muscat); tested as VPI32. Berry: large; skin reddish-black, adherent; ovate; flesh crisp, meaty, with a sweet and fruity flavor; ripens late midseason. Cluster: large. Vine: over-productive; above-average vigor; medium cold hardy.

Challenger.—Orig. in Mountain Grove, Mo., by Kenneth W. Hanson and Merlyn A. Haag, Missouri State Fruit Expt. Sta. Introd. in 1983. Parentage unknown. Selected in 1968; tested as G-2963. Berry: medium; round, tear-shaped; skin orange-red to deep red, thick, tough, adheres to pulp; flesh very meaty and firm with one or two soft, small seeds; high soluble solids; sweet, delicate flavor; ripens 10 days after Reliance and Himrod. Cluster: medium; loose, with almost no shoulders. Vine: moderately vigorous; winter hardiness comparable to Reliance. Susceptible to black rot, downy mildew, and powdery mildew.

Champagne.—See Adams.

Chancellor *(Seibel 7053)*.—Orig. in Aubenas, Ardeche, France, by Albert Seibel. Introd. in U.S. in the 1940s and in Canada in 1946. Seibel 5163 x Seibel 880. Cluster: medium; cylindrical; fairly compact; winged. Berry: round to slightly ovoid; skin jet black; flesh not juicy, very firm; ripens early, 2 weeks before Concord. Vine: vigor medium; productivity medium; good crop recovery after spring frost; upright in growth habit; leaf entire, rounded outline, petiolar sinus closed with overlapping lobes; buds-out early; very susceptible to powdery mildew early in the season.

Chardonel.—White wine grape with Chardonnay character. Orig. in Geneva, N.Y., by B.I. Reisch, R.M. Pool, W.B. Robinson, T. Henick-Kling, J.P. Watson,

K.H. Kimball, M.H. Martens, G.S. Howell, D.P. Miller, C.E. Edson, and J.R. Morris, New York State Agr. Expt. Sta., Cornell Univ. Introd. in 1990. Plant patent 7860, 5 May 1992; assigned to Cornell Res. Foundation. Seyval x Chardonnay; selected in 1960; tested as NY 45010. Berry: amber; medium; spherical; seeded. Cluster: shouldered; medium-large (200 g); average 1.6 clusters/shoot; cluster thinning occasionally required. Vine: productive; vigorous; moderately winter hard; moderately susceptible to powdery mildew, downy mildew, and botrytis bunch rot; foliage is not susceptible to sulfur injury. Produces high-quality wine with Chardonnay character.

Chelois.—Orig. in Aubenas, Ardeche, France, by Albert Seibel. Introd. in 1946 in Ontario, Canada; in the U.S. in 1948. Seibel 5163 x Seibel 5593. Cluster: medium; elongate; compact. Berry: small; round; skin black; flesh firm; ripens medium early, 1 week before Concord; makes a red wine of almost neutral character. Vine: buds-out late; healthy; productivity medium; vigorous; hardy; sometimes injured by drought.

Chief.—Orig. in Meridian, Miss., by N.H. Loomis, U.S. Dept. of Agriculture. Introd. in 1967. Creek x sibling of Topsail; selected in 1951; tested as Meridian 18. Muscadine type. Berry: slightly larger than Bountiful, about 140 berries per lb; skin darker than Bountiful; less tendency to shatter when ripe than Bountiful; flavor less pronounced than Bountiful; sugar content 19% to 20%. Vine: very vigorous; productive; has produced very well in the central and southern portions of the Gulf Coast states and Georgia; suggested for trial commercially, for home use, and as a pollinizer for cultivars having imperfect flowers.

Chilcott.—Orig. in Woodward, Okla., by L.F. Locke, U.S. Dept. of Agriculture, Southern Great Plains Field Sta. Introd. in 1959. Open-pollinated seedling of Volney; tested as Woodward 8556. Cluster: compact, medium to large. Berry: medium to large; skin dark red to nearly black at maturity; sugar and acid content high; flavor distinctly muscat; quality good to excellent; ripens in mid-August at place of origin. Vine: growth medium to vigorous; crops from light to heavy; less winter-hardy than Reating and Osborn; very drought resistant; probably best adapted to the southwest.

Chonkee.—Orig. in Brookings, S.D., by N.E. Hansen, South Dakota Agr. Expt. Sta. Introd. in 1925. Lady x a wild grape found in North Dakota. Berry: medium; skin yellowish-white; quality good. Vine: productive; vigorous.

Chontay.—Orig. in Brookings, S.D., by N.E. Hansen, South Dakota Agr. Expt. Sta. Introd. in 1925. Massasoit x Beta. Berry: very large; skin bluish-purple; flavor very good; seeds separate easily from flesh. Vine: vigorous.

Chowan.—Orig. in Raleigh, N.C., by the U.S. Dept. of Agriculture, and the North Carolina Agr. Expt. Sta. Introd. in 1962. Creswell x Burgaw; selected in 1952; tested as NC 29-193. Muscadine type. Cluster: loose. Berry: large; skin light brown to bronze, appearance attractive; flesh subacid, aromatic; flavor good; soluble solids about 19%, or 3% higher than Scuppernong; ripens in early midseason, about 1 week before Scuppernong, which it resembles. Vine: vigorous; productive, with good foliage; flower perfect, self-fertile; superior to Wallace and Willard in flavor, soluble solids, appearance, production, fruit size, and attractiveness. Suggested uses: as a bronze pollinator for commercial plantings or home gardens where a single vine is wanted.

Christmas.—Orig. in Santa Rosa, Calif., by Luther Burbank. First introd. in 1915 by the originator; reintroduced in 1926 by a California nursery. Claimed to be a seedling of Pierce. Berry: similar to Pierce; ripens late, over a long season. Vine: very vigorous; suitable for arbor planting; tetraploid.

Christmas Rose.—Orig. in Davis, Calif., by H.P. Olmo and Albert T. Koyama, California Agr. Expt. Sta. Plant patent 5056; 31 May 1983; assigned to Regents of Univ. of California, Berkeley. Introd. in 1980. S44-35c x 9117D. Of complex parentage involving four generations, derived from Emperor, Hunisa, Pirovano 75, and Nocera. Cross made in 1962; seedling vine selected in 1966; tested as c15-47. Cluster: long conical; large to very large. Berry: large, up to 9 g; ellipsoidal to ellipsoidal obovoid; skin dark red to reddish-black; flavor neutral. Believed superior to commercial Emperor in yield and regularity in cropping, and more uniform earlier coloring, larger berry, and improved palatability compared to Emperor.

Cimarron.—Orig. in Perkins, Okla., by Herman A. Hinrichs, Oklahoma Agr. Expt. Sta. Named and released in 1958. *Vitis cinerea* var. *canescens* x Seneca; cross made in 1944; tested as Okla. 23S52. Cluster: medium to large; moderately compact; slightly tapered. Berry: medium to large; roundish; skin blue-black with heavy bloom; flesh tender, juicy, sweet, acidity low; seed size medium; ripens evenly, in midseason, about 1 week before Concord; similar to Concord in shape and color, but somewhat smaller; for fresh, juice, and jelly purposes. Vine: healthy; vigorous; productive; drought resistant; very hardy; resistant to black rot.

Clarke.—Discovered about 1925 in the woods of Mississippi by M.D. Clarke and J.C.C. Price. Introd. in the early 1930s. Parentage unknown. Female muscadine. Berry: large; skin reddish-brown with large russet dots; season late.

Colobel.—Orig. in Aubenas, Ardeche, France, by Albert Seibel. Introd. into Canada and U.S. in 1946. Seibel 6150 x Seibel 5455. Cluster: long; medium to large; narrow; compact. Berry: spherical; skin bluish-black; pulp soft, acidity high; juice light red; ripens in Concord season; holds well on vine. Vine: growth upright; vigor medium; shoots with cobwebby hairs; young unfolding leaves bordered red, petiolar sinus V-shaped, teeth sharp, foliage reddening in autumn; tolerant of downy mildew and can be grown without spray control; subject to anthracnose; tendency to winter-kill until vine is well established; color of wine very intense, but acidic, bitter, and astringent.

Conquistador.—Orig. in Leesburg, Fla., by John A. Mortensen, Agr. Res. Ctr., Univ. of Florida. Introd. in 1983. (*V. smalliana* x Concord) x [(Norris x Concord) Ell-40]; selected in 1974; tested as Fla. L4-33. Berry: resembles Concord; quality good; for fresh fruit, wine, jelly, or juice. Vine: slightly delayed budbreak lessens frost damage in spring; productive when grafted on suitable rootstock; tolerant to Pierce's disease.

Cook *(Blue Boy, Davenport Blue)*.—Orig. in Shrewsbury, Mass., by H.A. Cook. Introd. in 1951 by S.L. Davenport, North Grafton, Mass. (Ring x Delaware) x Niagara. Cluster: large; shouldered. Berry: large; round; skin blue, tough; flesh excellent quality and flavor; ripens before Concord, 7–15 Aug. at Chekotah, Okla., hangs for long time on vine. Vine: vigorous; hardy; productive.

Cowart.—Orig. in Experiment, Ga., by B.O. Fry, Georgia Agr. Expt. Sta. Introd. in 1968. Higgins x Ga. 28; selected in 1960; tested as Ga. 12-2-2. Muscadine type. Cluster: size good. Berry: large; skin black; flavor good; ripens 12 Sept. in Experiment. Vine: vigorous; very productive; self-fertile; moderately tolerant of diseases and insects; adapted throughout southeastern U.S.

Creek.—Orig. in Experiment, Ga., by J.G. Woodroof, Georgia Agr. Expt. Sta. Introd. in 1938. Open-pollinated seedling of San Monta; selected in 1934. Muscadine type. Berry: skin very thin; very high percentage of free-run juice; high in sugar and acidity; ripens late.

Creswell.—Orig. in Creswell, N.C., on farm of F.R. White. Introd. in 1946. Parentage unknown; discovered about 1915. Female muscadine. Cluster: round, compact; one to 15 berries. Berry: large; round; good skin adherence; skin dark red, turning black when fully ripe, medium thick, tough; high sugar content; very good flavor; late-ripening. Vine: average vigor and productivity; growth drooping; average disease resistance; growing tips reddish-green.

Crimson Seedless.—Red, seedless, late-ripening table grape. Orig. in Fresno, Calif., by David W. Ramming and Ronald Tarailo, U.S. Dept. of Agriculture Horticultural Crops Res. Lab., and Sayed A. Badr, California State Univ. Not patented. C133-199 x Emperor; selected in 1983; tested as C102-26; introd. in 1989. Berry: medium (4.0 g, 20.8 mm long, 16.6 mm in diam.); cylindrical to oval; responds to gibberellic acid, although rates used for Thompson Seedless cause excessive thinning and a thickened, rigid rachis and pedicel resulting in excessive shatter; ripens in early October in Fresno, slightly before Emperor; has achieved 19% soluble solids and a sugar : acid ratio of >20:1; flesh is light yellow, translucent, meaty, firm; skin bright red with thick, white bloom, thick, medium tough, and does not separate from flesh; flavor sweet, neutral; contains two aborted, soft seeds that are very small and undetectable when eaten; holds on vine well and does not crack. Cluster: well-filled to slightly compact; medium (0.5 kg weight, length 20 cm); conical with a shoulder. Vine: very vigorous on its own roots; cane pruning advisable due to low production on spurs in initial trials; can be grown where other late-ripening *V. vinifera* cultivars are grown.

Dallas.—Orig. in Dallas County, Ala., on the farm of W.K. Campbell. Introd. in 1953 by Dalco Nursery, Selma, Ala. Parentage unknown; discovered in 1940 by a member of the Alabama Agr. Expt. Sta. and C.B. Neador. Muscadine type. Berry: extremely large, over 1 inch in diam.; skin reddish-black and dotted; quality good; matures in midseason; hangs well on vine, but leaves clean scar when harvested.

Darlene *(Darling)*.—Orig. in Brooks, Ga., by W.G. Ison. Introd. in 1988. 5-11-3 x Carlos. Plant patent 7314; 4 Sept. 1990. Female muscadine. Berry: 31 mm, consistently large throughout vine; skin bronze to pinkish-red; dry scar; melting pulp; 24% solids; excellent flavor; ripens early to midseason. Cluster: large. Vine: vigorous; disease resistant.

Darling.—See Darlene.

Dattier of St. Vallier.—See Seyve-Villard 20-365.

Dauphine.—Late-season, white, seeded table grape. Orig. in Stellenbosch, Republic of South Africa, by P.J.L. Ellis and E.P. Evans, Viticultural and Oenological Res. Inst. Almeria open-pollinated; tested as B16–32; introd. in 1983. Berry: large (9 g); long; oval; firm; sweet-neutral flavor. Cluster: large; oblong; well-filled to slightly compact; shouldered. Vine: vigorous; productive; similar to other cultivars for disease susceptibility. Developed as a late-season, white, crack-resistant variety.

Davenport Blue.—See Blue Boy, Cook.

Dawn.—Orig. in Experiment, Ga., by H.P. Stuckey, Georgia Agr. Expt. Sta. Introd. in 1938. Scuppernong x black male muscadine; selected in 1937. Muscadine type. Berry: resembles Scuppernong; early ripening.

Dawn Seedless.—Orig. in Davis, California, by H.P. Olmo and Albert T. Koyama, California Agr. Expt. Sta. Plant patent 4788; 10 Nov. 1981. Introd. in 1980. Gold x Perlette; cross made in 1966. Seedling vine first fruited in 1969 and selected and tested in trial blocks as g4-36. Cluster: similar to Perlette, but more uniform, less compact, and usually requires no berry thinning. Berry: larger than Perlette and more uniform; ripens 1 week later than Perlette in the Coachella Valley desert region. Vine: resembles Perlette, but less vigorous.

Daytona.—Orig. in Leesburg, Fla., by John A. Mortensen, Agr. Res. Ctr., Univ. of Florida. Introd. in 1983. Fla. 133-90 x Exotic. Selected in 1968; tested as Fla. F4-16. Berry: resembles Tokay; superior bunch and berry size; pulp easily chewed; tenacious berries permit handling. Vine: tolerant to Pierce's disease.

Dearing.—Orig. in Raleigh, N.C., by Charles Dearing, U.S. Dept. of Agriculture and the North Carolina Agr. Expt. Sta. Introd. in 1957. Luola x Burgaw; cross made in 1917; evaluated by N.H. Loomis and Carlos F. Williams; tested as US-B6-44. Muscadine type. Berry: size similar to Thomas; skin translucent green; sugar content averages 1% higher than Topsail; flesh crisper than most other muscadine cultivars; flavor sweet, superior to other perfect-flowered cultivars ripening at the same time; matures in late midseason, after Topsail but before Yuga; resembles Scuppernong. Vine: perfect-flowered; productive; adapted primarily for home gardens but suitable for use as a pollinator in commercial plantings of imperfect-flowered cultivars.

De Chaunac *(Seibel 9549)*.—Orig. in Aubenas, Ardeche, France, by Albert Seibel (1844–1936). Introd. into Canada in 1946. Seibel 5163 x Seibel 793. Cluster: medium to large; somewhat loose. Berry: blue; suitable for wine. Vine: vigorous; productive; hardier and more resistant to disease than many other French hybrids.

Delight.—Orig. in Davis, Calif., by H.P.

Olmo, California Agr. Expt. Sta. Introd. in 1947. Scolokertek kiralynoje (P.I. 103996) x Sultanina marble (P.I. 98388); cross made in 1936; first fruited in 1940; tested as Calif. 1220F21; sibling of Perlette. Cluster: large, 400–600 g; well-filled to compact; conical; shouldered. Berry: slightly larger than Sultanina (Thompson Seedless); uniform; oval; skin thick, dark greenish-yellow, lenticels very prominent, resistant to injury; flesh firm; flavor distinct; seedless; ripens early, just after Perlette; resembles Sultanina, but ripens 10 days earlier, shatters less, and in most localities has a slight muscat flavor; sometimes astringent; suitable for table and raisin use. Vine: vigor medium; fruitful when spur-pruned.

Desert Flame.—Orig. in Lodi, Calif., by Marvin L. Nies. Introd. in 1964. Plant patent 2345; 21 Jan. 1964. (Emperor x Calif. I8-70) x (Scolokertek x Muscat Hamburg); selected in 1956; tested near Thermal, Calif. Cluster: medium to small; averages 20–30 cm long; shoulder not pronounced on less-vigorous vines. Berry: 2.25 cm wide; ellipsoidal elongated to round; skin bright red, uniform, relatively thin; flesh clear, crisp, firm, not juicy; flavor unique with a very slight Muscat flavor; normal acidity; ripens early, with Perlette; intended principally as a table cultivar. Vine: moderately vigorous; leaf sinuses deep; habit with no particular distinguishing characteristics.

Digby.—Orig. in Brooks, Ga., by B.O. Fry, Griffin, Ga. Introd. in 1983. Jumbo x seedling 29-49; selected in 1971; tested as 19-6-1. Cluster: large. Berry: large; skin bronze; ripens evenly, midseason. Vine: vigorous; productive; tolerant to black rot.

Dix.—Orig. in Beltsville, Md., by C.A. Magoon and I.W. Dix, U.S. Dept. of Agriculture (USDA). Introd. in 1954 jointly by the USDA and the New Jersey Agr. Expt. Sta. America x Bailey; seed grown in 1938; selected in 1942; evaluated by D.H. Scott and E.G. Christ; tested as US-712-1. Berry: medium; skin black; flavor mild to neutral; ripens about the same season as Concord; soluble solids content ranging from 16% to 19%; acidity of 0.7% to 0.8%; intense red coloration in the juice and pulp valued for blending where more color is desired. Vine: vigorous and productive. Primarily of value as a juice grape rather than for dessert purposes.

Dixie.—Orig. in Clayton, N.C., by C.F. Williams, W.B. Nesbitt, and V.H. Underwood, U.S. Dept. of Agriculture and North Carolina Agr. Res. Serv., Dept. of Horticultural Science, North Carolina State Univ., and J.A. Mortensen, Univ. of Florida. Introd. in 1976. Topsail x NC 28-193 (Lucida x Wallace); cross made in 1953; selected in 1957; tested as N.C. 88-102. Muscadine type. Berry: large, 8/cluster; skin bronze; flesh mild, sweet; quality resembles Topsail; excellent wine quality. Vine: productive; vigorous; self-fertile; breaks dormancy after limiting chilling in Florida, hardy under North Carolina Coastal Plains conditions; susceptible to powdery mildew; highly resistant to Pierce's disease; resistant to leaf spot; practically immune to anthracnose and downy mildew; adapted to mechanical harvest.

Dixieland.—Orig. in Brooks, Ga., by B.O. Fry. Introd. in 1976. Plant patent 4771; 29 Sept. 1981. Fry x S. 29-49; cross made in 1969; selected in 1974; tested as S. 5-7-5. Muscadine type. Berry: large; oval; skin bronze, thickness medium; quality good. Vine: vigorous; susceptible to black rot.

Dixiered.—Orig. in Brooks, Ga., by B.O. Fry, Ison Nursery. Introd. in 1976. Plant patent 4770; 29 Sept. 1981. Seedling 44-6 x S. 44-7; cross made in 1969; selected in 1974; tested as S.11-10-1. Muscadine type. Berry: skin bronze to red; ripens midseason. Vine: vigorous; self-fertile; hardy; productive; tolerant to blackrot and ripe rot; most valuable characteristics are cluster size and yield.

Doreen.—Orig. in Clayton, N.C., by C.F. Williams, W.B. Nesbitt, and V.H. Underwood, U.S. Dept. of Agriculture and North Carolina State Univ. and developed by J.P. Overcash, Mississippi State Univ. Introd. in 1981. Higgins x Dixie. Tested as NC 276-108; cross made in 1959. Muscadine type. Berry: 5 g; 8–10/cluster; skin bronze, attractive, medium thick; flavor pleasing, typical muscadine aroma; ripens September in Mississippi and October in North Carolina. Vine: very productive; vigorous; hardy; self-fertile; resistant to berry shatter and most diseases (including Pierce's disease);

adapted to mechanical harvest. Produces a light, golden wine equal to Carlos, Scuppernong, and Sterling.

DOVine.—Early ripening raisin grape with potential to dry on the vine. Orig. in Fresno, Calif., by David W. Ramming and Ronald Tarailo, U.S. Dept. of Agriculture Horticultural Crops Res. Lab. Not patented. P79-101 x Fresno Seedless (using embryo rescue techniques); selected in 1987; tested as A25-58; introd. in 1995. Berry: oval to slightly rounded; medium (1.85 g fresh weight); resulting raisin weight 0.52 g and similar to Fiesta, but larger than Thompson Seedless; skin white, medium tough, medium thick, adheres to flesh; flavor sweet, neutral, good; one to two aborted seeds that are small and undetectable; soluble solids reach 22% by the second week of August in Fresno, allowing the canes to be cut and the fruit to dry on the vine; fruit dried on the vine consistently produced plumper raisins compared to drying on trays; ripens 7 days before Fiesta and 14–17 days before Thompson Seedless. Cluster: medium to large in size and length; conical; well-filled to slightly compact. Vine: vigorous; canes grow upright; "T" trellis recommended to spread fruiting canes and dense canopy; production good to high when cane-pruned.

Dulcet.—Orig. in Experiment, Ga., by J.G. Woodroof, Georgia Agr. Expt. Sta. Introd. in 1934. Open-pollinated seedling of Irene; selected in 1928. Muscadine type. Berry: excellent quality; persistent; resembles Thomas. Vine: foliage very resistant to black rot.

Dunkirk.—Orig. in Geneva, N.Y., by New York State Agr. Expt. Sta. Introd. in 1920. Brighton x Jefferson; cross made in 1899; tested as N.Y. 2569. Cluster: uniform; compact; resembles Delaware. Berry: larger than Delaware; skin dark red; good shipper; ripens late. Vine: vigorous; hardy; healthy; productive; very short internodes.

Duplin.—Orig. in Willard, N.C., by Charles Dearing, U.S. Dept. of Agriculture and North Carolina Agr. Expt. Sta. Introd. in 1946. Stanford x V10 R15 B4 [Eden x V23 R4 B2 (Eden x *Vitis munsoniana*)]; cross made about 1912. Muscadine type. Cluster: very loose; pendant; to 2 inches long. Berry: medium to large; oval; skin dull black, medium thick; pulp soft; average flavor; seeds large. Vine: vigorous; canes few; open growth habit; growing tips reddish-green; flower perfect, large.

Early Catawba.—Orig. in the area of Geneva, N.Y.; originator unknown. Introd. about 1945. Seedling of unknown origin sent about 1945 by F.E. Gladwin, New York State Agr. Expt. Sta., to K.F. Rynalski, Fredonia, N.Y., for planting. Cluster: medium; fairly compact. Berry: size average; heavy bloom; ripens about 2 weeks before Catawba; yields wine of good quality in northeastern U.S. Vine: vigorous; hardy; susceptible to mildew; only fair productivity.

Early Fry.—Female, early ripening, bronze muscadine for fresh market. Orig. in Brooks, Ga., by W.G. Ison. Introd. in 1993. Plant patent 9225; 1 Aug. 1995. Sweet Jenny x Ison. Berry: round; large (32 mm); dark bronze; seeded; dry scar; ripens 1–2 weeks earlier than Ison. Cluster: 5–15 berries/cluster. Vine: productive; very vigorous; tolerant to common diseases.

Early Giant.—Orig. in Altus, Ark., by Herman J.B. Wiederkehr. Introd. in 1932. Plant patent 42; 8 Nov. 1932; assigned to Stark Bros. Nurseries & Orchards Co., Louisiana, Mo. Parentage unknown. Berry: Concord type; larger and earlier than Campbell's Early, which it resembles; similar to Worden in flavor. Vine: thrifty; dark green foliage; the tetraploid form of Campbell's Early.

Early Muscat.—Orig. in Davis, Calif., by H.P. Olmo, California Agr. Expt. Sta. Introd. in 1958. Muscat Hamburg x Scolokertek kiralynoje; cross made in 1943; tested as Calif. K4-19. Cluster: loose; pyramidal. Berry: medium; shape uniform; skin dull yellow; flesh color white, texture firm; Muscat flavor; seeds small; ripens very early, just after Perlette and Pearl of Csaba; carries well in transit; principally for early table market, also sweet Muscat wine. Vine: vigor medium but very productive; should be girdled (ringed) to increase berry size; requires crop reduction by cluster thinning.

Early Niabell.—Orig. in Davis, Calif., by H.P. Olmo, California Agr. Expt. Sta. Introd. in 1958. Campbell 4n x Niagara 4n;

cross made in 1942; tested as Calif. L11-2. Cluster: medium; cylindrical; well-filled. Berry: large; slightly oval; skin purplish-black; flesh greenish; ripens in early midseason; useful for table fruit, fresh juice, semi-sweet wines. Vine: vigor medium to poor; very productive; tolerant to powdery mildew; requires some flower cluster thinning if long-pruned; tetraploid.

Early Superior Seedless™ *(Sugrafive)*.— Orig. in Riverside County, Calif, by Luther C. Hahn and Joseph Maranto, Superior Farming Co. Introd. in 1983 as Sugrafive. Plant patent 5151; 6 Dec. 1983; assigned to Superior Farming Co. Bud mutation of Superior Seedless; discovered in 1978. Berry: slightly smaller than Superior Seedless; similar in color and shape; ripens with Perlette, 7–10 days before Superior Seedless; develops a very mild muscat flavor when very ripe. Cluster: medium loose to compact; cylindrical to conical. Vine: very vigorous; medium to low productivity; breaks bud very early.

Edelweiss.— Orig. in Osceola, Wis., by E. Swenson, P. Pierquet, and C. Stushnoff. Introd. in 1980. MN 78 (Beta x Witt) x Ontario; selected in 1955; tested as E.S. 40. Seeded table grape. Berry: medium; round; skin green, slipskin, white bloom; flesh tender, melting; flavor fruity-labrusca, becoming strongly foxy at late maturity; early ripening. Cluster: large; conical; loosely formed; usually winged. Vine: vigorous; productive; resists foliage diseases similar to Beta; very cold hardy.

Eden.— Orig. in Geneva, N.Y., by Richard Wellington, New York State Agr. Expt. Sta. Introd. in 1938. Ontario x N.Y. 10085 (Triumph x Mills); cross made in 1923; selected in 1928. Cluster: small, to 300 g; long conical; poorly filled. Berry: medium; spherical; reddish-black; good quality. Vine: medium to poor vigor; low productivity; not sufficiently hardy.

Einset Seedless.— Orig. in Geneva, N.Y., by B.I. Reisch, G.W. Remaily, R.M. Pool, and J.P. Watson, New York State Agr. Expt. Sta., Cornell Univ. Introd. in 1985. Plant patent 6160; 26 Apr. 1988; assigned to Cornell Res. Foundation. Fredonia x Canner Seedless; selected in 1967; tested as NY 63.878.1. Berry: skin bright red with a light, waxy bloom; ovoid; medium-soft seed remnant; early ripening. Cluster: medium, 150 g; well-defined shoulder; moderately loose. Vine: productive; vigorous; moderately winter hardy; tolerant to botrytis bunch rot.

Eleven Point.— Orig. in Mountain Grove, Mo., by Paul H. Shepard, Missouri State Fruit Expt. Sta. Introd. in 1947. Captain x Terret Monstre; sibling of Bokay and Gagconade; selected in 1947; tested as G-1320. Cluster: compact; cylindrical; no shoulder; often over 1 ft long. Berry: large; skin black; flesh firm, tender, sweet, sprightly, separates readily from seeds; suitable for table or red wine; resembles Captain, but larger and better quality. Vine: vigorous; heavy producer; female flower.

Emerald Riesling.— Orig. in Davis, Calif., by H.P. Olmo, California Agr. Expt. Sta. Introd. in 1948. Muscadelle (of California) x White Riesling; cross made in 1936; first fruited in 1939; selected in 1939; tested as Calif. 1139E29. Cluster: large, 600–700 g; conical; well-filled; very symmetrical. Berry: round; skin dark bluish-green, thick, lenticels few; pulp bright green, soft, gelatinous, not very juicy, high in acidity; for production of white table wine of good acidity, of Chablis type. Vine: vigorous; spreading; heavy foliage; wood matures late in autumn; very productive.

Emerald Seedless.— Orig. in Davis, Calif., by H.P. Olmo, California Agr. Expt. Sta. Introd. in 1968. Emperor x Pirovano 75. Cross made in 1939; selected in 1950; tested as I7-11. Cluster: large; conical; loose to well-filled; uniform. Berry: medium-large; obovoid; skin greenish-yellow, tender; flesh moderately firm, light green; seedless; recommended for fancy raisins and as a table cultivar; ripens midseason, 2 weeks after Thompson Seedless. Vine: moderately productive; vigorous; leaves large, thick; canes large, few in number.

Empress *(Seedless Emperor)*.— Orig. in Visalia, Calif., by Vahan Mkhalian. Introd. in 1939. Plant patent 311; 17 Jan. 1939. Bud mutation of Emperor; discovered in 1928. Berry: seedless, but retains other characteristics of parent; much smaller than parent.

Eona.— Orig. in Brookings, S.D., by N.E. Hansen, South Dakota Agr. Expt. Sta. Introd. in 1925. Lady Washington x Beta;

selected in 1925. Berry: small; skin white; flesh sweet, quality fair for table use. Vine: fairly hardy; productive.

Erie.—Orig. in Fredonia, N.Y., by F.E. Gladwin, New York State Agr. Expt. Sta. Introd. in 1932. (Goff x Worden) x Worden; named in 1930; tested as Gladwin 100. Cluster: very small, less than 100 g; short conical; straggly or poorly filled. Berry: spherical, 16–20 mm; skin jet black; flesh tough, greenish; flavor of Concord; ripens very early, a few days before Van Buren. Vine: vigor medium; not as hardy as Concord; flower self-sterile, reflexed stamens; very irregular in productivity.

Esperanza.—Tropical, multi-use grape for home production. Orig. in Aibonito, Puerto Rico, by Francisco Watlington. Not patented. Native *Vitis caribaea* x Lakemont; selected in 1970; tested as 66H4. Berry: blue-black; seeded; spherical; small to medium size. Cluster: large; well filled. Vine: resistant to powdery and downy mildew. Adapted to tropical climate.

Esprit.—Orig. in Osceola, Wis., by Elmer Swenson. Introd. in 1984. Plant patent 5716; 15 Apr. 1986; assigned to Swenson Smith Vines. Edelweiss x Villard Blanc; tested as E.S. 422. Berry: large; skin white; suitable for both fresh and wine use. Cluster: large. Vine: productive; ripens in midseason, with Seyval; winter hardy to –28C.

Eureka.—Orig. in Stillwater, Okla., by Herman A. Hinrichs, Agr. Expt. Sta., Oklahoma State Univ. Introd. in 1975. America x Ontario; cross made in 1950; selected in 1955; tested as Okla. 229. Cluster: very large with medium-length shoulder; medium to long shank. Berry: medium; round; skin blue-black, attractive; brown-amber jelly; sugar 21%; tartaric acid 0.676%.Vine: very productive; foliage dense.

Everglades.—Orig. in Turrialba, Costa Rica, Central America, by Joseph L. Fennell, Lady Lake, Fla. Introd. in the U.S. in 1948. Fennell 18 x Fennell 105; selected in July 1945; *Vitis vinifera*. Berry: quality fair. Vine: able to grow and produce well under tropical, high-rainfall conditions; moderately resistant to rust, anthracnose, and black rot.

Excel.—Orig. in Brooks, Ga., by B.O. Fry, Ison Nursery. Introd. in 1983. Sugargate x open-pollinated seedling. Muscadine type. Berry: resembles Fry; large; skin bronze; flesh firm; flavor good. Vine: moderately vigorous; tolerant to diseases.

Exotic.—Orig. in Fresno, Calif., by Elmer Snyder and Frank N. Harmon, U.S. Dept. of Agriculture. Introd. in 1958. Flame Tokay x Ribier (Alphonse Lavallée); cross made in 1947; seedling budded to a growing rootstock in 1948; first fruited in 1949; tested as USDA G8-30. Cluster: very large; long with medium-long shoulders; well-filled. Berry: firm; medium-large; spherical; similar to Ribier; skin black; quality good; texture crisp; seed size medium, usually three/berry; shipping quality and market acceptance good; berries split badly in some localities; matures about 7–10 days later than Cardinal. Vine: vigorous, productive; leaves large, deeply lobed; blossoms self-fertile, with upright stamens.

Fairchild.—Orig. in Princeton, Fla., by J.L. Fennell, Lady Lake, Fla. Introd. in 1940. *Vitis tiliaefolia* x Alphonse Lavallée; selected in 1940; tested as Fennell 106. Cluster: large. Berry: small; skin black; quality fair. Vine: adaptable to wet, tropical conditions; immune to grape rust; reflexed stamens; has been a valuable parent in breeding.

Fantasy Seedless.—Black, seedless grape for fresh market. Orig. in Fresno, Calif., by David W. Ramming and Ronald Tarailo. B36-27 x C78-68; selected in 1981; tested as C23-92; introd. in 1994. Not patented. Berry: mostly oval; large (6.5 g); firm; sweet flavor; one or two soft, small, aborted seeds/berry; ripens slightly before Ribier. Cluster: medium; medium-compact to loose. Vine: vigorous; moderately productive when cane-pruned. Vines should not be sprayed with gibberellic acid due to resulting poor berry set, crop elimination the following year, and elongated vegetative growth. Recommended as a large-fruited, fresh-market grape.

Farmer Great Northern Seedless.—See Royal Seedless.

Farrer.—Orig. in Brooks, Ga., by B.O. Fry, Griffin, Ga. Introd. in 1983. Plant patent 7294; 14 Aug. 1990. Sugargate x o.p. seedling. Selected in 1971; tested as 20-17-2. Muscadine type. Berry: large; skin dark; ripens evenly, early. Vine: vigorous; productive.

Favorite.—Orig. in Brenham, Texas, by John Niederauer. Introd. about 1938. Probably Black Spanish x Herbemont; selected in early 1930s. Cluster: medium to large; similar to Black Spanish in appearance, but more compact. Berry: larger and less acidic than Black Spanish, of more pleasing taste; makes an attractive dark purple juice. Vine: very productive, 6 tons/acre or more in favorable seasons.

Fennell 113.—See Biscayne.

Festivee.—Orig. in Vineland Sta., Ontario, Canada, by O.A. Bradt, Hort. Res. Inst. of Ontario. Introd. in 1976. Alden x Verdelet; cross made in 1953; selected in 1961; tested as Vineland 53033. Berry: large, oval; skin reddish-black; flesh meaty; table quality good; ripens 27 Sept., 1 week before Concord. Cluster: medium-large; usually shouldered; medium to loose compactness. Vine: very vigorous; sprawling; very productive; moderately hardy; moderately susceptible to mildew.

Fiesta.—Orig. in Fresno, Calif., by U.S. Dept. of Agriculture. Introd. in 1973. (Calmeria x 43-13N) x [(Cardinal x Sultanina) x 43-13S]. 43-13N and 43-13S selections have the parentage: (Red Malaga x Tifafihi Ahmer) x (Muscat of Alexandria x Sultanina); cross made in 1965; first fruited in 1968; selection made by J.H. Weinberger and F.N. Harmon; tested as F18-94 in cooperation with California State Univ., Fresno. Cluster: medium-large; usually not compact. Berry: medium; oval, but rounder than Thompson Seedless; skin white, tender; flesh seedless; flavor good; ripens 10–14 days before Thompson Seedless; recommended for raisins, which are slightly larger, plumper, meatier, and more attractive than Thompson Seedless. Vine: vigor moderate; leaves darker color, tend to curl less and more serrated than Thompson Seedless; must be cane-pruned for maximum yield; berry set has been heavier on own-rooted vines, but vigor is greater on rootstock.

Flame Seedless.—Orig. in Fresno, Calif., by U.S. Dept. of Agriculture. Introd. in 1973. (Cardinal x Sultanina) x [(Red Malaga x Tifafihi Ahmer) x (Muscat of Alexandria x Sultanina)]; cross made in 1961; first fruited in 1964; selected by J.H. Weinberger and F.N. Harmon; tested as F32-137 in cooperation with California State Univ., Fresno Foundation. Cluster: medium; conical; berries well-spaced. Berry: small; round; skin bright red, attractive at Fresno and Bakersfield, did not develop color well at Indio, Calif., due to high-temperature conditions; flesh firm, crisp; flavor excellent; seedless; ripens before Thompson Seedless, about with Perlette. Vine: vigor moderate; very productive when cane-pruned and on Harmony rootstock; responds well to girdling and gibberellic acid treatment.

Flaming Red.—Orig. in Lodi, Calif., by Marvin L. Nies. Introd. in 1964. Plant patent 2344; 21 Jan. 1964. Queen x Cardinal; selected in 1956; tested near Thermal, Calif. Cluster: large, up to 1–2 lb; shoulders well-defined; well-filled; up to 26 cm long. Berry: size uniform, averaging 3 cm long, 2.2 cm wide; ellipsoidal elongated; skin bright red under desert conditions (Coachella Valley, Calif.), relatively tough; flesh very firm with some tendency to be pulpy and flaccid; flavor neutral; normal acidity; good shipping quality; seeds usually two/berry, about 60% well developed, 30% poorly developed; ripens early, with or before Cardinal, slightly later than Perlette; intended principally as a table variety; resembles Queen. Vine: moderately vigorous; leaf sinuses deep; habit with no particular distinguishing characteristics.

Flora.—Orig. in Davis, Calif., by H.P. Olmo, California Agr. Expt. Sta. Introd. in 1958. Semillon x Gewürztraminer; cross made in 1938; tested as Calif. H59-90. Cluster: small to medium; very compact. Berry: small; skin dark green; high in sugar and acid; ripens in early midseason. Vine: small; productivity medium; canes with very short internodes; buds small. Produces an aromatic white table wine of very high quality and flowery bouquet.

Florida Concord.—Orig. in Altoona, Fla., by Charles Demko. Introd. in 1963. (*Vitis simpsoni* x Carman) x Concord; cross made in 1946; tested as Demko 10-35. Cluster: small. Berry: medium; skin black; Concord type; early ripening. Vine: self-fertile; adapted to subtropical, humid climate.

Florida Fry.—Bronze, self-fertile muscadine. Orig. in Leesburg, Fla., by J.A. Mortensen,

J.W. Harris, and D.L. Hopkins. Triumph x Fla. AD3-42; selected in 1987; tested as AA7-44. Not patented. Berry: spherical; large (9.9 g); 3.5 seeds/berry; firm pulp, which, along with edible skin, breaks up easily when chewed; 75% dry scars; ripens over extended period. Cluster: moderately loose; 6.3 berries/cluster. Vine: less vigorous than Fry; vines can be spaced more closely than most muscadines; self-fertile; no Pierce's disease observed; more resistant to ripe rot, bitter rot, and black rot than Fry, Triumph, and Summit.

Foch.—See Maréchal Foch.

Four Five.—White, mostly seedless table grape. Orig. in Fresno, Calif., by John M. Garabedian. Plant patent 5908, 17 Mar. 1987. Almeria x unknown male parent; selected in 1968. Berry: medium; two to four vestigial seeds/berry that are not noticeable when eaten; mostly oval and elongated; soft; juicy; sweet; ripens 2 weeks before Delight. Cluster: large and long; cylindrical; medium density; double-shouldered. Vine: medium vigor to vigorous; productive; one or two clusters/shoot. Primary use is for fresh market but also suitable for raisin production.

Foxxylottie.—Orig. at Montgomery, Ala., by Booker T. Whatley, Whatley Farms. Introd. in 1982. Plant patent applied for. *Vitis vinifera* x *V. labrusca*. Selected in 1978. Berry: round; skin deep purple, moderately tough; flesh yellow-green, tender, juicy; moderately sweet, flavor excellent; dry stem scar. Cluster: large; open. Vine: vigorous; productive; resistant to Pierce's disease and other diseases common to the lower southern U.S.

Fredonia.—Orig. in Fredonia, N.Y., by F.E. Gladwin, New York State Agr. Expt. Sta. Introd. in 1927. Champion x Lucile; cross made in 1915. Berry: large; skin black, thick, tough; quality good; inclined to shatter badly at maturity; ripens early, 2 weeks before Worden. Cluster: medium; cylindrical; compact. Vine: susceptible to downy mildew; vigorous; hardy; fairly productive.

Frontenac™.—Red wine grape adapted to the midwestern U.S. Orig. in Excelsior, Minn., by P.R. Hemstad, J.J. Luby, and P.R. Pierquet, Univ. of Minnesota. Not patented. Landot 4511 x University of Minnesota *V. riparia* selection no. 89; selected in 1983; tested as MN 1047; introd. in 1995. Berry: small (1.1 g); bluish-black; juice characteristics include 21.6% soluble solids, pH 3.05, acidity 16.5 $g \cdot L^{-1}$; ripens 7–10 days after Foch, or about 20 Sept. at Excelsior; wine good quality with very little of the herbaceous character typical of many interspecific hybrids. Cluster: medium (130 g); somewhat loose. Vine: vigorous; productive with overcropping a possibility, requiring a small amount of cluster thinning in some years; open growth habit; has exhibited good disease resistance with high resistance to downy mildew and moderate resistance to powdery mildew and black rot; quite tolerant to adverse effects of 2,4-D herbicide drift; hardy enough for consistent production in central Minnesota, where winter minima reach –30C, and at least as hardy as Foch.

Fry.—Orig. in Experiment, Ga., by B.O. Fry. Introd. in 1970. Ga. 19-13 x USDA 19-11; cross made in 1959; tested as Ga. 20-55. Muscadine type. Cluster: large. Berry: large; skin bronze; quality good; soluble solids averaging 18.6%; ripens mid-September at Experiment. Vine: vigorous; productive; susceptible to cold injury below 10F; susceptible to black rot, but little damage done to fruit; flower pistillate.

Fry Seedless.—Seedless muscadine for fresh market. Orig. in Brooks, Ga., by W.G. Ison. Introd. in 1990. Plant patent 7296, 14 Aug. 1990; assigned to Ison's Nursery and Vineyards. Farrer x Redgate. Berry: seedless; small to medium (20 mm); pink-red; 20% solids; thin skin; round; pulp melting; ripens midseason. Vine: self-fertile; vigorous; low to medium production, erratic in some years; tolerant to common diseases.

Fuji Muscat.—Orig. in Del Rey, Calif., by Paul K. Fujihara. Introd. in 1954. Plant patent 1250; 9 Feb. 1954. Red Malaga x Muscat of Alexandria; selected in 1936. Cluster: large; long; shouldered; well-filled. Berry: medium; spherical; skin color dull green-yellow; flesh firm; mild Muscat flavor; matures early, 2 weeks before Muscat of Alexandria. Vine: more vigorous than either parent; tends to overbear; a table grape variety.

Gar Muscat.—White, seeded, table grape.

Orig. in Fresno, Calif., by John M. Garabedian. Plant patent 5907; 17 Mar. 1987. Lebanese No. 9 x Gar-Supreme; selected in 1978. Berry: light yellow; oblate; medium-large (4.5 g); one to four seeds/berry; soft and crisp; juicy; sweet, subacid, and spicy flavor; ripens 10–14 days before Muscat of Alexandria. Cluster: medium to large; conical; well-filled. Vine: vigorous; productive. Developed as an improved Muscat of Alexandria-type variety with better cluster fill, crisper texture, and earlier ripening.

Garlate.—White, late-ripening, seedless table grape for fresh market. Orig. in Fresno, Calif., by John M. Garabedian. Plant patent 5987; 19 May 1987. Seedling of two vines selected from a population of seedlings from a Lebanese variety; selected in 1977. Berry: large (3–4 g); oblate and cylindrical; soft and meaty; medium juicy; sweet; ripens several days after Garmeria (10–15 Oct. at Fresno). Cluster: large to very large; long; broad and tapering; medium in density. Vine: vigorous; productive.

Garonnet.—Orig. in Saint Vallier, Drôme, France, by Seyve-Villard. Introd. about 1932. Assumed to be from the cross Seibel 7053 x Seyve-Villard 12-375. Cluster: large; compact; short; conical. Berry: medium; oval; skin black; flesh soft; matures early; produces a neutral wine comparatively high in alcohol and of light color, but of only ordinary quality. Vine: vigorous; upright; buds-out early; leaf glabrous, deeply incised, lobes overlapping, surface undulated; petiolar sinus open, U-shaped; teeth prominent, sharp; poor crop recovery after spring frost; moderately resistant to downy and powdery mildew; heavy producer; short or spur pruning recommended; hardy in New York. Authorized for commercial use in many French districts; widely propagated on 3309 and St. George (du Lot) rootstocks.

Gasconade.—Orig. in Mountain Grove, Mo., by Paul H. Shepard, Missouri State Fruit Expt. Sta. Introd. in 1947. Captain x Terret Monstre; sibling of Bokay and Eleven Point; selected in 1947; tested as G-1013. Cluster: large; compact. Berry: medium; skin black; flesh tender; sweet, flavor good; suitable for juice, wine, or table use; ripens 2 weeks after Concord. Vine: very productive; flower perfect.

Georgia Red.—Orig. in Brooks, Ga., by B.O. Fry. Introd. in 1977. S. 42-28 x 46-32; cross made in 1970; selected in 1975; tested as S. 15-16-2. Muscadine type. Berry: small, 3/4 inch in diam.; round; skin red, thin, rough; quality good: ripens midseason. Vine: vigorous; tolerant to black rot.

Glenora.—Orig. in Geneva, N.Y., by R.M. Pool, J.P. Watson, K.H. Kimball, and J. Einset, New York Agr. Expt. Sta. Ontario x Russian Seedless; cross made in 1947; selected in 1952; tested as NY 35814. Berry: medium; blue-black; skin adherent; flesh melting; flavor delicate; ripens early; does not store well. Cluster: large; cylindrical, not excessively compact; responds favorably to gibberellic acid. Vine: vigorous; phylloxera resistant; not fully winter hardy, comparable to Himrod and Interlaken in hardiness; less resistant to powdery mildew and downy mildew than Concord, and extra sprays may be required to control these diseases; a seedless cultivar adapted to northeastern North America.

Gold.—Orig. in Davis, Calif., by H.P. Olmo, California Agr. Expt. Sta. Introd. in 1958. Calif. A3-94 (Muscat Hamburg x Sultanina) x Calif. K3-78 (Muscat Hamburg x Scolokertek kiralynoje); cross made in 1951; tested as Calif. Q17-16. Cluster: medium; short; conical; well-filled. Berry: large; oval; skin tender, with golden sheen; seeds few, small; flesh firm; mild muscat flavor; quality excellent for table use; ripens in early August; also used for production of a light Muscat wine. Vine: vigor medium; canes short, very leafy; very productive.

Golden City.—Orig. in Roodeplaat, Pretoria, South Africa, by Edward P. Evans, Hort. Res. Inst. Introd. in 1973. Queen of the Vineyards x Pearl of Csaba; cross made in 1959; selected in 1966; tested as P/Q89. Cluster: large; well-filled; needs thinning. Berry: 6.0 g; round-oval; skin straw-colored, very attractive, slight tendency to crack; flesh crisp; slight muscat flavor; ripens early, 10 days before Queen of the Vineyards and Seneca, mid-December in Roodeplaat. Vine: vigorous; productive; slightly susceptible to anthracnose.

Golden Isles.—Orig. in Griffin, Ga., by R.P.

Lane, Univ. of Georgia, and R.P. Bates, Univ. of Florida. Introd. in 1987. Fry x Ga. 19-6 (Creek x US 53-8b). Tested as Ga. 18-7-3. A muscadine type for less aromatic white wine of good quality. Berry: size of Dixie; skin bronze; dry stem scar; self-fertile; ripens 25 Sept. in Griffin, some unevenness of ripening may occur; degrees Brix usually 14.4, range 10.8 to 20.5, varying with season and picking date. Cluster: berries / cluster = 7.8. Vine: moderately vigorous and productive; not damaged by temperature of –21C in 1985; intermediate resistance to bitter rot, angular leaf spot, and powdery mildew; moderately resistant to black rot; has produced well without pest control program.

Golden Muscat.—Orig. in Geneva, N.Y., by R.D. Anthony, New York State Agr. Expt. Sta. Introd. in 1927. Muscat Hamburg x Diamond; cross made in 1915. Cluster: large; tapering; compact; subject to cluster spoilage. Berry: large; oval; flesh juicy, soft; sweet, aromatic; resembles Diamond in golden color; some aroma of European Muscat; ripens 2 weeks later than Concord; for home use. Vine: vigorous; productive; not as hardy as Concord.

Granny Val.—Orig. in Brooks, Ga., by B.O. Fry, Griffin, Ga. Introd. in 1983. Plant patent 5823; 9 Dec. 1986. Fry x Carlos. Selected in 1974; tested as 1-11-2. Muscadine type. Cluster: 10–15 berries / cluster. Berry: large; round; skin bronze; ripens early. Vine: vigorous; productive; tolerant to disease; self-fertile.

Great Northern Seedless.—See Royal Seedless.

Hanover.—Orig. in Geneva, N.Y., by New York State Agr. Expt. Sta. Introd. in 1928. Brighton x Niagara; selected in 1908 at Fredonia, N.Y., and named by F.E. Gladwin in 1924. Cluster: medium; conical; well-filled. Berry: variable in size; spherical; skin dark red, not attractive; average quality; ripens late; resembles Brighton. Vine: variable in vigor; poor in productivity; hardier than Niagara; flower self-fertile.

Hartnel.—Origin unknown, but perhaps Oregon. Introd. about 1941. Berry: large; skin blue; quality good; ripens about 1 week before Concord. Vine: considered to be a tetraploid form of Concord.

Hector.—Orig. in Geneva, N.Y., by Richard Wellington, New York State Agr. Expt. Sta. Introd. and named in 1937. Chagselas Rose x Brocton; cross made in 1923; selected in 1928. Cluster: short; conical; small; compact. Berry: dark red; spherical to slightly ellipsoidal, 15–18 mm, variable; pulp meaty, not foxy, acid center; ripens about 1 week before Catawba; seeds small. Vine: poor in vigor and tends to overbear; lacks hardiness; very susceptible to downy mildew; crop ripens unevenly.

Helena.—Orig. in Davis, Calif., by H.P. Olmo, California Agr. Expt. Sta. Introd. in 1958. Zinfandel x Refosco; cross made in 1937; tested as Calif. G22-6 and Oakville 6. Cluster: large; winged; compact. Berry: skin yellow-white; flesh soft, juicy; ripens in midseason. Vine: Zinfandel type, but more tolerant to red spider mite attack and fruit spoilage. Produces a distinctive white, dry table wine.

Henryetta.—Orig. in Henryetta, Okla., by Frank B. Cross, Oklahoma A&M College, Stillwater. Introd. in 1936. Probably a bud mutation of Brighton; discovered in 1933. Berry: skin darker red and ripens a few days earlier than Brighton; berries shrivel just before harvest at Stillwater. Vine: susceptible to droughty conditions.

Higgins.—Orig. in Experiment, Ga., by M.M. Murphy and B.O. Fry, Georgia Agr. Expt. Sta. Introd. in 1955. Yuga x a white male pollinator; selected in 1950; tested as Ga. 3. Muscadine type. Cluster: large; compact. Berry: large; skin bronze, moderately thick yet tender; flesh soft; flavor good; ripens in midseason, a few days before Hunt. Vine: moderately vigorous; yields outstanding; moderately resistant to black rot; self-unfertile, requiring a pollinator. Named in honor of B.B. Higgins, head, Botany Dept., Georgia Expt. Sta., for 42 years.

Himrod *(Himrod Seedless).*—Orig. in Geneva, N.Y., by A.B. Stout, New York State Agr. Expt. Sta., in cooperation with the New York Botanical Garden. Introd. in 1952. Ontario x Sultanina; cross made in 1928; selected in 1935; propagated in 1939; tested as N.Y. 15310. Cluster: large; very long; poorly filled. Berry: small; oval;

greenish-yellow, translucent; small, soft seeds; flesh tender, juicy, melting; vinous-flavored; quality good; ripens early, 1 week before Fredonia; resembles Interlaken Seedless. Vine: vigorous; low productivity; fairly hardy.

Himrod Seedless.—See Himrod.

Honey Seedless.—Triploid, seedless table grape. Orig. in Akitsu, Hiroshima, Japan, by Hiroyasu Yamane. Kyohou x Concord Seedless; selected in 1976; introd. in 1993. Berry: greenish-yellow; round; small (2 g); responds to gibberellic acid treatment; slipskin; mild foxy flavor. Cluster: small to medium size; cylindrical and shouldered; loose. Vine: vigorous; more resistant to downy and powdery mildew than *V. vinifera* varieties; medium cold hardiness.

Horizon.—Orig. at Geneva, N.Y., by New York State Agr. Expt. Sta., Cornell Univ. Introd. in 1982. Seyval x Schuyler. Sister seedling of Cayuga White. Cross made in 1945. Cluster: medium; moderately compact; cylindrical in shape; 2 clusters/shoot. Berry: medium; spherical; skin light green; flesh produces a white wine that is fruity with good body and balance; ripens midseason, 20–30 Sept. at Geneva. Vine: hardy; vigorous; productive; flower perfect with upright stamens; less resistant to powdery mildew than Concord.

Howard.—Orig. in Experiment, Ga., by H.P. Stuckey, Georgia Agr. Expt. Sta. Introd. in 1929. Scuppernong x black male muscadine; selected in 1926. Muscadine type. Berry: resembles Scuppernong.

Hunt.—Orig. in Experiment, Ga., by H.P. Stuckey, Georgia Agr. Expt. Sta. Introd. in 1920. Flowers x a white male muscadine; selected in 1918. Muscadine type. Berry: quality excellent; very even ripening, unusual in muscadine grapes; skin dull black with little bloom when prime ripe, thickness medium to thin with abundant pigment, prized by manufacturers of wine and frozen pulp; sugar content higher than Scuppernong, containing appreciable amount of sucrose. Vine: very prolific.

Imperial Cardinal *(Imperial Red)*.—Orig. in Fresno, Calif., by John M. Garabedian. Introd. in 1971. Plant patent 3336; 15 May 1973; assigned to Superior Farming Co., Bakersfield, Calif. Cardinal x Perlette; cross made in 1965; selected in 1970. Berry: medium to large; round to slightly elongated; skin dark red, medium thick, tough, adheres to pulp; flesh greenish, translucent, tender, juicy, texture medium; a dessert variety; ripens 3 weeks before Cardinal, 1 July in Fresno. Vine: open, upright; vigor medium; productivity excellent.

Imperial Red.—See Imperial Cardinal.

Interlaken Seedless.—Orig. in Geneva, N.Y., by A.B. Stout, G.D. Oberle, and R. Wellington, New York State Agr. Expt. Sta. Introd. in 1946. Ontario x Sultanina; selected in 1937. Cluster: medium; tapering. Berry: small; seedless; skin greenish-white, non-slipskin type; flesh crisp; sweet; quality excellent; resembles Sultanina, but has a slight Ontario flavor; ripens early. Vine: seems to be sufficiently hardy for the eastern U.S.; vigor moderate; productive; semihardy; long-cane pruning needed.

Irene.—Orig. in Experiment, Ga., by H.P. Stuckey, Georgia Agr. Expt. Sta. Introd. in 1920. Thomas x black male muscadine; selected in 1918. Muscadine type.

Ison.—Orig. in Brooks, Ga., by B.O. Fry and W.G. Ison. Introd. in 1986. Sugargate x Senoia. Plant patent 5822; 9 Dec. 1986. Self-fertile muscadine. Berry: large, 28 mm; skin black; dry scar; 20% solids; good flavor; ripens uniformly, early to midseason. Cluster: large. Vine: very productive; vigorous; disease resistant.

Italia.—Orig. in Rome, Italy, by A. Pirovano, Inst. of Fruit Culture. Introd. in California by California Agr. Expt. Sta., Davis, in 1946. Bicane x Muscat Hamburg; cross made in 1911. Cluster: loose; conical. Berry: very large; skin golden; mild muscat flavor; ripens in midseason. Vine: vigor medium; very productive.

Jakaranda.—Orig. in Roodeplaat, Pretoria, South Africa, by Edward P. Evans, Hort. Res. Inst. Introd. in 1973. Queen of the Vineyards x Pearl of Csaba; cross made in 1959; selected in 1966; tested as P/Q7. Cluster: smaller than Pearl of Csaba. Berry: slightly larger than Pearl of Csaba; round; skin green; flesh crisp; muscat flavor, pleasant; ripens early, 10 days before Pearl of Csaba (end of November in Roodeplaat). Vine: vigorous; productive; slightly susceptible to anthracnose.

Janebell.—Orig. in Brooks, Ga., by W.G. Ison. Introd. in 1988. Fry x Senoia. Plant patent 7268; 10 July 1990. Self-fertile muscadine. Berry: medium-large, 28 mm; skin bronze; dry scar; 22% solids; good flavor; ripens mid- to late season. Cluster: large. Vine: very productive; disease resistant.

Janet.—Orig. in Brooks, Ga., by W.G. Ison. Introd. in 1988. Fry x Senoia. Plant patent pending. Self-fertile muscadine. Berry: large, to 31 mm; skin bronze; wet stem scar; 21% solids; excellent flavor; ripens mid-season. Cluster: large. Vine: vigorous; very productive; disease resistant; cold hardy.

Jelly Bean.—Orig. in Reedley, Calif., by Vahan Apkarian. Introd. in 1963. Plant patent 2217; 15 Jan. 1963. Bud mutation of Emperor; discovered in 1950. Berry: shape of a kidney bean or candy jelly bean; 17 mm in diam., 25 mm long; skin purplish-red, with a small, light-green basal area; flavor sweet; ships very well; ripens about 2 weeks before Emperor; usually only one seed/berry; primarily for fresh table use. Vine: all characteristics the same as the parent.

Jingkejing.—Early ripening, seedless table grape. Orig. in Beijing, China, by Chen Zhenbang, Yang Meirong, Zhang Yingzhu, and Li Shengchen, Beijing Botanical Garden. French Blue x Black Monukka; tested as 60-34-9. Berry: 2.2 g; ovoid or ellipsoid; dark-purple, fragile skin; crisp flesh; medium juicy; seedless; sweet with muscat aroma when fully ripe; 15% to 19% soluble solids, 0.58% to 0.72% acidity; ripens about 25 July in Beijing area. Cluster: conic; 1.39/shoot; 385 g. Vine: high vigor; medium disease resistance.

Jingxiu.—Early ripening, seeded grape. Orig. in Beijing, China, by Chen Zhenbang, Yang Meirong, Zhang Yingzhu, and Li Shengchen, Beijing Botanical Garden. Pannonia x (Muscat Hamburg x Black Monukka); tested as 81-3-72. Berry: 5–6 g; ellipsoidal; dark purplish-red; medium-thick skin; thick, crisp flesh; sweet flavor; two to three small seeds/berry; 15% to 17.5% soluble solids; 0.46% acidity; ripens early, middle to end of July in Beijing. Cluster: 420 g; conic; 1.13/shoot. Vine: medium vigor; medium to high productivity; medium disease resistance.

Jingya.—Large-fruited, seeded grape. Orig. in Beijing, China, by Chen Zhenbang, Yang Meirong, Zhang Yingzhu, and Li Shengchen, Beijing Botanical Garden. Seedling of Black Olympia. Berry: large (10.8 g); ellipsoidal; dark-purple skin of medium thickness and thick bloom; flesh soft or medium-soft; juicy; sweet with somewhat sour and strawberry flavor; 13.5% to 18% soluble solids; 0.65% to 0.90% acidity; ripens early August in Beijing area; seedless berries can be produced after gibberellic acid treatment. Cluster: 486 g; cylindric-conic; 1.55/shoot. Vine: medium to slightly high vigor; high fruiting ability; high disease resistance.

Jingyou.—Large-fruited, seeded grape. Orig. in Beijing, China, by Chen Zhenbang, Yang Meirong, Zhang Yingzhu, and Li Shengchen, Beijing Botanical Garden. Seedling of Black Olympia. Berry: large (10.3 g); red-purple; nearly globose or ovoid; thick skin; crisp flesh; sweet flavor with a slight strawberry aroma; 14% to 19% soluble solids; 0.55% acidity; ripens in mid-August in Beijing area. Cluster: 580 g; conic; 1.15/shoot. Vine: high vigor; high disease resistance.

Jingyu.—Amber-yellow, seeded grape. Orig. in Beijing, China, by Chen Zhenbang, Yang Meirong, Zhang Yingzhu, and Li Shengchen, Beijing Botanical Garden. Italia x Queen of the Vineyards; tested as 60-36-4. Berry: 6.5 g; medium-thick, amber-yellow skin; ellipsoidal; thick, crisp flesh; juicy; sour-sweet flavor; one or two small seeds in most berries; 13% to 16% soluble solids; 0.48% to 0.55% acidity; ripens in early August in Beijing area. Cluster: 685 g; conic; 1.18/shoot. Vine: vigor medium to slightly high; medium to high productivity; downy mildew resistant with medium resistance to other diseases, except for high susceptibility to anthracnose.

Jingzaojing.—Seedless table grape. Orig. in Beijing, China, by Chen Zhenbang, Yang Meirong, Zhang Yingzhu, and Li Shengchen, Beijing Botanical Garden. Queen of the Vineyards x Thompson Seedless; tested as 60-27-13. Berry: 2.5–3.0 g; ovoid or ellipsoidal; thin, amber-yellow skin; flesh crisp; seedless; mild sweet-sour taste with a bit of muscat flavor when fully ripe; 16.4% to

20.3% soluble solids; 0.47% to 0.62% acidity; ripens in late July in Beijing area. Cluster: 518 g; conic; 1.09/shoot. Vine: vigor high; medium productivity; medium disease resistance.

Jolly.—Seedless table grape. Orig. in San Rafael, Mendoza, Argentina, by Angel A. Gargiulo. Plant patent 8547, 18 Jan. 1994; assigned to Milan M. Caratan and Luis M. Caratan. A hybrid of Lattuario Nero, Cardinal, and Thompson Seedless (complete pedigree not available); selected in 1969. Berry: pink-red; good size; good flavor. Cluster: loose.

Joyous.—Orig. in Cincinnati, Ohio, by Byron T. Johnson. Introd. in 1982. Mericadel x Delicatessen; tested as selection 64.ZZ-2. Berry: slipskin; seeded; red to dark red; medium-large; round; pulp tender, juicy; very late ripening, requires very long, hot growing season to ripen properly; produces a light muscat wine of golden color, resembling Gewurztraminer. Cluster: compact; medium-large; often with shoulder; borne three/shoot. Vine: very vigorous; growth habit open, with few laterals; moderately productive; drought tolerant; disease and insect resistant.

July Muscat.—Orig. in Davis, Calif., by H.P. Olmo, California Agr. Expt. Sta. Introd. in 1958. Calif. 126-11 (Muscat of Alexandria x Flame Tokay) x Calif. R4-41 (Muscat Hamburg x Scolokertek kiralynoje); cross made in 1950; tested as Calif. P61-12. Cluster: loose. Berry: medium; subovoid; skin resistant to weather, white; flesh greenish-yellow, firm; strong Muscat flavor; transports well; table variety. Vine: vigorous; medium productivity.

Jumbo.—Orig. in Experiment, Ga., by B.O. Fry. Introd. in 1970. Higgins x USDA 19-11; cross made in 1959; tested as Ga. 43-17. Muscadine type. Cluster: large. Berry: large; skin purple; quality good; soluble solids averaging 15.3%; ripens over several weeks, with largest harvest in late September in Experiment. Vine: very vigorous; productive; moderately resistant to common muscadine diseases; adapted to entire muscadine growing region.

Jumbo Red.—Orig. in Bridgman, Mich., by Ackerman Nurseries. Introd. in 1946. Considered to be a seedling of Concord; discovered in 1941 in vineyard of Concord. Cluster: large, compact. Berry: color dark when fully ripe; flesh amber-red; flavor very sweet; seeds very large; resembles *Vitis vinifera*.

Kat.E.Lin.—Seeded, multi-use grape. Orig. in Smithville, Ontario, Canada, by William L. Lounsbury, Sr., and Maurice W. Lounsbury. Plant patent 7644; 10 Sept. 1991. Mutation of Concord. Berry: red to dark-maroon; medium (2.7 g); round to oblate; slipskin; mild foxy flavor; two or three seeds/berry; high sugar (19–22°Brix); fruit ripens about 5 days before Concord. Cluster: small to medium; winged; conical; loose; three or four clusters/shoot. Vine: vigorous; productive; hardy to temperatures below –18C; resistant to mildew. Released as an earlier ripening, sweeter variety compared to Concord, for use as a fresh-market, jam, juice, or sweet, dessert-wine grape.

Kay Gray.—Orig. in Osceola, Wis., by Elmer Swenson. Introd. in 1981. Plant patent 4943; 16 Nov. 1982; assigned to Swenson Smith Vines. E.S. 217 (Minn. 78 x Golden Muscat) open-pollinated (Onaka is probable pollen parent); tested as E.S. 1-63. Berry: medium; skin white to golden; mild, fruity labrusca flavor; seeded; for fresh fruit use or for wine; early ripening. Cluster: small. Vine: excellent disease resistance, especially to downy mildew; productive, but limited by small cluster size; extremely hardy, better than Beta; primarily used as a home table grape.

Keating.—Orig. in Woodward, Okla., by L.F. Locke, U.S. Dept. of Agriculture, Southern Great Plains Field Sta. Introd. in 1959. Open-pollinated seedling of Last Rose; tested as Woodward 8493. Cluster: compact; medium to large. Berry: medium to large; skin yellowish-green; quality good to excellent; average ripening date 8 Aug. at Woodward; does not keep well in storage. Vine: growth medium to vigorous; crops medium to heavy; very drought resistant; somewhat lacking in winter hardiness; probably best adapted to the southwest.

Kee-Wah-Din.—Orig. in Cincinnati, Ohio, by Byron T. Johnson. Introd. in 1982. OSBU x Baco Noir; tested as selection 64.B-2.

Berry: medium-small; skin blue-black; round; produces red wine resembling Baco Noir. Cluster: medium-large; four/shoot. Vine: vigorous; productive; self-fertile; very cold hardy. Best adapted to Minnesota, Wisconsin, northern Michigan, and regions with similar climate.

Kendaia.—Orig. in Geneva, N.Y., by Richard Wellington, New York State Agr. Expt. Sta. Introd. in 1939. Portland x Hubbard; cross made in 1925. Cluster: medium, to 600 g; conical; shouldered; well-filled. Berry: spherical, 18–20 mm; slipskin; bluish-black; flavor slightly foxy, medium acid; tendency to crack; maturity 3 weeks before Concord; fair keeping quality. Vine: vigorous; productive; one of the hardiest of the New York introductions.

Keuka.—Orig. in Geneva, N.Y., by Richard Wellington, New York State Agr. Expt. Sta. Introd. in 1924. Chasselas Rose x Mills; cross made in 1913; seed germinated in 1914; first fruited in 1919. Cluster: medium; compact. Berry: dark red; flesh crisp and hardly slipskin; *V. vinifera*-type flavor; ripens about 1 week before Catawba; resembles Delaware in appearance.

Kilgore.—Orig. in Willard, N.C., by Charles Dearing, U.S. Dept. of Agriculture and North Carolina Agr. Expt. Sta. Introd. in 1946. Open-pollinated seedling of Labama; tested as V80 R3 B2. Muscadine type. Cluster: fairly compact; 6–10 berries. Berry: medium; round; skin green to reddish-bronze; pulp very juicy and soft, separating from seeds; acid; flavor good; ripens midseason; seeds pointed. Vine: vigor medium; leaves cordate-pointed; growing tips yellow-green; disease susceptible; flower pistillate. Named for W.B. Kilgore, former director, North Carolina Agr. Expt. Sta.

King Husainy.—Seedless, white table grape. Orig. in Zichron, Ya'acov, Israel, by Mordechai Karniel. Plant patent 6464, 13 Dec. 1988; assigned to Zanzivivai-Ferrara S.r.l. *Vitis vinifera* parentage. Berry: very large (40 x 18 mm); green-yellow; very firm; crisp; neutral flavor; no seed traces; fruit ripens 1 week after Superior Seedless, 2 weeks before Thompson Seedless. Cluster: medium; conical; very loose. Vine: vigorous; productive.

Kyoho.—Orig. in Nakaizu, Shizuoka, Japan, by Yasushi Oinoue, Oinoue Inst. for Agronomical and Biological Sci. Introd. in 1955. Ishiharawase x Centennial; cross made in 1935; selected in 1939. Cluster: large; loose. Berry: large; skin deep purple; flesh similar to *V. vinifera;* flavor foxy like Herbert; soluble solids content varies between 16% to 18%; ripens early September in Japan, with Campbell's Early; recommended as a table cultivar. Vine: very vigorous; productive; canes thick and vigorous; flower large, 50–60/cluster, sheds too much under bad climatic conditions; tetraploid.

Lacrosse.—Orig. in Osceola, Wis., by Elmer Swenson. Introd. in 1983. (Minn. 78 x Seibel 1000) x Seyval; tested as E.S.294. Berry: medium; skin white; produces excellent, fruity, nonlabrusca wine; ripens earlier than Seyval. Cluster: medium. Vine: vigor medium; very productive; good disease resistance; tolerates 2,4-D; very winter hardy.

Lady Patricia.—Orig. in Urbana, Ill., by H.C. Barrett, Univ. of Ill. Introd. in 1968. Seibel 14664 x Seyve Villard 20365. Cross made in 1951, selected in 1956, tested as Ill. 182-1. Cluster: large; loose; long-tapering; compound, long peduncle. Berry: large; very elongated, pointed, "lady finger" shape; flesh tender, sweet, slightly astringent, firm; skin golden yellow at maturity, attractive, thin but tough, adheres to flesh, no cracking; ripens about mid-September at Urbana. Vine: vigorous; very productive; over-bearing should be prevented by cluster thinning and pruning; bud burst after winter dormancy late; tolerant to downy mildew; best adapted from central Illinois southward, as well as in the grape-growing areas of the northwest.

Lake Emerald.—Orig. in Leesburg, Fla., by L.H. Stover, Watermelon and Grape Investigations Lab., Florida Agr. Expt. Sta. Introd. in 1954. *Vitis simpsoni* Pixiola x Golden Muscat; cross made in 1945; selected in 1951. Berry: up to 5/8 inch in diam., slightly smaller than desired for commercial table use; very few "shot" or unfertilized berries; skin emerald green to light golden, fairly tough; flesh soft, agreeably flavored, sweet, fragrant, and rather unique aroma; seed size medium, two to four/fruit; juice aromatic, sweet, well-flavored, light-colored; ripens during July at Leesburg; primarily for home gardens and local markets. Vine: high yields; canes often set four flower clusters, recommended to thin to two clusters/cane; cane size medium; leaf thick, firm; buds large, arising from prominent nodes; no symptoms of degeneration (due to Pierce's disease); more resistant to black rot and downy mildew than post-oak grape hybrids.

Lakemont.—Orig. in Geneva, N.Y., by John Einset, New York Agr. Expt. Sta. Introd. in 1972. Ontario x Sultanina; cross made in 1928; tested as New York 15305. Cluster: medium-large; medium compact to compact; wedge-shaped. Berry: medium-small; oval; skin yellowish-green; flesh tender, adheres to skin, juicy, sweet; quality good-excellent; seedless; ripens with Delaware; recommended for home garden and roadside market. Vine: vigor moderate; less hardy than Himrod, but more so than Interlaken Seedless; productive; flower perfect.

Largo.—Orig. in Turrialba, Costa Rica, Central America, by Joseph L. Fennell, Lady Lake, Fla. Introd. in 1942 in the U.S. Fennell 23 x Fennell 106; first fruited in U.S. in 1946. Berry: quality fair. Vine: grows and produces well under tropical, high-rainfall conditions.

Larson B-36.—Early ripening, black, seedless table grape. Orig. in Thermal, Calif., by Drake Larson. Plant patent 9039; 10 Jan. 1995. Cross of two unnamed seedlings originated by the developer. Berry: average 3.8 g, 1.5–1.8 cm in diam., 2.0–2.4 cm in length; no gibberellic acid required for berry sizing; black skin of medium thickness; flesh very firm, crisp, translucent, well attached to skin; flavor very sweet with slight plum flavor; seedless; ripens earlier than Beauty. Cluster: loose; medium length (20 cm); 0.3–0.8 kg; conical. Vine: very vigorous; moderate productivity.

Larson D-12.—See Mariah.

La Gloire de Seibel.—See Seibel 5409.

La Rochelle.—Early ripening, seeded table grape. Orig. in Stellenbosch, Republic of South Africa, by E.P. Evans and P.J.L. Ellis, Viticultural and Oenological Inst. Barlinka x Alphonse Lavallée; selected in 1979; tested as 11-13-96; introd. in 1986. Berry: large (9 g); blue-black; round; neutral flavor. Cluster: oblong; medium-large; well-filled to slightly compact. Vine: vigorous; productive; disease susceptibility similar to other cultivars. Developed as an early ripening, black variety with exceptional cold-storage capability.

Late Fry.—Late-ripening, self-fertile, bronze muscadine for fresh market. Orig. in Brooks, Ga., by W.G. Ison. Introd. in 1993. Plant patent 9224; 1 Aug. 1995. Fry x Granny Val. Berry: round; large (30 mm); 20% solids; fruit ripens 1 week later than Granny Val. Cluster: large. Vine: productive; vigorous; tolerant to common diseases.

Le Séquanis.—See Seibel 4643.

Le Subéreux.—See Seibel 6905.

Liberty.—Orig. in Leesburg, Fla., by J.A. Mortensen, Inst. of Food and Agr. Sci., Univ. of Florida. Introd. in 1976. W716 (43-47 x Golden Muscat) x Buffalo; tested as Fla. D4-84; selected in 1966. Cluster: 1/4-lb; moderately loose, open cluster allows good spray penetration. Berry: spherical; 3.2 g; slip-skin; good nonfoxy flavor; skin red. Vine: resistant to Pierce's disease and anthracnose; tolerant of powdery mildew and ripe-fruit rot.

Lollar.—Considered to have originated in Denison, Texas, by T.V. Munson, Munson Nurseries; however, Mrs. Minnie S. (Will) Munson stated in a personal letter dated 12 May 1955 that she was not aware that Mr. Munson so named a grape. Introd. about 1935. Parentage unknown; a *Euvitis* type. Berry: skin dark brown; ripens unevenly.

Loomis.—Orig. in Meridian, Miss., by N.H. Loomis. Introd. in 1989. Creek x US 15. Tested at Gay, Ga., as USDA 18-7B; selected by R.P. Lane and M.A. Owen in 1979. Female muscadine. Berry: larger than Cowart; skin burgundy-colored, thin; dry stem scar; very good quality; ripens with Nesbitt. Vine: vigorous; winter hardy to –21C; consistently productive; disease resistance superior to other dark-fruited muscadines.

Lucida.—Orig. in Experiment, Ga. by J.G. Woodroof, Georgia Agr. Expt. Sta. Introd. in 1933. Open-pollinated seedling of Irene;

selected in 1930. Muscadine type. Berry: very large; bronze-colored.

Magnolia.—Orig. in Raleigh, N.C., by the U.S. Dept. of Agriculture and the North Carolina Agr. Expt. Sta. Introd. in 1962. (A seedling derived from Hope, Thomas, and Scuppernong) x (Topsail x Tarheel); selected in 1954; tested as NC 60-60. Muscadine type. Berry: large; skin attractive, bronze, smooth; flavor good; soluble solids average 19% or about 3% higher than Scuppernong; ripens in late midseason or about 1 week later than Scuppernong. Vine: self-fertile; vigorous; very productive. Suggested uses: as a white pollinator for commercial plantings of white muscadine grape; for home gardens.

Magoon.—Orig. in Meridian, Miss., by N.H. Loomis, U.S. Dept. of Agriculture (USDA). Introd. in 1959. Thomas x Burgaw; first fruited and selected in 1946; distributed for trial in 1950; tested as muscadine grape selection 9. Muscadine type. Berry: medium; round; skin heavy, dull reddish-purple; flavor fine, pleasing and sprightly, aromatic; soluble solids content high, being above that of Scuppernong. Vine: productive; vigorous; flower perfect. Popular in the muscadine region, proving to be one of the best-quality, perfect-flowered cultivars as a pollinizer. Does well as far south as central Florida. Named in honor of C.A. Magoon (1883–1958), who was in charge of grape research for many years in the USDA.

Magruder.—See Pixiola.

Mandan.—Orig. in Brookings, S.D., by N.E. Hansen, Agr. Expt. Sta. Introd. in 1925. Wilder x North Dakota wild. Berry: 3/8 inch in diam.; skin black; flesh flavor fair; seeds separate readily from flesh; ripens early. Vine: heavy producer; vigorous.

Mantey.—Orig. in Eustis, Fla., by Karl Mantey. Introd. in 1951. Parentage unknown, believed to be open-pollinated seedling of *Vitis shuttleworthii*. Cluster: medium, larger than *V. shuttleworthii*. Berry: 16 mm in diam.; slightly ovoid; skin red, fairly tough, slight bloom; flesh slipskin; pulp medium tough, juicy; ripens 20 July in Central Florida. Vine: medium; robust; vigor moderate; moderately hardy; productivity medium to heavy; tolerant to Pierce's disease, downy mildew, and berry rot; recommended for breeding purposes.

Marco.—Orig. in Princeton, Fla., by J.L. Fennell, Lady Lake, Fla. Introd. about 1948. *Vitis rufotomentosa* 9 x Masters (*V. shuttleworthii* x Niagara) ?; selected in 1942; tested as Fennell 111. Berry: similar in flavor and quality to Delaware, which it resembles.

Maréchal Foch *(Foch)*.—Orig. at the Oberlin Inst., Colmar, Alsace, France, by Eugene Kuhlman. 101-14 x Goldriesling, but some authors maintain it is derived from Oberlin 595 x Pinot Noir; tested as Kuhlman 188-2. Cluster: loose; small, 8–10 cm long; cylindrical; winged. Berry: very small; skin blue-black; ripens very early; wine taste is herbaceous; violet coloration. Vine: very vigorous, but poor production; best adapted to difficult grape areas with short growing season; does not produce much crop after spring frost.

Mariah *(Larson D-12)*.—Early ripening, black, seedless table grape. Orig. in Thermal, Calif., by Drake Larson. Plant patent 9040; 10 Jan. 1995. Cross of two unnamed seedlings originated by the developer. Berry: average 4.8 g, 2.0–2.2 cm in diam., 2.0–2.4 cm in length; no gibberellic acid required for berry sizing; skin black to slight blue-red with medium thickness; flesh firm, translucent, well attached to skin; flavor very sweet with slight plum flavor; seedless; ripens earlier than Beauty. Cluster: well-filled; 0.4–0.9 kg; medium length (about 23 cm); elongated. Vine: very vigorous; high productivity; average resistance to leafhoppers and thrips; good resistance to nematodes; good resistance to disease (no disease observed).

Marroo Seedless.—Black, seedless table grape. Orig. in Merbein, Victoria, Australia, by Allan J. Antcliff, Commonwealth Scientific and Industrial Res. Organization. Plant patent 7377, 6 Nov. 1990; assigned to CSIRO. Carolina Blackrose x Ruby Seedless; selected in 1978; tested as MS23-7. Berry: blue-black; medium-large (4.5–6.0 g without gibberellic acid application); short elliptic; firm; juicy; neutral flavor; occasional lignified rudimentary seed; ripens about 2 weeks before Ruby Seedless. Cluster: large; loose; uniform. Vine: vigor similar to Ruby Seedless. Released as a naturally large-berried, seedless table grape.

Mars.—Orig. in Fayetteville, Ark., by James N. Moore, Univ. of Arkansas. Introd. in 1985. Plant patent 5680; 4 Mar. 1986. Island Belle x Ark. 1339 (a complex hybrid derived from Buffalo, Alden, Ontario, Ribier, Russian Seedless, Vergennes, and Zinfandel); selected in 1975; tested as Ark. 1508. Berry: seedless; medium-large; round; skin blue, nonadherent, medium thick; flavor resembles Campbell's Early. Cluster: medium; cylindrical; well-filled. Vine: productive; vigorous; may bear fruit precociously. Resistant to black rot, anthracnose, powdery mildew, and downy mildew. Recommended for home gardens and limited commercial use.

Marvel Seedless.—Orig. in Sanger, Calif., by Henry Folmar, Jr. Introd. in 1963. Plant patent 2335; 24 Dec. 1963. Bud mutation of Delight; apparently a tetraploid of this variety. Cluster: large; well-filled; less compact than Thompson or Delight and better adapted for ringing and hormone treatment to increase berry size than parent. Berry: large, averaging 17 mm in diam., 25 mm long; long, ovate; skin bright yellowish-green to yellow, tough; flesh color similar to skin color, firm, medium texture; flavor sweet; seedless; eating and shipping quality very good; ripens early, 1–12 Aug. in the central San Joaquin Valley of California; used for dessert, raisin, wine; resembles Delight. Vine: similar to Delight, except canes shorter with short internodes; leaf thicker and darker green than parent; productivity good, regular; vigorous; canes medium-brown.

Masters.—Orig. in Eustis, Fla., by Joseph L. Fennell, Lady Lake, Fla. Introd. about 1940. Exact parentage unknown but considered to be *Vitis shuttleworthii* x Niagara; discovered in the garden of Zack Savage in 1936. Berry: quality fair to poor; of little value except for breeding purposes; resembles certain *V. labrusca* clones. Vine: fair adaptability to subtropical humid conditions.

McCampbell.—Orig. in Holmdel, N.J., by T. Sara McCampbell. Introd. in 1961. Received by New York Agr. Expt. Sta., Geneva, N.Y., for trial in 1950. Bud sport of Fredonia. Cluster: very large, three to four times as large as Fredonia. Berry: identical to Fredonia. Vine: maximum of 1200 flowers as compared to maximum of 200 on inflorescence of Fredonia.

Meier Everbearing.—Orig. in Mt. Pleasant, Texas, by L.G. Meier. Introd. in 1922. Parentage unknown; possibly a hybrid of some American species; discovered in 1918. Cluster: compact. Berry: seedless; skin blue; flavor good; ripens evenly. Vine: very vigorous; roots readily from cuttings.

Melody.—Orig. in Geneva, N.Y., by B.L. Reisch, R.M. Pool, J.P. Watson, W.B. Robinson, and T.H.E. Cottrell, New York State Agr. Expt. Sta, Cornell Univ. Introd. in 1985. Plant patent 6159; 26 Apr. 1988; assigned to Cornell Res. Foundation. Seyval x GW5 (Pinot Blanc x Ontario); selected in 1972; tested as NY 65.444.4. Berry: medium; spherical; skin light green to amber, resistant to cracking. Cluster: compact; medium, 150 g; does not require cluster thinning like the Seyval parent. Vine: productive; vigorous; moderately winter hardy; moderately resistant to powdery mildew and botrytis bunch rot. Produces high-quality wine with floral overtones and slight hints of herbaceousness.

Melton.—Orig. in Geneva, N.Y., by R. Wellington, New York State Agr. Expt. Sta. Introd. in 1923. Triumph x N.Y. 4064 [(Winchell x Diamond) x Jefferson]. Cluster: long; attractive. Berry: quality and texture good; berries tend to crack during ripening.

Meteor.—Orig. in Stillwater, Okla., by Herman A. Hinrichs, Agr. Expt. Sta., Oklahoma State Univ. Introd. in 1975. Bailey x Keuka; cross made in 1950; selected in 1957; tested as Okla. 324. Cluster: medium to large; short shoulder; shank medium in length. Berry: medium; round; skin dark blue-black with heavy bloom; sugar 19%; tartaric acid 0.751%; ripens evenly, about 18 Aug. in central Oklahoma. Vine: productive.

Micala.—Dark-red, seedless table grape. Orig. in Fowler, Calif., by Alan J. Asadoorian. Plant patent 6348; 25 Oct. 1988. Ribier x Black Beauty Seedless; selected in 1984. Berry: dark red with bluish hue; medium (3.8 g); average of 1.5 seed traces/berry; oval; juicy; sweet flavor; ripens with

Ribier. Cluster: medium to large; tapering, with one or two shoulders; well-filled; two or three clusters / shoot. Vine: vigorous; productive.

MidSouth.—Orig. at Meridian, Miss., by N.H. Loomis and developed at Mississippi State Univ. by J.P. Overcash. Introd. in 1981. De Grasset (*Vitis champini*) x Galibert 255-5. Tested as M20-3C; cross made in late 1940s. Berry: 3.2 g; skin blue; juice red like Concord, makes good jelly; soluble solids 15.1%; flesh sweet to subacid mild, flavor foxy; uniformity of berry ripening on cluster has been better than Concord; ripens late July to August in Mississippi. Cluster: 122 g. Vine: vigorous, average pruning weight 4–10 lb / vine; self-fertile; resistant to Pierce's disease.

MissBlanc.—Orig. at Meridian, Miss., by N.H. Loomis and developed at Mississippi State Univ. by J.P. Overcash. Introd. in 1982. Galibert 261-12 x seedling of Extra x Marguerite. Tested as M26-4D. Cluster: averaged 5 oz with 189 berries per lb. Berry: white; flesh sweet, mild; soluble solids 19.4%; ripens mid-August in Mississippi. Vine: vigorous; survived 9 years, with yields up to 60 lb / vine, when many bunch grape cultivars (such as Concord, Baco Noir, and Chancellor) died from Pierce's disease.

MissBlue.—Orig. at Meridian, Miss., by N.H. Loomis and developed at Mississippi State Univ. by J.P. Overcash. Introd. in 1981. Dog Ridge (*Vitis champini*) x Moore Early (o.p. seedling of Concord). Tested as M18; cross made in early 1950s. Berry: 3.6 g; skin blue; flesh sweet to subacid, flavor mild foxy; juice red like Concord, makes good jelly; soluble solids 15.4%; ripens more evenly on the cluster than does Concord, from late July to mid-August in Mississippi. Cluster: 135 g. Vine: vigorous, average pruning weight 4–8 lb / vine; yields 9–12 tons / acre. Healthy vines survived 9 years, whereas Concord, Moored, Aurora, Chancellor, and other bunch grape cultivars died of Pierce's disease.

Monticello.—Orig. in Blacksburg, Va., by G.D. Oberle and R.C. Moore, Agr. Expt. Sta. Introd. in 1973. USDA 4606-5 (Fredonia x Niagara) x VPI 5-32 (Fredonia x Athens). Cross made in 1959 by R.C. Moore; first fruited 1963; selected in 1964; tested as VPI 31. Cluster: above average in size; compact; long cylindrical to tapering; single or double-shouldered. Berry: medium to small; spherical; skin blue-black, moderately thick, heavy, waxy bloom; uniform ripening; flesh moderately stringy, separates readily from seeds; flavor sweet, spicy, more fruity than foxy; quality very good; resembles Steuben; ripens 5 days after Fredonia, 2 weeks before Concord. Vine: moderately vigorous; productive; tolerant to black rot, mildew, and anthracnose with spray program; recommended for cooler areas of Virginia and possibly similar areas in nearby states. Named after President Thomas Jefferson's estate and home, where he showed great interest in grape culture and winemaking.

Moonbeam.—Orig. in Excelsior, Minn., by Univ. of Minnesota Fruit Breeding Farm. Introd. in 1944. Parentage unknown; selected in 1923; tested as Minn. 66. Berry: larger than Concord; skin greenish-yellow; flavor bland; ripens in early September in Excelsior. Vine: vigorous; hardy; easily propagated.

Moored.—Orig. in Blacksburg, Va., by R.C. Moore, Agr. Expt. Sta., Virginia Polytechnic Inst. Introd. in 1969. Fredonia x Athens. Cross made in 1949; selected in 1955; tested as V.P.I. 26. Cluster: above average in size; well-filled; compact. Berry: above average in size; round; skin red, medium bloom, which disappears rapidly after berry reaches maturity; flesh characteristically *V. labrusca* in texture, being somewhat rough, slightly acid near the seeds; flavor foxy; ripens about with Fredonia or 3 weeks before Concord; for fresh fruit; most nearly resembles Lucile. Vine: above average in vigor, productivity, and hardiness; no difficulty in controlling black rot, downy mildew, powdery mildew, and anthracnose.

Morrison.—Orig. in Willard, N.C., by Charles Dearing, U.S. Dept. of Agriculture and North Carolina Dept. of Agriculture. Introd. in 1946. Scuppernong x white male (AF7082); cross made in 1909; tested as V63 Rll B2. Muscadine type. Cluster: round; three to five berries. Berry: large; round; skin green with pink tinge; poor adherence;

pulp soft, smooth; flavor good; matures early; quality better than Scuppernong, which it resembles. Vine: weak grower; growing tips yellow-green; leaves with fine serrations; flower pistillate. Named for Cameron Morrison, former governor of North Carolina.

Muscat de Saint Vallier.—See Seyve-Villard 20-473.

Muscat Seedless.—White, muscat-flavored table grape. Orig. in Stellenbosch, Republic of South Africa, by E.P. Evans and P.J.L. Ellis, Viticultural and Oenological Res. Inst. Muscat of Alexandria x Perlette; selected in 1982; tested as 12-16-102; introd. in 1986. Berry: yellowish-green; small (3 g); roundish; firm; juicy. Cluster: medium; slightly oblong; well-filled to slightly compact; retains its flavor in storage. Vine: vigorous; productive.

Muska.—Orig. in Pretoria, South Africa, by E.P. Evans, Hort. Res. Inst. Introd. in 1966. Isabella x Pirovano 15; selected in 1961. Cluster: varies from 6–22 oz with average bunch weight of 16 oz; good shape; well-filled; fairly compact. Berry: averages 1/2 inch in diam.; resembles Prince; skin blue-black, medium bloom, not tough; flesh greenish, firm; soluble solids content varies between 16% to 18%; flavor foxy, similar to Isabella; ripens early, end of December in Pretoria, 2–3 weeks before Niagara. Vine: productive; tolerant to disease, especially anthracnose; vigor good; for home garden and commercial production.

Myakka.—Orig. in southern Florida by Joseph L. Fennell. Introd. in 1947. (*Vitis shuttleworthii* x *V. smalliana*) x *V. vinifera* Malaga; selected in 1943. Cluster: large. Berry: flavor sweet; skin purple; resembles Sultana, except for purple color. Vine: productivity high; vigorous; moderate disease resistance for semi-tropical conditions.

Napka.—Orig. in Brookings, S.D., by N.E. Hansen, Agr. Expt. Sta. Introd. in 1925. Salem x Beta. Cluster: compact. Berry: small; skin black; flesh flavor fair. Vine: vigorous; heavy producer; not reliably hardy.

Naples.—Orig. in Geneva, N.Y., by Richard Wellington, New York State Agr. Expt. Sta. Introd. in 1952. Delaware x N.Y. 8042 (Mills x Iona); cross made in 1929; selected in 1935; propagated in 1937; tested as N.Y. 17594. Cluster: medium (larger and less compact than Delaware); tapering; shouldered. Berry: medium; roundish-oval; skin red, tougher and less prone to crack than Delaware; flesh juicy, slightly tough; sweet, with Delaware aroma, quality good; resembles Delaware; desirable for white wine; season just after Concord. Vine: vigorous; tends to overbear.

Nava.—Large-fruited, seeded table grape for fresh market. Orig. in Bet Dagan, Israel, by P. Spiegel-Roy, R. Asaph, and I. Baron, Inst. of Horticulture, Agricultural Res. Organization, The Volcani Center. Dabouki x Alphonse Lavallée (Ribier); selected in 1975; tested as no. 101; introd. in 1988. Patented in Israel. Berry: black; roundish; large (8 g); contains two seeds; some tendency for skin to crack; fruit ripens during similar season as Alphonse Lavallée. Cluster: large (500–600 g); fairly loose; elongated. Vine: vigorous; adapted to spur pruning; productive; chilling requirement similar to Ribier; susceptibility to mildew less than Ribier.

Nesbitt.—Orig. in Raleigh N.C., by R.G. Goldy and W.B. Nesbitt, North Carolina State Univ. Introd. in 1985. Fry x Cowart. Self-fertile muscadine. Selected in 1971; tested as NC 67A013-12. Berry: size of Fry, 30–32 mm, about 12 g; skin black; tendency to wet scar; high quality for fresh consumption; extended ripening between early September and early October. Vine: productive, 40 kg/vine; moderately vigorous; more resistant than Carlos to ripe rot. For U-pick, fresh fruit, or dooryard plantings.

Nevermiss.—Orig. in Gay, Meriwether County, Ga., by M. Aubrey Owen. Introd. in 1945. Plant patent 692; 28 May 1946; assigned to H.G. Hastings Co. Muscadine type. Cluster: averages nine berries, much better than Scuppernong, which it resembles. Berry: averages 3/4 inch in diam.; round; skin white to bronze; pulp translucent, tender, very juicy, high sugar content; three or four seeds/berry; peak of ripening season first week in September. Vine: production regular; vigorous; flower pistillate; resistant to disease.

New River.—Orig. in Willard, N.C., by Charles Dearing, U.S. Dept. of Agriculture and North Carolina Dept. of Agriculture.

Introd. in 1946. Open-pollinated seedling of San Jacinto; 1908 seedling from seed sent by T.V. Munson, Denison, Texas; tested as V48 R9 B2. Muscadine type. Cluster: moderately loose; one to seven berries; to 2 inches long. Berry: mostly large; round; skin reddish-bronze, similar to Scuppernong, but earlier in maturity; flavor fair. Vine: vigor medium; open growth habit; fairly disease resistant; leaves curl inward; flower small, pistillate. Named after the New River in North Carolina.

New York Muscat.—Orig. in Geneva, N.Y., by Richard Wellington, New York State Agr. Expt. Sta. Named and introd. in 1961. Muscat Hamburg x Ontario; cross made in 1926; first described in 1932; tested as N.Y. 12997. Cluster: medium; loose to well-filled. Berry: oval; skin reddish-black with heavy bloom; muscat flavor; ripens early, in Delaware season. Vine: moderately vigorous and productive; may suffer winter injury below –15F. Promising as a cultivar for sweet red wine and possibly for dessert use.

Niabell.—Orig. in Davis, Calif., by H.P. Olmo, California Agr. Expt. Sta. Introd. in 1958. Campbell 4n x Niagara 4n; cross made in 1942; tested as Calif. L11-3. Cluster: well-filled; short; conical. Berry: uniformly very large; spherical; skin jet black; ripens in midseason; adapted for table fruit, fresh juice, semi-sweet wines. Vine: vigorous; canes long; productive; leaves very large, thick, slightly lobed; tolerant to powdery mildew; tetraploid.

Niagara rosada.—See Red Niagara.

Noble.—Orig. in Clayton, N.C., by W.B. Nesbitt, D.E. Carroll, Jr., and V.H. Underwood, North Carolina Agr. Expt. Sta. Introd. in 1973. Thomas x Tarheel. Cross made in 1946 by C.F. Williams (of above station); selected in 1951; tested as N.C. 20-119. Muscadine type. Cluster: 20–22 berries; loose. Berry: 4.3 g; round; skin black, medium thick, does not shell; flesh typical *V. rotundifolia*, quality good; ripens 18 Sept. in Clayton; recommended for wine, juice, jelly. Vine: vigorous; highly productive; tolerant to black rot; susceptible to powdery mildew, angular leaf spot, and bitter rot.

Norris.—Orig. in Leesburg, Fla., by J.A. Mortensen and L.H. Stover, Watermelon and Grape Investigations Lab. Introd. in 1967. Fla. W987 [(*V. smalliana* x *V. lincecumii*) x Cardinal] x Lake Emerald; cross made in 1956; selected in 1959; tested as FES A4-46. Cluster: moderately large. Berry: 3/4 inch in diam.; ellipsoidal; skin deep purple; for fresh fruit; ripens late July in Leesburg. Vine: vigorous; productive; resistant to Pierce's disease and downy mildew; requires a pollinator. Named for R.E. (Bob) Norris, county agent, Lake County, Fla., and secretary for 28 years of Florida Grape Growers Assn.

North Black.—Black, seeded table grape. Orig. in Akitsu, Hiroshima, Japan, by Hiroyasu Yamane. Introd. in 1991. Seneca x Campbell's Early. Berry: medium (4 g); violet-black; round; slipskin; foxy flavor. Cluster: medium; moderately compact; winged-cylindrical. Vine: somewhat low vigor; hardiness similar to Campbell's Early; fairly resistant to diseases.

North East.—Orig. in Fredonia, N.Y., by Fred E. Gladwin, New York State Agr. Expt. Sta. Named and released locally around North East, Pa., about 1925 by D.C. Paschke, who formerly worked for Gladwin. Champion x Lucile. Cluster: medium. Berry: not as large as Concord; skin black-blue; good quality; resembles Concord; too many small undeveloped green berries produced. Vine: vigorous; productive. Superseded by Fredonia.

North Fork.—Orig. in Mountain Grove, Mo., by Paul H. Shepard, Missouri State Fruit Expt. Sta. Introd. in 1947. Agawam x Early Daisy; selected in 1947; tested as G-2861. Cluster: large; compact. Berry: large; skin black with blue bloom, tough, does not crack from moisture at harvest time; pulp firm, crisp, sweet; ripens about 10 days before Concord. Vine: vigorous; very productive; perfect flower.

North Red.—Early ripening, hardy, seeded table grape. Orig. in Akitsu, Hiroshima, Japan, by Hiroyasu Yamane. Introd. in 1990. Seneca x Campbell's Early. Berry: medium (4 g); round; red-brown; slipskin; foxy flavor. Cluster: medium; moderately compact; winged-cylindrical. Vine: moderately vigorous; productive; hardiness similar to Campbell's Early; fairly resistant to diseases.

Nova.—Orig. in Ceres, Stanislaus County, Calif., by F.E. Triplett. Introd. in 1975. Plant patent 3989; 14 Dec. 1976. Ruby Cabernet x Barbera; cross made in 1961; selected in 1970. Cluster: medium to large; long stem; conical. Berry: black to deep blue-purple with thick gray-blue bloom; flesh greenish, soft, juicy; ripens 15–30 Sept. in Ceres. Vine: vigorous; highly productive; medium-length cane; flower fertile; tolerant to fruit spoilage. A table wine variety.

November.—Orig. in Experiment, Ga., by H.P. Stuckey, Georgia Agr. Expt. Sta. Introd. in 1920. Scuppernong x a black muscadine; selected in 1918. Muscadine type.

Nyora.—Orig. in Yanco, New South Wales, by W. Poggendorf, Dept. of Agriculture, Sydney, N.S.W. Introd. about 1956. Ohanez (Almeria) x Purple Cornichon. Cross made in 1929; selected in 1939. Table grape variety. Cluster: large; moderately loose; tapering. Berry: medium-large; ovate; skin black, heavy attractive bloom, tough; flesh flavor neutral; ripens 1–7 Mar., but can be harvested as late as May–June in New South Wales; resembles Purple Cornichon, but less prone to shatter when packed, less tendency toward flaccidity, and more attractive. Vine: moderately vigorous; productive, but some tendency toward millerandage; tolerant to bunch rot. Nyora, which is the language of the Australian aborigines, means "native berry."

Oconee.—Orig. in Clemson, S.C., by Harold J. Sefick, Agr. Expt. Sta. Introd. in 1970. S.C. 4707 (Alden x Ellen Scott o.p.) x Niagara. Cross made in 1954; selected in 1960; tested as SC 1. Cluster: medium; less compact and slightly longer than Concord. Berry: medium; round; skin blue, tolerant to cracking; flesh melting; *V. labrusca* flavor absent, softer than Concord; seeds separate from flesh when fully ripe; recommended for table and wine; ripens 10–14 days before Concord, 1–15 Aug. at Clemson. Vine: moderately vigorous; upright; hardiness and productivity equal to Concord; tolerance to black rot and downy mildew comparable to Concord. Oconee is the name of the westernmost county of South Carolina, where the cultivar was first observed.

Odem.—Late-ripening, large-fruited table grape for fresh market. Orig. in Bet Dagan, Israel, by P. Spiegel-Roy, R. Asaph, and I. Baron, Inst. of Horticulture, Agricultural Res. Organization, The Volcani Center. Introd. in 1991. Patented in Israel. Zeni x Toufahi; selected in 1975; tested as no. 110. Berry: red; ellipsoidal; large (8 g) with firm skin; fruit ripens about 14 days after Italia and Dabouki. Cluster: large (500 g); fairly loose; elongated. Vine: productive; adapted to spur pruning; vigorous.

Onslow.—Orig. in Willard, N.C., by Charles Dearing, U.S. Dept. of Agriculture and North Carolina Agr. Expt. Sta. Introd. in 1946. V22 R5 B4 (Scuppernong x male) x Burgaw; cross made about 1916; tested as 6B8. Muscadine type. Cluster: flat; loose, three to five berries. Berry: medium; slightly oval; dull black; adherence good, very small attachment area; skin thin, tough; seed small and broad; early ripening. Vine: vigor medium; canes short, growing tips reddish with red tendrils; flower pistillate. Named for Onslow County.

Orlando Seedless.—Orig. in Leesburg, Fla., by J.A. Mortensen and D.J. Gray, Central Florida Res. and Education Center. Introd. in 1986. Fla. D4-176 (Norris x Schuyler) x Fla F9-68 (Fla. A4-23 x Perlette). Selected in 1981; tested as Fla. BD8-77. The first seedless bunch grape with resistance to Pierce's disease. Berry: smaller than Lake Emerald, requiring gibberellic acid application after bloom to promote berry size and reduce occasional gritty seed remnants; skin light green. Vine: moderately vigorous and productive; self-fertile; budbreak early, sometimes leading to spring frost damage. Susceptible to anthracnose and leaf blight.

Orton.—Orig. in Willard, N.C., by Charles Dearing, U.S. Dept. of Agriculture and North Carolina Agr. Expt. Sta. Introd. in 1946. Latham x Burgaw; cross made about 1915; tested as 22B6. Muscadine type. Cluster: round; compact; 6–12 berries. Berry: medium; round; skin pearly bronze, thin, tough; ripens in midseason. Vine: vigorous; foliage dense; disease resistant; leaves shed late; flower pistillate. Named for Orton plantation.

Osborn.—Orig. in Woodward, Okla., by L.F.

Locke, U.S. Dept. of Agriculture, Southern Great Plains Field Sta. Introd. in 1959. Open-pollinated seedling of Armalaga; tested as Woodward 3894. Cluster: medium. Berry: medium; skin black; quality good; average ripening date mid-August at Woodward; stores well. Vine: growth medium to vigorous; very drought resistant; somewhat lacking in winter hardiness; probably best adapted to the southwest.

Ozark Prize.—Orig. in Mountain Grove, Mo., by Paul H. Shepard, Missouri State Fruit Expt. Sta. Introd. in 1947. Dr. Collier x Sheridan; selected in 1947; tested as G-1925. Cluster: medium to large; compact; similar to Sheridan. Berry: large; round; skin black; flesh firm; flavor good, sweet; good keeper; ripens almost with Concord at Mountain Grove; resembles Sheridan. Vine: moderately vigorous; good production; flower fertile.

Pam.—Orig. in Brooks, Ga., by W.G. Ison. Introd. in 1988. 5-11-3 x Senoia. Plant patent 7707; 12 Nov. 1991. Female muscadine. Berry: large, to 31 mm; skin superior bronze color; dry scar; 21% solids; very good flavor; uniform ripening in midseason. Cluster: large; long. Vine: vigorous; productive; disease resistant.

Pamlico.—Orig. in Raleigh, N.C., by the U.S. Dept. of Agriculture and the North Carolina Agr. Expt. Sta. Introd. in 1962. Lucida x Burgaw; selected in 1953; tested as NC 27-262. Muscadine type. Berry: large; skin light green, appearance attractive; flavor good; ripens in midseason, about 1 week before Dearing; superior to Willard and Wallace in flavor, size, soluble solids. Vine: vigorous; productive; foliage good; superior to Willard and Wallace in disease resistance; self-fertile. Suggested uses: as a white pollinator; for home gardens where a single vine is wanted.

Patricia.—Orig. in St. Catharines, Ontario, Canada, by W. Backus. Introd. in 1930 by E.D. Smith & Sons, Winona, Ontario. Trademarked by this company; when nursery department discontinued, assigned rights in the Canadian trademark to C.H. Prudhomme & Sons, Nurserymen, Beamsville, Ontario, Sept. 1947. Parentage unknown; chance seedling; discovered about 1915. Cluster: large; compact; often shouldered. Berry: skin black, tough; quality fair to poor; good shipping qualities; matures early, about Fredonia season at St. Catharines; resembles Concord. Vine: vigorous; very productive.

Pender.—Orig. in Willard, N.C., by Charles Dearing, U.S. Dept. of Agriculture and North Carolina Agr. Expt. Sta. Introd. in 1946. Latham x V20 R36 B4 [Kilgore x V19 R7 B2 (Scuppernong x male)]; cross made about 1917; tested as 25B6. Muscadine type. Cluster: small; loose; two to three berries. Berry: medium; oval; skin yellow-green; pulp very scanty; flavor apple-like; ripens later than Scuppernong. Vine: vigor very poor; susceptible to disease; leaves small; flower perfect. Named for Pender County.

Pêrel.—White, muscat-flavored, early ripening, seeded table grape. Orig. in Stellenbosch, Republic of South Africa, by E.P. Evans and P.J.L. Ellis, Viticultural and Oenological Res. Inst. Queen of the Vineyard x Jacaranda; selected in 1980; tested as 71-20; introd. in 1986. Berry; medium; round; yellow-green to straw-colored; soft; juicy; delicate muscat flavor. Cluster: small, conical; well-filled. Vine: moderate vigor; moderately productive; similar susceptibility to diseases as other varieties.

Perelli-101.—Orig. in Delano, Calif., by Antonio Perelli-Minetti. Introd. in 1964. Unknown seedling. Plant patent 2408; 2 June 1964. Berry: small, diam. about 0.5 inch; round; skin practically black, with a blue bloom, medium thick; juice gray-white, distinctive fresh fruit flavor; high sugar and high total acid content; maceration of the skins produces an intense red having substantially no blue. Vine: prolific; producing large number of side runners; main runners bearing leaves having four axially symmetrically disposed clefts, such leaves having sharp tips and serrated margins between clefts; secondary branches extending from the main runners, having substantially orbiculate leaves with pointed tips and serrated margins; bears numerous bunches on main and secondary runners; tolerant to bunch rot. A wine grape cultivar.

Perlette.—Orig. in Davis, Calif., by H.P. Olmo, California Agr. Expt. Sta. Introd. in

1946. Scolokertek kiralynoje (P.I. 103996) x Sultanina marble (P.I. 98388); cross made in 1936; first fruited in 1940; tested as Calif. 1253F21; sibling of Delight. Cluster: large, 450 to 700 g; compact to very compact; requires heavy thinning of berries after set. Berry: spherical; seedless; skin white to yellowish, translucent, thin, very tender; flesh translucent, crisp, juicy; flavor mild but distinctive, low in sugar and acidity; very resistant to high-temperature scald; exceptionally good handling qualities for an early variety; ripens 1 month before Sultanina at Davis and 2–3 weeks earlier than Sultanina in the desert regions of southern California. Vine: vigorous; more fruitful than Sultanina; spur-pruned on bilateral cordon.

Phil S. Taylor.—Orig. in Altoona, Fla., by Charles Demko. Introd. in 1957. (Edna x *Vitis simpsoni*) x (Seyve-Villard 12375 x Concord); cross made in 1947; tested as Demko 21-39. Cluster: medium. Berry: skin white; ripens in midseason. Vine: flower perfect. Named in honor of Phil S. Taylor, assistant commissioner of agriculture, state of Florida.

Pierrelle.—See Seyve-Villard 20-366.

Pineapple.—Orig. in Brooks, Ga., by W.G. Ison. Introd. in 1988. Fry x Senoia. Plant patent 7266; 10 July 1990. Self-fertile muscadine. Berry: medium-large, up to 28 mm; skin bronze; wet scar; 21% solids; good flavor, somewhat similar to a pineapple. Cluster: large. Vine: very vigorous; very productive; disease resistant.

Piney.—Orig. in Mountain Grove, Mo., by Paul H. Shepard, Missouri State Fruit Expt. Sta. Introd. in 1947. Open-pollinated seedling of Merrimac; selected in 1945; tested as G-135. Cluster: larger than Concord or Merrimac. Berry: large; skin black, bloom blue, thin, tough; flesh adheres somewhat to seeds; flavor good; ripens about with Concord at Mountain Grove; resembles Merrimac. Vine: vigorous; productive; flower fertile.

Piropella.—Orig. in Pretoria, South Africa, by E.P. Evans, Hort. Res. Inst. Introd. in 1966. Isabella x Pirovano 15; selected in 1961. Cluster: medium, average weight 16 oz; well-filled; slightly less compact than Muska; shoulder well-shaped. Berry: average 1/2 inch in diam.; skin blue-black with medium bloom, not tough; flesh greenish, firm with soluble solids content varying 16% to 18%; ripens mid-December, 4 weeks before Niagara. Vine: productive; tolerant to anthracnose; vigor good; for home garden and commercial production.

Pixiola *(Magruder)*.—Orig. in native woodlands near Leesburg, Fla., by Kenneth Loucks. Introd. in 1942. Open-pollinated seedling of *Vitis simpsoni*; discovered in 1939. Berry: small; skin greenish-white to emerald green; flesh very sweet; flavor good; sugar content unusually high; ripens in midseason. Vine: longevity excellent in native woodlands but variable under vineyard conditions. Pixiola is the female parent of Lake Emerald. Named after the location near Picceola Island where it was discovered growing wild in woodlands.

Plantet.—Orig. in Aubenas, Ardeche, France, by Albert Seibel. Introd. in U.S. and Canada about 1946. Parentage doubtful, but most often claimed as Seibel 4461 x (*Vitis berlandieri* x Jacquez). Cluster: large; compact; cylindrical. Berry: small; spherical; very solid; skin jet black, resistant to cracking; foxy flavor; ripens early. Vine: very highly productive and regularly so; adapted to wide range of climatic and soil conditions; buds-out late; grafts readily and well; shoots break off easily in wind; foliage tolerant to most diseases; difficult to root from cuttings. Produces wine of slight foxy taste.

Pontiac.—Orig. in Geneva, N.Y., by S.A. Beach, New York State Agr. Expt. Sta. Introd. in 1922. Herbert x Worden; cross made in 1903. Cluster: large; well-formed. Berry: large; skin dark purple, rarely cracks; flavor sweet and vinous; quality high; small number of seeds; long-keeping; good for home use. Vine: self-sterile.

PreCal™ *(Sugrasix)*.—Orig. in Bakersfield, Calif., by John Weinberger and Timothy Sheehan, Superior Farming Co. Introd. in 1987 as Sugrasix. Plant patent 6009; 8 Sept. 1987; assigned to Superior Farming Co. Ribier x Superior Seedless; tested as 20-5. Berry: size variable; skin purplish-black; seedless, with medium-size vestigial seed traces; ovate; flesh greenish, translucent, juicy, firm; most nearly resembles Ribier;

ripens about with Flame Seedless. Cluster: medium; double-shouldered, tapering; good berry attachment. Vine: medium vigor; medium productivity.

Price.—Orig. in Blacksburg, Va., by R.C. Moore, Agr. Expt. Sta. Introd. in 1973. VPI 4 (Hector x Seibel 13035) x VPI 5-7 (Fredonia x Athens). Cross made in 1959; first fruited in 1963; selected in 1964; tested as VPI 30. Cluster: medium; well-filled. Berry: medium; spherical; skin blue-black, heavy waxy bloom, uniform coloring; flesh firm, somewhat stringy, slightly acid near seeds, *V. labrusca* type texture and flavor; sweet; moderately foxy; quality comparable to Worden or Concord; ripens early with Van Buren or 4 weeks before Concord. Vine: moderately vigorous and productive; tolerant to black rot and powdery mildew; recommended for Mountain and Piedmont areas of Virginia and possibly areas in nearby states having similar soil and climatic conditions. Named in honor of Harvey L. Price, former head of VPI Dept. of Hort. and dean of College of Agr.

Pride.—Orig. in Experiment, Ga., by B.O. Fry, Agr. Expt. Sta., Univ. of Georgia. Introd. in 1972. Georgia 19-13 x USDA 19-11; cross made in 1959; tested as Georgia 23-61. Muscadine type. Berry: large; skin purple; quality good; average 17.9% soluble solids; ripens over 3–4 weeks, most abundantly last week of September in Experiment. Vine: vigorous; hardy; productive; moderate disease resistance.

Qualitas.—Orig. in Experiment, Ga., by H.P. Stuckey, Georgia Agr. Expt. Sta. Introd. in 1920. Thomas x black male muscadine; selected in 1918. Muscadine type.

Queen.—Orig. in Davis, Calif., by H.P. Olmo, California Agr. Expt. Sta. Introd. in 1954. Muscat Hamburg x Sultanina; cross made in 1931; first fruited in 1935; tested as Calif. A3-94. Cluster: large; loosely winged. Berry: large; ellipsoidal; uniform; skin dark red; flesh firm at maturity, but not as firm as Flame Tokay; medium sugar and acid; matures just after Red Malaga (Molinera); stores and ships well; competes successfully with Flame Tokay. Vine: medium vigor; flower cluster thinning required; fruitful when spur-pruned.

Queen Arlene™ *(Bianco One).*—Red, seedless, early ripening table grape. Orig. in Mecca, Calif., by Robert O. Bianco. Plant patent 8463, 16 Nov. 1993; assigned to Anthony Vineyards. Mutation of Flame Seedless. Berry: small (1 g); gibberellic acid treatment increases size to 3 g; oblate; slightly flat; neutral flavor; very crisp; fruit ripens about 15 days earlier and has smaller berries than Flame Seedless. Cluster: medium; compact; conical. Vine: medium vigor; productive.

Ravat 51.—See Vignoles.

Rayon d'or.—See Seibel 4986.

Red Amber.—Orig. in Excelsior, Minn., by Univ. of Minnesota Fruit Breeding Farm. Introd. in 1944. Parentage unknown; selected in 1923; tested as Minn. 45. Cluster: medium; compact. Berry: smaller than Concord; skin reddish-amber; flesh sweet, aromatic; quality very good; ripens early September in Excelsior. Vine: hardy.

Redgate.—Orig. in Brooks, Ga., by B.O. Fry, Ison Nursery. Introd. in 1974. Higgins x seedling 29-49. Crossed in 1968; selected in 1971; tested as S. 4-13-2. Muscadine type. Cluster: large. Berry: small; round; skin red. Vine: productive; vigorous; hardiness average; self-fertile.

Red Gem.—Orig. in Delano, Calif., by Luigi Chiarappa, Di Giorgio Fruit Corp., San Francisco, Calif. Introd. in 1964. Plant patent 2397; 28 Apr. 1964; assigned to Di Giorgio Fruit Corp. Queen x Emperor; selected in 1960. Cluster: large, averaging 9 inches long and 550 g, with a range of 6–13 inches; loose; conical; well-filled; with few or no shot berries; average of two clusters/cane. Berry: large, averaging 3/4 to 1 inch in diam.; a prolate ellipsoid; skin red in sun, sometimes greenish at stem end, tough, averaging 0.16 mm thick, not as easily separable from pulp as Emperor; seeds average three/berry, 7 mm long; flesh white, very firm, crisp, moderately juicy; averages 17° to 19°Brix, well-balanced sugar : acid ratio; very fine eating quality; primarily a table variety, but can be used for wine; ripens in Tulare County, Calif., about 1 Sept. if vines not girdled, about 1 month before Emperor and 1–2 weeks after Queen; storage quality good for 4 months under refrigeration; strongly attached to capstem. Vine: bears regularly, not as heavily as

Queen; no thinning required; growth habit like Queen; flower perfect; leaf large, up to 6–8 inches wide, elongated and lobed with petiole sinus open.

Redglobe.—Orig. in Davis, Calif., by H.P. Olmo and Albert T. Koyama, California Agr. Expt. Sta. Plant patent 4787; 10 Nov. 1981. Introd. in 1980. Cluster: large to very large; well-filled; peduncle very long (9 cm) and very thin in diam. Berry: very large, often over 13 g; spherical to oblate; skin dark red; flesh whitish, meaty, neutral flavor; seeds large, separating easily from pulp; pedicel with exceptionally large torus and very adherent. Vine: Emperor type, but fruit colors 2 weeks earlier and is more productive; excellent for long periods of cold storage.

Red Niagara *(Niagara rosada)*.—Orig. in the Louveira district, Jundiaí, São Paulo, Brazil, by Aurelio Franzini. Introd. in 1933; introd. in the U.S. in 1944 as P.I. 149001. A somatic mutation of a single cane of Niagara bearing red fruit; discovered in 1933 in the vineyard of Comendador Antonio Carbonari. Berry: skin entirely red; otherwise appears identical to Niagara.

Regale.—Orig. in Clayton, N.C., by C.F. Williams [U.S. Dept. of Agriculture–North Carolina State Univ. (NCSU)] and developed by W.B. Nesbitt (NCSU) and J.P. Overcash (Mississippi State Univ.). Introd. in 1981. Hunt x Magnolia. Tested as NC 153-1; cross made in 1957. Muscadine type. Berry: 5.0–5.5 g; skin blue-black, attractive; flesh somewhat tan, indicating a low sugar : acid ratio; ripens early September in Mississippi and 10–14 days later than Noble in North Carolina. Cluster: 5–10 berries. Vine: vigorous; hardy; long bloom period, self-fertile; productive; resistant to Pierce's disease; not adapted for mechanical harvest. Recommended for red wine, juice, or jelly.

Reliance.—Orig. in Clarksville, Ark., by J.N. Moore, Univ. of Arkansas. Introd. in 1982. Plant patent 5174; 10 Jan. 1984. Ontario x Suffolk Red; cross made in 1964; selected in 1967; tested as Ark. 1163. Cluster: medium (330 g); cylindrical; well-filled. Berry: round; 2.7 g; flesh pink, seedless; flavor sweet, delicate *V. labrusca*; skin tender; stores well for up to 3 months. Vine: very hardy, enduring temperatures to –34C; vigorous; productive; tolerant to black rot, anthracnose, powdery mildew, and downy mildew. Has performed well from central Texas to Wisconsin.

Remaily Seedless.—Orig. in Bucks County, Pa., by G.W. Remaily, Hammondsport, N.Y. [retired, New York State Agr. Expt. Sta. (NYSAES), Geneva]. Introd. in 1980. Lady Patricia (Seyve Villard 20-365 x Seibel 14665) x NY 33979 (Russian Seedless x Ontario). Cross made in 1965; selected in 1974 by Robert M. Pool, NYSAES, as a dessert selection; tested as N.Y. 65.085.1. Cluster: large; tapered; attractive. Berry: medium; oval; skin light green, adherent, slightly tough, resistant to cracking; flavor neutral; ripens with Concord or about 10 Oct. at Geneva; good holding quality. Vine: large, rampantly expansive; very vigorous; hardy to –27C; very productive, must be cluster-thinned; resistant to botrytis bunch rot; tolerant to downy mildew and powdery mildew; proper growth control must be exercised to retain berry size.

Roanoke.—Orig. in Raleigh, N.C., by U.S. Dept. of Agriculture and North Carolina Agr. Expt. Sta. Introd. in 1962. Lucida x (Topsail x Tarheel); selected in 1954; tested as NC 59-32. Muscadine type. Cluster: loose; medium-large. Berry: medium; skin attractive, white with a golden-yellow tint; ripens in early midseason, about 1 week before Scuppernong. Vine: self-fertile; vigorous; very productive. Superior to Wallace and Willard in flavor, attractiveness, vigor, and production; superior to Dearing in production and attractiveness. Suggested uses: as a white pollinator; especially for home gardens where a single vine is desired.

Robin Cardinal *(Thornburgs Robin)*.—Orig. in Litchfield Park, Ariz., by Wayne Thornburg. Introd. in 1960. Plant patent 1895; 5 Jan. 1960. Bud mutation of Cardinal; discovered in 1952. Berry: ripens about 5–10 days before Cardinal; otherwise same characteristics as Cardinal.

Roi des blancs.—See Seibel 4986.

Roi des Noirs.—See Seibel 4643.

Romulus *(Romulus Seedless)*.—Orig. in Geneva, N.Y., by New York State Agr. Expt. Sta. in cooperation with A.B. Stout,

New York Botanical Garden. Introd. in 1952. Ontario x Sultanina; cross made in 1928; selected in 1936; propagated for trial in 1937; tested as N.Y. 15291. Cluster: large; shouldered; compact. Berry: small but variable in size, roundish; skin greenish-yellow; flesh melting, juicy, vinous-flavored; quality good; seedless; resembles Sultanina; ripens with Concord. Vine: very vigorous; very productive; fairly hardy.

Romulus Seedless.—See Romulus.

Rosa.—Orig. in Brooks, Ga., by B.O. Fry and W.G. Ison. Introd. in 1988. Higgins x Granny Val. Plant patent applied for. Female muscadine. Berry: medium-large, 28 mm; skin pinkish-red; wet scar; 22% solids; ripens midseason to late. Cluster: medium. Vine: very vigorous; moderately productive.

Rosebelle.—Orig. in Beltsville, Md., by D.H. Scott, U.S. Dept. of Agriculture and Maryland Agr. Expt. Sta. Introd. in 1967. Fredonia x Niagara. Cross made in 1946 by N.H. Loomis; selected in 1950 by Scott; tested as U5-4606-87. Cluster: large; compact; attractive. Berry: medium; skin red; flavor pleasantly foxy; soluble solids content varying between 17% to 21%; ripens just before Concord; ripens more evenly than Fredonia and Concord; for roadstand sales and home gardens. Vine: vigor good; foliage excellent; moderately productive and more consistently productive than Fredonia.

Rosette *(Seibel 1000).*—Orig. in Aubenas, Ardeche, France, by Albert Seibel. Introd. in U.S. in the early 1900s and in Ontario, Canada, in 1946. Parentage unknown, but derived from Jaeger 70 (*Vitis rupestris-lincecumii*) and *V. vinifera*. Cluster: medium; compact; long conical or cylindrical; often setting irregularly and interspersed with small, green shot berries. Berry: round to slightly oval; small; black with bluish cast; juice white; ripens in early midseason. Vine: very vigorous; bushy; can withstand very low winter temperatures, to –35F; breaks bud in midseason; moderate to low productivity; susceptible to anthracnose and only fairly tolerant to downy mildew, but commercially resistant to powdery mildew; shoot tips downy and carmine-tipped; shoots reddish with glabrous and shiny leaves. Produces a light-bodied, poorly colored wine, slightly bitter, but used for local consumption in parts of France. Planted in New York and Canada.

Roubidoux.—Orig. in Mountain Grove, Mo., by Paul H. Shepard, Missouri State Fruit Expt. Sta. Introd. in 1947. Open-pollinated seedling of Prune de Cazouls; selected in 1945; tested as G-194. Cluster: medium; loose. Berry: large; slightly oval, resembling *Vitis vinifera*; skin blue; flesh firm; ripens late, with Catawba. Vine: *V. labrusca* type; hardy; extremely vigorous; productive; flower perfect.

Roucaneuf.—Orig. in Saint Vallier, Drôme, France, by Seyve-Villard. Introd. in the 1930s. From the cross Seibel 6468 x Seibel 6905, according to Galibert. Cluster: large; shouldered; cylindrical. Berry: medium; long ellipsoidal; about 4/5 inch long; skin suffused with pink; flesh tender, crisp; ripens midseason to late; sheds badly at maturity; good to eat fresh; wine alcoholic, aromatic, somewhat bitter. Vine: vigorous; very productive; buds-out late; young shoots with few scattered hairs; moderately tolerant to downy mildew; grafts well on most common rootstocks, except 3309; young leaves shiny, reddish-bronze; petiolar sinus wide, open, and U-shaped.

Rouge.—Red, seeded table grape, an early ripening sport of Emperor. Orig. in Delano, Calif., by George L. Caratan. Plant patent 6421, 29 Nov. 1988; assigned to Anton George Caratan. Berry: large (9.9 g); two or three seeds/berry; oval or occasionally roundish; sweet and subacid flavor; ripens 2–3 weeks before Emperor. Cluster: very large; long; conical; double-shouldered; well-filled but not compact. Vine: average vigor to vigorous; productive; one cluster/shoot.

Rougeon *(Seibel 5898).*—Orig. in Aubenas, Ardeche, France, by Albert Seibel. Introd. in the U.S. about 1927; named by the Great Lakes Grape Nomenclature Committee in 1972. Seibel 880 x Seibel 4202(?); cross made in late 1800s. Cluster: medium to small; cylindrical; compact. Berry: oblate; skin dark blue; flesh meaty, high acidity; ripens midseason; wine grape useful for blending. Vine: vigorous; hardy; productivity irregular, tends to overbear; upright in growth, open.

Royal Seedless ***(Farmer Great Northern Seedless, Great Northern Seedless).***—Orig. in Mankata, Minn., by Frank M. Schwab, Schwab Fruit Farm & Nursery. Introd. in 1956 by this nursery. Plant patent 1681; 11 Feb. 1958. Parentage unknown; discovered in 1940. *Vitis labrusca* type. Berry: seedless; smaller than Concord, which it resembles; appears to be sweeter and with higher aroma than Concord. Vine: more hardy than Concord; bears heavily; vigorous.

Royalty.—Orig. in Davis, Calif., by H.P. Olmo, California Agr. Expt. Sta. Introd. in 1958. Alicante Ganzin x Trousseau; cross made in 1938; tested as Calif. 526. Cluster: medium; loose to well-filled. Berry: small; ellipsoidal; juice dark red; high in sugar and acid; very resistant to spoilage; ripens in midseason; selected for production of port-type wine. Vine: very productive; foliage heavy; canes semi-erect to drooping; tolerant to powdery mildew.

Rubaiyat.—Orig. in Stillwater, Okla., by Herman A. Hinrichs, Agr. Expt. Sta., Oklahoma State Univ. Introd. in 1975. Seibel 5437 x Bailey; cross made in 1952; selected in 1957; tested as OK 308. Cluster: medium with a shoulder; shank medium to short. Berry: medium; round; skin dark blue-black with bloom; juice very dark red; sugar 22%; tartaric acid 0.63%; ripens 15 Aug. in Stillwater. Vine: productive; self-fertile; vigor medium to strong. Uses: juice, jelly, red wine.

Rubilande.—Orig. in Aubenas, Ardeche, France, by Albert Seibel. Seibel 2859 x Seibel 4643. Cluster: medium; compact; often winged. Berry: medium; ovoid; skin lavender-pink; flesh hard, crisp; ripens in midseason; some tendency to rot; only useful for pink wine production, but wine oxidizes readily and bitterness is objectionable. Vine: young shoots with downy tips; expanding leaves deeply bronzed; buds-out late, but not hardy; production irregular in many localities because of shedding flowers and poor set; vine becoming weak; not resistant to summer drought; susceptible to downy mildew.

Rubired.—Orig. in Davis, Calif., by H.P. Olmo, California Agr. Expt. Sta. Introd. in 1958. Tinto Cão x Alicante Ganzin; cross made in 1938; tested as Calif. 58. Cluster: medium; loose. Berry: small; ellipsoidal; very intense red juice; ripens in midseason; very resistant to spoilage; for production of sweet red wine, concentrate, or blending where intense and stable color is desired. Vine: very productive; very vigorous; all shoots fruitful; habit semi-upright; leaves cover fruit; tolerant to powdery mildew.

Rubistar.—Dark red, early–ripening, seeded table grape. Orig. in Roodeplat, Republic of South Africa, by E.P. Evans and P.J.L. Ellis, Viticultural and Oenological Res. Inst. Queen of the Vineyard x Muscat Hamburg; selected in 1973; tested as CC 65; introd. in 1986. Berry: medium to large; ovoid truncate; firm; crisp; slight muscat flavor. Cluster: medium-large; round; well-filled; numerous shot berries. Vine: vigorous; productive.

Ruby.—Orig. in Geneva, N.Y., by Richard Wellington, New York State Agr. Expt. Sta. Introd. and named in 1938. Keuka x Ontario; cross made in 1923; selected in 1930. Cluster: medium; long, conical; shouldered; compact. Berry: variable size, to 20 mm; skin dark red, slipskin; quality good; ripens 1 week before Catawba. Vine: lacks vigor; overbears; not sufficiently hardy at Geneva.

Ruby Cabernet.—Orig. in Davis, Calif., by H.P. Olmo, California Agr. Expt. Sta. Introd. in 1948. Carignane x Cabernet Sauvignon; cross made in 1936; first fruited in 1940; tested as Calif. 234F2. Cluster: medium; loose; peduncle woody. Berry: skin black; pulp juicy; resistant to spoilage; produces table wine of high quality, similar to Cabernet Sauvignon but of more intense color and earlier maturation; especially adapted to regions too hot for best quality of Cabernet Sauvignon. Vine: vigorous; upright habit; more productive than Cabernet Sauvignon.

Ruby Seedless.—Orig. in Davis, Calif., by H.P. Olmo, California Agr. Expt. Sta. Introd. in 1968. Emperor x Pirovano 75. Cross made in 1939; selected in 1950; tested as I8-17. Cluster: very large; conical and shouldered; well-filled. Berry: medium; ovoid; skin reddish-black to dark red, tender; flesh firm, tender; seedless; good eating quality, stores well; a dessert cultivar; ripens late midseason. Vine: very vigorous; very productive.

Saint Croix.—Orig. in Osceola, Wis., by Elmer Swenson. Introd. in 1981. Plant patent 4928; 9 Nov. 1982; assigned to Swenson Smith Vines. E.S. 283 (Minn. 78 x Seibel 1000) x E.S. 193 (Minn. 78 x Seneca); tested as E.S. 2-3-21. Berry: medium; skin blue; very juicy. Cluster: medium. Useful for neutral, high-quality red wine. Vine: very hardy, but not as hardy as Kay Gray; vigorous; good resistance to black rot; leaves susceptible to downy mildew; very precocious.

Saint Pepin.—Orig. in Osceola, Wis., by Elmer Swenson. Introd. in 1983. Plant patent 5771; 22 July 1986; assigned to Swenson Smith Vines. (MN78 x Seibel 1000) x Seyval; tested as E.S.282. Female bunch grape. Berry: medium; slipskin; skin white; flesh tender; pink juice; resembles Lacrosse, but earlier ripening; produces an excellent German-style white wine; also has good dessert quality. Cluster: large; conical; loose. Vine: very vigorous; very productive; slightly less hardy than Lacrosse.

Salba.—Orig. in Keerweder, Franschhoek, South Africa, on the farm of P.K. Siebritz by C.J. Coetzee. Introd. and named in 1965. Barlinka x Alphonse Lavallée; cross made in 1955. Cluster: medium; pyramidal; compact to very compact, requiring berry thinning; resembling Alphonse Lavallée; peduncle reddish. Berry: weight 63.5 per lb, smaller than Barlinka; ovoid truncate to oblong, with apical grooves; skin black, purplish bloom, resistant to splitting; flesh greenish, firm, attractive; flavor more acid than parents; ripens in early March in Keerweder, between season of parents; subject to sunburn when exposed. Vine: vigorous; productive, produces many small secondary clusters; susceptible to anthracnose. Leaf: medium, five-lobed with broad sinuses very deeply cut, so that all lobes overlap; teeth ovate, very large; petiolar sinus very wide, open U-shaped; indumentum similar to Alphonse Lavallée; leaves turn dark red in autumn.

Sanderson Special.—Orig. in Glendale, Ariz., by Chester A. Sanderson. Introd. in 1948. Plant patent 782; 20 Jan. 1948. Parentage unknown; discovered in 1942 as a chance seedling. Cluster: medium, about 7 inches long; oblong; not winged; compact. Berry: 3/4 inch in diam.; spherical; skin dark reddish-black, slipskin; flesh soft; flavor mild; matures in Concord season. Vine: very vigorous; very productive; foliage dense, leaf smaller than Concord; *V. labrusca* type.

Sanger Sweet.—Orig. in Sanger, Calif., by E.L. Magnone. Introd. in 1942. Plant patent 509; 14 Apr. 1942; assigned to Ernest Magnone. Claimed to be an open-pollinated seedling of Muscat of Alexandria; selected about 1924. Cluster: short; pyramidal; slightly compact. Berry: medium; spherical; skin dark green to amber; flesh soft, very juicy, low acid; mild Muscat flavor; ripens in early July. Vine: vigor medium; very productive; leaves small, dark green. Apparently similar to or identical to Pearl of Csaba.

Sanoma.—See Sonona.

Saturn.—Orig. in Fayetteville, Ark., by James N. Moore, John R. Clark, and Justin R. Morris. Introd. in 1989. Dunstan 210 (Blackrose x Aurelia) x NY 45791 (Bath x Himrod); selected in 1974; tested as Ark. 1448. Plant patent 670; 28 Mar. 1989. Berry: seedless; large; oval; skin bright red, adherent; texture firm, crisp; sweet and fruity, mild flavor; good in storage; ripens midseason. Cluster: medium-large; conical. Vine: productive, may require cluster-thinning; medium vigor; moderate resistance to black rot, powdery mildew, and anthracnose; susceptible to downy mildew; vines have withstood –23C with minimal injury in Arkansas.

Scarlet.—Orig. in Davis, Calif., by H.P. Olmo, California Agr. Expt. Sta. Introd. in 1946. Golden Muscat x Teinturier; cross made in 1935; first fruited in 1939; tested as Calif. 294E7. Cluster: small; compact; cylindrical. Berry: medium; skin black; juice abundant, dark red, high sugar and acid; mild Concord-type flavor; resembles Teinturier in color of berry and juice. Vine: vigorous; resistant to powdery mildew; foliage blood red in autumn. For home and commercial production of grape juice.

Schuyler.—Orig. in Geneva, N.Y., by Richard Wellington and G.D. Oberle, New York State Agr. Expt. Sta. Introd. in 1946. Zinfandel x Ontario; selected in 1932. Cluster: medium; long; shouldered; well-filled. Berry: has much of the flavor and flesh characteristics of *V. vinifera*; juicy, soft; skin blue-black, tough, astringent; ripens in midseason; resembles Zinfandel more than Ontario. Vine: appears to be sufficiently hardy and disease resistant to be adapted to the climate of eastern U.S.; vigor moderate; productive.

Scioto.—Orig. in Cincinnati, Ohio, by Byron T. Johnson. Introd. in 1982. America x Westfield; tested as selection 64.AA-1. Berry: medium; oblate; skin dark blue-black; flavor resembles raspberries; *V. labrusca* type; ripens late midseason. Cluster: medium; with shoulder; medium shank. Vine: vigorous; productive; healthy. Introduced as juice and jelly grape for southern Ohio.

Seedless Emperor.—See Empress.

Seedless Muscat.—See Thomuscat.

Seedless Tokay.—Orig. in Lodi, Calif., by Walter A. Perrin. Introd. in 1965 by Stribling's Nurseries, Merced, Calif. Plant patent 2340; 14 Jan. 1964. Bud mutation of Flame Tokay; discovered about 1951. Cluster: smaller than parent; loose to well-filled. Berry: average 9/16 inch in diam.; ovoid-truncate; seedless with no trace of seedcoats; gibberellic acid treatment necessary to obtain adequate berry size. Vine: like the parent in all details.

Seibel 1000.—See Rosette.

Seibel 4643 *(Le Séquanis, Roi des Noirs)*.—Orig. in Aubenas, Ardeche, France, by Albert Seibel. Introd. in U.S. in 1940s. Seibel 29 x *Vitis vinifera* Danugue. Cluster: medium to large; cylindrical; compact. Berry: medium; round; skin dull black; pulp soft; spoilage severe in wet localities or seasons. Vine: vigorous; breaks bud early; a very heavy producer; many vines infected with virus degeneration; mature leaf rounded in outline, three-lobed with only shallow superior sinuses, petiolar sinus very narrow and closed by overlapping lobes; only fair tolerance to anthracnose; recovers poorly after frost. Wine: very heavy-bodied, intense color, foxy and disagreeably bitter.

Seibel 4986 *(Rayon d'or, Roi des blancs)*.—Orig. in Aubenas, Ardeche, France, by Albert Seibel. Introd. in U.S. in the 1940s. Seibel 405 x Seibel 2007. Cluster: medium; cylindrical; winged; compact. Berry: round; skin dull golden with lilac bloom; flesh pulpy; spoils quickly at maturity. Vine: vigor medium; buds-out late; good regrowth of fruitful shoots after frost; shoot tip felted with carmine-edged young leaves; leaf five-lobed, dark green, shiny surface, sinuses deep, narrow, and closed by overlapping lobes; teeth large, acute, and few; satisfactory tolerance to phylloxera in most soils; tolerant to both downy mildew and black rot. Wine: white, highly aromatic, with a berry-like flavor, but without the bitterness of many similar hybrids; used with some success in production of sparkling wines. Authorized for commercial use in Loiret and Loire-et-Cher in France.

Seibel 5279.—See Aurore.

Seibel 5409 *(La Gloire de Seibel)*.—Orig. in Aubenas, Ardeche, France, by Albert Seibel. Introd. in U.S. in the 1950s. Seibel 867 x Seibel 452. Cluster: medium; cylindrical; very compact. Berry: small; spherical; very juicy; skin white with pink tinge; ripens early. Vine: only fair vigor; bushy growth habit; buds-out late; production below average; foliage healthy, resistant to downy mildew, only slightly susceptible to powdery mildew; does not recover with a second crop after spring frost; young shoot tips cottony with carmine edges; leaf rounded in outline, large obtuse teeth. Produces a white wine of pronounced berry-like flavor.

Seibel 5437.—Orig. in Aubenas, Ardeche, France, by Albert Seibel. Introd. in U.S. and Canada about 1946. Seibel 867 x Seibel 2512. Cluster: medium; cylindrical. Berry: small; ovoid; skin bluish-black; some shelling off at maturity; ripens before Concord; pulp with deep-red juice; foxy flavor. Vine: very vigorous and highly productive; buds-out early; susceptible to downy mildew, but resistant to powdery mildew; leaf large, dark green, bullated surface, petiolar sinus closed by overlapping lobes; canes very long. Wine with accentuated *V. labrusca* flavor, used only for blending.

Seibel 5813.—Orig. in Aubenas, Ardeche, France, by Albert Seibel. Introd. in U.S. about 1949. Seibel 175 x Seibel 3012. Cluster: very large, to 26 cm long; conical; shouldered; loose. Berry: large; ovoid; elongated, resembles Dattier; skin thick, dull black; pulp juicy; flavor neutral and flat; matures late. Vine: buds-out very late; shoot glabrous, bright green, very shiny; unfolding leaves tinged bronze; leaf reniform, with very wide, open petiolar sinus; blade thick, dark green; very susceptible to anthracnose and downy mildew; injured by high summer temperatures. Wine is sometimes only pink, with bitterness apparent. Should be heavily cluster-thinned; needs a long, warm growing season and good soil moisture for best results.

Seibel 5898.—See Rougeon.

Seibel 6905 *(Le Subéreux)*.—Orig. in Aubenas, Ardeche, France, by Albert Seibel. Introd. in U.S. in the 1940s. Seibel 4595 x Seibel 4199. Cluster: medium; conical; compact. Berry: medium; ovoid; skin jet black, with corky blotches on surface, which is the best mark of identification; texture meaty; flavor distinct but not foxy; ripens early. Vine: often tends toward overproductivity; shoot tips glabrous, shiny, red; fair resistance to downy mildew, but very susceptible to anthracnose.

Seibel 7053.—See Chancellor.

Seibel 8229.—Orig. in Aubenas, Ardeche, France, by Albert Seibel. Introd. into Canada in 1946 and into U.S. in the early 1940s. Seibel 5163 x Seibel 4989. Cluster: medium to small; compact. Berry: small; slightly oval; skin yellowish-white. Vine: upright grower; vigorous and productive; fair tolerance to powdery mildew; well adapted to heavy soils; not fully hardy in Ontario, Canada.

Seibel 8365.—Orig. in Aubenas, Ardeche, France, by Albert Seibel. Introd. in the U.S. in the early 1940s. Seibel 5410 x Seibel 4643. Cluster: medium; elongated; winged with many green shot berries, often poorly set. Berry: medium; spherical; skin bluish-black; pulp firm, flavor foxy; matures in midseason. Vine: vigor poor; tolerant to fungal diseases; productivity much below average. The poor fruit set is a serious defect in many localities.

Seibel 9110.—See Verdelet.

Seibel 9549.—See De Chaunac.

Seibel 10096.—Orig. in Aubenas, Ardeche, France, by Albert Seibel. Introd. in the U.S. in 1940s. Seibel 5455 x Seibel 5163. Cluster: large to very large; long; narrow; winged; very compact. Berry: small; spherical to oval; skin bluish-black, thick, leathery; pulp juicy, very sweet; flavor almost neutral or slightly herbaceous. Vine: very irregular in time of flowering and set; vigor medium; fair resistance to downy and powdery mildew, susceptible to black rot; tends to weaken with age. Produces a wine of deep red color, well-balanced, and without objectionable flavor.

Seibel 10868.—Orig. in Aubenas, Ardeche, France, by Albert Seibel. Introd. in 1946 into Canada, in the U.S. in the 1940s. Seibel 5163 x Seibel 5593. Cluster: short conical; medium-large; compact. Berry: small; skin white with pinkish tinge; flesh soft, with foxy flavor; subject to botrytis spoilage; ripens in Concord season. Vine: vigor medium; stocky; hardy; susceptible to anthracnose; leaf size medium, with deeply indented lobes, teeth obtuse, very large. Considered promising on the Niagara, N.Y., peninsula for wine making, of foxy taste.

Seibel 13047.—Orig. in Aubenas, Ardeche, France, by Albert Seibel. Introd. in 1946 into Canada, in the U.S. in late 1940s. Seibel 5658 x Seibel 4995. Cluster: very large; shouldered; compact. Berry: medium; skin yellow-white; flesh firm; flavor poor; ripens early; mediocre as a table grape. Vine: productivity irregular; hardy; vigorous; susceptible to black rot in wet seasons. Wine quality poor.

Seibel 13053.—See Cascade.

Seibel 14596.—Orig. in Aubenas, Ardeche, France, by Albert Seibel. Introd in the U.S. in 1940s. Seibel 6468 x Seibel 5455. Cluster: medium; cylindrical; winged; fairly compact. Berry: small; round; skin blue-black; pulp firm, meaty, tannic, flavor neutral; defect of shedding at maturity; ripens in midseason. Vine: vigor fair; medium productivity; buds-out late; leaf with very wide, shallow, petiolar sinus; susceptible to downy mildew and powdery mildew, often subject to cluster spoilage by botrytis bunch rot.

Seibel 15051.—Orig. in Aubenas, Ardeche, France, by Albert Seibel. Introd. in the U.S. about 1950. Seibel 5455 x Seibel 8216. Cluster: large; narrow; to 10 inches in length; double-shouldered; compact. Berry: spherical; skin light amber; flesh firm; maturity in early midseason. Vine: young shoot tip downy, white; leaves with roughened surface; extremely vigorous; very erratic in production. Produces a wine of yellow color with an exotic aroma and taste.

Seinoir.—Orig. in Aubenas, Ardeche, France, by Albert Seibel. Introd. in the U.S. in the 1940s. Seibel 5163 x Seibel 880. Cluster: small; conical; compact. Berry: small; ovoid; skin black; pulp with deep red juice; flavor almost neutral; matures in early midseason. Vine: buds-out late; shoots glabrous, bronzed tips; leaf rounded in outline, petiolar sinus narrow-V or closed, apical teeth very large and acute; vigor moderate; recovers with some crop after spring frost, but productivity only average; susceptible to downy mildew. Wine quality fair, often used in blends to add color.

Seminole.—Orig. in southern Florida by Joseph L. Fennell, Lady Lake, Fla. Introd. in 1947. (*Vitis shuttleworthii* x *V. rofotomentosa)* x (*V. candicans* x Rommel); selected in 1945. Vine: flower perfect; vigorous; productive; disease resistant in semitropical climate; resembles Catawba or Extra.

Seneca.—Orig. in Geneva, N.Y., by R.D. Anthony, New York State Agr. Expt. Sta. Introd. in 1930. Lignan Blanc x Ontario; cross made in 1917. Cluster: medium; loose; tapering; medium compact. Berry: small to medium; oval; skin thin, tender, yellowish-green becoming a translucent golden; flesh greenish, translucent, firm, tender, juicy, very sweet, melting; flavor rich, aromatic; quality very good to best; ripens early to midseason, with Ontario, shortly after Perle de Csaba; almost *V. vinifera* type; hangs well on vine until end of season. Vine: vigorous; productive; hardy; susceptible to powdery mildew; requires long-cane pruning; canes long, flattened; flower stamens upright; blooms 1–5 days after Concord.

Senoia.—Orig. in Brooks, Ga., by B.O. Fry, Ison Nursery. Introd. in 1982. Plant patent 4772; 6 Oct. 1981. Higgins x Carlos; cross made in 1971; selected in 1976; tested as 22-1-1. Muscadine type. Cluster: large; long stem. Berry: medium; ripens evenly; skin red; flesh quality good; recommended for wine. Vine: productive; vigorous; self-fertile; tolerant of diseases.

Seyval.—Orig. in Saint Vallier, Drôme, France, by Seyve-Villard. Introd. about 1930; in the U.S. during the 1940s. Assumed to be Seibel 4995 x Seibel 4986. Cluster: small; cylindrical. Berry: small; round; skin golden; ripens early to midseason; subject to spoilage; wine light, yellow, aromatic. Vine: vigor medium; hardy; fairly resistant to downy mildew; productivity very high but very variable in different localities; short pruning recommended; shoot tip white woolly with carmine-edged leaves, young leaves bronzed; leaf small, rounded outline, shiny, petiolar sinus with overlapping lobes, teeth small, rounded, surface with light cobwebby hairs.

Seyve-Villard 1-72.—Orig. in Saint Vallier, Drôme, France, by Seyve-Villard, and the first hybrid introduced commercially by this breeder, about 1932. Imported from France into the U.S. as P.I. 181602. Cluster: small; cylindrical; compact. Berry: round; skin light pink on a yellow ground color, almost lavender; flesh pulpy; slight foxy taste; sometimes shelling before full maturity; massed at base of shoots; ripens in early midseason. Vine: spreading; very vigorous at first, but tends to overbear and often weakens; short pruning recommended; tolerant to downy mildew; shoot tip downy, bordered with carmine, later retaining a purplish cast.

Seyve-Villard 23-18.—Orig. in Saint Vallier Drôme, France, by Seyve-Villard. Introd. about 1932. Probably Seibel 11803 x Seibel 7053, according to Galibert. Cluster: conical; long, up to 10 inches; two short wings; well-filled. Berry: medium; slightly ovoid; skin dull black; ripens in midseason; wine neutral, bright red, improves considerably on aging. Vine: vigorous, but weakening after a few years; fruiting irregular and sets poor in some localities; buds-out fairly late; shoots with only few loose hairs; leaf glabrous, shiny, petiolar sinus bounded by unequal basal lobes; satisfactory resistance to downy mildew.

Seyve-Villard 12-303.—Orig. in Saint Vallier, Drôme, France, by Seyve-Villard. Introd. about 1930. Probably from the cross Seibel 6468 x Seibel 6905, according to Galibert. Cluster: medium; long conical; compact. Berry: ovoid; very juicy; dull rose; wine almost neutral, flat, sometimes bitter. Vine: vigorous; shoots easily broken off by wind; recovers poorly after spring frost; productivity medium; leaf undulated, very shiny, deeply lobed and overlapping lateral sinuses; petiolar sinus wide, open, and U-shaped.

Seyve-Villard 20-365 *(Dattier of St. Vallier)*.—Orig. in Saint Vallier, Drôme, France, by Seyve-Villard. Introd. in the early 1930s, in the U.S. before 1960. Panse x Seyve-Villard 12-375; cross made in 1931; sibling of Seyve-Villard 20-366. Cluster: large; short conic; loose. Berry: large, 1/2 x 1 inch; ovoid truncate; elongated; skin greenish-yellow, thin; flesh tender, meaty; flavor bland, neutral, sweet; sheds easily after rain and subject to rot. Vine: vigorous; very productive; buds-out late; shoots glabrous and the unfolding leaf red, sinuses deep, well marked, open; high tolerance to downy mildew and moderate tolerance to powdery mildew; not hardy. One of the best varieties of this breeder. Must not be allowed to overbear or quality will be indifferent. A dessert grape.

Seyve-Villard 20-366 (Pierrelle).—Orig. in St. Vallier, Drôme, France, by Seyve-Villard. Introd. in 1946; imported into the U.S. as P.I. 255591. Parentage not available. Cluster: large; short conical; loose to poorly filled; often with shot berries. Berry: medium to large; ovoid; skin yellowish-white; flesh meaty, with faint flavor of Muscat; maturity early. Vine: vigorous; productivity variable; shoots erect; young, expanding leaves maroon, glabrous, shiny; mature leaf large, three-lobed, reniform; petiolar sinus broad, open; fairly resistant to downy mildew and anthracnose.

Seyve-Villard 20-473 *(Muscat de Saint Vallier)*.—Orig. in Saint Vallier, Drôme, France, by Seyve-Villard. Introd. about 1942. Parentage unknown. Cluster: medium; long; cylindrical; very loose to poorly set; does not keep long on vine; tendency to dry and blacken at maturity. Berry: medium to large, to 22 mm in length; elliptical; skin dull whitish-yellow, high in tannin; flesh soft, Muscat flavor; ripens in early midseason; a table grape type for local use. Vine: vigor medium to weak; young shoot tips with cobwebby hairs, unfolding leaves reddish; mature leaf deeply lobed with sinuses narrow and straight, teeth sharp; petiolar sinus very wide, lyre-shaped, open; very susceptible to powdery mildew and moderately susceptible to downy mildew.

Shakoka.—Orig. in Brookings, S.D., by N.E. Hansen, Agr. Expt. Sta. Introd. in 1925. Lady x North Dakota wild. Berry: large; round; skin black; flesh quality fair; seeds separate readily from pulp. Vine: vigorous; heavy producer.

Shani.—Red, noncracking table grape to replace Cardinal for fresh market. Orig. in Bet Dagan, Israel, by P. Spiegel-Roy, R. Asaph, and I. Baron, Inst. of Horticulture, Agricultural Res. Organization, The Volcani Center. Introd. in 1981. Patented in Israel. Dabouki x Cardinal; selected in 1974; tested as no. 7. Berry: red; large (7 g); ovoid; less cracking than Cardinal; ripening period similar to that of Cardinal. Cluster: long to very long; large (600 g); shouldered; not compact. Vine: vigorous; adapted to spur pruning; productive.

Shauman.—Orig. in Peru, Ind., by J.F. Wittel. Introd. in the 1930s. Believed to be an interspecific hybrid of *Vitis labrusca* x *V. vinifera;* probably selected during 1920–25.

Sheridan.—Orig. in Geneva, N.Y., by S.A. Beach, New York State Agr. Expt. Sta. Introd. in 1921. Herbert x Worden; cross made in 1903. Cluster: compact; large. Berry: large; round; with good quality; skin black, tough; stores exceptionally well; matures about 1 week after Concord. Vine: hardy; fairly productive; vigor moderate.

Simone.—Orig. in Summerland, B.C., by A.G. Reynolds, L.G. Denby, and M. Bouthillier, Agriculture Canada, Res. Sta. Patricia x Himrod; selected in 1971; tested as Summerland Sel. 375 and under the provisional name Sovereign Charter. Berry: medium; orbicular; skin dark blue, thin; nonslipskin; seedless; mildly *V. labruscana* flavor; moderately crisp texture; ripens midseason. Cluster: medium-large (265–300 g); well-filled; conical; winged; slight tendency to shatter in some years. Vine: very vigorous; recommended for divided canopy culture; winter hardiness similar to Okanagan Riesling; resistant to powdery mildew.

Sivan.—Yellow, seeded table grape for fresh market. Orig. in Bet Dagan, Israel, by P. Spiegel-Roy, R. Asaph, I Baron, and N. Sahar, Inst. of Horticulture, Agricultural Res. Organization, The Volcani Center. Introd. in 1981. Patented in Israel. Dabouki x Cardinal; selected in 1968; tested as no. 42. Berry: yellow; ellipsoidal; medium (5 g); containing one or two small seeds; skin fairly tough; some tendency for shot berries and sunburn of berries; slight muscat flavor; ripens at similar time as Queen of Vineyards. Cluster: medium (400 g); not compact; short, conical shape. Vine: vigorous; adapted to spur pruning; productive.

Sonoma.—See Sonona.

Sonona *(Sanoma, Sonoma)*.—Orig. in Brookings, S.D., by N.E. Hansen, Agr. Expt. Sta. Introd. in 1925. Lady x North Dakota wild. Berry: medium to small; skin white with light-pink tinge turning to light red with white bloom at maturity; flesh sweet with trace of wild grape flavor; seeds separate readily from pulp. Vine: very productive.

Southern Home.—Black interspecific hybrid introd. for ornamental planting in the southern U.S. Orig. in Leesburg, Fla., by J.A. Mortensen, J.W. Harris, D.L. Hopkins, and P.C. Andersen. Summit (*Vitis rotundifolia*) x P9-15 (a hybrid of *V. rotundifolia, V. popenoei, V. munsoniana,* and *V. vinifera*); selected in 1984; tested as Fla. AA12-3. Plant patent applied for. Berry: oval; medium-large (6.8 g); 3.4 seeds/berry; pulp breaks up when chewed; scar moderately dry; soluble solids similar to other black muscadines; unique aromatic flavor; berries adhere to vine longer than other muscadines. Cluster: moderately loose; 12.2 berries/cluster. Vine: vigorous; self-fertile; unique leaf shape with deep-cut sinuses resembling those of maple leaf; no Pierce's disease has been observed; highly resistant to ripe rot, bitter rot, black rot, anthracnose, and downy mildew.

Southland.—Orig. in Meridian, Miss., by U.S. Dept. of Agriculture. Introd. in 1967. Thomas x seedling of Topsail; first fruited in 1950; tested as Meridian 13; selected and evaluated by N.H. Loomis. Muscadine type. Berry: large, about 100 berries per lb; skin purple, nonglossy; flavor good; quality high; sugar content 18% to 18.8% or about 3% higher than Scuppernong; ripens midseason, 10 Sept. at place of origin. Vine: moderately vigorous; productive; perfect-flowered; has produced very well in the central and southern portions of the Gulf Coast states and Georgia; suggested for trial commercially, for home use, and as pollinizer for cultivars having imperfect flowers.

Sovereign Coronation.—Orig. in Summerland, B.C., Canada, by L.G. Denby, Agriculture Canada, Res. Sta. Introd. in 1977. Lady Patricia x Himrod; cross made in 1966; selected in 1971; tested as Sel. 361. Cluster: averages 118 g (about 4 bunches per lb); cylindrical; tight; shatter-resistant. Berry: medium, averages 2.8 g; slip-skin; seedless; dark purple to black, with light bloom giving bluish overcast; skin split-resistant, disintegrates in mouth; flesh crisp, juicy; mild Concord flavor; 20.0°Brix soluble solids; matures early to mid-September in Summerland. Vine: vigor medium-weak, 1.8-m spacing in row on sandy soil; uniform fruitfulness on cane averages three clusters/shoot; productive when pruned to cordon system or long or short canes on side-by-side parallel wires, pruning formula 60 + 10 good with cluster thinning; cluster progressively smaller on shoot so tertiary cluster must be removed early; leaf large, resembling *Vitis labrusca* L., three-lobed, margins irregular dentate to rounded serrate; lower surfaces pubescent, whitish-green; yield 18–27 kg or 160–240 bunches/plant with cluster thinning. Recommended for table grape of Concord type, full-bodied jams, juice. Easy handling; storage at 1–3C up to 6 months with SO_2 storage pads.

Sovereign Opal.—Orig. in Summerland, B.C., by A.G. Reynolds, L.G. Denby, G.E. Strachan, and M. Bouthillier, Agriculture Canada, Res. Sta. Introd. in 1988. Golden Muscat x Marechal Foch; tested as

Summerland Selection 166. Berry: medium; round; skin green, thick, astringent; produces high-quality wine described as citrus, vegetative, herbaceous, and grassy; ripens late midseason. Cluster: medium-large; cylindrical. Vine: moderately vigorous; reliably hardy to –20C; resistant to powdery mildew.

Sovereign Rose.—Orig. in Summerland, B.C., Canada, Agriculture Canada, Res. Sta. by L.G. Denby and D.F. Wood. Introd. in 1977. Bath x Pearl of Csaba; cross made in 1966; selected in 1971; tested as Sel. 80. Cluster: averages 133 g; shouldered; cylindrical; well-filled; medium tight; shatter-resistant; averages two clusters/shoot. Berry: round; average 2.50 g; skin pink or light red, heavy bloom, tough, split-resistant; flesh meaty, firm, very sweet mild muscat flavor; seeds small, 2–3/berry; soluble solids 19.0° to 21.0°Brix; total acidity 0.7% to 1.0%, malic acid low (20% of total); tannins in juice 60–90 mg/100 ml. Vine: medium vigor; uniform fruitfulness along cane; cane or spur pruning; basal cluster largest, second and third 25% to 30% smaller; for market, remove small clusters; bud retention formula 60 + 10, or downward if cluster thinning not practical; leaf three-lobed, similar to Marechal Foch in size, margins deeply and irregularly serrated; yield 18–36.4 kg, average 27 kg/plant, depending on cluster thinning. Makes excellent muscatel wine (flowery bouquet and mild muscat flavor like Pearl of Csaba) if grapes are pressed before fermentation to minimize effect of high tannin in skin; wine best appreciated with 2% to 5% sugar; wine color light amber to rose, sometimes requires blending to improve color.

Sovereign Tiara.—Orig. in Summerland, B.C., by A.G. Reynolds, L.G. Denby, M. Bouthillier, and G.E. Strachan, Agriculture Canada, Res. Sta. Introd. in 1989. Golden Muscat x Marechal Foch; tested as Selection 123. Berry: medium; orbicular; seeded; skin gold; ripens late-midseason; produces high-quality wines described as fruity, herbaceous, vegetative, and candy-like. Cluster: medium, about 100 g; cylindrical; shouldered; slightly loose. Vine: highly productive; vigor moderate to high; best when grown with cordon training on a divided canopy; winter cold hardiness better than Okanagan Riesling, but less than Marechal Foch; resistant to powdery mildew.

Spalding.—Orig. in Experiment, Ga., by H.P. Stuckey, Georgia Agr. Expt. Sta. Introd. in 1920. Flowers x white male muscadine; selected in 1918. Muscadine type. Cluster: medium; compact. Berry: smaller than Hunt; skin jet black, medium thick; ripens over a long season, persistent; quality fair. Vine: vigor medium; productivity fair; flower pistillate.

Spurbeck.—Orig. in Charles City, Iowa, by a Mr. Spurbeck. Introd. in 1939. Chance seedling of *Vitis labrusca* type; first fruited in 1934. Berry: white. Vine: not hardy.

Stanford.—Orig. in Willard, N.C., by Charles Dearing, U.S. Dept. of Agriculture and North Carolina Agr. Expt. Sta. Introd. in 1946. Open-pollinated seedling of San Jacinto from Denison, Texas; selected about 1910; tested as V47 R9 B2. Muscadine type. Cluster: round; usually two to three berries. Berry: very large; slightly oblong; skin greenish-pearl, with russet specks and blotches; seeds large. Vine: foliage only fair in disease resistance; flower pistillate. Named for J.A. Stanford of Willard, N.C.

Sterling.—Orig. in Clayton, N.C., by C.F. Williams, W.B. Nesbitt, and V.H. Underwood, U.S. Dept. of Agriculture and North Carolina Agr. Res. Serv., Dept. of Horticultural Science, North Carolina State Univ., Raleigh. Introd. in 1981. NC 50-55 x Magnolia; selected in 1958. Muscadine type. Berry: large; resembles Magnolia. Vine: productive; vigor moderate; self-fertile. Uses: fresh fruit and wine.

Steuben.—Orig. in Geneva, N.Y., by Richard Wellington and Geo. D. Oberle, New York State Agr. Expt. Sta. Introd. in 1946. Wayne x Sheridan; selected in 1937. Cluster: medium; cylindrical; compact; resembles Sheridan. Berry: resembles Wayne; medium; skin bluish-black, slightly tough; flesh with high sugar content; aromatic flavor of Eumelan; good keeping quality; ripens in Concord season. Vine: very productive; hardy; vigorous; tolerant to downy mildew and black rot. Makes an aromatic white wine of American type.

St. Francis.—Orig. in Mountain Grove, Mo., by Paul H. Shepard, Missouri State Fruit Expt. Sta. Introd. in 1947. Muench x *Vitis vinifera* Gros Guillaume; selected in 1947; tested as G-2496. Cluster: large; compact. Berry: large; slightly ovoid; skin black; flesh firm, tender, separates from seed easily; flavor sweet, sprightly; ripens 2 weeks after Concord. Vine: vigorous; productive; flower perfect.

Stout Seedless.—Orig. in Geneva, N.Y., by New York State Agr. Expt. Sta. in cooperation with A.B. Stout, New York Botanical Gardens. Introd. in 1930. (Triumph x Dutchess) x Sultanina rose; cross made in 1921; first fruited in 1926. Cluster: large; medium compact. Berry: small; oval; soft seeds; skin greenish-yellow; flesh juicy, soft, sweet. Vine: not hardy and fails to ripen fruit well at Geneva.

Stover.—Orig. in Leesburg, Fla., by Loren Stover and J.A. Mortensen, Florida Agr. Expt. Sta. Introd. in 1968. Mantey x Seyve-Villard 12-309; cross made in 1956; tested as FES A4-43. Cluster: medium; frequently shouldered. Berry: 9/16 inch in diam., ellipsoidal; skin translucent light green to golden at maturity; sugar content at maturity averages 17% to 18%; flavor mild and free from excessive sweetness or tartness; dessert or table cultivar; ripens late June or early July; refrigeration prolongs shelf life and seems to improve flavor; keeping quality inferior to Thompson Seedless but superior to Lake Emerald. Vine: yield and vigor are significantly improved by grafting on a compatible rootstock; grows vigorously and produces well on Lake Emerald and Dog Ridge; on these rootstocks, canes grow 5–10 ft in length and 3/16 to 3/8 inch in diam., upright; leaves heart-shaped, with regular serrate margins; tolerant to Pierce's disease and downy mildew, susceptible to anthracnose and powdery mildew; flower self-fertile.

Stuckey.—Orig. in Experiment, Ga., by H.P. Stuckey, Georgia Agr. Expt. Sta. Introd. in 1920. Scuppernong x a black male muscadine; selected in 1918. Muscadine type. Cluster: small; berry adherence good. Berry: medium; spherical; skin medium thick, russet bronze; pulp soft; fair quality; ripens after Scuppernong. Vine: vigor very poor; low yield; foliage very susceptible to black rot; flower pistillate.

Suffolk Red.—Orig. in Geneva, N.Y., by John Einset, Agr. Expt. Sta. Introd. in 1972. Fredonia x Black Monukka; cross made in 1935; selected in 1941; tested as New York 21572. Cluster: medium-large; long cylindrical; loose to scraggly when no ringing or gibberellic acid treatment applied. Berry: medium; round; skin bright red, attractive; flesh melting, meaty, sweet; quality very good; seedless; ripens early, before Delaware. Vine: not sufficiently hardy for colder areas; does well in Suffolk County on Long Island.

Sugargate.—Orig. in Brooks, Ga., by B.O. Fry, Ison Nursery. Introd. in 1974. Plant patent 4056; 31 May 1977; assigned to B.O. Fry, Griffin, Ga. Fry x S. 29-49; cross made in 1968; selected in 1971; tested as S. 5-5-5. Muscadine type. Berry: large; skin dark color; early ripening. Vine: pistillate; vigorous; productivity low.

Sugar Pop.—Orig. in Brooks, Ga., by W.G. Ison. Introd. in 1988. Fry x 8-16-1. Plant patent 7591; 16 July 1991. Female muscadine. Berry: medium-large, 28 mm; skin bronze; dry scar; 21% solids; ripens midseason to late. Cluster: large. Vine: very vigorous; very productive.

Sugrathree.—See Black Imperial™.

Sugrafive.—See Early Superior Seedless™.

Sugrasix.—See PreCal™.

Sugratwelve.—Seedless, white table grape to replace Superior Seedless. Orig. in Mecca, Calif., by H. Newby, Jr., D. Cain, and K.S. Andrew, Sunworld International. Introd. in 1993. Plant patent 8298, 13 July 1993; assigned to Sunworld International. Mutation of Superior Seedless. Berry: elongated, similar to Thompson Seedless; crisp texture; white; no gibberellic acid needed for sizing; ripens at similar time as Superior Seedless.

Summit.—Orig. in Experiment, Ga., by Georgia Agr. Expt. Sta. by R.P. Lane. Introd. in 1977. Fry x Ga. 29-49; cross made in 1965; selected in 1971; tested under Ga. 37-40. Muscadine type. Berry: large; bronze to medium-brown; flavor mild, sweet; 20.5% soluble solids; quality good to very good; separates with dry stem scar. Vine: flower pistillate; vigorous; productive, 50 lb per vine; moderate resistance to the more serious diseases of muscadine; cold hardy at place of origin (5-year trial).

Sunbelt.—Even-ripening juice grape for southern areas where Concord is not adapted. Orig. in Clarksville, Ark., by J.N. Moore, J.R. Morris, and J.R. Clark, Univ. of Ark. Introd. in 1993. Plant patent 8511; 21 Dec. 1993. Concord open-pollinated; selected in 1971; tested as Ark. 1335. Berry: blue; large; seeded; round. Cluster: small; even ripening in high-temperature climates. Vine: moderately productive; vigorous; moderately resistant to black rot and anthracnose and highly resistant to powdery mildew and downy mildew. Fruit and vine characteristics are similar to Concord, except that Sunbelt ripens its fruit evenly at high temperatures.

Sunset.—Orig. in Stillwater, Okla., by Herman A. Hinrichs, Agr. Expt. Sta., Oklahoma State Univ. Introd. in 1975. Bailey x Keuka; cross made in 1950; selected in 1957; tested as OK 303. Cluster: medium. Berry: medium; skin red; juice straw-colored; sugar 20%; tartaric acid 0.67%; ripens about 18 Aug. in central Oklahoma. Vine: medium; vigorous; self-fertile. Excellent table grape.

Superior Seedless.—Orig. in Madera, Calif., by John M. Garabedian. Introd. in 1971. Plant patent 3106; 11 Apr. 1972; assigned to Superior Farming Co., Bakersfield, Calif. Cardinal x unnamed seedless selection; cross made in 1963; selected in 1967. Berry: large; oval to slightly obovoid; skin greenish-yellow, thick, tough; seedless; flesh firm; slight Muscat flavor, slightly acidic, crisp; quality good; good shelf life; ripens 7–10 days before Thompson Seedless. Vine: large; vigorous; productive. Recommended for fresh-market growing areas of California and Arizona.

Supreme.—Orig. in Brooks, Ga., by W.G. Ison. Introd. in 1988. Black Fry x Dixieland. Plant patent 7267; 10 July 1990. Female muscadine. Berry: large, 31 mm; skin black, edible; dry scar; 24% solids; ripens midseason to late; fruit shatters. Cluster: large. Vine: vigorous; productive.

Suwanee.—Orig. in Leesburg, Fla., by John A. Mortensen, Agr. Res. Ctr., Univ. of Florida. Introd. in 1983. C5-50 (W1521 x Villard Blanc) x Fla. F8-35 (Norris x Alden); selected in 1974; tested as Fla. H15-13. Berry: resembles Golden Muscat; larger than Stover; quality good; fresh fruit and wine have slight muscat flavor; early ripening. Vine: blooms later than Stover; satisfactory on most soils without grafting; tolerant to Pierce's disease.

Sweet Jenny.—Orig. in Brooks, Ga., by W.G. Ison. Introd. in 1986. 11-2-2 x 12-12-1. U.S. plant patent 7265; 10 July 1990. Female muscadine. Berry: large, 30–35 mm; skin bronze; 24% solids; ripens early to midseason. Cluster: large. Vine: vigorous and productive; susceptible to angular leaf spot.

Symphony.—Orig. in Davis, Calif., by H.P. Olmo, California Agr. Expt. Sta. Plant patent 5013; 29 Mar. 1983; assigned to the Regents of Univ. of California, Berkeley. Introd. in 1981. Muscat of Alexandria x Grenache Gris; cross made in 1940; selected as seedling J5-58 in 1945; tested as Oakville 919 and Fresno 250. Wine tests began in 1948. Cluster: medium; compact; short conical with shoulder or wing; peduncle very short and thick. Berry: spherical; 3 g; skin yellow-green, thick, dull aspect; flesh soft; expressed juice clear; medium acidity, with mild Muscat flavor; matures in mid-September at Davis. Vine: canes with very short basal internodes; vigorous; production 7 to 10 tons / acre; widely adapted. Makes wide range of white-wine types, from dry Muscat table wine, natural sweet wines, to sparkling wines; wines retain Muscat flavor and aroma for many years in the bottle.

Tamiami.—Orig. in Princeton, Fla., by Joseph L. Fennell, Lady Lake, Fla. Introd. in 1957. Fennell 6 (a wild species) x Malaga; first fruited in 1943. Berry: quality good; sugar content high; resembles a medium-sized berry of *Vitis vinifera*; tends to ripen unevenly on the cluster and produces over 50 lb / vine in tests (1967) at Watermelon and Grape Investigations Lab., Leesburg, Fla. Vine: extremely vigorous, with some canes 40–50 ft long; large trunk and arms; resistant to downy and powdery mildew, susceptible to anthracnose and moderately susceptible to Pierce's disease.

Tara.—Self-fertile, large-fruited bronze muscadine for fresh market and to use as pollinizer. Orig. in Griffin, Ga., by R.P. Lane, Univ. of Ga. Introd. in 1993. Not patented. Summit x Triumph; selected in 1979; tested as GA 33-8-2. Berry: large (11.4 g); bronze; dry stem scar; similar to Fry; ripens early in season. Vine: vigorous; productive; tolerant to most diseases affecting muscadines except black rot; escaped cold damage at –15C in 1989.

Tarheel.—Orig. in Willard, N.C., by Charles Dearing, U.S. Dept. of Agriculture and North Carolina Agr. Expt. Sta. Introd. in 1946. Luola x V36 R15 B4 [Eden x V23 R4 B2 (Eden x Munsoniana)]; cross made about 1912; tested as 42B6. Muscadine type. Cluster: large; loose; open; generally 8–10 berries. Berry: small; round; good adherence; skin shiny black, thin; flavor distinctive; ripens in midseason; resembles Eden. Vine: very vigorous and productive; foliage dense; resistant to black rot but susceptible to mildew; flower perfect, serves as a good pollinator.

Tetra.—Orig. in Mountain Grove, Mo., by Paul H. Shepard, Missouri State Fruit Expt. Sta. Introd. in 1947. Herbert x Worden; tested as G-2918. Cluster: medium to large; loose. Berry: very large, equal to Columbian and Golden Giant; flesh juicy; sweet, very fine flavor; ripens with Concord. Vine: growth fair; production good; flower with recurved stamens; considered to be tetraploid.

Thomuscat (Seedless Muscat).—Orig. in Fresno, Calif., by W.H. Henderson, Henderson's Experimental Gardens. Introd. in 1949. Muscat of Alexandria x Thompson Seedless. Cluster: rather scraggly, not uniform. Berry: tends to be small; mostly seedless; Muscat flavor when fully ripe; sets irregularly. Vine: vigor low; tends to over-produce; primarily for home gardens.

Thornburgs Robin.—See Robin Cardinal.

Topsail.—Orig. in Willard, N.C., by Charles Dearing, U.S. Dept. of Agriculture and North Carolina Agr. Expt. Sta. Introd. in 1946. Latham x Burgaw; selected from a cross made in 1915; tested as 15B6. Muscadine type. Cluster: round; three to five berries. Berry: large; round; shells off when ripe; skin light green but splotched with bronze; seeds broad, stubby; high sugar content; eating quality considered better than most standard cultivars; ripens midseason. Vine: very vigorous; drooping, canes long and thin; foliage abundant and disease resistant; basal tendrils often forked and fruit-bearing; flower pistillate.

Toscha.—Orig. in Brookings, S.D., by N.E. Hansen, South Dakota Agr. Expt. Sta. Introd. in 1925. Lady x a North Dakota wild species; selected in 1925. Cluster: compact. Berry: averages 5/8 inch in diam.; skin white; flesh firm, sweet, meaty; quality fair as a table grape in relation to other cultivars grown in area of origin. Vine: vigorous; moderately hardy.

Tropico.—Orig. in Princeton, Fla., by J.L. Fennell, Lady Lake, Fla. Introd. in 1943. *Vitis shuttleworthii* 5 x Lomanto; selected in 1943; tested as Fennell 240. Berry: large; flesh tender; seeds small; resembles Worden. Vine: adaptable to subtropical conditions.

Triumph.—Self-fertile, bronze muscadine for fresh market and for use as pollinizer. Orig. in Griffin, Ga., by R.P. Lane, Univ. of Georgia. Introd. in 1980. Not patented. Fry x Ga. 29-49; cross made by B.O. Fry. Selected in 1971; tested as Ga. 9-6-2. Berry: large (8 g); green to bronze; thin skin; thick pulp, nonmelting; 3.2 seeds; dry scar; 18% solids. Vine: vigorous; hardy; productive; moderate resistance to common diseases.

Tudor Premium Red.—Red, seedless table grape, an early ripening sport of Emperor. Orig. in Delano, Calif., by John A. Buska. Plant patent 6760, 25 Apr. 1989; assigned to Dan Tudor and Sons. Berry: two naturally occurring seed traces per berry that are visually detectable, but following treatment with gibberellic acid and girdling are substantially undetectable; berry large (naturally 4.4 g; treated with gibberellic acid and girdling 5.7 g); neutral flavor; crisp; matures 10–14 days before Emperor. Cluster: large; broad and tapering; well-filled but not compact. Vine: vigorous; productive to very productive; one cluster/shoot usually.

Tzori.—Triploid, seedless, white table grape, a sport of Perlette. Discovered in 1978 in Nir-Banim, Israel, by Moshe Tzori. Plant patent 6295; 20 Sept. 1988. Berry: round;

medium; seedless; responds to gibberellic acid treatment; green-yellow to yellow; slight muscat flavor. Cluster: medium; less dense than Perlette. Vine: fertile; fast-growing. Use is intended for fresh market or raisins.

Urbana.—Orig. in Geneva, N.Y., by S.A. Beach, New York State Agr. Expt. Sta. Introd. for trial in 1912. Ross x Mills; cross made in 1899. Berry: red; flesh crisp; quality good; keeps well; matures late; resembles Catawba. Vine: subject to mildew.

Valiant.—Orig. in Brookings, S.D., by R.M. Peterson, South Dakota State Univ. Introd. in 1982. Fredonia x SD S9-39 (*Vitis riparia* from northeastern Montana). Cross made in 1967; selected in 1972; tested as SD 72S15 or SD7 121. Berry: attractive; round; skin blue, slip skin; matures early. Cluster: compact; well-filled; average length 4 inches. Vine: very hardy; productive; excellent for juice.

Valplatinta.—Tropical juice or wine grape. Orig. in Aibonito, Puerto Rico, by Francisco Watlington. Introd. in 1993. Plant patent 8434, 26 Oct. 1993; assigned to the Commonwealth of Puerto Rico. A complex hybrid of *V. caribaea*, *V. vinifera*, and *V. labruscana* x Esperanza. Berry: medium (2 g); ellipsoidal; dark blue-black with light bloom; 2 or 3 seeds; skin medium thick and highly pigmented; juicy pulp; flavor of slight tartness but sweet; 19% solids. Cluster: long (18 cm); conical; loose; medium (250 g); 100 berries / cluster; three clusters / cane usually. Vine: medium vigor—performance may vary; medium productivity; moderate to good resistance to powdery mildew and downy mildew; possible resistance to anthracnose. Ripens uniformly with high sugar and acidity in tropical conditions.

Van Buren.—Orig. in Fredonia, N.Y., by F.E. Gladwin, New York State Agr. Expt. Sta. Introd. in 1935. Fredonia x Worden. Cluster: small to medium; well-filled; cylindrical; shouldered. Berry: medium; skin jet black; flesh juicy, sweet, foxy; quality of unfermented juice not equal to Concord; table quality good, poor keeping quality, tender, like Worden, which it resembles in all fruit characteristics except berry size; ripens very early. Vine: fairly vigorous, drooping; hardy; moderately productive; susceptible to downy mildew; leaf large, three-lobed; petiolar sinus wide, U-shaped. Recommended as an early maturing Concord type for home and local use.

Vanessa Seedless.—Orig. at Vineland Sta., Ontario, by K.H. Fisher and O.A. Bradt, Hort. Res. Inst. Introd. in 1983. Seneca x NY 45910; cross made in 1965; selected in 1972; tested as V.65164. Cluster: medium; conical with small shoulder; well-filled. Berry: spherical; larger than eastern seedless cultivars; flesh crisp; neutral flavor; seed traces green and noticeable, but rarely lignified; skin bright solid red; matures 20 Sept. at Vineland. Vine: aspect and foliage resembles Canadice; hardier than standard eastern seedless cultivars (Himrod, Interlaken Seedless, Lakemont). Recommended as a fresh market variety.

Varousset.—Orig. in Saint Vallier, Drôme, France, by Seyve-Villard. Introd. about 1932. Assumed to be from the cross Seibel 7053 x Seyve-Villard 12-375. Cluster: very large, to 15 inches long; tapering; loose. Berry: medium; slightly ovoid; skin dull black; flesh firm, crisp; sugar content very high; ripens in midseason; can be used as wine or table grape; produces red wine, light in color, highly alcoholic, often flat, but of neutral and clean taste; somewhat subject to cluster rot. Vine: habit open, spreading; extremely vigorous; very productive; hardy in central Illinois; leaf large, dark green, surface shiny; only medium tolerance to downy and powdery mildew. Authorized for commercial use in many regions of southwestern France.

Veeblanc.—Orig. at Vineland Sta., Ontario, Canada, by O.A. Bradt, Hort. Res. Inst. of Ontario. Introd. in 1977. Cascade x Seyve-Villard 14-287; cross made in 1953; selected in 1961; tested as Vineland 53263. Berry: medium-large; round; pulp soft, juicy; skin white. Cluster: medium; medium compactness; usually shouldered. Vine: good vigor and productivity; open short arms; similar to De Chaunac in hardiness; tolerant to downy mildew. Recommended for the production of dry white table wine.

Veeport.—Orig. in Vineland Sta., Ontario, Canada, by O.A. Bradt, Hort. Expt. Sta., Ontario Dept. of Agriculture. Introd. in

1961. Wilder x Winchell; selected in 1938; tested as Ontario 29143. Cluster: medium; well-filled. Berry: spherical; skin black. Vine: vigorous; many long canes; leaf trilobed, strongly concave. Chief value is for wine, where it has consistently received good ratings as a dessert wine; resembles Wilder.

Ventura.—Orig. in Ontario, Canada, by O.A. Bradt, Hort. Res. Inst., Vineland Sta. Introd. in 1974. Chelois (Seibel 10878) x Elvira; cross made in 1951; selected in 1959; tested as Vineland 51061. Cluster: medium; medium compact; usually shouldered. Berry: small; round; skin white, resistant to cracking; flesh somewhat astringent; ripens 30 Sept. at Vineland (few days before Concord); soluble solids 18.1; titratable acidity 1.37; pH 3.01; recommended for a quality dry white wine. Vine: vigorous; hardy; very productive; tolerant to mildew.

Venus.—Orig. in Clarksville, Ark., by J.N. Moore and E. Brown, Univ. of Arkansas. Introd. in 1977. Alden x NY 46000; cross made in 1964; selected in 1967; tested as Ark 1196. Berry: blue-black with whitish bloom; round; seedless; 3.0 g; ripens 7–10 days before Fredonia; responds well to girdling and gibberellin treatment for increasing berry size. Vine: vigorous; very productive; tolerant to black rot, anthracnose, powdery mildew; susceptible to downy mildew; hardy to –10F.

Verdelet *(Seibel 9110)*.—Orig. in Aubenas, Ardeche, France, by Albert Seibel. Introd. into Canada in 1946, into the U.S. in 1949. Seibel 5455 x Seibel 4938. Cluster: medium; cylindrical; well-filled to compact; appearance very attractive. Berry: about 1/2 inch in diam.; ovoid-truncate; skin golden-yellow, beautiful aspect; flesh texture firm and crisp; flavor bland but pleasing. Vine: vigor medium; highly fertile; shoot tips downy with young leaves carmine-edged; leaf with petiolar sinus open lyre; teeth acute and sharp, hairs cobwebby on under surface. Sometimes recommended as a table grape; produces wine with almost perfumed odor.

Vidal 256.—Orig. at Foundation Fougerat, Bois-Charentes, France, by J.L. Vidal. Introd. in 1952 in the U.S. as PI200684. St. Emilion x Seibel 4986. Cluster: medium; cylindro-conical; compact. Berry: small; round; greenish white; high in sugar; wine-perfumed, like Seibel 4986. Vine: vigorous; hardy; in general aspect like St. Emilion; shoots downy white; leaves rough, undulated, teeth sharp, petiolar sinus closed by overlapping lobes, arachnoid slightly hairy; good resistance to downy mildew; fruitful recovery after spring frost; fruit set sometimes deficient.

Vignoles (Ravat 51).—Orig. in France, by J.F. Ravat. Introd. in the U.S. as PI178547 and 181481 in 1949; named by the Great Lakes Grape Nomenclature Committee in 1972. Seibel 6905 x Pinot. Cross made about 1930. Cluster: very small; very compact; occasionally with small wing. Berry: oblate with contortion due to compactness; skin light green with pinkish tinge, tendency to cracking; flesh juicy; high sugar content; wine cultivar; ripens early midseason. Vine: vigorous; low in productivity; very hardy; upright, open; susceptible to downy mildew.

Villard Blanc.—Orig. in Saint Vallier, Drôme, France, by Seyve-Villard. Introd. and first cultivated in France about 1937; P.I. 157,518; received in 1947 in the U.S. Assumed to be from the cross Seibel 6468 x Seibel 6905. Cluster: large, to 1 1/2 lb; widely branched; loose; stems long. Berry: above medium; ovoid; skin golden-yellow; flesh meaty; matures in late midseason, 10 days before Seyve-Villard 12-309; wine light-bodied, straw-colored, highly aromatic, only faintly bitter. Vine: young shoots with cobwebby hairs, new leaves bronzed, leaf medium, surface very shiny; teeth very long, sharp; growth habit open; canes long, straight, few in number, with clusters borne well out from the trunk and hanging free; very productive after the first few years; in good soils can be grown as a direct producer as it has high tolerance to phylloxera; sufficient resistance to downy and powdery mildew in all except the most severe seasons; long pruning recommended in localities where shoots can be easily broken by wind and when vines are young and too vigorous; grafts well on many stocks. Considered to be the best all-around French hybrid.

Vincent.—Orig. in Vineland Sta., Ontario, Canada, by O.A. Bradt, Ontario Dept. of Agriculture, Hort. Res. Inst. Introd. in 1967. V 370628 (Lomanto x Seneca) x Seibel 10878; selected in 1958; tested as V 49431. Cluster: medium to large; compact to very compact. Berry: medium; skin dark blue to black; juice very dark; wine cultivar; ripens during Concord season. Vine: vigorous; productive.

Vinered.—Orig. in Vineland Sta., Ontario, Canada, by O.A. Bradt, Ontario Dept. of Agriculture, Hort. Res. Inst. Introd. in 1964. Brocton x self; cross made in 1929; tested as Ontario 29186. Cluster: large to very large; medium compactness. Berry: slightly larger than Catawba; skin tender, an attractive red when well-matured; ripens shortly before Catawba, requiring a lengthy growing season to produce best quality; flesh with a pleasant vinous flavor; slightly higher sugar and lower acid content than Catawba; a promising dessert type; probable wine use for blending as it produces a large amount of juice with a neutral flavor, with little *V. labrusca* flavor in the wine. Vine: vigorous; productive, up to 6–7 tons/acre; performs very well in Ontario and southward.

Vinok.—Orig. in Stillwater, Okla., by Herman A. Hinrichs, Agr. Expt. Sta., Oklahoma State Univ. Introd. in 1975. Bailey x Keuka; cross made in 1954; selected in 1958; tested as OK 353. Cluster: very long with long shank. Berry: medium to large; round; skin blue-black; sugar 19%; tartaric acid 0.67%; ripens evenly about 18 Aug. at Stillwater. Vine: vigorous; productive; drought and winter-hardy.

Vivant.—Orig. at Vineland Sta., Ontario, by K.H. Fisher and O.A. Bradt, Hort. Res. Inst. Introd. in 1983. V.50154 x NY.25681; cross made in 1963; selected in 1971; tested as V.63331. Cluster: long; cylindrical with one or two shallow shoulders; well-filled to compact. Berry: small; spherical; flesh soft; skin yellow-tan at maturity; ripens 3–5 Oct. at Vineland; produces a light, crisp, German-type wine. Vine: vigorous; productive; requires spray program; hardier than Seyval and Vidal 256, but not as hardy as Ventura or Concord; adapted to milder regions in Ontario.

Wachula.—Orig. in Princeton, Fla., by J.L. Fennell, Lady Lake, Fla. Introd. about 1948. Open-pollinated seedling of *Vitis rufotomentosa* 9; selected in 1942; tested as Fennell 126.

Wallace.—Orig. in Willard, N.C., by Charles Dearing, U.S. Dept. of Agriculture and North Carolina Agr. Expt. Sta. Introd. in 1946. V26 R5 B4 x Willard; cross made about 1915; tested as 2B8. Muscadine type. Cluster: large, 2–15 berries; loose. Berry: medium; slightly oval; skin yellow to bronze, thin, tender; seeds small; ripens late; resembles Scuppernong. Vine: vigorous; foliage abundant but subject to black rot and other diseases; flower perfect. Named for the town of Wallace.

Watergate.—Orig. in Brooks, Ga., by B.O. Fry, Ison Nursery. Introd. in 1974. Seedling 2-3-1 x S. 19-6-1; cross made in 1968; selected in 1971; tested as seedling 1-13-1. Female muscadine. Berry: large; round; bronze; pulp melting; ripens midseason. Vine: hardy; productive; vigorous.

Watkins.—Orig. in Geneva, N.Y., by New York State Agr. Expt. Sta. Introd. and named in 1930. Mills x Ontario; cross made in 1911; first fruited in 1915; selected in 1916. Cluster: smaller than Concord; cylindrical; loose. Berry: to 18 mm in diam.; ellipsoidal; skin reddish-black, slipskin; flavor very good, foxy; ripens after Concord. Vine: vigor only fair; productivity unreliable; lacks hardiness.

Wayne.—Orig. in Geneva, N.Y., by Richard Wellington, New York State Agr. Expt. Sta. Introd. in 1927. Mills x Ontario; first fruited in 1916. Berry: resembles Mills in flavor, color, texture, quality; more nearly *V. vinifera* than most varieties of its period of introduction. Cultivar not planted in New York because of its late season, tendency to shell, unreliable productivity.

Welder.—Orig. in Tavares, Fla., by Henry M. Welder. Introd. in 1972. Open-pollinated seedling of unknown muscadine; discovered in 1957. Muscadine type. Berry: resembles Dearing but ripens earlier; quality good; used as fresh fruit and for making wine; wet stem scar. Vine: more vigorous, hardier, productive, and tolerant to leafhoppers than Dearing; self-fertile; precocious.

Westfield.—Orig. in Fredonia, N.Y., by F.E. Gladwin, New York State Agr. Expt. Sta. Introd. in 1935. Herbert x Concord Seedless; tested as Gladwin 200. Cluster: 100–300 g; cylindrical; compact. Berry: *Vitis labrusca* type; skin blue, pigment content high, considered valuable for use in blending in the making of wines and unfermented juices; ripens with Concord. Vine: hardy; moderately productive.

Willard.—Orig. in Willard, N.C., by Charles Dearing, U.S. Dept. of Agriculture and North Carolina Agr. Expt. Sta. Introd. in 1946. Stanford x V19 R7 B2 (Scuppernong x New Smyrna); cross made about 1910; tested as V36 R B4. Muscadine type. Cluster: loose, about 2 inches long. Berry: medium; slightly oval; skin greenish to bronze; seeds difficult to separate from pulp. Vine: vigor medium; drooping growth habit; foliage sparse and subject to disease, shedding early and exposing fruit; flower perfect, useful as a pollinizer. Named for the town in which the experimental vineyard is located.

Yates.—Orig. in Geneva, N.Y., by Richard Wellington, New York State Agr. Expt. Sta. Introd. and named in 1937. Mills x Ontario; cross made in 1923; selected in 1928. Cluster: medium; conical; compact. Berry: spherical, to 19 mm in diam.; slipskin, pulp meaty; skin dark red, very tough, adherence good; excellent keeping quality; ripens late, 1 week after Concord. Vine: vigorous; hardy; productive.

Yuga.—Orig. in Experiment, Ga., by J.G. Woodroof, Georgia Agr. Expt. Sta. Introd. in 1934. Open-pollinated seedling of San Monta; selected in 1932. Muscadine type. Berry: skin reddish-bronze, thin; flavor sweet; quality excellent; ripens late and irregular; for the home vineyard.

not named.—Orig. in Athenia, N.J., by T.C. Kevitt. Introd. in 1936. Plant patent 195; 15 Sept. 1936; assigned to Meyer Aronowitz, Barclay Nursery, New York. Seedling of Niagara; selected about 1929. Berry: slightly larger and sweeter than Niagara; skin amber color when ripe; ripens 4–5 days earlier than Niagara. Vine: more thrifty and productive, and hardier than Niagara.

Rootstocks for Grape

Florilush.—Improved bunch grape rootstock for Florida. Orig. in Leesburg, Fla., by J.A. Mortensen, J.W. Harris, and D.L. Hopkins. Dogridge x Tampa; selected in 1984; tested as CD9-81. Not patented. Vine (compared to Dog Ridge): higher percentage of successful grafts; lower tendency to sprout below the graft union; more tolerant to grape root borer than Tampa; resistant to Pierce's disease, anthracnose, downy mildew, powdery mildew, and Isariopsis leaf blight; highly tolerant to various nematodes; adapted to well-drained soils of Florida.

Freedom.—Orig. in Fresno, Calif., by U.S. Dept. of Agriculture in cooperation with California State Univ., Fresno Foundation. Introd. in 1974. Couderc 1613 x seedling (5) of Dog Ridge; cross made in 1956; selected in 1957 by J.H. Weinberger and F.N. Harmon; tested as 91-102. Vine: more vigorous than Harmony but less vigorous than Ramsey (Salt Creek) and Dog Ridge; tolerance to rootknot nematode and phylloxera same as Harmony, neither immune.

Harmony.—Orig. in Fresno, Calif., by J.H. Weinberger and F.N. Harmon, U.S. Hort. Field Sta. Seedling 39 (Couderc 1613, o.p.) x seedling 5 (Dog Ridge, o.p.), both seedling selections from open-pollinated seed; cross made in 1955; selected in 1956; tested as US16-154. Vine: more vigorous than Couderc 1613, but less so than Dog Ridge or Salt Creek; greater resistance to phylloxera and rootknot nematode than Couderc 1613; Thompson Seedless tends to overbear on this stock, Emperor has been vigorous and productive; propagates readily from cuttings and bench grafts; flowers have reflexed stamens. Recommended as a rootstock for Thompson Seedless for wine or raisin production in soils where nematode or phylloxera is a problem.

La Pryor.—Orig. in La Pryor, Texas, by the Texas Agr. Expt. Sta. Introd. in 1934. Parentage unknown, probably a hybrid of *Vitis candicans* x *V. rupestris*; discovered in 1933. Cluster: small. Berry: medium; skin black; flesh sweet. Vine: light producer; used only as a clonal rootstock: roots readily from cuttings; resistant to a root rot (*Phytomatotrichum* sp.); vigorous.

Leverkuhn.—Orig. near Houston, Texas; first called to attention by the late J.T. Patterson of that city. Introd. about 1940. Originated as a spontaneous hybrid of *Vitis candicans* and an unknown American variety. Berry: acidity very high. Vine: very vigorous and long-lived in the southern states; medium to heavy producer; being tried for rootstock purposes in some areas.

Tampa.—Clonal rootstock to replace Dog Ridge. Orig. in Leesburg, Fla., by J.A. Mortensen and L.H. Stover. Introd. in 1982. Not patented. Fla. 43-47 (*Vitis aestivalis* ssp. *smalliana* open-pollinated) x Niagara. Selected in 1961; tested as Fla. 48-1-26. Vine: low tendency to sprout from below the graft union; produces very vigorous scions; very compatible; resistant to Pierce's disease, anthracnose, downy mildew, powdery mildew; drought tolerant; tolerant to nematodes. Berry: unmarketable; small; purple.

VR 039-16.—Rootstock for sites with grape fanleaf/*Xiphinema index* complex. Orig. in Davis, Calif., by M.A. Walker, L.A. Lider, A.C. Goheen, and H.P. Olmo, Univ. of California. Introd. in 1991. Plant patent 6166, 3 May 1988; assigned to the Regents of the Univ. of California. Almeria x Male No. 1 (*Vitis rotundifolia);* cross made in 1948. Vine: sterile; has resistance to phylloxera; shows some resistance to fan leaf virus/X. *index* complex compared to susceptible rootstocks, but possibly not total resistance. Recommended for use only on sites with the grape fan leaf/*X. index* complex.

VR 043-43.—For sites with grape fanleaf virus. Orig. in Davis, Calif., by L.A. Lider, H.P. Olmo, and A.C. Goheen. Plant patent 6319, 4 Oct. 1988; assigned to the Regents of the Univ. of California. Hunisa x *Vitis rotundifolia* male No. 2. Vine: immune to feeding by *Xiphinema index*, the nematode vector of grape fanleaf virus; tolerant to phylloxera; tolerant to most strains of root knot nematode. Asexual propagation by cuttings is improved by application of rooting hormones. Distinguished from its companion rootstock, VR 039-16, in that VR 043-43 is not immune, but only tolerant, to phylloxera.

GRAPEFRUIT

Burgundy Red.—Orig. in Fort Pierce, St. Luice County, Fla., by Oliver L. Peacock of Fort Pierce and Hudson J. McReynolds of Orlando. Introd. in 1956. Plant patent 1276; 11 May 1954. Bud mutation of Thompson (Pink Marsh); discovered in 1943. Fruit: rind thin, no blush; flesh deep red, high acidity, nearly seedless, juice content fairly low in early and midseason; juice much deeper red than that of Ruby Red; harvested in May. Tree: round-topped, foliage dense; resembles parent.

Cecily.—Orig. in Uitenhage, Cape Colony, Republic of South Africa, by Sir Percy Fitzpatrick. Introd. into the U.S. in 1928; by U.S. Dept. of Agriculture as P.I. 58457. Bud mutation of Walters; trees of Walters from Florida imported into South Africa by Fitzpatrick; discovered in 1922. Fruit: seedless; otherwise similar to parent; resembles Marsh.

Flame.—Orig. in 1973 as a seedling of Henderson grapefruit at A.H. Whitmore Foundation Farm, Leesburg, Fla. Selected by C.J. Hearn, U.S. Dept. of Agriculture (USDA), Orlando, in 1980. Propagated by C.O. Youtsey, Citrus Budwood Registration Bureau. Introd. by USDA in 1987. Fruit: globose; some external blush; flesh red, darker than Ray Ruby but not as dark as Star Ruby; color holds later than Ray Ruby; seedless. Tree similar to Henderson.

Frost Marsh.—Orig. in Riverside, Calif., by Howard B. Frost, California Citrus Expt. Sta. Introd. in 1952. A nucellar seedling of Marsh, from seed planted in 1916. Fruit: resembles the old-line Marsh, but perhaps lower in acid. Tree: greater vigor and better yield than the old-line Marsh.

John Garner.—Orig. in Texas by George W. Baylor. Introd. in 1934. Open-pollinated seedling of Duncan. Fruit: size similar to Duncan; nearly seedless; ripens about 2 weeks later than Duncan; quality good.

Jordan.—Orig. in Phoenix, Ariz., by Mr. and Mrs. S.E. Jordan. Introd. about 1925. Considered to be a bud mutation of Marsh; discovered about 1920. Fruit: color and shape resemble Marsh, size is more uniform; quality very good; two to three seeds; ripens 2 weeks before Marsh.

Oroblanco.—Orig. in Riverside, Calif., by R.K. Soost and J.W. Cameron, Univ. of California at Riverside. Plant patent 4645; 10 Feb. 1981; assigned to Regents, Univ. of California. Introd. in 1980. A triploid hybrid of*Citrus grandis* (CRC 2240) x *Citrus paradisi* (tetraploid) cross made in 1958. Fruit: size and shape similar to Marsh; peel color paler than Marsh and thicker; flesh paler than Marsh and larger hollow core, tender, juicy, lacks bitterness, separates well from segment membranes, seedless. Tree; large; rounded to drooping; vigorous; hardy to about 30F; productivity moderate to heavy; may alternate.

Pink Blush *(Prothero Pink)*.—Orig. in El Toro, Calif., by R.H. Prothero. Introd. about 1953. Bud mutation of Marsh; discovered in the 1930s. Fruit: size runs smaller than most varieties; pink skin color breaks with high heat at maturity and color does not hold too long; practically no pink flesh color; sweeter than most varieties; few to no seeds.

Rio Red.—Orig. in 1963 as a bud sport of A&I-1-48, a selection from irradiated budwood of Ruby Red. Introd. by R.A. Hensz, Texas A&I, in 1984. Fruit: globose; good external blush; flesh red; five to seven seeds; five times the lycopene content of Ruby Red. Tree: vigorous; precocious; productive; not sensitive to herbicide and foot rot like Star Ruby.

Ruby *(Glenred, Henninger Ruby, Henninger's Ruby, Redblush, Ruby Red)*.—Orig. in McAllen, Texas, by Albert E. Henninger. Introd. in 1934 or 1935. Plant patent 53; 24 Jan. 1933. Bud mutation of Thompson; discovered in 1929. Fruit: rind blushed pink to red; flesh pink, deeper red than Thompson, fine-grained; few to no seeds. Ruby and Redblush are considered identical, for all practical purposes (see Webb grapefruit, below).

Shambar.—Orig. in Corona, Calif., by the Chase Plantation, by A.D. Shamel, California Citrus Expt. Sta., Riverside. Introd. in 1945. Bud mutation of Marsh; discovered in 1936. Fruit: rind and flesh pink; considered by some to be superior to Marsh in juiciness, flavor, and long storage life. Tree: same as Marsh or Ruby.

Stanfield Red.—Orig. in Elsa, Texas, by J.H. Stanfield. Not introd. commercially; included here because it has been used in experimental work and its name has appeared in publication. Bud mutation of Marsh; discovered about 1945. Fruit: flesh red; resembles Webb, but not superior to it. Tree: contains many viruses.

Thompson *(Pink Marsh)*.—Orig. in Oneco, Fla., in a grove of W.B. Thompson. Introd. in 1924 by Reasoner Brothers, Royal Palm Nurseries, Oneco. Bud mutation of Marsh; discovered in 1913 by S.A. Collins of Oneco. According to Webber and Batchelor (*The Citrus Industry*, 1943, p. 582), a similar pink-fleshed bud mutation of Marsh occurred in Riverside, Calif. Fruit: size medium; rind light yellow, with no pink; flesh pink, juice not colored, abundant; seeds few; ripens before Marsh.

Vaughan Early Seedless.—Orig. in Fort Myers, Fla., by Alan W. Vaughan. Introd. in 1968. Parentage unknown; discovered in 1966. Fruit: size medium; round; rind yellow, very thin; flesh darker than Marsh, extremely sweet and juicy; none to five seeds; ripens early December; resembles Marsh. Tree: large; vigorous; productivity moderate.

Warren.—Orig. in Piru, Calif., by C. Mavro Warren. Introd. in 1958. Chance seedling; discovered about 1946. Fruit: nearly seedless, none to three seeds per fruit; a little less yellow than Marsh.

Webb *(Redblush, Webb Redblush, Webb's Redblush, Webb's Redblush Seedless)*.—Orig. in Donna, Texas, by J.B. Webb, Redblush Nursery. Introd. in 1934. Bud mutation of Thompson (Pink Marsh); discovered in 1931. Fruit: skin blushed; flesh a deep red; average three seeds per fruit. Tree: prolific bearer. Ruby and Redblush are considered identical for all practical purposes (see Ruby grapefruit, above).

GUAVA

Allahabad Safeda.—Indian origin. White dessert guava. Weight 173.06 g; diam. 7.5 cm; cavity width 5.0 cm; seeds 3.2%; 12.5% soluble solids, 0.6% total acid. Sweet, thick flesh, has a pleasant flavor but lumpy surface. Somewhat resistant to red-banded thrips and Mucor fruit rot.

Beaumont.—Orig. in Hawaii, by F.A.I. Bowers and H.Y. Nakasone, Univ. of Hawaii. Introd. in 1960. Also known as B-30. Open-pollinated seedling of a selection from a wild population; discovered in 1954–55. Fruit: 2.5–3.0 inches in diam., 3.5–4.5 inches long, slightly oval; skin yellow, slightly rough; flesh medium-pink, 1.0% to 1.2% acid, 9% to 10% soluble solids; ripens year-round with peaks in April–May and September–October. Pulp in locules has a tendency to be loose and puffy; puree recovery 78%. Yield from 3-year-old trees was 24 t/ha/year. Vitamin C 143.8 to 154 mg/100 g fruit. Tree: vigor excellent; 12–15 ft high; productive; susceptible to red-banded thrips, fruit flies. The first commercial guava cultivar in Hawaii for processing.

Blitch.—Selected by D. Sturrock, West Palm Beach, Fla. Fruit oval, 85–142 g. Skin slightly rugose, yellow to greenish yellow. Light-pink flesh, tart, pleasant; 0.74% total acidity and 0.12% vitamin C. Seeds small and numerous.

Donrom.—Thailand origin. Other names: Thailand apple, Bangkok apple. White dessert guava, also used for canning and juice; triploid, seedless. Short to medium tree with large, dark-green oval leaves with wavy margins. Fruit: oblate to globose, ridged, and frequently asymmetrical, 7–8.5 cm in diam., average 250–350 g. At firm-ripe, flesh is white with a tinge of pink, frequently has no seed core but a hollow carpellary cavity; 10% to 11% soluble solids and 0.55% total acid. Poor yield (15–20 t/ha/year) and heavy fruit drop (50%). Susceptible to anthracnose and root-knot nematode, but somewhat resistant to Phomopsis stylar and ring rot.

Fan Retief.—South Africa origin, probably a hybrid from Madeira guava. Fruit ripens from May to September, may extend up to November in S. Africa. Slightly pear-shaped with a tendency to be round. Average fruit weight 116.4 g, 5.25 cm in diam., seed cavity 3.5 cm in diam.; seed is 2.3% of fruit weight; 12.6% soluble solids and 3.9% acid; vitamin C 248 mg/100 g. Pulp is light pink, pleasant guava flavor. Fruit are reasonably resistant to bruising. Tree is vigorous, prolific, and ripe fruit tends to hold onto trees.

Glom Toon Klau.—Thailand origin. White dessert guava, early crop, high yield (40–50 t/ha/year), up to three croppings per year in Malaysia. Fruit: round to oblate, 8–9 cm in diam., rough skinned with no groove or line on fruit surface. Average 400–500 g to as much as 700 g. Flesh smooth, crunchy, and sweet at firm-ripe. Seed core is large but few seeds 2% to 3% by weight). Vitamin C 92 mg//100 g; 8% to 10% soluble solids and 0.36% total acid. Short postharvest shelf life of 2–4 days.

Glom Um-porn.—Thailand origin. White dessert guava, round to oblate, 10 cm equatorial diam. and 8.5 cm polar diam. Rough skin with slight lines radiating from stem end. Weight 400–500 g, 7% soluble solids at green firm-ripe.

Hart.—Orig. in New Smyrna, Fla., by J.Y. Detwiler. Introd. in 1945. Seedling of Detwiler guava; parent variety was grown by Detwiler from seed reported to have come from India; named after W.S. Hart of New Smyrna. Fruit: large; flesh white to pale yellow; sugar content about 8%. Tree: vigorous; hardy in southern California.

Homestead.—Florida origin. Crosses made by G.D. Ruehle in 1945, between Ruby and Supreme and was designated as Ruby x Supreme 6-29. Released by Carl W. Campbell, Tropical Research and Education Center, Univ. of Florida, IFAS, 1989. Fruit: 150 g, bright yellow, thin skin, smooth to slightly undulated. Pulp is red, smooth, and uniform 6.8–7.7 degrees Brix, 0.29% acid, and

0.12% vitamin C. A good variety for the fresh market and for processing into guava shells.

Hong Kong Pink.—Seeds from Hong Kong. A pink dessert guava selected at Poamoho farm, Hawaii. Fruit: texture smooth, average 218.5 g, thick flesh, few seeds, 10% soluble solids and 0.2% total acid. Flavor pleasant. Vitamin C 81.6 mg/100 g when mature and 113.6 mg/100 g when ripe. Spreading tree.

Indonesian Seedless.—Indonesian origin. White dessert guava. Fruit: weight 173 g, 7.3 cm in diam., seed cavity 4.3 cm in diam. with 0.7% seed; 12.5% soluble solids and 0.4% acid. Sweet, thick fleshed, none to few seeds. Reported to be a triploid.

Ka Hua Kula.—Orig. in Honolulu, Hawaii, selection No. 097. Released by H.Y. Nakasone and P.J. Ito, 1978, Univ. of Hawaii. Introd. in 1972. Open-pollinated seedling of Beaumont progeny; discovered in 1962. Fruit: 2.3–3.0 inches in diam., 5.0–8.5 oz, oblong, rough exterior, yellow skin, mild flavor; flesh 1.0% to 1.3% total acidity, 8% to 9.5% soluble solids, dark pink; recommended for processing; higher acidity and darker pink flesh than Beaumont. Puree recovery 78%, yield at 20 t/ha/year from 3-year-old trees. Somewhat resistant to red-banded thrip. Tree: 12–15 ft; crown round; very vigorous; productive.

Kampuchea.—Other name: Vietnam. A white dessert guava Kampuchea/Vietnam origin. Diploid, seedy (80–240/fruit). Early crop, high yield (60 t/ha/year), average three crops per year in Malaysia. Medium to short spreading tree. Oblong to pear-shaped fruit 400–600 g, thick flesh 14–16 mm, crunchy, and sweetish; 6.5% to 7% soluble solids, 0.5% acid. Vitamin C 55–180 mg/100 g. Also used for juice and canning. Susceptible to anthracnose, Phomopsis stylar end rot, and root-knot nematode.

Lucknow 49.—White dessert guava. Indian origin. Smooth texture fruit average 226 g, few seeds; 11% soluble solids and 0.4% acid. Flavor musky when fully ripe. Tree: spreading; somewhat resistant to red-banded thrips.

Miami Red.—Orig. in Miami, Fla., by the Expt. Farm, Univ. of Miami. Introd. in 1959. Open-pollinated seedling of Red Indian; seed planted in 1948; first fruited in 1950. Fruit: large; skin yellow; flesh red, thick, odor mild, flavor good; shipping and keeping qualities good. Plant: productive; relatively disease resistant.

Miami White.—Orig. in Miami, Fla., by the Expt. Farm, Univ. of Miami. Introd. in 1959. Open-pollinated seedling of Red Indian; seed planted in 1948; first fruited in 1951. Fruit: large; skin light yellow; flesh white, thick, flavor good; shipping quality good. Plant: productive; relatively disease resistant.

Patillo.—Original seedling population was planted by A.J. Patillo in De Land, Fla., in 1920s. One plant was brought to G.D. Ruehle's attention in 1952. Pink processing guava, ovate to obovate, with smooth, yellow skin, small and numerous seeds. Flavor is subacid, mild aroma. Average weight 113.5 g, 11.9% soluble solids, 1.23% acid, and 1.7% vitamin C. Tree: upright. Low incidence of Algae disease (*Cephaleuros* sp.).

Pink Acid.—A selected seedling from Speer and Stone Acid. Fruit round, average 167.4 g, suitable for processing. Pulp strong pink, 11.7% soluble solids and 1.7% acid; few seeds. Trees are upright.

Puerto Rico No. 2.—Puerto Rico origin. Processing variety also known as Trujillo #2. Round fruit, medium flesh, average 184.5 g. Seed cavity meaty. Salmon flesh color and has a distinct guava flavor, 11.8% soluble solids. A spreading tree with good yield.

Red Indian.—Orig. in Dade County, Fla., by Fred Lenz. Introd. in 1946. Open-pollinated seedling of P.I. 57828; discovered in 1940. Fruit: large; flesh red, sweet, quality good; ascorbic acid content averages 195 mg/100 g fresh fruit; total sugars average 7% to 10%. Tree: productive; not vigorous.

Redland.—Orig. in Homestead, Fla., by S.J. Lynch and H.S. Wolfe, Florida Subtropical Expt. Sta. Introd. in 1941. Parentage unknown; seed obtained from Atkins Inst., Arnold Arboretum, Cuba; selected in 1938. Fruit: large; pyriform; no musky guava odor; flavor mild; matures in winter. Tree: vigorous; heavy cropper; susceptible to red algae spot.

Ruby.—Orig. in Homestead, Fla., by George

D. Ruehle, Univ. of Florida, Subtropical Expt. Sta. Introd. in 1946. Open-pollinated seedling of P.I. 81849; selected in 1944. Fruit: large; flesh red, sweet, thick, mild, quality good; seeds few; ascorbic acid averages 180 mg/100 g fresh fruit; total sugars average 9% to 10%. Tree: vigor medium; productive; tolerant to algal spotting.

Supreme.—Orig. in Homestead, Fla., by George D. Ruehle, Univ. of Florida, Subtropical Expt. Sta. Introd. in 1946. Parentage unknown; seed planted in 1936; selected in 1940. Fruit: white flesh, large, averaging 6–10 oz; flesh thick, flavor mild; seed cavity small; seeds few; ascorbic acid 247 mg/100 g fresh fruit; total sugars 7% to 8%. Tree: productive; thrifty; vigorous; tolerant to algal spotting.

Thai Maroon.—Thailand origin. Ornamental. Tree medium and slender with oblong greenish-maroon leaves. Young leaves are distinctly maroon colored. Flowers are pink to pale maroon, clustered in twos or threes. Fruit: oblong, smooth, deep-maroon skin and flesh, and acid.

Waiakea.—HAES 11-26. Pink processing guava, released by P.J. Ito, Univ. of Hawaii. Weight 246.1 g; fruit 7.33 x 7.0 cm; cavity 4.16 x 4.03 cm; exterior L: 67.24, a: –8.707, b: 43.96; pink flesh, L: 44.19, a: 22.04, b: 10.97, 7.9% soluble solids. Somewhat resistant to Mucor rot.

Hazelnut

(Filbert)

Barcelona *(Castanyera (eastern Spain), Fertile de Coutard (France), Grande (northern Spain)).*—Old cultivar widely distributed in western Europe; introd. to the U.S. by Felix Gillet about 1885; trees first offered for sale in 1895. Barcelona has been the standard for the U.S. industry for the past century. Nut: large (3.2 g), brown; 43% kernel by weight, kernel fibrous, pellicle half removed by dry heat; husk 50% longer than nut, free-husking. Tree: very vigorous, more productive than most European cultivars. Highly resistant to big bud mites, intermediate response to eastern filbert blight. Incompatibility alleles $S_{\underline{1}}$ S_2. Female inflorescences emerge early; midseason pollinizers recommended.

Bawden.—Orig. in the Saanichton District of Vancouver Island, British Columbia, Canada, by A.W. Bawden. Introd. in 1937. Parentage unknown; discovered in 1936. Nut: large; husk short; matures early. Now obsolete.

Bixby.—Orig. in Lancaster, Pa., by J.F. Jones. Introd. in 1937. *Corylus americana* Rush x *C. avellana* Italian Red; first fruited in 1924; tested as Jones 200. Nut: long, pointed; appearance attractive; kernels sweet, clean. Tree: very hardy, prolific.

Brag.—Orig. in Westbank, British Columbia, Canada, by J.U. Gellatly. Introd. in 1928. F_2 open-pollinated seedling of Kentish Cob; selected in 1926. Nut: large; free-husking, clean kernels; three to four nuts per cluster; cracks easier than Craig, which it resembles. Now obsolete.

Brixnut.—Orig. in McMinnville, Ore., by C.T. Brixey. Introd. in 1919. Parentage unknown, appears to be seedling of Barcelona. Nut: large (3.8 g), round at base, tapering to apex, color dull brown because of light pubescence; 60% free-husking, husk slightly longer than nut; matures one week after Barcelona; 44% kernel by weight, kernel round, slight fiber on pellicle, pellicle half removed by dry heat. Tree: moderately vigorous, spreading and becoming drooping with heavy crop, productive. Resistant to big bud mite, susceptible to eastern filbert blight. Planted commercially in the 1930s and 1940s but now obsolete. Incompatibility alleles $S_{\underline{1}}$ $S_{\underline{14}}$.

Buchanan.—Orig. in Lancaster, Pa., by J.F. Jones. Introd. in 1930. Cross of the Rush hazel by one of the European filberts; cross made in 1920. Nut: kernel 18 x 22 mm, ovoid pointed, medium brown; flavor good; nut in husk pale, brownish-gray; buds 1/8 inch long, rounded, blunt; catkins greenish, single or in clusters of one to four, about 1 inch long in early October, resembles Bixby, except longer and more slender, and shell thicker. Tree: hardy, productive, reaching height of 10 ft; broad, rounded top, outline oval; younger wood greenish-brown; lenticels whitish, somewhat broader than long; leaf 8.4 x 9.4 cm, apex missing or blunt, glabrous, shiny, medium dark green on top, lighter green and dull on under surface; named in honor of President Buchanan, who spent his last years in Lancaster and is buried there.

Butler.—Orig. in Wilsonville, Ore., by Harry B. Lagerstedt, U.S. Dept. of Agriculture and Agr. Res. Sta., Corvallis, Ore. Introd. in 1980 as a pollinizer to replace Daviana. Seedling appears to be Barcelona x Daviana; discovered in 1958 by Joseph C. Butler. Nut: slightly larger than Daviana; free husking; nut drop occurs over a long period, beginning slightly before Barcelona and ending with Ennis; kernel slightly longer than wide. Tree: vigorous and erect; tendency toward biennial bearing; high yields; pollinates Barcelona. Susceptible to eastern filbert blight and bacterial blight. Incompatibility alleles S_2 $S_{\underline{3}}$.

Caldwell.—Orig. in Steamboat Rock, Iowa, by Willard Caldwell. Introd. in 1934. Seedling of the native hazelnut discovered in 1930. Nut: very large. Now obsolete.

Carlola.—Orig. in River Falls, Wis., by Carl Weschcke. Introd. in 1941. *Corylus americana* x Brag. Nut: matures early; shell very thin; large. Tree: very prolific. Resistant to wild hazel blight. Now obsolete.

Casina.—Orig. in Asturias, Spain, where it is a minor cultivar of no commercial importance. Introd. to the U.S. by Q. Zielinski about 1960 and propagated by nurseries in the late 1980s. Nut: small (1.8 g), round-oval, dull brown; 56% kernel by weight, pellicle not removed by dry heat; flavor and texture very good, suitable for the unblanched kernel market; husk 50% longer than nut, about 75% free-husking; matures 3 days before Barcelona. Tree: vigorous, productive, strong tendency to alternate bearing; kernels fill shells poorly with heavy crop. Susceptible to eastern filbert blight, intermediate susceptibility to big bud mite. Incompatibility alleles $S_{\underline{10}}$ $S_{\underline{21}}$. Suitable as mid-late shedding pollinizer.

Castanyera.—See Barcelona.

Comet.—Orig. in Westbank, British Columbia, Canada, by J.U. Gellatly. Introd. in 1928. Seedling of unknown parentage. Nut: very attractive; long, thin shell; kernel clean, smooth, plump. Tree: produces good crops if cross-pollinated. Now obsolete.

Corabel.—Orig. in Pont-de-la-Maye, near Bordeaux, France, by Eric Germain, INRA. Introd. in 1988. Barcelona (Fertile de Coutard) o.p.; pollen parent appears to be Daviana. Cross made in 1969; selected in 1975; tested as G1FC9. Nut: very large (4.2 g); color light brown, attractive; kernel well-filled, 46% by weight, blanches readily. Tree: vigorous, late-leafing, few suckers. Moderately susceptible to big bud mites.

Craig.—Orig. in Westbank, British Columbia, by J.U. Gellatly. Introd. in 1928. F_2 open-pollinated seedling of Kentish Cob; selected in 1926. Nut: large, oval; medium shell; large kernel. Tree: produces good crop if cross-pollinated. Now obsolete.

Daviana.—Orig. in England. Introd. to the U.S. by Felix Gillet in 1870s; trees first offered for sale in 1888. Nut: medium-large (2.9 g), long, light brown, striped; 56% kernel by weight, pellicle not removed by dry heat; husk longer than nut, mostly free-husking; matures one week before Barcelona. Tree: vigorous, erect, not productive. Highly susceptible to eastern filbert blight and big bud mites. Widely planted as pollinizer for Barcelona, now obsolete. Incompatibility alleles $S_{\underline{3}}$ S_{11}.

Dolores.—Orig. in River Falls, Wis., by Carl Weschcke. Introd. in 1941. *Corylus americana* x Brag. Nut: kernel has very thin pellicle; light-colored meats; shell thin; matures early. Now obsolete.

DuChilly *(Kentish Cob)*.—Orig. in Kent, England. Introd. to the U.S. by Felix Gillet in 1870s, trees first offered for sale in 1887. Nut: medium-large (3.4 g), long, brown; 48% kernel by weight, pellicle not removed by dry heat; husk longer than nut, 60% free-husking; matures after Barcelona. Tree: low vigor, semi-erect, productive. Susceptible to eastern filbert blight, moderately resistant to big bud mites. Incompatibility alleles $S_{\underline{10}}$ $S_{\underline{14}}$.

English Henneman.—See Henneman.

Ennis.—Orig. in Scholls, Ore., by Harry B. Lagerstedt, U.S. Dept. of Agriculture and Agr. Res. Sta., Corvallis, Ore. Introd. in 1980 as a main crop cultivar to replace Barcelona for the in-shell trade. Selected in 1960. Parentage unknown, appears to be Barcelona x Daviana. Nut: very large (4.2 g); round-oblong with a distinct point. Shell: medium to light brown; moderately pubescent; 46% kernel by weight; kernel large, slightly longer than wide and somewhat pointed, plump and less wrinkled than most large nuts; matures about same time as Barcelona, but tends to stay in husk longer, causing nuts to drop 1 week later, by early October. Tree: more branched and compact than Barcelona and slower growing. Resistance to *Xanthomonas corylina* Dowson is equal or better than Barcelona; more susceptible to big bud mite (*Phytoptus avellanae* Nal.); highly susceptible to *Anisogramma anomala* (Peck) Müller. Not an effective pollinizer for Barcelona. Incompatibility alleles $S_{\underline{1}}$ S_{11}. Female inflorescences emerge very late; late pollinizer required.

Farris' Bellhusk.—Orig. in Lansing, Mich., by Cecil Farris. Introd. in 1971. Chance seedling of *Corylus avellana*. Nut: medium; shell thin, round, striped; husk half as long as nut and recurved, resembling a bell; matures in early September. Resistant to big bud mite. Produces average amounts of pollen, but with low germination rates.

Fertile de Coutard.—See Barcelona.

Firstoka.—Orig. in Westbank, British Columbia, Canada, by J.U. Gellatly, Gellatly Nut Nursery. Introd. in early 1930s. Kentish Cob o.p. Nut: matures early; flavor fine; falls free from husk. Tree: vigorous, produces good crops. Now obsolete.

Fitzgerald.—Orig. in Washougal, Wash., by D. Fitzgerald. Introd. in 1936. Parentage unknown; open-pollinated seedling; discovered in 1936. Nut: round type, very large, uniform in size and shape; base flattened to concave, wide; apex rounded and grooved on sides; shape round, oval in cross section; color tan to light brown, striped; kernel rough, plump, angular, pointed; flavor sweet, moderately oily; appearance poor due to pellicle adherence; 39% kernel. Now obsolete.

Fitzgerald 18.—See Gem.

Freehusker.—See Self Husker.

Gasaway.—Orig. in Battle Ground, Wash., by Richard Gasaway. Introd. in 1926 by Carroll D. Bush Nurseries, Barton, Ore. Nut: small (1.5 g), long, light gray, pubescent; matures after Ennis; 47% kernel by weight, kernel fibrous, pellicle not removed by dry heat; husk half as long as nut, free-husking. Immune to eastern filbert blight, resistant to big bud mite. Sets many catkins, which shed pollen late in the season. Incompatibility alleles $S_{\underline{3}}$ S_{26}.

Géant de Halle.—See Hall's Giant.

Gellatly.—Orig. in Westbank, British Columbia, Canada, by J.U. Gellatly, Gellatly Nut Nursery. Introd. in 1928. F_2 open-pollinated seedling of Kentish Cob; selected in 1926. Nut: matures early; cracks out about 41% kernel; size averages 148 per lb; resembles Du Chilly. Tree: produces heavy crops and numerous catkins; will pollinate Craig and Holder varieties. Now obsolete.

Gem *(Fitzgerald 18)*.—Orig. in Washougal, Wash., by Dell M. Fitzgerald. Introd. in 1937 as Fitzgerald 18; then in 1947 as Gem. Barcelona o.p.; selected in 1931. Nut: large, long, resembles Daviana; kernel averages 40.8% of nut; quality superior. Tree: vigorous, upright; holds catkins and sheds pollen over a long period of time. Susceptible to big bud mite (*Phytoptus avellanae* Nal.), resistant to eastern filbert blight [*Anisogramma anomala* (Peck) E. Müller] under field conditions. Incompatibility alleles S_2 $S_{\underline{14}}$.

Graham *(Longwink)*.—Orig. in Ithaca, N.Y., by Samuel H. Graham. Introd. in 1950. Winkler x Longfellow; selected in 1940. Nut: large; quality good. Tree: hardy; annual bearer. Blight resistant.

Grande.—See Barcelona.

Grand Traverse.—Orig. in Lansing, Mich., by Cecil Farris. Introd. in 1989. Faroka x Royal; cross made in 1977; selected in 1981; tested as 81C. Nut: nearly round, brown; 51% kernel by weight; ripens with Ennis; kernel plump, clean, excellent flavor. Introduced for the in-shell trade. Tree: vigorous, productive, not precocious, winter hardy. Resistant to big bud mites.

Hallesche Riesen.—See Hall's Giant.

Hall's Giant *(Géant de Halle, Hallesche Riesen, Wunder aus Bollwiller)*.—Orig. in Germany or Alsace. Introd. to the U.S. by Felix Gillet about 1890. Nut: large (4.0 g), round-pointed, brown; shell thick; matures after Barcelona; 38% kernel by weight, pellicle removed by dry heat; husk longer than nut, 80% free-husking. Tree: vigorous, erect, not productive. Moderately resistant to eastern filbert blight, highly resistant to big bud mites. Recommended as late pollinizer, compatible with most cultivars. Incompatibility alleles $S_{\underline{5}}$ $S_{\underline{15}}$.

Hardin.—Orig. in Hardin County, Iowa, by E.F. Huen, Eldora, Iowa. Introd. in 1923. Chance seedling selected from the wild. Nut: large; shell fairly thick. Now obsolete.

Henneman *(English Henneman)*.—Orig. in Salem, Ore., by H.A. Henneman. Introd. in 1927. English Cob x Barcelona; direction of cross unknown. Nut: shape similar to that of Alpha, larger than that variety. Tree: good pollinator for Du Chilly. No longer propagated.

Holder.—Orig. in Westbank, British Columbia, Canada, by J.U. Gellatly. Introd. in 1928. F_2 open-pollinated seedling of Kentish Cob; selected in 1926; tested as G2. Nut: quality high; shell thin; kernel 50% to 53% of dried nut, smooth, high oil content. Tree: late-blossoming. Now obsolete.

Jemtegaard #5 *(J5)*.—Orig. in Boring, Ore., by Olger Jemtegaard. Introd. in 1960s. Parentage believed to be Nonpareil x Gem. Nut: large (3.4 g), round-oval; 50% kernel by weight, slight fiber on pellicle, pellicle not removed by dry heat; free-husking, husk slightly shorter than nut; matures late (with Ennis). Tree: vigorous, moderately productive. Moderately resistant to big bud mites, susceptible to eastern filbert blight. Sheds pollen late, used as pollinizer for Ennis. Incompatibility alleles S_2 $S_{\underline{3}}$.

J5.—See Jemtegaard #5.

Kentish Cob.—See DuChilly.

Latson.—Orig. in Ann Arbor, Mich., by Merle D. Latson. Introd. in 1963. May be an open-pollinated seedling of Italian Red, Medium Long, or Cosford; discovered in 1958. Nut: large, averaging 29.5 mm long, 19.6 mm wide; averages 49.5% kernel; oil content averages 47.9%; ripens late, with Cosford; resembles Medium Long. Tree: bears annually and well. Now obsolete.

Lisa.—Orig. in Lansing, Mich., by Cecil Farris. Introd. in 1990. Faroka x OSU 18-114 (Barcelona x Royal). Nut: slightly elongated, well-suited to marketing in the shell; shell lightly striped; husk open, recurved; kernel slick, smooth, light-colored; excellent flavor; ripe first week of September in central Michigan. Resistant to big bud mites.

Longwink.—See Graham.

Magdalene.—Orig. in River Falls, Wis., by Carl Weschcke. Introd. in 1941. *Corylus americana* x Brag. Nut: large; thin shell; matures early. Now obsolete.

Morningside.—Orig. in Ithaca, N.Y., by Samuel H. Graham. Introd. in 1945. Rush x Du Chilly; selected in 1945. Nut: quality very good. Tree: bears heavily. Variety lost because of blight.

Nonpareil.—Orig. in Washougal, Wash., by D. Fitzgerald, Fitzgerald Nursery. Introd. in 1938. Parentage unknown. Nut: very large; base flattened and wide; apex rounded and grooved; shape long, oval, round in cross-section and grooved; surface color medium tan, distinctly striped; shell is checked or scarred, color uniform with a darker-colored base; kernel light type, skin round, medium striped, plump, pointed and angular; kernel flavor sweet, oily, quality good, but kernel has a large hollow center; 37% kernel. Tree: bears moderately well; blooms midseason, sheds pollen over a very long period. Incompatibility alleles S_1 $S_{\underline{3}}$.

Nooksack.—Orig. in Nooksack, Wash., by H.E. Altman. Introd. in 1928. Chance seedling. Nut: large, like Du Chilly; drops freely from husk; kernel fills shell well, cracks out about 40% more kernel than Barcelona. Tree: productive; habit of growth similar to Du Chilly, but more spreading; pollinates Du Chilly and the late bloom of Barcelona. Incompatibility alleles $S_{\underline{6}}$ S_{14}.

O'Rourke.—Orig. in Lansing, Mich., by Cecil Farris. Introd. in 1985. Morrisoka x Royal; cross made in 1976; selected in 1981. Nut: very large; 41% kernel by weight; kernel development is very good for so large a nut; husk short, open. Tree: precocious.

Pearson Early Ped.—Orig. unknown. Introd. in the 1920s by David Gellatly, Rosefield Nursery, Westbank, British Columbia, Canada. Nut: appearance fine; quality good. Tree: hardy. Now obsolete.

Potomac.—Orig. in Beltsville, Md., by J.W. McKay, H.L. Crane, and C.A. Reed, U.S. Dept. of Agriculture. Introd. in 1951. Rush x Du Chilly; selected in 1943; tested as USDA 2336. Nut: averages 185–195 per lb; yields 50% kernels; resembles Barcelona. Tree: vigorous, hardy, productive; suited to areas in which varieties of the European species cannot be grown due to lack of hardiness.

Reed.—Orig. in Beltsville, Md., by J.W. McKay, H.L. Crane, and C.A. Reed, U.S. Dept. of Agriculture. Introd. in 1951. Rush x Bolwyller; selected in 1943; tested as USDA 1667. Nut: averages 180–190 per lb; yields 48% kernels; resembles Red Aveline. Tree: vigorous, productive; suited to areas where varieties of the European species cannot be grown successfully due to lack of hardiness.

Royal.—Orig. in Stayton, Ore., by E. Roy. Introd. in 1934 by H.L. Pearcy Nursery, Salem, Ore. Barcelona x Daviana; discovered in 1930. Nut: larger than either parent; shell thin; kernel large; color and markings similar to Daviana; ripens about same time as Barcelona. Tree: early production; sheds pollen medium to early. Susceptible to big bud mite and eastern filbert blight. Incompatibility alleles S_1 $S_{\underline{3}}$.

Scherf.—Orig. in Newberg, Ore., by A.B. Scherf. Introd. in 1936. Chance seedling discovered about 1930. Nut: very large, ovate to long, round in cross-section and grooved; shell reddish-tan, striped, often checked or scarred; kernel rough, dark, collapsed, sweet, oily, quality fair, retains considerable pellicle; attractive; Daviana type. Tree: blooms in midseason; pollinated by Clackamas and probably by other late-blooming varieties; productivity light. Now obsolete.

Self Husker *(Freehusker)*.—Orig. in Newberg, Ore., by A.B. Scherf. Introd. in 1940. Chance seedling; discovered in late 1930s. Nut: falls free from husk; shell very thin; resembles Du Chilly. Tree: moderately vigorous, upright grower; bears annually, produces one to two nuts per cluster. Now obsolete.

Skinner.—Orig. in Heuvelton, N.Y., by Fred L. Ashworth, St. Lawrence Nurseries. Introd. in 1967. Dropmore x seedling of Italian Red; selected in 1966. Nut: large, a little wider than long; shell medium thickness, chestnut color; kernel plump, flavor good; pellicle about average in thickness; resembles Barcelona; ripens 24 Sept. at Heuvelton. Tree: very hardy. Tolerant to blight.

VR 4-31.—Orig. in Corvallis, Ore. Introd. in 1991 by S.A. Mehlenbacher and Maxine M. Thompson, Oregon State Univ., as a pollinizer for Barcelona in areas exposed to eastern filbert blight. Montebello x Gasaway; cross made in 1976; selected in 1987. Nut: small (1.9 g), round compressed, light brown, pubescent; ripens with Barcelona; kernel 51% by weight, very fibrous, pellicle not removed with dry heat; husk slightly shorter than nut, free-husking. Tree: sets many catkins; sheds abundant pollen at same time or slightly later than Daviana. Immune to eastern filbert blight, intermediate response to big bud mite. Incompatibility alleles S_1 $S_{\underline{3}}$.

VR 11-27.—Orig. in Corvallis, Ore. Introd. in 1991 by S.A. Mehlenbacher and Maxine M. Thompson, Oregon State Univ., as a pollinizer for Barcelona and other early flowering varieties in areas exposed to eastern filbert blight. Montebello x Gasaway; cross made in 1976; selected in 1987. Nut: small (2.3 g), round compressed, light brown, pubescent at apex; kernel 46% by weight, very fibrous, pellicle not removed with dry heat; husk same length as nut, free-husking; ripens with Barcelona. Tree: sets many catkins; sheds abundant pollen at same time or slightly later than Daviana. Immune to eastern filbert blight, moderate resistance to big bud mite. Incompatibility alleles S_1 $S_{\underline{3}}$.

VR 20-11.—Orig. in Corvallis, Ore. Introd. in 1991 by S.A. Mehlenbacher and Maxine M. Thompson, Oregon State Univ., as a pollinizer for Barcelona in areas exposed to eastern filbert blight. OSU 10-68 (Barcelona x Compton) x Gasaway; cross made in 1976; selected in 1987. Nut: small to medium (2.3 g), nearly round, light brown, attractive; kernel 51% by weight, little fiber on pellicle, pellicle not removed with dry heat; husk slightly shorter than nut, free-husking; ripens with Ennis. Tree: sets many catkins; sheds abundant pollen at same time as Daviana. Immune to eastern filbert blight, moderately susceptible to big bud mite. Incompatibility alleles S_2 $S_{\underline{3}}$.

VR 23-18.—Orig. in Corvallis, Ore. Introd. in 1991 by S.A. Mehlenbacher and Maxine M. Thompson, Oregon State Univ., as a pollinizer for Barcelona in areas exposed to eastern filbert blight and also for late-flowering varieties. OSU 13-19 (Barcelona x Lansing) x Gasaway; cross made in 1976; selected in 1987. Nut: small to medium (2.2 g), oval, light brown with stripes and pubescence; kernel 51% by weight, much fiber on pellicle, pellicle not removed with dry heat; husk slightly shorter than nut, free-husking; ripens with Ennis. Tree: sets many catkins; sheds abundant pollen over long period, starting slightly later than Daviana. Immune to eastern filbert blight, intermediate response to big bud mite. Incompatibility alleles S_1 $S_{\underline{3}}$.

Willamette.—Orig. in Corvallis, Ore., by Shawn A. Mehlenbacher and Maxine M. Thompson, Oregon State Univ. Introd. in 1990. Montebello o.p. The female parent is Montebello; the pollen parent is believed to be OSU 18.114 (Barcelona x Royal) rather than Compton as published previously. Cross made in 1973; selected in 1979; tested as OSU 43-58. Nut: medium (2.8 g), dark brown, attractive; kernel 50% by weight, pellicle easily removed by dry heat; husk 50% longer than nut, about 60% free-husking; flavor and texture very good. Moldy kernels noted in some years. Matures 1 week later than Barcelona. Introduced for the blanched kernel market. Tree: vigorous, productive. Moderately resistant to eastern filbert blight, intermediate susceptibility to big bud mites. Incompatibility alleles S_1 $S_{\underline{3}}$.

Woodford.—Orig. in Forest Grove, Ore., by E.W. Woodford. Introd. in 1936. Parentage unknown. Nut: large, round, oval in cross section, slightly grooved; 34% kernel; kernel rough, striped, dark, plump, pointed and ridged; flavor sweet, medium dry, quality good; ripens late. Tree: sheds pollen very early, but light in quantity. Incompatibility alleles S_1 $S_{\underline{3}}$. Now obsolete.

Wunder aus Bollwiller.—See Hall's Giant.

ROOTSTOCKS FOR HAZELNUT (FILBERT)

Dundee.—Orig. in Corvallis, Ore., by Harry B. Lagerstedt, U.S. Dept. of Agriculture, Agr. Res. Serv. Introd. in 1993. *Corylus colurna* o.p.; appears to be *C. colurna* x *C. avellana* hybrid. Selected in 1971; tested as USOR 15-71. Ennis trees grafted to Dundee are vigorous and productive, produce very few suckers, and have smooth unions. Dundee bark is silver-colored and smooth. Propagation is by tie-off layerage with girdling; propagation by semi-hardwood cuttings or simple layerage is less successful. Highly susceptible to eastern filbert blight (*Anisogramma anomala*).

Newberg.—Orig. in Corvallis, Ore., by Harry B. Lagerstedt, U.S. Dept. of Agriculture, Agr. Res. Serv. Introd. in 1993. *Corylus colurna* o.p.; appears to be *C. colurna* x *C. avellana* hybrid. Selected in 1971; tested as USOR 7-71. Ennis trees grafted to Newberg are vigorous, productive, and produce very few suckers; the rootstock slightly overgrows the scion. Newberg bark is intermediate in appearance between *C. colurna* and *C. avellana*. Propagation is by tie-off layerage with girdling; propagation by semi-hardwood cuttings or simple layerage is less successful. Highly susceptible to eastern filbert blight.

HEARTNUT

The Japanese walnut variety heartnut was indigenous only to the Japanese Islands. In older references, it was named *Juglans sieboldiana* and the heartnut (*J. sieboldiana* var. *cordiformis*; common name Siebold walnut). Currently, the correct name is *Juglans ailantifolia* var. *cordiformis*; common name "heartnut."

The heartnut grows best in the northeastern United States and Canada, where lakes moderate the climate. Farther west, severe autumn freezes kill-back one or more year's growth. Farther south, late-spring frosts are a problem.

Few commercial nurseries sell heartnut cultivars. The following list includes the more popular cultivars grown currently.

Bates—Orig. by a Mr. Bates of Jackson, S.C. Introd. by J.F. Jones, Lancaster, Pa. Yields 50 nuts per lb; 25% kernel.

Canoka—Orig. by J.U. Gellatly, Westbank, British Columbia, Canada. Foliates late. Yields 50 nuts per lb; 25% kernel.

Etter—Orig. by Fayette Etter, Lancaster, Pa. Yields 78 nuts per lb; 33% kernel.

Fodermaier.—Orig. by John J. Fodermaier in Dover Plains, N.Y. Yields 55 nuts per lb; 37% kernel. Rated the best.

Marvel.—Introd. by Gilbert L. Smith & Sons Nursery, Millerton, N.Y., in 1948. Fodermaier x Wright cross.

Rhodes.—Orig. by George Rhodes Covington in the 1940s. Yields 65 nuts per lb; 28% kernel.

Schubert.—Orig. by Kenneth Schubert, Millstadt, Ill., in the 1960s. Introd. by Louis Gerardi. Yields 53 nuts per lb; 33% kernel.

Walters.—Orig. by J.U. Gellatly and introd. by him in the late 1930s. Yields 61 nuts per lb; 26% kernel.

Wright.—Orig. by Ross Pier Wright in Erie, Pa. Introd. by Gilbert L. Smith & Sons Nursery, Millerton, N.Y. Yields 54 nuts per lb; 31% kernel. Rated similar to Fodermaier.

HICAN

Baress.—Orig. in Freeburg, Ill., by William Schmisseur on his "Bar S" ranch. Introd. in 1953 by the Louis Gerardi Nursery, Caseyville, Ill. *Carya* x*nussbaumeri*; chance seedling, considered to be a natural pecan–shellbark hickory cross; selected in 1953. Nut: shell thin; larger than and ripens later than Burlington, which it resembles.

Burlington.—Orig. in Des Moines County, Iowa. Introd. in 1940 by Benton & Smith Nut Tree Nursery, Millerton, N.Y. *Carya* x*nussbaumeri*; discovered in 1915. Nut: large, averaging 55 per lb; quality good; ripens early. Tree: productive; comes into bearing slowly. No longer propagated.

Burton.—Orig. in Masonville, Ky., by Emmett Burton. Introd. in 1930 by J. Ford Wilkinson, Indiana Nut Nursery, Rockport, Ind. Chance seedling of *Carya illinoinensis* x *C. ovata*, discovered in 1880. Nut: size medium; shell thin with slight shagbark ridges; flavor blend of pecan and shagbark hickory; harvested in October. Tree: large; vigor good; productive, more so than most hicans, more annual than shagbark hickory; no evidence of pecan scab, susceptible to pecan weevil; protogynous, pollinated by protandrous pecans.

Clarksville.—Orig. in Clarksville, Mo., by Arthur L. Norton. Introd. in 1936. Parentage unknown; *Carya* x*nussbaumeri*. Nut: large; shell fairly thin, very much like that of the less productive Rockville; matures early. Tree: bears well; vigorous; protogynous.

Des Moines.—Orig. in Burlington, Iowa, near the Des Moines River. Introd. in 1924 by Snyder Bros., Center Point, Iowa. *Carya* x*nussbaumeri*. Nut: resembles Burlington. Tree: grows satisfactorily in southern Illinois.

Gerardi.—Orig. in New Memphis, Clinton County, Ill., by Joseph Gerardi, Louis Gerardi Nursery, O'Fallon, Ill. Introd. in 1931 by J.G. Duis; named in 1927 by Arthur S. Colby, Illinois Agr. Expt. Sta., Urbana. *Carya* x*nussbaumeri*; pecan x shellbark hickory; direction of cross unknown; discovered in 1893. Nut: matures earlier than most northern pecans; more productive than McCallister, less than Burlington, which is smaller but earlier; to produce a plump meat, good culture and protection from shuckworm and curculio are necessary. Tree: protogynous, apparently pollinated by various hickories and McCallister hican; flowers early at Urbana.

Henke.—Orig. in southern Iowa by a Mr. Henke. Introd. in 1932. (*Carya ovata* x *C. illinoinensis*); shagbark hickory x pecan; direction of cross unknown; discovered in 1928. Nut: small; high-quality kernel, fills shell well; ripens early. Tree: bears precociously and well.

Jay Underwood *(Underwood)*.—Orig. in Ft. Scott, Kan., by Jay S. Underwood, Uniontown, Kan. Introd. in 1956 by the Louis Gerardi Nursery, O'Fallon, Ill. Native (wild) seedling: (*Carya* x*brownii*) x *C. laciniosa*; discovered in 1952; named by Joseph C. McDaniel, Illinois Agr. Expt. Sta., Urbana. Nut: large; shell thin; kernel plump, fine flavor. Tree: protogynous; vigorous; production light to medium.

Pixley.—Orig. in Wabash County, Ill., by E. Guy Pixley, West Salem, Ill. Introd. in 1962. (*Carya illinoinensis* x *C. ovata*). Chance seedling from seed produced in Louisiana from an unknown pecan variety; apparently crossed with shagbark hickory; selected about 1950; named in 1957. Nut: quality similar to Burton; smaller than Burton; ripens early. Tree: produces regular crops; successfully grafted to both parental species, pecan and shagbark hickory; vigorous; flowering protogynous; ornamental.

Radcliff.—Orig. in Vandalia, Ill., by W.F. Sonneman; named after Mr. Radcliff, who brought the original tree to Sonneman's attention. Introd. in 1954 by the Louis Gerardi Nursery, O'Fallon, Ill. *Carya* x*nussbaumeri*. Chance seedling, considered to be a natural pecan–shellbark hickory cross; selected in 1951. Nut: shell thin; shorter than Burlington; kernel well-filled; resembles Des Moines.

Underwood.—See Jay Underwood.

Wright.—Orig. in Missouri by C.C. Wright. Introd. about 1926–28 by John W. Hershey, Nut Tree Nurseries, Downington, Pa. *Carya* x*nussbaumeri*; considered to be a natural cross of pecan with shellbark hickory. Nut: kernel does not fill shell.

HICKORY

Aber.—Orig. in Jacksonville, Cherokee County, Texas, by Edwin A. Aber. Introd. in 1946. Discovered as a very old tree. *Carya texana* (named as *C. ovalis*). Nut: shell quite thin; kernel very sweet; poor cracking quality. Tree: very prolific. Primarily for home gardens; mostly a novelty.

Abscoda.—Orig. in Union City, Mich., discovered by Harry Burgart in 1934. Introd. in 1944. *Carya ovata.* Nut: size medium; shell average thickness; kernel plump, sweet. Tree: annual bearer, but heaviest crops come on alternate years. Original tree was 30–35 ft tall and about 50 years old when discovered.

Abundance.—Orig. in Franklin County, Pa., by Fayette Etter. Introd. in 1947. *Carya* x*dunbarii.* A native shellbark hickory x a native shagbark hickory; tested as Etter 54. Nut: shagbark; size medium; kernel white, cracks out readily; shell thin. Tree: bears annually; top-worked trees often bear second year after grafting.

Adelhurst.—Orig. in New Milford, Conn., by Henry C. Beeman. Introd. in 1935. *Carya ovata.* Shagbark tree native to the region; discovered in 1935. Nut: kernel superior.

Anthony.—Orig. in Ogle County, Ill., by A.B. Anthony, Sterling, Ill. Introd. in 1932. *Carya ovata.* Seedling selected in 1932; tested as Anthony 1. Nut: shagbark; large; matures very early; kernel cracks out well; quality very good. Tree: usually a prolific bearer; found relatively resistant to hickory leaf spot at Beltsville, Md.

Beeman.—Orig. in New Milford, Conn., by Henry C. Beeman. Introd. about 1935. Native; discovered in 1935. *Carya ovata.* Nut: kernel superior, cracks out well. Tree: annual bearer.

Bradley.—Orig. in Franklin County, Pa., by Fayette Etter. Introd. in 1932. Native shellbark hickory. *Carya laciniosa.* Nut: large; kernel large, never tight in shell; cracking quality excellent; shell thickness medium. Tree: bears annually; bears precociously after grafting; requires good soil.

Cranz.—Orig. in Ira, Ohio, by Eugene F. Cranz. Introd. in 1934. *Carya ovata;* native seedling; selected in 1934.

Creager.—Orig. in Sumner, Iowa, by W.O. Creager; introd. by D.C. Snyder, Snyder Bros., Center Point, Iowa, in 1925. *Carya* x*laneyi.* Nut: smooth, white, fine quality, but too small for commercial use. Tree: vigorous and rapid grower; very prolific.

Davis.—Orig. in Duchess County, N.Y., by Gilbert L. Smith, Gilbert L. Smith & Sons Nursery. Introd. in 1934. Seedling selected in 1934. *Carya ovata.* Nut: size medium; kernel quality high, plump, bright; cracking quality excellent; ripens at point of origin a little later than Fox; resembles Wilcox. Tree: vigorous. Won first prize in the New York and New England Nut Contest of 1934.

De Acer *(White)*.—Orig. in Urbana, Iowa, by Linn County Nurseries, Center Point, Iowa. Introd. in 1950 by this nursery. Shagbark hickory (*Carya ovata*) x bitternut hickory (*C. cordiformis*); direction of cross not known; discovered about 1948 by Delbert White on farm owned by a Mr. De Acer; tested and developed by this nursery. *Carya* x*laneyi.* Nut: large; shell thin; flavor like that of shagbark hickory, pleasant; of best quality of this parentage. Tree: bears very well.

Fairbanks.—Orig. in Anamosa, Jones County, Iowa, on the farm of C.A. Fairbanks. Discovered by A.B. Dennis in 1910; introd. by S.W. Snyder, Linn County Nurseries, Center Point, Iowa, in 1920. Shagbark hickory (*Carya ovata*) x bitternut hickory (*C. cordiformis*). *Carya* x*laneyi.* Nut: large; shell smooth, fairly thin, attractive, white; kernel quality poor. Tree: very vigorous; foliage very dark green, stiff; heavy bearer.

Fayette.—Orig. in Franklin County, Pa., by Fayette Etter. Introd. in 1932. Native seedling; tested as Etter 9. *Carya laciniosa.* Nut: shellbark; large; shell thin; kernel cracks out well, flavor good.

Fox.—Orig. in Fonda, N.Y., by Roland Fox. Introd. about 1934. Native selection from tree growing in the Mohawk River Valley.

Carya ovata. Nut: kernel color light, flavor fine, large percentage crack out whole; ripens early. Tree: bears consistently. First prize winner in the Northern Nut Growers Assn. contest of 1934.

Goheen.—Orig. in Pennsylvania Furnace, Pa., by Mrs. Martha Goheen. Introd. in 1934. Chance seedling of shagbark hickory; discovered 1926. *Carya ovata*. Nut: hulled nuts very large, averaging 1 1/4 x 1 inch; flavor fair. No longer propagated since 1962.

Grainger.—Orig. on the farm of J.N. Stapleton, Washburn, Grainger County, Tenn., by the Tennessee Valley Authority. Introd. in 1935 by John W. Hershey, Downingtown, Pa. Chance seedling discovered in 1934. *Carya ovata*. Nut: large; shell thin; kernel averages high percentage of nut; cracks well. Tree: vigorous.

Hadley.—Orig. in Leominster, Mass., by George C. Hadley. Introd. in 1934. Chance seedling of shagbark hickory; discovered in 1934. *Carya ovata*. Nut: quality very uneven. Now obsolete.

Haviland.—Orig. in Waterford, Ontario, Canada, on the farm of Ira Haviland by Bion T. Forse, Simcoe, Ont. Introd. in 1934. Seedling of shagbark hickory; discovered about 1931. *Carya ovata*. Nut: large; shell thin; kernel plump, good flavor; ripens early. Tree: productive.

Hefty.—Orig. in Evansville, Wis., by Thomas C. Hefty. Introd. in 1934. Seedling of shagbark hickory; selected in 1933. *Carya ovata*. Nut: size medium; shell well-filled; flavor good.

Heibner.—Orig. in Danville, Iowa, by Harvey Heibner. Introd. in 1934. Seedling of shagbark hickory; selected in 1932. *Carya ovata*. Nut: shell thin; kernel plump, rich in quality, flavor sweet.

Hoffeditz.—Orig. in Franklin County, Pa., by Fayette Etter. Introd. in 1930. Open-pollinated seedling of a native shellbark hickory; discovered in 1914. *Carya laciniosa*. Nut: size medium; kernel flavor good; cracking quality good; shape not always uniform. Tree: bears annually.

Jolliffe.—Orig. in Uffington, W.Va., by George C. Jolliffe. Introd. in 1933. Selected seedling of shellbark hickory. Nut: very large; shell paper-thin. Tree: bears heavily annually.

Keystone.—Orig. in Franklin County, Pa., by Fayette Etter. Introd. in 1955. Open-pollinated seedling of a native shellbark hickory; discovered in 1955. *Carya laciniosa*. Nut: shellbark; large; best cracking shellbark ever found by Etter, as kernel always falls from shell. Tree: bears annually.

Last.—Orig. in Steamboat Rock, Hardin County, Iowa, by E.F. Huen, Eldora, Iowa. Introd. in 1934. Chance seedling of a native hickory discovered about 1933 on the Last farm; original tree still alive in 1963. *Carya ovata*. Nut: shell thin; flavor good.

Lingenfelter.—Orig. in Blair County, Pa., by Jesse B. Lingenfelter. Introd. in 1943. Discovered in Spring 1925, as a result of high ratings received from Northern Nut Growers Assn. in selecting choice hickory nuts. *Carya ovata*. Nut: averaging about 60 per lb; shell medium-thick, dark; cracks reasonably well; kernel dark, texture medium, quality and flavor fair. Tree: precocious; prolific and regular bearer; early harvest.

Murdock.—Orig. in Crown Point, N.Y., by John Murdock. Introd. about 1940. Parentage unknown; shagbark, probably discovered as a chance seedling in 1935. *Carya ovata*. Nut: quality high. Tree: hardy; bears early.

Peck.—Orig. in Duchess County, N.Y., by Gilbert L. Smith, Gilbert L. Smith & Sons Nursery, Millerton, N.Y. Not introduced commercially but included here because name has been publicized. Shagbark x bitternut; selected in 1934. *Carya xlaneyi*. Nut: quality good, sweet; shell thin, cracks well; only about one-half of the nuts fill well, the reason for it not being introduced to the trade.

Porter.—Orig. in Mackeyville, Clinton County, Pa., on farm owned by Ralph R. Porter. Introd. in 1960. Chance seedling selected by Mrs. Delmont Williams, whose husband rented the farm; selected as promising in 1958 at a judging contest. *Carya ovata*. Nut: large; shell thin; yields high percentage of kernel (46.7%); cracking quality excellent.

Stauffer.—Orig. in Franklin County, Pa., by Fayette Etter. Introd. in 1934. A native shellbark hickory x a native shagbark hickory; selected in 1930. *Carya xdunbarii*. Nut: large; shell white, thickness average;

kernel tight in shell. Tree: bears annually.

Stocking.—Orig. in Owosso, Mich., by Eugene Stocking. Introd. in 1936. Shagbark hickory (*Carya ovata*) x bitternut hickory (*C. cordiformis*); discovered in 1932. *Carya xlaneyi*. Nut: fairly large; shell thin; cracking quality very good. Tree: symmetrical.

Stratford.—Orig. in Stratford, Iowa. Introd. about 1926. Chance seedling of shagbark hickory (*Carya ovata*), although some persons consider it a hybrid between shagbark hickory and bitternut hickory (*C. cordiformis*); first propagated by Simon Snyder, Linn County Nurseries, Center Point, Iowa; discovered between 1920 and 1925. Nut: shell thin, smooth, near white, cracks easily, resembling bitternut hickory; kernel hard, keeps very well, quality high, no bitter taste, quite smooth, oil content quite high, typically shagbark hickory. Tree: buds, flaking bark, and leaves typically shagbark hickory; productive.

Weschcke.—Orig. in Fayette, Iowa, by Carl Weschcke. Introd. in 1936. Selected about 1928. *Carya xlaneyi*. Nut: thin shell; kernel full and rich, cracking out in entire halves; matures 1–15 Sept. Tree: grafts well on wild bitternut hickory (*C. cordiformis*); produces consistently.

Westpha.—Orig. in Kendall, Wis., by Otto Westphal. Introd. in 1935. Chance seedling of shagbark hickory; discovered in 1921. *Carya ovata*. Nut: large; shell thin; flavor very good. Tree: yields well.

White.—See De Acer.

Wilcox.—Orig. in Geneva, Ohio, by C.F. Walker, H.L. Jacobs, and W.R. Fickes. Introd. about 1936. Seedling of a wild shagbark hickory; discovered in 1934. *Carya ovata*. Nut: size medium; flavor good; cracking quality excellent; keeps well. Tree: bears early.

JOSTABERRY

Jogranda.—See Jostaki.

Josta.—A vigorous, disease-resistant thornless hybrid of blackcurrant x gooseberry. Orig. in Germany by Rudolf Bauer. Cross made in 1955; selected in 1970s; introd. to Europe in 1977. F_2 open-pollinated selection of an amphidiploid F_1 hybrid (*R. nigrum* L. 'Langtraubige Schwarze' x *R. divaricatum* Dougl.). Fruit: resembles a black currant more than a gooseberry in size and color but is much milder in flavor; persistent with a very long shelf-life; high vitamin C content. Plant: vigorous, spreading, thornless, without oil glands like those found in black currants. Resistant to American powdery mildew, white pine blister rust, and gall mites. Not suitable to mechanical harvest. Fruit good for processing.

Jostagranda.—See Jostaki.

Jostine.—A vigorous, disease-resistant thornless hybrid of blackcurrant x gooseberry. Orig. in Germany by Rudolf Bauer. German plant breeder rights. Cross made in 1959, selected in 1970s; introd. to Europe in 1985. F_2 open-pollinated selected of an amphidiploid F_1 hybrid (*R. nigrum* L. 'Silvergieters Schwarze' x *R. grossularia* 'Grune Hansa'). Fruit: large to medium size berries, one to three per truss, uneven ripening, very good flavor, keeps well on the bush, slow picking rate by hand. Plant: very strong growth, spreading bushes; midseason; yield high; resistant to American powdery mildew, white pine blister rust, leaf spot, and gall mite. Valuable for home gardens because of resistance and suitability for jams, jellies, and pies.

Jostaki (*Jogranda, Jostagranda*).—A vigorous, disease-resistant thornless hybrid of blackcurrant x gooseberry. Orig. in Germany by Rudolf Bauer. U.S. Plant Patent 5918; issued on 31 Mar. 1987 to Rudolf Bauer, now deceased, previously of Breitbrunn, Germany; Annelise Bauer, heir; cross made in 1959; selected in 1970s, introd. in 1987. F_2 open-pollinated selection of an amphidiploid F_1 hybrid (*R. nigrum* L. 'Langtraubige Schwarze' x *R. divaricatum* Dougl.). Fruit: Berries larger than Josta; two to four per truss, violet-black, matures about 1 July in Switzerland; uneven ripening, very good flavor, keeps well on the bush, good for hand picking, not for machine harvest. Plant: medium growth, spreading, slightly drooping habit, needs trellis; midseason ripening; yield high, resistant to American powdery mildew, white pine blister rust, leaf spot, and gall mite. Valuable for home gardens because of resistance and suitability for jams, jellies, and pies.

ORUS 1.—A vigorous, disease-resistant jostaberry. Orig. in Corvallis, Ore., by George F. Waldo. Cross made about 1966. *R. nigrum* x *R. grossularia*. Fruit: black about 1.5 g/berry, about as productive as Josta, ripening about 1 week after Josta. Plant: vigorous, spreading (unsuitable for machine harvest), resistant to winter cane dieback, American powdery mildew, and white pine blister rust.

ORUS 10.—A vigorous, disease-resistant jostaberry. Orig. in Corvallis, Ore., by George F. Waldo. Cross made about 1966. *R. nigrum* x *R. grossularia*. Fruit: black, about 1.7 g/berry, less productive than Josta, ripens 2 days after Josta. Plant: vigorous, spreading (not suitable for machine harvest), resistant to winter can dieback, American powdery mildew, and white pine blister rust.

JUJUBE

Lang.—Orig. in China; brought into the U.S. in 1908 by Frank N. Meyer, U.S. Dept. of Agriculture; P.I. 22686. Introd. in 1924. Fruit: small oblong; skin red; flesh melting, sweet. Tree: very large, spreading.

Li.—Orig. in China; brought into the U.S. in 1914 by Frank N. Meyer, U.S. Dept. of Agriculture; P.I. 38249. Introd. in 1924. Fruit: very large, up to 2 inches long, round to ovoid; skin mahogany brown; primarily for eating fresh.

Mu Shing Hong.—Orig. in China; brought into the U.S. in 1908 by Frank N. Meyer, U.S. Dept. of Agriculture; P.I. 22684. Introd in 1924. Fruit: size medium, up to 1 3/4 inches in diam.; oblong, tapering toward stylar end; skin red; processes well. Tree: large; when old, very few side branches on main branches.

Shui men.—Orig. in China; brought into the U.S. in 1914 by Frank N. Meyer; P.I. 38245. Introd. in 1924. Fruit: size medium; elongated; good only when fresh.

So.—Orig. in China; brought into the U.S. in 1914 by Frank N. Meyer, U.S. Dept. of Agriculture; P.I. 37484. Not introd. but included because it has been used in research studies. Fruit: size medium; round; skin brown-red; flesh sweet. Tree: very gnarled with zigzag growth.

Yu.—Orig. in China; brought into the U.S. in 1913 by Frank N. Meyer, U.S. Dept. of Agriculture; P.I. 36854. Not introd. but included because it has been used in research studies. Fruit: large; elongated, tapering toward stylar end; skin reddish-brown; flesh sweet, firm; keeping quality rather good. Tree: tall; main branches with few side branches; leaves large.

• *Revisions for this commodity were not received before the book went to press; therefore, the above variety descriptions are virtually identical to those published previously.*

Kiwifruit (Chinese Gooseberry)

Actinidia deliciosa* (A. Chev.) C.F. Liang et A.R. Ferguson var. *deliciosa

syn. *A. chinensis* Planch. var. *deliciosa* (A. Chev.) A. Chev.

A. chinensis Planch. var. *hispida* C.F. Liang

Abbott.—Female. Orig. in Auckland, New Zealand, probably by Hayward R. Wright (deceased). Introd. in the 1930s; named in 1958 by Harold M. Mouat (deceased), New Zealand Dept. of Scientific and Industrial Res., Auckland. Parentage unknown, chance seedling. U.S. PI 257464, PI 295683. Fruit: very similar to Allison; mediocre quality; small-medium size, squat oblong; skin brownish-green; hairs dense, long, and silky; flesh light green; flavor sweet; harvested before Hayward; poor storage life. Vine: medium vigor; productive; flowers earlier than Hayward. Now only of historical interest or for districts having inadequate winter chilling for Hayward. Named for owner of plant from which propagation material was taken.

Allison.—Female. Orig. in Palmerston North, New Zealand, by Bruno H. Just (deceased). Introd. in the 1930s. Chance seedling; discovered in the 1920s; named in 1958 by Harold M. Mouat (deceased), New Zealand Dept. of Scientific and Industrial Res., Auckland. U.S. PI 257465. Fruit: mediocre; medium size, oblong; very similar to Abbott, but somewhat larger, slightly broader in proportion to length, and more rounded at stem end; skin brownish-green; hairs dense, stiff; flesh light green; flavor sweet; harvested before Hayward; medium storage life. Vine: medium vigor; productive; flowers earlier than Hayward. Now only of historical interest or for districts having inadequate winter chilling for Hayward. Named for Alexander Allison, probably the first to grow kiwifruit in New Zealand.

Autari.—Male. Orig. in Udine, Italy, by Raffaele Testolin, Università di Udine. Introd. 1996. Seedling from a cross made in 1986 of Hayward by fruiting male, M3h. For use as a pollinizer for Hayward, with which it is coincident under northern Italy conditions. Floriferous, high pollen germinability, good seed setting and fruit sizing capability. Named for ancient Longobard king.

Bien Donné.—See Donné.

Blake.—Female. Orig. in Berkeley, Calif. Chance seedling, origin unknown. Distributed from 1970s. Fruit: not of commercial quality; very small; tear-shaped, sloping inwards toward stylar end; hair easily brushed off; matures very early. Vine: very productive; flowers early; requires less winter chilling than Hayward. Claimed to be self-fertile.

Bruno.—Female. Probably orig. in Palmerston North, New Zealand, by Bruno H. Just (deceased). Introd. in the 1930s. Chance seedling; discovered in the 1920s; named in 1958 by Harold M. Mouat (deceased), New Zealand Dept. of Scientific and Industrial Res., Auckland. U.S. PI 257466, PI 295684. Fruit: medium-large; elongated cylindrical, distinct from all other cultivars; darker brown skin; hairs dense, bristly; flesh green; very good flavor; small core; high vitamin C content; harvested earlier than Hayward; medium storage life. Vine: vigorous; productive; flowers earlier than Hayward. No longer widely grown. Named for Bruno Just.

California male *(Chico male)*.—Male. Probably orig. from seeds collected in China by Ernest H. Wilson (deceased) and sent from Ichang, U.S. PI 21781; selected and later renumbered in 1958 as U.S. PI 252006 by Robert L. Smith, U.S. Dept. of Agriculture Plant Introduction Sta., Chico, Calif. Chosen as pollinizer for Hayward in California. Floriferous. (Other males of unstated origin selected in California include **Cal Chico No. 3, Chico No. 3, Chico Early, Chico Extra Early**.)

Chico.—See Hayward.

Chico male.—See California male.

Chieftain *(M59).*—Male. Orig. near Kumeu, Auckland, New Zealand, by Ronald M. Davison, New Zealand Dept. of Scientific and Industrial Res., Auckland. Origin unknown, chance seedling selected in early 1970s as pollinizer for Hayward. Renamed and released in 1989. Very vigorous, highly floriferous; high pollen germinability; good seed setting capability; flowering consistently coincident with Hayward under New Zealand conditions.

Constricted.—See Greensill.

Derek.—Male. Orig. in Waiuku, near Auckland, New Zealand, by Donald A. Skelton. Origin unstated. Early flowering pollinizer selected for Skelton.

Dexter.—Female. Orig. in Coffs Harbour, New South Wales, Australia, by J. Dexter. Origin unknown, rogue plant in a shipment of Hayward plants sent from New Zealand. Now thought to be Abbott or very similar to Abbott. Named for selector.

Donné *(Bien Donné).*—Female. Orig. in Pickstone Nursery, Groot Drakenstein, South Africa. Chance seedling, origin unknown. Selected by Fruit and Food Technology Res. Inst., Stellenbosch, South Africa, in 1982; released in 1990. Fruit: medium size, thick cylindrical; stylar end sometimes protruding slightly; skin brown; hairs short, bristly, easily lost; flesh dark green; matures late; good storage life. Vine: very vigorous; vegetative growth continues late into autumn; lower winter chilling requirement; flowers late. Considered more suitable than other cultivars for South African conditions. Named for Bien Donné Experimental Farm, where it was evaluated.

Elmwood.—Female. Orig. in Tauranga, New Zealand, by Harold M. Mouat (deceased), New Zealand Dept. of Scientific and Industrial Res., Auckland. Origin unknown; selected about 1960. U.S. PI 300973. Fruit: medium-large; elongated, cylindrical, round to elliptical in cross section; skin brownish-green; hairs stiff, easily removed; flesh bright green; good flavor; poor storage life. Vine: medium vigorous; productive. Flowering and fruit maturity about same time as for Hayward.

Gracie *(Jones).*—Female. Orig. in Avondale, Auckland, New Zealand, by Hayward R. Wright (deceased) in 1930s. Chance seedling, probably a sibling of Hayward; later named and distributed by John Gracie, assistant to Wright. Fruit: medium to large; longitudinal ellipsoid, apex rounded, elliptical in cross section; skin greenish-brown; hairs long, soft, easily removed; flesh green; flavor average; poor storage life. Vine: moderately vigorous and not very productive. Named for John Gracie. The original plant, Jones, was probably a grafted plant of Gracie.

Greensill *(Constricted).*—Female. Orig. probably in Palmerston North, New Zealand, by Bruno H. Just (deceased). Introd. in 1930s. Origin unknown; named in 1960s by Harold M. Mouat (deceased), New Zealand Dept. of Scientific and Industrial Res., Auckland. U.S. PI 257467, PI 276401. Fruit: medium size; broadly oblong, often lopsided, broader at stalk end, elliptical in cross section, often constricted at waist (hence the name Constricted); skin greenish-brown; stylar end often depressed; hairs short; flesh dark green; good flavor; restricted storage life. Vine: reasonably productive. Named for L. Greensill, Otumoetai, Tauranga, New Zealand, from whose property vine was selected.

Hayward *(Chico).*—Female. Orig. in Auckland, New Zealand, by Hayward R. Wright (deceased). Introd. in late 1930s. Chance seedling; selected in the 1920s; named in 1958 by Harold M. Mouat (deceased), New Zealand Dept. of Scientific and Industrial Res., Auckland. Introd. to U.S. in 1935 as U.S. PI 112053; subsequently named Chico, until it was recognized as being identical to Hayward; reintrod. as PI 295685. Fruit: large; distinctly broad oval, slightly flattened laterally, elliptical in cross section; hair fairly fine, silky; skin greenish-brown; flesh light green; core elliptical, often markedly fluted; good flavor; harvested later than almost all other cultivars; exceptional storage life. Vine: moderate vigor and productivity; flowers late. Currently, only important fruiting cultivar in commercial orchards throughout world. Common faults are misshapen, flat fruit and Hayward mark. Named for Hayward Wright.

Hayward Clone 8.—A clone of Hayward; selected at Università di Udine, Italy, by Raffaele Testolin from many Hayward variants collected from the main kiwifruit areas in Italy. Released in 1993. Produces essentially only single fruits, very few misshapen fruit or fruit with Hayward mark.

Hayward Clone K.—A clone of Hayward; orig. in Cesena, Italy, by G. Zuccherelli. Selected from an individual plant of French provenance. Distributed since 1985 by Vitroplant, Italy. Considered to be less vigorous, more productive, and more consistent in fruit shape and size than Hayward, and matures earlier.

Hayward Clone Kramer.—A clone of Hayward; orig. in Te Puke, Bay of Plenty, New Zealand, by John Kramer (deceased). Distributed since about 1985. Considered to be less vigorous and more consistent in fruit shape.

Hayward Clone Maeba®.—Orig. in Italy by Maurizio Sciutti. Distributed by Vivai Piante Battistini since 1989. Considered to be more productive with fewer malformed fruit than other Hayward strains.

Jones.—See Gracie.

Katiuscia.—Female. Orig. in Cesena, Italy, by Giuseppe Zuccherelli. Seedling from controlled cross using Hayward as mother. Distributed in 1991. Fruit: exceptionally large; elliptical tending to rounded, rounded to elliptical in cross section; skin brown; hairs easily shed at maturity; flesh light green; core elliptical; flavor similar to Hayward. Vine: medium vigor; produces exclusively single flowers; flowers about 1 week before Hayward; matures about 2 weeks before Hayward.

Koryoku.—Female. Orig. in Kagawa Agr. Expt. Sta., Kagawa Prefecture, Japan. Seedling of Hayward; selected in 1967. Japan plant patent 1446. Fruit: medium-large, long cylindrical; skin brown; hairs dense, stiff, easily removed; flesh bright green; high sugar content but low acidity; harvested earlier than Hayward; medium storage life. Vine: vigorous; floriferous; productive.

M3.—See P1 Vitroplant®.

M51.—Male. Orig. in Te Puke, Bay of Plenty, New Zealand, by Ronald M. Davison, New Zealand Dept. of Scientific and Industrial Res., Auckland. Origin unknown; chance seedling selected in early 1970s as pollinizer for Hayward under New Zealand conditions. Small, flat flowers; not abscising; highly floriferous; high pollen germinability.

M52.—Male. Orig. in Te Puke, Bay of Plenty, New Zealand, by Ronald M. Davison, New Zealand Dept. of Scientific and Industrial Res., Auckland, and Graham R. Bayliss. Origin unknown; chance seedling selected in early 1970s as pollinizer for Hayward under New Zealand conditions. Medium-sized, cupped flowers; not abscising; only moderately floriferous; high pollen germinability; good seed set and fruit weight.

M54.—Male. Orig. in New Zealand by Ronald M. Davison, New Zealand Dept. of Scientific and Industrial Res., Auckland. Origin unknown; chance seedling selected in early 1970s as pollinizer for Hayward under New Zealand conditions. Medium-sized, cupped flowers, abscising; very floriferous.

M56.—Male. Orig. in Te Puke, Bay of Plenty, New Zealand, by Ronald M. Davison, New Zealand Dept. of Scientific and Industrial Res., Auckland. Origin unknown; chance seedling selected in early 1970s as pollinizer for Hayward. Small, cupped flowers; abscise after flowering; very floriferous; high pollen germinability; consistently good seed set and fruit weight; coincident with Hayward under New Zealand conditions. Widely grown.

M59.—see Chieftain.

Matua.—Male. Orig. in Te Puke, Bay of Plenty, New Zealand, by William A. Fletcher (deceased), New Zealand Ministry of Agr. and Fisheries, Auckland, and Harold M. Mouat (deceased), New Zealand Dept. of Scientific and Industrial Res., Auckland. Named by W.A. Fletcher in 1968. Parentage unknown; chance seedling; collected in 1962. For use as a pollinizer, particularly for Hayward, with which it is coincident under New Zealand conditions. Vine: large, very vigorous; highly floriferous; extended flowering season. "Matua" means "father" in Maori.

McLean.—Male. Orig. in Victoria, Australia. Parentage unknown; chance seedling. Used as a pollinizer in Victoria, but earlier than most female cultivars.

Montcap®.—Female. Orig. in Montauban, France, by Patrice Blanchet, Lycée agricole de Montauban-Capou. Second-generation Hayward o.p. seedling. Released in 1995. Fruit: medium-large, elongated oblong (shape intermediate between Bruno and Hayward); fruit matures about 1 month ahead of Hayward; limited storage life. Vine: medium vigorous; flowers at same time as Hayward. Commercial potential only at beginning of season.

Monty.—Female. Orig. in Te Puke, Bay of Plenty, New Zealand, by William A. Fletcher (deceased), New Zealand Ministry of Agr. and Fisheries, Auckland, and Harold M. Mouat (deceased), New Zealand Dept. of Scientific and Industrial Res., Auckland. Parentage unknown; chance seedling; raised by Bruno H. Just (deceased) of Palmerston North, New Zealand; collected in early 1960s. U.S. PI 255681, PI 255682. Fruit: mediocre; medium size; oblong, but often wider at stylar end, which is often depressed; skin brown; hair dense; flesh light green; good flavor; harvested about same time as Hayward; medium storage life. Vine: very vigorous; very productive with many lateral flowers and tendency to overcrop; flowers late. Now only of historical interest. Named for C. Montgomery, original owner of vine.

Moonya.—Male. Orig. in Tamborine Mountain, Queensland, Australia. Parentage unknown; chance seedling. Prolific flowerer, coinciding with earlier flowering female cultivars.

P1 Vitroplant® *(M3)*.—Male. Orig. in Cesena, Italy, by Gilberto Zuccherelli (deceased). Parentage unknown; chance seedling; distributed in 1985. Plant: medium vigorous; very floriferous; extended flowering period; good production of pollen. Selected as pollinizer for Hayward, especially in Verona district.

Saanichton 12.—Female. Orig. by Henry Bailey and supplied to Saanichton Agriculture Canada, Res. Sta., Sidney, British Columbia. Chance seedling, origin unspecified. Fruit: medium large; oblong-round, very square shoulders; often pronounced vertical ribs; inverted stylar end; core markedly fluted.

Skelton.—Female. Orig. in Waiuku, near Auckland, New Zealand, by Donald A. Skelton. Arose from controlled cross of two kiwifruit selections; raised in 1980; introd. in 1992. New Zealand Plant Variety Rights 638 (1991); U.S. plant patent 8344 (1993). Fruit: large; very long, narrowly obovoid; distinctly sloping inwards at fruit stalk end; stylar periphery rounded; skin light- to mid-brown; hair similar to Hayward; flesh light green; soft texture; sweeter, less acid than Hayward; harvested earlier than Hayward; medium storage life. Vine: medium vigor; productive; precocious; flowers early. Named for raiser.

Tewi.—Female. Orig. in Tenerife, Canary Islands. Chance seedling from Hayward o.p. Fruit: oblong, slightly fan-shaped; hairs stiff. Vine: very vigorous; precocious; productive; flowers early; requires little winter chilling. "Tewi" from Tenerife kiwi.

Tomua.—Female. Orig. in Te Puke, Bay of Plenty, by Kiwifruit Breeding Group, HortResearch. Seedling from a cross between Hayward and an early flowering male; selected in 1990 and released in 1995. Fruit: large; broadly oval, slightly flattened laterally, elliptical in cross section; tapering at stylar end; similar to Hayward but generally flatter and longer; hairs fairly fine, brittle, easily rubbed off; skin reddish-brown; flesh deep green; core elliptical; good flavor; harvested at least 1 month before Hayward; medium storage life. Vine: moderate vigor and productivity; budbreak about 2 weeks before Hayward; flowers 1 week before Hayward. Very similar in most attributes to Hayward, except for earlier flowering and fruit maturity. "Tomua" means "early" in Maori.

Tomuri.—Male. Orig. in Te Puke, Bay of Plenty, New Zealand, by William A. Fletcher (deceased), New Zealand Ministry of Agr. and Fisheries, Auckland, and Harold M. Mouat (deceased), New Zealand Dept. of Scientific and Industrial Res., Auckland. Named by W.A. Fletcher in 1968. Parentage unknown; chance seedling; collected in 1962. For use as a pollinizer, particularly for Hayward, as it flowers later than most other pollinizers. Vine: large, vigorous; moderately floriferous, with tendency for flowers to be in larger cymes; flowers late. "Tomuri" means "late" in Maori. Widely used in Italy.

Top Star®.—Female. Orig. in Azienda Vantini, Verona, Italy. Obtained in 1985 from a bud mutation on Hayward. Patented by Con.E.Co.Vi. s.c.r.l., Lugo, Ravenna, Italy. Distributed in 1990. Fruit: large; broadly oval, slightly flattened laterally, elliptical in cross section; essentially glabrous; skin brown with numerous and obvious suberose lenticels; flesh light green; core elliptical; good flavor; harvested about same time as Hayward; good storage life, similar to Hayward. Vine: compared with Hayward, less vigorous, shorter internodes, winter buds more exposed; leaves curled; no hairs on buds or canes; moderate productivity; flowers late.

Vincent.—Female. Orig. in Yorba Linda, Orange County, Calif., by Raymond Vincent. Chance seedling from Hayward; origin unspecified; raised about 1968. Fruit: small, oval; skin brown; flesh yellowish-green; matures very early. Vine: requires little winter chilling; productive; flowers early.

Wilkins Super.—Female. Orig. in Pongakawa, Bay of Plenty, New Zealand, by Marcus J. Wilkins and Jan Wilkins. Sport of Hayward; collected in 1979. New Zealand Plant Variety Rights 288 (1986); U.S. patent 4,686,318 (1987); U.S. plant patent 6815 (1989). Fruit: large; elongated cylindrical, much longer than Hayward; prominent shoulders; other characteristics, such as hair density, color, and elliptical shape in cross section, similar to Hayward; hairs longer than Hayward; flesh medium green; core elliptical; flavor and long storage life similar to Hayward. Vine: habit similar to Hayward.

Rootstocks for Kiwifruit

D1 *(D1 Vitroplant®)*.—Male. Orig. in orchard of Forlivesi Giacomo, Italy. Chance seedling from Bruno o.p.; raised in 1975; distributed in 1983. Particularly suitable for heavy or waterlogged soils. Promoted by Gilberto Zuccherelli (deceased) for its good tolerance of calcareous soils.

***D1 Vitroplant®*.**—See D1.

Kaimai *(TR 2)*.—Male. Orig. in Te Puke, Bay of Plenty, New Zealand, by Kiwifruit Breeding Group, Dept. of Scientific and Industrial Res. Fruit and Trees (now HortResearch). Chance seedling; origin unknown; male plant of (probably) *A. hemsleyana*; released in 1992. New Zealand Plant Variety Rights 727 (1992). Chosen as a rootstock for Hayward under New Zealand conditions, where it reduces vegetative growth and encourages precocity and heavy flowering.

***TR 2*.**—See Kaimai.

KUMQUAT

Centennial.—Orig. as a spontaneous sport on a twig of US1515 (a seedling of Nagami), at A.H. Whitmore Foundation Farm, Leesburg, Fla. Selected by H.C. Barrett, U.S. Dept. of Agriculture, Orlando. Fruit: ovate; skin variegated grey-yellow / green-yellow when immature, yellow-orange with darker orange regions when mature; moderately acid; juicy. Fruit produced year-round. Tree: small, shrubby; upright; dense; thornless; short internodes; leaf variegation variable, tricolored pale yellow / grey-green / dark green.

LEMON

Allen Eureka.—Orig. in Pasadena, Calif., by a Mr. Allen. Introd. about 1925. Parentage unknown; selected by Fruit Growers Supply Co., Los Angeles, Calif.; tested as 6408. Fruit: typical Eureka type. Tree: very vigorous; high production.

Armstrong Seedless.—Orig. in Riverside, Calif., by Sanford Johnson. Introd. in 1939. Plant patent 342; 10 Oct. 1939. Considered to be a bud mutation of Eureka; year of discovery unknown. Fruit: similar to Eureka, but entirely seedless. Tree: flower does not produce pollen; more vigorous and more hardy than Eureka, but less fruitful.

Avon.—Orig. in Arcardia, Fla., by J.H. Jones, Avon Park, Fla. Introd. in 1940 by Ward's Nursery, Avon Park. Parentage unknown; discovered about 1934. Fruit: larger and more nearly round than Eureka; well-defined, broad nipple demarcated by a depressed ring; juice suitable as a frozen product; satisfactory peel oil. Tree: productive; vigorous.

Cavers Lisbon.—Orig. in Upland, Calif., by W.B. Cavers. Introd. in 1920. Probably a bud mutation of Lisbon on a tree planted about 1892. Fruit: resembles Lisbon. Tree: vigorous; productive; free of shellbark.

Cuban *(Cuban shaddock)*.—Orig. in Cuba. Imported into the U.S. in 1923 and in 1930 as P.I. 84565 by the U.S. Dept. of Agriculture; more closely related to lemon or citron than to shaddock. Fruit: large to very large, 3 7/8 to 6 7/8 inches in diam. and length; rind rough, lemon-yellow to light orange-yellow, not nippled, to 1 1/4 inches thick, tender, sweet, oil glands medium in size and number, oil content low; inner rind thick, white, spongy; pulp light greenish-yellow, sour; seeds numerous, small. Tree: large; upright; vigorous; productive; foliage dense.

Frost Eureka.—Orig. in Riverside, Calif., by Howard B. Frost, Univ. of California Citrus Expt. Sta. Introd. in 1948. A nucellar seedling of Eureka; selected in 1918. Fruit: resembles Eureka. Tree: has added vigor and disease resistance in comparison with other clones of Eureka.

Frost Lisbon.—Orig. in Riverside, Calif., by Howard B. Frost, California Citrus Expt. Sta. Introd. in 1952. A nucellar seedling of Lisbon, from seed planted in 1917. Typical Lisbon, but tree has greater vigor than its seed parent line.

Harvey.—Orig. in Clearwater, Fla., by Harvey Smith on the home grounds of George James. Introd. in 1943 by the Glen Saint Mary Nurseries Co., Winter Haven, Fla. Parentage unknown; discovered in 1939; this nursery purchased and propagated trees from the original stock. Fruit: main crop ripens August to October; resembles Eureka. Tree: hardy; upright; thornless.

Perrine.—Orig. in Eustis, Fla., by Walter T. Swingle, U.S. Dept. of Agriculture. Introd. in 1931. Genoa lemon x Mexican lime; selected in 1909. Fruit: size and shape of ordinary lemon; rind pale yellow, thin, tough, somewhat corrugated or wrinkled; 10–12 segments; pulp translucent, greenish-yellow, lime type, high acid with flavor of lemon; usually four to six up to 12 seeds, slender, long, pointed. Tree: evergreen, vigorous, productive, resembling lime; leaf unifoliate, 3–5 inches long, oval, pointed, petiole wingless; hardier than lime; resistant to scab and withertip, but highly susceptible to gummosis caused by *Diplodia natalensis*. Variety gone out of commercial use.

Variegated Pink *(Pink-Fleshed, Variegated)*.—Orig. in San Fernando, Calif., by a Mr. Field of Brocton, Mass., and first propagated in 1932 by A.D. Shamel, U.S. Dept. of Agriculture, Riverside, Calif.; listed as C.E.S. 2367. Introd. in 1937. Bud mutation of Eureka; discovered before 1932. Fruit: when immature has abundant, irregular, thin, longitudinal, white streaks on normal green rind, sometimes with wider sectors; yellow and slightly ridged when mature; both flesh and juice pink. Tree: leaf with incomplete and irregular white margins; variegated pink, white, and green; flowers pink. A novelty for home gardens.

ROOTSTOCKS FOR LEMON

Estes.—Orig. in southern Polk County, Fla., as reported by Lawrence Estes; propagated, evaluated, and released by Harry W. Ford, Florida Citrus Expt. Sta., Lake Alfred, and William A. Feder, U.S. Dept. of Agriculture, Orlando, Fla. Introd. in 1964. Similar to Rough lemon; tested as Rough lemon-B. Used only as a clonal rootstock; tolerant to the burrowing nematode [*Radopholus similis* (Cobb) Thorne]; tolerant to tristeza, exocortis, and xyloporosis virus diseases.

Milam.—Orig. in Polk County, Fla., in an orchard owned by Mrs. Muriel Milam; propagated, evaluated, and released by Harry W. Ford, Florida Citrus Expt. Sta., Lake Alfred, and William A. Feder, U.S. Dept. of Agriculture, Orlando, Fla. Introd. in 1964. A citrus hybrid of unknown parentage, but considered to be a lemon for practical trade purposes; discovered in 1954 as a rootstock of a Parson Brown sweet orange tree, 15 years of age; tested as Clone X-hybrid; tree subsequently moved to Lake Alfred. Used only as a clonal rootstock: resistant to the burrowing nematode [*Radopholus similis* (Cobb) Thorne]; not resistant to the citrus nematode (*Tylenchulus semipenetrans* Cobb); appears to be tolerant to tristeza virus, anthracnose, xyloporosis, and exocortis; propagated by seeds and cuttings.

Lime

Ballard.—Orig. in Weslaco, Texas, by E.B. Ballard and C.A. Ballard, Ballard's Nurseries. Introd. in 1936. Bud mutation of a Mexican thornless lime, discovered in 1931. Fruit: characteristics of a lime when green and a lemon when ripe; acidity higher than ordinary limes or lemons; seedless; harvests easily. Tree: produces well; thornless; everbearing.

Giant Key.—Orig. as a spontaneous autotetraploid seedling of Key lime. Selected by H.C. Barrett, U.S. Dept. of Agriculture, Orlando, in 1973. Fruit: size large, 47 mm, twice as large as Key lime; 7.3% acid; self-fertile. Tree: shrub-like; semi-upright; continuous flowering and multiple fruit crops; not cold-tolerant.

Idemor.—Orig. in Homestead, Fla., by George L. Polk. Introd. in 1941. Plant patent 444; 28 Jan. 1941. Bud mutation of Tahiti (Persian) lime (*Citrus aurantifolia*); discovered before 1934. Fruit: smaller than Tahiti, averaging 1 3/4 inches in diam.; round, uniform, symmetrical; stem short, stout; rind texture grained, adherent to pulp, green; flesh very light green, tender, vesicles long and narrow, abundant juice, very good quality; seedless; ripens early, 10 June to last of September. Tree: regular and productive bearer; vigorous grower; resistant to disease.

Newell *(Newell's Thornless Key)*.—Orig. in Orlando, Fla., on the John Buchanan place. Introd. in 1945. Parentage unknown; discovered about 1940. Fruit: seedless; smooth-skinned; flavor like Mexican lime; resembles Mexican or "Key" lime. Tree: more frost resistant than Mexican lime; thornless; everbearing; pollen sterile.

Thornless Mexican *(Mexican Thornless)*.—Orig. in Weslaco, Texas, by E.B. Ballard and C.A. Ballard, Ballard's Nurseries. Introd. in 1930. Bud mutation of a Mexican thorny lime; discovered in 1925. Fruit: medium to small; skin thin; seedless. Tree: thornless and everbearing.

USDA 1.—Orig. in Lake Placid, Fla., by J.F.L. Childs, U.S. Hort. Sta., Orlando, Fla. Introd. in 1963. Nucellar seedling of Persian lime, *Citrus aurantifolia* (Christm.) Swing.; selected in 1960. Fruit and tree: resembles commercial Persian lime; shows leaf blotch disease, otherwise considered identical with USDA 2.

USDA 2.—Orig. in Lake Placid, Fla., by J.F.L. Childs, U.S. Hort. Sta., Orlando, Fla. Introd. in 1963. Nucellar seedling of Persian lime, *Citrus aurantifolia* (Christm.) Swing.; selected in 1960. Fruit and tree: same as USDA 1, but selected free of leaf blotch (wood pocket) disease.

Limequat

Eustis.—Orig. in Eustis, Fla., by Walter T. Swingle, U.S. Dept. of Agriculture. Introd. in 1923. Mexican lime (*Citrus aurantifolia*) x round kumquat (*Fortunella japonica*); selected in 1910; sibling of Lakeland. Fruit: size medium; nearly oval; rind light yellow, very sweet, smooth, thin; flesh juicy, fine quality; main crop borne in late fall and winter. Tree: resistant to withertip; hardy; small thorns abundant.

Lakeland.—Orig. in Eustis, Fla., by Walter T. Swingle, U.S. Dept. of Agriculture. Introd. in 1923. Mexican lime (*Citrus aurantifolia*) x round kumquat (*Fortunella japonica*); selected in 1910; sibling of Eustis. Fruit: larger than Eustis; shape similar to Eustis; rind deep yellow, smooth, thin; flesh juicy, fine quality; seeds few. Tree: resistant to withertip; hardy; nearly thornless.

Loquat

MacBeth.—Orig. in Carlsbad, Calif., by A.R. MacBeth. Plant patent 2591; 1 Feb. 1966; assigned to Select Nurseries, Whittier, Calif. A chance seedling; discovered in 1948. Fruit: borne in clusters of three to eight, on husky stems; very large, 1 1/2 inch x 2 inches; ovoid; skin saffron yellow, smooth, tough; flesh cream with yellow streaks, has texture of peach, flavor pleasing and mild, juicy, low acidity; seeds small, two to four per fruit, dark brown. Tree: spreading, crown dome-shaped, maximum overall diam. 20 ft; leaves large, elliptic in form, serrate, acute apex with a substantially attenuate base, Y-shaped ventation characteristic on fifth vein from tip on left-hand side.

Oliver.—Orig. at the U.S. Dept. of Agriculture Plant Introduction Garden, Miami, Fla., about 1910. From cross between Olivier and Tanaka made by George W. Oliver. Fruit: up to 5 cm long and 3.5 cm in diam.; ovoid to slightly pyriform; skin orange, firm; flesh orange, thick, juicy, firm; flavor excellent; four to six seeds; ripens February through April in Florida. Tree: up to 9 m tall; dense, rounded canopy.

Wolfe.—Orig. at Univ. of Florida Subtropical Expt. Sta., Homestead. Described in 1965 by Carl W. Campbell. Open-pollinated seedling of Advance; planted in 1936 by H.S. Wolfe of Homestead station; tested as Fla. SES 4. Fruit: up to 5.5 cm long and 3.2 cm in diam.; obovoid to slightly pyriform; skin yellow, resistant to bruising; flesh pale yellow, thick, juicy, firm; flavor excellent; one to three seeds; ripens March through April in Florida, several days later than Oliver; suitable for all purposes. Tree: up to 9 m tall; dense, upright canopy.

Lychee

Bengal.—Orig. in Calcutta, India. Introd. in 1929 into the U.S. by the Plant Introduction Garden, U.S. Dept. of Agriculture, Coconut Grove, Fla., as P.I. 94066; propagated and distributed in 1948. Parentage unknown; first fruited in the U.S. in 1940. Fruit: produced in large clusters of 20 or more fruit; large, 1 inch in diam. and up to 1 3/4 inches long, 22 g in weight; round; deep pink when ripe, rose-scented; matures in May and June in Florida; firm flesh, TSS 17%, very rough peel; flavor excellent. Tree: spreading; vigorous, up to 40 ft high; foliage dark green; apparently well adapted to alkaline soil.

Bosworth 3.—Named in Australia, probably a variant of Kwai mi. Other names: Kwai may pink, Kwai mi pink (Australia). Season: mid-June to early July (Northern Hemisphere). Fruit: almost round or spherical, 15 to 28 g, yellow-pink to orange-red at harvest, thick peel with sharp-pointed protuberance, suture not obvious. Flesh: firm, sweet, edible portion 65% to 82%. Tree: moderately vigorous and distinctively upright. A consistent bearer in a wide range of climates.

Brewster.—Collected by W.N. Brewster from Hengwha in the Fukien province, China, between 1903 and 1906. Introd. into Hawaii in 1907 as PI 21204. Brewster was suggested to be the same as the Chinese variety Chen-tze or Chen Family Purple. A midseason variety available in June-July. Fruit: large, average 45 fruit/kg, conical or wedge-shaped, bright purplish-red, protuberance short and sharp-pointed, endocarp distinctly pink in color. Flesh: soft, slightly fragrant, juicy, sweet when full-ripe; edible portion 74%. Seeds: 30% to 50% shriveled. Tree: large, vigorous, upright with sharp crotches. Leaves: dark green, with six leaflets. Requires a distinctive cold period for flower initiation.

Hak-ip.—China origin. Widely distributed. Other names: Haak-yip, Hei-ye, Woo-yip (China), Baidum, Oh Hia (Thailand). Season: early to mid-June in Northern Hemisphere. Fruit: oval or uneven cordate, weight 16 to 32 g; 3.1 x 3.2 cm, thin smooth skin dull-red in color. Flesh: milky white, tender and very juicy; edible portion 63.5% to 73.3% of the fruit, 16.5% to 20% soluble solids; vitamin C 22.31 to 45.7 mg/100 ml juice, 0.37% to 0.39% acid. Large to medium seed. Tree moderately vigorous, long, thin branches susceptible to wind damage, no reported on resistance to insect pests. High yield, good quality as fresh fruit, commonly used for canning in China and Taiwan.

Groff.—Orig. in Poamoho Branch Sta., Hawaii Agr. Expt. Sta., Honolulu. Introd. in 1952. Probably an open-pollinated seedling of Hak-ip; selected by members of the Horticulture Dept. of this station; tested as Hawaii A.E.S. 18-3. Fruit: small, 76–84 count/kg; 97% to 100% of fruit have small, shriveled seeds; higher proportion of flesh to seed than other clones growing in Hawaii; matures between August and September. Tree: more reliable in bearing than other clones tested.

Kaimana *(Poamoho)*.—Released in 1982 by R.A. Hamilton, P.J. Ito, and H. Rapoza, Univ. of Hawaii. Selected from a population of Hak-ip seedlings planted at the Poamoho Expt. Sta., Univ. of Hawaii, in 1955. "Kaimana" means "diamond" in the Hawaiian language and was named to commemorate the 75th anniversary of the Hawaii Agricultural Expt. Sta. Tree: low spreading, wide crotches, light-gray bark. Leaves: large, shiny, and dark green, with six leaflets. Fruit: season from June to July, 3 to 4 cm in diam., 20 to 26 g, deep maroon in color, smooth leathery skin with distinctive rows of spines across the blossom end. Soluble solids 20% to 24% and pH 4.5. Edible portion is 71% to 75% of the fruit weight, 40% of the fruit have shriveled seeds.

Kwai mi.—China origin. One of the best lychee cultivars. Different forms and variants present. Other names: Gui Wei, Kwai mei, Kwai May (China). Season: late June to early July (Northern Hemisphere). Fruit: round, 15–22 g, thin skin, light red color, very spiny or sharp protuberances. Flesh: milky white, firm, sweet, very juicy, with cinnamon flavor or osmanthus aroma, edible portion 75% to 81%, TSS 18% to 21%. Most seeds are small and aborted. Tree: upright, with scattered branches, light green elongated-elliptic leaflets with inward folded margins. Strong tendency for biennial bearing.

No mai t'sz.—China origin. One of the best lychee cultivars. Many variants and clones are present. Other names: No mai chee, Nuo mi ci (China). Season: late June to early July (Northern Hemisphere). Fruit: narrow-bodied cordate, rounded blossom end, 20 to 27 g, bright red or with yellow blotches, smooth protuberances. Flesh: milky white, soft, very juicy, very sweet, slightly fragrant; edible portion 78% to 84%, TSS 18% to 21%; vitamin C 20.4–36 mg/100 ml juice, acid 0.18% to 0.26%. Almost all seeds are small and shriveled. Tree: vigorous, compact. Strong tendency for biennial bearing, requires distinct period of low temperatures for flower initiation.

Peerless *(Ware Seedless)*.—Orig. in Oneco, Fla., by C.E. Ware, Clearwater, Fla. Introd. in 1956. Probably an open-pollinated seedling of Brewster; selected in 1938. Fruit: large, averaging 16–18 per lb; seeds are aborted, small, and shriveled. Tree: bears regularly.

Poamoho.—See Kaimana.

Sha keng.—Origin from Taiwan. Fruit: dull-red in color, smooth skin, average 24.9 g, edible portion 73.1%, 17.7% soluble solids, 42.1% aborted seed. Flavor acceptable, good yield.

Tai so.—China origin. Widely distributed. Other names: Kwai mi (misidentified), Da zao, Qiu zao, Wu yue li (China), Mauritius (S. Africa), Charley Tong, Wailua (Hawaii), Hong huey and Maw Mong (Thailand). Different forms and variants are present. Season: early to mid-June (Northern Hemisphere). Fruit: elongated oval or elliptical, 19–27 g; skin, thin; bright but turns dull-red when fully mature; sharp-pointed protuberance, suture not obvious. Flesh: milky white, rather coarse and chewy; sweet acidic flavor, moderately juicy; edible portion 65% to 75%; 15.3% to 18% soluble solids; vitamin C 11.7–41.8 mg/100 ml juice; acid 0.27%. Most seeds are large. Tree: very vigorous with open crown, crotch angle very narrow weak. High yield and relatively consistent.

Wai chee.—China origin. Other names: Huai zhi (China), Kim chen (Thailand). Season: late June to early July (Northern Hemisphere). Fruit: round or sub-global, 15–28 g, skin thick, dull-red in color, smooth protuberances with obvious suture. Flesh: waxy white, soft, very juicy; 17% to 21% soluble solids, vitamin C 17.3–42.2 mg/100 ml juice; acid 0.15% to 0.36%. Edible portion 69% to 77%. Most seeds are large, about 35% shriveled seeds. Tree: lacks vigor when young, branches short with dense foliage; crown dome-shaped, compact. Regular bearing and long harvesting period, fruit stores well on tree, up to 14 days after maturity. This is also a suitable variety for canning and drying.

Ware Seedless.—See Peerless.

MACADAMIA

Beaumont.—Orig. in Highfields, New South Wales, Australia, on the Goswell property by R.G. Kebby, N.S.W. Dept. of Agriculture. Named and introd. in 1965 by the California Macadamia Society, Carlsbad, Calif. *Macadamia tetraphylla* x *M. integrifolia*; discovered in 1954; tested in Australia as NSW 44; imported into Hawaii in 1954 and tested as HAES 695; brought to California in 1959 by the Dept. of Horticultural Science, Univ. of California, Riverside. Fruit: husk (pericarp) round, dull green, thin, slightly rough, apical point sharp; tends to split on tree at maturity before fruit drops. Nut: medium to large, range 65–80 per lb, 7/8 to 1 1/2 inches in diam.; dark brown, round, shell; thin on side, 1/32 to 1/16 inch, thicker toward base, kernel 40% of nut, round, slightly flattened, white, texture and flavor very good; oil content 70% to 75% at 3% kernel moisture; high percentage (85%) of grade A kernels (no culls); fruiting season January to March in California. Tree: also introduced for ornamental purposes, handsome; upright, branches spreading somewhat; new leaves with reddish-bronze blades and bright red midribs and veins; mature leaves green; flowers numerous, 300–500, bright pink, racemes up to 14 inches long. Named to commemorate the sojourn to Australia in 1953 of Dr. and Mrs. J.H. Beaumont, horticulturist, Hawaii Agr. Expt. Sta., who selected and imported the clone.

Burdick.—Orig. in Encinitas, Calif., by Ernest S. Burdick and Rodney B. Taylor. Introd. in 1958. Plant patent 1868; 22 Sept. 1959; by Burdick, assignor of one-half to Taylor. Seedling of unknown origin; *Macadamia tetraphylla*; planted in 1949; selected in 1956; tested at the Citrus Expt. Sta., Riverside, Calif., as CES 288. Nut: large, averaging 40 per lb; shell thin, averaging 1/16 inch thick; kernel averages 4 g; shell well-filled; cracks out about 34% of total nut weight; October harvest season. Tree: bears annually.

Dennison.—*Macadamia integrifolia*. HAES 790. Univ. of Hawaii, CTAHR, R.A. Hamilton and P.J. Ito, 1989. A Keauhou seedling planted at Waimanalo Expt. Sta. in 1953, selected in 1963. Adapted to low elevations, better yield, kernel quality, and percent recovery of grade 1 kernels than Keauhou at low elevations. Tree: medium, upright, strong, wide crotches and rounded canopy. Nut: 143/kg (65/lb), 39% kernel, and 99% grade 1.

Elimbah.—Orig. in Caboolture, Queensland, Australia, on the R. Markwell property by A.A. Ross and J. McG. Wills, Dept. of Agriculture and Stock, Brisbane, Queensland. Introd. in 1961. Seedling of *Macadamia tetraphylla*; selected in 1948; tested as F-1 in Australia. Nut: fairly large; shell rough; quality good. Tree: vigorous; bears well; some plantings in California.

Faulkner.—Orig. in Santa Paula, Calif., on G.S. Faulkner property. Introd. about 1954. Chance seedling. Nut: shell averages 7/8 inch in diam.; kernel averages 33% of nut; quality good; ripens during November to June. Tree: tendency toward biennial bearing.

Ikaika.—Orig. in the Nutridge Orchard, Oahu, Hawaii, by R.A. Hamilton, W.B. Storey, and E.T. Fukunaga, Hawaii Agr. Expt. Sta., Honolulu. Introd. in 1951. Open-pollinated seedling of *Macadamia integrifolia*; selected in 1936; tested as Hawaii A.E.S. 333. Nut: size medium, averaging 9/10 inch in diam.; kernel averages 31% to 35%. Tree: produces heavily; vigorous; extremely hardy; very wind-resistant. Highly susceptible to Macadamia Quick Decline disease.

Jordan.—Orig. in Keauhou, Hawaii, by W.B. Storey, Hawaii Agr. Expt. Sta., Honolulu. Introd. in 1963. Open-pollinated seedling of unknown parentage; discovered in 1936; tested as HAES 462. Nut: large, 9.2 g, spherical; shell thin, 4.5 mm at base, 2.2 mm on side; kernel large, average 3.4 g = 37%. Harvesting date: California, February–April; Hawaii, August–November. Tree: 40 ft; conical; vigorous; slightly cold tolerant; good productivity; fair quality; not known to be susceptible to any disease. Nut considered too large for commercial operations, propagated and planted only for home use; obsolete. Named for brothers R.A. and E.W. Jordan, who introduced the seeds from Australia.

Kakea.—Orig. in Honolulu, Hawaii, by W.B. Storey, Hawaii Agr. Expt. Sta. Introd. in 1948. Open-pollinated seedling of *Macadamia integrifolia*; selected in 1936; tested as Hawaii A.E.S. 508. Nut: size medium, averaging about 60 per lb; desirable shape; 35% to 40% kernel. Tree: produces heavy crops; vigorous; hardy; good resistance to anthracnose. Suitable for areas with deep well-drained soil and wind protection.

Kau.—*Macadamia integrifolia*. HAES 344. Released in 1971 by R.A. Hamilton and M. Nakamura. Original seedling from the Hawaiian Macadamia Co. orchard on Mt. Tantalus, Honolulu, selected in 1936. Adapted better to windy and dry areas than other cultivars. Tree: upright, stong, and well-formed ascending branches. Nut: 132/–154/kg (60–70/lb), 36% to 40% kernel, 95% grade 1. Harvest: September to December in Hawaii. Resistance to anthracnose.

Keaau.—Orig. in Lawai Valley, Kalaheo, Kauai, Hawaii, on the Deschwanden orchard, operated by A. Glaisyer, Jr. Introd. in 1967. Unknown seedling; selected in 1948; 1953 testing began by the Univ. of Hawaii at Poamoho Expt. Farm, Oahu, and Makawao Expt. Farm, Maui; first grafted trees fruited in 1958; tested as HAES 660. Nut: size medium, averaging about 80 nuts/lb, about 0.9 inch in diam.; shell smooth, medium brown with variable amounts of lighter speckling; kernel 42% to 46% of nut, color light cream; harvest August–November. Tree: moderately vigorous, with well-formed ascending branch structure so that the tree does not take up as much space as more-spreading varieties; strong, medium-wide crotch angles; productive, kernel yield greater than Keauhou, Ikaika, and Kakea in replicated field testing over a 4 years; not as tolerant to anthracnose as Keauhou, but superior in tolerance to Ikaika or Kakea. Named after Keaau Orchard of the Royal Hawaiian Macadamia Nut Co., near Hilo, Hawaii, where the first comprehensive yield test was conducted.

Keauhou.—Orig. in the Keauhou section, Kona, Hawaii, by W.B. Storey, Hawaii Agr. Expt. Sta., Honolulu. Introd. in 1948. Open-pollinated seedling of *Macadamia integrifolia*; selected in 1936; tested as Hawaii A.E.S. 246. Nut: fairly large, averaging about 54 per lb, average diam. about 1 inch; 37% to 40% kernel; shell very slightly pebbled; harvest season comparatively short, major part of crop maturing within about 3 months. Tree: vigorous; yields well; extremely resistant to anthracnose.

Kohala.—Orig. in the Nutridge Orchard, Oahu, Hawaii, by W.B. Storey, Hawaii Agr. Expt. Sta., Honolulu. Introd. in 1948. Open-pollinated seedling of *Macadamia integrifolia*; selected in 1936; tested as Hawaii A.E.S. 386. Nut: size medium, averaging about 70 per lb; 38% to 42% kernel.

Makai.—*Macadamia integrifolia*. HAES 800. Selected in 1967 from 4000 seedlings of Keauhou at the Univ. of Hawaii, Waiakea Research Station, outside of Hilo, Hawaii. Better adaptation to lower elevations than most cultivars. Tree: broad, spreading, high anthracnose resistance. Consistent bearer of large, uniform nuts and high-quality kernels. Nut: medium large, 130/kg (59/lb), 40% kernel and 98% grade 1. Excellent kernel quality after cooking. Resistance to southern green stink bug (*Nezara viridula* L.) and tropical nut borer (*Hypothenemus obscurus* F.).

Mauka.—*Macadamia integrifolia*. HAES 741. Selected in 1957 from unknown seedling at the Glaisyer/Deschwanden orchard, Lawai Valley, Kauai. Performed well between 549 to 670 m (1800 to 2200 ft) elevation. Trees: vigorous, upright, oval crown. Nut: 143/kg (65/lb), 43% kernel with 96% grade 1. Better yeild than Keauhou or Ikaika at higher elevation.

Nuuanu.—Orig. in the Nutridge Orchard, Oahu, Hawaii, by W.B. Storey, Hawaii Agr. Expt. Sta., Honolulu. Introd. in 1948. Open-pollinated seedling of *Macadamia integrifolia*; selected in 1936; tested as Hawaii A.E.S. 336. Nut: size medium, averaging about 70 per lb; 38% to 43% kernel; shell smooth; inclined to crack when husked by machinery. Tree: vigorous;

somewhat susceptible to anthracnose. Named after the valley on Oahu where the first macadamia trees were planted when introduced from Australia in 1892.

Pahala.—Orig. in Honolulu, Hawaii, at Poamoho Expt. Farm by R.A. Hamilton, P.J. Ito, and C.G. Cavaletto, Univ. of Hawaii. Introd. in 1981. Open-pollinated seedling of HAES 462 planted in 1953. Tested as HAES 788; selected in 1963. Shell: slightly pebbled. Kernel: light chocolate-brown without spots or mottling, attractive, uniform in shape, size, and color; 41.2% kernel, quality excellent; harvest season extends from mid-October to March, with 40% maturing from January to March, requiring two or more shaking operations if shake-harvested. Tree: narrow and upright, requiring less space, up to 120% more trees per acre can be planted; moderately resistant to anthracnose. Named for the Pahala orchard of the Mauna Loa Macadamia Nut Corp.

Pahau.—Orig. in the Nutridge Orchard, Oahu, Hawaii, by W.B. Storey, Hawaii Agr. Expt. Sta., Honolulu. Introd. in 1948. Open-pollinated seedling of *Macadamia integrifolia*; selected in 1936; tested as Hawaii A.E.S. 425. Nut: size medium, averaging about 60 per lb; 37% to 40% kernel; shell smooth, fairly dark; inclined to crack when husked by machinery. Tree: fairly vigorous; somewhat susceptible to anthracnose. Named in memory of Robert K. Pahau, superintendent, Kona Branch Sta., 1930–47.

Purvis.—Orig. in the Nutridge Orchard of the Hawaiian Macadamia Co. on Mount Tantalus near Honolulu by R.A. Hamilton, P.J. Ito, and C.G. Cavaletto, Univ. of Hawaii. Unknown seedling. Selected in 1936; tested as HAES 294. Named in honor of the late William Purvis to commemorate the centennial of his introduction of macadamia nuts into Hawaii from Queensland in 1882. Nut: resembles Keauhou, except kernel quality and percent kernel are notably better and more consistent. Tree: resembles Keauhou, but has better climatic adaptation.

Stephenson.—Orig. in Gilston, Queensland, Australia, in the orchard of W. Hill, by A.A. Ross and J. McG. Wills, Dept. of Agriculture and Stock, Brisbane, Queensland. Introd. in 1961 in Australia. Seedling of *Macadamia tetraphylla*; selected in 1948; tested in Australia as H3; tested in Hawaii as HAES 732. Nut: high quality. Tree: large, open; vigorous; good producer; fairly widely distributed in California.

Wailua.—Orig. in the Nutridge Orchard, Oahu, Hawaii, by W.B. Storey, Hawaii Agr. Expt. Sta., Honolulu. Introd. in 1951. Open-pollinated seedling of *Macadamia integrifolia*; selected in 1934; tested as Hawaii A.E.S. 44 and HAES 475. Nut: size medium, averaging about 60–68 per lb; shell smooth; kernel percentage high, 37% to 42%; resembles Keauhou. Tree: bears heavily; comes into bearing at an early age; vigorous.

MANDARIN

Bower.—Orig. in Texas, by U.S. Dept. of Agriculture (USDA) and Texas Agr. Expt. Sta., Weslaco, Texas. Introd. in 1973. Clementine mandarin x Orlando tangelo; cross made in 1942 in Orlando, Fla.; tested since 1955 in Crystal City, Texas, and Edcouch, Texas; selected by H.K. Wutscher (USDA) and N.P. Maxwell (Texas Agr. Expt. Sta.). Fruit: medium-large, about 3 inches in diam.; base and apex slightly depressed; rind deep orange at maturity, thin, slightly bumpy, peels easily, severe creasing at the apex in Florida, but not in Texas.; flesh deep orange, juicy, high acidity, good blending qualities, flavor pleasant, 13–15 segments, 25–40 monoembryonic seeds. Tree: upright, spreading; vigorous; hardy; highest yields have been on Morton citrange rootstock; needs pollinator.

Encore.—Orig. in Riverside, Calif., by J.W. Cameron, R.K. Soost, and H.B. Frost, California Agr. Expt. Sta. Introd. in 1965. F_1 of King (tangor?) x Willowleaf (Mediterranean) mandarin; selected in 1957; evaluated at Riverside. Fruit: fairly large; oblate; rind yellow-orange, thin, nearly smooth, peels easily with little albedo; flesh deep orange, flavor rich and sprightly; seeds numerous and monoembryonic; unusually late-ripening and late-holding on tree; best in Riverside from June through August. Tree: somewhat alternate-bearing habit.

Fallglo.—Orig. from a cross of [*Citrus reticulata* x (*C. paradisi* x *C. reticulata*)] Bower x (*C. reticulata* hybrid) Temple. Cross made by P.C. Reese in 1962; introd. in 1987 by C.J. Hearn, U.S. Dept. of Agriculture, Orlando. Fruit: large, 76–82 mm; flattened; rind smooth, reddish-orange; resistant to plugging; easily peeled; high juice content (65.8%); 20–40 monoembryonic seeds. Matures early to midseason. Tree: vigorous; thornless; foliage light green; more upright than Temple; moderately cold hardy; resistant to citrus scab fungus (*Sphaceloma fawcetti*).

Fairchild.—Orig. in Indio, Calif., by J.R. Furr, U.S. Date and Citrus Sta., U.S. Dept. of Agriculture. Introd. in 1964. Clementine tangerine x Orlando tangelo; cross made in 1948; selected in 1959; tested as Indio C48-14-39. Fruit: tangerine type; size medium; oblate, typical of the species; peel smooth, peels freely, dark orange; flesh orange-colored, firm, yet tender, juicy, flavor rich, sweet; retains good quality through December; 20–30 small to medium polyembryonic seeds; best suited as fresh fruit; usually ripens in early November, with Clementine, but in some seasons may be harvested before Clementine because its color changes from green to orange earlier. Tree: hardy; heat tolerant; vigorous; productive; nearly thornless; adapted especially to Coachella and Imperial valleys of California, may not require pollinators, but such should be provided until proved otherwise. Named in memory of David Fairchild (1869–1954), plant explorer and writer.

Fortune.—Orig. in Indio, Calif., by J.R. Furr, U.S. Date and Citrus Sta., U.S. Dept. of Agriculture. Introd. in 1964. Clementine x Dancy; cross made in 1948; selected in 1959; tested as Indio C489-6. Fruit: tangerine type; medium to large, 2 1/4 to 3 inches in diam.; oblate, typical of the species; peel attractive, slightly bumpy, orange to reddish-orange, only moderately easy to remove from flesh; flesh sprightly, subacid, pleasing, firm, but tender, juicy, orange-colored; seed medium, monoembryonic, numerous (20–30) in mixed plantings; best suited as fresh fruit; quality satisfactory on trees on Rough lemon rootstock; for highest quality and late-holding on tree, it is recommended to use sour orange, Cleopatra mandarin, or Troyer citrange rootstock; harvested late, from late January through March. Tree: productive; vigorous; spreading, dense; hardy; heat tolerant; adapted especially to Coachella and Imperial valleys of California; pollinator probably necessary. Named for Robert Fortune (1812–1880), English plant explorer who introduced the kumquat into Europe and America.

Fremont.—Orig. from a cross Clementine x Ponkan made in 1954 in Orlando, Fla., by Philip C. Reece, U.S. Dept. of Agriculture; seed planted at Indio, Calif.; tree grown to fruiting at the Southwestern Irrigation Field Sta., Brawley, Calif., where Fremont was selected in 1959 by J.R. Furr, U.S. Date and Citrus Sta., Indio; tested as Indio C54-3-1. Introd. in 1964. Fruit: size medium, 2 1/2 inches in diam.; oblate; peel smooth, bright reddish-orange, easy to remove; flesh deep orange, tender; juicy, free from granulation even when grown on Rough lemon rootstock in light sandy soil; flavor excellent, sprightly; quality excellent, retaining good quality even when overripe in January or later; about 15 small to medium seeds per fruit, about half being monoembryonic; best suited as fresh fruit; ripens early, just before Dancy and after Clementine; sunburns easily because of tree's upright growth. Tree: hardy; moderately vigorous, precocious, productive; branches upright, nearly thornless; heat tolerant; adapted to Coachella and Imperial valleys of California; pollinator probably needed. Named in memory of John Charles Fremont (1813–90), explorer and scientific observer in western North America.

Frost Dancy.—Orig. in Riverside, Calif., by Howard B. Frost, California Citrus Expt. Sta. Introd. in 1952. Nucellar seedling of Dancy seed planted in 1916. A vigorous line of Dancy, initially free of several viruses.

Frost Satsuma.—Orig. in Riverside, Calif., by Howard B. Frost, California Citrus Expt. Sta. Introd. in 1956. Nucellar seedling of Satsuma; seed planted in 1916. Fruit: typical Satsuma, but has higher soluble solids than its seed parent budline. Tree: initially free of several viruses; greater vigor and better yields than its seed parent budline.

Frua.—Orig. in Riverside, Calif., by Howard B. Frost and James W. Cameron, California Citrus Expt. Sta. Introd. in 1950. King tangor x Dancy tangerine; selected in 1925. Fruit: large, 4–6 1/2 per lb, averaging 16% larger than Dancy; oblate, neck usually absent, more angled than Dancy; skin pebbled, deep orange to red-orange; rind about 2 to 3 1/2 mm thick, peels easily, oil glands numerous with pleasant aroma; core slender; puffs earlier and more rapidly than Dancy; segments separate easily; pulp yellow-orange, vesicles stout, usually tender, very juicy, aroma of juice pleasant; flavor and texture very good; seeds small, 9-13 per fruit; ripens early, 3–4 weeks before Dancy, with a short picking season of December and January; resembles Dancy. Tree: vigor rather low; nonproductive under hot desert conditions, e.g., Coachella Valley in southern California; best adapted to southern California navel orange districts. Frua is the Esperanto word for "early" taken from the German *früh.*

Honey.—Orig. in Riverside, Calif., by Howard B. Frost, California Citrus Expt. Sta. Described in 1943. Not introduced commercially; propagated on a small scale in Coachella Valley of California. King tangor x Willowleaf mandarin. Fruit: small; rind orange; pulp deep orange, flavor exceptionally sweet and rich; seedy; resembles Willowleaf; ripens early. Tree: vigorous; extremely alternate bearer. Value of variety is for breeding purposes.

Kara.—Orig. in Riverside, Calif., by Howard B. Frost, California Citrus Expt. Sta. Introd. in 1935. Owari satsuma (considered to be this variety) x King mandarin; selected in 1925. Fruit: large, up to 3 inches in diam., averaging three to five per lb; slightly to medium oblate, often somewhat globose, rather variable; skin deep orange, somewhat wrinkled and bumpy; rind thin, peels easily and cleanly, rather tough; oil glands many, odor pleasant; pulp deep, rich yellowish-orange, tender, very juicy with a unique aroma, sugar high, flavor rich, excellent; seeds 12–20, plump, medium to large; ripens from January or February to April or May. Tree: vigorous; moderately open, usually spreading.

Kimbrough.—Orig. in Port Sulphur, La., by R.T. Brown, P.L. Hawthorne, J.E. Love, D.W. Newsom, and T.J. Raiford, Louisiana Agr. Expt. Sta., Baton Rouge. Introd. in 1976. Open-pollinated seedling of Owari. Selected in 1962; tested as P-62-7-3. Fruit: medium to large; oblate to spherical; skin orange or orange-chrome; rind thinness equal to Owari; peel separates easily from flesh; flesh color and taste equal to Owari;

seedless; ripens early to mid-November or with Owari. Tree: larger and hardier than Owari and 40% more productive. Named for W.B. Kimbrough, professor emeritus, Dept. of Horticulture, Louisiana State Univ., in recognition of his work in citrus research.

Kincy.—Orig. in Riverside, Calif., by H.B. Frost, Univ. of California, Citrus Res. Center. Introd. in 1943. King tangor x Dancy mandarin; cross made in 1914; selected in 1925. Fruit: 7 cm in diam.; deep oblate, small neck; skin orange, medium to thin; flesh deep orange, sugar medium, acidity medium or more; ripens February to March at Riverside. Tree: 15 to 18 ft; globose; vigor moderate; not frost hardy. A mandarin hybrid of fair eating quality and for use as a zygotic parent.

Kinnow.—Orig. in Riverside, Calif., by Howard B. Frost, California Citrus Expt. Sta. Introd. in 1935. King x Willowleaf; selected in 1925. Fruit: large, up to 3 inches in diam., averaging three to five per lb; medium to slightly oblate, uniform; skin yellow-orange, larger oil glands very conspicuous; rind thin, smooth, peels poorly, rather tough; pulp segments 9–12, separating fairly easily, deep yellowish-orange, moderately tender, very juicy with a unique and very pleasant aroma, sugar very high, flavor very rich, excellent; 12–24 medium size, polyembryonic seeds; ripens December to February in California desert areas, February to May in California coastal areas. Tree: tends to bear alternately; vigorous; tall, erect.

Ledar.—Orig. in Gayndah, Bowen County, Queensland, Australia, by N.M. Darrow and R.J. Darrow. Introd. in 1965. Plant patent 2714; 14 Feb. 1967. Bud mutation of Ellendale Beauty (*Citrus sinesis* x *Citrus reticulata*). Fruit: resembles Ellendale Beauty; skin adheres to fruit better than parent variety; ripens later and fruit hangs on tree longer. Tree: more precocious than Ellendale Beauty; leaders develop in recumbent condition; can be pruned to form weeping or espaliered tree.

Lee.—Orig. in Orlando, Fla., by the U.S. Hort. Field Sta. Introd. in 1960. Clementine x Orlando tangelo; cross made in 1942. Fruit: size medium-large; round; rind color orange, loose, texture smooth, easily peeled; flavor sweet; seeds many; ripens early (October); resembles a tangelo. Tree: vigorous; branches rather upright, ascending.

Osceola.—Orig. in Orlando, Fla., by the U.S. Hort. Field Sta. Introd. in 1960. Clementine x Orlando tangelo; cross made in 1942. Fruit: size small; slightly flat to round; rind red-orange, thin, texture smooth, peels easily; pulp quality high, flavor rich; ripens early (November), before Dancy; seeds many. Tree: vigor moderate; rounded.

Pitman's Chinese Ponkan.—Orig. in Apopka, Fla., by Robert G. Pitman. Introd. in 1949. Plant patent 863; 9 Aug. 1949. Bud mutation of Ponkan. Fruit: larger than Dancy and without its characteristic neck; skin light orange; seeds few; flesh soft, melting, quality very good. Tree: more upright in growth than Dancy; vigor fair.

Pixie.—Orig. in Riverside, Calif., by J.W. Cameron, R.K. Soost, and H.B. Frost, California Agr. Expt. Sta. Introd. in 1965. Open-pollinated seedling of F_1 of Kincy (King x Dancy); selected about 1948; evaluated at Riverside. Fruit: small to medium; shape variable; rind yellow-orange, peels easily without dripping juice; flesh medium-orange, flavor mild and pleasant, essentially seedless, regardless of exposure to other pollen sources; matures late, April to July. Tree: somewhat alternate-bearing habit; not successfully grown in the hot desert.

Robinson.—Orig. in Orlando, Fla., by the U.S. Hort. Field Sta. Introd. in 1960. Clementine x Orlando tangelo; cross made in 1942. Fruit: size medium; shape slightly flat; rind loose, texture smooth, orange, peels easily; pulp quality high, rich, sweet; many monoembryonic seeds; ripens early (October), before Dancy. Tree: very vigorous; rounded.

Silverhill.—Orig. in Mary, Fla., by Walter T. Swingle, U.S. Dept. of Agriculture. Introd. in 1931. Seedling of Owari Satsuma x sweet orange pollen, but apparently not a hybrid; selected in 1909. Fruit: large; quality good. Tree: extra vigorous; hardy.

Sunburst.—Orig. from a 1961 cross of [*Citrus reticulata* x (*C. paradisi* x *C. reticulata*)] Robinson x [*C. reticulata* x (*C. paradisi* x *C.

reticulata)] Osceola. Introd. in 1979 by C.J. Hearn, U.S. Dept. of Agriculture, Orlando. Fruit: size medium, 64–76 mm; oblate, flattened at stem end; depressed navel; rind thin, dark orange, smooth, easily peeled; flesh dark orange; 10–20 polyembryonic seeds; midseason. Tree: moderately vigorous; upright; thornless; foliage dark green. Moderately cold tolerant; tolerant to snow scale; resistant to Alternaria and to scab; susceptible to rust mites.

Sunset.—Orig. from a cross of *Citrus reticulata* Imperial x (*C. reticulata* x *C. sinensis*) Ellendale. Cross made in 1963 by L. Stafford, Irymple, Victoria, Australia; tested as C034; introd. in 1991 by I. Thornton, M. Edwards, and S. Sykes. Fruit: oblate; slightly larger and higher juice content than Imperial; occasional navel; rind thin, bumpy, easily peeled; flesh orange; few seeds. Early to midseason. Tree: spreading, less upright than Imperial; thornless; medium lanceolate leaves.

Wilking.—Orig. in Riverside, Calif., by Howard B. Frost, California Citrus Expt. Sta. Introd. in 1935. King x Willowleaf; selected in 1925; sibling of Kinnow. Fruit: large, up to 3 1/8 inches in diam., averaging three to five per lb; medium oblate, fairly uniform; skin pebbled to nearly smooth, glossy, deep yellowish-orange; rind thin to medium, peels fairly well, oil glands many; flesh segments 9–13, separating easily, firm, rich yellow-orange, moderately tender, very juicy, aroma of juice unique, sugar content high, flavor rich and very good, being very sweet late in season; 10–17 monoembryonic seeds; some granulation under California desert conditions; ripens January to March in California desert, February to May in California coastal areas. Tree: vigorous; alternate bearer.

Mango

The following cultivar descriptions have been excerpted, with permission, from Mangos: A Guide to Mangos in Florida *[R.J. Campbell (ed.), 1992, Fairchild Tropical Garden, Miami, Fla., USA].*

Alice.—Seedling of Saigon from the grove of Fred Herman, South Miami, Fla. Planted in 1935; first fruited in 1944. Tree: vigorous; large and spreading canopy. Fruit: eating quality good. Shape: oval; base slightly flattened; slender stem inserted squarely in an elevated manner; apex rounded with no beak; surface smooth. Size: length 10.5–14 cm; breadth 8.5–10 cm; thickness 7–8 cm. Weight: 300–520 g. Skin: ground color yellow-orange; crimson blush; numerous large white dots; skin thick, tough, and easily separating. Flesh: soft and juicy orange flesh; flavor rich, aromatic and sweet with a strong, pleasant aroma; no fiber. Stone and seed: thick and woody stone with a monoembryonic seed filling 70% to 90% of the stone. Season: late June to July. Consistent bearer, but light crops; not extensively propagated in Florida; probably a Southeast Asian (Saigon) / Indian hybrid.

Anderson.—Seedling of Sandersha that was planted at the residence of Mrs. L.F. Anderson, Miami, Fla. Seed sent from Jamaica in 1926 and bore fruit in 1931. Named in 1948. Tree: vigorous; large and spreading, semi-open canopy. Fruit: eating quality fair. Shape: long and slender; base rounded; stout stem inserted squarely in an elevated manner; apex bluntly pointed, often with a small lateral beak; surface undulating. Size: length 26–34 cm; breadth 9.5–12.5 cm; thickness 8.5–12 cm. Weight: 1300–1700 g. Skin: ground color greenish-yellow; crimson or dark red blush; few large white dots; skin thick, tough, and easily separating. Flesh: soft and juicy pale yellow flesh; flavor mild and acidulous with a weak, pleasant aroma; no fiber. Stone and seed: thick and woody stone with a monoembryonic seed filling 40% to 60% of the stone. Season: July to August. Mainly of interest due to its large size; good cooking mango used for chutney, mango pickles, etc.; tendency to split on the tree.

Baptiste.—Introd. to Florida from Haiti by Lawrence H. Zill, Boynton Beach, Fla. Tree: low vigor; small, spreading, open canopy. Fruit: eating quality fair to good. Shape: oval; base rounded; slender stem inserted obliquely in a shallow, grooved cavity; apex rounded with a small beak; surface smooth. Size: length 9–11.5 cm; breadth 7.5–8.5 cm; thickness 7–7.5 cm. Weight: 250–450 g. Skin: ground color bright yellow; no blush; few small russet dots; skin thin, tender, and adherent. Flesh: very firm and dry deep yellow flesh; flavor mild and sweet with a weak, pleasant aroma; no fiber. Stone and seed: thick and woody stone with a polyembryonic seed filling 70% to 85% of the stone. Season: June to July. Extremely firm at maturity with no fiber in flesh; mild carrot flavor; one of the most popular cultivars in Haiti; sometimes used as a rootstock.

Becky.—Seedling of Sunset from J.D. Ensey, Merritt Island, Fla. Tree: vigorous; dense, rounded canopy, branching near the ground. Fruit: eating quality fair. Shape: ovate to oblong; base rounded; stout stem inserted obliquely in a deep, grooved cavity; apex bluntly pointed without a beak; surface undulating. Size: length 11.5–15.5 cm; breadth 8.5–10 cm; thickness 7.5–8.5 cm. Weight: 510–850 g. Skin: ground color greenish-yellow; pink or crimson blush; numerous small white dots; skin thick, tender, and easily separating. Flesh: soft, melting, and juicy lemon-yellow flesh; flavor somewhat acidulous and spicy with a weak, pleasant aroma; little fiber. Stone and seed: thick and woody stone with a monoembryonic seed filling 50% to 75% of the stone. Season: July to August. Combination of pink, crimson, and yellow make this an unusually attractive fruit; may be too delicate for commercial shipping and handling.

Big Yellow.—See Cushman.

Brooks.—Seedling of Sandersha planted in 1910 on the property of a Mr. Brooks, Miami, Fla. First fruited in 1916; propagated in 1924. Second mango cultivar to be named in Florida. Tree: low vigor; small, open canopy. Fruit: eating quality fair. Shape: oblong; base rounded; thick stem inserted squarely in a shallow, slightly grooved cavity; apex rounded to bluntly pointed with no beak; surface smooth. Size: length 12.5–15.5 cm; breadth 8.5–10 cm; thickness 8–9 cm. Weight: 480–830 g. Skin: ground color greenish-yellow; no blush; few small white dots; skin medium thick, medium tough, and adherent. Flesh: firm and juicy deep yellow flesh; flavor mild and sweet with a weak, pleasant aroma; medium fiber. Stone and seed: thin and papery stone with a monoembryonic seed filling 70% to 85% of the stone. Season: August to early October. Also called Brooks Late; replaced long ago by Keitt as the important commercial late-season cultivar in Florida; fruit are "starchy" until fully ripe; favored by Asians in Florida for pickling when green.

Brooks Late.—See Brooks.

Cambodiana.—Best of a group of seedlings grown from some Saigon seeds sent by David Fairchild from Indo-China to Miami, Fla., in 1902. Tree: moderately vigorous; dense, rounded canopy. Fruit: eating quality good. Shape: oblong to ovate; base rounded; slender stem inserted squarely in a level manner; apex bluntly pointed usually with no beak; surface smooth. Size: length 9–11.5 cm; breadth 6.5–7.5 cm; thickness 5–6 cm. Weight: 220–340 g. Skin: ground color greenish-yellow; no blush; few small white dots; skin thin, tender, and adherent. Flesh: soft, tender, melting, and juicy lemon-yellow flesh; flavor sweet and mildly subacid with a weak, pleasant aroma; little fiber. Stone and seed: thick and woody stone with a polyembryonic seed filling 75% to 95% of the stone. Season: June to early July. Limited by its small size, lack of color, and poor storage characteristics; frequently planted as a dooryard mango in Florida; may be the same as Rockdale Saigon.

Carabao.—Introd. to Florida from the Philippines in 1909 and 1930 by the U.S. Dept. of Agriculture, Miami, Fla. Tree: vigorous; large, dense canopy. Fruit: eating quality good to excellent. Shape: long and slender; base rounded to slightly flattened; slender stem inserted obliquely in a shallow cavity; apex bluntly pointed with no beak; surface smooth. Size: length 11–13 cm; breadth 6.5–7 cm; thickness 6–6.5 cm. Weight: 270–440 g. Skin: ground color greenish to bright yellow; no blush; few small green dots; skin thin, medium tough, and easily separating. Flesh: soft, tender, melting, and juicy lemon-yellow flesh; flavor mild, spicy, and sweet with a weak, pleasant aroma; no fiber. Stone and seed: thin and papery stone with a polyembryonic seed filling 50% to 60% of the stone. Season: June to July. Smoke has been used in the Philippines to induce flowering, but this practice has now been replaced by potassium nitrate sprays; flavor enjoyed by many, but a poor producer in Florida, and has poor storage characteristics.

Carrie.—Seedling of Sophie Fry from the residence of Lawrence H. Zill, Boynton Beach, Fla. First fruited in 1940; named in honor of Zill's mother. Propagated since about 1949. Tree: vigorous; dense, rounded, compact canopy. Fruit: eating quality excellent. Shape: oblong; base rounded; stout stem inserted obliquely in a slightly raised or level manner; apex rounded with no beak; surface slightly undulating. Size: length 10–13 cm; breadth 6–7 cm; thickness 5.5–6 cm. Weight: 280–440 g. Skin: ground color greenish-yellow; no blush; numerous small yellow dots; skin thin, tender, and easily separating. Flesh: extremely soft, tender, melting, and juicy orange flesh; flavor rich, sweet, and spicy with a strong musky aroma; no fiber. Stone and seed: thick and woody stone with a monoembryonic seed filling 60% to 90% of the stone. Season: June and July. Recommended as a dooryard mango in Florida; regular producer in Florida without fungicide application; considered one of the most exotic-flavored mangos in Florida; lack of color and firmness has prevented commercial use.

Chino.—Introd. to Florida from Cuba in 1931 by the U.S. Dept. of Agriculture, Miami, Fla. This seedling race probably originated from Number 11 in the vicinity of

Cienfuegos, Cuba. Tree: vigorous; large, spreading, open canopy. Fruit: eating quality poor to fair. Shape: oval; base flattened; slender stem inserted squarely in a deep cavity; apex rounded, sometimes with a small lateral beak; surface smooth. Size: length 9.5–11 cm; breadth 7.5–8.5 cm; thickness 6–6.5 cm. Weight: 360–440 g. Skin: ground color greenish-yellow; pink blush; numerous small white dots; skin thin, tender, and easily separating. Flesh: soft, tender, and juicy orange flesh; flavor rich, aromatic, and sweet with a weak, pleasant aroma; abundant fiber. Stone and seed: thick and woody stone with a polyembryonic seed filling 70% to 90% of the stone. Season: late May to June. Taste is quite pleasant, but there is an abundance of fiber; highly susceptible to anthracnose.

Cogshall.—Of unknown parentage from Pine Island, Fla. Selected in the 1940s for inclusion in the Univ. of Florida germplasm collection at Homestead. Tree: moderately vigorous; dense, semi-dwarf canopy. Fruit: eating quality good to excellent. Shape: oblong; base rounded; slender stem inserted squarely in a shallow grooved cavity; apex bluntly pointed with no beak; surface undulating. Size: length 11–13 cm; breadth 7.5–8.5 cm; thickness 6–6.5 cm. Weight: 280–450 g. Skin: ground color greenish-yellow to yellow-orange; crimson blush; numerous medium-yellow dots; skin thin, tender, and somewhat adherent. Flesh: soft, melting, and juicy lemon-yellow flesh; flavor rich, aromatic, and spicy with a weak, pleasant aroma; little fiber. Stone and seed: thick and woody stone with a monoembryonic seed filling 60% to 85% of the stone. Season: June to July. Semi-dwarf habit due to short internodes; may be a good dooryard mango, but too soft for commercial use.

Cowasji Patel.—Introd. to Florida from the state of Bombay, India, in 1910 by the U.S. Dept. of Agriculture, Miami, Fla. Tree: moderately vigorous; large, dense canopy. Fruit: eating quality fair. Shape: oval; base rounded; slender stem inserted squarely in a shallow cavity; apex rounded with no beak; surface undulating. Size: length 10.5–13.5 cm; breadth 8.5–10 cm; thickness 8–9 cm. Weight: 500–830 g. Skin: ground color greenish-yellow; pink blush; numerous small russet dots; skin thick, tough, and adherent. Flesh: firm and dry deep yellow flesh; flavor acidulous and resinous with a weak, pleasant aroma; little fiber. Stone and seed: thick and woody stone with a monoembryonic seed filling 60% to 75% of the stone. Season: late June to July. Often used for chutney, mango pickles, etc.; has a strong resinous aroma before ripening; tends to ripen unevenly.

Cushman.—From a seed of unknown parentage planted in 1936 at the residence of E.L. Cushman, Miami, Fla. Tree: vigorous; large, spreading, medium-dense canopy. Fruit: eating quality excellent. Shape: round; base flattened; stout stem inserted squarely in a shallow cavity; apex rounded with a small lateral beak; surface smooth. Size: length 10.5–13 cm; breadth 10.5–13 cm; thickness 10–11 cm. Weight: 700–950 g. Skin: ground color bright yellow; rarely a pink blush; numerous large white and russet dots; skin thick, tough, and easily separating. Flesh: soft, melting, and juicy pale yellow flesh; flavor sweet, rich, and aromatic with a strong, pleasant aroma; no fiber. Stone and seed: thick and woody stone with a monoembryonic seed filling 90% to 100% of the stone. Season: July to August. Distinctive appearance due to its shape; long shelf life when kept at ambient temperature; seeds of late-season fruit sometimes sprout within the fruit; previously known in Florida as Big Yellow.

Dot.—Seedling of Carrie from the residence of Lawrence H. Zill, Boynton Beach, Fla. Paheri may have been the pollinating parent. Named after Dorothy Zill. Tree: moderately vigorous; medium, open canopy. Fruit: eating quality excellent. Shape: oval to ovate; base flattened to slightly rounded; stout stem inserted squarely in a level manner; apex bluntly pointed or rounded without a beak. Size: length 7.5–11.5 cm; breadth 6.5–8 cm; width 5.5–7 cm. Weight: 260–450 g. Skin: ground color bright yellow; sometimes a slight pink blush; numerous small russet dots; skin medium thick, tough, and easily separating. Flesh: medium-firm, juicy, deep yellow flesh; flavor rich, aromatic, and sweet with a strong, pleasant aroma; little fiber. Stone and seed: thick and woody stone with a monoembryonic seed filling 50% to 70% of the stone. Season: June to August. Flavor is exceptionally rich and aromatic; good dooryard mango because the fruit ripen over an extended length of time; tendency to produce many halfsized fruit, but these are of good eating quality.

Duncan.—Of unknown parentage, but Saigon probably one of its parents. Patented by David Sturrock, West Palm Beach, Fla. Tree: vigorous; medium-sized, open canopy. Fruit: eating quality good to excellent. Shape: oblong; base flattened or slightly rounded; slender stem inserted obliquely in a deep, grooved (variable) cavity; apex rounded without a beak; surface smooth. Size: length 12.5–14.5 cm; breadth 8.5–10 cm; thickness 7–8 cm. Weight: 450–670 g. Skin: ground color bright yellow; no blush; few large yellow dots; skin medium thick, tender, and easily separating. Flesh: soft, tender, and juicy orange flesh; flavor mild and sweet with a weak, pleasant aroma; no fiber. Stone and seed: thick and woody stone with a monoembryonic seed filling 50% to 70% of the stone. Season: July to August. Eating quality excellent, but fruit easily damaged by handling when ripe; seed often sprouts within the overmature fruit; resistant to anthracnose.

Edward.—Thought to be a Haden x Carabao cross made by Edward Simmonds of the Plant Introduction Garden, Miami, Fla., in the 1920s; however, recent isozyme studies have not supported this. Original tree grown at the Simmonds residence in South Miami; named and described in the late 1930s by Tom and David Sturrock. Tree: vigorous; large, compact, dense canopy. Fruit: eating quality excellent. Shape: oval to oblong; base rounded; stout stem inserted squarely in a shallow cavity; apex rounded without a beak; surface undulating. Size: length 10–15 cm; breadth 8–10 cm; thickness 7–8 cm. Weight: 450–620 g. Skin: ground color bright yellow; pink or reddish blush; numerous small white dots; medium thick, tender, and adherent. Flesh: tender, melting, and juicy deep yellow or orange flesh; flavor sweet, spicy, and rich with a weak, pleasant aroma; no fiber. Stone and seed: thin and papery stone with a monoembryonic seed filling 35% to 50% of the stone. Season: June and July. Considered by many as the finest tasting of the Florida mangos; consistent bearer, usually producing only a few fruit in a single year; often the seed is poorly developed, but the fruit develops to full size.

Fascell.—Probably a cross between Haden and Brooks made by Michael Fascell, Miami, Fla. Seeds planted in 1929; first fruited in 1936. Patented in 1941. Tree: moderately vigorous; medium to large, rather open canopy. Fruit: eating quality good. Shape: oval; base somewhat flattened; stout stem inserted squarely in a shallow cavity; apex rounded with no beak; surface smooth. Size: length 9.5–11.5 cm; breadth 6–8.5 cm; thickness 5–6 cm. Weight: 300–450 g. Skin: ground color bright yellow; pink to crimson blush; numerous small white dots; skin thin, tough, and adherent. Flesh: medium-firm, tender, and juicy deep yellow flesh; flavor mild and sweet with a weak, pleasant aroma; little fiber. Stone and seed: medium-thick and woody stone with a monoembryonic seed filling 60% to 80% of the stone. Season: late June to July. First patented mango cultivar in Florida; often difficult to determine maturity; fruit can be twisted apart into two halves when fully ripe because there is so little fiber.

Florigon.—Seedling of Saigon planted on the property of John C. Kaiser, Fort Lauderdale, Fla. Seed planted in 1932; first fruited in 1938. Tree: vigorous; large, upright, open canopy. Fruit: eating quality good to excellent. Shape: oval; base flattened; slender stem inserted squarely in a shallow, narrow cavity; apex bluntly pointed with a small lateral beak; surface smooth. Size: length 9.5–13 cm; breadth 7.5–9 cm; thickness 7–7.5 cm. Weight: 270–450 g. Skin: ground color bright yellow; rarely a faint pink blush; numerous small yellow dots; skin thin, tough, and medium adherent. Flesh: medium-soft, melting, and juicy deep yellow flesh; flavor mild and sweet with a weak, pleasant aroma; little fiber. Stone and seed: thick and woody stone with a polyembryonic seed filling 90% to 100% of the stone. Season: June to July. Recommended as a dooryard cultivar in Florida because is resistant to anthracnose, and a consistent producer.

Francis.—See Madame Francis.

Francisque.—See Madame Francis.

Gaylour.—Introd. to Florida from Israel about 1970 by Lawrence H. Zill, Boynton Beach, Fla. Orig. in Egypt. Tree: vigorous; large, dense canopy. Fruit: eating quality fair to good. Shape: oblong; base rounded; stout stem inserted obliquely in a level manner; apex rounded with a large lateral beak; surface undulating. Size: length 12–14.5 cm; breadth 7–8 cm; thickness 5.5–6 cm. Weight: 300–490 g. Skin: ground color greenish-yellow; no blush; few small white dots; skin thin, tender, and easily separating. Flesh: soft and juicy deep yellow flesh; flavor acidulous, aromatic, and spicy with a strong somewhat repugnant aroma; little fiber. Stone and seed: thick and woody stone with a monoembryonic seed filling 50% to 75% of the stone. Season: July to August. Fruit remains green when ripe; limited evaluation has shown it to be a good producer in Florida.

Glenn.—Seedling of Haden planted on the property of Roscoe E. Glenn, Miami, Fla. Planted in 1940; first fruited in 1945. Tree: moderately vigorous; small to medium, dense, rounded canopy. Fruit: eating quality excellent. Shape: oval to oblong; base rounded; slender stem inserted obliquely in a shallow, grooved cavity; apex bluntly pointed with no beak; surface smooth. Size: length 9.5–12.5 cm; breadth 7.5–8.5 cm; thickness 7–8 cm. Weight: 400–620 g. Skin: ground color bright yellow; orange-red blush; numerous small yellow and white dots; skin thin, tough, and easily separating. Flesh: soft and juicy deep yellow flesh; flavor rich and spicy with a strong, pleasant aroma; little fiber. Stone and seed: thick and woody stone with a monoembryonic seed filling about 50% to 75% of the stone. Season: June to July. Recommended for dooryard planting in Florida due to excellent eating quality and consistent production; fruits with a minimum of care; poor storage characteristics.

Golden Lippens.—Seedling of Lippens planted in 1942 by Peter and Irene Lippens, Miami, Fla. First fruited in 1951. Tree: vigorous; medium-sized, upright, open canopy. Fruit: eating quality excellent. Shape: oblong; base rounded; stout stem inserted obliquely in a shallow cavity; apex rounded sometimes with a small lateral beak; surface smooth. Size: length 12.5–14.5 cm; breadth 9–10 cm; thickness 8–8.5 cm. Weight: 550–950 g. Skin: ground color bright yellow; sometimes a pink blush; numerous small yellow or white dots; skin medium thick, tender, and adherent. Flesh: medium-firm, melting, and juicy lemon-yellow flesh; flavor rich, aromatic, and sweet with a weak, pleasant aroma; no fiber. Stone and seed: medium-thick and woody stone with a monoembryonic seed filling 40% to 60% of the stone. Season: June to July. Heavy, regular producer in Florida; tendency for limb breakage in some years due to heavy crop load.

Golden Nuggets.—Patented in Feb. 1990 by Edward Mitchell, Miami, Fla. Saigon may be one of its parents. Plant patent 7158. Tree: vigorous; large, upright, open canopy with well-spaced branching. Fruit: eating quality good to excellent. Shape: oval; base irregularly rounded; stout stem inserted obliquely in a shallow cavity; apex rounded with no beak; surface undulating. Size: length 9–11 cm; breadth 6.5–7.5 cm; thickness 5.5–6 cm. Weight: 340–550 g. Skin: ground color yellow-orange; pink blush; numerous small russet dots; skin medium thick, tender, and adherent. Flesh: firm, melting, and juicy lemon-yellow flesh; flavor mild and sweet with a weak, pleasant aroma; medium fiber. Stone and seed: thin and papery stone with a monoembryonic seed filling 75% to 90% of the stone. Season: late July to August. A consistent, heavy producer in limited commercial production in Florida; good harvesting and handling characteristics, but sometimes damaged by hot-water treatment for anthracnose control.

Graham.—Seedling of Julie introduced to Florida from Trinidad in 1932 by the U.S. Dept. of Agriculture, Miami, Fla. Tree: moderately vigorous; medium-sized, dense, rounded canopy. Fruit: eating quality good to excellent. Shape: oval; base rounded; slender stem inserted squarely in a shallow grooved cavity; apex rounded with a small lateral beak; surface undulating and irregular. Size: length 11.5–13.5 cm; breadth 9.5–11 cm; thickness 7–8 cm.

Weight: 570–820 g. Skin: ground color yellow; rarely a pink blush; few small white or russet dots; skin medium thick, tough, and easily separating. Flesh: soft and juicy deep yellow to orange flesh; flavor rich, aromatic, and medium sweet with a strong resinous aroma; little fiber. Stone and seed: medium-thick and woody stone with a monoembryonic seed filling 30% to 50% of the stone. Season: June to August. Strong, aromatic flavor reminiscent of Julie; well known throughout the Windward Islands.

Haden.—Seedling of Mulgoba planted in 1902 by Capt. John J. Haden of Coconut Grove, Fla. Taste, bearing habit, and isozyme analysis suggest that Turpentine may have been the pollinating parent. First fruited in 1910; first superior mango cultivar selected and named in Florida. Tree: vigorous; large and spreading canopy. Fruit: eating quality good to excellent. Shape: oval; base rounded; stout stem inserted squarely in a shallow cavity; apex rounded without a beak; surface smooth. Size: length 10.5–14 cm; breadth 9–10.5 cm; thickness 8.5–9.5 cm. Weight: 510–680 g. Skin: ground color bright yellow; crimson or red blush; numerous large yellow dots; skin thick, tough, and adherent. Flesh: firm and juicy deep yellow flesh; flavor rich and sweet with a weak, pleasant aroma; abundant fibers in flesh. Stone and seed: medium-thick and woody stone with a monoembryonic seed filling 80% to 95% of the stone. Season: June to mid-July. Haden dominated the commercial mango industry of Florida for 25 years; due to unreliable production, anthracnose susceptibility, and internal breakdown, it has been replaced by other varieties in Florida, but is grown on a commercial scale in many other countries in the world; parent of many Florida mangos; often produces small seedless fruits (nubbins); grows best in tropical climates.

Hansen.—Of unknown parentage from Judge Ware of Clearwater, Fla. Selected in 1943 for inclusion in the Univ. of Florida germplasm collection at Homestead. Tree: vigorous; large, dense, rounded canopy. Fruit: eating quality fair. Shape: oblong; base rounded and irregular; slender stem inserted obliquely in a shallow cavity; apex rounded without a beak; surface undulating. Size: length 11.5–13.5 cm; breadth 9.5–10.5 cm; thickness 8.5–9 cm. Weight: 450–740 g. Skin: ground color bright yellow; crimson blush; numerous small russet dots; skin thick, tough, and easily separating. Flesh: soft, melting, and juicy deep yellow flesh; flavor rich and sweet with a weak, pleasant aroma; no fiber. Stone and seed: thick and woody stone with a monoembryonic seed filling 50% to 75% of the stone. Season: June to July. With limited evaluations, production appears to be consistent and heavy in Florida.

Hatcher.—Of unknown parentage from the residence of Mr. Hatcher, Boynton Beach, Fla. Tree: vigorous; medium to large, dense canopy. Fruit: eating quality good. Shape: oval; base rounded and irregular; stout stem inserted squarely in a deep, grooved cavity; apex rounded without a beak; surface smooth. Size: length 12–15 cm; breadth 10.5–13 cm; thickness 9.5–10.5 cm. Weight: 600–850 g. Skin: ground color dark yellow; crimson or red blush; numerous medium-white dots; skin thick, tough, and easily separating. Flesh: firm and juicy deep yellow to orange flesh; flavor mild and sweet with a weak, pleasant aroma; little fiber. Stone and seed: medium-thick stone with a monoembryonic seed filling 60% to 85% of the stone. Season: July to August. Little is known about this cultivar, but the color, size, and eating quality suggest promise.

Hodson.—Of unknown parentage from Coral Gables, Fla. Selected in 1950 for inclusion in the Univ. of Florida germplasm collection at Homestead. Tree: vigorous; medium to large, dense canopy. Fruit: eating quality good. Shape: oblong; base flattened; stout stem inserted squarely in a shallow cavity; apex rounded without a beak; surface smooth. Size: length 11.5–14 cm; breadth 8.5–9.5 cm; thickness 7.5–8 cm. Weight: 440–650 g. Skin: ground color yellow-orange; dark red blush; numerous small russet dots; skin thin, tender, and moderately adherent. Flesh: soft, tender, and juicy orange flesh; flavor rich, aromatic, spicy, and very sweet with a strong, pleasant aroma; little fiber. Stone and seed: medium-thick and woody stone with a polyembryonic seed filling 60% to 90% of the stone. Season: July to August. Attractive, good-flavored mango that may have promise as a dooryard cultivar.

Irwin.—Seedling of Lippens planted in 1939 by F.D. Irwin, Miami, Fla. First fruited in 1945; named in 1949; propagated since 1949. Tree: moderately vigorous; small to medium, open canopy. Fruit: eating quality good. Shape: ovate; base rounded; rather slender stem inserted obliquely in a shallow cavity; apex pointed with no beak; surface smooth. Size: length 11.5–13 cm; breadth 8–9 cm; thickness 6.5–7.5 cm. Weight: 340–450 g. Skin: ground color bright yellow; crimson or dark red blush; numerous large white dots; skin medium thick, tender, and adherent. Flesh: soft, tender, melting, and juicy lemon-yellow flesh; flavor sweet and mild with a weak, pleasant aroma; no fiber. Stone and seed: thin and papery stone with a monoembryonic seed filling 70% to 90% of the stone. Season: June to early July. Consistent and heavy producing cultivar with outstanding color; fruit generally soft with a short postharvest life, but often exported from Tropical America to Europe; often produces fruit in clusters, and has a tendency to produce seedless fruit.

Jacquelin.—Of unknown parentage; named after the daughter of Jack Faircloth of Lake Worth, Fla. Propagated since about 1950; named and described in 1954. Tree: vigorous; large, upright, dense canopy. Fruit: eating quality good to excellent. Shape: round; base flattened; stout stem inserted squarely in a shallow cavity; apex rounded with no beak; surface smooth. Size: length 11–14 cm; breadth 9.5–11 cm; thickness 8–8.5 cm. Weight: 450–730 g. Skin: ground color greenish to bright yellow; orange-red blush; numerous medium-yellow dots; skin thick, tough, and easily separating. Flesh: tender, melting, and juicy orange flesh; flavor rich, spicy, and sweet with a strong, pleasant aroma; no fiber. Stone and seed: thick and woody stone with a monoembryonic seed filling about 60% to 80% of the stone. Season: June to July. Has the appearance of a large Zill mango; fruit tender and easily bruised during handling or shipping; attracted attention in the Caribbean due to its size and color.

Jakarta.—Seedling of Paheri from the residence of Lawrence H. Zill, Boynton Beach, Fla. Tree: vigorous; medium to large, rounded, dense canopy. Fruit: eating quality good. Shape: round to oblong; base flattened; stout stem inserted obliquely in a shallow cavity; apex rounded with a small lateral beak; surface smooth. Size: length 12.5–14.5 cm; breadth 10–11.5 cm; thickness 8–9 cm. Weight: 550–840 g. Skin: ground color orange-yellow; orange-red blush; numerous large, white dots; skin thin, tender, and medium adherent. Flesh: tender and juicy deep yellow flesh; flavor rich, spicy, and slightly resinous with a weak, pleasant aroma; little fiber. Stone and seed: thick and woody stone with a monoembryonic seed filling 75% to 90% of the stone. Season: June to August. Orange-red color gives this cultivar a distinctive beautiful appearance on the tree; heavy producer in Florida that may have promise as a commercial cultivar in Tropical America.

Jehangir.—Brought to Florida from India in 1979 by Frank Smathers, Jr., Miami, Fla. Tree: moderately vigorous; small to medium, spreading, dense canopy. Fruit: eating quality fair to good. Shape: oval; base slightly flattened; stout stem inserted squarely in a level manner or in a slight cavity; apex rounded with no beak; surface undulating. Size: length 8–11.5 cm; breadth 7.5–8.5 cm; thickness 7–7.5 cm. Weight: 400–625 g. Skin: ground color greenish; no blush; numerous large, russet dots; skin thick, medium tough, and adherent. Flesh: firm, tender, and juicy light yellow flesh; flavor mild, sweet, aromatic, and slightly resinous with a weak, resinous aroma; no fiber. Stone and seed: thick and woody stone with a monoembryonic seed filling 80% to 95% of the stone. Season: July to September. Considered one of the choicest mangos of southern India; flesh color almost white in Florida; the few trees grown in Florida have been poor producers.

Jewel.—Seedling of Lippens from the residence of Lawrence H. Zill, Boynton Beach, Fla. Tree: moderately vigorous; small, open, and spreading canopy. Fruit: eating quality good to excellent. Shape: oval, oblique; base rounded; slender stem inserted squarely in a level manner; apex bluntly pointed with no beak; surface smooth. Size: length 10.5–11.5 cm; breadth 8–9 cm; thickness 7–7.5 cm. Weight: 275–440 g. Skin: ground color yellow-orange to orange; dark red blush; numerous small white dots; skin thin, tender, and medium adherent. Flesh: soft, tender, and juicy deep yellow flesh; flavor rich and sweet with a weak, pleasant aroma; no fiber. Stone and seed: medium-thick and woody stone with a monoembryonic seed filling 50% to 75% of the stone. Season: June to July. Clusters of highly colored fruit hanging within the typically small, open canopy is impressive; fruit color early on the tree, before physiological maturity; internal breakdown a problem.

Jubilee.—Of unknown parentage promoted by Edward Mitchell of Miami, Fla. Tree: moderately vigorous; large, spreading canopy. Fruit: eating quality fair to good. Shape: rounded, sigmoid; base rounded; slender stem inserted obliquely in a shallow cavity; apex rounded with a small lateral beak; surface smooth. Size: length 11–15 cm; breadth 9–10.5 cm; thickness 8–8.5 cm. Weight: 400–500 g. Skin: ground color light red; purplish-red blush; few small, russet dots; skin thin, tough, and easily separating. Flesh: firm and juicy deep yellow flesh; flavor mild and sweet with a weak, pleasant aroma; medium fiber. Stone and seed: thick and woody stone with a monoembryonic seed filling 50% to 75% of the stone. Season: July to early August. Entire fruit purplish-red when ripe; fruit color extremely attractive, but flavor watery and insipid.

Keitt.—Seedling of Mulgoba planted in 1939 on the property of Mrs. J.N. Keitt, Homestead, Fla. Tree: moderately vigorous; medium-sized, upright, open canopy. Fruit: eating quality good to excellent. Shape: oval; base rounded; stout stem inserted slightly obliquely either in a level or a slightly raised manner; apex rounded with no beak; surface smooth. Size: length 13–15 cm; breadth 9–11 cm; thickness 8.5–10 cm. Weight: 510–2000 g. Skin: ground color greenish-yellow; red or pink blush; numerous small white or yellow dots; skin thick, tough, and adherent. Flesh: firm and juicy lemon-yellow flesh; flavor sweet and mild with a weak, pleasant aroma; little fiber. Stone and seed: thick and woody stone with a monoembryonic seed filling 30% to 50% of the stone. Season: August to early October. Precocious, regular, and heavy producer; leading late-season commercial mango in Florida, and a good dooryard tree; produced commercially in many regions of the world; good anthracnose resistance; excellent handling and shipping characteristics; often difficult to manage in humid climates due to vigorous "leggy" growth, but an easier-managed, small, compact tree under arid conditions.

Kent.—Seedling of Brooks planted in 1932 on the property of Leith D. Kent, Coconut Grove, Fla. First fruited in 1938; named and described in 1945. Tree: vigorous; large, compact, upright canopy. Fruit: eating quality excellent. Shape: oval; base rounded; stout stem inserted slightly obliquely in a level or slightly raised manner; apex rounded with no beak; surface smooth. Size: length 11–13 cm; breadth 9.5–11 cm; thickness 9–9.5 cm. Weight: 600–750 g. Skin: ground color greenish-yellow; red or crimson blush; numerous small yellow dots; skin thick, tough, and adherent. Flesh: firm, tender, melting, and juicy deep-yellow to orange-yellow flesh; flavor rich, medium aromatic, and sweet with a weak, pleasant aroma; little fiber. Stone and seed: thick and woody stone with a monoembryonic seed filling 70% to 90% of the stone. Season: July to August. Considered one of the best-tasting Florida mangos; still grown on a small commercial scale in Florida due to its excellent eating quality and late season; poor storage characteristics and susceptible to internal breakdown.

Langra.—Introd. to Florida from India in 1960 by the U.S. Dept. of Agriculture, Miami, Fla. Tree: moderately vigorous; medium-sized, upright, open canopy. Fruit: eating quality fair to good. Shape: oblong; base rounded; stout stem inserted obliquely in a level manner; apex rounded with a small lateral beak; surface undulating. Size: length 8.5–9.5 cm; breadth 7–7.5 cm; thickness 6–6.5 cm. Weight: 220–350 g. Skin: ground color greenish-yellow; pink blush; numerous small white dots; skin thick, tough, and easily separating. Flesh:

firm and juicy lemon-yellow flesh; flavor mild and sweet with a very weak, pleasant aroma; medium to abundant fibers in flesh. Stone and seed: thick and woody stone with a monoembryonic seed filling 80% to 100% of the stone. Season: July to August. Good flavor, but fruit small and fibrous; inconsistent, light production in Florida, although a commercial cultivar in India.

Lily.—Seedling of unknown parentage grown by William W. Carmichael of Miami, Fla. Named after his wife, Lily Carmichael. Tree: vigorous; large, open, spreading growth. Fruit: eating quality fair to good. Shape: oval; base rounded; stout stem inserted squarely or slightly obliquely in a shallow cavity; apex rounded with no beak; surface smooth. Size: length 11.5–13 cm; breadth 8.5–9.5 cm; thickness 7.5–8.5 cm. Weight: 530–750 g. Skin: ground color yellow-orange; predominant dark red blush; numerous small white dots; skin thin, tough, and easily separating. Flesh: soft, tender, and juicy lemon-yellow flesh; flavor mild and spicy with almost no aroma; little fiber. Stone and seed: thick and woody stone with a monoembryonic seed filling 40% to 60% of the stone. Season: July and August. Attractive fruit, flavor mild and watery; never widely tested in Florida for commercial use.

Lippens.—Seedling of Haden planted on the property of Peter Lippens, Miami, Fla., in 1931. First fruited in 1938. Propagated commercially since 1945. Tree: vigorous; medium-sized, compact, rather dense canopy. Fruit: eating quality good to excellent. Shape: ovate to oblong; base rounded to slightly flattened; medium-stout stem inserted obliquely in a slight cavity; apex rounded with no beak; surface smooth. Size: length 9.5–11 cm; breadth 8–9 cm; thickness 7.5–8 cm. Weight: 350–520 g. Skin: ground color yellow; pink or crimson blush; numerous small yellow dots; skin medium thick, tough, and adherent. Flesh: firm, melting, and juicy deep yellow flesh; flavor mild, sweet, and aromatic with a weak, pleasant aroma; no fiber. Stone and seed: thick and woody stone with a monoembryonic seed filling 85% to 100% of the stone. Season: late June and July. Heavy producer in cool, dry climates like that of the Canary Islands; flavor often likened to that of pineapple; susceptibility to anthracnose has hampered its use as a commercial cultivar in humid climates.

Madame Francis.—Of unknown parentage from Haiti. Tree: moderately vigorous; medium-sized, open canopy. Fruit: eating quality fair to good. Shape: oblong, sigmoid; base rounded; stout stem inserted slightly obliquely in a shallow cavity; apex rounded with a medium-size lateral beak; surface undulating. Size: length 15–17 cm; breadth 8.5–11 cm; thickness 5.5–7.5 cm. Weight: 370–520 g. Skin: ground color greenish to bright yellow; no blush; few large russet dots; skin thin, tender, and adherent. Flesh: soft and juicy orange flesh; flavor rich, spicy, and sweet with a weak, pleasant aroma; little fiber. Stone and seed: thin and papery stone with a polyembryonic seed filling 25% to 40% of the stone. Season: June to July. Also called Francis and Francisque in the Caribbean; trees often grown from seed in Haiti and the Caribbean; poor color and anthracnose susceptibility limit its commercial use.

Magshamim.—Introd. to Florida from Israel by Lawrence H. Zill, Boynton Beach, Fla. Tree: vigorous; medium-sized, upright, dense canopy. Fruit: eating quality good. Shape: oblong; base rounded; medium-slender stem inserted slightly obliquely in a level manner; apex rounded or bluntly pointed with a small lateral beak. Size: length 11.5–13 cm; breadth 7.5–8.5 cm; thickness 6.5–7 cm. Weight: 400–520 g. Skin: ground color bright yellow; crimson or dark red blush; numerous small white dots; skin thick, tough, and easily separating. Flesh: firm, tender, and juicy lemon-yellow flesh; flavor mild and sweet with a weak, pleasant aroma; little fiber. Stone and seed: medium-thick and woody stone with a monoembryonic seed filling 60% to 90% of the stone. Season: late June to early August. Color, shape, and size desirable, but flavor a disappointment to many in Florida; somewhat cold-tolerant in Florida.

Mallika.—Introd. to Florida from India in 1978 by Frank Smathers, Jr. Hybrid between Neelum and Dasheri. Tree: moderately vigorous; dwarf, dense canopy. Fruit:

eating quality excellent. Shape: oblong; base rounded; slender stem inserted obliquely in a raised manner; apex rounded with a small lateral beak; surface undulating. Size: length 10–12 cm; breadth 6.5–7.5 cm; thickness 5–5.5 cm. Weight: 280 to 450 g. Skin: ground color bright yellow; no blush; numerous small yellow dots; skin thick, tough, and easily separating. Flesh: soft, tender, and juicy deep yellow to orange flesh; flavor rich, aromatic, and sweet with a strong, pleasant aroma; little fiber. Stone and seed: medium-thick and woody stone with a monoembryonic seed filling 40% to 60%. Season: late June to July. Of considerable interest due to its dwarf nature; irregular producer in Florida, but good eating quality; aroma is one of the strongest and most pleasant.

Manila.—Philippine-type seedling race taken from the Philippines to Chiapas State, Mexico, and later to Florida. Tree: vigorous; large, upright, open canopy. Fruit: eating quality good to very good. Shape: long and slender; base rounded; slender stem inserted squarely in a level or elevated manner; apex bluntly pointed sometimes with a small beak; surface smooth. Size: length 12.5–14 cm; breadth 5.5–6 cm; thickness 5–5.5 cm. Weight: 180–260 g. Skin: ground color bright yellow; sometimes a light pink blush; few small reddish dots; skin thin, medium tough, and easily separating. Flesh: mediumfirm and juicy deep yellow flesh; flavor rich, spicy, and sweet with a weak, pleasant aroma; little to abundant fiber. Stone and seed: medium-thick and woody stone with a polyembryonic seed filling 50% to 60% of the stone. Season: late June to July. Used for fresh fruit and canned slices in Mexico; trees in Mexico often grown from seed.

Manzano Colorado.—Introd. to Florida from the Virgin Islands in 1976 by the U.S. Dept. of Agriculture, Miami, Fla. Orig. in Puerto Rico. Tree: vigorous; medium-sized, dense, rounded canopy. Fruit: eating quality fair. Shape: round to oval; base flattened; stout stem inserted squarely in a deep cavity; apex rounded with a small lateral beak; surface undulating. Size: length 9.5–11.5 cm; breadth 10–11 cm; thickness 8.5–9 cm. Weight: 450–630 g. Skin: ground color greenish-yellow; dark red blush; numerous small white dots; skin thin, tough, and easily separating. Flesh: soft and dry lemon-yellow flesh; flavor rich, spicy, and resinous with a strong resinous aroma; little fiber. Stone and seed: thick and woody stone with a polyembryonic seed filling 70% to 90% of the stone. Season: June to early August. Common cultivar throughout the Caribbean; resinous aroma quite distinctive before ripening.

Martin.—Of unknown parentage, selected by D.H. Martin, West Palm Beach, Fla. Tree: moderately vigorous; large, open, spreading canopy. Fruit: eating quality fair. Shape: oblong; base rounded; slender stem inserted obliquely in a level manner; apex rounded with no beak; surface undulating. Size: length 17–20.5 cm; breadth 12–15 cm; thickness 9–11 cm. Weight: 650–980 g. Skin: ground color greenish-yellow; pink blush; few medium-white dots; skin thick, tough, and easily separating. Flesh: firm and melting orange flesh; flavor mild and sweet with a weak, pleasant aroma; little fiber. Stone and seed: thick and woody stone with a monoembryonic seed filling 60% to 75% of the stone. Season: July to August. Nicknamed "potbelly" because of unique shape.

Marty.—Of unknown parentage, perhaps a cross between Tommy Atkins and Keitt. Patented in Feb. 1990 by Neal P. Brooks, Homestead, Fla. Plant patent 7140. Tree: vigorous; large, upright, dense canopy. Fruit: eating quality good. Shape: oval; base flattened; medium-stout stem inserted squarely in a shallow cavity; apex rounded with no beak; surface smooth. Size: length 11.5–13 cm; breadth 8.5–9.5 cm; thickness 7.5–8.5 cm. Weight: 560–750 g. Skin: ground color greenish-yellow; prevalent dark reddish-purple blush; few small white dots; skin medium thick, tough, and easily separating. Flesh: medium-firm and juicy lemon-yellow flesh; flavor mild and sweet with a weak, pleasant aroma; little fiber. Stone and seed: thick and woody stone with a monoembryonic seed filling 50% to 75% of the stone. Season: July to August. Reddish-purple blush gives cultivar a unique look when ripe; tastes sour until fruit fully ripe; poor internal color.

Micongensis.—Introd. to Florida from the Philippines by the U.S. Dept. of Agriculture, Miami, Fla. Originally said to be from Vietnam. Tree: vigorous; medium-sized, open canopy. Fruit: eating quality good to excellent. Shape: ovate to long and slender; base rounded; slender stem inserted obliquely in a level manner; apex sharply pointed with no beak; surface smooth. Size: length 11.5–13 cm; breadth 7.5–8.5 cm; thickness 6–6.5 cm. Weight: 200–410 g. Skin: ground color greenish to bright yellow; slight pink blush; few small white dots; skin thin, tender, and easily separating. Flesh: soft, tender, melting, and juicy lemon-yellow flesh; flavor rich, spicy, and sweet with a weak, pleasant aroma; no fiber. Stone and seed: thin and papery stone with a polyembryonic seed filling about 40% to 60% of the stone. Season: July to August. Considered one of the best mangos in Vietnam.

Mulgoba.—Introd. to Florida from India in 1889 by the U.S. Dept. of Agriculture, Miami, Fla. First fruited in 1902 on Prof. Gale's property near Lake Worth, Fla. Tree: vigorous; large, spreading, open canopy. Fruit: eating quality good to excellent. Shape: oval to ovate; base flattened; stout stem inserted squarely in a shallow cavity; apex rounded to bluntly pointed with a small beak; surface smooth. Size: length 8.5–10.5 cm; breadth 6.5–7.5 cm; thickness 5–6 cm. Weight: 340–450 g. Skin: ground color bright yellow; pink blush; numerous, large white dots; skin thick, medium tough, and adherent. Flesh: soft, tender, melting, and juicy lemon-yellow flesh; flavor rich, spicy, and sweet with a strong, pleasant aroma; little to medium fiber. Stone and seed: thick and woody stone with a monoembryonic seed filling 90% to 100% of the stone. Season: July and August. Considered one of the finest flavored mangos in Florida; highly susceptible to anthracnose; inconsistent, low production in Florida.

Nam Doc Mai.—Introd. to Florida from Thailand in 1973 by the U.S. Dept. of Agriculture, Miami, Fla. Multiple introductions of Nam Doc Mai have been made to Florida, resulting in two recognizable types, Nam Doc Mai and Mun. These may be different cultivars. Tree: vigorous; medium-sized, upright, dense canopy. Fruit: eating quality excellent. Shape: long and slender, sigmoid; base rounded; medium-stout stem inserted squarely in a level or slightly raised manner; apex sharply to bluntly pointed with no beak. Size: length 17–19 cm; breadth 7.5–8.5 cm; thickness 6.5–7.5 cm. Weight: 340–580 g. Skin: ground color greenish to bright yellow; slight pink blush; numerous small green dots; skin medium thick, tender, and easily separating. Flesh: soft, tender, and juicy lemon-yellow flesh; flavor rich, spicy, and very sweet with a weak, pleasant aroma; no fiber. Stone and seed: thin and papery stone with a polyembryonic seed filling 25% to 50% of the stone. Season: mid-June to July. Low-acid fruit often eaten while still green; regular producer that occasionally has multiple crops in a single season.

Nam Tan Teen.—Introd. to Florida from Thailand in 1973 by Robert J. Knight for the Rare Fruit Council International of Miami, Fla. Tree: vigorous; large, upright, somewhat open canopy. Fruit: eating quality fair to good. Shape: oblong; base rounded; slender stem inserted squarely in a slightly elevated manner; apex rounded to bluntly pointed sometimes with a small lateral beak; surface smooth. Size: length 12–13.5 cm; breadth 7.5–8.5 cm; thickness 6.5–7 cm. Weight: 320–450 g. Skin: ground color greenish-yellow; no blush; no dots; skin thick, medium tough, and medium adherent. Flesh: mediumfirm and juicy pale yellow flesh; flavor mild and insipid with no aroma; little fiber. Stone and seed: medium-thick and woody stone with a polyembryonic seed filling 90% to 100% of the stone. Season: late May to June. Name means "Chinese sugar" in the Thai language; consistent, heavy producer in Florida.

Neelum.—Introd. to Florida from India in 1979 by Frank Smathers, Jr., Miami, Fla. Tree: moderately vigorous; small, compact canopy. Fruit: eating quality good to excellent. Shape: ovate; base flattened to slightly rounded; stout stem inserted squarely in a shallow cavity; apex bluntly pointed with a small lateral beak; surface undulating. Size: length 9.5–11 cm; breadth 7.5–8.5 cm; thickness 6–6.5 cm. Weight: 230–300 g. Skin:

ground color bright yellow; no blush; numerous small white dots; skin thick, tender, and easily separating. Flesh: soft, melting, and juicy deep yellow flesh; flavor mild and sweet with a weak, pleasant aroma; no fiber. Stone and seed: medium-thick and woody stone with a monoembryonic seed filling 90% to 100% of the stone. Season: September to early October. Late maturation season makes this cultivar a valuable source of germplasm; crop can often be sold at a high price due to the scarcity of fruit at this time of year.

Number 11.—Came to Florida in the 1880s as a shipment of seeds from Jamaica sent to the Reasoner Brothers, who grew and distributed trees around South Florida, particularly along Lake Worth and Biscayne Bay. Tree: vigorous; medium-sized, upright, dense canopy. Fruit: eating quality poor to fair. Shape: ovate, reniform; base flattened or slightly rounded; slender stem inserted squarely in a shallow cavity; apex bluntly pointed, sometimes with a small lateral beak; surface smooth. Size: length 9–10 cm; breadth 5.5–6.5 cm; thickness 5–5.5 cm. Weight: 200–300 g. Skin: ground color greenish-yellow; sometimes an orange-red blush; few small russet dots; skin thin, tough, and adherent. Flesh: firm and dry lemon-yellow flesh; flavor mild, sweet, insipid with a very weak, pleasant aroma; abundant fiber. Stone and seed: thick and woody stone with a polyembryonic seed filling 95% to 100% of the stone. Season: June to August. Used as a rootstock in Florida in the past; produces fruit in clusters.

Nuwun Chan.—Introd. to Florida from Thailand in 1973 by Robert J. Knight for the Rare Fruit Council International, Miami, Fla. Tree: moderately vigorous; small, upright, dense canopy. Fruit: eating quality good. Shape: long and slender, sigmoid; base flattened; slender stem inserted obliquely in a shallow cavity; apex bluntly pointed with no beak; surface undulating. Size: length 16–18 cm; breadth 7–8 cm; thickness 6–6.5 cm. Weight: 340–500 g. Skin: ground color greenish-yellow; pink to red blush; numerous small green dots; skin thick, tough, and easily separating. Flesh: soft, melting, and juicy pale yellow flesh; flavor mild, sweet, and insipid with a weak, pleasant aroma; little fiber. Stone and seed: thick and woody stone with a polyembryonic seed filling 60% to 85% of the stone. Season: late May to June. In some years fruit have a strong pink to red blush, and in some years there is little or no blush; fruit often eaten green.

Okrung.—Introd. to Florida from Thailand in 1973 by Robert J. Knight for the Rare Fruit Council International, Miami, Fla. Tree: moderately vigorous; medium-sized, upright, dense canopy. Fruit: eating quality good. Shape: oblong, sigmoid; base rounded; slender stem inserted obliquely in a grooved cavity; apex rounded with no beak; surface smooth. Size: length 11–13 cm; breadth 5–6 cm; thickness 4.5–5.5 cm. Weight: 160–240 g. Skin: ground color green to greenish-yellow; no blush; numerous small white dots; skin thick, tough, and medium adherent. Flesh: soft and juicy pale yellow or greenish flesh; flavor mild, insipid, and very sweet with a weak, pleasant aroma; abundant fiber. Stone and seed: thick and woody stone with a polyembryonic seed filling 20% to 40% of the stone. Season: June to August. Low-acid fruit, good for eating while still green; heavy production, occasionally producing multiple crops within a single year; grooved cavity extends the length of the fruit, giving it a unique appearance.

Ono.—Introd. to Florida from Hawaii in 1927 by the U.S. Dept. of Agriculture, Miami, Fla. Supposedly a seedling of Paheri from R.W. Smith, Makiki Heights, Honolulu. Tree: vigorous; large, upright, somewhat open canopy. Fruit: eating quality fair. Shape: oblong; base rounded; stout stem inserted squarely in a shallow cavity; apex bluntly pointed with no beak; surface undulating. Size: length 9.5–11 cm; breadth 7–8 cm; thickness 6–6.5 cm. Weight: 290–340 g. Skin: ground color bright yellow; dark red blush; numerous medium-yellow dots; skin thin, tender, and adherent. Flesh: soft and juicy deep yellow flesh; flavor rich and sweet with a strong, pleasant aroma; little fiber. Stone and seed: thick and woody stone with a polyembryonic seed filling 70% to 85% of the stone. Season: June to July. Regular producer, but highly susceptible to anthracnose; fruit very soft at maturity; was an important cultivar in Hawaii.

Osteen.—Seedling of Haden planted in 1935 on the property of S.A. Osteen, Merritt Island, Fla. First fruited in 1940. Tree: vigorous; medium-sized, dense canopy. Fruit: eating quality good. Shape: oblong; base rounded; stout stem inserted squarely in an elevated manner; apex rounded sometimes with a small beak; surface smooth. Size: length 12–15.5 cm; breadth 9–10.5 cm; thickness 8.5–9.5 cm. Weight: 500–760 g. Skin: ground color yellow-orange; purple or lavender blush; numerous small white dots; skin thick, tough, and easily separating. Flesh: firm and juicy lemon-yellow flesh; flavor mild and sweet with a weak, pleasant aroma; little fiber. Stone and seed: thick and woody stone with a monoembryonic seed filling 40% to 60% of the stone. Season: July to early September. Fruit turns lavender long before physiological maturity; seriously considered as a commercial cultivar, but never widely planted.

Palmer.—Of unknown parentage from a seed planted about 1925 on the property of Mrs. Victor Mell, Miami, Fla. First propagated in 1945; named and described in 1949. Tree: moderately vigorous; large, upright, tight canopy. Fruit: eating quality good. Shape: oblong; base rounded; stout stem inserted obliquely in a level or slightly raised manner; apex rounded with no beak; surface smooth. Size: length 12–15 cm; breadth 8.5–10 cm; thickness 6.5–7.5 cm. Weight: 510–850 g. Skin: ground color yellow-orange; dark red to crimson blush; few small white dots; skin medium thick, tough, and adherent. Flesh: firm and melting orange-yellow flesh; flavor mild and aromatic with a weak, pleasant aroma; little fiber. Stone and seed: medium-thick and woody stone with a monoembryonic seed filling 70% to 85% of the stone. Season: July to early September. Fruit turns purple early, long before physiological maturity; still grown on a commercial scale in Florida; often harvested immature due to early coloring.

Parvin.—Seedling of Haden planted at the residence of Clint Parvin, Bradenton, Fla. Tree: vigorous; rounded, spreading, dense canopy. Fruit: eating quality good to excellent. Shape: oval to oblong; base rounded; stout stem inserted squarely in a slightly grooved cavity; apex rounded with no beak; surface smooth but irregular near stem. Size: length 11.5–14 cm; breadth 9–10.5 cm; thickness 8–9 cm. Weight: 450–690 g. Skin: ground color greenish-yellow; dark red to crimson blush; numerous small yellow dots; skin thick, tough, and easily separating. Flesh: firm and juicy deep yellow flesh; flavor mild and sweet with a strong, pleasant aroma; little fiber. Stone and seed: thick and woody stone with a monoembryonic seed filling 60% to 80% of the stone. Season: July to August. Excellent storage and handling characteristics; responds well to the induction of off-season bloom by chemical means; mature fruit have a waxy feel due to the heavy bloom; tendency in some years to produce seedless fruit.

Philippine.—Seedling race probably originating from a shipment of seeds sent to Florida from Cuba in the 1910s. Seed source in Cuba probably came from Mexico under the name of Manila. Tree: moderately vigorous; large, upright, dense canopy. Fruit: eating quality good to excellent. Shape: long and slender; base rounded; slender stem inserted obliquely in a shallow grooved cavity; apex bluntly pointed with no beak; surface smooth. Size: length 10.5–14.5 cm; breadth 6–7.5 cm; thickness 5–6 cm. Weight: 200–340 g. Skin: ground color bright yellow; no blush; numerous small white dots; skin thin, tough, and medium adherent. Flesh: soft, melting, and juicy yellow-orange flesh; flavor rich, aromatic, and medium sweet with a weak, pleasant aroma; little fiber. Stone and seed: thin and papery stone with a polyembryonic seed filling 70% to 85% of the stone. Season: June to July. Excellent eating quality, but fruit are easily damaged by handling and will not store very long.

Pillsbury.—Of unknown parentage from the property of C.T. Earle, Bradenton, Fla. Seed planted in the late 1930s; first fruited in 1943. Tree: vigorous; large, rounded, open canopy. Fruit: eating quality good. Shape: oval; base flattened; medium-stout stem inserted obliquely in a shallow cavity; apex rounded with no beak; surface smooth. Size: length 9.5–11 cm; breadth 8.5–9.5 cm;

thickness 8–8.5 cm. Weight: 400–750 g. Skin: ground color greenish-yellow; crimson blush; numerous small white dots; skin thin, medium tough, and adherent. Flesh: tender, melting, and juicy lemon-yellow flesh; flavor mild, aromatic, and sweet with a weak, pleasant aroma; little fiber. Stone and seed: thick and woody stone with a monoembryonic seed filling 60% to 80% of the stone. Season: July to August. Consistent, but light production in Florida with limited evaluation.

Pruter.—Of unknown parentage from H.C. Pruter, Pine Island, Fla. Tree: vigorous; large, upright, dense canopy. Fruit: eating quality good. Shape: oval to oblong; base rounded; stout stem inserted squarely in a shallow cavity or in a level manner; apex rounded with no beak; surface smooth. Size: length 11–13 cm; breadth 9.5–11 cm; thickness 9–10 cm. Weight: 510–920 g. Skin: ground color bright yellow; crimson or dark red blush; few small white dots; skin thick, tough, and easily separating. Flesh: soft, tender, and melting deep yellow flesh; flavor mild and sweet with a weak, pleasant aroma; no fiber. Stone and seed: thick and woody stone with a monoembryonic seed filling 60% to 80% of the stone. Season: July to August. Good color and size, but the soft flesh and poor storage characteristics have discouraged commercial use.

Ruby.—Of unknown parentage from the residence of E.P. Davis, Miami, Fla. Tree: moderately vigorous; large, upright, somewhat open canopy. Fruit: eating quality good. Shape: oblong; base rounded; stout stem inserted squarely in an elevated manner; apex rounded with no beak; surface smooth. Size: length 10.5–12.5 cm; breadth 6.5–7.5 cm; thickness 5.5–6 cm. Weight: 200–300 g. Skin: ground color yellow-orange; red or crimson blush; numerous small white dots; skin thick, tough, and easily separating. Flesh: firm and juicy deep yellow flesh; flavor rich, aromatic, spicy, and sweet with a strong, pleasant aroma; little fiber. Stone and seed: thick and woody stone with a monoembryonic seed filling 70% to 85% of the stone. Season: July to August. Tendency to produce seedless fruit in some years; often fruits in clusters; internal breakdown sometimes a problem, but usually not severe.

Saber.—Introd. to Florida from Israel in 1953 by the U.S. Dept. of Agriculture, Miami, Fla. Originally from East Africa. Tree: moderately vigorous; small, open canopy. Fruit: eating quality fair. Shape: long and slender, sigmoid; base rounded; stout stem inserted squarely in an elevated manner; apex sharply pointed with a large lateral beak; surface smooth. Size: length 11–13 cm; breadth 6.5–7.5 cm; thickness 5.5–6.5 cm. Weight: 300–440 g. Skin: ground color bright yellow; dark red blush; numerous small white dots; skin thick, tough, and easily separating. Flesh: firm and juicy orange flesh; flavor rich and sweet with a weak, pleasant aroma; abundant fiber. Stone and seed: thick and woody stone with a polyembryonic seed filling 80% to 90% of the stone. Season: mid-June to July. Very attractive color; previously used as a rootstock in Israel.

Saigon.—Seedling race with considerable variation in size and form. David Fairchild sent seeds of Saigon from Indo-China in 1904 to the U.S. Dept. of Agriculture's Introduction Garden in Miami, Fla., where they were grown by the garden's director, Edward Simmonds. Tree: vigorous; medium-sized, open, rounded canopy. Fruit: eating quality good to excellent. Shape: oval to ovate; base rounded; stout stem inserted squarely in an elevated manner; apex bluntly pointed with a small lateral beak; surface smooth. Size: length 7.5–10 cm; breadth 6.5–7.5 cm; thickness 5–6 cm. Weight: 250–350 g. Skin: ground color greenish to bright yellow; rarely a pink blush; few small russet dots; skin thin, tender, and adherent. Flesh: soft and juicy pale yellow flesh; flavor sweet and mild with a weak, pleasant aroma; no fiber. Stone and seed: thick and woody stone with a polyembryonic seed filling 90% to 100% of the stone. Season: June and July. Consistent, heavy producer; good resistance to anthracnose.

Sandersha.—Introd. to Florida from Bangalore, India, in 1901 by the U.S. Dept. of Agriculture, Miami, Fla. Tree: vigorous; large, spreading, open canopy. Fruit: eating quality poor to fair. Shape: oblong; base rounded; stout stem inserted squarely in an elevated manner; apex rounded to bluntly pointed with a large lateral beak; surface smooth. Size: length 17.5–20 cm; breadth 9–11.5 cm; thickness 8.5–10.5 cm. Weight: 800–1100 g. Skin: ground color greenish-yellow; pink blush; few small white dots; skin thick, tough, and adherent. Flesh: firm and medium-juicy lemon-yellow flesh; flavor rich and spicy with a weak, somewhat repugnant aroma; abundant fiber. Stone and seed: thin and papery stone with a monoembryonic seed filling 40% to 60% of the stone. Season: July to August. Often used for cooking or processing, such as chutney; important as a parent for many Florida cultivars.

Santaella.—Introd. to Florida from Puerto Rico by the U.S. Dept. of Agriculture, Miami, Fla. Tree: moderately vigorous; semi-dwarf, dense canopy. Fruit: eating quality fair to good. Shape: oblong; base rounded; stout stem inserted obliquely in a deep cavity; apex rounded with a small lateral beak; surface smooth. Size: length 12–13.5 cm; breadth 7–8.5 cm; thickness 6–7 cm. Weight: 330–500 g. Skin: ground color yellow-orange to orange; no blush; few small white dots; skin thick, tough, and easily separating. Flesh: firm and dry orange flesh; flavor rich, aromatic, and sweet with a strong, slightly repugnant aroma; little fiber. Stone and seed: thick and woody stone with polyembryonic seed filling 95% to 100% of the stone. Season: June to July. Firm flesh texture makes this cultivar a valuable germplasm resource; consistent, heavy production with limited evaluation in Florida.

Sensation.—Of unknown parentage from North Miami, Fla. Seed planted in 1935; first fruited in 1941. Original tree was moved to W.W. Carmichael's nursery, Perrine, Fla., and was first propagated there. Tree: vigorous; moderately open, symmetrical canopy. Fruit: eating quality fair to good. Shape: oval; base rounded; stout stem inserted squarely to slightly obliquely in a shallow grooved cavity; apex rounded, sometimes with a small beak; surface undulating. Size: length 9–11.5 cm; breadth 7–8 cm; thickness 6.5–7 cm. Weight: 280–340 g. Skin: ground color dark yellow; prominent dark red to purple blush; few small yellow dots; skin medium thick, tough, and easily separating. Flesh: firm and medium-juicy deep yellow flesh; flavor mild and sweet with a weak, pleasant aroma; no fiber. Stone and seed: thick and woody stone with a monoembryonic seed filling 70% to 80% of the stone. Season: mid-July to September. Consistent and heavy producer, but eating quality is only fair; well adapted to cool subtropical climates like the Canary Islands.

Smith.—Seedling of Haden planted by C.W. Smith of Perrine, Fla. First fruited in 1939; propagated since 1946. Named and described in 1948. Tree: moderately vigorous; large, upright, symmetrical, and dense canopy. Fruit: eating quality fair to good. Shape: oblong; base rounded; slender stem inserted obliquely in a level manner; apex rounded with no beak; surface smooth. Size: length 12.5–15.5 cm; breadth 9–10.5 cm; thickness 8–9 cm. Weight: 580–950 g. Skin: ground color yellow-orange; crimson blush; few large yellow dots; skin thin, tough, and adherent. Flesh: firm, medium-juicy deep yellow flesh; flavor mild and sweet with little aroma; medium fiber. Stone and seed: thick and woody stone with a monoembryonic seed filling 60% to 75% of the stone. Season: July to August. Consistent, relatively heavy producer; highly susceptible to anthracnose.

Spirit of '76.—Seedling of Zill from the residence of Lawrence H. Zill, Boynton Beach, Fla. Pollinating parent may have been Haden. Tree: moderately vigorous; medium-size, spreading canopy. Fruit: eating quality good to excellent. Shape: oblong; base rounded; medium-stout stem inserted squarely in a slightly elevated manner; apex rounded, sometimes with a small lateral beak; surface smooth. Size: length 11.5–13.5 cm; breadth 7.5–9 cm; thickness 7–8 cm. Weight: 360–450 g. Skin: ground color bright yellow; dark red blush; numerous small white dots; skin thick, tough, and easily separating. Flesh: soft and juicy deep yellow flesh; flavor rich, aromatic, and sweet with a weak, pleasant aroma; no fiber. Stone and seed: thick and woody stone with a monoembryonic seed filling 40% to 60% of the stone. Season: June to July. Fruit can be cut and twisted apart in two halves when fully ripe because there is so little fiber; productive tree with beautiful fruit, but the soft flesh will probably discourage commercial use.

Springfels.—Seedling of Haden, perhaps with Sandersha as the pollinating parent. Seed planted at the residence of Charles Springfels, West Palm Beach, Fla. First fruited in 1925. Tree: moderately vigorous; spreading, semi-dwarf canopy. Fruit: eating quality fair to good. Shape: oblong; base rounded; stout stem inserted squarely in a shallow cavity or in a raised manner; apex rounded, sometimes with a small lateral beak; surface undulating. Size: length 19–22 cm; breadth 10–12.5 cm; thickness 10–11 cm. Weight: 850–1350 g. Skin: ground color bright yellow; dark red blush; numerous large white dots; skin thick, tough, and easily separating. Flesh: firm and juicy deep yellow flesh; flavor rich, aromatic, and sweet with a strong, pleasant aroma; little fiber. Stone and seed: medium-thick and woody stone with a monoembryonic seed filling 40% to 50% of the stone. Season: July to August. Color is similar to Haden, and is called Super Haden in some Latin American countries.

Stevens.—Seedling of Keitt planted in 1982 at the residence of Robert H. Stevens, Perrine, Fla. First fruited in 1989. Tree: vigorous; medium-sized, rounded, open canopy. Fruit: eating quality good to excellent. Shape: ovate to oblong; base rounded; stout stem inserted obliquely in a level manner; apex rounded with a small lateral beak; surface undulating. Size: length 14–16.5 cm; breadth 9.5–11 cm; thickness 8–9 cm. Weight: 620–970 g. Skin: ground color bright yellow; crimson or dark red blush; numerous large white dots; skin thick, tough, and easily separating. Flesh: soft, melting, and juicy lemon-yellow flesh; flavor mild and sweet with a weak, pleasant aroma; no fiber. Stone and seed: medium-thick and woody stone with a monoembryonic seed filling 60% to 75% of the stone. Season: July to August. Information limited on this cultivar; trees appear productive; fruit are attractive, but the flesh is soft.

Stringless peach.—Of unknown parentage from the area of Homestead, Fla. Tree: vigorous; large, upright, dense canopy. Fruit: eating quality fair. Shape: oval; base flattened; slender stem inserted obliquely in a level manner; apex rounded with no beak; surface smooth. Size: length 8.5–9.5 cm; breadth 7.5–8 cm; thickness 6.5–7 cm. Weight: 170–240 g. Skin: ground color bright yellow; no blush; numerous large yellow dots; skin thin, tough, and easily separating. Flesh: soft, tender, and juicy deep yellow flesh; flavor rich, aromatic, and sweet with a weak, pleasant aroma; medium fiber. Stone and seed: thick and woody stone with a polyembryonic seed filling 90% to 100% of the stone. Season: July to August. Consistent and heavy producer; fruit are usually produced in clusters.

Sunset.—Of unknown parentage from the grove of Frank Adams, Pine Island, Fla. Tree: vigorous; medium-sized, thin, and spreading canopy. Fruit: eating quality good. Shape: oblong; base rounded; slender stem inserted obliquely in a level manner; apex rounded usually without a beak; surface smooth. Size: length 10.5–12 cm; breadth 7.5–9 cm; thickness 7–7.5 cm. Weight: 200–370 g. Skin: ground color bright yellow; dark red blush; numerous small white dots; skin thick, tough, and easily separating. Flesh: soft, tender, and juicy orange flesh; flavor rich, aromatic, and sweet with a strong, pleasant aroma; little to medium fiber. Stone and seed: thick and woody stone with a monoembryonic seed filling 65% to 75% of the stone. Season: late June to July. Consistent, heavy producer, but fruit easily damaged by handling and storage procedures.

Super Haden.—See Springfels.

Tahar.—Seedling of Irwin introduced to Florida from Israel by the U.S. Dept. of Agriculture, Miami, Fla. Tree: vigorous; medium-sized, upright, dense canopy. Fruit: eating quality fair to good. Shape: ovate; base flattened; slender stem inserted obliquely in a level manner; apex rounded with a small lateral beak; surface undulating. Size: length 11.5–13 cm; breadth 8.5–9.5 cm; thickness 7.5–8 cm. Weight: 360–520 g. Skin: ground color bright yellow; dark red blush; numerous small white dots; skin thick, tough, and easily separating. Flesh: soft and juicy deep yellow flesh; flavor mild, aromatic, and slightly insipid with a strong repugnant odor; little fiber. Stone and seed: medium-thick and woody stone with a monoembryonic seed filling 50% to 70% of the stone. Season: July to August. Good color and size, but flavor insipid.

Tommy Atkins.—Seedling of Haden planted about 1922 in Broward County, Fla. First fruited in the early 1940s; trees first sold in 1948. Tree: vigorous; dense, rounded canopy. Fruit: eating quality fair to good. Shape: oval to oblong; base broadly rounded; stout stem inserted squarely in a slightly raised manner; apex rounded with a small lateral beak; surface smooth. Size: length 12–14.5 cm; breadth 10–13 cm; thickness 8.5–10 cm. Weight: 450–700 g. Skin: ground color orange-yellow; crimson or dark red blush; numerous small white dots; skin thick, tough, and adherent. Flesh: firm and medium-juicy lemon to deep yellow flesh; flavor mild and sweet with a strong, pleasant aroma; moderate fine fiber. Stone and seed: thick and woody stone with a monoembryonic seed filling 50% to 75% of the stone. Season: June to early August. Along with Keitt, one of the two most important cultivars in Florida; highly resistent to anthracnose; consistent, heavy producer; good shipping and handling characteristics; susceptible to internal breakdown.

Tong Dam.—Introd. to Florida from Thailand in 1973 by Robert J. Knight for the Rare Fruit Council International, Miami, Fla. Tree: vigorous; large, rounded, dense canopy. Fruit: eating quality good to excellent. Shape: oblong to long and slender; base rounded; slender stem inserted squarely in a shallow cavity; apex bluntly pointed with no beak; surface undulating. Size: length 14.5–18 cm; breadth 8.5–10.5 cm; thickness 7–8 cm. Weight: 380–500 g. Skin: ground color greenish-yellow; no blush; few small white dots; skin thick, tender, and medium adherent. Flesh: soft and juicy lemon to deep yellow flesh; flavor rich, spicy, and sweet with a weak, pleasant aroma; little or no fiber. Stone and seed: medium-thick and woody stone with a polyembryonic seed filling 60% to 70% of the stone. Season: late May to June. Remains green when ripe; consistent, heavy producer.

Torbet.—Of unknown parentage planted on the property of Nellie Torbet, Goulds, Fla. Tree: moderately vigorous; medium-sized, rounded canopy. Fruit: eating quality fair to good. Shape: round; base flattened; stout stem inserted squarely in a deep cavity; apex rounded without a beak; surface undulating. Size: length 9–10.5 cm; breadth 8.5–10 cm; thickness 9–10.5 cm. Weight: 360–440 g. Skin: ground color yellow-orange; orange-red blush; numerous large russet dots; skin thick, tough, and medium adherent. Flesh: firm and medium-juicy lemon-yellow flesh; flavor mild and insipid with a weak, pleasant aroma; abundant fiber. Stone and seed: thick and woody stone with a polyembryonic seed filling 90% to 100% of the stone. Season: June to July. Misidentified in some places as Tolbert; shape and color give this cultivar a unique look.

Turpentine.—Seedling race that came to South Florida as seed probably in the 1870s from the Caribbean and South America. The first seeds may have been planted near Coconut Grove, Fla. Tree: vigorous; large, spreading, rounded canopy. Fruit: eating quality poor to fair. Shape: oval; base flattened; medium stem inserted squarely in a shallow cavity; apex rounded without a beak; surface smooth. Size: length 7.5–8.5 cm; breadth 6.5–7.5 cm; thickness 6–6.5 cm. Weight: 140–200 g. Skin: ground color bright yellow; sometimes a pink blush; few large white dots; skin thick, tough, and easily separating. Flesh: firm and juicy lemon-yellow flesh; flavor rich, aromatic, spicy, resinous, and sweet with a strong, pleasant aroma; abundant coarse fiber. Stone and seed: thick and woody stone with a polyembryonic seed filling nearly 100% of the stone. Season: June to August. Commonly used as a rootstock in Florida and other countries; although small and fibrous, has a pleasing, quite desirable taste; may have more than one crop in a single season, and fruit often produced in clusters.

Valencia Pride.—Seedling of Haden planted at the residence of Mr. and Mrs. Zapiain, Miami, Fla. Seed planted in 1937; first fruited in 1941. Tree: vigorous; large, spreading, open canopy. Fruit: eating quality good. Shape: long and slender, sigmoid;

base rounded; stout stem inserted squarely usually in an elevated manner; apex rounded with a large apical beak; surface undulating. Size: length 18–21 cm; breadth 7.5–10 cm; thickness 7.5–8.5 cm. Weight: 600–900 g. Skin: ground color yellow; pink to dark red or crimson blush; numerous large yellow dots; skin thin, tender, and easily separating. Flesh: firm, melting, and juicy deep yellow flesh; flavor mild and sweet with a weak, pleasant aroma; little fiber. Stone and seed: thick and woody stone with a monoembryonic seed filling 50% to 70% of the stone. Season: July to August. Exceptionally attractive mango due to shape, size, and color; recommended for dooryard planting because of good eating quality and consistent production.

Van Dyke.—Seedling of unknown parentage planted in the 1930s at the residence of Mrs. Madelin Van Dyke, Miami, Fla. First propagated on a commercial scale in the 1950s and 1960s. Tree: moderately vigorous; large, open canopy. Fruit: eating quality good to excellent. Shape: oval; base rounded; slender stem inserted squarely in a level or slightly elevated manner; apex bluntly pointed with a small lateral beak; surface smooth. Size: length 9–11.5 cm; breadth 7.5–9.5 cm; thickness 7–8 cm. Weight: 250–520 g. Skin: ground color bright yellow; bright red or crimson blush; numerous small yellow dots; skin thick, tough, and easily separating. Flesh: firm, melting, and juicy orange-yellow flesh; flavor rich, spicy, and sweet with a strong, pleasant aroma; little fiber. Stone and seed: medium-thick and woody stone with a monoembryonic seed filling 60% to 80% of the stone. Season: July to August. Regular, heavy producer; good resistance to anthracnose; exceptional color and storage characteristics; small size fits well with marketing in Europe, but has limited its acceptance in the U.S. market; internal breakdown a limitation of this cultivar.

Wallace.—Seedling of Haden planted in 1935 by George R. Wallace, Bradenton, Fla. Tree: vigorous; large, rounded, dense canopy. Fruit: eating quality good. Shape: oblong; base rounded; stout stem inserted squarely in a level manner; apex rounded with no beak; surface smooth. Size: length 11–13.5 cm; breadth 7.5–9 cm; thickness 6–7.5 cm. Weight: 320–430 g. Skin: ground color bright yellow; crimson blush; numerous small yellow dots; skin thick, tough, and adherent. Flesh: medium-firm, and juicy lemon-yellow flesh; flavor rich, acidulous, and spicy with a strong, pleasant aroma; little fiber. Stone and seed: thick and woody stone with a monoembryonic seed filling 70% to 80% of the stone. Season: late June to July. Consistent producer with attractive and good-flavored fruit, but has received only limited interest in Florida for dooryard or commercial use.

Zill.—Seedling of Haden planted in 1922 by Carl King at Lake Worth, Fla. Lawrence Zill now has the original tree on his property in Boynton Beach, Fla. First fruited in 1930; propagated in 1940 by Zill. Named and described in 1945. Tree: vigorous; large, spreading, open canopy. Fruit: eating quality good to excellent. Shape: oval to ovate; base slightly flattened; fairly stout stem inserted obliquely in a narrow, shallow cavity; apex rounded to bluntly pointed with a small lateral beak; surface undulating. Size: length 8.5–10 cm; width 7.5–8.5 cm; thickness 6–7 cm. Weight: 230–370 g. Skin: ground color greenish-yellow to yellow; dark red or crimson blush; numerous small white dots; skin thin, tender, and adherent. Flesh: soft and juicy pale yellow flesh; flavor mild and sweet with a strong, pleasant aroma; no fiber. Stone and seed: thick and woody stone with a monoembryonic seed filling 70% to 90% of the stone. Season: June to early July. Regular producer; good color and flavor, but does not hold up well in storage or shipping.

MAYHAW

Big Red (*No. 1 Big*).—Chance seedling of *Crataegus aestivalis* discovered in Pearl River swamps near Sandy Hook Mississippi, by T.O. Warren. Introduced in 1969. Fruit: large to very large, about 2.5 cm; oval; skin brilliant red, glistening finish; flesh mixed white and red; high soluble solids; ripens late April to early May in southern U.S. Tree: very vigorous; upright spreading; not precocious; very productive; susceptible to rust, hardy in zone 6.

No. 1 Big.—See Big Red.

Super Spur.—Chance seedling of *Crataegus opaca* discovered in the swamps near Sibley, La., by J.S. Akin. Fruit: more red color than average for this species, good for jelly and syrup. Tree: spur-type bearer; productive, hardy in zone 6.

Superberry—See Texas Superberry.

Texas Superberry (*Superberry, T.O.'s Superberry*).—Chance seedling of *Crataegus aestivalis* found in the wild by T.O. Warren near Buna, Texas, in early 1980s. Fruit: large, shiny, red and purple, attractive, makes good preserves. Tree: early blooming, hardy in Zone 6.

T.O.'s Superberry.—See Texas Superberry.

MULBERRY

Illinois Everbearing.—Orig. in White County, Ill., on the bottom-lands of the Wabash River, by Peter Glaser of Evansville, Ind. Introd. in 1958 by Burgess Seed & Plant Co., Galesburg, Mich., and by Louis Gerardi Nursery, O'Fallon, Ill. Considered to be a polyploid derived from *Morus rubra*; discovered in 1947. Fruit: large; flavor good; nearly seedless; matures over a long season. Tree: extremely hardy; very productive; young grafts extremely vigorous.

• *Revisions for this commodity were not received before the book went to press; therefore, the above variety description is virtually identical to that published previously.*

NECTARINE

***Anderbrite*.**—See Stark Sweet Melody.

Annabella.—Orig. in Madera, Calif., by John M. Garabedian. Introd. in 1954. Plant patent 1287; 29 June 1954. Tioga o.p. Fruit: skin yellow, deeply blushed on sunny side with red; flesh yellow, mottled very red next to pit, texture firm, crisp, meaty, flavor mild, aroma distinct, quality good, clingstone; matures about 7 Sept. at place of origin or about 4 weeks later than Tioga. Tree: size medium; bears regularly. Leaves: glandular.

Apache.—Orig. in Wheeler Ridge, Calif., by J.J. Kovacevich, Arvin, Calif. Introd. in 1977. Plant patent 4042; 19 Apr. 1977; assigned to J.J. Kovacevich, Jr. Mutation of Sunrise nectarine, which it resembles, but ripens 10–14 days earlier.

April Glo.—Orig. by C.F. Zaiger, G.N. Zaiger, L.M. Gardner, and G.G. Zaiger, Modesto, Calif. Plant patent 7305; 21 Aug. 1990. (Ruby Gold x peach of unknown parentage) x (Sunred x Royal). Fruit: medium; globose with tip and pronounced suture; red to dark red over yellow to yellow-orange ground color; flesh medium firm, light yellow, melting, semi-freestone; matures 10–14 days before May Glo. Flowers medium, nonshowy, pink; leaf glands globose. Tree: large; vigorous; upright; productive. Chilling requirement 200 h.

Arctic Pride.—Orig. by C.F. Zaiger, G.N. Zaiger, L.M. Gardner, and G.G. Zaiger, Modesto, Calif. [(Ruby Gold x Redwing) x (O'Henry x Giant Babcock)] F_4. Plant patent 8450; 9 Nov. 1993. Fruit: large; nearly globose; skin attractive red; flesh white, firm, melting; freestone; flavor excellent; high soluble solids. Ripens in late season. Flowers: large, showy, pink; leaf glands reniform. Tree: large; vigorous; upright; productive.

Armking.—Orig. in Wasco, Calif., by David L. Armstrong. Introd. in 1969. Plant patent 2943; 25 Nov. 1969; assigned to Armstrong Nurseries, Ontario, Calif. Unnamed peach x unnamed peach; selected in 1963. Fruit: 2 1/4 to 2 1/2 inches in diam.; skin color good, attractive; flesh yellow, firm, pronounced aroma; semi-freestone; resembles Armred, but ripens 2–4 days earlier; of commercial quality. Tree: normal size; vigor good; hardiness untested; leaf glands globose, sometimes weakly reniform; flowers nonshowy, dark pink.

Armqueen.—Orig. in Wasco, Calif., by David L. Armstrong. Introd. in 1970. Plant patent 2958; 23 Dec. 1969; assigned to Armstrong Nurseries, Ontario, Calif. Panamint nectarine x unnamed peach seedling (Palomar x Springtime). Cross made in 1957; selected in 1961; tested as 57205-2. Fruit: 2 1/2 to 2 5/8 inches in diam.; uniform, nearly globose, no beak; skin smooth, no pubescence, thin but tough, highly colored with dark red, occasional mottling and flecking of yellow ground color showing through, particularly on side away from sun; flesh yellow, minimal amount of red at pit cavity at tree-ripe stage, quality, flavor, and aroma excellent, medium firm, texture fine, melting, freestone; ripens early and uniformly, 11–13 days after Mayred or about 6–7 June at Wasco; most nearly resembles Sunrise (Merrill Sunrise) nectarine. Tree: vigorous; upright-spreading; productive; flowers showy, light pink; leaf glands weakly reniform; chilling requirement nearly same as Springtime peach. Recommended for commercial planting.

Armred.—Orig. in Northridge, Calif., by David L. Armstrong and Herbert C. Swim, Armstrong Nurseries, Ontario, Calif. Introd. in 1964. Plant patent 2580; 21 Dec. 1965; assigned to Armstrong Nurseries. Parentage from an extended series of crosses on one side going back three generations to Babcock and July Elberta peaches and Boston and Goldmine nectarines; on the other side side, going back two generations to Goldmine and July Elberta; selected in 1957. Fruit: size medium; round; skin attractive, smooth, heavily blushed with little or no speckling; flesh yellow, semi-freestone; ripens early, generally several days after Robin peach and a few days before Sunrise nectarine; particularly good quality and nectarine flavor for such an early ripening variety. Tree: vigorous; produces heavy crops; chilling requirement medium, between June Gold and Robin peaches.

Arctic Glo.—Orig. by C.F. Zaiger, G.N. Zaiger, L.M. Gardner, and G.G. Zaiger, Modesto, Calif. Plant patent 7884; 16 Jan. 1992. [(Carnival o.p.) x (Fayette o.p.)] x white-fleshed nectarine seedling. Fruit: large; nearly globose; red over pale yellowish-white ground color; flesh white with pinkish streaks near pit cavity, firm, meaty, clingstone. Ripens 9 June at Modesto. Flowers large, showy, pink; leaf glands globose. Tree: large; vigorous; upright.

Arctic Queen.—Orig. by C.F. Zaiger, L.M. Gardner, G.N. Zaiger, and G.G. Zaiger, Modesto, Calif. Plant patent 8094; 12 Jan. 1993. Complex origin. Fruit: large; nearly globose; firm; flesh melting, white, freestone with bleeding of red from pit cavity; orange-red over yellowish-white ground color. Ripens late July at Modesto. Flowers large, showy, pink; leaf glands reniform. Tree: large; vigorous; upright; productive.

Arctic Rose.—Orig. by C.F. Zaiger, G.N. Zaiger, L.M. Gardner, and G.G. Zaiger, Modesto, Calif. Plant patent 7889; 23 June 1992. (Ruby Gold x Red Wing) o.p.2. Fruit: large; globose; medium red over chalk-white ground color leaving mottled red skin; flesh white with pale red streaks from pit cavity, firm, meaty, clingstone. Ripens 1 July at Modesto. Flowers large, showy, pink; leaf glands reniform. Trees: large; vigorous; upright.

Arctic Show.—Orig. by C.F. Zaiger, G.N. Zaiger, L.M. Gardner, and G.G. Zaiger, Modesto, Calif. Plant patent 7920; 21 July 1992. [(Sun Grand x Merrill Gem) o.p. x (late nectarine x Autumn Grand) o.p.] x (Redwing x nectarine seedling) o.p. Fruit: large; globose; high percentage of red over white to pinkish-white ground color; flesh white with red at cavity, firm, freestone. Ripens 10 days after Autumn Grand. Flowers large, showy, pink; leaf glands primarily globose with some reniform. Tree: large; vigorous; upright.

Autumn Free.—Orig. in Le Grand, Calif., by F.W. Anderson. Introd. in 1970. Plant patent 2976; 30 June 1970; assigned to Reedley Nursery, Reedley, Calif. Sun Grand x Gold King. Tested as 9E845. Fruit: large, round; skin yellow, substantially overspread with red; flesh yellow, with red next to stone, firm, freestone; ripens about 3 weeks after Golden Grand and Late Le Grand. Tree: large; spreading; productive and regular bearer; flowers large, showy; leaf glands reniform.

Autumn Grand.—Orig. in Le Grand, Calif., by F.W. Anderson, Merced, Calif. Introd. in 1969. Plant patent 2894; 10 June 1969; assigned to Reedley Nursery, Reedley, Calif. Late Le Grand x unnamed seedling of Gold King. Tested as Anderson 23H890. Fruit: large, uniform, globose; skin yellow, partially overspread with red; flesh yellow, clingstone; resembles Gold King, but distinguished by the absence of the red suture line, which ripens before the remainder of the fruit, has better shipping quality and longer shelf life, ripens 1 week later. Tree: large to medium; vigorous and spreading; productive and regular bearer; flowers large; leaf glands reniform.

Autumn Red.—Orig. in Kingsburg, Calif., by Yoshio Yamada. Plant patent 3915; 15 June 1976; assigned to Vaughn Girazian, Fresno, Calif. Probable mutation of Autumn Grand. Fruit: large; semi-cling; yellow fleshed; yellow skinned with light-red blush; excellent eating quality; ripens about 1 Oct., 3–4 weeks after Autumn Grand. Tree: large; vigorous; spreading; productive.

Banquet.—Orig. in Los Angeles, Calif., by Walter E. Lammerts, Dept. of Subtropical Hort., Univ. of California. Introd. in 1945. Babcock peach x Quetta; selected in 1945. Fruit: 2 1/4 to 2 1/2 inches in diam.; skin bright red; flesh yellow with a rich subacid blend, semi-freestone; pit small; cans well. Tree: low chilling requirement, 100 h at or below 40F.

Big Jim.—Orig. by J.W. Taylor, Dinuba, Calif. Plant patent 8021; 3 Nov. 1992. Assigned to Ito Packing Co., Reedley. (Red Jim x Summer Grand) o.p. Fruit: large; nearly round; very firm and meaty; flesh yellow, clingstone; 90% shiny red over yellow ground color. Ripens mid-June. Flowers large, showy, pink; leaf glands globose. Tree: large; vigorous; spreading; productive but low percentage of flower buds leads to less than normal fruit thinning.

Big Juan.—Orig. by L.G. and N.G. Bradford, LeGrand, Calif. Plant patent 8197; 6 Apr. 1993. Red Diamond x nectarine. Similar to May Diamond but larger in size, has no red in yellow flesh, and ripens 7 days earlier; freestone. Flowers small, pink; leaf glands reniform. Tree: large; vigorous; spreading and dense; productive.

Bim.—Orig. in Le Grand, Calif., by F.W. Anderson. Introd. in 1944. Plant patent 575; 13 Apr. 1943. (J.H. Hale peach x Lippiatt)F_2; selected in 1934. Fruit: medium to large; skin bright red; flesh yellow, red at pit; texture medium firm; flavor slightly subacid, mild; quality good; freestone; ripens 7–10 days before J.H. Hale. Tree: large, spreading, resembles Elberta; leaf glands reniform; flowers large, pink.

Bob Grand.—Orig. in Le Grand, Calif., by F.W. Anderson. Introd. in 1970. Plant patent 3000; 17 Nov. 1970; assigned to Reedley Nursery, Reedley, Calif. Red Grand x unnamed never ripening seedling; tested as 37P657. Fruit: resembles Late Le Grand, but more uniform in size, skin smoother and more overspread with brighter red, more attractive, flesh more melting and better quality, ripens same time; clingstone. Tree: resembles Late Le Grand; flowers large, showy; leaf glands reniform.

Bradcrim.—See Stark Crimson Snow™.

Brendavilha.—Orig. in New South Wales, Australia, by John Christie. Introd. in 1985. Sport of Maravilha peach. Registered in New South Wales. Fruit: smaller, higher percent red blush, and rounder than Maravilha. Fruit subject to surface cracks. Other fruit, tree, and foliage characters resemble Maravilha.

Bright.—See Stark Sweet Melody.

Burbank Flaming Gold.—See Flaming Gold.

Burbank's Fuzzless.—See Flaming Gold.

California Gold.—Orig. in Santa Rosa, Calif., by Luther Burbank. Introd. in the late 1920s. Parentage unknown; selected about 1925. Fruit: skin yellow, blushed with crimson; flesh yellow, firm, very juicy; pit small, freestone. Tree: tendency to set double fruit.

Carlson.—Orig. by H.D. Carlson, Yakima, Wash. Plant patent 7527; 21 May 1991. Chance seedling. Fruit: medium; symmetrical globose; red over yellow-orange skin; flesh yellow, firm, crisp, semi-freestone. Ripens 14 days before Red Gold. Flowers medium, pink; leaf glands reniform. Tree: medium size and vigor; spreading.

Carolina.—Orig. at Agr. Expt. Sta., Univ. of Florida, Gainesville. Bred in 1976, selected in 1979, and tested as Fla. 9-9N by W.B. Sherman. Introduced in Spain by Viveros Orero, S.A. in 1990. Complex parentage of Florida selections. Fruit: medium, round, skin 90% to 100% red blush over yellow ground color; flesh yellow, medium firm, melting, semi-freestone. Ripening about 90 days after bloom. Flowers pink, showy; leaf glands reniform. High resistance to bacterial spot. Tree: vigorous; productive. Chilling requirement about 325 chill units, blooming with Flordagold.

Casa Linda *(Merrill Casa Linda)*.—Orig. in Red Bluff, Calif., by Grant Merrill. Introd. in 1953. Seedling of J.H. Hale x Quetta o.p.3; selected in 1952. Fruit: large; appearance attractive; flesh white, freestone, flavor very good; resembles Quetta; for home orchards, as too soft for long-distance shipment. Tree: large, moderately productive. Variety now obsolete.

Cashmere.—See Sunfre.

Cavalier.—Orig. in Blacksburg, Va., by George D. Oberle and R.C. Moore, Virginia Agr. Expt. Sta. Introd. in 1952. (P.I. 43143 x unnamed V.P.I. peach seedling) x self; cross made in 1936 by Moore; selected in 1940; tested as V.P.I. 17. Fruit: 2 inches in diam., slightly smaller than Redchief; ovate, about 2 1/4 inches long; skin orange-yellow, splashed and mottled rather dark red, attractive, good finish; flesh yellow, freestone, firm, aromatic, slightly bitter, quality better than average; ripens about 8 days before Elberta peach; under conditions at Blacksburg, it has shown outstanding ability to escape infection by brown rot; resembles Garden State. Tree: vigorous; productive; flowers showy, hardy.

Cherokee.—Orig. in Blacksburg, Va., by George D. Oberle, Virginia Agr. Expt. Sta. Introd. in 1965. (Sunhigh x V.P.I. 40N nectarine) o.p.; tested as V.P.I. 59N. Fruit: about 2 1/4 inches in diam.; ovate, symmetrical; skin highly colored with fine finish; flesh yellow, texture noticeably stringy, flavor rich and above-average quality, semi-freestone; ripens 13 days before Lexington. Tree: vigor medium; productive; leaf glands mostly reniform, but difficult to classify; flowers large, showy; above-average tolerance to brown rot infection and spring frost.

Chiyodared.—Orig. by M. Yoshida, H. Kyotani, M. Yamaguchi, T. Kozono, T. Nishida, Y. Ishizawa, K. Kanato, and K. Nishimura of Japan. Plant patent 8116; 26 Jan. 1993. Assigned to Fruit Tree Res. Sta., Tsukuba. Hiratsukared x Nectared 5; first fruited in 1976 from seedling no. N-45-10; selected as Tsukuba no. 76 in 1980. Fruit: medium; round to elliptic; medium firm; flesh yellow, melting, clingstone; solid deep red over yellow ground color. Ripens early season in mid-July at Tsukuba, about 2 weeks before Hiratsukared. Flowers large, pink. Tree: medium vigor; spreading.

Clinton.—Orig. in Visalia, Calif., by Clinton L. Hagler. Introd. in 1960. Plant patent 1935; 19 Apr. 1960. Bud mutation of Le Grand; discovered in 1953. Fruit: large, up to 3 inches in diam.; skin ground color yellow overlaid with red to a full blush, more than average in toughness and thickness; flesh yellow with some red around pit, flavor very mild, aroma very slight, slightly tart, quality very good, clingstone; ripens about 10–14 days later than parent, about 7–10 days before Late Le Grand, or about 21 July for first picking for shipping at place of origin; good keeping and handling qualities; resembles Le Grand. Tree: leaves more crinkly than those of Le Grand, having a tendency to roll along the edges; flowers self-fruitful, showy, petal edges slightly ruffled.

Columbia.—Orig. in Blacksburg, Va., by G.D. Oberle, Virginia Agr. Expt. Sta. and Virginia Polytechnic Inst. & State Univ. Introd. in 1974. Cavalier x unnamed yellow-fleshed, freestone nectarine developed by U.S. Dept. of Agriculture, Beltsville, Md.; cross made in 1954; first fruited in 1958; seedling I-56 tested as VPI 90N. Fruit: 2 1/4 inches in diam.; slightly ovate; skin 75% medium-red over orange-yellow ground color, tendency to russet in years having blossoming season frosts; flesh yellow-orange, firmer than average, moderately stringy, flavor rich, quality above average; ripens 3 days before Cavalier, 10 days before Elberta; for use as fresh fruit. Tree: productive, hardiness of fruit buds and blossoms nearly that of Madison peach.

Columbina.—Orig. in Agr. Expt. Sta., Univ. of Florida, Gainesville, by R.H. Sharpe. Complex parentage of Florida selections; selected in 1969; tested as Fla. 19-37S. Named in southern Brazil. Fruit: medium; skin 50% red blush; flesh yellow, semi-freestone. Ripens 85 days from bloom. Flowers showy; leaf glands reniform. Resistant to bacterial leafspot. Chilling requirement: 350 chill units.

Compton.—Orig. in Ivanhoe, Calif., by Verna A. Compton. Introd. in 1958. Plant patent 1674; 14 Jan. 1958. Parentage unknown; discovered in 1953. Fruit: skin yellow ground color blushed with pink; flesh yellow with some red at the pit, texture firm, quality good; freestone when fully ripe, semi-clingstone at early stages and even into early picking; cooking quality very good, being equal in tests to Elberta peach. Tree: size controlled by pruning, tends to be medium, spreading; productive; flowers small to medium.

Crimson Gold.—Orig. in Modesto, Calif., by C. Floyd Zaiger. Introd. in 1967. Plant patent 2825; 6 Aug. 1968; assigned to Dave Wilson Nursery, Modesto. Seedling of unknown parentage. Fruit: resembles Early Sungrand as to color, firmness, shape, and size, but early maturing; ripens about 14 days earlier than Grand River and Sunrise. Tree: medium to large; productivity heavy; flowers large to medium, pink; leaf glands reniform; chilling requirement 400–600 h at or below 45F.

CrimsonStar *(Sunecteleven).*—Orig. in Fresno, Calif., by John H. Weinberger. Introd. in 1982. Plant patent 4942; 16 Nov. 1982; assigned to Superior Farming Co. Unnamed late-ripening nectarine o.p.; selected in 1977; tested as 87-101. Fruit: compared to Fantasia, slightly smaller, more ovate, with less-prominent apex tip; skin bright red on yellow ground color, medium thick; flesh yellow with red streaks next to pit; freestone. Ripens 1 week after Summer Grand. Tree: vigor medium; upright-spreading; productive. Flower: large, showy, pink.

Davis.—Orig. in Austin, Texas, by Walter E. Davis. Introd. in 1922 by Austin Nursery Co., Austin. Chance seedling; discovered in 1916. Fruit: large; skin red over pale yellow; flesh yellow, Honey type; freestone. Tree: very productive.

Del Rio Rey.—Orig. near Le Grand, Merced County, Calif., by F. Anderson. Plant patent 4081; 19 July 1977; assigned to Reedley Nursery, Reedley, Calif. May Grand x Red Free. Fruit: large to medium; symmetrical; globose; freestone; resembles May Grand in red skin and juicy flesh; resembles Red Free in other respects, but ripens 1 week later; kernel sweet. Tree: medium; medium vigor; spreading; productive; flowers large; leaf glands reniform.

Desert Delight.—Orig. by C.F. Zaiger, G.N. Zaiger, L.M. Gardner, and G.G. Zaiger, Modesto, Calif. Plant patent 7891; 23 June 1992. (Ruby Gold o.p.) x May Glo. Fruit: medium; globose; flesh light yellow, firm, clingstone. Ripens early in subtropical climates such as Palm Desert, Calif. Flowers large, showy, pink; leaf glands globose. Tree: large; vigorous; upright. Chilling requirement 200 h.

Diamond Jewel.—Orig. by N.G. and L.G. Bradford, LeGrand, Calif. Plant patent 7050; 24 Oct. 1989. Red Diamond x unnamed seedling. Fruit: large; round; dark red over entire surface; flesh brilliant yellow to pit, extremely firm, crisp, clingstone. Ripens 22 days before Red Diamond. Flowers small, moderate pink; leaf glands reniform. Tree: medium size and vigor; semi-spreading.

Diamond Jim.—Orig. by J.W. Taylor, Dinuba, Calif. Plant patent 6471; 20 Dec. 1988. Assigned to Ito Packing Co., Reedley, Calif. Red Jim x May Glo. Fruit: uniform and large; nearly round; reddish-purple skin; flesh yellow with red throughout but more intense at pit and suture, firm, meaty, clingstone. Ripens 10 days before May Grand. Flowers large, showy, red anthers; leaf glands reniform. Tree: large; vigorous; spreading; very productive.

Durbin.—Orig. in Byron, Ga. Selected by V.E. Prince in cooperation with U.S. Dept. of Agriculture. Introd. in 1980. (Flaming Gold x Candoka) o.p. x (Garden State x Tenn. Natural seedling) o.p. from 1964 cross. Fruited for first time in 1967; tested as BY 5-6930. Fruit: size medium, but larger than Nectared 2; round to slightly ovate; 75% covered with bright red, undercolor yellow with slight green tint; flesh yellow, firm but melting, medium in texture, flavor good, semi-freestone; ripens early June at Byron, with Nectared 2, slightly earlier than Coronet peach. Tree: vigorous and productive for early maturing variety; less susceptible to brown rot than Nectared 2; blossoms small, nonshowy, self-fertile; leaf glands reniform; chilling requirement 850 h below 45F.

Earliglo.—Orig. by C.F. Zaiger, G.N. Zaiger, L.M. Gardner, and G.G. Zaiger, Modesto, Calif. Plant patent 7402; 25 Dec. 1990. (Ruby Gold x peach of unknown parentage) x May Glo. Fruit: medium; globose; red over yellow ground color; flesh yellow, firm, meaty, clingstone; maturing 5 days before May Glo. Tree: large; vigorous; upright; productive. Chilling requirement 200 h.

Earliscarlet.—Orig. at U.S. Dept. of Agriculture Appalachian Fruit Res. Sta., Kearneysville, W.Va., by R. Scorza, H.W. Fogle, and L.E. Gilreath. Introd. in 1985. Parentage unknown; selected in 1971; tested as B68598. Fruit: skin 80% scarlet over bright yellow ground color; somewhat elongate without prominent suture; flesh yellow, firm, semi-freestone; excellent flavor. Ripens 3–7 days before Redhaven. Tree: vigorous; productive. Abundance and hardiness of flower buds usually makes thinning necessary. Chilling requirement about 800 h. Winter hardiness similar to Redhaven. Blossom: showy; self-fertile. Leaf glands globose. Resistant to bacterial leafspot.

Early Le Grand.—Orig. in Merced, Calif., by F.W. Anderson. Introd. in 1950. Plant patent 980; 19 Sept. 1950; assigned to Reedley Nursery, Reedley, Calif. Le Grand o.p.; selected in 1947. Fruit: resembles Le Grand but ripens from 2 to 3 weeks earlier. Tree: large, spreading; leaf glands reniform; flowers large, showy, pink.

Early Maygrand.—Orig. near Le Grand, Merced County, Calif., by F.W. Anderson. Plant patent 4435; 10 July 1979; assigned to Reedley Nursery, Reedley, Calif. Red Diamond x Early Sungrand. Fruit: size medium, uniform, symmetrical; skin yellow, substantially overspread with red; flesh yellow, firm; freestone; ripens early June, between Aurelio Grand and May Grand. Tree: large; vigorous; spreading; productive; flowers large, pink; leaf glands reniform.

Early Red.—Orig. in Le Grand, Calif., by F.W. Anderson, Merced, Calif. Introd. in 1974. Plant patent 3633; 8 Oct. 1974; assigned to Reedley Nursery, Reedley, Calif. Red June o.p.; tested as SP730. Fruit: size medium; globose to oblong; skin yellow, overspread with deep red; flesh yellow, freestone, excellent for tis season; for long-distance market; ripens 1–10 June. Tree: large; vigorous; requires heavy thinning; flower large, showy; leaf glands reniform.

Early Red Grand.—See Royal Grand.

Early Red Jim.—Orig. by J.W. Taylor, Reedley, Calif. Plant patent 7186; 13 Mar. 1990. Assigned to Ito Packing Co., Reedley. Red Jim x May Grand. Fruit: large; round but compressed at base; high percentage of dark red over yellow skin; flesh very firm, meaty, clingstone. Ripens 20 July at Reedley. Flowers large, showy, pink, red anthers; leaf glands reniform. Tree: large; vigorous; spreading.

Early Sun.—Orig. in Del Rey, Calif., by Tom Chamberlin, H.P. Metzler & Sons. Introd. in 1990. Unknown parentage. Fruit: large; round; skin bright red on yellow ground color; flesh yellow, firm; ripens with May Grand. Tree: vigorous; productive.

Early Sungrand.—Orig. in Le Grand, Calif., by F.W. Anderson. Introd. in 1955. Plant patent 1420; 20 Sept. 1955; assigned to Reedley Nursery, Reedley, Calif. Sun Grand o.p.; selected in 1953; tested as Anderson 2B49. Fruit: large; skin highly colored; flesh yellow, firm, semi-freestone; ripens about with John Rivers or about 2 weeks before Sun Grand. Tree: large, upright; productive; leaf glands reniform; flowers large.

Early Sunrise.—See Sunbright II.

Eastern Glo.—Orig. by C.F. Zaiger, G.N. Zaiger, L.M. Gardner, and G.G. Zaiger, Modesto, Calif. Plant patent 7890; 23 June 1992. (Red Grand x Crimson Gold) o.p. Fruit: large; firm; nearly globose; excellent flavor; high percentage of red over yellow ground; flesh yellow, melting, clingstone. Ripens 7 days after Crimson Gold. Flowers pink; leaf glands reniform. Tree: large; vigorous; upright; productive. More resistant than Red Grand to bacterial spot.

Elbertarine.—Orig. in Farmersville, Calif., by Harold D. Dougherty. Introd. in 1958. Plant patent 1753; 23 Sept. 1958. F_1 Elberta peach seedling x (Le Grand) sp.2. Fruit: medium; elongated rather than round; skin pale yellow, blushed with red, slight tendency to crack; flesh yellow, flavor mild, resembling that of Elberta; aroma pronounced; freestone; ripens with Sun Grand. Tree: leaf glands reniform; flowers large.

Fairlane.—Orig. in Fresno, Calif., by U.S. Dept. of Agriculture. Introd. in 1973. (Le Grand x Sun Grand) x Fantasia; cross made in 1966; first fruited in 1969; selected by J.H. Weinberger; tested as F94-98. Fruit: large; oval; skin bright yellow ground color, slightly redder than Flamekist; flesh yellow, firm, smooth texture, flavor good, clingstone; ripens immediately after Flamekist, fruit sometimes drop before maturity. Tree: vigorous; productive; flowers self-fertile, large; leaf glands reniform.

Fantasia.—Orig. in Fresno, Calif., by U.S. Dept. of Agriculture. Introd. in 1969. Gold King x seedling of Red King; cross made in 1961; tested as F58-40; first fruited in 1964; selected in 1964 by John H. Weinberger. Fruit: larger than Red Grand; ovate; skin bright yellow with 1/3 to 2/3 covered with bright red blush, more highly colored than Le Grand; flesh yellow, firm, texture smooth, quality good, freestone; shipping quality good; ripens immediately after Flavortop, or a few days earlier than Red Grand. Tree: vigorous; productive; flowers large, self-fertile; leaf glands reniform; short chilling requirement.

Firebrite.—Orig. in Fresno, Calif., by U.S. Dept. of Agriculture. Introd. in 1974.

Flavortop x 66-90 (F2 seedling of Red King, o.p.). Cross made in 1966; selected in 1969 by J.H. Weinberger; tested as F103-32. Fruit: large; ovate; skin blushed bright red over 80% to 100%, attractive; flesh yellow, firm, flavor excellent, texture smooth; ripens with Early Sungrand, a few days before Independence. Compared to Early Sungrand: more attractively colored, slightly larger, smoother shape. Tree: moderately vigorous and productive; flowers large, self-fertile; leaf glands reniform; medium chilling requirement.

Fire Diamond.—Orig. by Norman G. and Lowell G. Bradford, Le Grand, Calif. Introd. in 1989. Early Diamond x unnamed genetic dwarf peach. Plant patent 7064; 5 Dec. 1989; assigned to Bright's Nursery, Le Grand. Fruit: most nearly resembles Early Diamond; small, uniform; globose; skin dark red 75% to 90% on orange-yellow ground color; flesh yellow, medium firm; semi-freestone; mild subacid. Susceptible to skin cracking and to blossom brown rot in wet years. Ripens with Mayfire at Le Grand, 3 weeks before Early Diamond. Tree: vigor medium; size medium; spreading; very productive; regular production. Blossoms purplish-red. Leaves mostly reddish-brown from midsummer to dormancy.

Fireglobe.—Orig. in Winters, Calif., by Guy L. Philp and Claron O. Hesse, California Agr. Expt. Sta., Davis. Introd. in 1954. (Fay Elberta peach x Lippiatt seedling 29F3)] o.p. selfed; cross made in 1943 by Philp; selected in 1946; since 1947 tested by Hesse, as Calif. C5-27B. Fruit: medium-large, 2 1/4 to 2 1/2 inches in diam.; nearly round; skin develops a high blush before maturity; flesh a bright clear yellow, pit cavity red, but color does not extend into flesh more than a fraction of an inch, firm, melting, freestone; quality very good; ripens with Quetta and Elberta peach. Tree: vigorous; good commerical crops, even after mild winters that produced some bud drop; chilling requirement more than Quetta or Stanwick, less than John Rivers or Philp; flowers abundantly, flowers showy, self-fruitful; leaf glands globose.

Flame Glo.—Orig. by C.F. Zaiger, G.N. Zaiger, L.M. Gardner, and G.G. Zaiger, Modesto, Calif. [Flamekist x (SunGrand x Merrill Gem) F2] F3. Plant patent 8441, 2 Nov. 1993. Fruit: large; nearly globose with high degree of attractive red skin; flesh light yellow, firm; clingstone with good flavor. Ripens 45 days before FlameKist. Flower: large, showy, pink; leaf glands reniform. Tree: large; vigorous; upright; productive.

Flamekist.—Orig. in Fresno, Calif., by Crops Res. Div., Agr. Res. Serv., U.S. Dept. of Agriculture. Introd. in 1968. Gold King selfed. Planted in 1959; first fruited and selected in 1962 by John H. Weinberger; tested as F113-87. Fruit: large; ovate; skin attractive yellow with red blush; heavily shaded fruits do not attain satisfactory color; flesh yellow, firm, texture smooth, quality excellent, clingstone; has less growth-cracking than other varieties of same ripening time; ripens 3 weeks after Elberta and a few days earlier than September Grand; good shipping quality. Tree: vigor moderate, should be controlled to obtain good fruit color; productive; flowers large, self-fertile; leaf glands reniform; moderately low chilling requirement; recommended for areas of low rainfall.

Flaming Gold (*Burbank Flaming Gold, Burbank's Fuzzless, Fuzzless*).—Orig. in Sebastopol, Calif., by Luther Burbank. Introd. in 1928. Parentage unknown; selected about 1916. Fruit: skin bright gold ground cover blushed with bright red, smooth, waxy; flesh yellow, freestone; ripens before Elberta. Tree: vigorous; medium-large, spreading; medium hardy; productive.

Flaming Red.—Orig. by G.T. Nakagawa, Fresno, Calif. Plant patent 5480; 28 May 1985. Entire tree mutant of Flamekist. Similar to parent, but ripens 12 days later.

Flavortop.—Orig. in Fresno, Calif., by U.S. Dept. of Agriculture. Introd. in 1969. Fairtime peach o.p. Selected in 1964 by John H. Weinberger; first fruited in 1964; tested as F58-80. Fruit: large; ovate; skin mostly red over attractive undercolor at maturity; flesh yellow, firm, texture smooth, quality excellent, freestone; shipping quality good; ripens just after Independence and with Sun Grand, second week in July. Tree: vigorous; productive; flowers large, self-fertile; leaf glands reniform; moderate chilling requirement.

Fox.—Orig. in Ann Arbor, Mich., on the farm of Grant Fox by Clarence J. Kirby of Ilgenfritz Nurseries. Introd. about 1938. Parentage unknown; discovered before 1933. Fruit: skin red; flesh yellow, freestone; resembles Halehaven peach. Variety apparently obsolete.

Francesco.—Orig. in Hammonton, N.J., by Anthony C. Curcio and Nunzio C. Curcio. Introd. in 1968. Plant patent 2839; 15 Oct. 1968; assigned to Bountiful Ridge Nurseries, Princess Anne, Md. Parentage unknown. Seedling discovered in 1965. Fruit: large; round; skin yellow ground color covered with bright red blush, slightly russeted, tough; flesh yellow, red next to pit, moderately juicy, subacid, quality good, freestone; keeping qualities good; resistant to bruising; ripens 1 Sept. with Blake and Redskin.Tree: medium-large; upright, round-topped; vigorous; hardy; productive, annual bearer; flowers medium-large, peach blossom; leaf glands mixed.

Freedom.—Orig. in Le Grand, Calif., by F.W. Anderson. Introd. in 1951. Plant patent 1161; 13 Jan. 1953; assigned to Thomas B. Stribling, Jr., Stribling's Nurseries, Merced, Calif. Le Grand x (Bim x Kim); first fruited in 1948; tested as Anderson 9K66. Fruit: very large; skin highly blushed; flesh golden yellow, freestone, firm, juicy, flavor excellent; good for shipping, canning, and dessert use; ripens a few days earlier than Quetta and Le Grand, which it resembles. Tree: size medium; spreading; productive; leaf glands globose; flowers large, pink.

Fresno Sun.—Orig. in Del Rey, Calif., by Tom Chamberlin, H.P. Metzler & Sons. Introd. in 1990. Unknown parentage. Fruit: large; round; skin bright red on yellow ground color, smooth, waxy; flesh yellow, firm; ripens just before Firebrite. Tree: vigorous; productive.

Fuzzless.—See Flaming Gold.

Fuzzless-Berta.—Orig. in Blacksburg, Va., by Frederick W. HofMann. Introd. in 1942. Plant patent 479; 22 July 1941. Parentage unknown, but probably an Elberta o.p.2. Fruit: flesh yellow; freestone; similar to Elberta, except for glabrous skin; quality fair; ripens late. Tree: only moderate vigor; blossoms large, showy, light pink; susceptible to brown rot.

Garden Delight.—Orig. in Modesto, Calif., by Floyd Zaiger. Introd. in 1975. (Red Grand nectarine x Flory Dwarf peach) F_2. Plant patent 4155; 25 Nov. 1975; owned by Zaiger's Genetics. Fruit: medium; globose; skin yellow overspread with red; flesh yellow, firm; flavor good, subacid. Tree: nonvigorous; spreading; productive.

Garden State.—Orig. in New Brunswick, N.J., by M.A. Blake, New Jersey Agr. Expt. Sta. Introd. in 1934. Plant patent 92; 3 Apr. 1934. Elberta o.p.2; selected in 1922. Fruit: large; round oval; skin greenish-yellow ground color, almost completely covered with red; flesh firm, juicy, yellow, freestone; ripens 21–30 Aug. Tree: vigorous; medium-spreading; leaf glands reniform; flowers large, showy, light pink, self-fertile.

Giant Free.—Orig. in Merced, Calif., by T.B. Stribling, Stribling's Nurseries. Introd. in 1964. Freedom o.p.2; selected in 1962; tested as Stribling-136-57. Fruit: large; skin blushed; flesh clear yellow, freestone; resembles Freedom. Tree: crop production more uniform than Freedom.

Globerta (*Glogold*).—Orig. in Omaha, Neb., by Frederick Wenzel HofMann, Henry Field Seed and Nursery Co. Introd. in 1954. Plant patent 1230; 17 Nov. 1953. Fuzzless-Berta peach selfed; selected in 1943. Fruit: smaller than Elberta; skin yellow, highly red-blushed, waxy, smooth; flesh yellow, firm, flavor good, freestone; ripens in Elberta season; ships well. Tree: large, spreading; flowers large, pinkish-light orchid; high tolerance to causticity of lead arsenical spray and brown rot.

Glogold.—See Globerta.

Gold (*Yellow*).—Commercial synonym of Gialla di Padova. Imported from Padova, Italy, by the U.S. Dept. of Agriculture (USDA) in 1926 as P.I. 65977. Distributed by Plant Introduction Sta., USDA, Chico, Calif., to Idaho and California; trees also propagated in Kansas; renamed Gold in 1932. Fruit: large; skin and flesh clear yellow; flavor good. Apparently now obsolete in U.S.

Golden Free.—Orig. in Le Grand, Calif., by

F.W. Anderson. Introd. in 1956 by Reedley Nursery, Reedley, Calif. (Le Grand x Kim) F_2; selected in 1953; tested as Anderson 6K32. Fruit: flesh yellow, quality good, freestone; ripens with and resembles Golden Grand, but not as large, nor as firm, and less attractive.

Golden Grand.—Orig. in Le Grand, Calif., by F.W. Anderson. Introd. in 1954. Plant patent 1324; 23 Nov. 1954; assigned to Reedley Nursery, Reedley, Calif. Le Grand x (Le Grand x Kim); selected in 1950; tested as Anderson 4K30A. Fruit: large; skin with an attractive red color; flesh yellow, freestone; ripens in Stanwick season. Tree: medium to large, spreading; productive; vigorous; leaf glands reniform; flowers large, showy, pink.

Golden Prolific.—Orig. in Le Grand, Calif., by F.W. Anderson, Merced, Calif. Introd. in 1963. Plant patent 2193; 20 Nov. 1962; assigned to L.E. Cooke Co., San Gabriel, Calif. F_2 (Late Le Grand x Flory Dwarf peach); selected in 1961. Fruit: large, averaging 2 1/2 inch in diam.; globose to slightly oblate, symmetrical; skin yellow, partially mottled orange-red, thickness and texture medium; flesh yellow, red next to the pit, flavor subacid to mild, soft, ripens evenly, eating quality fair to good, freestone; shipping quality medium to poor; keeping quality medium. Tree: dwarf, average height of 6–8 ft, the first dwarf nectarine to be named and introduced; very ornamental; vigor medium; spreading; productive, leaf glands reniform; flowers large, self-fruitful, anthers red.

Goldenshade.—Orig. in Oakland, Calif., by Henry G. Wong. Plant patent 3964; 26 Oct. 1976. Chance seedling of unknown parentage. Fruit: globose, uniform, medium size; medium thick skin, partly blushed red, smooth texture; juice mildly subacid; clingstone; yellow flesh, good eating quality. Tree: medium; vigorous; spreading; upright; productive.

Gold King.—Orig. in Visalia, Calif., by L.A. Hagler. Introd. in 1956. Plant patent 1424; 4 Oct. 1955. Bud mutation of Le Grand; discovered in 1952. Fruit: large; skin highly blushed; flesh yellow, firm, flavor fine, texture fine, clingstone; ripens about 4 weeks later than Le Grand or about 1 Sept. at place of origin; ships well; canning qualities good. Tree: medium to large; upright; hardy; productive and regular bearer; flowers pink; leaf glands reniform.

Gold Nugget (*Rio Grand*).—Orig. in Le Grand, Calif., by F.W. Anderson. Introd. in 1953. Plant patent 1249; 2 Feb. 1954; assigned to Reedley Nursery, Reedley, Calif. (Kim x July Elberta peach) F_2; tested as Anderson 6K10. Fruit: size medium; skin yellow overspread with red, appearance attractive; flesh yellow, firm, freestone; ripens between John Rivers and Gower; resembles Sun Grand. Tree: large, spreading; medium vigor; productive; leaf glands globose; flowers large.

Grandandy.—Orig. in Le Grand, Calif., by F.W. Anderson. Introd. in 1957. Plant patent 1545; 1 Jan. 1957; assigned to Reedley Nursery, Reedley, Calif. Le Grand x Sun Grand; selected in 1953; tested as Anderson 1B41. Fruit: large; skin yellow, partially covered with red; flesh yellow, flavor subacid, freestone; ripens 1 week later than Sun Grand, which it resembles, but flesh not quite as firm and is of better quality. Tree: large to medium; spreading; vigorous; productive; leaf glands globose; flowers large.

Granderli.—Orig. in Le Grand, Calif., by F.W. Anderson. Introd. in 1957. Plant patent 1645; 24 Sept. 1957; assigned to Reedley Nursery, Reedley, Calif. Le Grand x Star Grand; selected in 1953; tested as Anderson 1B715. Fruit: skin light yellow overspread with red; flesh yellow, firm, subacid, clingstone; ripens with Early Le Grand, which it resembles; split pits and growth cracks fewer than in Early Le Grand. Tree: lower chilling requirement and produced crops more regularly than Early Le Grand.

Grandeur.—Orig. in Le Grand, Calif., by F.W. Anderson. Introd. in 1953. Plant patent 1265; 23 Mar. 1954; assigned to Reedley Nursery, Reedley, Calif. Le Grand o.p.; selected in 1948; tested as Anderson 477. Fruit: flesh yellow, clingstone; very similar to Le Grand, but is more productive, has more external red color, and ripens about 1 week later. Tree: resembles Le Grand.

Grand Diamond.—Orig. near Le Grand,

Merced County, Calif., by F.W. Anderson. Plant patent 4095; 30 Aug. 1977; assigned to Reedley Nursery, Reedley, Calif. Red Diamond x Early Sungrand. Fruit: large; uniform; symmetrical, globose to oblong; freestone; yellow flesh, skin red, uniformly colored; good long-distance shipping and keeping quality; similar to Red Diamond, but ripens 2 weeks later. Tree: large; vigorous; productive; fruit kernel sweet; flower medium; leaf glands reniform.

Grand Harry.—Orig. in Le Grand, Calif., by F.W. Anderson, Merced, Calif. Introd. in 1969. Plant patent 2896; 17 June 1969; assigned to Reedley Nursery, Reedley, Calif. Late Le Grand x unnamed seedling of Gold King. Tested as Anderson 48H335. Fruit: large; globose; skin yellow, substantially overspread with red at maturity; flesh yellow, clingstone; resembles September Grand, but less susceptible to growth cracks and decay, ripens in the same season. Tree: size medium; vigorous; productive and regular bearer; flowers large, showy, pink; leaf glands reniform.

Grand Haven.—Orig. in Le Grand, Calif., by F.W. Anderson, Merced, Calif. Introd. in 1954. Plant patent 1326; 30 Nov. 1954; assigned to Reedley Nursery, Reedley, Calif. (Halehaven peach x Le Grand) F_2; selected in 1950; tested as Anderson B72E. Fruit: flesh yellow, freestone; ripens with John Rivers, but is larger and firmer with more red skin color. Tree: large; upright to spreading; vigorous; productive; leaf glands reniform; flowers large, showy, pink.

Grand Melody.—Orig. near Le Grand, Merced County, Calif., by F.W. Anderson. Plant patent 4079; 19 July 1977; assigned to Reedley Nursery, Reedley, Calif. Golden Prolific x Early Sungrand. Fruit: large; uniform; globose; clingstone; resembles Red Grand nectarine size and appearance; eating quality fair to good. Tree: genetically dwarf; vigorous; upright; productive.

Grandoso.—Orig. in Le Grand, Calif., by F.W. Anderson. Introd. in 1955. Plant patent 1427; 1 Nov. 1955; assigned to John M. Garabedian, Fresno, Calif. (Le Grand x seedling of Bim) F1; selected in 1950; tested as Andergon 10K6. Fruit: size medium; skin yellow overspread with red; flesh yellow, texture medium to soft, flavor delicate, vinous, eating quality good, freestone; ripens between John Rivers and Sun Grand; resembles Sun Grand. Tree: size medium; spreading; vigorous; productive; flowers large.

Grandoso.—Orig. in Le Grand, Calif., by F.W. Anderson. Introd. in 1955. Plant patent 1427; 1 Nov. 1955; assigned to John M. Garabedian, Fresno, Calif. (Le Grand x seedling of Bim) F1; selected in 1950; tested as Andergon 10K6. Fruit: size medium; skin yellow overspread with red; flesh yellow, texture medium to soft, flavor delicate, vinous, eating quality good, freestone; ripens between John Rivers and Sun Grand; resembles Sun Grand. Tree: size medium; spreading; vigorous; productive; flowers large.

Grand Prize.—Orig. in Le Grand, Calif., by F.W. Anderson. Introd. in 1955. Plant patent 1344; 25 Jan. 1955; assigned to Reedley Nursery, Reedley, Calif. Le Grand x Kim x Le Grand; selected in 1948; tested as Anderson 4K30B. Fruit: larger than Gower; skin yellow, overlaid with red; flesh yellow, freestone; ripens with Gower. Tree: large; upright; spreading; vigorous; productive; leaf glands globose; flowers large, showy, pink.

Grand River.—Orig. in Le Grand, Calif., by F.W. Anderson. Introd. in 1953. Plant patent 1248; 2 Feb. 1954; assigned to Reedley Nursery, Reedley, Calif. (Le Grand x John Rivers) F_2; selected in 1951. Fruit: flesh yellow, semi-freestone; ripens 1 week earlier than John Rivers or about 15 June at place of origin.

Grand River B3.—Orig. in Le Grand, Calif., by F.W. Anderson. Introd. in 1957. Plant patent 1543; 1 Jan. 1957; assigned to Reedley Nursery, Reedley, Calif. (Le Grand x John Rivers) F_2; selected in 1951; tested as Anderson B3. Fruit: flesh yellow, freestone to semi-freestone; resembles Grand River, having better quality; ripens with Grand River; primary difference between the two varieties: seed bitter in Grand River; seed sweet in Grand River B3. Tree: resembles Grand River, but more vigorous and productive.

Gransun.—Orig. by J.L. and L. Jackson of Kingsburg, Calif. Plant patent 8255; 15 June 1993. Assigned to Kings Gate Ranch,

Kingsburg. Sport of May Grand, ripens about 2 weeks earlier.

Grand Time.—Orig. in Le Grand, Calif., by F.W. Anderson. Introd. in 1978. Plant patent 4318; 17 Oct. 1978; assigned to Reedley Nursery, Reedley, Calif. Niagara x Autumn Gold. Fruit: medium to large; globose, symmetrical; flesh yellow with red at pit cavity; eating quality good; skin yellow overspread with red; clingstone; ripens a few days after Late Le Grand and a few days earlier than Regal Grand. Tree: medium to large; upright to spreading; productive and regular bearer; leaf glands reniform; flowers large, pink.

Harblaze.—Orig. in Harrow, Ont., by R.E.C. Layne, Agriculture Canada, Res. Sta. Introd. in 1989. Stark Delicious x Hardired; cross made in 1973; selected in 1978; tested as HW104. Fruit: skin 90% bright red blush over yellow-orange ground color; larger and rounder than Nectared 2; flesh yellow, firm, freestone; good flavor. Ripens 3 days after Nectared 2. Tree: vigorous; productive; equally bud-hardy and wood-hardy with Redhaven. Flower: showy, pink. Leaf glands reniform. Resistant to bacterial spot, brown rot, leucostoma canker, and powdery mildew.

Hardired.—Orig. in Harrow, Ontario, Canada, by R.E.C. Layne and G.M. Weaver, Agriculture Canada, Res. Sta. Introd. in 1974. Lexington x NJN32; cross made in 1963; selected in 1968; tested as HW102. Fruit: 2 1/4 inches in diam.; ovate; skin 90% red with greenish-yellow ground color, attractive; flesh bright yellow, moderately firm, quality good, freestone; ripens 5 days after Redhaven, 20 Aug. at Harrow. Tree: spreading; very hardy, like Reliance peach; very productive; tolerant of bacterial spot and brown rot; flower pink, large, showy; leaf glands reniform.

Harko.—Orig. in Harrow, Ontario, Canada, by R.E.C. Layne and G.M. Weaver, Agriculture Canada, Res. Sta. Introd. in 1974. Lexington x NJN32; cross made in 1963; selected in 1968; tested as HW101. Fruit: 2 1/4 to 2 3/4 inches in diam.; ovate; skin 90% red with greenish-yellow ground color, attractive; flesh bright yellow, red at pit, moderately firm, quality good, freestone; ripens 1–2 days after Redhaven peach, 16 Aug. at Harrow. Tree: spreading; hardier than Redhaven, like Madison; productive; tolerant of bacterial spot and brown rot; flower pink, large, showy; leaf glands reniform.

Harvest Sun.—Orig. in Del Rey, Calif., by Tom Chamberlin, H.P. Metzler & Sons. Introd. in 1988. Parentage unknown. Plant patent 7012; owned by Metropolitan Insurance Co.; licensed exclusively to Agri Sun Nursery, Selma, Calif. Fruit: large; slightly elongated; skin red on yellow ground color, smooth; flesh yellow, firm; ripens just before Fairlane. Tree: vigorous; productive.

Hayes Late (*Strawberry*).—Orig. in Fresno, Calif., by Mrs. Signa Larsen Hayes. Introd. in 1944. Plant patent 587; 8 June 1943. Stanwick x a late peach; selected in 1939. Fruit: large, up to 2 1/2 inches in diam.; flesh white, aromatic; cans well because of firmness; clingstone; pit large, conspicuously corrugated and pointed; ripens late, about 1 Sept., some 3 weeks after Quetta and 1 week before Stanwick; good keeping and shipping quality; resembles Quetta. Tree: size medium; leaves with globose glands.

Hi-Red (*Sunectsix*).—Orig. in Fresno, Calif., by John Garabedian. Introd. in 1970. Plant patent 3133; 25 Apr. 1972; assigned to Superior Farming Co. June Belle o.p. Fruit: larger than LeGrand, which it most nearly resembles. Ripens earlier than LeGrand. Now surpassed by better cultivars.

Honey.—Orig. in Comal County, Texas, by Comal Springs Nursery, New Braunfels, Texas. Introd. about 1920. Honey peach o.p.; selected about 1918. Fruit: flesh white, very tender, as tender as known parent; ripens in July. Variety now obsolete.

How Red (*Sunectnineteen*).—Orig. by J.H. Weinberger, Fresno, Calif. Plant patent 8336; 10 Aug. 1993. Assigned to SunWorld, Indio, Calif. Sunecteight o.p. Fruit: large; oblong; melting, yellow-orange flesh, clingstone. Ripens in late July, about 2 weeks before Red Jim. Flowers large, red-purple; leaf glands globose. Tree: medium size and vigor; semi-upright; productive. Isoenzyme patterns for MDH are unique and present in about 2% of all peach and nectarine varieties.

Hunter.—Orig. in Geneva, N.Y., by U.P. Hedrick, New York State Agr. Expt. Sta. Introd. for trial in 1924. Hunter 1 o.p.; seed planted in 1915; tested as N.Y. 50; first fruited in 1920; named in 1922. Fruit: good size; flesh yellow, quality good, freestone; ripens midseason, 7–10 days later than Hunter 1. Tree: mildews badly; foliage eglandular. Variety supplanted by parent Hunter 1; no longer propagated.

Hunter 1.—Orig. in White Plains, N.Y., by Harry Hunter. Introd. for trial in 1930. Supposedly Elberta o.p.; seed planted in 1910; first fruited in 1913; bud sticks received by U.P. Hedrick, New York State Agr. Expt. Sta., for testing in 1914. Fruit: good size; attractive; flesh yellow, flavor and quality good, freestone; ripens midseason. No longer propagated.

Improved Sunrise.—Orig. in Red Bluff, Calif., by Grant Merrill. Introd. in 1961. Plant patent 2145; 1 May 1962. Sunrise o.p.; discovered in 1957; selected in 1960. Fruit: skin very red; flesh yellow, flavor sweet; slightly freestone when ripened off the tree; resembles Sunrise, but is sweeter, skin brighter red with less russet, shape slightly longer; ripens early; ships very well. Tree: good producer; bears regularly.

Independence.—Orig. in Fresno, Calif., by John H. Weinberger, U.S. Hort. Field Sta. Introd. in 1965. Red King o.p.; selected in 1960; tested as US F100-37. Fruit: size medium; ovate; skin brilliant cherry-red, attractive, some tendency toward roughness of suture and tip; flesh yellow, firm, good texture and flavor, freestone; suitable for long-distance shipments; ripens early, in late June or early July, about with Early Sungrand, with an average ripening date of 4 July. Tree: good productivity; moderately vigorous; leaf glands reniform; petals large; chilling not adversely affected by warm winter of 1963 in the San Joaquin Valley, Calif.; self-fruitful.

Jolly Red Giant.—Orig. by C.F. Zaiger, L.M. Gardner, G.N. Zaiger, and G.G. Zaiger, Modesto, Calif. Plant patent 7947; 18 Aug. 1992. {Ruby Gold x [Flame Kist x (Red Grand x Royal Gold peach)]} o.p. Fruit: very large; firm; nearly globose; excellent flavor; attractive red skin over golden-yellow ground color; flesh yellow, melting, freestone. Ripens with Red Grand. Flowers large, showy, and pink; leaf glands reniform. Tree: large; vigorous; upright; very productive.

Jordana.—Orig. in Madera, Calif., by William T. Kirkman, Jr. Introd. in 1950. Plant patent 1096; 13 May 1953; assigned to John M. Garabedian, Fresno, Calif. Parentage unknown. Fruit: large to medium; broadly ovoid; skin yellowish ground color, blushed with red on the sunny side, medium thickness, tough, somewhat astringent, smoother finish than Quetta; ripens 4–5 weeks later than Quetta, which it resembles. Tree: large; upright; spreading; vigorous; medium productivity; leaf glands reniform; flowers small, dark pink.

July Red.—Orig. by N.G. Bradford, LeGrand, Calif. Plant patent 5663; 18 Feb. 1986. [(Red Free x Autumn Gold) x Red Diamond] o.p. Resembles Red Grand, but ripens 3 days later, has larger fruit, brighter solid red color, more attractive and globose form, and firmer flesh equaling that of Red Diamond. Melting flesh; clingstone.

June Belle.—Orig. in Fresno, Calif., by John M. Garabedian of Peters and Garabedian, Peach and Willow Farms. Introd. in 1964. Plant patent 2323; 26 Nov. 1963; assigned to Peach and Willow Farms. Bud mutation of Sun Grand o.p. Fruit: size medium, uniform; skin orange-red to red; flesh yellow with some red at pit, firm, slightly acid but good balance with sugar, eating quality good, semi-freestone; keeping and shipping qualities good; ripens 2 weeks before Sun Grand; resembles Sun Grand. Tree: large; vigorous; spreading; productive; leaf glands globose; blooms in midseason; flowers large, petals pink, anthers light yellow, pollen sterile.

June Brite.—Orig. by C.F. Zaiger, G.N. Zaiger, L.M. Gardner, and G.G. Zaiger, Modesto, Calif. Plant patent 6078; 5 Jan. 1988. (Early Sungrand x Zee Gold) o.p. Fruit: large; nearly globose; high percentage of red over yellow ground color; flesh yellow, firm, meaty, clingstone. Ripens 3 days before Zee Gold. Flowers large, showy, pink; leaf glands reniform. Tree: medium; upright; vigorous; productive.

June Glo.—Orig. in Modesto, Calif., by Floyd Zaiger. Introd. in 1984. (Zee Gold x

Early Sungrand) F_2. Plant patent 5228; 8 May 1984; assigned to Zaiger's Genetics. Fruit: large; nearly globose; skin yellow overspread with red; flesh dark yellow, firm; flavor mild. Tree: vigorous; upright; productive.

June Grand.—Orig. in Le Grand, Calif., by F.W. Anderson, Merced, Calif. Introd. in 1958. Plant patent 1723; 8 July 1958; assigned to Kim Bros., Reedley Nursery, Reedley, Calif. Early Sungrand o.p.; selected in 1956. Fruit: 2 3/8 inches in diam., not quite as large as seed parent; skin dark red over an orange-yellow ground color; flesh yellow, melting, moderately juicy, subacid, aroma not pronounced, quality good, semi-clingstone; ripens extremely early, about 7–10 days before Early Sungrand and John Rivers. Tree: bears well and regularly; leaf glands reniform; flowers large, showy, anthers red.

Kay Grand.—Orig. in Le Grand, Calif., by F.W. Anderson. Introd. in 1970. Plant patent 2993; 20 Oct. 1970; assigned to Reedley Nursery, Reedley, Calif. Red Grand x Stark SunGlo. Tested as 37P577. Fruit: resembles Sun Grand; large, spherical; skin yellow overspread with red at maturity; flesh yellow, flecked with red, firm; recommended as a late freestone shipping variety; ripens with Le Grand, 2–3 weeks later than Sun Grand. Tree: large; spreading; productive and regular bearer; flowers large, showy; leaf glands reniform.

Kism Grand.—Orig. by N.G. Bradford, LeGrand, Calif. Plant patent 5666; 18 Feb. 1986. Red Free x Sparkling Red. Fruit: large; globose; attractive red blush over yellow skin; flesh yellow, firm, melting clingstone ripening 5 days before Late LeGrand. Flowers large, pink; leaf glands reniform. Tree: vigorous; spreading; very productive.

Kent Grand.—Orig. in Cutler, Calif., by Leonard K. Wileman. Introd. in 1969. Plant patent 2937; 11 Nov. 1969; assigned to Reedley Nursery, Reedley, Calif. Bud mutation of Red Grand. Discovered in 1960. Fruit: yellow-fleshed, clingstone; resembles Red Grand and Royal Grand (Early Red Grand); ripens 10 days earlier than Red Grand and is more highly colored, and about 4 days later than Royal Grand; more uniform ripening period, more uniformity in bright red overcolor, smoother and larger than Royal Grand. Tree: similar to Red Grand and Royal Grand; flowers large, showy, pink; leaf glands reniform.

Kim.—Orig. in Merced, Calif., by F.W. Anderson. Introd. in 1938. Plant patent 173; 7 Apr. 1936. (J.H. Hale x Lippiatt) F_2; selected in 1935. Fruit: large; skin yellow background blushed with red; flesh yellow, firm, flavor acid, quality good, freestone; ripens 2 weeks earlier than Lippiatt; suitable for drying, canning, and shipping. Tree: large; spreading; vigorous; productive; flowers large, pink.

Krasvynos.—Orig. in the Fergana Valley, western Asiatic Russia. Introd. in 1944 as P.I. 119844 by M.A. Blake and J.S. Joffe, New Jersey Agr. Expt. Sta., New Brunswick, N.J. Parentage unknown; tested as Ferganensis 0932. Fruit: 2 inches in diam.; flesh white, slightly stringy, quality fair to good, semi-freestone; ripens 7–10 days before Elberta; subject to cracking, brown rot, and internal breakdown when fully ripe in some seasons. Tree: productive; hardier than Elberta; flowers showy. Used as a parent for hardiness.

Lafayette.—Orig. in Blacksburg, Va., by George D. Oberle, Virginia Polytechnic Inst. Introd. in 1967. V.P.I. 22N x V.P.I. 41N. V.P.I. 22N is derived from Muir's Seedling, a nectarine introduced from New Zealand, and V.P.I. 41N is derived from the cross (Newboy x Golden Jubilee). Cross made in 1952; first fruited in 1955; selected in 1956; released by above institution in 1967. Fruit: average 2 1/4 inches in diam.; skin nearly solid bright red with smooth finish, very attractive; flesh white with attractive pink adjacent to pit cavity, firmness above average, texture good, flavor mild, quality good, freestone; ripens 10 days before Elberta. Tree: vigor average; hardiness above average; heavy producer of fruit buds; flowers large, showy, self-fruitful; leaf glands weakly reniform.

Late Le Grand.—Orig. in Merced, Calif., by F.W. Anderson. Introd. in 1951. Plant patent 1035; 11 Sept. 1951; assigned to Reedley Nursery, Reedley, Calif. Le Grand o.p.; selected in 1950. Fruit: very similar to Le Grand, but ripens about 2 weeks later. Tree: more vigorous and more productive than Le Grand.

Late Red Grand.—Orig. in Le Grand, Calif., by F.W. Anderson. Introd. in 1957. Plant patent 1544; 1 Jan. 1957. Red Grand o.p.; selected in 1954; tested as Anderson 2L7D. Fruit: resembles Red Grand, but ripens about 2 weeks later and has more attractive red skin color. Tree: size medium; vigor medium; spreading; very productive; leaf glands reniform.

Late Red Jim.—Orig. by J.W. Taylor, Reedley, Calif. Plant patent 7176; 6 Mar. 1990. Assigned to Ito Packing Co., Reedley. Red Jim x May Grand. Fruit: very large; round; high percentage of dark red over yellow skin; flesh yellow with much red streaking at pit into flesh, very firm, meaty, clingstone. Ripens mid-August at Reedley. Flowers large, showy, pink, red anthers; leaf glands reniform. Tree: large; vigorous; spreading.

Late Red Jim II.—Orig. by J.W. Tos, Hanford, Calif. Plant patent 7505; 23 Apr. 1991. Assigned to Tos Farms, Hanford. Bud mutation of Red Jim ripening 3 weeks after Red Jim.

Le Grand (*Pot of Gold*).—Orig. in Le Grand, Calif., by F.W. Anderson. Introd. in 1942. Plant patent 549; 13 Oct. 1941. (J.H. Hale x Quetta) F_2; selected in 1936. Fruit: large; flesh yellow, firm, clingstone; resembles Quetta. Tree: large, spreading; productive; leaf glands reniform; flowers large, showy, pink.

Leprechaun.—Orig. at Univ. of Arkansas, selected in 1983, tested as Ark. 327, and released by J.N. Moore, R.C. Rom, S.A. Brown, and G.L. Klingaman. Patent pending. [(Flowery dwarf x unnamed nectarine) F_2 x Ark. 164] selfed. Fruit: small; firm; good flavor, freestone, ripens with Loring peach. Flowers pink and showy. Ornamental brachytic dwarf with bright-green leaves. Plants susceptible to damage from peach-tree borer.

Lexington.—Orig. in Blacksburg, Va., by George D. Oberle and R.C. Moore, Virginia Agr. Expt. Sta. Introd. in 1957. V.P.I. 13 selfed, which is a white-fleshed, freestone peach selected from seedling peaches grown from open-pollinated seed from a tree developed by crossing Lippiatt's Late Orange with Golden Jubilee peach; selected in 1950; tested as V.P.I. 34N. Fruit: slightly larger than Cavalier, which it resembles; skin deep yellow overlaid with medium red; flesh deep yellow, medium firm, texture medium, quality good, freestone; ripens 7–10 days before Cavalier. Tree: extremely vigorous; considerable ability to escape brown rot and blossom-season frost; leaf glands reniform; flowers showy.

Liberty.—Orig. in Le Grand, Calif., by F.W. Anderson. Introd. in 1952. Plant patent 1160; 13 Jan. 1953; assigned to Kim Brothers, Reedley Nursery, Reedley, Calif. F2 [Kim x July Elberta]; first fruited in 1948; tested as Anderson 7K62. Fruit: skin almost completely red at shipping ripeness; flesh yellow, freestone, firm; resembles Sun Grand, but ripens about 1 month later, around 10 Aug. at place of origin. Tree: large, spreading; vigorous; productive; leaf glands globose; flowers large, pale pink.

Mabel.—Orig. in Winters, Calif., by Guy L. Philp, California Agr. Expt. Sta., Davis. Introd. in 1948. Humboldt x Burbank; selected in 1942; tested as Calif. 27-12a and Calif. 27-13; sibling of Philp. Fruit: medium or larger when well thinned, 5.8 cm in cheek diam.; shape variable, round-oval to round, often very slightly ovate; skin rich yellow, deeply blushed; flesh clear yellow, sweet, mildly subacid, flavor excellent, firm to firm-melting, texture fine-grained; stone medium or smaller, broad-oval, plump, free; ripens in early midseason, about with Philp, before Gower; dessert and local-market uses. Tree: very vigorous; very productive; leaf glands reniform; flowers large, heavy to medium bloom; moderate chilling requirement.

Marigold.—Orig. in Le Grand, Calif., by F.W. Anderson. Introd. in 1956. Plant patent 1724; 8 July 1958; assigned to the Burchell Nursery Co., Modesto, Calif. (Le Grand x Royal Hale peach) F_1; selected in 1955; tested as Anderson 10L9a. Fruit: large; skin well blushed; flesh yellow, firm, freestone; ships very well; resembles Golden Grand and Late Le Grand; matures late, 10–14 days after Freedom and about with Late Le Grand. Tree: large; vigorous; productive; leaf glands reniform; flowers large, showy.

Martha Jane.—A red-leaved, double-flowering nectarine, with 18–26 petals/flower. Orig. at Florida Agr. Expt. Sta., Gainesville.

Fla. 15-103 x Fla. 8-23; tested as Fla. 0-20R. Selected in 1980; named in 1991 by G. Freeman. High flower bud set. Flower: showy, pink, double; bright red leaves in spring fade to dull red in summer. Fruit small, ripening in mid-July at Gainesville. Resistant to bacterial leafspot. Tree: vigorous, spreading; few fruits produced. Chilling requirement 475 chill units.

May Belle.—Orig. near Fowler, Fresno County, Calif., by Sam Parnagian. Plant patent 4084; 26 July 1977. Discovered in 1973 as a bud mutation of Armking. Similar to Armking, ripening about 10 days earlier; fruit more spherical; tree more vigorous.

Mayfair.—Orig. in Le Grand, Calif., by F.W. Anderson, Merced, Calif. Introd. in 1965. Plant patent 2532; 8 June 1965; assigned to George Abe, Selma Nursery, Selma, Calif. Grandandy x Grand Haven; selected in 1962; tested as Anderson 50H725. Fruit: size medium, average 2 1/8 inches in suture diam.; globose to slightly oblong; skin bright red, medium thickness; flesh yellow with no red at pit cavity, firm, melting, subacid, vinous, juicy, aroma very slight, quality good; stone size medium, free to semi-free; ripens early, 2 weeks before John Rivers, 10 days before Sunrise and Grand River, 2–3 days before Red June; resembles Red June. Tree: size medium; vigorous; spreading; bears regularly and very well; leaf glands reniform; flowers large.

Mayfire.—Orig. in Fresno, Calif., by Agr. Res. Serv., U.S. Dept. of Agriculture (USDA). Introd. in 1983. Armking o.p. Selected in 1979 by David W. Ramming, USDA; tested as P56-17. Fruit: matures in 65 days with average diam. of 2 1/4 to 2 3/8 inches, oval; skin has brief red blush over 70% to 90%; flesh yellow, firm, texture smooth, quality good. Tree: very productive, thinning necessary; vigor moderate; blossoms large, showy, self-fertile; leaf glands globose; Chilling requirement 650 h; blooms with Springcrest at Fresno.

May Glo.—Orig. C.F. Zaiger, Modesto, Calif. Plant patent 5245; 12 June 1984. (Fayette x May Grand) o.p. Fruit: medium; globose; firm; high percentage of red over yellow ground color; flesh yellow, firm, melting, semi-clingstone; ripens 3 days before ArmKing. Flowers large, showy, pink; leaf glands reniform. Tree: large; vigorous; spreading; productive. Chilling requirement 300 h.

May Grand.—Orig. in Le Grand, Calif., by F.W. Anderson, Merced, Calif. Introd. in 1967. Plant patent 2794; 6 Feb. 1968; assigned to Reedley Nursery, Reedley, Calif. (Red Grand x Early Sungrand) F_2; tested as Anderson 24P1000. Fruit: large; globose to oblong; skin yellow, most overspread with red; flesh yellow, freestone; resembles Early Sungrand; ripens early, about 10 days earlier than Grand River and Sunrise, 2–3 weeks earlier than Early Sungrand, 3 weeks before John Rivers. Tree: size medium; productive and regular bearer; flowers large, showy; leaf glands reniform.

May Jim.—Orig. by J.W. Taylor, Dinuba, Calif. Plant patent 8084; 5 Jan. 1993. Assigned to Ito Packing Co., Reedling. (Red Jim x May Glo) o.p. Fruit: medium; round; firm; melting; flesh yellow, clingstone, solid light red over greenish-yellow ground color. Ripens late May at Reedley. Flowers large, showy, light pink; leaf glands reniform. Tree: large; vigorous; upright; productive.

May Kist.—Orig. by C.F. Zaiger, G.N. Zaiger, L.M. Gardner, and G.G. Zaiger, Modesto, Calif. Plant patent 7976; 15 Sept. 1992. (May Grand x Ruby Gold) x Armking mutation. Fruit: medium; nearly globose; firm; good flavor; medium red over yellow ground; flesh pale yellow, melting, clingstone. Ripens 3 weeks before May Grand. Flowers nonshowy, pink; leaf glands globose. Tree: large; vigorous; upright; productive.

May Lion.—Orig. by D.M. and L.M. Serimian, Selma, Calif. Plant patent 6542; 17 Jan. 1989. ArmKing x Summer Grand. Fruit: large; uniformly broad ovate with variable but dark red-purple suture; high percentage of red over yellow skin; flesh firm, melting, semi-clingstone. Ripens 18 May in central part of San Joaquin Valley. Flowers small, light pink; leaf glands mostly reniform but some globose. Tree: medium size and vigor; upright; very productive.

May Queen.—Orig. in Exeter, Calif., by Grant Merrill. Introd. in 1971. Plant patent 3581; 6 Aug. 1974. Scarlet Queen o.p. Fruit:

1 3/4–2 1/4 inches in diam.; globose; skin bright red, waxy, no russet; flesh yellow, firm, clingstone; excellent shipper; ripens late May to early June. Tree: large, vase-shaped; vigorous; productive; low chilling requirement; flowers small to medium, pink; leaf glands reniform.

Mayred.—Orig. in Wasco, Calif., by David L. Armstrong. Introd. in 1969. Plant patent 2758; 8 Aug. 1967; assigned to Armstrong Nurseries, Ontario, Calif. Unnamed nectarine x unnamed peach. Parentage going back through series of crosses to peach varieties Babcock, July Elberta, Swatow, and nectarine varieties Boston and Gold Mine. Pollen parent derived from Panamint x Springtime. Fruit: size small to medium; globose to broadly ovoid, no beak; skin ground color yellow covered with heavy, bright red blush, smooth, lack of pubescence, attractive; flesh yellow, no red at pit, flavor sweet; freestone when fully ripe; ripens a few days before Robin, 1 week before Armred, 1 week after Springtime peach; resembles Armred. Tree: vigorous; chilling requirement equal to Springtime peach; very productive; leaf glands mixed, weakly reniform, with a few globose; flowers 1 1/2 inches in diam., pink, showy.

Mericrest.—Orig. in Durham, N.H., by E.M. Meader (retired), Agr. Expt. Sta. Introd. in 1971. Meredith peach x Nectacrest nectarine; cross made in 1958; selected in 1962; tested as N.H. 62N2. Fruit: size medium; skin dark red, attractive; flesh yellow, quality good; freestone; ripens 5–10 Aug. in central Missouri. Tree: medium large; hardy; productivity heavy; susceptible to bacterial spot.

Merrill Princess.—See Princess.

Merrill Queen.—Orig. in Red Bluff, Calif., by Grant Merrill. Introd. in 1961. Plant patent 2010; 10 Jan. 1961. J.H. Hale x Flaming Gold, o.p.2; selected in 1958. Fruit: large, skin highly colored red; flesh yellow, very firm, well flavored, partly freestone; resembles Princess, but ripens earlier, larger and sets more heavily and more regularly than that variety. Tree: large, medium spread; vigorous; productive and regular bearer; leaf glands reniform; flowers large, showy, pink. Variety now obsolete.

Merrill Sunrise.—See Sunrise.

Mid Glo.—Orig. in Modesto, Calif., by Floyd, Gary, and Grant Zaiger and Leith Gardner. Introd. in 1989. (May Grand x unknown late-maturing nectarine) F_2. Plant patent 6753; 25 Apr. 1989; assigned to Zaiger's Genetics. Fruit: large; globose; yellow ground color almost covered by red blush; flesh yellow, firm; subacid. Tree: vigorous; upright; productive.

Mike's Grand.—Orig. by R. Peters and E. Wuhl, Fresno, Calif. Plant patent 5587; 3 Dec. 1985. Bud sport of Early Sungrand. Similar to May Grand, but fruit has more yellow flesh, skin is less bright, fruit has flatter configuration and ripens 2 or 3 days later.

Moon Grand.—Orig. in Le Grand, Calif., by F.W. Anderson, Merced, Calif. Introd. in 1967. Plant patent 2878; 1 Apr. 1969; assigned to Reedley Nurseries, Reedley, Calif. Red Grand nectarine x seedling of Early Sungrand; tested as Anderson 24PA55. Fruit: large; slightly elongated; skin yellow, substantially covered with red; flesh yellow, freestone; resembles Early Sungrand, but ripens immediately after that variety, with harvest complete before the start of Star Grand II. Tree: large; vigorous; productive and regular bearer; flowers large, showy, pink; leaf glands reniform.

Morton.—Orig. in Geneva, N.Y., by H.O. Bennett and R.C. Lamb, New York Agr. Expt. Sta. Introd. in 1965. Self of N.Y. 302 (Newboy x Schumaker); selected in 1945; tested as N.Y. 884. Fruit: size medium, about 2 inches in diam.; spherical; skin white, about 80% red-blushed; semi-clingstone; quality good; ripens early mid-season. Tree: vigorous; hardy; flowers showy, self-fertile; leaf glands reniform.

Nectacrest.—Orig. in New Brunswick, N.J., by M.A. Blake. New Jersey Agr. Expt. Sta. Introd. in 1947. Garden State x [(Gold Mine x Belle selfed) o.p.]. Fruit: large; flesh white, fairly firm, flavor fine, freestone; ripens about 10 days before Elberta. Tree: hardier than Elberta; vigorous; flowers showy.

Nectaheart.—Orig. in New Brunswick, N.J., by M.A. Blake, New Jersey Agr. Expt. Sta. Introd. in 1947. Garden State x [(Gold Mine x Belle selfed) o.p.]. Fruit: flesh white, freestone; ripens about 2 weeks before

Elberta. Tree: hardier than Elberta; flowers showy.

Nectalate.—Orig. in New Brunswick, N.J., by M.A. Blake, New Jersey Agr. Expt. Sta. Introd. in 1947. Garden State x N.J. 25032 (a nectarine seedling). Fruit: flesh white, freestone; ripens 1 week after Elberta. Tree: not as hardy as Elberta; flowers showy.

Nectared 1.—Orig. in New Brunswick, N.J., by L.F. Hough and Catherine H. Bailey, New Jersey Agr. Expt. Sta. Introd. in 1962. [N.J. 53739 (Candoka x Flaming Gold)] o.p. x N.J. N17 [open-pollinated seedling of (Garden State x N.J. 25032, nectarine)]; selected in 1960; tested as N.J. N49. Fruit: averages 2 1/4 to 2 1/2 inches in diam.; skin attractive, smooth, up to 3/4 red over a yellow ground color; flesh yellow, no red at pit, semi-clingstone; ripens 7–10 days before Jerseyland peach, 6 1/2 weeks before Elberta. Tree: moderately productive; flowers nonshowy; leaf glands reniform.

Nectared 2.—Orig. in New Brunswick, N.J., by L.F. Hough and Catherine H. Bailey, New Jersey Agr. Expt. Sta. Introd. in 1962. [N.J. 53939 (Candoka x Flaming Gold) o.p. x N.J. N14 (Nectalate o.p.)]; selected in 1959; tested as N.J. N34. Fruit: averages 2 1/4 to 2 1/2 inches in diam.; skin attractive, smooth, 3/4 to a full blush over a yellow ground color; flesh yellow, some red at pit, semi-clingstone; ripens with, or a few days before, Jerseyland peach. Tree: productive; appears resistant to brown rot; flowers showy; leaf glands reniform.

Nectared 3.—Orig. in New Brunswick, N.J., by L.F. Hough and Catherine H. Bailey, New Jersey Agr. Expt. Sta. Introd. in 1962. [N.J. 53939 (Candoka x Flaming Gold) o.p. x N.J. N14 (Nectalate o.p.)]; selected in 1959; tested as N.J. N42. Fruit: averages 2 3/8 to 2 3/4 inches in diam.; skin attractive, smooth, three-fourths to full blush over yellow; flesh yellow, semi-clingstone; ripens 7 days after Nectared 2, just after Redhaven peach. Tree: moderately productive; appears to be resistant to brown rot; flowers nonshowy; leaf glands reniform.

Nectared 4.—Orig. in New Brunswick, N.J., by L.F. Hough and Catherine H. Bailey, New Jersey Agr. Expt. Sta. Introd. in 1962. N.J. 53939 (Candoka x Flaming Gold) x N.J. N14 (Nectalate o.p.); selected in 1955; tested as N.J. N23. Fruit: averages 2 3/8 to 2 1/2 inches in diam.; skin smooth, three-fourths to full blush over yellow; flesh yellow, semi-freestone; ripens about with Triogem peach, or about 3 1/2 weeks before Elberta. Tree: productive; flowers showy; leaf glands reniform.

Nectared 5.—Orig. in New Brunswick, N.J., by L.F. Hough and Catherine H. Bailey, New Jersey Agr. Expt. Sta. Introd. in 1962. N.J. 53939 (Candoka x Flaming Gold) x N.J. N14 (Nectalate o.p.); selected in 1955; tested as N.J. N24. Fruit: averages 2 3/8 to 2 3/4 inches in diam.; skin smooth, three-fourths to full blush over yellow; flesh yellow, semi-freestone to full freestone when fully ripe; ripens with Sunhigh peach, 3 weeks before Elberta. Tree: productive; flowers nonshowy; leaf glands reniform.

Nectared 6.—Orig. in New Brunswick, N.J., by L.F. Hough and Catherine H. Bailey, New Jersey Agr. Expt. Sta. Introd. in 1962. N.J. N14 (Nectalate o.p.) x Nectaheart; selected in 1955; tested as N.J. N25. Fruit: 2 1/4 to 2 1/2 inches in diam; skin smooth, three-fourths to full red over yellow; flesh yellow, freestone; ripens with Goldeneast peach.Tree: productive; flowers showy; leaf glands reniform.

Nectared 7.—Orig. in New Brunswick, N.J., by L.F. Hough and Catherine H. Bailey, New Jersey Agr. Expt. Sta. Introd. in 1962. N.J. 53939 (Candoka x Flaming Gold) x N.J. N14 (Nectalate o.p.); selected in 1955; tested as N.J. N26. Fruit: averages 2 1/2 to 2 3/4 inches in diam.; skin one-half to three-fourths red over yellow; flesh yellow, quality good, freestone; ripens about with Summercrest peach. Tree: moderately productive; flowers showy; leaf glands reniform.

Nectared 8.—Orig. in New Brunswick, N.J., by L.F. Hough and Catherine H. Bailey, New Jersey Agr. Expt. Sta. Introd. in 1962. N.J. 53939 (Candoka x Flaming Gold) x N.J. N14 (Nectalate o.p.); selected in 1955; tested as N.J. N27. Fruit: 2 3/8 to 2 1/2 inches in diam.; skin smooth, three-fourths to full blush over yellow; flesh yellow, freestone; ripens with Elberta peach.

Tree: productive; flowers showy; leaf glands reniform.

Nectared 9.—Orig. in New Brunswick, N.J., by L.F. Hough and Catherine H. Bailey, New Jersey Agr. Expt. Sta. Introd. in 1962. N.J. 67239 (Candoka peach x N.J. 25032) x N.J. N14 (Nectalate o.p.); selected in 1955; tested as N.J. N28. Fruit: 2 3/8 to 2 3/4 inches in diam.; skin up to three-fourths red over yellow; flesh yellow, freestone; pit size medium; ripens 1 week after Elberta peach. Tree: moderately vigorous; moderately productive; flowers showy; leaf glands reniform.

Nectared 10.—Orig. in New Brunswick, N.J., by L.F. Hough and Catherine H. Bailey, New Jersey Agr. Expt. Sta. Introd. in 1962. Nectalate o.p.; selected in 1948; tested as N.J. N53. Fruit: 2 1/2 to 2 3/4 inches in diam.; skin one-half to three-fourths red over yellow; flesh yellow, firm, freestone; ripens 2 weeks or more after Elberta. Tree: productive; flowers showy, self-unfruitful, pollen aborted; leaf glands reniform.

Nectarina.—Orig. in Wasco, Calif., by David L. Armstrong. Introd. in 1968. Plant patent 2929; 7 Oct. 1969; assigned to Armstrong Nurseries, Ontario, Calif. Seedling of unnamed dwarf nectarine o.p.; selected about 1963. Fruit: 2 to 2 1/4 inches in diam.; globose to broadly ovoid; skin ground color yellow, blushed with red; flesh yellow, red at pit cavity, quality good, freestone; ripens early July at Wasco, probably late July in cooler areas; resembles Panamint. Tree: dwarf; leaf glands weakly reniform; flowers 1 1/2 inches in diam., showy.

Nectarose.—Orig. in New Brunswick, N.J., by M.A. Blake, New Jersey Agr. Expt. Sta. Introd. in 1947. Garden State x [(Gold Mine x Belle selfed) o.p.]. Fruit: large; skin nearly covered with a dark red blush; flesh white, quality good, freestone; ripens 2 1/2 weeks before Elberta and about with Sunhigh. Tree: hardier than Elberta; flowers showy; vigorous; productive.

Necta Zee.—Orig. in Modesto, Calif., by Floyd, Gary, and Grant Zaiger and Leith Gardner. Introd. in 1988. (Early Sungrand x a dwarf nectarine) o.p. Plant patent 6283; 31 Sept. 1988; assigned to Zaiger's Genetics. Fruit: medium to large; globose; skin dark yellow overspread with red; flesh yellow, firm; good flavor, subacid. Tree: medium vigor; spreading; very productive. For home-garden use.

New Yorker.—Orig. in Geneva, N.Y., by New York Agr. Expt. Sta. Introd. in New Zealand in 1954, in U.S. in 1967. (Livingston x Greensboro) o.p.; selected in 1945 by Henry Bennett, New York Fruit Testing Cooperative Assn.; tested as N.Y. 1017 since 1955. Fruit: round to oblate; skin greenish-white ground color, 80% covered with a dark red blush; flesh juicy, slightly coarse, medium firm; flavor sweet, rich; quality high; ripens about 20 Aug. at Geneva, just after Redhaven. Tree: vigorous and productive; has tendency toward narrow crotches; blossom buds are slightly more resistant than Redhaven to injury by low winter temperatures; flowers large, showy; leaf glands reniform. Planted in New Zealand.

Niagara.—Orig. in Le Grand, Calif., by F.W. Anderson, Merced, Calif. Introd. in 1969. Plant patent 2883; 6 May 1969; assigned to Reedley Nursery, Reedley, Calif. Red Grand x Early Sungrand; tested as Anderson AlH416. Fruit: large; globose to slightly oblong; skin yellow, substantially overspread with red at maturity; flesh yellow, freestone; ripens about with Le Grand; resembles Sun Grand, but larger and ripens about 2 weeks later. Tree: large; vigorous; productive and regular bearer; flowers large, pink; leaf glands reniform.

Olympia.—Orig. by W.D. Krause, Squaw Valley, Calif. Plant patent 6248; 16 Aug. 1988. Royal Giant o.p. Fruit: medium large; uniform globose; bright red over yellow skin; flesh yellow, uniformly firm, clingstone. Ripens 7 days after Red Jim. Flowers large, showy, pink; leaf glands reniform. Tree: medium to large; very productive; half hardy—not for cold winter areas.

PacificStar (*Sunectseven*).—Orig. in Fresno, Calif., by John H. Weinberger. Introd. in 1982. Plant patent 4893; 5 Oct. 1982; assigned to Superior Farming Co. May Grand o.p.; selected in 1977; tested as 69-128. Fruit: most nearly resembles May Grand, but rounder and smoother; size medium, 60 mm; skin highly colored, crimson on yellow ground color; flesh yellow with slight red at apex, firm, melting, excellent quality. Ripens 1 week after May Grand. Tree: medium vigor; upright-spreading; medium productivity; regular bearing; flowers midseason. Well-suited for commercial shipping.

Palomar.—Orig. in Ontario, Calif., by Herbert C. Swim. Introd. in 1956. Plant patent 1652; 8 Oct. 1957; assigned to Armstrong Nurseries, Ontario, Calif. Unnamed peach o.p.; selected in 1949. Fruit: large; skin highly blushed red; flesh yellow, flavor subacid, pleasing; resembles Panamint, but ripens later, early to mid-August at place of origin. Tree: medium; produces heavy crops. Particularly suited to southern California and central California inland valley areas, as chilling requirement medium.

Panamint.—Orig. in Ontario, Calif., by Herbert C. Swim, Armstrong Nurseries. Introd. in 1952. Plant patent 1100; 25 May 1952; assigned to Armstrong Nurseries. (Babcock peach x Boston) x (Gold Mine x Rio Oso Gem peach); selected in 1945. Fruit: high red skin color; flesh yellow, freestone; ripens during third week of July; resembles Pioneer. Tree: vigorous; productive; very short chilling requirement, adapted to southern California conditions.

Philp.—Orig. in Winters, Calif., by Guy L. Philp, California Agr. Expt. Sta., Davis. Introd. in 1948. Humboldt x Burbank; selected in 1942; tested as Calif. 27-12; sibling of Mabel. Fruit: medium to large, up to 5.7 cm in diam.; round to round-oval; skin light lemon-yellow, deeply blushed, nearly full-colored, surface slightly irregular; flesh light apricot-yellow, clear, firm to firm-melting, fine-grained, excellent texture, flavor excellent, subacid, sweet; stone small, oval, free, ripens in early midseason, about with Mabel, before Gower. Tree: extremely productive; very vigorous; flowers large, heavy bloom; moderate chilling requirement; leaf glands globose.

Pinta.—Orig. by C.B. Bubani, Fresno, Calif. Larry's open-pollinated. Plant patent 8486; 7 Dec. 1993. Similar to Red Jim, but ripens 5 weeks later with brighter color and richer flavor. Fruit: large; ovate; skin 60% to 90% red blush over bright yellow ground color; flesh yellow, melting, firm; freestone; red at pit. Flower: large, showy; leaf glands reniform. Tree: upright-spreading; very productive.

Pioneer.—Orig. in Ontario, Calif., by Herbert C. Swim. Introd. in 1949. Plant patent 787; 9 Mar. 1948; assigned to Armstrong Nurseries, Ontario, Calif. (Gold Mine nectarine x Rio Oso Gem peach) selfed; selected in 1943. Fruit: skin red, thin, sufficiently crisp to be eaten with ease; flavor rich; distinctive; flesh yellow with red coloring, especially around pit, freestone; ripens in same season as Gold Mine (last week in July); resembles Lippiatt's Late Orange nectarine. Tree: resistant to delayed foliation, being similar to Redwing peach; flowers large, pink, very showy; leaf glands reniform.

Pocahontas.—Orig. in Blacksburg, Va., by George D. Oberle, Virginia Agr. Expt. Sta. Introd. in 1965. (Sunhigh x V.P.I. 40N nectarine) o.p.2; tested as V.P.I. 57N. Fruit: medium large, 2 1/8 inches in diam.; ovate; skin highly colored; flesh yellow, slightly stringy, quality good, semi-freestone; ripens 19 days before Lexington. Tree: vigorous; productive; blossom large, nonshowy; leaf glands mostly reniform, but difficult to classify.

Pot of Gold.—See Le Grand.

Princess (*Merrill Princess*).—Orig. in Red Bluff, Calif., by Grant Merrill. Introd. in 1954. Plant patent 1410; 9 Aug. 1955; assigned to Grant Merrill Orchards, Red Bluff, Calif. (J.H. Hale peach x Flaming Gold) F2; discovered in 1949; selected in 1951. Fruit: large; skin mahogany-red nearly all over; flesh yellow, firm-melting when ripe, freestone, ripens with John Rivers. Tree: bears regularly and heavily. Variety now obsolete.

P-R Red.—Orig. by P. Ricchiuti, Clovis, Calif. Plant patent 6488; 27 Dec. 1988. Bud sport of September Grand, ripens 5–7 days later.

Redbud.—Orig. in Blacksburg, Va., by George D. Oberle and R.C. Moore, Virginia Agr. Expt. Sta. Introd. in 1957. Sister seedling of Lexington; selected in 1951; tested as V.P.I. 47N. Fruit: size about 2 inches in diam.; skin bright, solid red with a smooth finish, very attractive; flesh white, freestone, texture medium, firmness above average, quality near-good to good; matures about 27 days before Elberta, or about 14 days before Redchief; resembles Redchief. Tree: vigor moderate; produces heavy sets of buds; leaf glands reniform; flowers showy.

Redchief.—Orig. in Blacksburg, Va., by George D. Oberle and R.C. Moore, Virginia Agr. Expt. Sta. Introd. in 1952. (P.I. 43143 x unnamed V.P.I. peach seedling) x (P.I. 63973 x an unnamed V.P.I. peach seedling); cross made in 1936 by R.C. Moore; first fruited in 1940; tested as V.P.I. 19. Fruit: about 2 inches in diam. and length, nearly round, with halves tending to be unequal; skin bright red, attractive finish; flesh white, moderately firm, flavor mild, quality good, freestone; keeps well in cold storage; ripens 10–14 days before Elberta peach; resembles Nectaheart. Tree: vigorous; productive; flowers showy.

Red Delight.—Orig. by C.F. Zaiger, Modesto, Calif. Plant patent 6024; 29 Sept. 1987. (May Grand x Red Grand o.p.) o.p. Fruit: large; globose; high percentage of red blush over bright yellow ground color; flesh yellow, very firm, meaty, clingstone; ripens 6 days before May Grand. Flowers large, showy; leaf glands reniform. Tree: large; vigorous; productive.

Red Diamond.—Orig. near Le Grand, Calif., by F.W. Anderson, Merced, Calif. Introd. in 1972. Plant patent 3165; 9 May 1972; assigned to Reedley Nursery, Reedley, Calif. Early Sungrand x Red Grand; tested as Anderson 25P360. Fruit: size medium, globose-oblong; skin bright red, tough; flesh yellow, firm, melting, freestone, red at pit; flavor subacid, very good; unusually long harvest period, end of June through first half of July. Tree: large, open, spreading; vigorous; very productive; flowers medium size; leaf glands reniform.

Red Flame.—Orig. near Dinuba, Tulare County, Calif., by J.H. Enns. Plant patent 3952; 14 Sept. 1976. Mayred mutation. Fruit: medium; semi-clingstone; yellow skin overspread with red; yellow flesh; larger fruit size than Mayred and ripens 3 days earlier; good eating quality. Tree: medium to large; vigorous; spreading; adapted to moderate climate.

Red Fred.—Orig. by F.K. Nagare, Reedley, Calif. Plant patent 7821; 10 Mar. 1992. Early May Grand mutation, ripens 7 days after May Grand with semi-freestone fruit of superior firmness, color, and shape.

Red Free.—Orig. in Le Grand, Calif., by F.W. Anderson, Merced, Calif. Introd. in 1973. Plant patent 3328; 8 May 1973; assigned to Reedley Nursery, Reedley, Calif. Red Grand x Early Sungrand. Selected in 1970; tested as Anderson 25P 190. Fruit: medium; round to oblong; skin completely covered with red blush at maturity, medium thick, no cracking; flesh yellow with red at pit cavity streaked into flesh, firm, freestone; compared to Sun Grand, larger, brighter red, ripening 10 days later, extending the shipping period; good shipping qualities. Tree: medium, spreading-open; vigorous; very productive; leaf glands reniform; flowers large, showy.

Red Glen.—Orig. by N.G. and L.G. Bradford, LeGrand, Calif. Plant patent 7193; 13 Mar. 1990. Red Diamond x unnamed seedling. Fruit: large; globose; very dark red over deep red background; flesh brilliant yellow with strong red next to stone, very firm, very crisp, clingstone, stays firm 3 weeks on tree after first picking stage. Ripens 28 July at LeGrand. Flowers small, moderate pink; leaf glands reniform. Tree: large; vigorous; hardy; very productive.

Red Glo.—Orig. by C.F. Zaiger, G.N. Zaiger, L.M. Gardner, and G.G. Zaiger, Modesto, Calif. Plant patent 7828; 17 Mar. 1992. (Ruby Gold x low-chill peach) x (Sun Red x June Glo). Fruit: large; globose; attractive red over dark-yellow to light-orange ground color; flesh yellow, firm, meaty, clingstone. Ripens 1 month before Sun Red. Tree: large; vigorous; productive; chilling requirement 250 h.

Red Grand.—Orig. in Merced, Calif., by F.W. Anderson. Introd. in 1952. Plant patent 1060; 1 Jan. 1952; assigned to Reedley Nursery, Reedley, Calif. Le Grani x (Kim x Le Grand); selected in 1948. Fruit: very similar to Le Grand, but ripens about 1 week earlier and has more red surface color; flesh yellow, clingstone; larger than Quetta and about the same size as Le Grand. Tree: large, spreading; vigorous; productive; leaf glands reniform; flowers large, showy, pink.

Red Jewel.—Orig. by L.G. and N.G. Bradford, LeGrand, Calif. Plant patent 8013; 27 Oct. 1992. (Red Diamond x Autumn Free) o.p. Fruit: large; slightly oblong; very firm; very good flavor; 100% dark red over vivid reddish-orange ground color. Flesh, brilliant yellow, melting, freestone with sweet kernel. Ripens in early July, 65 days before Autumn Free. Flowers small, pink; leaf glands reniform. Tree: large; vigorous; spreading and dense; very productive.

Red June.—Orig. in Le Grand, Calif., by F.W. Anderson, Merced, Calif. Introd. in 1961. Plant patent 2044; 4 Apr. 1961; assigned to Reedley Nursery, Reedley, Calif. (Le Grand x July Elberta peach) F_2. Fruit: large to medium, suture diam. average 2 1/4 inches; globose, uniform; skin red to maroon, which almost completely covers a yellow ground color, smooth; flesh yellow, firm, meaty, moderately juicy, subacid, very little aroma, eating quality good, freestone; ripens about 2 weeks before John Rivers or Early Sungrand and 1 week before Grand River or Sunrise, with a ripening period of 5–12 June; market and dessert use; keeping and shipping qualities good. Tree: large, spreading; vigorous; very productive; bears regularly.

Red King.—Orig. in Visalia, Calif., by Lisle A. Hagler. Introd. in 1958. Plant patent 1718; 10 June 1958. Le Grand mutation. Fruit: skin predominantly red on tree-ripe fruit; flesh yellow, slightly red around the pit, firm, melting, flavor sweet and slightly tart, slightly aromatic, quality very good, freestone; ripens at place of origin about 4–6 July, being about midway between Sun Grand and Le Grand. Tree: large; vigorous; open, vase-formed; very productive; regular bearer; leaf glands globose; flowers self-fertile.

Red River.—Orig. in Le Grand, Calif., by F.W. Anderson. Introd. in 1954. Plant patent 1430; 1 Nov. 1955; assigned to Reedley Nursery, Reedley, Calif. (Le Grand x John Rivers) F_2; selected in 1951; tested as Anderson 6K52W. Fruit: medium to small; flesh white, texture medium to soft, melting, flavor subacid to mild, aroma pronounced, semi-freestone; ripens about 1 week earlier than John Rivers, which it resembles, but has more red skin color. Tree: medium; upright to spreading; productive; leaf glands reniform; flowers large.

Red Sunset.—Orig. by N.G. Bradford, LeGrand, Calif. Plant patent 6982; 15 Aug. 1989. (Dwarf seedling x May Grand) o.p. Fruit: medium; slightly oblong with protruding suture area. Flesh yellow with pink at cavity, soft, melting, freestone. Ripens 8 days after May Grand. Flowers large, pink; leaf glands reniform. Tree: genetic dwarf; average size.

Regal Grand.—Orig. in Le Grand, Calif., by F.W. Anderson. Introd. in 1959. Plant patent 1751; 16 Sept. 1958; assigned to Reedley Nursery, Reedley, Calif. Cross between two late-ripening, clingstone nectarine seedlings; selected in 1957; tested as Anderson 18K18. Fruit: skin yellow, partly overspread with red, thickness medium; flesh yellow, firm, flavor subacid, aroma distinct, quality good, clingstone; extends ripening and marketing season of Le Grand-type nectarines by maturing 7–10 days after Late Le Grand. Tree: large to medium; spreading; vigorous; productive; leaf glands reniform; flowers large.

Richards Grand (*Richards Pride*).—Orig. in Fresno, Calif., by John M. Garabedian, Peach & Willow Farms. Introd. in 1964. Plant patent 2419; 7 July 1964; assigned to Peach & Willow Farms. Chance seedling. Fruit: large, average 3 3/8 inches in suture diam.; globose; distinct deep suture at apex, cavity rounded, apex short and slightly depressed; skin yellow, highly blushed, medium thick, tenacious to flesh; flesh yellow, red next to pit, firm, slightly acid, distinct aroma, eating quality good; pit size medium, cling; ripens 3–4 weeks after Gold King; resembles Gold King. Tree: size medium; medium vigor; leaf glands reniform; flowers large, bloom in midseason.

Richards Pride.—See Richards Grand.

Rio Grand.—See Gold Nugget.

Rose (*White Rose*).—Orig. in Le Grand, Calif., by F.W. Anderson. Introd. in 1955. Plant patent 1421; 27 Sept. 1955; assigned to Burchell Nursery, Modesto, Calif. (Le Grand x John Rivers) F_2; tested as Anderson 6K50W. Fruit: flesh white, freestone; resembles John Rivers, but is larger, firmer, and has more attractive red skin; ripens during John Rivers season. Tree: chilling requirement lower and a heavier and more regular producer than John Rivers.

Roseprincess.—Orig. at Byron, Ga., by W.R. Okie, U.S. Dept. of Agriculture Southeast Fruit & Tree Nut Res. Lab. Introd. in 1989. Red King nectarine o.p.3e; seed collected in 1979 by V.E. Prince; selected in 1982; tested as BY80N384. Fruit: medium; skin attractive red blush on greenish-white ground color; round; flesh white, melting, freestone; flavor very good, high acidity. Ripens 2 weeks after Snow Queen, 1 week after Redhaven. Released to provide a firmer white nectarine for local sales and mid-range shipping. Flowers showy, self-fertile; leaf glands reniform; moderately resistant to bacterial spot. Chilling requirement about 850 h.

Royal Flame.—Orig. in Fowler, Calif., by H.E. and R.K. Hiraoka. Introd. in 1951. Flaming Gold o.p.; selected in 1949. Fruit: medium to large; flesh yellow, freestone, firmer than Flaming Gold; resembles Flaming Gold, but with a little less color; ripens at place of origin about 10 July, about 1 week earlier than Flaming Gold.

Royal Giant.—Orig. by C.F. Zaiger, Modesto, Calif. Plant patent 4107; 20 Sept. 1977. (Red Grand x late nectarine seedling) o.p. Fruit: very large; globose; high percentage of red over yellow ground color, coloring 2–3 weeks before ripening; flesh yellow, firm, clingstone. Ripens with Late Le Grande. Flowers large, showy, pink; leaf glands reniform. Tree: large; vigorous; upright; very productive—large crops of large-sized fruits.

Royal Glo.—Orig. by C.F. Zaiger, L.M. Gardner, G.N. Zaiger, and G.G. Zaiger, Modesto, Calif. Plant patent 8281; 29 June 1993. [(Dwarf nectarine x Ruby Gold) o.p.] x May Glo. Fruit: large; globose; firm, meaty, yellow flesh, clingstone; high percentage of attractive red over yellow ground color. Ripens 6 days after May Glo. Flowers large, showy, pink; leaf glands reniform. Tree: large; vigorous; upright; productive. Chilling requirement 450 h.

Royal Grand (*Early Red Grand*).—Orig. in Le Grand, Calif., by F.W. Anderson, Merced, Calif. Introd. in 1957. Plant patent 1634; 20 Aug. 1957. Red Grand o.p.; selected in 1954; tested as Anderson 2L7A2. Fruit: large, averaging 3 inches in diam.; skin yellow but almost completely covered red, thick, tenacious to flesh; flesh yellow with some red at pit, firm, meaty, aroma wanting, quality good, subacid, clingstone; ripens 12–22 July at place of origin, about 14 days before Red Grand and about with Early Le Grand; resembles Red Grand. Tree: size medium; vigor medium; bears regularly, very productive; leaf glands reniform; flowers large, bloom in midseason, about with Elberta peach.

Ruby Diamond.—Orig. by L.G. and N.G. Bradford, LeGrand, Calif. Plant patent 7918; 21 July 1992. Red Diamond x unnamed nectarine seedling. Fruit: large; very symmetrical; globose; deep red over entire surface; flesh bright orange-yellow with moderate red streaking near pit, very firm, tough, crisp, freestone. Ripens 6 days after Red Diamond. Flowers large, light pink; leaf glands reniform. Tree: large; vigorous; spreading; hardy; very productive.

Ruby Gold.—Orig. in Modesto, Calif., by Floyd Zaiger. Introd. in 1969. Parentage unknown. Fruit: medium to large; generally oval; skin highly colored; flesh yellow, semi-freestone. Tree: low-chilling requirements (400–600 h). This variety is cited as a parent of Stark Gulf Queen peach.

Santa Maria.—Orig. by C.B. Bubani, Fresno, Calif. Larry's o.p. Plant patent 8485; 7 Dec. 1993. Similar to Fantasia, but ripens 2 weeks later with a brighter red blush, firmer fruit, and enhanced flavor. Fruit: large; globose to oval; 25% to 60% red blush with some streaking; flesh light yellow, with red cavity radiating into flesh; melting; freestone. Flower: large, showy, pink; leaf glands reniform. Tree: hardy; upright-spreading; productive.

Scarlet Queen.—Orig. in Exeter, Calif., by Grant Merrill. Plant patent pending. Introd. in 1969. Merrill Queen o.p.; discovered in 1966; tested as C13-6S. Fruit: large for season; oval; skin highly colored, 50% covered with red or scarlet blush; flesh yellow, firm; clingstone, but semi-cling if tree-ripened; good shipping qualities; ripens first week in June in San Joaquin Valley. Tree: large, vase-shaped; vigorous; productivity good for the season; flowers large, showy; leaf glands reniform.

Scarlet Red.—Orig. by N.G. Bradford, LeGrand, Calif. Plant patent 5665; 18 Feb. 1986. (Red Free x Autumn Gold) x Rio Del Ray. Most nearly resembles Regal Grant, but ripens 2 days earlier and has brighter red skin and more flesh firmness. Fruit: globose; firm, crisp; clingstone; flesh yellow with red next to stone. Flowers large, pink; leaf glands reniform. Tree: large; vigorous; hardy; very productive.

September Grand.—Orig. in Le Grand, Calif., by F.W. Anderson. Introd. in 1958. Plant patent 1755; 23 Sept. 1958; assigned to Reedley Nursery, Reedley, Calif. Late Le Grand x peach seedling (which had as the female parent Late Le Grand); selected in 1957; tested as Anderson 11G22. Fruit: large, averaging 3 inches in diam.; skin yellow, partly overspread with red; flesh yellow, texture firm, flavor subacid, aroma distinct, clingstone; matures second week in September at place of origin, thus extending the marketing season about 3 weeks after Late Le Grand. Tree: large, spreading; productive; leaf glands reniform; flowers large.

September Red.—Orig. by N.G. and L.G. Bradford, LeGrand, Calif. Plant patent 5664; 18 Feb. 1986. (Red Free x Tom Grand) x Sparkling Red. Fruit: medium; globose; rose-beige blush over yellow ground color; flesh yellow, firm, crisp, clingstone with red next to stone throughout. Ripens 5 days later than September Grand. Flowers very late, large size, pink; leaf glands reniform. Tree: medium size; very productive; dense.

Sequoia.—Orig. in Porterville, Calif., by F.D. Williams. Introd. in 1942. Plant patent 496; 23 Dec. 1941. Hutchison peach mutation; discovered in 1937. Fruit: medium to large; skin smooth with deep red color; flesh nearly white with slight greenish tint, bright red at pit, firm, flavor good, slightly acid, freestone; ripens 15–25 Aug. Tree: large, upright; vigorous; productive.

Sierra Star (*Sunecthirteen*).—Orig. in Fresno, Calif., by John H. Weinberger and Timothy Sheehan. Introd. in 1988. Plant patent 6158; 26 Apr. 1988; assigned to Superior Farming Co. S#3 x Flamekist; selected in 1978; tested as 181-119. Fruit: very large,77 mm in diam.; round; skin dark red on yellow ground color; flesh yellow with red at pit, firm, melting, freestone; ripens 7–10 days after Flamekist, which it most nearly resembles. Tree: medium vigor; upright-spreading; very productive; regular bearing; flowers midseason. Flowers large, showy, pink.

Silver Lode.—Orig. in Ontario, Calif., by Herbert C. Swim, Armstrong Nurseries. Introd. in 1951. Plant patent 1023; 24 July 1951; assigned to Armstrong Nurseries, Ontario, Calif. (Gold Mine x Rio Oso Gem peach) x (Gold Mine x July Elberta peach); selected in 1944. Fruit: skin red; flesh white, sweet, texture good, freestone. Tree: adaptable to southern California's climate.

Silver Prolific.—Orig. in Le Grand, Calif., by F.W. Anderson, Merced, Calif. Introd. in 1964. Plant patent 2325; 3 Dec. 1963; assigned to L.E. Cooke Co., San Gabriel, Calif. (Late Le Grand x Flory Dwarf peach) F_2; sibling of Golden Prolific. Fruit: large, averaging 2 1/2 inches in diam.; globose to slightly oblate, uniform, symmetrical; skin pale yellow, slightly mottled with red, texture and thickness medium; flesh soft, subacid to mild, eating quality fair to good, ripens evenly, aroma pronounced, white with very pale greenish cast, red next to pit; stone free, size medium; seed bitter; harvest season second week in August; keeping quality medium; not firm enough for commercial long-distance shipping; resembles Golden Prolific, but is white instead of yellow-fleshed, ripens about 10 days before it, with more aroma. Tree: dwarf with very short internodes; vigorous; flowers large, anthers red, self-fruitful, bloom in midseason; leaf size medium, glands reniform, petiole short, blade 5 1/2 inches long; ornamental with fruit production.

Snow Queen.—Orig. in Wasco, Calif., by David L. Armstrong, Armstrong Nurseries, Ontario, Calif. Introd. in 1975. Plant patent applied for; to be assigned to Armstrong Nurseries. Unnamed peach selection o.p.; cross made in 1957; selected in 1964; tested as 57278-4. Fruit: 3 inches in diam.; slightly oblong; skin fair blush with russeting; flesh white, melting, texture fine, very juicy, freestone, quality very good; keeps well; recommended for market, shipping, dessert, home gardens; ripens mid-to late June. Tree: vase-shaped; very vigorous; very productive, regular bearer; flower small, deep pink, non-showy; leaf glands reniform.

Southern Belle.—Orig. in Le Grand, Calif., by F.W. Anderson. Introd. in 1975. Plant patent 3758; assigned to L.E. Cooke Co., Visalia, Calif. Empress peach x Bonanza peach. Fruit: large; skin yellow overspread with red; flesh yellow, freestone; ripens early August. Tree: dwarf, 5 ft at maturity; productive; chilling requirement about 300 h.

Sparkling June.—Orig. by N.G. Bradford, LeGrand, Calif. Plant patent 6981; 15 Aug. 1989. (Red Diamond x Early Sungrand) o.p. Fruit: medium; uniform globose; 95% deep red over yellow ground color; flesh yellow, extremely firm, crisp, melting, freestone. Ripens 14 days before Early Sungrand. Flowers large, pink; leaf glands reniform. Tree: medium size; spreading.

Sparkling May.—Orig. by N.G. and L.G. Bradford, LeGrand, Calif. Plant patent 6675; 14 Mar. 1989. Red Diamond x Aurora Grand. Fruit: uniform medium size; slightly oblong; full red blush; flesh light yellow, very firm, melting, freestone. Ripens 29 days before Red Diamond, 3 days before Aurora Grand. Flowers small, moderate pink; leaf glands reniform. Tree: medium size; vigorous; spreading; productive.

Sparkling Red.—Orig. by F.W. Anderson, Merced, Calif. Plant patent 4314; 10 Oct. 1978. Assigned to Burchell Nursery, Modesto, Calif. Summer Grand x May Grand. Fruit: medium large; globose; mottled and streaked red over golden glow yellow; flesh yellow, firm, melting with red next to freestone. Ripens a few days after LeGrand, which the fruit strikingly resembles. Flowers large; reniform leaf glands. Tree: medium; vigorous; productive.

Spring Bright.—Orig. by N.G. and L.G. Bradford, LeGrand, Calif. Plant patent 7507; 23 Apr. 1991. May Diamond x unnamed seedling. Fruit: large; oval; high percentage of deep red over light orange-yellow; flesh vivid yellow with deep pink at pit, very sweet, extremely firm, crisp, clingstone. Ripens 16 days after May Diamond. Flowers large, light pink; leaf glands globose. Tree: large; spreading; hardy; very productive.

Spring Grand.—Orig. in Livingston, Calif., by Sam Maeda. Introd. in 1966. Plant patent 2767; 12 Sept. 1967; assigned to Reedley Nursery Co., Reedley, Calif. Early Sungrand mutation; discovered in 1964. Fruit: flesh yellow, freestone; resembles Early Sungrand, but ripens 10 days earlier. Tree: vigor medium; moderately hardy; medium productivity, bears regularly; leaf glands reniform; flowers large, pink.

Spring Red.—Orig. in Le Grand, Calif., by F.W. Anderson, Merced. Introd. in 1974. Plant patent 3619; 17 Sept. 1974; assigned to Reedley Nursery, Reedley, Calif. Summer Grand o.p.; discovered in 1970. Fruit: size medium; globose to oblong symmetrical; skin yellow, entirely overspread with red at maturity; flesh yellow, freestone; resembles Early Sungrand; ripens between May Grand and Early Sungrand. Tree: large; spreading; vigorous; productive, bears regularly.

Star Bright (*Sunecten*).—Orig. in Fresno, Calif., by John H. Weinberger. Introd. in 1982. Plant patent 4927; 9 Nov. 1982; assigned to Superior Farming Co. Summer Grand o.p.; selected in 1977; tested as 83-50. Fruit most nearly resembles Summer Grand; compared to Early Sungrand, rounder, smoother, with less prominent tip; medium; skin scarlet on yellow ground color; medium thickness; flesh yellow, mottled with red next to pit, fine texture, melting, excellent eating quality; freestone. In central California, ripens about with Early Sungrand. Tree: vigor medium; upright-spreading; productive. Flower: large, showy, pink.

Star Brite.—Orig. by R. Peters and E. Wuhl, Fresno, Calif. Plant patent 5586; 3 Dec. 1985. Early Sungrand mutation. Similar to May Grand, but fruits are larger, redder, more pointed, and ripen 5 days earlier.

Star Grand.—Orig. in Le Grand, Calif., by F.W. Anderson, Merced, Calif. Introd. in 1954. Plant patent 1327; 30 Nov. 1954; assigned to Reedley Nursery, Reedley, Calif. Le Grand x Gold Nugget; selected in 1950; tested as Anderson 6K98. Fruit: large; skin highly colored; flesh yellow, freestone; ripens between John Rivers and Gower. Tree:

large, spreading; vigorous; productive; leaf glands reniform; flowers medium, non-showy, pink.

Star Grand II.—Orig. in Le Grand, Calif., by F.W. Anderson. Introd. in 1956. Plant patent 1646; 8 Oct. 1957; assigned to Reedley Nursery, Reedley, Calif. Le Grand x Sun Grand; selected in 1954; tested as Anderson 1B15N. Fruit: seed sweet; otherwise similar to Star Grand. Tree: chilling requirement less than that of Star Grand.

Stark Crimson Snow™ (*Bradcrim*).—Orig. by L.G. Bradford and N.G. Bradford, LeGrand, Calif. (Sun Grand x Robin) o.p. Plant patent 8461, 16 Nov. 1993. Fruit: size medium; globose, with inconspicuous shallow suture line; skin nearly full-red blush; flesh white, medium firm, melting, non-acidic; freestone; pronounced aroma with very good eating quality. Ripens 1 month after Robin. Flower: late, large, medium pink; leaf glands globose. Tree: large; vigorous; spreading; very productive.

Stark Delicious.—Orig. in Merced, Calif., by Fred W. Anderson. Introd. in 1962. Plant patent 1921; 15 Mar. 1960; assigned to Stark Bros. Nurseries & Orchards Co., Louisiana, Mo. Late Le Grand x Star Grand; selected in 1958; tested as Anderson 7B94. Fruit: large, averaging about 2 1/4 inches or more in diam.; globose to oblong; skin thick to medium, yellow mottled with red, with full all-over red in Midwest and eastern U.S.; flesh yellow with considerable red around the pit, firm, crisp, melting, subacid, flavor delicate, quality good, freestone; shipping and keeping qualities good; ripens first of July at place of origin. Tree: vigorous; productive; bears regularly; leaf glands reniform; flowers large, showy.

Stark EarliBlaze.—Orig. in Le Grand, Calif., by F.W. Anderson, Merced, Calif. Introd. in 1962 by Stark Bros. Nurseries & Orchards Co., Louisiana, Mo. Plant patent 1949; 7 June 1960; assigned to Stark Bros. Nurseries & Orchards Co. Sun Grand x Le Grand x Merrill Gem peach; tested as Anderson 14K54. Fruit: large to medium; form uniform, symmetrical, globose; skin yellow, largely overspread with red, thickness medium; flesh yellow, firm, meaty, eating quality fair to good, flavor subacid, ripens evenly; semi-freestone to freestone when fully ripe; ripens early, about 2 weeks before Redhaven peach; shipping quality good, for early season; resembles Grand River. Tree: moderately vigorous; upright to spreading; productive; bears regularly; leaf glands globose; flowers large, showy.

Stark Early Bird (*Anderbird*).—Orig. in Merced, Calif., by Fred Anderson. Plant patent 4076; assigned to Stark Bros. Nurseries, Louisiana, Mo. Introd. in 1980. Red Diamond x May Grand. Selected about 1977. Fruit: size medium; skin 100% bright red; flesh yellow, firm, flavor excellent; semi-freestone; good shipping qualities; ripens early, 1 week ahead of Redhaven. Tree: bud hardiness better than Loring, but slightly less than Redhaven; vigorous.

Stark Early Flame.—Orig. in Sheffield, Ala., by R.H. King. Introd. in 1946. Plant patent 759; 2 Sept. 1947; assigned to Stark Bros. Nurseries & Orchards Co., Louisiana, Mo. Flaming Gold o.p.; discovered in 1942. Fruit: appearance, size, and flavor resemble Flaming Gold, but ripens 14 days earlier, around 1 Aug., stone semi-free. Tree: large, spreading; vigorous; hardy; productive and regular bearer; flower size medium, pink; leaf glands reniform.

Stark Gulf Pride (*Zaipride*).—Orig. in Modesto, Calif., by Floyd Zaiger. Introd. in 1983. Plant patent 5461; 7 May 1985.; Assigned to Stark Bros. Nurseries & Orchards Co., Louisiana, Mo. Red Grand nectarine x Desertgold peach; selected in 1976; tested as 44G1235. Fruit: 2–2 1/2 inches in diam.; skin covered with 50% red blush; flesh yellow, firm, flavor excellent. Tree: very low chilling requirement (about 350 h); matures early.

Stark HoneyGlo (*Arderhone*).—Orig. in Merced, Calif., by Fred Anderson. Plant patent 4789; 10 Nov. 1981; to be assigned to Stark Bros. Nurseries, Louisiana, Mo. Introd. in 1980. Parentage unknown. Discovered and selected in 1970. Fruit: size medium; skin 80% bright red, ground color yellow; flesh firm, yellow, quality fair to good; freestone; ripens 2 weeks before Loring. Tree: miniature or genetic dwarf, brachytic growth; medium height 6 ft at maturity; productive in Midwest; slightly less hardy than Redhaven, more bud hardy than Loring.

Stark LateGold.—Orig. in Merced, Calif., by F.W. Anderson. Introd. in 1972. Plant patent 2793; 30 Jan. 1968; assigned to Stark Bros. Nurseries & Orchards Co., Louisiana, Mo. Late Le Grand x Early Sungrand; cross made between 1960 and 1965; tested as 18K84B. Fruit: 2 3/4 inches in diam.; globose; skin yellow, 70% overspread with red, medium thick; flesh yellow, firm, quality excellent, freestone; for shipping and dessert; ripens slightly later than Elberta peach. Tree: large; spreading; vigorous; hardy in north central Missouri; flowers large; leaf glands reniform.

Stark RedGold.—Orig. in Le Grand, Calif., by F.W. Anderson, Merced, Calif. Introd. in 1962. Plant patent 1947; 24 May 1960; assigned to Stark Bros. Nurseries & Orchards Co., Louisiana, Mo. (Sun Grand) o.p.2; tested as Anderson 13H1138. Fruit: large to medium; form symmetrical, globose; skin smooth, yellow, overspread with red, medium to thick; flesh yellow, streaked with red next to pit, firm, meaty, ripens evenly, flavor subacid, eating quality good to best, freestone; shipping quality very good; ripens immediately after Sun Grand; resembles Sun Grand. Tree: vigorous; very productive; leaf glands reniform; flowers large.

Stark SunGlo.—Orig. in Le Grand, Calif., by F.W. Anderson, Merced, Calif. Introd. in 1962 by Stark Bros. Nurseries & Orchards Co., Louisiana, Mo. Plant patent 1929; 5 Apr. 1960; assigned to Stark Bros. Nurseries. Sun Grand x (Le Grand x Halehaven peach); tested as Anderson 14K14. Fruit: large; symmetrical, globose; skin yellow, partly overspread with red, thickness medium; flesh yellow, shading to red near pit, firm, meaty, melting, subacid, ripens evenly, quality good, freestone; keeping and shipping quality good; ripens 4–5 days after Redhaven peach, about 7 days before Sun Grand, which it resembles. Tree: large to medium; moderately vigorous; productive; bears regularly; leaf glands reniform; flowers nonshowy, bloom in midseason.

Stark Sweet Melody (*Anderbrite, Bright*).—Orig. in Le Grand, Calif., by F.W. Anderson, Merced, Calif. Introd. in 1974. Originally introduced as Bright under plant patent 3329; 8 May 1973; assigned to Arthur Bright, Le Grand. Reintroduced by 1 Aug. 1978; reassigned to Stark Bros. Nurseries, Louisiana, Mo., in 1980 as Stark Sweet Melody. Cross of two unnamed dwarf seedlings. Fruit: large-medium; globose; skin yellow, blushed with red over about 70%; flesh yellow with red at pit cavity, freestone; good shipping qualities, for long distances, intended to compete with fruit from standard-size trees; resembles Golden Prolific, ripens 1 month before, about with Red Haven peach. Tree: dwarf, 5–8 ft; vigor moderate; productive; leaf glands reniform; flowers large, showy.

Strawberry.—See Hayes Late.

Summer Beaut.—Orig. near Le Grand, Merced County, Calif., by F.W. Anderson. Plant patent 4093; 23 Aug. 1977; assigned to Burchell Nursery, Modesto, Calif. From F_1 cross of Red Diamond x Sun Grand. Fruit: large; symmetrical; globose; freestone; yellow skin overspread with red; flesh yellow and firm; ripens 10 days earlier than Sun Grand; good eating quality. Tree: large; vigorous; upright to spreading; blooms midseason with large flowers; leaf glands globose.

Summer Beaut (*Anderbeaut*).—Orig. in Merced, Calif., by Fred Anderson. Plant patent 4093; 23 Aug. 1977; assigned to Stark Bros. Nurseries, Louisiana, Mo., and Burchell Nurseries, Modesto, Calif. Introd. in 1979. Parentage unknown. Discovered and selected about 1970. Fruit: large; skin 100% bright red; flesh yellow, firm, good shipper; flavor good; freestone; ripens 1 week after Redhaven; hangs well on tree. Tree: bud hardiness better than Loring, similar to Redhaven.

Summer Bright.—Orig. in LeGrand, Calif., by Norman G. and Lowell G. Bradford. Introd. in 1987. Red Diamond nectarine x unnamed peach seedling. Plant patent 7049; 24 Oct. 1989; assigned to Bright's Nursery, LeGrand. Fruit most nearly resembles Red Diamond. Fruit: large; globose to slightly oblong; skin 75% to 90% deep red on reddish-orange ground color; finish very smooth; flesh yellow, very firm; clingstone; excellent flavor, good sugar/acid balance. In central California, ripens 18 days after Red Diamond, 5 days before Fantasia. Tree: very vigorous; spreading;

very productive; hardy in central California. For long-distance shipping.

Summer Diamond.—Orig. in LeGrand, Calif., by Norman Bradford. Introd. in 1984. Red Diamond o.p. Plant patent 5654; 4 Feb. 1986; assigned to Bright's Nursery, LeGrand. Fruit resembles Red Diamond. Fruit: medium; slightly oblong; skin red on orange-yellow ground color; finish smooth and shiny; flesh yellow, very firm; flavor good; freestone; pit cavity larger than desired. In central California, ripens with Fantasia. Tree: medium vigor; spreading; production medium to heavy; hardy in central California. For long-distance shipping.

Summer Fire.—Orig. by N.G. and L.G. Bradford, LeGrand, Calif. Plant patent 7506; 23 Apr. 1991. Red Diamond x unnamed seedling. Fruit: large; uniform; oval; mostly dark red over orange background; flesh light yellow with strong red mottling toward the stone, extremely firm, crisp, clingstone. Ripens 9 days after Summer Bright. Flowers small, moderate pink; leaf glands reniform. Tree: large; vigorous; spreading; productive.

Summer Grand.—Orig. in Le Grand, Calif., by F.W. Anderson, Merced, Calif. Introd. in 1969. Plant patent 2879; 1 Apr. 1969; assigned to Reedley Nursery, Reedley, Calif. Late Le Grand nectarine x Early Sungrand nectarine; tested as Anderson 18K84a. Fruit: large; globose; skin yellow, substantially overspread with red; flesh yellow, freestone; resembles Sun Grand, but is characterized by larger fruit and more attractive appearance, ripens about 3 weeks before Le Grand and about with Sun Grand. Tree: resembles Sun Grand, but distinguished by reniform leaf glands.

Summer Lion.—Orig. by D.M. and L.M. Serimian, Selma, Calif. Plant patent 6543; 17 Jan. 1989. Red Lion x May Grand. Fruit: very large; uniform and slightly asymmetrical; 85% to 95% bright red with dark-red streaks on shoulder over yellow skin; firm, melting clingstone. Ripens 1 July in central part of San Joaquin Valley. Flowers large, showy; leaf glands reniform. Tree: average size and vigor; semi-spreading.

Summer Lion Two.—Orig. by D.M. and L.M. Serimian, Selma, Calif. Plant patent 6544; 17 Jan. 1989. May Grand x Red Lion. Fruit: very large; uniform; nearly globose; 90% red blush over yellow ground color; flesh amber-yellow with red at pit cavity, crisp, firm, melting, freestone. Ripens 10 July in central part of San Joaquin Valley. Flowers large, showy, pink; leaf glands predominately reniform with some globose. Tree: average size; vigorous; upright.

Summer Lion Three.—Orig. by D.M. and L.M. Serimian, Selma, Calif. Plant patent 6541; 17 Jan. 1989. May Grand x Red Lion. Fruit: very large; moderately asymmetrical ovate; 80% to 95% red with dark-red mottling over yellow skin; flesh yellow-amber with dark red at pit, crisp but melting, freestone. Ripens 15 days earlier than Red Lion. Flowers large, showy, pink; leaf glands reniform. Tree: moderate in size and vigor; upright.

Summer Star *(Sunectwelve)*.—Orig. in Fresno, Calif., by John H. Weinberger. Introd. in 1985. Plant patent 5473; 21 May 1985; assigned to Superior Farming Co. Fantasia o.p.; selected in 1977; tested as 90-110. Fruit: most nearly resembles Fantasia, with which it ripens, but skin has more red blush and b yellow ground color; rounder and more uniform; fewer cleft sutures; large, 75 mm; flesh yellow, with some red near stone; firm, melting; good eating quality. Tree: vigor medium; upright-spreading; productive.

Sunblaze.—Orig. at Agr. Expt. Sta., Univ. of Florida, Gainesville, from a 1977 cross; introd. in 1990 by W.B. Sherman, P.M. Lyrene, and F.G. Gmitter. Complex parentage of Florida selections. Selected in 1979 and tested as Fla. 9-15N. Fruit: medium; slightly oval; skin 90% to 100% red blush over yellow ground color; flesh yellow, firm, melting, freestone. Ripens with Sunred, about 90 days from bloom. Flowers small, deep pink, nonshowy; leaf glands reniform. High resistance to bacterial spot. Tree: vigorous; productive. Chilling requirement about 250 chill units.

Sunbob.—Orig. at Agr. Expt. Sta., Univ. of Florida, Gainesville. Introduced in New South Wales, Australia, in 1989. Parentage unknown. Fruit: medium large, long ovate; attractive 60% to 80% bright red over bright-yellow ground color; skin tough and very waxy. Flesh yellow, firm, melting, freestone. Ripens 100 days after bloom. Flowers large, showy, light pink; leaf glands reniform. Chilling requirement about 200 chill units.

Sun Bonnet.—Orig. in Le Grand, Calif., by F.W. Anderson, Merced, Calif. Introd. in 1974. Plant patent 3325; 1 May 1973; assigned to L.E. Cooke Co., Visalia, Calif. Empress peach o.p.; selected in 1969; tested as Anderson 9E725. Fruit: resembles Sunburst; medium to small; globose, uniform; symmetrical; skin yellow, substantially overspread with red; flesh yellow with red flecks, red at pit, firm, crisp, clingstone; subacid to mild; ripens last half of July, 1 month before Golden Prolific; recommended for long-distance shipping. Tree: genetically dwarf; spreading; vigorous; very productive; flower large; leaf glands reniform.

Sunbright.—See Sunbright I.

Sunbright I *(Sunbright)*.—Orig. in Red Bluff, Calif., by Grant Merrill. Introd. in 1960. Plant patent 2138; 17 Apr. 1962. Merrill Sunrise o.p.; discovered in 1957; selected in 1960. Fruit: size medium, average 2 1/4 inches in diam.; skin medium thick, bright orange-yellow overlaid with medium orange-red, brighter than Merrill Sunrise, similar to Sunbright II; flesh yellow, firm, moderately juicy, mild, subacid; partly freestone when ripened off the tree; ripens early, about 5 days before Sunrise; shipping quality good; very similar to Sunbright II and Sunrise. Tree: very strong, vigorous; produces well; very similar to Sunrise.

Sunbright II *(Early Sunrise)*.—Orig. in Red Bluff, Calif., by Grant Merrill. Introd. in 1961. Plant patent 2240; 12 Mar. 1963. Sunrise o.p.; discovered in 1957; selected in 1960. Fruit: average 2 5/16 inches in diam.; oblong to globose; skin over-color more red than brown; flesh yellow, very firm, mild, very sweet, flavor good; slightly freestone when ripened off the tree; ships well; ripens about 5 days before Sunrise; resembles Sunbright I, colors slightly later, sugars earlier. Tree: large; vigorous; medium upright; productive.

Sunbrite.—See Sun Grand.

Sunburst.—Orig. in Le Grand, Calif., by F.W. Anderson, Merced, Calif. Introd. in 1973. Plant patent 3330; 8 May 1973; assigned to Stark Bros. Nurseries and Orchards Co., Louisiana, Mo. Cross between two unnamed genetic dwarf seedlings; selected in 1967. Fruit: size medium; uniform, symmetrical, globose; skin yellow, substantially overspread with red, glossy; flesh firm, crisp, yellow with red at pit cavity; good shipping quality, recommended for long distances, intended to compete with fruit from standard-size trees; resembles Golden Prolific, differentiated by its clingstone fruit and globose leaf glands, ripens 1 month sooner. Tree: dwarf, short internodes; vigorous; spreading; regular and productive bearer; flowers large, self-fertile.

Sunburst.—Orig. by C.J. Riano, Dinuba, Calif. Plant patent 7364; 23 Oct. 1990. Parentage unknown. Fruit: medium to large; asymmetrical; deep red over yellow skin; flesh yellow, firm, crisp, clingstone, ripens 5 weeks after Sun Grand. Flowers medium to large, pink; leaf glands reniform with some globose types. Tree: average size; vigorous; highly productive.

Suncoast.—Orig. at Florida Agr. Expt. Sta., Gainesville. Complex parentage; tested as Fla. 85-6N; released by P.C. Andersen and W.B. Sherman in 1994. Released for large size early in season. Fruit: oblong; large; 80% red blush over yellow ground color; skin tough, with some sugar speckles; flesh firm, sweet, yellow, melting; semi-clingstone at soft ripe. Ripens 7 days after Flordaking, about 75 days after full bloom. Flower: small, deep pink, nonshowy; leaf glands reniform. Leaves medium small; tend to roll in autumn. Good resistance to bacterial leafspot. Chilling requirement 400 chill units.

Sundollar.—Orig. in Agr. Expt. Sta., Univ. of Florida, Gainesville, by P.C. Andersen, W.B. Sherman, and R.H. Sharpe. Introd. in 1982. Sunlite x Armqueen; cross made in 1978; selected in 1982; tested as 2-4N. Fruit: medium when properly thinned; round, with no significant tip and smooth suture; skin 80% red blush; flesh yellow, melting, good firmness; relatively nonbrowning; semi-clingstone. Ripens 70 days from bloom, early May in northern Florida. Flowers showy, medium-pink; leaf glands reniform. Tree: moderately vigorous; productive; semi-upright; sets moderate to high number of fruit buds; self-fertile. Resistant to bacterial leafspot. Chilling requirement: 400 chill units.

Sundowner.—Orig. in Agr. Expt. Sta., Univ. of Florida, Gainesville, by W.B. Sherman and R.H. Sharpe and named by New South Wales Dept. of Agriculture, Australia. Open-pollinated seedling of a Columbina x Sunred selection. Selected in 1976; tested as Fla. 6-3N. Fruit: medium; skin 90% red blush; flesh yellow, medium firm, semi-freestone; ripens 90 days after bloom. Flowers showy; leaf glands reniform. Resistant to bacterial leafspot. Chilling requirement: 250 chill units.

Sunectsix.—See Hi-Red.

Sunectseven.—See Pacific Star.

Sunecteight.—See Superstar.

Sunecten.—See Star Bright.

Sunecteleven.—See Crimson Star.

Sunectwelve.—See Summer Star.

Sunecthirteen.—See Sierra Star.

Sunectnineteen.—See How Red.

Sun Flame.—Orig. in Le Grand, Calif., by F.W. Anderson. Introd. in 1955. Plant patent 1429; 1 Nov. 1955; assigned to Beck and Hiraoka. (Le Grand x Halehaven peach) F_2; selected in 1952; tested as Anderson 4K20. Fruit: size medium; flesh yellow, firm, melting, with red mottling next to the pit, flavor rich with some acid, freestone; ripens between Sun Grand and Freedom; resembles Sun Grand. Tree: large; upright to spreading; vigorous; productive; leaf glands globose; flowers large, pink.

Sunfre *(Cashmere)*.—Orig. in Fresno, Calif., at U.S. Hort. Field Sta. by U.S. Dept. of Agriculture and released jointly with Florida Agr. Expt. Sta., Monticello. P42-81 [P60-38 (LeGrand x Sun Grand) x Bonanza] x P42-91 [P60-38 (LeGrand x Sun Grand) x Bonanza]. Introd. in 1981. Cross made in 1969; first fruited in 1973 and selected by John H. Weinberger. Tested as C73-40. Fruit: large, 2 3/8 to 2 1/2 inches in diam., round; flesh yellow, firm, smooth texture, quality good; semi-freestone; good storage qualities; skin in Florida has yellow ground color with bright red blush covering 75% of surface; in California not as attractive as May Grand, having many lenticel markings and russeting. Tree: vigorous; highly productive; chilling requirement about 500 h at or below 45F; blooms self-fertile, with large, showy petals; leaf glands reniform. Bacterial spot susceptibility similar to June Gold peach.

Sungem.—Orig. in Agr. Expt. Sta., Univ. of Florida, Gainesville by W.B. Sherman, P.M. Lyrene, P.C. Andersen, T.E. Crocker, and R.H. Sharpe. Introd. in 1977. Complex parentage of Florida selections x Armking; included in the pedigree are Okinawa, Panamint, and Columbina. Tested as Fla. 7-4N. Fruit: small to medium; skin 90% blush, russeting rare; flesh yellow, melting, firm, semi-clingstone; ripens 70–75 days after bloom, 5–12 days before Armking. Flower: large, showy, medium-pink, self-fruitful; leaf glands reniform. Sets unusually high amount of flower buds. Resistant to bacterial leafspot. Chilling requirement: 450 chill units.

Sungold.—Orig. in Gainesville, Fla., by Agr. Expt. Sta., Univ. of Florida. Introd. in 1969. NJ5107397 derived from (Candoka x Flaming Gold) x [(Garden State x NJ25032) o.p.] crossed with (Okinawa x Panamint); cross made in 1958 at Agr. Expt. Sta., New Brunswick, N.J., by L.F. Hough and Catherine Bailey; seed started by R.H. Sharpe at Gainesville, Fla.; seedlings planted at North Florida Expt. Sta., Quincy, in 1959; selected in 1961 by H.W. Young and R.H. Sharpe; tested as Q303-4 and Q45. Fruit: size medium; roundish; skin 80% to 100% covered with bright red blush; suture not prominent; flesh golden-yellow, quality excellent, firm, freestone when fully ripe; ripening date varies from 9 to 29 June. Tree: considerable resistance to brown rot; 550 h; flowers nonshowy, small; leaf glands reniform.

Sun Grand *(Sunbrite)*.—Orig. in Merced, Calif., by F.W. Anderson. Introd. in 1950. Plant patent 974; 22 Aug. 1950; assigned to Reedley Nursery, Reedley, Calif. F_2 of (Kim x July Elberta); selected in 1947. Fruit: large, about 2 1/2 inches in diam.; flesh yellow, firm, freestone; ripens between John Rivers and Gower, about 5 July. Tree: large, spreading; productive; leaf glands globose; flowers large.

Sunhome.—Orig. Agr. Expt. Sta., Univ. of Florida, Gainesville by W.B. Sherman, P.M. Lyrene, and R.H. Sharpe. Introd. in 1985.

Sunred x complex parentage of Florida selections; included in the pedigree are Elberta, J.H. Hale, and Southland. Selected in 1979; tested as Fla. 9-13NR. Fruit: medium-small; nearly round without prominent suture or tip; skin 100% red blush; flesh yellow, firm, smooth-textured, relatively nonbrowning, semi-freestone; quality medium; ripens 90 days from bloom. Petals large, pink, showy; leaves dark red; leaf glands reniform. Intended for dooryard and landscape purposes. Chilling requirement: 300 chill units. Leaves and fruit highly resistant to bacterial leafspot.

Sunking.—Orig. in Merced, Calif., by Walter D. Krause, Stribling's Nurseries. Introd. in 1965. Plant patent 2775; 7 Nov. 1967. Seedling of Sun Grand; selected in 1960; tested as S-137-123163-61. Fruit: resembles Red June; skin brillant red; flesh firm for early shipping, flavor excellent; semi-freestone, free when soft-ripe; ripens late May and early June. Tree: vigorous, sturdy; productivity medium to good, requires very little thinning; chilling requirement 700 h; leaf glands reniform; flowers medium size, pink to dark pink, showy.

Sunlite.—Orig. in Gainesville, Fla., by R.H. Sharpe, Univ. of Florida. Introd. in 1974. (Okinawa peach x Panimint) x NJN21; cross made in 1959; selected in 1966; tested as 44-28. Fruit: small to medium; round; skin 2/3 covered with light blush; flesh yellow, medium firm, flavor and texture good, freestone at maturity; ripens 1 June; recommended for fresh shipping. Tree: vigor good; productivity medium to good; chilling requirement 450 h; flowers large, showy, self-fruitful; leaf glands reniform.

Sunmist.—An early white nectarine with low chilling requirement. Released in 1994 by W.B. Sherman, P.C. Andersen, and P.M. Lyrene, Florida Agr. Expt. Sta., Gainesville, in 1994. Flordaglo peach x Mayfire; tested as Fla. 88-11NW. Plant patent applied for. Fruit: round; 90% red blush over creamy white ground color; skin moderately tough; flesh firm, white, melting, nonbrowning on cut surfaces; semi-clingstone. Ripens about 80 days after full bloom, about 7 days before Sunblaze. Flower: showy, pink; leaf glands globose. Resistant to bacterial leafspot. Chilling requirement 300 chill units.

Sunny Red.—Red Jim mutation, found by P. Chiamari and S. R. Hirasuna, Fresno, Calif. Plant patent 8534; 11 Jan. 1994. Fruit: similar to Red Jim but ripening 10 days later; large; ovate with broad groove suture; skin mostly dark red with striping; flesh yellow, melting, firm, dark red at pit; freestone. Flower: large, showy, pink; leaf glands reniform.

Sunraycer.—Orig. at Florida Agr. Expt. Sta. Cross of complex parentage, tested as Fla. 84-16N. Released in 1994 by W.B. Sherman, P.C. Andersen, and P.M. Lyrene for its large, attractive fruit. Fruit: elongated oval; large; 90% red blush over bright yellow ground color; flesh firm, yellow, melting, nonbrowning; semi-cling when soft ripe. Few split pits. Ripens 5 days after Sunred, 85 days after bloom. Flower: numerous, nonshowy; high fruit set. Leaf glands reniform. Very resistant to bacterial leafspot. Chilling requirement 275 chill units.

Sunred.—Orig. in Gainesville, Fla., by R.H. Sharpe, Florida Agr. Expt. Sta. Introd. in 1964. [Panamint x (Southland x Hawaiian) F_2] o.p.; cross made in 1959; first fruited in 1961; selected in 1961; tested as Fla. 6872. Fruit: small; round; skin bright red over 90% to 100% of surface; flesh yellow, firm, dessert quality excellent, semi-freestone when fully ripe; ripens during the first or second week in May in central Florida. Tree: requires about 250 h of winter chilling at or below 45F to break the rest period.

Sunrich.—Orig. in Gainesville, Fla., by H.W. Young and R.H. Sharpe, Univ. of Florida Agr. Expt. Sta. Introd. in 1971. NJ 5107397 [(Candoka Flaming Gold) x (Garden State x NJ 25032)] x (Okinawa x Panamint). Cross made in 1958 by L.F. Hough and Catherine Bailey, Rutgers Univ., Agr. Expt. Sta., New Brunswick, N.J. Seedling grown, tested, and selected at Gainesville. Tested as Q303-24 and Q42; selected in 1961. Fruit: 2–2 1/4 inches in diam.; round; skin covered with 80% to 100% red blush; flesh yellow, freestone; ripens second week of June. Tree: size medium; vigorous; moderately productive; chilling requirement 550 h; flowers nonshowy; leaf glands reniform.

Sunripe.—Orig. in Gainesville, Fla., by Dept. of Fruit Crops, Univ. of Florida. Introd. in

1979. Flordawon x Merril Princess. Cross made in 1967; selected in 1970; tested as Fla. 7E-62. Fruit: medium-large, roundish-oblong; skin deep yellow ground color with 80% bright red blush; flesh clear yellow with trace of red near pit, high quality, firm, very sweet; freestone; ripens after Sunlite. Tree: vigorous; upright-spreading; susceptible to bacterial spot; flowers large, showy, medium-pink; leaf glands reniform; chilling requirement 350 h.

Sunrise *(Merrill Sunrise)*.—Orig. in Red Bluff, Calif., by Grant Merrill. Introd. in 1953. Plant patent 1256; 16 Feb. 1954. (J.H. Hale peach x Flaming Gold) F_2 o.p.; selected in 1949. Fruit: large for its season, average 2 1/2 inches in diam.; globose, symmetrical; skin thickness medium, medium tough, undercolor yellow but nearly completely overlaid with mahogany-red; flesh yellow, firm, crisp, moderately juicy, mild, subacid, semi-freestone when soft-ripe; ships well; ripens 10 days before John Rivers. Tree: size medium; vigor medium; upright, open; productive and regular bearer; leaf glands reniform; flowers large, pink.

Sunsnow.—Orig. at Agr. Expt. Sta., Univ. of Florida, Gainesville. Bred in 1981, selected in 1985, and tested as 83-5NW. Introduced in Spain by Viveros Orero, S.A. in 1990. Complex parentage of Florida selections. Fruit: medium size; slightly oval; skin 70% red blush with a moderate amount of sugar speckles; flesh white, firm, melting, semi-freestone. Ripens about 90 days after bloom. Flowers large, light pink, showy; leaf glands reniform. Very resistant to bacterial spot. Tree: vigorous; productive. Chilling requirement about 250 chill units.

Sunsplash.—From 1975 cross made at Florida Agr. Expt. Sta., Gainesville; selected in 1978. Sunlite x ArmKing; tested as Fla. 8-2N. Released in 1993 by G. Krewer, T. Beckman, and W.B. Sherman for its early, attractive fruit. Fruit: oblong; medium size; 60% red blush over bright yellow ground color; flesh medium firm, yellow, melting, nonbrowning; semi-freestone when soft ripe. Ripens with Sungem, about 80 days after bloom; high fruit set and fewer split pits than Sungem. Flower: showy, medium, pink; pollen bright yellow; leaf glands globose. Medium-high resistance to bacterial leafspot. Chilling requirement 400 chill units.

SunWright.—Orig. at Agr. Expt. Sta., Univ. of Florida, Gainesville. Bred in 1979, selected in 1981 by W.B. Sherman, and tested as Fla. 81-17N. Introduced in New South Wales, Australia, in 1991. Complex parentage of Florida selections. Fruit: medium small; round; skin 90% red blush over yellow ground color; flesh yellow, medium firm, melting, semi-clingstone. Ripens about 78 days after bloom. Flowers pink, showy; leaf glands reniform. Highly resistant to bacterial spot. Tree: vigorous; productive. Chilling requirement about 200 chill units.

Superstar (*Sunecteight*).—Orig. in Fresno, Calif., by John H. Weinberger. Introd. in 1981. Plant patent 4894; 12 Oct. 1982; assigned to Superior Farming Co. Summer Grand o.p.; selected in 1978; tested as 84-55. Fruit: large; color brighter than either Sun Grand or Flavortop, with which it ripens; ovate; flesh yellow streaked with red at pit, firm, melting; excellent flavor; freestone. Most nearly resembles Summer Grand. Tree: medium vigor; productive. Flower: large, showy, pink.

Supreme Red.—Orig. by F. Paganelli, Wapato, Wash. Plant patent 4639, 3 Feb. 1981. Red Gold mutation, similar in all respects except ripens 2 or 3 days earlier.

Tastyfree.—Orig. in Le Grand, Calif., by F.W. Anderson, Merced, Calif. Introd. in 1974. Plant patent 3759; 5 Aug. 1975; assigned to Burchell Nursery, Modesto, Calif. Red Free x Autumn Gold. Fruit: medium; globose; skin yellow overspread with red, partially mottled; flesh yellow, firm, crispy, meaty; pit red streaked into flesh; subacid; vinous; freestone; ripens mid- to late August, 3 weeks after Red Free. Tree: upright to spreading; very productive; leaf glands reniform; large flowers.

Tasty-Gold.—Orig. in Modesto, Calif., by Floyd Zaiger. Introd. in 1986. Ruby Gold x (Early Sungrand x Royal Gold). Plant patent 5623; 7 Jan. 1986; assigned to Zaiger's Genetics. Fruit: large; globose; skin yellow overspread with red; flesh dark yellow, firm; mild subacid. Tree: vigorous;

productive; chilling requirement 800–1000 h.

Tiger.—Orig. in Talent, Ore., by David B. Lowry, Medford, Ore. Introd. in 1967. Plant patent 2776; 14 Nov. 1967. Fireglobe x Sun Grand. Fruit: size medium, uniform; skin yellow ground color shading more to orange and substantially overspread with a diffused red having distinct spotting, smooth, thin, highly waxed; flesh yellow with red next to pit, firm, moderately juicy, flavor mild, quality good; good shipping quality; stone free, small; ripens 3 Sept. Tree: size medium; open, spreading; vigorous; productive; hardy; leaf glands reniform; flower size medium, pink.

Tioga.—Orig. in Palo Alto, Calif., by W.F. Wight, U.S. Dept. of Agriculture. Introd. in 1941. (Salwey x Quetta) x Lippiatt; tested as USDA 429-40B. Fruit: large; broad oval; skin greenish-yellow to deep yellow, blushed or streaked with deep red when exposed, mottled with lighter dots toward apex, thick and tough, surface sometimes rough; flesh pale yellow, streaked or flecked with red at suture and near stone cavity, texture firm to firm-melting, slightly fibrous, freestone; season 27 days after Elberta. Tree: vigorous; productive; leaf glands globose, small; flowers nonshowy, medium, pink.

Tom Grand.—Orig. in Le Grand, Calif., by F.W. Anderson, Merced, Calif. Introd. in 1969. Plant patent 2895; 10 June 1969; assigned to Reedley Nursery, Reedley, Calif. Red Grand x Gold King o.p.; tested as Anderson 3E950. Fruit: large; uniform, globose; skin smooth, yellow, substantially overspread with bright red at maturity; flesh yellow, clingstone; has the same skin characteristics as Red Grand, but ripens about 1 month later. Tree: size medium; spreading; productive and regular bearer; flowers large, showy, pink; leaf glands reniform.

Violet.—Orig. in Hughson, Calif., by George Edward Alexander. Introd. in 1937. Plant patent 828; 25 Apr. 1939; assigned to Karl K. Snyder, Hughson, and A.D. Di Grazia, San Francisco. Chance seedling; discovered in 1929. Fruit: moderately large; skin mottled purple, tough, thick, sour, tenacious to flesh, slight tendency to crack; flesh whitish-cream, mottled red next to skin and pit, texture fine and melting, flavor subacid to mild, aroma distinct; freestone; ripens evenly, late, 12 Aug. to 1 Sept. Tree: size medium; upright, vase-formed; productive and regular bearer; flower size medium, deep pink. Appears to be similar if not identical to P.I. 65979, Vineuse Henri Monicourt.

Western Red.—Orig. by L.G. and N.G. Bradford, Le Grand, Calif. Plant patent 8196; 6 Apr. 1993. Red Diamond x May Diamond. Fruit: large; globose; firm; excellent flavor; crisp; yellow flesh; freestone; very dark red over deep reddish-orange background color. Ripens 7–10 days before August Red. Flowers small, pink; leaf glands reniform. Tree: large; vigorous; upright and dense; very productive.

White Glory.—A white-flowering, weeping ornamental. Orig. at Agr. Expt. Sta., North Carolina; selected by F.E. Correll in early 1960s; and introduced in 1985 by D.J. Werner, P.R. Fantz, and J.C. Raulson for ornamental qualities. S-37 o.p. Fruit: small, subglobose; flesh white, melting, subject to skin cracking, freestone, ripens with Elberta. Flowers large, showy, subsessile, pure white, numerous, single to sparsely semi-double. Leaves larger than for most standard cultivars. Fruit set is minimal. Tree: vigorous; extremely weeping (pendant), suggest grafting 6–7 feet above ground. Chill requirement about 900 h.

White Rose.—See Rose.

Yellow (*Gold*).—Orig. in Santa Rosa, Calif., by Luther Burbank (1849–1926). Introd. in 1920; possibly renamed Gold and re-introduced in 1927. Muir peach x New White. Fruit: large; skin deep golden-yellow, blushed; flesh yellow, firm, subacid, rich, freestone; ripens 10–25 Aug. Tree: bears well and consistently.

Zee Glo.—Orig. by C.F. Zaiger, G.N. Zaiger, L.M. Gardner, and G.G. Zaiger, Modesto, Calif. Plant patent 6408; 22 Nov. 1988. [Red Grand o.p. x (Sun Grand x Merrill Gem)] o.p. Fruit: large; slightly elongated; high percentage of red over yellow skin; flesh orange-yellow, firm, clingstone. Ripens 4 days before Late Le Grand. Flowers large, showy, pink; leaf glands reniform. Tree: large; vigorous; upright; productive.

Zee Gold.—Orig. in Modesto, Calif., by Chris Floyd Zaiger, Zaiger's Nursery. Introd. in 1971. Plant patent 3087; 28 Mar. 1972; assigned to Dave Wilson Nursery, Modesto. (Early Sungrand x July Elberta) o.p. Fruit: large to medium; uniform, nearly globose; skin yellow ground color, almost entirely covered with red blush; flesh yellow, clingstone; ripens about 10 days earlier than Grand River and Sunrise. Tree: medium to large; very productive; flowers large, pink, blooms 4–12 Mar.; leaf glands reniform.

Zee Grand.—Orig. by C.F. Zaiger, L.M. Gardner, G.N. Zaiger, and G.G. Zaiger, Modesto, Calif. Plant patent 7475; 19 Mar. 1991. (Red Grand x Royal Gold) o.p.2. Fruit: medium large; globose; overspread with red on bright-yellow ground color; flesh yellow, firm, meaty, clingstone. Ripens 3 days before May Grand. Flowers large, showy, pink; leaf glands reniform. Tree: large; vigorous; upright; productive.

Rootstocks for Nectarines

Damas GF 1869.—See under Peach. Incompatible with most nectarines. Compatible with Apache, Aurora Grand, Early Maygrand, Firegold, Fireking, John Rivers, July Red, June Star, Maria Emilia, Maygrand, Maystar, Merrill X, Mike Star, Moongrand, Nectared 1, Nectarose, Niagara Grand, Pacific Star, Red Diamond, Red Free, Ruby Grand, Snow Queen, Spring Star, Stark Early Bird, Summer Beaut, Summer Red, Summer Star, Sun Star, and Zeegold.

Flordaguard.—See under Peach. Compatible with all nectarine varieties tested.

GF 53.7.—See under Peach. Compatible with many, but not all, nectarine varieties.

Hansen 536.—See under Almonds.

Hansen 2168.—See under Almonds.

Ishtara®.—See under Plums. Compatible with Summergrand; further compatibilities not yet determined. Semi-dwarfing.

Jaspi®.—See under Plums. Compatible under Fuzalode and Nectared 6; not tested under other varieties. Reduces tree by about 20% compared to peach seedling.

Julior®.—See under Peach. Compatible with nectarines. Medium vigor.

Mr.S.2/5.—See under Peach. Compatible with all varieties tested.

Myran®.—See under Peach. Compatible with nectarine.

Nemared.—See under Peach. Compatible with nectarine.

OLIVE

Bonita.—Orig. in Ontario, Calif., by Herbert C. Swim, Armstrong Nurseries. Introd. in 1954. Open-pollinated seedling of Ascolano; selected in 1950. Tree: very vigorous; well-shaped; foliage abundant; unproductive; recommended as an ornamental.

Kadesh.—Orig. at the Volcani Center, Bet Dagan, Israel, by Shimon Lavee. Of unknown origin from trees collected throughout the Mediterranean basin, in the Kadesh oasis; selected in 1968; tested as K12. Fruit: oval, length 24 mm, width 19 mm; fruit 6 g at green-mature stage, pit large; extremely low oil content; tasty and firm; for table use as a pickled green olive; 10 days earlier than Manzanillo. Tree: strong; tendency to widen; relatively easy rooting.

Minerva.—Orig. in Ivanhoe, Calif., by Lloyd Main of Visalia, Calif. Introd. in 1956. Plant patent 1521; 25 Sept. 1956. Open-pollinated seedling of Manzanillo; selected in 1949. Fruit: large; skin tough, not easily bruised; appearance good; takes a rich black color in processing for ripe pickling; somewhat intermediate between Manzanillo and Sevillano. Tree: crops annually.

Pollen King.—Orig. in Ontario, Calif., by Herbert C. Swim, Armstrong Nurseries. Introd. in 1967; named in 1971. Plant patent 2957; 23 Dec. 1969; assigned to Armstrong Nurseries. Open-pollinated seedling of Ascolano. Seed from parent tree planted in 1947; selected in 1950; tested as A-2. Fruit: size medium; oblong to heart-shaped; skin tender, lenticels relatively inconspicuous at beginning of ripening and becoming prominent as ripening progresses; flesh between near white and ivory-yellow when ripe, tender, flavor strong when pickled; ripens early to mid-November. Tree: very large; spreading; extremely vigorous; highly productive; flowers produced regularly with high pollen viability, making it a suitable pollinizer.

Swan Hill—Introd. from Australia in 1961 and patented in the United States by the Univ. of California. Produces primarily staminate flowers and thus sets few fruit, making it a desirable ornamental olive. However, it is difficult to root and is therefore propagated by grafting onto Oblonga rootstock.

Orange

Altoona.—Orig. in Altoona, Fla., by Bertman V. Johnson. Introd. in 1953. Plant patent 1170; 24 Feb. 1953; assigned to J.A. Schiruska, Lady Lake, Fla. Parentage unknown. Fruit: sweet; resembles Parson Brown, but larger and matures earlier. Tree: upright, spreading; vigorous; resembles Parson Brown.

Ambersweet.—Orig. from a 1963 cross of [*Citrus reticulata* Clementine x (*C. paradisi* x *C. reticulata*) Orlando] x *C. sinensis*. Introd. by C.J. Hearn, U.S. Dept. of Agriculture, Orlando, Fla., in 1989. Fruit: size medium, 63–95 mm; slightly pyriform; rind moderately thick, pebbly, easily peeled; 0–30 seeds; rind and flesh bright orange; early to midseason. Tree: moderately vigorous; upright; dense foliage; some thorns; moderately cold-tolerant.

Armstrong Seedless Valencia.—Orig. in Anaheim, Calif., by Pearl C. Mohn. Introd. in 1939. Plant patent 124; 5 Mar. 1935; assigned to Armstrong Nurseries, Ontario, Calif. Bud mutation of Valencia; discovered about 1928. Fruit: seedless; quality equal to or slightly better than Valencia. Tree: does not produce as well as Valencia.

Atwood Early.—Orig. in Lemon Cove, Calif., by Frank Atwood. Introd. in 1945. Bud mutation of Washington Navel; discovered in 1935. Fruit: navel type; skin deep orange-red, of a more intense color than that of Washington Navel, which it resembles; matures earlier than Washington Navel, remaining firm very late in the season; responds to ethylene treatment at a lesser color intensity than standard varieties.

Barrington Navel.—Orig. unknown. First introd. by B.A. Barrington, who observed it over the years as budded trees in his grove at Melrose, Fla. Introd. in 1965. Fruit: reported to have excellent quality; navel not protruding; medium-late maturing, consistent yields. Entered in the Florida Budwood Registration Program in 1955 by D.W. Tredinick of Jacksonville, Fla. Not widely grown.

Best Seedless *(Best's, Best Navel)*.—Orig. in Avondale, Auckland, New Zealand, by a Mr. Best. Introd. about 1935. Open-pollinated seedling of Cook; discovered about 1910. Fruit: often small and misshappen; flesh juicy, flavor rich; ripens early midseason; resembles Washington Navel. Tree: greater vigor and heavier yields than Washington Navel.

Burton *(Helen Burton)*.—Orig. in Clanwilliam, Western Cape Province, on the farm of Helen Burton.Selected in 1954 by G. Joubert. Unknown seedling. Fruit: size 7 cm in diam.; shaped like Valencia; skin coarser than Valencia; flesh texture fine, quality good, few or no seeds; ripens June to July or a few weeks before Valencia. Tree: resembles Valencia; productivity good. Recommended for areas normally producing Valencias and with high acid content.

Butler.—Orig. in Phoenix, Ariz., by Robert Butler. Introd. about 1920. Probably a nucellar seedling of Valencia brought in from Florida during the 1890s. Fruit: resembles Valencia; disadvantage is seediness, 12–20 seeds per fruit; midseason, January through March. Tree: vigorous; medium to heavy production.

Campbell Valencia.—Orig. in the Earl Campbell orchard, Santa Ana, Calif. Introd. in 1942. Considered to be a seedling of a Valencia tree planted in 1871; a nucellar line planted in 1942. Fruit: appears to be a Valencia type; large; holds late on tree. Tree: free of recognized virus diseases; extremely vigorous, thorny and somewhat slower in coming into bearing than standard varieties.

Cara Cara.—Orig. as a limb sport of Washington navel orange at Hacienda Cara Cara, Valencia Province, Venezuela. Introd. to Florida in 1977 by A.H. Krezdorn through Div. of Plant Industry, Gainesville. Fruit: size and shape comparable to Washington; no red pigmentation in rind; navel sometimes small and inconspicuous; flesh dark red (from lycopene, not anthocyanin as in blood oranges); seedless. Tree: vigorous; some faint variegation on young foliage; cambial tissues may have red pigmentation. Occasional reversion on limbs, producing normal, orange-fleshed fruit.

Choate Navel.—Orig. unknown. Brought to Florida from California many years ago by S.S. Sadler; in 1939 E.S. Choate of Eustis, Fla., propagated this variety on Cleopatra root; in 1948 John H. Kauffman, Sr. of Grand Island Nurseries, Eustis, propagated the variety, introduced it in 1950. Bud mutation of Washington Navel. Fruit: size average; rind light color; navel not protruding; quality good; can be held until after Christmas without drying out; ripens fairly late October to early November.

Clanor.—Orig. in Clanwilliam, Cape Province, South Africa, by P. Nortier. Introd. in 1954. Parentage unknown; discovered in 1930. Fruit: resembles Valencia, maturing 30 days earlier; quality good. Tree: similar to nucellar Valencia; productive, but with degree of instability.

Cornell.—Orig. in Winter Haven, Fla., by Harold E. Cornell. Introd. in 1945. Plant patent 657; 17 Apr. 1945; assigned to Thelma H. Cornell. Bud mutation of Parson Brown; discovered in 1938. Fruit: medium to large; oblong-rounded; rind deep yellow to orange at maturity; seeds 0–5; flesh juicy, sweetness and acidity well blended; very early maturity. Tree: new growth has a tendency to be thorny.

Cutter Valencia.—Orig. in Riverside, Calif., by H.S. Fawcett, California Citrus Expt. Sta. Introd. in 1952. A nucellar seedling; seed selected from the J.C. Cutter orchard in Riverside, by Fawcett; planted in 1935. Fruit: a Valencia type; large; slightly more round than standard varieties. Tree: initially free of recognized virus diseases; exceptionally vigorous; slightly slower coming into bearing than standard varieties.

Delta *(Delta Valencia)*.—Orig. in Rustenburg, Transvaal, South Africa, by A.J. Smith. Introd. in 1958. Bud mutation of Valencia; discovered in 1952. Fruit: size same as Valencia, but more elongated; skin thicker and coarser than Valencia; flesh texture fine, seedless, with lower sugar and acid content; ripens 15 June to 15 July, shortly before Valencia. Tree: same as Valencia; very productive; recommended for areas normally producing Valencias and with high acid content.

Diller.—Orig. in Phoenix, Ariz., by Daniel Diller. Introd. in 1920. Probably originated as a seedling from a shipment of trees purchased in Florida and fruited in Arizona; planted in the early 1890s; selected in 1910. Fruit: only a few seeds; resembles Parson Brown; highly frost resistant.

Dream *(Dream Navel)*.—Orig. in Seminole County, Fla., by Donald J. Nicholson. Introd. in 1944. Plant patent 625; 25 Apr. 1944. Parentage unknown, but considered to be a bud mutation of Washington Navel; discovered in 1939 when the tree was 25 to 30 years old. Fruit: navel type; medium to large; seedless; appearance fine; flavor fine; juice content high; matures in early October at Orlando, Fla., but holds on the tree better than most navels. Tree: medium vigor; productive.

Frost Navel.—Orig. in Riverside, Calif., by Howard B. Frost, California Citrus Expt. Sta. Introd. in 1952. Nucellar seedling of Washington Navel, from seed planted in 1917. Fruit: resembles Washington Navel, having greater vigor than its seed parent line.

Frost Valencia.—Orig. in Riverside, Calif., by Howard B. Frost, California Citrus Expt. Sta. Introd. in 1952. Nucellar seedling of Valencia, from seed planted in 1915. Fruit: typical Valencia, but has greater vigor and better yield than its seed parent. Tree: initially free of psorosis virus carried by seed parent tree.

Fukuhara.—Orig. in Chiba, Japan, by Shuhei Fukuhara. Introd. in 1930. Supposed to be bud mutation of Joppa; discovered about 1915. Fruit: size medium; skin reddish-orange, rough, medium thick; flesh juicy, flavor sprightly, seeds few; ripens late. Tree: vigorous; hardy; productive; leaf large and thick.

Gillette.—Orig. in Hemet, Calif., by A.F. Gillete. Introd. about 1945. Presumed bud mutation of Washington Navel; discovered about 1940. Fruit: navel type; ripens slightly before Washington Navel, which it resembles. Tree: distinctive; heavily foliaged with large, thick leaves.

Gisler.—Orig. in Saticoy, Calif., by Paul S. Gisler, Clyde J. Bradley, and Joe Gisler, Jr., Camarillo, Calif. Introd. in 1966. Plant patent 2610; 22 Mar. 1966. Apparently a mutation of Valencia. Fruit: about one-third larger than ordinary Valencia, having very few seeds; oval or pear shape; skin yellow-orange; quality good; grows in large clusters; flowers self-fertile. Tree: very vigorous; leaves very large, dark green, thick texture.

Glen Improved.—Orig. in the grove of William G. Roe, Winter Haven, Fla., by the Glen Saint Mary Nursery Co., Glen Saint Mary, Fla. Introd. in 1934. Bud mutation of Washington Navel; discovered in 1932. Fruit: navel; resembles Washington Navel. Tree: apparently bears larger crops than Washington Navel under Florida conditions.

Glen Summer *(Pope Summer, Pope's Summer)*.—Orig. in Lakeland, Fla., by Fred W. Pope. Introd. in 1938. Parentage unknown; discovered about 1934 in a 40-acre grove of Pineapple. Fruit: skin thin, texture smooth, color deep; flesh juicy, seedless; season very late in Florida, from June to September; keeps well on tree; resembles Valencia. Tree: resembles Valencia.

Harward Late.—Orig. in Tauranga, New Zealand, by G.F. Joseph. Orchard later owned by a Mr. Harward. Introd. about 1935. Open-pollinated seedling of Valencia; discovered about 1915. Fruit: larger than Valencia; resembles Valencia; possibly nucellar strain of Valencia. Tree: large; vigorous; productive.

Heard Early *(Heard Sweet)*.—Orig. in Arcadia, DeSoto County, Fla., by J.J. Heard. Introd. in 1924. Chance seedling; discovered in 1916 on "the old Willingham Place," 4 miles southwest of Arcadia. Fruit: quality good, soluble solids high; has seeds; ripens in October; holds well on the tree; resembles Parson Brown. Tree: very large; very productive.

Jordan Sweet.—Orig. in Phoenix, Ariz., by Mrs. S.E. Jordan. Introd. in 1930. Bud mutation of a rootstock, probably Valencia; discovered in 1925. Fruit: carpels separate easily; seeds few, five to six per fruit; resembles Valencia; early midwinter ripening; holds well on tree.

Koethen.—Orig. in an old sweet-orange seedling orchard of E.L. Koethen. Introd. in 1943. Seedling of Washington Navel; selected about 1915 for its vigor by H.B. Frost, California Citrus Expt. Sta., Riverside; tested as la4-24; selected in 1922 for rootstock purposes by H.J. Webber of this station. Fruit: size medium; round to slightly elliptical; seeds 10–25 per fruit; ripens during January to March. Tree: vigorous. Seedlings used as a sweet-orange-type rootstock for orange, grapefruit, lemon, and mandarin.

Lemor.—Orig. in Chula Vista, Calif., by James M. Ater. Introd. about 1935. Plant patent 142; 1 Oct. 1935. Probably an open-pollinated seedling of Valencia; discovered in 1930. Fruit: shape similar to Lisbon lemon; rind and flesh orange in color, similar to Valencia; flavor tart, somewhat intermediate between orange and lemon. Tree: using sour orange as rootstock are medium-sized; vigorous; leaves more slender and pointed than lemon or orange; flowers pure white, smaller and not as sweetly scented as the orange.

Marrs *(Marrs Early)*.—Orig. in Donna, Texas, by O.F. Marrs. Introd. in 1940. Bud mutation of a navel orange received in 1925 from Nuisbickel-Warren Nursery in California. Fruit: large; flesh sweet, low acid content; 5–15 seeds per fruit; marketed from early to midseason, November to January in Texas. Tree: high yields.

Newhall Navel.—Orig. in Duarte, Calif., by Newhall Land and Cattle Co., Newhall, Calif. Introd. in 1955. Bud mutation of Washington Navel; discovered in 1950. Fruit: resembles Washington Navel, having a deeper orange color to rind and flesh, more oval in shape, and earlier maturity. Tree: budline carries exocortis virus.

Oasis (*Casa Grande Seedling, Wood's Seedling*).—Orig. in Casa Grande, Ariz., by C.J. Wood. Introd. in 1935. Parentage unknown, probably a nucellar seedling; discovered before 1928. Fruit: very small; resembles Hamlin; disadvantage is its seediness, 8–15 seeds per fruit; juice very high in solids; early season. Tree: vigorous; production high.

Olinda Valencia.—Orig. in Olinda, Orange County, Calif., in the garden of Ollie Smith. Introd. in 1957. Chance seedling; discovered in 1939. Fruit: large; skin thin; flesh sweet, quality good; Valencia type but matures slightly earlier than standard varieties. Tree: vigorous; productive; free of recognized virus diseases.

Paradise *(Paradise Navel)*.—Orig. in Lake County, Fla., by Donald J. Nicholson. Introd. in 1944. Plant patent 548; 13 Oct. 1942. Parentage unknown; discovered in 1934. Fruit: large, shape is variable; rind deep orange to reddish tinge when fully mature, very smooth, satin-like, glossy, tough; flesh yellowish, tender, habit of slow and gradual filling of juice cells, generally remains juicy in Florida; ripens October to January. Tree: large; vigorous; thornless.

Parson Brown (not the Parson Brown described in Citrus Industry by Webber and Batchelor.).—Orig. in Phoenix, Ariz., by R.H. Fuqua. Introd. about 1920. Parentage unknown; discovered before 1915. Fruit: similar to Diller, but contains more seeds. Tree: more tolerant to cold than Valencia and Washington Navel.

Plaquemines.—Orig. in Diamond, La., by P.O. Hawthorne, R.T. Brown, L.G. Jones, D.W. Newson, and J.C. Miller, Louisiana Agr. Expt. Sta., Baton Rouge. Introd. in 1960. Bud mutation of Pineapple; discovered in 1948 in the Pineapple variety section of Magnolia Orange Grove near Port Sulphur, Plaquemines Parish, Diamond. Fruit: seedless; round to slightly oblate; medium-large, 2 3/8 to 3 inches in diam.; rind smooth, thick, glossy, deep orange; flesh yellowish-orange, juicy, medium grain, flavor good, acidity and sugars well balanced, none to one seed per fruit; season January and February. Tree: as hardy as parent and other sweet orange varieties; produces satisfactory crops; recommended for home gardens along the lower Gulf Coast of Louisiana or in areas where other varieties of round sweet oranges do not winter kill.

Robertson Navel.—Orig. in Redlands, Calif., by Roy Robertson. Introd. in 1936. Plant patent 126; 28 May 1935; assigned to Armstrong Nurseries, Ontario, Calif. Bud mutation of Washington Navel; discovered in 1925. Fruit: resembles Washington Navel, in some areas ripens 2–4 weeks earlier than Washington Navel. Tree: precocious in bearing habit, but somewhat lacking in vigor on sweet and sour orange rootstock.

Rohde Red Valencia.—Orig. near Sebring, Fla., by the late Paul Rohde, Sr. Original budwood was discovered in 1955 and budded to sour orange seedlings. Fruit: medium-large; typical of Valencia in appearance, time of maturity, and flavor; but the juice color is much deeper orange than that of known types of this cultivar, due to the orange pigment cryptoxanthin. Tree: large on rough lemon rootstock.

Rufert.—Orig. in Riverside, Calif., by Howard B. Frost, California Citrus Expt. Sta. Described in 1943. A fertile, nucellar seedling variant of Ruvel, a pollen-sterile navel cultivar; from seed planted in 1915. Fruit: similar to Ruvel, but with more seeds and no navel; early maturity in desert areas. Tree: yields well in desert areas. Not grown commerically.

Salustiana.—Orig. in Enova, Valencia, Spain. Introd. in 1950 by Don Salustiano Pallas. Bud mutation on a "communa" tree discovered in the yard of a convent. Fruit: medium-large; subglobose to spherical; skin well-covered at maturity, medium thick; flesh juicy, flavor sweet; seedless; ripens mid-season. Tree: medium-large; vigorous; productive.

Sarah *(Pink Shamouti)*.—Orig. in Petah-Tiqvah, Israel, by B. Ben-Ezer (Raab). Introd. there in 1945. Bud mutation of Shamouti; discovered in 1939 on a tree planted in 1909. Fruit: pulp and juice pink (from lycopene, not anthocyanin as in blood oranges); ripens somewhat earlier than Shamouti; otherwise, characteristic of parent. Tree: leaf slightly variegated, being more or less distinctly veined.

Saxon Early Sweet.—Orig. in Arcadia, Fla., by John Franklin Saxon, Sunshine Nurseries (no longer in business). Introd. about 1920. Chance seedling; discovered in 1913. Fruit: size tends to be small; rind color attractive; ripens very early.

Seedless Pineapple.—Orig. in Shiloh, Fla., by Robert S. Ragin. Introd. in 1941. Plant patent 477; 24 June 1941; assigned to Clapp and Clapp, Orlando, Fla. Bud mutation of Pineapple; discovered in 1934. Fruit: seeds few; in other respects, identical with parent. No longer planted commercially because of low fruit yields.

Skaggs Bonanza.—Orig. in Strathmore, Calif., by Everett Skaggs. Introd. in 1964 by John Walker, Porterville, Calif. Bud mutation of Washington Navel. First propagated in 1959 by Leo Weisenberger, nurseryman, Porterville. Fruit: resembles Washington Navel, but colors and matures earlier. Tree: precocious; very productive. Budline apparently free of psorosis, exocortis, stubborn, and cachexia.

Summerfield Navel.—Orig. in Weirsdale, Fla., by W.J. Lyles, Summerfield Nursery Co. Introd. about 1931. Parentage unknown; discovered in a grove planted by Charlie Waterman before the 1895 freeze; two trees were different from all others in that grove and these were propagated as this variety. Fruit: size above average; rind fine-textured; flavor excellent, flesh juicy; matures early; external navel medium-sized. Tree: vigorous; slightly taller than it is wide.

Summernavel *(Workman Navel)*.—Orig. in Riverside, Calif., by John Albert Workman. Introd. in 1942. Plant patent 347; 21 Nov. 1939; assigned to Armstrong Nurseries, Ontario, Calif. Bud mutation of Washington Navel; discovered about 1934. Fruit: large, 3 1/4 inches in diam.; roundish, navel type; rind smooth to slightly pebbled, 1/8 to 1/4 inch thick, separates easily from flesh; 8–12 segments, juicy; no seeds; ripens later than its parent; resembles Washington Navel. Tree: upright; moderately vigorous; semi-dwarf; practically seedless; no pollen produced.

Tarocco.—Introd. in Riverside, Calif., in 1951 by L.D. Batchelor, California Citrus Expt. Sta. Budwood of *Citrus sinensis* Osbeck sent in 1930 by Howard S. Fawcett of the above Citrus Expt. Sta. from the orchard of Brogna Guisepee of Lentini, Sicily. Fruit: about the size of Valencia; elongated; frequent navel structure; skin with slight red blush; flesh red, commercially seedless, juice deep red, flavor good, with slight grape flavor; ripens in April. Tree: characteristics like sweet orange, more open, sensitive to twig dieback.

Tomango.—Orig. in Nelspruit, South Africa, by Hugh Hall, H.L. Hall & Sons, Mataffin E. Tul, S. Africa. Introd. in 1930. Bud mutation of Joppa; planted in 1903; selected in 1910. Fruit: resembles Joppa; flesh melting; seedless; quality high. Tree: similar to Shamouti; productive; incompatible on rough lemon rootstock.

Trovita.—Orig. in Riverside, Calif., by Howard B. Frost, California Citrus Expt. Sta. Introd. in 1935. Three seedlings, doubtless nucellar, all Trovita, were grown from seeds found in one fruit in the main Washington Navel orange crop. The parent tree was not located, but the record of the fruit does not mention any peculiarity. Probably a bud mutation had occurred in an ordinary tree of navel orange; seed from crop of 1914–15; selected in 1928. Fruit: navel structure usually absent or rudimentary; flesh somewhat more tender and juicy than Washington Navel, which it resembles; two to six seeds, in a mixed varietal planting. Tree: more vigorous than Washington Navel; otherwise, resembles Washington Navel. The name is Esperanto for "found," suggested by the accidental finding of the seeds that produced Trovita.

Tulegold Early Navel *(Tule Gold)*.—Orig. in Porterville, Calif., by Arthur D. Mabs. Introd. in 1966. Plant patent 2818; 16 July 1968; assigned to Central Valley MM Farms, Strathmore, Calif. Bud mutation of Washington Navel; discovered in 1955. Fruit: resembles parent variety Washington Navel; differentiated by its rapid accelerated growth of fruit and earlier coloring habits; ripens 10 days to 3 weeks earlier than Washington Navel. Tree: resembles parent, but foliage is somewhat smaller; precocious, bearing fruit at 1 or 2 years old when field-grown; stout fruit stems hold the fruit more firmly than parent variety.

Wood Valencia.—Orig. in Mesa, Ariz., by C.J. Wood. Introd. about 1944. Parentage unknown, probably a nucellar seedling of Valencia; planted about 1930; discovered in 1940. Fruit: typical Valencia type; skin has tendency to be rough and "sheepnose" at stem end in off-crop years; ripens early. Tree: extremely vigorous; production high, tends to alternate bearing.

Zellwood Satin.—Orig. in Zellwood, Orange County, Fla., by K.C. Moore. Introd. in 1945. Parentage unknown. Fruit: small; skin smooth, satiny, glossy; flesh very juicy, rich flavor, nearly seedless; resembles Hamlin. Tree: vigorous; dense; prolific.

Barnes.—Orig. in Gainesville, Fla., and later sent to Riverside, Calif., where it was developed by H.J. Webber, Univ. of California Citrus Expt. Sta. Introd. in 1957. Open-pollinated seedling of trifoliate orange (*Poncirus trifoliata*); selected in 1932. Used as a clonal rootstock; highly resistant to phytophthora root rots.

Pomeroy.—Orig. in Riverside, Calif., by C.S. Pomeroy, U.S. Dept. of Agriculture. Introd. in 1951. Open-pollinated seedling of trifoliate orange (*Poncirus trifoliata*); selected in 1925. Tree: trifoliate; more vigorous than some other trifoliate oranges; flowers large. Used only as a clonal rootstock.

areas where other varieties of round sweet oranges do not winter-kill.

Ridge Pineapple.—Orig. in Orlovista, Fla., by Harry W. Ford, Florida Citrus Expt. Sta., Lake Alfred, and William A. Feder, U.S. Dept. of Agriculture, Orlando, Fla. Introd. in 1964. A sweet orange; tested as Pineapple-156; resembles Pineapple orange. Used only as a clonal rootstock: resistant to the burrowing nematode [*Radopholus similis* (Cobb) Thorne]; tolerant to tristeza, anthracnose, and scab diseases.

Rubidoux.—Orig. in Riverside, Calif., probably by Herbert John Webber, at the Rubidoux Laboratory, which later moved to the California Citrus Expt. Sta., Riverside. Introd. about 1920. Seedling of sour orange (*Citrus aurantium*); selected 1910–15 from seed brought from Florida; propagated and grown in the grove of Elijah Hawkins, Riverside. Used only as a rootstock; used extensively in the past for orange and grapefruit varieties; propagated by seed, which are highly polyembryonic and hence quite uniform, with most of the viable embryos being from the nucellus; no longer recommended because of susceptiblity to quick decline strains of citrus tristeza virus.

Standard.—Orig. in Riverside, Calif., by H.J. Webber, California Citrus Expt. Sta. Introd. in 1932. Seedling from mixed sour orange seed from Florida; selected in 1915. Used for experimental rootstock purposes. Fruit: sour; medium large; seeds numerous, averaging 51 per fruit. Tree: large; vigorous; thorny; small twigs nearly thornless; similar to Brazilian.

ORANGEQUAT

Nippon.—Orig. in Washington, D.C., by Eugene May, U.S. Dept. of Agriculture. Introd. in 1932. Unshiu Satsuma orange (*Citrus reticulata*) x Meiwa Kumquat (*Fortunella japonica* x *F. margarita*). Fruit: rind sweet, mild, edible, very attractive; flesh acid; excellent for marmalade; ripens October to February. Tree: slow growing; hardy; productive.

PAPAYA

Betty.—Orig. in Miami, Fla., by Bronson Bayliss. Introd. about 1934. Parentage unknown; selected about 1933. Fruit: rounded-oblong, 2–3 lb; flesh 3/4 inch thick; excellent dessert quality; rind and pulp soft, not a good shipper. Tree: semi-dwarf. Variety now obsolete.

Cariflora.—Florida origin. Released in 1986 by R.A. Conover, R.E. Litz, and S.E. Malo of IFAS, Univ. of Florida. Dioecious, round fruit 13.5–14.5 cm in diam., weight 0.5–0.75 kg; moderately firm, yellow flesh with an agreeable taste and aroma. Seed cavity is small and star shaped, 10.7% soluble solids. Yield about 35 t/ha/year. Tolerant to papaya ringspot virus and develops little distortion or mottling on leaves or fruit when infected. Well adapted to warm lowland condition.

Eksotika.—Released by Y.K. Chan, Malaysian Agricultural Research and Development Institute, Malaysia, in 1987. Developed from a cross between Subang 6 and Sunrise solo, followed by repeated backcrosses to Sunrise. Fruit weight 650 g, 13% to 14% soluble solids. Flesh reddish pink, flavor pleasant; yeild 45 t/ha/year at 18 months.

Higgins.—Orig. in Honolulu, Hawaii, by H.Y. Nakasone, Univ. of Hawaii. Introd. in 1974. Complex parentage involving Betty, Kapoho Solo, Line 5, and Line 10. Cross made in 1960; selected in 1964; tested as line 17a. Fruit: 14–18 oz, pyriform; skin smooth, orange-yellow when ripe; flesh deep yellow, mild flavor; 12% to 17% soluble solids; small-fruited (1 lb). Tree: bearing height 30–36 inches; moderately vigorous; produces 100–140 lbs per tree per year; susceptible to phytophthora root rot. Variety now obsolete.

Kapoho *(Puna Solo)*.—Orig. in Hilo, Hawaii, by growers. Date of introd. unknown. Parentage unknown; discovered before 1950. Also known as Matsumoto solo. Selected for the Puna, Kapoho, area with annual rainfall of more than 250 cm. Fruit: 14–28 oz, pyriform; skin fairly smooth, yellow when ripe; flesh pale orange; 13% to 16% soluble solids; fruit size 400–800 g, small (1 lb), firm, and good quality for shipping. The most widely grown export cultivar. Tree: bearing height 45–60 inches; vigorous; produces 90–120 lbs per tree per year; susceptible to phytophthora root rot; adapted to high rainfall areas.

***Puna Solo*.**—See Kapoho.

Sunrise.—Orig. in Honolulu, Hawaii, by the Univ. of Hawaii, Poamoho Expt. Farm. Introd. in 1968. Line 9 x Kariya solo; cross made in 1961; tested as HAES 63-22. Fruit: averages 22–26 oz, depending on location; TSS 15.5%; pyriform with slight neck; skin generally smoother with less ribbing; flesh reddish-orange, sweet, high sugar content, firmer than parent varieties; quality and flavor similar to solo, but superior to Line 9; seed cavity not as deeply indented as other solo strains, making it easier to remove seeds; firmness, texture, and size vary considerably with location; shelf life and shipping quality similar to solo, similar care must be maintained in picking, handling, and marketing. Tree: precocious, maturing first fruit about 9 months after transplanting at a height about 3 feet; carpellody (cat-facing) almost completely absent at elevations below 700 ft and little or no sterility, depending on season and growing conditions.

Sunset.—Released in 1993 by R.A. Hamilton, P.J. ito, and R.E. Paull. Sunset (HAES 72-12) is an inbred sister line of Sunrise, selected from a population of seedlings from Line 9 solo crossed with Kariya in 1961. This hermaphrodite line is precocious, has low bearing height averaging 80 cm from ground level, and low incidence of sterility or carpelloidy. Fruit are pyriform and slightly necked, weight 400 to 600 g, 12% to 17% soluble solids, and firmer than Sunrise. Skin has a tendency to freckle in warm weather. Fruit picked at color break take about 10 days to reach 70% yellow and 15 days to reach 100% yellow. Production is 49 to 100 t/ha/year. Somewhat resistant to anthracnose.

Waimanalo.—Orig. in Honolulu, Hawaii, by H.Y. Nakasone, Univ. of Hawaii. Introd. in 1970. Complex parentage involving Betty, Line 5, and Line 8. Selected in 1960; tested as line 77. Fruit: 20–30 oz; roundish-pyriform; skin smooth, yellow when ripe; firm; flesh bright yellow, flavor mild, 12% to 16% soluble solids; firmness and shelf life superior to older cultivars; recommended for use as fresh fruit or for processing. Tree: bearing height 30–34 inches; vigorous; produces 90–120 lbs / tree / year; slightly tolerant to phytophthora root rot.

Wilder.—Line 25. Orig. in Honolulu, Hawaii, by H.Y. Nakasone, Univ. of Hawaii. Introd. in 1974. A sister line of Higgins. Complex parentage involving Betty, Line 5 solo, and Kapoho. Tree produces well with superior fruit qualities in a wide range of environments. Fruit pyriform, 468 g; flesh orange-yellow, 15.9% soluble solids. Variety now obsolete.

Passion Fruit

Noel's Special.—Orig. in Hilo, Hawaii. Released by P.J. Ito, Univ. of Hawaii, CTAHR. Selected in 1968 from open-pollinated seedlings of a single plant growing since the early 1950s on an abandoned farm in Hilo. Fruit: round, average 90 g; cavity packed with dark-orange pulp yielding 43% to 56% juice by weight; refractometer readings range from 15% to 19.8% total soluble solids; flowers are self-incompatible and need other cultivars for adequate cross-pollination. Vine: vigorous; produces fruit after 1 year; dark-green leaves and fruit are tolerant to alternaria brownspot (*Alternaria passiflorae*). Named in honor of Noel Fujimoto, a deceased graduate student who discovered the cultivar. The original clone of Noel's Special was lost for more than 15 years. in 1995, this cultivar was re-released after a vine with almost identical fruit qualities was discovered in Puueopaku, Hawaii. Fruit: oval, 6.1 x 5.5 cm; orange-yellow exterior, average Hunter L, a, b were 69.89, 7.44, and 61.16, respectively. Weight ranged between 56–160 g, average 119 g. Juice recovery average 39%. Juice: bright orange, average L, a, b values of 42.66, 11.46, and 30.7. Total SSC ranged between 13% to 15.5%, average 14.4%. Total acidity averaged 2.61%. Not self-compatible.

PAWPAW

Asimina triloba (L.) Dunal

Davis.—Selected from the wild in Illinois by Corwin Davis in 1959. Introd. in 1961 from Belleview, Mich. Fruit: 115 g, up to 12 cm long; green skin; yellow flesh; large seed; ripens first week of October in Michigan; keeps well in cold storage.

Ford Amend.—Selected from wild seedling of unknown parentage by Ford Amend around 1950. Introd. from Portland, Ore. Fruit: slightly smaller and earlier than Sunflower; ripens late September in Oregon; greenish-yellow skin; orange flesh.

G-2.—Selected from G.A. Zimmerman seed by John W. McKay, College Park, Md., in 1942.

Glaser.—Selected by P. Glaser of Evansville, Ind.

Kirsten.—Hybrid seedling of Taytwo x Overleese; selected by Tom Mansell, Aliquippa, Pa.

Little Rosie.—Selected by P. Glaser of Evansville, Ind. Fruit: small in size. Reported to be an excellent pollinator.

M-1.—Selected from G-2 seed by John W. McKay, College Park, Md., in 1948.

Mango.—Selected from the wild in Tifton, Ga., by Major C. Collins in 1970.

Mary Foos Johnson.—Selected from the wild in Kansas by Milo Gibson. Seedling donated to North Willamette Expt. Sta., Aurora, Ore., by Mary Foos Johnson. Fruit: similar to Sunflower. May be same as Sunflower.

Mason/WLW.—Selected from the wild in Mason, Ohio, by Ernest J. Downing in 1938.

Middletown.—Selected from the wild in Middletown, Ohio, by Ernest J. Downing in 1915.

Mitchell.—Selected from the wild in Jefferson Co., Ill., by Joseph W. Hickman in 1979. Fruit: large to medium size; slightly yellow skin; golden flesh; few seeds; excellent flavor.

NC-1.—Hybrid seedling of Davis x Overleese; selected by R. Douglas Campbell, Ontario, Canada, in 1976. Fruit: 340 g; few seeds; yellow skin and flesh; thin skin; early ripening, 15 Sept. in Ontario.

Overleese.—Selected from the wild in Rushville, Ind., by W.B. Ward in 1950. Fruit: 340 g; few seeds; bears in clusters of three to five; ripens first week of October in Michigan.

PA-Golden.—Selected as seedling from seed originating from George Slate collection by John Gordon, Amherst, N.Y. Fruit: yellow skin; golden flesh; matures mid-September in New York.

Prolific.—Selected by Corwin Davis, Belleview, Mich., in mid-1980s. Fruit: 200–225 g; yellow flesh; ripens first week of October in Michigan.

Rebecca's Gold.—Selected from Corwin Davis seed, Belleview, Mich., by J.M. Riley in 1974. Fruit: 85–170 g; kidney-shaped; yellow flesh.

SAA-Overleese.—Selected from Overleese seed by John Gordon, Amherst, N.Y., in 1982. Fruit: 285 g; rounded shape; green skin; yellow flesh; few seeds; matures in mid-October in New York.

SAA-Zimmerman.—Selected as seedling from seed originating from G.A. Zimmerman collection by John Gordon, Amherst, N.Y., in 1982. Fruit: 170–225 g; yellow skin and flesh; few seeds. May be same as Zimmerman.

Silver Creek.—Selected from the wild in Millstedt, Ill., by K. Schubert. Fruit: medium size.

Sunflower.—Selected from the wild in Chanute, Kan., by Milo Gibson in 1970. Tree reported to be self-fertile. Fruit: up to 225 g; yellow skin; butter-color flesh; few seeds; ripens first week of October in Michigan. May be same as Mary Foos Johnson.

Sweet Alice.—Selected from the wild in West Virginia by Homer Jacobs of the Holden Arboretum, Mentor, Ohio, in 1934.

Taylor.—Selected from the wild in Eaton Rapids, Mich., by Corwin Davis in 1968.

Fruit: small; bears up to seven fruit in a cluster; green skin; yellow flesh; ripens first week of October in Michigan.

Taytwo.—Selected from the wild in Eaton Rapids, Mich., by Corwin Davis in 1968. Sometimes spelled Taytoo. Fruit: up to 285 g; light-green skin; yellow flesh; ripens first week of October in Michigan. Taylor and Taytwo both originated in the same wild patch and should be very closely related.

Wells.—Selected from the wild in Salem, Ind., by David Wells in 1990. Fruit: 340–400 g; green skin; orange flesh.

Wilson.—Selected from the wild on Black Mountain, Harlan Co., Ky., by John V. Creech in 1985. Fruit: medium size; yellow skin; golden flesh.

Zimmerman.—Selected from G.A. Zimmerman seed by George Slate. May be same as SAA-Zimmerman.

PEACH

Adria.—Orig. in New Brunswick, N.J., by Catherine H. Bailey and L. Fredric Hough, Agr. Expt. Sta. Introd. in 1974. 492012 (Sunhigh x Blake) x 5110043 [Jerseyland x NJ156 (Raritan Rose x Duke of York)]; cross made in 1955; selected in 1960; tested as NJ224. Fruit: small to medium; round to slightly oval; skin 60% red striped on light yellow; flesh yellow, moderately firm; ripens 7 to 10 days before Collins. Tree: size slightly substandard with relatively short internodes resistant to Fusicoccum canker under artificial inoculation; flower large, showy; leaf glands reniform. Named in cooperation with Ing. Milena Leksan, Agriculture Inst. of Slovenia, Ljubljana, Yugoslavia, because of the value of this variety to their peach growing region.

Afterglow.—Orig. in New Brunswick, N.J., by M.A. Blake, New Jersey Agr. Expt. Sta. Introd. in 1938. J.H. Hale x N.J. 27116; cross made in 1923; tested as N.J. 84. Fruit: large; round, uniform; skin medium to light blush, not attractive; flesh yellow, firm-melting, freestone; ripens 1 week after Elberta; processing quality fair. Tree: only moderate in vigor and productivity; flower small, nonshowy; flower buds not hardy; chilling requirement 750 h.

Alabama.—Orig. in Mississippi by J.C.C. Price. Introd. about 1930. Carman o.p. Fruit: large; round ovate; skin cream, partially overlaid with red; flesh white, streaked with red, firm, juicy, mildly subacid, quality poor, semi-clingstone; ripens about 10 June. Apparently obsolete.

Alamar.—Orig. in Winters, Calif., by Allen L. Martin. Introd. in 1949. Plant patent 930; 4 Apr. 1950. Parentage unknown; discovered in 1940. Fruit: large; globose; skin yellow with a very high blush; flesh yellow, streaked with red next to stone; stone small, free; ripens 1 week earlier than Rio Oso Gem, 5 to 6 days after J.H. Hale. Tree: size medium; vigor medium; pollen-sterile; leaf glands reniform; flower large, pink.

Albatros.—Orig. in Roodeplaat, Pretoria, South Africa, by Edward P. Evans, Hort. Res. Inst. Introd. in 1973. [Bells November x Culemborg 388] x Robin = [9/72] sp.; cross made in 1965; selected in 1967, tested as 15/7. Fruit: large; round; skin green ground color with red blush over 60% to 90%, pubescence medium; flesh white, melting, flavor excellent, freestone; ripens early November. Tree: medium-large; open vase-shaped; vigorous; productive; must be well-thinned; leaves develop dark red color in autumn; flower showy light rose; leaf glands reniform; low chilling requirement, 300 to 500 h.

Albru.—Orig. in Port Clinton, Ohio, by Lawrence Satterfield. Introd. in 1948. Plant patent 813; 7 Dec. 1948; assigned to French Nursery Co., Clyde, Ohio. J.H. Hale mutation; discovered in 1930. Fruit: large; skin highly colored; flesh yellow, freestone; equal halves that hold their shape well when canned; free from undeveloped fruit or "buttons" characteristic of J.H. Hale. Tree: stronger growth habit than J.H. Hale; flower smaller than J.H. Hale and darker pink.

Alcorn.—Orig. in Modesto, Calif., by T.M. Alcorn. Introd. in 1950. Probable seedling of Lovell; discovered about 1943. Fruit: flesh clear yellow; commercial canning clingstone; ripens in early midseason. Tree: sets heavily; vigorous; leaves with globose glands.

Allgold.—Orig. at Arkansas Agr. Expt. Sta., Clarksville, by James N. Moore, Roy C. Rom, Stanley A. Brown, and William A. Sistrunk. Introd. in 1984. NJ554367 x G17-5E; cross made by Catherine Bailey and L.F. Hough of New Jersey Agr. Expt. Sta.; selected in 1971; tested as A-142. Complex pedigree includes J.H. Hale, Slappey, Admiral Dewey, Orange Cling Seedling, Alameda, and St. Helena. Fruit: 60 to 80 mm when well-thinned; round, without prominent suture or tip; skin 25% red on yellow ground color; flesh orange-yellow, very firm, nonmelting, clingstone. Ripens 7 to 10 days before Babygold 5 in Arkansas; uniform maturity within tree. Tree: vigorous; spreading, with good crotch angles. Winter hardiness above average. Leaves and fruit resistant to bacterial spot. Flowers nonshowy; self-fertile; blossom period early.

Almena Hart.—Orig. in Weatherford, Texas, on the property of D.A. Hart, Jr., by the late R.B. Alexander. Introd. in the 1920s. Parentage unknown; discovered in 1918. Fruit: flesh white, with some red next to the pit, clingstone; ships well; ripens just after Elberta, or about 25 Aug. at place of origin. Tree: winter hardy; vigorous; spreading.

Altair.—Orig. in Ontario, Calif., by Herbert C. Swim, Armstrong Nurseries. Introd. in 1951. Plant patent 1022; 24 July 1951; assigned to Armstrong Nurseries, Ontario, Calif. (Swatow P.I. 41395 x Rio Oso Gem) x [Babcock x (Babcock x Swatow P.I. 41395)]; selected in 1947. A fruiting-flowering variety. Fruit: uniform in form but slightly unsymmetrical; skin yellow ground color, highly blushed, color lightened by heavy pubescence; flesh white, quality high for this type, freestone; ripens during the second week in August at Ontario, Calif.; picking period extends for 10 days. Tree: winter rest requirement about the same as that of Babcock; flower large, ornamental, double, with as many as 14 petals per flower, closely arranged on the stem, pink; leaf glands reniform.

Alton (*Johnson 2*).—Orig. in Yuba City, Calif., by Carl Johnson. Introd. in 1946. Lovell o.p.; selected in 1942. Fruit: flesh yellow, commercial canning clingstone; ripens in early midsummer; similar to Johnson but ripens 1 week earlier, with narrower pit. Not the Alton variety introduced about 1887.

Amador.—Orig. in Palo Alto, Calif., by W.F. Wight, U.S. Dept. of Agriculture. Introd. in 1942. Elberta x Ontario; tested as USDA W24-42. Fruit: medium to large; oval-round to ovate; skin medium yellow with dull red blush; flesh light yellow, some red at pit, firm, melting, fairly sweet, quality good, aroma moderate, freestone; ripens 1 week before Elberta. Tree: vigorous; productive; leaf large, glands globose; flower nonshowy, small.

Amber Crest (*Supechfour*).—Orig. by J.H. Weinberger, Fresno, Calif. Plant patent 5503; 2 July 1985. Assigned to Superior Farming Co., Bakersfield, Calif. Flavorcrest x Springcrest. Fruit: medium large, globose, very symmetrical; more red blush than Flavorcrest, over bright-yellow ground color with short pubescence; flesh yellow, firm, melting, freestone. Ripens 3 days before Flavorcrest. Flowers pink, showy; leaf glands globose. Tree: large and productive.

Ambergem.—Orig. in New Brunswick, N.J., by M.A. Blake, New Jersey Agr. Expt. Sta. Introd. in 1934. Belle selfed; cross made in 1914; tested as N.J. 46 B. Fruit: size medium; oval to round, irregular, suture well-marked; skin dark red, blushed or mottled over most of fruit; pubescence medium; flesh golden yellow, red around pit extending into flesh, firm, nonmelting; canning clingstone, pit very large; ripens in early midseason, 3 weeks before Elberta; quality average for processing. Tree: vigor medium; very fruitful; flower medium, nonshowy, very fertile, even setting well in adverse weather; leaf with reniform glands.

Ames (*J.L. Ames*).—Orig. in Live Oak, Calif., by J.L. Ames. Introd. in 1946 by Fowler Nurseries, Newcastle, Calif. July Elberta mutation; discovered in 1943. Fruit: large; skin highly blushed; flesh yellow, firm, juicy, freestone; ripens 7 to 10 days earlier than parent, which it resembles.

Amparo.—Orig. in Modesto, Calif., by Floyd, Gary, and Grant Zaiger and Leith Gardner, Modesto, Calif. Introd. in 1988. O'Henry x (May Grand nectarine o.p. x Sam Houston o.p.). Plant patent 6472; 20 Dec. 1988; owned by Zaiger's Genetics. Fruit: large; nearly globose; skin yellow ground color overspread with red; flesh golden yellow, medium firm; mild flavor. Tree: vigorous; productive.

Amrein.—Orig. in Noble, Okla., by C.E. Garee, Noble Nursery. Introd. about 1945. Elberta o.p. Fruit: flesh yellow, clingstone; ripens with known parent. Tree: somewhat more productive than Elberta Cling.

Anaheim.—Orig. in Anaheim, Calif., by a Mr. Schaffer. Introd. in 1946, by the Orange County Nursery Co., Norwalk, Calif. Parentage unknown; selected in 1944. Fruit: medium to large; flesh white, freestone;

matures late October, about 10 days after Millers Late, which it resembles. Variety now obsolete.

Andergen.—See Stark Sensation.

Andora.—Orig. in Palo Alto, Calif., by W.F. Wight, U.S. Dept. of Agriculture. Introd. in 1941. Libbee x Lovell; tested as USDA W3525. Fruit: medium to large; oval-round to oblate-round; skin blushed or streaked red over half of surface; flesh clear yellow, more meaty texture than usual, light pink to red at pit; commercial canning clingstone; ripens few days after Paloro, coloring before full maturity; fruit sometimes drops before harvest. Tree: vigorous; productive; leaf glands reniform, large; flower nonshowy, size medium, light pink.

Andross.—Orig. in Davis, Calif., by L.D. Davis, California Agr. Expt. Sta. Introd. in 1964. Dix 5A-l [(Paloro x Round Tuscan) x (Orange Cling seedling x Australian Muir)] x Fortuna; cross made in 1953; selected in 1957; tested as Calif. GHA 5-18. Fruit: medium to large; skin blush medium with excellent harvest color; flesh clear yellow; pit pink to red; ripens in Paloro season. Tree: consistently sets heavy crops; leaf with globose glands; flower large, nonshowy type.

Angelus (*Merricle Y-105*).—Orig. in Red Bluff, Calif., by Grant Merrill. Introd. in 1958. Plant patent 1966; 9 Aug. 1960. J.H. Hale o.p., believed to be crossed with Fortyniner; discovered in 1955; selected in 1957; first fruited in 1958. Fruit: size variable, large; skin highly colored red, medium thick; flesh yellow, red near pit, flavor mild, texture firm, freestone; ripens fairly evenly, with some tendency to ripen first at the base; ripens with Elberta and Fortyniner; resembles Fortyniner. Tree: vigorous; medium large; productive; leaves with reniform glands; flower small, dark pink.

Anza.—Orig. in Riverside, Calif., by J.W. Lesley, California Citrus Expt. Sta. Introd. in 1948. Prenda x Golden State; selected in 1941. Fruit: skin yellow, blushed; flesh yellow, firm, sweet, juicy, quality good, freestone; ripens in midseason, the second week in August in southern California. Tree: rather upright; moderately vigorous; leaf glands reniform; chilling requirement fairly long, but less than J.H. Hale.

Arctic Gem.—Orig. in Modesto, Calif., by Floyd, Gary, and Grant Zaiger and Leith Gardner. Introd. in 1988. O'Henry x Giant Babcock. Plant patent 7065; 12 May 1989; assigned to Zaiger Genetics. Fruit: medium; skin 75% attractive red, medium pubescence; flesh white, very firm, flavorful, freestone. Ripens about 15 days after Redhaven in New Jersey. Tree: vigorous; upright-spreading. Chilling requirement 800 to 1000 h.

Arkansas Black (*Early Red Cling*).—Orig. in Bentonville, Ark., by a Mr. Cotton. Introd. about 1920. Parentage unknown; discovered about 1920. Fruit: averages 2 1/4 inches in diam.; skin very dark red; flesh white, quality fair to good, clingstone; matures early, about 7 1/2 weeks before Elberta; resembles Erly-Red-Fre. Apparently obsolete.

Arkansas Seedling.—See McGraw.

Arline.—Orig. in Fort Valley, Ga., by John Allen. Introd. in 1924. Parentage unknown. Fruit: very large; skin yellow, highly blushed; flesh white, some red around pit, semi-freestone; ripens just before Elberta; resembles Carman. No longer propagated.

Armgold.—Orig. in Wasco, Calif., by David L. Armstrong, Armstrong Nurseries, Ontario, Calif. Introd. in 1964. Plant patent 2399; 5 May 1964; assigned to Armstrong Nurseries. (Flamingo x Springtime) F_2; selected in 1959. Fruit: small to medium, averaging 2 to 2 3/8 inches in diam.; globose, nearly symmetrical, relatively uniform; skin lemon yellow, mottled red, pubescence moderate to scant, short, thickness medium to thin; flesh yellow, no red at pit, medium firm, melting, pronounced aroma, good balance between sugar and acid; semi-freestone to freestone when fully ripe, size medium; ripens very early, just a few days after Springtime. Tree: chilling requirement, similar to June Gold; vigorous grower, upright-spreading; produces large crops, considerable thinning necessary; flower nonshowy; leaf glands reniform.

August Sun.—Orig. in Del Ray, Calif., by Tom Chamberlain, H.P. Metzler & Sons. Introd. in 1983. Plant patent 5140; 15 Nov. 1983; licensed exclusively to Agri Sun Nursery. O'Henry o.p. Fruit about 75 mm;

skin color red on yellow ground color; round; flesh yellow, freestone. Ripens 55 days after Redhaven in Selma, Calif. Tree: vigorous; very productive, chilling requirement 350 to 500 h.

Aurora (*Merrill Aurora*).—Orig. in Red Bluff, Calif., by Grant Merrill. Introd. in 1952. Plant patent 1245; 19 Jan. 1954. (J.H. Hale x Early Wheeler) F2 selfed; selected in 1951. Fruit: flesh yellow, freestone; has much of the firmness and high skin color of Redhaven but sized much more easily; has many J.H. Hale characteristics; takes on a polish when brushed; ripens with or shortly after Redhaven. Tree: lower chilling requirement than Redhaven and larger; leaf glands reniform; flower size medium, pink to light red.

Autumn.—Orig. in New Brunswick, N.J., by M.A. Blake, New Jersey Agr. Expt. Sta. Introd. in 1947. (J.H. Hale x Eclipse) x Late Crawford; tested as N.J. 145. Fruit: medium to large; skin half red with greenish undercolor until soft, dull; flesh yellow, firm, quality good for a late peach; ripens 3 to 4 weeks after Elberta. Tree: hardier than Elberta; flower nonshowy.

Autumn Beauty.—Orig. in Modesto, Calif., by Alexander Kojakanian. Introd. in 1959. Plant patent 1795; 13 Jan. 1959; one-half interest assigned to Aram Kojakanian, Modesto. Chance seedling; discovered in 1931. Fruit: large; skin yellow, blushed with red; flesh yellow, red next to pit, nonmelting, flavor mild, aroma distinct, quality good, clingstone; harvest season extends through the last 3 weeks of October in place of origin. Tree: size medium; vigor medium; upright, spreading; productive; leaf glands reniform; flower small, pink; blooms with Merrill Fiesta.

Autumn Gem.—Orig. in Exeter, Calif., by Grant Merrill. Introd. in 1971. Fiesta x unnamed seedling Cross made in 1960, selected in 1966, tested as 36-86AB. Fruit: large, globose; skin ground color light yellow, one-half to three-fourths covered with red blush, pubescence short, light; flesh yellow, considerable red at stone, firm, freestone; stem long, reducing stem crease and drop; ripens late September or early October with Halloween l; recommended for market and long-distance shipping, also as replacement for Halloween due to brighter color and less drop. Tree: size medium, vigorous, productive; flower pink, small to medium, midseason; leaf glands reniform.

Autumnglo.—Orig. at New Jersey, Agr. Expt. Sta., by C.H. Cook and L.F. Hough. Introd. in 1977. (Candoka x sel. Nect. 25032) x Merrill Fiesta; cross made in 1961; selected in 1965; tested as N.J. 232. Fruit: large; round; yellow flesh, firm, melting; freestone; ripens 1 week after Rio Oso Gem; fresh-market use. Tree: average size; spreading; vigorous; productive; susceptible to bacterial leaf spot; introduced due to its hardiness, high quality and late ripening.

Autumn Gold.—Orig. in Hotchkiss, Colo., by Mrs. Arley Sylvester. Introd. in 1973. Plant patent 3388; 21 Aug. 1973; assigned to Stark Bros. Nurseries & Orchards Co., Louisiana, Mo. Elberta o.p., discovered in 1958, selected in 1961. Fruit: resembles Elberta but flavor less bitter, less red in flesh, ripens 10 days later, 22–25 Sept. in central Missouri. Tree: medium large; broadly ovoid; hardy; leaf glands reniform.

Autumn Lady.—Orig. by G. Merrill, Red Bluff, Calif. Plant patent 4398; 20 Mar. 1979. Autumn Gem o.p. Fruit: large; slightly unsymmetrical; spotted with short red stripes on yellowish-green ground color; flesh yellow, firm, melting, freestone with red pit cavity. Ripens mid-October at Red Bluff. Flowers large, showy, pink; leaf glands reniform. Tree: medium vigor and very productive.

Autumn Rose.—Orig. by R.D. Richards, Bakersfield, Calif. Plant patent 7990; 29 Sept. 1992. Autumn Gem mutation, which it strongly resembles, but ripens 4 weeks later.

Aztecgold.—Orig. by Florida Agr. Expt. Sta. in 1984 cross, selected in 1986 by W.B. Sherman, tested as Fla. 86-31c. Named by J. Rodriguez at Chapingo, Mexico, in 1992. [(Mexican cling x Sunred nectarine) F_2] o.p. Released for fresh fruit market. Fruit: medium size; oval; 60% attractive red blush over bright yellow ground color; short pubescence; flesh yellow, nonmelting, good flavor; clingstone. Ripens about 5 days after Diamante, 15 days after TropicSweet. Flowers: nonshowy; leaf glands reniform.

Moderately resistant to bacterial leafspot and powdery mildew. Chilling requirement 300 chill units.

Babcock.—Orig. in Berkeley, Calif., by E.B. Babcock and C.O. Smith. Selected in 1923 at the California Citrus Expt. Sta. by H.B. Frost and J.W. Lesley. Introd. in 1933 by George P. Weldon, Ontario, Calif. (Strawberry x Peento) F_2. Fruit: small to medium, up to 2 1/2 inches in diam.; round to ovoid-spherical, ventral suture prominent; skin light pink with delicate blush to deep red on sun-exposed portions, attractive, pubescence light, thin but fairly tough, separates easily from flesh when fully ripe; flesh nearly pure white, red near pit, tender, very juicy, quality good, very sweet, mild, low acid; pit small, free; ripens 2 weeks before Elberta. Tree: medium to large; vigorous; spreading; leaf glands globose; leafs out early; very productive; requires little winter chilling.

Babdon.—Orig. in Ontario, Calif., by George P. Weldon, Chaffey College. Introd. in 1948. Babcock x Weldon; selected in 1945. Fruit: skin color similar to Babcock; flesh white, red at pit, fairly firm, quality fair; adheres well to tree. Variety now obsolete.

Babygold 5.—Orig. in New Brunswick, N.J., by L.F. Hough and Catherine H. Bailey, New Jersey Agr. Expt. Sta. Introd. in 1961. P.I. 35201 x [(J.H. Hale x Goldfinch) o.p.]; cross made in 1951; selected in 1954; tested as N.J. C3. Fruit: large; skin one-half red; flesh yellow, nonmelting, very slight red at pit; clingstone; ripens 1 week before Ambergem. Tree: productive; flowers nonshowy; leaf glands reniform.

Babygold 6.—Orig. in New Brunswick, N.J., by L.F. Hough and Catherine H. Bailey, New Jersey Agr. Expt. Sta. Introd. in 1961. [(J.H. Hale x Bolivian Cling P.I. 36126) o.p.] x [(J.H. Hale x Goldfinch) o.p.]; cross made in 1951; selected in 1954; tested as N.J. C15. Fruit: large; skin one-half red; flesh yellow, nonmelting, very little red at pit; canning clingstone; ripens with Ambergem, 2 weeks before Elberta. Tree: productive; flower nonshowy; leaf glands reniform.

Babygold 7.—Orig. in New Brunswick, N.J., by L.F. Hough and Catherine H. Bailey, New Jersey Agr. Expt. Sta. Introd. in 1961. (Lemon Free x P.I. 35201) x [(J.H. Hale x Goldfinch) o.p.]; cross made in 1951; selected in 1954; tested as N.J. C19. Fruit: large; skin up to one-half red; flesh yellow, very slight red at pit, nonmelting, quality good; pit small, cling; ripens 1 week before Elberta. Tree: productive; flowers nonshowy; leaf glands reniform.

Babygold 8.—Orig. in New Brunswick, N.J., by M.A. Blake, New Jersey Agr. Expt. Sta. Introd. in 1961. P.I. 35201 x Ambergem; cross made in 1938; selected in 1951; tested as N.J. C64 and at Michigan Agr. Expt. Sta., South Haven. Fruit: large; skin one-half red; flesh yellow, very little red at pit, nonmelting; clingstone; ripens with Elberta. Tree: productive; flowers nonshowy; leaf glands reniform.

Babygold 9.—Orig. in New Brunswick, N.J., by M.A. Blake, New Jersey Agr. Expt. Sta. Introd. in 1961. P.I. 35201 x Up-to-Date, P.I. 43137; cross made in 1938; selected in 1957; tested as N.J. C66. Fruit: large; skin one-half red; flesh yellow, very slight red at pit, nonmelting; clingstone; ripens 1 week after Elberta and Babygold 8. Tree: productive; flower showy; leaf glands reniform.

Barden Tilford.—See Telford.

Bar-Don.—Orig. in Rockport, Ind., by J. Ford Wilkinson, Indiana Nut Nursery. Introd. in 1951. Parentage unknown; discovered in 1930. Fruit: skin red; flesh white, flavor excellent, freestone; ripens early; obsolete.

Barry Cling.—Orig. in Dallas, Texas, by B.T. Barry. Introd. about 1925 by Foster Nurseries, Denton, Texas. Parentage unknown. Fruit: large; skin white, blushed; flavor fine, clingstone; ripens in latter part of May.

Bates.—Orig. in Delhi, Calif., by W.H. Bates. Introd. in 1939. Plant patent 604; 19 Oct. 1942. Parentage unknown; discovered in 1929; selected in 1935. Fruit: resembles Lovell freestone variety but distinguished therefrom by a cling stone that is relatively small, and later ripening, in Phillips Cling season; commercial canning clingstone. Tree: large; vigorous; productive; leaf glands reniform; flower size medium, salmon. No longer commercially planted.

Beardon.—See Carson.

Beauty Flame.—See Royal June.

Beauty-Gem.—See Jerseyland.

Bellaire.—Orig. near Martinsburg, W.Va., by Freddie Blair. Introd. in 1985. Discovered in 1977; tested as FB-244. Loring mutation. Similar to Loring except for ripening 10 days earlier. Fruit: medium to large; skin brilliant red blush on yellow-orange ground, little pubescence; round; flesh clear yellow, nonbrowning; very firm; excellent quality. Tree: vigorous; easily trained; leaves resistant to bacterial spot; hardiness similar to that of Redhaven.

Bella Rosa.—Orig. in Acampo, Calif., by R.R. Macoubrie. Introd. in 1950. A flowering peach o.p.; discovered in 1943. Fruit: skin well-colored; flesh yellow, very firm, does not discolor when frozen, freestone; ripens about Rio Oso Gem time; hangs well on tree during ripening. Tree: self-fruitful; moderately vigorous.

Bells November (*King Edward*).—Orig. unknown. Introd. about mid-1930s. Parentage unknown. Fruit: size medium; round; skin greenish ground color with 60% covered with red blush; flesh melting, texture too soft for good handling, freestone; ripens early, end of October in South Africa. Tree: large, spreading: vigorous; productive, needs thinning; low chilling requirement; flowers from April to end of July, nonshowy.

Bennett's Perfection.—Introd. in 1923 from Avondale, New Zealand, by U.S. Dept. of Agriculture as P.I. 57912. Fruit: large; flesh yellow, freestone; matures very late. Tree: productive.

Ben Owens.—Orig. in Plains, Mont., by David A. Lawyer. Introd. in 1959 by David A. Lawyer Nursery. Parentage unknown; discovered in 1945. Fruit: small to medium; skin yellow, blushed, tender; flesh yellow, some red at pit, freestone; ripens in a short growing season. Tree: some hardiness; bears most years. Home garden variety.

Berenda Sun.—Orig. in Del Ray, Calif., by Tom Chamberlain, H.P. Metzler & Sons. Introd. in 1984. Plant patent 5297; 23 Nov. 1984; licensed exclusively to Agri Sun Nursery. O'Henry o.p. Fruit: medium-large, about 75 mm; round; skin dark red; flesh yellow, freestone. Ripens 45 days after Redhaven in Selma, Calif. Tree: moderately vigorous; productive. Chilling requirement about 350 to 500 h.

Bicentennial.—Orig. in Calhoun, La., by J.C. Taylor, P.L. Hawthorne, A.C. Miller, J.E. Love. R.J. Constantin, N.L. Horn, and D.W. Newsom, Louisiana State Univ., Baton Rouge. Introd. in 1977. LaGold x Redglobe. Cross made in 1965; selected in 1966; tested as L-66-1-1. Fruit: medium, slightly oval with a small tip; flesh medium firm, golden-yellow, free of red; semi-freestone; skin covered with 60% bright red; pubescence very short; ripens 14–25 May. Tree: vigorous; hardy; productive; flowers showy, rose-colored; leaf glands globose; chilling requirement 700 to 750 h below 45F; resistant to bacteriosis (*Xanthomonas pruni*) and brown rot (*Sclerotinia fructicola*).

Big Red.—See Shippers Late Red.

Biscoe.—Orig. in Jackson Springs, N.C., by North Carolina Agr. Expt. Sta. Introd. in 1968. Raritan Rose x Redskin; cross made in 1955; first fruited in 1958; tested as NC 4639. Fruit: medium to large; round; skin attractive, one-half or more of surface covered with red blush overbright yellow ground color; flesh deep yellow to orange, firm, holding well after picking, texture fine-melting with very little fiber, more resistant to browning than Elberta, flavor good, freestone; ripens about same time as Redskin. Tree: vigorous; productive; flowers small, nonshowy, self-fertile; leaf glands reniform; chilling requirements and blooming time about same as Elberta; highly resistant to bacterial spot and rated better than Redskin.

Black.—Orig. in Stellenbosch, South Africa, by P.A.L. Steyn, Fruit and Food Technology Res. Inst. Introd. in 1961. Walgant x (Kakamas x Early Dawn); selected in 1960; sibling of Malherbe and Neethling. Fruit: flesh yellow, no red in pit cavity, clingstone; excellent canning quality; ripens early, 36 days before Kakamas, with a statistical mean ripening date of 14 Jan.; resembles Kakamas. Tree: very resistant to delayed foliation. Named in memory of M.W. Black, Professor in Pomology, Univ. of Stellenbosch.

Blake (*M.A. Blake*).—Orig. in New Brunswick, N.J., by M.A. Blake, New Jersey Agr. Expt. Sta. Introd. in 1953. J.H. Hale x Primrose; selected in 1953; tested as N.J. 117. Fruit: large, round; skin an attractive red, pubescence slight; flesh yellow, red around pit and suture, firm-melting, quality good, sometimes too acid, freestone; ripens 1 week before Elberta; freezing quality good; excellent for home canning; resembles J.H. Hale. Tree: productive; vigorous; chilling requirement 750 h; flower buds tender in winter but resistant to spring frost; susceptible to bacteriosis; flower small, nonshowy; leaf glands reniform. Named in memory of Professor M.A. Blake (1882–1947).

Blazing Gold.—Orig. in Merced, Calif., by T.B. Stribling, Jr., Stribling's Nurseries. Introd. in 1951. Plant patent 1127; 9 Sept. 1952. July Elberta o.p.; selected in 1947; tested as S-47-3. Fruit: skin blushed; flesh yellow with a few reddish streaks next to pit, flavor slightly acid, freestone; ripens very early, with Florence, 3 to 5 days before Gold Dust; resembles Gold Dust. Tree: vigorous; very productive; bears regularly; leaf glands reniform; flower size medium, pink.

Bobolink.—Orig. in Ontario, Calif., by Herbert C. Swim, Armstrong Nurseries. Introd. in 1953. Plant patent 1150; 20 Nov. 1951. (Gold Mine nectarine o.p. x Rio Oso Gem) x (Gold Mine nectarine o.p. x July Elberta); selected in 1944. Fruit: size medium; broadly ovoid; skin yellow ground color, blushed with red; pubescence moderate, medium to short; flesh yellow, medium firm, flavor fine, good balance of acid and sugar, freestone; ripens 4 to 7 days before July Elberta. Tree: a shorter winter chilling requirement than most yellow-fleshed freestone peaches having a similar ripening period to that of July Elberta, making it especially adapted for southern California conditions; leaf glands globose to reniform; flowers large, showy, color reddish-pink, blooms 5 to 10 days earlier than July Elberta.

Bodine Elberta.—See July Elberta.

Boland.—Orig. in Elsenburg, South Africa, by A.F. de Wet, Stellenbosch-Elsenburg College of Agr., Dept. Agr., Stellenbosch. Introd. in 1944. Parentage unknown; selected in 1943. Fruit: shape similar to Elberta; large; flesh white, good flavor, good dessert quality; stone small, semi-free; ripens later. Tree: vigorous; productive; low winter-chilling requirement.

Bonanza.—Orig. in Wasco, Calif., by David L. Armstrong, Armstrong Nurseries, Ontario, Calif. Introd. in 1962. Plant patent 2213; 8 Jan. 1963; assigned to Armstrong Nurseries. Complex parentage involving Mayflower, Babcock, Rio Oso Gem, Golden Blush, July Elberta, Socala, Coolidge Double Red, Early Imperial and Swatow peaches and Gold Mine nectarine. Fruit: size medium; skin moderately red; flesh yellow, freestone; ripens early, a few days before Coronet and a few days after June Gold. Tree: dwarf; bushy; internodes short, about 1/4 inch apart; chilling requirement moderate, similar to June Gold; flower light pink, semi-double, showy, ornamental.

Bonette.—Orig. in Red Bluff, Calif., by Charles A. Bonetti. Introd. in 1948. Elberta mutation; discovered in an orchard planted in 1923. Fruit: matures later than Elberta; otherwise similar to Elberta.

Bonita.—Orig. in Riverside, Calif., by J.W. Lesley, California Citrus Expt. Sta. Introd. in 1943. Prenda x Golden State; selected in 1940. Fruit: medium to large; skin light yellow with a deep blush, attractive; flesh yellow, dark pink around pit, contains some fiber, sweet, subacid, quality good; freestone; midseason, ripening in late July. Tree: vigorous; upright; foliage dark green, leaf glands reniform; produces very well; requires less winter chilling than Babcock.

Bonfire.—Dwarf red-leaved ornamental orig. at Univ. of Arkansas from 1984 seed, selected in 1988, tested as Ark. 84186-T3, and released in 1992 by J.N. Moore, R.C. Rom, S.A. Brown, and G.L. Klingaman. Plant patent 8509; 21 Dec. 1993. Mistakenly issued patent under name Tom Thumb. Tsukuba no. 2 o.p. Plant patent pending. Fruit: round; small; poor eating quality. Flowers double, showy, pink. Ornamental brachytic dwarf with dark-red leaves holding color late in summer with good resistance to bacterial spot. Trees susceptible to peach-tree borer.

Bonjour.—Orig. in Red Bluff, Calif., by Grant Merrill, Exeter, Calif. Introd. in 1966. Plant patent 2694; 27 Dec. 1966. Merrill Gold Rush x Gemfree. Fruit: resembles Gemfree being distinguished by a shallower, smaller-lipped suture that presents a more uniformly smooth outer skin surface having a brighter red coloring; flesh is distinguished by its more brilliant yellow coloring with a minimum of red flecking when firm-ripe, in having a better flavor, more melting consistency when soft-ripe, and a reduced tendency to oxidize or brown when exposed to air; freestone when soft-ripe and permitted to ripen off tree; for fresh market; shipping quality good; ripens a few days earlier than Gemfree. Tree: large; vigorous; leaf glands reniform; blooming about the same time as July Elberta; flowers medium, dark pink; annually productive.

Bonnie.—Orig. in Burlington, Ont., Canada, by Wallace V. Hopkins. Introd. about 1928. Chance seedling; discovered about 1920. Fruit: skin yellow, blushed; flesh yellow, semi-freestone. Obsolete.

Bounty.—Orig. at U.S. Dept. of Agriculture Appalachian Fruit Res. Sta., Kearneysville, W.Va., by R. Scorza, H.W. Fogle, L.E. Gilreath, K. Patten, E.W. Neuendorff, and W.R. Okie. Introd. in 1988. [(Halberta o.p.) x Redskin] x {Loring x [(Hiley x Fireglow) x Fireglow]}. Fruit: large (usually 65 to 75 mm, but can reach 100 mm); round; skin 30% to 80% red blush on yellow ground color; flesh yellow, nonbrowning, firm, freestone; flavor excellent. Ripens uniformly, about 16 days after Redhaven. Good cropping after natural freezes of –21 to –28C. Tree: vigorous; flower bud production 1.25 times that of Redhaven. Resistant to bacterial leafspot under normal disease pressure. Chilling requirement –800 h.

Bowen.—Orig. in Davis, Calif., by L.D. Davis, Agr. Expt. Sta., Univ. of California, Davis. Introd. in 1971. Dix 22A-S (unnamed seedling) x Str 14-25B (Halford x Dixon 1); cross made in 1955, first fruited in 1959, tested as D20-3E. Fruit: size medium; truncated, symmetrical; skin yellow ground color with slight blush, excellent harvest color, pubescence moderate; flesh clear yellow, becoming slightly darker when fully mature, flavor excellent, clingstone; introduced because of superior flavor when canned; ripens late Dixon or early Cortez season. Tree: size medium, upright; vigorous; highly productive; flower nonshowy; leaf glands globose. Named after the late Logan Bowen, a grower in Stanislaus County who had a long and distinctive association with the California canning peach industry.

Boyd Elberta.—Orig. in Cobden, Ill., by J.J. Boyd and Sons Fruit Farm. Introd. in 1953. Parentage unknown; discovered in 1930. Fruit: skin averages three-fourths red; flesh yellow, firm, freestone; ripens 1 week before Elberta.

Brandywine.—Orig. in Tulsa, Okla., by Carl Burton Fox. Introd. in 1943. Plant patent 580; 4 May 1943. J.H. Hale x pollen parent unknown; seed planted in 1930. Fruit: large, 3 1/2 inches in diam.; skin yellow, marked with a deep shade of pink (usually on one side) that extends through the flesh to the seed; flesh yellow with pink tinge, firm, freestone; develops a characteristic flavor when canned; stone small, deeply embedded. Tree: resembles J.H. Hale.

Brayberta.—Orig. in Bedford, Ky., by Joe S. Bray. Introd. in 1944. Elberta or Elberta Queen mutation; discovered in 1938. Fruit: larger than Elberta; skin nearly solid red; flesh yellow, freestone; resembles a large-sized Elberta fruit except with more of a velvety red skin.

Brentwood Beauty.—See July Elberta.

Brighton.—Orig. in Geneva, N.Y., by Robert C. Lamb, Agr. Expt. Sta. Introd. in 1972. Sunhigh x Redhaven; cross made in 1949, selected in 1953, tested as NY 2622. Fruit: size medium; roundish; skin 90% bright red over yellow ground color, attractive; flesh yellow, medium firm, juicy, sweet, smooth, high quality; ripens uniformly, 8 Aug. at Geneva. Tree: size medium; round, spreading; vigorous; moderately hardy, about the same as Sunhaven; productive; tolerant to perennial canker; flower nonshowy; leaf glands reniform.

Brilliant (*Merrill Brilliant*).—Orig. in Red Bluff, Calif., by Grant Merrill. Introd. in 1950. July Elberta x Florence; selected in 1945. Fruit: small; skin very brilliant color, tough; flesh white, clingstone; ripens very early; resembles Alexander, but firmer and with less melting flesh. Tree: large; vigorous; hardy; regular bearer.

Burbank Elberta.—See July Elberta.

Burbank Giant (*Giant*).—Orig. in Santa Rosa, Calif., by Luther Burbank. Introd. in

1931 by Stark Bros. Nurseries & Orchards Co., Louisiana, Mo. Muir o.p.; selected about 1928. Fruit: large; flesh yellow, quality good, softens too fast for long shipment, freestone; subject to brown rot. No longer propagated.

Burbank July Elberta.—See July Elberta.

Burbank's Orchid.—Orig. in Santa Rosa, Calif., by Luther Burbank. Introd. in 1939. Plant patent 290; 6 Sept. 1938; assigned to Stark Bros. Nurseries & Orchards Co., Louisiana, Mo. Parentage unknown. Most valuable characteristic is the large, double, pink flowers. Fruit: edible; small; flesh white, quality poor, freestone; resembles Stark Summer Heath in appearance. Tree: hardy, strong growing.

Burbank's Santa Rosa.—Orig. in Santa Rosa, Calif., by Luther Burbank. Introd. in 1939. Plant patent 291; 6 Sept. 1938; assigned to Stark Bros. Nurseries & Orchards Co., Louisiana, Mo. Parentage unknown. Most valuable characteristic is large, double, crimson flowers. Fruit: edible; small; flesh white, freestone. Tree: strong grower; hardy.

Burggraff.—Orig. in Live Oak, Calif., by Floyd Leslie Burggraff. Introd. in 1928. Possibly Lovell o.p.; planted in 1923. Fruit: large; flesh yellow, firm; commercial canning clingstone; ripens in Paloro time. Tree: leaf nonglandular. No longer commercially planted.

Burnett Elberta.—See July Elberta.

Burton.—Orig. in Chowchilla, Calif., by Burton P. Jackson. Introd. in 1950. Assigned to R.C. Randolph, Chico, Calif. Parentage unknown; discovered in 1943. Fruit: flesh yellow, flavor good, freestone; ripens very late, 20 Oct.–15 Nov. Tree: bears well.

Buttercup.—Orig. in New Brunswick, N.J., by M.A. Blake, New Jersey Agr. Expt. Sta. Introd. in 1925. Lola x Arp; cross made in 1916. Fruit: medium to small, round, slight suture; skin mottled red over half of surface, pubescence medium; flesh yellow, soft-melting, semi-clingstone, some red at pit; ripens 1 month before Elberta; quality fair, too soft for processing. Tree: vigor medium, sets very heavily; flower large, showy, light pink; leaf with reniform glands.

Calred.—Orig. in Fresno County, Calif., by C.O. Hesse, Univ. of California Kearney Horticultural Field Station. Plant patent 3977; 16 Nov. 1976; assigned to The Regents of the Univ. of California, Berkeley. B4-59 X B4-59; B4-59 = Fireglobe x 34M; cross made in 1962; selected in 1967; on trial in 1968 as 6,11-12B. Fruit: globose to broad ovate; uniform, smooth, golden ground color; medium to dark red blush; sparse pubescence; freestone; flesh firm, rich flavor, low acid; ripens end of August before Fairtime and Summerset. Tree: medium vigor; upright; spreading; leaf glands weakly reniform; flowers large, showy, light pink.

Camden.—Orig. in Columbia, S.C., by South Carolina Agr. Expt. Sta., Clemson Univ., Clemson, S.C., and U.S. Dept. of Agriculture. Introd. in 1972. FV 89-14 [(Fireglow x Hiley) x Fireglow] x Springtime; cross made in 1958 by V.E. Prince, at the former U.S. Hort. Field Lab., Fort Valley, Ga., tested as FV 9-144 at Univ. Sandhill Expt. Sta., Columbia, S.C.; first fruited in 1965. Fruit: small to medium; ovate, indistinct suture line; skin ground color yellow, covered with 60% bright red blush at maturity, pubescence light; flesh yellow, melting, medium textured, flavor good; clingstone, up to 25% split pits; can be harvested in one picking; ripens with Springtime or 3 to 4 days before Springold. Tree: vigorous; productive for an early maturing variety; tolerant to bacterial leaf spot; chilling requirement 750 h below 45F; flower large, showy, self-fertile; leaf glands reniform.

Cameron.—Orig. in Yakima Valley, Wash., by a Mr. Cameron. Introd. about 1943. Possibly Elberta o.p. Fruit: flesh yellow, firm, freestone; ripens about 5 days later than Elberta, which it resembles. No longer propagated.

Canadian Harmony.—Orig. in Harrow, Ont., Canada, by Gerald M. Weaver, Canada Dept. of Agr., Res. Sta. Introd. in 1968. Redskin x Sunhaven. Cross made in 1961, selected in 1964, tested as Harrow 1748. Fruit: 2 3/4 to 3 inches in diam.; round; skin 80% blushed red over a yellow ground cover, pubescence scant; flesh yellow, slightly red at pit, firm but melting, quality good, freestone; ripens 31 Aug., about 2 weeks before Elberta. Tree: large; upright-spreading; exceptional vigor; hardiness equivalent to Redhaven; very productive; moderately tolerant to bacterial leaf and fruit spot *Xanthomonas pruni*; flower nonshowy, salmon, late blooming; leaf glands globose. Recommended for southwestern Ontario.

Canadian Queen.—Orig. in Xenia, Ill., by J.E. Markham, Markham's Nursery. Introd. in 1932. Elberta x Canadian Banner. Fruit: very large; shape uniform; skin yellow, blushed red; flesh yellow, highly flavored, freestone; ships well; closely resembles Early Elberta. Tree: self-fruitful.

Canadian Wonder.—Orig. near Sardis, B.C., Canada, by Henry Krahn. Plant patent 4051; 10 May 1977; assigned to May Nursery Co., Yakima, Wash. Pacific Gold mutation. Fruit: similar to Pacific Gold; large; attractive; flesh juicy; flavor sweet; semifreestone; ripens 7 to 10 days after Redhaven. Tree: very small; full dwarf, height 3 feet.

Candoka.—Orig. in Okanogan, Wash., by William F. Ramsey, Okanogan, and Andrew T. Gossman, Wenatchee, Wash. Introd. in 1932. Plant patent 51; 20 Dec. 1932; assigned to Gossman by Ramsey. Parentage unknown; discovered in 1927. Fruit: size medium; round, regular; skin deep red over yellow, rather thick, pubescence very slight, very attractive; flesh yellow, red around pit, very firm, quality very good, freestone; ripens a few days before Elberta; resembles J.H. Hale; good for canning and freezing. Tree: vigor medium, spreading; fruitfulness erratic, often poor; flower very small, nonshowy, pollen-sterile; leaf glands reniform; chilling requirement 850 h at or below 45F.

Candor.—Orig. in Candor, N.C., at the Sandhills Res. Sta., by F.E. Correll and C.N. Clayton, North Carolina Agr. Expt. Sta., Raleigh, N.C. Introd. in 1965. Redhaven x Erly-Red-Fre; cross made in 1957; selected in 1960; tested as N.C. 7422. Fruit: size uniformly good; round, only a slight tendency to be pointed; skin with a bright red blush over about three-fourths of a rich yellow ground color, very attractive, pubescence medium short; flesh yellow, texture very fine, very firm with no tendency to soften at tip or along suture, flavor unusually good, highly resistant to browning; semifreestone; ripens just before Earlired and Dixired, about 10–15 June in the Sandhills area. Tree: moderately vigorous; producing medium to very heavy crops; flower nonshowy; blooms in Redhaven season; self-fruitful; chilling requirement similar to Redhaven; foliage moderately resistant to bacterial spot; leaf glands reniform.

Cardinal.—Orig. in Fort Valley, Ga., by J.H. Weinberger, U.S. Dept. of Agriculture Hort. Field Sta. Introd. in 1951. Halehaven selfed; cross made in 1941; selected in 1945; tested as USDA FV-101. Fruit: size medium; round; skin about three-fourths of the surface covered with a bright, attractive blush over a yellow ground color, with very light pubescence; flesh yellow, clingstone, firm but melting, medium-textured, and of good flavor; approaches Dixired in size, almost identical with it in color, appearance and firmness; ripens 4 days ahead of Dixired, or 6 weeks earlier than Elberta. Tree: productive, but only moderately vigorous; flowers medium-sized and self-fertile; chilling requirement is about 900 h or slightly more than Elberta and slightly less than Dixired. Resembles Dixired in fruit and tree habits.

Carnival (*Merrill Carnival*).—Orig. in Red Bluff, Calif., by Grant Merrill. Introd. in 1960. Plant patent 2144; 1 May 1962. Rodeo x [(July Elberta x Maxine) x unknown], discovered in 1955; selected in 1959; tested as Merrill 61-167. Fruit: skin highly colored red, being more attractive than Pageant, which it resembles, but larger; flesh yellow, firm, quality very good, freestone; ripens 5 to 10 days after Pageant. Tree: larger and more productive than Pageant; chilling requirement lower than that of Pageant.

Carol.—Orig. in Sigourney, Iowa, by John P. Tremmel and Minnie K. Tremmel. Introd. in 1959. Plant patent 1901; 2 Feb. 1960. Tremmel o.p.; selected in 1954, first fruited in 1954. Fruit: averages 2 1/2 to 3 inches in diam.; skin yellow with a red blush; flesh yellow, nonmelting, with some red around the pit, flavor mild, freestone; claimed to be good for canning and freezing; resembles Rochester. Tree: very winter hardy.

Carolina Belle.—Orig. at Agr. Expt. Sta.,

North Carolina. Bred by J.R. Ballington in 1975, selected in 1980, tested as NCT1333, and introduced in 1987 by D.J. Werner, J.R. Ballington, and D.F. Ritchie. Biscoe x Starlite. Fruit: large; round; skin 50% to 70% red over cream-white ground color; flesh white, medium firm, melting, freestone. Ripens 7 to 10 days after Redhaven. Flowers nonshowy; leaf glands reniform. Tree: growth habit similar to Redhaven; moderate resistance to bacterial spot. Chilling requirement 750 h.

Carolyn.—Orig. in Palo Alto, Calif., by W.F. Wight, U.S. Dept. of Agriculture and the California Agr. Expt. Sta. Introd. in 1942. Libbee x Lovell; tested as USDA W35-37. Fruit: medium to large, highly uniform, truncate-round to oval-round; skin blushed or streaked red over half of surface, medium pubescence; flesh clear yellow, red at pit but not in flesh, firm, texture uniform; commercial canning clingstone; ripens between Guame and Sims; some tendency to drop at maturity. Tree: vigorous, highly fruitful, leaf glands globose; flower nonshowy, small.

Carrie.—Orig. near Mexia, Texas, by J.W. Stubenrauch. Named and introd. in 1925 by Munson Nurseries, Denison, Texas. Frank o.p.; selected in 1923. Fruit: larger than Frank; round; skin yellow, blushed, clingstone. Tree: productive.

Carson (*Beardon, Encinas, Percoca*).—Orig. in Palo Alto, Calif., by W.F. Wight, U.S. Dept. of Agriculture. First commercial planting in 1943. Maxine x Leader; selected in 1935; tested as USDA-W33-5B. Fruit: flesh yellow, being more pale and with poorer flavor than Fortuna; sizes well; conventional canning clingstone. Leaf glands reniform; flowers showy.

Cary Mac.—Orig. near McBee, S.C., by A.C. McLeod. Plant patent 3965; 26 Oct. 1976. Loring mutation. Fruit: medium to large; fairly uniform; skin yellow with red cheek; flesh yellow with tendency of nonbrowning; subacid with slight soluble tannin; excellent eating quality; semi-cling to freestone; market and dessert use; keeping and shipping quality excellent; early ripening, 7 to 10 days before Loring. Tree: medium, slightly more spreading than upright; average hardiness and productivity.

Casellaqueen.—Orig. in Swedesboro. N.J., by Samuel Casella, Casella Bros. Sons, Swedesboro, and Earl E. Slaybaugh, Slaybaugh Bros. Nursery, Aspers, Pa. Plant patent 4406; 2 June 1978; assigned to Casella Bros. Sons with propagation and sales rights to Slaybaugh Bros. Nursery. Introd. in 1979. M.A. Blake mutation. Fruit: resembles M.A. Blake, but stronger-blushed red color, about 80% bright red; skin does not crack; freestone; good storage qualities; ripens 7 to 10 days before parent variety. Tree: similar to M.A. Blake, but hardier; resistant to canker and bacterial leaf spot.

Cassie.—Orig. in Modesto, Calif., by Floyd Zaiger. Introd. in 1980. O'Henry open-pollinated. Plant patent 4571; 15 July 1980; assigned to Zaiger Genetics. Fruit: medium to large; pubescence medium; flesh yellow, firm, good flavor, freestone. Ripens 35 days after Redhaven at Modesto, 25 days after Redhaven in New Jersey. Tree vigorous; upright. Chilling requirement 800 to 1000 h.

Chadon.—Orig. in Ontario, Calif., by George P. Weldon, Chaffey Junior College. Introd. in 1948. Chaffey x Weldon; selected in 1945. Fruit: large; skin greenish-yellow with a blush; flesh white, firm, quality good, similar to Chaffey, freestone; ripens mid-August to September in southern California. Tree: resistant to delayed foliation. Variety now obsolete.

Chaffey.—Orig. in Ontario, Calif., by George P. Weldon, Chaffey Junior College. Introd. in 1939. Lukens Honey x Elberta. Fruit: skin attractive, developing a pink blush on parts exposed to light when ripe, pubescence light; flesh yellowish-white, slightly red at pit, juicy, firm, quality excellent, semi-sweet; freestone; halves slightly uneven; ripens a little earlier than Elberta. Tree: fairly resistant to delayed foliation; dependable producer; vigorous.

Chan (*Extra, Extra Fancy*).—Orig. in Sutter Co., Calif., by Ronald Chan, Yuba City. Introd. in 1984. Andross mutation. Plant patent 5710; 8 Apr. 1986; assigned to Fowler Nurseries, Newcastle, Calif. Fruit: most resembles Halford; large; symmetrical; skin color golden yellow, almost luminous; flesh golden yellow; firm; sweet but mild flavor; clingstone. For commercial canning. Ripens 5 days before Loadel. Tree: moderate vigor; highly productive; spreading; hardy in southern California.

Chandler.—Orig. in Yuba City, Calif., by W. Ray Chandler. Introd. in 1953. Possibly Lovell o.p.; discovered in 1943. Fruit: large; flesh yellow, commercial canning clingstone; pit small; resembles Halford 2. No longer commercially planted.

Charlotte (*Charlotte Giant*).—Orig. in Arlington, Texas, by J.A. Evans. Introd. about 1935. Parentage unknown; discovered about 1922. Fruit: larger than Elberta; stylar end somewhat pointed; flesh yellow, flavor good, freestone; ripens 10 to 14 days after Elberta. Tree: similar to Elberta; vigorous grower; heavy bearer.

Charlotte Giant.—See Charlotte.

Cherryred.—Orig. in New Brunswick, N.J., by M.A. Blake, New Jersey Agr. Expt. Sta. Introd. in 1947. (J.H. Hale x Goldfinch) o.p.; tested as N.J. 129. Fruit: medium; ovate; skin mottled red on yellow-green ground color, pubescence above average; flesh yellow, heavily streaked with red, firm, nonmelting, clingstone; ripens 5 days before Red Bird. Tree: vigorous, hardy, over productive; flower small, nonshowy; leaf glands reniform; chilling requirement 900 h.

Cheyenne Indian.—Orig. in Wheeler County, Texas, by Paul Bruce, Bruce and Sons Nurseries, Alanreed, Gray County, Texas. Introd. in 1926. Chance seedling discovered in 1924. Fruit: flesh yellow, mild, subacid, soft, flavor excellent, freestone.

Chinese Beauty.—Orig. in Clarendon, Texas, by A.L. Bruce & Sons Nursery. Introd. about 1925. Parentage unknown; discovered about 1920. Fruit: size medium; skin white with a blush; flesh white, freestone; for eating fresh or for freezing. Tree: very prolific.

Chinese Dwarf.—See Swatow.

Chinese Dwarf Evergreen.—See Swatow.

Chinese Dwarf Mandarin.—See Swatow.

Christensen Early Elberta.—Orig. in Brigham City, Utah, by C.W. Christensen. Introd. in 1929. Elberta o.p.; selected about 1920. Fruit: skin color high, pubescence heavy; flesh considerable red coloring, freestone; ripens 2 to 3 weeks earlier than known parent; ships well. Tree: buds only moderately hardy.

Clark.—Orig. in Bristol, Ind., by Virgil Clark, Bristol Orchards. Introd. in 1946. Halehaven o.p.; discovered in 1942. Fruit: ripens 2 weeks earlier than Halehaven and resembles that variety. Probably no longer propagated.

Clasen Hale.—Orig. in Twin Falls County, Idaho, by Kimberly Nursery, Kimberly, Idaho. Introd. about 1930. Parentage unknown; discovered about 1924; may be a renamed variety. Fruit: large; flesh dark yellow, firm, aroma fine, quality good, freestone; good fresh, excellent for home canning; resembles Candoka.

Clayton.—Orig. in North Carolina by F.E. Correll and C.N. Clayton, North Carolina State Univ. Pekin x Candor; selected in 1965; tested as NCA 2679. Fruit: medium; round; attractive; yellow ground color with 75% red blush; flesh deep yellow with red around the pit, melting with good texture and excellent flavor; freestone; highly resistant to flesh browning and bacterial spot; ripens 3 to 5 days after Redhaven; fruit set is outstanding and heavy thinning is required most years. Tree: medium vigor; well-formed; productive. Flowers nonshowy, self-fertile, blooms with Redhaven.

Clayton.—Orig. at Sandhills Res. Sta., Jackson Springs, N.C., by F.E. Correll and C.N. Clayton, North Carolina State Univ., Raleigh. Introd. in 1975. Pekin x Candor. Tested as NCA 2679. Fruit: medium: round; very attractive; flesh firm, melting, texture good, flavor excellent, deep yellow with red at pit cavity; freestone; skin medium yellow with 75% red blush; ripens 25 June–10 July, about 3 to 5 days after Redhaven. Tree: productive; heavy thinning required; flowers nonshowy, self-fertile; chilling requirement appears similar to that of Redhaven; highly resistant to bacterial spot (*Xanthomonas pruni*).

Clendining.—See Jon Boy.

Clocolan.—Orig. in Roodeplaat, Republic of South Africa, by Edward P. Evans, Hort. Res. Inst., Pretoria. Introd. in 1975. Moreira Jubilee o.p., in 1963, selected in 1969, tested as 542. Fruit: large; ovate; skin yellow with attractive red blush; flesh yellow, melting, freestone; ripens mid-December. Tree: medium large, 3 m; open vase-shaped; productivity moderate; flower showy, blooms after mid-August in Southern Hemisphere; leaf glands reniform.

Clyde Wilson.—Orig. by W.J. Wilson, Fort Valley, Ga. Plant patent 4916; 2 Nov. 1982. Assigned to Bountiful Ridge Nursery, Princess Anne, Md. Loring mutation. Similar to Loring in tree and fruit characters except ripening begins at end of Loring harvest.

Collins.—Orig. in New Brunswick, N.J., by L.F. Hough and Catherine H. Bailey, New Jersey Agr. Expt. Sta. Introd. in 1959. Jerseyland x [Raritan Rose x (J.H. Hale x Goldfinch) o.p.]; selected in 1955; tested as NJ 200. Fruit: size medium; skin an attractive yellow, blushed; flesh yellow, firm, high quality, semi-clingstone but a freestone when fully ripe, no split pit; ships well; ripens 3 to 7 days before Sunrise. Tree: vigorous; very productive.

Collins Kakamas.—See Kakamas.

Colora.—Orig. in Colora, Md., by Lloyd Balderson III. Introd. in 1936. Brackett mutation; selected in 1933. Fruit: skin yellow with blush; flesh yellow, quality high, texture fine, freestone; ripens after Golden Jubilee. Tree: some bud hardiness, producing crop after –17F; resembles Vedette.

Comanche.—Orig. in Olney, Ill., by the Illinois Agr. Expt. Sta., Urbana, Ill. Introd. in 1958. Vedette x Halehaven; cross made in 1937 under the direction of M.J. Dorsey of this station; selected in 1941; tested as Illinois K76. Fruit: size medium to large; round; skin 75% red over a yellow ground color; flesh yellow, firm, texture medium, mildly subacid, melting, quality good; semi-freestone; resembles Redhaven but with more smoothly shaped fruit, more red flesh at the suture, a less acid flavor, and matures 3 days before Redhaven and 4 weeks before Elberta in southern Illinois. Tree: vigorous; very productive annually; buds hardy; more so than Redhaven; self-fruitful; blooms with Redhaven; petals showy; foliar glands small, reniform; moderately resistant to bacterial spot.

Compact Flavorette.—Orig. by C.F. Zaiger, L.M. Gardner, G.N. Zaiger, and G.G. Zaiger, Modesto, Calif. Plant patent 8071; 22 Dec. 1992. (Unknown x Amparo) o.p. Fruit: large; globose; very firm; attractive red skin color over yellow with moderate short pubescence; flesh yellow, melting, freestone. Ripens just after mid-July at Modesto. Flowers large, showy, pink; leaf glands reniform. Tree: semi-dwarf, about 60% of normal size; nonvigorous; upright; productive. Chilling requirement 450 h.

Compact Redhaven.—Orig. in Orondo, Wash., by Lester Pratt, Van Well Nursery, Wenatchee, Wash. Introd. in 1971. Plant patent applied for, to be assigned to Van Well Nursery. Redhaven mutation, discovered in 1963. Fruit: resembles Redhaven, ripening same time. Tree: dwarf, compact and full; vigorous and productive.

Constitution.—Orig. in New Brunswick, N.J., by M.A. Blake, New Jersey Agr. Expt. Sta. Introd. in 1947. (J.H. Hale x Eclipse) x Late Crawford; tested as N.J. 161. Fruit: skin color good; flesh yellow, freestone; quality good; ripens 7 to 10 days after Elberta. Tree: hardier than Elberta; flower nonshowy.

Contender.—Orig. at Agr. Expt. Sta., North Carolina. Bred by F.E. Correll in 1974, selected in 1977, tested as NCT 544, and introduced in 1988 by D.J. Werner, J.R. Ballington, and D.F. Ritchie. Winblo x complex parentage of North Carolina selections. Fruit: large; round; skin 70% bright red over yellow ground color; flesh yellow and resistant to browning, firm, melting, freestone. Ripens 21 days after Redhaven. Flowers nonshowy; leaf glands reniform. Tree: superior flower bud hardiness; large number of flower buds; moderate resistance to bacterial spot; more vigorous than Redhaven but similar growth habit. Chilling requirement 1050 h.

Corona.—Orig. in Palo Alto, Calif., by W.F. Wight, U.S. Dept. of Agriculture and the California Agr. Expt. Sta. Introd. in 1942. Libbee x Lovell; tested as USDA W35-34. Fruit: large, up to 3 1/3 inches in diam.; nearly round; skin golden yellow, blushed or streaked red; flesh yellow, red next to pit, fine-grained, firm; commercial canning clingstone; late (after Phillips by 3 to 4 days). Tree: vigorous, productive; leaf glands globose; flower nonshowy, small.

Coronado.—Orig. in Palo Alto, Calif., by W.F. Wight, U.S. Dept. of Agriculture and the California Agr. Expt. Sta. Introd. in 1950. (Pratt Low x Tuscan) x Leader Seedling Cling; selected about 1936; tested as USDA W38-30B. Fruit: medium to large; round, somewhat ovate; suture shallow, well-marked; skin yellow overlaid with lightly streaked red up to 25% of surface, pubescence average; flesh clear yellow, very little red at cavity, firm, fine-textured, nonmelting, flavor mild; commercial canning clingstone; ripens very early, about 10 days before Fortuna. Tree: vigorous; productive; leaf glands globose, small to medium; petals nonshowy, size medium.

Coronet.—Orig. in Fort Valley, Ga., by John H. Weinberger, U.S. Dept. of Agriculture Hort. Field Sta. Introd. in 1953. USDA FV 5-56 (Halehaven x self) x Dixigem; cross made in 1945; selected in 1948; tested as USDA FV 126-79. Fruit: size medium; ovate; skin yellow ground color, three-fourths covered with a bright, attractive red blush, pubescence very light; flesh yellow, firm but melting, smooth textured and of good but mild flavor; near-freestone when fully ripe; ripens 2 to 3 days earlier than Dixigem and about 4 days earlier than Redhaven at Fort Valley. Tree: vigorous and productive; susceptibility to bacterial spot disease is evidently about the same as Elberta; flowers small-petaled, self-fruitful; leaf glands reniform; chilling requirement is slightly higher than Hiley and less than Elberta, or about 800 h.

Correll.—Orig. in North Carolina by F.E. Correll and C.N. Clayton, North Carolina State Univ. Pekin x Candor; selected in 1966; tested as NCA 2699. Fruit: medium; round; attractive; bright yellow ground color covered with 80% to 90% medium red blush; flesh medium yellow, firm, melting, good texture and flavor; clingstone until soft-ripe; highly resistant to flesh browning and bacterial spot; thinning necessary most years; ripens 2 1/2 to 3 weeks before Redhaven. Tree: medium vigor; well-formed; productive. Flowers small, nonshowy, self-fertile, blooms with Red-haven and Candor.

Correll.—Orig. at Sandhills Res. Sta., Jackson Springs, N.C., by F.E. Correll and C.N. Clayton, North Carolina State Univ., Raleigh. Introd. in 1975. Pekin (Summercrest x Redhaven) x Candor (Redhaven x Erly-Red-Fre); tested as NCA 2699. Fruit: medium; round; flesh medium yellow, with trace of red, texture and flavor good; clingstone until soft-ripe; skin bright yellow covered with 80% to 90% medium red blush, very attractive; ripens 3 to 5 days before Candor. Tree: productive; thinning required; flowers small, nonshowy, self-fertile; chilling requirement appears similar to that of Redhaven; highly resistant to both flesh browning and bacterial spot (*Xanthomonas pruni*).

Cortez.—Orig. in Palo Alto, Calif., by W.F. Wight, U.S. Dept. of Agriculture. Introd. in 1944. Paloro x Halford #1; tested as USDA W22-38. Fruit: medium to large; oblate-round; skin golden yellow, washed and striped with red blush over one-fourth of surface, very light pubescence; flesh clear yellow, some pink at pit, flavor mild, quality good; commercial canning clingstone; ripens just before Paloro. Tree: vigorous; productive; leaf glands globose; flower nonshowy, very small, light pink, early blossoming.

Corvallis.—Orig. in Corvallis, Ore., by D.M. John. Introd. about 1945. Parentage unknown; discovered in 1880. Fruit: skin thick; flesh yellow, somewhat coarse in texture, quality fair, freestone; ripens with J.H. Hale, which it resembles, but is smaller in size.

C.O. Smith.—Orig. in Whittier, Calif., by E.B. Babcock and C.O. Smith, California Agr. Expt. Sta. Introd. in 1933. (Strawberry x Peento) F_2; selected in 1930. Fruit: similar to Lukens Honey, but much larger; too soft for a commercial variety. Tree: resistant to delayed foliation; vigorous; heavy producer.

Crawford Special.—Orig. in the Kurtz orchard, The Dalles, Ore. Introd. in 1935 by the Carlton Nursery Co., Forest Grove, Ore., in 1933. Fruit: skin red; flesh yellow, freestone; ripens 2 weeks before Early Crawford, which it resembles. Propagation discontinued about 1943.

Cresthaven.—Orig. in South Haven, Mich., by Stanley Johnson and J.E. Moulton, Michigan Agr. Expt. Sta. Introd. in 1963. Kalhaven x South Haven 309 (SH 50 x Redhaven); cross made in 1952; tested as SH 385. Fruit: medium to large, uniform; nearly round; skin bright red over gold, very attractive, moderately thick and tough; flesh clear yellow, too much red in pit cavity for commercial canning but satisfactory for home canning, freezes unusually well, firm, very resistant to browning; freestone, easily pitted by mechanical means; remains on tree when mature; ripens 1 week or slightly more before Elberta, in a very favorable season. Tree: medium to large, vigorous, productive; self-fruitful; above average in hardiness; flower medium in size; leaves with reniform glands.

Crimson Cascade.—A weeping, red-leaved ornamental orig. at Univ. of Arkansas from 1978 seed. Selected in 1981 and released in 1992 by J.N. Moore, R.C. Rom, S.A. Brown, and G.L. Klingaman. (New Jersey selection x Rutgers Redleaf) selfed. Fruit: small, poor eating quality. Flowers double, showy, dark red. Leaves dark red fading to greenish-red by midsummer. Extreme weeping growth habit; vigorous.

Crimson Lady.—Orig. by L.G. and N.G. Bradford, LeGrand, Calif. Plant patent 7953; 25 Aug. 1992. (Red Diamond x unknown) x (Springcrest x unknown). Fruit: large; round; very uniform; very firm; dark red blush smoothly blended into dark yellowish-pink ground color with very short, scant pubescence; flesh brilliant yellow, nonmelting, clingstone. Ripens in early June about 28 days before Red Diamond. Fruits store on the tree 10 to 14 days. Flowers large, pale pink; leaf glands globose. Tree: large; vigorous; very productive.

Crown Princess.—Orig. in Le Grand, Calif., by Norman G. and Lowell G. Bradford, 11875 E. Savana Rd. Introd. in 1988. Red Diamond nectarine x unnamed peach seedling. Plant patent 7070; 12 Dec. 1989; licensed exclusively to Bright's Nursery, Le Grand, Calif. Fruit most nearly resembles Merrill Gemfree. Fruit: large, uniform; skin 50% dark red-streaked on dark pink ground color; attractive; globose, slightly oblate shape; flesh yellow, nonmelting, very firm, semi-clingstone to clingstone; mild subacid, very good flavor. Ripens with Merrill Gemfree in California. Tree: large; very vigorous; productive; regular bearing; spreading; adapted to central California.

Crystal.—Orig. in Modesto, Calif., by G.C. Kircher. Introd. in 1966. Hale Early x Salwey. Plant patent 2671; 13 Sept. 1966. Fruit: large, 3 1/2 inches in diam.; round, uniform; skin light yellow, blush on side exposed to sun, pubescence firm or wiry more widely spaced hairs than usual peach down, freely peelable flesh white, yellowish near pit, firm, juicy, aromatic; texture fine; stone free; quality excellent; ripens late August. Tree: vigorous, inclined to be bushy; productive, requires thinning each year; regular bearer; rapid grower producing in 3 years; flower large, showy.

Culemborg.—Orig. in Stellenbosch, Republic of South Africa, by P.A.L. Steyn, Western Province Fruit Res. Sta., Dept. of Agr. Introd. there in 1952; imported into U.S. in 1954 as P.I. 219558. Babcock x Early Dawn. Fruit: flesh white, freestone; excellent storage quality; ripens early. Tree: very resistant to delayed foliation.

Cullinan.—Orig. in Beltsville, Md., by H.W. Fogle, Agr. Res. Serv., U.S. Dept. of Agriculture. Introd. in 1977. McNeely x Goldenred; cross made in 1963; selected in 1966, tested as Beltsville 64019 or 6419. Fruit: large; round; 80% bright red blush; lightly pubescent; yellow ground color; flesh yellow without red streaks, firm; freestone; ripens 7 to 10 days after Redhaven. Tree: average size and vigor; productive; leaf glands reniform; moderate bud hardiness; tolerant to bacterial spot (*Xanthomonas pruni*), vegetative growth moderately resistant to brown rot (*Monolinia fructicola*). Flower medium, nonshowy, and self-fertile. Recommended for mid-Atlantic and Northeastern states.

Cumberland.—Orig. in New Brunswick, N.J., by M.A. Blake, New Jersey Agr. Expt. Sta. Introd. in 1925. Belle x Greensboro; cross made in 1914. Fruit: medium to large, oval with pointed apex; skin dull red over greenish-white ground, not attractive, pubescence long; semi-freestone; flesh white with red at pit, soft, melting, texture coarse; ripens 3 weeks before Elberta, persists on tree. Tree: vigorous; medium productivity; resistant to bacterial spot; flower large, showy, light pink; leaf glands reniform; chilling requirement 850 h.

Curlew.—Orig. in Long Beach, Calif., by John D. Davis. Introd. in 1945. Plant patent

651; 27 Feb. 1945; assigned to Armstrong Nurseries, Ontario, Calif. Possibly Salwey o.p.; selected in 1935. Fruit: large to medium, up to 2 7/8 inches in diam.; some fruit with a pronounced beak at apical end; skin light yellow, blushed, moderately tough, pubescence short and scant; flesh yellow, considerable red at pit, medium firm, meaty, moderately juicy, mild, subacid, aroma slight; freestone, pit fairly large; ripens late, 20 Sept. to 10 Oct. at Ontario. Tree: size medium; vigorous; productive, bears regularly; low chilling requirement; flower small, petals pink.

Daily News Four Star.—See Four Star.

Daily News Three Star.—See Three Star.

Daily News Two Star.—See Two Star.

Daily News One Star.—See One Star.

Dandy (*Merrill Dandy*).—Orig. in Red Bluff, Calif., by Grant Merrill. Introd. in 1950. Candoka o.p.; selected in 1943. Fruit: flesh yellow, freestone; ripens 3 weeks earlier than Condoka, which it resembles. Tree: vigorous; heavy bearer; self-fertile. No longer propagated because of tendency to develop split pits.

Daroga Ped.—Orig. in Orondo, Wash., by Auvil Bros. Introd. in 1957 by Van Well Nursery, Wenatchee, Wash. July Elberta mutation; discovered in 1948. Fruit: skin color with extremely high blush; flesh yellow, quality good, freestone; ripens uniformly and a few days after Redhaven, which it resembles. Tree: upright-spreading; vigorous.

Davidson.—See Kingston Redleaf.

David Sun.—Orig. in Del Ray, Calif., by Tom Chamberlain, H.P. Metzler & Sons. Introd. in 1985. Plant patent 5874; 10 Feb. 1987; assigned to Metropolitan Ins. Co.; licensed exclusively to Agri Sun Nursery. Parentage unknown. Fruit: large, 70 to 75 mm; round; skin fully red on yellow ground; flesh yellow, semi-freestone; no split pits; ripens with Gemfree, 10 days before Redhaven in Selma, Calif. Good shipper. Tree: very vigorous; very productive. Chilling requirement about 300 h.

Dawne.—Orig. in Ashfield, Carbon County, Pa., by J. Robert Stetler. Introd. in 1960 by Bountiful Ridge Nurseries, Princess Anne, Md. Plant patent 1955; 28 June 1960; by James Robert Stetler, assignor to Edgar H. Kemp and Homer S. Kemp, Bountiful Ridge Nurseries. Chance seedling; discovered in 1952; tested in Princess Anne. Fruit: medium to large; skin brilliant red appearing early over yellow ground color; flesh yellow, streaked with red, flavor mild, subacid, nonbrowning for 24 h when cut flesh exposed to air, freestone; ripens before Cardinal, Dixired, June Gold, Redcap, and some 2 weeks before Erly-Red-Fre. Tree: vigorous; hardy; above average in bud hardiness; consistent heavy bearer; chilling requirement about 750 h below 45F; leaf glands reniform; flowers showy, petals pale pink; self-fruitful.

Dean.—Orig. in Newcastle, Calif., by E.F. Fowler, Fowler Nurseries. Introd. in 1960 by Fowler Nurseries. Fay Elberta o.p.; discovered in 1950; tested as Fowler 46-7. Fruit: medium to large; shape similar to Fay Elberta; skin yellow with a decided blush, attractive; flesh yellow, firm, moderately juicy, freestone; very highly flavored and of good quality when canned commercially; ripens about 4 days before July Elberta. Tree: vigorous; productive; leaf glands reniform.

DeBerard.—Orig. in Ontario, Calif., in the A.M. DeBerard orchard. Introd. in 1948. Sims mutation; discovered about 1940. Fruit: flesh yellow, clingstone. Tree: as early leafing as Babcock; otherwise, strikingly similar to Sims; slight chilling requirement, being well-adapted to southern California climatic conditions.

Delicious.—Orig. in New Brunswik, N.J., by M.A. Blake, New Jersey Agr. Expt. Sta. Introd. in 1925. Belle x Greensboro; cross made in 1914. Fruit: small to medium; ovate, apex depressed; skin dull red over greenish-white ground, fair appearance, pubescence heavy; flesh white, heavily marked red at pit, soft, melting, lacks flavor, semi-freestone to free; ripens 2 weeks before Elberta, quality fair. Tree: vigorous, productivity medium; flower medium, nonshowy; leaf glands reniform.

Delight.—Orig. in Newcastle, Calif., by E.F. Fowler, Fowler Nurseries. Introd. in 1960. Parentage unknown; discovered in 1954. Fruit large, up to 3 1/2 inches in diam.; skin well-blushed, attractive; flesh yellow, firm, flavor good, freestone; ripens about 10 days before Elberta; resembles Rio Oso Gem.

Delmass.—Orig. by J.D. Masso, Solvang, Calif. Plant patent 6122; 8 Mar. 1988. Unknown parentage. Fruit: large; very firm; suborbicular; flesh white, melting, freestone, ripening in late July through August. Flowers small, pale orange; leaf glands reniform.

Delp Early Hale.—Orig. in Wapato, Wash., by Mr. and Mrs. Ervin Delp. Introd. in 1963. Plant patent 2365; 18 Feb. 1964; Ervin Delp and Julia M. Delp. Possibly J.H. Hale o.p.; however, this orchard also contained bearing trees of Candoka, Carmen, Golden Elberta, Golden Jubilee, and Slappey; discovered in 1954. Fruit: medium-large; skin lightly pubescent, appearance excellent, deeply blushed on one side, golden yellow on other side, tough; flesh yellow, red around pit, firm, melting, juicy, tender, slightly acid, eating quality good; freestone, does not tend to split; keeps and ships well; very little preharvest drop; ripens 10 to 14 days before J.H. Hale, which it resembles. Tree: large; vigorous; very productive; bears regularly; flower size medium; self-fruitful; blooms with J.H. Hale.

Delta.—Orig. by Idlewild Expt. Sta., Clinton, La. Selected in 1969 by P.L. Hawthorn, tested as L-9-A47-33, and released by C.E. Johnson in 1992. [LeGold x (Redskin x Cardinal)] o.p. Fruit: large; ellipical; 40% to 50% red blush over yellow ground color; flesh yellow, nonmelting, clingstone. Ripens at end of June Gold harvest, about 1 June. Flowers showy; leaf glands reniform. Self-sterile. Trees: vigorous; moderate resistance to bacterial spot. Chilling requirement 550 h.

Derby.—Orig. at Sandhills Res. Sta., Jackson Springs, N.C., by D.J. Werner, F.E. Correll, J.R. Ballington, C.N. Clayton, and D.F. Ritchie, North Carolina State Univ., Raleigh. Introd. in 1978. NC 7130 (Halehaven x Redskin) x Candor. Cross made in 1962; selected in 1967; tested as NCA 2183. Fruit: large; round with slight bulge at suture; flesh yellow with trace of red, firm, flavor good; semi-clingstone; skin covered with 70% light-red blush; ripens early 1 to 3 days before Candor, 17 to 20 days before Redhaven. Tree: self-fertile; productive; leaf glands globose; flowers nonshowy; tolerant to bacterial spot (Xanthomonas pruni).

Desertgold.—Orig. in Fresno, Calif., by U.S. Dept. of Agriculture. Introd. in 1969. [(Newday x Southland) x (Southland x Hawaiian)] x [(Sunhigh x Southland) x (Southland x Hawaiian)]. Cross made in 1958, selected in 1961 by John H. Weinberger, tested as F81-79. Fruit: size medium; round; skin yellow with red blush, attractive, resembling, but superior to Rochon; flesh yellow, moderately firm, quality good, semi-cling; ripens early June at Fresno a few days ahead of Early Amber and Cardinal, at Indio in late April and early May a week or more ahead of Springtime. Tree: vigor moderate; productive; flower small, self-fertile; leaf glands reniform. Chilling requirement 300 h.

Desertred.—Orig. in Agr. Expt. Sta., Univ. of Florida, Gainesville, by W.B. Sherman and named in northwest Mexico. Complex parentage of Florida selections. Selected in 1979; tested as Fla. 9-10. Fruit: medium; skin 90% dark red blush; flesh yellow, firm, semi-freestone. Ripens 90 days after bloom. Flowers showy; leaf glands reniform. Highly susceptible to fruit twinning and to bacterial leaf spot. Chilling requirement 250 chill units.

Diamond King.—Orig. in Farina, Ill., by Lincoln Graves, Egyptian Nursery & Landscape Co. Introd. in 1920. Chance seedling discovered about 1910. Fruit: flesh yellow, freestone; matures about the same time as Elberta. Tree: hardy.

Diamond Princess.—Orig. in LeGrand, Calif., by Norman G. and Lowell G. Bradford. Introd. in 1988. Red Diamond nectarine x unnamed peach seedling. Plant patent 7066; 5 Dec. 1989; licensed exclusively to Bright's Nursery, Le Grand, Calif. Fruit most nearly resembles RedTop. Fruit: large, uniform; globose; skin color 75% to 95% deep solid red on yellowish pink ground color; flesh yellow, very firm, freestone; light pubescence; excellent flavor with good sugar/acid balance. Ripens 3 days after RedTop in California. Tree: very vigorous; very productive; regular bearing; spreading; hardy in central California. For long-distance shipping.

Dickey.—Orig. in Coloma, Mich., by Erich L. Kerlikowske, Coloma Nursery. Introd. in 1955. Red Hale mutation, discovered in

1952. Fruit: size medium; round; skin bright red; underlaid with slight goldish-yellow ground cover; flesh clear yellow with slight red at pit, texture fine, freestone; ripens 10 days before Elberta; recommended for fresh market and canning; varieties most valuable characteristic is its high color in season following Redhaven. Tree: strongly vigorous; hardiness comparable to Redhaven; requires heavy thinning for productivity; fruit hangs well without dropping.

Dillon.—Orig. in Norco, Calif., by Glen O. Dillon. Introd. in 1943. Plant patent 572; 9 Mar. 1943. Unknown parentage. Fruit: apex pointed and protruding; skin greenish-yellow; flesh yellow, red at pit, firm; ripens in late November; halves unequal. Tree: vigor and growth moderate; one of last varieties to lose its leaves; flowers nonshowy.

Dinkle.—Orig. in Sparta, Mich., by a Mr. Dinkle. Introd. about 1939. Parentage unknown; discovered about 1930. Fruit: skin a dull green, unattractive; flesh yellow, flavor fair for such a late variety in Michigan; ripens 7 days or more after Elberta. Tree: very hardy; productive.

Dixigem.—Orig. in Fort Valley, Ga., by John H. Weinberger, U.S. Dept. of Agriculture Hort. Field Sta. Introd. in 1944. (Admiral Dewey x St. John) x South Haven; seedling planted in 1937. Fruit: medium, round, uniform; skin bright yellow ground color, light blush, pubescence light; flesh light yellow, nonbrowning, firm, melting, texture fine, flavor mild, low tannin; freestone when ripe but sometimes clings; matures early, 4 to 5 weeks before Elberta. Tree: vigorous, highly productive; flower medium, nonshowy; leaf glands reniform; chilling requirement 950 h.

Dixigold.—Orig. in Fort Valley, Ga., by L.M. Hutchins, U.S. Dept. of Agriculture Hort. Field Sta. Introd. in 1939. Hiley o.p.; probably selected in 1931. Fruit: medium large, oblong, apex apparent; skin mottled red over greenish-yellow ground, pubescence heavy; flesh yellow, firm, melting, freestone; ripens 2 1/2 weeks before Elberta; resembles Hiley. Tree: vigorous, open; medium fruitfulness; flowers large, showy; leaf glands reniform; tolerant to bacteriosis; chilling requirement 950 h.

Dixiland.—Orig. in Fort Valley, Ga., by John H. Weinberger and Victor E. Prince, U.S. Dept. of Agriculture Hort. Field Sta. Introd. in 1962. FV 5-56 (Halehaven selfed) x Dixigem; cross made in 1945 by Weinberger; early testing by Weinberger; later testing and final selection by Prince; tested as USDA FV 127-1. Fruit: large; ovate; skin blushed red over a yellow ground color, pubescence light; flesh yellow, firm, melting, flavor and texture good, freestone; matures in late midseason, with Redskin and 2 to 3 days before Elberta in central Georgia; superior to Redskin in attractiveness and size. Tree: chilling requirement about 750 h below 45F, similar to Southland; resistant to bacterial spot; vigorous; productive; self-fruitful; leaf glands globose; flower petals large, showy. Appears superior to Elberta in attractiveness, flavor and bacterial spot resistance; less color than Blake.

Dixired.—Orig. in Fort Valley, Ga., by J.H. Weinberger, U.S. Dept. of Agriculture Hort. Field Sta. Introd. in 1945. Halehaven x self; selected in 1939. Fruit: size medium; round; skin bright red, attractive, light pubescence; flesh yellow, medium firm; clingstone; ripens early, about 10 to 12 days before Redhaven, 5 to 6 weeks before Elberta; good quality for very early shipping and local markets. Tree: vigor medium; highly productive; blooms late; tolerant to leaf curl; flower small, nonshowy; leaf glands globose; chilling requirement 1050 h at or below 45F.

Dixon 1.—Orig. in Linden, Calif., by F.A. Dixon, Canners League of California, San Francisco. Introd. in 1956. Orange Cling x Australian Muir; sibling of Dixon 2; cross made in 1933; tested as Dix 6-6. Fruit: flesh yellow, turning red if overripe; flavor superior to conventional canning clingstone varieties; season extra early, filling the gap between Fortuna and Cortez; indistinguishable from Dixon 2. Tree: very high yields; indistinguishable from Dixon 2. No longer commercially planted.

Dixon 2.—Orig. in Linden, Calif., by F.A. Dixon, Canners League of California, San Francisco. Introd. in 1956. Orange Cling x Australian Muir; sibling of Dixon l; cross made in 1933; tested as Dix 6-8. Fruit: flesh yellow, turning red if overripe; flavor superior to conventional canning clingstone varieties; some fragments in pit; season extra early, filling the gap between Fortuna and Cortez; indistinguishable from Dixon 1. Tree: very high yields; indistinguishable from Dixon 1. No longer commercially planted.

Donwel.—Orig. in Ontario, Calif., by George P. Weldon, Chaffey College. Introd. in 1948. Babcock x [Elberta x (Elberta x Peento)]; selected in 1945. Fruit: skin yellow with a blush; flesh yellow, quality excellent, freestone; ripens with Weldon. Tree: early foliation; bloom showy; recommended for home and semi-commercial plantings. Variety now obsolete.

Double Delight.—Orig. in Ontario, Calif., by H.C. Swim, Armstrong Nurseries. Introd. in 1957. Plant patent 1787; 30 Dec. 1958; assigned to Armstrong Nurseries. Parentage complex, involving many parents, such as Rio Oso Gem, Swatow, Socala, Early Imperial, Coolidge Double Red, Lukens Honey, Florence; selected in 1949. Fruit: large; skin with moderate red blush; flesh yellow, firm, flavor sweet, freestone; ripens early midseason, before Saturn. Tree: vigorous; chilling requirement higher than that of Saturn. Flower: large; petals pale pink, double, 14 to 18 per flower, blooms after Saturn; leaf glands reniform; ornamental.

Double Jewel.—Orig. in Modesto, Calif., by Floyd, Gary, and Grant Zaiger and Leith M. Gardner, Modesto, Calif. Introd. in 1988. Red Grand x genetic dwarf peach. Plant patent 6410; 22 Nov. 1988; owned by Zaiger's Genetics. Fruit: large; globose; skin red over yellow ground color; flesh light, pale yellow, firm; flavor mild, subacid. Tree: vigorous; upright; productive.

Douthit.—Orig. in Pendleton, S.C., by J.B. Douthit. Introd. about 1921. Parentage unknown; discovered during the period 1910–12. Fruit: resembles Lizzie in appearance; flesh yellow with no red, nonbrowning, with very high retention of yellow color when processed, nonmelting, clingstone; ripens 7 days after Elberta.

Dow Special.—Orig. in Modesto, Calif., by Neil L. Dow. Introd. in 1950. Parentage unknown; discovered about 1927 in a Muir orchard. Fruit: flesh yellow, firm; commercial canning clingstone; resembles Phillips Cling, but with more skin blush and does not produce split pits or gum, as does that variety; ripens in advance of Phillips Cling, about the Corona. No longer commercially planted.

Dr. Davis.—Orig. by L.D. Davis, Davis, Calif. Plant patent 4861; 29 June 1982; assigned to Regents of Univ. of California, Berkeley. Hybrid of L.D. Davis selections D25-9E x G40-5E. Tested as 7-7-52. Fruit: firm; large; globose; yellow skin over golden yellow flesh; flesh nonmelting, clingstone, brown stone. Ripens late, 1 day before Carolyn. Medium-short pubescence. Flowers pink, medium, nonshowy; leaf glands reniform. Tree: medium size and vigor; very productive.

Dripstone.—Introd. by the U.S. Dept. of Agriculture in 1932 from Australia as P.I. 99243. Fruit: skin red, more so than Elberta; flesh yellow, freestone; ripens about with Elberta. No commercial value.

Duke of Georgia.—Orig. in Fort Valley, Ga., by John F. Duke. Plant patent 1236; 29 Dec. 1953; by John F. Duke, Sr., John F. Duke, Jr., and Leman P. Duke, Fort Valley. Possibly Erly-Red-Fre mutation. Fruit: average 2 to 2 1/4 inches in cheek diam.; round, apex somewhat pointed; skin light green to cherry red, thin, tough; flesh white, blended with pink and red, juicy, melting, firm, subacid; freestone at maturity, split-pits common; resembles Erly-Red-Fre very closely but matures about 10 days earlier. Tree: low winter chilling requirement; leaf glands globose.

Dwarf Elberta.—See Gage.

Earlibelle.—Orig. in Pretoria, South Africa, by E.P. Evans, Hort. Res. Inst. Introd. in 1966. C.O. Smith Seedling 212 x Culemborg; selected in 1963. Fruit: size medium; flesh white, firm, flavor good, semi-clingstone; resembles Culemborg; ripens very early. Tree: productive; flowers after danger of late frost is past; flower showy, 1 inch in diam., pink; leaf glands reniform.

Earligold.—Orig. in Ontario, Calif., by Herbert C. Swim, Armstrong Nurseries. Introd. in 1958. Plant patent 1883; 8 Dec. 1959. Complex parentage; selected in 1952. Fruit: size medium; skin color good if not picked too early, for it colors late; flesh yellow, semi-freestone; maturity period 5 to 7 days longer than for Springtime. Tree: vigorous; chilling requirement low, equal to or slightly less than for Springtime; yield good for its season.

EarliGrande.—Orig. in College Station. Texas, by H.H. Bowen. Texas Agr. Expt. Sta. Introd. in 1979. Fla. 5-58 x Early Amber. Seed obtained in 1966 from R.H. Sharpe and W.B. Sherman, Univ. of Florida, Gainesville. Seedling planted at College Sta. in 1967; selected in 1970; tested as A1-70. Fruit: medium to large; skin varies from 25% to 75% red blush, pubescence medium; flesh yellow with slight red next to pit, fine texture, good flavor, moderately firm; semi-freestone; ripens early, in southern Texas before any other commercial variety. Tree: very vigorous, large, upright; moderately resistant to bacterial spot. Chilling requirement 275 h.

Earlihale.—Orig. in Beltsville, Md., by Leon Havis, U.S. Dept. of Agriculture, and by Harold W. Fogle, Washington Agr. Expt. Sta. at Prosser, Wash., where it was tested as B2-2221. Introd. in 1958. July Elberta selfed; cross made in 1941; selected in 1944. Fruit: size medium to large; flesh yellow, firm, texture medium, flavor good, freestone; ripens with Halehaven, about 2 weeks before J.H. Hale and 3 to 4 days after Redglobe; similar to J.H. Hale; ships well. Tree: vigorous; pollen-sterile; petals small; blooms with parent; chilling requirement is similar to parent; tree and flower buds hardy to low winter temperatures in tests at Prosser, Wash.; susceptible to bacterial spot but not nearly as susceptible as Sunhigh.

Earlired.—Orig. in Beltsville, Md., by the U.S. Dept. of Agriculture. Introd. in 1960. Redhaven x B3-674 [Halehaven x B3293 (Halehaven x Oriole)]; cross made in 1948; first fruited in 1951; tested as US B9-134. Fruit: size medium; requires early and heavy thinning for best size; round; skin attractive, about 85% of surface covered a bright, attractive red over yellow, pubescence light; flesh yellow, firm, melting, texture medium, flavor good; semi-freestone when soft-ripe; develops a prominent apex and suture when grown as far south as the central parts of Georgia, Alabama, and Texas; ripens early, usually 2 to 5 days before Cardinal, finishing before Sunhaven begins, up to 6 weeks before Elberta. Tree: vigorous; bears annually; productive; moderately susceptible to bacterial spot disease; chilling requirement about the same as for Elberta; flower small; leaf glands globose.

Earli Rio.—Orig. in 1961 by D.A. Diebold, Kelso, Mo. Plant patent 4171; 27 Dec. 1977. Assigned to Stark Brothers Nursery and Orchards Co., Louisiana, Mo.; Rio Oso Gem mutation. Fruit: identical to Rio Oso Gem except ripens 17 or 18 days earlier, about 8 days earlier than Early Rio Oso Gem. Tree: more vigorous with smaller flowers than Rio Oso Gem.

Early Amber.—Orig. in Balm, Fla., by Tom Huston and H.G. Sweat, Peaches of Florida. Introd. in 1964. Plant patent 2458; 1 Dec. 1964; by Tom Huston, Miami, Fla., assignor to Peaches of Florida. Fla. 3-5 (Southland x Hawaiian) F_2 x Springtime; cross made in 1958 in Miami, by Huston. Fruit: size medium, 1 7/8 to 2 3/8 inches in diam.; oval; skin dark red over yellow, thickness medium; flesh orange yellow, slightly reddish at pit, firmness medium, flavor and quality good, clingstone; ripens early, about 80 days from bloom, shipping ripe 30 Apr. into early May in central Florida. Tree: vigorous; productive; flower very small, blooms in early February in central Florida; winter chilling requirement of about 350 h below 45F. Obsolete.

Early Babcock.—Orig. in Ontario, Calif., by George P. Welda Chaffey College. Introd. in 1948. Babcock x Nectar; selected in 1945. Fruit: larger than Babcock, which it resembles; skin color deep red; flesh white, quality good, semi-freestone; ripens 10 days before Babcock. Tree: subject to some delayed foliation.

Early Champion.—Orig. in Troy, Kan., by Fred C. Baker. Introd. in 1963. Champion mutation; discovered in 1954. Fruit: characteristics of its parent except it ripens 10 to 14 days before its parent. Tree: extremely hardy.

Early Coronet.—Orig. in Bakersfield, Calif., by Glen A. Moody. Introd. in 1964. Plant patent 2336; 31 Dec. 1963. Coronet mutation; discovered in 1960. Fruit: size medium, uniform, averaging 2 3/4 inches in cheek diam.; ovate, nearly round; skin yellow, brightly blushed, some red striping;

definite suture line white or light yellow; flesh yellow, with small pink areas, firm, melting, flavor mild, quality good, moderately juicy; freestone when fully ripe; ripens about 8 days before Coronet, which it resembles. Tree: vigorous; productive; bears regularly; blooms early; leaf glands reniform; flower size medium, pink.

Early Dawn.—Orig. in Goede Rust, French Hoek, South Africa, by a Mr. Louw. Introd. by H. Page in 1953. Duke of York mutation; discovered about 1928. Fruit: similar to parent but ripening 2 to 3 weeks earlier. Tree: requires less winter chilling than parent.

Early East.—Orig. in New Brunswick, N.J., by M.A. Blake, New Jersey Agr. Expt. Sta. Introd. in 1946. J.H. Hale o.p. x (Slappey x Dewey); tested as N.J. 134. Fruit: size medium; ovate, elongate, apex nippled; skin an attractive solid red, pubescence heavy; flesh yellow, firm to soft; semi-clingstone, freestone when tree-ripe, pits often split at maturity; ripens 10 days before Golden Jubilee. Tree: vigorous; hardier than Elberta; highly productive; flower nonshowy; leaf gland reniform; chilling requirement 850 h.

Early Elberta (*Early Elberta Stark, Early Improved Elberta, Gleason, Gleason Early Elberta, Gleason Elberta, Golden Elberta, Improved Early Elberta, Improved Elberta, July Elberta, Stark Early Elberta, Sumner Gleason*).—Orig. in Kaysville, Utah, by Sumner Gleason. Introd. in 1907. Parentage unknown; discovered in 1900. Fruit: flesh yellow, quality very high, freestone; matures 10 days before Elberta, which it resembles; flavor much better than that of Elberta; excellent for canning and freezing.

Early Elberta Stark.—See Early Elberta.

Early Elegant Lady.—Orig. by M.D. Lewis, Reedley, Calif. Plant patent 7169; 27 Feb. 1990. Probable Elegant Lady mutation; ripening 7 days earlier.

Early Fair.—See Early Fair Beauty.

Early Fair Beauty (*Early Fair, King's Delicious*).—Orig. in Nashville, Ark., by J.T. Warren. Introd. in 1946. Fair Beauty mutation; discovered in 1942. Fruit: flesh yellow, freestone; ripens 2 to 3 weeks before Fair Beauty. A second Fair Beauty mutation, orig. in Johnson County, Ark., by Hiram Taylor. Introd. in 1953; discovered in 1950. Fruit: flesh yellow, freestone; ripens 2 to 3 weeks before Fair Beauty. Tree: hardy. Both are similar in all respects.

Early Fairtime.—Orig. in Bakersfield, Calif., by Lewis B. Sherrill. Introd. in 1973. Plant patent 3603; 27 Aug. 1974. Fairtime mutation; discovered in 1968. Fruit: large, average diam. 3 to 3 1/8 inches; globose; skin deep yellow overspread with red; flesh texture medium to firm, freestone; suitable for canning, freezing and fresh; ripens 5–10 Aug. in Bakersfield. Tree: size medium; vase-shaped; vigor moderate; hardy for Calif.; productivity excellent; flower light pink, showy; leaf glands reniform; chilling requirement 750 h.

Early Gold Dust.—Orig. in Merced, Calif., by T.B. Stribling, Stribling's Nurseries. Introd. in 1960. Seedling of 5-50-29 (Gold Dust o.p.); selected in 1956. Fruit: resembles Gold Dust, ripening 10 days earlier; fruit too small for commercial purposes. Tree: vigorous, sturdy; leaf glands reniform; flowers pink, large, showy. For home orchard.

Early Halehaven.—Orig. in Benton Harbor, Mich., by John Ugene Nametz. Introd. in 1939. Plant patent 325; 5 Apr. 1936; assigned to Greening Nursery Co., Monroe, Mich. Halehaven mutation; discovered in 1935. Fruit: ripens 7 to 10 days before Halehaven; all other characteristics similar to Halehaven. Tree: resembles Halehaven.

Early Hiley.—Orig. in Fort Valley, Ga., by A.J. Evans. Introd. in 1931. Probable Hiley mutation; selected about 1921. Fruit: medium to large, oblong, irregular, suture well-marked; skin with attractive red blush over half of surface, pubescence heavy but short; flesh white, melting, firm, tinged red near pit, freestone; pit large; good quality, suitable as frozen product; ripens 3 weeks before Elberta; resembles Hiley. Tree: vigorous and highly productive; flower large, showy; leaf glands reniform; chilling requirement 750 h.

Early Improved Elberta.—See Early Elberta.

Early Kirkman.—Orig. in Madera, Calif., by William T. Kirkman. Introd. in 1952. Plant patent 1147; 18 Nov. 1952. Kirkman Gem mutation. Fruit: skin yellow but washed and dotted medium red; flesh yellow, but red next to stone, freestone; ripens 2 to 3 weeks earlier than parent, or about 20 Aug., between the seasons of Rio Oso Gem and Kirkman Gem. Tree: resembles Kirkman Gem; leaf glands reniform; flower large, pink.

Early Loring.—Orig. by F. Beyer, Paducah, Ky. Plant patent 4170; 27 Dec. 1977. Assigned to Stark Brothers Nursery and Orchards Co., Louisiana, Mo.; Loring mutation. Tree and fruit similar to Loring except fruit ripens 10 to 14 days earlier with more intense red blush.

Early Maycrest.—Orig. in Reedley, Calif., by Ron Toeus. Plant patent 6229; 26 July 1988; licensed exclusively to Riesner Nursery. Maycrest mutation. Fruit: medium to large; skin blushed red on golden yellow ground; flesh yellow, firm, clingstone; globose; most nearly resembles Queencrest. Ripens 20 days before Redhaven in California.

Early O'Henry.—Orig. in Exeter, Calif., by Grant Merrill. Plant patent pending. Introd. in 1971. Parentage unknown, discovered in 1960, selected in 1966, tested as E3-12. Fruit: size medium globose; skin ground color light yellow, covered with one-half to three-fourths red blush, pubescence very short, light; flesh rich yellow, very little red, firm, freestone; ripens mid-July before Elberta; very similar to O'Henry but ripening 3 weeks earlier; recommended for shipping, canning, and freezing. Tree: size medium; vigorous; productive; flower pink, showy, late; leaf glands reniform.

Early Perfection.—Orig. in Muncie, Ind., by Louis Cowing. Introd. about 1935. Parentage unknown; first fruited about 1928. Fruit: flesh white, soft, quality very good, semi-clingstone; matures very early; regembles Champion. Obsolete.

Early Profit.—Orig. in Bridgman, Mich., by the Ackerman Nurseries. Introd. in 1947. Parentage unknown; discovered in 1940. Fruit: nearly round; skin blush bright red; flesh yellow, firm, juicy, very sweet; freestone to semi-clingstone, depending upon the season; ripens in July at place of origin. Tree: exceptionally hardy in Michigan; produces crops annually. Now superseded by other and better varieties.

Early Raven.—Orig. in Perry, Ohio, by Paul Secor, Secor's Nursery. Introd. in 1956. Halehaven mutation; discovered in 1954. Fruit: large; round; skin yellow, blushed over three-fourths of cheek; flesh yellow, freestone; ripens about 2 weeks before Halehaven. Tree: vigorous; bears annually and well.

Early Red Cling.—See Arkansas Black.

Early Redhaven.—Orig. in Palisade, Wash., by Homer Patrick Montgomery. Introd. in 1961 by Columbia & Okanogan Nursery Co. and Van Well Nursery, both of Wenatchee. Plant patent 2343; 14 Jan. 1964; by H.P. Montgomery, assignor of one-half to Columbia & Okanogan Nursery Co. Redhaven mutation; discovered in 1955. Fruit: medium to large; apex less pointed than Redhaven; skin red; flesh yellow, freestone; ripens 12 to 14 days before Redhaven; otherwise resembles Redhaven in appearance, color and quality. Tree: resembles Red-haven.

Early Rochester.—Orig. in Yakima, Wash., by V.C. Campbell. Introd. in 1940. Plant patent 351; 26 Dec. 1939. Rochester o.p.; selected about 1932. Fruit: clingstone; ripens 18 to 21 days before Rochester; other characteristics similar to Rochester. Tree: resembles Rochester.

Early Shinn.—Orig. in Pope County, Ark., by Truman Shinn. Introd. in 1933. Bud mutation of Shinn; discovered about 1930. Fruit: large; flesh white, clingstone; resembles Shinn except 2 weeks earlier. Tree: yields well; hardy.

Early Sunhigh.—See Garnet.

Early Top.—Orig. in Fowler, Calif., by Hitoshi P. Nakayama. Plant patent 4072; 5 July 1977. Redtop mutation. Similar to Redtop in appearance and quality except the fruit matures 7 to 10 days earlier. Fruit: large; uniform, and nearly round; flesh yellow color, good eating quality; freestone. Tree: large; productive; regular bearing.

Early Triogem.—Orig. in the orchard of Ed Harvey, Cream Ridge, N.J., by M.A. Blake, New Jersey Agr. Expt. Sta., New Brunswick. Introd. in 1942. Triogem mutation; discovered in 1940. Fruit: smaller than parent; flesh yellow, semi-clingstone; ripens about 1 week before Triogem, which it resembles. Tree: hardier than Elberta, similar to Triogem.

Early White Giant.—Orig. in Arkadelphia, Ark., by Frank Geheb. Introd. in 1949. Plant patent 878; 11 Nov. 1949; assigned to Stark Bros. Nurseries & Orchards Co., Louisiana, Mo. Parentage unknown; discovered in 1937. Fruit: large; skin attractive; flesh white, quality good, semi-freestone; ripens early, 1 week before Early Red Bird. Tree: responds to early and heavy thinning, similar to Redhaven; bears fairly well; leaf glands reniform; flower size medium, pink.

Eclipse.—Orig. in New Brunswick, N.J., by M.A. Blake, New Jersey Agr. Expt. Sta. Introd. in 1925. Belle selfed; seed gathered in 1914. Fruit: size medium; flesh yellow, soft, freestone; ripens same season as Sunhigh. Tree: very hardy; flowers showy. Formerly planted to replace Hiley in New Jersey.

Eden.—Orig. in Geneva N.Y., by Robert C. Lamb, Agr. Expt. Sta. Introd. in 1972. Champion x Raritan Rose; cross made in 1940, selected in 1949, tested as NY 1466. Fruit. size medium; roundish; skin 60% bright red on creamy white ground color; flesh white, moderately firm, sweet, quality good; ripens 25 Aug. at Geneva. Tree: large; round-spreading; vigorous; hardiness equal to Redhaven; very productive; tolerant to perennial canker; flower nonshowy; leaf glands reniform.

Edna Elliott.—Orig. in Modesto, Calif., by Lawrence Elliott. Introd. in 1952. Halford mutation; discovered in 1950. Fruit: large; flesh yellow; commercial canning clingstone; ripens 1 week earlier than parent. Tree: Halford type.

Elberta Queen.—Orig. in Louisiana, Mo., by Gus Jordan. Introd. in 1925. Elberta mutation; discovered in 1902. Fruit: somewhat larger than its parent; round; flesh yellow, freestone. Tree: identical to Elberta.

Elbertita.—Orig. in New Brunswick, N.J., by M.A. Blake, New Jersey Agr. Expt. Sta. Introd. in 1958 by Del Rancho Fortuna, McFarland, Calif. A dwarf mutation of Elberta o.p.; selected in 1930. Fruit: flesh yellow, clingstone; quality and appearance similar to Elberta; ripens about with, to 1 week after, Elberta. Tree: semi-dwarf, one-third to one-half the size of Elberta; internodes short; foliage very dense, dark green; not as productive as Elberta; flowers nonshowy. An unusual home garden variety.

Elcling.—Orig. in Arlington, Texas, by J.A. Evans, Evans Specialty Nursery. Introd. apparently in 1922. Probable Elberta o.p.; discovered during the 1910s. Fruit: characteristics of Elberta except pit is clingstone.

Eldorado.—Orig. in Orange, Calif., by Fedor Wassiliew, deceased. Introd. in 1980. Parentage unknown. Cross made in 1967, selected in 1974. Plant patent applied for, to be assigned to Jackson & Perkins Co., Medford, Ore. Fruit: 2 1/2 to 3 inches in diam., round, shallow suture; flesh firm, yellow; freestone; ripens 20–30 July at Irvine, Calif. Tree: dwarf, 3 to 5 feet, broad as tall, good vigor, productive; flower semi-double; recommended for home gardens.

Elegant Lady.—Orig. by G. Merrill, Red Bluff, Calif. Plant patent 4399; 20 Mar. 1979. Early O'Henry x July Lady. Fruit: medium large; globose; blushed red over yellow skin; flesh yellow, firm, melting, freestone with bright red at pit cavity, ripens late July at Red Bluff. Flowers large, nonshowy; leaf glands reniform. Tree: medium vigor and productive.

Ellerbe.—Orig. in North Carolina by F.E. Correl and C.N. Clayton, North Carolina State Univ. Pekin x Candor; selected in 1966; tested as NCA 2705. Fruit: medium-large; round; ground color bright yellow with 60% to 70% bright red blush; flesh firm, melting, with good texture and flavor; flesh deep yellow with red in pit cavity; freestone; highly resistant to bacterial spot; ripens 14 days after Redhaven.

Ellerbe.—Orig. at Sandhills Res. Sta., Jackson Springs, N.C., by F.E. Correll and C.N. Clayton, North Carolina State Univ., Raleigh. Introd. in 1975. Pekin x Candor. Tested as NCA 2705. Fruit: medium to large; round; flesh firm, melting, texture and flavor good; deep yellow with red at pit; freestone; skin bright yellow with 60% to 70% bright red blush; browning is considered intermediate; ripens with Winblo (10–21 July). Tree: flowers nonshowy, self-fertile; chilling requirement appears to be similar to that of Redhaven; highly resistant to bacterial spot (*Xanthomonas pruni*).

Elliott Special.—Orig. in Modesto, Calif., by William C. Elliott. Introd. about 1940. Plant patent 166; 18 Feb. 1936. Lovell o.p. Fruit: averages 2 3/4 inches in diam.; round; skin yellow, blushed; flesh yellow, red at pit, moderately juicy, firm, mild, sweet, freestone; matures late, about 3 weeks after Paloro, Elberta, and J.H. Hale but before Salwey; keeps well; resembles Paloro. Tree: vigorous; size medium; hardy; productive; flower size medium, pink.

Ellis.—Orig. in Palo Alto, Calif., by W.F. Wight, U.S. Dept. of Agriculture. Introd. in 1935. Phillips Cling x Linden; cross made in 1925; came into bearing in 1929. Fruit: large, averaging 7.6 cm in diam.; apex rounded, either with or without a short point; skin yellow, sometimes mottled red; flesh yellow, firm, fine-grained, no red at pit cavity, subacid, quality good; commercial canning clingstone; ripens before Phillips Cling, in late midseason. Tree: vigorous; productive; leaf glands globose; flower with small petals. No longer commercially planted.

Emery.—Orig. in Jackson Springs, N.C., by North Carolina Agr. Expt. Sta. Introd. in 1968. Rochester x Redskin; cross made in 1958; first fruited in 1961; tested as NC 9062. Fruit: size medium; round; skin smooth, one-half of surface covered with red blush over a dull yellow ground color; flesh yellow, less red pigment than Redskin, firm and stays firm well after harvest, fine-melting texture, less juicy than Elberta, flavor excellent, fully freestone; ripens 5 to 7 days later than Elberta. Tree: vigorous; productive; flowers large, showy, self-fertile; leaf glands reniform; chilling requirements about same as Redhaven; equally as resistant to bacterial spot as Redskin.

Empress.—Orig. in Le Grand, Calif., by F.W. Anderson, Merced, Calif. Introd. in 1964 by L.E. Cooke Co., San Gabriel, Calif. Plant patent 2533; 8 June 1965; assigned to L.E. Cooke Co. F_2 of (Flory x Red Grand nectarine); selected in 1962; tested as Anderson 9H15. Fruit: size medium, averages about 2 1/4 inches in suture diam.; globose, uniform; skin bright red over yellow; flesh yellow, firm, melting, subacid to mild, very well-flavored, quality good; ripens evenly, about 3 weeks before Golden Glory; resembles Golden Glory and Rio Oso Gem. Tree: genetically dwarf, averages 5 to 8 feet; internodes short; productive, bearing regularly; leaf large, 6 1/2 inches long, glands reniform; flower large, self-fruitful; suitable also as an ornamental.

Encinas.—See Carson.

Envoy.—Orig. in New Brunswick, N.J., by M.A. Blake, New Jersey Agr. Expt. Sta. Introd. in 1949, jointly with Hort. Res. Inst. Ontario Dept. Agr., Vineland, Ont., Canada. J.H. Hale x Sunbeam; tested as N.J. 102. Fruit: small to medium, round ovate; skin blushed and striped red over most of surface, attractive, slightly pubescent; flesh yellow, melting, almost firm; quality very good, average for canning; freestone; pit small; ripens 4 weeks before Elberta. Tree: vigorous; crotches weak; hardier than Elberta; very productive; flower large, showy; leaf glands reniform.

Erli-Berta.—Orig. in Fowler, Calif., by H.E. and R.K. Hiraoka. Introd. in 1951. Early Elberta mutation; discovered in 1948. Fruit: large; flesh yellow, clings slightly to pit, firm; ripens about 4 weeks before Early Elberta. Tree: bears well, requires heavy thinning.

Erly-Red-Fre.—Orig. in Chase City, Va., by Willie Marsden Perry. Introd. in 1938. Plant patent 320; 28 Mar. 1939; assigned to Bountiful Ridge Nurseries, Princess Anne, Md. Chance seedling discovered in 1936. Fruit: large; round; skin attractive red, thick; flesh white, greenish near pit, firm, melting; pit small, very slight clingstone to freestone; ripens 5 to 6 weeks before Elberta and with or slightly after Red Bird, which it resembles. Tree: vigorous; hardy; average productivity; blossom small, nonshowy; leaf glands reniform; chilling requirement 1150 h.

Erlyvee.—Orig. in Vineland, Ont., Canada, by O.A. Bradt, Ontario Dept. Agr., Hort. Res. Inst. Introd. there in 1949; introd. in U.S. in 1950; P.I. 188908. Golden Jubilee o.p.; selected in 1941; tested as Ontario 350113. Fruit: size small-medium; skin yellow, highly blushed, tender; flesh yellow, quality fair, semi-freestone; matures early, 4 to 5 days after Mayflower; resembles Golden Jubilee. Tree: vigorous; productive; leaf glands reniform; flowers small, similar to Golden Jubilee; tolerant to bacterial leaf spot.

Ernie's Choice.—Orig. in New Brunswick by L.F. Hough and C.H. Bailey, New Jersey Agr. Expt. Sta.; distributed for testing in 1973 by E. Christ as NJ 275; introduced in 1990 by J.C. Goffreda, A.M. Voordeckers, and J.L. Frecon. Parentage unknown. Fruit:

large; globose to ovate-round; attractive 65% mottled red blush on a bright-yellow ground color; flesh yellow, firm, melting, freestone. Ripens 10 days after Redhaven, 21 days before Rio Oso Gem. Flowers large, showy, light pink, small calyx cup; blooms 3 days before Redhaven; leaf glands globose. Leaves and fruit similar to Redhaven in bacterial spot susceptibility. Tree: vigorous; spreading; productive.

Evalyn Gem.—Orig. in Yuba City, Calif., by Perry M. Reedy. Introd. in 1951. Plant patent 971; 8 Aug. 1950. Chance seedling; discovered in 1945; first fruited in 1948. Fruit: large; globose; skin deep yellow ground color blushed with red, thin; flesh yellow, mottled with red, medium firm, subacid, freestone; low tannin content when ripe; ripens between J.H. Hale and Rio Oso Gem; resembles J.H. Hale. Tree: large; vigorous; productive; leaf glands reniform; flowers small, generally salmon turning light pink near base, shading to red margins and tips.

Eva's Pride.—Orig. by C.F. Zaiger, G.N. Zaiger, L.M. Gardner, and G.G. Zaiger, Modesto, Calif. Plant patent 7751; 17 Dec. 1991. [Red Grand o.p. x Desert Gold] o.p.2. Fruit: large; globose; firm; red blush over yellow ground color with moderate pubescence; flesh yellow, melting, clingstone. Ripens in mid-June at Modesto. Flowers large, showy, pink; leaf glands reniform. Tree: large; vigorous; productive.

Eve.—Orig. in Red Bluff, Calif., by Dareld R. Vestal. Introd. in 1957. Plant patent 1716; 10 June 1958; assigned to Reedley Nursery, Reedley, Calif. Redhaven mutation; discovered in 1954. Fruit: large; flesh yellow, firm, subacid; semi-freestone when firm ripe, freestone when soft-ripe; ripens about 2 weeks before parent; resembles parent. Tree: resembles parent, characterized by blooming period 3 to 4 days earlier.

Everett.—Orig. in Noble, Okla., by H.H. Everett. Introd. about 1920. Parentage unknown; discovered in 1910. Fruit: size medium; flesh deep yellow, quality very good, clingstone; ripens very late, hangs well on tree. Apparently now obsolete.

Everts.—Orig. in Davis, Calif., by L.D. Davis, California Agr. Expt. Sta. Introd. in 1962. Dix 5A-1 x Dix 22A-5; Dix 5A-1: (Paloro x Round Tuscan) x Dixon #2; Dix 22-A-5: (Orange Cling o.p. x Alameda) x (Goodman Choice x Transvaal); cross made in 1950; selected in 1954; tested as GH 11-38. Fruit: commercial canning clingstone; medium to large; skin an excellent harvest yellow with medium blush; flesh clear yellow; outstanding flavor quality when canned; ripens just after Halford. Tree: sets and sizes satisfactory crops; leaf glands reniform; nonshowy flower.

Extra.—See Chan.

Extra Fancy.—See Chan.

Fair Beauty (*Murry Beauty*).—Orig. in Arp, Texas, by W.V. Henso and R.W. Fair. Introd. in 1925. Elberta o.p.; discovered in 1920. Fruit: small to medium, round-ovate; skin bright red, over dark yellow, attractive; flesh yellow, red at pit, firm, melting, semi-clingstone, quality very good, low in tannin; ripens 3 1/2 weeks before Elberta; resembles Elberta. Tree: vigorous; heavy bearer; 850 h chilling requirement during dormant season, not fruitful in areas with less than this amount.

Fairberta.—Orig. in Tyler, Texas, by Robert W. Fair. Not introd. commercially. Plant patent 283; 23 Aug. 1938. Chance seedling. Fruit: size medium; somewhat pointed at the stylar end; skin highly blushed; flesh yellow, quality excellent, freestone. Tree: size medium; vigor medium; productive; flower size medium, salmon.

Fairhaven.—Orig. in South Haven, Mich., by Stanley Johnston, Michigan Agr. Expt. Sta. Introd. in 1946. J.H. Hale x South Haven; cross made in 1930; selected in 1935. Fruit: medium-large; nearly round; skin bright yellow with an attractive bright red cheek, smooth, very tough; flesh medium-yellow with some red at the pit, very firm, moderately fine-textured, flavor good, resists browning when cut; freestone; ripens about 3 weeks before Elberta, end of Redhaven season. Tree: vigorous; scaffold limbs strong; bears well annually; fruit buds as hardy as Halehaven; sets heavily.

Fairtime.—Orig. in Fresno, Calif., by Crops Res. Serv., U.S. Dept. of Agriculture. Introd. in 1968. Rodeo x Kirkman Gem o.p.; cross made in 1958; selected in 1961 by John H. Weinberger; tested as F102-13. Fruit: large; round; suture and apex may be slightly raised; skin attractive yellow with one-fifth to one-half covered by a bright red blush, depending on tree vigor and shading, very light pubescence; flesh yellow, considerable red around pit, firm but melting, texture smooth, flavor good, freestone; suitable for fresh market and freezing; ripens about second week in September at Fresno, with Merrill Fiesta and 1 week later than Summerset. Tree: vigorous; productive, heavy cropping is desirable to prevent excessive fruit size; flower large, self-fertile; leaf glands reniform; tree blooms with standard varieties having a low chilling requirement.

Fairway.—Orig. in Fresno, Calif., by Agr. Res. Serv., U.S. Dept. of Agriculture. Introd. in 1974 [Kirkman Gem x (J.H. Hale x Rio Oso Gem) x (Le Grand x Dripstone). Cross made in 1965, selected in 1968 by John H. Weinberger, tested as 51-20. Fruit: large; round; skin bright yellow with two-thirds to three-fourths covered by bright red, attractive, pubescence light; flesh yellow, firm, texture smooth, flavor good, freestone; ripens in early July at Fresno. Tree: moderately vigorous; productive; flower large, self-fertile, full bloom 6 days earlier than Springcrest, 9 days later than Desertgold; leaf glands reniform; chilling requirement 550 h.

Fallate.—Orig. in New Brunswick, N.J., by M.A. Blake, New Jersey Agr. Expt. Sta. Introd. in 1947. (J.H. Hale x Eclipse) x Berks; tested as N.J. 183. Fruit: flesh white, firm, freestone; ripens 7 to 10 days after Elberta. Tree: hardier than Elberta; flower nonshowy.

False Fireglow.—See Southern Glow.

False Sunhigh.—See Fireball.

Fancy Lady.—Orig. by W.K. Mizuno and W.T. Mizuno, Reedley, Calif. Plant patent 7023; 12 Sept. 1989. Sparkle mutation; ripening 1 month earlier with more red color and more symmetrical shape.

Fantastic Elberta.—Orig. near Hickman, Calif., by Merle M. Fountain. Plant patent 3806; 18 Nov. 1975. Fay Elberta mutation. Fruit: medium; globose; skin medium yellow with substantial overspreading with red shading to darker red; flesh firm, abundant juice, mild flavor, yellow color; good eating quality; freestone; good keeping and shipping quality; good resistance to insects and medium resistant to diseases; ripens 12 Aug. Tree: medium; spreading; very productive; regular bearer; medium vigor; flowers double with wavy petals, attractive as ornamental.

Farida.—Orig. in Palo Alto, Calif., by W.F. Wight, U.S. Dept. of Agriculture. Introd. in 1938. Leader Seedling 26-13 x (Paloro x Tuscan); cross made in 1931; first fruited in 1935; tested as US 33-21A. Fruit: large, averaging 7.6 cm in diam.; nearly round, symmetrical; skin yellow, blushed, pubescence medium; flesh yellow, nonmelting, slightly red in pit cavity, firm, fine-grained, flavor and quality good; clingstone; free from split pits; recommended for shipping, but may be canned; ripens early, Tuscan season. Tree: produces well; leaf glands reniform; flower with small petals.

Fayette.—Orig. in Fresno, Calif., by J.H. Weinberger, U.S. Hort. Field Sta. Introd. in 1966. Fay Elberta x [(Fireglow x Hiley) x Fireglow]; cross made in 1957; first fruited in 1961; tested as F5-8. Fruit: large; round; skin yellow with red blush, lightly pubescent; flesh yellow, very firm, smooth textured, flavor fair to good, freestone; smoother shaped, more highly colored and firmer fleshed than Fay Elberta; ripens between Fay Elberta and Rio Oso Gem. Tree: vigorous; productive; blossoms self-fertile and large petaled; leaf glands reniform; not susceptible to bacterial spot; greater vigor than Rio Oso Gem.

Fertile Hale (*Perfect Hale, Le Duke*).—Orig. in Lawrence, Mich., by Lawrence B. Le Duke. Introd. in 1935. Plant patent 175; 28 Apr. 1936; assigned to Greening Nursery Co., Monroe, Mich. J.H. Hale mutation; discovered in 1927. Fruit: size of J.H. Hale; globular; skin an attractive yellow with a deep carmine blush, moderately thick, pubescence slight; flesh yellow, red at pit, juicy, tender, moderately firm, slightly stringy, quality pleasing, freestone. Tree: self-fruitful; productive; vigorous; hardier than parent or Elberta; leaf glands reniform.

Ficksburg.—Orig. in Roodeplaat, Republic of South Africa, by Edward P. Evans, Hort. Res. Inst., Pretoria. Introd. in 1975. Moreira Jubilee o.p.; cross made in 1963, selected in 1969, tested as 14/52. Fruit: medium to large; round; skin yellow with red blush; flesh yellow, slight red at pit, clingstone; ripens early December. Tree: 3 to 4 m; open vase-shaped; productivity moderate; flower showy, blooms after mid-August; leaf glands reniform; recommended for areas where late frost is a problem.

Fidler.—Orig. in Fresno or Tulare County, Calif., by a Mr. Fidler. Introd. about 1944 by the California Nursery Co., Niles, Calif. Bud mutation of a chance seedling. Fruit: flesh yellow, freestone, firm; ripens 10 to 15 days after Salwey, in late September or early October; resembles Krummel.

Fiesta.—Orig. in St. Joseph, Mich., on the farm of a Mr. Brown. Introd. in 1948. Parentage unknown. Fruit: resembles J.H. Hale; ripens with Elberta. Obsolete.

Fillette.—Orig. in New Brunswick, N.J., by Catherine H. Bailey and L.F. Hough, New Jersey Agr. Expt. Sta. Introd. in 1966. (Jerseyland x Mayflower) o.p.; selected in 1960; tested as NJ 221. Fruit: size medium; skin yellow splashed with red; flesh yellow, soft, good quality; clingstone; ripens very early, 3 weeks before Collins. Tree: late blooming; flowers showy, light pink; leaf glands reniform.

Fireball (*False Sunhigh*).—Orig. unknown. Introd. in 1952 by Peach Ridge Farms & Nursery, Clemson, S.C. Parentage unknown; discovered about 1948. Fruit: large; round, apex tipped; skin blushed and red striped over most of surface, ground color greenish-yellow, appearance fair; flesh yellow, red at pit, freestone, firm, melting, very good quality; ripens 2 weeks before Elberta, with Sunhigh. Tree: vigor medium, low productivity in some areas; flower medium, nonshowy; leaf glands globose; chilling requirement 850 h.

Fireglow.—Orig. in New Brunswick, N.J., by M.A. Blake, New Jersey Agr. Expt. Sta. Introd. in 1939. J.H. Hale x Marigold; cross made in 1923; tested as N.J. 71. Fruit: large; round; skin red, attractive; flesh yellow, medium firm, melting, quality good; small pit, free; ripens 17 days before Elberta. Tree: vigorous; very sensitive to environment, hence not widely adapted; susceptible to brown rot and fruit cracking; flower large, showy; leaf glands reniform; chilling requirement 850 h.

Fireprince.—Orig. at U.S. Dept. of Agriculture S.E. Fruit & Tree Nut Res. Lab., Byron, Ga. Introd. by W.R. Okie. [(Hal-Berta x Fireglow) x Redglobe] x [(Sunhigh x Southland) x Redcap]; cross made in 1964 by V.E. Prince; selected in 1966; tested as FV5-1942 and BY5-1942. Fruit: large; skin attractive red on yellow ground color, light pubescence; flesh yellow, melting; freestone. Ripens about 10 to 12 days after Redhaven; ripens between Harvester and Redglobe. Flowers showy, self-fertile; leaf glands reniform; moderately resistant to bacterial spot. Chilling requirement about 850 h.

Firered.—Orig. in Davis, Calif., by C.O. Hesse, Univ. of California at Davis Pomology Dept. Plant patent 3973; 16 Nov. 1976; assigned to The Regents of the Univ. of California, Berkeley. Introd. in 1977. B 8-12 x (Fireglobe x 14-14M); cross made in 1960; selected in 1965; tested as 4,9-25A. Fruit: medium or larger; globular; skin has yellow ground color with bright red blush; flesh yellow with red at cavity; freestone; large showy pink flowers; ripens early August, just before Fay Elberta. Tree: vigorous; upright; spreading; good productivity.

Fisher.—See Howard Fisher.

Flamecrest.—Orig. in Fresno, Calif., by U.S. Dept. of Agriculture. Introd. in 1973. Fayette x {Southland x [(Fireglow x Hiley) x Fireglow] x (J.H. Hale x Rio Oso Gem)}. Cross made in 1965, selected in 1968 by J.H. Weinberger. Fruit: large, but not as large as Suncrest; round; skin bright yellow covered with 50% red blush, pubescence light, more attractive and smoother than Suncrest; flesh yellow, firm, smooth-textured, flavor good, freestone; ripens early July, latter part of Suncrest season. Tree: vigorous; productive; flower self-fertile, large; leaf glands reniform; medium chilling requirement.

Flameprince.—Orig. at U.S. Dept. of Agriculture Southeastern Fruit and Tree Nut Res. Lab., Byron, Ga.; selected in 1983; tested as BY81P584; and released by W.R. Okie in 1993. Complex parentage; sibling of Summerprince. Fruit: large; round; 70% bright red over an attractive golden-yellow ground with short pubescence; flesh yellow with some red in the pit cavity, very firm, melting, freestone, excellent eating quality and texture. Ripens with O'Henry at Byron, 2 weeks after Elberta. Flowers large, showy, pink; leaf glands reniform. Tree: vigorous; productive; moderately resistant to bacterial spot. Chilling requirement 850 h.

Flamingo.—Orig. in Ontario, Calif., by Walter E. Lammerts. Introd. in 1948. Plant patent 661; 20 Nov. 1945; assigned to Armstrong Nurseries, Ontario, Calif. Rio Oso Gem o.p.; selected in 1940. Fruit: more uniform and more smoothly shaped than Rio Oso Gem, which it resembles; ripens 5 to 10 days earlier than Rio Oso Gem and in same season as J.H. Hale. Tree: chilling similar to that of Socala; begins to leaf and flower earlier than Rio Oso Gem; flower large, pink with reddish base.

Flavorcrest.—Orig. in Fresno, Calif., by U.S. Dept. of Agriculture. Introd. in 1974. (110-47 x 109-89) x FV89-14 [(Fireglow x Hiley) x Fireglow] . 110-47 is Kirkman Gem x Dripstone; 109-89 is Kirkman Gem x (J.H. Hale x Rio Oso Gem). Cross made in 1966, selected in 1969 by J.H. Weinberger. Fruit: large, round; skin bright red over a bright yellow ground color; flesh yellow, firm, flavor excellent, texture smooth; ripens with Coronet, 1 to 2 days before June Lady. Compared with Coronet: firmer, better flavor, red exterior does not become dark; larger than June Lady. Tree: moderately vigorous and productive; flower large, self-fertile; leaf glands reniform; medium chilling requirement. Ripens between Springcrest and Redtop.

Flavor Queen.—Orig. by J.F. Doyle, Clovis, Calif. Plant patent 6759; 25 Apr. 1989. Either Redtop or Regina x Snowflame. Fruit: medium; uniform; nearly globose; attractive 60% to 80% red blush with streaking and mottling over bright-yellow ground color; flesh firm, crisp, melting, semi-clingstone. Ripens with Gemfree. Flowers large, nonshowy; leaf glands predominately reniform. Tree: average size and vigor; productive.

Flordabeauty.—Orig. in Agr. Expt. Sta., Univ. of Florida, Gainesville, by R.H. Sharpe and named in Central America. Introd. in 1972. Complex parentage of Florida selections. Selected in 1967; tested as Fla. 26-31. Fruit: medium large; skin 40% red blush on greenish-yellow ground color; flesh yellow, firm, semi-freestone. Ripens 100 days after bloom. Flowers showy; leaf glands reniform. Moderately susceptible to bacterial leaf spot. Chilling requirement 150 chill units.

Flordabelle.—Orig. in Gainesville, Fla., by R.H. Sharpe, Fla. Agr. Expt. Sta., Gainesville. Introd. in 1970. (Southland x Hawaiian, F_2) x Flordawon. Cross made in 1960 selected in 1963, tested as 'W' and W68-1. Fruit: medium to large; nearly round; skin 40% to 80% medium red blush, with dull greenish ground color at picking stage, changing to bright yellow at edible stage; flesh yellow, firm, freestone; ripens mid-May in central Florida. Tree: vigorous, but moderately compact; chilling requirement of 150 h; flower showy; leaf glands reniform.

Flordacrest.—Orig. in Agr. Expt. Sta., Univ. of Florida, Gainesville, by W.B . Sherman, P.M. Lyrene, and T.E. Crocker. Introd. in 1989. Fla. 5-15N x Flordaking. Seed parent has complex parentage including Maygold, Okinawa, Southland, Hawaiian, Merrill Sunrise, Palomar, and Panamint. Selected in 1982; tested as Fla. 82-19. Fruit: medium; skin 70% red blush; flesh yellow, flesh, firm, semi-clingstone; some split pits when crop is light. Ripens 74 days after bloom, after Flordaking and before June Gold. Showy flower; leaf glands globose. Highly resistant to bacterial leaf spot. Chilling requirement 400 chill units.

Flordadawn.—Orig. in Agr. Expt. Sta., Univ. of Florida, Gainesville, by P.C. Andersen and W.B. Sherman. Introd. in 1989. Flordagold x EarliGrande. Selected in 1979; tested as Fla. G9-1. Fruit: medium small; skin striped red with 60% blush; flesh yellow, softening first at tip, semi-clingstone. Ripens 60 days from bloom. Flowers showy; leaf glands reniform. Resistant to bacterial leaf spot. Chilling requirement 250 chill units.

Flordaglo.—Orig . in Agr. Expt. Sta., Univ. of Florida, Gainesville, by W.B. Sherman and P.M. Lyrene. Introd. in 1989. Sundowner nectarine x Maravilha. Selected in 1982, tested as Fla. 82-9W. Fruit: medium large; skin 90% red blush; flesh white, firm, melting, semi-clingstone. Ripens 85 days after bloom. Flower showy; sets high amount of flower buds; leaf glands reniform. Moderately resistant to bacterial leaf spot. Chilling requirement 150 chill units.

Flordaglobe.—Orig. in Agr. Expt. Sta., Univ. of Florida, Gainesville, by P.C. Andersen, W.B. Sherman, and R.H. Sharpe. Introd. in 1989. Redglobe x EarliGrande. Selected in 1982; tested Fla. M2-2. Fruit: medium small; skin 80% red blush; flesh yellow, firm, semi-clingstone. Ripens 63 days after bloom. Flowers nonshowy; sets large number of flower buds, reniform leaf glands. Resistant to bacterial leaf spot. Chilling requirement 475 chill units.

Flordagold.—Orig. in Gainesville, Fla., by W.B. Sherman and R.H. Sharpe, Univ. of Florida. Rio Grande o.p.; orig. in 1965; selected in 1967; tested as Fla. 15-39. Fruit: large, round-oblong; attractive with red over two-thirds of the surface; firm, texture medium coarse; pubescence medium; flesh is clear yellow with a trace of green near the pit; flesh clings to the pit but becomes semi-free on softening; ripens 10–20 May at Gainesville. Tree: round; vigorous; requires 300 h; leaf glands globose; blossoms large, showy, pink; pollen bright yellow and abundant; trees are self-fertile.

FlordaGrande.—Orig. in Agr. Expt. Sta., Univ. of Florida, Gainesville, Fla., by R.H. Sharpe, W.B. Sherman, and R.E. Rouse as a joint release of Florida and Texas Agr. Expt. Sta. Introd. in 1984. Complex parentage of Florida selections. Cross made in 1963; selected in 1965; tested as Fla. 10-64. Fruit: medium but large for season; skin 60% red overcolor with bright yellow ground color; flesh yellow, firm; freestone. Tree: hardy; ripens 103 days from bloom; flower large, showy, self-fertile; leaf glands reniform; chilling requirement 50 h; tolerant to bacterial spot.

Flordahome.—Orig. in Gainseville, Fla., by R.H. Sharpe, Florida Agr. Expt. Sta. Introd. in 1960. Chico 11 (P.I. 146130) F_2 x *Prunus davidiana* (C-26712); original cross made by Lloyd E. Joley, U.S. Plant Introduction Sta., Chico, Calif., in 1949; Chico 11 is a seedling selected from the Shau Thai peach (P.I. 65821); Flordahome tested as US H 97. Up to 1.9 inches in diam. under good conditions; oval; skin one-half blushed; flesh white, soft, quality fair, freestone; resembles Jewel; ripens early. Tree: ornamental; flower about 2 inches in diam., number of petals average 32 to 36 per flower; chilling requirement 400 h; fairly resistant to root-knot nematode.

Flordaking.—Orig. in Gainesville, Fla., by Univ. of Florida, Dept. of Fruit Crops. Introd. in 1978. {Early Amber x [(Southland x Hawaiian) o.p.]. Cross made in 1966; selected in 1969; tested as Fla. 15-34. Fruit: large, roundish with a slight tip and suture bulge; skin yellow ground color with 70% red overblush; flesh yellow, nonbrowning, medium firm; semi-cling; ripens early, 10 May in North Florida. Tree: slightly upright. vigorous, moderately productive; resistant to bacterial spot; flowers nonshowy; leaf glands globose; chilling requirement 450 h.

FlordaMex.—Orig. at Agr. Expt. Sta., Univ. of Florida, Monticello, by W.B. Sherman and R.H. Sharpe. Redglobe x EarliGrande. Selected in 1983; tested as Fla. M3-4; named in Mexico in 1987. Fruit: medium; nearly round; skin 70% red blush with dark-red stripes; flesh yellow, firm, melting, semi-clingstone. Ripens about 75 days after bloom, 5 days after Flordaking. Flowers nonshowy; leaf glands globose. Moderately resistant to bacterial leaf spot and powdery mildew. Chilling requirement 450 chill units.

FlordaMex 1.—Orig. at Agr. Expt. Sta., Univ. of Florida, Monticello, by W.B. Sherman and R.H. Sharpe. Harken x EarliGrande. Selected in 1982; tested as M2-9; named in Mexico in 1989. Fruit: medium large; skin 50% red blush; flesh yellow, firm, melting, semi-freestone. Ripens about 95 days after bloom. Flowers showy; leaf glands reniform. Highly resistant to bacterial leaf spot and powdery mildew. Chilling requirement 450 chill units.

Flordaprince.—Orig. in Gainesville, Fla., by R.H. Sharpe and W.B. Sherman, Univ. of Florida. Introd. in 1982. Fla. 2-7 (complex parentage) x Maravilha. Cross made in 1972; selected in 1975; tested as Fla. 5-2. Fruit: large for season of maturity; round; attractive; skin covered with 80% red blush, pubescence medium; flesh yellow, firm, high quality, semi-clingstone; susceptible to bacterial spot; ripens 80 days from bloom or 20 days before Flordabelle. Tree: vigorous, upright productive; leaf glands reniform; flowers large, showy, medium pink; chilling requirement 150 h.

Flordaqueen.—Orig. in Gainesville, Fla., by R.H. Sharpe, Florida Agr. Expt. Sta. Introd. in 1961. (Southland x Jewel) o.p.; tested as Fla. 4-26. Fruit: about 2 1/2 inches in diam., sizing well under heavy crops; round, no beak or rough suture; skin yellow with light red blush; flesh yellow, moderately firm, flavor only fair and rather bland, clingstone; ripens about 10 days after Maygold or earlier when rest period is broken satisfactorily. Tree: blooms well; bud-set heavy; chilling requirment 550 h. Obsolete.

Flordared.—Orig. in Gainesville, Fla., by R.H. Sharpe, Florida Agr. Expt. Sta. Introd. in 1970. (Southland x Hawaiian) o.p.4 made by J.H. Weinberger. Selected in 1962, tested as L27-12 and Mored. Fruit: small to medium; round; skin 30% to 50% light red blush; flesh white, moderately firm, freestone; ripens late April to early May in central Florida. Tree: vigorous; chilling requirement of 50 to 100 h; flower showy; leaf glands reniform. Obsolete.

FlordaRio.—Orig. by Florida Agr. Expt. Sta. in 1981; selected in 1983; tested as Fla. M3-2; released in 1994 by P.C. Andersen, W.B. Sherman, and R.H. Sharpe. Harken x EarliGrande. Fruit: large; round, with bulging suture; 70% red blush over bright yellow ground color; flesh firm, melting, yellow; semi-clingstone when soft ripe. Ripens with June Gold, about 92 days after full bloom. Flowers: nonshowy; globose petiolar leaf glands. Resistant to bacterial leafspot. Chilling requirement 400 chill units.

Flordastar.—Orig. in Agr. Expt. Sta., Univ. of Florida, by W.B. Sherman, P.M. Lyrene, and T.E. Crocker. Introd. in 1989. Flordagold x EarliGrande. Selected in 1978, tested as Fla. 8-1. Fruit: medium small; nearly round; skin 60% red blush; smooth suture; pubescence medium short; flesh yellow, some red in flesh when full-ripe; firm, melting, semi-clingstone. Ripens with Flordaprince, 72 days after bloom. Flowers showy; leaf glands globose. Highly resistant to bacterial leaf spot. Chilling requirement about 225 chill units.

Flordasun.—Orig. in Gainesville, Fla., by R.H. Sharpe, Florida Agr. Expt. Sta. Introd. in 1964; distributed by Florida Foundation Seed Producers, P.O. Box 14006, University Station, Gainesville, Fla. 32603. [(Southland x Hawaiian) F_3] x Springtime; cross made in 1959; top-worked trees first fruited in 1961, tested as Fla. 16-33W. Fruit: small, averaging about 2 inches in diam.; nearly round; skin yellow, faintly blushed; flesh yellow with little red at pit, medium-firm, ripens uniformly; clingstone until fully ripe, being nearly freestone at the soft-ripe stage; dessert quality fair to good, with some bitterness in overripe condition; ripens uniformly in late April, with or up to 6 days before Flordawon. Tree: needs to be well-grown and properly thinned for desirable commercial fruit size; chilling requirement 300 h; fruit sets medium to heavy, some thinning required; somewhat more spreading and open than Flordawon; leaf glands reniform; flower large, showy, self-fruitful.

Flordawon.—Orig. in Gainesville, Fla., by R.H. Sharpe, Florida Agr. Expt. Sta. Introd. in 1961. (Southland x Hawaiian) o.p.3; selected in 1959; tested as Florida L1-58. Fruit: small, 1 3/4 to 2 inches in diam. under favorable growing conditions and proper thinning; skin slightly blushed; flesh an attractive yellow, nonbrowning, moderately firm, tendency to be soft at tip, freestone; dessert quality very good; ripens in late April, 3 weeks before Jewel. Tree: chilling requirement 150 h. Obsolete.

Florence.—Orig. in Vacaville, Calif., by Millard Sharpe. Introd. in 1920 by Silva-Bergtholdt Nurseries, Newcastle, Calif. Alexander o.p.; discovered about 1915. Fruit: large; round; skin more highly colored red and quality a bit better than Alexander; flesh whitish-yellow, firm, clingstone; ships well; matures with Mayflower or about 10 days earlier; resembles Alexander.

Floretta.—Orig. in Le Grand, Calif., by F.W. Anderson. Introd. in 1926. Quetta nectarine x Florence; selected in 1925. Fruit: flesh white, quality good, clingstone; ripens following Florence, which it resembles. Variety now obsolete.

Flory Dwarf.—Orig. in Modesto, Calif., by C.R. Flory. Introd. in 1945. Seed brought from northern China in 1938. Fruit: flesh white, tender, fine aroma, bitter, freestone; resembles White Heath. Tree: dwarf, growing about 2 to 3 feet tall; flower very large, showy, bright red, very ornamental, similar to a flowering peach; high chilling requirement. Primarily for home gardens.

Fontana.—Orig. in Ontario, Calif., by George P. Weldon, Chaffey Junior College. Introd. in 1939. Sims x (Feicheng x Bolivian Cling). Fruit: large; 2 1/2 inches diam. and over; flesh quality good, subacid, deep yellow while skin still appears green, no red at pit; conventional canning clingstone, pit small; hangs well on tree; ripens moderately late, last of August and first of September; resembles Sims. Tree: more resistant to delayed foliation than Sims, but needs slightly more winter chilling than Babcock; good producer. No longer commercially planted.

Ford.—Orig. in Merced, Calif., on the Fancher Ranch of the California Packing Corp., Berkeley, Calif. Introd. in 1957. Bud mutation of Sims; discovered in 1948; tested as CPC-114. Fruit: size medium; slightly truncated; skin yellow; flesh yellow, some pink at pit cavity; a commercial canning clingstone; ripens with Peak, about the second week of August; resembles Paloro.

Forestgold.—Orig. at Agr. Expt. Sta., Univ. of Florida, Gainesville, by W.B. Sherman and R.H. Sharpe. Selected in 1977; tested as Fla. 7-11; named in New South Wales, Australia, in 1991. (Sunnyside x Flordawon) o.p. Fruit: medium large; nearly round with slight uneven suture and small point; skin 50% red blush over yellow ground color; flesh yellow, firm, melting, freestone. Ripens about 95 days after bloom, about 5 to 7 days after Flordagold. Flowers showy, medium pink; leaf glands reniform. Moderately susceptible to bacterial leaf spot. Chilling requirement 350 chill units.

Fortuna.—Orig. in Palo Alto, Calif., by W.F. Wight, U.S. Dept. of Agriculture and California Agr. Expt. Sta. Introd. in 1941. A clingstone seedling of Leader x (Tuscan x Paloro); tested as USDA W33-21B; sibling of Shasta. Fruit: medium to large, averaging 7.2 cm in diam.; almost round, slightly variable; suture well marked; skin deep yellow, washed with red up to one-fourth of surface, pubescence average; flesh clear yellow, slightly red next to pit, subacid, firm, fine-grained, nonmelting, flavor mild, aroma slight; commercial canning clingstone; quality excellent; ripens early, just before Tuscan, about 18 days before Paloro. Tree: vigorous; heavy producer; leaf glands reniform, large; flower non-showy, rather large; sets heavily, requires thinning.

Fortyniner (*Merrill 49'er*).—Orig. in Red Bluff, Calif., by Grant Merrill. Introd. in 1950. J.H. Hale o.p.; selected in 1943. Fruit: large; skin brightly blushed over yellow, very attractive; flesh yellow, quality excellent, freestone; ripens 1 week before J.H. Hale and Elberta; resembles J.H. Hale. Tree: vigorous; fruitful; subject to bacterial spot. Tested in Kentucky (1966), considered one of best tested, but tender in bud, needing further testing. Variety replaced by the originator with Angelus.

Four Star (*Daily News Four Star*).—Orig. in La Canada, Calif., by W.E. Lammerts, Descanso Distributors. Introd. in 1951. Plant patent 1093; 6 May 1952. [(Swatow x Rio Oso Gem) selfed] x [(Swatow x Rio Oso Gem) x (Babcock x Mayflower)] selected in 1948. Fruit: 2 1/2 inches in diam.; skin color high; fiesh white, quality high, freestone; ripens middle of June; resembles Robin. Tree: vigorous; short chilling requirement; flowers large, 2 inches in diam., double, light pink, very abundant.

Fowler.—Orig. in Marlboro, N.Y., by George R. Fowler. Introd. in 1943. Plant patent 567; 23 Feb. 1943. Parentage unknown. Fruit: large; skin yellow, highly colored, tough; flesh yellow, firm, flavor sweet with slight acid undertone, freestone; Elberta type; ripens about same time as Elberta in New York. Tree: hardy.

Fragar.—Orig. in Bathurst, New South Wales, Australia, by F.A. Fragar. Introd. about 1949. Chance seedling. Fruit: large; skin white with dark red blush; flesh white, red at pit, juicy, firm, flavor excellent, clingstone; ripens 9–21 Feb. at Bathurst; recommended for coast and tablelands area of N.S.W. and areas with similar conditions. Tree: large; vigor moderate; productivity good, regular; chilling requirement moderate.

Franciscan (*Merricle L-152*).—Orig. in Red Bluff, Calif., by Grant Merrill. Introd. in 1961. Plant patent 2239; 12 Mar. 1963. (J.H. Hale x Flaming Gold nectarine) o.p.2, discovered in 1956; selected in 1960; tested as L-152. Fruit: large; skin more brilliant red than Gemfree; flesh yellow with red near pit, very firm, flavor fine, freestone; ripens about 45 days later than Gemfree, which it resembles, and about 5 days before J.H. Hale and with Suncrest; shipping quality good. Tree: size medium; productive; chilling requirement moderately low; leaf glands reniform; flower pink, self-fertile.

Frankie.—Orig. in Marble Falls, Texas, by Kirk Schroeter. Introd. in 1938. Probable Frank mutation; discovered in 1936. Fruit: flesh yellow, slightly red around the pit, freestone. Tree: prolific bearer, adapted to mild winter climate. Closely resembles Frank.

Freeland.—Orig. in Clarinda, Iowa, by E.W. Freeland. Plant patent 623; 11 Apr. 1944; assigned to Stark Bros. Nurseries & Orchards Co., Louisiana, Mo. Parentage unknown; discovered in 1932. Fruit: large to medium; globose; skin highly blushed red, thin, colors before ripening; flesh yellow, freestone; ripens in Elberta season; an Elberta type. Tree: large; vigorous; hardy; leaf glands reniform; flowers very large, pink.

Freestone.—Orig. by D.L. Armstrong, Orange, Calif. Plant patent 4865; 13 July 1982. Assigned to Armstrong Nurseries, Ontario, Calif. Rio Oso Gem x Redwing. Fruit: medium large; slightly unsymmetrical globose; moderate red over greenish-yellow ground color; flesh white, firm, freestone. Ripens in mid- to late August with a strong brown stone. Flowers small and nonshowy; leaf glands reniform. Tree: medium size with round shape.

FrostKing.—Orig. in Danby, Tompkins County, N.Y., by Joseph J. Ponchalek, Ithaca, N.Y. Introd. in 1969. Plant patent 2941; 18 Nov. 1969; assigned to Stark Bros. Nurseries & Orchards Co., Louisiana, Mo. Parentage unknown; selected in 1955 from seedlings originating from a large number of peach pits planted from unknown parentage. Fruit: resembles Blake; medium large; uniform, symmetrical, broadly ovoid; skin highly colored red, very attractive, coloring about 1 week before maturity, pubescence moderate; flesh deep yellow, streaked with red near pit, moderately firm, flavor mild, quality good, freestone; ripens about the same season as Early Elberta; recommended for market, dessert, and canning. Tree: medium large; vigorous; very hardy; productive and regular bearer; flower medium small, pale pink, blooms late in comparison with other varieties; leaf glands globose.

Frostqueen.—Orig. in New Brunswick, N.J., by M.A. Blake, New Jersey Agr. Expt. Sta. Introd. in 1947. (J.H. Hale x Eclipse) x Berks; tested as N.J. 159 and N.J. 185. Fruit: flesh white, firm, freestone; ripens up to 3 weeks after Elberta. Tree: hardier than Elberta; flower nonshowy.

Frye.—Orig. in Yuba City, Calif., by Lloyd Frye. Introd. in 1953. Plant patent 1210; 11 Aug. 1953. Paloro mutation; discovered in 1933. Fruit: size medium; flesh yellow, fine-grained, clingstone; pit small; matures about 10 to 12 days before Paloro, which it resembles; subject to mildew. Tree: resembles Paloro; leaf eglandular; flowers small, nonshowy, salmon.

Fujihara Babcock.—Orig. in Del Rey, Calif., by Paul K. Fujihara. Introd. in 1963. Plant patent 2202; 18 Dec. 1962. Giant Babcock x Merrill Gem; cross made in 1955. Fruit: size uniform, up to 3 1/4 inches in diam.; slightly asymmetrical; skin deeper red than Giant Babcock; flesh creamy white, streaked with pink, firm, quite juicy, very fragrant, flavor good, relatively low acidity, freestone; keeps very well; ripens during the period 15–30 June at place of origin, being 2 to 4 weeks earlier than Giant Babcock; resembles Babcock. Tree: vigorous; bears regularly and heavily; leaf glands reniform.

Fulmur.—Orig. in Modesto, Calif., by Carroll M. Fulkerth and Willette B. Murray. Introd. in 1954. Plant patent 1172; 3 Mar. 1953. Elberta o.p. Fruit: averages 2 7/8 inches in diam.; skin attractive, red blushed and mottled over entire fruit; flesh yellow, firm, melting, aroma distinct, flavor mild and subacid, freestone; ripens 2 weeks before Fay Elberta at place of origin or about 24 July. Tree: large; vigorous; very productive;

leaf glands reniform; flower size medium, pink.

Gage (*Dwarf Elberta, Gage Elberta*).— Orig. in Texico, Ill., by John A. Gage, Gage & Hawkins Orchards. Introd. in a limited way in 1915 and more generally in 1926. Probable Elberta mutation; discovered about 1901. Fruit: large; round; skin highly blushed; flesh yellow; freestone; ships well; a little more pointed and more pubescent fruit than Elberta, which it resembles; ripens 2 to 7 days after Elberta; believed to be less subject to dropping than Elberta. Tree: exceptionally low, spreading; framework branches form strong, wide-angled crotches; blooms 2 to 3 days after Elberta; has shown some resistance to defoliation by bacterial spot; buds hardier than Elberta.

Gage Elberta.—See Gage.

Gaiety (*Merricle P-75*).—Orig. in Red Bluff, Calif., by Grant Merrill. Introd. in 1958. Plant patent 1872; 13 Oct. 1959. J.H. Hale F4; selected in 1955. Fruit: large; skin highly colored red; flesh yellow, with red near skin, firm, freestone; ripens a few days later than Redhaven. Tree: size medium; vigor medium; very productive; leaf glands reniform; flower small, salmon-pink.

Gaiety 2 (*Merricle P-77*).—Orig. in Red Bluff, Calif., by Grant Merrill. Introd. in 1961. Plant patent 2295; 12 Nov. 1963. J.H. Hale F4; selected in 1960. Fruit: skin very red; flesh yellow, firm, freestone; ripens with Redhaven; resembles Gaiety but larger, firmer, skin with more red. Tree: more productive than Gaiety. No longer being propagated.

Gala.—Orig. at Louisiana Agr. Expt. Sta., Calhoun. Selected by J.C. Taylor in 1977, tested as L72-3-8, and jointly released by C.E. Johnson with the U.S. Dept. of Agriculture Southeastern Fruit and Tree Nut Res. Lab., Byron, Ga. Harvester o.p. Fruit: large; round; 70% bright red over yellow ground color; short pubescence; flesh yellow, melting, freestone. Ripens 34 days before Elberta. Flowers small, nonshowy; leaf glands reniform. Moderate resistance to bacterial spot. Trees: vigorous and productive. Chilling requirement 750 h.

Garabedian Hale.—Orig. in Madera, Calif., by John Garabedian of Fresno, Calif. Introd. in 1958. Kirkman Gem o.p.; selected in 1954. Fruit: skin highly colored; flesh yellow, freestone; matures 10 days before J.H. Hale.

Garnet (*Early Sunhigh*).—Orig. in East Wenatchee, Wash., by Ruth D. Reeves. Introd. in 1957 by Van Well Nursery, Wenatchee, Wash. Sunhigh mutation; discovered in 1951. Fruit: skin bright garnet color over most of surface at harvest time; flesh yellow, texture good, flavor fine, freestone; matures about 10 days ahead of Sunhigh; qualities similar to Sunhigh plus the features of earlier ripening and more red skin color; dessert qualities excellent; ships, cans, and freezes well. Tree: vigorous; spreading; very productive.

Garnet Beauty.—Orig. in Ruthven, Ont., Canada, by Garnet Bruner. Introd. in 1958; propagated and distributed by George P. Whaley, Ruthven. Redhaven mutation; discovered in 1951. Canadian Hort. Council registration no. 220. Fruit: medium to large; oval in cross section; flesh yellow, streaked with red, firm, melting, flavor excellent; semi-freestone; ripens 8 to 10 days earlier than parent; hangs on tree until overripe; resembles Redhaven except for earliness and slight elongation of fruit. Tree: vigor above average; heavy crops produced, which size and color well even inside the tree; slightly susceptible to bacterial spot; petals small; resembles Redhaven.

Gemcrest.—Orig. in Davis, Calif., by the Univ. of California. Stribling's Nurseries, Merced, Calif., obtained budwood from the Univ. as UCD S45-8. Named and introd. by Chas. Preuss, Clovis, Calif. Parentage unknown. Fruit: medium large; slightly oblong; skin yellow with attractive blush; fresh yellow, firm, flavor and eating quality excellent, freestone; ripens between Fay Elberta and Rio Oso Gem; for fresh market and shipping. Tree: vigorous; productive.

Gemfree (*Merrill Gemfree*).—Orig. in Red Bluff, Calif., by Grant Merrill. Introd. in 1954. Plant patent 1409; 9 Aug. 1955; assigned to Grant Merrill Orchards. July Elberta x Merrill Gem; selected in 1952; tested as Merricle 27-24. Fruit: large; skin red; flesh yellow, when held until soft, meaty to melting and freestone; when firm ripe, has the firmness of Merrill Gem; resembles Merrill Gem; ripens a few days later than Cardinal and a few days earlier than Merrill Gem. Tree: blooms and sets more heavily than Merrill Gem; leaf glands reniform; flowers medium size to small, pink.

Gemmers Late Elberta.—Orig. in Sussex, N.J., in the orchard of Conrad Gemmer, by M.A. Blake, New Jersey Agr. Expt. Sta. Introd. in 1941 by Bountiful Ridge Nurseries, Princess Anne, Md. Parentage unknown; discovered in 1932 by Blake. Fruit: flesh yellow, freestone; a late Elberta type, ripening 3 to 7 days after Elberta. Tree: not as hardy as Elberta.

Gene Elberta.—Orig. in Newcastle, Calif., by E.F. Fowler, Fowler Nurseries. Introd. in 1960. Parentage unknown; selected in 1954; tested as 33-10 Elberta. Fruit: large, up to 3 inches in diam.; skin blushed; flesh yellow, good flavor, nonoxidizing type; ripens during 20–25 July at place of origin; resembles Elberta. Tree: vigorous; bears well.

Ghoosen.—See Goosen.

Giant Babcock.—Orig. in Del Rey, Calif., by Paul K. Fujihara. Introd. in 1955. Plant patent 1353; 22 Feb. 1955. Babcock x July Elberta; selected in 1950. Fruit: large; skin predominantly red; flesh ivory yellow, streaked outwardly from the stone with shades of red, freestone; resembles Babcock but larger; matures about 12 to 14 days after Babcock or about 7 to 10 days before Elberta at place of origin. Tree: similar to July Elberta; medium to large; vigorous; round topped; productive; leaf glands reniform; flowers large, more colorful than Babcock.

Giant Elberta.—Orig. in Paonia, Colo., by Gilbert Hice. Introd. in 1947; assigned to Carlton Nurseries, Forest Grove, Ore. Parentage unknown; discovered in 1946. Fruit: larger than Elberta; occasionally irregular in shape; skin highly blushed; flesh yellow, freestone; ships well; ripens 7 days before Elberta; resembles July Elberta, but with more skin color. Tree: vigorous and strong growing; when mature is tall, wide-spreading, open, and well-branched; strong, wide crotch angles; leaf glands mixed in shape; flower medium-small, reddish, resembling Elberta and opening about the same time.

Giant Snowball.—Orig. in Star City, Ind., by F.A. Ogle. Introd. in 1930. Parentage unknown. Fruit: flesh white, freestone pit small; size of J.H. Hale; matures in midseason. Probably no longer propagated.

Giblin.—Orig. in Yuba City, Calif., by Thomas Giblin. Introd. in the early 1920s. Parentage unknown; discovered about 1915. Fruit: flesh yellow; commercial canning clingstone. No longer commercially planted.

Glamar.—Orig. in Osoyoos, B.C., Canada, by John Stuart Leckie. Introd. in 1942. Plant patent 592; 15 June 1943. Rochester mutation; discovered in 1939. Fruit: equal to or larger than Rochester; round, apex pointed; skin smooth, bright yellow, highly blushed, medium pubescence; flesh yellow, firm, very sweet; semi-freestone to freestone; ripens 14 to 20 days before parent; resembles Rochester. Tree: resembles Rochester.

Gleason.—See Early Elberta.

Gleason Elberta.—See Early Elberta.

Glenalton.—Orig. in Concord West, Sydney, New South Wales, Australia, by R. Anderson. Chance seedling; discovered about 1949, named in 1958. Fruit: size medium; skin yellow blushed with red; flesh yellow, quality good; ripens last half of February, 2 weeks after J.H. Hale at Bathurst; a dessert variety. Tree: vigor good; productivity excellent; flower small, red or dark pink; leaf glands reniform; chilling requirement low.

Globehaven.—Orig. in Xenia, Ill., by J.E. Markham, Markham's Nursery. Introd. about 1930. Probable cross of Vivid Globe x South Haven. Fruit: medium to large; round, uniformskin half covered bright red; flesh yellow, very firm, melting, freestone; ripens about 10 days before Elberta; ships well; resembles Halehaven, but ripens earlier. Earlier-ripening than Hale-haven. Tree: vigorous, low to good productivity; flower size medium, showy; leaf glands globose; chilling requirement 850 h at or below 45F.

Glohaven.—Orig. in South Haven, Mich., by Stanley Johnston and J.E. Moulton, Michigan Agr. Expt. Sta. Introd. in 1963. South Haven 20 (J.H. Hale o.p.) x Kalhaven; cross made in 1949; tested as SH 363. Fruit: large; nearly round; skin mostly red over deep yellow, pubescence little, attractive, tough; flesh clear yellow, almost no red in pit cavity, firm, very resistant to browning; freestone, easily pitted by mechanical means; ships unusually well; cans and freezes well; remains on tree when mature; ripens about 2 weeks before Elberta. Tree: medium to large; vigorous; productive; self-fruitful; average, or somewhat above in hardiness; flower medium large, deep pink; leaves with reniform glands.

Gloribloom.—Orig. in Ontario, Calif., by George P. Weldon, Chaffey College. Introd. in 1948. Babcock x [Elberta x (Elberta x Peento)]; selected in 1945. Fruit: skin light yellow, blushed; flesh light yellow, freestone, quality fair; ripens with Weldon. Flower: large, pink, showy.

Glory.—Orig. at Louisiana Agr. Expt. Sta., Calhoun; selected in 1977, tested as L73-1-32, and released by C.E. Johnson in 1992. LaPremiere x (Redglobe x Cardinal). Fruit: nearly round; 50% attractive red blush over golden-yellow ground color with short pubescence; flesh deep yellow with red near pit, melting, firm, freestone. Ripens 5 to 7 days before Elberta. Flowers showy, rose-colored; leaf glands globose. Trees: moderately high resistance to bacterial spot. Chilling requirement 800 h.

G.M.—See Stark Compact Elberta.

Goldcrest.—Orig. in Fresno, Calif., by Agr. Res. Serv., U.S. Dept. of Agriculture. Introd. in 1983. [(Fireglow x Hiley) x Fireglow] x Springtime o.p. Selected in 1976 by David W. Ramming, U.S. Dept. of Agriculture. Fruit: ripens in only 60 days and averages 2 1/2 inches in diam.; round; skin covered by 70% bright red blush, undercolor bright yellow, pubescence light; semi-freestone; flesh yellow, firm but melting, good flavor. Tree: moderately vigorous, moderately productive, self-fertile; blossoms large, showy; leaf glands reniform; chilling requirement not yet determined; blooms with Springcrest at Fresno; ripens before Maycrest and Springcrest; tolerant to Prunus ring spot and prune dwarf virus.

Gold Dust.—Orig. in Merced, Calif., by T.B. Stribling, Jr. Stribling's Nurseries, Merced, Calif. Introd. in 1950. Plant patent 1144; 11 Nov. 1952. July Elberta o.p.; selected in 1947; tested as Stribling S-47-4. Fruit: skin yellow ground color, mottled and streaked in red; flesh yellow, freestone; ripens 30 days earlier than July Elberta, 2 weeks earlier than Redhaven, 2 to 3 days later than Florence; being distinguished from July Elberta by rounder form and legg down; from the Redhaven by rounder form; from Florence by larger size, rounder, less down, yellow instead of white flesh. Tree: large; vigorous; upright; dense; round topped; hardy; productive; regular bearer; leaf glands reniform; flower small, pink.

Golden Amber.—Orig. in Pretoria, South Africa, by E.P. Evans, Hort. Res. Inst. Introd. in 1966. Moreira Jubilee selfed. Selected in 1963. Fruit: size 2 1/2 to 3 inches in diam.; oval like Moreira Jubilee but not pointed and without suture; skin light orange-yellow with attractive red blush varying from 10% to 80% of skin surface; flesh light orange-yellow, with slight red around pit cavity, semi-cling; resembles Golden Jubilee; ripens early mid-November. Tree: medium to large; vigorous; leaf glands reniform; flower nonshowy, flowers profusely, but a high percentage of fruit abort.

Golden Babcock.—Orig. in Del Rey, Calif., by Paul K. Fujihara. Introd. in 1958. Plant patent 1743; 19 Aug. 1958. Giant Babcock x Redhaven; selected in 1953. Fruit: medium to large; skin rough, almost entirely red; flesh yellow, some red around the pit, very firm, flavor sweet with low acid content, freestone; shipping quality good; can remain on tree 5 to 7 days at maturity without becoming soft; ripens 1–8 Aug. at place of origin, or about 1 month after Redhaven, about 2 weeks after Giant Babcock, and about with J.H. Hale.

Golden Beauty.—Orig. in Smith County, Texas, by R.W. Fair, of Tyler, Texas. Introd. in 1933. Fair Beauty mutation; discovered in 1929. Fruit: round-ovate, medium to large; flesh yellow, freestone; resembles Fair Beauty in appearance but round, having a distinct yellow skin color with a red blush; ripens about 3 days later than Fair Beauty, 2 weeks before Elberta. Tree: vigorous, globular in outline; productive, very little fruit thinning needed; flower medium, showy, dark pink; leaf glands reniform.

Golden Blaze.—Orig. in the Victoria Dept. of Agr. Res. Sta., Tatura, Victoria, Australia. Bred by A. Czerkaskyj, selected in 1984, tested as T241, and introduced in 1991 by L.G. Issell and P.H. Jerie. Babygold 7 x Oom Sarel. Fruit: medium large; nearly round with slightly uneven suture and small point; skin 40% to 60% scarlet maroon blush in streaks over golden ground color; medium pubescence; flesh light orange with light red at pit surface, moderately fine texture, firm, nonmelting, clingstone with good separation of flesh and pit when twisted. Ripens 6 days after Redhaven. Flowers showy and medium pink; leaf glands round to reniform. Tree: medium vigor; relatively upright; productive. International rights licensed to Australia Nursery Fruit Improvement Co.

Golden Blush.—Orig. in Lakeside, Calif., by Benjamin Harrison Haley. Introd. in 1938. Plant patent 473; 3 June 1941; assigned to Armstrong Nurseries, Ontario, Calif. Parentage unknown; selected in 1925. Fruit: large; somewhat elongated, ventral suture slightly raised; skin yellow, blushed, thickness medium, tough; flesh yellow, red around pit, firm, Juicy, high flavor, freestone; ripens in midseason, with Elberta; resembles Elberta. Tree: more resistant to delayed foliation in southern California than Elberta; very vigorous; bears very well; petals large, showy; leaf glands reniform.

Golden Charm.—Orig. in the Victoria Dept. of Agr. Res. Sta., Tatura, Victoria, Australia. Bred by A. Czerkaskyj in 1973, selected in 1985, tested as T244, and introduced in 1991 by L.G. Issell and P.H. Jerie. Golden Queen x Babygold 5. Fruit: medium large; oblique with an uneven suture and a moderate point; skin 40% light-maroon blush over a golden ground color. Flesh yellow to orange with slight red at the pit surface, firm, nonmelting, clingstone with good separation of flesh and pit when twisted, sweet flavor. Ripens 30 days after Redhaven. Flowers nonshowy; leaf glands round to reniform. Tree: medium vigor; moderately upright; productive. Chilling requirements similar to Golden Queen. International rights licensed to Australia Nursery Fruit Improvement Co.

Golden Crest (*Supechthree*).—Orig. by J.H. Weinberger, Fresno, Calif. Plant patent 4903; 26 Oct. 1982. Assigned to Superior Farming Co., Bakersfield, Calif. Flavorcrest x Springcrest. Fruit: medium; globose; medium light red over yellow ground color; flesh yellow, firm, melting, semi-freestone, ripens 7 days after Springcrest. Flowers large, showy, and pink; leaf glands globose. Tree: medium size and vigor; productive.

Golden Dawn.—Orig. in Bolton, Mass., by Jonathan Davis of Sterling Junction, Mass. Introd. in 1958. Plant patent 1746; 2 Sept. 1958. Possible Lovell o.p.; discovered in 1953. Fruit: skin yellow, overlaid with red where exposed to the sun; flesh yellow with no red around the pit, very juicy, melting, quality high, freestone; starts to ripen about with Erly-Red-Fre. Tree: leaves with sharply serrate margins; flowers nonshowy.

Golden Early Bind.—Orig. in Downers Grove, Ill., by R.H. Christofferson. Introd. in 1949. Plant patent 970; 1 Aug. 1950; assigned to Stark Bros. Nurseries & Orchards Co., Louisiana, Mo. Crawford o.p.; discovered in 1937. Fruit: 2 3/4 inches in diam.; skin highly blushed red, pubescence heavy; flesh yellow, semi-freestone, quality good; ripens very early, 6 days before Mikado. Tree: size medium; vigor medium; open, upright; productive; leaf glands reniform; flower size medium, pink.

Goldeneast.—Orig. in New Brunswick, N.J., by M.A. Blake, New Jersey Agr. Expt. Sta. Named and introd. in 1937. Elberta x New Jersey 38 E.G.; cross made in 1923; tested as N.J. 87. Fruit: large; oval; skin yellow, deeply blushed, appearance especially attractive; flesh yellow, red streaked, sometimes greenish near pit, firm, melting, low acid, high quality; freestone; ripens in midseason, in mid-August in New Jersey, about 2 weeks before Elberta and 10 to 14 days after Jubilee; hangs well on tree. Tree: vigorous, dormant fruit buds medium hardy, productive; flower large, showy; leaf glands reniform; chilling requirement 1050 h at or below 45F.

Golden Elberta.—See Early Elberta.

Golden Elberta Cling.—Orig. in Fort Smith, Ark., by Will R. Gaunaway. Introd. in 1925 by Stark Bros. Nurseries & Orchards Co., Louisiana, Mo. Parentage unknown; discovered in 1911. Fruit: medium to large; ovate, flattened, apex tipped; skin brightly blushed, pubescence light; flesh yellow, firm, juicy, fine texture, clingstone; ripens in midseason, 5 days before Elberta; resembles Early Elberta. Tree: globe shaped; flower large, showy; leaf glands reniform; chilling requirement 950 h at or below 45F.

Golden Flame.—Orig. in Fowler, Calif., by H.E. and R.K. Hiraoka. Introd. in 1951. Howard Fisher mutation; discovered in 1950. Fruit: medium to large; round; skin yellow, highly blushed, very little pubescence; flesh yellow, semi-clingstone, firm; ripens in early June, with Florence.

Golden Globe.—Orig. at New Brunswick, N.J., by M.A. Blake, New Jersey Agr. Expt. Sta. Introd. in 1937. J.H. Hale x Marigold; cross made in 1923; selected in 1933; tested as N.J. 73. Fruit: very large; round; skin blushed dark red with some streaking; flesh yellow, sometimes red streaked, red at pit, firm, quality very good, freestone; ripens somewhat later than Jubilee, 2 weeks before Elberta. Tree: sometimes low in fruitfulness, lacks hardiness, very susceptible to bacteriosis; flower medium, nonshowy; leaf glands globose.

Golden Glory.—Orig. in Le Grand, Calif., by F.W. Anderson, Merced, Calif. Introd. in 1964. Plant patent 2418; 7 July 1964; by F.W. Anderson, assignor to L.E. Cooke Co., San Gabriel, Calif. (Late Le Grand nectarine x Flory Dwarf) F_2; selected in 1960. Fruit: large, averaging 2 3/8 inches in diam.; globose to slightly oblate, uniform, symmetrical, suture distinct; skin thickness and texture medium, yellow, partially mottled orange-red, pubescence moderate; flesh yellow, red next to stone, soft, melting, subacid to mild, eating quality fair to good; ripens evenly; stone size medium, free, seed bitter; keeping quality medium; primarily for home use and local market; ripens during last week of August; resembles Golden Prolific nectarine, being a sibling of this variety. Tree: genetically dwarf, averaging 6 to 8 feet tall; vigorous, spreading; productive, bears regularly; trunk stocky; leaf large, glands reniform; flower large, blooms in midseason, anthers red; self-fruitful; an attractive ornamental.

Golden Honey.—Orig. in Tahlequah, Okla., by John Willescraft. Introd. in the early 1940s by Ben Davis, Ozark Nursery Co., Tahlequah. Parentage unknown. Seed planted in 1910. Fruit: round with distinct suture; flesh yellow, firm, quality good, clingstone; ripens 10 days after Elberta. Tree: vigorous; upright, similar to Elberta; leaves have more blue sheen than Elberta; flowers medium size.

Golden Jubilee.—Orig. in New Brunswick, N.J., by M.A. Blake, New Jersey Agr. Expt. Sta. Introd. in 1926. (Elberta x Greensboro) o.p.; the o.p. seed was harvested in 1920. Fruit: medium to large; oval to oblong, flattened; skin mottled bright red over one-third of surface, pubescence light; flesh yellow, red at pit, firm, melting, texture coarse, freestone; very good quality, average for canning and freezing; ripens 3 weeks before Elberta, first along suture, drops early from tree. Tree: sets heavily but is self-thinning, flower nonshowy, small; leaf glands reniform.

Golden Lady.—Orig. by G. Merrill, Red Bluff, Calif. Plant patent 4295; 29 Aug. 1978. Parentage unknown. Fruit: medium to large; firm; nearly globose; red blush over yellow; flesh firm, melting, yellow, semi-freestone. Ripens with Merrill Gem Free; few split pits. Flowers pink, nonshowy; leaf glands reniform. Tree: medium vigor; very productive.

Golden Logan.—Orig. in Cedar Ridge, Nevada County, Calif., by Mabel A. Logan. Introd. in 1965. Plant patent 2807; 23 Apr. 1968. Parentage unknown, discovered in 1960. Fruit: size medium; skin and flesh yellow; freestone; ripens late (early September); flavor and eating quality very good. Tree: vigor medium; productive; globose leaf glands; tends to not require heavy thinning.

Golden Masterpiece.—Orig. in Eikenhof, near Johannesburg, South Africa, by J.A. Herholdt, Herholdt's Nurseries, Johannesburg. Introd. in 1967. Kakama x Amsterdam nectarine. Cross made in 1960, tested as H1779/60. Fruit: large; roundish; skin orange-yellow, with red cheek; flesh yellow, red next to stone, fine-grained, sweet, aromatic; ripens 1 week later than Elberta; home garden and commercial variety. Tree: large, upright, well-formed; vigor moderate; hardy; productive and regular bearer; tolerant to mildew; flower resembles Elberta.

Golden Monarch.—Orig. in Wenatchee, Wash., by Wilfred Pratt. Introd. in 1971. Plant patent applied for, to be assigned to Van Well Nursery, Wenatchee. Redhaven mutation, discovered in 1964. Fruit: large, round; skin characteristics similar to Redhaven; flesh very firm, yellow, freestone; good shipping quality; ripens 10 days before Redhaven. Tree: size medium; vigorous and productive.

Golden Pineapple.—Orig. in Hollister, Calif., by Raymond Cheek. Introd. in 1957. Chance seedling; discovered in 1954. Fruit: flesh yellow; flavor very good, resembling pineapple; clingstone.

Golden Promise.—Orig. in Chico, Calif., by G.C. Banta. Introd. in 1954. Chance seedling discovered about 1949. Fruit: flesh yellow flavor fine, clingstone; ripens end of July. Tree: bears well.

Golden Queen.—Orig. in Tauranga, New Zealand. Introd. in 1908; purchased, named and propagated by David Hay, David Hay & Sons Nurserymen; introd. into Australia shortly thereafter and into the U.S. in 1923 under P.I. 68353 by U.S. Plant Introduction Sta., Chico, Calif. Fruit: medium large; round-oblate; suture medium depth, halves almost equal; apex variable, may or may not terminate with a nipple; skin greenish-yellow, changing to deep yellow, very slight red wash in some localities, appearance good, pubescence scant; flesh deep yellow, no red at pit, firm, juicy, good flavor, quality excellent; pit relatively small, clingstone; season first 2 weeks in March in Southern Hemisphere; very suitable for canning. Tree: good bearer; must be judiciously thinned; a very adaptable variety.

Goldenred.—Orig. in Stillwater, Okla., by Herman A. Hinrichs, Oklahoma Agr. Expt. Sta. Released in 1958. Early Elberta x Halehaven; cross made in 1946; selected in 1951. Fruit: skin smooth, pubescence short, color intermediate between Elberta and Redhaven; resembles Halehaven in shape; flesh yellow, good quality, freestone. Tree: very productive.

Golden State.—Orig. in Riverside, Calif., by J.W. Lesley, California Citrus Expt. Sta. Introd. in 1942. Paragon x (Elberta x a peento) F_2; selected in 1933. Fruit: large; skin lemon-yellow with a marbled red blush on one side, pubescence dense, medium thick, appearance not too attractive; slightly oblong, halves nearly equal; flesh light yellow, light pink next to stone, texture coarse, flavor fair, slightly subacid, firm, freestone; ripens first week in August, a few days before Elberta; used fresh and for home canning. Tree: vigorous, upright; leaf glands reniform, large, not numerous; irregular producer; subject to preharvest drop; short winter-chilling requirement; flower very small, petals pink, self-fruitful. Obsolete.

Golden Sun.—Orig. in the Victoria Dept. of Agr. Res. Sta., Tatura, Victoria, Australia. Bred by A. Czerkaskyj and selected in 1985, tested as T242, and introduced in 1991 by P.H. Jerie and L.G. Issell. Fruit: medium; round; skin slight red blush on yellow-orange ground color; flesh yellow to orange, strong sweet flavor, firm, nonmelting, clingstone. Ripens 11 days after Redhaven. Flowers showy. International rights licensed to Australia Nursery Fruit Improvement Co.

Golden Supreme.—Orig. in Modesto, Calif., by C. Floyd Zaiger. Introd. in 1968. Plant patent pending, to be assigned to Dave Wilson Nursery, Modesto. Sun Grand nectarine x Merrill Gem F2. Fruit: skin highly colored; flesh very firm; shipping qualities good; ripens slightly earlier than Early Gold Dust. Tree: very productive.

Golden Treasure.—Orig. in Wasco, Calif., by David L. Armstrong. Introd. in 1968. Parentage unknown. Plant patent 2930; 7 Oct. 1969; assigned to Armstrong Nurseries, Ontario, Calif. Fruit: about 2 1/2 inches in diam.; globose; skin yellow ground color overlaid wlth red blush, pubescence moderate; flesh yellow, firm, melting, fine texture, aroma pronounced, freestone; ripens 6 to 8 weeks after Bonanza; resembles a lightly colored July Elberta. Tree: dwarf; leaf glands weakly reniform; flower pale to light plnk, showy, 1 1/2 inches in diam.

Goldfinch.—Orig. in New Brunswick, N.J., by M.A. Blake, New Jersey Agr. Expt. Sta. Introd. in 1925. Slappey x Admiral Dewey; cross made in 1916. Fruit: medium to large; ovate, apex pointed; skin mottled bright red on yellow ground; flesh deep yellow, melting, firm, streaked red at pit, tannin low, quality very good, semi-freestone; rlpens 2 weeks before Elberta. Tree: vigorous, average productivity; flower small, nonshowy; leaf glands reniform.

Goldgem.—Orig. in New Brunswick, N.J., by M.A. Blake, New Jersey Agr. Expt. Sta. Introd. in 1957. J.H. Hale x Cumberland; selected in 1930; tested as N.J. 197. Fruit: large; skin color slightly dull; flesh yellow, only moderately firm, freestone; ripens with Golden Jubilee, between Jerseyland and Triogem. Tree: probably as hardy as Elberta.

Gold Hale.—Orig. in Farmersville, Calif., by Harold D. Dougherty. Introd. in 1956. Plant patent 1696; 15 Apr. 1958. Parentage unknown; selected in 1951. Fruit: large, averaglng 2 1/2 inches in diam.; uniformly round; skin deep red overlaylng yellow, thick, tenacious to flesh; flesh yellow, mottled orange or red, red at pit cavity, firm, flavor mild, quallty good, freestone; shipping quality very good; matures in early July at place of origin. Tree: similar to Halehaven and Johnson in growth, spread and vigor; hardy; flower large; leaf glands globose; regular bearer.

Goldilocks.—Orig. at Arkansas Agr. Expt. Sta., Clarksville, James N. Moore, Roy C. Rom, Stanley A. Brown, and William Sistrunk. Introd. in 1984. NJ554367 x G17-5E; cross made by Catherine Bailey and L.F. Hough of New Jersey Agr. Expt. Sta.; selected in 1966; tested as A-15. Complex pedigree includes J.H. Hale, Slappey, Admiral Dewey, Orange Cling Seedling, Alameda, and St. Helena. Fruit: about 6 cm when well-thinned; round, without prominent suture or tip; skin yellow with light splash of red; flesh yellow-orange with; red pigment, very firm, nonmelting, clingstone. Ripens about 3 days before Babygold 5. Tree: medium vigor; upright-spreading with good crotch angles. Considerable tolerance to bacterial spot. Flowers nonshowy; self-fertile. Leaf glands reniform. Chilling requirement about 750 h.

Goldprince.—Orig. at Byron, Ga., by W.R. Okie, U.S. Dept. of Agriculture S.E. Fruit & Tree Nut Res. Lab. Introd. in 1989. Loring x [(Fairhaven x FV89-14) x (FV89-14 x Duke of Georgia)]; FV89-14 is an F_2 of (Hiley x Fireglow) x Fireglow. Cross made in 1966 by V.E. Prince; selected in 1970; tested as FV7-1240 and BY7-1240. Fruit: medium; skin attractive red, light pubescence; flesh melting, clingstone. Ripens with Junegold, 3 weeks before Redhaven. Better shape and color and few split pits than Junegold, which it replaces . Flowers showy, self-fertile; leaf glands reniform; moderately resistant to bacterial spot. Chilling requirement about 650 h.

Goldray.—Orig. in Lancaster, S.C., by J. Roy Cunningham. Introd. in 1950. Bud mutation of Golden Jubilee; selected in 1939. Fruit: identical in color, size and quality wlth Golden Jubilee; semi-freestone; used for commercial canning; ripens 10 to 14 days earlier than Golden Jubilee. Tree: hardy; thrifty; growth similar to Golden Jubilee.

Gold Rush (*Merrill Gold Rush*).—Orig. in Red Bluff, Calif., by Grant Merrill. Introd. in 1950. J.H. Hale o.p.; selected in 1942. Fruit: flesh yellow, freestone; ripens 3 weeks earlier than J.H. Hale, which it resembles. Tree: very vigorous; self-fertile.

Gomes.—Orig. in Modesto, Calif., by Felix Gomes. Introd. in 1936. Parentage unknown; selected in 1935. Fruit: skin yellow; flesh yellow, firm; ripens in late Phillips Cling season; commercial canning clingstone; resembles Stuart.

Goodcheer.—Orig. in New Brunswick, N.J., by M.A. Blake, New Jersey Agr. Expt. Sta. Introd. in 1947. (J.H. Hale x Eclipse) x Laterose; tested as N.J. 152. Fruit: small; round, apex tipped; skin blushed dark red; flesh yellow, melting, firm, fibrous, dry, red at pit, color extending into flesh, fair quality; small pit, free; ripens about 1 week after Elberta. Tree: hardier than Elberta; flower nonshowy, small; leaf glands reniform; chilling requirement 750 h at or below 45F.

Goosen (*Ghoosen*).—Orig. in Prince Alfred's Hamlet, Cape Town, South Africa, by a Mr. Goosen. Introd. in 1950 in the U.S. as P.I. 220045 and in 1956 as P.I. 234193. Chance seedling discovered in the early 1940s; propagation rights granted to a Mr. G. Malherbe. Fruit: large; flesh pale yellow, texture fine-grained, quality good; canning clingstone; early ripening. Tree: vigorous; not a consistent or heavy cropper in all areas.

Granite Supreme.—Orig. in Applethorpe, Queensland, Australia, by C.W. Winks, A.M. Hibberd, and B.L. Topp, Granite Belt Hort. Res. Sta. Introd. in 1988. Selfed seedling of a Goldmine nectarine x Early Elberta selection. Seed collected in 1974 by A.M. Hibberd; selected in 1980; tested as N1T6. Fruit: medium; skin 75% red blush; flesh medium yellow with red flecks near pit; exceptionally firm, melting, clingstone; flavor well-balanced sub-acid; texture medium. Ripens days before Redhaven. Flower nonshowy. Tree: vigorous, highly productive; chilling requirement about 800 chill units. Moderate susceptible to bacterial leaf spot.

Grebing.—Orig. in Perryville County, Mo., on the farm of a Mr. Grebing. Introd. about 1925, apparently by H.J. Webster and Sons Nursery Co., Afton, Mo. Parentage unknown. Fruit: medium to large; nearly round; skin white, blushed; flesh white, melting, juicy; stone small, free; ripens in July. Tree: productive; hardy.

Gurney's Dakota (*Miss South Dakota*).—Orig. in Yankton, S.D., by A.C. Ellerman, Gurney Seed & Nursery Co. Introd. about 1928. Elberta o.p.; discovered in 1921. Fruit: large; flesh yellow, firm; freestone; resembles Elberta. Now obsolete.

Gypsy.—Orig. ln Red Bluff, Calif., by Grant Merrill, Exeter, Calif. Introd. ln 1960. Seedling of Splendor; discovered in 1956; tested as seedling 48-89. Fruit: large; skin highly colored; flesh yellow, firm, freestone; resembles Splendor; ripens with Franciscan; tends to become soft on one side and has a tip on the end, tip does not soften; quality good. Tree: size medium; productive; leaf glands reniform; flower size medium.

Haig Arakelian, Sr.—Orig. near Modesto, Calif., by Haig Arakelian, Sr. Unknown parentage. Plant patent 6230; 26 July 1988; assigned to Burchell Nursery. Fruit: medium to large, typically about 65 mm; globose, symmetrical; skin light red over 15% of surface on light yellow ground color; flesh bright yellow with no red around pit; nonmelting; flavor good, well-balanced; clingstone. Ripens week of July at Modesto. Tree: vigorous; upright-spreading; very productive; regular production; hardy under typical San Joaquin Valley conditions. Leaf glands globose, small, light green, darken with age. Flowers small, nonshowy, abundant. Released for commercial canning.

Halate.—Orig. in Louisiana, Mo., by Stark Bros. Nurseries & Orchards Co. Introd. in 1937. Parentage unknown; discovered in 1924 in a test orchard of the above nursery. Fruit: large; flesh yellow, freestone; ripens 10 to 14 days after Elberta; resembles J.H. Hale. Tree: vigorous; very productive.

Hal-Berta Giant (*Hale Berta*).—Orig. in Xenia, Clay County, Ill., by James E. Markham, Markham's Nursery. Introd. in 1932. Plant patent 7; 16 Feb. 1932; assigned to Stark Bros. Nurseries & Orchards Co., Louisiana, Mo. Parentage unknown; selected in 1928. Fruit: very large; variable, round, apex pointed; skin thin, mottled and blushed red over yellow, pubescence moderate; flesh golden yellow, red at pit, firm, melting, freestone; ripens a few days after Elberta; quality very good. Tree: vigorous; low in productivity; flower small, nonshowy, pollen-sterile; chilling requirement 850 h at or below 45F.

Hale Berta.—See Hal-Berta Giant.

Halegold.—Orig. in Commerce, Ga., by Early Stark. Introd. about 1930. Parentage unknown. Fruit: large; round; skin bright red blushed and mottled on attractive yellow ground; flesh yellow, melting, firm, red at pit, freestone; ripens a few days after Elberta. Tree: vigorous and productive; flower small, nonshowy, leaf glands reniform; chilling requirement 950 h at or below 45F.

Hale Harrison Brillant.—Orig. in Crozet, Albemarle County, Va., by Oscar F. Jones. Introd. in 1949. Plant patent 814; 14 Dec. 1948; by Oscar F. Jones, Crozet, Va., and Wilbur C. Jones, Hagerstown, Md.; assigned to Harrison Brothers Nurseries, Berlin, Md. Possibly Elberta x Belle; discovered in 1936. Fruit: large; oblong, apex tipped, suture striped russet raised and slightly prominent; skin bright red, mottled only one-third of surface, pubescent light; flesh yellow, red at pit, firm, melting, quality very good, freestone; ripens with Elberta; drops at maturity. Tree: vigorous; productive; flower small, nonshowy; leaf glands reniform; susceptible to bacteriosis.

Halehaven.—Orig. in South Haven, Mich., by Stanley Johnston, Michigan Agr. Expt. Sta. Introd. in 1932. J.H. Hale x South Haven; cross made in 1924. Fruit: medium to large; round, uniform; skin dark red over most of surface; flesh yellow, red at pit, firm, melting, quality excellent, freestone; ripens 2 weeks before Elberta; resembles South Haven, but better colored. Tree: vigorous and productive; flower medium, nonshowy; leaf glands globose; susceptible to brown rot.

Halford (*Halford 2*).—Orig. in Modesto, Calif., by John Halford. Introd. in 1921. Parentage unknown; discovered in 1919. Fruit: large; flesh yellow, nonmelting, clingstone; canning quality excellent. Tree: produces heavy crops.

Halford 1.—See McKnight.

Halford 2.—See Halford.

Halford 3.—Orig. in Modesto, Calif., by John Halford. Introd. in 1921. Parentage unknown; discovered about 1919. Fruit: skin yellow, slight blush; flesh yellow, firm, commerical canning clingstone; ripens about with Halford. Tree: leaf eglandular. Now obsolete.

Halloween I (*Merrill Halloween I*).—Orig. in Red Bluff, Calif., by Grant Merrill, Grant Merrill Orchards. Introd. in 1954. Plant patent 1473; 1 May 1956. Rodeo o.p.; discovered in 1951; selected in 1953. Fruit: large; flesh yellow, firm, freestone; ripens extremely late, about 70 days after Elberta or about 10 Oct. at place of origin; resembles Merrill Fiesta. Tree: size medium; vigor medium; productive; leaf glands reniform; flower size medium, pink.

Halloween II (*Merrill Halloween II*).—Orig. in Red Bluff, Calif., by Grant Merrill, Grant Merrill Orchards. Introd. in 1954. Plant patent 1472; 1 May 1956. (Candoka x Rio Oso Gem) o.p.; discovered in 1952; selected in 1953. Fruit: flesh yellow, firm, freestone; ripens very late, 80 days after Elberta or about 20 Oct. at place of origin; resembles Merrill Fiesta. Tree: size medium; vigor medium; productive; leaf glands reniform; flower size medium, pink with salmon tinge.

Halo.—Orig. in Clarkston, Wash., by H. Lynn Tuttle, Lynn Tuttle Nursery. Introd. in 1942. Tuscan o.p.; selected about 1936. Fruit: fairly large; skin highly blushed 1 or 2 weeks before ripening; flavor excellent; freestone; resembles Tuscan. Tree: resistant to spring frosts; bears early and heavily.

Hamlet.—Orig. in North Carolina by F.E. Correll and C.N. Clayton, North Carolina State Univ. Pekin x Candor; selected in 1966; tested as NCA 2684. Fruit: medium; round; skin color is medium yellow with 70% bright red blush; flesh yellow, firm, melting, separates from pit only when fruit is soft ripe; highly resistant to bacterial spot; thinning required for proper sizing of fruit; ripens 4 weeks before Redhaven. Tree: medium vigor; well-formed; productive. Flowers nonshowy, self-fertile, blooms with Redhaven.

Hamlet.—Orig. at Sandhills Res. Sta., Jackson Springs, N.C., by F.E. Correll and C.N. Clayton, North Carolina State Univ., Raleigh. Introd. in 1975. Pekin x Candor; tested as NCA 2684. Fruit: medium; round; flesh firm, melting, texture and flavor good, medium yellow with little or no red; clingstone; skin medium yellow with 70% bright red blush ripens 3 to 5 days later than Whynot (23 May–10 June). Tree: productive; flower nonshowy, self-fertile; chilling requirement appears similar to that of Redhaven; highly resistant to bacterial spot (*Xanthomonas pruni*).

Harbelle.—Orig. in Harrow, Ont., Canada, by G.M. Weaver, Canada Dept. of Agr., Res. Sta. Introd. in 1968. Sunhaven selfed Cross made in 1961, selected in 1964, tested as Harrow 429. Fruit: 2 3/4 inches in diam.; round; skin yellow ground color covered with 60% bright red blush, pubescence short; flesh yellow, melting, firm, quality good, freestone; ripens early iust ahead of Sunhaven, about 4 Aug. at Harrow. Tree: small, round; vigor moderate; hardy; productivity moderate, but consistent; tolerant to bacterial spot; flower nonshowy, salmon; leaf glands globose; recommended for Northeast and Midwest peach growing regions of Ontario.

Harbinger.—Orig. in Harrow, Ont., Canada, by G.M. Weaver and R.E.C. Layne, Canada Dept. Agr., Res. Sta. and L. F. Hough and Catherine H. Bailey, Rutgers Univ., Agr. Expt. Sta., New Brunswick, N.J. Introd. in 1971. Cherryred x NJ 560519 Jerseyland x Mayflower; cross made in 1962, selected in

1966, tested as Harrow 0640 (H-0640). Fruit: 21/4 inch in diam., oval to round; skin attractive, 80% to 90% red blush on yellow ground color, pubescence scant; flesh yellow, trace of red some seasons, firm, melting, quality good; few split pits, clingstone; ripens 19 July at Harrow, 1 week before Earlired, 1 month before Redhaven. Tree: large, more upright than spreading; vigorous; hardier than Redhaven; productive; tolerant to bacterial spot and brown rot; flower large, showy, pink; leaf glands globose; recommended for southwestern Ontario and midwestern United States.

Harbrite.—Orig. in Harrow, Ont., Canada, by Gerald M. Weaver, Canada Dept. of Agr., Res. Sta. Introd. in 1969. Redskin x Sunhaven. Cross made in 1961, selected in 1964, tested as Harrow 430. Fruit: 2 1/2 inches in diam.; ovate; skin covered with 80% brilliant red blush over yellow ground cover, attractive; flesh yellow, some red at pit, firm, melting, quality good, freestone; ripens 18 Aug., 4 weeks before Elberta. Tree: size medium; upright-spreading; vigor medium; hardiness is greater than Redhaven; exceptionally productive; highly tolerant to bacterial spot *Xanthomonas pruni*; flower nonshowy, salmon; leaf glands globose; considerable dwarfing and early ripening obtained with propagation to Rutgers Red Leaf rootstock. Recommended for southwestern Ontario.

Harcrest.—Orig. in Harrow, Ont., by R.E.C. Layne, Res. Sta., Agr. Canada. Introd. in 1983. Redskin x H 4219 (NJ 155 x Madison); selected in 1975; tested as HW 225. Fruit: resembles Cresthaven, ripening 3 days later; flesh firm; freestone; for fresh market, home canning, and freezing. Tree: bud and wood hardy; tolerant to perennial canker, bacterial spot, brown rot .

Hardee.—Orig. in Sundusky County, Ohio, by Donald S. Byers, Clyde, Ohio. Introd. in 1936. Plant patent 120; 15 Jan. 1935. Possibly Elberta o.p; discovered in 1925. Fruit: size of Elberta; elongated, apex pointed, ventral suture very pronounced; skin yellow, blushed, thick; flesh yellow, pronounced flavor, freestone; ripens with or somewhat after Elberta; in other characteristics resembles Elberta. Tree: growth intermediate between J.H. Hale and Elberta; hardier than Elberta; leaf glands reniform.

Hardy-Berta.—Orig. in Mount Vernon, Jefferson County, Ill., by Marsh Harpole. Introd. in 1934. Plant patent 271; 15 Feb. 1938; assigned to Stark Bros. Nurseries & Orchards Co., Louisiana, Mo. Parentage unknown; discovered in 1932. Fruit: Elberta type; large, averaging 3 inches in diam.; slightly oblong, apex short and pointed; skin almost completely red; flesh yellow, some red around pit, firm, moderately juicy, some acidity, eating quality good with only a slight suggestion of Elberta bitterness; pit extremely long, very large, freestone; shipping quality average; ripens late, 7 to 10 days after Elberta. Tree: very hardy; vigorous.

Harken.—Orig. in Harrow, Ont., Canada, by G.M. Weaver and R.E.C. Layne, Canada Dept. of Agr., Res. Sta. Introd. in 1970 by the above station and the Univ. of Kentucky, Princeton, Ky. Redskin x Sunhaven. Cross made in 1961, selected in 1965, tested as H 2066. Fruit: 2 1/2 inches in diam., oval to round; skin covered with 80% bright red on yellow ground color, attractive, very smooth; flesh bright yellow, nonbrowning, firm, freestone, quality good; ripens 3 days after Redhaven, hangs well. Tree: size medium, spreading; vigor moderate; very hardy, better than Redhaven; productive; tolerant to bacterial spot and brown rot; flower nonshowy, medium pink; leaf glands globose.

Harland.—Orig. in Harrow, Ont., Canada, by G.M. Weaver, Agr. Canada, Res. Sta. Introd. in 1979. [(J.H. Hale x Vedette) x (V37016l x Earlired) o.p.] Cross made in 1961; selected in 1967; tested as H614 and H2049. Evaluated in Canada, United States, and New Zealand. Fruit: medium small (larger in New Zealand than Ontario), round to oblate; skin attractively colored with 70% red-striped blush, medium tough, pubescence light and short; subject to split pits some seasons; flesh bright yellow with trace of red pigment near pit, flavor good; ripens mid-July at Harrow. Tree: medium vigor, open, spreading; hardiness intermediate, between Loring and Redhaven; productivity moderate; resistant to bacterial spot and brown rot, susceptible to perennial canker.

Harmony (*Merricle 0-99*).—Orig. in Red Bluff, Callf., by Grant Merrill. Introd. in 1960. Plant patent 2152, 12 June 1962. Fortyniner x Gemfree; selected in 1959. Fruit: averages 2 1/2 inches or more in diam.; round; skin medium thick, smooth, dull orange with dark red cheek; flesh yellow, toward apical end to bright orange-red at stem end, juicy, very firm when shipping ripe, flavor vinous, aroma distinct, freestone; keeping and shipping qualities good; matures about 15 July at place of origin, or about 2 weeks after Gemfree and before Redhaven; resembles Gemfree. Tree: medium small; productive; leaves with reniform glands; flowers bloom early, about with July Elberta, small, pink.

Harriet.—Orig. in South Portland, Maine, by a Miss Coe. Introd. before 1945. Chance seedling; P.I. 151655. Tree: very hardy.

Harris.—See Harris Elberta.

Harris Elberta (*Harris*).—Orig. in Parlier, Calif., by David B. Harrls, Jr. Introd. in 1952. July Elberta mutation; discovered in 1951. Fruit: skin with higher blush and less pubescence than parent; flesh yellow, freestone; matures 5 days before parent. Tree: moderately vigorous; productive.

Harrison.—Orig. in Blacksburg, Va., by G.D. Oberle, Virginia Agr. Expt. Sta., Virginia Polytech. Inst. & State Univ. Introd. in 1972. Veteran x Jerseyland. Cross made in 1951, first selected in 1954, tested as VPI 66. Fruit: size medium or above; round ovate; skin two-thirds bright medium; red over yellow ground color, pubescence light; flesh orange yellow, texture smooth, freestone; ripens early, 3 1/2 weeks before Elberta. Tree: size average or above; spreading; vigor and hardiness average or above; productivity above average; flower small, nonshowy, self-fertile; leaf glands reniform.

Harrow Beauty.—Orig. in Harrow, Ont,. by R.E.C. Layne, Res. Sta., Agr. Canada. Introd. in 1983. Cresthaven x Harken: selected in 1976: tested as HW 231. Fruit: resembles Harken; skin well-colored, attractive; flesh very firm; freestone. Tree: more cold hardy than Loring or Canadian Harmony, ripening same season; tolerant to perennial canker, bacterial spot, brown rot.

Harrow Blood.—Orig. in Harrow, Ont., Canada, by G.M. Weaver, Canada Dept. of Agr., Res. Sta. Introd. in 1967. Parentage unknown. Fruit: small, 1 3/4 inches in diam.; round; skin has 75% red blush on heavy green ground cover; flesh white, with blood red color next to pit, freestone; ripens 1 Oct. Tree: small and compact, upright; vigor moderate; hardy; productive; tolerant to peach tree borer and bacterial spot; flower pink, large, showy; leaf glands small, reniform; recommended as a dwarfing rootstock for peach in areas where root-knot nematode is not a problem; compatible with peach.

Harrow Diamond.—Orig. at Harrow, Ont., Canada, by R.E.C. Layne, Agriculture Canada Res. Sta. Introd. in 1984. Redskin x Harbinger; tested as HW 213. Fruit: medium, to 60 mm when properly thinned; skin color solid, bright red blush over 60% to 70% of surface on bright yellow ground color; pubescence short and sparse; flesh yellow, nonbrowning, melting, medium firm, moderately sweet and juicy; freestone when fully ripe; free of split pits. Fruit ripens 1 day before Candor and Earlired. Pit deeply sculptured; medium; plump; without wings. Flowers showy, pink mid-season. Tree: vigor medium; productive; open-spreading, with medium to wide crotch angles; tolerant to Leucostoma canker; resistant to bacterial spot and brown rot. Wood and fruit buds are winter hardy like Redhaven; full crops were produced after exposure to –26 and –28C.

Harson.—Orig. in Harrow, Ont., Canada, by Richard E.C. Layne and G.M. Weaver. Harrow Res. Sta., Agr. Can. Introd. in 1982. Redskin x Sunhaven. Selected in 1965; tested as H781. Fruit: resembles Redhaven; size uniform; flesh yellow, melting; ripens uniformly 1 day before Redhaven; good quality fresh and processed. Tree: hardiness similar to that of Redhaven; good resistance to bacterial spot and brown rot; moderate tolerance to perennial canker.

Harvester.—Orig. in Calhoun, La., by John C. Taylor and P.L. Hawthorne, La. Agr. Expt. Sta., Louisiana State Univ. Introd. in 1973. Redskin x Southern Glow, cross made in 1961, selected in 1964, tested as L-61-242. Fruit: medium to large; round; skin yellow, washed red over 50% to 75%, pubescence short, attractive; flesh yellow with slight red near pit, firm, no split pits observed, freestone, quality good; ripens 3 weeks before Elberta in north Louisiana, with Keystone; for fresh market and canning. Tree: large; upright; vigorous; very productive; flower dark rose, ample viable pollen; leaf glands globose, good resistance to bacteriosis (*Xanthomonas pruni*), moderately resistant to brown rot (*Schlerotinia fructicola*); chilling requirement 750 h below 45F; recommended for north Louisiana and areas of comparable growing conditions.

Havis.—Orig. in Beltsville, Md., by H.W. Fogle, Agr. Res. Serv., U.S. Dept. of Agriculture. Introd. in 1977. Dixiland x Sentinel; cross made in 1962 by A. Leon Havis; tested as Beltsville 63221. Fruit: large, round; 80% to 90% bright red blush; short and sparce pubescence; yellow fleshed; redness at pit; freestone; moderately firm; ripens 5 days after Redskin. Flower: nonshowy and self-fertile. Tree: average size and vigor; productive; leaf glands reniform; good bud hardiness; tolerant to bacterial spot (*Xanthomonas pruni*), moderately resistant to brown rot (*Monolinia fructicola*). Recommended for mid-Atlantic and northeastern states.

Hawaiian.—Orig. in Fort Valley, Ga., by U.S. Dept. of Agriculture. Not introd.; included because it is one parent of Flordawon. Hawaii seedling from S. China o.p.; selected about 1945. Fruit: small; flesh white; worthless. Tree: low-chilling requirement to break rest period of its buds; flower large, showy. Used as a parent in breedlng low-chilling varieties.

Hawkeye.—Orig. in Shenandoah, Iowa, by Henry Field, Henry Field Seed and Nursery Co. Introd. in 1937. Parentage unknown; selected about 1930. Fruit: size similar to Elberta; skin pale yellow with a faint blush; flesh white; freestone. Now obsolete.

Hawkins.—Orig. in Altoona, Fla., by Edna Hawkins. Introd. in 1950. Parentage unknown; selected in 1944. Fruit: small; skin yellow, blushed; flesh yellow, good flavor, freestone; ripens late. Tree: long-lived.

Hawthorne.—Orig. at Calhoun, La., by C.E. Johnson, W.A. Young, J.E. Boudreaux, W.J. Bourgeois, F.J. Peterson, and P.W. Wilson, Louisiana Agr. Expt. Sta. Introd. in 1990. Southern Glow o.p.3; selected in 1976; tested as L73-A10-10. Fruit: large, 70 mm; skin 90% red over medium yellow ground; flesh yellow, with some red near pit, firm, good texture, freestone; nearly round, with nonprominent suture. Ripens in late June at Calhoun, midway between Harvester and LaFeliciana. Flowers large, showy, self-fertile. Leaf glands reniform. Tree: vigorous, productive; no bacterial spot has been observed on fruit and relatively little on leaves. Chilling requirement 500 to 600 h.

Heap Big Indian.—Orig. in Hulbert, Cherokee County, Okla., by J.L. Sullivan. Introd. in 1950 by Ben Davis, Ozark Nursery, Tahlequah, Okla. Parentage unknown; seed planted in 1919. Fruit: flesh firm, red, but not as red as Indian Cling; resembles Indian Cling but larger; ripens with Elberta or about 1 Aug. in Okla. Tree: medium size, spreading; sets heavy crop; requires thinning of fruit; leaves have reddish veins.

Herb Hale.—Orig. in Yakima, Wash., by Herbert W. Donahey. Introd. in 1944. Plant patent 588; 8 June 1943; assigned to Columbia & Okanogan Nursery Co., Wenatchee, Wash. Chance seedling. Fruit: large, up to 3 1/4 inches in diam.; round, suture deep; skin bright red over half of surface, thick, tough, pubescence medium; flesh orange-yellow, some red next to pit, juicy, firm, melting, texture fine, acidity high, freestone, quality good; ripens 1 1/2 weeks before Elberta; resembles J.H. Hale and Halehaven. Tree: vigorous; average productivity; flower medium, showy; leaf glands reniform; chilling requirement 950 h.

Herholdt's Summer Pride.—Orig. in Eikenhof, near Johannesburg, South Africa, by J.A. Herholdt, Herholdt's Nurseries, Johannesburg. Introd. in 1968. Sheasby nectarine x Jubilee. Cross made in 1962, tested as H2708. Fruit: large; elongated; skin white with red on side exposed to sun; flesh white, with red next to stone, sweet, flavor excellent, freestone; ripens 2 weeks before Elberta. Tree: medium large, well-branched; vigor moderate; hardy; productive and regular bearer; tolerant to mildew; flower resembles Elberta.

Hermosa.—Orig. in Riverside, Calif., by J.W. Lesley, California Citrus Expt. Sta.

Introd. in 1942. J.H. Hale x Babcock; selected in 1937. Fruit: medium large; slightly oblong and cordate; skin creamy white, some pink to solid red, attractive, thick, tough, semi-adherent to flesh, pubescence medium in amount; flesh white, pink next to stone, sweet, mild like Babcock, nonacid, juicy; freestone; ripens in midseason, in mid-August; resembles Babcock. Tree: fairly vigorous; spreading, resembling J.H. Hale; bears heavily; flower large, showy, self-fruitful; leaf glands reniform; large; short chilling requirement, being similar to Babcock; little grown now.

Hermosillo.—Orig. in Agr. Expt. Sta., Univ. of Florida, Gainesville, by W.B. Sherman and named in northwest Mexico. Introd. in 1984. Complex parentage of Florida selections. Selected in 1981; tested as Fla. 81-30. Fruit: large; skin 90% purplish-red blush; flesh yellow, firm, freestone. Ripens 108 days after bloom. Flower showy; leaf glands reniform. Moderately resistant to bacterial leaf spot. Chilling requirement 300 chill units.

Herrington.—Orig. in Davis, Calif., by L.D. Davis, Agr. Expt. Sta., Univ. of California, Davis. Introd. in 1971. Everts selfed. Cross made in 1956, first fruited in 1960, tested as G44-3W. Fruit: size medium; truncated; symmetrical; skin yellow ground color with slight to moderate blush, excellent harvest color, pubescence moderate; flesh clear yellow, flavor excellent, clingstone; introduced because of superior flavor when canned; ripens in McKnight season, 2 to 3 days before Halford. Tree: size medium; upright; vigorous; productivity high; flower nonshowy; leaf glands reniform.

Hesse.—Orig. in 1973 at Univ. of California, Davis. Plant patent 8332; 3 Aug. 1993, assigned to Univ. of California Regents. Selected in 1978 by R. Fenton; tested as 18,6-60; and released in1993 by T.M. Gradziel, W. Beres, J. Doyle, and C. Weeks. Everts o.p. Fruit: medium; globose; moderate pubescence; flesh uniform yellow to yellow-orange throughout, nonmelting, clingstone with a smaller than average, light-brown pit. Ripens with Starn. Flowers large, nonshowy; leaf glands globose. Tree: medium size and vigor; productive; slightly upright.

Hickman's Elberta.—Orig. in Palisade, Colo., by Virgil Hickman. Introd. in 1967 by Carlton Nursery Co., Forest Grove, Ore. Elberta mutation, discovered in 1960. Fruit: resembles Elberta, but larger and more highly colored; ripens 10 days earlier than Elberta; recommended for fresh fruit and canning. Tree: size medium; vigor and hardiness moderate.

Hiland.—Orig. in Fort Valley, Ga., by J.H. Weinberger, U.S. Dept. of Agriculture Hort. Field Sta. Introd. in 1952. Southland x (Hiley x Halehaven); cross made in 1945; fruited first in 1948; tested as USDA FV 123-41. Fruit: size medium; ovate; skin has light red blush covering about one-half the surface at the shipping-ripe stage, pubescence very light; flesh yellow, fairly firm, medium textured, flavor good, clingstone; ripens 1 day earlier than Cardinal, or 5 days before Dixired, up to 6 weeks before Elberta. Tree: vigorous and productive; chilling requirement of 750 h at or below 45F, being about the same as for Hiley; flower large, showy, self-fruitful.

Hinners (*Hinners Hale*).—Orig. in Metropolis, Ill., by Ed Hinners. Introd. about 1932 by Vienna Nursery, Vienna, Ill. Seedling tree mutation; discovered in 1932; first propagated by the Vienna Nursery. Fruit: large; skin attractive, about 75% red over yellow; flesh yellow, texture fine, firm, quality good, freestone; matures after Elberta; hangs well after maturity; resembles Elberta. Tree: very productive; moderately hardy; vigorous.

Hinners Hale.—See Hinners.

Hiraoka Flame.—Orig. in Fowler, Calif., by Ross and Harry Hiraoka. Introd. in 1950. Fay Elberta o.p.; selected in 1939. Fruit: large as J.H. Hale; flesh yellow, firm, good flavor, freestone; keeps well in storage; quality good for shipping, freezing or drying; resembles J.H. Hale. Tree: low growing; self-fruitful.

Hoffman.—Orig. in Live Oak, Calif., by Phillip B. Hoffman. Introd. in 1937. Plant patent 593; 22 June 1943. Chance seedling; first fruited in 1935; selected in 1935. Fruit: size medium, uniform; symmetrical, globose; skin deep yellow overspread with a light blush, thin, tender, tenacious; flesh yellow, juice moderate, texture firm, fine, crisp, flavor mild; commercial canning clingstone; ripens in Phillips Cling season. Tree: medium size; vigorous; upright; very productive; leaf glandless; flower size medium, salmon. No longer commercially planted.

Home Canner (*Merrill Home Canner*).—Orig. in Red Bluff, Calif., by Grant Merrill. Introd. in 1950. J.H. Hale F2; selected in 1944. Fruit: flesh yellow, clingstone; 4 weeks later than J.H. Hale, which it resembles. Tree: vigorous. Variety now obsolete.

Hon-ee-gold.—Orig. in Michigan by Stahelin's Nursery, Bridgman, Mich. Introd. in 1954. Chance seedling; discovered in 1952. Fruit: size medium; pubescence moderate; halves unequal; flesh yellow, with some red next to pit; semi-clingstone; ripens after Elberta.

Honey Bee.—Orig. in Modesto, Calif., by Floyd, Gary, and Grant Zaiger and Leith Gardner. Introd. in 1984. Springcrest x genetic dwarf peach. Plant patent 5276; 28 Aug. 1984. Fruit: large; slightly asymmetrical; skin bright yellow ground color almost covered by red blush; flesh yellow, moderately firm; subacid. Tree: relatively low vigor; spreading; productive.

Honeyberta.—Orig. in Ontario, Calif., by Geo. P. Weldon, Chaffey College. Introd. in 1948. Lukens Honey x Elberta; selected in 1945. Fruit: skin yellow with an attractive blush; flesh yellow, fairly firm, quality good; ripens in Elberta time. Tree: early foliation; production good.

Honey Dew Hale.—Orig. in New Cumberland, Pa., by Charles O. Grissinger. Introd. in 1949. Plant patent 888; 1 Nov. 1949; assigned to Stark Bros. Nurseries & Orchards Co., Louisiana, Mo. J.H. Hale mutation; discovered in 1936. Fruit: large; skin white with pink blush; flesh white, marbled with yellow, with thin yellow segment from pit through skin along suture; freestone; good shipping qualities; resembles J.H. Hale. Tree: vigor medium; open, spreading; leaf glands reniform; flower small, pink.

Honeygem.—Orig. in New Brunswick, N.J., by M.A. Blake, New Jersey Agr. Expt. Sta. Introd. in 1947. (J.H. Hale x P.I. 55564) x Estella; tested as N.J. Low Acid 8. Fruit: flesh yellow, firm, attractive, low in titratable acidity; freestone; ripens about a week after Elberta. Tree: not as hardy as Elberta; flower showy.

Hope Farm.—Orig. in Westwood, Bergen County, N.J., by Herbert W. Collingwood. Introd. in 1925 by Bountiful Ridge Nurseries, Princess Anne, Md. Parentage unknown; discovered in 1919 on the Hope Farm. Fruit: large; flesh white, quality good, freestone; matures about 5 days after Elberta and follows Belle. Tree: apparently self-unfruitful; hardy.

Howard.—Orig. in Merced, Calif., on the Fancher Ranch of the California Packing Corp. Introd. in 1949. Parentage unknown; discovered in 1941. Fruit: sizes fairly well; flesh yellow, very little red at pit; commercial canning clingstone; matures with Paloro; unusually free of defects such as gum, blossom-end breakdown, etc.; resembles Paloro. Tree: very vigorous; sets well. No longer commercially planted.

Howard Fisher (*Fisher*).—Orig. in Queenston, Ont., Canada, by C. Howard Fisher. Introd. in 1936. Plant patent 233; 9 Feb. 1937. Valiant mutation; discovered in 1934. Fruit: large; oblate; skin butter-yellow, brightly blushed, light pubescence; flesh yellow, coarse texture, melting, slightly soft, flavor good; semi-freestone; ripens 5 weeks before Elberta; does not drop from tree even when fully ripe; ships well; resembles Valiant. Tree: vigorous; average productivity; susceptible to brown rot; blossom medium, nonshowy; leaf glands reniform.

Hudelson.—Orig. in Hughson, Calif., by Warren E. Hudelson. Introd. in 1938. Parentage unknown; discovered in 1933. Fruit: round; flesh yellow; commercial canning clingstone; ripens in Gaume season; resembles Halford. Tree: produces good crops consistently. No longer commercially planted.

Hutchison.—Orig. in Woodville, Calif., by Perry Hutchison. Introd. in 1938. Parentage unknown; selected in 1934. Fruit: large; texture smooth; ripens well when picked hard; good for frozen pack. Tree: vigorous.

Idlewild.—Orig. in Idlewild Expt. Sta., Clinton, La., by James E. Boudreaux, C.E. Johnson, P.L. Hawthorne, W.A. Young, R.L.

Cunningham, T.J. Raiford, and D.W. Newsom of Louisiana Agr. Expt. Sta., Baton Rouge. Introd. in 1983. La Feliciana o.p.; discovered in 1973; tested as L71-A73-30. Fruit: medium to large; round to oval with slight suture, halves almost equal; semi-freestone; skin medium-yellow with 75% to 80% overcolored with dark red, pubescence medium to long, heavy; flesh flecked throughout with red; quality, firmness, and texture good for ripening season; ripens 2nd week of June, about 36 days before Elberta, 12 to 15 days after Bicentennial, and 4 to 6 days before Harvester. Tree: vigorous, strong, productive, heavy thinning required; blossoms nonshowy, light pink; leaf glands reniform; chilling requirement 500 to 600 h; tolerant to bacteriosis (*Xanthomonas campestris* sp. *pruni*).

Imperani.—Orig. in Roodeplaat, Republic of South Africa, by Edward P. Evans, Hort. Res. Inst., Pretoria. Introd. in 1975. F_2 seedling of Jubilee x Keimoes, cross made in 1962, selected in 1969, tested as 8/51. Fruit: medium to large; round; skin yellow with attractive red blush; flesh yellow, slight red at pit, firm, clingstone; ripens 1 Jan.; for dessert. Tree: medium large, 34 m; vase-shaped; very productive; flower showy, blooms about 20 Aug.; leaf glands reniform.

Impon.—Orig. in Ontario, Calif., by George P. Weldon, Chaffey College. Introd. in 1948. Imperial x Paragon; selected in 1945. Fruit: large as Elberta; skin yellow; flesh yellow, subacid; freestone; ripens end of Babcock season; quality superior to Imperial, which it resembles. Variety now obsolete.

Improved Elberta.—See Early Elberta.

Improved Early Elberta.—See Early Elberta.

Improved Mardigras.—Orig. in Red Bluff, Calif., by Grant Merrill, Exeter, Calif. Introd. in 1968. Plant patent pending. [J.H. Hale x (J.H. Hale x Levi)] selfed. Cross made in 1953, selected in 1956, tested as 39-55. Fruit: large; globose to cordate; skin covered with one-fourth to two-thirds bright red blush, attractive, pubescence short; flesh yellow, very little red at pit, very firm, freestone; excellent freezing qualities; ripens late August for shipping; early September for freezing; variety may be picked in one picking if desired. Tree: large, vase shaped; vigorous; productivity heavy; flower medium size, blooms with Scarlet Lady; leaf glands reniform.

Improved Pacifica.—Orig. in Red Bluff, Calif., by Grant Merrill, Exeter, Calif. Introd. in 1964. Plant patent 2425; 14 July 1964. J.H. Hale x (July Elberta x Maxine); discovered in 1955; selected in 1963; tested as Merrill Y-25. Fruit: large, averaging 3 inches in diam.; form variable, globose to slightly flattened; suture small, distinct; skin deep yellowish-orange overlaid 50% to 90% bright red, pubescence short, medium tough, medium thick, tenacious to flesh; flesh deep yellow, streaked and mottled red throughout, particularly near base with red next to pit cavity, meaty, melting, flavor delicate, subacid, distinct aroma, eating quality good; stone free, size medium, with fewer split-pits than Pacifica; keeping and shipping qualities good; ripens about with Rio Oso Gem and Pacifica; resembles Pacifica. Tree: large, vigorous; bears regularly; flower size medium, blooms about with Elberta.

Improved Pallas.—See Melba.

Improved Rochester.—See Western Pride.

Ingwe.—Orig. in Roodeplaat, Pretoria, South Africa, by Edward P. Evans, Hort. Res. Inst. Introd. in 1972. Oom Sarel x Moreira Jubilee; cross made in 1964, selected in 1967, tested as 16/15. Fruit: large; round-oval; skin yellow ground color, with attractive red blush over 40% to 60%; flesh yellow, firm, melting, freestone, flavor excellent; ripens mid-November. Tree: large; open vase-shaped; vigorous; moderately productive; slight thinning required; flower nonshowy, light rose; leaf glands reniform; chilling requirement 300 to 500 h.

Inkoos.—Orig. in Groot Drakenstein Valley, Paarl, South Africa, by a Mr. Gray on the Bellingham farm. Introd. about 1923. Duke of York mutation. First propagated by a Mr. Arton of neighboring Packwood farm. Fruit: ripens about 2 weeks earlier than parent.

Ito Red.—Orig. by J.W. Taylor, Dinuba, Calif. Plant patent 7248; 19 June 1990. Assigned to Ito Packing Co., Reedley, Calif. Red Jim x late peach. Fruit: large; round; mottled rose blush over yellow ground color with very short pubescence; flesh yellow with red flecks and dark red at pit, melting, semi-clingstone. Ripens late August to early September. Flowers with Red Jim, large, pink, showy, anthers red; leaf glands small reniform but some globose. Tree: large; vigorous; productive.

Jane's Favorite.—Orig. in Birmingham, Mich., by Robert A. Nitschke, Southmeadow Fruit Gardens. Introd. in 1962. Parentage unknown; discovered in 1958. Fruit: medium large; round; skin yellow, blushed, attractive; flesh yellow, firm, flavor fine; freestone; ripens in mid-September. Tree: bears regularly and heavily; valuable for the home garden.

Jayhaven.—Orig. at South Haven, Mich., by Stanley Johnston, Michigan State Univ. Introd. in 1976 by R.L. Andersen, J.E. Moulton, T.M. Thomas, and H.J.Belter. SH333 (Redhaven x Halehaven o.p.) x Ambergem o.p.; selected in 1965; tested as SH467. Fruit: medium; round, uniform; skin 80% red blush over golden-yellow ground color, pubescence sparse; flesh clear yellow, very firm, freestone; quality commercially acceptable. Ripens with Glo-haven, 10 days after Redhaven. Tree: vigor medium; upright-spreading; productive. Flower bud hardiness similar to that of Redhaven. Blossoms nonshowy. Tolerant of bacterial spot.

Jefferson.—Orig. in Blacksburg, Va., by George D. Oberle and R.C. Moore, Virginia Agr. Expt. Sta. Introd. in 1960. J.H. Hale x Valiant; tested as V.P.I. 48. Fruit: large; skin bright red; flesh yellow, very firm, texture fine, flavor good; fully freestone; cans and freezes well; resembles J.H. Hale. Tree: vigorous; bud set heavy; buds and flowers tolerant to blossoming-season frost; reliable annual production; pollen fertile.

Jefferson Sun.—Orig. by R.H. Metzler, Del Ray, Calif. Plant patent 7003; 29 Aug. 1989. Assigned to Metzler Investments and Metropolitan Life Insurance Co., Del Ray. Berenda Sun mutation. Similar to parent but ripens 3 to 4 weeks earlier.

Jerseydawn.—Orig. in 1965 in New Brunswick by L.F. Hough and C.H. Bailey, New Jersey Agr. Expt. Sta. Selected in 1968, tested as NJ 246, and introduced in 1984 by S.A. Mehlenbacher, L.F. Hough, and C.H. Bailey. Solo x (Jerseyland x Mayflower). Fruit: medium; round; skin 40% to 70% red blush on a dull-yellow ground color; flesh yellow, firm, melting, semi-freestone; few split pits. Ripens 10 to 14 days before Redhaven. Flowers large, showy, pink; blooms 1 to 3 days after Elberta; leaf glands reniform. Leaves and fruit resistant to bacterial spot. Tree: medium vigor; productive.

Jerseyglo.—Orig. in New Brunswick, N.J., by Catherine H. Bailey and L.F. Hough in cooperation with New Jersey Agr. Expt. Sta. Dept. of Hort. & Forestry, Rutgers Univ., New Brunswick, N.J. Introd. in 1979. Jefferson x Loring. Cross made in 1964; selected in 1968; tested as NJ 244. Fruit: large, round; skin three-fourths full red on yellow; flesh yellow, firm, freestone, ripens with Elberta. Tree: average size, vigor good, hardy, productive, resistant to bacterial leaf spot.

Jerseyland (*Beauty-Gem*).—Orig. in New Brunswick, N.J., by M.A. Blake, New Jersey Agr. Expt. Sta. Introd. in 1946. J.H. Hale o.p. x (Slappey x Admiral Dewey); tested as N.J. 135. Fruit: very large; ovate, halves unequal, suture and tip prominent; skin dark red over greenish-yellow ground, attractive, rather heavy pubescence; flesh yellow, not very firm, melting, semi-freestone; pit long, sharp apex; ripens 1 month before Elberta; rough and unattractive in southern Georgia. Tree: vigorous and very productive; some split pit tendency; flower small, nonshowy; leaf glands reniform; chilling requirement 950 h at or below 45F.

Jerseypink.—A double-flowered ornamental orig. at New Jersey Agr. Expt. Sta., New Brunswick. Selected in 1985 by A.M. Voordeckers, tested as NJ 308, and introduced in 1990 by J.C. Geffreda, A.M. Voordeckers, and S.A. Mehlenbacher for ornamental qualities. Complex parentage of New Jersey selections. Fruit: medium; oblong-round; 40% to 50% scarlet-red blush over a greenish-yellow ground color; flesh yellow, nonmelting, clingstone. Flowers numerous, large, showy, rich pink, extra petals (double); late blooming; leaf glands reniform. Trees: tolerant to bacterial spot, vigorous, moderately spreading, slightly productive.

Jerseyqueen.—Orig. in New Brunswick, N.J., by Catherine H. Bailey and L.F. Hough, New Jersey Agr. Expt. Sta. Introd. in 1964; trees available from New Jersey Peach Council, Princeton. (Candoka x NJ 25032 selected nectarine) x (Garden State x NJ 25032) o.p.]; cross made in 1951; selected in 1954; tested as N.J. 216. Fruit: large; oval; skin brighter red than Elberta, attractive; flesh yellow, firmer and of better quality than Elberta; freestone; ripens about with Elberta; excellent dessert quality. Tree: productive; flower showy; leaf glands reniform. Suggested planting regions: where an Elberta substitute is wanted.

Jewel.—See July Elberta.

Jim Bowie.—Orig. in Montague, Texas, by U.A. Randolph, Texas Agr. Expt. Sta. Introd. in 1968. Parentage unknown. seedling discovered in 1945. Fruit: 2 1/2 inches in diam.; truncate to round; skin greenish-yellow ground cover, 40% blushed with light red; flesh light to medium yellow, tinted red around pit, flavor sweet, subacid; ripens 5–20 Sept.; for local markets. Tree: size medium; spreading; vigorous; hardy; fairly resistant to bacterial spot; flower standard Elberta type; chilling requirement probably 850 to 900 h.

J.L. Ames.—See Ames.

J.M. Mack.—Origin unknown, believed to be Australia. Introd. into the U.S. in 1926; introd. into New Zealand from Australia in 1940. Parentage unknown. Fruit: under New Zealand conditions: large; round, one-half bulging, unequal; skin ground color cream lightly streaked and mottled pink on very limited areas, attractive, smooth, thick, pubescence very short, peels easily; flesh creamy-white, firm but very tender, very juicy, mild subacid flavor, quality very good; freestone; very good dessert variety; brown rot susceptibility seems high. Tree: size medium; spreading; flowers small, dark pink. Under U.S. conditions: Fruit: size medium-large; flesh creamy-white, flavor and aroma distinctive, quality very good; ripens a few days before Elberta. Tree: hardier than Elberta. Obsolete in Australia.

Jody Gaye.—Orig. in Fresno, Calif., for trial only by U.S. Dept. of Agriculture. Tested as F-11. Variety was not introduced U.S. Dept. of Agriculture. About 1957, Stribling Nursery Co., Merced, Calif., propagated the selection for a grower Alamar x Goldust. Fruit: size medium; round to elongate; skin mostly red; flesh firm, melting; freezing qualities good; ripens 24 days after Rio Oso Gem. Tree: size medium; vigor moderate; productive; susceptible to brown rot.

Johnson 2.—See Alton.

Johnson Early Elberta.—Orig. in Brigham City, Utah. Introd. in 1935. Probably an Elberta or Early Elberta o.p.; discovered in 1922 on the Johnson farm. Fruit: ripens with Early Elberta; firmer and better color than Early Elberta; resembles J.H. Hale. Tree: very hardy; buds only moderately hardy.

John S. Shaw.—Orig. in Somerset, Texas, by John S. Shaw, New Braunfels Nursery and Floral Co., New Braunfels, Texas. Introd. in 1925. Fruit: size of Carman; skin yellow, blushed; flesh white, firm, flavor pronounced, freestone; ripens in mid-June; resembles Mamie Ross and Honey. Tree: bears well; winter-chilling requirement slight.

John G.—See Late John Gee.

Jon Boy (*Clendining*).—Orig. in Richwood, N.J., by Edward Clendining. Introd. in 1981. Plant patent 6827, 6 June 1989; assigned to Adams Co. Nursery. Loring mutation. Resembles Loring, but ripens 10 days earlier with much more red skin color. Fruit: large; firm; semi-freestone. Tree: vigorous; productive; resistant to bacterial leaf spot.

Jubilant.—Orig. in Danville, Morgan County, Ala., by Penn-Orr-McDaniel Orchards (Wm. Arthur Penn, Lovic Orr, and Jos. C. McDaniel). Introd. in 1947. Golden Jubilee mutation; discovered and first propagated in 1945. Fruit: round; skin yellow, more highly blushed than Golden Jubilee; flesh yellow, as firm as Golden Jubilee; freestone; ripens 7 to 10 days before Golden Jubilee, which it resembles. Tree: growth moderately vigorous; productive; flower medium, nonshowy; leaf glands reniform; chilling requirement 850 h at or below 45F. Useful as a similar variety to precede Golden Jubilee.

July Elberta (*Bodine Elberta, Brentwood Beauty, Burbank Elberta, Burbank July*

***Elberta, Burnett Elberta, Early Elberta, Jewell, Kim Early Elberta, Kim Elberta, Mulberry, Smith Beauty, Smith Mountain*).**—Orig. in Sebastopol, Calif., by Luther Burbank. Introd. in 1930. Plant patent 15; 15 Apr. 1932; assigned to Stark Bros. Nurseries & Orchards Co., Louisiana, Mo. Parentage unknown; selected about 1925. Fruit: size medium; round; skin dull red, blushed and streaked on greenish-yellow ground, pubescence heavy and dense; flesh yellow, only slightly red at pit, firm, melting, very good quality; pit small, free; ripens about 2 weeks before Elberta; ships very well. Tree: vigorous; highly fruitful; flower size medium, nonshowy; leaf glands reniform; susceptible to bacteriosis; chilling requirement 750 h.

July Gold.—Orig. in Sebastopol, Calif., by Luther Burbank. Introd. in 1930 by Stark Bros. Nurseries & Orchards Co., Louisiana, Mo. Parentage unknown; selected in 1928. Fruit: large, up to 3 inches in diam.; skin orange yellow; flesh golden, no red at pit, flavor good, freestone; ripens about 20 days before Elberta. Tree: very hardy; productive; vigorous.

July Hale.—Orig. in Caldwell, Idaho. Introd. in 1940. Parentage unknown; propagated in 1939 by Lafayette Nursery Co., Lafayette, Ore. Fruit: skin highly colored; ripens just before July Elberta; resembles Rochester.

July Heath (*Othelle*).—Orig. in Paducah, McCracken County, Ky., by J.W. Harris. Introd. in 1925 by O. Pipers, Clinton Nursery, Clinton, Ky. Parentage unknown; discovered in 1923. Fruit: large; ovate; skin blushed rose over greenish-white ground; flesh white to greenish-white, red at pit, firm, nonmelting, quality fair to poor, clingstone; ripens 10 to 14 days before Elberta. Tree: average fruitfulness; flower medium, nonshowy; leaf glands reniform.

July Lady.—Orig. in Red Bluff, Calif., by Grant Merrill, Exeter, Calif. Introd. in 1970. Plant patent 3023, 2 Feb. 1971. J.H. Hale x Merrill Gem. Cross made in 1957, selected in 1969, tested as 28-28E. Fruit: large; variable, generally globose; skin yellow ground cover, red blush over 50% to 60%, medium thick, pubescence scant; flesh white to light yellow, specked with red, flavor subacid and mild, firm; shipping quality excellent; freestone; recommended for fresh market and freezing; ripens mid-July a few days earlier than Angelus, which it resembles. Tree: vigorous; productive; flower small, pinkish-salmon; leaf glands reniform.

July Lady Improved.—Orig. in Red Bluff, Calif., by Grant Merrill (deceased). Introd. in 1972. Plant patent 3590; 13 Aug. 1974. Red Lady o.p. Fruit: size medium to large; generally cordate, some globose, with slight tip; skin medium thin, tenacious, sweet, no blush unless very ripe; flesh yellow, red at pit, moderate amygdalin, texture generally firm, freestone; for short and long distance shipping; ripens first week of July. Tree: size medium; vase-shaped; more vigorous than Red Lady; bears regularly; flower small to medium, pink to pinkish salmon; leaf glands reniform.

July Sun.—Orig. in Del Ray, Calif., by Tom Chamberlain, H.P. Metzler & Sons. Introd. in 1983. Plant patent 5139; 15 Nov. 1983; assigned to Metropolitan Insurance Co.; licensed exclusively to Agri Sun Nursery. July Lady o.p. Fruit: very large, up to 100 mm; skin medium red on bright yellow ground; round; flesh yellow, freestone. Ripens 12 days after Redhaven in New Jersey, 20 days after Redhaven in Selma, Calif. Tree: very vigorous; very productive. Chilling requirement about 350 h.

Jumbo.—Orig. in St. Joseph, Ill., by Walter Ewbank. Introd. in 1974 by Herschel L. Boll, Champaign, Ill. Unknown seedling; discovered about 1968 in a fence row. Fruit: large, round-oblong and flattened, 50% to 75% covered with red blush; flesh yellow, firm; freestone; ripens about with Elberta. Tree: highly productive.

Jun-Berta.—Orig. in Dover, Ark., by Henry Frank Smith. Introd. in 1951. Plant patent 1021; 10 July 1951; assigned to Stark Bros. Nurseries & Orchards Co., Louisiana, Mo. Fair Beauty mutation; discovered in 1944. Fruit: flesh yellow, semi-clingstone; resembles parent, except for earlier maturing, 2 weeks earlier than Fair Beauty, about 8 weeks before Elberta. Tree: same as parent.

June Bride.—Orig. in Kingfisher, Okla., by Paul M. Goodwin, Goodwin Nursery. Introd. in 1965. Sunhigh o.p. Discovered in

1962. Fruit: size medium; round; skin highly blushed with red, pubescence very slight; flesh amber, quality very good, flavor and aroma good, semi-clingstone; ripens 15 June; for home or commercial use. Tree: medium to large; vigorous; hardy; productive, tends to overbear.

Junecrest.—Orig. in Modesto, Calif., by Floyd, Gary, and Grant Zaiger and Leith Gardner. Introd. in 1986. Fayette x a Mexican cling peach. Plant patent 5641; 21 Jan. 1986; assigned to Zaiger Genetics. Fruit: medium to large; skin well-colored; pubescence medium; flesh yellow, semi-freestone; round with firm tip; very good flavor. Ripens 15 days before Redhaven at Modesto. Tree: vigorous; upright. Chilling requirement 800 to 1000 h.

June Gold.—Orig. in Ontario, Calif., by Herbert C. Swim, Armstrong Nurseries. Introd. in 1958. Plant patent 1884; 8 Dec. 1959. Flamingo x Springtime; selected in 1952. Fruit: very large; skin highly colored red; flesh yellow, firm, freestone; matures 7 to 10 days before Redhaven and 7 to 10 days after Robin; suitable for shipping; J.H. Hale type. Tree: chilling requirement, slightly higher than for Springtime; leaf glands globose; flower size medium, showy.

June Lady.—Orig. in Red Bluff, Calif., by Grant Merrill, Exeter, Calif. Introd. in 1968. Plant patent 3022; 2 Feb. 1971. Fortyniner x Gemfree. Cross made in 1956, selected in 1959, tested as 0-46A. Fruit: considered large for the season; globose to cordate; skin smooth, very little pubescence, one-fourth to three-fourths covered with red blush; flesh firm, clingstone; may be picked in one picking; for local and distant markets; ripens in San Joaquin Valley in mid-June with Coronet. Tree: size medium, vase shaped; vigorous; productivity considered good for the season; flower small, early about with July Elberta; leaf glands reniform.

June Pride.—Orig. in Modesto, Calif., by Floyd, Gary, and Grant Zaiger and Leith Gardner. Introd. in 1989. O'Henry x June Lady. Plant patent 6747; 18 Apr. 1989; assigned to Zaiger Genetics. Fruit: medium large; skin dark, full red, attractive; pubescence medium; flesh yellow, very firm, good flavor; semi-freestone. Ripens with Redhaven at Modesto. Tree: vigorous; upright. Chilling requirement 800 to 1000 h.

Juneprince.—Orig. at Byron, Ga., by W.R. Okie, U.S. Dept. of Agriculture S.E Fruit & Tree Nut Res. Lab. [(Sunhigh x Southland) x Redcap] x Junegold; cross made in 1963 by V.E. Prince; selected in 1966; tested as FV4-3751 and BY4-3751. Fruit: large; skin attractive red on yellow ground color, pubescence light; flesh melting, freestone. Ripens 5 to 7 days before Redhaven; released to replace Coronet. Flowers showy self-fertile; leaf glands absent; leaf margin serrate. Moderately resistant to bacterial spot. Chilling requirement about 650 h.

June Sun.—Orig. in Del Ray, Calif., by Tom Chamberlain, H.P. Metzler & Sons. Introd. in 1987. Plant patent 5873; 3 Feb. 1987; owned by Metropolitan Insurance Co.; licensed exclusively to Agri Sun Nursery. Unknown parentage. Fruit: large, up to 85 mm; round skin dark red on yellow ground; flesh yellow, freestone. Ripens with or just ahead of Redhaven in California. Tree: very vigorous; very productive; upright. Chilling requirement about 300 h.

June Wax.—Orig. in Arlington, Texas, by J.A. Evans, Evans Specialty Nursery. Introd. about 1925. Parentage unknown. Fruit: size medium; skin appears waxy, almost entirely white with an occasional touch of red near the stem end; flesh white, sweet, pleasing, too tender for shipping; semi-clingstone, but flesh separates readily from stone when fully ripe; ripens in June. Obsolete.

Jungerman.—Orig. in Davis, Calif., by L.D. Davis, California Agr. Expt. Sta. Introd. in 1964. Dix 22A-5 [(Orange Cling seedling x Alameda) x (Goodman's Choice x Transvaal)] x Dixon 1; cross made in 1950; selected in 1954; tested as Calif. GH 1-3. Fruit: medium to large; skin highly blushed with reasonably good harvest color; flesh a clear yellow but may be flecked with red if allowed to become overripe; pit red, somewhat elongated; ripens about with Cortez, 2 or 3 days before Paloro. Tree: consistently sets heavy crops; leaf with globose glands; flower large, showy, pink.

Kakamas (*Collins Kakamas*).—Orig. in Kakamas, Orange River, South Africa, by

O.S.H. Reinecke, Stellenbosch-Elsenburg College of Agr., Dept. Agr., Stellenbosch. Introd. there in 1935 , imported into U.S. as P.I. 227402 and in 1956 as P.I. 234194; the former was indexed and found to be virus-positive in 1965 and P.I. 234194 virus-positive in 1967. St. Helena o.p.; selected by A.D. Collins in 1933. Fruit: size medium; rounded with small apical beak, ridged with shallow suture at the stem end, shallow depression on side opposite suture; skin yellow; flesh deep orange-yellow to pit, very firm, texture fine, flavor distinctive; stone small, cling; excellent for canning; ripens late. Tree: vigorous; twigs willowy; narrow lanceolate leaves; very resistant to delayed foliation; high and regular production. A leading commercial variety in South Africa.

Kalhaven.—Orig. in South Haven, Mich., by Stanley Johnston, Michigan Agr. Expt. Sta. Introd. in 1936. J.H. Hale x Kalamazoo; cross made in 1924. Fruit: size medium; round; skin bright red over half of surface; flesh yellow, melting, firm, medium acidity, semi-freestone; ripens just before Elberta; quality very good; resembles J.H. Hale, but smaller. Tree: vigor moderate, highly productive; flower small, non-showy; leaf glands reniform.

Kearney.—Orig. in Fresno County, Calif., by C.O. Hesse, Univ. of California Kearney Horticulture Field Station. Plant patent 3976; 16 Nov. 1976; assigned to The Regents of the Univ. of California, Berkeley. Fay Elberta x Lippiatt seedling 29 F3; cross made in 1956; selected in 1960; tested as E3-46. Fruit; medium to large; broad elliptic; yellow ground color with red blush; flesh yellow, red at cavity; freestone; average ripening date 10 Aug. Tree: medium; upright; spreading; vigorous; good productivity. Flower large, showy pink; leaf glands globose; chilling requirement about equal to Fay Elberta.

Keimoes.—Orig. in Keimoes, South Africa, by A.F. de Wet, Stellenbosch-Elsenburg College of Agr., Dept. Agr., Stellenbosch. Introd. there in 1950; imported into U.S. in 1954 as P.I. 219559 and in 1956 as P.I. 234195. Transvaal o.p.; discovered in 1933. Fruit: skin yellow; flesh deep yellow, firm; commercial canning clingstone; resembles Kakamas. Tree: vigorous; heavy bearer; low chilling requirement; indexed and found virus-positive in 1965.

Kenlate Elberta.—Orig. in Clinton, N.C., by John H. Rendall Introd. in 1969 by Bountiful Ridge Nurseries, Princess Anne, Md. Elberta mutation; discovered in 1962. Fruit: large, round; skin highly blushed with red, distinctive red stripe on suture when fully ripe; flesh yellow, melting; freestone; ripens 2 weeks after Elberta, recommended for fresh market. Tree: large, spreading; vigorous; hardy; productive; flower small, deep pink; leaf glands weakly reniform.

Kern Sun.—Orig. in Del Ray, Calif., by Tom Chamberlain, H.P. Metzler & Sons. Introd. in 1983. Plant patent 4980; 8 Feb. 1983; owned by Metropolitan Insurance Co.; licensed exclusively to Agri Sun Nursery. June Lady o.p. Fruit: medium, 65 to 75 mm; round; skin dark red on yellow ground; flesh yellow, semi-freestone, firm; early; attractive. Ripens 15 days before Redhaven in Selma, Calif. Tree: medium vigorous; very productive; upright. Chilling requirement 300 to 350 h.

Kern Triumph.—Orig. in McFarland, Kern County, Calif., by W.C. Hanawalt. Introd. in 1929. Parentage unknown. Fruit: large; skin purplish-red; flesh yellow, freestone; ripens in late October and November.

Kette.—Orig. in Dayton, Ohio, by A.E. Kette. Introd. in 1929 by W.N. Scarff & Sons Nursery, New Carlisle, Ohio. Elberta o.p.; discovered in 1915. Fruit: very large; skin golden yellow, lightly blushed; flesh yellow, fine-grained, freestone; resembles a very large Elberta. Tree: heavy annual producer. Now obsolete.

Keystone.—Orig. in Fort Valley, Ga., by John H. Weinberger, U.S. Dept. of Agriculture Hort. Field Sta. Introd. in 1954. Newday x Southland; cross made in 1946; tested as USDA FV 177-17. Fruit: about the same size as its parents; round; skin with a light attractive blush over bright golden ground cover, pubescence light; flesh yellow, firm, smooth textured, flavor good; pit very nearly free, resembling Newday and slightly less free than Southland; ripens with Early Hiley, slightly ahead of Triogem, or 3 to 3 1/2 weeks ahead of Elberta. Tree: vigorous; unproductive in some areas; susceptible to late spring frost; less bacterial spot than Southland and Elberta; flower small, self-fruitful, blooms about the same time as Newday and Southland in the Fort Valley area; leaf glands globose; chilling requirement 750 h.

Kimbo.—Orig. in Sandy Spring, Md., by William W. Moore. Introd. in 1959. Plant patent 1647; 8 Oct. 1957; assigned to Buntings' Nurseries, Selbyville, Del. Possible Elberta mutation. Fruit: large; flesh yellow, juicy; freestone when ripe; ripens about 10 days later than Elberta, or during the first 2 weeks of September at place of origin. Tree: very large; vigorous; hardy; very productive; bears regularly; self-fruitful.

Kim Earling.—Orig. in Farmersville, Calif., by H.D. Dougherty. Introd. in 1952. Plant patent 1216; 8 Sept. 1953. Johnson x Halehaven; selected in 1950. Fruit: medium to large; skin highly colored; flesh yellow, very firm, semi-freestone; matures 5–10 June in the San Joaquin Valley of California; resembles Redhaven; ships well. Tree: vigorous; bears well; flower showy, large, pink; leaf glands none.

Kim Early Elberta.—See July Elberta.

Kim Elberta.—See July Elberta.

King Edward.—See Bells November.

King's Delicious.—See Early Fair Beauty.

Kingston Redleaf (*Davidson, Redleaf, Tennessee Redleaf*).—Orig. in Davidson County, Tenn., by Joseph C. McDaniel, Tennessee Dept. Agr., Nashville, Tenn. Introd. in 1946, by Peach Ridge Farm, Clemson, S.C. Parentage unknown; propagated in 1938. Fruit: yellow fleshed, freestone; poor edible quality; pits give high germination without special treatment, many double kernels. For use as a rootstock: about 75% of its seedlings are red-leafed.

Kirkman Gem (*Late Rio Oso Gem*).—Orig. in Madera County, Calif., by Wm. T. Kirkman, Jr., Tracy, Calif. Introd. in 1942. Plant patent 506; 17 Mar. 1942; assigned to Kirkman Corp., Tracy, Calif. Rio Oso Gem mutation; discovered in 1937. Fruit: large; round; truncate; skin yellow, blushed, tough, ventral suture with a deep reddish streak about 1/4 inch wide, which appears in advance of rest of color; flesh yellow, streaked with red next to stone, juice moderate, firmer than Rio Oso Gem, flavor mildly acid, ripens evenly; freestone; ripens in mid-September, 4 to 5 weeks after parent; resembles Rio Oso Gem. Tree: medium vigor and productivity; flower large, showy, light pink; leaf glands reniform.

Klamt.—Orig. in Davis, Calif., by L.D. Davis, California Agr. Expt. Sta. Introd. in 1964. Dixon 1 x Wiser; cross made in 1950; selected in 1954 tested as Calif. Da 20-29. Fruit: medium to large; skin blush medium, with an excellent harvest yellow; flesh clear yellow; pit brown; when canned, flavor superior to existing conventional commercial canning clingstone varieties; maintains its superior flavor quality; it appears to process satisfactorily; ripens 2 or 3 days after Paloro. Tree: consistently sets good crops; leaf with globose glands; flower like that of Dixon 1, being of the larger nonshowy type.

Klondyke Early Elberta.—Orig. in Tremonton, Utah. Introd. in 1930 by Moore's Nursery, Ogden, Utah. Parentage unknown; discovered in 1925. Fruit: roundish; skin highly blushed over yellow; flesh firm, yellow, freestone; good shipping and dessert quality; ripens early. Tree: vigorous; productive; moderately hardy.

Kurakata Yellow.—Orig. near Setagaya-Ku, Tokyo, Japan, by Elizo Kurakata. Plant patent 4423; 5 June 1979. Tusberta x Ban Kitou. Fruit: large, round; skin yellow, partially tinged with red where exposed to sun; flesh yellow, delicate texture, slightly sweet-sour; clingstone; ships well for market; ripens August in Tokyo. Tree; size medium, vigorous, productive, good resistance to disease; flowers small, rose-pink, self-fertile; leaves lanceolate.

Kuster Fay.—Orig. in Yuba City, Calif., by Fred Kuster. Introd. in 1958. Fay Elberta mutation; discovered in 1950. Fruit: flesh yellow, freestone; ripens about 10 days after Fay Elberta, which it resembles; ships well. Tree: self-fruitful.

Lady Lou.—Orig. in Parlier, Calif., by Kengo Osumi. Plant patent 6940; 25 July 1989. Licensed exclusively to Riesner Nursery. June Lady mutation. Fruit: medium; globose; skin fully red-blushed; flesh yellow, firm, clingstone. Productive. Ripens 16 days before Redhaven in California.

Lady Nancy.—Orig. near Hammonton,

N.J., by Medio DeMarco, De Marco Orchards. Introd. in 1989. Plant patent 7069; 12 Dec. 1989; assigned to Adams Co. Nursery and Burchell Nursery. Jerseyqueen mutation. Fruit: large to very large; shape globose to ovate, not symmetrical; skin 75% scarlet red, on greenish-white ground color; flesh white, firm, with yellow suture line through the flesh; excellent quality; ripens with Jerseyqueen, before Rio Oso Gem. Excellent storage characteristics. Tree: similar to Jerseyqueen; very vigorous; large; open-spreading; production regular.

La Feliciana.—Orig. in Calhoun, La., by P.L. Hawthorne, W.A. Young, and J.C. Taylor, Louisiana State Univ., Baton Rouge. Introd. in 1980. L5-20-18 o.p. Cross made in 1960: selected in 1964: tested as L1-81-23. Fruit: medium to large: round; equal halves: flesh yellow with red flecks throughout. firmness and texture excellent, quality good; skin 60% covered with dark red, very little pubescence; freestone; ripens late June, after Harvester. Tree: large, upright, excellent vigor, productivity outstanding; tolerant to bacteriosis (*Xanthomonas pruni*) and brown rot; flower nonshowy; leaf glands reniform; chilling requirement 550 to 660 h at 45F.

La Festival.—Orig. at Idlewild Research Station, Clinton, La., by J.E. Boudreaux, C.E. Johnson, P.L. Hawthorne, W.A. Young, R.L. Cunningham, M.G. Lartigue, T.J. Raiford, F.J. Peterson, and P.W. Wilson. Introd. in 1985. La Feliciana o.p.; selected in 1983; tested as L71-A73-3. Fruit: medium to large; skin 80% bright red over medium yellow ground; spherical; flesh yellow with red flecks; firm; freestone; very good flavor. Ripens midway between Harvester and La Feliciana. Tree: vigorous; productive. Leaves and fruit resistant to bacterial spot. Chilling requirement 400 to 500 h.

LaGem.—Orig. in Calhoun, La., by J.C. Taylor, P.L. Hawthorne, A.C. Miller, J.C. Miller, and R.S. Woodward, Louisiana Agr. Expt. Sta. Introd. in 1965. Redglobe x Redskin; tested as L-9-8-10. Fruit: medium-large, up to 2 1/2 inches in diam.; oval to round; nonprominent suture; halves equal; skin yellow, brightly blushed, attractive, pubescence medium short, resistant to cracking; flesh yellow, very firm, texture good, quality very good; pit nearly free; ripens evenly, about with Dixigem; in many characteristics better than Dixigem in areas tested. Tree: vigorous; very productive; leaf glands globose; hardier than Dixigem; flower large, showy; rest period of leaf and fruit buds broken by 850 to 900 h at or below 45F; resistant to bacterial spot and brown rot.

LaGold.—Orig. in Calhoun, La., by J.C. Taylor, P.L. Hawthorne, A.C. Miller, N.L. Horn, D.W. Newsom, and R.S. Woodward, Louisiana Agr. Expt. Sta. Introd. in 1967. Redhaven o.p. seedling, probably pollen parent Early Elberta. Tested as L-3-16-16. Fruit: medium-large; slightly pointed oval; skin yellow ground cover with 15% to 25% red blush, attractive, pubescence medium, nonprominent suture, equal halves; flesh yellow with some red near pit, firm, quality and texture good, freestone; ripens 15–20 June at place of origin, 25 to 28 days before Elberta; good for canning and freezing. Tree: strong, vigorous; very productive; hardy; chilling requirement of 700 to 750 h at 45F; leaf glands globose; flower large, showy, rose colored; tolerant to *Xanthomonas pruni* and *Sclerotinia fructicola.*

LaJewel.—Orig. in Calhoun, La., by C.E. Johnson and W.A. Young. Introd. in 1988. Redglobe x Prairie Rose. Cross made in 1959; selected in 1962; tested as L-9-6-4. Fruit: medium to large, up to 76 mm; skin medium red over 40% of surface, ground color deep yellow; pubescence short; flesh yellow; firm; freestone; occasionally flesh at stem end separates from stone. Ripens about 15 days after Elberta. Flowers: showy; rose; self-fertile; tolerant of spring frosts. Leaf glands reniform. Tree: vigorous; productive; resistant to bacterial spot; no tendency to sun scald or cold damage. Chilling requirement about 850 h.

La Pecher.—Orig. by Louisiana Agr. Expt. Sta. at Baton Rouge in 1971. Selected in 1973 by P.L. Hawthorn, tested as L71-A72-23, and released by J.E. Boudreaux et al. in 1984. La Feliciana open-pollinated. Fruit: medium large; round; 85% bright red over medium yellow; flesh yellow with red flecks throughout, firm, melting, semi-freestone. Ripens 9 days before Harvester. Flowers numerous, nonshowy, light pink with rose margins; leaf glands reniform. Good resistance to bacterial spot. Tree: vigorous; highly productive. Chilling requirement 450 h.

LaPremiere.—Orig. in Calhoun, La., by P.L. Hawthorne, J.C. Taylor, A.C. Metler, J.C. Metler, and R.S. Woodward, North Louisiana Agr. Expt. Sta. Introd. in 1965. Redglobe x Prairie Rose; cross made in 1958; tested as L-9-5-92. Fruit: large, up to 2 3/4 inches in diam.; round; halves equal; nonprominent suture; skin yellow, highly blushed, pubescence short, very attractive; flesh clear yellow, firm, texture good, quality very good, freestone; cans and freezes very well; ripens about 12 days before Elberta. Tree: strong, vigorous; hardy; very productive; flower large; leaf glands globose; rest period of fruit and leaf buds broken by 900 to 950 h at or below 45F; resistant to bacteriosis and brown rot.

LaRed.—Orig. in Calhoun, La., by P.L. Hawthorne, J.C. Taylor, A.C. Miller, N.L. Horn, D.W. Newsom, and R.S. Woodward, Louisiana Agr. Expt. Sta. Introd. in 1967. Red Skin x Southern Glo. Cross made in 1960; tested as L-1-2-39. Fruit: large, 2 1/4 to 2 1/2 inches in diam.; round; skin yellow ground color covered with 90% to 95% red blush, attractive, pubescence medium, may have a slight suture some seasons; flesh yellow with some red, very firm, texture and quality good, freestone; ripens 12 days before Elberta. Tree: strong, vigorous; very productive; hardy; leaf glands globose; flowers large, showy, dark rose color; chilling requirement 850 to 900 h; tolerant to *Xanthomonas pruni* and *Sclerotinia fructicola.*

Last Chance Number One.—Orig. by J.W. Spraque, Lancaster, Calif. Plant patent 6119; 1 Mar. 1988. Last Chance mutation, blooms and ripens later; very small leaves.

Late Alamar.—Orig. in Newcastle, Calif., by E.F. Fowler, Fowler Nurseries. Introd. in 1960. Parentage unknown; selected in 1954. Fruit: large, up to 3 1/2 inches in diam.; skin blushed, attractive; flesh yellow, well flavored, quite firm, freestone; ripens in mid-September; resembles Alamar.

Late Babcock.—Orig. in Strathmore, Calif., by Samuel F. Molfino. Introd. in 1951. Plant patent 2014; 24 Jan. 1961; assigned to Kenneth R. Clifford, Strathmore. Heath Cling o.p.; discovered in 1925. Fruit: large; skin medium thick, light yellow with greenish cast, partially blushed red, pubescence slight, sometimes glabrous; flesh white to cream, mottled with red next to pit, subacid, moderately juicy, freestone; ripens about 10–26 Aug. Tree: large; vigorous; upright; bears regularly; very productive; flowers large; leaf glands reniform.

Late Bess.—See Swetzer.

Late Canner (*Merrill Late Canner*).—Orig. in Red Bluff, Calif., by Grant Merrill. Introd. in 1950. J.H. Hale o.p.; selected in 1947. Fruit: flesh yellow, clingstone; 8 weeks later than J.H. Hale, which it resembles. Tree: size medium; regular bearer. Variety now obsolete.

Late Elberta.—See Shippers Late Red.

Late John Gee (*John G.*).—Orig. in Madera, Calif., by John G. Garabedian. Introd. in 1952. Plant patent 1141; 28 Oct. 1952. Kirkman Gem o.p.; discovered in 1949. Fruit: resembles Late Kirkman being distinguished therefrom by a rounder shape, more yellow, ground color, and ripening 2 weeks later; fruit can remain on tree for a considerable time without over-ripening, permitting longer harvest period. Tree: large; vigor medium; productive; leaf glands reniform; flowers large, pink.

Late Kirkman.—Orig. in Madera County, Calif., by William T. Kirkman. Introd. in 1951. Plant patent 920; assigned to Kirkman Corp., Tracy, Calif.; 7 Feb. 1950. Kirkman Gem o.p. Fruit: flesh yellow, freestone; yellow suture line ripens evenly with the fruit as distinguished from the reddish early ripening suture line of Kirkman Gem, which it resembles, but has a later ripening period, mid-September to mid-October; hangs well on tree. Tree: vigorous grower; leaf glands reniform; flowers large, pink.

Late Rio (*Merrill Late Rio*).—Orig. in Red Bluff, Calif., by Grant Merrill, Grant Merrill Orchards. Introd. in 1951. Plant patent

1098; 20 May 1952. Kirkman Gem o.p.; selected in 1950. Fruit: skin highly colored; flesh yellow, freestone, does not darken upon exposure to air, sweeter and firmer than known parent; no soft suture; same season as Kirkman Gem, 5 1/2 weeks later than Elberta. Tree: size medium; vigor medium; productive; leaf glands reniform; flowers large, pink.

Late Rio Oso Gem.—See Kirkman Gem.

Laterose.—Orig. in New Brunswick, N.J., by M.A. Blake, New Jersey Agr. Expt. Sta. Introd. in 1945; J.H. Hale x Delicious; tested as N.J. 109. Fruit: medium to large; ovate; skin with attractive blush; flesh white, red at pit, fine-textured, melting, firm, freestone; ripens late, 10 days after Elberta; excellent for home canning and freezing. Tree: more hardy than Elberta; vigorous and productive; susceptible to brown rot; flower size medium, nonshowy; leaf glands reniform; chilling requirement 850 h.

Late Sunhaven.—Orig. in South Haven, Mich., by South Haven Expt. Sta. Introd. in 1960. Sunhaven mutation; discovered in 1959. Fruit: resembles Sunhaven with more red color and ripening 1 week after Sunhaven; ripens 3 days before Redhaven, which it also resembles, but flesh characteristics not as good. Tree: resembles Sunhaven exactly except Sunhaven has small, globose leaf glands and late Sunhaven has large, reniform leaf glands.

Laurol.—Jerseyqueen mutation, found by D. Reuter in New Jersey. Plant patent 8558; 25 Jan. 1994; assigned to New Jersey Peach Council, Clayton. Fruit very large; skin attractive 75% red skin on orange ground color; slight pubescence; slightly ovate with distinct suture; flesh yellow, firm, melting; freestone, with red streaks radiating from red pit cavity. Ripens 14 days after Jerseyqueen. Flowers: late, large, showy, pink; leaf glands reniform. Tree: very large, vigorous; spreading, willowy; moderately hardy, similar to Jerseyqueen; moderately tolerant to bacterial leafspot.

La White.—Orig. by Louisiana Agr. Expt. Sta. at Baton Rouge in 1971. Selected in 1973 by P.L. Hawthorn, tested as L71-A64-42W, and released by J.E. Boudreaux et al. in 1984. Nectar o.p.2. Fruit: medium large; round; 75% bright red overcolor; flesh white with red flecks throughout, firm, melting, semi-freestone. Ripens 27 days before Elberta. Flowers numerous, nonshowy, light pink with rose margins; leaf glands reniform. Tree: vigorous and productive. Chilling requirement 650 h.

Lawrence.—Orig. in Puyallup, Wash., by Alex Lawrence. Introd. in 1941. Parentage unknown. Fruit: semi-freestone; resembles Rochester in fruit characteristics and quality, but ripens 2 weeks earlier. Tree: bears early and regularly; hardy to spring frosts.

Le Duke.—See Fertile Hale.

Leeton.—Orig. in Palo Alto, Calif., by W.F. Wight, U.S. Dept. of Agriculture. Introd. in 1935. Leader o.p. Fruit: large, averaging 6.3 to 8.0 cm in diam.; skin golden yellow, mottled, and splashed with bright red, pubescence medium; flesh lemon yellow, fairly firm, juicy, ripens from outside in, mildly subacid; stone small, semi-cling; a good dessert type; not adapted to canning or drying; ripens early, with Triumph. Tree: vigorous; leaf long, 6 to 7.5 inches, glands reniform; flower showy, petals large; not adapted to areas where delayed foliation is a problem.

Legacy.—Orig. in Princess Anne, Md., by Homer S. Kemp, Bountiful Ridge Nurseries. Introd. in 1971. Parentage unknown, discovered in 1960, selected in 1964, tested as BRN 6 and H.K. 6. Fruit: size medium to large; round; skin yellow overspread with bright red; flesh yellow, freestone; for fresh market; ripens 2 to 3 weeks before Redhaven. Tree: large; shape similar to Redhaven; vigor excellent; hardy; very productive; flower large, showy.

Legend.—Orig. at North Carolina Agr. Expt. Sta. Bred by F.E. Correll in 1974, selected in 1977, tested as NCT 516, and introduced in 1991 by D.J. Werner. Winblo x complex parentage of North Carolina selections. Fruit: large; round; skin dull red over yellow ground color; flesh yellow, very firm, melting, freestone. Ripens 7 to 10 days after Emery. Flowers nonshowy; leaf glands reniform. Tree: moderate resistance to bacterial spot; very bud hardy; high number of flower buds. Chilling requirement 950 to 1000 h.

Lena.—Orig. in Parkersburg, W.Va., by S.S. Shimer, Rose Hill Nursery, Annamoriah, W.Va. Introd. about 1922. Parentage unknown; discovered about 1915. Fruit: very large; flesh firmer but less juicy than Elberta, which it resembles. Now obsolete.

Levi 5.—Origin obscure. Probably synonymous with or a seedling of Levy.

Licking.—Orig. in Neward, Ohio, by the Price Fruit Farm. Introd. in 1930. Parentage unknown; discovered in 1930. Fruit: size medium; skin highly colored; flesh yellow, freestone; ripens about 10 Aug. in place of origin. Variety obsolete.

Lindy.—Orig. in Kingsburg, Calif., by Helge W. Lindholm. Introd. in 1962. Plant patent 2135; 17 Apr. 1962; assigned to Reedley Nursery, Reedley, Calif. Chance seedling; discovered in 1955. Fruit: large, averaging 3 inches in diam.; skin yellow, partially blushed, pubescence moderate; flesh yellow, streaked with red near pit, firm, moderately juicy, slightly acid, quality good, freestone; ships, cans, and freezes well, tends to be nonbrowning; hangs well on tree; ripens about 2 days after Elberta and before Rio Oso Gem. Tree: self-fruitful; large; vigorous; very productive; bears regularly.

Lizbeth.—Orig. in Red Bluff, Calif., by Grant Merrill (deceased). Introd. in 1972. Plant patent 3589; 13 Aug. 1974. Fortyniner x Gemfree. Fruit: 2 3/8 inch axial diam., 2 1/2- to 3-inch transverse diam.; globose to cordate with some tip; skin redder than Harmony; flesh moderate amygdalin, moderate juice, fine, mild, delicate, freestone; a shipping var.; ripens about 25 June. Tree: size medium to large; vase-shaped; vigorous; heavy bearer for its season; flower medium small, pink to pinkish-salmon, abundant pollen; leaf glands reniform; with proper handling may be mechanically harvested in one picking.

Loadel.—Orig. in Yuba City, Calif., by Howard H. Harter, Harter Packing Co. Introd. about 1950. Probable offshoot of the Lovell seedling rootstock; discovered in 1946. Fruit: flesh yellow; commercial canning clingstone; of good shape; same season as that of Fortuna; leaf glands globose; flowers nonshowy. Tree: bears heavily; vigorous grower.

Loring.—Orig. in Mountain Grove, Mo., by Paul H. Shepard, Missouri State Fruit Expt. Sta. Introd. in 1946. Frank x Halehaven; selected in 1943; tested as PH-811. Fruit: large; round to ovate; skin blushed over most of surface, yellow ground color, pubescence light, very attractive; flesh yellow, firm, melting, medium texture, red at pit, quality very good, freestone; ripens 10 days before Elberta. Tree: very vigorous; regular production even in adverse spring weather; tolerant to bacteriosis; flower large, showy; leaf glands globose; chilling requirement 950 h.

Los Angeles.—Orig. in Norwalk, Calif., by M. Veyna, Orange County Nursery. Introd. in 1950. Parentage unknown; selected in 1947. Fruit: resembles Elberta in size, but rounder and of lighter skin color; flesh yellow, freestone; high canning and shipping quality. Obsolete.

Lovell Cling.—See Wiser.

Lucas (*Lucas Beauty*).—Orig. in Cucamonga, Calif., by Vincent G. Lucas. Introd. in 1933. Plant patent 81; 14 Nov. 1933; assigned to the Lucas Ranching Co., Cucamonga. Parentage unknown; discovered about 1927. Fruit: nearly round; skin yellow, blushed; flesh yellow, red at pit, firm, quality good, freestone; ripens late in October at Ontario, Calif. Tree: resistant to delayed foliation; leaves narrow, bears very well.

Lucas Beauty.—See Lucas.

Lyman Late.—Orig. in St. Helena, Calif., by W.W. Lyman. Introd. about 1925. Parentage unknown; discovered in 1920. Fruit: large; flesh yellow, freestone; ripens 10 to 14 days later than Krummel at St. Helena.

M.A. Blake.—See Blake.

Magraw.—See McGraw.

Madeleine Pouyet (*Nivolet, Primissima Delbard, May Flower précoce Madeleine Pouvet*).—Orig. in Saint-Paul-de-Romans (26), France, in the orchard of Mr. Aimé Pouyet by a Mr. Noir, Pepinieres du Val d'Or, Manthes (26), France. Introd. in 1950 in France, in 1961 in the U.S. as P.I.

277702. Presumably a Mayflower mutation; discovered in 1935. Fruit: size 60 to 90 g, maximum 125 g; round; skin three-fourths covered with red blush; suture quite deep, apex nippled; flesh white, no red at pit, tender, moderately sweet, same quality as Mayflower; stone tends to split, cling; one of earliest ripening varieties, 9 days before Mayflower. Tree: size moderate; vigor good; susceptible to bud drop in some seasons; productivity good; flower small, leaf glands globose. Recommended for early ripening regions.

Madison.—Orig. in Blacksburg, Va., by George D. Oberle, Virginia Agr. Expt. Sta. Introd. in 1963. Ideal x Redhaven; cross made in 1952; selected in 1955; tested as V.P.I. 54. Fruit: averages 2 5/8 inches in diam.; skin bright red over a bright orange-yellow ground color, very attractive; flesh orange-yellow, bright red next to pit, exceedingly firm, texture fine, free from stringiness; flavor mild and rich, quality considered as good as Redhaven; ripens 7 days before Elberta. Tree: vigor average; fruit buds set heavily; very tolerant to blossom-season frost; flower nonshowy, large, self-fruitful.

Maffey.—Orig. in Modesto, Calif., by Albert W. Maffey. Introd. in 1958. Plant patent 1713; 27 May 1958. Parentage unknown; seedling discovered in 1956. Fruit: medium to large; skin light green, with rose blush at stem end; flesh white, red next to pit, does not oxidize readily, texture medium, flavor subacid, good, freestone; freezes well; resembles Nectar, but ripens end of August. Tree: large; vigorous; spreading; leaf glands globose; flower size medium, salmon.

Magnolia.—Orig. in Fort Valley, Ga., by Wm. J. and E. Clyde Wilson, Magnolia Farms. Introd. in 1970. Coronet mutation; discovered in 1961, tested as W100. Fruit: 2 to 2 1/2 inches in diam.; round; skin ground color yellow, 80% blushed bright red, attractive; suture has a narrow green hairline dividing both halves; flesh yellow, melting; ripens 3 days later than Redcap; semi-clingstone. Tree: vigor moderate; productivity similar to Coronet; susceptible to brown rot; flower small, dark pink; leaf glands reniform.

Maiden (*Merricle B-108*).—Orig. in Red Bluff, Calif., by Grant Merrill. Introd. in 1958. Plant patent 1870; 6 Oct. 1959. J.H. Hale F4; discovered in 1954; selected in 1956; first fruited in 1958. Fruit: large; skin highly colored red; flesh yellow, red next to pit, firm, freestone; ripens uniformly over the tree, about 5 days before Elberta; resembles Gemfree. Tree: size medium; vigor medium; productive; leaf glands reniform; flower small, pink. No longer propagated.

Majestic.—Orig. in Calhoun, La., by P.L. Hawthorne, J.C. Taylor, T.J. Raiford, W.A. Young, N.H. Horn, R.S. Woodward, and D.W. Newsom. Introd. in 1979. Meadowlark o.p.; selected in 1963. Fruit: large; slightly oval; almost equal halves; attractive; excellent quality, firmness and texture; flesh golden-yellow with red flecks throughout; skin 60% medium red over-color; freestone; ripens 3 to 4 days before Redglobe (22 June–1 July), between Harvester and Redglobe. Tree: productive; vigorous; hardy; moderately resistant to bacteriosis (*Xanthomonas pruni*) and brown rot; flower large, showy, rose-colored; 800 to 850 h.

Malherbe.—Orig. in Stellenbosch, South Africa, by P.A.L. Steyn, Fruit and Food Technology Res. Inst. Introd. in 1961. Walgant x (Kakamas x Early Dawn); selected in 1960; sibling of Black and Neethling. Fruit: flesh yellow, no red at pit cavity, clingstone; excellent canning quality; ripens very early, 55 days before Kakamas, with a statistical mean ripening date of 20 Dec.; resembles Kakamas. Tree: very resistant to delayed foliation.

Maluti.—Orig. in Elsenburg, South Africa, by A.F. de Wet, Stellenbosch-Elsenburg College of Agr., Dept. Agr., Stellenbosch. Introd. there in 1950; imported into U.S. in 1954 as P.I. 219560 and in 1956 as P.I. 234196, distributed in U.S. in 1965. Transvaal o.p.; selected in 1942. Fruit: skin yellow; flesh deep yellow, firm; commercial canning clingstone; resembles Kakamas. Tree: vigorous; prolific cropper; low winter-chilling requirement.

Mamacita.—Orig. in Sebastopol, Calif., by James Robinson. Introd. in 1968. Parentage unknown; seed planted in 1963; discovered in 1966; selected in 1967. Fruit: size medium; very pointed at lower end, resembling a nipple; flesh yellow, medium juicy, firm, flavor and quality good, freestone; resembles Honey; ripens late July; needs thinning. Tree: size medium; vigor medium; precocious and very productive; tolerant to leaf curl; flowers deep rose, double, blooms 2 months; leaf glands globose.

Maravilha.—Orig. in Agr. Expt. Sta., Univ. of Florida, Gainesville, by R.H. Sharpe and named in southern Brazil. Complex parentage of Florida selections; heterozygous for nectarine. Selected in 1966; tested as Fla. 13-72. Fruit: medium small; skin 90% red blush; flesh white; softens first at tip; semicling. Ripens 80 days after bloom. Flower showy; reniform leaf glands. Moderately resistant to bacterial leaf spot. Chilling requirement 250 chill units.

Marcus.—Orig. in Jackson Springs, N.C., by G.W. Schneider, C.N. Clayton, and F.E. Correll, Sandhills Peach Res. Sta., North Carolina Agr. Expt. Sta. Introd. in 1959. Erly-Red-Fre selfed; first fruited and selected in 1956; tested as N.C. 1179. Fruit: averages 2 to 2 1/4 inches in diam.; round , pointed on tip, being similar to its parent; skin surface shows 60% to 70% red over yellow ground color when fully ripe; flesh yellow, relatively firm for an early maturing variety; suture rough, sometimes softening early; clingstone; ripens very early, 2 to 7 days before Mayflower or 10 days before Cardinal; recommended for local use rather than shipping purposes. Tree: moderately vigorous, productive; flowers showy; moderately susceptible to bacterial spot.

Mardigras.—Orig. in Red Bluff, Calif., by Grant Merrill. Introd. in 1968. Plant patent 2965; 27 Jan. 1970. J.H. Hale x (J.H. Hale x Levy) F_3. Cross made in 1957; selected in 1967; tested as 39-56. Fruit: 2 1/2 to 3 1/2 inches in diam.; skin highly colored; flesh yellow; firm, freestone; excellent for freezing; good shipping qualities; resembles Parade; ripens with Parade in early September. Tree: large; vigorous; productive; leaf glands reniform; flower small, pink.

Margaret Kane. —Orig. in Middletown, N.J., by William Kane. Introd. in late 1930s. Bud mutation of Golden Jubilee; discovered in 1934. Fruit: flesh yellow, freestone, too soft for shipping; ripens 10 to 14 days earlier than parent, which it resembles. Obsolete.

Marglow.—Orig. in College Park, Md., by I.C. Haut, Maryland Agr. Expt. Sta. Introd. in 1966. Redskin o.p. Fruit: similar to Redskin, including high red color and firmness, but ripening 3 days after Redskin. Tree: lacks willowy type of growth characteristic of Redskin.

Marhigh (*Supernuts*).—Orig. in College Park, Md., by I.C. Haut, Maryland Agr. Expt. Sta. Introd. in 1966. (J.H. Hale x Carmen selection) o.p. Fruit: excellent size; skin has good red color; good shipping firmness; resembles Redskin, but ripens 5 to 7 days after Redskin.

Marigold.—Orig. in New Brunswick, N.J., by M.A. Blake, New Jersey Agr. Expt. Sta. Introd. in 1925. Lola x Arp; cross made in 1916. Fruit: size medium; ovate; skin blushed and mottled bright red overlight yellow ground, tender, peels easily, attractive; flesh deep yellow, red at pit, melting, soft, very good quality for its season, semiclingstone; ripens early, with Arp and 1 mo. before Elberta; difficult to pick because of strong attachment and softening at maturity. Tree: medium vigor; high productivity, requires heavy thinning; flowers large, showy; leaf glands reniform.

Marilyn.—Orig. in Beamsville, Ont., Canada, by T. Konkle. Introd. about 1940. Chance seedling; discovered in 1935. Fruit: skin more highly colored than Elberta; flesh yellow, freestone; ripens with Elberta. Tree: unproductive.

Marina.—Orig. in Goede Rust, French Hoek, South Africa, by H.S.H. Page. Introd. about 1935. Duke of York mutation. Fruit: small to medium; flattish; skin three-fourths to full red, has so little fuzz it appears smooth; flesh greenish-white, juicy; stone white, large, semi-free, inclined to split; ripens with Duke of York. Tree: vigorous; upright; subject to delayed foliation.

Marion.—Orig. in Yuba City, Calif., by C.E. Sullivan. Introd. in 1964 by Sierra Gold Nurseries, Yuba City. Gaume mutation; discovered in 1957. Fruit: larger than Gaume; no blossom-end breakdown nor splitting of pit; flesh yellow, firm; a commercial canning clingstone; ripens with Gaume and about 10 days before Halford;

resembles Gaume. Tree: sets heavier crops than Gaume. No longer commercially planted.

Mark-Berta.—Orig. in Xenia, Ill., by J.E. Markham, Markham Nurseries. Introd. about 1932. Hal-Berta Giant x Canadian Queen. Fruit: size medium; skin yellow with some blush; flesh yellow, firm, freestone; ripens about Elberta season. Obsolete.

Markham's Jewel.—Orig. in Xenia, Ill., by J.E. Markham, Markham's Nursery. Introd. about 1930. Parentage unknown. Fruit: skin with high blush; flesh yellow, freestone.

Mark-Late.—Orig. in Xenia, Ill., by J.E. Markham, Markham's Nursery. Introd. in the 1930s. Canadian Queen x Mark-Berta. Fruit: large; flesh yellow, freestone; ripens about 5 days after Elberta.

Marland.—Orig. in College Park, Md., by I.C. Haut, Maryland Agr. Expt. Sta. Introd. in 1966. Schooldays o.p. Fruit: large; skin high red color; flesh firm, quality excellent; resembles Redhaven; ripens 5 to 8 days after Redskin.

Marpride.—Orig. in College Park, Md., by I.C. Haut, Agr. Expt. Sta., Univ. of Maryland. Introd. in 1971. Schooldays o.p. Discovered in 1963, selected in 1965, tested as Md. 3-39. Fruit: large; very slightly oval; skin brightly colored; flesh firm highly flavored for season of ripening, freestone; ripens 1 to 2 days before Marqueen. Tree: average size; vigor good; productive; flower pink, showy; leaf glands reniform. Recommended for mid-Atlantic region.

Marqueen.—Orig. in College Park, Md., by I.C. Haut, Maryland Agr. Expt. Sta. Introd. in 1966. Schooldays o.p. Fruit: large; skin color high; flesh yellow, firm, freestone; resembles Redhaven; ripens 12 to 15 days after Redskin; has good commericial qualities.

Marsun.—Orig. in College Park, Md., by I.C. Haut, Maryland Agr. Expt. Sta. Introd. in 1966. Schooldays o.p. Fruit: good size, quality and color for the season; flesh yellow, freestone; ripens very late, 28 to 31 days after Redskin.

Mary Jane.—Orig. in Corsicana, Texas, by J. Walter Little, Little Nursery. Introd. in 1935. Chance seedling; discovered in 1931. Fruit: large; round; flesh white, some red at pit, very sweet; semi-clingstone; ripens 15 June at place of origin; resembles Early Wheeler. Tree: productive; vigorous; hardy.

Massasoit.—Orig. in New Brunswick, N.J., by M.A. Blake, New Jersey Agr. Expt. Sta. Introd. in 1925. Slappey x Admiral Dewey; cross made in 1916. Fruit: skin tender, dark red; flesh yellow, medium firm, melting, juicy, freestone; ripens with Triogem and Carman, just after Oriole. Tree: hardy; late blooming.

Maxine.—Orig. in Palo Alto, Calif., by W.F. Wight, U.S. Dept. of Agriculture. Introd. in 1935. Mich. No. 1 Late x Lemon Free; cross made in 1919 by Michigan Agr. Expt. Sta., South Haven; seed taken to California; resultant seedling planted in 1922 at Palo Alto; first fruited in 1925. Fruit: large, about 7.3 cm in diam.; nearly round, sides equal, symmetrical; suture well-defined; skin lemon yellow, somewhat mottled and splashed red where exposed to sun, fairly tough, pubescence medium; flesh clear yellow with no red at pit, fine-textured, juicy, flavor mild, quality excellent; stone small, free; ships and cans well; sets heavily, need thinning; ripens in early midseason. Tree: very heavy producer; flower showy, petals large; leaf glands globose; similar to Elberta in amount of cold necessary to break the rest period.

Maybelle.—Orig. in New Brunswick, N.J., by M.A. Blake, New Jersey Agr. Expt. Sta. Introd. in 1948. Raritan Roge x [(J.H. Hale x Goldfinch) o.p.]; tested as N.J. 164. Fruit: size medium; ovate, unequal halves; skin bright red over greenish-white ground, pubescence fairly heavy; flesh white, clingstone, melting, firm, quality fair; ripens 7 weeks before Elberta. Tree: medium vigor, hardier than Elberta, high productivity; flower medium, nonshowy; leaf glands reniform; chilling requirement 800 h.

May Crest.—Orig. in Reedley, Fresno County, Calif., by W.Y. Minami. Plant patent 4064; 21 June 1977. Discovered in 1973; Springcrest mutation. Tree and fruit similar to parent but ripens 7 to 10 days earlier.

May Flower précoce Madeleine Pouvet.—See Madeleine Pouyet.

Maydon.—Orig. in Ontario, Calif., by George P. Weldon, Chaffey College. Introd. in 1948. Mayflower x Weldon; selected in

1945. Fruit: very large;skin light green or yellow with a blush; flesh white, juicy, quality good, freestone; ripens with Babcock. Tree: bears well; recommended for planting in southern California. Variety now obsolete.

Maygold.—Orig. in Fort Valley, Ga., by John H. Weinberger, U.S. Dept. of Agriculture Hort. Field Sta. Introd. in 1953. Sunhigh x Southland; cross made in 1945; selected in 1948; tegted as USDA FV 13212. Fruit: size medium; ovate; skin has a light red blush covering about half the fruit, over an attractive yellow ground color, pubescence very light; flesh yellow, firm but melting, medium textured, of good flavor, clingstone; ripens 5 to 6 weeks before Elberta; for early shipping. Tree: vigorous and productive; leaves and fruit subJect to bacterial spot disease; flowers small petaled and self-fruitful; leaf glands reniform; outstanding characteristic is its relatively low chilling requirement, about 650 h; blooms 2 to 15 days earlier than Hiley and as much as 5 weeks earlier than Elberta at Fort Valley.

May Lady.—Orig. in Red Bluff, Calif., by Grant Merrill, Exeter, Calif. Introd. in 1968. Gemfree x Blazing Gold. Cross made in 1956, selected in 1959, tested as 57-131. Fruit: size medium; generally globose; skin smooth, pubescence very short, one-half to three-fourths covered with red blush; flesh firm for the geagon, clingstone; shipping qualities good; has few split pits; ripens late May and early June in San Joaquin Valley. Tree: size medium, vase shaped; vigor moderate; productivity considered good for its season; flower small, blooms early to midseason, abundant pollen; leaf glands reniform.

May Pride.—Orig. by C.F. Zaiger, C.N. Zaiger, L.M. Gardner, and G.G. Zaiger, Modesto, Calif. Plant patent 7775; 28 Jan. 1992. [(Red Grand x Desert Gold) F_2] x (Ruby Gold x unknown low-chill peach). Fruit: medium; firm; attractive red skin; fresh yellow, melting, clingstone. Flowers large, showy; leaf glands globose. Tree: large; vigorous; upright; productive. Chilling requirement 250 h.

May Sun.—Orig. in Del Ray, Calif., by Tom Chamberlain, H.P. Metzler & Sons. Introd. in 1989. Plant patent 6992; 22 Aug. 1989; owned by Metropolitan Insurance Co.; licensed exclusively to Agri Sun Nursery. Parentage unknown. Fruit: large, about 3.5 inches; round; skin medium red on bright yellow ground; flesh yellow, clingstone; ripens 25 days before Redhaven in Selma, Calif. Tree: very vigorous; very productive. Chilling requirement about 300 h.

Maytime.—Orig. in Le Grand, Calif., by F.W. Anderson. Introd. in 1958. Plant patent 1725; 8 July 1958; assigned to Kim Bros., Reedley, Calif. Sun Grand nectarine x Robin; selected in 1956; tested as 3B100. Fruit: skin bright red over a pale greenish-yellow ground color; flesh white, clingstone; matures very early, about 1 week before Mayflower; resembles Springtime. Tree: size medium; vigorous; spreading; leaf glands reniform; flower large, showy.

Maywel.—Orig. in Ontario, Calif., by George P. Weldon, Chaffey College. Introd. in 1948. Mayflower x Weldon; selected in 1945. Fruit; large; shape excellent; gkin yellow or greenish with attractive blush; flesh white, freestone; ripens about 5 days ahead of Babcock. Tree: very fruitful; early foliation. Obsolete.

McGraw (*Arkansas Seedling, Magraw*).—Orig. in Republic, Mo., by W.T. Barron, W.T. Barron & Son Nursery. Introd. about 1920. Chance seedling; discovered about 1910. Fruit: size medium; skin blushed on one side; flesh white, firm, very sweet, clingstone; keeps well in cool storage; ripens in early September; resembles Heath Cling. Tree: often propagated by seed; buds quite resistant to frost.

McGuigan.—Orig. in Cedar Springs, Ont., Canada, by V.G. McGuigan. Introd. in 1935. Plant patent 624; 18 Apr. 1944; assigned to E.D. Smith & Sons, Limited, Winona, Ont., Canada. Possibly Elberta x Chili; selected in 1922. Fruit: resembles Elberta in size and shape; skin yellow with red cheeks; flesh yellow, freestone; ripens before Elberta, 1–10 Sept.; for dessert and canning. Tree: large; vigorous; productive; leaf glands reniform; flowers medium large, pink, self-fertile.

McKay.—Orig. in Waterloo, Wis., by the McKay Nursery Co. Introd. in 1945. Chance seedling; discovered in 1942. Fruit: size medium; flesh yellow, freestone; resembles Rochester. Tree: some hardiness, survives some winters, but cannot maintain long life or sustained yields.

McKnight (*Halford 1*).—Orig. in Modesto, Calif., by John Halford. Introd. in 1921. Parentage unknown. Fruit: flesh yellow; ripens 3 to 4 days before Halford; commercial canning clingstone. Tree: leaf glands reniform.

McKune.—Orig. in Davis, Calif., by L.D. Davis, California Agr. Expt. Sta. Introd. in 1964. [(Paloro x Round Tuscan) x Dixon 2] [(Orange Cling seedling x Alameda) x (Goodman Choice x Transvaal)]; cross made in 1950; selected in 1954; tested as Calif. GH 11-49; sibling of Everts. Fruit: medium to large; skin blush medium, with an excellent harvest color; flesh clear yellow; pit pink; when canned, flavor superior to existing conventional commercial canning clingstone varieties; maintains its superior flavor quality; it appears to process satisfactorily; ripens about 5 days after Everts, which it resembles. Tree: consistently sets good crops; leaf glands globose; flower like that of Dixon 1, being of the larger nonshowy type.

McNeely.—Orig. in Beltsville, Md. Introd. in 1972 by U.S. Dept. of Agriculture and South Carolina Agr. Expt. Sta., Clemson Univ., Clemson, S.C. Kalhaven x (Halberta x Goldenglobe). Planted at Beltsville in 1943; fruited at Sandhill Expt. Sta., Columbia, S.C. since 1962. Fruit: medium to large; skin attractive, 50% covered with light red blush at maturity, in western U.S. does not color well; pubescence light; flesh yellow, firm quality good, freestone; ripens with Keystone. Tree: vigor moderate; more productive than Keystone; leaves susceptible to bacterial spot (*Xanthomonas pruni*); chilling requirement 900 h below 45F; flower salmon color, small nonshowy, self-fertile; leaf glands reniform.

Meadow Lark.—Orig. in Ontario, Calif., by Walter E. Lammerts. Introd. in 1947. Plant patent 528; 30 June 1942; assigned to Armstrong Nurseries, Ontario, Calif. (Early Imperial x Coolidge Double Red) x Socala; selected in 1941. Fruit: medium large, 2 to 2 1/8 inches in diam.; broadly ovoid, unsymmetrical; skin thin, free, yellow, blushed, dots numerous, pubescence moderate; flesh yellow, juicy, medium firm, melting, subacid, mild, eating quality good, freestone; keeping quality good; ripens early, mid-June in southern California, about 28 days before Early Imperial; somewhat resembles Elberta. Tree: large; vigorous; very productive; bears regularly; leaf glands globose; flower large, blooms early; low winter-chilling requirement.

Melba (*Improved Pallas*).—Orig. in San Antonio, Texas, by a Mr. Yost. Introd. in 1936. Parentage unknown; propagated as Improved Pallas; Wolfe Nursery, Stephenville, Texas, obtained buds from F.P. Wittman, Horticulturist, Missouri Pacific Lines, Dilley, Texas, and renamed it Melba. Fruit: size of Belle of Georgia; pit small; honey flavor; ripens 25 days ahead of Elberta. Tree: adapted to mild winters.

Melvin.—Orig. in Marshalltown, Iowa, by Wheelock Wilson. Introd. in 1947 by Wheelock Wilson Nursery, Marshalltown. Bailey mutation; discovered about 1938. Fruit: small, 1 1/4 to 1 1/2 inches in diam.; flesh yellow, flavor sweet, freestone. Tree: very hardy; bears prolifically.

Meredith.—Orig. in New Brunswick, N.J., by M.A. Blake, New Jersey Agr. Expt. Sta. Introd. in 1955 by the New Hampshire and New Jersey Agr. Expt. Sta. Slappey x Dewey; selected in 1924; tested as N.J. 41SD. Fruit: large; flesh yellow, freestone, ripens 1 Sept. in Durham, N.H. Tree: hardy in southern New Hampshire; flower nonshowy; has value as a parent in breeding for winter hardiness.

Merriam.—Orig. in Davis, Calif., by L.D. Davis, Agr. Expt. Sta., Univ. of California, Davis. Introd. in 1971. Everts selfed. Cross made in 1956, first fruited in 1960, tested as G44-1E. Fruit: size medium; truncate, symmetrical; skin yellow ground cover with slight blush, excellent harvest color, pubescence moderate; flesh clear yellow, flavor excellent, clingstone; introduced because of superior flavor when canned; ripens 34 days before Corona. Tree: size medium; upright; productivity high; flower nonshowy; leaf glands reniform.

Merricle B-108.—See Maiden.

Merricle L-14.—See Pacifica.

Merricle L-75.—See Merrill Gem II.

Merricle L-120.—See Toreador.

Merricle L-129.—Orig. in Red Bluff, Calif.,

by Grant Merrill. Introd. in 1958. Plant patent 1818; 10 Mar. 1959. J.H. Hale F_4; selected in 1954. Fruit: large; variable; flesh yellow, streaked with red, freestone; ripens about 5 days later than Splendor, which it resembles, but is more highly colored than that variety. Tree: bears more regularly than Splendor; leaf glands reniform; flower small; pink; blooms 5 days later than Splendor.

Merricle L-130.—See Scarlet Lady.

Merricle L-152.—See Franciscan.

Merricle 0-99.—See Harmony.

Merricle P-75.—See Gaiety.

Merricle P-77.—See Gaiety 2.

Merricle Y-105.—See Angelus.

Merricle 61-167.—See Parade.

Merricle Schooldays.—See Schooldays.

Merrill Beauty.—Orig. in Red Bluff, Calif., by Grant Merrill. Introd. in 1947. Plant patent 905; 6 Dec. 1949; J.H. Hale o.p.2; selected in 1943. Fruit: large; round; skin highly colored; flesh yellow, freestone; ripens 5 weeks before Elberta and 1 week before Redhaven; resembles J.H. Hale. Tree: strong; vigorous; large; productive and regular bearer; leaf glands globose; flower small, light pink, self-fertile.

Merrill Bonanza.—Orig. in Red Bluff, Calif., by Grant Merrill, Grant Merrill Orchards. Introd. in 1953. (J.H. Hale x Levi F_3) o.p.; selected in July 1948; tested as Merrill C16-51. Fruit: round; skin dull color; flesh yellow, very firm, freestone; ripens during July Elberta season, 15–25 July at Modesto, Calif.; resembles Merrill Gem. Tree: regular bearer.

Merrill Carnival.—See Carnival.

Merrill Dandy.—See Dandy.

Merrill Delicious.—Orig. in Red Bluff, Calif., by Grant Merrill. Introd. in 1947. J.H. Hale o.p.2; selected in 1943. Fruit: small; flesh yellow, flavor very good, freestone; tends to ripen on one side; ripens 5 weeks before Elberta. Tree: size medium; regular bearer; resembles Merrill Beauty but smaller and legs uniform in ripening.

Merrill Fiesta.—Orig. in Red Bluff, Calif., by Grant Merrill, Grant Merrill Orchards. Introd. in 1952. Plant patent 1099; 20 May 1952. Kirkman Gem o.p.; selected in 1950. Fruit: large; more elongated than parent; skin highly colored red; flesh yellow, firm, freestone; no soft suture; season late, being a few days after Kirkman Gem but not as late as Rodeo. Tree: resembles parent, being stronger and more vigorous; leaf glands reniform; flowers large, pink.

Merrill 49'er.—See Fortyniner.

Merrill Gem.—Orig. in Red Bluff, Calif., by Grant Merrill. Introd. in 1947. Plant patent 868; 30 Aug. 1949. J.H. Hale o.p.2; selected in 1943; tested as C5-57. Fruit: large; averaging 2 3/4 inches in cheek diam.; round, uniform, suture and tip well-marked; skin bright red mottled over 90% of surface, coloring early; flesh yellow, some red flecks, very firm, nonmelting, clingstone; ripens 6 weeks before Elberta. Tree: vigorous rather variable in production from year to year; flower small, nonshowy; leaf glands reniform.

Merrill Gem II (*Merricle L-75*).—Orig. in Red Bluff, Calif., by Grant Merrill. Introd. in 1958. Plant patent 1869; 29 Sept. 1959. [July Elberta x (J.H. Hale x Red Bird)] o.p.; selected in 1956; first fruited in 1958. Fruit: skin highly colored red; flesh yellow, very firm, clingstone; ripens about 1 week before Redhaven and 10 days after Gemfree, which it resembles. Tree: size medium; vigor medium; productive; leaf glands reniform; flower small, pink.

Merrill Gemfree.—See Gemfree.

Merrill Gold Rush.—See Gold Rush.

Merrill Hale.—Orig. in Red Bluff, Calif., by Grant Merrill, Grant Merrill Orchards. Introd. in 1953. Plant patent 1246; 19 Jan. 1954. J.H. Hale o.p.; selected in 1942. Fruit: larger, firmer and skin better colored than July Elberta but not as large or as firm as J.H. Hale; flesh yellow, freestone; ships well; season between July Elberta and Elberta; resembles July Elberta and J.H. Hale. Tree: very large; very vigorous; very productive; leaf glands reniform; flower size medium, pink to light red, self-fertile.

Merrill Halloween I.—See Halloween I.

Merrill Halloween II.—See Halloween II.

Merrill Home Canner.—See Home Canner.

Merrill June.—Orig. in Red Bluff, Calif., by Grant Merrill. Plant patent 869; 30 Aug. Introd. in 1947. J.H. Hale o.p.2; selected in 1943. Fruit: resembles Arp, but is firmer, semi-freestone and more pointed; skin highly colored, yellow with red overcolor and splashed in red; ripens with Redbird, 6 weeks before Elberta. Tree: resembles Arp, but less vigorous; productive, regular bearer; leaf glands reniform; flower size medium, pink.

Merrill Late Canner.—See Late Canner.

Merrill Late Gold.—Orig. in Red Bluff, Calif., by Grant Merrill. Introd. in 1950. Kirkman Gem o.p.; selected in 1946. Fruit: flesh yellow, freestone; no soft stripe on suture; resembles Kirkman Gem. Tree: self-sterile.

Merrill Late Rio.—See Late Rio.

Merrill Necta-Heath.—See Necta-Heath.

Merrill Pageant.—See Pageant.

Merrill Rodeo.—See Rodeo.

Merrill Splendor.—See Splendor.

Merrill Surprise.—Orig. in Red Bluff, Calif., by Grant Merrill. Introd. in 1950. July Elberta o.p.; selected in 1944. Fruit: flesh yellow, freestone; ripens 3 weeks earlier than July Elberta, which it resembles. Tree: size medium; sets rather heavily. Obsolete.

Merrill Yellow King.—See Yellow King.

Merritt.—Orig. in Madera, Calif., by Wm. T. Kirkman, Kirkman Peach Co. Introd. about 1948. J.H. Hale x John Rivers nectarine; selected in 1945. Fruit: flesh white, rather firm, freestone; resembles Nectar.

Merritt Gold.—Orig. in Madera, Calif., by Wm. T. Kirkman, Kirkman Peach Co. Introd. about 1950. Kirkman Gem mutation; discovered in 1947. Fruit: flesh yellow, quality good, freestone; resembles Rio Oso Gem; ripens between the seasons of Rio Oso Gem and Kirkman Gem.

Meteor.—Orig. in New Brunswick, N.J., by M.A. Blake, New Jersey Agr. Expt. Sta. Introd. in 1925. Belle selfed; seed gathered in 1914. Fruit: flesh yellow, freestone; ripens just before Elberta. Tree: very hardy; flowers showy.

Michigold.—Orig. in the sand dunes of Michigan by Stahelin's Nursery, Bridoman, Mich. Introd. in 1945. Parentage unknown; discovered in 1935. Fruit: larger than Halehaven; round; skin color poor; flesh yellow, tinged with green, dry, quality poor. Tree: not hardy and not productive at South Haven Expt. Sta., Mich.

Midway.—Orig. in New Brunswick, N.J., by M.A. Blake, New Jersey Agr. Expt. Sta. Introd. in 1939. J.H. Hale x N.J. 27116; cross made in 1923. Fruit: large; round, deep suture; skin blushed dark red; flesh yellow, melting, soft, fibrous, freestone; ripens just after Goldeneast. Tree: vigorous, productivity sometimes low; flower large, nonshowy; leaf glands reniform; chilling requirement 1050 h at or below 45F.

Milam.—Orig. at College Station, Texas, by A.H. Krezdorn, Texas A&M Univ. Seedling derived from Halehaven selfed; selected by J.B. Storey in 1957; tested as A181-4 and released as Milam in 1975 by H.H. Bowden. Fruit: truncate, up to 7.5 cm in diam.; medium yellow color, 60% to 90% solid dark to mottled, light-red blush; pubescence medium; firm, moderately good flavor; storage life relatively long; fruit matures at about same time as Loring; equal to Redglobe in bacterial spot resistance.

Milam.—Orig. in College Station, Texas, by Texas A&M Univ. Introd. in 1977. FV5-56 x (Halehaven selfed). Cross made by A.H. Krezdorn. Selected in 1957 by J.B. Storey and tested as A181-4. Evaluated by H.H. Bowen, J.B. Storey, T.E. Denman. U.A. Randolph, and B.D. Reeder. Fruit: truncate; skin medium yellow with 60% to 90% solid dark to mottled, light-red blush; pubescence medium; flesh uniformly yellow, some red at cavity, firm, fine texture, flavor moderately good. Storage life long; ripens with Loring. Tree: more productive than Loring; equal to Redglobe in bacterial spot resistance; flower red, showy; requires 650 to 750 h.

Missouri.—Orig. in Mountain Grove, Mo., by Paul H. Shepard, Missouri State Fruit Expt. Sta. Introd. in 1946. Sunbeam o.p.; selected in 1942; tested as PH-127. Fruit: larger than Sunbeam; round; resembles Tuscan in appearance; skin blushed, pubescence short; flesh yellow, firm, texture fine, flavor good, quality fair; pit small, round, freestone when ripe; ships well; freezes well; ripens 4 weeks before Elberta. Tree: vigorous; productive; very hardy.

Miss South Dakota.—See Gurney's Dakota.

Moffett.—Orig. in Ceres, Calif., by John Moffett, Superior Fruit Ranch. Introd. in 1942. Lovell o.p.; discovered in 1939. Fruit: flesh yellow, firm, canning clingstone; ripens early for a canning variety. No longer commercially planted.

Monaco.—Orig. in Empire, Calif., by James Monaco, Modesto, Calif. Introd. in 1948. Lovell mutation; discovered in 1937. Fruit: flesh yellow; commercial canning clingstone; leaf glands globose; flowers showy; ripens in Gaume time; resembles Halford.

Monroe.—Orig. in Blacksburg, Va., by George D. Oberle, Virginia Agr. Expt. Sta. Introd. in 1966. Rio Oso Gem x (Shippers Late Red o.p. x Sunhigh). First fruited in 1958; selected in 1959; tested as V.P.I. 58. Fruit: large, 2 1/2 to 3 inches in diam.; skin rich orange-yellow ground cover with 60% bright medium-red blush, pubescence short, attractive; flesh orange-yellow with moderate pinkish-red color adjacent to the pit cavity, firm, smooth, flavor mild, quality well above other varieties ripening in its season; ripens about 11 days after Elberta and 7 days after Rio Oso Gem; tolerant to diseases. Tree: very vigorous; moderately productive; flowers large, pale pink, showy, self-fruitful; leaf glands predominately reniform.

Mountaingold.—Orig. in Blacksburg, Va., by Virginia Agr. Expt. Sta. Introd. in 1968. P.I. 35201 x N.J. 196; cross made in 1951 by L.F. Hough, Rutgers Univ., New Brunswick, N.J.; selected in 1954 by G.D. Oberle, Virginia Polytechnic Inst.; tested as W2-7-22. Fruit: size above average, 2 3/4 inches; skin bright orange-yellow with two-thirds an attractive red blush; flesh firm, nonmelting, yellow with essentially no red pigment near pit; clingstone; ripens 14 days before Elberta; resembles Babygold 5; recommended for commercial processing (canning). Tree: size average; vigor moderate; productivity moderate; flower small, nonshowy, hardiness of fruit buds above average; leaf glands reniform.

Mulberry.—See July Elberta.

Murry Beauty.—See Fair Beauty.

My-O-My.—Orig. in Fresno, Calif., by John M. Garabedian. Introd. in 1970. Plant patent 2963; 20 Jan. 1970; assigned to Peach & Willow Farms, Fresno. Late John Gee o.p. Fruit: 2 1/2 to 3 inches in diam.; unsymmetrical, round ovate; skin yellow, slightly mottled and streaked with red, pubescence moderate; flesh pale yellow, red next to stone and apex, flavor mellow, quality good, firm, freestone; good shipping quality; ripens early September, 1 week to 10 days later than Rio Oso Gem. Tree: large, open-spreading, vase-shaped, vigorous; hardy; regular and productive bearer; flower large, showy, pink; leaf glands reniform.

Nectar.—Orig. in Bakersfield, Calif., by Oliver P. Blackburn. Introd. in 1935. Plant patent 86; 6 Feb. 1934; assigned to Calif. Nursery Co., Fremont, Calif. Stanwick nectarine x unknown peach. Fruit: large; round-ovate, apex pointed; skin blushed pink to red; flesh white tinged with red, juicy, soft-melting, sweet, aromatic, very good quality; pit size medium, free, some split pits; ripens 3 weeks before Elberta; susceptible to brown rot; resembles Early Wheeler. Tree: vigorous; foliage heavy, leaf with reniform glands; flower nonshowy, size medium; chilling requirement 1050 h.

Nectar-Florence.—Orig. in Di Giorgio, Calif., by Edward Champness, Earl Fruit Co., now Di Giorgio Fruit Corp. Not introd. commercially as a salable commodity. Plant patent 161; 17 Dec. 1935; assigned to the Earl Fruit Co. Parentage unknown; selected about 1929. Fruit: size medium; skin deep red, smooth, pubescence slight; flesh reddish-yellow, firm, flavor pleasant, semiclingstone; matures with Florence, which it resembles. Tree: shy bearer, obsolete.

Neethling.—Orig. in Stellenbosch, South Africa, by P.A.L. Steyn, Fruit and Food Technology Res. Inst. Introd. in 1961. Walgant x (Kakamas x Early Dawn); sibling of Black and Malherbe. Fruit: flesh yellow, no red in pit cavity, clingstone; excellent canning quality; ripens 33 days before Kakamas, with a statistical mean ripening date of 17 Jan.; resembles Kakamas. Tree: very resistant to delayed foliation.

Nell.—Orig. in East Berlin, Pa., by Chester B. Worley, York Springs, Pa., in the yard of a Mr. Nell. Introd. in 1965 by Worley's Nurseries, York Springs. Possibly Halehaven o.p., discovered in 1963. Fruit: medium large, round; skin highly colored, resembling Halehaven, medium tough; flesh yellow, smooth, freestone; for canning and roadside market; ripens 25 Sept. at York Springs. Tree: size medium; upright; vigorous; moderately hardy; productive.

Nemaguard.—Orig. in Fort Valley, Ga., by

the U.S. Dept. of Agriculture, Hort. Field Laboratory. Introd. in 1961. A selected seedling from seed obtained in 1949 from a commercial importer as *Prunus davidiana;* tree and fruit characters resemble peach (*P. persica*); may be of hybrid origin; tests of seed germination, compatibility with various peach varieties, nematode resistance, and top growth of scions have been made principally at U.S. Dept. of Agriculture Stations at Fort Valley, Ga.; Fresno, Calif.; Beltsville, Md.; released for trial in 1959 as US FV 234-1. Rootstock: percentage of seed germination high; peach buds compatible, showing very satisfactory growth, with good fruit yields; resistance to root-knot nematode species equal to any other stock tested at Beltsville and Fort Valley; tests of seedlings inoculated with *Meloidogyne incognita* variety *acrita,* root-knot nematode, showed some root damage on only about 25% of seedlings, but reproduction of these nematodes was uncertain; similar tests with *M. javanica,* root-knot nematode, showed slight galling, but nematode reproduction was doubtful. Tree: blooms in Elberta season; chilling requirement for fruit and leaf buds same as for Elberta; trees grown at Fort Valley have shown some trunk damage usually associated with winter injury in that area.

Nestor.—Orig. in Palo Alto, Calif., by W.F. Wight, U.S. Dept. of Agriculture. Introd. in 1938. Muir x Paloro; cross made in 1923; fruited in 1928; tested as US 26-53. Fruit: large; averaging 7.6 cm in diam.; nearly round; skin golden yellow, slightly blushed, pubescence medium; flesh yellow slightly pink near pit, quite juicy, flavor good; stone small, free; primarily for drying purposes; ripens first week in August. Tree: vigorous; productive; leaf glands globose; flower with small petals; not suited for growing in warm-winter climates.

Newbelle.—Orig. in Agr. Expt. Sta., Univ. of Florida, Gainesville, by R.H. Sharpe and named by local growers. Introd. in 1984. Complex parentage of Florida selections. Selected in 1968; tested as Fla. 1E-138. Fruit: large; skin 60% red blush; flesh yellow, firm, freestone. Ripens 105 days after bloom. Flower showy; leaf glands reniform. Resistant to bacterial leaf spot. Chilling requirement 150 chill units.

Newcheer.—Orig. in New Brunswick, N.J., by M.A. Blake, New Jersey Agr. Expt. Sta. Introd. in 1947. (J.H. Hale x Mexican Honey) x (J.H. Hale x Iron Mountain). Tested as N.J. Low Acid 5. Fruit: small; ovate, apex tipped; skin red over half of surface; freestone; flesh white, low in titratable acidity; ripens a few days after Elberta. Tree: vigor medium; hardier than Elberta; productivity medium; flower small, nonshowy; leaf glands reniform.

New Covenant.—Orig. in Chico, Calif., by G.C. Banta. Introd. in 1954. Parentage unknown; discovered about 1948. Fruit: flesh yellow, clingstone; ripens at the end of the usual commercial canning clingstone season.

Newday.—Orig. in New Brunswick, N.J., by M.A. Blake, New Jersey Agr. Expt. Sta. Introd. in 1938; tested as N.J. 79. J.H. Hale x N.J. 40 C.S.; cross made in 1923. Fruit: large; ovate; skin blushed bright red, attractive, rough; flesh yellow, semi-clingstone, soft, melting; ripens 2 weeks before Elberta. Tree: vigorous, good to low productivity; flower size medium, nonshowy; leaf glands reniform.

New Generation.—Orig. in Yakima, Wash., by Leroy W. Brackett. Plant patent 4418; 22 May 1979; assigned to May Nursery Co., Yakima, Wash. Early Elberta mutation. Fruit: similar to parent, but appears more winter hardy; matures 7 to 10 days earlier than parent in same growing conditions. Tree: similar to parent, but more upright in growth habit and with smaller scaffolding.

Newhaven.—Orig. at South Haven, Mich., by Stanley Johnston, Michigan State Univ. Introd. in 1978 by R.L. Andersen, J.E. Moulton, T.M. Thomas, and H.J. Belter. (Redhaven x Fairhaven) o.p.; selected in 1962; tested as SH415. Fruit: medium-large; roundish oblong; skin 70% bright red over golden-yellow ground color; flesh clear yellow, nonbrowning, very firm, freestone; flavor excellent. Ripens 5 to 7 days after Redhaven. Tree: vigor medium; spreading; productive; leaf glands reniform. Flower bud hardiness similar to that of Redhaven. Blossoms nonshowy. Considerably more tolerant than Redhaven to bacterial spot.

Nivolet.—See Madeleine Pouyet.

Norman.—Orig. in Jackson Springs, N.C., by North Carolina Agr. Expt. Sta. Introd. in 1968. Sunhigh x Redskin; cross made in 1957; first fruited in 1960; tested as N.C. 7244. Fruit: medium to large; round; skin smooth, most of surface covered early with a dark red blush over a medium yellow ground; flesh deep yellow with small amount of red pigment, very resistant to browning, very firm, texture fine and melting, flavor good when ripe, freestone; ripens 3 to 5 days earlier than Ranger. Tree: vigor medium; productive; flowers large, showy, fruitful; leaf glands reniform, medium size; chilling requirements and bloom date about the same as Redhaven; moderately resistant to bacterial spot, rating better than Redhaven.

North Greenleaf.—See Rancho Resistant.

Norwalk.—Orig. in Los Nietos, Calif., by William Cole. Introd. in 1951, by the Orange County Nursery Co., Norwalk, Calif. Parentage unknown; selected in 1946. Fruit: flesh yellow, clingstone; resembles Sims; good canner. Variety now obsolete.

Norwalk Free.—Orig. in Buena Park, Calif., by Henry Ramsey. Introd. in 1950 by the Orange County Nursery Co., Norwalk, Calif. Parentage unknown; selected in 1945. Fruit: large; flesh yellow; freestone; high shipping and canning qualities; resembles Sims. Variety now obsolete.

Nuevo.—Orig. in Riverside, Calif., by J.W. Lesley and M.M. Winslow, California Citrus Res. Center and Agr. Expt. Sta. Introd. in 1961. Parentage includes P.I. 32374, Peak, Elberta, Peento, etc.; cross made in 1947; first fruited in 1951; tested as U.C. 335-10. Fruit: large, with proper thinning; nearly round, quite symmetrical; skin yellow, exposed fruit about 30X splashed red, tough, pubescence short; flesh yellow, red next to pit, sweet, acidity medium, nonmelting, firm, texture fine; canning clingstone; keeps well; no preharvest drop; ripens in late August to early September. Tree: vigorous; winter-chilling requirement slight, similar to Rubidoux and Meadow Lark, intermediate between Elberta and Bonita; leaf glands globose; flower small, nonshowy, petals often waxy; produces good crops in most years. No longer commercially planted.

O'Boy.—Orig. in Watervliet, Mich., by George Schairer. Introd. in 1936. Parentage unknown; discovered in 1933. Fruit: round; skin blushed dark red; flesh yellow, red around the pit, firm, freestone with excellent flavor. Tree: seems to be stronger than J.H. Hale; self-fruitful. Tested at South Haven Expt. Sta., Mich., and resembles J.H. Hale so closely as to appear to be that variety, unproductive and appeared self-sterile.

O'Henry.—Orig. in Red Bluff, Calif., by Grant Merrill. Introd. in 1968. Plant patent 2964; 27 Jan. 1970. Merrill Bonanza o.p.; apparently a cross with a nectarine variety; discovered in 1960; selected in 1967. Fruit: 2 1/2 to 3 inches in diam.; skin brightly colored, pubescence short; flesh yellow streaked with red, very firm, quality good, freestone; ships well; resembles Improved Pacifica; ripens about with Improved Pacifica to slightly later. Tree: size medium; vigor moderate; productive; leaf glands reniform; flower pink, large, showy.

O.K.—See Oklaberta.

Okinawa.—Orig. in Gainesville, Fla., by R.H. Sharpe, Florida Agr. Expt. Sta. Introd. in 1957. Seed importation from Okinawa in 1953. Fruit: edible; small; flesh white, freestone. Tree: resistant to the root-knot nematodes (*Meloidogyne javanica and M. incognita*); appears to be a compatible rootstock for Florida peach varieties; chilling requirement 150 chilling units. Good germination can be obtained with carefully harvested seed and removal of seeds from pit.

Oklaberta (*O.K.*).—Orig. in Wynnewood, Okla., by Otis Warren, Warren and Sons Nursery. Introd. in 1938. Parentage unknown; discovered in 1927. Fruit: roundish-oval; skin yellow, blushed with red; flesh yellow, freestone; finer-grained than Elberta; ripens with Elberta; develops a high percentage of split pits. Tree: vigorous; upright; blooms with Elberta.

Old Boy.—See Shippers Late Red.

Olinda.—Orig. in Ruthven, Ont., Canada, by George P. Whaley. Introd. in 1962. Chance seedling; discovered in 1945. Fruit: round; skin attractively red, pubescence medium; flesh yellow, firm, quality good, freestone; ripens 5 days before Elberta; hangs well on tree. Tree: vigorous; moderately resistant to bacterial leaf spot. Tests conducted in Kentucky in 1966 showed susceptibility to leaf spot, resistance to browning.

One Star (*Daily News One Star*).—Orig. in La Canada, Calif., by W.E. Lammerts, Descanso Distributors. Introd. in 1951. St. Helena seedling x [(Swatow x Rio Oso Gem) x (Babcock x Mayflower)]; selected in 1948. Fruit: 2 1/2 inches in diam.; flesh yellow, quality high, semi-clingstone; ripens middle of June; resembles Meadowlark. Tree: vigorous; short chilling requirement; flower large, very abundant.

Oom Sarel.—Orig. in Stellenbosch, South Africa, by P.A.L. Steyn Western Province Fruit Res. Inst., Dept. Agr. Introd. there in 1961. (Kakamas x Early Dawn) F_2; selected in 1960. Fruit: flesh orange, no red around pit, clingstone; excellent commercial canning quality; ripens 61 days before Kakamas, which it resembles. Tree: very resistant to delayed foliation.

Oriole.—Orig. in New Brunswick, N.J., by M.A. Blake, New Jersey Agr. Expt. Sta. Introd. in 1924. Slappey x Admiral Dewey; cross made in 1916 in Vineland, N.J. Fruit: large; roundish oval; skin yellow, deeply blushed with irregular stripes of dull red, thick, medium tough, pubescence medium heavy; flesh yellow, juicy, pleasantly flavored, too tender for shipping but fine for local markets; high quality; stone free, medium large; ripens early, just before Golden Jubilee, 5 to 7 days before Carman and Rochester. Tree: large, vigorous, upright-spreading; very productive; hardy; flower large, blooms in midseason.

Orion.—Orig. in Roodeplaat, Pretoria, South Africa, by Edward P. Evans, Hort. Res. Inst. Introd. in 1973. [Bells November x Culemborg 388] x Robin; cross made in 1963; selected in 1965; tested as 71. Fruit: large, 6.5 cm; round; skin green ground color, blushed with 60% to 80% red; flesh white, melting, firm, flavor good, freestone; ripens early, 20 Oct. Tree: large, open vase-shaped; very vigorous and productive; leaves develop dark red color in autumn; needs thinning; flower showy, light rose; leaf glands reniform; low chilling requirement 300 to 500 h.

Oro-A.—Orig. in Agr. Expt. Sta., Univ. of Florida, Gainesville, by W.B. Sherman and released jointly with J. Rodriguez, Centro de Fruticultura, Colegio de Postgraduates, Chapingo, Mexico. Introd. in 1990. Diamante o.p. Selected in 1984; tested as Fla. 84-12C. Fruit: medium small; skin yellow; flesh yellow, nonmelting, clingstone; devoid of red in skin and flesh. Ripens 85 days after bloom. Flower small, showy; leaf glands reniform. Resistant to bacterial leaf spot and powdery mildew. Chilling requirement 275 chill units.

Osage.—Orig. in Mountain Grove, Mo., by Paul H. Shepard, Missouri State Fruit Expt. Sta. Introd. in 1946. Alton o.p.; selected in 1945; tested as PH-631. Fruit: large; round; skin red over greenish-yellow under color; flesh creamy-white, streaked with red, fine-grained, firm, tender, sweet, juice an attractive pink, clingstone; resembles Indian in color and flavor; high quality canner; ripens 2 weeks before Elberta; hangs well on tree. Tree: vigorous; resistant to low temperatures.

Oso Grande.—Orig. in Fresno, Calif., by F.J. Harkness, United Packing Co. Introd. in 1955. Chance seedling; discovered in 1953. Fruit: flesh yellow, freestone; ripens 1 week later than Elberta; resembles Candoka. Tree: self-fruitful; resembles Rio Oso Gem.

Othelle.—See July Heath.

Ouachita Gold.—Orig. in Calhoun, Louisiana, by Charles E. Johnson. N. Louisiana Expt. Sta. Introd. in 1981. Redhaven o.p. x L61-2-26 (Redskin x Southern Glo). Cross made in 1965: tested as L65-10-63. Fruit: large. 3 1/4 inches in diam.; flesh excellent quality, firmness, and texture; freestone; resembles LaGold; late maturity, 15 to 20 days after Elberta. Tree: vigorous, strong. consistent yielder; resistant to bacterial spot; flowers large, showy; leaf glands reniform; chilling requirement 80 to 850 h at or below 45F.

Ozark.—Orig. in Mountain Grove, Mo., by Paul H. Shepard, Missouri State Fruit Expt. Sta. Introd. in 1946. Frank x Halehaven; selected in 1943; tested as PH-820. Fruit: size medium; round ovate; skin full red; flesh yellow, freestone, firm, fine-grained, excellent flavor; ships well; ripens 2 weeks before Elberta; resembles J.H. Hale. Tree: hardy; very productive; vigor medium;

flower large, showy; leaf glands reniform; chilling requirement 1050 h.

Pacemaker.—Orig. in New Brunswick, N.J., by M.A. Blake, New Jersey Agr. Expt. Sta. Introd. in 1939; tested as N.J. 99. J.H. Hale x Marigold; cross made in 1923. Fruit: large; round; skin bright red over half of surface, attractive; flesh yellow, red at pit, firm, melting, semi-clingstone; extends the Goldeneast season, ripens 1 week before Elberta. Tree: vigor medium, very productive; flower small, nonshowy, pollen-sterile; leaf glands reniform; chilling requirement 750 h.

Pacifica (*Merricle L-14*).—Orig. in Red Bluff, Calif., by Grant Merrill. Introd. in 1958. Plant patent 1954; 21 June 1960. [J.H. Hale x (July Elberta x Maxine)] o.p.; selected in 1954. Fruit: skin highly colored red; flesh yellow with red at stone, firm, flavor mild, moderately juicy, quality good, freestone; ripens with Rio Oso Gem; resembles Rio Oso Gem, suture less pronounced. Tree: size medium; vigorous; productive; leaf glands reniform; flower small, pink, self-fruitful.

Pacific Gold.—Orig. in Buckley, Wash., by W.M. Schwab, Buckley Nursery Co. Introd. in 1933. Rochester o.p.; discovered in 1929. Fruit: large; flesh firm; resembles Rochester. Tree: long-lived, productive. Similar to and perhaps identical with Rochester.

Pageant (*Merrill Pageant*).—Orig. in Red Bluff, Calif., by Grant Merrill, Grant Merrill Orchards. Introd. in 1954. Plant patent 1338; 29 Dec. 1954. Kirkman Gem o.p.; discovered in 1948; selected in 1953. Fruit: large; flesh yellow, freestone; ripens very late, late September to early October, or 50 days later than Elberta; resembles Merrill Fiesta but 10 days later and with more color. Tree: size medium; vigor medium; productive, self-fertile; leaf glands reniform; flower size medium, light pink.

Palmer White.—Orig. in Montrose, Colo., by Corwin Palmer. Introd. in 1966. Parentage unknown. Plant patent 2665; 23 Aug. 1966; discovered 1958. Eruit: medium, 2 1/8 to 2 3/8 inches in diam.; oblong to round-ovate; skin greenish-cream to cream-white ground cover, washed and blushed with a dull cherry red, streaked and mottled with darker oxblood red; pubescence light, giving a smooth appearance; flesh white with streaks of red close to stone, melting, juicy, sweet, texture medium to firm, flavor mild, subacid; quality as dessert or canned, good to very good; stone moderately free; ripens late August and first week of September. Tree: medium to small; moderately vigorous; moderately productive; annual bearer; tolerant to high summer temperature and hardy to –24F; has not been afflicted by blight, mildew, or mosiac.

Parade (*Merricle 61-167*).—Orig. in Red Bluff, Calif., by Grant Merrill. Introd. in 1960. Plant patent 2153; 12 June 1962. Parentage unknown; selected in 1959. Fruit: large, averaging 2 5/8 inches or more in diam.; globose to slightly compressed; skin medium thick, undercolor bright yellow, light red near pit, moderately juicy, firm, flavor mild, subacid, vinoug, aroma moderate, eating quality good, not darkening when exposed to air; freestone, but adhering some on ventral suture near base; ripens very late, 15–30 Sept. at place of origin, with Merrill Fiesta, or 2 weeks after Splendor, which it resembles, or 1 month after J.H. Hale. Tree: size medium, smaller than Splendor; productive; leaf glands reniform; flower salmon pink, small, blooming about midseason with Fay Elberta; leaf wavy, distinctly crinkled; self-fruitful.

Paradise.—Orig. in Paradise, Calif., by J.E. Fisher. Introd. about 1946. Chance seedling. Fruit: skin reddish-white; flesh white, with some red at pit, freestone; matures in midseason. Tree: bears well.

Parkhill.—Orig. in Hollister, Calif., by Raymond Cheek. Introd. in 1957. Chance seedling; discovered in 1950. Fruit: flesh yellow, flavor good, freestone; resembles Muir. Tree: considered to be somewhat resistant to leaf curl (*Taphrina deformans*) fungus.

Pat's Pride.—Orig. in Bakersfield, Calif., by Lewis B. Sherrill, Patterson & Hale, Bakersfield. Introd. in 1960. Plant patent 2048; 18 Apr. 1961; 75% assigned to Patterson & Hale. Possible Springtime mutation; discovered in 1959 in an orchard of Springtime in the Wheeler Ridge area. Fruit: large; globose; skin yellow with attractive blush; flesh creamy white, firmer than Springtime, clingstone; less subject to split pits than Springtime; ripens 4 to 6 days after Springtime; resembles Robin in shape and Springtime in flesh texture and flavor. Tree: size medium; vigorous; regular bearer; leaf glands globose; flower size medium, salmon.

Pat's Redhaven.—Orig. in the Mettler Station area, Kern County, Calif., by Lewis B. Sherrill, Bakersfield, Calif. Introd. in 1961. Plant patent 2102; 7 Nov. 1961; 75% assigned to Patterson & Hale, Bakersfield, Calif. Redhaven mutation. Fruit: large; skin yellow overspread with red; flesh yellow to orange yellow, firm, clingstone; excellent eating quality; resembles Redhaven but ripens about 10 days earlier. Tree: resembles Redhaven.

Peach-Arine.—Orig. in Santa Rosa, Calif., by Luther Burbank. Introd. in 1930 by Stark Bros. Nurseries & Orchards Co., Louisiana, Mo. Muir x unknown white nectarine. Fruit: large; oval; skin yellow, no blush, nearly fuzzless; flesh creamy yellow, firm, fine-grained, juicy, stone small and free; for dessert and canning; ripens late August in north central Missouri. Tree: vigorous; very productive. Apparently no longer propagated.

Pearson Hiley.—Orig. in Lee Pope, Ga., by John W. Pearson. Introd. in 1946. Plant patent 760; 9 Sept. 1947. Early Hiley mutation; discovered in 1941. Fruit: skin and flesh full red at maturity; freestone; ripens 10 days before Early Hiley; resembles parent. Tree: same as parent.

Pedersen.—Orig. in Modesto, Calif., by L.E. Pedersen. Introd. in 1937. Parentage unknown; selected in 1935. Fruit: flesh yellow, firm; ripens in Phillips Cling season; commercial canning clingstone.

Pekin.—Orig. in Jackson Springs, N.C., by North Carolina Agr. Expt. Sta. Introd. in 1968. Summercrest x Redhaven; cross made in 1957; first fruited in 1960; tested as NC 7366. Fruit: size medium; round or slightly elongated; skin smooth, light yellow ground color, 75% to 85% covered with bright red blush; pubescence short; flesh yellow, more resistant to browning than Redhaven, smooth melting texture, pleasing flavor; freestone when soft-ripe; ripens 2 to 3 days earlier than Redhaven. Tree: vigorous; productive; flowers nonshowy, self-fruitful; leaf glands reniform; chilling requirement about same as and blooms just after Redhaven; highly resistant to bacterial spot, rated better than Redhaven.

Penryn.—Orig. in Palo Alto, Calif., by W.F. Wight, U.S. Dept. of Agriculture. Introd. in 1938. Maxine x Leader; tested as US 32-11A. Fruit: large, averaging 7.6 cm in diam., halves unequal; flesh light yellow with no red at pit, juicy, mildly subacid, quality good, freestone; skin yellow, highly blushed, pubescence short, appearance attractive; for dessert purposes; ships well; ripens early, with or slightly before Maxine. Tree: vigorous; very productive; leaf glands reniform; flower with large petals; not suited for growing in warm-winter climates.

Percoca.—See Carson.

Perfect Hale.—See Fertile Hale.

Perkins.—Orig. in Arlington, Texas, by J.A. Evans. Introd. about 1947. Open-pollinated seedling of Charlotte. Fruit: flesh yellow, freestone; ripens between Elberta and Charlotte; superior in flavor to Charlotte, otherwise very similar to Charlotte. Tree: vigorous; chilling requirement 850 to 900 h; subject to bacterial spot.

Peterson Elberta.—Orig. in Reedley, Calif., by Howard C. Peterson. Introd. in 1953. Plant patent 1228; 4 Dec. 1953. Early Elberta mutation; discovered in 1944. Fruit: large; resembles Rio Oso Gem but skin more highly colored; flesh yellow, firm, freestone; ships well; matures before July Elberta. Tree: resembles July Elberta; glands few; flowers small, pink.

Piedmontgold.—Orig. in Blacksburg, Va., by Virginia Agr. Expt. Sta. Introd. in 1968. P.I. 35201 x N.J. 196. Cross made in 1951 by L.F. Hough, Rutgers Univ., New Brunswick, N.J.; selected in 1954 by G.D. Oberle, Virginia Polytechnic Inst.; tested as W2-7-28. Fruit: size above average, 2 3/4 inches; skin bright orange-yellow with two-thirds an attractive red blush; flesh firm, nonmelting, yellow with essentially no red pigment near pit, clingstone; ripens 9 days before Elberta. Tree: size average; vigor moderate; hardiness average or above; productivity average; flower small, nonshowy; leaf glands reniform.

Pink Cascade.—Double-flowered, weeping ornamental orig. at Univ. of Arkansas from 1978 seed, selected in 1981, and released in 1992 by J.N. Moore, R.C. Rom, S.A. Brown, and G.L. Klingaman. (New Jersey selection x Rutgers Redleaf) selfed. Fruit: small, poor eating quality. Flowers double, showy, pink. Leaves dark red fading to greenish-red by midsummer. Ornamental with extreme weeping growth habit; vigorous.

Pioneer.—Orig. in New Brunswick, N.J., by M.A. Blake, New Jersey Agr. Expt. Sta. Introd. in 1925. Belle x Greensboro; cross made in 1915. Fruit: size medium; oval; skin bright red, pubescence heavy, attractive; flesh white, coarse, soft, melting, semiclingstone to almost freestone; ripens 3 weeks before Elberta; resembles Cumberland. Tree: vigorous; productive; hardy; flower small, nonshowy; leaf glands reniform; chilling requirement 800 h.

Plaisir de Merle.—Orig. near Simondium, Groot Drakenstein, South Africa, by a Mr. Gray. Duke of York mutation; discovered before 1945. Fruit: ripening before Early Dawn; size often too small for commercial use. Tree: requires less winter chilling than parent.

Plantz.—Orig. in Marysville, Calif., by William Plantz. Introd. in 1936. Plant patent 262; 29 Dec. 1936. Parentage unknown; selected in 1935. Fruit: large; oblate spheroid, slightly irregular; skin apricot-yellow, red blush at top, tough, tenacious, lightly pubescent; flesh firm, meaty, juicy, yellow with no red at pit, texture fine, few fibers, flavor sub-acid; clingstone, small; ripens with Phillips Cling (early September); used essentially for canning. Tree: slightly weeping; average vigor; leaves large; flowers small, deep rose pink, blooms 26 Feb.–9 Mar.

Polar.—Orig. in Osage, Iowa, by Farmer Seed & Nursery Co., Faribault, Minn. Introd. in 1962. Plant patent pending. Chance seedling; discovered in 1956. Fruit: large, averaging 3 inches in diam.; skin highly blushed; flesh yellow, firm, freestone; resembles Elberta. Tree: hardy; blooms late; leaf glands reniform.

Polly.—Orig. in Glenwood, Iowa, by S.A. Beach, Iowa State Agr. Expt. Sta. Introd. in 1934. (Hill's Chili)F_2 o.p.; selected in 1916; first fruited in 1920. Fruit: flesh white, high quality; resembles Champion. Tree: very hardy. Not entirely desirable for commercial planting; well adapted for home orchards in Iowa.

Pomeroy.—Orig. in Merced, Calif., by the Calif. Packing Corp. Introd. in 1934. Parentage unknown; selected in 1932. Fruit: flesh yellow; commercial canning clingstone; ripens in Peak season.

Poppy.—Orig. in Mountain Grove, Mo., by Paul H. Shepard, Missouri State Fruit Expt. Sta. Introd. in 1947. Frank x Halehaven; selected in 1947; tested as PH-812. Fruit: large; shape resembles Elberta; skin pronounced red; flesh yellow, fine texture and flavor, freestone; ripens 8 days before Elberta, but appears to be ripe 7 to 10 days before it actually reaches the hard-ripe stage; good shipper; resembles Elberta. Tree: vigor and productivity average; flower small, nonshowy; leaf glands globose.

Powell Gem.—Orig. in Winton, Calif., by L. Thornton Powell, Powell Orchards. Introd. in 1968. Plant patent 3015; 12 Jan. 1971. Kirkman Gem mutation; discovered in 1961. Fruit: medium-large; flesh yellow, firm, freestone; resembles Rio Oso Gem; ripens about with Rio Oso Gem. Tree: size medium; vigor comparable to Rio Oso Gem; productive; leaf glands reniform; flowers large, pink, showy.

Prairie Clipper.—Orig. at the Hort. Farm, Olney, Ill., by M.J. Dorsey, Illinois Agr. Expt. Sta., Urbana, Ill. Introd. in 1946. J.H. Hale x Gage; cross made in 1933; tested as Ill. K47. Fruit: very large; skin medium yellow with dark red blush, dull appearance; flesh yellow, deeper color at stone, firm, J.H. Hale type of flesh, quality good, flavor sharp, freestone; ripens 3 days before Elberta.

Prairie Dawn.—Orig. at the Hort. Farm, Olney, Ill., by M.J. Dorsey, Illinois Agr. Expt. Sta., Urbana. Introd. in 1946. Valiant x Halehaven; cross made in 1937; tested as Ill. K73. Fruit: medium to large; skin medium yellow, light blush, appearance excellent; flesh medium yellow, soft at maturity, tender, juicy; clingstone, tending to be freestone when mature; ripens 35 to 40 days before Elberta; good for fresh market because of early maturity and good appearance. Tree: fruit buds with moderate degree of hardiness, hardier than Elberta; flower small, nonshowy; leaf glands reniform; chilling requirement 850 h.

Prairie Daybreak.—Orig. at the Hort. Farm, Olney, Ill., by M.J. Dorsey, Illinois Agr. Expt. Sta., Urbana, Introd. in 1946. Halehaven x Sun Glo; cross made in 1937; tested as Ill. K69. Fruit: large; tipped, oval; skin medium yellow, dark blush, dull to attractive appearance; flesh light yellow, tender, firm, moderately juicy, quality poor; seed sometimes splits, freestone; ripens 37 days before Elberta. Tree: vigor medium, high productivity; flower small, nonshowy; leaf glands reniform; chilling requirement 750 h.

Prairie Rambler.—Orig. at the Hort. Farm, Olney, Ill., by M.J. Dorsey, Illinois Agr. Expt. Sta., Urbana, Ill. Introd. in 1946. Elberta x Gage; cross made in 1933; tested as Ill. K43. Fruit: very large; skin greenish-yellow, light to dark red blush, attractive; flesh yellow with some red at stone, firm, juicy, subacid, spicy flavor, freestone; ripens 3 days after Elberta.

Prairie Rose.—Orig. at the Hort. Farm, Olney, Ill., by M.J. Dorsey, Illinois Agr. Expt. Sta., Urbana. Introd. in 1946. Gage x Halehaven; cross made in 1937; tested as Ill. K80. Fruit: medium to large; round; skin medium yellow, blushed and splashed with dark red, light pubescence, very attractive; flesh medium yellow with occasional trace of red through flesh and at stone, medium firm, texture coarse, juicy; semi-cling to freestone; ripens 25 days before Elberta; too soft in Georgia. Tree: vigor and productivity average; flower large, showy, light pink; leaf glands reniform; chilling requirement 950 h.

Prairie Schooner.—Orig. at the Hort. Farm, Olney, Ill., by M.J. Dorsey, Illinois Agr. Expt. Sta., Urbana, Ill. Introd. in 1946. Elberta x South Haven; cross made in 1933; tested as Ill. K40. Fruit: large; oval; skin light yellow, light red blush; flesh light yellow, lighter color at stone, tender, juicy; semi-cling until fully mature; inclined to overset.

Prairie Sunrise.—Orig. at the Hort. Farm, Olney, Ill., by M.J. Dorsey, Illinois Agr. Expt. Sta. Introd. in 1946. Valiant x Halehaven; cross made in 1937; tested as Ill. K74. Fruit: large; round; skin medium yellow, dark red blush, heavy pubescence; flesh entirely yellow, tender, firm, melting, juicy; semi-cling until fully mature; ripens 37 days before Elberta. Tree: moderate vigor and highly productive; flower large, showy; leaf glands reniform; chilling requirement 850 h.

Prenda.—Orig. in Riverside, Calif., by J.W. Lesley, California Citrus Expt. Sta. Introd. in 1942. J.H. Hale x Columbiana-Peento-Elberta hybrid; selected in 1935. Fruit: medium to large; skin light cream, blushed on one side near stem end, smooth; flesh white, texture smooth, juicy, mild, sweet, pleasing aroma, freestone; ripens with Golden State; subject to preharvest drop in some localities. Tree: vigorous; upright; produces well; leaf dark green, glands reniform; short chilling required. Originally introduced under the name Rosy, now supplanted by the name Prenda due to prior usage of the name Rosy for an established brand name.

Pressnell.—Orig. in Modesto, Calif., by Earl Pressnell. Introd. in 1938. Parentage unknown. Fruit: round; flesh yellow; commercial canning clingstone; ripens after Halford.

Preuss Suncrest.—Orig. in Clovis, Fresno County, Calif., by Charles Preuss. Introd. in 1968. Plant patent 2797; 5 Mar. 1968. Suncrest mutation; discovered in 1962; first fruited in 1966. Fruit: resembles Suncrest but ripens about 10 days later, darker colored red suture, and sweeter flavor; flesh yellow, freestone. Tree: vigorous; regular and heavy bearer; leaf glands reniform; flower large, light pink, self-fertile, blooms with Fay Elberta.

Primissima Delbard.—See Madeleine Pouyet.

Primrose.—Orig. in New Brunswick, N.J., by M.A. Blake, New Jersey Agr. Expt. Sta. Introd. in 1925. Belle x Elberta; cross made in 1915. Fruit: medium to large; ovate, suture well-marked; skin yellow with bright blush, attractive; flesh yellow, tinged red, melting, soft, very good quality, high tannin, freestone; ripens with Elberta. Tree: very vigorous, spreading; sets heavily and requires considerable thinning; flower small, nonshowy; leaf glands reniform.

Princess Anne.—Orig. in Princess Anne, Md., by Robert Reynolds. Introd. in 1970 by Bountiful Ridge Nursery, Princess Anne, Md. Chance seedling; discovered in 1965. Fruit: large, round, skin blushed with about 85% red, tough; flesh yellow, firm, freestone, quality good; ripens 20 July. Tree: size medium; upright to spreading; vigorous, hardy and productive; flower pink, showy; leaf glands globose.

Queencrest.—Orig. in Reedley, Calif., by Leo and Rosalie Balakian, Reedley, Calif. Introd. in 1986. Plant patent 6025; 29 June 1987. Maycrest mutation; tested as B41871. Fruit: medium to large; round; skin red-blushed on golden ground color; flesh yellow amber, firm, clingstone; flavor sweet, not acid. Ripens 20 days before Redhaven in California. Good shipper. Tree: vigorous and productive for an early variety.

Radiance.—Orig. in New Brunswick, N.J., by M.A. Blake, New Jersey Agr. Expt. Sta. Introd. in 1925. Belle x Greensboro; cross made in 1914. Fruit: large; oval, being similar to Belle; skin greenish-white with a slight blush; flesh white, quality fair to good; almost a freestone; ripens with Carman. Tree: moderately hardy; very vigorous and productive; sets well in adverse spring weather; flowers large, showy; leaf glands reniform.

Rainbow.—Orig. in Parlier, Calif., by Frank R. Coddington. Introd. in 1956 by Parlier Nursery. Plant patent 1406; 19 July 1955. July Elberta mutation; discovered in 1949. Fruit: flesh yellow, freestone; matures about 1 week earlier than parent or about 20 June at place of origin; flavor similar to July Elberta, but flesh is firmer. Tree: medium large; vigorous, stout branching with well-knit forks of divergent habit; productive; leaf glands small, varying from globose to reniform; flower similar to July Elberta but with more pink and blossoming a few days later.

Ramona.—Orig. in Riverside, Calif., by J.W. Lesley, California Citrus Expt. Sta. Introd. in 1943. (Peak x P.I. 32374) F_2; selected in 1933. Fruit: large; slightly oblate, halves equal or nearly so; skin yellow, blushed, pubescence medium dense, medium thick, tough, adherent to flesh; flesh yellow, some red next to pit, nonmelting, sweet, subacid; clingstone; ripens late, after Fontana, about the second week in September. Tree: large; vigorous; upright to slightly spreading; produces many twin fruit; short chilling requirement; leaf glands globose; flower small, pink, self-fruitful. Obsolete.

Ramos.—Orig. in Ceres, Calif., by C.M. Ramos, Hughson, Calif. Introd. in 1965. Possibly Lovell o.p.; discovered in 1940; selected in 1950. Fruit: sizes better than Cortez; flesh yellow, firm; commercial canning clingstone; ripens in Cortez season; holds well on tree; resembles Paloro. Tree: sets well but not heavily. No longer commercially planted.

Ramsey.—Orig. in Okanogan, Wash., by W.F. Ramsey. Introd. about 1948. J.H. Hale x Candoka. Fruit: large; skin red; flesh yellow, red around pit, slightly coarse, astringency high, firm, quality good, freestone; ripens about 10 days later than J.H. Hale; Candoka type. Tree: bears heavily; strong grower. Obsolete.

Ranchero.—Orig. in Red Bluff, Calif., by Grant Merrill, Exeter, Calif. Introd. in 1968. Plant patent 3002; 17 Nov. 1970. (Kirkman Gem) F_3 o.p.; discovered in 1959, tested as Merrill 77-110. Fruit: large; generally globose; skin smooth one-fourth to one-half covered with red blush; flesh yellow, buttery, freestone; good freezing qualities; recommended for local or distant markets; ripens mid-late September in San Joaquin Valley. Tree: large, vase shaped; vigorous; productivity heavy; flower large, showy, pink, pollen moderate; leaf glands reniform.

Rancho Redleaf.—Orig. in Bakersfield, Calif., by Grant Merrill, Red Bluff, Calif. Introd. in 1956. Plant patent 1440; 20 Dec. 1955; assigned to Del Rancho Fortuna, McFarland, Calif. Parentage unknown; discovered in 1949; selected in 1954. Tree: most seedlings highly resistant to nematodes; foliage red, which results in faster and more accurate suckering of nursery stock from its seed. Three-fourths of seedlings with red leaves; flowers pink and double; leaf glands globose; attractive as an ornamental. Introduced as a nematode-resistant rootstock for commercial peaches.

Rancho Resistant (*North Greenleaf*).—

Orig. in McFarland, Calif., by T.A. Sand, Rancho Fortuna. Introd. in 1958. Plant patent 1941; 10 May 1960. S-37 o.p.; selected in 1953. Fruit: 2 1/16 inches average diam.; uniform; skin greenish-yellow background, thick, tough; flesh white to greenish-white, juicy, texture coarse, flavor poor, quality poor, freestone. Tree: used as a rootstock; seedlings very uniform; seeds germinate very well; consistent and heavy bearer; resistant to rootknot nematode (*Meloidogyne incognita* variety *acrita*); flower large, single, white.

Rand.—Orig. in Marysville, Calif., by O.B. Rand. Introd. in 1940. Gaume mutation; discovered in 1934. Fruit: flesh yellow, commercial canning clingstone; same season as that of Phillips Cling and serves as a replacement for that variety; resembles Corona.

Ranger.—Orig. in Beltsville, Md., by U.S. Dept. of Agriculture. Introd. in 1952. Raritan Rose selfed; cross made in 1940; selected in 1944; tested as USDA B12160. Fruit: medium to large; almost round; skin with light to medium pubescence, red blotched, mottled, striped over 90% of greenish-yellow to yellow ground; flesh yellow, red at pit, some red flecking throughout, medium firm, melting, medium textured, flavor good; pit size medium, free; ripens 3 to 4 weeks before Elberta; ships only moderately well; cans and freezes well. Tree: productive, moderately vigorous; flower size medium, nonshowy, self-fruitful, usually opening shortly after Elberta; leaf glands reniform; chilling requirement 900 h; relatively high bacterial spot resistance.

Raritan Rose.—Orig. in New Brunswick, N.J., by M.A. Blake, New Jersey Agr. Expt. Sta. Introd. in 1936. J.H. Hale x Cumberland; cross made in 1926; tested as N.J. 97. Fruit: medium to large; round, ridged along suture; skin attractive red; flesh white, streaked red, fine-textured, soft, melting, watery, freestone; ripens 4 weeks before Elberta; quality good. Tree: vigorous, spreading, productive, hardy, tolerant to bacteriosis; flower small, fairly showy; leaf glands reniform; chilling requirement 1150 h.

Ray Crest.—Orig. by R. Goosen, Dinoba, Calif. Plant patent 5388; 15 Jan. 1985. Springcrest mutation; ripening 7 days earlier, but otherwise resembling Springcrest.

Rayon.—Orig. in Agr. Expt. Sta., Univ. of Florida, Gainesville, by R.H.Sharpe and named in northwest Mexico. Introd. in 1982. Complex parentage of Florida selections. Selected in 1971; tested as Fla 1-11. Fruit: large; skin 40% red blush; flesh yellow, medium firm, freestone. Ripens 102 days after bloom. Flower showy; leaf glands reniform. Moderate resistance to bacterial leaf spot. Chilling requirement 200 chill units.

Red Baron.—Orig. in Wasco, Calif., by D.L. Armstrong, Ontario, Calif. Introd. in 1977. Plant patent to be assigned to Armstrong Nurseries, Ontario, Calif. Seedling of advanced generations of Rio Oso Gem x Coolidge Double Red Flowering; selected in 1963. Fruit: round, 3 inches; flesh yellow, flavorful; freestone; ripens early July; resembles Flordabelle or Early Amber. Flower: petals red double, profuse. Backyard cultivar.

Redcal.—Orig. in Davis, Calif., by C.O. Hesse, Univ. of California, Davis, Pomology Dept. Plant patent 3978; 16 Nov. 1976 assigned to The Regents of the Univ. of California, Berkeley. Introd in 1977. Da 31-24 x Da 31-24; cross made in 1960; selected in 1965 tested as 4,9-98A. Fruit: medium or larger; globular skin has yellow ground color with bright red blush; flesh yellow, red at cavity; ripen early August just before Fayette and 7 to 8 days after Firered. Tree: medium upright; spreading vigorous; productivity good.

Redcap.—Orig. in Fort Valley, Ga., by J.H. Weinberger, U.S. Dept. of Agriculture Hort. Field Sta. Introd. in 1952. Southland x Dixired; cross made in 1945; fruited first in 1948; tested as USDA FV 121-58. Fruit: size medium; round; skin yellow, covered with a bright, attractive blush; flesh yellow, firm but melting, medium textured, flavor good, clingstone; ripens 1 to 2 days earlier than Dixired; resembles Dixired. Tree: moderately vigorous and productive; chilling requirement of 750 h at or below 45F, being about the same as that for Hiley and Southland; flower small, self-fruitful; less chilling than Dixired; susceptibility to bacterial spot disease is similar to Elberta.

Redcrest.—Orig. in New Brunswick, N.J., by M.A. Blake, New Jersey Agr. Expt. Sta. Introd. in 1946. Parentage unknown; tested as N.J. 126. Fruit: large; ovate; skin very attractive blush over yellow; flesh yellow, firm, melting, good flavor, freestone; excellent for freezing and canning; ripens 1 week before Elberta. Tree: vigorous and highly productive; not as hardy as Elberta; flower small, nonshowy; leaf glands reniform; chilling requirement 650 h at or below 45F. Obsolete.

Reddawn.—Orig. in Benton Harbor, Mich., by Mr. Molter, Molter Nurseries. Discovered and introd. about 1960. Parentage unknown. Fruit: size medium; round; skin yellow ground color covered with bright red, medium thick; flesh melting, firm; semi-clingstone, excessive split pits; for fresh market; ripens 15 days before Redhaven. Tree: large; spreading; vigorous; flower buds susceptible to winter cold injury.

Redelberta (*Scarlet Elberta*).—Orig. in Kennewick, Wash., by Jay Perry. Introd. in 1936. Plant patent 232; 2 Feb. 1937; assigned to Columbia & Okanogan Nursery Co., Wenatchee , Wash. Elberta mutation; discovered in 1928. Fruit: slightly smaller than Elberta; skin more highly colored than parent; flesh yellow, with some red throughout, freestone, quality excellent; ripens 10 days earlier than Elberta, which it resembles. Tree: resembles Elberta.

Redglobe.—Orig. in Beltsville, Md., by the U.S. Dept. of Agriculture. Introd. in 1954. (Admiral Dewey x St. John) x Fireglow; cross made in 1940; selected in 1944; tested as USDA B7398. Fruit: medium to large; round, halves equal; skin color especially bright and attractive, at maturity about three-fourths covered with bright red over yellow, pubescence heavy; flesh yellow, especially free from any red through the flesh, but with some red around the stone, very firm melting, fine-textured, flavor fair to good; ships well; pit size medium, free; acceptable for canning, attractive as a frozen product; well-suited for long-distance shipment; ripens about 15 days before Elberta. Tree: productive; vigorous; susceptibility to bacterial spot disease seems similar to Elberta; buds relatively tender to frost; flower large, showy, deep pink, self-fruitful; leaf glands reniform; chilling requirement of 850 h.

Red Gold.—Orig. in Blount County, Ala., by Charles Thomason of Bangor, Ala. Introd. in 1956. Plant patent 1329; 14 Dec. 1954; assigned to Stark Bros. Nurseries & Orchards Co., Louisiana, Mo. Hal-Berta Giant x Sunrise; selected in 1946. Fruit: skin highly blushed with crimson; flesh yellow, quality high, freestone; highly resistant to *Xanthomonas pruni* and brown rot; ripens about 15 days earlier than Elberta. Tree: size medium; vigorous; open, spreading; productive; leaf glands reniform; flower small, pink.

Red Gold Giant.—Orig. in Bridgman, Mich., by Ackerman Nurseries. Introd. in 1951. J.H. Hale o.p. Fruit: large; round; skin heavily blushed; flesh yellow, fine-grained, juicy, firm, flavor good, freestone; ripens 4 days after J.H. Hale; ships well; resembles Halberta. Tree: yields better when cross-pollinated with such varieties as Elberta, Halehaven, etc.; hardy.

Red Hale.—Orig. is unknown. Introd. by Stahelin Nursery, Detroit, Mich., in 1948. Parentage unknown. Fruit: large; skin attractive, blushed; flesh characteristics poor. Tree: vigorous; large; branches tend to break down rather easily.

Red Jack.—Orig. by C.F. Zaiger, L.M. Gardner, C.N. Zaiger, and G.G. Zaiger, Modesto, Calif. Plant patent 7194; 13 Mar. 1990. [(Ruby Gold x peach) F_2] x [(Sunred x Royal April) F_2]. Fruit: medium; globose; very firm; high percentage of red over yellow ground cover with moderately heavy pubescence; flesh yellow, melting, semi-clingstone, ripens 20 May at Modesto. Flowers medium, pink, nonshowy; predominantly globose leaf glands but some reniform. Tree: large; vigorous; productive. Chilling requirement 150 h.

Redhaven.—Orig. in South Haven, Mich., by Stanley Johnston, Michigan Agr. Expt. Sta. Introd. in 1940. Halehaven x Kalhaven; cross made in 1930. Fruit: size medium; round, suture somewhat prominent, apex distinct; skin yellow overlaid red to deep red, attractive, tough, pubescence light to medium; flesh yellow, red at pit, nonbrowning, very firm, melting, quality good; pit large, usually free when ripe; ships very well; ripens very early, 30 days before Elberta. Tree: spreading, vigorous; leaf glands reniform; flower size medium, nonshowy, petals red, self-fruitful; highly productive; nearly always requires thinning; chilling requirement of 950 h.

Redkist.—Orig. in Watervliet Township, Berrien County, Mich., by A.C. Wendzel. Plant patent 4055; 17 May 1977; assigned to Hilltop Orchards & Nurseries, Hartford, Mich. Redskin mutation; asexual propagation began in 1967. Fruit: rounder than Redskin, bright yellow ground color bright red over-color as compared with Redskin; skin smooth; light pubescence; fruit size averages 2 1/2 inches; pit small; excellent quality mild subacid flavor; matures 7 to 10 days earlier than Redskin; shipping and holding quality equal to Redskin. Tree: resembles Redskin.

Red Lady (*Scarlet Lady II*).—Orig. in Red Bluff, Calif., by Grant Merrill. Introd. in 1965. Plant patent 2601; 8 Feb. 1966. Gold Rush o.p.; selected at Red Bluff in 1958 and tested at Lamont and Exeter, Calif. Fruit: about size of J.H. Hale; cordate; apex tipped; skin bright to dark red over most of surface, tightly adherent; flesh firm, bright yellow, red near pit extending some into flesh, texture fine; pit large, freestone; ripens with Gold Rush. Tree: vigorous; size medium to small, but large with controlled pruning, spreading; very productive; leaf glands reniform; leaves on short internodes; flower small, showy.

Redleaf.—See Kingston Redleaf.

Redqueen.—Orig. in New Brunswick, N.J., by Catherine H. Bailey and L.F. Hough, New Jersey Agr. Expt. Sta. Introd. in 1964; (J.H. Hale x Eclipse) x Blake; cross made in 1951; selected in 1954; tested as N.J. 212. Fruit: large when thinned; skin three-fourths to full red, slightly dark, attractive; Blake type; flesh yellow, as firm as Elberta; excellent dessert quality; freestone; ripens about with Goldeneast and Loring, 2 weeks before Elberta. Tree: extremely hardy; flower showy; sets heavily; leaf glands reniform; suggested test area: where hardiness is a problem.

RedRan.—Orig. in McFarland, Calif., by Harold R. Martin. Introd. in 1964. Plant patent 2588; 18 Jan. 1966; assigned to Fortuna Trees, McFarland, Calif. Rancho Redleaf F_3. Fruit: 2 inches in diam.; unusually uniform; skin reddish-brown to yellowish-orange, thick, tough; flesh light yellowish-green, juicy, flavor poor, texture coarse, freestone. Tree: for rootstock use, greater production of germinable seeds than Lovell or Rancho Redleaf; selected for resistance to nematodes, *Meloidogyne incognita acrita* and *Meloidogyne javanica*.

Red Robin.—Orig. in Ontario, Calif., by Herbert C. Swim, Armstrong Nurseries. Introd. in 1959. Plant patent 1976; 18 Oct. 1960. Parentage unknown; selected in 1951. Fruit: size medium; shape almost round, variable; skin highly colored red; flesh white, flecked with considerable red near skin area, flavor well balanced, subacid, freestone; ripens early, a few days after Robin; resembles Robin but more highly colored. Tree: vigorous; chilling requirement less than that of Robin and almost as low as that of Springtime; leaf glands reniform; flower size medium, showy, pink.

Redrose.—Orig. in New Brunswick, N.J., by M.A. Blake, New Jersey Agr. Expt. Sta. Introd. in 1940. J.H. Hale x Delicious; cross made in 1925; tested as N.J. 98. Fruit: large; round, uniform; skin bright red on white ground, attractive, pubescence medium; flesh creamy white, red at pit, melting, moderately firm, fine texture, good quality, freestone; ripens 2 weeks before Elberta. Tree: vigorous, globular, very productive; flower large, showy, petals pink, self-fruitful, blooms with Elberta; leaf glands reniform; chilling requirement 850 h.

Redskin.—Orig. in College Park, Md., by A.L. Schrader and I.C. Haut, Maryland Agr. Expt. Sta. Introd. in 1944. J.H. Hale x Elberta; selected in 1931. Fruit: medium to large, good size even with heavy production; nearly round; skin deep red over yellow, attractive, pubescence light to medium; flesh yellow, red at pit, melting, nonbrowning, firm, quality somewhat better than Elberta, fine flavor; shipping and keeping, canning and freezing qualities good; pit size small to medium, free; ripens with or slightly ahead of Elberta. Tree:

spreading; vigorous; very productive; shorter chilling requirement than Elberta, 750 h; flowers large, showy, petals pink, self-fruitful, blooms with Elberta; resistant to bacteriosis.

Red Sun.—Orig. by T.O. Chamberlin, Sr., Visalia, Calif. Plant patent 7829; 17 Mar. 1992. Assigned to Metropolitan Life Insurance Co., Overland, Kan. Unknown parentage. Fruit: large; round; dark-red blush over faint yellow ground color with medium pubescence; flesh yellow, melting, freestone, ripens 7 days before Ryan's Sun. Flowers large, showy, pink; leaf glands reniform. Tree: upright; productive.

Redtop.—Orig. in Fresno, Calif., by John H. Weinberger, U.S. Dept. of Agriculture and Fresno State College Foundation. Introd. in 1961. Sunhigh x July Elberta o.p.; cross made in 1949 at Beltsville, Md.; seedling grown and selected at Fresno; tested as U.S. F-15. Fruit: larger than Coronet, usually not as large as Redglobe; nearly round, apex slightly pointed; skin nearly covered with an attractive blush, pubescence slight; flesh yellow, unusually firm, fine-grained, quality high, freestone; product good when canned or frozen; moderately susceptible to bacterial spot disease; ripens about 3 1/2 weeks before Elberta, between the seasons of Regina and Coronet and Redglobe. Tree: productive; moderately vigorous; moderately susceptible to bacterial spot disease; flower with large petals, self-fruitful, blooms with July Elberta; leaf glands reniform.

Redwin.—Orig. in White Horse Pike, Hammonton, N.J., by Joseph G. Battaglia. Introd. in 1957, by Bountiful Ridge Nurseries, Princess Anne, Md. Plant patent 1661; 3 Dec. 1957; assigned to Edgar H. Kemp and Homer S. Kemp, Princess Anne, Md., copartners, trading as Bountiful Ridge Nurseries. Erly-Red-Fre mutation; discovered in 1950. Fruit: cheek diam. about 2 1/4 inches; flesh white, mottled with red next to skin on the sunny size, quality good, semi-freestone; ripens before Mayflower and about 3 weeks before parent. Tree: vigorous; blooms before most other varieties; short chilling requirement; leaf glands globose; flower size medium, salmon.

Redwing.—Orig. in Ontario, Calif., by Walter E. Lammerts, Armstrong Nurseries. Introd. in 1944. Plant patent 621; 28 Mar. 1944; Walter E. Lammerts, assignor to Armstrong Nurseries. Babcock x Stensgaard July Elberta; selected in 1939. Fruit: size medium, up to 2 5/8 inches in diam., larger than Babcock; globose, most fruit unsymmetrical; skin yellow, blushed, medium, thick, tough, pubescence moderate and short; flesh yellowish-white, medium firm, melting, fibers few and tender, flavor mild, subacid, delicate, aroma distinct; freestone, small; ripens early mid-season, usually shortly before Babcock, which it resembles. Tree: medium to large; vigorous; very productive annually; slightly more winter chilling required than Babcock; flower small, petals deep pink.

Regal.—Orig. at Louisiana Agr. Expt. Sta; outstanding for attractiveness of large, early season fruit. Harvester x Surecrop; crossed in 1977; tested as L77-2-79. Introd. in 1993 by C.E. Johnson, J.T. Payne, W.A. Young, and E.W. Bush. Fruit: medium size; oval, slightly pointed; 65% smooth red blush over yellow ground color; short pubescence; flesh yellow, firm, melting; semifree when soft-ripe. Ripens 3 days before Bicentennial. Flowers: nonshowy with rose-colored petals; leaf glands reniform. Highly resistant to bacterial leafspot. Chilling requirement 700 h.

Regina.—Orig. in Beltsville, Md., and Fresno, Calif., by the Crops Research Division, Agr. Res. Serv., U.S. Dept. of Agriculture. Introd. in 1958. Sunhigh x (Admiral Dewey x St. John) selfed]; cross made in 1946 at Beltsville; seed planted in Fresno in 1950; selected as promising in 1953 by Leon Havis; tested as USDA F-14. Fruit: size medium; ovate; skin pubescence light, three-fourths of surface covered with a bright, attractive, red blush over a yellow ground color; flesh yellow, texture smooth, flavor good, firm but melting; freestone; when fully mature, similar to Redhaven and Coronet; firmer, more highly colored and smoother in shape than Redhaven; does not tend to soften first on suture and apex; resembles Coronet, differing primarily in ripening about 3 days later and in being somewhat more acid. Tree: above average in vigor; productive, though tending to set less heavily than Redhaven; petals small, opening later than Elberta but before Redhaven.

Reliance.—Orig. in Fonthill, Ont., Canada, by Jesse H. McCombs, Union Nurseries. Introd. about 1927. Fruit: large; skin yellow overlaid with dark red; flesh yellow, medium quality, freestone; ripens about 10 to 14 days before Elberta; resembles South Haven. Very cold hardy.

Reliance.—Orig. in Durham, N.H., by E.M. Meader, New Hampshire Agr. Expt. Sta. Introd. in 1964. (Minn. PH04559 x Meredith) o.p.2; selected in 1959; tested as NH 59E. Fruit: size medium; round; suture slightly prominent; skin dull red splashed over yellow, pubescence moderate; flesh bright yellow, medium firm, slightly stringy, ripens evenly, good flavor, freestone; ripens about with Golden Jubilee and Oriole. Tree: productive as fruit buds very hardy; flowers showy; leaf glands globose; adapted to northern states and for trial where peach culture is an uncertain venture due to cold winters.

Renton.—Orig. in Coquitlam, B.C., Canada, by Jack Renton. Introd. in 1940. Parentage unknown; discovered in 1926. Fruit: flesh yellow, semi-freestone; resembles Howard Fisher. Tree: more resistant to leaf curl than are other varieties.

Richhaven.—Orig. in South Haven, Mich., by Stanley Johnston, Michigan Agr. Expt. Sta. Introd. in 1955. Redhaven x SH 50 (J.H. Hale x Halehaven); cross made in 1944; selected in 1949. Fruit: large to very large; nearly round, uniform; skin blushed, with stripes of red and gold, thick, tough, appearance very attractive, pubescence almost negligible; flesh yellow with some red near the pit, texture medium, firm, melting, resistant to browning; high quality for shipping, commercial canning and freezing; matures about with Halehaven, 14 days before Elberta; pit small, free, even when firm ripe; quality excellent; resembles Halehaven but larger, skin color brighter, flesh firmer. Tree: large; vigorous; highly productive; susceptible to bacteriosis; hardy; leaf glands reniform; flower very small, nonshowy, pink, self-fruitful, blooms with Elberta; chilling requirement of 950 h.

Rich Lady.—Orig. by C.F. Zaiger, G.N. Zaiger, L.M. Gardner, and G.G. Zaiger, Modesto, Calif. Plant patent 7290; 7 Aug. 1990. Amparo o.p. Fruit: large; globose; red over golden yellow with moderately heavy pubescence; flesh yellow, very firm, melting, semi-clingstone, ripens 17 days before Amparo. Flowers large, light pink, showy; leaf glands reniform. Tree: large; vigorous.

Rich May.—Orig. by C.F. Zaiger, L.M. Gardner, G.N. Zaiger, and G.G. Zaiger, Modesto, Calif. Plant patent 7432; 29 Jan. 1991. [(May Grand o.p. x peach) x Sam Houston o.p.] x (Tasty Gold x May Crest). Fruit: large; nearly globose; firm; liquid red blush over yellow ground color with moderate pubescence; flesh yellow, melting, semi-clingstone. Ripens 5 days before May Crest. Flowers large, showy, pink; leaf glands reniform. Tree: large; vigorous; productive.

Rio Grande.—Orig. in Hawthorne, Fla., by Tom Huston, Miami, Fla. Introd. in 1969. Plant patent 2888; 20 May 1969; assigned to Peaches of Florida, Balm, Fla. Blazing Gold x (Southland x Hawaiian) F_2. Fruit: medium to large; shape variable; skin yellow, covered with red blush, pubescence moderate; flesh yellow, firm, freestone; shipping quality good; for fresh market and dessert; ripens midseason, 10 June. Tree: size medium; vigor medium; flower large, showy, pink; chilling requirement 400 to 450 h.

Rio Oso Gem.—Orig. in Rio Oso, Calif., by William F. Yerkes. Introd. in 1933. Plant patent 84; 28 Nov. 1933. Parentage unknown; selected in 1926. Fruit: large, averaging between 2 3/4 to 3 3/4 inches in diam.; round to slightly elongated, slight to prominent suture ridge; skin with more red than J.H. Hale, very attractive; flesh yellow, firm, fine texture, good quality, melting, freestone; ripens 1 week after Elberta, about 2 weeks after H.J. Hale, which it resembles. Tree: productive; susceptible to bacterial spot; leaf glands reniform; flower showy, very large, light pink, self-fruitful, chilling requirement of 850 h.

Riverdale.—Orig. in Riverdale, Ga., by Emory Travis, Riverdale Nurseries. Introd. in 1930. Parentage unknown; discovered in 1930. Fruit: flesh yellow, quality good, clingstone; ripens in Georgia about the first week of July. Probably obsolete.

Rizzi.—Orig. in 1973 at Univ. of California, Davis. Plant patent 8394; 28 Sept. 1993 to Univ. of California Regents. Selected in 1978 by R. Fenton; tested as 19, 2-72; and released in 1993 by T.M. Gradziel, W. Beres, J. Doyle, and C. Weeks. Everts o.p. Fruit: large; globose; 20% to 60% red blush over yellow-orange with moderate pubescence; flesh uniform yellow throughout, nonmelting, clingstone with a smaller than average, light-brown pit. Ripens late season with Ross and Dr. Davis. Fruit have potential for extended storage beyond existing processing cultivars. Flowers small, nonshowy; leaf glands reniform. Tree: vigorous; slightly upright.

R.L. Stoner.—Orig. near Dayton, Ohio, by R.L. Stoner. Introd. in 1948. Plant patent 798; 11 May 1948. Early Elberta mutation; discovered in 1929. Fruit: size medium, averages 2 1/2 inches in diam.; roundish oblong, halves unequal, suture prominent, apex rounded, tip pointed; skin deep red, thick, tough, finely pubescent; flesh white to cream, red at pit, melting, sometimes mealy, fine-textured, freestone; shipping and canning quality good; ripens 7 days before Elberta; resembles Belle. Tree: vigorous; consistent and high bearer; hardy; leaf glands reniform; flower small, nonshowy, petals pink, self-fruitful; chilling requirement of 750 h.

Roberta.—Orig. in Villa Ridge, Ill., in the orchard of R.B. Endicott. Introd. in 1933. Elberta mutation; discovered in 1928, by J.W. Lloyd, Illinois Agr. Expt. Sta., Urbana, Ill. Fruit: resembles Elberta, but round and ripens 1 week earlier, skin deeper red and more fully colored before maturity than parent.

Robin.—Orig. in Ontario, Calif., by Walter E. Lammerts, Armstrong Nurseries. Introd. in 1944. Plant patent 529; 30 June 1942; Walter E. Lammerts, assignor to Armstrong Nurseries. Babcock x Mayflower; selected in 1941. Fruit: medium to small, up to 2 3/8 inches in diam., larger than Mayflower, fairly uniform; globose, symmetrical; skin yellow, highly blushed, thin, tender, free when fully ripe, pubescence short and abundant; flesh yellowish white, firm, melting, fibers few and tender, flavor mild, sweet with slight acid, aroma pronounced; semi-clingstone, small; ripens very early, with Mayflower, about 1–10 June at place of origin; resembles Babcock. Tree: resistant to delayed foliation in mild winter climates, being similar to Babcock; bears annually and well; medium large, medium vigorous; flower very large, petals pink.

Robison.—Orig. in Greer, S.C., by John B. Robison. Introd. in 1962. Plant patent 2163; 14 Aug. 1962; assigned to Haley Nursery Co., Smithville, Tenn. Chance seedling. Fruit: medium to large, averaging about 2 3/4 inches in diam.; shape symmetrical, suture line smooth; skin yellow with cheek bright red; flesh light yellow, not excessively red around the pit, tends to be nonbrowning, firm, flavor sweet, subacid, mild, delicious, freestone; ripens late, after Elberta and J.H. Hale and before Rio Oso Gem. Tree: size medium; bears regularly; vigor average; flower large, showy, pink, blooming in midseason; medium chilling requirements.

Rochon.—Orig. in Riverside, Calif., by J.W. Lesley, California Agr. Expt. Sta. Introd. in 1958. Peento hybrid x (Rochester x Peento); selected in 1953. Fruit: rather small, distinct beak; skin an attractive red over yellow; flesh yellow, juicy, soft, quality fair; semi-clingstone; matures early; resembles Tejon. Tree: short winter-chilling requirement; leaf glands globose; flower large, showy, blooms early; fruit set medium to heavy after a warm winter, producing a medium-sized crop.

Rodeo (*Merrill Rodeo*).—Orig. in Red Bluff, Calif., by Grant Merrill, Grant Merrill Orchards. Introd. in 1951. Plant patent 1097; 20 May 1952. Kirkman Gem o.p.; selected in 1950. Fruit: skin more highly colored than Kirkman Gem; flesh yellow, freestone, does not darken upon exposure to air, sweeter and firmer than known parent; no soft suture; season very late, 7 weeks after Elberta and 10 days later than Kirkman Gem; resembles Kirkman Gem. Tree: more vigorous and larger than parent; leaf glands reniform; flowers large, light pink.

Rogue Rio.—Orig. in Talent, Ore., by Donald Charles Korth. Introd. in 1958. Parentage unknown; discovered in 1954. Fruit: skin with a high blush; flesh yellow, very firm; freestone, pit small; ripens between J.H. Hale and Rio Oso Gem. Tree: very vigorous; self-fruitful.

Romance.—Orig. in Mountain Grove, Mo., by Paul H. Shepard, Missouri State Fruit Expt. Sta. Introd. in 1947. Wilma x Halehaven; selected in 1947; tested as PH-1132. Fruit: large; skin becomes full red while still firm; flesh yellow with no red, firm, fine-grained, flavor good; good shipper; ripens 16 days before Elberta; resembles Wilma. Tree: buds and flowers more hardy than Elberta.

Rosebud.—Orig. in New Brunswick, N.J., by M.A. Blake, New Jersey Agr. Expt. Sta. Introd. in 1925. Carman x Slappey; cross made in 1916. Fruit: small; oval, apex sharply pointed; skin greenish-white overlaid with red, pubescence short, medium; flesh white, red at pit, firm, melting, texture medium fine; quality fair, semi-clingstone; ripens 30 days before Elberta. Tree: very hardy; prolific, requires severe thinning; vigorous, spreading; leaf glands reniform; flowers small, nonshowy, self-fruitful.

Roslyn.—Orig. in Monticello, Ga., by Roslyn Benton Fitzpatrick. Introd. about 1947. Possibly Elberta o.p. Fruit: very large; elongated, with a very decided tip; flesh yellow, firm, clingstone; ripens about 2 days after Elberta, which it resembles; shipping quality excellent. Tree: large; vigorous.

Ross.—Orig. by L.D. Davis, Davis, Calif. Plant patent 4863; 29 June 1982; assigned to Regents of Univ. of California, Berkeley. Hybrid of L.D. Davis selections GH,8-14 x D,30-3E. Tested as N30-12E. Fruit: firm; large; globose; yellow skin over golden yellow flesh; flesh nonmelting, clingstone, brown stone. Ripens late, 2 days before Carolyn. Medium-short pubescence. Flowers pink, showy; leaf glands globose. Tree: semi-upright, medium size and vigor; very productive.

Rosy Dawn.—Orig. in Buckley, Wash., by Buckley Nursery Co. Introd. in 1961. Pacific Gold x Redhaven; cross made in 1955, selected in 1958. Covered by trademark. Fruit: medium to large; oval; skin yellow ground color with heavy red blush; flesh deep yellow, flavor good, freestone; ripens about 20 July; for fresh market and canning. Tree: 10 to 12 feet; oval shape; vigorous; hardy to –10F; productive; more tolerant to leaf curl than other varieties in the area.

Royal (*Royal Hale*).—Orig. in Le Grand, Calif., by F.W. Anderson. Introd. in 1954. Plant patent 1247; 26 Jan. 1954; assigned to Reedley Nurseries, Reedley, Calif. Le Grand nectarine x Elberta; selected in 1949; tested as Anderson lB36. Fruit: more red skin color than J.H. Hale, which it resembles; quality better than J.H. Hale; ripens 3 days earlier than J.H. Hale. Tree: leaf glands reniform; flower large, self-fruitful. Since the introduction of Royal Hale peach in 1954, confusion has resulted because of its being called "Hale" in the trade, which is J.H. Hale. Thus, the name has been changed to Royal.

Royal April.—Orig. in Modesto, Calif., by C. Floyd Zaiger, Zaiger's Nursery. Introd. in 1972. Plant patent 3411; 20 Nov. 1973, assigned to Dave Wilson Nursery, Hughson, Calif. Royal Gold mutation; discovered about 1965. Fruit: size medium; round; skin heavy red blush when mature; flesh melting, fine, clingstone; a dessert variety; ripens 7 to 10 days before Royal Gold. Tree: flower nonshowy; leaf glands globose.

Royal Cocktail.—Orig. in Sacramento, Calif., by Edwin M. Wagner. Introd. in 1957. Plant patent 1623; 30 July 1957. Chance seedling; discovered in 1948. Fruit: 2 1/2 to 3 inches in diam.; pronounced apex; flesh yellow, firm, tough; commercial canning clingstone; matures about 10 days later than Phillips Cling, which it resembles, or about 9–20 Sept. at place of origin; lasts well on tree. Tree: large; upright; vigorous; productive; flower very small, pink; leaf glands mostly globose. No longer commercially planted.

Royal Elberta.—See Royal Haven.

Royal Fay.—Orig. in Poplar, Tulare County, Calif., by Fred D. Williams, Porterville, Calif. Introd. in 1945. Plant patent 795; 21 Mar. 1948. Fay Elberta mutation; discovered in 1941. Fruit: flesh yellow, firm, freestone; pubescence sparse and short; similar to parent but with higher color. Tree: large; vigorous; upright; leaf glands reniform; flower size medium, salmon.

Royal Gem.—Orig. in Le Grand, Calif., by F.W. Anderson. Introd. in 1959. Plant patent 1754; 23 Sept. 1958; assigned to Reedley Nursery, Reedley, Calif. Le Grand nectarine x late-ripening freestone peach; selected in 1955; tested as Anderson llL9A. Fruit: large, averaging 3 1/4 inches in diam.; skin yellow, overspread with mottled red; flesh yellow, reddish next to stone, firm, flavor subacid, aroma distinct, eating quality good to best, freestone; J.H. Hale type, ripening about 1 month later and with more red skin color and less acidity than that variety; resembles Fiesta. Tree: large; vigorous; spreading; productive; leaf glands reniform; flower large.

Royal Gold.—Orig. in Modesto, Calif., by C.F. Zaiger, Zaiger's Nursery. Introd. in 1967. Plant patent 2663; 23 Aug. 1966; assigned to Dave Wilson Nursery and Buller Farms, Modesto, Calif.; discovered in 1961; named in 1967. Springtime mutation. Fruit: medium; 2 1/4 to 3 inches in diam.; globose to slightly flattened at stem in large size, apex projecting, oblong in smaller sizes; skin bright yellow ground cover, covered with red blush; flesh faintly yellow to darker yellow at apex, firm, crisp, mildly acid to sweet, aromatic, quality good; stone free but tight when not fully ripe, frequently retains thin filaments of flesh along ridges near base; ripens 2 to 4 days earlier than Springtime. Tree: medium to large; vigorous; productive; regular bearer; leaf glands globose; flower size medium, pink.

Royal Hale.—See Royal.

Royal Haven (*Royal Elberta*).—Orig. in Fowler, Calif., by Harry E. Hiraoka. Introd. in 1955. Plant patent 1499; 24 July 1956. July Elberta mutation; discovered in 1952. Fruit: large; shspe round; skin deep, glossy red, tough; flesh yellow streaked with red, firm, sweet, quality very good, freestone; matures before Redhaven, 30 days earlier than July Elberta. Tree: large; vigorous; upright; productive; leaf glands reniform; flower size medium, salmon.

Royal June (*Beauty Flame*).—Orig. in Fowler, Calif., by Harry E. Hiraoka. Introd. in 1955. Plant patent 1426; 25 Oct. 1955. July Elberta mutation. Fruit: large; skin blushed a deep red; flesh yellow, firm, fibers few, flavor sweet, freestone; shipping qualities good; ripens evenly, 1 week before Gold Dust and 30 to 35 days before July Elberta. Tree: large; vigorous; upright; productive; flower size medium, salmon.

Royal Lady.—Orig. by F.J. Janzen, Reedley, Calif. Plant patent 8211; 27 Apr. 1993. Elegant Lady mutation, ripens 2 weeks later.

Royal May.—Orig. in Le Grand, Calif., by F.W. Anderson. Introd. in 1956. Plant patent 1516; 11 Sept. 1956; assigned to Reedley Nursery, Reedley, Calif. (Le Grand nectarine x July Elberta seedling) F_2; selected in 1954; tested as Anderson 10K99. Fruit: large to medium, 2 1/2 inches in diam.; skin light yellow almost completely overspread with bright red, pubescence short and inconspicuous; flesh yellow, quality good, flavor subacid, aroma wanting, semi-freestone to clingstone; ripens about 3 to 4 days earlier than Blazing Gold and Robin and about 3 to 4 days later than Mayflower, or 10–19 June. Tree: large to medium; vigorous; upright to spreading; productive; leaf glands reniform.

Royal Red.—Orig. in Reedley, Calif., by Nickolai J. Friesen. Not introd., no trees sold. Plant patent 3273; 19 Dec. 1972. Fortyniner mutation, discovered in 1961. Fruit: about 2 1/2 inch diam.; generally globose, uniform; skin yellow, spotted with red, medium thick; flesh bright yellow, generally firm, fine, flavor sweet, quality excellent, freestone; can be harvested in two pickings; ripens 7 to 10 days before Fortyniner, 15 July, hangs well; for local and distant fresh market, keeping and shipping qualities excellent. Tree: size and vigor medium; productivity regular and heavy; flower deep purplish pink, blooms early, with Fortyniner.

Royal Redleaf.—Orig. in Bridgman, Mich., by the Ackerman Nurseries. Introd. about 1940. Chance seedling, discovered in 1938. Fruit: size medium; skin pinkish-red; flesh white, quite juicy, fine flavor. Tree: foliage red; flowers pink. Primarily an ornamental variety, but the fruit is edible.

Royalvee.—Orig. in Vineland Station, Ont., Canada, by the Ontario Dept. of Agr., Hort. Res. Inst. Introd. and named in 1959. (Halehaven x Vedette) x Veteran; cross made in 1946; tested as Ontario 46071. Fruit: size medium; skin streaked red over orange ground color; flesh bright yellow, fairly firm, quality good, oxidizes slowly; semi-freestone, but reasonably free at full maturity; ripens 2 weeks before Redhaven or about Goldray season, averaging 6 Aug. at Vineland. Tree: very productive; requires heavy thinning in most seasons, being similar to Veteran in this respect; hardy; self-fruitful, setting fruit under adverse conditions; flower large, showy, being similar to Veteran; self-fruitful; leaf glands reniform.

Roza.—Orig. in Prosser, Wash., by Wash. State Agr. Expt. Sta. and U.S. Dept. of Agriculture. Introd. in 1971. J.H. Hale x July Elberta o.p. Cross made in 1949 at Beltsville, Md., by A.L. Havis; planted at Prosser and selected in 1953 by H.W. Fogle; tested as Prosser 0-1384 by T.K. Toyama, Prosser (U.S. Dept. of Agriculture). Fruit: 2 3/4 to 3 inches in diam.; round; skin light yellow ground cover, faintly streaked with medium red blush over 75% of surface; flesh deep yellow, very firm, texture slightly coarse, flavor excellent, freestone; good shipping quality; ripens with Redglobe and about 2 weeks before J.H. Hale; recommended for fresh market and canning. Tree: vigorous; productive; flower small, self-fertile; susceptible to bacterial spot (*Xanthomonas pruni*).

Rubidoux.—Orig. in Riverside, Calif., by J.W. Lesley, California Citrus Expt. Sta. Introd. in 1949. Parentage includes Elberta, Late Champion, Lukens Honey, Peento; selected in 1940. Fruit: medium large; skin greenish-yellow with a dull red blush; flesh light yellow, deep red around pit extending into flesh, some fiber, sweet, well-flavored, freestone; keeps very well in cold storage; ripens last of August, about 10 days after J.H. Hale. Tree: vigorous; moderately spreading; retains foliage into late fall; leaf glands reniform; production good; short winter-chilling requirement. Obsolete.

Rubired.—Orig. in Jackson Springs, N.C., by F.E. Correll and C.N. Clayton, North Carolina Agr. Expt. Sta. Introd. in 1972. Redhaven x Dixired. Cross made in 1955, selected in 1959, tested as NC 4911. Fruit: size medium to large; round; skin light yellow ground color, three-fourths covered with red blush with darker blotches, attractive; flesh medium yellow with very little or no red pigment, firm, fine texture, melting, semi-freestone when soft-ripe, highly resistant to browning; ripens 6 days earlier than Redhaven. Tree: spreading; vigor moderate; productive; flower small, nonshowy, self-fertile; leaf glands reniform; highly resistant to bacterial spot disease. Chilling requirement 800 to 900 h.

Ruby May.—Orig. by J.S. Takeda, Orosi, Calif. Plant patent 7170; 27 Feb. 1990. Springcrest mutation; ripens earlier, about 3–10 May at Orosi.

Ruston Red.—Orig. in Calhoun, La., by C.E. Johnson, Louisiana Agr. Expt. Sta., Baton Rouge. Introd. in 1982. LaPremiere [Prairie Rose x (Redglobe o.p.)]. Cross made in 1971; selected in 1974; tested as L71-5-4. Fruit: large, 2 1/2 inches in diam.; round, slight point, equal halves; skin yellow, 70% red over-color, pubescence medium; flesh yellow with red flecks near pit, firm, excellent texture and quality; freestone; ripens 17–20 July, 2 to 5 days after Elberta. Tree: vigorous, productive, resistant to bacteriosis (*Xanthomonas pruni*); chilling requirement of 800 to 900 h; flower medium size, light-pink, nonshowy; leaf glands reniform.

Rutgers Green Leaf.—Orig. in New Brunswick, N.J., by M.A. Blake, New Jersey Agr. Expt. Sta. Introd. in 1947. Parentage unknown; tested as N.J.E. Fruit: very small; flesh white, freestone; of no value. Tree: suggested as a source of vigorous hardy seedling rootstocks; flowers showy.

Rutgers Red Leaf.—Orig. in New Brunswick, N.J., by M.A. Blake, New Jersey Agr. Expt. Sta. Introd. in 1947. Parentage unknown. Fruit: very small; flesh white, freestone; of no value. Tree: suggested as a source of hardy seedling rootstocks; may have slight dwarfing effect; readily identified in nursery because of red color of young leaves; flowers showy.

Ryan Sun.—Orig. at Del Ray, Calif., by Tom Chamberlain, H.P. Metzler & Sons. Introd. in 1983. Plant patent 5103; 13 Sept. 1983; owned by Metropolitan Insurance Co.; licensed exclusively to Agri Sun Nursery. O'Henry o.p. Fruit: large, up to 100 mm; round; skin dark red on yellow ground;

flesh yellow, freestone. Ripens 60 days after Redhaven in Selma, Calif. Tree: very vigorous; upright. Chilling requirement about 500 h.

S-37.—Orig. in Atwater, Calif., by Thomas B. Stribling, Jr., Stribling's Nurseries, Merced, Calif. Introd. in 1947. Plant patent 904; 6 Dec. 1949. ornamental flowering peach o.p.; seed planted in 1935; selected about 1937; tested as Stribling 37-18. Rootstock for fruit trees: resistant to *incognita* root-knot nematode.

S-60.—Orig. in Merced, Calif., by Walter D. Krause, Stribling's Nurseries. Introd. in 1960. S-37 o.p.; selected in 1960. Used only as a rootstock: seedlings very uniform- resistant to root-knot nematode (*Meloidogyne incognita*), being similar to S-37; more compatible with scions of almond and plum varieties than S-37, well-suited for peach and nectarine; roots deeply.

Safari.—Orig. in Roodeplaat, Pretoria South Africa, by Edward P. Evans, Hort. Res. Inst. Introd. in 1914. Moreira Jubilee selfed. Cross made in 1960, selected in 1963, tested as 6-62. Fruit: large; round-oval; skin yellow ground color, 60% covered with red blush, very attractive, pubescence slight; flesh yellow, melting, firm, flavor slightly tart; freestone; ripens early December. Tree: large; vase-shaped; vigorous; moderately productive; thinning required; late flowering; chilling requirement about 500 h; flower nonshowy; leaf glands reniform.

Salberta.—Orig. in Ottawa County, Ohio, by George Rofkas, Clinton, Ohio. Introd. in the early 1920s. Possibly Elberta x Salwey. Fruit: flesh yellow, quality only fair, freestone; ripens about 2 weeks after Elberta; resembles Elberta.

Salem.—Orig. by C.W. Haines, Elmer, N.J. Plant patent 6267; 30 Aug. 1988; assigned to Adams County Nursery, Aspers, Pa. Unknown parentage. Fruit: large; globose; high percentage of red blush; moderate pubescence; flesh yellow, melting, semi-freestone. Ripens 5 to 10 days before Loring. Flowers pink, nonshowy; leaf glands reniform. Tree: large; spreading; vigorous; very productive.

Sam Houston.—Orig. in College Station, Texas, by J.B. Storey, Texas Agr. Expt. Sta. Introd. in 1965. July Elberta x (Early Elberta x Floriberta); cross made in 1949; tested as A-162-4. Fruit medium to large, 2 1/4 to 2 1/2 inches in diam.; skin attractive, red, coloring before harvest time; flesh yellow, fine texture, low in acid, flavor mild, remains firm; pit unusually small, freestone; ripening about 23 June, with Sunhigh and Goldenred. Tree: a low chilling variety adapted to regions of southern U.S. with about 500 to 700 h 45F. Sets heavily and requires much thinning, full bloom 6 days before Maygold; susceptible to bacterial spot, *Xanthomonas pruni*.

San Pedro.—Orig. in Agr. Expt. Sta., Univ. of Florida, Gainesville, by R.H. Sharpe and named in northern Argentina. Complex parentage of Florida selections. Selected in 1964; tested as Fla. 16A-33. Fruit: medium large; skin 30% red blush; flesh yellow, medium-firm, semi-freestone. Ripens 85 days after bloom. Flower showy; leaf glands reniform. Moderate resistant to bacterial leafspot. Chilling requirement 300 chill units.

Saturn.—Orig. in Ontario, Calif., by Herbert C. Swim, Armstrong Nurseries. Introd. in 1955. Plant patent 1485; 19 June 1956. Parentage traces to Rio Oso Gem, Swatow, Socala, Early Imperial, Coolidge Double Red, Lukens Honey, Florence; selected in 1949. An ornamental fruiting variety. Fruit: large; shape of Babcock; skin color of Elberta; flesh yellow, freestone; ripens 7 to 10 days later than Babcock. Tree: large; vigorous; upright; flower large, double, pink, petals 14-16; leaf glands reniform; low winter chilling requirement.

Scarlet Elberta.—See Redelberta.

Scarlet Lady (*Merricle L-130*).—Orig. in Red Bluff, Calif., by Grant Merrill. Introd. in 1958. Plant patent 1785; 23 Dec. 1958; assigned to Grant Merrill Orchards. J.H. Hale F4; selected in 1954. Fruit: large; round; skin highly colored red; flesh yellow, firm, freestone; ripens about 10 days be fore Elberta; resembles Gemfree. Tree: size medium; vigor medium; productive; leaf glands reniform.

Scarlet Lady II.—See Red Lady.

Scarletpearl.—Orig. at Byron, Ga., by W.R. Okie, U.S. Dept. of Agriculture S.E. Fruit & Tree Nut Res. Lab. Introd. in 1989. (Biscoe x Redgold nectarine) o.p. seed collected in 1980; selected in 1983; tested as BY81P1411. Fruit: medium; skin attractive red blush on greenish-white ground color, light pubescence; shape round; flesh white, melting; clingstone. Flavor is typical of other early-season peaches. Ripens with June Gold, 3 weeks before Redhaven. Flowers showy, self-fertile; leaf glands globose; moderately resistant to bacterial spot. Chilling requirement about 750 h, making it adapted to most of the Southeast.

Scarlet Snow.—Orig. by C.F. Zaiger, L.M. Gardner, G.N. Zaiger, and G.G. Zaiger, Modesto, Calif. [Golden Supreme o.p. x (O'Henry x Giant Babcock)] x (unknown nectarine x Golden Supreme o.p.). Plant patent 8439; 26 Oct. 1993. Fruit: large; globose; attractive red blush with short pubescence; flesh white, very firm, melting; freestone. Ripens 35 days after O'Henry. Flowers: medium, nonshowy, pink; leaf glands reniform. Tree: large; vigorous; upright; productive.

Schooldays (*Merricle Schooldays*).—Orig. in Red Bluff, Calif., by Grant Merrill. Introd. in 1950. J.H. Hale o.p.; selected in 1942. Fruit: flesh yellow, freestone; ripens 3 weeks later than J.H. Hale, which it resembles. Tree: very vigorous; self-fertile. Obsolete.

Scott Early Elberta.—See Scott Elberta.

Scott Elberta (*Scott Early Elberta*).—Orig. in Sandhill Expt. Sta. of Clemson College, Columbia, S.C., by L.E. Scott, South Carolina Agr. Expt. Sta., now of the Maryland Agr. Expt. Sta., College Park. Introd. in 1958. Elberta mutation; selected in 1937; P.I. 151650; tested as Sandhill I. Fruit: size medium-ovate to round; halves equal; skin greenish-yellow overlaid with red, striped, pubescence light to medium; flesh yellow, red at pit, sometimes flecked red throughout, firm, to soft, melting, quality fair to good; pit size medium, free; ripens 7 to 10 days before Elberta depending upon where grown; requires thinning to achieve good commercial size. Tree: highly productive; moderately vigorous; leaf glands reniform; flower small, nonshowy, self-fruitful, blooms 2 days after Elberta; chilling requirement of 850 h.

Sensation.—See Stark" Sensation.

Sentinel.—Orig. in Fort Valley, Ga., by U.S. Hort. Lab. Introd. in 1966. FV 5-56 (Halehaven selfed) x Dixigem; cross made in 1946; selected in 1951; first fruited in 1956; tested as FV 17347. Fruit: size medium; round; skin yellow ground color with 75% red blush, pubescence medium; flesh yellow, firm but melting; flavor good, texture good; freestone when fully matured. Tree: vigorous; productive; self-fertile; good resistance to bacterial spot disease; leaf glands globose; blossoms small petaled; chilling requirement 850 h.

Sentry.—Orig. in Beltsville, Md., by Harold W. Fogle in cooperation with U.S. Dept. of Agriculture. Introd. in 1980. Loring x Sentinel. Planted in 1964; selected in 1966; tested as Beltsville 64302. Fruit: large, attractive red blush covering 80%; flesh bright, clear yellow, firm, good quality for season; semi-freestone; ripens 12 July at Beltsville, 10 to 14 days before Redhaven and 6 weeks before Elberta. Tree: vigorous, productive, resistant to bacterial spot (*Xanthomonas pruni*); blossoms nonshowy and self-fertile; leaf glands small, globose. Chilling requirement 850 to 900 h of chilling.

September Lady.—Orig. in Red Bluff, Calif., by Grant Merrill, Exeter, Calif. Introd. in 1970. Plant patent pending. Seedling of J.H. Hale x (Hale x Levi) F2; discovered in 1967, tested as E287. Fruit: large, round; skin highly colored; flesh yellow, little red, freestone; recommended for fresh use, freezing and shipping; ripens mid- to late September in Fresno. Tree: large; vigorous, productivity heavy and regular. Recommended as an improvement over Carnival by ripening a little later and being a better commercial freezer variety.

September Sun.—Orig. at Del Ray, Calif., by Tom Chamberlain, H.P. Metzler & Sons. Introd. in 1987. Plant patent 6008; 8 Sept. 1987; assigned to Metropolitan Insurance Co.; licensed exclusively to Agri Sun Nursery. Parentage unknown. Fruit: large, up to 100 mm; round; skin medium red on yellow ground; flesh yellow, freestone. Ripens 90 days after Redhaven in Selma, Calif. Tree: very vigorous; very productive; upright. Chilling requirement >500 h.

Sessen Cling.—Orig. in West Plains, Mo., by Geo. L. Sessen. Introd. in 1922. Elberta o.p.; discovered in 1910. Fruit: flesh deep yellow with some red around the pit, very sweet, clingstone; ripens with and resembles Elberta.

Shalil.—Orig. in the Kurram Valley of northwestern India. Introd. into the U.S. in 1924 by U.S. Dept. of Agriculture as P.I. 36485 and later as P.I. 63850. Fruit: flesh yellow, freestone. seedlings of it are being used as rootstocks because of their resistance to root-knot nematode (*Meloidogyne incognita* var. *acrita*).

Shamel.—Orig. in Merced, Calif., on the Fancher Ranch by A.D. Shamel, U.S. Dept. of Agriculture, Riverside, Calif. Introd. in 1935. Probable Paloro o.p.; discovered in 1931. Fruit: flesh yellow; commercial canning clingstone; tendency to develop excessive blossom-end breakdown and split pits; resembles Paloro. Tree: tendency toward faulty fruit set. No longer commercially planted.

Shannon.—Orig. in Modesto, Calif., by the Hume Cannery. Introd. in 1939. Parentage unknown; selected in 1935. Fruit: ripens in Phillips Cling season; commercial canning clingstone. No longer commercially planted.

Sharon.—Orig. in McFarland, Calif., by R.E. Armantrout and E.W. Root. Introd. in 1943. Plant patent 540; 30 Sept. 1940. Parentage unknown; selected in 1930. Fruit: 3 1/4 inches in diam.; roundish-oval, halves slightly unequal; skin medium thick, medium tough, pubescence medium in length and thickness; flesh firm, white, flavor distinctive; freestone, medium-large, flattened and obovate; ripens 15–30 Aug. Tree: size medium; vigorous, upright-spreading; productive;leaf glands reniform; flower size medium, pink.

Shasta.—Orig. in Palo Alto, Calif., by W.F. Wight, U.S. Dept. of Agriculture and California Agr. Expt. Sta. Introd. in 1941. A clingstone seedling of Leader x (Tuscan x Paloro); tested as USDA W36-37A; sibling of Fortuna. Fruit: medium to large, averaging 7.2 cm in diam.; nearly round, uniform; suture well-marked, shallow; skin deep golden yellow, blushed up to one-fourth of surface, pubescence average; flesh clear yellow, pink to reddish at pit, fine-grained, nonmelting, flavor mild, sweet, aroma very slight, quality good; commercial canning clingstone; ripens early, with Fortuna, just before Tuscan; resembles Fortuna. Tree: vigorous; heavy producer; leaf glands globose, small to medium; flower nonshowy, size medium. No longer commercially planted.

Shepard Beauty.—Orig. in Crawford, County, Ga., by John F. Duke, John F. Duke, Jr., and Leman P. Duke, Fort Valley, Ga. Introd. in 1964. Plant patent 2439; 25 Aug. 1964; assigned to Charles L. Shepard, Jr., Fort Valley, Ga. Coronet mutation; discovered in 1958; first fruited in 1960. Fruit: resembles Coronet; average size 2 1/8 inches in diam.; 70% less fuzz than Coronet; skin yellow-gold with red flaking; ripens 16 days before Coronet; superior taste to Coronet, and does not deteriorate at maturity as early as Coronet; semi-freestone. Tree: productive; normally requires heavy thinning to make 2-inch fruit; chilling requirement 850 to 950 h.

Shermans Early.—Orig. in Agr. Expt. Sta., Univ. of Florida, Gainesville, by R.H. Sharpe and named in New South Wales, Australia. Introd. in 1985. Complex parentage of Florida selections. Selected 1973; tested as Fla. 3-1. Fruit: small; skin 60% red blush; flesh yellow, medium-firm but softening first at tip, semi-clingstone. Ripens 60 days from bloom. Flower nonshowy; leaf glands reniform. Resistant to bacterial leaf spot. Chilling requirement 400 chill units.

Shermans Red.—Orig. in Agr. Expt. Sta., Univ. of Florida, Gainesville, by R.H. Sharpe and named in New South Wales, Australia. Introd. in 1985. Sunred x Springbrite. Selected in 1972; tested as Fla. 2-2. Fruit: medium small; skin 90% red blush; flesh yellow, firm, semi-clingstone. Ripens 70 days from bloom. Flowers showy; leaf glands reniform. Resistant to bacterial leaf spot. Chilling requirement 300 chill units.

Shifflet.—Orig. in Gridley, Calif., by Harry Shifflet. Introd. in 1927. Parentage unknown; discovered in 1924. Fruit: flesh yellow, firm, flavor good; commercial canning clingstone. No longer commercially planted.

Shinn (*Shinn Cling, Shinn's*).—Orig. by J.C. Shinn, London, Ark.; place of origin unknown, but probably Pope County, Ark. Introd. in 1920. Chance seedling. Fruit:

large; flesh white, clingstone; resembles Uneeda, except larger and superior. Tree: yields very well; hardy.

Shinn Cling.—See Shinn.

Shinn's.—See Shinn.

Shinn's Delicious.—Orig. in Russellville, Arkansas, by C.T. Shinn and L.C. Smith. Introd. in 1951. July Elberta mutation; discovered in 1946. Fruit: small; round, tip pointed; skin yellow overlaid with dull red, moderately attractive, pubescence medium to heavy; flesh yellow, red at pit, firm; pit small, free; identical with parent except that it ripens 3 weeks earlier than July Elberta; closely resembles Starking Delicious. Tree: vigorous; quite susceptible to brown rot; leaf glands reniform; flowers small, nonshowy, self-fruitful; chilling requirement of 750 h at or below 45F.

Shippers Late Red (*Big Red, Late Elberta, Old Boy, Yates Late Elberta*).—Orig. unknown. Introd. in early 1930s. Probably J.H. Hale o.p., being propagated from trees grown by H.W. Miller, Paw Paw, W.Va. Fruit: size medium; ovate, halves somewhat unequal, suture and tip prominent; skin yellow, blushed, pubescence medium; flesh yellow, with more red around pit than J.H. Hale, firm, stringy, somewhat astringent; pit size medium, free; ripens about 3 days before Elberta; ships very well. Tree: vigorous; leaf glands reniform; flower small, nonshowy, self-fruitful; highly productive; chilling requirement of 850 h; susceptible to bacterial leaf spot.

Shoji (*Shoji Early Redhaven*).—Orig. in Livingston, Calif., by Frank Shoji. Introd. in 1955. Redhaven mutation; discovered about 1947. Fruit: similar to Redhaven but ripens 1 week earlier; has the green suture line characteristic of early ripening varieties. Tree: moderately vigorous; spreading.

Shoji Early Redhaven.—See Shoji.

Short.—Orig. in Arlington, Texas, by the late J.A. Evans. Introd. about 1937. Possibly Elberta o.p.; discovered about 1930. Fruit: large; round, with no prominent ridge or tip; flesh yellow, streaked with red, flavor acid, freestone; ripens 2 to 3 days after Elberta, which it resembles. Tree: growth habit low and spreading.

Shotz Late Hale.—Orig. in Sherwood, Ore., by Tualatin Valley Nurseries. Introd. about 1950. Parentage unknown. Fruit: size medium; round; skin yellow, dull, thick; flesh yellow, with little red at the pit, coarse, firm, very sweet, quality fair, freestone; ripens 2 weeks later than J.H. Hale; resembles Muir. Tree: upright; moderately vigorous.

Siberian.—Orig. in Harrow, Ont., Canada, by G.M. Weaver, Canada Dept. Agr. Introd. in 1965. Parentage unknown. Fruit: flesh white, freestone. Tree: chief characteristic is its value as a rootstock, hardy, early fall wood maturation and semi-dwarfing size controlling properties, frost tolerant.

Sierra.—Orig. in Escalon, Calif., by Carl T. Miller. Introd. in 1955; assigned to Burchell Nursery, Modesto, Calif. Parentage unknown; discovered in 1949. Fruit: flesh yellow, a commercial canning clingstone; round and more highly colored than Palora, which it resembles; ripens at Paloro time. Tree: not as subject to mildew as Paloro. No longer commercially planted.

Sierra Lady.—Orig. in Reedley, Calif., by Michael Jost. Plant patent 6045; 27 Oct. 1987; licensed exclusively to Sierra Gold Nursery. O'Henry mutation discovered in 1981. Fruit: medium-large to large; globose to slightly irregular; skin red to red-orange on yellow ground; flesh yellow with moderate amount of red flecking; pit cavity dark red; very firm, melting, fine-textured; freestone; flavor very good, moderate acid, well-balanced. Ripens 35 days after Redhaven in California weeks before O'Henry. Tree: vigor medium; upright spreading; productive. Flowers large, pink, showy, slightly later than most varieties.

Silverado.—Orig. in Orange, Calif., by Fedor Wassiliew, deceased. Introd. in 1981 by Jackson & Perkins Co., Medford, Ore. Plant patent applied for. Parentage unknown. Selected in 1974. Fruit: 2 1/2 to 3 inches in diam.; round; shallow suture; flesh firm, white with red streaks, freestone; ripens 20–30 July at Irvine, Calif. Tree: genetic dwarf, 3 to 5 feet, broad as tall; vigor good; productivity fair; flower semi-double.

Silver Logan.—Orig. in Cedar Ridge, Nevada County, Calif., by Mabel A. Logan. Introd. in 1965. Plant patent 2808; 23 Apr. 1968. Parentage unknown; discovered in 1960. Fruit: ripens late August; quality fine;

white-fleshed freestone. Tree: vigor medium; globose leaf glands; productive; moderate cropping.

Sixty-six.—Orig. in New Brunswick, N.J., by M.A. Blake, New Jersey Agr. Expt. Sta. Introd. in 1945. J.H. Hale x (Carman x Slappey); tested as N.J. 66. Fruit: large; skin red all over; flesh white, freestone, fairly firm; ripens about with Carman. Tree hardy as Elberta; flowers nonshowy. Obsolete.

Skibbe's Elberta.—Orig. in Eau Claire, Mich., by a Mr. Skibbe. Introd. in 1948 by Ackerman Nurseries, Bridgman, Mich. Chance seedling. Fruit: large; shape similar to J.H. Hale; skin bright red with occasional striping; flesh yellow, flavor very good, freestone; ripens shortly after Elberta. Tree: vigorous; quite hardy; self-fruitful; fruit-bud set usually light. Obsolete.

Smith Beauty.—See July Elberta.

Smith Mountain.—See July Elberta.

Snowball.—Orig. in Guilford County, N.C. Introd. in 1926, by Lindley Nurseries, Greensboro, N.C. Parentage unknown. Fruit: skin yellowish-white; flesh clear white, flavor good, juicy, semi-clingstone. Not the Snowball variety, which apparently originated near Austin, Texas, in 1900. Obsolete.

Snowbrite.—Orig. by C.F. Zaiger, G.N. Zaiger, L.M. Gardner, and G.G. Zaiger, Modesto, Calif. Plant patent 8195; 6 Apr. 1993. [(O'Henry x Giant Babcock) x (May Grand nectarine x Sam Houston)] o.p. Fruit: large; globose; very firm; attractive red skin coloring 6 or 7 days before maturity over yellowish-white ground color with moderate pubescence; flesh white, melting, freestone. Ripens 34 days before Giant Babcock. Flowers large, showy, pink; reniform leaf glands. Tree: large; vigorous; upright; productive.

Snow Diamond.—Orig. by Mitchell, Rose, and Vernon Langford, Woodlake, Calif. Plant patent 7336; 25 Sept. 1990; assigned to Plum-Cot. Unknown parentage. Fruit: large; globose; dark red over pale-yellow ground color with very short pubescence; flesh white, melting, freestone. Ripens late July at Fresco, Calif. Flower large, light pink; leaf glands globose. Tree: large; vigorous; very productive.

Snow Giant.—Orig. by C.F. Zaiger, G.N. Zaiger, L.M. Gardner, and G.G. Zaiger, Modesto, Calif. Plant patent 8085; 5 Jan. 1993. (Unknown nectarine seedling x Royal Gold) x {(O'Henry o.p.) x [(May Grand nectarine) o.p. x Sam Houston]}. Fruit: large; nearly globose; firm; attractive red skin color over yellowish-white ground color with moderate pubescence; flesh white, melting, freestone. Ripens 1 month after Giant Babcock in the late maturity season. Flowers large, showy, pink; leaf glands globose. Tree: large; vigorous; upright; productive.

Snow King.—Orig. by C.F. Zaiger, L.M. Gardner, G.N. Zaiger, and G.G. Zaiger, Modesto, Calif. (Unknown nectarine x Redwing) x (O'Henry o.p.). Plant patent 8415; 12 Oct. 1993. Fruit: large; nearly globose; nearly covered with attractive red blush; moderate pubescence; flesh white, very firm, melting; excellent flavor; freestone. Ripens 5 or 6 days after O'Henry. Flowers: large, showy, pink; leaf glands reniform. Tree: large, vigorous; upright; productive.

Snyder Delicious.—Orig. in Campbell, Mo., by Lawrence Snyder. Introd. about 1925. Parentage unknown. Fruit: flesh yellow, freestone; apparently identical to July Elberta. Tree: flowers showy.

Socala.—Orig. in Ontario, Calif., by Mrs. C.C. Barnes. Introd. about 1942. Parentage unknown; discovered about 1935. Fruit: flesh yellow, subacid, freestone; matures earlier than July Elberta, which it resembles in southern California; ripens in early midseason. Tree: requires considerably more winter chilling than Babcock.

Sogood.—Orig. in College Park, Md., by I.C. Haut, Maryland Agr. Expt. Sta. Introd. in 1961. Parentage unknown. Fruit: large; globe-type; skin highly colored; flesh yellow, firm, freestone; resembles Halehaven; ripens 10 to 12 days after Redskin, slightly later than Rio Oso Gem in Maryland. Tree: productive; vigorous.

Solo.—Orig. in Summerland, B.C., Canada, by J.E. Britton, Canada Dept. Agr., Res. Sta. Introd. in 1949. J.H. Hale x Veteran; cross made in 1933; selected in 1941 by A.J. Mann of the above station; tested as Summerland S-11-3. Fruit: large; flesh yellow, juicy, firm, freestone; easily handles with minimum amount of bruising; good for canning; ripens with Valiant and Veteran; resembles J.H. Hale. Tree: medium size; compact; moderately hardy; heavy annual bearer; requires heavy thinning. No longer recommended for commercial plantings in British Columbia, fruit has a tendency toward bald patches, nonbrowning, flower self-sterile.

Somerset.—Orig. in Princess Anne, Md., by Homer S. Kemp, Bountiful Ridge Nurseries. Introd. in 1966. Chance seedling; discovered in 1956. Fruit: large; skin golden yellow ground color, covered with a red blush, tough, thick; flesh texture coarse, but melting, flavor fine, freestone; ripens with or iust after Sunhaven, but before Red-haven. Tree: vigorous; hardy; productive; leaf glands globose.

Somervee.—Orig. in Vineland Station, Ont., Canada, by E.F. Palmer and O.A. Bradt, Ontario Dept. Agr., Hort Res. Inst. Introd. in 1951. Halehaven x Oriole; selected in 1944. Fruit: size medium; skin streaked with red; flesh yellow, melting, quality good, clingstone; matures early, 1 week after Mayflower. Tree: moderately vigorous; very productive; leaf glands globose; flowers small; requires heavy thinning.

Som-Mor.—Orig. in Woolwine, Va., by Murray Harmon. Intro in 1988. Plant patent 6828; 6 June 1989; assigned to Adams Co. Nursery. Chance seedling. Fruit: large; skin three-fourths red blush over yellow grow color; flesh yellow, firm, freestone. Ripens with and similar to Rio Oso Gem. Buds above average cold hardiness and frost tolerance. Blossoms large, pink, showy, usually profuse. Some brown rot tolerance.

Southern Flame.—Orig. in Le Grand, Calif., by F.W. Anderson, Merced, Calif. Introd. in 1975. Plant patent 3620; 17 Sept. 1974; assigned to L.E. Cooke Co., Visalia, Calif. Empress x Bonanza, selected in 1973. Fruit: large; round; skin yellow overspread with red, pubescence scant- flesh yellow flecked with red, red at pit, firm, crisp, melting, freestone; recommended for local market; ripens 14–23 July. Tree: dwarf; spreading; vigor medium; productive; flower large; leaf glands globose.

Southern Glow (*False Fireglow*).—Origin unknown; discovered in 1948. Introd. in 1951, by John T. Bregger, Peach Ridge Farms & Nursery, Clemson. Fruit: medium to large; round ovate, skin yellow, blushed, pubescence medium; flesh yellow, soft, watery, melting, quality good, freestone; ripens 3 weeks before Elberta; resembles Fireglow. Tree: vigor medium; chilling requirement of 700 h; flower small, nonshowy, self-fruitful.

Southern Rose.—Orig. near Le Grand, Calif., by F.W. Anderson, Merced, Calif. Introd. in 1974. Plant patent 3476; 12 Feb. 1974; assigned to L.E. Cooke Co., Visalia, Calif. Golden Glory x Bonanza, selected in 1971, tested as Anderson 50P457. Fruit: size medium, globose to slightly oblate, uniform, symmetrical; skin yellow, partially overspread with red; flesh yellow, melting, mild, red at pit streaked into flesh; freestone; ripens early August. Tree: genetically dwarf; spreading; productive; vigorous; flower large, blooms earlier than Bonanza and most other dwarf varieties; leaf glands globose, less chilling required than either Golden Glory or Bonanza.

Southern Sweet.—Orig. in Le Grand, Calif., by F.W. Anderson. Introd. in 1975. Plant patent 3757; assigned to L.E. Cooke Co., Visalia, Calif. Empress x Bonanza; tested as Anderson 5G25. Fruit: medium: skin yellow with slight overspread of red; flesh yellow; freestone; ripens early to mid-June. Tree: dwarf, about 4 feet at maturity; chilling requirement 500 h.

South Haven (*Sun Glo*).—Orig. in South Haven, Mich., by A.G. Spencer, Greening Nursery Co., Monroe, Mich. Introd. in 1920. St. John mutation; discovered in 1911. Fruit: medium to large; roundish-oblate, halves somewhat unequal, apex mucronate; skin light yellow, blushed or striped, pubescence medium, tender; flesh yellow, traces of red at pit, somewhat soft, tender, melting, good to very good; stone large, free; ripens 14 days before Elberta; Crawford type. Tree: vigorous; leaf glands globose; flower small, nonshowy, petals red, self-fruitful; productivity very high; resistant to bacteriosis.

Southland.—Orig. in Fort Valley, Ga., by J.H. Weinberger, U.S. Dept. of Agriculture Hort. Field Sta. Introd. in 1946. Halehaven selfed; selected in 1939. Fruit: medium to large; round; skin yellow, overlaid red to deep red, striped, pubescence light to medium, attractive; flesh yellow, red at pit, firm, nonbrowning, melting, slower softening than Halehaven, very good to excellent; stone size medium, free; ripens 18 days before Elberta; resembles Halehaven. Tree: vigorous; chilling requirement of about 750 h; quite susceptible to bacteriosis; flower large, showy, self-fruitful; leaf glands globose.

Sowell.—Orig. in Modesto, Calif., by W.J. Sowell. Introd. in 1938. Chance seedling discovered in 1927. Fruit: sizes well; flesh yellow, commercial canning clingstone; does not gum as badly as Phillips Cling, matures same season. No longer commercially planted.

Sparkle.—Orig. in Exeter, Calif., by Grant Merrill. Introd. in 1971. Plant patent applied for. Franciscan o.p. Discovered in 1965, tested as E12-3NW. Fruit: large; round to heart-shaped; skin bright yellow undercolor, covered with one-half to three-fourths bright red blush, nearly fuzzless; flesh yellow, firm, freestone; shipping quality good; ripens late July to early August. Tree: large; vigorous and productive; flower large, showy, pink; leaf glands reniform.

Special Red July.—Spring Red nectarine mutation, found by F.E. Viscarra, Readley, Calif. Plant patent 8597; 15 Feb. 1994. Similar to Elegant Lady, but ripens 2 weeks later, and to O'Henry, but ripens 1 week earlier. Skin: 70% to 95% red blush; blush slightly duller than Elegant Lady; more speckling and russeting than occurs on Elegant Lady or O'Henry. Fruit: large; flesh yellow, with red radiating from pit; melting; freestone. Flowers: medium, non-showy, bright pink; leaf glands reniform. Tree: vigorous; upright-spreading; productive.

Splendor (*Merrill Splendor*).—Orig. in Red Bluff, Calif., by Grant Merrill. Introd. in 1952. Plant patent 1257; 16 Feb. 1954. (J.H. Hale x Levi) o.p.3; selected in 1951. Fruit: large, uniform; unsymmetrical, globose; skin bright yellow ground color with dark red splotches; flesh yellow, pit dark red, firm, freestone; ripens immediately following Rio Oso Gem; resembles Candoka. Tree: flower small, dark pink, needs cross-pollination; leaf glands reniform.

Spotlight.—Orig. in Summerland, B.C., by J.E. Britton, Canada Dept. Agr., Res. Sta. Introd. in 1946. Veteran x Rochester; cross made in 1934; original tree planted in 1937; selected in 1942 by A.J. Mann, of the above station. Fruit: medium to large; skin attractive high color; flesh yellow, rather soft, quality good, semi-clingstone; matures 1 week earlier than Rochester. Tree: spreading; vigorous; hardy; quality good. Home garden variety.

Spoto Gem.—Orig. in Yuba City, Calif., by Henry R. Spoto, Sr., Spoto Nursery. Introd. in 1967. Plant patent 2901; 8 July 1969. Parentage unknown; seedling selected in 1957. Fruit: large, 2 3/4 to 3 inches in diam.; variable, globose, slightly compressed axially; skin yellow, mottled and streaked with red, thin; pubescence moderate; flesh very firm, yellow, with red next to pit, freestone; ripens late, the first week in September shortly before Kirkman Gem; fruit does not drop; recommended for shipping, freezing or canning. Tree: large, spreading and round topped; vigorous, hardy; productive and regular bearer; flower large, pink; leaf glands globose.

Spraque.—Orig. by J.W. Spraque, Ontario, Calif. Plant patent 7337; 25 Sept. 1990. Last Chance Number One mutation. Dwarf tree. Distinctive in its short bushy tree shape with small-diameter branches; tree height rarely exceeds 5 feet.

Springbrite.—Orig. in Fort Valley, Ga., by the U.S. Dept. of Agriculture. Introd. in 1972. (Sunhigh x Southland) x Springtime. Cross made in 1958 at U.S. Hort. Field Lab., Fort Valley, Ga.; first fruited in 1961, tested as FV 9-266; selected and tested by Victor Prince. Fruit: small to medium; round with a slight tip; skin ground color yellow, 70% covered with bright red blush, more attractive than Springcrest; pubescence fine and short; flesh yellow, firm but melting, texture medium, flavor good; semi-freestone when fully ripe; ripens with Springcrest or 1 week before June Gold. Tree: vigor moderate; productivity moderate; moderately susceptible to bacterial leaf Spot; chilling requirement 650 h flower small, self-fertile, leaf glands reniform.

Springcrest.—Orig. in Fort Valley, Ga., by the U.S. Dept. of Agriculture. Introd. in 1969. [(Fireglow x Hiley) x Fireglow] x Springtime. Cross made in 1958, first fruited in 1961 and selected that year by Victor E. Prince of the above Sta., tested as FV9-170; has also been tested and fruited at U.S. Hort. Field Sta., Fresno, Calif. Fruit: small to medium; round with slight tip; skin has a bright red blush over a yellow ground color in California, very attractive; in the Southeast, skin tends to have a slightly dark blush unless the trees are growing vigorously; pubescence is fine and short; flesh yellow, firm, but melting, medium in texture, flavor good; semi-freestone when fully ripe; ripens about 1 week later than Springold and 1 week earlier than Cardinal; percentage of fruit with split pits have not been excessively high. Tree: vigor moderate; productivity moderate; susceptible to bacterial spot disease; chilling requirement about 700 h; flower large, self-fertile; leaf glands globose.

Springold.—Orig. in Fort Valley, Ga., by U.S. Hort. Lab. Introd. in 1966. [(Fireglow x Hiley) x Fireglow] x Springtime; cross made in 1959; first fruited and selected in 1961; tested as FV 9-149. Fruit: small; round with a slight tip; 80% of surface covered with bright, attractive, red blush over yellow, ground color; pubescence light; flesh yellow, firm, but melting, texture medium, flavor good; matures early, 2 weeks before Cardinal, 3 days after Springtime. Tree: vigorous; productive; self-fertile; susceptible to bacterial spot disease; leaf glands reniform; blossoms large petaled; chilling requirement is 850 h below 45F.

Spring Lady.—Orig. by G. Merrill, Red Bluff, Calif. Plant patent 4661; 3 Mar. 1981. Unknown parentage. Fruit: medium; very firm; globose; blushed mottled red over yellow skin; flesh yellow, freestone, melting, beige stone, ripening 5 or 6 days after Springcrest. Flowers pink, medium; leaf glands reniform. Tree: large and vigorous.

Springtime.—Orig. in Ontario, Calif., by Herbert C. Swim, Armstrong Nurseries. Introd. in 1953. Plant patent 1268; 20 Apr. 1954; (Luken's Honey o.p. x July Elberta) x Robin; selected in 1949. Fruit: ripens about 2 weeks earlier (late May to early June) than Robin, which it resembles; flesh white, semi-freestone. Tree: particularly suited for regions having a mild climate as it has a short winter chilling requirement; leaf glands globose; flowers medium to small, nonshowy, color pink.

Stanford.—Orig. in Palo Alto, Calif., by W.F. Wight, U.S. Dept. of Agriculture. Introd. in 1935. Hauss x Phillips Cling; cross made in 1925; first fruited in 1929. Fruit: large, averaging 8.5 cm in diam., halves usually equal; skin golden-yellow splashed with red, pubescence short; flesh deep yellow, firm, fine-grained, subacid, flavor good; commercial canning clingstone; hangs well on tree; ripens in late midseason, with or iust after Halford. Tree: heavy producer; upright; leaf glands globose; flower petals small; unsuited for growing in a warm-winter climate.

StanisLaus.—Orig. by C.F. Zaiger, L.M. Gardner, G.N. Zaiger, and G.G. Zaiger, Modesto, Calif. Carolyn x Loadel. Plant patent 8376, 14 Sept. 1993. Fruit: large; nearly globose; moderate pubescence; flesh yellow, very firm, nonmelting; clingstone. Ripens 7 or 8 days after Loadel. Flowers: small, nonshowy, pink; leaf glands globose. Tree: vigorous; semi-upright; productive. Fresh market and processing.

Starcrest.—Orig. by P. Chapus and H. Veauvy in France. Plant patent 6387; 8 Nov. 1988. Assigned G.I.E. Star Fruits, Vedene, France. Springcrest mutation; matures 15 days before Springcrest.

Stardust.—Orig. in Merced, Calif., by Stribling's Nurseries. Introd. in 1959. S-53-23 o.p.; selected in 1956; tested as S-55-32. Fruit: uniformly large; round; skin with a sheen and a distinctive orange ground color, highly blushed; flesh yellow, red at pit, very firm, flavor very good; freestone, pit small; ships very well; ripens before Summertime and about with Redhaven and Regina; resembles Redhaven and Redtop.

Stark Compact Elberta (*G.M.*).—Orig. in Moses Coulee, Douglas County, Wash., by Pete Van Well, Wenatchee, Wash. Plant patent 4370; 23 Jan. 1979; licensed to Stark Bros. Nurseries, Louisiana, Mo. Introd. in 1980. Whole tree mutation of Fay Elberta. Fruit: identical to Fay Elberta. Tree: compact, similar to Com-Pact Redhaven: many lateral shoots on first year's growth.

Stark EarliGlo.—Orig. in Simcoe, Ont., Canada, by John M. Roxburgh. Introd. in 1957 by Stark Bros. Nurseries & Orchards Co., Louisiana, Mo. Plant patent 1632; 13 Aug. 1957; assigned to Stark Bros. Nurseries & Orchards Co. Redhaven mutation; discovered in 1950. Fruit: flesh yellow, mottled throughout with carmine, semifreestone; matures 7 to 10 days earlier than parent, which it resembles. Tree: size medium; vigor medium; open, spreading; productive; leaf glands mostly reniform, some globose.

Stark Early Elberta.—See Early Elberta.

Stark Encore.—Orig. in Cream Ridge, N.J., by L. Fredric Hough and C.H. Bailey, in cooperation with Rutgers Univ., N.J . Plant patent 4572; 15 July 1980; assigned to Stark Bros Nurseries, Louisiana, Mo. Introd. in 1980. NJ 585414 x Autumnglo. Tested as NJ 260. Fruit: large 21/2 to 3 inches in diam.; skin greenish-yellow under color, 75% covered with crimson red blush; flesh firm to very firm, yellow, quality good; freestone; ripens 5 1/2 weeks after Redhaven. Tree: vigorous; productive, similar to Redhaven; tolerant to bacterial spot.

Stark Finale (*Tyson 820*).—Orig. in Monroe, N.C., by J.D. Tyson. Introd. in 1983. Plant patent applied for; to be assigned to Stark Bros. Nurseries & Orchards Co., Louisiana, Mo. Chance seedling; discovered in 1978. Fruit: 3 inches in diam.; skin covered with 50% red blush; flesh yellow, high quality, meaty and nonbrowning; late maturing, 10 days after Rio Oso Gem. Tree: productive and regular bearer.

Stark Gulf Queen (*Zaiqueen*).—Orig. in Modesto, Calif., by Floyd Zaiger. Introd. in 1983. Plant patent applied for; to be assigned to Stark Bros. Nurseries & Orchards Co., Louisiana, Mo. Ruby Gold nectarine o.p.. Selected in 1976; tested as 37G804. Fruit: 2 3/4 inches in diam.; skin nearly full red; flesh yellow, firm, good shipping and handling characteristics; flavor very good. Tree: chilling requirement about 300 h; matures early.

Starking Delicious.—Orig. in Dover, Ark., by Henry Frank Smith. Introd. in 1949. Plant patent 803; 27 July 1948; assigned to Stark Bros. Nurseries & Orchards Co., Louisiana, Mo. July Elberta mutation; discovered in 1944. Fruit: flesh yellow, freestone when fully ripe; identical to parent except for ripening period, ripens 18 days before July Elberta and 6 days before Redhaven. Tree: same as parent; leaf glands reniform; flower small, pink.

Stark LateGlo.—Orig. in New Park, Pa., by Robert L. Brown. Introd. in 1961. Plant patent 2124; 30 Jan. 1962; assigned to Stark Bros. Nurseries & Orchards Co., Louisiana, Mo. Elberta mutation; discovered in Fawn Township, York County, Pa., in 1938. Fruit: large, averaging 2 3/4 to 3 inches in diam.; skin highly blushed, medium thin; flesh yellow with some red around pit, firm, tender, flavor mild, aromas distinct, juice quantity moderate, quality high, freestone; resembles Elberta. Tree: medium large, spreading; vigorous; hardy; productive; leaf glands reniform.

Stark Late Gold.—Orig. in Bangor, Ala., by Charles Thomason. Introd. in 1955 by Stark Bros. Nurseries & Orchards Co., Louisiana, Mo. Plant patent 1321; 16 Nov. 1954; assigned to Stark Bros. Nurseries & Orchards Co. Hal-Berta Giant x Elberta o.p.; selected in 1940. Fruit: large; skin attractive, yellow overlaid with bright carmine-red; flesh yellow, quality very good, freestone; ripens 7 days after J.H. Hale, which it resembles. Tree: resistant to brown rot; very productive; very hardy.

Stark® Sensation (*Andergen, Sensation*).—Orig. in Le Grand, Calif., by Norman G. Bradford. Introd. in 1983. Plant patent 5124, 18 Oct. 1983; assigned to Stark Bro's Nurseries. Unnamed nectarine seedling 16K275 x unnamed peach seedling 54P105. Fruit: medium to large; globose; skin 85% red on yellow ground color; flesh yellow; medium firm; flavor mild, very good. Ripens late June in central California, 7 weeks before Elberta. Tree: genetic dwarf; medium productivity; spreading and drooping limbs; leaves large; hardy in zones 5 to 8.

Stark SunBright.—Origin unknown. Introd. in 1963 by Stark Bros. Nurseries & Orchards Co., Louisiana, Mo. Parentage unknown. Fruit: large; round; skin nearly solid red over an orange to red ground, attractive; flesh yellow, red around pit, somewhat stringy, juicy, flavor slightly strong, quality very good; freestone, pit size medium; ships well; ripens in midsummer, 12 days before Elberta and with July Elberta; resembles Redglobe and J.H. Hale. Tree: vigorous self-fruitful; highly resistant to bacterial leaf spot; leaf glands round.

Stark Sure-Crop.—Orig. in Ash Grove, Mo., by John W. Nicholson. Introd. in 1946. Plant patent 670; 15 Jan. 1946; assigned to Stark Bros. Nurseries & Orchards Co., Louisiana, Mo. Greensboro o.p. Fruit: size medium to large, average 2 1/2 inches in diam.; round, halves unequal; skin fairly thick, tough, blushed over yellow; flesh white, firm, moderately juicy, clingstone; ripens early, about 5 days after Early Wheeler (Red Bird) and 7 to 10 days before Golden Jubilee; primarily for home and roadside orchard use. Tree: hardy; produces well under adverse conditions.

Starlet.—Orig. in Wasco, Calif., by David L. Armstrong, Armstrong Nurseries, Ontario, Calif. Introd. in 1969. Plant patent 2944; 25 Nov. 1969; assigned to Stark Bros. Nurseries & Orchards Co., Louisiana, Mo. Parentage unknown; selected in 1965. Fruit: resembles Golden Treasure; quality and color good; flesh yellow; freestone when fully ripe; ripens 2 weeks after Bonanza, 4 to 6 weeks before Golden Treasure. Tree: dwarf; productivity heavy under California conditions; flowers light pink, showy; leaf glands mixed, weakly reniform.

Starlite.—Orig. in Fort Valley, Ga., by V.E. Prince. of U.S. Dept. of Agriculture. [(Fireglow x Hiley) x Fireglow] x Springtime. Introd. in 1980. First fruited in 1961; tested as FV9-239. Fruit: small to medium, round with slight tip; skin under color white with green tint, 70% covered with bright red blush at maturity, pubescence light; flesh white, firm but melting, texture medium, flavor good, semi-clingstone; ripens with Springold or 3 days after Springtime. Tree: vigorous, productive; susceptible to bacterial spot, but because of early maturity fruit usually escapes damage; chilling requirement 650 h; blossoms showy, large. self-fertile; leaf glands globose.

Starn.—Orig. in Hughson, Calif., by J.E. Starn. Introd. in 1950. Chance seedling planted in 1936; discovered about 1938. Fruit: flesh yellow; commercial canning clingstone; leaf glands reniform; matures in Phillips Cling season.

St. Clair.—Orig. in Belleville, Ill., by Louis Dintelman, Louis Dintelman Nursery. Introd. about 1934. Parentage unknown. Fruit: large; flesh yellow, flavor fair, clingstone.

Stephenson.—Orig. in Dallas, Texas, by J.A. Evans, Evans Specialty Nursery, Arlington, Texas. Introd. in the 1920s. Parentage unknown. Fruit: extremely large; skin deep yellow, with only a slight blush; flesh yellow, juicy, flavor pleasing, clingstone. Tree: large; produces heavy crops.

Stuart.—Orig. in French Camp, Calif., by J.F. Stuart, Ripon Nursery and Improvement Co. Introd. in 1927. Lovell o.p.; selected in 1925. Fruit: flesh yellow, firm, commercial canning clingstone; ripens in Phillips Cling season; resembles Gomes.

Sugar Giant.—Orig. by C.F. Zaiger, G.N. Zaiger, L.M. Gardner, and G.G. Zaiger, Modesto, Calif. (Unknown nectarine x Royal Gold) x [O'Henry o.p. x (MayGrand o.p. x Sam Huston)]. Plant patent 8442; 2 Nov. 1993. Fruit: large; globose; light red blush over white ground; medium pubescence; flesh white, firm, melting; freestone. Ripens 1 week before Giant Babcock. Flowers: large, showy, pink; leaf glands reniform. Tree: large; vigorous; upright; productive.

Sugar Lady.—Orig. in Modesto, Calif, by Floyd, Gary, and Grant Zaiger and Leith Gardner. Introd. in 1989. Plant patent pending; owned by Zaiger Genetics. Fruit: large; skin dark red blush on cream background; flesh white, very firm, sweet, low acid, freestone. Ripens 25 days after Redhaven in California.

Sullivan 1.—Orig. in Tudor, Calif., by C.E. Sullivan. Introd. in 1936. Plant patent 186; 23 June 1936. Parentage unknown; selected in 1928. Fruit: large, uniform, averaging 3 3/8 inches in diam.; globose, oblate, suture distinct, shallow; skin yellow, blushed, tough, pubescence moderate; flesh yellow, commercial canning clingstone; ripens in late midsummer, 5 days before Phillips; resembles Johnson. Tree: large, vigorous;

very productive; leaf glands reniform; flower size medium; hardy.

Sullivan 2.—Orig. in Yuba City, Calif., by C.E. Sullivan. Introd. in 1939 by Sierra Gold Nurseries, Yuba City. Lovell o.p.; selected in 1936. Fruit: sizes well; skin with a high blush; flesh yellow; commercial canning clingstone; pit small; stem long; hangs on tree better than other commercial canning clingstone; ripens just before Halford and Sullivan 3 and after Sullivan 1. Tree: upright; vigorous; easily trained by pruning.

Sullivan 3.—Orig. in Yuba City, Calif., by C.E. Sullivan. Introd. in 1940 by Sierra Gold Nurseries, Yuba City. Lovell o.p.; selected in 1937. Fruit: skin with a high blush; flesh yellow, no red around pit; commercial canning clingstone; ripens with Halford, before Sullivan 4 and after Sullivan 2. Tree: moderate grower; productive; tends to put more than usual growth into fruiting rather than tree growth.

Sullivan 4.—Orig. in Tudor, Calif., by C.E. Sullivan. Introd. in 1940. Parentage unknown; selected in 1929. Fruit: ripens in Phillips Cling season; leaf glands globose; commercial canning clingstone.

Sullivan Early Elberta (*Sullivans Elberta*).—Orig. in Zebulon, Ga., by P.M. Sullivan. Introd. in 1938. Elberta mutation; discovered in 1933. Fruit: size medium; ovate, halves somewhat unequal, tip prominent, suture medium prominent; skin greenish-yellow partially overlaid with red; flesh yellow, red at pit, firm, melting, quality good; ships well; ripens 7 days before Elberta, which it resembles. Tree: vigorous; leaf glands reniform; flower small, nonshowy, petals red, self-fruitful; chilling requirement of 850 h at or below 45F; more susceptible to bacteriosis than Elberta.

Sullivans Elberta.—See Sullivan Early Elberta.

Summercrest.—Orig. in New Brunswick, N.J., by M.A. Blake, New Jersey Agr. Expt. Sta. Introd. in 1938. J.H. Hale x Cumberland; cross made in 1926; tested as N.J. 94. Fruit: large; ovate; skin predominantly red over yellow, dull, pubescence medium; flesh yellow, quality good, melting, firm, freestone; ripens with or just before Elberta; popular for local market in New Jersey; canning quality only fair. Tree: very vigorous; as hardy as J.H. Hale; susceptible to brown rot; flower size medium, nonshowy, self-fruitful; leaf glands reniform; chilling requirement 950 h.

Summerglo.—Orig. in New Brunswick, N.J., by L.F. Hough and Catherine H. Bailey. New Jersey Agr. Expt. Sta. Introd. in 1978. Collins x Red Slovenia. Cross made in 1960; selected in 1966: tested as NJ 233. Fruit: large; flesh yellow; melting; freestone; ripens 1 week after Redhaven. Tree: flower buds cold tender; tolerant to bacterial spot (*Xanthomonas campestris* sp. *pruni*).

Summergold.—Orig. in Fort Valley, Ga., by U.S. Dept. of Agriculture. Introd. in 1970. V 370224 [(J.H. Hale x Valiant) o.p.] x Redglobe. Cross made in 1955, first fruited in 1958, tested as FV 6256. Fruit: medium to large; round with slight tip; skin yellow undercolor with 60% covered with a bright red blush, attractive; flesh yellow, firm but melting, texture fine, flavor good, freestone; ripens about same time as Loring and 4 days after Redglobe; in 5-year tests by the Food Science Dept., Georgia Agr. Expt. Sta., Experiment, Ga., variety was rated excellent for processing; also recommended for fresh fruit. Tree: vigor moderate; productive; susceptibility to bacterial spot about the same as Redglobe; flower small, self-fertile; leaf glands reniform; chilling requirement 750 h.

Summerlong.—Orig. in Dayton, Ohio, by Clara V. Benjamin. Introd. in 1955. Plant patent 1503; 31 July 1956; assigned to Greening Nursery Co., Monroe, Mich. Parentage unknown; discovered in 1951. Fruit: size medium; skin yellow blushed with red, pubescence moderate; flesh yellow with reddish tinge, quality good, freestone; ripens at intervals of up to 5 weeks from late July to late September; resembles Crawford. Tree: size medium; vigor medium; spreading; productive; flower small, pink.

Summer Pearl.—Orig. in New Brunswick, N.J. by L. Fredric Hough and Catherine H. Bailey, New Jersey Agr. Expt. Sta., Cook College, New Brunswick, N.J. Introd. in 1979 by Stark Bros. Nurseries, Louisiana, Mo. NJ 554039 x NJ 585399. Cross made in 1965; selected in 1972; tested as NJ 252. Fruit: 2 3/8 to 2 5/8 inches in diam., globose; skin one-half to one-fourth red on cream white; flesh white with red at pit. quality good; freestone; ripens 3 1/2 weeks after Redhaven. Tree: standard, vigorous, productive; self-fertile; bud hardiness better than Loring, slightly less than Red-haven; moderately tolerant to bacterial leaf spot.

Summerprince.—Orig. at U.S. Dept. of Agriculture Southeastern Fruit and Tree Nut Res. Lab., Byron, Ga. Summercrest cross of complex parentage; selected in 1983, tested as BY81P512, and released by W.R. Okie in 1992. Fruit: medium large; very round; 80% bright red on attractive yellow with very short pubescence; flesh yellow with red at pit cavity, firm, melting, semi-clingstone, good texture, and high flavor. Ripens 14 days before Redhaven. Flowers large, showy, pink; blooms with Redglobe; leaf glands reniform. Tree: vigorous; productive; moderate resistance to bacterial spot. Chilling requirement 850 h.

Summerqueen.—Orig. in New Brunswick, N.J., by M.A. Blake, New Jersey Agr. Expt. Sta. Introd. in 1957. J.H. Hale x Cumberland; selected in 1930; tested as N.J. 157. Fruit: flesh yellow, firm, freestone; ripens with Sunhigh. Tree: flower showy; pollen-sterile; less susceptible to bacterial leaf spot than Sunhigh.

Summerrose.—Orig. in New Brunswick, N.J., by M.A. Blake, New Jersey Agr. Expt. Sta. Introd. in 1947. J.H. Hale x Delicious; tested as N.J. 101. Fruit: large; oval, tip and suture somewhat prominent; skin greenish-white overlaid with dull rose, blushed and striped, pubescence light; flesh white, red at pit, firm, melting, quality fair to good; stone size medium, free; ripens 10 to 16 days before Elberta depending upon locality; resembles Belle in shape and appearance. Tree: hardier than Elberta; flowers nonshowy, small, petals red, self-fruitful; medium to high productivity; chilling requirement of 750 h.

Summerset.—Orig. in Fresno, Calif., by John H. Weinberger, U.S. Hort. Field Sta. Introd. in 1965. Kirkman Gem x (J.H. Hale x Rio Oso Gem); cross made in 1955; selected in 1958; tested as F110-28. Fruit: large; round; 20% to 30% bright blush over an attractive yellow, pubescence light; suture raised slightly; flesh yellow, firm, but melting, good texture and flavor, quality high, freestone; ripens 5 to 6 weeks after Rio Oso Gem, about 1 week before Merrill Fiesta, in early September at Fresno and about 1 Oct. at Beltsville, Md. Tree: vigorous, more so than Rio Oso Gem; productive; leaf glands reniform; petals large.

Summer Sun.—Orig. in Del Rey, Calif., by Tom Chamberlain, H.P. Metzler & Sons. Introd. in 1986. June Lady o.p. Plant patent 6007; 8 Sept. 1987; owned by Metropolitan Insurance Co.; licensed exclusively to Agri Sun Nursery. Fruit: most resembles June Lady; medium; skin bright red on yellow ground; flesh yellow, crisp, excellent flavor, freestone. Ripens 10 days after Redhaven in Selma, Calif., 3 days after June Lady. Tree: vigorous; productive.

Summer Sweet.—Orig. by C.F. Zaiger, G.N. Zaiger, L.M. Gardner, and G.G. Zaiger, Modesto, Calif. Plant patent 8070; 22 Dec. 1992. Complex parentage. Fruit: large; globose; very firm; attractive red skin color over yellowish-white ground color with short pubescence; flesh white, melting, freestone. Ripens early July at Modesto. Flowers large, showy, pink; leaf glands reniform. Tree: large; upright; vigorous; productive.

Summertime.—Orig. in Merced, Calif., by Stribling's Nurseries. Introd. in 1959. S-53-23 o.p.; selected in 1956; tested as S-57-2. Fruit: uniformly large; round; skin bright golden, highly blushed; flesh yellow, very fine flavor, no oxidation when frozen; freestone; good used fresh or frozen; ripens after Stardust and about with Fortyniner; resembles Suncrest and Fortyniner.

Sumner Gleason.—See Early Elberta.

Sunapee.—Orig. in Durham, N.H., by A.F. Yeager and E.M. Meader, New Hampshire Agr. Expt. Sta. Introd. in 1955. (Oriole x P.I. 104315) F_2; selected in 1949. Fruit: oval; flesh yellow, freestone; ripens last of August for northernmost limits of peach culture. Tree: hardy.

Sunbeam.—Orig. in New Brunswick, N.J., by M.A. Blake, New Jersey Agr. Expt. Sta. Introd. in 1925. Slappey x Admiral Dewey; cross made in 1916. Fruit: small to medium; round ovate; skin bright red over orange ground; flesh yellow, red at pit, melting, firm, fine texture, good quality; semi-clingstone, partially free when ripe; ripens 3 weeks before Elberta; good for processing, free from catechol tannin. Tree: vigor medium; rounded; productivity high; flower small, nonshowy; leaf glands reniform; chilling requirement 950 h.

Sunbrite.—Orig. in Fort Valley, Ga., by the U.S. Dept. of Agriculture. [(Southland x Sunhigh) x Coronet] x [(Sunhigh x Southland) x Springtime]. Cross made in 1962; first fruited in 1965 and selected that year by V.E. Prince; tested as FV3-381. Fruit: small to medium; round with a slight point; skin has a bright red blush over 70% of the surface; pubescence is fine and short; flesh is firm but melting, medium in texture, flavor is good, semi-freestone when fully ripe; ripens 1 week before Red Cap and 3 days after Springcrest; fruit with split pits have not been excessively high. Tree: vigorous; productive moderately susceptible to bacterial spot; chilling requirement about 750 h; flower has small petals, self-fertile; leaf glands reniform.

Suncling.—Orig. in South Haven, Mich., by Stanley Johnston and J.E. Moulton, Michigan Agr. Expt. Sta. Introd. in 1961. P.I. 35201 x New Jersey 196; cross made in 1951 by New Jersey Agr. Expt. Sta.; selected at South Haven Expt. Sta. in 1957; tested as SH 382. Fruit: medium large; nearly round; skin golden with some blush where exposed to sun, pubescence short, nearly unnoticeable; flesh orange yellow, very firm, texture fine-grained; pit small, nonmelting clingstone; ripens about with Ambergem or 10 to 12 days before Elberta. Tree: vigorous; productive; fruit buds unusually hardy; flowers very frost resistant.

Suncloud.—Orig. in Dinuba, Calif., by Robert A. Callison. Introd. in 1963. Plant patent 2166; 14 Aug. 1962. J.H. Hale x Blazing Gold nectarine mutation; selected in 1958; tested as Callison 6. Fruit: large, up to 3 1/4 inches in diam.; symmetrical, rounded; skin thickness medium, yellow overlaid with bright red blush, moderately pubescent; flesh yellow, pale pink next to pit, medium firm, subacid, moderately juicy, dessert quality very good, freestone; ventral suture distinct; ripens in same general season as Merrill Hale, Redglobe, Redtop; resembles Redtop. Tree: large; vigorous; consistently good bearer; bloom time average; chilling requirement of about 900 h; self-fruitful.

Suncrest.—Orig. in Fresno, Calif., by the Crops Res. Division, U.S. Dept. of Agriculture. Introd. in 1959. Alamar x Gold Dust; cross made in 1952; first fruited in 1956; tested as USDA F-69. Fruit: large; round; at maturity two-thirds of skin covered with a bright, attractive red over a yellow ground color, pubescence light; flesh yellow, firm but melting, texture and flavor good, freestone; ripens about 10 days before Elberta and about 10 days after Redglobe and July Elberta; suitable for distant shipment. Tree: vigorous, productive; flower large, self-fruitful, opening about with Elberta; leaf glands reniform; susceptible to bacterial spot disease.

Sundance.—See Sun Lady.

Sundar.—Orig. in Riverside, Calif., by J.W. Lesley, California Agr. Expt. Sta. Introd. in 1965. Parentage involved Prenda, Late Champion, Elberta, and Peento; selected in 1947; tested as U.C. 254-2. Fruit: large; skin white, red blushed, slightly pubescent, semi-adherent, tough; flesh white, heavily colored red at pit, juicy, freestone; flavor sweet, aromatic, slightly bitter; ripens late, in mid-Sept. at place of origin. Tree: vigorous, crops regularly; flower large, showy, fertile; leaf glands globose, chilling requirement 700 h.

Sunday (*Sunday Elberta*).—Orig. in Baroda, Mich., by George P. Sunday. Introd. in 1940. Plant patent 418; 27 Aug. 1940; assigned to Greening Nursery Co., Monroe, Mich. Chance seedling; selected in 1927. Fruit: size medium; shape similar to Elberta; skin bright yellow, blushed smooth, pubescence slight; flesh orange-yellow with some red around pit, firm, quality similar to J.H. Hale; freestone; suitable for commercial or home freezing; ships well; ripens just after Elberta; requires thinning. Tree: very hardy in bud; very vigorous; self-fruitful.

Sunday Elberta.—See Sunday.

Sun Glo.—See South Haven.

Sunglow.—Orig. in Riverside, Calif., by J.W. Lesley, California Citrus Expt. Sta. Introd.

in 1942. J.H. Hale x Bolivian Cling o.p.; selected in 1936. Fruit: medium large; slightly oblong; skin golden yellow, blushed, appearance attractive, ventral suture often red, pubescence short, not dense; flesh light to orange yellow; pit cavity medium pink; texture smooth, juicy, firm; flavor good, sweet; subacid; aroma slight, richly flavored; freestone; quality good; ripens in midseason, 1 Aug. at Riverside; subject to preharvest drop. Tree: vigorous, upright to somewhat spreading; foliage dark green, leaf glands reniform; flower small, pink, self-fruitful; produces well; blooms slightly after Babcock; obsolete.

Sungold.—Orig. in Des Moines, Iowa. Introd. in 1936; assigned to Interstate Nurseries, Hamburg, Iowa. Professor T.J. Maney, Iowa State College, selected it about 1924. Parentage unknown; selected about 1928. Fruit: large; flesh yellow, flavor fine, freestone; ripens 1 week before Elberta; resembles J.H. Hale. Tree: productive.

Sunhaven.—Orig. in South Haven, Mich., by Stanley Johnston, Michigan Agr. Expt. Sta. Introd. in 1955. Redhaven x (J.H. Hale x Halehaven); cross made in 1944; selected in 1949; tested as SH 232. Fruit: medium large; nearly round, uniform; skin very attractive, largely bright red over golden, pubescence very short and soft as to be almost unnoticeable; flesh yellow, flecked with red, texture fine, firm, flavor very good to excellent, very resistant to browning; freestone when fully ripe, partially cling when flesh is firm; ripens very early, 10 days before Redhaven, 40 days before Elberta; primarily for fresh market; resembles Redhaven. Tree: large; vigorous; consistently productive; leaf glands globose; flower small, nonshowy, self-fruitful; chilling requirement of 900 h.

Sunhigh.—Orig. in New Brunswick, N.J., by M.A. Blake, New Jersey Agr. Expt. Sta. Introd. in 1938. J.H. Hale x N.J. 40 C.S.; cross made in 1923; tested as N.J. 82. Fruit: medium to large; oval to round, ribbed, apex pointed; skin bright red over yellow, pubescence light, attractive; flesh yellow, red at pit, firm, melting, fine-textured, quality good; stone small to medium, free; ripens 16 days before Elberta. Tree: very susceptible to bacteriosis; vigorous, spreading; leaf glands reniform; flower nonshowy, size medium, petals red, self-fruitful; chilling requirement of 750 h; fairly hardy.

Sun Lady (*Sundance*).—Orig. in Red Bluff, Calif., by Grant Merrill. Introd. in 1966. Plant patent 2676; 11 Oct. 1966; J.H. Hale x (Maxine x July Elberta) o.p.; selected in 1962; first fruited 1964; tested as Y-22E. Fruit: medium, averaging 2 5/8 inches in diam.; generally globose, frequently flattened toward base on one side, uniform; skin dark yellow with varying amounts of dark reddish-orange blush in stripes covering about 50% to 80% of surface, thick, medium tough; flesh brilliant yellow to vivid yellow, some dark reddish-orange near pit, moderately juicy, aromatic; stone generally free, sometimes adhering on less mature fruit; texture firm; flavor mild, delicate; eating quality good; keeping and shipping quality good; recommended for fresh market; ripens 2 to 3 weeks earlier than Carnival; resembles Carnival being distinguished therefrom by its smaller fruit, highly colored skin, firmer flesh; fruit can be harvested in one picking. Tree: large, upright, vigorous; regular and heavy bearer; leaf glands reniform; flowers salmon pink, small; blooms with July Elberta.

Sunland.—Orig. in Byron, Ga., by V.E. Prince of U.S. Dept. of Agriculture. Introd. in 1980. [(Southland x Sunhigh) o.p.] x (Dixiland x Keystone). Cross made in 1963; first fruited in 1966; tested as BY 4-7140. Fruit: large, round with a slight tip, 75% of surface bright red, under color yellow; flesh yellow with some red, firm, but melting, medium texture, flavor good; freestone; ripens 3 to 5 days after Harvester. Tree: vigorous, productive, moderately susceptible to bacterial spot, chilling requirement 750 h; blossoms nonshowy, small petaled, self-fertile; leaf glands globose.

Sunnyside.—Orig. in Fresno, Calif., by Agr. Res. Serv., U.S. Dept. of Agriculture. Introd. in 1974. [(Sunhigh x Southland) x (Southland x Hawaiian)] x [(Newday southland) x (Southland x Hawaiian)] . Cross made in 1958, selected in 1961 by John H. Weinberger, tested as 83-16. Fruit: size medium; round; skin yellow with three-fourths covered by red, attractive, light pubescence; flesh yellow, firm, texture smooth, flavor good; ripens in late June at Fresno. Tree: moderate vigor; productive; flower large, self-fertile, full bloom about 28 Feb. at Fresno; leaf glands reniform; chilling requirement 550 h.

Sunprince.—Orig. in Byron, Ga., by W.R. Okie and V.E. Prince, S.E. Fruit & Tree Nut Res. Lab and U.S. Dept. of Agriculture. [Redglobe x (Dixieland x FV240-1)] x [(Hiley x Fireglow) x B2832]. Introd. in 1981. Cross made in 1963; first fruited 1966. Fruit: large; round; attractive: better color and shape than those of Redskin and Blake; skin 50% to 70% bright red with attractive yellow ground color, pubescence light; flesh yellow with some red at pit, firm, but melting, texture and flavor good; freestone; ripens 2 days before Elberta. Tree: vigorous, productive; blossoms large, pink, showy, self-fertile, leaf glands globose, chilling requirement 800 h; good resistance to bacterial spot, more so than Blake, but more susceptible than Redskin.

Sunqueen.—Orig. in New Brunswick, N.J., by Catherine H. Bailey and L.F. Hough, New Jersey Agr. Expt. Sta. Introd. in 1966. (J.H. Hale x Meredith) x [(J.H. Hale x *P. kansuensis*) o.p.]; selected in 1954; tested as N.J. 210. Fruit: large; skin bright yellow ground color with one-half blushed with carmine, very attractive; flesh yellow, not as firm as Sunhigh; freestone; ripens in Sunhigh season. Tree: hardier than Sunhigh; very productive; flowers nonshowy; leaf glands reniform.

Sunray.—Orig. in Pretoria, South Africa, by E.P. Evans, Hort. Res. Inst. Introd. in 1966. Moreira Jubilee selfed; 1963. Fruit: size medium; skin attractive, light orange-yellow with slight red around pit cavity, semiclingstone; resembles Golden Jubilee; ripens early mid-November. Tree: medium to large; vigorous; leaf glands reniform; flower nonshowy, flowers profusely but fairly high percentage of fruit abort.

Sunrise.—Orig. in New Brunswick, N.J., by M.A. Blake, New Jersey Agr. Expt. Sta. Introd. in 1953. [J.H. Hale x (Slappey x Dewey)] o.p.; selected in 1953; tested as N.J. 133. Fruit: small to medium; elongate, oval, apex mucronate; skin dull red over yellow; flesh firm, melting, semi-clingstone; ripens 2 weeks before Jerseyland, 41 days before Elberta. Tree: hardy; vigorous; must be well thinned; flower small, nonshowy, self-fruitful; leaf glands reniform; chilling requirement 850 h; highly productive.

Sunshine.—Orig. in Hammonton, N.J., by John Errera. Introd. in 1962 by Bountiful Ridge Nurseries, Princess Anne, Md. Plant patent 2360; 11 Feb. 1964; assigned to Edgar H. Kemp and Homer S. Kemp, Bountiful Ridge Nurseries. Redhaven mutation; discovered in 1958. Fruit: large, averaging about 2 1/2 inches in diam.; skin yellow, almost solidly overlaid with red, thick, tough, no tendency to crack, pubescence short; flesh yellow, mottled red next to stone, firm, crisp, melting, subacid, mild, delicate, quality good for fresh use, canning, freezing; ripens evenly; stone small, free; ripens 3 to 5 days before Redhaven, which it resembles. Tree: vigorous; very productive; hardy; leaf glands reniform; flower size medium, blooms medium early.

Supechfour.—See Amber Crest.

Supechthree.—See Golden Crest.

Superior.—Orig. in Summerland, B.C., by J.E. Britton, Canada Dept. Agr., Expt. Sta. Introd. in 1946. J.H. Hale x Veteran; cross made in 1933; original tree planted in 1936; selected in 1941 by A.J. Mann, of the above station. Fruit: size above medium; flesh yellow, moderately firm, quality above fair, practically freestone (similar to Vedette & Valiant); fair as a canning variety; similar season to that of Valiant and Veteran. Tree: compact; medium hardiness.

Supernuts.—See Marhigh.

Super Precoce.—This is the name under which it was introduced in France; not introduced in the United States. Orig. in Marges, France, by Raymond Deroux. Introd. commercially in France in 1957. Plant patent 1668; 10 Dec. 1957; assigned to mesne assignments, to Stark Bros. Nurseries & Orchards Co., Louisiana, Mo. Mayflower mutation; discovered in 1955. Fruit: rather small, being similar to Mayflower; skin with more red coloration than that of parent; flesh white, sometimes slightly red underneath the skin and around the pit, tender, abundant juice, fibers few, flavor slightly acid, medium sweet, slightly more than that of parent, semi-freestone; ripens a few days before Mayflower. Tree: medium tall; vigorous; semi-erect; hardy; very productive; leaf glands globose; flower medium large, rosy color, similar to Mayflower.

Surabian.—Orig. in Reedley, Calif., by Surabian Bros. Intod. in 1960. Honey Dew Hale o.p.; discovered in 1956. Fruit: flesh white, very soft; ripens latter part of June; unsatisfactory as a commercial shipping or market variety; suitable for home gardens.

Surecrop.—Orig. in Calhoun, La., by P.L. Hawthorne and J.C. Taylor, Louisiana Agr. Expt. Sta., Louisiana State Univ. Introd. in 1 973. Redglobe x Dixired. Cross made in 1959; selected in 1962; tested as L-59-12-1. Fruit: medium to large, round oval; skin yellow, overcolor 80% to 90% red, attractive, pubescence medium to long and heavy; flesh bright yellow with slight red throughout, firm, good texture and quality; equal halves; very few split pits; ripens 45 days before Elberta or Dixired. Tree: strong, medium vigor, productive; flower nonshowy. medium rose color; leaf glands reniform; good resistance to bacteriosis (*Xanthomonas pruni*); medium resistance to brown rot (*Sclerotinia fructiocola*); chilling requirement 950 to 1000 h .

Suwanee.—Orig. in Fort Valley, Ga., by John H. Weinberger and Victor E. Prince, U.S. Dept. of Agriculture Field Laboratory. Introd. in 1962. J.H. Hale x FV 14-86 (Hiley x Halehaven); cross and selection made by Weinberger; later testing by Prince as USDA FV 166-79. Fruit: large; round; pubescence light, about 80% blushed red over an attractive yellow ground color; flesh yellow, firm, melting, texture medium fine, flavor good; freestone; ripens with Golden Jubilee and Keystone, about 10 days after Maygold season begins. Tree: moderately vigorous; productive; chilling requirement about 650 h; blooms with Maygold; flowers with small petals, self-fruitful; leaf glands reniform; susceptible to bacterial spot.

Swatow (*Chinese Dwarf, Chinese Dwarf Mandarin, Chinese Dwarf Evergreen*).—Seeds brought from Kiayingchow, Swatow, China, by George Campbell through George C. Hanson, American Consul, Swatow, China. Introd. in 1935 by Walter Lammerts, Armstrong Nurseries, Ontario, Calif. Parentage unknown. Received in the U.S. by U.S. Dept. of Agriculture in 1915 as P.I. 41395. Fruit: small; flesh white, fragrant, inferior quality, clingstone. Tree: dwarf; foliage dense, evergreen, ornamental; low chilling requirement. Used for breeding.

Sweeney.—Orig. in Davis, Calif., by L.D. Davis, Agr. Expt. Sta., Univ. of California, Davis. Introd. in 1971. Early Muir o.p. x Jungerman. Cross made in 1955, first fruited in 1959, tested as D-8-13W. Fruit: size medium; somewhat elongated; skin yellow ground cover with slight blush, excellent harvest color, pubescence moderate; flesh clear yellow, flavor excellent, canning clingstone; introduced because of superior flavor when canned; ripens with Fortuna or 2 to 3 days after. Tree: size medium; upright; productive; flower large, pink, showy; leaf glands reniform.

Sweet Gem.—Orig. by C.F. Zaiger, G.N. Zaiger, L.M. Gardner, and G.G. Zaiger, Modesto, Calif. Plant patent 7952; 25 Aug. 1992. [(May Grand x peach) x (Sam Houston o.p.)] o.p. Fruit: large; very firm; nearly globose; attractive red blush over bright-yellow ground color, with moderately short pubescence; flesh yellow, clingstone. Ripens early mid-June at Modesto. Fruit stores on tree 10 to 14 days. Flowers large, showy, pink; leaf glands reniform. Tree: large; vigorous; upright; productive.

Sweethaven.—Orig. at South Haven, Mich., by Stanley Johnston, Michigan State Univ. Introd. in 1976 by R.L. Andersen, J.E. Moulton, T.M.Thomas, and H.J.Belter. [Redhaven x (Halehaven o.p.)] x Crosby; selected in 1965; tested as SH448. Fruit: medium-small, requiring heavy thinning; round; skin brilliant 90% red-striped blush over bright yellow ground color; pubescence short; flesh clear yellow, semi-cling; few split pits; flavor superior for its season. Ripens with Garnet Beauty, 14 days before Redhaven. Tree: vigor medium; spreading; leaf glands globose. Flower buds hardier than those of Redhaven. Blossoms nonshowy. Slightly more tolerant than Redhaven to bacterial spot.

Sweet Sue.—Orig. near Glassboro, N.J., by

J. Wilson Hughes, Broad Acres Fruit Farm. Parentage unknown. Introd. in 1979. Plant patent 4451; 21 Aug. 1979; assigned to Newark Nurseries. Fruit: large; skin bright red blush over strong yellow ground; smooth suture; flesh deep yellow, very firm, melting, pleasant aroma, excellent flavor. Ripens uniformly with or slightly later than Rio-Oso-Gem; excellent storage and shipping quality. Tree vigorous, spreading; very productive; moderate resistance to bacterial spot.

Swellengrebel.—Orig. in Stellenbosch, South Africa, by P.A.L. Steyn, Western Province Fruit Res. Sta., Dept. of Agr. Introd. there in 1953. Babcock x Early Dawn. Fruit: flesh white, freestone; excellent storage quality; ripens early. Tree: very resistant to delayed foliation.

Swetzer (*Late Bess*).—Orig. in Wheatland, Calif., by Peter H. Swetzer. Introd. in 1940. Chance seedling; discovered about 1932. Fruit: sizes well; flat, blocky; flesh yellow, flavor good; canning clingstone; pit small; ripens in Phillips Cling season, about 5 days after Halford; mildews if not sprayed early. Tree: produces crops consistently above average; leaf eglandular.

TAMUDenman.—Orig. at Stephenville, Texas, by J.W. Worthington, J.S. Newman, and J. Lasswell, Texas A&M Univ. Res. & Ext. Center. Introd. in 1984. Redglobe o.p.; selected in 1962. Fruit: mostly 55 to 70 mm if well-thinned; skin color attractive red blush on 75% to 100% of surface; medium pubescence; round; flesh yellow, fine-textured, excellent flavor; very firm; freestone. Small pit. Harvest about 1 week after Loring. Tree: vigorous; upright-spreading, easily trained to open center; highly branched, producing large quantities of fruiting wood. More resistant to bacterial spot than is Redglobe. Abundant bud set and extended bloom period.

Tasty Zee.—Orig. in Modesto, Calif., by Floyd, Gary, and Grant Zaiger and Leith Gardner, Modesto, Calif Introd. in 1988. O'Henry x Giant Babcock. Plant patent 6409, 22 Nov. 1988; owned by Zaiger Genetics. Fruit: medium to large; globose; attractive, 80% red; pubescence medium; flesh white, very firm, good flavor, free to semi-freestone. Ripens about 18 days after Redhaven in New Jersey. Tree: vigorous; upright-spreading. Chilling requirement 800 to 1000 h.

Tatura Aurora.—Orig. in Tatura, Australia, by L.A. van Heek, Tatura Hort. Res. Sta. Introd. in 1960. Yellow Transvaal x Pullar's Cling; cross made by P.F. Marriott in 1938; seed planted in 1939; original seedling no. S68; station designation Code 58. Fruit: tends to be small, symmetrical; skin rich yellow; flesh yellow, very firm, fairly juicy, clingstone; canning quality very good; ripens late, about with Pullar's Cling; needs medium to heavy thinning.

Tatura Dawn.—Orig. in Tatura, Australia, by L.A. van Heek, Tatura Hort. Res. Sta. Introd. in 1960. Probably Levis sp.; seed collected in 1936 by P.F. Marriott; planted in 1937; original seedling no 15-43; station designation Code 37. Fruit: large; round, slightly lipped; skin light to midyellow with about one-third blushed red; flesh medium orange-yellow, being similar to Levis; texture fairly firm, juicy, flavor good, clingstone; pit size average, round; ripens early, 7 days before Levis; picking range of about 6 days; good uniformity of ripening of individual fruit; slight susceptibility to preharvest drop; canning quality fair. Tree: vigorous; light to medium amount of thinning required.

Tatura Sunrise.—Orig. in Tatura, Australia, by L.A. van Heek, Tatura Hort. Res. Sta. Introd. in 1960. Levis x Young's Cling; cross made by P.F. Marriott in 1937; seed sown in 1938; original seedling no. 4024; station designation Code 40. Fruit; size medium; round, slightly lipped; skin yellow, slightly blushed; flesh medium orange-yellow, texture fairly firm, juicy, clingstone; canning quality good; stone average to small; ripens early, 7 days before Levis; picking range of about 6 days; good uniformity of ripening of individual fruit; slight susceptibility to preharvest drop. Tree: yield good; vigorous; medium amount of thinning required.

Tatura Sunset.—Orig. in Tatura, Australia, by L.A. van Heek, Tatura Hort. Res. Sta. Introd. in 1960. Pullar's Cling x Golden Queen; cross made by P.F. Marriott in 1936; seed planted in 1938; original seedling no. 19-95; station designation Code 45. Fruit:

large, good uniformity in size; skin with red blush; flesh yellow, very firm, fairly juicy, slight red tinge around stone but not extending into flesh, clingstone; canning quality good; ripens in late season, slightly before Pullar's Cling.

Taylor.—Orig. in Placer County, Calif., by E.G. Taylor. Introd. in 1936. Parentage unknown; selected in 1930. Fruit: ripens in Phillips Cling season; commercial canning clingstone. Tree: leaves eglandular. No longer commercially planted.

Taylor Queen.—Orig. in Griffith, New South Wales, by Hort. Res. Sta, Dept. of Agriculture, New South Wales. Introd. in 1968. Golden Queen o.p. Fruit: large; ovate, long tip; skin golden with little or no red blush, pubescence moderate; flesh golden, no red at pit, no green shoulder, flavor and texture good, clingstone; processes well, recommended for canning; ripens about 4 Apr. Tree: size medium, erect; very productive; flower large, showy, pale pink.

Tejon.—Orig. in Riverside, Calif., by J.W. Lesley, California Citrus Expt. Sta. Introd. in 1958. Complex parentage involving a Peento hybrid, Columbiana, Strawberry, Rochester, and Early Imperial; crosses completed in 1949; selected in 1952. Fruit: flesh yellow, semi-clingstone; matures early; resembles Maygold. Tree: short winter-chilling requirement; bears very well; leaf eglandular; blooms early and heavily.

Telford (*Barden Tilford, Telford Elberta, Tilford, Tilford Early Elberta, Tilford Elberta*).—Orig. in Centralia, Ill., in the J.W. Telford orchard by Roy E. Gibson, Greening Nursery Col, Monroe, Mich. Introd. in 1947. Elberta mutation; discovered in 1939. Fruit: flesh yellow, freestone; resembles Elberta but with somewhat better skin color, flavor, and quality, ripening 3 to 7 days earlier. Tree: resembles Elberta.

Telford Elberta.—See Telford.

Tennessee Redleaf.—See Kingston Redleaf.

Terry.—Orig. in Strathmore, Calif., by D.M. Terry. Introd. in 1950. Chance seedling; discovered in 1940. Fruit: very large, symmetrical; flesh yellow, firm; commercial canning clingstone; ripens in Phillips and Sullivan season and a few days before Giblin; resembles Halford. Tree: sets and matures a good crop with low cullage; leaf glands globose. No longer commercially planted.

Texaberta.—Orig. in Belton, Texas, by E.E. Griffith. Introd. in 1938. Parentage unknown; discovered in 1936. Fruit: small to medium; round-ovate, tip and suture somewhat prominent; skin dull red over yellow, pubescence short, not too attractive; flesh yellow, slightly red at pit, firm, melting, somewhat tart, quality fair to poor, freestone; ripens with Elberta, which it resembles. Tree: productivity medium; vigor low; leaf glands globose; flower very small, nonshowy, self-fruitful; chilling requirement of 750 h.

TexRoyal.—Orig. at New Jersey Agr. Expt. Sta., Cream Ridge. Complex parentage of New Jersey selections x Early Amber. Bred and selected in 1975 by H.H. Bowen, Texas Agr. Expt. Sta., Yoakum; tested as Y18-101; introduced in 1990 by D.H. Byrne and T.A. Bacon. Fruit: medium large; round with slight suture and tip; rich red ground color over yellow ground color; flesh yellow, firm, melting, freestone. Ripens with Juneprince, 10 days after June Gold. Flowers small, nonshowy, pink; leaf glands globose. Tree: vigorous; productive; intermediate spreading; moderate resistance to bacterial leaf spot. Chilling requirement 600 chill units. Superior fruit set and yield; tolerant to reduced chilling.

Texstar.—Orig. in College Station, Texas, by D.H. Byrne and J.S. Kamas. Introd.in 1984. Parentage unknown; selected in 1973 by H.H. Bowen; tested as TAES M9. Distributed by Foundation Seed Services, Texas Agr. Expt. Sta. Fruit: moderately large (57 to 68 mm); skin attractive red blush over 40% of surface; pubescence moderate; flesh yellow, firm, good flavor and texture, semi-freestone; round to oblong with no prominent suture or tip when sufficient chilling has been received; good storage life. Ripens about with Junegold, 45 days before Elberta. Tree: vigorous; productive; heavy thinning required to size fruit. Susceptible to peach leaf curl; no observation of bacterial leaf spot. Leaf glands reniform. Flowers: showy; bloom date 6 days before Junegold. Chilling requirement about 550 h.

Thiara.—Chance seedling found by S.

Thiara, Yuba City, Calif. Plant patent 8744; 24 May 1994. Similar to Loadel, but ripening 8 days earlier, with longer fruit and more oblate shape. Fruit: blush has 5% to 40% red, with light striping common; medium large; medium pubescence; flesh yellow; clingstone. Flowers: nonshowy; leaf glands globose. Tree: vigorous; upright-spreading; productive.

Thomson Early Elberta.—Orig. in Toppenish, Wash., by Morris A. Thomson. Plant patent 2180; 9 Oct. 1962; assigned to CaliforniaPacking Corp., San Francisco. J.H. Hale x Halehaven. Fruit: large; oblate to round, base end flattened, apex slightly depressed; skin yellow with red blush; flesh yellow, freestone; ripens 1 1/2 weeks before Halehaven, 3 1/2 weeks before J.H. Hale; good for canning. Tree: vigor medium; productive; leaf glands reniform; flower large, showy, light pink to white, self-fertile.

Three Star (*Daily News Three Star*).—Orig. in La Canada, Calif., by W.E. Lammerts, Descanso Distributors. Introd. in 1951. Plant patent 1092; 6 May 1952; assigned to Descanso Distributors. (Swatow x Rio Oso Gem) selfed] x (Swatow x Rio Oso Gem) selfed; selected in 1948. Fruit: 2 1/2 to 2 3/4 inches in diam.; skin color good; flesh yellow, quality high, freestone; resembles Crawford. Tree: vigorous; short chilling requirement; flowers large, 2 inches in diam., double, deep pink, very abundant.

Tilford.—See Telford.

Tilford Early Elberta.—See Telford.

Tilford Elberta.— See Telford.

Tiz-A-Jewel.—Orig. in Monetta, S.C., by Jerrold A. Watson. Plant patent 2890; 10 June 1969. Introd. in 1967. Chance seedling, discovered in 1961. Fruit: size medium, uniform; skin yellow ground color, with red blush; flesh yellow, medium firm, medium subacid, medium juiciness; freestone when fully ripe; keeping and shipping quality good, for local and distant markets; ripens 6 June, earlier than Coronet and slightly later than Cardinal. Tree: large; upright, spreading; vigorous, hardy; productivity heavy, bears regularly; flower small, nonshowy; leaf glands reniform.

Tokane.—Orig. in Elsenberg, South Africa, by A.F. de Wet, Stellenbosch-Elsenburg College of Agr., Dept. Agr. Introd. there in 1950; imported into U.S. in 1954 as P.I. 219561 and P.I. 220051, in 1956 as P.I. 234197. Transvaal Cling o.p.; selected in 1942. Fruit: large; skin yellow; flesh deep yellow, firm; commercial canning clingstone; ripens late; resembles Kakamas. Tree: strong grower; heavy bearer; low chilling requirement; P.I. 219561 indexed and found virus-positive in 1965; indexing of P.I. 220051 and P.I. 234197 found virus-positive in 1967.

Tomlinson October.—Orig. in Clarendon, Texas, by a Mr. Tomlinson. Introd. about 1922 by A.L. Bruce, nurseryman, of Clarendon. Parentage unknown; discovered about 1915. Fruit: flesh white, clingstone; matures very late.

Tom Thumb.—Orig. in Argentonsur-Creuse (Indre), France, by Leon Maillochon and Ernest Marlaud. Introd. about 1938 by Stark Bros. Nurseries & Orchards Co., Louisiana, Mo. Plant patent 306; 29 Nov. 1938; assigned to Stark Bros. Nurseries & Orchards Co., Louisiana, Mo., on 13 July 1937. Parentage unknown; discovered during the late 1920s. Fruit: edible, of poor quality; resembles Mayflower. Tree: once considered valuable as a low-growing ornamental with rather attractive flowers.

Topcrest.—Orig. in Modesto, Calif., by Floyd, Gary, and Grant Zaiger and Leith Gardner, Modesto, Calif. Introd. in 1986. Fayette x a Mexican seedling. Plant patent 5745; 13 May 1986. Owned by Zaiger Genetics. Fruit: medium; pubescence medium; flesh yellow, firm, semi-freestone; very good flavor. Ripens 35 days before Redhaven in Modesto. In New Jersey, smaller than Redhaven, clingstone; ripens 40 days before Redhaven.Tree: vigorous; upright-spreading. Chilling requirement about 800 h.

Toreador (*Merricle L-120*).—Orig. in Red Bluff, Calif., by Grant Merrill. Introd. in 1958. Plant patent 2071; 11 July 1961. J.H. Hale F_4; discovered in 1954; selected in 1956. Fruit: large; round; skin red over about half the surface; flesh yellow, red near pit, firm, freestone; good for freezing; ripens about 5 days later than Splendor, which it resembles and about 20 days after Elberta. Tree: size medium; vigor medium; productive; leaf glands reniform; flower small, pink, self-fruitful; fruit easily thinned.

Tra-zee.—Orig. by C.F. Zaiger, L.M. Gardner, G.N. Zaiger, and G.G. Zaiger, Modesto, Calif. Plant patent 6347; 25 Oct. 1988. Late-ripening nectarine x O'Henry o.p. Fruit: large; globose; light-red blush over bright-yellow ground color; flesh dark yellow with intense red at pit, ultra firm, melting, freestone. Ripens 21 days after O'Henry. Flowers large, pink, showy; leaf glands reniform. Tree: large; vigorous; productive.

Treasure.—Orig. in Exeter, Calif., by Grant Merrill. Introd. in 1971. Plant patent applied for. J.H. Hale x (J.H. Hale x Levi) F_3. Cross made in 1961, selected in 1969, tested as E8-7W. Fruit: large; round to heart-shaped; skin bright yellow undercolor with red blush, nearly fuzzless, flesh yellow, firm, flavor sweet, freestone; shipping quality good; ripens mid- to late August. Tree: large; vigorous; productivity heavy; flower large, showy; leaf glands reniform.

Tremmel.—Orig. in Sigourney, Iowa, by John P. Tremmel and Minnie K. Tremmel. Introd. in 1959. Plant patent 1925; 29 Mar. 1960. Chance seedling; discovered in 1936. Fruit: averages 2 1/2 to 3 inches in diam.; skin yellow, blushed to dark red, color appearing early, pubescence heavy; flesh yellow with some red around the pit, texture firm, quality fair to good, freestone; resembles Rochester; suitable for canning, freezing and fresh market; requires thinning; ripens about 15–25 Aug. at place of origin. Tree: size medium; precocious; productive; very hardy; flowers large, pink, fertile; leaf glands reniform; tolerant to leaf curl.

Triogem.—Orig. in New Brunswick, N.J., by M.A. Blake, New Jersey Agr. Expt. Sta. Introd. in 1938. J.H. Hale x Marigold; cross made in 1923; tested as N.J. 70. Fruit: medium to large; round to oval, suture prominent; skin yellow overlaid with dull red, attractive, pubescence light to medium; flesh yellow, streaked with red, firm, melting, high quality; stone large, free; ripens slowly and early, 19 days before Elberta; ships well. Tree: vigorous; upright; leaf glands reniform; flower size medium, nonshowy, petals red, self-fruitful; chilling requirement of 850 h; sets good crops, requires thinning.

Tropi-Berta.—Orig. in Long Beach, Calif., by Mrs. Alexander B. Hepler, Jr. Plant patent 4014; 1 Feb. 1977; assigned to L.E. Cooke Co., Visalia, Calif. Chance seedling grown from seed of Elberta peach. Fruit: large; uniform; globose; medium thick skin, partly red; flesh yellow, red at pit extending into flesh; subacid flavor; freestone Tree: large; vigorous; spreading. Characterized by tree's ability to set fruit with less winter chilling requirement than other similar varieties.

Tropi-Berta.—Orig. in Long Beach, Calif., by Mrs. Alexander B. Hepler, Jr. Introd. in 1971. Plant patent 4014; assigned to L.E. Cooke Co., Visalia, Calif. Chance seedling. Fruit: resembles Elberta; large; skin blushed red; flesh yellow, juicy, firm; freestone; ripens early to mid-August. Tree: chilling requirement about 600 h.

TropicBeauty.—Orig. in Agr. Expt. Sta., Univ. of Florida, Gainesville, by W.B. Sherman. Introd. in 1988 by R.E. Rouse, Texas Agr. Expt. Sta. and W.B. Sherman. Complex parentage of Florida selections x Flordaprince. Selected in 1982; tested as Fla. 82-10. Fruit: medium; skin 80% red blush on bright yellow ground color; pubescence short; round; flesh bright yellow, very firm, semi-clingstone. Ripens 90 days after bloom. Flowers showy with small calyx cup; leaf glands reniform. Heterozygous for nectarine. Moderately susceptible to bacterial leaf spot. Chilling requirement 150 chill units.

Tropico.—Orig. in Fullerton, Calif., by Lawrence W. Sherwood, Sherwood Speciality Nursery. Introd. in 1951. Plant patent 924; 14 Feb. 1950. Parentage unknown; discovered about 1930. Fruit: resembles J.H. Hale in size and shape; skin cream with blush; flesh white, juicy, firm, quality good, flavor slightly tart, distinctive, freestone; seed small; nonsplitting; ripens about 1 Oct. Tree: heavy bearer; vigorous; dense growth; flowers rich pink color.

TropicSnow.—Orig. in Agr. Expt. Sta., Univ. of Florida, Gainesville, by W.B. Sherman. Introd. in 1988 by R.E. Rouse, Texas Agr. Expt. Sta and W.B. Sherman. Complex parentage of Florida selections x Maravilha. Selected in 1982; tested as Fla. 82-24W. Fruit: large; skin 25% red blush; flesh white, medium firm, melting, freestone; pronounced suture bulge; flavor tart but sweet. Ripens 95 days after bloom, 18 to 20 days after Flordaprince. Flower showy; leaf glands reniform. Resistant to bacterial leaf spot. Chilling requirement 200 chill units.

TropicSweet.—Orig. in Agr. Expt. Sta, Univ. of Florida, Gainesville, by R.H. Sharpe and W.B. Sherman. Introd. in 1986 by R.E. Rouse of Texas Agr. Expt. Sta. and W.B. Sherman. Complex parentage of Florida selections. Selected in 1975 tested as Fla. 5-12. Fruit: medium large; skin 70% red blush on dull yellow ground color; flesh orange-yellow, firm, semi-freestone. Ripens 98 days after bloom. Flowers showy; small calyx cup; leaf glands reniform. Moderately susceptible to bacterial leaf spot. Chilling requirement 175 chill units .

Troy.—Orig. in Jackson Springs, N.C., by North Carolina Agr. Expt. Sta. Introd. in 1968. Raritan Rose x Redskin; cross made in 1955; first fruited in 1958; tested as N.C. 4640. Fruit: medium to large; round; skin covered with 60% to 70% bright red blush over yellow ground color; flesh bright orange-yellow, moderately resistant to browning, remains firm well after picking, texture fine with very little fiber, melting, flavor excellent, freestone; ripens 2 to 3 days later than Ranger. Tree: vigor medium; productive; leaf glands reniform; flower small, nonshowy, self-fertile; chilling requirements and blooming date about the same as Elberta; moderately resistant to bacterial spot or equal to Redskin.

Tucker's.—Orig. by J.H. Tucker, Kingsburg, Calif. Plant patent 8237; 25 May 1993. Chance seedling. Fruit: medium size; oblate; very firm; 50% red blush over golden-yellow ground color with light pubescence; flesh yellow, melting, freestone. Ripens late October and hangs well on the tree into November. Flowers large, showy, pink; leaf glands mixed globose and reniform. Tree: medium; open; vigorous.

Tudor.—Orig. in Palo Alto, Calif., by W.F. Wight, U.S. Dept. of Agriculture and the California Agr. Expt. Sta. Introd. in 1941. Libbee x Newkom; tested as USDA W22-27. Fruit: medium to large; truncate-round, halves equal; skin golden yellow with deep red blush; flesh clear yellow with red at pit; commercial canning clingstone; season between Guame and Sims; drops occasionally at maturity. Too much red color at pit. Tree: vigorous; productive; leaf glands reniform; flower nonshowy, small.

Tufts.—Orig. in Davis, Calif., by L.D. Davis, Agr. Expt. Sta., Univ. of California, Davis. [Goodman Choice o.p. x Dixon 1] x Everts. Cross made in 1956, first fruited in 1960, tested as G19-14E. Fruit: size medium; symmetrical and truncated; skin yellow ground color with slight blush, excellent harvest color, pubescence moderate; flesh clear yellow, flavor excellent, clingstone; introduced because of superior flavor when canned; ripens in Dixon season. Tree: size medium; upright; productivity high; flower nonshowy; leaf glands reniform.

Tulip.—Orig. in Mountain Grove, Mo., by Paul H. Shepard, Missouri State Fruit Expt. Sta. Introd. in 1947. Sunbeam o.p.; selected in 1947; tested as PH-128. Fruit: medium to large; nearly round; skin well-colored before picking time, pubescence short; flesh firm, texture fine, yellow throughout, flavor good, semi-freestone; good shipper; ripens very early, 6 to 7 weeks before Elberta, before Marigold and Mikado; resembles Missouri. Tree: vigorous; leaf glands reniform; flower small, nonshowy, petals red, self-fruitful; chilling requirement of 950 h; moderately productive.

Two Star (*Daily News Two Star*).—Orig. in La Canada, Calif., by W.E. Lammerts, Descanso Distributors. Introd. in 1951. Plant patent 1156; 23 Dec. 1962; assigned to Descanso Distributors. (Babcock x Quetta nectarine) x [(Swatow x Rio Oso Gem) x (Babcock x Mayflower)]; selected in 1948. Fruit: 2 1/2 inches in diam.; flesh white, quality high, semi-clingstone; as early as Robin, which it resembles. Tree: vigorous; short chilling requirement; flowers single, large and very abundant.

Tyler.—Orig. in Blacksburg, Va., by G.D. Oberle, Agr. Res. Sta., Virginia Polytechnic Inst. Introd. in 1970. Rio Oso Gem x (Shippers Late Red x Sunhigh). Cross made in 1956, selected in 1960, tested as V.P.I. 60. Fruit: large, 3 inches in diam.; round, slightly flattened halves; skin rich yellow groundcover with two-thirds covered with bright red blush, pubescence short; flesh orange-yellow with attractive pinkish-red near pit, firm, texture smooth; flavor mild, superior to Rio Oso Gem; ripens 6 days after Elberta or with Rio Oso Gem. Tree: size average or above; round to spreading; vigor and productivity average or above; hardiness same as Rio Oso Gem; flower small, medium red; leaf glands reniform. Variety is similar to Rio Oso Gem, but has better tree characteristics and is more productive. May replace Rio Oso Gem in Virginia.

Tyson 820.—See Stark Finale.

Valdero.—Orig. in Parlier, Calif., by James Valdero. Introd. in 1954. Plant patent 1205; 14 July 1953. Chance seedling. Fruit: skin dark red; flesh white, red around the pit, acid low, freestone; ripens about 10 days before Elberta; can remain on tree 7 days or more after ripening without deterioration; can be shipped. Tree: medium large; open; vigor medium; foliage bright green; flower large, bright pink.

Valentine.—Orig. in Lakewood, Colo., by William M. Bunger. Introd in 1969. Parentage unknown. First fruited in 1968. Fruit: large to 3 1/4 inches in diam.; conic to roundish; skin color yellowish 30% to 50% blushed red; flesh yellow, firm, medium tender, juicy, quality and flavor good, freestone; recommended for fresh market or canning; ripens late August or early September. Tree: medium to large, rounded; vigorous and productive; very hardy.

Valiant.—Orig. in Vineland Station, Ont., Canada, by the Ontario Dept. Agr., Hort Res. Inst. Introd. in 1925. Elberta o.p., probably self-pollinated in 1917; selected in 1922; tested as Ontario 174770. Fruit: large; roundish oblate; skin thick, tough, adherent to flesh, yellow, attractively blushed, pubescence short and fine; flesh deep yellow, firm, tender, juicy, sweet; best of quality; stone free, medium large; ripens in midseason, about 2 1/2 weeks before Elberta, which it resembles. Tree: size medium; vigorous, upright-spreading; very productive, self-fruitful; hardy; flower small, nonshowy, petals red, blooms in midseason; leaf glands reniform; chilling requirement of 850 h; quite susceptible to brown rot.

Valigold.—Orig. in Ontario, Calif., by Herbert C. Swim, Armstrong Nurseries. Introd. in 1956. Plant patent 1649; 8 Oct. 1957. Parentage unknown; selected in 1949. Fruit: very large; skin blushed red; flesh yellow, flavor good, subacid, freestone; ripens during the July Elberta season; resembles July Elberta, but skin more highly colored. Tree: vigorous; very productive; bears regularly; leaf glands reniform; flower small, nonshowy.

ValleGrande.—Orig. in Weslaco, Texas, by R.E. Rouse and D.H. Byrne. Introd. in 1990. EarliGrande mutation; discovered in 1983; tested as RGV-83-1. Fruit: compared to EarliGrande, larger (120 g), more highly blushed (60% to 75% red) and 5 to 7 days later ripening; round; flesh yellow, moderately firm, fine-textured, slightly red next to pit, resistant to browning; semi-freestone. Tree: vigorous; productive; leaf glands globose; flowers nonshowy. Susceptible to rust; assumed to be moderately tolerant to bacterial leafspot as is EarliGrande. Chilling requirement 250 chilling units.

Valley Gem.—Orig. by Paul E. Hansche, Univ. of California, Davis. Introd. in 1989. Cal Red x A 54P455 (a brachytic dwarf selection from the late Fred Anderson). Plant patent 6976; 8 Aug. 1989; assigned to Regents of Univ. of California. Fruit: medium; fairly regular in shape, slightly oblong; skin covered 80% to 100% with deep red blush on bright golden ground; flesh medium to moderate orange yellow with occasional short radiating red lines near pit; very firm; freestone; sweet with mild subacid flavor; excellent eating quality. Ripening relatively uniform, 10 to 12 days after O'Henry at Davis. Tree: brachytic dwarf, not expected to exceed 2 m on standard rootstock; very precocious and very productive; blooms about 1 week earlier than standard cultivars. Flowers showy, profuse.

Valley Red.—Orig. by Paul E. Hansche, Univ. of California, Davis. Introd. in 1989. Standard peach selection, 11,7-15, x A54P455 (a brachytic dwarf selection from the late Fred Anderson). Plant patent 6671. Assigned to Regents of Univ. of California. Fruit: medium; nearly spherical; skin covered 80% to 100% with deep red blush on bright golden ground; flesh medium to golden yellow with little or no red; firm; sweet with mild subacid flavor; very good eating quality. Ripening relatively uniform, about 2 weeks before O'Henry at Davis. Tree: brachytic dwarf, not expected to exceed 2 m on standard rootstock; very precocious and very productive; blooms about 1 week earlier than standard cultivars . Flowers showy, profuse.

Valley Sun.—Orig. by Paul E. Hansche, Univ. of California, Davis. Introd. in 1989. 11,5-19 (standard peach selection) x A50P620 (a brachytic dwarf selection from the late Fred Anderson). Plant patent 6912; 11 July 1989; assigned to Regents of Univ. of California. Fruit: medium; nearly spherical; skin covered 80% to 100% with deep red blush on bright golden ground; pubescence very fine, light; flesh medium to golden yellow with little or no red; very firm; freestone; sweet with mild subacid flavor; very good eating quality. Ripening relatively uniform, slightly later than O'Henry at Davis. Tree: brachytic dwarf, not expected to exceed 2 m on standard rootstock; very precocious and very productive; blooms about 1 week earlier than standard cultivars. Flowers showy, profuse.

Vanderpoole.—Orig. in Council Bluffs, Iowa, by Mrs. Raymond Smith. Introd. in 1946. Elberta o.p.; discovered in 1945. Fruit: large; skin golden yellow with 25% of area blushed; apex very prominent; flesh yellow, quality good, freestone; ripens 1 week before Elberta. Tree: not hardy. Obsolete.

Vanguard.—Orig. in Vineland Station, Ont., Canada, by the Ontario Dept. Agr., Hort. Res. Inst. Introd. in 1941. Vaughan x Valiant; cross made in 1925; selected in 1930; tested as Ontario 250512. Fruit: size medium; round to oval, tip somewhat elongated; skin yellow, attractive, pubescence medium; flesh yellow, firm, melting, quality good, semi-freestone; ripens 5 to 6 weeks before Elberta; resembles Valiant. Tree: moderatly vigorous; leaf glands reniform; flower small, nonshowy, petals red, self-fruitful; chilling requirement of 950 h.

Vanity.—Orig. in Vineland Station, Ont., Canada, by E.F. Palmer and O.A. Bradt, Ontario Dept. Agr., Hort. Res. Inst. Introd. in 1965. (J.H. Hale x Valiant) o.p.; selected in 1942; tested as Vineland 370244. Fruit: skin an all-over red, attractive; flesh yellow, firm, juicy, freestone; ripens 7 to 10 days before Elberta; for fresh market. Tree: moderately vigorous; productive; leaf glands reniform; flowers large, showy, self-fruitful.

Van Riebeeck.—Orig. in Stellenbosch, South Africa, by P.A.L. Steyn, Western Province Fruit Res. Inst., Dept. Agr. Introd. there in 1952; imported into U.S. in 1954 by U.S. Dept. of Agriculture as P.I. 219562. Babcock x Early Dawn. Fruit: flesh white, freestone; excellent dessert type; excellent storage quality; ripens in midseason. Tree: resistant to delayed foliation.

Vaughan.—Orig. in Vineland Station, Ont., Canada, by F.S. Reeves and J.R. Haarlem, Ontario Dept. Agr., Hort. Res. Inst. Introd. in 1925. Leamington selfed; cross made in 1913; selected in 1919. Fruit: skin appearance dullish; flesh yellow, freestone; ripens early; resembles Rochester. Tree: unusually hardy in fruit bud.

Vedette.—Orig. in Vineland Station, Ont., Canada, by the Ontario Dept. Agr., Hort. Res. Inst. Introd. in 1925. Elberta o.p.; cross made in 1915; selected in 1921; tested as Ontario 150369. Fruit: size medium; round to oval; skin yellow splashed dark red, attractive, pubescence light; flesh yellow, firm, melting, somewhat stringy; stone size medium, free; quality good to very good; matures 3 weeks before Elberta, a few days before Halehaven; resembles Elberta. Tree: vigorous; chilling requirement of 1050 h at or below 45F; flower large, showy, petals pink, self-fruitful; leaf glands reniform.

Vedoka.—Orig. in Vineland Station, Ont., Canada, by J.R. Van Haarlem and O.A. Bradt, Ontario Dept. Agr., Hort Res. Inst. Introd. in 1961 by O.A. Bradt and D.V. Fisher, Canada Dept. Agr., Expt. Sta., Summerland, B.C. J.H. Hale x Vedette; cross made in 1929 in Vineland; tested as Vineland 290147; sent to Summerland in 1946 and other areas for trial. Fruit: round; skin with attractive red blush; flesh yellow, firmer than Valiant, quality good to high; freestone, but slightly clingstone in some seasons in Ontario; ripens with Valiant; canning tests consistently satisfactory. Tree: vigorous; very susceptible to bacterial leaf spot, being discarded in 1953 at Vineland Station because of this; since Summerland is a low-precipitation area where this disease is unknown, Vedoka is released for limited commercial trial there and in similar areas; produces heavily; not quite as hardy as Veteran.

Veecling.—Orig. in Vineland Station, Ont., Canada, by O.A. Bradt, Hort. Res. Inst. of Ontario. Introd. in 1974. Babygold 6 o.p., cross made in 1958; selected in 1965; tested as V 58011. Fruit: size medium to large; round; skin red; clingstone; for processing (canning); ripens 2 weeks before Babygold 5, 4 to 5 days after Redhaven. Tree: size medium; vigorous; hardiness similar to Babygold series; very productive, requires thinning; some resistance to brown rot and bacterial leaf spot; flower large, showy; leaf glands reniform.

Veeglo.—Orig. in Ontario by S.J. Leuty and O.A. Bradt, Hort. Res. Inst. of Ontario, Vineland Sta. Introd. in 1981. Sunhigh x Royalvee. Cross made in 1955; selected in 1962; tested as V55123. Fruit: attractive: skin medium-bright red blush over a light, golden-yellow background; flesh yellow, melting, slow to oxidize; freestone: ripens 26 Aug. (9 days after Redhaven) between Redhaven and Loring. Tree: vigorous, spreading; productive; moderately susceptible to perennial canker and moderately resistant to bacterial spot; flower small, nonshowy; leaf glands reniform.

Veefreeze.—Orig. in Vineland Station, Ont., Canada, by the Ontario Dept. Agr., Hort. Res. Inst. Introd. in 1940. Elberta o.p. x Arp Beauty; cross made in 1920; selected in 1926; tested as 202339. Fruit: large; round-ovate; skin dull dark red over yellow, not too attractive; pubescence medium; flesh yellow, red at pit, some red throughout, texture coarse, freestone; retains flavor in frozen pack; nonbrowning flesh; ripens 1 day before Elberta; which it resembles. Tree: moderately vigorous; leaf glands reniform; flower small, nonshowy, petals red, self-fruitful; highly productive.

Velvet.—Orig. in Vineland Station, Ont., Canada, by O.A. Bradt, Ontario Dept. Agr., Hort. Res. Inst. Introd. in 1965. (Halehaven x Vedette) x Vesper; selected in 1951; tested as Vineland 46108. Fruit: skin an attractive all-over red; flesh yellow, firm, juicy, freestone; ripens 1 week after Redhaven, which it resembles; for fresh markets or processing. Tree: productive; bud hardiness similar to Redhaven; flower large, showy, self-fruitful; leaf glands reniform.

Ventura.—Orig. in Riverside, Calif., by J.W. Lesley, California Citrus Expt. Sta., Riverside, and C.C. Delphey, California Agr. Ext. Serv., Ventura. Introd. in 1951. Hermosa x 71-9; cross made in 1938; first released in 1946 as seedling 202-10. Fruit: small; oblate; skin yellow, wine-red blush, attractive; flesh yellow, firmer than Babcock, subacid, quality fairly good, freestone; ripens early in July in southern California. Tree: vigorous, upright; short chilling requirement, less than Babcock; productive; leaf glands reniform.

Vesper.—Orig. in Vineland Station, Ont., Canada, by the Ontario Dept. Agr., Hort. Res. Inst. Introd. in 1949. J.H. Hale x Vedette; selected in 1934; tested as Ont. 290159. Fruit: medium large; round, uniform; skin yellow-green, blushed deep red over most of surface, pubescence light, attractive; flesh yellow, red at pit, melting, firm, texture coarse, freestone; good shipping quality; ripens 1 week after Elberta; resembles Vedette. Tree: vigor medium, very productive; susceptible to bacterial leaf spot; flower large, showy; light pink; leaf glands reniform.

Veteran.—Orig. in Vineland Station, Ont., Canada, by the Ontario Dept. Agr., Hort. Res. Inst. Introd. in 1928. Vaughan x Early Elberta; cross made in 1919; selected in 1925; tested as Ont. 1940208. Fruit: medium to large; round-oblate; skin yellow, splashed red, attractive, pubescence short, amount medium; flesh yellow, melting, soft, quality good, low in tannin; semi-freestone to freestone; matures 8 to 10 days before Elberta, or about St. John time; Elberta type. Tree: vigorous; highly productive; flower small nonshowy; leaf glands reniform; chilling requirement 1150 h; self-fruitful.

Vetter Elberta.—Orig. in Reedley, Calif., by Joseph J. Vetter. Introd. in 1948. Plant patent 966; 18 July 1950; one-half assigned to Roy R. Huth, West Visalia Nursery, Visalia, Calif. July Elberta mutation; discovered in 1940. Fruit: resembles parent, having better sizing and color, more roundness in shape; flesh yellow; stone semi-free; ripens about 10 June, some 3 weeks before July Elberta; stores well. Tree: strong grower; yields well; leaf glands globose; flower size medium, salmon, late blooming.

Viceroy.—Orig. in Vineland Station, Ont., Canada, by the Ontario Dept. Agr., Hort. Res. Inst. Introd. in 1929. Vaughan x Early Elberta; cross made in 1919; selected in 1925; tested as Ont. 194055. Fruit: size medium; round, apex slight; skin yellow with dark red blush over half of surface, attractive; flesh yellow, light red at pit, firm, melting, juicy; semi-freestone; Elberta type, but fuller, better quality, and matures 3 weeks earlier than Elberta. Tree: hardy, vigorous, productivity high; flower showy, large; leaf glands reniform; self-fruitful.

Vimy.—Orig. in Vineland, Ont., Canada, by the Ontario Dept. Agr., Hort. Res. Inst. Introd. in 1924. Elberta x Arp. Fruit: large; flesh yellow, quality fair to good, freestone; not favored for processing. Tree: overly vigorous; shy bearer. Obsolete.

Vivian.—Orig. in Palo Alto, Calif., by W.F. Wight, U.S. Dept. of Agriculture and the California Agr. Expt. Sta. Introd. in 1950. (Maxine x Leader) x [(Tuscan x Paloro) x (Paloro x Pratt Low)]; selected about 1936; tested as USDA W21-19C. Fruit: medium or larger, averaging 7.6 cm in diam.; nearly round to broad-oval, symmetrical; skin yellow with some red streaking, covering up to 40% of surface, pubescence average; flesh yellow with very little red at pit, flavor mild, subacid, slight aroma, non-melting, firm, fine-grained; flavor good when canned; commercial canning clingstone; ripens late, about 5 days after Fortuna, about 13 days before Paloro. Tree: quite vigorous; productive; leaf glands globose; petals nonshowy, medium.

Vivid.—Orig. in Vineland Station, Ont., Canada, by O.A. Bradt, Hort. Res. Inst. of Ontario. Introd. in 1974. Sunhigh x (Early Halehaven x Envoy), cross made in 1955, selected in 1961, tested as V 55115. Fruit: large; slightly oblong; skin red; flesh yellow, freestone, oxidizes slowly; for canning, freezing and as fresh fruit; ripens 3 to 4 days after Redhaven. Tree: size medium; vigorous; hardiness similar to Redhaven; productive, requires thinning; flower small; leaf glands reniform.

Vivid Globe.—Orig. in Xenia, Ill., by J.E. Markham, Markham's Nursery. Introd. in 1932. Probably J.H. Hale x Yellow Globe. Fruit: large; skin with a high blush; flesh yellow, freestone; resembles J.H. Hale. Tree: self-unfruitful.

Wahlbert.—Orig. in Clarkston, Wash., by Albert W. Wahl. Introd. in 1942. Plant patent 520; 26 May 1942; assigned to H. Lynn Tuttle, Lynn Tuttle Nursery, Clarkston. Chance seedling; parentage unknown; first observed in 1918; first crop produced in 1921 propagated in 1935. Fruit: large, up to 3 inches in diam.; round, uniform, halves equal; skin deep red shading to orange, colors well, thick, pubescence short; flesh deep orange to yellow, red around pit, firm, meaty, medium sweet, sprightly, very good flavor; juice reddish; pit small, red, free, does not crack; handles and ships well; ripens in early midseason, 2 weeks before Elberta; holds well on tree; resembles J.H. Hale. Tree: large; vigorous; hardy; self-fruitful; flower showy; leaf glands globose.

Walgant.—Orig. in Stellenbosch, South Africa, by P.A.L. Steyn, Western Province Fruit Res. Inst., Dept. Agr. Introd. there in 1958. Kakamas selfed; selected in 1957. Fruit: flesh orange, no red around pit, excellent canning quality; clingstone; ripens 14 days before Kakamas, which it resembles. Tree: very resistant to delayed foliation.

Walter.—Orig. in Marshalltown, Iowa, by Wheelock Wilson. Introd. in 1947 by Wheelock Wilson Nursery, Marshalltown. Parentage unknown; discovered about 1933. Fruit: large; flesh yellow, freestone; matures early October in central Iowa. Tree: hardy.

Washington.—Orig. in Blacksburg, Va., by George D. Oberle, Virginia Agr. Expt. Sta. and Virginia Polytechnic Inst. Introd. in 1959. V.P.I. 15 x Sunhigh; selected in 1954; cross made in 1951; tested as V.P.I. 49. Fruit: large; flesh yellow, firm, quality above average, freestone; skin an attractive red; ripens about 3 weeks before Elberta; superior to Elberta for canning and freezing; resembles Sunhigh. Tree: flowers hardy, showy; pollen viable; productive; vigorous; somewhat resistant to bacteriosis.

Watson.—See What-a-Flavour.

Watt's Early.—Introd. from New Zealand in 1922 by the U.S. Dept. of Agriculture as P.I. 55741 and from Australia at a later date as P.I. 111834. Fruit: size medium; skin blushed red; flesh white, freestone; ripens early. Tree: rest period extremely short.

Welberta.—Orig. in Ontario, Calif., by George P. Weldon, Chaffey College. Introd. in 1948. Weldon x Elberta; selected in 1945. Fruit: skin yellow with red streaks; flesh yellow, firm, quality good, freestone; ripens 10 to 14 days ahead of Elberta, which it resembles. Tree: resistant to delayed foliation; adapted to southern California conditions. Obsolete.

Welcome Hale.—Orig. in Monroe, Mich., by Clarence Kirby, Ilgenfritz Nurseries. Introd. about 1930. J.H. Hale mutation. Fruit: large; skin attractive; flesh yellow, freestone; ripens last week of August; ships well. Tree: vigorous; hardy. Tested at South Haven Expt. Sta., Mich., and variety appeared to be the same as J.H. Hale and was self-sterile.

Weldon.—Orig. in Ontario, Calif., by George P. Weldon, Chaffey Junior College. Introd. in 1939. Babcock x [Elberta x (Elberta x Peento)]. Fruit: large, 2 3/8 to 3 inches in diam.; skin tough, yellow with reddish-pink blush, pubescence rather fine; flesh yellow, high quality, juicy, sweet, firm, some red near the pit, fine-grained, flavor good, freestone; ripens about 1 Aug., just after Babcock and just before Elberta. Tree: highly resistant to delayed foliation; dependable producer. Variety now obsolete.

Western Pride (*Improved Rochester*).—Orig. in Fresno County, Calif., by Carlton Nursery Co., Forest Grove, Ore. Introd. in 1943. Possibly Rochester mutation; discovered about 1904; selected in 1935. Fruit: large; skin highly blushed; flesh yellow, some red around pit, fir freestone, very few split pits; primarily for fresh use and home canning; early midseason, ripening 2 to 3 days after Rochester and about 15 days before Elberta; hangs well on tree; resembles Halehaven. Tree: consistent and heavy producer.

What-a-Flavour (*Watson*).—Orig. in Forrest, Ont., Canada, by Leslie B. Watson. Trademark 50643 of the Canadian Hort. Council; registered 8 Oct. 1930. Possibly Early Crawford o.p. Fruit: flesh yellow, flavor excellent, freestone; pit small; season last of September in Ontario.

White Hale.—Orig. in New Brunswick, N.J., by M.A. Blake, New Jersey Agr. Expt. Sta. Introd. in 1932. Plant patent 31; 11 Oct. 1932; Maurice A. Blake, assignor to New Jersey Agr. Expt. Sta. J.H. Hale x Belle; cross made in 1922; tested as N.J. 63. Fruit: large; round; uniform, apex slightly protruded; skin greenish-white to cream, overspread with red, pubescence short, fairly attractive; flesh white, red around pit, firm, melting, juicy, quality good; ripens with J.H. Hale; a midseason freestone for districts south of New Brunswick. Tree: vigorous, upright, very productive; blossom size medium, nonshowy, petals red; leaf with reniform glands; chilling requirement 750 h; self-fruitful.

White Knight 1.—Orig. in Miami, Fla., by Tom Huston, Peaches of Florida, Balm, Fla. Introd. in 1963 by this firm. Okinawa x Palomar nectarine; cross made in 1958; selected in 1961. Fruit: oval, apex beaked; skin light cream to white, 50% to 75% of surface blushed; flesh white, soft, quality fair, clingstone; browning flesh ripens early, in late April. Tree: winter-chilling requirement about 250 to 300 h. Obsolete.

White Knight 2.—Orig. in Miami, Fla., by Tom Huston, Peaches of Florida, Balm, Fla. Introd. in 1963 by this firm. Okinawa x Springtime; cross made in 1958; selected in 1961. Fruit: oval, apex beaked; skin cream to white, 50% to 75% of surface blushed; flesh white, soft, browning flesh; clingstone; quality fair; ripens in early May; resembles Springtime. Tree: winter-chilling requirement about 225 to 250 h. Obsolete.

White Lady.—Orig. in Modesto, Calif., by Floyd, Gary, and Grant Zaiger and Leith Gardner, Modesto, Calif. Introd. in 1986. [(O'Henry x Giant Babcock) x May Grand] x Sam Houston. Plant patent 5821; 9 Dec. 1986; owned by Zaiger Genetics. Fruit: medium to large; skin highly colored; pubescence medium; flesh white, very firm, high quality, low acid, mild-flavored. Ripens 5 days after Redhaven in New Jersey. Tree: vigorous; upright-spreading. Chilling requirement 800 to 1000 h.

White Rose.—Orig. in Cutler, Tulare County, Calif., by Stella O. Vuich. Introd. in 1959. Plant patent 1831; 5 May 1959. Chance seedling. Fruit: large; skin cream, blushed with red; flesh white, red near the pit, flavor mild, very sweet, freestone; ripens about 1 month later than Babcock; resembles Babcock. Tree: large; vigorous; productive; leaf glands reniform; flower large, pink.

Whynot.—Orig. in Jackson Springs, N.C., by North Carolina Agr. Expt. Sta. Introd. in 1968. Erly-Red-Fre x o.p.; seeds collected from center of large orchards in 1956; first fruited in 1959; tested as N.C. 6743. Fruit: small, 1 3/4 to 2 inches in diam.; round; skin yellow, three-fourths covered with bright red, lightly pubescent; flesh clear yellow, smooth melting texture, flavor good for early peach; clingstone; ripens 5 days before Marcus, is the earliest ripening peach tested in North Carolina. Tree: vigor medium; production moderate; leaf glands reniform; chilling h and blooming time about same as Redhaven; flower small, nonshowy, self-fruitful; moderately resistant to bacterial spot.

Wickersham.—Orig. in Yakima, Wash., by J.A. Wickersham. Introd. in 1949. Plant patent 821; 1 Feb. 1949; assigned to May Nursery Company, Yakima, Wash. Parentage unknown; discovered in 1938. Fruit: skin highly colored; flesh yellow streaked with red next to stone, flavor mild, freestone; sugar content high; good keeping quality; ripens 1 week later than Elberta; resembles J.H. Hale. Tree: vigorous, spreading; resembles Elberta; leaf glands reniform; flower size medium, salmon-pink.

Wight.—Orig. in New South Wales, Australia, by the Dept. of Agr. Introd. about 1940. Transvaal Yellow o.p.; imported in 1934 from the U.S. Fruit: size medium, being difficult to size; skin, average to rich yellow; flesh yellow, very firm, no red at pit; clingstone, pit average to small; canning quality very good; ripens in late March; very slight susceptibility to preharvest drop.

Wildrose.—Orig. in New Brunswick, N.J., by M.A. Blake, New Jersey Agr. Expt. Sta. Introd. in 1947. J.H. Hale x Delicious; tested as N.J. 118. Fruit: medium to large; round-ovate; skin yellow-green, mottled dull red over most of surface, pubescence light; flesh white, red at pit, firm, melting, quality good; pit small, free; ripens 3 weeks before Elberta. Tree: vigorous; hardier than Elberta; productivity medium; flower nonshowy, self-fruitful; leaf glands reniform; chilling requirement of 750 h.

Williams.—Orig. in Merced, Calif., by the California Packing Corp. Introd. about 1936. Parentage unknown; selected aboout 1932. Fruit: flesh yellow, nonmelting, commercial canning clingstone; ripens in Halford time.

Williams Gem.—Orig. in Porterville, Calif., by Horace J. Williams. Introd. in 1958. Plant patent 1839; 9 June 1959; assigned to Williams & Sons. Rio Oso Gem mutation; discovered in 1949. Fruit: large, averaging 2 7/8 to 3 1/8 inches in cheek diam.; flesh yellow, streaked with red around the pit, melting, flavor subacid, moderately sweet, mild, very good, freestone; ripens about 2 weeks after Rio Oso Gem; resembles that variety. Tree: resembles parent; flowers large, showy, pink.

Wil-Rai (*Wilray*).—Orig. in Green Bay, Va., by R.B. Wilson. Introd. about 1935. Parentage unknown; discovered about 1929. Fruit: averages about 2 1/4 inches in diam.; round; skin blushed red; flesh yellow, freestone. Tree: fairly hardy. Obsolete.

***Wilray*.**—See Wil-Rai.

Wilson.—Orig. in Aledo, Ill., by Dave Wilson. Introd. about 1930. Chance seedling; first fruited in 1923; selected in 1926. Fruit: large; skin yellow with a deep blush; flesh yellow, juicy, soft, freestone; ripens 10 days before Champion. Obsolete.

Winblo.—Orig. in Jackson Springs, N.C., by F.E. Correll and C.N. Clayton, North Carolina Agr. Expt. Sta. Introd. in 1972. Redskin selfed. Cross made in 1958, first fruited 1961, tested as NC 8969. Fruit: large, 2 1/2 x 2 3/4 inches in diam.; round, symmetrical; skin light yellow, three-fourths covered with bright red blush, attractive; flesh bright yellow, resistant to browning, very little red pigment except in pit cavity, firm, texture; fine, melting, flavor excellent; freestone, for freezing, canning and fresh market; ripens with Loring, 1 1/2 weeks before Elberta. Tree: spreading; vigor moderate; productive; flower large, showy, self-fertile; leaf glands reniform; moderate resistance to bacterial spot. Chilling requirement 800 to 850 h.

Windsor.—Orig. in Red Bluff, Calif., by Grant Merrill. Introd. in 1968. Plant patent 2966; 27 Jan. 1970. Rodeo x (July Elberta x Leeton). Cross made in 1960; selected in 1967. Fruit: 2 1/2 to 3 1/2 inches in diam.; skin bright red; flesh yellow, streaked with red, quality good, freestone; firm for mechanical handling and for shipping; resembles Improved Pacifica; ripens between Rio Oso Gem and Merrill Fiesta and about with Toreador. Tree: size medium; vigor medium; productive; leaf glands reniform; flower size medium, pinkish-salmon.

Wintercheek.—Orig. in Arlington, Texas, by J.A. Evans, Evans Specialty Nursery. Introd. about 1930. Parentage unknown. Fruit: large; skin highly colored red; flesh white, firm, red around pit, fragrant, good quality, clingstone; ripens from 15 Oct.–15 Nov. Tree: large; vigorous; productive. Obsolete.

Wiser (*Lovell Cling*).—Orig. in Gridley, Calif., by Ray B. Wiser. Introd. in 1943. Plant patent 507; 17 Mar. 1942. Lovell x Sims; selected in 1933. Fruit: flesh yellow, free from red at pit; commercial canning clingstone; ripens late Halford season, 3 to 4 days before Phillips Cling; free from red at pit; resembles Lovell. Tree: growth willowy; vigorous; bears heavily; leaf glands globose; flowers small, nonshowy, pale pink.

Woltemade.—Orig. in Weltevrede, Barrydale, South Africa, by Nel Brothers, Montagu, South Africa. Introd. there in 1950; in the U.S. in 1956 as P.I. 234198. Probably Kakamas o.p. Selected in 1948. Fruit: resembles Kakamas; earlier ripening than Kakamas; good canning qualities. Tree: very low chilling requirement.

World's Earliest.—Orig. in Moscow, Tenn., by E.L. Morris. Introd. in 1936 by Stark Bros. Nurseries & Orchards Co., Louisiana, Mo. Parentage unknown; discovered in 1933. Fruit: small to medium; round; skin greenish-white, blushed, pubescence light; flesh white, melting, soft, clingstone; ripens early, 7 weeks before Elberta, tends to fall before full maturity; quality poor; resembles Red Bird. Tree: vigorous, very productive; flower large, showy, petals light pink; chilling requirement 850 h; leaf glands reniform; self-fruitful.

X77.—First called to attention as a young planting in Sutter County, Calif., about 1952. Propagated from topworked trees that appear to be Dixon 1 or Dixon 2.

Yakima Hale.—Orig. in Red Bluff, Calif., by Grant Merrill. Introd. in 1955. Plant patent 1538; 11 Dec. 1956. Parentage unknown; discovered in 1949; selected in 1954. Fruit: skin as highly colored as J.H. Hale, which it resembles; flesh yellow, freestone; matures 10 days before J.H. Hale and after July Elberta. Tree: larger and more vigorous than J.H. Hale; self-fruitful; leaf glands reniform; flower small, pink with salmon tinge.

Yanco Queen.—Orig. in Yanco, New South Wales, by New South Wales Dept. of Agr., Agr. College and Res. Sta. Introd. in 1968. Golden Queen mutation; discovered in 1960. Fruit: large; round, short tip, skin golden, slightly blushed, pubescence moderate; flesh golden, no red at pit, no green shoulder, flavor and texture good; processes well; recommended for canning; clingstone; ripens 22 Feb. Tree: size medium, erect; very productive; flesh large, showy, pale pink.

Yates Late Elberta.—See Shippers Late Red.

Yellow Hiley.—Orig. in Fort Valley, Ga., by Feston Fagan. Introd. about 1922. Hiley o.p.; discovered about 1916. Fruit: flesh yellow, freestone; ripens with Hiley, which it resembles. Obsolete.

Yellow Hobson.—Orig. in Stephenville, Texas. Introd. in 1936. Chance seedling; planted in 1923; first fruited in 1927. Fruit: skin red; flesh yellow, red at pit, clingstone; matures last of July at Stephenville.

Yellow King (*Merrill Yellow King*).—Orig. in Red Bluff, Calif., by Grant Merrill. Introd. in 1950. J.H. Hale o.p.; selected in 1945. Fruit: flesh yellow, freestone; ripens 1 week later than J.H. Hale, which it resembles. Tree: self-fertile. Variety now obsolete.

Yelo.—Orig. in Ontario, Calif., by George P. Weldon, Chaffe College Introd. in 1948.

(Lukens Honey x Elberta) x Weldon; selected in 1945. Fruit: skin yellow with a blush; flesh light yellow, quality high, freestone; ripens in mid-August, about the same time as Elberta. Obsolete.

Young's Cling.—Orig. in Australia by W. Young. Introd. about 1945. Paloro x Pratt-Low; seed secured from the U.S. in 1935. Fruit: size fair to small; symmetrical, round; skin yellow with deep-red blush; flesh medium to orange-yellow, slightly soft, no red around pit, clingstone; canning quality fair; ripens in late January; considerable susceptibility to preharvest drop.

Zachary Taylor.—Orig. in Blacksburg, Va., by G.D. Oberle, Virginia Agr. Expt. Sta., Virginia Polytech. Inst. & State Univ. Introd. in 1972. Freelman's Free o.p. x Sunhigh. Cross made in 1951, first selected in 1 954, tested as VPI 51. Fruit: large; round; skin three-fourths bright medium red over orange yellow ground color, attractive; flesh deep orange yellow, pinkish red adjacent to pit, very firm, texture smooth; freestone; ripens 2 weeks before Elberta or 2 weeks after Redhaven. Tree: size below average; upright to fairly spreading; productivity above average; vigor and hardiness average; flower large, showy, pale pink, self-fertile; leaf glands reniform.

Zaiqueen.—See Stark Gulf Queen.

Zee Lady.—Orig. in Modesto, Calif., by Floyd, Gary, and Grant Zaiger and Leith Gardner, Modesto, Calif. Introd in 1986. O'Henry x June Lady. Plant patent 5832; 16 Dec. 1986; owned by Zaiger Genetics. Fruit: large; skin full red; pubescence medium; round; flesh yellow, very firm, very good flavor; freestone. Ripens 35 days after Redhaven in California, 30 days after Redhaven in New Jersey. Tree: vigorous; upright-spreading. Chilling requirement 800 to 1000 h.

Zee Red.—Orig. in Modesto, Calif., by Floyd, Gary, and Grant Zaiger and Leith Gardner, Modesto. Calif. Introd. in 1986. Fayette x May Grand nectarine. Plant patent 5831; 16 Dec.1986; owned by Zaiger Genetics. Fruit: large; skin even dark red; pubescence medium; flesh yellow, very firm, good flavor, freestone. Ripens 20 days after Redhaven in New Jersey. Tree: vigorous; upright-spreading. Chilling requirement 800 to 1000 h.

Zolezzi.—Orig. in Linden, Calif., by Zolezzi Bros. Introd. in 1952. Parentage unknown; found in 1940. Fruit: flesh yellow, firm, canning clingstone; ripens late, with Halford; resembles Stanford. Tree: very heavy production.

not named.—Orig. in Centralia, Ill., by D.B. Perrine. Plant patent 1449; 17 Jan. 1956; discovered in 1946. Probable Redhaven mutation. Fruit: flesh yellow, firm, freestone; resembles Redhaven, but is larger, rounder and with a slightly more subacid flavor. Tree: identical to Redhaven. Considered by some people as apparently similar, if not identical to Redhaven.

not named.—Orig. in Reedley, Calif., by Albert A. Boos. Not introd. commercially. Plant patent 1626; 6 Aug. 1957. Parentage unknown. Fruit: large; skin color similar to that of Merrill Gem; flesh yellow, usually freestone when fully ripe; matures 7 to 10 days earlier than Redhaven and with Merrill Gem. Obsolete.

not named.—Orig. in Merced, Calif., by F.W. Anderson. Not commercially introduced. Plant patent 1738; 29 July 1958; assigned to Kim Bros., Reedley Nursery, Reedley, Calif. Le Grand nectarine x Royal Hale. Fruit: large; skin yellow overspread with orange-red, pubescence moderate; flesh yellow, red near pit, firm, flavor delicate, freestone; ripens 1 week later than Royal Hale and J.H. Hale, ripening with Rio Oso Gem. Tree: large; vigorous; productive; leaf glands reniform; flower large, self-fertile. Included here because of plant patent.

not named.—Orig. in Merced, Calif., by F.W. Anderson. Not introduced. Plant patent 1946; 24 May 1960; assigned to Stark Bros. Orchards & Nurseries Co., Louisiana, Mo. Sun Grand nectarine x Merrill Gem. Fruit: yellow fleshed, free to semi-freestone; tests conducted by Stark Bros. showed variety to be susceptible to bacterial leaf spot.

not named.—Orig. in Davisboro, Ga., by Herman J. Heidrich, Orlando, Fla., and Charles F. Windham, Davisboro, Ga. Introd. in 1964. Plant patent 2459; 1 Dec. 1964; assinged to Herman J. Heidrich &

Sons, Orlando, Fla. Fruit : variable, averages 2 inches in diam.; globose, oblate; skin ground color cream yellow with high red blush, color resembles Early Rose, medium thick, tenacious to flesh, no tendency to crack during wet or dry seasons; flesh yellow, firm, juicy, flavor mild, clingstone; ripens 30 Apr.–7 May. Tree: medium; vigorous; spreading; open; hardy; resembles Hiley; leaf glands reniform; flower medium, pink.

not named.—Orig. in Hughson, Calif., by John R. Borghello and Angelo Ertassi, Modesto, Calif. Introd. in 1965. Plant patent 2497; 20 Apr. 1965; assigned to R.M. Parker, Modesto, Calif. Chance seedling. Fruit: large; globose; skin yellow overspread with red, medium tough, tenacious; flesh yellow, firm, flavor mild; clingstone, without split pits; ripens after Paloro and before Gaume. Tree: medium, upright; productive; leaf eglandular; flowers small, pink.

Rootstocks for Peaches

Adarcias.—Orig. by M.A. Moreno and R. Cambra, Est. Exp Aula Dei (CSIC), Zaragoza, Spain. [Almond (*Prunus dulcis*) x peach (*P. persica*)] o.p. Selected in 1970. Roots by hardwood cuttings and can be produced by in vitro techniques. As rootstock, produces less vigor (two-thirds size) than similar peach–almond hybrids (e.g., GF 677 and Adafuel), but shows equally high productivity index in tests with peach cultivars. Adapted to light and calcareous soils, but requires good drainage. Graft compatible with peach, almond, and Japanese plum (*P. salicina*). Variable compatibility with European plum (*P. domestica*). Leaves are peach-like, internodes are intermediate. One-year-old shoots green in summer and intensely red in winter. Flowers: large, showy, light pink. Fruit is intermediate between peach and almond.

BY520-9.—Orig. by U.S. Dept. of Agriculture, Byron, Ga.; released jointly with Clemson Univ. in 1994 for extending tree life in situations where peach tree short-life syndrome has been a major problem. {[S-37 o.p. x Nemaguard] o.p.} o.p. produced siblings that are planted and harvested in bulk for the BY520-9 rootstock line. Seedlings appear to have resistance to *Criconemella xenoplex*, *Meloidogyne incognita*, and *M. javanica* nematodes. Plant variety protection applied for.

Avimag.—See Cadaman®.

Cadaman® (*Avimag*).—Orig. jointly at INRA Fruit Res. Sta., Pont-de-la-Maye, France, and GDFVEA, Budapest, Hungary. Introd. in 1989. Selected from a cross of *Prunus persica* x *P. davidiana* made by I. Gergely (GDFVEA); tested as No. 41-4-21. A clonal rootstock propagated by softwood and semi-hardwood cuttings and by in vitro techniques. Very vigorous in early years, then stabilizing when production begins. Compatible with all varieties of peaches and almonds. Tolerance to waterlogging better than GF 677; similar productivity and tolerance to root-knot nematodes. On the unbudded tree, flower showy, light pink; fruit small, freestone.

Citation.—Orig. in Modesto, Calif., by Floyd Zaiger. Introd. in 1982 . Red Beaut o.p. (presumed plum x peach hybrid); tested as 4-G-816. Plant patent 5112; 29 Apr. 1982; owned by Zaiger Genetics . A red-leaved, precocity-inducing clonal rootstock. Reduces tree size by about 40% to 50%. Readily propagated by hardwood and softwood cuttings. Recommended for use only with virus-free scions. Induces very early maturation of wood and very early leaf fall. The unbudded tree is an attractive ornamental, with showy pink flowers that are mostly female-sterile. Leaf glands globose.

Damas GF 1869.—Orig. at Pont-de-la-Maye, near Bordeaux, France, at INRA Fruit Res. Sta. Damas de Toulouse o.p.; selected in 1945; introd. in 1955. Putative parent *P. spinosa*. A clonal rootstock compatible with most peach varieties, but incompatible with Babcock, Fairway, Redwing, Summer Pearl, Sunnyside, and several European varieties, Semi-vigorous, producing tree somewhat smaller than one on peach rootstock. Intermediate productivity index. More tolerant than peach roots on poorly drained soils. Propagated readily by softwood and hardwood cuttings. Suckers profusely.

Chanturgue.—See Montclar®.
Ferdor.—See Julior®.
Fereley.—See Jaspi®.

Flordaguard.—A seed-propagated, nematode-resistant rootstock orig. at Agr. Expt. Sta., Univ. of Florida, Gainesville, and introduced in 1990 by W.B. Sherman, P.M. Lyrene, and R.H. Sharpe. {[(Shau thai-P.I. 65821) o.p. x *P. davidiana*] F3 x (either Okinawa or Rancho 23/32) o.p.}o.p. Selected in 1976 and tested as Fla. 14-11. Percentage of seed germination high; peach and nectarine buds show very satisfactory growth; resistant to root-knot nematode *Meloidogyne javanica* and *M. incognita* races 1 and 3. Tree: blooms with Flordagold and requires about 300 chill units; long whippy growth; no blind nodes; high flower bud set, precocious. Leaves: homozygous for red-leaf; reniform petiolar glands. Flowers deep pink and showy. Fruit: flesh yellow; deep red at cavity and on pits; freestone; ripens about 130 days from bloom. Susceptible to lime-induced chlorosis.

GF 43.—See under Plums. Compatible with many peach varieties; very vigorous.

GF 53.7.—Orig. at Pont-de-la-Maye, near Bordeaux, France, by Ch. Grasselly, INRA Fruit Res. Sta. Introd. in 1975 . St. Julien d'Orleans o.p.; selected in 1960. A pentaploid, probably after natural cross with *Prunus spinosa.* A clonal rootstock easily propagated by hardwood cuttings. Compatible with all peach and nectarine varieties tested. Tolerant to waterlogging. Considerable suckering after 5 to 6 years.

GF 305.—Orig. at Pont-de-la-Maye, near Bordeaux, France, by R. Bernhard, INRA Fruit Res. Sta. Selected in 1940 from seedling population of local peach variety of Montreuil near Paris. Introd. in 1950. A high-vigor seedling rootstock. Used as plant indicator for viruses. Susceptible to crown gall, root-knot and lesion nematodes, and waterlogging. Flower pink, nonshowy; fruit white-fleshed, freestone.

GF 557.—Orig. at Pont-de-la-Maye, near Bordeaux, France, by R. Bernhard, INRA Fruit Res. Sta. Introd. in 1955. Shalil o.p.; selected in 1945. A presumed peach x almond hybrid. A very vigorous clonal rootstock rarely producing suckers. Ease of propagation similar to GF 677. Resistant to chlorosis and *Meloidogyne incognita.* Susceptible to crown gall and very susceptible to waterlogging. On the unbudded tree, flower pale pink, showy; fruit freestone, dehiscent at maturity.

GF 677.—Orig. near Lot-et-Garonne in southwestern France. Peach x almond; discovered in 1938 by Siloret and Souty; selected in 1939 by R. Bernhard and Ch. Grasselly. Intro 1965; UPOV no. 692; assigned to Agri Obtention, Guyancourt, France. Very vigorous, making a tree 10% to 20% larger than tree on peach rootstock; high productivity index. A clonal rootstock difficult to propagate by conventional methods; readily amenable micro-propagation. Susceptible to *Meloidogyne,* but high vigor makes it suitable for replant situations. Tolerant of high lime concentration in soil. Sensitive to root asphyxia. As unbudded tree, flower small, showy, pale pink; fruit freestone.

Hansen 536.—Compatible; see under Almonds.

Hansen 2168.—Compatible; see under Almonds.

Higama.—Orig. at Pont-de-la-Maye, near Bordeaux, France, by Ch. Grasselly, INRA Fruit Res. Sta. Introd. in 1970. Selected in 1960 from mixed lot of peach seedlings received from Japan. Tested as S-2543. Seed-propagated; very vigorous rootstock. Tolerant to *Meloidogyne incognita.* Very susceptible to lime-induced chlorosis. Flowers pale pink, showy; leaf glands reniform; fruit white-fleshed, freestone.

Ishtara®.—See under Plums. Semi-dwarfing. Compatible with Springcrest and Merrill O'Henry; further compatibilities not yet determined. Productivity and fruit size are enhanced.

Jaspi® *(Fereley)*.—See under Plums. Compatible with Redhaven, Merrill O'Henry, Springtime, and a number of French varieties. Reduces tree size by about 20% compared with peach seedling stocks. Induces very good productivity. Excellent anchorage; no suckers

Julior® *(Ferdor)*.—Orig. at Pont-de-la-Maye, near Bordeaux, France, by Ch. Grasselly, INRA Fruit Res. Sta. Selected in 1965 from St. Julien (*Prunus insititia*) x

Pershore (*P. domestica*). Introd. in 1988. A clonal rootstock moderately easy to propagate by hardwood cuttings; very amenable to micropropagation. Confers very high vigor. Tolerant of waterlogging and crown gall. Induces medium vigor; almost no suckers. Susceptible to chlorosis at pH above 8. On unbudded tree, flower white, small, showy; fruit yellow skin, freestone.

Mirabi®.—See under Plums. Compatible with some peach varieties.

Montclar®**(*Chanturgue*).**—Orig. at Pont-de-la-Maye, near Bordeaux, France, by Ch. Grasselly, INRA Fruit Res. Sta. Local peach seedline, tested as S-2489. Seedling rootstocks are very vigorous, very uniform. More resistant to chlorosis than most peach seedling lines. Susceptible to root-knot nematodes. As unbudded tree, flowers pink, showy; fruit white-fleshed freestone.

Mr.S.2/5.—Orig. at Pisa, Italy, by F. Scaramuzzi, Univ. of Pisa. Myrobalan o.p.; pentaploid, presumed to be *Prunus cerasifera* x *P. spinosa*. Readily propagated by cuttings and by layering. Compatible with all peach and nectarine varieties tested. Compared to Damas GF 1869, which it most resembles, similarly tolerant of waterlogging, similar vigor, more productive; produces very few suckers. As unbudded tree, flowers white, mostly female sterile; no thorns; vigorous.

Myran (*Yumir, PPH 1*).—Orig. in Pont-de-la-Maye, near Bordeaux, France, by R. Bernhard and Ch. Grasselly, INRA Fruit Res. Sta. Introd. in 1988. Belsiana plum (P322, probably *Prunus cerasifera* x *P. salicina*) x Yunnan peach S1058. UPOV registration #2116. When used as a rootstock under peach, slightly more vigorous than peach seedlings. Compatible with all peach varieties tested. Productivity moderate during early years, improving as tree matures. Propagated by semi-hardwood cuttings. More resistant than peach and peach x almond to *Armillaria mellea;* tolerant to *Meloidogyne arenaria, M. javanica,* and *M. incognita;* susceptible to *Pratylenchus vulnus.*Tolerant of alkaline soils to pH 8. More resistant to root anoxia than are peach roots. As an unbudded tree, flower pale pink, nonshowy, sterile; vigorous.

Nemared.—Orig. at Fresno, Calif., by J.H. Weinberger, U.S. Dept. of Agriculture; introd. in 1983 by D.W. Ramming. (Nemaguard x a red-leaved seedling of Tennessee Natural) F_3. Tested as P115-95 very red-leaved seedling rootstock. More tolerant of *Meloidogyne incognita* and *M. javanica* than is Nemaguard. In the nursery, produce fewer side branches and can be budded earlier than Nemaguard.

Pixy.—Not compatible under peaches.

***PPH 1*.**—See Myran.

Rubira.—Orig. at Pont-de-la-Maye, near Bordeaux, France, by Ch. Grasselly, INRA Fruit Res. Sta. Introd. in 1980. (Red-leaved peach from the United States) F_2; selected in 1960. Tested as S-2605. A semi-dwarfing, precocity-inducing, red-leaved, seed-propagated rootstock. Susceptible to root-knot nematodes. As an unbudded tree, flowers pale pink, showy; leaf glands reniform; fruit white-fleshed freestone.

St. Julien GF 655-2.—Orig. at Pont-de-la-Maye, near Bordeaux, France, by Ch. Grasselly, INRA Fruit Research Station. Introd. in 1965. St. Julien d'Orleans (*Prunus insititia*) o.p. UPOV #699; assigned to Agri Obtention, Guyancourt, France. Clonal rootstock for peaches and prunes. Readily propagated by cuttings and by suckers. Induces medium level of tree vigor, good productivity index with concentrated fruit maturity. Few suckers produced under peaches, very few under prunes. Fair resistance to *Phytophthora* and to bacterial canker. As unbudded tree, flower white, showy; produces typical blue-skinned fruit with green flesh; freestone.

St. Julien No. 1.—Orig. in Pont-de-la-Maye, near Bordeaux, France, by Ch. Grasselly, INRA Fruit Res. Sta. Introd. in 1960. Self-sterile selection of St. Julien d'Orleans (*Prunus insititia*) x Common Mussel (*P. domestica*). Produces tree similar to that on Lovell. Resistant to lime-induced chlorosis. More tolerant to waterlogging than peach stocks. More uniform than most St. Julien seedling lines, but less uniform than St. Julien No. 2. Less winter hardy than St. Julien A. Moderate suckering. As un-budded tree, flowers white, self-sterile; fruit purple, semi-freestone, 2 cm.

St. Julien No. 2.—Orig. in Pont-de-la-Maye, near Bordeaux, France, by Ch. Grasselly, INRA Fruit Res. Sta. Introd. in 1960. A controlled F_1 seedling rootstock produced by field hybridization of a self-sterile selection of St. Julien d'Orleans (*Prunus insititia*) x Brompton (*P. domestica*). Produces tree about 75% of standard size. Moderately productive. Moderately resistant to *Phytophthora*. More uniform than most St. Julien seedling lines. More tolerant to waterlogging than peach stocks. Less winter hardy than St. Julien A. Moderate suck-ering. As unbudded tree, flowers white, self-sterile; fruit purple, semi-freestone, 2 cm.

Yumir.—See Myran.

Peach x Plum Hybrid

Tri-Lite.—Orig. by C.F. Zaiger, G.N. Zaiger, L.M. Gardner, and G.G. Zaiger, Fresno, Calif. O'Henry x [mixture of three pollens from (Red Beaut plum x unknown peach)]. Plant patent 8393; 28 Sept. 1993. A regular and productive bearer of round, medium-size, very firm, clingstone, very short pubescent, white-fleshed fruit with good flavor and ability to hang on the tree 12 to 15 days in late June. Flesh clearing when canned. Flowers: showy, pink. Tree: large; vigorous; upright; leaf glands reniform.

PEAR

Andrew.—Orig. in Saskatoon, Saskatchewan, Canada, by C.F. Patterson, Univ. of Saskatchewan. Introd. in 1960 for home gardens. *Pyrus ussuriensis* Maxim. x Bartlett; selected in 1958; tested as Sask. PR 1. Fruit: 2 inches long by 2 inches in diam. under nonirrigated field conditions at Saskatoon; skin thin; flesh quality fair to good, but breaks down quickly; ripens in early September. Tree: hardy.

Andy.—Orig. in Le Grand, Calif., by F.W. Anderson, Merced, Calif. Introd. in 1970. Plant patent 2975; 30 June 1970, assigned to Fowler Nurseries, Newcastle, Calif. Bartlett x Rose of July (Rosee de Juillet). Tested as Anderson 19K7. Fruit: large to medium; resembles Bartlett; skin yellow-green when hard-ripe, to predominately yellow when fully ripe; flesh white, streaked with yellow, texture fine, juicy, subacid; recommended for dessert, local and long distance market, processing; ripens 1 week to 10 days earlier than Bartlett. Tree: size medium; vigorous; spreading; flower large, self-fertile as well as interfertile with Bartlett.

Arganche (*Klementinka, Mustafabey, Zaharoasa de Vara*).—A small, early ripening pear from Macedonia. Introduced to U.S. from Belgrade, Yugoslavia, in 1960 (PI 264694). Fruit: small (about 50 g) like Seckel, pyriform, skin yellow with red blush and no russet; flesh fine-textured, sweet, juicy, firm; ripe early, about 16 July in western Oregon. Tree: naturally compact due to relatively short internodes, abundant fruiting spurs, consistently productive, resistant to scab. Apparently identical to Mustafabey from Turkey, Zaharoasa de Vara from Romania, and Klementinka from Bulgaria.

Aurora.—Orig. in Geneva, N.Y., by New York State Agr. Expt. Sta. Named and introd. in 1964. Marguerite Marillat x Bartlett; selected in 1950; propagated since 1954 by the New York State Fruit Testing Assn., Geneva; tested as N.Y. 7620. Fruit: large; pyriform; skin bright yellow, slightly russeted, sometimes blushed, very attractive; flesh melting, smooth, juicy, sweet, aromatic, high quality for dessert purposes; longer storage and shelf life than Bartlett; ripens with or just after Bartlett; very well suited for the home gardener and fresh fruit market; keeps well in cold storage until December. Tree: vigorous; spreading; productive; fire blight (*Erwinia amylovora*) resistance similar to Bartlett; has performed well in New York, Ohio, Pennsylvania, and Ont., Canada.

Ayers.—Orig. in Knoxville, Tenn., by Brooks D. Drain, Tennessee Agr. Expt. Sta. Introd. in 1954. Garber x Anjou; tested as Tenn. 37S21. Fruit: skin golden russet with a rose tint flesh juicy, sweet; good for eating fresh and average for canning; first picking in mid-August. Tree: resistant to fire blight, pollen-sterile.

Bantam.—Orig. in Excelsior, Minn., by Univ. of Minnesota Fruit Breeding Farm. Introd. in 1940. Parentage unknown; seed planted 1914; tested as Minn. 3. Fruit: small; ovate; skin green, blushed red; flesh tender, melting, juicy, sweet, quality good; ripens in early September. Tree: very hardy; resistant to fire blight; upright-spreading; productive.

Beierschmidt.—Orig. in Fairbanks, Fayette County, Iowa, by J.A. Beierschmidt. Introd. in 1927. Considered to be a seedling of Bartlett; seed planted by Marie Beierschmidt, mother of J.A. Beierschmidt, about 1900; first fruit borne about 1908 to 1910; original tree died when about 15 years old, but many suckers had been transplanted from it; first called to attention of S.A. Beach in 1921. Fruit: medium to large; broader than, and not as necked as Bartlett; skin thin and tender, greenish-yellow to clear pale yellow when ripened, with slight russet; flesh firm, tender, very juicy, highly aromatic, of high quality.

Berger.—Orig. in San Diego County, Calif., by an unknown woman. First announced in print in 1926 by Luther Burbank. Bud mutation discovered in 1925. Fruit: large;

apple-shaped; skin greenish-yellow, slightly tough, numerous russet dots; flesh pure white, firm, texture fine, very good; ripens last of September and into October.

Big Productive.—Orig. in Santa Rosa, Calif., by Luther Burbank. First announced in print in 1920. Considered to be a hybrid between Bartlett and Le Conte. Fruit: ripens in the fall. No longer propagated.

Bristol Cross.—Orig. in Bristol, England, by G.T. Spinks, Univ. of Bristol Res. Sta., Long Ashton. Introd. in 1933. Williams Bon Chretien (Bartlett) x Conference; selected in 1926. Fruit: large; quality good; characteristics intermediate between the parents.

Butirra Precoce Morettini.—An early season, high-quality dessert variety developed in Florence, Italy, by A. Morettini. Introd. in 1956. Coscia x Bartlett; ("butirra" = buttery, "precoce" = early). Fruit: medium to large; pyriform, but not as uniform as Bartlett; skin green-yellow with red blush, thin; flesh white, melting, juicy, sweet. Ripens 20 days before Bartlett; storage 1 to 2 months. Tree: vigorous; productive. Graft compatible on quince.

Butirra Rosata Morettini.—Orig. in Firenze, Italy, by Alessandro Morettini. Selected in 1940. Introd. in 1960; in the U.S. as P.I. 282935 in 1962. Coscia x Beurré Clairgeau. Fruit: large; skin yellow with red blush; flesh white, juicy, flavor excellent; ripens 6 to 7 days before Bartlett. Tree: very vigorous; self-incompatible and considerably parthenocarpic; scarcely compatible with quince rootstock; peduncle thick and short.

California.—Orig. in Davis, Calif., by William H. Griggs and Ben T. Iwakiri, Agr. Expt. Sta., Univ. of California. Introd. in 1975. Plant patent 3599; 27 Aug. 1974; assigned to the Regents of the Univ. of California. Max-Red Bartlett x Comice, cross made in 1959, selected in 1968. Fruit: average length 76.5 mm, average width 76.2 mm; ovate-obtuse-pyriform in longitudinal section, angular in transverse section; skin yellowish-green ground color, bright red blush on exposed side, yellow lenticels, attractive; flesh white, texture smooth, fine, melting, juicy, tender, flavor sweet, subacid, quality excellent; ripens first 3 weeks of August at Davis; keeps well, prime quality when removed from storage at 32F and ripened in late October and November; recommended as fresh fruit. Tree: medium to large; upright, spreading; vigorous; productive, precocious, regular bearer; susceptibility to fire blight between Bartlett and Comice.

Campas.—Orig. in St. Petersburg, Fla., on the farm of Martin Campas. Introd. in 1935 by Stanley Johnston, Michigan Agr. Expt. Sta., South Haven, Mich. Parentage unknown; discovered in 1923 by the U.S. Dept. of Agriculture, which propagated and distributed trees for trial purposes to many experiment stations; it proved of sufficient value for canning in comparison with Kieffer at the Michigan Sta. to warrant its introduction and naming. Fruit: size medium; skin yellow, somewhat russeted; flesh with fewer grit cells, whiter, and softer than Kieffer, which it resembles; ripens late. Tree: high degree resistance to fire blight.

Canal Red (*Reimer Canal*).—Orig. in Medford, Ore., by F.C. Reimer. Introd. in 1974 by Stark Bros. Nurseries & Orchards, Co., Louisiana, Mo. Forelle x Max-Red Bartlett, discovered about 1955, selected in 1966, tested as Reimer Canal. Fruit: most nearly resembles Comice; skin yellow ground color with bright red blush, attractive; flesh smooth, melting, sweet, quality excellent; a dessert variety; ripens late August in central Missouri. Tree: large; upright; vigor good, hardy; productivity moderate.

Carrick.—Orig. in Knoxville, Tenn., by Brooks D. Drain and Lawson M. Safley, Tennessee Agr. Expt. Sta. Introd. in 1957. Seckel x Garber; cross made in 1934. Fruit: medium to large, 2 3/4 inches in diam.; oblong pyriform; calyx open, large; skin thickness medium, tender, yellow, very russeted, slightly blushed; dots many, russeted; core small; flesh white, tinged with yellow, firm, tender, juicy, crisp; flavor sweet-subacid, sprightly, good; ripens in September. Tree: large; vigorous; spreading; some resistance to fire blight. Named in honor of the Reverend Samuel Carrick, President of the Univ. of Tennessee, 1794–1809.

Cascade.—Orig. at Medford, Ore., by F.C. Reimer, Southern Oregon Expt. Sta. Max

Red Bartlett x Comice. Plant patent 6245; 9 Aug. 1988; assigned to Carlton Plants. Fruit: large; skin dark red blush on yellow ground; texture similar to Bartlett; excellent dessert quality; susceptible to limb rub. Matures late in Bartlett season; keeps well in refrigerated storage for 6 to 8 months. Tree: vigorous; upright; mature trees crop regularly. Blooms just after Bartlett; Bartlett, Bosc, and Comice are suitable pollinizers.

Cashman.—Orig. in Ortonville, Minn., by Francis Black. Introd. in 1942 by Cashman Nurseries, Owatonna, Minn. Parentage unknown; possibly a chance seedling of Flemish Beauty; discovered about 1940. Fruit: identical in shape and color but a trifle larger than Parker; resembles Parker and Flemish Beauty. Tree: more hardy in northern climates than Parker.

Cayuga.—Orig. in Geneva, N.Y., by U.P. Hedrick, New York State Agr. Expt. Sta. Introd. in 1920. Seckel o.p.; seed borne in 1906. Fruit: of Bartlett size; similar in shape to Seckel; color of Clairgeau; mediocre quality, shy bearer at Southern Oregon Expt. Sta. Tree: has some fire blight resistance.

Caywood.—Orig. in Geneva, N.Y., by U.P. Hedrick, New York State Agr. Expt. Sta. Introd. in 1938. Seckel o.p.; seed collected in 1908; first crop in 1922. Fruit: larger, rounder, and more russeted than Seckel; flesh sweet, highly aromatic, Seckel flavor; season just before Seckel. Tree: round topped, spreading; vigorous and productive.

Chang.—Orig. in Brookings, S.D., by N.E. Hansen, South Dakota Agr. Expt. Sta. Introd. in 1926. seedling of *Pyrus ussuriensis* Maxim.; seed from Arnold Arboretum, Boston, Mass.; first fruited in 1923. Fruit: up to 2 1/4 inches in diam.; oblong pyriform; skin yellow; flesh white, firm, juicy, quality fair. Tree: leaves red in autumn, attractive.

Chapin.—Orig. in Geneva, N.Y., by U.P. Hedrick, New York State Agr. Expt. Sta. Introd. in 1945. Seckel o.p.; from seed borne in 1908; first full crop in 1920. Fruit: slightly larger than Seckel, little longer in shape; skin greenish-yellow with considerable russet; flesh richly aromatic, very sweet; season with Clapp Favorite; desirable for home use; resembles Seckel. Tree: vase form, open; vigor medium; productive.

Christmas Holiday.—Orig. in Sebastopol, Calif., by Luther Burbank. Introd. in 1940 by Stark Bros. Nurseries & Orchards Co., Louisiana, Mo. Parentage unknown. Fruit: large; skin russeted appearance attractive; flesh quality good; resembles Bosc in shape and flavor; matures late. Tree: produces well.

Clyde.—Orig. in Geneva, N.Y., by U.P. Hedrick, New York State Agr. Expt. Sta. Introd. in 1932. Seckel o.p.; seed collected in 1908; first full crop in 1921. Fruit: larger, less obovate, longer necked than Seckel; skin light greenish-russet over entire surface; flavor not as rich as Seckel; develops best quality after a period of cold storage. Tree: round topped, spreading; vigorous and productive.

Colette.—Orig. in Freeport, Ill., by Marie E. Dreyer. Introd. in 1953. Plant patent 1401; 28 June 1955; assigned to Henry Field Seed & Nursery Co., Shenandoah, Iowa. Parentage unknown; discovered in 1941. Fruit: resembles Bartlett; flavor fine; matures over a long period of time; suitable for canning. Tree: flowers bloom over a long period of time.

Columbia Red Anjou.—A long-storing red sport of Anjou. Orig. near Hood River, Ore., in 1976 by Eugene C. Euwer. Plant patent 6194; issued 7 June 1988 to Eugene C. Euwer. Discovered as a sport in the top of a 60-year-old Beurre d'Anjou tree. Fruit: similar to standard Anjou except skin is overall bright red; lower rate of respiration and ethylene production result in longer storage and shelf life than standard Anjou; higher titratable acids and soluble solids than standard Anjou and Gebhard Red Anjou. Tree: size, shape, and growth characteristics same as parent; leaves, bark, and flower petals darker than parent Anjou.

Concorde.—A precocious, midseason variety similar to Conference. Orig. at East Malling Research Station, Kent, U.K. Conference x Doyenne du Comice. Plant patent 9192; issued 11 July 1995 to Plant Breeding International, Cambridge, England. Introd. by F.H. Alston in 1987; U.S. propagation rights assigned to TRECO, Woodburn, Ore.

Fruit: long-pyriform, large, similar to Conference; excellent skin finish with less russet than Conference; good, fresh eating quality. Harvest maturity mid-September, 2 weeks after Conference. Stores in air at –1C for 6 months; storage life better than Comice, not as good as Conference. Blooms between Conference and Doyenne du Comice; both parents and Bartlett are suitable pollinizers. Tree: spreading, with wide crotch angles; precocious; early yields superior to either parent. Produces few secondary blossoms.

Conference.—A standard for midseason dessert pears in western Europe. Orig. at Sawbridgeworth, England, by Thos. Rivers and Sons and introd. in 1894. Leon LeClerc de Laval o.p. Exhibited at the National British Pear Conference in 1885, for which it was later named. Fruit: medium to large size; long pyriform; skin smooth, green, unevenly russeted; flesh pale yellow with pink tinge, tender, melting, very juicy, sweet. Season October and November; good storage life. Cropping early, regular, heavy; tends to produce misshapen parthenocarpic fruit. Self-fertile, diploid; good pollen. Tree: moderately vigorous. Susceptible to fire blight.

Cope's Seedless.—Orig. in Columbiana, Ohio, by Joe J. Cope, Cope Bros. & Fultz Nursery, Salem, Ohio. Introd. in 1935. Parentage unknown; discovered in 1928. Fruit: seedless except when planted with varieties other than Bartlett, which it resembles; ripens 2 weeks later. Tree: closely resembles Bartlett, to the extent that they appear identical.

Covert.—Orig. in Geneva, N.Y., by Richard Wellington, New York State Agr. Expt. Sta. Introd. in 1935. Bartlett x Dorset; cross made in 1921; first full crop in 1921. Fruit: one of largest of all pears; similar to Bartlett, but not as pale yellow; flesh tender, granular; develops best quality after a period of cold storage; season December to January. Tree: spreading, dense; vigorous and productive.

Cow.—See Texking.

Crimson Gem Comice.—Orig. in Phoenix, Ore., by Lyle Kinney. Introd. in 1977 by Sierra Gold Nurseries, Yuba City, Calif. Plant patent 4263. Bud mutation of Regal Red Comice. Discovered in 1965. Fruit: resembles Comice; skin thick, smooth, glossy, solid red throughout growing season, with little or no purplish tint and having slight yellowish undercolor when ripe; flesh white, fine, tender, juicy, sweet; ripens about 22 Sept. in southern Oregon; hangs well, keeping quality good to 15 Jan. Tree: spur-type, large, roundish.

Dabney.—Orig. in Knoxville, Tenn., by Brooks D. Drain, Tennessee Agr. Expt. Sta. Introd. in 1954. Seckel x Garber; crossed in 1935; tested as Tenn. 35583. Fruit: size medium; oblong obovate, pyriform, sides unequal; skin thick, medium in toughness, smooth, waxen and dull, greenish; dots many, medium in size, russeted and conspicuous; core large; flesh yellowish-white, melting, tender, juicy, quality very good; flavor sprightly, sweet-subacid and very good dessert quality; picked late July and early August, ripening rapidly in summer temperatures; scored low for canning. Tree: small; spreading, becoming drooping with loads of fruit; comes into bearing at five years; productive; moderately resistant to fire blight.

Daves's Delight.—Orig. at Kentville, Nova Scotia, by A.D. Crowe, Agriculture Canada Res. Sta. Beierschmidt x Bartlett. Fruit: medium to below medium; pyriform, slightly irregular to ribbed; skin yellow, up to 50% washed with attractive bright red; many inconspicuous lenticels; flesh light cream, very fine and buttery, nearly free of grit cells; flavor aromatic, equal to or better than Bartlett. Early maturing, with Buerre Giffard; holds well on or off tree with good resistance to core breakdown. Tree: vigor moderate; relatively precocious and productive. Released as a dessert pear; processing qualities not known.

David.—Orig. in Saskatoon, Saskatchewan, Canada, by C.F. Patterson, Univ. of Saskatchewan. Introd. in 1960 for home gardens. *Pyrus ussuriensis* Maxim. x Bartlett; selected in 1958; tested as Sask. PR 3. Fruit: up to 2 3/4 inches long and 2 1/4 inches in diam. under nonirrigated field conditions at Saskatoon; skin thin; flesh quality good, not breaking down quickly. Tree: extremely hardy.

Dawn.—Orig. in Beltsville, Md., by U.S.

Dept. of Agriculture. Introd. for trial in 1960. Michigan 437 (Barseck x Bartlett) x Comice; tested as US 570. Fruit: slightly smaller than Bartlett; pyriform, resembling Bartlett in shape but with a somewhat smoother surface; flesh sweet with a trace of acid, spicy, very juicy, quality very good, aromatic; almost entirely free of grit cells; ripens for prime eating in 12 to 14 days when held at 70F; ripens about 14 days earlier than Bartlett at Beltsville. Tree: upright; moderately vigorous; bears mainly on short terminals; not highly resistant to fire blight, probably no more so than Bartlett; flowers contain moderate quantities of pollen. Recommended for trial in areas where Bartlett can be grown, as a high-quality variety ripening 2 weeks before Bartlett.

Delbard Precoce.—An early ripening pear from France. Orig. by George Delbard Nurseries, Malicorne-Commentry, France, in 1958. Released in 1975. Seedling of Max Red Bartlett. Fruit: long conic, uniform; size medium, about 140 g; skin yellow, slightly marbled with russet; flesh fine, melting, juicy, tart; similar in appearance and maturity with Dr. J. Guyot, about 15 days before Bartlett. Tree: vigorous, branching habit similar to Bartlett; bloom mid-late, pollinated by Bartlett, Conference, Dr. J. Guyot, Beurre Hardy.

Delbard Premiere.—A very early ripening pear from France. Orig. by George Delbard Nurseries, Malicorne-Commentry, France, in 1955. Released in 1975. Akca x Dr. J. Guyot. Fruit: conic-round, uniform; size medium, about 140 g, 65 mm in diam.; skin green, becoming yellow when ripe; flesh crisp, juicy, sweet; mature in early July about 20 days before Dr. Jules Guyot, about 35 days before Bartlett; keeps about 30 days in cold storage. Tree: vigorous, medium branching, spreading; bloom early and abundant, pollinated by Conference, Bartlett; regular cropping, requires early thinning for good fruit size.

Devoe.—Orig. in Marlboro, N.Y., by Charles A. Greiner. Introd. in 1947. Plant patent 728; 25 Mar. 1947. Parentage unknown, but thought to beClapp Favorite o.p. Fruit: shape similar to Bosc; coloring similar to Clapp Favorite. Tree: vigorous; tolerant to fire blight and pear psylla.

De Zahar.—See Zaharoasa de Vara.

Dropmore.—Orig. in Dropmore, Manitoba, Canada, by F.L. Skinner, Skinner's Nursery, Ltd. Introd. in 1958. Pioneer x *Pyrus ussuriensis.* Tree: very large, upright; considered to be of value in breeding work as the tree is extremely hardy; also, of value as an ornamental.

Dymond.—Orig. in Tunkhannock, Pa., by Lawrence B. Dymond. Introd. in 1954. Plant patent 1806; 3 Feb. 1959; assigned to Kelly Brothers Nurseries, Dansville, N.Y. Chance seedling; discovered in 1950. Fruit: 2 1/2 to 2 3/4 inches in diam.; skin color very similar to that of Bartlett; flesh juicy, firm, tender, melting, sprightly subacid, quality good; ripens 10 days later than Bartlett; keeps well in cold storage; resembles Clapps Favorite. Tree: tolerant to fire blight and scab; hardy; productive.

Early Faulkner.—Orig. in Knoxville, Tenn., by J.A. McClintock, Tennessee Agr. Expt. Sta., who secured propagating material from the farm of Arthur Faulkner. Not introd.; used for breeding material. Considered to be *Pyrus pyrifolia* x *P. communis*; direction of cross unknown; discovered about 1933. Fruit: ripens early. Tree: resistant to fire blight. A local variety.

Early Seckel.—Orig. in Geneva, N.Y., by U.P. Hedrick, New York State Agr. Expt. Sta. Introd. in 1935. Seckel o.p.; seed collected in 1906; first full crop in 1915. Fruit: resembles Seckel but ripens 2 to 3 weeks earlier; keeps in storage longer than Seckel; recommended for local and roadside markets. Tree: round topped, spreading; vigor medium.

El Dorado.—Orig. in Placerville, El Dorado County, Calif., by Robert Patterson. Introd. in 1945. Considered to be an open-pollinated seedling of Bartlett; discovered in 1931; Patterson secured grafting wood from the original tree during 1935–40. Fruit: large, sometimes larger than Bartlett; skin very smooth, greenish-yellow, not changing to yellow as does Bartlett upon ripening; flesh quality fair to excellent, depending upon time of harvest, storage, and ripening conditions; sweet, with fair balance of sugar to acid; texture smooth, melting, with few grit cells; ripens very late, beginning about the third week in September;

keeps well in cold storage at 30 to 32F until May and June; home cans very well; considered as a possible replacement for Anjou. Tree: growth habit similar to Bartlett.

Eller Bartlett.—Orig. in Yakima, Wash., by Walter H. Eller. Introd. in 1954. discovered in 1942. Plant patent 1184; 12 May 1953; assigned to Eller Pear Co. Yakima. Bud mutation of Bartlett. Fruit: flavor good; resembles Bartlett; good keeping qualities; claimed to ripen 4 to 5 weeks before Bartlett. Tree: claimed to have some fire blight resistance; productive; said to bloom earlier and have less second bloom than Bartlett. Appears to be no different than other Bartlett clones in bloom and ripe dates at the pear germplasm collection in Corvallis, Ore.

Elliot.—Orig. at Univ. of California, Davis, by Kay Ryugo; released in 1988. Elliot #4 x Vermont Beauty; cross made in 1964. Plant patent 6452; 6 Dec. 1988. Fruit: 50 to 60 mm in diam.; shape pyriform to conical, similar to Comice; skin yellow-green, 50% russeted. Buttery texture; flavor similar to Bosc; soluble solids to 18%. Ripens 2 to 4 weeks after Bartlett; stores 16 weeks at 0C. Tree: upright habit; no fire blight during 20 years of observation, apparently resistant.

Elwert.—Orig. in Sherwood, Ore., by the late Fred Elwert of Tualatin Valley Nurseries, Sherwood. Introd. about 1943. Parentage unknown; discovered about 1930. Fruit: round; flesh firm, flavor fine; good for eating fresh and for home canning; ripens in September. Tree: upright like a Bartlett.

Enie.—Orig. in Ottawa, Ont., Canada, by Canada Dept. Agr., Res. Sta. Introd. in 1948. Zuckerbirne x Clapp Favorite; selected in 1930. Fruit: size medium; quality good. Tree: vigorous; hardy at Ottawa. Recommended for home gardens in eastern Ontario and Quebec.

Etrusca.—A very early ripening pear from Italy. Orig. at the Univ. of Florence, Florence, Italy, by Elvio Bellini. Coscia x Gentile. Cross made in 1971. Evaluated as selection FI-71.PC.526. Released and patented in 1992. Italian patent number 186/NV/92. Fruit: large, elongate, very regular, attractive; skin green with 30% to 40% red blush; flesh firm, white, fine texture, resistant to core breakdown; matures early July in Florence, with Delbard Premiere, about 37 days before Bartlett. Tree: vigorous on pear seedling rootstock, compatible with quince, abundant bloom, good fruit set, self-incompatible.

Ewart.—Orig. in East Akron, Ohio, by Mortimer Ewart. Introd. in 1928. Parentage unknown. Fruit: size medium; skin greenish-yellow, russeted, appearance fair; flesh fine, melting, tender, juicy, flavor excellent; quality excellent; season 10 to 20 days after Bartlett, which it resembles; keeps well into December under refrigerated storage. Tree: somewhat more resistant to fire blight than Bartlett; very productive; effectively pollinates Bartlett.

Finland.—Orig. in Brookings, S.D., by N.E. Hansen, South Dakota Agr. Expt. Sta. Introd. in 1933. Open-pollinated seedling of Finland Early Yellow. Fruit: small, about 2 inches in diam.; flesh yellow; stem long. Tree: very hardy; quite resistant to fire blight

Finsib.—Orig. in Brookings, S.D., by N.E. Hansen, South Dakota Agr. Expt. Sta. Introd. in 1939. Finland Early Yellow (from Finland) x *Pyrus ussuriensis* 'Saponsky' (P.I. 20336; from Siberia); selected in 1939. Fruit: globular, 2 x 2 inches; skin yellow with small russet dots; flesh juicy; ripens in late summer; resembles Finland Early Yellow. Tree: extremely hardy. Name is a contraction of Finland and Siberia.

Finsib Sister.—Orig. in Brookings, S.D., by N.E. Hansen, South Dakota Agr. Expt. Sta. Introd. in 1939. Finland Early Yellow x *Pyrus ussuriensis* 'Saponsky' (P.I. 20336); selected in 1939. Fruit resembles Finsib. Tree: extremely hardy; more vigorous than Finsib.

Flamingo.—An attractive, early midseason dessert variety introduced in Republic of South Africa in 1993. Bon Rouge (a red sport of Bartlett) x Forelle; released by Infruitec. Fruit: size medium-small, 150 g; oblong-ovate; skin light yellow with distinctive red blush and conspicuous lenticels; flesh creamy white, soft texture, good flavor. Harvest maturity early February in the Western Cape. Tree: vigorous; productive; tendency to produce a few thorny shoots; moderate chilling requirement. Packhams Triumph and December are suitable pollinizers.

Flordahome.—Orig. in Gainesville. Fla., by Univ. of Florida, Dept. of Fruit Crops. Introd. in 1981. Tenn (Tennessee 37-20) x Hood. Cross made in 1954; selected in 1970; tested as Fla. 41-116. Fruit: large, round-ovate; skin green, tender; flesh white, melting (buttery texture); early ripening, mid- to late July, needs to be harvested at hard green stage for maximum aroma and flavor; stores up to 6 weeks at 40 to 45F without significant loss in quality. Tree: upright, semi-compact; highly productive; shorter and more compact than Hood; resistant to fire blight, moderately resistant to leaf spot; low chilling requirement, adapted to north and central Florida; requires cross-pollination. Hood and Pineapple good pollinizer.

Funks Colorado.—Orig. in Fort Morgan, Colo., on the Funk farm by the Fort Morgan Nursery. Introd. in 1942 by this nursery. Parentage unknown; discovered during the early 1920. Fruit: resembles Bartlett. Tree: hardy; resistant to fire blight.

General Leclerc.—Orig. near Paris, France, by Nomblot Nursery and Institut National de la Recherche Agronomique, Angers, France. Parentage unknown, but probably open-pollinated seedling of Doyenne du Comice; discovered in 1950. Fruit: size medium to large; short pyriform; skin thick, fairly tough, deep yellow almost overlain with russet; flesh texture fine, melting, juicy, flavor sweet-subacid, quality outstanding; ripens a few days before Doyenne du Comice; recommended for fresh market. Tree: upright, then spreading; moderately vigorous; very productive; tolerant to scab, drought and high temperatures: strong tendency to parthenocarpy; compatible with quince rootstocks.

Gilmore Pride.—Orig. near Siler City, N.C., by Glenn G. Gilmore, North State Nursery Co., Julian, N.C. Introd. in 1946. Chance seedling, parentage unknown. Fruit: size medium; round to turbinate shape with no neck; skin clear yellow with large russet dots and splashes of russet over much of surface when mature; flesh white, coarse texture, some grit cells near core and skin; ripens from late August to early October in North Carolina. Tree: vigorous; upright; little branching and very open when spread with fruit; leaves large, thick, dark green; blooms early or about the same time and as resistant to fire blight as Kieffer.

Golden Russet Bosc.—A russeted sport of Bosc. Discovered in Hood River, Ore., by Sumio Fukui in 1940. Bud mutation of Beurre Bosc. Plant patent 5243; 4 June 1984; assigned to Carlton Plants, Dayton, Ore. Fruit: skin uniformly orange russeted; otherwise identical to ordinary Beurre Bosc in fruit quality, tree habit, phenology, etc.

Golden Spice.—Orig.in Excelsior, Minn., by Univ. of Minnesota Fruit Breeding Farm. Introd. in 1949. Parentage unknown; seed planted in 1914; selected about 1924; tested as Minn. 4. Fruit: small; skin clear yellow with blush; flesh light yellow, juicy, flavor pleasant, tart, spicy; ripening in midseason; resembles Seckel in size, but not as sweet. Tree: very hardy; productive; a good parent for further breeding.

Gorham.—A high-quality, very productive dessert pear. Orig. in Geneva, N.Y., by Richard Wellington, New York State Agr. Expt. Sta. Introd. in 1923. Bartlett x Josephine de Malines; cross made in 1910. Fruit: quality excellent; ripens 10 days later than Bartlett, which it resembles; keeps 6 to 8 weeks longer than Bartlett. Tree: upright, dense; vigorous and productive; moderately susceptible to fire blight.

Gourmet.—A high-quality, cold-hardy dessert pear, resistant to fire blight. Orig. at South Dakota State Univ., Brookings, by R.M. Peterson and J.R. Waples; released in 1988. Not patented. South Dakota F15 x Ewart; cross made in 1954. Background includes *Pyrus ussuriensis* and *P. pyrifolia*, as well as *P. communis*. Tested as South Dakota 69-S-1 and South Dakota 55-8-22. Fruit: 56 to 74 mm in diam., 60 to 76 mm long; turbinate; skin green-yellow with numerous russet dots; thick skin. Flesh firm, yellow, coarse, crisp, and juicy like Asian pear; European pear flavor. Ripens third week of September at Brookings. Tree: upright; medium vigor; moderately productive.

Grand Champion.—Orig. in Hood River, Ore., by W.F. Shannon. Introd. in 1943. Plant patent 585; 18 May 1943; assigned to

Stark Bros. Nurseries & Orchards Co., Louisiana, Mo. A russet sport of Gorham; discovered in 1936. Fruit: skin overspread with attractive golden russet, does not show bruises; flesh white, juicy, flavor spicy; dessert quality good; keeping quality good, storing well until late December; ripens about 2 weeks after Bartlett. Tree: blooms late; identical to Gorham.

Guraly Jr.—Orig. in Fairport Harbor, Ohio, by Joseph T. Guraly, Jr. Introd. in 1940. Plant patent 435; 17 Dec. 1940. Parentage unknown. Fruit: round; skin brownish-green, thick; flesh white with strong bright red streaks radially from core, coarse-grained, sweet, juicy; ripens early August at place of origin. Tree: tall; foliage dark green; leaves wide and short.

Hansen Seedless.—Orig. in Brookings, S.D., by N.E. Hansen, South Dakota Agr. Expt. Sta. Introd. in 1945. *Pyrus sinensis* x Marguerite Marillat; selected in 1945. Fruit: supposed to be seedless, but in some years contains seeds.

Harbin.—A cold-hardy, fire-blight-resistant pear line used as a rootstock or windbreak. Orig. in Brookings, S.D., by N.E. Hansen, South Dakota Agr. Expt. Sta. Introd. in 1926. *Pyrus ussuriensis*; original seed gathered in 1924 about 50 miles east of Harbin, North China, by Professor N.E. Hansen. The Harbin pear is usually propagated from seed, so the name is given to the seedlings. Fruit: quality poor. Tree: hardy; used in South Dakota as a rootstock; some seedlings show fire blight resistance; good windbreak tree since the foliage and branching are dense.

Harper.—Orig. in southwest Georgia by a Mr. Harper. Introd. in 1933 by Wight Nursery & Orchard Co., Cairo, Ga. Chance seedling discovered about 1919. Fruit: skin russeted; flesh very juicy; three to four seeds per fruit; keeps well; quality good. Tree: productive.

Harrow Delight.—Introd. for early fresh-market and home-garden use. Orig. at Res. Sta., Harrow, Ont., Canada, by H.A. Quamme, Agr. Can. Introd. in 1982. Purdue 80-51 (Old Home x Early Sweet) x Bartlett. Cross made by R.E.C. Layne, Res. Sta., Harrow; selected in 1973; tested as HW-603. Fruit: 5% smaller than that of Bartlett; ovate-pyriform, shallow, broad basin; flesh quality high, juicy, grit equal to that of Bartlett, flavor as good as that of Bartlett but distinctly different; skin light-green to yellow-green ground color with 20% to 30% covered with a light blush, no russeting; processed fruit inferior to that of Bartlett and only a little better than that of Kieffer; ripens 2 weeks before Bartlett. Tree: spreading; vigor moderate; productive; leaves ovate with rounded base, leaf serrations indistinct; flowers white; resistance to fire blight (*Erwinia amylovora*) slightly less than Old Home. Cross fertile with Bartlett, Bosc, Anjou, and Harvest Queen.

Harrow Sweet.—A high-quality, late-season pear for the fresh market, selected for precocity, productivity, cold hardiness, and fire blight resistance. Orig. at Agriculture Canada Research Station, Harrow, Ont., by D.M. Hunter, P. Pinsonneault, Frank Kappel, H.A. Quamme, W.G. Bonn, and R.E.C. Layne. Bartlett x Purdue 80-51. Cross made in 1965 by R.E.C. Layne; selected by H.A. Quamme in 1980; tested as HW-609. Fruit: medium to large; pyriform; skin yellow with red blush; some russeting of lenticels. Flesh white, sweet, juicy, flavorful; stores 10 weeks at 10C. Tree: medium size; upright to spreading; productive. Fire blight resistance similar to Harrow Delight. Pollen cross-compatible with Bartlett. Graft compatible on quince.

Hartmann.—Orig. in Talent, Ore., by F.C. Reimer, Southern Oregon Branch Expt. Station. Comice x Max Red Bartlett tested as OSU 3-11-50. Fruit: medium large, 2 to 3 inches in diam.; pyriform, resembles Bartlett in form; skin dark red, tough, when ripe can be peeled like an avocado; subject to friction bruising like Comice; flesh white, juicy, sweet, but often insipid, aromatic, often gritty near core; storage comparable to Comice; shelf life outstanding; ripe with Comice in mid-September in southern Oregon. Tree: only fair in vigor; upright; similar to Bartlett in structure; foliage quite red, leaves stiff and brittle; blooms with Comice; production moderate. Named for the late Prof. Henry Hartmann of the Dept. of Horticulture, Oregon State Univ. Still growing at S. Oregon Expt. Sta. in 1996; not propagated commercially.

Harvest Queen.—Introd. for early fresh-market and home-garden use. Orig. at Res. Sta., Harrow. Ont., Canada, by H.A. Quamme, Agr. Can. Introd. in 1982. Michigan 572 (Barseck x Bartlett) x Bartlett. Selected in 1972 from cross made by L.F. Hough and Catherine H. Bailey, Rutgers Univ.; tested as HW-602. Fruit: ovate-pyriform with shallow broad basin; flesh quality high, equal to that of Bartlett but less gritty, flavor almost identical; skin yellow resembling Bartlett, no blush or russet; fresh and processing quality equal to Bartlett, but 10% less in size for processing; ripens 1 week before Bartlett. Tree: upright; vigor moderate; productive; leaves elliptical, acuminate tips and tapering base, leaf serrations distinct; flowers white; resistance to fire blight similar to Kieffer; cross fertile with Bosc, Anjou, and Harrow Delight, but not with Bartlett.

Henderson Special.—Orig. in Athens, Texas, by E.M. Henderson, Henderson's Nursery. Introd. in 1927. Open-pollinated seedling of Bartlett; first fruited in 1922. Fruit: large; flesh yellow, quality good; tendency for flesh to grow up one side of stem; ripens last of August and first of September Tree: relatively resistant to fire blight; comes into bearing early.

Highland.—Orig. in Geneva, N.Y., by Robert C. Lamb, Agr. Expt. Sta. Introd. in 1974. Bartlett x Comice, cross made in 1944, selected in 1956, tested as NY 10274. Fruit: 2 1/2 inches in diam.; pyriform; skin yellow with irregular light russet; flesh melting, juicy, few grit cells, flavor sweet, fruity, quality very good; ripens 28 Sept. at Geneva, 4 weeks after Bartlett, a dessert variety. Tree: size medium; somewhat upright; vigor medium; productive; hardiness about like Bartlett; susceptible to fire blight.

Honeysweet.—Orig. in New Hampshire, Seckel x U.S. 220 (Vermont Beauty x Roi Charles de Wurtemburg); cross made in 1955 by L.F. Hough, Rutgers Univ., N.J., seedling screened for fire blight at Lafayette, Ind.; seedling first fruited in 1967; tested as 117-1 and TH7-230; selected in 1969 by Jules Janick, Purdue Univ., West Lafayette, Ind. Released in 1977, plant patent 4499; 13 Feb. 1979 to Stark Bros. Nurseries. Fruit: pyriform-turbinate; ripens to golden russet; flesh smooth, buttery, no detectable grit; flavor rich, sweet, resembles Seckel; cultivar sets without pollination, but fruit size is reduced; pollen is fertile; local market use; ripens 1–10 Sept.; expected to be adapted to the Midwest. Tree: spreading; does not defoliate even without spraying for leaf-spotting diseases; resistance to fire blight rated with Kieffer.

Hood.—A low-chilling, fire blight resistant pear that can be grown where warm climate precludes culture of higher quality varieties. Origin unknown, probably from Florida. Asian x European hybrid that has been grown successfully in Florida and southern California. Fruit: large (240 to 280 g); skin smooth, yellow-green; flesh white, firm, buttery texture, mild flavor; susceptible to core breakdown. Ripens mid-July in northern Florida. Tree: upright-spreading; requires 160 to 260 h chilling. Highly resistant to fire blight; moderately resistant to *Fabrea maculata* leaf spot. Hood is parent of Flordahome.

Hoskins.—Orig. in Knoxville, Tenn., by Brooks D. Drain, Tennessee Agr. Expt. Sta. Introd. in 1954. Seckel x Late Faulkner; crossed in 1938; tested as Tenn. 38S10. Fruit: size medium; roundish obovate, pyriform, sides unequal; skin medium thick and medium tough, dull, light yellow blushed and mottled with russet; dots many, large, russeted and conspicuous; core size medium; flesh white often tinged with pink, fine-grained, melting, juicy; flavor subacid to sour, sprightly, good; ripens 25 Sept.–7 Oct. Tree: large; vigorous; spreading; fire blight resistant.

Illinois Bartlett.—Orig. in Farina, Ill., by A.M. Whitford. Introd. in 1953. Probably Kieffer x Duchess d'Angouleme; discovered in 1932. Fruit: flesh firm, with no grit cells, flavor good; matures in late September. Tree: upright; spreading, similar to Garber; much more resistant to fire blight than Bartlett.

Ilya.—Orig. in Brookings, S.D., by N.E. Hansen, South Dakota Agr. Expt. Sta. Introd. in 1940. Vinnaja Selenaja x an unknown pear from southern Missouri. Fruit: diam. about 2 1/4 inches; globular, somewhat irregular and angular; skin yellow with minute russet dots; flesh pleasantly subacid, melting, grit-free; a good summer table and culinary pear.

Improved Kieffer.—Orig. in Somerset, Texas. Introd. in 1956 by Aldridge Nursery, Von Ormy, Texas. Kieffer o.p. discovered in 1950. Fruit: softer and earlier than Kieffer, which it resembles. Tree: very vigorous.

James.—Orig. in Saskatoon, Saskatchewan, Canada, by C.F. Patterson, Univ. of Saskatchewan. Introd. in 1960 for home gardens. *Pyrus ussuriensis* Maxim. x Aspa; selected in 1959; tested as Sask. PR 16. Fruit: averages 2 5/8 inches long and 2 1/4 inches in diam. under nonirrigated field conditions at Saskatoon; skin yellowing well before flesh breakdown occurs; flesh texture very fine; ripens in late September; quality good; resembles Bartlett. Tree: hardy.

John.—Orig. in Saskatoon, Saskatchewan, Canada, by C.F. Patterson, Univ. of Saskatchewan. Introd. in 1960 for home gardens. *Pyrus ussuriensis* Maxim. x Aspa; selected in 1958; tested as Sask. PR 15. Fruit: up to 3 inches long and 2 1/2 inches in diam. under nonirrigated field conditions at Saskatoon; skin thin, yellowing well before flesh becomes soft; quality good; ripens in late September. Tree: hardy.

Johnson.—Orig. in Waynesville, Mo., by the late father of Alfred Johnson. Apparently introd. for trial in 1948 by the Missouri State Fruit Expt. Sta., Mountain Grove. Chance seedling; discovered in 1942. Fruit: large; skin golden; quality fair, being better than most fire blight resistant varieties; ripens about 15 Sept.; resembles Maxine. Tree: fire blight resistant.

Kalle **.**— See Starkrimson.

Klementinka (*Arganche*).—A small, early ripening pear from Bulgaria. Local Bulgarian variety of unknown parentage, imported to U.S. from Kustendil, Bulgaria, in 1974 (PI 392320). One of the parents of Ubileen Gift. Apparently identical to Arganche from Macedonia and Mustafabey from Turkey. See Arganche for description.

Lacock.—Orig. in Plains, Mont., by David A. Laywer & Sons Nurseries. Introd. in 1959. Parentage unknown, but two trees have been known since 1900. Fruit: size medium to small; skin russeted and not too attractive; flesh firm, sweet, juicy; cans well; resembles Flemish Beauty. Tree: hardy.

Lady Marion.—Orig. in Tacoma, Wash., by Thomas F. Wells. Introd. in 1971. Plant patent 2985; 28 July 1970. Chance seedling, discovered in 1921, selected in 1958, first propagated in 1968. Fruit: size variable, rounded pyriform slightly flattened at calyx end, shape distinctly different than other varieties ripening same season; skin yellowish-green ground color with reddish blush; flesh cream-colored, high sugar content; recommended for canning, requires less sugar than Bartlett, possibly has dietetic value for low sugar diets. Tree: size medium prolific and annual bearer; hardy.

Late Faulkner.—Orig. in Knoxville, Tenn., by J.A. McClintock, Tennessee Agr. Expt. Sta., who secured propagating material from the farm of Arthur Faulkner. Not introd.; used for breeding material. Considered to be *Pyrus pyrifolia* x *P. communis*; direction of cross unknown; discovered about 1933. Fruit: resembles Mooers. Tree: resistant to fire blight. A local variety.

Laxton's Early Market.—Orig. in Bedford, England, by Laxton Bros. Nursery. Introd. in 1927 in England; in the U.S. for trial in 1938. Marie Louise x Doyenne d'Ete. Fruit: size medium; flesh yellow, subacid, dessert flavor fair; harvested first week of August at Wooster, Ohio.

Laxton's Foremost.—Orig. in Bedford, England, by Laxton Bros. Nursery. Introd. in 1939 in England; in the U.S. in 1989 (P.I. 541542). Marechal du Cour x Fertility. Fruit: large; skin yellow, often blushed with red; flavor very good; harvested during the Bartlett season, last week of August, at Wooster, Ohio; resembles Bartlett. Tree: quite susceptible to fire blight.

Laxton's Progress.—Orig. in Bedford, England, by Laxton Bros. Nursery. Introd. in England in 1933; in the U.S. in 1938 as P.I. 127039. Marie Louise x Bartlett. Fruit: large; skin yellow with a pink blush; flesh yellow, juicy, subacid, flavor good to very good, high quality; harvested first week in September in Wooster, Ohio, about 1 week after Bartlett; resembles Bartlett. Tree: susceptible to fire blight.

Laxton's Record.—Orig. in Bedford, England, by Laxton Bros. Nursery. Introd. in 1936 in England; in the U.S. for trial in 1938 and again in 1989 (P.I. 541540). Marie Louise x Doyenne du Comice. Fruit: size medium to above medium; skin yellow blushed with red, with some fine russet; flesh tender, juicy, sweet, flavor fair to good; harvested during second week in October at Wooster, Ohio; resembles Bartlett.

Laxton's Satisfaction.—Orig. in Bedford, England, by Laxton Bros. Nursery. Introd. in 1929 in England; in the U.S. for trial in 1938. Bartlett x Beurre Superfin. Fruit: size medium to above medium; skin greenish-yellow; flesh juicy, sweet to subacid, flavor good; season of Bartlett at Wooster, Ohio; resembles Bartlett. Tree: quite susceptible to fire blight.

Laxton's Superb.—Orig. in Bedford, England, by Laxton Bros. Nursery. Introd. in 1913 in England; in the U.S. in 1937. Beurre Superfin x Bartlett. Fruit: flesh sweet, juicy, dessert quality high; season of harvest precedes Bartlett by 1 week at Wooster, Ohio; softens very quickly; primarily for home gardens as it is of little commercial value. Tree: very susceptible to fire blight.

Laxton's Victor.—Orig. in Bedford, England, by Laxton Bros. Nursery. Introd. in 1933 in England; in the U.S. in 1938 as P.I. 127042. Marie Louise x Seckel. Fruit: small to medium; skin greenish-yellow, blushed with red, with considerable russet; flesh juicy, sweet, flavor fair to good; harvested during the first week of October at Wooster, Ohio; resembles Seckel. Tree: quite susceptible to fire blight.

Lee.—Orig. in New Brunswick, N.J., by L. Fredric Hough and Catherine H. Bailey, New Jersey Agr. Expt. Sta. Introd. in 1968. Beierschmidt x N.J. 1. Cross made in 1950; first fruited in 1958; selected in 1958; tested as N.J. 4. Fruit: size medium; short-pyriform; skin light yellow when ripened on tree; flesh cream colored, flavor nearly comparable to Bartlett and pleasant to eat when tree ripened, quality good; ripens just after Bartlett; as blight resistant as Kieffer, but usually does not transmit a high level of resistance to its seedlings Tree: size moderate; central leader with fairly open branching; productive, good pollen, compatible with other varieties; later blooming than Bartlett.

LeRoi.—See Red Anjou.

Luscious.—Orig. in Brookings, S.D., by Ronald M. Peterson, Agr. Expt. Sta. Introd. in 1973. SD E31 x Ewart; cross made in 1954, selected in 1967, tested as South Dakota 67Sll. Fruit: size medium; pyriform with broad neck; skin thick, tender, attractive rich yellow with occasional small scattered brown russeted areas, sometimes with a pink blush; flesh light yellow, firm, fine texture, melting, very juicy, flavor similar to Bartlett, quality good; ripens 25 Sept. at Brookings; recommended as a dessert variety. Tree: size medium; broad-oval; vigorous; moderately productive and moderately hardy at Brookings; shows more tolerance to fire blight than most varieties, adapted to parts of the northern Great Plains.

Mac.—Orig. in New Brunswick, N.J., by L. Fredric Hough and Catherine H. Bailey, New Jersey Agr. Expt. Sta. Introd. in 1968. Gorham x N.J. 1. Cross made in 1950; first fruited in 1958; selected in 1958; tested as N.J. 6. Fruit: size medium; acute-pyriform; skin straw yellow when tree ripened; flesh creamy white, texture fine, nearly buttery, no indication of astringency, quality good, comparable to Gorham; ripens with Gorham or 2 weeks after Bartlett; as resistant to fire blight as Kieffer, the best quality blight resistant variety selected so far. Tree: vigor below medium; central leader with open branching; not as productive as Lee or Star, pollen good, compatible with other varieties.

Magness.—Orig. in Beltsville, Md., by U.S. Dept. of Agriculture. Introd. for trial in 1960. Released in 1968 by H.J. Brooks. Seckel seedling x Comice; tested as US 3866-E. Fruit: size medium; oval; skin lightly covered with russet, relatively tough, somewhat resistant to insect puncture and decay; flesh soft, very juicy, almost free of grit cells, flavor sweet, highly perfumed, aromatic; ripens at Beltsville about 1 Sept., being a week later than Bartlett; ripens for prime eating in about 10 days when held at 70F; can be held in cold storage up to 3 months, then ripens with good quality. Tree: very vigorous and spreading for a pear; original tree and first trees propagated from it have some thorns, which may be expected to decrease with additional repropagations; begins bearing at about 6 years; early fruiting mainly on medium-long terminals; entirely pollen-sterile, but sets well with all varieties that have been tested; very resistant to fire blight. Recommended for general trial because of high degree of apparent blight resistance and high quality of fruit. Named in honor of John R. Magness who retired in 1959 as chief of the fruit and nut crops section at U.S. Dept. of Agriculture, Agricultural Research Center, in Beltsville, Md.

Manning-Miller.—Orig. in Potsdam, N.Y., on farm of Samuel Miller, but previously owned by the Manning family. Introd. in 1974 by St. Lawrence Nurseries, Heuvelton, N.Y. Parentage unknown, believed to have originated from seed sent by Robert Manning of Massachusetts. Fruit: 3 1/2 inches long x 3 inches wide; pyriform; skin smooth, green to yellow, no russeting or blush; flesh white, smooth, sweet, juicy, flavor good; ripens 1–7 Sept. in New York. Tree: medium-large; vigorous; hardy, withstanding -40F; annual bearer; shows tolerance to fire blight and scab.

Marks (*Marks Seedling*).—Orig. in Geneva, N.Y. Introd. in 1973 by George D. Oberle, Virginia Polytechnic Inst. & State Univ., Agr. Expt. Sta. Parentage unknown. Seed sent to New York State, Agr. Expt. Sta. (NYS, AES), in 1928 by a Mr. Marks, from an area near the Hudson River, for identification; seed planted and germinated by Richard Wellington, NYS, AES, and tested as Marks l; selected in 1940 by George D. Oberle then of NYS, AES, and tested by him in Virginia since 1948. Fruit: small; ovate, short pyriform; skin olive green to yellowish ground color, unattractive, smooth to slightly russeted; flesh creamy white, tender, juicy, very few grit cells, flavor and quality good, flesh qualities resemble Sheldon; ripens 1 Sept. or 1 week before Seckel. Tree: size average or below; upright with profuse branching; moderate vigor and productivity; shows more tolerance to fire blight than most *P.communis* types; recommended for cooler areas of Virginia.

Max-Red Bartlett (*Red Bartlett*).—Orig. in Zillah, Wash., by Arthur Moritz. Introd. in 1945. Plant patent 741; 1 July 1947; assigned to Arthur D. MacKelvie, Zillah. Bud mutation of Bartlett; discovered in 1938 on a tree planted in 1913. Fruit: skin dark cranberry red, turning to an attractive bright red color after picking; flesh finer grained and higher sugar content than parent variety; ripens with Bartlett. Tree: shoots and leaves reddish. Mutation genetic as it transmits the reddish color to a percentage of its seedlings

McIlhenny.—Orig. in Avery Island, La., by E.A. McIlhenny. Introd. in 1933. Bud mutation of Le Conte; discovered about 1917; P.I. 46566. Fruit: large, from 2 1/2 to 4 inches in diam. and 2 1/2 to 3 1/2 inches long; rounded; skin greenish-yellow, medium to thick, waxen, smooth, numerous large russet dots; flesh white, texture medium, flavor mild, moderately juicy, slightly aromatic, sweet to slightly acid, quality fair—better than Le Conte and Kieffer, a few large grit cells near the rather large core. Tree: shows more limb resistance to fire blight than do Le Conte and Kieffer, but resistance at bloom time is similar to these varieties.

Mendel.—Orig. in New Ulm, Minn., by William Pfaender, Jr. Introd. in 1920. Parentage unknown. Fruit: large to medium; pyriform; skin yellow, inclined to russet, some blemish; flesh juicy, firm, yet medium tender, quality good; ripens in mid-September. Tree: upright; vigorous; moderately hardy.

Menie.—Orig. in Ottawa, Ont., Canada, by Canada Dept. Agr., Res. Sta. Introd. in 1947. Kurskaya x Flemish Beauty; selected in 1930. Fruit: large; flavor pleasant; quality good. Tree: hardy; vigorous; slow to come into bearing. Recommended for home gardens in eastern Ontario and Quebec.

Mericourt.—Orig. in Clarksville, Tenn., by Tennessee Agr. Expt. Sta. at Mericourt Expt. Sta., Clarksville. Introd. in 1966. Seckel x Late Faulkner. Cross made in 1938; first fruited in 1947; tested as Tenn. 38S63. Fruit: 2 3/4 x 2 3/8 inches in diam.; short pyriform, necked; skin green or greenish-yellow, occasionally blushed with dark red, dull, waxy in texture, smooth; dots small and brownish; flesh creamy-white, buttery, almost completely lacking in stone cells, abundantly juicy, flavor sprightly subacid, sweet, quality excellent; stem short, 3/4 inch long, moderately thick, well attached in deep, abrupt basin; basin cavity shallow, acute and medium-broad; core very small; calyx open and medium; ripens 25 Aug.–9 Sept. at Highland Rim Expt. Sta., Springfield, Tenn.; recommended for fresh dessert and canning; when canned tends to soften at edges of slices. Tree: vigorous; hardy, withstanding temperatures from a low of –23F to 70s during month of January and –7F to above 75F during February; tolerant to fire blight and leaf spot.

Merton Pride.—Orig. in Bayfordbury, Hertford, Herts, England, by the John Innes Hort. Institution. Introd. in 1959. Glou Morceau x Williams Bon Chretien (tetraploid); selected in 1941. Fruit: large; skin greenish-yellow covered with fine russet; flesh quality excellent; ripens in early midseason. Tree: strong, being capable of carrying heavy crops. Merton Pride is a triploid variety.

Miney.—Orig. in Ottawa, Ont., Canada, by Canada Dept. Agr., Res. Sta. Introd. in 1947. Zuckerbirne x Clapp Favorite; selected in 1930. Fruit: large; flavor pleasing; quality good. Tree: fairly vigorous; hardy at Ottawa; resistant to fire blight. Recommended for home gardens in eastern Ontario and Quebec.

Ming.—Orig. in Brookings, S.D., by N.E. Hansen, South Dakota Agr. Expt. Sta. Introd. in 1927. *Pyrus ussuriensis* Maxim. x *P. communis* 'Louise Bonne de Jersey'; selected in 1917. Fruit: pyriform; about 2 inches in diam.; quality poor. Tree: very hardy; apparently quite resistant to fire blight.

Mock's William.—See Sensation Red Bartlett.

Moe.—Orig. in Ottawa, Ont., Canada, by Canada Dept. Agr., Res. Sta. Introd. in 1947. Zuckerbirne x Clapp Favorite; selected in 1930. Fruit: size medium; quality good; early ripening. Tree: very hardy at Ottawa; quite resistant to fire blight. No longer recommended for home gardens in eastern Ontario and Quebec.

Monterrey.—Orig. in Monterrey, Nuevo Leon, Mexico. Introd. in 1952 by Aldridge Nursery, Von Ormy, Texas. Chance seedling, probably a hybrid of 'Pineapple' with a European pear (*Pyrus communis*); discovered about 1947. Fruit: quite large; round to pyriform; large calyx giving a distinctive appearance; skin yellow; flesh soft, nearly free of grit cells, quality very good for fresh market and canning; ripens in late August at Von Ormy. Tree: resembles pineapple.

Mooers.—Orig. in Knoxville, Tenn., by Brooks D. Drain, Tennessee Agr. Expt. Sta. Introd. in 1954. Duchesse d'Angouleme x Late Faulkner; hybridized in 1934; tested as Tenn. 34S272. Fruit: large; skin golden russet; ripening extends from late September into October; keeps until December under refrigeration; fine for canning. Tree: large; upright; vigorous; resistant to fire blight and leaf spot, pollen sterile.

Moonglow.—Orig. in Beltsville, Md., by U.S. Dept. of Agriculture. Introd. in 1960. US-Mich. 437 x Roi Charles de Wurtemburg; tested as US 353. Fruit: large; attractive; flesh rather soft, moderately juicy, nearly free of grit cells, flavor mild, subacid, rated good; for processing as well as being of good quality for fresh use; ripens at Beltsville in mid-August, about 7 days earlier than Bartlett; ripens for prime eating and processing in 10 days when held at 70F. Tree: very upright; vigorous; heavily spurred; fruits heavily at an early age; appears to be very resistant to fire blight; flowers contain abundant pollen. Recommended for trail in areas where fire blight is a major problem.

Morgan.—Orig. in Knoxville, Tenn., by Brooks D. Drain and Lawson M. Safley, Tennessee Agr. Expt. Sta. Introd. in 1957. Bartlett x Late Faulkner. Fruit: medium to large, 2 3/4 inches in diam.; round oblong, slightly pyriform; calyx open, large; skin tough, thick, light yellow, russeted, slight blush; dots many, large, russeted; core large, closed; flesh white, tinged with yellow, moderately juicy, tender, melting, flavor sweet-subacid, sprightly, quality good; ripens in late August. Tree: large, upright, vigorous; under favorable conditions produces good annual crops; considered resistant to fire blight. Named in honor of Harcourt A. Morgan, President, Univ. of Tennessee, from 1919 to 1933.

Mustabej.—See Mustabey.
Mustabey.—See Mustafabey.
Mustafabey (*Mustabej, Mustabey*).—A small, early ripening pear from Turkey. Introd. into U.S. in 1960 from Belgrade, Yugoslavia, as "Mustabey" (PI 264697), from the Instituto di Frutticoltura in Rome, Italy, in 1967 as "Musta Bey" (PI 324134), collected in Macedonia by T. van der Zwet in 1978 as "Mustabej." The name apparently refers to its origin in the garden of a man named Mustafa. Identical to Arganche from Macedonia and Zaharoasa de Vara from Romania. See Arganche for description.
Mustafabey.—See Arganche.
Nikto.—Orig. in Brookings, S.D., by N.E. Hansen, South Dakota Agr. Expt. Sta. Introd. in 1923. Marillot x *Pyrus ussuriensis* Maxim. Fruit: small, up to 1 1/2 inches in diam.; obtuse pyriform; skin yellow; quality fair. Tree: some susceptibility to fire blight. Now obsolete.
Nixon.—Orig. in Pennsylvania by Richard Nixon. Introd. for trial about 1924. Probably a seedling of Kieffer; discovered about 1924. Fruit: round at apex, tapering abruptly to a point at the stem end; skin green; flesh smooth, quality very good; ripens in late August and early September in Columbia, Mo.; keeping quality poor. Tree: vigorous; bears alternately; resistant to fire blight. Recommended for home planting in Missouri.
Nye Russet Bartlett.—Orig. in Talent, Ore., by Stephen G. Nye of Medford, Ore. Introd. experimentally in 1957. Bud mutation of Bartlett; discovered in 1924. Fruit: skin deep yellow overlaid with a very attractive smooth light golden russet; resembles Bartlett, but ripens 1 week later, firmer, somewhat more spicy than that variety. Tree: similar in performance and fire blight susceptibility to Bartlett.
Okolo.—Orig. in Brookings, S.D., by N.E. Hansen, South Dakota Agr. Expt. Sta. Introd. in 1940. Open-pollinated seedling of *Pyrus ussuriensis* Maxim. Fruit: diam. about 2 1/4 inches; pyriform; skin light yellow with minute russet dots; stem long and stout; flesh white, firm, juicy, with delicious flavor. Tree: bears well.
Olia.—Orig. in Habarovsk, Maritime Province, Russia, by A.V. Lukasheff. Introd. into Manitoba, Canada, in 1940. *Pyrus ussuriensis* x Finland Early Yellow; cross made in 1910; sister seedling of Tioma. Fruit: somewhat smaller than Tioma but sweeter and more aromatic than that variety; quality too low for commercial value; as in Tioma, chief value lies in being a parent for development of large fruit and a hardy tree. Tree: very similar to Tioma; hardy.
Orcas.—A fall pear suitable for fresh consumption, canning, and drying. Discovered on Orcas Island, Wash., by Joseph C. Long in 1966. Roadside seedling of unknown parentage. Fruit: medium to large, 65 to 80 mm in diam., elongated, with broad base, slightly lumpy; short stem; skin yellow with green dots and red-orange blush. Flesh creamy white, juicy, mild flavor, small core. Ripens 2 days after Bartlett in western Washington; stores 11 weeks at 1C. Tree: upright, becoming spreading; blooms 5 days after Bartlett; partially self-fruitful; pollen cross-compatible with Bartlett.
Orcross.—Orig. in Medford, Ore., by F.C. Reimer, Oregon Agr. Expt. Sta. Introd. in 1945. Oregon 18 x Farmingdale. Used occasionally as fire blight resistant rootstock.
Oregon 18.—Orig. in Medford, Ore., by F.C. Reimer, Oregon Agr. Expt. Sta. Introd. in 1939. Open-pollinated seedling of a French seedling (*Pyrus communis*); selected in 1926; tested as Ore. P-18. Fruit: small; flesh hard, astringent; of no commercial value. Rootstock: fire blight resistant; used as parent for rootstock breeding studies.
Orient.—Orig. in Chico, Calif., by Walter Van Fleet, Plant Introd. Garden, U.S. Dept. of Agriculture. Introd. in 1945 through the Tennessee Agr. Expt. Sta. *Pyrus communis* x *Pyrus* sp. from China; direction of cross unknown; P.I. 64224. Fruit: nearly round, averaging 3 inches long and 2 3/4 inches in diam.; flesh firm, juicy, slightly sweet, lacking in flavor; good for canning; ripens 15–18 Aug. at Knoxville, Tenn. Tree: produces annual medium-sized crops; resistant to fire blight; mostly of interest in Tennessee and southward.
Ovid.—Orig. in Geneva, N.Y., by Richard Wellington, New York State Agr. Expt. Sta. Introd. in 1931. Bartlett x Dorset; cross made in 1912; sibling of Willard. Fruit:

large; pyriform, neck rather long; skin thick, medium tender, smooth, deep yellow with broad bright blush resembling Bartlett but with russet patches, attractive; flesh fine-grained, tender, nearly white, juicy, melting, sweet, agreeably flavored, quality good; core small; ripens in December at Geneva; keeps late in storage, resembles Dorset. Tree: large, vigorous, upright spreading; hardy; productive; blooms early.

Packham's Triumph.—Orig. in Molong, New South Wales, Australia, by Charles Henry Packham (1832–1909). Introd. at the turn of the century; introd. into the U.S. as P.I. 43185 in 1916 from Auckland, New Zealand; as P.I. 155945 in 1945 from Buenos Aires, Argentina; as P.I. 157506 in 1946 from Angers, France; and as P.I. 280405 in 1965 from France; first commercial planting in Oregon made in 1950. Uvedale St. Germain x Williams Bon Chretien (Bartlett); the selection from the cross was made in 1896 (1897?). Fruit: large; usually obtuse pyriform; surface slightly uneven; skin thin, lemon-yellow when ripe; flesh texture very fine, very juicy, melting, flavor particularly rich and pleasing; one of best varieties for cool storage, keeping 4 to 5 months in fine condition; stem length medium, moderately stout, usually curved, set in a shallow depression; ships extremely well; matures about 30 days after Bartlett. Tree: upright; moderately vigorous; good, consistent bearer; highly susceptible to fire blight; susceptible to stony-pit virus.

Para de Zahar.—See Zaharoasa de Vara.

Parker.—Orig. in Excelsior, Minn., by the Univ. of Minnesota Fruit Breeding Farm. Introd. in 1934. Open-pollinated seedling of a Manchurian cultivated pear (probably *Pyrus communis*); selected in 1920; tested as Minn. 1. Fruit: medium to large; pyriform, rounded; skin yellow, blushed red; flesh whitish, juicy, pleasant, sweet, quality good; season late September. Tree: fairly hardy; upright, vigorous; susceptible to fire blight.

Patten.—Orig. in Charles City, Iowa, by the late C.G. Patten. Named and introd. in 1922 by H.L. Lantz, Iowa State College. Orel 15 x Anjou; selected probably about 1915. Fruit: large; juicy; if picked 7 to 10 days before maturity, it ripens well and has excellent quality for dessert; fair as a canned product. Tree: among the hardiest of large-fruited varieties; valuable only in the colder areas of the Upper Mississippi Valley where Bartlett, Anjou, and other standard varieties cannot be grown; moderately resistant to fire blight. A good variety for the home orchard and for local markets.

Peter.—Orig. in Saskatoon, Saskatchewan, Canada, by C.F. Patterson, Univ. of Saskatchewan. Introd. in 1960 for home gardens. *Pyrus ussuriensis* Maxim. x Aspa; selected in 1958; tested as Sask. PR 6. Fruit: 2 1/8 inches long and 2 inches in diam. under nonirrigated field conditions at Saskatoon; skin moderately thin; quality very fair; ripens the last 2 weeks in September. Tree: hardy.

Phelps.—Orig. in Geneva, N.Y., by Richard Wellington, New York State Agr. Expt. Sta. Introd. in 1925. Winter Nelis x Russet Bartlett; cross made in 1912. Fruit: skin color duller than Bartlett; flesh juicy, tender, pleasant vinous flavor; ripens 1 month later than Bartlett, and keeping until Christmas. Tree: spreading; vigorous and productive.

Phileson.—Orig. in Ottawa, Ont., Canada, by Canada Dept. Agr., Res. Sta. Introd. in 1947. Kurskaya x Flemish Beauty; selected in 1930. Fruit: large; attractive; quality good. Tree: hardy at Ottawa; vigorous; slightly susceptible to fire blight. Recommended for home gardens in eastern Ontario and Quebec.

Philip.—Orig. in Saskatoon, Saskatchewan, Canada, by C.F. Patterson, Univ. of Saskatchewan. Introd. in 1960 for home gardens. *Pyrus ussuriensis* Maxim. x Aspa; selected in 1958; tested as Sask. PR 4. Fruit: 2 3/4 inches long and 2 1/2 inches in diam. under nonirrigated field conditions at Saskatoon; skin tender; very fair quality; ripens during the last 2 weeks of September. Tree: hardy.

Pioneer 2.—Orig. in Brooks, Alberta, Canada, by A.L. Young. Introd. in 1936. Parentage unknown. Fruit: small, pyriform, shape slightly irregular; flesh firm, rather coarse and granular, flavor mild, sweet, slightly aromatic. Tree: tall, upright; moderately hardy.

Pioneer 3.—Orig. in Brooks, Alberta, Canada, by A.L. Young. Introd. in 1934. Parentage unknown; fruited first in 1931. Fruit: small; oval to pyriform; skin greenish-yellow, blushed dull red; flesh whitish, medium firm, granular, juicy, sweet, quality fair, spicy as dessert fruit, fair as sauce; season mid-September. Tree: upright; vigorous; hardy; productive.

Pontotoc.—Orig. in Pontotoc, Miss., by J.C.C. Price, Mississippi Agr. Exp. Sta., State College, Miss. Introd. in 1926. Parentage unknown; discovered in 1922. Fruit: size medium, round; flesh tender, fair for canning. Tree rest period short; field resistance to fire blight very good.

Potomac.—A fire blight-resistant mid-season pear with good fruit quality. Released in 1993 jointly by R.L. Bell and T. van der Zwet, U.S. Dept. of Agriculture, Appalachian Fruit Research Station, Kearneysville, W.Va., and The Ohio State Univ. Moonglow x Beurre d'Anjou; cross made by H.J. Brooks in Beltsville, Md., in 1961; tested as U.S. 62537-048. Not patented. Fruit: medium size, 65 mm in diam.; ovate-pyriform; skin light green, glossy. Flesh moderately fine, buttery with some small grit under skin; flavor subacid and mild, similar to Anjou. Ripens 2 weeks after Bartlett; stores for 8 weeks or less. Tree: medium size; precocity and productivity similar to Anjou; fire blight resistance greater than Seckel.

Pulteney.—Orig. in Geneva, N.Y., by Richard Wellington, New York State Agr. Expt. Sta. Introd. in 1925. Winter Nelis x Russet Bartlett; cross made in 1912. Fruit: ripens 4 weeks later than Bartlett with fruit a little more regular in outline, skin smoother, but not as attractive in color as Bartlett. Tree: open, spreading; vigor medium; productive.

Pushkin.—Orig. in Brookings, S.D., by N.E. Hansen, South Dakota Agr. Expt. Sta. Introd. in 1920. *Pyrus ovoidea* x R K553. Fruit: small, 1 1/2 inches in diam.; roundish; skin green, blushed dull red, thick, tough; flesh yellow, coarse, juicy, firm, quality fair; ripens in early September.

Red D'Anjou (*Gebhard Red Anjou, LeRoi*).—Variety name changed from LeRoi to Red D'Anjou, and plant patent rights assigned to Columbia & Okanogan Nursery, Wenatchee, Wash. Orig. in Medford, Ore., by Harold L. Gebhard and Edward W. Gebhard. Plant patent 1992; 6 Dec. 1960; H.L. Gebhard and E.W. Gebhard; by Lucille M. Gebhard, Central Point, Ore., and Marjorie A. Grigsby, Medford, Ore. (executrices). Red bud mutation of Anjou. Fruit: overall dark red skin; otherwise similar to parent. Tree: leaf veins and bark with reddish overcast.

Red Doyenné du Comice.—See Regal Red Comice.

Red Hardy.—See Royal Red.

Redspire.—Ornamental *Pyrus calleryana* cultivar. Orig. in South Brunswick Township, N.J., by William Flemer, III, Princeton, N.J. Plant patent 3815; 16 Dec. 1975; assigned to Treesearch, Kingston, N.J. Open-pollinated seedling of Bradford. Fruit: very small, not abundant, self-fruitful, egg-shaped, very astringent. Tree: small pyramidal; hardy, survived winters in central Vermont.

Reed.—Orig. in Shenandoah, Iowa, by a Mr. Reed. Introd. in 1937. Considered to be *Pyrus communis* x some oriental species; discovered about 1937. Fruit: of no value. Tree: possibly some fire blight resistance, but this is now considered doubtful.

Regal Red Comice (*Red Doyenné du Comice*).—Orig. in Phoenix, Ore., by Charles Lyle Kinney, Medford, Ore. Introd. in 1965. Plant patent 2527; 1 June 1965; assigned to Bear Creek Orchards, Medford, Ore. Bud mutation of Doyenné du Comice; discovered in 1960. Fruit: similar in size, shape, conformation and character to that of Doyenné du Comice; color overall red with slight purplish tint when ripe, has a mottled greenish-yellow undercoat when it approaches ripeness; flesh white, fine, melting when ripe; eating quality excellent. Tree: identical to Comice, except for darker foliage.

Reimer Canal.—See Canal Red.

Reimer Red.—Orig. in Medford, Ore., by Frank C. Reimer. Plant patent 2054; issued 25 Apr. 1961 to Frank C. Reimer. Named and introduced by Stark Bros. Nurseries. Comice x Max-Red Bartlett. Fruit: large; skin yellow ground color, overspread with red; flesh white, fine-grained; resembles Max-Red Bartlett ripening about 3 weeks later. Tree medium to large; vigor moderate.

Rescue.—An attractive, early ripening, large-fruited pear suitable for canning and fresh use. Orig. near Vancouver, B.C., by a Mr. Abrey. *Pyrus communis* type of unknown parentage. K. Nomura rescued scions from the original tree in 1976 before it was destroyed, after death of its owner. Fruit: large to very large, 75 to 85 mm in diam., often 110 mm long; pyriform, symmetrical, with wide base and small core; short, fleshy stem. Skin bright yellow with orange-red blush; flesh cream-colored, smooth, melting, juicy, and mild, similar to Clapp Favorite. Ripens 6 days before Bartlett; stores about 12 weeks at 1C. Tree: upright; moderately vigorous; few branches and spurs. Annual bearer; pollen cross-compatible with Bartlett; blooms about 3 days after Bartlett.

Richard Peters.—Orig. in State College, Pa., by E.L. Nixon, Pennsylvania Agr. Expt. Sta. Introd. in 1927. Probably an open-pollinated seedling of Kieffer; selected in 1924. Fruit: quality better than Bartlett, which it resembles; ripens early. Tree: practically immune to fire blight; vigorous; self-unfruitful, productive when properly cross-pollinated. Named in honor of Richard Peters, founder and former president of Philadelphia Society for Promoting Agriculture.

Rogue Red.—Orig. in Medford, Ore., by F.C. Reimer, E. Degman, and V. Quackenbush, S. Oregon Expt. Sta. Introd. in 1969. Comice x [Seckel x Farmingdale seedling 122]; cross made in 1947; selected in 1955; tested as 5-235. Fruit: similar in shape to Comice; skin is covered with 60% to 80% red blush; flesh quality excellent; late winter variety to supplement Anjou, Packham's Triumph and El Dorado; harvest period coincides with Comice and Bosc, but may be ripened any time up to May; following cold storage, variety is ripe and ready to eat after 7 to 10 days at room temperature; storage scald has not been observed, and virtually no bruises or skin abraisions appear. Tree: unusually upright; vigorous; productivity moderate, but regular; foliage light green and rolled; compatible with French and quince rootstocks; moderately susceptible to fire blight; symptoms of stony pit have been found when grafted on infected trees; blooming period overlaps Anjou and Bartlett; pollen of Bartlett, Bosc, and Comice twice as effective as Anjou.

Rosemarie.—The first pear cultivar bred in the Republic of South Africa. Released in 1990 by P.R. Jolly, J.D. Stadler, P.F. Louw, W.J.C. Smith, T.R. Visagie, and P.C. Fourie, Stellenbosch. Bon Rouge (a red-skinned Bartlett) x Forelle; cross made in 1974 by N. Hurter, P. Louw, and C. Smith at Simondium, Cape Province. Red-leaved seedlings were selected from the progeny and planted in Bien Donne in 1977. Tested as selection OR8-25-38. Propagation restricted by South African plant breeder's rights. Fruit: small to medium; skin yellow with bright orange-red blush, very attractive; shape similar to Bartlett; early ripening. Tree: vigorous; semi-spreading.

Rosired Bartlett.—Orig. in Placerville, Calif., by James W. Sweeney. Introd. in 1963. Plant patent 2012; 24 Jan. 1961; assigned to Fowler Nurseries, Newcastle, Calif. Bud mutation of Bartlett; discovered about 1948. Fruit: size comparable to Bartlett and Max-Red Bartlett; oblong, obtuse, pyriform, tapering toward apex, symmetrical, uniform; skin thin, tender, maroon red at maturity, dots smaller and much less conspicuous than on Bartlett or Max-Red Bartlett (this characteristic especially distinguishing it from Max-Red Bartlett); flesh white, similar to Bartlett, juicy, fine-grained, melting, buttery, flavor very good, aromatic, mild; dessert and canning quality good. Tree: size medium, generally similar to Bartlett; vigorous; hardy; upright; regular bearer; very productive.

Royal Red (*Red Hardy*).—Orig. in Milpitas, Calif., by Victor A. Silvera. Introd. in 1940. Plant patent 380; 23 Apr. 1940. Bud mutation of Beurre Hardy; discovered in 1934. Fruit: identical with parent, except for red skin. Tree: new shoot growth dark reddish-brown; older wood-tinged brown.

Russet Bartlett.—Orig. in Jordan Harbor (now Vineland Station), Ont., Canada, by the late N.P. Moyer. Introd. in 1927. Bud mutation of Bartlett; discovered in 1918. Fruit: skin russetted; may be held somewhat longer than Bartlett in ordinary storage; in CA storage, it does not bruise as much as Bartlett; resembles parent. Tree: blooms 1 to 2 days later than Bartlett, flowers smaller with more pink; defoliates earlier than Bartlett.

Saponsky.—Orig. in Saponsky, Siberia; brought to U.S. by Frank N. Meyer, agricultural explorer, in 1907. Introd. in 1924 by N.E. Hansen, South Dakota Agr. Expt. Sta., Brookings, as P.I. 20336. *Pyrus ussuriensis.* For rootstock purposes only: seedlings hardy; tolerant to fire blight.

Sakarnaia.—See Zaharoasa de Vara.

Sensation Red Bartlett (*Mock's William*).—Believed to have originated in East Burwood, Victoria, Australia, by W. Mock. Introd. from Australia in 1959 by H. E. Franklin, Pozieros, Queensland, Australia. Parentage unknown; possibly bud mutation of Bartlett; received in the U.S. in 1959 as P.I. 255600 and P.I. 292507; discovered about 1940. Fruit: 3 x 2 5/8 inches in diam.; ovate pyriform; skin yellow, 80% to 95% dark red blush, rough and bumpy; dots few, medium, obscure; flesh creamy white, melting, tender, moderately juicy, sweet, quality good. Tree: size medium; upright; somewhat less vigorous than Bartlett; branches stocky; possibly heavier than Bartlett, but less linear growth, smooth reddish-brown, some scarf skin; branchlets short, glabrous, smooth, glossy.

Shea.—Origin unknown. Introd. about 1928 by the Cole Nursery, Painesville, Ohio. *Pyrus communis*; apparently a seedling, possibly of Sheldon, obtained by Cole Nursery before 1928. Fruit: size above medium; obovate to obtuse pyriform; skin yellow with a fine russet similar to Sheldon; flesh white, soft to firm, tender, juicy, flavor good to very good, sweet, stone cells confined to core area and not objectionable; ripens during third week of September, about 3 weeks after Bartlett; keeps well for short period; resembles Sheldon somewhat in appearance although surface is roughened. Tree: some susceptibility to fire blight.

Sierra.—Orig. in Summerland, B.C., Canada, by Canada Dept. of Agr. Introd. in 1969. Bartlett x Marguerite Marillat. Cross made in 1947, selected in 1956, tested as 9R-1-13. Fruit: large; long, pyriform; skin smooth, thin, very tender, easily develops russeting from bruises or frost; green when picked, yellowish green when ripe; flesh very fine and smooth, very sweet, juicy, excellent quality; ripens with Anjou, or third week of September, keeps in good condition at 31F until February. Tree: large; irregular shape, spreading; vigorous; hardier than Bartlett; very productive; recommended for home gardens or commercial plantings, requiring the best care in growing and handling of fruit.

Simon.—Orig. in Saskatoon, Saskatchewan, Canada, by C.F. Patterson, Univ. of Saskatchewan. Introd. in 1960 for home gardens. *Pyrus ussuriensis* Maxim. x Aspa; selected in 1958; tested as Sask. PR 9. Fruit: 2 1/2 inches long and 1 7/8 inches in diam. under nonirrigated field conditions at Saskatoon; skin moderately thin; quality fair; ripe during the last 2 weeks of September. Tree: hardy.

Sirrine.—Orig. in Riverhead, Long Island, N.Y., by F. Atwood Sirrine. Introd. in 1969. Open-pollinated seedling of Bartlett. Discovered in 1954, tested as Sirrine 1 by the New York State Agr. Expt. Sta., Geneva. Fruit: size medium, 2 3/8 to 2 1/2 inches in diam.; pyriform, short; skin bright yellow, with an occasional bronzy blush, smooth, thin; flesh smooth, melting, juicy, flavor sweet and rich; ripens about first week in September, with Bartlett; will store in 32F until November. Tree: large; round-topped; vigor good; productive; slightly more tolerant to blight than Bartlett. Recommended for home orchard and local market.

Sodak.—Orig. in Brookings, S.D., by N.E. Hansen, South Dakota Agr. Expt. Sta. Introd. in 1945. *Pyrus ussuriensis* Maxim. x *Marguerite Marillat*. Fruit: 2 x 2-1/2 inches; oblong-obovate-pyriform; flesh white, very juicy, melting, buttery. Tree: vigorous; probably obsolete.

Southworth.—Scionwood brought from near Duluth, Minn., by a Great Lakes captain sometime between 1900–10 and grafted on two trees at his home near Massena, N.Y. Introd. in 1968 by Fred L. Ashworth, St. Lawrence Nurseries, Heuvelton, N.Y. Brought to Ashworth's attention in 1967 by Frank Southworth of Massena, N.Y., present owner of the property. Parentage unknown. Fruit: medium large, as large as Bartlett; shaped like Duchesse d'Angouleme; skin green, rarely blushed; flesh has very little grit cells, flavor good, not as sweet as Bartlett; seeds per fruit average about three; ripens 20–30 Sept. in New York; some fire blight resistance. Tree: medium size; upright, vigorous; extremely hardy; productive; self-fertile. Original trees are still strong and vigorous, showing no sign of winter injury.

Spalding.—Orig. in Spalding County, Calif., by J.W. Daniell, R.P. Lane, W.A. Chandler, and Gerard Krewer, Univ. of Georgia, Experiment. Introd. in 1982. Probably Pineapple x unknown. Selected in 1957. Fruit: resembles that of Magness; large; round-pyriform; flesh creamy-white, fine texture, not buttery, juicy with grit cells only around the core; flavor good, subacid, light aroma; skin yellow, lightly russet with brown dots, no blush, smooth. Tree: size similar to that of Magness; flowers self-fertile; productive; resistant to fire blight.

Spartlet.—Orig. in Farmington, Mich., by Edward Palacky. Introd. in 1964. Plant patent 3705; issued 15 Apr. 1975 to Hilltop Orchards and Nurseries, Hartford, Mich. Chance seedling of Bartlett. Discovered in 1963, tested since 1964 as Bartlett-x by Michigan State Univ. and Gerber Products Co. Fruit: larger than Bartlett, with large basal end; skin smooth, green-yellow at maturity, often with a bright scarlet blush on exposed side, dotted with numerous pin-pointed russet lenticels; flesh aromatic, fine-grained to granular in nonripened state, smooth to buttery and slightly fibrous when fully ripened, quality rated closely to Bartlett but slightly coarser; ripens mid-September, 12 days after Bartlett; has stored well until mid-February at 31F, makes acceptable packs of puree and sliced canned halves. Tree: large, spreading to upright; moderately vigorous; hardy; productive; shows more tolerance to fire blight than Bartlett. Recommended as a complementary choice to Bartlett not as an alternate. The name is a concatenation of Bartlett and Spartan, the Michigan State Univ. athletic teams.

Stanley.—Introd. into the U.S. in 1939 from Australia as P.I. 133594. Josephine de Malines x Winter Nelis. Fruit: above medium to large; shape irregular, obovate pyriform to turbinate; skin greenish-yellow, russetted; flesh white to yellow, crisp to tender, medium dry, sweet to subacid; stone cells not objectionable, confined to core area; flavor fair to good; keeps well into December; harvested during last week of September in midwest.

Star.—Orig. in New Brunswick, N.J., by L. Fredric Hough and Catherine H. Bailey, New Jersey Agr. Expt. Sta. Introd. in 1968. Beierschmidt x N.J. 1. Cross made in 1950; first fruited in 1958; selected in 1958; tested as N.J. 3. Fruit. size above medium; pyriform; skin light green; flesh white, juicy, fine-grained, but not buttery, quality good; pleasant to eat as it ripens on the tree; internal breakdown observed some seasons; ripens with Clapp Favorite; the earliest ripening, good quality variety presently available with any degree of resistance to fire blight, but less resistant than Kieffer. Tree: upright; moderately strong; productive, good pollen, compatible with other varieties.

Starking Delicious.—Orig. in Miami County, Ohio, by Marvin A. Cook of Tipp City, Ohio. Introd. in 1953 by Stark Bros. Nurseries & Orchards Co., Louisiana, Mo. Plant patent 1044; 9 Oct. 1951; assigned to Stark Bros. Nurseries & Orchards Co. Parentage unknown; discovered in 1930. Fruit: large; appearance attractive; flesh firm, white, sweet, juicy, quality excellent. Tree: displays resistance to fire blight. Apparently identical to Maxine.

Stark Jumbo.—Orig. in Ukiah, Calif., by Louis F. Johnson. Introd. in 1975. Plant patent 3928; issued 29 June 1976 to Stark Bros. Nurseries & Orchards Co., Louisiana, Mo. Limb mutation of Bartlett. Fruit: large, 4 1/2 inches; oblong, obtuse, pyriform; flesh juicy, melting, sweet; recommended for dessert or canning; ripens 15–20 Aug. Tree: semi-dwarf; low spreading, compact; vigor and productivity medium; hardy.

Starkrimson (*Kalle*).—Orig. in South Haven, Mich., by Adrian G. Kalle. Introd. in 1956 by Stark Bros. Nurseries & Orchards Co., Louisiana, Mo. Plant patent 1095; 13

May 1952; assigned to Stark Bros. Nurseries & Orchards Co. Bud mutation of Clapp Favorite; discovered in 1939. Fruit: skin color solid red, otherwise resembles Clapp Favorite; ripens about 26 Aug. Tree: identical to parent variety.

Stewart Bartlett.—Orig. in Wenatchee, Wash., by Howard M. Stewart. Introd. in 1961 by Van Well Nursery, Wenatchee. Bud mutation of Bartlett; discovered in 1952. Fruit: resembles Bartlett. Tree: the wood has some resistance to fire blight, however, considered susceptible under eastern U.S. conditions.

Summercrisp.—An early season pear with good resistance to fire blight; among the hardiest tested at Univ. of Minnesota. Originally received by Univ. of Minnesota Fruit Breeding Farm, Excelsior, in 1933 from John Gaspard, Caledonia, Minn. Released in 1987 by J.J. Luby, D.S. Bedford, E.E. Hoover, S.T. Munson, W.H. Gray, D.K. Wildung, and C. Stushnoff. Parentage unknown; ancestry includes *P. ussuriensis* and *P. communis*. Tested as Gaspard #5 and N33201. Not patented. Fruit: pyriform; 60 to 80 mm diam., 80 to 100 mm long; skin green with red blush and prominent red lenticels. Flesh crisp, juicy, mild flavor. Stores for 6 weeks under refrigeration; develops internal browning when allowed to ripen before storage. Not suitable for canning. Tree: medium size; conical shape; blooms with Patten and Parker; requires cross-pollination.

Sungari.—Orig. in Brookings, S.D., by N.E. Hansen, South Dakota Agr. Expt. Sta. Introd. in 1940. Vinnaja Selenaja x *Pyrus ussuriensis* Maxim. Fruit: 2 to 2 1/2 inches in diam.; oblong, pyriform; stem long; skin yellow with faint russet dots; flesh tender, very juicy when fully ripe, pleasant.

Sure Crop.—Orig. in Bentonville, Ark., by Ben Davis, Ozark Nursery Company, Tahlequah, Okla. Introd. in 1930. Chance seedling of unknown parentage; selected in 1917; one of Davis' close friends, Dr. Cotton of Bentonville, was riding on the railroad at the turn of the century; he ate a pear, put the seeds of this pear in his pocket, and later planted them; this variety is the result. Fruit: large; resembles Bartlett. Tree: annual bearer; blooms late; more resistant to fire blight than many other varieties.

Tait 1.—Orig. in Cartertown, St. Joseph, Ont., Canada, by David Tait. Introd. about 1920. Open-pollinated seedling of Kurskaya; selected about 1905. Fruit: small, up to 2 inches in diam.; pyriform; skin greenish-yellow; flesh firm, sweet, tender, juicy, quality good although some noticeable stone cells; season mid-September. Tree: hardy, withstanding –42F; vigorous; upright-spreading; usually productive.

Tait 2.—Orig. in Cartertown, St. Joseph, Ont., Canada, by David Tait. Introd. about 1920. Open-pollinated seedling of Warner. Fruit: similar to Tait 1, but larger in diam.; flavor rich. Tree: extremely hardy, but less so than Tait 1.

Tait-Dropmore.—Orig. in Cartertown, Ont., Canada, by David Tait. Introd. in 1928 by Skinner's Nursery, Dropmore, Manitoba, Canada. Open-pollinated seedling of Patten; discovered about 1935. Fruit: small, up to 2 inches in diam.; pyriform; irregular; skin greenish-yellow, blushed dull carmine, thick; flesh dull yellow, coarse, gritty; flavor sweet, pleasant as dessert fruit, fair as sauce; season early to mid-September. Tree: upright-spreading; vigorous; very hardy; partially resistant to fire blight; productive.

Tanya.—Orig. in Brookings, S.D., by N.E. Hansen, South Dakota Agr. Expt. Sta. Introd. in 1938. Ideal x *Pyrus ussuriensis*. Fruit: small; skin deep red; flesh quality poor; ripens late; keeps well. Tree: hardy; vigorous; appears to be resistant to fire blight. Now obsolete.

Texking (*Cow*).—Orig. in Ellis County, Texas, by a Mr. Weaver. Introd. in 1945. Parentage unknown; discovered about 1938 as a rootstock shoot. Fruit: large; skin color solid russet; similar to Delicious apple in shape; flesh sweet, not soft, free from stone cells; bears in clusters. Tree: heavy and early producer; good keeper; fire blight resistant.

Thomas.—Orig. in Saskatoon, Saskatchewan, Canada, by C.F. Patterson, Univ. of Saskatchewan. Introd. in 1960 for home gardens, *Pyrus ussuriensis* Maxim. x Bartlett; selected in 1959; tested as Sask. PR 17. Fruit: 2 1/4 inches long and 2 1/4 inches in diam. under nonirrigated field conditions at Saskatoon; flesh very firm, quality fair; keeps well in cold storage; ripens during late September to early October. Tree: hardy.

Tioma.—Orig. in Habarovsk, Russia, by A.V. Lukasheff. Introd. into Manitoba, Canada, in 1940. *Pyrus ussuriensis* x Finland Early Yellow; cross made in 1910; sister seedling of Olia. Fruit: large, 2 3/4 to 3 inches long and 2 1/2 to 2 3/4 inches in diam.; pyriform to oval; skin light green to creamy green, blush faint; flesh white to creamy, coarse, tender, medium juicy, flavor moderately acid, lacking the sweet pear flavor, lacks quality; ripens in early September, season short; storage life up to 2 months. Tree: hardy; similar to Olia but perhaps more productive.

Ubileen Gift.—A large-fruited, early ripening pear from Bulgaria. Orig. Institute for Fruit Growing, Kustendil, Bulgaria, by Vasil Georgiev in 1957. Released in 1984, introduced into the U.S. in 1974 (PI 392323). Clapp Favorite x Klementinka. Fruit: large to very large (about 230 g), pyriform, ripening in late July; skin yellow with red blush; flesh yellow, fine-textured, buttery, sweet, juicy, subacid, and aromatic; can be stored for 4 to 6 days at room temperature. Similar to Butirra Precoce Morettini in appearance and ripening season. Tree: large, vigorous, resistant to scab.

Valya.—Orig. in Brookings, S.D., by N.E. Hansen, South Dakota Agr. Expt. Sta. Introd. in 1938. Lincoln x Russian sand pear. Fruit: up to 2 inches in diam.; round, tapering toward stem; skin yellow, dots small and russetted; quality fair.

Vistica Nectar.—Orig. in Yuba City, Calif., by Louis Vistica. Introd. in 1952. Plant patent 1153; 9 Dec. 1952; assigned to Louis Vistica Nurseries. Farmingdale x Burkett. Fruit: larger, more symmetrical, smoother flesh than Burkett; matures between 1 and 15 Aug. at place of origin. Tree: highly resistant to fire blight; vigorous; productive.

Waite.—Orig. in Arlington, Va., by M.B. Waite, U.S. Dept. of Agriculture. Introd. in 1938. Parentage unknown; selected about 1920; tested as USDA 66131. Fruit: almost as large as Bartlett, resembling it in shape; flesh smooth, almost free of grit cells, more acid than Bartlett; apparently lacks quality; good for cooking and canning, fairly good for dessert; ripens about Kieffer time. Tree: good fire blight resistance; growth rather weak; not very productive; cross-pollination essential.

Warren.—Orig. at Hattiesburg, Miss., by T.O. Warren as chance seedling. Introd. in 1976. Fruit: medium to large; shape variable; skin dull brown, sometimes with red blush, smooth; flesh whitish, buttery, smooth, moderately firm; flavor said to be comparable to Magness or Comice; ripens about with Magness; stores much better than Bartlett. Tree: vigorous; pyramidal, with flat crotch angles; cold hardy; tolerant of high summer temperatures; resistant, but not immune, to fire blight; disease-free foliage.

Willard.—Orig. in Geneva, N.Y., by Richard Wellington, New York State Agr. Expt. Sta. Introd. in 1981. Bartlett x Dorset; cross made in 1921; sibling of Ovid. Fruit: large; oblong pyriform, uniform, similar to Bartlett; skin thick, tough, surface uneven, clear yellow, fine-grained, tender, juicy, sweet, with a piquant flavor; quality good; core small; ripens late, 2 weeks after Ovid. Tree: large; vigorous; upright-spreading; hardy; productive.

Zachman.—Orig. in Rogers, Minn., by L.M. Zachman. Introd. in 1929. Considered to be an open-pollinated seedling of Flemish Beauty. Fruit: large; skin green, blushed; flavor good; flesh with few grit cells; similar to Parker; resembles Flemish Beauty. Tree: hardy; productive. No longer available.

Zaharoasa de Vara.—See Arganche.

Zaharoasa de Vara (*Para de Zahar or De Zahar in Romania, Sakarnaia in Russia*).—A small, early ripening pear from Romania. Introd. by L.F. Hough to U.S. in 1970 (PI 352661). Orig. possibly in Ukraine. Name refers to the sweet early season fruit (zahar = sugar, vara = summer), apparently identical to Arganche from Macedonia, and Mustafabey from Turkey. See Arganche for description.

not named.—Orig. in northern Wisconsin by Rudolf N. Ruedlinger. Not commercially introd. Plant patent 309; 17 Jan. 1939; assigned to Ruedlinger Nursery, St. Louis Park, Minn. This tree has not proved worthy; it will not be introduced to the trade. Information is recorded here because this pear tree was patented.

not named.—Orig. in Youngstown, Ohio, by Roy C. Bowman. Plant patent 757; 2 Sept. 1947. Probably Kieffer x Anjou. Fruit: shape similar but smaller than Anjou; keeping quality good; faint core; seeds few; thick, permanent stem. Tree: long, slender, tough stems; little suckering; somewhat fire blight resistant; blossoming is in two cycles about 2 weeks apart. Included here because of plant patent.

Rootstocks for Pears

BP-1.—A semi-dwarfing, precocious rootstock from Republic of South Africa. Grown from seed of a wild pear tree at Elsenberg in 1928. Selected and evaluated by A.F. deWet at Fruit and Fruit Technology Research Inst., Stellenbosch; released in 1974. Propagated with some difficulty by hardwood cuttings. Susceptible to fire blight.

BP-2.—A clonal rootstock introd. in Republic of South Africa in 1974. From seed taken from wild pear tree in Elsenberg in 1928. Selected and evaluated by A.F. deWet, Fruit and Fruit Technology Research Inst., Stellenbosch. Trees on BP-2 are standard size, precocious.

BP-3.—A clonal pear rootstock introd. in Republic of South Africa in 1979. From seed taken from wild pear tree in Elsenberg in 1928. Selected and evaluated by A.F. deWet, Fruit and Fruit Technology Research Inst., Stellenbosch. Trees on BP-3 are standard size, precocious and very productive. Adaptable to variety of soil types, from sandy loam to clay loam. Hardwood cuttings propagate satisfactorily with bottom heat. Susceptible to fire blight.

***Brooks Selections*®.**—See OHxF Series.

Daygon.—See OHxF 40.

Daynir.—See OHxF 69.

Dayre.—See OHxF 282.

Daytor.—See OHxF 87.

OHxF Series (*Brooks Selections*®).—Orig. near Forest Grove, Ore., by Lyle A. Brooks from seed obtained at Agriculture Canada Res. Sta., Summerland, B.C. Old Home x Farmingdale. From several thousand seedlings, Brooks, M.N. Westwood, and P. Lombard selected and evaluated in the orchard >500 candidates. Thirteen have been introduced commercially; all are resistant to fire blight, pear decline, and woolly pear aphids; they produce no or few suckers; propagation (with difficulty) is by hardwood cuttings. The Brooks Selections are compatible with *Pyrus communis* varieties and appear to be compatible with Asian pear varieties. Patented clones are assigned to Carlton Plants; all carry the "Brooks Selections®" trademark.

OHxF 18.—Vigorous.

OHxF 40 (*Daygon*).—Plant patent 5412; 5 Mar. 1985. Semi-dwarfing.

OHxF 51.—Dwarfing, but less so than EM Quince A. Only moderately winter hardy.

OHxF 69 (*Daynir*).—Plant patent 5559; 14 Sept. 1985. Intermediate vigor. Induces early, heavy production, especially of Bartlett.

OHxF 87 (*Daytor*).—Plant patent 6362; 1 Nov. 1988. Semi-vigorous.

OHxF 97.—Vigorous. Induces early, heavy production.

OHxF 217.—Plant patent 5468; 14 May 1985. Semi-vigorous.

OHxF 266.—Semi-dwarfing.

OHxF 267.—Semi-vigorous.

OHxF 282 (*Dayre*).—Plant patent 5573; 15 Oct. 1985. Nearly as vigorous as seedling but more productive.

OHxF 288.—Semi-vigorous.

OHxF 333.—Intermediate vigor. Induces early production. Fruit may be slightly smaller than from trees on OHxF 69.

OHxF 514.—Vigorous.

OHxF 516.—Vigorous.

Pear, Asian

Chojuro.—Orig. in Kawasaki, Kanagawa Prefecture, Japan, in 1895 by Chojuro Toma. Chance seedling of *Pyrus pyrifolia*. Introd. to U.S. in 1939 by U.S. Dept. of Agriculture. Evaluated at Univ. of California, Davis, beginning in 1955. Fruit: medium, about 66 mm in diam., 55 mm long; oblate; skin thick, russeted, green to orange-brown. Flesh white, mildly sweet and bland; firm; coarse; distinctive aroma. Ripens mid-August in Oregon; stores 20 weeks. Tree: medium vigor; spreading. Resistance to an outbreak of scab (*Venturia nashicola*) in 1897, high sugar content, high productivity, and resistance to blackspot (*Alternaria kikuchiana*) helped Chojuro become an important variety in Japan.

Daisui Li.—A large, smooth-skinned, pyriform Asian pear orig. at Univ. of California, Davis. Kikusui x Tse Li (Tsu Li); tested as selection 12-44 near Winters, Calif. Introd. by Ben Iwakiri in 1988. Plant patent 6075, Jan. 1988. Fruit: large to very large; obovate to globular, more pear-shaped than most Asian pears; skin smooth, thick, light green; flesh white, firm, coarse, crisp, juicy, sweet, subacid; distinct aroma, similar to Tse Li. Ripens in early September in Davis. Fruit stores 5 to 6 months at 0C. Tree: large; more vigorous than other Asian varieties; upright to slightly spreading; open; hardy. Blooms with Chojuro and Kikusui, before Bartlett, slightly after Ya Li and Tse Li. Resistant to fire blight.

Gold Nijisseiki.—A mutation of Nijisseiki resistant to black spot disease. Developed at the Inst. of Radiation Breeding at the National Inst. of Agrobiological Resources, Ohmiya-machi, Ibaraki, Japan, from gamma-irradiated buds of Nijisseiki. Registered as Pear Norin 15; named and released in 1991 by T. Sanada, K. Kotubuki, T. Nishida, H. Fujita, and F. Ikeda. Similar in habit, flowering season, and fruit characteristics to Nijisseiki, except fruit ripens a few days later and tree is resistant to the Japanese pear pathotype of *Alternaria alternata*; not as resistant as Chojuro.

Hosui (English translation = *Much Water*).—Orig. at the Natl. Hort. Res. Sta., Tsukuba, Japan. Cross of Ri-14 (Kikusui x Yakumo) x Yakumo made in 1954; selected in 1963; introd. in 1972 by M. Kajiura, K. Kanato, Y. Machida, M. Maeda, I. Kozaki, T. Tashiro, O. Kishimoto, and K. Seike. Fruit: large, 300–350 g; globose to oblate; skin russeted, golden to golden brown, enlarged lenticels. Flesh off-white, sweet, mild, crisp, juicy; finer texture than Chojuro and higher quality; ripens with Chojuro, mid-August in Oregon; stores 4 weeks. Tree: medium to large; vigorous; willowy; productive.

Hougetsu.—A late-ripening, highly productive, russet-skin Japanese pear with good resistance to black spot disease. Orig. at the Fruit Tree Research Station, MAFF, in Hiratsuka, Kanagawa, Japan, by K. Kotobuki, Y. Sato, K. Abe, T. Saito, M. Omura, I. Kajiura, T. Ogata, T. Kozono, K. Seike, Y. Machida, A. Kurihara, and I. Shimura. Named and released in Japan in 1992 under registration number 16. U.S. plant patent 9179; issued 27 June 1995. Okusankichi x 75-23 [Nijisseiki x Ri-14 (Kikusui x Yakumo)]. Cross made in 1974, first fruited in 1981, selected in 1982. Tested as Tsukuba No. 40. Fruit: large (600 g), round (108 mm in diam., 92 mm height), similar size and shape to Okusankichi; skin yellow-reddish-brown; deciduous calyx, thick peduncle about 3.7 cm long; flesh white, soft and crisp, sweet (about 11.7% sugar) similar to Kosui and Hosui, moderate acidity (pH about 4.4), more acid than Hosiu, less than Okusankichi, no astringency, no aromatic flavor; late ripening about 2 weeks before Okusankichi; keeps about 20 days at 25C. Tree: vigorous, open, low pruning requirement, late flowering with Okusankichi, highly resistant to black spot disease (*Alternaria* sp. and pear necrotic spot virus). Suitable pollinators include Kosui, Shinsei, Niitake.

Kosui (English translation = *Good Water*).—A high-quality, early ripening Japanese pear. Orig. at the Natl. Hort. Res. Sta., Tsukuba, Japan. Kikusui x Wasekozo; cross made in 1941; first fruited in 1947; released in 1959. Fruit: medium size; skin mostly russeted, yellow to golden brown. Flesh: crisp, juicy, very sweet; ripens 2 weeks before Hosui and Chojuro; stores 8 weeks. Tree: medium size; vigorous; spreading; resistant to blackspot; moderately resistant to scab.

New Century.—See Shinseiki.

Nijisseiki (English translation =*Twentieth Century*).—The standard against which other Asian pears are compared. Orig. in Matsudo City, Chiba Prefecture, Japan, in 1888 by Kakunosuke Matsudo. Chance seedling; introd. in 1898. Fruit: medium-large, 66 mm in diam., 55 mm long; globose-oblate; skin smooth, greenish-yellow to yellow; semi-glossy; inconspicuous lenticels. Flesh: crisp, white, juicy, sweet, bland. Ripens mid-August in Oregon, just after Chojuro; stores 20 weeks. Tree: medium size; upright, becoming spreading; productive. Must thin for good fruit size. Susceptible to blackspot.

Shin Li.—A fine-textured, fire-blight-resistant Asian pear bred at Univ. of California, Davis. A cross between the Japanese pear, Kikusui, and the Chinese variety, Tse Li (Tsu Li); tested near Winters, Calif. as selection 12-43. Plant patent 6076; Jan. 1988; introd. by Ben Iwakiri in 1988. Fruit: medium to large; round-oblate, slightly flatter than Daisui Li. Skin thick, smooth, light green to yellow-green; flesh firm, tender, crisp, juicy, sweet; texture finer than Daisui Li. Ripens early September at Davis; stores 5 to 6 months at 0C. Tree: vigorous; large to medium; upright to slightly spreading, open; hardy; resistant to fire blight. Leaves large, wide, leathery. Blooms early, with Chojuro, slightly after Ya Li and Tse Li.

Shinseiki (English translation = *New Century*).—Orig. at Okayama Prefecture Agr. Expt. Sta., Japan, by Teiji Ishikawa. Nijisseiki x Chojuro. Named and released in 1945. Fruit: medium size; globose to oblate; skin smooth, yellow. Flesh: white, crisp, juicy, bland. Ripens before Nijisseiki; stores 12 weeks. Tree: medium size; dense; spreading; precocious and productive.

Shinsui (English translation = *New Water*).—A russeted, early midseason Japanese pear selected in 1956 at Yatabe, Japan. Kikusui x Kimizukawase. Introd. in 1967 by M. Kajiura, K. Kanato, Y. Machida, and I. Kozaki at the Horticultural Res. Sta., Yatabe. Fruit: medium size, 250 g; skin yellow-brown, russetted; globose-oblate; flesh crisp, juicy, fine texture, very sweet; not as firm as other Asian pears; excellent eating quality. Ripens mid-August in Oregon and Washington. Susceptible to blackspot (*Alternaria*). A superior replacement for Ishiiwase.

Tse Li (*Tsú Li*).—Grown in Shantung Province of northern China for thousands of years. Fruit: large, 75 mm in diam., 90 mm long; ovate-pyriform with no distinct neck; irregular, lumpy. Skin yellow with large prominent tan lenticels. Flesh crisp, juicy, sweet, trace of tartness, distinct aroma. Ripens mid- to late September in Oregon, 4 weeks after Nijisseiki; stores 25 weeks at 0C. Tree: large; upright; cold hardy; very early bloom, requires early blooming pollinizer such as Ya Li. Low chilling, about 350–500 h.

Tsú Li.—See Tse Li.

Twentieth Century.—See Nijisseiki.

Ya Li (English translation = *Duck Pear*).—An old variety from northeastern China. Fruit: large, 70 mm in diam., 80 mm long; globular-pyriform; neck may be obscure. Skin light green to yellow, clean, waxy, small lenticels, free of russet. Stem curved, often fleshy and off-center. Flesh white, crisp, juicy, sweet, fragrant. Ripens mid- to late September in Oregon; stores 24 weeks. Tree: large; upright; vigorous; blooms very early, requires early pollinizer. Cold hardy. Low chilling requirement, about 350–500 h.

PECAN

Aggie.—Seedling selection orig. by N.H. Hander, Tex-Han Nursery, Belton, Texas. Discovered in 1953, introd. in 1960. Nut: elliptic with obtuse apex and base, laterally compressed (taller than wide) to round in cross section; 69 nuts/lb, 63% kernel; kernels golden in color with smooth texture, wide dorsal grooves, and very tight basal cleft; shells easily, quality good. Up to eight nuts per cluster. Tree vigorous, prolific.

Alley.—Orig. in Scranton, Miss., by Mrs. C.H. Alley, from seed planted in 1871; bore first fruit about 1880. Introd. in 1896 by F.H. Lewis, Scranton, Miss. Widely planted in the early 1900s. Known as one of the "big four" in the southeastern U.S., along with Stuart, Schley, and Pabst. Male parent of Caddo. Nut: elliptic, often asymmetric, with obtuse apex and base; round in cross section; moderate stripes on shell; 75 nuts/lb, 52% kernel; kernel light colored, with broad dorsal grooves, wrinkled texture. Protandrous bloom pattern, with early pollen shed and midseason receptivity. Ripens late midseason. Very susceptible to scab [*Cladosporium caryigenum* (Ell. et Lang) Gottwald] and black pecan aphid (*Melanocallis caryaefoliae* Davis).

Apache.—Orig. by controlled cross (Burkett x Schley) made in 1940 by L.D. Romberg, U.S. Dept. of Agr.–Agr. Res. Serv., Brownwood, Texas. Seedling recorded as 40-4-17; first fruited in 1945; selected in 1947; tested as T-110, released in 1962. Nut: oval elliptic to elliptic, with acute apex and obtuse base; round in cross section; 45 nuts/lb, 59% kernel; kernels golden in color, with prominent basal cleft. Protogynous, with mid- to late-season pollen shed and midseason receptivity. Precocious and prolific. Ripens late midseason. Susceptible to scab. Used as a seedstock in Texas, New Mexico, Arizona.

Arthur.—See Colby.

Barton.—Orig. by controlled cross (Moore x Success) made by L.D. Romberg, U.S. Dept. of Agr.–Agr. Res. Serv., Brownwood, Texas. Cross made in 1937 in the orchard of John Barton, Sr., of Utley, Texas. Seedling recorded as 37-3-20; first fruited in 1944; tested as T-15, released in 1953. Nut: elliptic with obtuse apex and acute base; round in cross section; shell suture usually dark at base; 48 nuts/lb, 57% kernel; kernels golden in color, with deep secondary dorsal and ventral grooves. Late to break buds in spring, with Stuart. Protandrous, with midseason pollen shed and mid- to late-season receptivity. Precocious and prolific, with a tendency to overbear with maturity. Ripens early midseason. Resistant to scab. Recommended (1990) for homeowner plantings in Mississippi.

Best's Early.—A native selection from the Mississippi River valley orig. in Eldred, Greene County, Ill., by Richard B. Best. Introd. in 1978 by R.D. Campbell. Nut: oblong with acute apex and base: round in cross section; few markings on shell; 203 nuts/lb; 39% kernel. Weak crotches. Early harvest.

Bradley.—See Ideal.

Brake.—Orig. as seedling selection in Rocky Mount, N.C., by W.H. Brake. Grown from seed from "Williamson tree," Raleigh, N.C., planted in 1910. Plant patent 47; issued 29 Nov. 1932. Introd. in 1934. Nut: elliptic with obtuse apex and obtuse to rounded base; round in cross section; 55 nuts/lb, 60% kernel; kernels golden in color, with wide dorsal grooves, deep basal cleft and ventral groove. Protandrous. Extremely thin shell. Very susceptible to scab.

Brooks.—Seedling selection made by B.W. Stone of Brooks County, Ga. Introd. prior to 1913. Female parent of Caddo. Nut: oblong with acuminate apex and base; slightly compressed in cross section; markings on shell dark and rough; 87 nuts/lb, 52% kernel; kernels cream to golden in color, with wide dorsal grooves. Protogynous bloom pattern, with midseason pollen shed and midseason receptivity. Scab resistant. Large leaflets. Recommended (1990) for planting in Arizona.

Burkett.—Native selection made by J.H. Burkett from Battle Fish Creek, near Putnam, Callahan County, Texas. Discovered 1900, first propagated in 1901. Introd. in 1911. Nut: orbicular, with obtuse apex and rounded base; round in cross section; 43 nuts/lb, 55% kernel; kernel golden to light brown in color with prominent dark brown speckles. Protogynous, with midseason pollen shed and receptivity. Ripens late midseason. Very susceptible to scab and downy spot (*Mycosphaerella caryigena* Demaree and Cole). Prone to preharvest germination. Widely used as a nursery seedstock for the western pecan growing region. Female parent of Apache and Comanche.

Caddo.—Orig. by controlled cross (Brooks x Alley) made in 1922 or 1923 by C.A. Reed, U.S. Dept. of Agr.–Agr. Res. Serv., Philema, Ga. Tested as Philema 1175, released in 1968. Nut: oblong elliptic to oblong, with acuminate apex and base; round in cross section; 60 nuts/lb, 56% kernel; kernels golden in color with shallow dorsal grooves, wide and extended dorsal ridge; excellent shelling characteristics. Protandrous, with early pollen shed and midseason pistillate receptivity. Precocious and prolific. Ripens midseason. Moderate scab resistance. Recommended (1990) for commercial plantings in Arizona, Louisiana, and Texas.

Caloro.—See Govett.

Candy.—Orig. as chance seedling on property of W.B. Schmidt, Ocean Springs, Jackson County, Miss. Introd. by T. Bechtel in 1913. Nut: oval, with obtuse apex and base; round in cross section; 66 nuts/lb, 48% kernel; kernels golden in color, smooth textured, with wide dorsal grooves; excellent shelling quality. Initiates growth early. Protogynous, with midseason pollen shed and receptivity. Bears at 5–6 years, prolific, ripens early. Resistant to scab. Recommended (1990) for planting in Arizona, Louisiana, Mississippi.

Cape Fear.—Orig. in Willard, N.C., at the Coastal Plain Branch Sta., North Carolina Agr. Expt. Sta. From Schley o.p. planted in 1912; reported first in 1937 by Robert Schmidt. Nut: elliptic with obtuse apex and base; shell dark striped; round in cross section; 45 nuts/lb, 54% kernel; kernels cream to golden in color, with wide dorsal grooves, deep secondary dorsal grooves. Protandrous, with early pollen shed and mid- to late-season pistillate receptivity. Very precocious and prolific, with a tendency to overbear with maturity. Ripens late midseason, about 3 days after Stuart. Susceptible to fungal leaf scorch complex (caused by *Alternaria, Pestolotia* and other fungi) which can defoliate trees. Variable in scab susceptibility by location, being moderately resistant in some orchards, very susceptible in others. Recommended (1990) for planting in Alabama, Arizona, Florida, Georgia, Louisiana, Mississippi, North Carolina, South Carolina, and Texas.

Carmichael.—Native seedling found by John Haschke in Harwood, Gonzales County, Texas, at Indian campsite on land owned by John R. Carmichael. Introd. in 1930. Pollen parent of Sioux. Nut: oblong elliptic with acute apex and obtuse base; round in cross section; 74 nuts/lb, 60% kernel; kernels cream to golden, with wide dorsal grooves. Excellent quality. Protogynous.

Cherokee.—Orig. by controlled cross (Schley x Evers) made by L.D. Romberg, U.S. Dept. of Agr.–Agr. Res. Serv., Brownwood, Texas. Cross made in 1948 and tested as 48-22-27 by Romberg and G.D. Madden. Scions budded into mature tree first fruited in 1952. Released in 1971. Nut: oblong elliptic with acute apex and obtuse base; flattened (wider than tall) to round in cross section; 48 nuts/lb, 55% kernel; kernels golden to light brown in color, but darkening quickly with age. Pro-tandrous, with early to midseason pollen shed and mid- to late-season pistillate receptivity. Very precocious and prolific, with a tendency to alternate bear, often at an early age. Nutrient deficiencies in N, P, and K associated with heavy production. Susceptible to scab. Recommended (1990) for planting in Arizona.

Chestnut.—See Mahan.

Chetopa (*KS112*).—Orig. as native seedling growing at the Kansas Agr. Expt. Field, Chetopa, Kan. Discovered by F. Brewster, former land owner and identified as tree #112. After donation of land to Kansas

State Univ., the tree was known as KS112. Introd. by Wm. Reid in 1995. Nut: oblong, with acute apex and obtuse base; flattened in cross section; 69 nuts/lb, 54% kernel; narrow dorsal grooves, thick. Protogynous, with late pollen shed and early to mid-season receptivity. Resistant to scab. Moderately precocious, with regular production. Ripens with Giles, 10–14 Oct. in Chetopa. Good branch angles

Cheyenne.—Orig. by controlled cross (Clark x Odom) made by L.D. Romberg, U.S. Dept. of Agr.–Agr. Res. Serv., Brownwood, Texas. Cross made in 1942; scion budded into mature tree first fruited in 1948; tested as 42-13-2 by Romberg and G.D. Madden; released in 1970. Nut: oval elliptic to elliptic, with acute apex and obtuse base; round in cross section; 48 nuts/lb, 58% kernel; kernels light cream to cream in color; very wrinkled texture; wide, shallow dorsal grooves. Protandrous, with midseason pollen shed and mid- to late season pistillate receptivity. Very precocious and prolific. Ripens in late mid-season, shortly after Stuart. Slow growing, "dwarf"-type tree. Very susceptible to yellow aphids. Susceptible to scab in south Louisiana; somewhat resistant in Texas. Recommended (1990) for planting in Arizona, California, Florida, New Mexico, Oklahoma, South Carolina, and Texas.

Chickasaw.—Orig. by controlled cross (Brooks x Evers) made by L.D. Romberg, U.S. Dept. of Agr.–Agr. Res. Serv., Brownwood, Texas. Cross made in 1944; scion budded into mature tree first fruited in 1949; tested as 44-4-101 by Romberg and G.D. Madden; released in 1972. Nut: elliptic with obtuse apex and base; flattened in cross section; 55 nuts/lb, 54% kernel; kernels flattened in cross section. Protogynous, with midseason pollen shed and early pistillate receptivity. Precocious and prolific, with a tendency to alternate bear with maturity. Ripens early midseason, about 18 days before Stuart. Nutrient deficiencies in N, P, and K associated with heavy production. Resistant to scab in Texas, but susceptible in south Louisiana and Georgia. Recommended (1990) for planting in Arizona and South Carolina.

Chief.—Native seedling selected in 1942 by H.C. Reeder, Ridgway, Ill. Introd. in 1953 by J.F. Wilkinson, Indiana Nut Nursery, Rockport, Ind. Nut: elliptic with acute apex and base; laterally compressed in cross section; ridges on shell; 54 nuts/lb, 45% kernel; kernels with tight dorsal grooves and very tight ventral groove, breaking up badly at shelling. Protandrous, with very late pistillate receptivity.

Chiles AAA.—See Owens.

Choctaw.—Orig. by controlled cross (Success x Mahan) made by L.D. Romberg, U.S. Dept. of Agr.–Agr. Res. Serv., Brownwood, Texas. Cross made in 1946. Scion budded into mature tree in 1947; produced first fruit in 1952. Tested as 46-15-276 from 1954 until released by Romberg in 1959. Nut: oval elliptic to elliptic, with obtuse apex and acute base; round in cross section; weak suture on shell; 37 nuts/lb, 58% kernel; kernels cream to golden in color with darker veins; wide, shallow dorsal grooves. Protogynous, with mid- to late-season pollen shed and early to midseason pistillate receptivity. Prolific. Resistant to scab in most locations, although very susceptible at Hanna, La. Requires good site and management to produce large crops of well-filled nuts. Recommended (1990) for planting in Arizona, Mississippi, Oklahoma, South Carolina, and east Texas.

Clark.—Native selection made by W.J. Millican, from the San Saba River near Bend, San Saba County, Texas. Nut: oval elliptic, with obtuse apex and base; round in cross section; 64 nuts/lb, 56% kernel; kernel cream in color, wrinkled, with broad dorsal grooves. Tree finely branched. Prolific. Susceptible to scab. Female parent of Cheyenne.

Colby (*Arthur, Orrell*).—Native seedling selection made from Wash Orrell farm, Clinton County, Ill., by Wm. W. and J.W. Lawrence, Fayette County, Ill., in the early 1940s. Selected by A.S. Colby and J.C. McDaniel, Univ. of Illinois, and tested as Illinois 1-19A. Released in 1957, named in honor of A.S. Colby. Nut: oblong, with acute apex and base; round in cross section; 66 nuts/lb, 44% kernel; kernels golden in color, with wide dorsal grooves and deep basal cleft. Protogynous, with mid- to late-season pollen shed and early pistillate receptivity. Medium precocity and prolificacy, with regular production. Matures in 160 day growing season; grown as far north as Niagara on Lake, Ontario, Canada. Used as a seedstock in Kansas and north. Recommended (1990) for planting in Arizona, Kansas, Kentucky, Oklahoma, and Tennessee.

Comanche.—Orig. by controlled cross (Burkett x Success) made by L.D. Romberg, U.S. Dept. of Agr.–Agr. Res. Serv., Brownwood, Texas, in orchard of John Barton, Sr., Utley, Texas. Cross made in 1937; scions budded in 1940 into mature bearing tree; first fruited in 1944; tested as 37-8-22; released in 1955. Nut: round with obtuse apex and base; round or slightly flattened in cross section; very similar in shape to Burkett, but with darker shell markings; 37 nuts/lb, 56% kernel; kernels golden in color, with very wide dorsal grooves; lacking the dark flecks that mark Burkett kernels. Protogynous, with mid- to late-season pollen shed and early to midseason pistillate receptivity. Very susceptible to scab in South Louisiana. Once recommended in area of Burkett culture, but no longer being planted.

Cowley.—Seedling selection from O.E. Cowley farm, Perkins, Payne County, Okla. Grown from seed of unknown parentage planted in 1951; brought to attention by Dept. of Hort., Oklahoma Expt. Sta., Stillwater, in 1960; introd. in 1966. Nut: oblong, with acute apex and base; flattened in cross section; 60 nuts/lb, 52% kernel. Protogynous. Six to seven nuts per cluster; tendency to overbear; preharvest germination problems in Shreveport, La. Resistant to scab.

Creek.—Orig. from a controlled cross (Mohawk x Starking Hardy Giant) made in Brownwood, Texas, by L.D. Romberg. Cross made in 1961. Tested as 61-6-67 by T.E. Thompson, W.D. Goff, M.L. Nesbitt, R.E. Worley, R.D. O'Barr, and B.W. Wood and released in 1996. Nut: elliptical, with an acute apex and base; round in cross section; 54 nuts/lb, 49% kernel; kernels cream to golden in color, with wide dorsal grooves. Released for its precocity, productivity, and disease resistance for use as a temporary cultivar in the early establishment phase of orchards in the southeast. Mature trees tend to overbear. Protandrous, with midseason pollen shed and mid-to late-season pistillate receptivity. Moderately scab resistant. Highly susceptible to pecan phylloxera (*Phylloxera devastatrix* Pergande).

Curtis.—Seedling selection grown from seed of Turkey Egg obtained from Arthur Brown, Bagdad, Fla., and planted by J.B. Curtis, Orange Heights, Fla. Nut was planted 1886, bore first crop in 1893 and introd. in 1896. Nut: oblong elliptic with acute apex and base; slightly compressed in cross section; shell with few markings; 89 nuts/lb, 57% kernel; kernels cream to golden in color, but often with darker speckles that detract from appearance; tight dorsal grooves. Protogynous, with late-season pollen and midseason pistillate receptivity. Resistant to scab. Susceptible to powdery mildew (*Microsphaera alni* de Candolle ex Winter). Used as a nursery seedstock in the southeastern U.S. Male parent of Houma. Curtis has a unique isozyme genotype: *bc* for phosphoglucomutase. Recommended (1990) for homeowner plantings in Florida, Georgia, and Louisiana.

Davis.—Native selection from Ocean Springs, Jackson County, Miss., made by Davis Nursery, Vancleave, Miss. Discovered in 1918; introd. in 1921. Nut: oblong with acute apex and base; flattened in cross section; shell smooth, with few markings; 53 nuts/lb, 46% kernel; kernels with wide dorsal grooves and narrow dorsal ridge. Excellent scab resistance. Recommended for homeowners in Alabama and Mississippi.

Delmas.—Seedling selection from orchard of A.G. Delmas, planted about 1877 in Scranton, Jackson county, Miss. Tree began bearing in 1884, was named in 1885, and was introd. commercially in 1890. Nut: elliptic, with obtuse apex and base; round in cross section; prominent ridges on shell, which is smooth, with few stripes; 50 nuts/lb, 48% kernel; kernels golden, with wide dorsal grooves; wrinkled on ventral surface. Protogynous with mid- to late-season pollen shed and midseason pistillate receptivity. Susceptible to scab. A prominent cultivar in Israel. Known as Nellie in Australia.

Desirable.—Orig. by controlled cross (unknown parents) made by C. Forkert, Ocean Springs, Jackson County, Miss. Parentage of Forkert's first introduction, Dependable, was Jewett x Success. Similarity in appearance led to speculation that Desirable is a sibling. Another controlled cross by Forkert, Admirable, was Russell x Success. That parentage has also been suggested for Desirable. Isozyme analysis is consistent with Russell x Success parentage, while the genotype of Jewett remains undetermined. Cross made in 1903. Introd. by Forkert as Desirable in 1914, but not widely distributed until after Forkert's death in 1928, when U.S. Dept. of Agr.–Agr. Res. Serv., Philema, Ga., tested it as #7191, beginning about 1930. Identified by U.S. Dept. of Agr. as Forkert's "Desirable" in 1936. Nut: elliptic with obtuse apex and obtuse to rounded base; round in cross section; shell rough, with elevated suture; 38 nuts/lb, 54% kernel; kernels golden in color, with wide dorsal grooves. Protandrous, with abundant early pollen shed and mid- to late-season pistillate receptivity. Bears in about 6 years and makes consistent, moderate production of high quality nuts, due in part to self-thinning fruit drop that reduces the number of nuts per cluster. Ripens in late midseason, shortly after Stuart. Susceptible to scab. Recommended for use in commercial orchards from Georgia west to Texas. Recommended (1990) for planting in Alabama, Arizona, Florida, Georgia, Louisiana, Mississippi, North Carolina, South Carolina, and Texas.

Dependable.—Orig. by controlled cross (Jewett x Success) made by C. Forkert, Ocean Springs, Jackson County, Miss., in early 1900s. Introd. prior to 1914. Nut: oblong, with obtuse apex and base; round in cross section; 50 nuts/lb, 51% kernel; Protogynous. Good resistance to scab at Elgin, Texas, while Desirable is very susceptible. Dependable has *bb* genotype for phosphoglucose isomerase, while Desirable has *bc*.

Devore.—Seedling selection made by R.B. Best, Wapello, Louisa County, Iowa. Introd. in 1978. Nut: elliptic with obtuse apex and acute base; oval in cross section; 143 nuts/lb, 43% kernel; kernels golden in color, with shallow dorsal grooves and deep basal cleft. Very early nut maturity. Evaluated for production in the far north.

Dooley.—Native seedling discovered about 1940 by Earl Dooley, Okmulgee, Okmulgee County, Okla., and introd. about 1945. Nut: oblong elliptic to oblong, with obtuse apex and base; round in cross section; 80 nuts/lb, 47% kernel; kernels cream to golden in color, with deep, wide basal cleft; high quality. Protandrous, with midseason pollen shed and mid- to late-season pistillate receptivity. Precocious. Ripens almost 3 weeks after Colby. Very susceptible to scab. Recommended (1990) in Kansas.

Elliott.—Seedling selection by H. Elliot, Milton, Santa Rosa County, Fla. Original tree was 2.5 ft in diam. and making large crops of high-quality nuts in 1912, when Elliot purchased the property where it stood. First propagated in 1919, with the establishment of Harlan Farms Nursery, Paxton, Fla. Nut: ovate, with acute apex and rounded base; round in cross section; 67 nuts/lb, 53% kernel; kernels golden in color, with wide dorsal grooves and a deep basal cleft. Protogynous, with midseason pollen shed and early pistillate receptivity. Early to leaf out in spring, making it (and its seedlings) susceptible to late spring freezes. Slow to bear, and not a heavy producer, but known for excellent quality nuts. Excellent scab resistance. Very susceptible to black aphid. Used as a seedstock in the southeast. Recommended (1990) in Alabama, Arizona, Florida, Georgia, Louisiana, Mississippi, and South Carolina.

Evers.—Seedling tree purchased from J.A. Evans Nursery, Arlington, Texas, by W.T. Evers and grown on his farm in Denton County, Texas. Introd. about 1950. Nut: oval elliptic to elliptic, with obtuse apex and rounded base; 104 nuts/lb, 54% kernel; kernels with wide, shallow dorsal grooves. Protogynous. Tree finely branched. Prolific. Nuts mature early, in September at Brownwood. Susceptible to scab. Resistant to downy spot. Used as pollen parent for Cherokee, Chickasaw, Shoshoni, and Osage to whom it conferred precocity and prolificacy.

Farley.—Seedling selection from Jackson County, Fla., discovered in 1918 by B.W.

Stone, Stone Nursery Co., Thomasville, Ga. Introd. about 1925. Nut: oval, with oblique (obtuse and asymmetric) to truncate apex and rounded base; cross section often distinctly four-sided due to flattened adaxial and abaxial surfaces of nut; 60 nuts / lb, 53% kernel; kernels wrinkled in texture, with wide dorsal grooves and basal cleft. Slow to bear, low in production, late fruit maturity. Resistant to scab.

Florida Giant.—See Mahan.

Forkert.—Orig. from a controlled cross (Success? x Schley) made by C. Forkert of Ocean Springs, Jackson County, Miss. Nuts from the cross were planted about 1913. Introd. by Forkert's nursery in the 1920s and used as a yard tree in south Mississippi. Potential as commercial cultivar rediscovered in the 1960s. Nut: oblong elliptic with acuminate apex and obtuse base, surface very rough, with prominent dark stripes; round in cross section; 49 nuts / lb, 62% kernel; kernels cream to golden in color, with deep, relatively narrow dorsal grooves. Protogynous, with mid- to late-season pollen and early pistillate receptivity. Susceptible to scab in Louisiana. Recommended (1990) in Alabama, Arizona, Louisiana, Mississippi, and South Carolina.

Frotscher.—Seedling selection made by Oscar Olivier from Bayou Teche, near Olivier, Iberia Parish, La. Seed was planted prior to the Civil War. First propagated by William Nelson in 1882. Nut: oblong elliptic, with obtuse apex and rounded base; flattened in cross section; 57 nuts / lb, 49% kernel; kernels golden in color, with darker veins; kernels wrinkled in texture, with wide dorsal grooves and a prominent basal cleft. Protogynous, with mid- to late-season pollen shed and early pistillate receptivity. Good resistance to scab and downy spot, but susceptible to vein spot (*Gnomonia nerviseda* Cole).

F.W. Anderson.—Orig. in Le Grande, Calif., by F.W. Anderson, Merced, Calif. Seedling grown from Caloro o.p. Selected in 1961; introd. in 1963. Plant patent 2392; issued 21 Apr. 1964, assigned to Stribling Nurseries, Merced. Nut: oblong with acute apex and base; round in cross section; slight ridges on shell, few dark markings; 50–60 nuts / lb. Pollen shed and pistillate receptivity overlap. Precocious. Ripens three weeks before Stuart in Central Valley of California.

Garner.—See John Garner.

Georgia Giant.—See Mahan.

Giles.—Orig. as native seedling on the Neosho River near Chetopa, Kan., on the property of A.E. Giles. Discovered about 1927 by J. Ford Wilkinson, Indiana Nut Nursery, Rockport, Ind. Intro. in 1930 by Wilkinson. Nut: oblong with acute, asymmetric apex and rounded base; laterally compressed in cross section; shell darkly marked with stripes, raised at suture; 74 nuts / lb, 53% kernel; kernels golden, with wide dorsal grooves, narrow dorsal ridge, deep, but wide basal cleft. Protandrous, with midseason pollen shed and pistillate receptivity. Recommended for commercial orchards in Arizona, Kansas, Missouri, Oklahoma, and Tennessee. Used as a seedstock in parts of Oklahoma and Kansas.

Gloria Grande.—Orig. as seedling selection (possibly from Stuart seed) growing in Elloree, S.C., discovered in 1920 by W.W. Watson, Watson's Pecanwood Nurseries, Orangeburg, S.C. Intro. in 1923. Nut: elliptic with obtuse often asymmetric apex and obtuse to rounded base; round in cross section; shell thick, smooth, darkly striped; 44 nuts / lb, 48% kernel; kernels golden, with shallow, wide dorsal grooves and a deep basal cleft. Protogynous with mid- to late-season pollen shed and early to midseason pistillate receptivity. Not precocious. Resistant to scab, fungal leaf scorch and downy spot. Susceptible to black pecan aphids. Recommended for use by homeowners in Alabama, Arizona, Georgia, Louisiana, and South Carolina. Recommended for commercial growers in Alabama, Arizona, Florida, and South Carolina.

Gormely.—Orig. as native seedling discovered in 1946 on the G. G. Gormely farm in Okemah, Okla., and brought to attention by Dept. of Hort., Oklahoma Agr. Expt. Sta., Stillwater. Intro. in 1966. Nut: oval elliptic with obtuse apex and base; round in cross section; 103 nuts / lb, 46% kernel; kernels golden with wide, shallow dorsal grooves and very wide, deep basal cleft;

finely wrinkled in texture. Resistant to scab. Recommended for homeowners and commercial growers in Oklahoma.

Govett (*Calaro*).—Orig. as native selection in Seguin, Texas, by R.C. Govett. Intro. in 1922. Original tree estimated to be 100 years old in 1957. Nut: oblong with acute apex and obtuse base; flattened in cross section; 55 nuts/lb, 57% kernel; kernels golden in color with wide dorsal grooves and deep secondary dorsal grooves. Protogynous. Not seed parent of F.W. Anderson.

GraBohls.—Orig. as seedling selection made by H.C. Bohls, Austin, Texas. Plant patent 3434; 15 Jan. 1974 assigned to O.S. Gray, Arlington, Texas. Parentage unknown, but suspected to be Mahan x Texas Prolific, which is consistent with isozymes , dichogamy, and appearance; tested as Mary. Nut: oblong with obtuse apex and right angled to acute base; nuts often asymmetric, with prominent suture stripe; round in cross section; 42 nuts/lb, 59% kernel; kernels golden with wide dorsal grooves, deep secondary dorsal grooves, deep basal cleft, and wrinkled texture. Protogynous with late-season pollen shed and early to midseason pistillate receptivity. Very precocious and prolific, with severe alternate bearing tendencies. Ripens midseason. Not recommended for planting in any state.

GraCross.—Seedling selection in orchard of Harry V. Cross , South Sabine River near Greenville, Hunt County, Texas. Discovered in 1950; Plant patent 4236, issued 4 Apr. 1978; assigned to O.S. Gray, Jr., Arlington, Texas. Nut oblong elliptic, with obtuse apex and base; shell with prominent dark stripes; 42 nuts/lb, 59% kernel; kernels cream to golden in color, with prominent veins; dorsal grooves medium wide, deep. Protogynous.

GraKing.—Orig. in Hugo, Okla. Introd. in 1959 by O.S. Gray Nursery & Landscape, Arlington, Texas. Chance seedling grown from Texas seed; original tree shown to Mr. Gray at the residence of Mrs. King in Hugo sometime during the period 1946–49, when the tree was 12 inches in diam. and 40 feet tall. Nut: oblong with obtuse apex and base; flattened in cross section; often with weak suture; 35 nuts/lb, 53% kernel; kernels with wide dorsal grooves, deep ventral groove. Protandrous, with midseason pollen shed and mid- to late-season pistillate receptivity in New Mexico. Ripens early. Recommended (1990) in Oklahoma.

GraPark Giant.—Orig. in Grand Prairie, Texas, by M.K. Parker. Introd. in 1974. Plant patent 3687; issued 11 Feb. 1975; assigned to O.S. Gray, Arlington, Texas. Chance seedling, from seed of unknown parentage planted in 1947. Nut: oblong, with obtuse apex and obtuse to rounded base; nuts often asymmetric, shells rough; 38 nuts/lb, 60% kernel;

GraTex.—Orig. by controlled cross (Ideal x Success) by O.S. Gray, Jr., and J.A. Evans, O.S. Gray Nursery, Arlington, Texas. Selected in 1945. Introd. in 1962. Nut: elliptic, with obtuse apex and base; round in cross section; 45 nuts/lb, 64% kernel; kernels golden in color, with deep, narrow dorsal grooves. Protogynous. Good resistance to scab.

GraZona.—Orig. as chance seedling from unknown parents, discovered in 1952 in Mesa, Ariz., by Lyman Coe on the ranch of J.W. May. Introd. in 1968 by the O.S. Gray Nursery, Arlington, Texas. Nut: orbicular, with obtuse, almost truncate apex and obtuse base; nuts often asymmetric; laterally compressed in cross section; 40 nuts/lb, 55% kernel; kernels golden in color with darker specks; wide dorsal grooves. Protogynous.

Greenriver.—Native selection made from woods on the L.B. Major farm, at the mouth of the Green River near Henderson, Henderson County, Ky. Other selections from the same woods include Major and Hinton. Introd. by T.P. Littlepage in 1911. Nut: oval elliptic with obtuse apex and base; laterally compressed in cross section; ridges on shell; 70 nuts/lb, 49 % kernel; kernels golden in color, with wide dorsal grooves and deep, wide basal cleft. Protogynous, with midseason pollen shed and early pistillate receptivity. Not appropriate for the far north, due to late nut maturity. Scab resistant in north. Greenriver has a unique isozyme genotype: *cc* for phosphoglucose isomerase. Recommended (1990) in Arizona, Kentucky, and Tennessee.

Halbert.—Native selection discovered in

1886 by H.A. Halbert in a grove near Coleman, Coleman County, Texas. Introd. by Halbert in 1901. Nut: oval elliptic with obtuse apex and rounded base; 60 nuts/lb, 60% kernel; kernels golden in color, with narrow dorsal grooves and a deep basal cleft. Protogynous. Precocious and prolific. Trees are finely branched and slow growing due to the early production. Very susceptible to scab. Female parent of Wichita.

Halsly.—Orig. by controlled cross (Schley x Halbert) made in Mason, Texas, by Frank J. Willmann, LaGrange, Texas. Seed planted in 1914. Introd. in 1924. Nut: resembles Halbert, but longer and larger; 45 nuts/lb, 56% kernel. Sibling of Slybert.

Harper.—Orig. by controlled cross (Mahan x Western) made by Roy E. Harper, New Mexico Agr. Expt. Sta., Las Cruces. Introd. in 1967. Selected and evaluated by Roy M. Nakayama of above station. Nut: obovate oblong with acute asymmetric apex and acute base; flattened in cross section; shell rough, darkly striped; 53 nuts/lb, 61% kernel; kernels golden in color with tight dorsal grooves and deep secondary dorsal grooves. Nut resembles Western, but larger, smoother, and longer. Protogynous with flowering period comparable to Ideal and Wichita. Ripens 10 days earlier than Western, reducing problems with sticktights at harvest.

Harris Super (*Schley-Harris*).—Orig. as chance seedling in Gunnison, Miss., discovered in 1952 by Edward Harris. Introd. in 1960. Plant patent 2051; issued 25 Apr. 1961; assigned to Simpson Nursery Co., Monticello, Fla. Parentage unknown; once thought to be Stuart x Schley, but that parentage is inconsistent with isozyme patterns (Schley and Stuart are *aa* for phosphoglucose isomerase, while Harris Super is *ab*). Nut: oblong with long acute apex and obtuse base; round in cross section; shell heavily specked, prominently ridged; 50 nuts/lb, 53% kernel; kernels with wide dorsal grooves, deep secondary dorsal grooves, wrinkled texture. Protogynous. Precocious and prolific. Some scab resistance in Georgia, but very susceptible in Florida. Not recommended for planting in any state.

Hastings.—Orig. as open-pollinated Stuart seedling in Monticello, Ga., selected by G.T. Chaffin in 1945. Introd. in 1955. Plant patent 1399; 14 June 1955; assigned to the H.G. Hastings Co., Atlanta, Ga. Nut: oblong elliptic, with acute apex and obtuse base; 68 nuts/lb, 48% kernel; kernels golden in color, with wide dorsal grooves, deep and wide basal cleft, wrinkled texture. Protogynous.

Hirschi.—See Steuck.

Hodge (*Hodge's Favorite, Illinois Mammoth*).—Orig. as native seedling in York, Clark County, Ill., discovered by H.G. Hodge about 1890. Specimens sent to U.S. Dept. of Agr. 20 Feb. 1893. First northern pecan cultivar to be discovered and described. Introd. in 1954 by the Gerardi Nursery, Caseyville, Ill. Nuts distributed as Hodge's Favorite and Illinois Mammoth. Cultivar described in 1908 as Hodge. Nut: oblong, with acute apex and obtuse asymmetric base; shell rough, with prominent ridge along suture; laterally compressed in cross section; 69 nuts/lb, 59% kernel; kernels with shallow, wide dorsal grooves, deep, narrow basal cleft, and deep ventral groove; kernels break up badly in shelling. Ripens early. Weak wood subject to splitting on windy sites. Protogynous, with late pollen shed. Recommended (1990) for planting in Kentucky.

Hodge's Favorite.—See Hodge.

Houma.—Orig. by controlled cross (Desirable x Curtis) made by L.D. Romberg, U.S. Dept. of Agr.–Agr. Res. Serv., Brownwood, Texas. Cross made in 1958; scion budded into bearing tree in 1959; tested as 58-4-61 by T.E. Thompson, E.F. Young, Jr., J.E. Boudreaux, R.D. O'Barr, R.S. Sanderlin; released in 1989. Nut: elliptic, with obtuse apex and base, often asymmetric; round in cross section; 55 nuts/lb, 51% kernel; kernels golden in color, with wide, shallow dorsal grooves and a deep, wide basal cleft. Protandrous, with early midseason pollen shed and midseason receptivity, similar to Desirable.

Humble.—Orig. as native seedling growing near the Nueces River, Zavala County, Texas, discovered in the 1920s by James A. Simpson, Uvalde, Texas. First propagated in 1928. Plant patent 73; issued 25 July 1932; assigned to Humble Oil & Refining Co.,

Houston, Texas. Nut: oblong elliptic, with obtuse apex and obtuse base; shell rough, heavily marked with dots; laterally compressed in cross section; 45 nuts/lb, 55% kernel; kernels rough textured, with medium dorsal grooves, deep secondary dorsal grooves, a deep ventral groove, and conspicuous secondary ventral grooves. Late to bloom. Many nuts (6–10) per cluster. Precocious and prolific. Ripens early.

Ideal (*Bradley*).—Orig. as native seedling in San Saba, Texas, selected about 1925 by J.A. Evans, Arlington, Texas. Introd. during the period 1930–35. Nut: oblong elliptic, with obtuse apex and rounded base; shell smooth in texture with slight ridge at suture; flattened in cross section; 67 nuts/lb, 59% kernel; kernels with wide dorsal grooves, deep basal cleft, and deep ventral groove. Protogynous, with mid- to late-season pollen shed and early to midseason receptivity. Often used as pollinizer for 'Western' in New Mexico, where it is sometimes called Bradley. Ripens early. Very susceptible to scab. Recommended (1990) in Arizona and New Mexico.

Illinois Mammoth.—See Hodge.

Imperial.—Orig. as chance seedling from unknown parents, selected in 1958 by E.W. Ross, San Antonio, Texas. Introd. in 1962. Plant patent 2211; issued 8 Jan. 1963; assigned to Aldridge Nursery, Von Ormy, Texas. Nut: elliptic or ovate elliptic with obtuse apex and base, base often with short tip; round in cross section; shell rough, with very weak suture; 32 nuts/lb, 56% kernel; kernels golden, with wide, shallow dorsal grooves beside wide, triangular dorsal ridge; deep basal cleft and ventral groove.

Indiana.—Native seedling discovered in the early 1900's by M.J. Niblack and W.C. Reed. Original tree stood west of Oaktown, Knox County, Ind., and was thought to be a natural seedling of the nearby 'Busseron' ortet. Introd. in 1909. Nut: oblong, with obtuse apex and obtuse to rounded base; laterally compressed in cross section; 75 nuts/lb, 47% kernel. Protogynous, with midseason pollen shed and midseason pistillate receptivity. Used as a seedstock in the 1960s by Linwood Nursery, Calif., under the name Linwood, but abandoned due to slow growth of seedlings.

James.—Limb sport of a seedling of unknown parentage, selected by G.W. James, Brunswick, Mo. Plant patent 2607; granted 15 Mar. 1966. Selection based on atypically abundant, dark glossy green, unusually large foliage, coupled with abundant production of nuts. Considered suitable for ornamental purposes. Nut; oblong elliptic with acute apex and obtuse to rounded base; 70–74 nuts/lb, 53% kernel.

James Early.—Seedling selection made by G.W. James from orchard near Brunswick, Chariton County, Mo. Discovered in Fall 1984. Plant patent 7419; received 15 Jan. 1991; assigned to James Pecan Farms. Nut: oblong with acute apex and base; round in cross section; 104 nuts/lb; kernels with very prominent, wide basal cleft. Very early nut maturity, ripening 10–15 days before Starking Hardy Giant.

John Garner (*Garner*).—Orig. as seedling of San Saba Improved growing on the J.W. Terry farm, selected in 1932 by E.E. Risien, San Saba, Texas. Introd. in 1934 by Wolfe Nursery, Stephenville, Texas. Named after Vice President John Garner, of Uvalde, Texas. Onliwon reported as pollen parent, discounted on the basis of isozymes. Nut: elliptic with obtuse asymmetric apex and rounded base; shell smooth, with dark suture line at base; laterally compressed in cross section; 52 nuts/lb, 61% kernel; kernels golden in color, with dorsal grooves flared at base, dorsal ridge narrow at apex. Protandrous. Tree: vigorous, overbears, often showing severe Zn deficiency in west Texas.

Johnson.—Native seedling discovered by Charles Stephens, Columbus, Kan., in 1931. Nut: oblong, with acuminate apex and base, round in cross section; 84 nuts/lb, 47% kernel; kernels with tight dorsal grooves. Protogynous. Early nut maturity.

Jubilee.—Orig. as chance seedling of unknown parentage selected by Melvin B. Clemons, Foley, Ala. Plant patent 5485; issued 4 June 1985. Nut: oblong with acute apex and base; round in cross section; 45 nuts/lb, 53% kernel; kernels golden in color, with deep, tight dorsal grooves, secondary dorsal grooves, deep basal cleft, deep ventral and secondary ventral grooves. Protogynous. Precocious and prolific. Resistant to scab. Recommended (1990) for trial plantings in Alabama.

Kallus.—See Prilop of Lavaca.

Kanza.—Orig. from a controlled cross (Major x Shoshoni) made in Brownwood, Texas, by L.D. Romberg. Cross made in 1955. Tested as 55-11-11 by T.E. Thompson, L.J. Grauke, Wm. Reid, M.W. Smith, and S.R. Winter and released in 1996. Nut: ovate, with an acute apex and obtuse to rounded base; round in cross section; 77 nuts/lb, 54% kernel; kernels golden in color, with dorsal grooves that do not trap packing material. Released for its ability to produce good yields of high quality nuts in the northern pecan producing regions (Zone 6b). Protogynous, with late pollen shed and early pistillate receptivity. Kanza has excellent scab resistance.

Kernodle.—Orig. as chance seedling in Camp Hill, Ala., discovered in 1948 by Julius A. Kernodle. Introd. in 1957. Plant patent 1744; issued 26 Aug. 1958; assigned to Simpson Nursery Co., Monticello, Fla. Nut: blocky elliptic, with truncate apex and base; laterally compressed in cross section; 57 nuts/lb, 55% kernel; kernels with deep secondary dorsal grooves, wide dorsal grooves. Protogynous. Ripens late.

Kibler.—Orig. as chance seedling in Coloma, Mich., selected in 1940 by the Burgess Seed & Plant Co., Galesburg, Mich. Introd. in 1959. Nut: oblong, with obtuse, asymmetric apex and acute base; round in cross section; 102 nuts/lb, 48% kernel; kernels with shallow dorsal grooves that do not trap shell; thick shell. Very early nut maturity, occasionally ripening in southern Michigan.

Kiowa.—Orig. by controlled cross (Mahan x Desirable) made by L.D. Romberg, U.S. Dept. of Agr.–Agr. Res. Serv., Brownwood, Texas. Cross made in 1953; scion grafted into a bearing tree in 1954 and first fruited in 1958. Tested as 53-9-191 by G.D. Madden, E.J. Brown, and H. Malstrom. Released in 1976. Nut: oblong elliptic, with obtuse apex and base; round in cross section; 38 nuts/lb, 58% kernel; kernels golden in color, with wide dorsal grooves, narrow dorsal ridge, deep basal cleft and ventral groove. Protogynous, with midseason pollen shed and early to midseason receptivity. Kiowa cross pollinates well with Desirable. More precocious than Desirable. Ripens in late midseason, with Desirable. Susceptible to scab. Recommended (1990) for use in commercial orchards in Alabama, Florida, Louisiana, Mississippi, Oklahoma, South Carolina, and Texas.

KS112.—See Chetopa.

Mahan (*Chestnut, Florida Giant, Georgia Giant, Masterpiece, Mayhan, Mississippi Giant*).—Orig. as chance seedling from seed of unknown parentage planted about 1910 by J.M. Chestnutt. Propagation rights sold to F.A. Mahan, Monticello Nursery Co., Monticello, Fla., in 1927. Nut: oblong, with acute apex and base; nut often asymmetric, appearing "pinched" in middle due to flattening of abaxial and adaxial surfaces; flattened in cross section; 32 nuts/lb, 58% kernel; kernels with deep secondary dorsal grooves and basal cleft, often poorly filled to base, woody in texture. Homozygous for protogyny (*PP*), which is a dominant allele, making all Mahan progeny protogynous. Mid- to late-season pollen shed, with early to mid-season receptivity. Very precocious and prolific, with a strong tendency to overbear as a mature tree. Ripens late, about 12 days after Stuart. Very susceptible to scab. Mahan is the female parent of Tejas, Kiowa, Harper, Mahan-Stuart, and Maramec and the pollen parent of Choctaw, Wichita, and Mohawk. Recommended (1990) for planting only in Arizona.

Mahan-Stuart.—Orig. by controlled cross (Mahan x Stuart) made by F.A. Mahan, Monticello Nursery Co., Monticello, Fla. Selected in 1948, first propagated in 1949. Plant patent 1532; issued 4 Dec. 1956. Nut: oblong with obtuse apex and base; flattened in cross section; 30 nuts/lb, 52% kernel; kernels with extended dorsal ridge, tight dorsal grooves, deep secondary dorsal grooves, and wrinkled texture. Protogynous. Precocious. Recommended (1990) for planting only in Arizona.

Major.—Native seedling discovered about 1907 by Mrs. L.B. Major, from woods on the Major farm near the confluence of the Green and Ohio Rivers, Henderson County, Ky. Introd. in 1908. Nut: oval elliptic to orbicular, with obtuse apex and base; few stripes, abundant spots, and ridges on the shell surface; 78 nuts/lb, 49% kernel; kernels cream to golden in color, with broad, shallow dorsal grooves. Protandrous, with midseason pollen shed and midseason pistillate receptivity. Good resistance to scab, but susceptible to vein spot. A standard cultivar for the northern pecan region of the U.S. Female parent of Osage and Kanza. Recommended (1990) for planting in Arizona, Kansas, Kentucky, Missouri, and Tennessee.

Maramec.—Orig. as open-pollinated seedling of Mahan by Emma Charlton, Maramec, Okla. Nut planted in 1933, selected in 1963, tested as 63M. Introd. in 1969. Nut: oblong with obtuse apex and base; prominent basal tip; shell rough, ridged, pinched in middle; round in cross section; 44 nuts/lb, 59% kernel; kernels with medium dorsal grooves, deep secondary dorsal grooves, dorsal ridge thickened at base. Protogynous with late pollen shed and early to midseason receptivity. Precocious. Ripens midseason, shortly before Stuart. Susceptible to scab in Louisiana, but considered somewhat resistant in Oklahoma and Georgia. Recommended (1990) for planting in Oklahoma and north Texas.

Masterpiece.—See Mahan.

Mayhan.—See Mahan.

McLeon.—Seedling selection grown from open-pollinated Moore seedstock planted in 1961 by Gilbert McDowell, Nevada, Mo. Seedling began to bear in 1972, was first propagated in 1974, and received Plant patent 5911; 24 Mar. 1987. Nut: oblong elliptic with acute apex and obtuse base; 55 nuts/lb, 55% to 60% kernel. Leaves are larger and lighter green in color than Stark Hardy Paper Shell. Late nut maturity (1 Nov. at Nevada, Mo.).

Melrose.—Orig. as chance seedling in the orchard of John Crow, Hanna, La. Selections from that orchard were numbered as CS (Crow Seedling)1 to 226. Propagated by Guidry Nursery, St. Martinsville, La., as CS 23. Collected by Louisiana Agr. Expt. Sta. in 1964 as L15. Tested by W.A. Young and W.A. Meadows in plantings at Melrose, La., and released under that name in 1979. Nut: oblong with acute apex and base; round in cross section; 53 nuts/lb, 57% kernel; kernels golden to light brown, with wide dorsal grooves. Protogynous with midseason pollen shed and midseason receptivity. Ripens late midseason, shortly after Stuart. Resistant to scab. Recommended (1990) for planting or trial in Alabama, Arizona, Louisiana, and Mississippi.

Mississippi Giant.—See Mahan.

Mobile.—Seedling selection made by F.H. Lewis and I.P. Delmas from Bayou LaBatre, Ala. Nut planted about 1887. First propagated by Lewis and Delmas about 1900. Introd. by Stone Nursery as Mobile in 1904. Nut: oblong with right angled apex and obtuse base; round in cross section; shell surface with ridges; 53 nuts/lb, 43% kernel; kernels wrinkled, with very broad, shallow dorsal grooves, and a prominent basal cleft. Protandrous, with midseason pollen shed and pistillate receptivity. Resistant to scab. Produces a high percentage of very poor quality nuts early in the life of the tree. Considerable numbers of old trees remain in Georgia and South Carolina. Considered one of the worst of all pecan cultivars.

Mohawk.—Orig. by controlled cross (Success x Mahan) made by L.D. Romberg, U.S. Dept. of Agr.–Agr. Res. Serv., Brownwood, Texas. Cross made in 1946 (full sibling of Choctaw), first fruited in 1953; selected for testing in 1954; tested as 46-15-195. Released in 1965. Nut: oblong with obtuse apex and base; slightly flattened in cross section; shell rough, dark striped; 32 nuts/lb, 59% kernel; kernels golden to light brown in color with narrow dorsal grooves when well-filled. Protogynous, with midseason pollen shed and early to midseason pistillate receptivity. Very precocious and prolific with a tendency to alternate bear on mature trees. Freeze susceptible, especially after large crop. Ripens midseason, about 9 days before Stuart. Susceptible to scab in Louisiana and some Georgia locations, but generally resistant in Texas and Oklahoma. Recommended (1990) in Arizona, Mississippi, Oklahoma, South Carolina, and Tennessee.

Moore.—Seedling selection owned by Miss Annie Wooten of Waukeenah, Jefferson County, Fla. Introd. in 1908. Known as Bester in South Africa. Nut: oval elliptic to oblong elliptic with obtuse apex and base; slightly flattened to round in cross section; 67 nuts/lb, 49% kernel; kernels golden to light brown, with tight dorsal grooves and a narrow dorsal ridge. Protandrous, with early pollen and midseason pistillate receptivity. Precocious, becoming an alternate bearer with maturity. Moore and its sibling Waukeenah have been used as seedstocks by the pecan nursery industry, primarily in the southeastern pecan region of the U.S. Moore is the female parent of Barton and of Selection 48-13-311 (the female parent of Navaho). Recommended (1990) for planting in Arizona.

Moneymaker.—Seedling selection by S.H. James, Mound, La., from an orchard planted about 1885 using nuts purchased in New Orleans. The nuts supposedly were collected from South Louisiana, between New Orleans and the Texas border. Named in 1896 and commercially propagated in 1898. Nut: oval elliptic with oblique rounded apex and rounded base; round in cross section; 62 nuts/lb, 50% kernel; kernels light brown in color with shallow dorsal grooves, prominent secondary dorsal grooves, and wrinkled texture. Protogynous, with midseason pollen and early pistillate receptivity. Precocious and prolific. Trees have strong structure, due to good limb angles. Resistant to scab. Early nut maturity. Substantial acreage in Georgia and Alabama. Recommended (1990) for planting in Arizona.

Mount.—Orig. as native seedling discovered in 1949 on farm of E.E. Mount, Okmulgee, Okla., and brought to attention by Dept. of Hort., Oklahoma Agr. Expt. Sta., Stillwater. Introd. in 1966. Nut: oval elliptic, with obtuse apex and base; round in cross section; 75 nuts/lb, 52% kernel; excellent shelling pecan. Protogynous. Hardy. Ripens early, 2 weeks before Stuart. Resistant to scab. Recommended (1990) for planting in Oklahoma.

Navaho.—Orig. from a controlled cross [48-13-311 (=Moore x Schley) x Wichita] made in Brownwood, Texas, by E.J. Brown and G.D. Madden in 1974. Tested as 74-1-11 by T.E. Thompson, L.J. Grauke, and J.B. Storey and released in 1994. Nut: oblong elliptic with an acuminate apex and obtuse base; round in cross section; 62 nuts/lb, 61% kernel; kernels golden to light brown in color, with deep, relatively narrow dorsal grooves and a prominent basal cleft. Protandrous, with early pollen shed and midseason pistillate receptivity. Susceptible to scab and vein spot. Trees are vigorous with strong structure, and begin growth early in the spring. Very precocious and prolific. Nut maturity mid-October in College Station, Texas.

Nugget.—Native selection from the Leon River near Gustine, Comanche County, Texas, discovered by J.A. Evans in the early 1930s. Nut: oval elliptic with acute apex and base; 86 nuts/lb, 57% kernel; kernels with very shallow dorsal grooves. Protogynous. Susceptible to scab. Consistent in production. Unusual in having *d* allele for malate dehydrogenase, as does its progeny Sullivan.

Oconee.—Orig. by controlled cross (Schley x Barton) made by L.D. Romberg, U.S. Dept. Agr.–Agr. Res. Serv., Brownwood, Texas. Cross made in 1956. Selected and tested as 56-7-72 by T.E. Thompson, E.F. Young, Jr., R.E. Worley, R.D. O'Barr, and R.S. Sanderlin; Released in 1989 for use in the southern U.S. pecan belt. Nut: elliptic, with obtuse apex and base; round in cross section; 48 nuts/lb, 56% kernel. Protandrous, with early to midseason pollen shed and mid- to late-season pistil receptivity, similar to Desirable. Good precocity and yield potential. Nut matures with Cheyenne, about 21 Oct. at Tifton, Ga. Moderate scab resistance, fair resistance to downy spot and vein spot. Recommended (1990) for planting in Louisiana.

Odom.—Orig. as chance seedling grown from MS nut in Toledo, Newton County, Texas, by R.L. Odom. Discovered about 1920. Introd. in 1923. Nut: oblong elliptic with obtuse asymmetric apex and rounded base; round in cross section; shell smooth, with few dark markings; 43 nuts/lb, 56% kernel; kernels cream to golden in color, with wide dorsal grooves and a deep, wide basal cleft; unusual in having an "inner shell" formed by packing material, as in Woodard. Female parent of Shoshoni. Pollen parent of Cheyenne. Susceptible to scab.

Orrell.—See Colby.

Osage.—Orig. by controlled cross (Major x Evers) made by L.D. Romberg, U.S. Dept. of Agr.–Agr. Res. Serv. Brownwood, Texas. Tested as 48-15-3 by T.E.Thompson and E.F. Young, Jr. Released in 1989 for use in the northern U.S. pecan production area. Nut: oval elliptic with obtuse apex and acute base; round in cross section; 81 nuts/lb, 55% kernel; excellent shelling characteristics. Protandrous, with early to mid-season pollen shed and mid- to late-season pistil receptivity, similar to Pawnee and Peruque. Good precocity and yield potential. Ripens very early, about 11 Sept. at Brownwood, Texas. Resistant to scab, downy spot and vein spot.

Owens (*Chiles AAA*).—Orig. as chance seedling among 2000 FL nursery seedlings planted in 1900 on Cuba Island in Moon Lake, Coahoma County, Miss., by F.M. Owens. Discovered in 1914. Introd. in 1930. Nut: oblong elliptic with obtuse apex and base; round in cross section; shell rough, ridged, with silver sheen; 48% kernel; kernels with wide dorsal grooves, narrow dorsal ridge, deep basal cleft; cracks out well. Protandrous, with early to midseason pollen shed and mid- to late-season receptivity. Strong tree structure withstands wind and sleet. Annual bearer of moderate crop. Resistant to scab. Resistant to black pecan aphid. Recommended (1990) for planting in Alabama, Arizona, and Mississippi .

Pabst.—Seedling selection from nuts obtained in New Orleans by W.B. Schmidt and planted in 1875 at Ocean Springs, Jackson County, Miss. First propagated by C.E. Pabst in 1890, and named in his honor in 1893 by B.M. Young of Morgan City, La. Nut: Oblong to oblong elliptic, with obtuse apex and obtuse to rounded base; round in cross section; 59 nuts/lb, 47% kernel; kernels golden to light brown, with wide, shallow dorsal grooves, pronounced secondary dorsal grooves, and a prominent basal cleft. Shells are thick, with good sutures that protect nuts after drop. Protandrous, with midseason pollen shed and late pistillate receptivity. Very susceptible to scab. Trees are vigorous and attain exceptional height with maturity.

Patrick.—Orig. in Rogers County, Okla., as a result of a survey of native trees in northeast Oklahoma. Discovered 1952. Introd. in 1954 by the Dept. of Hort., Oklahoma A&M College, Stillwater. Nut: oblong with obtuse apex and rounded, apiculate base; round in cross section; 115 nuts/lb, 69% kernel; narrow dorsal grooves and tight basal cleft. Produces well, bears biennially. Resistant to scab.

Pawnee.—Orig. by controlled cross (Mohawk x Starking Hardy Giant) made by L.D. Romberg, U.S. Dept. of Agr.–Agr. Res. Serv., Brownwood, Texas. Cross made in 1963; selected and tested by T.E. Thompson and R.E. Hunter as 63-16-125. Released in 1984 to be grown throughout the U.S. pecan belt. Nut: elliptical with obtuse apex and rounded base; laterally compressed in cross section; 44 nuts/lb, 58% kernel; kernels golden in color, with wide dorsal grooves and deep basal cleft. Protandrous, with early to midseason pollen shed and mid- to late season receptivity. Precocity and production medium, with some tendency to biennial bearing. Nut matures early, latter half of September in Brownwood, Texas. Medium susceptibility to scab, fair resistance to downy spot, outstanding resistance to yellow aphids. Recommended (1990) for planting across Texas, Oklahoma, Kansas, to Alabama and Arizona.

Pensacola Cluster.—Orig. as chance seedling in Pensacola, Fla., discovered by E W. Moring in 1958. Introd. in 1960. Plant patent 2099; issued 31 Oct. 1961; assigned to Simpson Nursery Co., Monticello, Fla. Nut: oblong with obtuse apex and base; round in cross section; resembles Van Deman; 32 nuts/lb. Protogynous, with late pollen shed and midseason receptivity. Nuts borne in clusters of six to 10. Nuts mature in early Oct. Very susceptible to scab in Louisiana.

Peruque.—Orig. as native seedling near Mississippi River, in St. Charles, Mo., on property owned by George Hunn and known locally in 1918 as Hunn. Tree first propagated in 1935 by R. Richterkessing, with permission of new owner Wm. Krause who suggested the name Peruque. Introd. in 1953. First commercially propagated by Gerardi Nursery, O'Fallen, Ill. Nut: ovate, with obtuse apex and rounded base; laterally compressed in cross section; dark suture line at base; apex slightly furrowed on abaxial and adaxial surfaces; 81 nuts/lb, 59% kernel; kernels golden, with tight dorsal grooves and a deep basal cleft. Protandrous, with early to midseason pollen shed, and midseason receptivity. Ripens very early. Good scab resistance. Recommended (1990) for commercial plantings in Arizona, Kansas, Kentucky, Missouri, Oklahoma, and Tennessee and for homeowners in all of those states except Missouri.

Posey.—Native selection from Gibson County, Ind., introd. in 1911 by W.C. Reed. Nut: oval elliptic with an acute apex and obtuse base; laterally compressed in cross section with raised suture; 63 nuts/lb, 54% kernel; kernels light brown in color, with prominent secondary dorsal grooves. Protogynous bloom pattern. Late to break buds in the spring. Resistant to scab. Early nut maturation. Recommended (1990) for planting in Kansas, Kentucky, Missouri, and Tennessee.

Price-Fleming.—Seedling selection of unknown parentage discovered in the 1950s by J.W. Price on his property in North Central Nash County, N.C. Plant patent 7228; issued 15 May 1990. Nut: oblong elliptic with obtuse apex and base; asymmetric; 44 nuts/lb, 55% kernel; kernels cream to golden in color. Resistant to scab.

Prilop of Lavaca (*Steffek, Kallus*).—Native selection from the Lavaca River, originally growing 2 1/2 miles south of Halletsville, in Lavaca County, Texas. The original tree grew near a river crossing, on land owned by G.W. Prilop, and was called Prilop by the first grafters. The cultivar is also called Steffek due to the work of F.T. Steffek in propagating the tree. Steffek collected graftwood from the ortet and propagated an orchard in 1939, before the destruction of the original tree due to a flood in 1940. The cultivar has also been called Kallus, after Louis Kallus, Sr., an early proponent of the cultivar who sent wood to the U.S. Dept. of Agr.–Agr. Res. Serv. Pecan Station in 1976. Nut: oblong with acute apex and obtuse base; round in cross section; shell smooth, with few markings; 78 nuts/lb, 57% kernel; kernels golden, with very thin dorsal ridge, especially at apex. Protandrous. Resistant to scab. Texas State Champion Native in 1991. One of the most widely propagated Texas natives in recent times, with more than 3000 acres in Lavaca County.

Pseudo-Desirable.—See Superdesirable.

Risien 1.—Orig. in San Saba, Texas, as open-pollinated seedling of unknown parentage, by Guy Risien. Propagated at U.S. Dept. of Agr. Pecan Field Sta., Brownwood, Texas, in 1938. Nut: ovate elliptic, with acute apex and base; round in cross section; 80 nuts/lb; kernels with wide, shallow dorsal grooves. Protandrous. Tree open, weeping habit. Very susceptible to scab. Pollen parent of Tejas.

Royal.—Orig. as open-pollinated seedling of Schley in Riverside, Calif., by Robert A. Harris, Arlington, Calif. Introd. in 1949. Plant patent 833; issued 26 Apr. 1949; assigned to Lawrence Sherwood, Sherwood Specialty Nursery, Fullerton, Calif. Nut: oblong, slightly obovate, with acute apex and obtuse, asymmetric base; round in cross section; 60 nuts/lb; shell dense but thin, well-sealed; cracks easily; kernel smooth, oily, fills well, keeps well without becoming rancid. Precocious and prolific. Ripens early.

San Saba.—Native selection from the San Saba River, San Saba County, Texas, made by E.E. Risien in 1882. Tree discovered when nuts were entered in a contest sponsored by Risien in an effort to find valuable trees. Risien purchased the land to obtain the tree and found the top had been cut out to harvest the crop. Introd. by Risien in 1893. Nuts from this tree were planted in 1895 by Risien to form his orchard, from which were selected San Saba Improved, Sovereign, Onliwon, Jersey, and Texas 60. Western Schley was once thought to be a progeny of San Saba, but could not be based on genotypes for malate dehydrogenase and leucine aminopeptidase. Nut: oval to orbicular, with obtuse apex and obtuse to rounded base; round in cross section; 103 nuts/lb, 64% kernel; kernels golden to light brown in color, with wide dorsal grooves and prominent secondary dorsal grooves. Protandrous. Susceptible to scab.

San Saba Improved.—Seedling selection grown from seed of San Saba planted by E.E. Risien, in his orchard on the San Saba River, San Saba County, Texas. Introd. 1912. Nut: oblong elliptic with acute apex and obtuse base; asymmetric; prominent dark stripes on shell apex; round in cross section; 61 nuts/lb, 59% kernel; kernels golden to light brown in color, with dorsal ridge pinched at apex and flared at base above a prominent basal cleft. Protandrous with early pollen shed and midseason pistillate receptivity. Susceptible to scab.

Schley.—Seedling selection from nut planted about 1881 by A.G. Delmas at Scranton, Jackson County, Miss. Once thought to be from Stuart seed, but isozymes negate that relationship (Schley has *aa* genotype for phosphoglucose isomerase, and *bb* for malate dehydrogenase, while Stuart is *bb* for the former and *ac* for the latter). Named by Delmas in 1898 in honor of Admiral Winfield Scott Schley, Commander of the U.S. Naval Forces in the Spanish American War. First commercially propagated under the name Admiral Schley in 1902 by D.L. Pierson, Monticello, Fla. Nut: oblong with acute apex and base; asymmetric;54 nuts/lb, 62% kernel; kernels golden with narrow dorsal grooves. Protogynous, with midseason pollen shed and midseason pistillate receptivity. Long considered the standard of nut quality for the pecan industry. Used extensively in breeding. Female parent of Sioux, Shawnee, Cherokee, Oconee, Woodroof, and Cape Fear. Male parent of Apache. Susceptible to scab, black pecan aphid, and a trunk disorder described as phomopsis canker. Recommended (1990) for planting in Arizona.

Schley-Harris.—See Harris Super.

Select.—Orig. as open-pollinated seedling of Altman in Riverside, Calif., by Robert A. Harris. Introd. in 1943. Plant patent 510; issued 28 Apr. 1942; assigned to Lawrence Sherwood, Sherwood Specialty Nursery, Fullerton, Calif. Nut: obovate, with obtuse apex and base; laterally compressed in cross section; 46 nuts/lb; shell thin, fills well; kernels with deep, tight dorsal grooves and deep secondary grooves. Tree: vigorous and pendulous growth; hardy; wide climatic adaptability. Precocious and productive. Ripens early in Riverside, Calif.

Seminole.—Orig. as chance seedling in Waukeenah, Fla. Selected in 1905. Introd. in 1923. Nut: oblong with long acuminate apex and acute base; often asymmetric; round in cross section; 60 nuts/lb, 53% kernel; kernels with narrow dorsal ridge, wide dorsal grooves, deep basal cleft. Protogynous. Bears regularly. Susceptible to vein spot in Brownwood, Texas. Recommended (1990) for planting in Arizona.

Shawnee.—Orig. by controlled cross (Schley x Barton) made by L.D. Romberg, U.S. Dept. of Agr.–Agr. Res. Serv., Brownwood, Texas. Cross made in 1949. Scion budded into bearing tree in 1950, bore first fruit in 1955; tested as 49-17-166 by Romberg and G.D. Madden. Released in 1968. Nut: oblong, with obtuse apex and base; round in cross section; shell light brown with few dark stripes; 48 nuts/lb, 58% kernel; kernels golden in color, with very narrow dorsal ridge. Protogynous, with midseason pollen shed and early to midseason receptivity. Medium precocity, productive. Ripens early midseason, one week before Stuart. Susceptible to scab, but manageable with fungicides. Recommended (1990) for commercial and homeowner plantings in Oklahoma, and for trial by commercial growers in Alabama and Louisiana.

Shoshoni.—Orig. by controlled cross (Odom x Evers) made by L.D. Romberg, U.S. Dept. of Agr.–Agr. Res. Serv., Brownwood, Texas. Cross made in 1944, first fruited in 1950, tested as 44-15-59 by Romberg and G.D. Madden. Released in 1972. Nut: oval elliptic with obtuse apex and rounded base; laterally compressed in cross section; 41 nuts/lb, 53% kernel; kernels wrinkled with very wide dorsal grooves and deep basal cleft. Protogynous, with midseason pollen shed and early to midseason receptivity. Very precocious and prolific with a severe tendency to alternate bear. Ripens in early midseason, over a week before Stuart. Susceptible to scab. Freeze injury to bearing trees in Oklahoma. Recommended (1990) in Arizona, South Carolina, and Tennessee.

Sioux.—Orig. by controlled cross (Schley x Carmichael) made by L.D. Romberg, U.S. Dept. of Agr.–Agr. Res. Serv., Brownwood, Texas. Cross made in 1943 in H.G. Lucas orchard, Brownwood, Texas. First fruited in 1948; selected in 1949; tested as 43-4-6. Released in 1962. Nut: oblong, with acute apex and right angled base; round in cross section; 60 nuts/lb, 60% kernel; kernels cream to golden in color, with narrow dorsal grooves, smooth texture, high oil. Protogynous, with mid season pollen shed and early to midseason receptivity. Medium precocity, good productivity. Susceptible to scab in Georgia and Louisiana. Recommended (1990) for commercial and homeowner plantings in Oklahoma and Texas.

Slybert.—Orig. by controlled cross (Schley x Halbert) made in Mason, Texas, by Frank J. Willmann. Seed planted in 1914, along with sibling Halsly. Introd. in 1924. Nut: resembles Schley, shell thin, shells easily; 60 nuts/lb, 59% kernel. Tree semi-dwarf. Precocious, very prolific. Initiates growth early in spring, susceptible to late spring frost.

Sovereign.—See Texas Prolific.

Squirrel.—See Squirrel's Delight.

Squirrel's Delight (*Squirrel*).—Seedling selection made by E.E. Risien, San Saba, San Saba County, Texas. Once thought to be a progeny of San Saba, but mdh-1 genotype is inconsistent (*aa* for San Saba, *bb* for Squirrel's Delight). Named for squirrel predation that accompanies the cultivar's early nut ripening. Nut: oval elliptic with obtuse apex and obtuse to rounded base; laterally compressed in cross section; 47 nuts/lb, 55% kernel; kernels golden, with deep dorsal grooves, well developed secondary dorsal grooves and a deep ventral groove. Susceptible to scab. Sometimes called Squirrel in Oklahoma, where there is still significant acreage. Recommended (1990) for planting in Oklahoma.

Starking Hardy Giant.—Orig. as native seedling in Brunswick, Mo., by George James. Discovered in 1947. Introd. in 1954. Plant patent 1361; issued 15 Mar. 1955; assigned to Stark Bros. Nurseries & Orchards Co., Louisiana, Mo. Nut: oblong elliptic with obtuse apex and base; round in cross section; 78 nuts/lb, 58% kernel; kernels finely wrinkled with narrow dorsal grooves and a deep, narrow basal cleft. Protandrous. Ripens very early, first week in September. Tree vigorous, hardy, medium productivity. Recommended (1990) for planting in Arizona.

Stark Surecrop.—Orig. as chance seedling in Calhoun County, Ill., by W.T. Osterman, Nebo, Ill. Discovered in 1959. Introd. in 1970. Plant patent 3158; issued 9 May 1972; assigned to Stark Bros. Nurseries & Orchards Co., Louisiana, Mo. Nut: oval elliptic with acute apex and rounded base; nuts often asymmetric; round in cross section; shell with ridges; kernels with flared dorsal grooves, deep secondary dorsal grooves. Ripens mid-October in west central Illinois. Tree large, vigorous, hardy to –18F, productive.

Steffek.—See Prilop of Lavaca.

Steuck (*Hirschi*).—Native seedling discovered in 1922 by J.F. Tiedke on bottom land of the Osage River near Papinsville, Bates County, Mo., on land owned by G.H. Steuck. Introd. in 1929. Popularized in 1940 by A.G. Hirschi of Oklahoma City, Okla., as Hirschi and propagated in 1954 by Inter-State Nursery, Hamburg, Iowa, under that name. The name Hirschi is much more widely known than Steuck. Nut: oblong, with acute apex and base; round in cross section; shells with prominent dark stripes; 72 nuts/lb, 49% kernel; kernels golden to light brown in color, with deep, narrow dorsal grooves; tight ventral groove and basal cleft. Protandrous , with midseason pollen shed and mid- to late-season pistillate receptivity. Initially considered scab resistant, but now quite susceptible. Recommended (1990) (as Hirschi) for planting in Kansas, Missouri, and Oklahoma.

Stuart.—Seedling selection from the orchard of J.R. Lassabe at Pascagoula, Jackson County, Miss. Lassabe planted the orchard about 1874 using nuts of unknown parentage obtained from Mobile, Ala. After the orchard was purchased by Captain E. Castanera, the tree gained fame for its high yields. First propagated by A.G. Delmas in 1886 as Castanera. Propagated about 1890 by J. Keller and Col. W.R. Stuart of Ocean Springs, Miss., under the name Stuart. Nut:

oblong elliptic with obtuse apex and rounded base; round in cross section; dark stripes on shell; 51 nuts / lb, 49% kernel; kernels golden to light brown in color with wide, shallow dorsal grooves, deep secondary dorsal grooves and a pronounced basal cleft. Protogynous with late pollen shed and midseason pistillate receptivity. Resistant to scab. Susceptible to downy spot, black pecan aphids, and yellow aphids. Late to begin growth in spring, making the cultivar hardy in North Georgia, AK, and Oklahoma. Slow to bear. There are more acres of Stuart trees than any other cultivar. Recommended (1990) for planting in Alabama, Arizona, Florida, Georgia, Mississippi , North Carolina, Oklahoma, South Carolina, and Tennessee.

Success.—Seedling selection from orchard planted about 1890 by W.B. Schmidt, Ocean Springs, Miss. Selected for excellent kernel quality by T. Bechtel in 1901; propagated in 1902; introd. in 1903. Nut: oval elliptic with obtuse asymmetric apex and obtuse to rounded base; round in cross section; shell dark striped at apex; 48 nuts / lb, 50% kernel; kernels golden to light brown in color, with wide, shallow dorsal grooves, prominent ventral groove and wide, secondary ventral grooves. Protandrous, with midseason pollen shed and mid- to late-season receptivity. Female parent of Choctaw and Mohawk. Male parent of Barton, Comanche, and GraTex. Very susceptible to scab, although once quite resistant. Very susceptible to shuck disorders "shuck dieback" (stress related) and "stem end blight" [possibly caused by *Glomerella cingulata* (Ston.) Spa. and V. Sch.]. Still a common cultivar found in old orchards, especially in Texas and Alabama, often planted with Stuart. Recommended (1990) for planting in Arizona and South Carolina.

Sullivan.—Orig. by controlled cross (Stuart? x Nugget) made at New Mexico State Univ. Agr. Expt. Sta., Albuquerque. (Sullivan has *bd* genotype for malate dehydrogenase, as does Nugget, while Stuart is *ac*). Tested as N2-G. Introd. in 1983. Nut: oblong, with acute apex and base, round in cross section; 65 nuts / lb, 55% kernel; kernels with very tight, deep dorsal grooves beside straight, projecting dorsal ridge; shallow basal cleft. Protogynous, 'Peruque' recommended as pollinizer. Ripens early, needing 175-day season. Named for D.T. Sullivan, Professor Emeritus, Hort. Dept. New Mexico State Univ.

Sumner.—Orig. as chance seedling in Tift County, Ga., discovered by W.E. Sumner about 1932. Nut: oblong, with obtuse apex and base; round to flattened in cross section; 39 nuts / lb, 61% kernel; kernels with wide dorsal grooves and deep secondary dorsal grooves. Protogynous, with mid- to late season pollen shed and early to midseason receptivity. Precocious and prolific, with consistent production. Ripens late, almost two weeks after Stuart. Generally resistant to scab, although recently susceptible in some locations. Susceptible to black pecan aphids. Recommended (1990) for planting in Alabama, Florida, Georgia, and Louisiana.

Superdesirable (*Pseudo-Desirable*).—Orig. in Ocean Springs, Miss., by Charles Forkert. Apparently a chance seedling discovered in 1930 in a variety trial, when one tree of six sent to the U.S. Dept. of Agr. Pecan Laboratory as Desirable by the nurseryman, T. Bechtel, proved to be different. Nut: oblong, with obtuse apex and obtuse to rounded base; round in cross section; kernels with wide, flared dorsal grooves and deep secondary dorsal grooves. Protogynous (while Desirable is protandrous). Moderately susceptible to scab. Resistant to bunch disease. Tree vigorous and erect, with bark and foliage darker than Desirable; fall foliage retention better than Desirable.

Surprize.—Orig. as chance seedling grafted to Pioneer in 1963 in Baldwin County, Ala., in orchard of L.V. Underwood. Scion died and stock grew to "surprise" the orchardist and become the Surprize tree. Discovered in 1968. Seedling grew from unknown seedstock (Success suspected based on nut shape; consistent with isozyme genotypes). Tested by Gulf Coast Substation, Fairhope, Ala., from 1983. Introd. commercially by Dellwood Nursery, Foley, Ala., about 1983. Nut: ovate elliptic, with acute apex and rounded base; flattened in cross section; 31 nuts / lb, 51% kernel; kernels with wide dorsal grooves, concave ventral surface. Initiates growth late, mid-April in Fairhope. Protandrous. Moderate precocity, with consistent production on mature trees. Ripens mid- to late-season, about 6 days after Stuart. Good disease resistance in south Alabama. Recommended for trial planting in Alabama.

Teche.—Sent by mistake from R. Frotscher to nurseryman W. Nelson in about 1885, in a shipment of Frotscher wood. Distributed by Nelson along with orders for Frotscher. Originally called several names referring to mixup: Fake Frotscher, False Frotscher, Frotscher #2, Spurious Frotscher, Duplicate Frotscher. Named Teche by National Nut Growers to clarify the problem and to note the origin of the tree from Bayou Teche, La. In 1990, there were still more than 1000 acres of this cultivar in Georgia. Nut: elliptic to oblong, with acute apex and base; few stripes; 65 nuts/lb, 45% kernel; kernels light brown in color, with very prominent apex and deep basal cleft. Trees have twisted pattern of growth in trunks and major limbs that keep the tree small, compared to Frotscher. Susceptible to scab, vein spot, downy spot, and yellow aphids. Late to mature nuts.

Tejas.—Orig. by controlled cross (Mahan x Risien 1) made by L.D. Romberg, U.S. Dept. of Agr.–Agr. Res. Serv., Brownwood, Texas. Cross made in 1944; scion budded into bearing tree in 1945; first fruited in 1949; tested as 44-10-293. Released in 1973. Nut: oblong, with acute apex and base; round in cross section; 54 nuts/lb, 54% kernel; kernels with shallow, wide dorsal grooves; excellent shelling characteristics. Late to initiate spring growth. Protogynous, with midseason pollen shed and early to midseason receptivity. Precocious and very productive. Ripens about 10 Oct. at Brownwood. Tree vigorous, with strong crotch angles. Very susceptible to scab.

Texas Prolific (*Sovereign*).—Seedling selection made by E.E. Risien from his orchard of San Saba seedlings. Nuts were planted in 1895 and buds from promising seedlings were budded into the tops of mature trees in 1897. One such bud bore fruit in 1898. Named Sovereign in 1899. Risien changed the name to Texas Prolific in 1900, and most trees of the cultivar are now called by that name. Nut: oblong elliptic with right angled apex and obtuse base; rough shell; 51 nuts/lb, 51% kernel; kernels golden in color, with deep dorsal grooves, secondary grooves, ventral groove and basal cleft. Protandrous, with early pollen shed and midseason receptivity. Mature trees have shaggy bark. Susceptible to stem phylloxera.

Tucker's Favorite.—Orig. by controlled cross (Stuart x Schley) made by William R Tucker, Dawsonville, Dawson County, Ga. Cross made in 1940, selected in 1956. Plant patent 2156; issued 17 July 1962. Nut: elliptic, with obtuse apex and obtuse to rounded base; round in cross section; shell ridged, striped as in Stuart; shape similar to Stuart, but more pointed at apex; 44 nuts/lb, well filled; kernels with wide, flared dorsal grooves and a deep, very wide basal cleft. Heavy production, ripens late, last half of November, or 15 days after Stuart. Dense foliage, resistant to scab.

Van Deman.—Seedling grown from a nut planted in 1836 by Duminie Mire, Union, St. James Parish, La. The nut was collected from an outstanding tree growing on nearby land owned by Mr. Gravois. Grafted by Emil Bourgeois as Duminie Mire in 1877 at Rapidan Plantation. Bourgeois later sold grafted seedlings. Renamed Van Deman by W.R. Stuart, Ocean Springs, Miss., in honor of H.E. Van Deman, U.S. Dept. of Agr. pomologist. Sold by H. Post, Fort Worth, Texas, in 1900 under the name Paragon. In 1990, there were more than 1000 acres of Van Deman planted in Georgia, and about 800 in Texas. Nut: oblong with acute apex and obtuse base; flattened in cross section; 67 nuts/lb, 44% kernel; kernels cream to golden in color, with deep, narrow dorsal grooves and a deep basal cleft. Protogynous, with late pollen shed and midseason pistillate receptivity. Susceptible to scab, vein spot and downy spot.

Western (*Western Schley*).—Seedling selection made by E.E. Risien, San Saba, San Saba County, Texas; introd. around 1924. Named in relation to Schley, which it resembles in nut shape. Once thought to be a progeny of San Saba, but could not be based on genotypes for malate dehydrogenase and leucine aminopeptidase. There are more trees of Western than any other cultivar in the western pecan growing region. Nut: oblong elliptic to oblong, with right angled apex and acute base; asymmetric; round in cross section; shell surface rough; 57 nuts/lb, 58% kernel; kernels golden to light brown, with deep, tight dorsal grooves that trap packing material and cause kernels to break during shelling. Very susceptible to scab and downy spot. Recommended (1990) for planting in Arizona, California, New Mexico, Oklahoma, and west Texas.

Western Schley.—See Western.

Wichita.—Orig. by controlled cross (Halbert x Mahan) made by L.D. Romberg, U.S. Dept. of Agr.–Agr. Res. Serv., Brownwood, Texas. Cross made in 1940, scion budded into bearing tree in 1941, first fruited in 1947, tested as 40-9-193. Released in 1959. Nut: oblong, with acute to acuminate, asymmetric apex and rounded apiculate base; round in cross section; 43 nuts/lb, 62% kernel; kernels golden to light brown in color with narrow dorsal grooves and a wide, shallow basal cleft. Protogynous, with midseason pollen shed and early to midseason receptivity. Precocious and prolific. Ripens in midseason, with Western, 4 to 20 days before Stuart, depending on location (earlier in SW). Tree moderately upright, vigorous, often with a late flush of growth. Very susceptible to scab. Recommended (1990) in Arizona, California, New Mexico, Oklahoma, and Texas.

Willman.—Orig. by controlled cross (Commonwealth x Schley) made by Frank J. Willmann, La Grange, Texas. Selected in 1937. Introd. in 1940. Nut: oblong with right angled apex and obtuse base; round in cross section; 48 nuts/lb, 58% kernel; kernels with wide, flared dorsal grooves, deep secondary dorsal grooves, and shallow basal cleft. Protogynous. Tree vigorous with heavy and regular production.

Witte.—Orig. as native seedling in Burlington, Iowa, selected by John H. Witte. Introd. about 1925. Discovered by pioneer woodsman who brought it to the attention of Witte. Nut: oval elliptic with obtuse apex and rounded, often asymmetric base; 82 nuts/lb, 52% kernel; kernels with wide dorsal grooves, deep secondary dorsal grooves, and wide, shallow basal cleft. Protandrous. Starts bearing very late. Very early nut maturity, being about 3 weeks ahead of Colby.

Woodard.—Orig. as chance seedling on Garrison farm near Ty Ty, Ga. Discovered by R.W. Patrick. Original tree was 75-80 years old in 1982. Grafted trees included in cultivar trials at Univ. of Georgia Coastal Plain Expt. Sta., Tifton, Ga., in 1955. Locally known as GCPES Seedling One or Patrick. Selected by R.E. Worley and O.J. Woodard of the above station. Introd. in 1982. Nut: obovate oblong, with obtuse apex and acute base; flattened in cross section; 60 nuts/lb, 54% kernel; kernels cream to golden in color, wrinkled, with wide, shallow dorsal grooves. Unusual in having "inner shell" formed by packing material in the dorsal grooves. Protandrous, with early to midseason pollen shed and midseason receptivity. Medium precocity, and moderate production, comparable to Desirable. Ripens late, about 10 days after Stuart. Susceptible to scab. Very susceptible to powdery mildew. Nuts damaged by mechanical harvest. Recommended (1990) for planting in Arizona.

Woodroof.—Orig. as chance seedling (progeny of A-93) selected by J.W. Daniell, B.B. Brantley, and K.E. Heaton, Univ. of Georgia, Experiment, Ga. Reported as F_2 progeny of A-93, which is supposedly a Schley x Moneymaker cross. However, Woodroof has a *c* allele for malate dehydrogenase that is not present in any of its putative ancestors. Introd. in 1983. Nut: elliptic, with obtuse, often asymmetric apex and obtuse base; flattened in cross section, with prominent suture; 48 nuts/lb, 45% kernel; kernels wrinkled, with wide, flared dorsal grooves and deep secondary dorsal grooves. Protogynous. Resistant to scab.

PERSIAN WALNUT

Aco.—Orig. in Dayton, Ohio, by Arthur C. Orth. Introd. in 1964 by the Gilbert L. Smith Nursery, Millerton, N.Y. Open-pollinated seedling of a Crath Carpathian type; seed planted in 1937; first propagated by Homer G. Bowser, Xenia, Ohio; tested as Orth 2 and Bowser. Nut: large; shell thin, good cracking quality; kernel well developed, flavor good. Tree: vigorous; medium hardy.

Adams.—Orig. in Salem, Ore., by Moses P. Adams. Introd. in 1952; assigned to Oregon State College, Corvallis. Open-pollinated seedling of P.I. 18256; selected in 1940. Nut: similar to Franquette but darker colored and matures 7 days earlier. Tree: hardy; may have some resistance to blight.

Amigo.—Orig. in Davis, Calif., by E.F. Serr and H.I. Forde, California Agr. Expt. Sta. Introd. in 1968. Sharkey x Marchetti; sibling of Chico; cross made in 1955; selected in 1963; tested as UC 56-226. Nut: large; round; shell seal poor in some years; kernel 54% of nut, quality high; 80% of those tested had light colored kernels; matures early to midseason. Tree: productive, many clusters of four nuts; 80% of lateral buds producing pistillate flowers; leafs out 14 days after Payne making it a valuable pollenizer for Hartley, Midland, Pioneer, Tehama, and Pedro. The name was chosen because of its meaning "friend" and because it was decided to give Spanish names to new varieties especially valuable as pollenizers.

Ashley.—Orig. in Woodland, Calif., by Pallas Neal Ashley. Introd. in 1960. Plant patent 2168; 4 Sept. 1962; assigned to Stuke Nursery Co., Gridley, Calif. Parentage unknown; discovered about 1945. Nut: large; ovoid; kernel averages 54% of nut, large, plump, light tan, flavor good, quality high; shell seal fair but adequate; ripens early; resembles Payne. Tree: small; early bearing; leafs out early, 4 days after Payne; fruitfulness on lateral buds 90% or more; extremely heavy yielding; chilling requirement medium; pollen shedding overlaps pistillate bloom; needs consistent pruning; often planted as filler trees.

Ashworth.—Orig. in Heuvelton, St. Lawrence County, N.Y, by Fred L. Ashworth, St. Lawrence Nurseries. Named and introd. in 1948 by John Hershey, Downingtown, Pa. Seed secured in 1937 by Ashworth from the Wisconsin Hort. Soc., imported from the Carpathian Mountains of Poland; selected in 1947; Carpathian type. Nut: rather small; well-shaped; kernel sweet; ripens early; seeds germinate well. Tree: very hardy, down to –36F; strong, vigorous.

Ata.—Orig. in Sangamon County, Ill., by Herman C. Helmle. Introd. in the early 1950s. Carpathian type; seed secured about 1935 from Wisconsin Hort. Soc.; seed originally from Carpathian Mountains in Poland; tested as Helmle 9. Nut: kernel flavor good; rather difficult to extract kernel as it fills the convolutions at base of shell; quality fair. Tree: produces a high percentage of perfect flowers. The name Ata was suggested by the word "atavism."

Bandoni.—Orig. in Merced, Calif., by Guido Bandoni. Introd. in 1963. Bud mutation of Eureka; discovered in orchard of Jack Muratore, Merced; tested and propagated by G. Bandoni. Nut: averages 64% large by weight; averages 51% kernel; harvested third week of September in 1962; sometimes borne in clusters; resembles Eureka in shape and other characters; appears to have higher color and less kernel shriveling than Eureka. Tree: averages about 60% of lateral buds producing pistillate flowers; starts growth about Payne time, 10 days before Eureka.

Baum (*Improved Franquette*).—Orig. in San Jose, Calif., by J.G. Baum. Introd. in 1920. Open-pollinated seedling of Franquette; selected in 1918. Nut: quality excellent; shell strong, well-filled; kernel light colored, fills shell; ripens fairly early, before Franquette and Mayette; resembles Franquette but much heavier. Tree: growth willowy; blooms later than other varieties;

production average; performs better in cooler districts such as Santa Clara Valley and Lake County than in the Sacramento Valley, Calif.

Beck.—Orig. in Hamilton, Mich., by Daniel Beck. Introd. about 1931. Chance seedling. Seed apparently came from England and is thought to have been planted about 1918. Nut: small; cracking quality good; kernel bright, flavor very good. Apparently no longer being propagated.

Bedford.—Orig. in Bedford, Va., by H.F. Stoke of Roanoke, Va. Introd. in 1940. Parentage unknown; discovered about 1930. Nut: resembles Mayette but smaller; shell medium thick, well-filled, well-sealed; quality very good. Tree: bears regularly; hardy; blooms in late midseason; self-fruitful; bears at an early age; grows moderately.

Bel.—Orig. in Doetinchem, The Netherlands, by C.J. Gerritsen, Inst. of Hort. Plant Breeding, Wageningen, The Netherlands. Introd. in 1954. Parentage unknown; selected in 1953. Nut: large; kernel color and flavor very good; shell opens easily.

Boone.—Orig. in Erlanger, Kenton County, Ky., by Philip Taliaferro. Named, propagated, and introd. in 1953 by Sterling Rouse, Florence, Ky. Parentage unknown; from a Carpathian type. Nut: flavor rich and sweet with no bitter taste; shell medium thickness; 40 per lb. Tree: bears well and early; self-fruitful; very hardy.

Breslau (*Russ*).—Orig. in Halsey, Ore., by E. Russ. Introd. in 1935. Chance seedling from seed imported about 1917 by Russ from Breslau, Germany. Nut: Bijou type; very large; nearly round; shell medium bright, well-sealed, attractive, thin; kernel medium plump, quality good, sometimes flavor mildly astringent, moderately rich, extracted easily from shell. Tree: precocious bearer; heavy crops; apparently hardier than Thomas eastern black walnut; twigs slender; vigorous.

Broadview.—Orig. in Westbank, British Columbia, Canada, by J.U. Gellatly. Introd. in 1930. Seed brought from the ancestral home of the Utki family in Odessa, Russia, to the North Okanogan Valley, British Columbia; dissemination of trees took place, and Gellatly selected this one; Carpathian type. Nut: large to medium; shell soft, well-sealed; cracking quality good; ripens a little before Franquette, beginning in early September; keeps well. Tree: very productive; very hardy. Both seeds for planting and scionwood distributed under this varietal name.

Broe.—Orig. in Westbank, British Columbia, Canada, by J.U. Gellatly, Gellatly Nut Tree Nursery. Introd. about 1930. Chance seedling from seed supplied by John Broe, Chilliwack, British Columbia. Nut: size medium; shell medium thick; well-filled; flavor good; season late.

Bruce.—See Spurgeon.

Burtner.—Orig. in Mercerburg, Pa., by Fayette Etter of Lemasters, Pa. Introd. in 1951. Parentage unknown; discovered in 1928; selected in 1945; named in 1949. Nut: medium size; well-filled with light-colored kernel; shell well-sealed, of average thickness. Tree: bears regularly; hardy; late leafing, escaping frosts; self-fruitful.

Caesar.—Orig. in Roanoke, Va., by Harvey F. Stoke. Introd. in 1938. Seedling selected in 1934 from Carpathian-type seed imported from Poland by Paul Crath. Nut: very large; kernel quality good; sometimes kernel not fully developed; shell very thin, resulting in some breakage. Tree: bears regularly.

Carmelo.—Orig. in Yuba City, Calif., by C.E. Sullivan. Introd. in 1957. Plant patent 1476; 8 May 1956; assigned to Sierra Gold Nurseries, Yuba City. Open-pollinated seedling of Payne; selected in 1948. Nut: bijou-type; low kernel percent; very large; good quality, little blight or sunburn; ripens about with Payne. An excellent home garden and novelty variety. Tree: protogynous; precocious; leafs out with Payne; productive.

Carpathian D (*Crath D*).—Orig. in the Carpathian Mountain Region in Poland. Scions cut by Paul C. Crath and shipped to James A. Neilson at Michigan State College, East Lansing. Introd. in 1934. Nut: size medium; odd appearance, equilateral triangle; kernels white, sweet; kernel 34.5% of nut; cracking quality by machine good, by ordinary means poor; shape of nut commercially undesirable. Tree: vigorous; slender growth; bark brown colored; easily grafted;

hardy; one of the latest to leaf out in spring; fair producer; consistent pollenizer.

Cascade.—Orig. from a selection made by William Schildgen from the Okanagan Valley, Loomis, WA, being one progeny of a series of crosses of a Russian, perhaps Carpathian, walnut with a Manchurian type, probably Manregian, which he obtained from Oregon. Nut: large, dark-shelled with a large kernel. Tree: leafs out with Harley and is late harvesting; yield is moderate.

Chandler.—Orig. in Davis, Calif., by Harold I. Forde, Univ. of California, Davis. Introd. in 1979. Plant patent 4388; assigned to Regents, Univ. of California, Davis. Pedro x UC 56-224; selected in 1968; tested as 64-172. Nut: resembles Payne; oval, well-sealed; kernel excellent quality, 48% to 53% of nut. Tree: precocious; highly productive; late midseason; leafs out 1 to 4 days after Hartley; Franquette or Cisco as pollenizer may increase yields in some years.

Chico.—Orig. in Davis, Calif., by E.F. Serr and H.I. Forde, California Agr. Expt. Sta. Introd. in 1968. Sharkey x Marchetti. Cross made in 1955; selected in 1961; tested as UC 56-206. Nut: Persian; round; 33 mm in cheek diam.; shell seal fair to good; kernel 49% of nut, quality excellent; 90% of those tested had light colored kernels; size of nut will be too small if pruning neglected, because trees will overbear; matures early to midseason. Tree: leafs out 6 days after Payne; 80% of lateral buds producing pistillate flowers; sheds pollen late, after the peak of its pistillate flower receptivity, making it a suitable pollenizer for moderately early leafing varieties that shed their pollen, early such as Ashley, Marchetti, Lompoc, Serr, Gustine, and Vina. This selection is mainly a pollenizer variety, but may be suitable for a main producing variety in some districts. Named after the city of Chico, near which it has produced well.

Cisco.—Orig. at Davis, Calif., by Gale H. McGranahan, David E. Ramos, Harold I. Forde, and Ronald G. Snyder. Introd. in 1990 pollenizer for Chandler and Howard. Meylan and Pedro, cross made in 1965 by Forde. Moderate pollen crop is produced during period of peak receptivity of Chandler and Howard, beginning by seventh leaf, with light yields as early as second leaf. Nuts: shell medium to light in color; texture medium; slightly larger than Franquette; kernel color variable. Tree: intermediate vigor, making tree moderate size; semi-upright; laterally fruitful; late-leafing.

Clarence.—Orig. in Wassaic, N.Y, by Gilbert L. Smith and William A. Benton, Benton Smith Nut Tree Nursery, Millerton, New York Introd. in 1959. Parentage unknown; selected about 1950; tested as S-25. Nut: Carpathian type; large; quality good. Tree: a late pollinator.

Clinton.—Orig. in Wassaic, N.Y, by Gilbert L. Smith and William A. Benton, Benton Smith Nut Tree Nursery, Millerton, N.Y. Introd. in 1961. Parentage unknown; Carpathian type; selected about 1950; tested as S-5. Nut: large; quality excellent. Tree: self-pollenizer.

Colby.—Orig. in Urbana, Ill., by A.S. Colby, Illinois Agr. Expt. Sta. Introd. in 1952; named in 1951. Open-pollinated seedling of Crath 10; Carpathian type seed secured from Cosseev, Poland, in 1934; first fruited in 1942 in Illinois; tested as Ill. 10. Nut: medium size; thin shell, well-sealed; kernel plump, good flavor; early maturity. Tree: hardy and fruitful in central Illinois; probably best adapted to areas north of the Ohio River; flowers early; protandrous.

Concha.—An introduction from Chile, selected from the orchard of Alfredo Concha by Forde in 1979. Nut: seal is good, kernel wight is 46% of nut weight, and kernel color is variable. Tree: leafs out with Payne, is laterally fruitful, and is harvested with Hartley.

Conway Mayette (*Lindale*).—Orig. in Los Molinos, Calif. Introd. in 1935. Considered to be a bud mutation of Mayette; discovered in 1935. Nut: shell sealed well; matures early; resembles Mayette. Tree: bears regularly; blooms late; less susceptible to blight than early blooming varieties in northern California.

Countryman 1.—Orig. in Ontario, Ore., by P.F. Countryman. Introd. in 1950. Carpathian type; selected in 1950 from nuts secured from the Wisconsin Hort. Soc. in 1937; originally secured from the Carpathian Mountains of Poland. Nut: medium to small; round; not quite as thin-shelled nor as sweet as Franquette; quality good; kernel cracks out well. Tree: large, medium spreading; yields as well as Countryman 3, or 100 to 150 lb per tree; hardy.

Countryman 2.—Orig. in Ontario, Ore., by P. F. Countryman. Introd. in 1950. Carpathian type; selected in 1950 from nuts secured from the Wisconsin Hort. Soc.; originally imported from the Carpathian Mountains of Poland. Nut: large; elongated; shell thin; kernel cracks out well, not as sweet as Countryman 1. Tree: large; more upright than Countryman 1; yield comparable to that of Countryman 1; hardy.

Countryman 3.—Orig. in Ontario, Ore., by P.F. Countryman. Introd. in 1950. Carpathian type; selected in 1950 from nuts secured from Wisconsin Hort. Soc. in 1937; originally from Carpathian Mountains of Poland. Nut: size medium; shell thin, but not as thin as Franquette; kernel quality good; cracks well. Tree: large; very hardy; used for seed propagation as they germinate well; crop averages 100 to 140 lb per tree.

Countryman 4.—Orig. in Ontario, Ore., by P.F. Countryman. Introd. in 1950. Carpathian type; seed originally imported from the Carpathian Mountains of Poland. Nut: size medium; slightly elongated; shell not quite as thin as Countryman l; kernel sweet, cracks out well. Tree: more upright than Countryman 1, with a comparable yield, 100 to 150 lb per tree; hardy.

Countryman 6-X.—Orig. in Ontario, Ore., by P.F. Countryman. Introd. in 1957. Chance seedling of Countryman 1, 2, 3, or 4; Carpathian type. Nut: medium to large; shape like Franquette; shell thinner than Countryman 3; kernel sweet, cracks out well. Tree: hardy; large; upright.

Crath 1 (*Stoke*).—Orig. in Toronto, Ontario, Canada, by P.C. Crath. Introd. about 1933. Seed imported by Crath from the Carpathian Mountains of Poland; planted in 1923; first fruited in 1931; imported in the U.S. between 1931 and 1935. Nut: size medium; excellent quality. Tree: extremely hardy. Apparently the first of the Carpathian-type Persian walnut varieties to be introduced.

Crath D.—See Carpathian D.

Crath-McIntyre.—Orig. in Ontario, Canada. Introd. in 1932. Seed given to a Mr. McIntyre by Paul C. Crath, who collected it in the Carpathian Mountains of Poland; sent to James Neilson, Michigan Agr. Expt. Sta., East Lansing, in 1932. Nut: medium to large; broad; blunt at apex; kernel 45.6% of nut, plump, white, somewhat bitter. Tree: bears well; needs pollenizer, Carpathian D a good pollen source.

Deming.—Orig. in Wassaic, Dutchess County, N.Y, by Gilbert L. Smith and William A. Benton, Gilbert L. Smith & Sons Nursery, Millerton, N.Y. Introd. in 1950. Seedling selected in 1949 from seeds imported from Poland in 1935. Nut: large; quality excellent. Tree: very vigorous; productive; medium late in leafing out; self-pollenized. Placed fifth in Northern Nut Growers' Assn. contest of 1949 to 1950.

Dependahl.—Orig. in Delhi, Ill., by W. Dependahl. Introd. about 1935. Chance seedling; discovered about 1929. Nut: quality fair. Now obsolete.

Drummond.—Orig. in Gustine, Calif., by A.L. Drummond. Introd. in 1955. Plant patent 1861; 1 Sept. 1959. Open-pollinated seedling of Eureka; selected in 1947. Nut: large; long cylindrical; shell light tan, medium thick, seal fair to good, cracks out about 48% kernel; kernels average 90% light color; shell larger and softer than Eureka; ripens 7 days earlier than Eureka. Tree: bears heavily; harvested with light shaking; catkins dehisce about 2 weeks after Eureka.

Early Ehrhardt (*Rutherford Ehrhardt*).—Orig. in Refugio Canyon, Goleta, Santa Barbara County, Calif., by George Rutherford. Introd. in 1942. Bud mutation of Ehrhardt; discovered about 1940 as a large limb. Soft-shelled Santa Barbara type. Nut: resembles parent, but lower quality. Kernel wight is 55% of the nut weight, but the kernel color is poor. Tree: low winter-chilling requirement; starts growth early, even following warm winters; leafs out almost two weeks earlier than Payne. It is protogynous and bears ony on the terminals. Subject to sunburn. No longer grown commercially, but walnut researchers often use it to signal the beginning of the breeding season.

Ehrhorn.—Orig. in Mountain View, Calif., by Adolph Ehrhorn. Introd. in 1943. Considered to be a seedling from the cross Payne x Concord; discovered about 1938 in the orchard of Peter Lusich; propagated and tested by Ehrhorn. Nut: shell rough and more pitted than Payne, seals fairly well; quality of kernel very good.

Equinox.—See Gent.

Etter.—Orig. in Mercerburg, Pa., by Fayette Etter, Etter's Nut Tree nursery. Introd. in 1935. Parentage unknown. Nut: shell thin, well-sealed; kernel sweet, plump.

Eureka.—Originated about 1903 as a seedling tree on the Stone Ranch near Fullerton, Calif., from nuts obtained from the Meek estate in Hayward, Calif. The parent tree was thought to be a Persian or Kaghazi type from Iran. Nut: elongated and rounded at both ends, shell quality excellent, seal good, but kernel color is poor. Tree: leafs out one week and harvests one to two weeks after Payne, little fruitfulness on lateral buds. Yield is moderate. Not highly susceptible to sunburn, but it is late maturing and susceptible to frost. If selfed, Eureka give one-quarter albinos.

Fateley.—Orig. in Franklin, Ind., by Nolan W. Fateley. Introd. in 1956 by J.F. Wilkinson, Indiana Nut Nursery, Rockport, Ind. Seedling from a Carpathian type; discovered about 1940. Nut: very large; shell rough; kernel good. Tree: very hardy; vigorous; bears well.

Fickes.—Orig. in Wooster, Ohio, by W.R. Fickes. Introd. in 1951. Parentage unknown; tested as Fickes 22 and Fickes 122. Nut: Carpathian type; round; size slightly above average; shell very thin, not well sealed; kernel white, very sweet; ripens early. Tree: comes into bearing early; prolific; easy to graft.

Firstling.—Orig. in River Falls, Wis., by Carl Weschcke, Hazel Hills Nursery Co. Introd. in 1942. Seedling selected in 1936; from seed imported from the Carpathian Mountains in Poland by Paul C. Crath, Toronto, Ontario, Canada. Nut: Carpathian type; medium to small; shell thin. Tree: extremely hardy; difficult to graft on native black walnut (*Juglans nigra*).

Gent (*Equinox*).—Orig. in Eugene, Ore., by Edward A. Gent. Introd. in 1963. Considered to be a bud mutation of Baum; discovered in 1930. Nut: very large; kernel good; harvesting begins about 21 Sept.; resembles Franquette. Tree: hardy.

G.O.K. 16.—See Wallick 16.

Grande.—Orig. in Mesilla Park, N.M., by Fabian Garcia, New Mexico. Agr. Expt. Sta. Introd. in 1932. Open-pollinated seedling of Franquette; selected in 1928. Nut: quality fine; resembles Mayette. Tree: prolific bearer.

Gratiot.—Orig. in Perrinton, Mich., by Lee Somers. Introd. in 1962; approval given by Mich. Nut Growers' Assn. and Northern Nut Growers' Assn. Open-pollinated seedling of a Carpathian-type Persian walnut; grown from seed distributed by the Wisconsin Hort. Soc. about 1937; from nuts collected in Poland by Paul C. Crath; tested as Somers NC4. Nut: averages 12.4 g; 38.9 mm long, 30.9 mm wide at cheek, and 31.5 mm wide at suture; kernel averages 50.5% of nut, color light, flavor sweet, oil content 65.2%, ripens in midseason, usually late September. Tree: annual bearer. Variety ranked second in 1959 Michigan contest and fifth in the 1960 Michigan evaluation.

Greenhaven.—Orig. in Perrinton, Mich., by Lee Somers at Greenhaven Farm. Introd. in 1962; approval given by Mich. Nut Growers' Assn. and Northern Nut Growers' Assn. Open-pollinated seedling of a Carpathian type; grown from seed distributed by the Wisconsin Hort. Soc. about 1937; from nuts collected in Poland by Paul C. Crath; tested as Somers NlW4. Nut: averages 12.73 g; 34.5 mm long, 34.8 mm wide at cheek, 33.3 mm wide at suture; kernel color light, flavor sweet, oil content 61.0%, averages 45.8% of nut; ripens in midseason, usually late in September. Tree: consistent annual bearer. Variety rated eighth in 1960 in Michigan evaluation; won first place in the 1961 Tri-State Nut Show, New Haven, Ill.; rated third in the 1959 Michigan contest.

Gustine.—Orig. in Davis, Calif., by E.F. Serr and H.I. Forde, California Agr. Expt. Sta. Introd. in 1968. Waterloo x Payne. Cross made in 1951; selected in 1957; tested as UC 52-61. Nut: long, Eureka type, pointed at the apex, 33 mm in cheek diam.; shell seal good; kernel 53% of nut, high quality; 80% of those tested had light colored kernels;

matures midseason. Tree: very productive; needs heavy heading back to prevent breakage when young; leafs out 6 days after Payne; 80% of lateral buds producing pistillate flowers. Performance has been good in the Gustine district, from which the name was obtained.

Hansen.—Orig. in Clay Center, Ohio, by the father of Awalt Hansen. Introd. in 1950. Chance seedling; selected in 1934; original tree considered to be more than 100 years old; seed that produced this tree considered to have come from Germany. Nut: small to medium; round; shell thin, smooth; flavor mild, sweet, good; kernel percentage high. Tree: very productive; self-fruitful; hardy; resistant to disease and to husk maggot; about 16 inches in diam. in 1961 and about 40 ft tall.

Harriette.—Orig. in San Jose, Calif., by Harry Pyle. Introd. about 1941. Parentage unknown. Nut: large; quality good; resembles Mayette and Meylan.

Hartley.—Orig. in Napa Valley, Calif., by Mr. and Mrs. John Hartley. Introd. about 1925. Parentage unknown; selected in 1909 from seedling trees grown by William Hunter; seed planted in 1892. Nut: large; broad, flat base and pointed tip; shell light colored, thin, seals fairly well; kernel light colored, not tight in shell; flavor mild; matures with the Mayette group and earlier than Franquette. Tree: leafing out medium, 2 weeks after Payne; very productive; higher yields than Franquette; size of tree not as large as Franquette but fruiting area larger; chilling requirement medium; pollinator for Payne. One of the most popular commercial varieties in California.

Helmle.—Orig. in Breckenridge, Sangamon County, Ill., by Herman C. Helmle, Springfield, Ill. Introd. in 1954. Open-pollinated seedling from seed of Carpathian type walnut imported from Poland; selected in 1953 by Geo. L. Hockenyos of Springfield, Ill., and J.C. McDaniel, Illinois Agr. Expt. Sta., Urbana; tested as Helmle 2. Nut: small. Tree: hardy; very productive; valued chiefly for its protogynous flowering as a pollenizer for protandrous Carpathian-group varieties.

Henry.—Orig. in Bluffs, Ill., by Royal Oakes. Introd. in 1955. Open-pollinated seedling of Lake, a Carpathian type; selected in 1948. Nut: medium to large; shell sometimes does not seal well; kernel of very good quality. Tree: fairly productive.

Himalaya.—Orig. in Kalispell, Mont., on the Pinkerman ranch. Introd. in 1968 by David A. Lawyer, David A. Lawyer & Sons Nursery, Plains, Mont. Parentage unknown; nuts brought back from northeast India; discovered about 1960. Nut: medium to large; slightly oval; shell thin; kernel good quality, except when there is a cool summer; harvest in September. Tree: upright, slightly spreading; vigorous; productive; not injured at –42F.

Hodges.—Orig. in Biggs, Calif., by D.L. Hodges. Named and introd. in 1956. Discovered in 1946 as a 25-year-old topworked tree, not as a chance seedling. Nut: bijou-type length 45 mm, diam. 38 mm on young trees; not as large as Willson Wonder, but better sealed than that variety; can be hulled by machinery without excessive breakage; resembles Willson Wonder and other Bijou-type nuts; ripens about 1 week before Payne. Tree: yields very well; original tree 45 ft tall with an 80-ft spread in 1965.

Houston.—Orig. in Byron, Calif., by Alba C. Houston. Introd. in 1958. Parentage unknown; appeared in shipment of Payne trees secured from a San Jose nursery in 1933; discovered in 1947. Nut: shell color similar to Blackmer, and similar to it in size and shape but better sealed; kernel color intermediate between Payne and Franquette, with slight tendency to speckle; kernel averages 52% of nut, being only slightly more than Payne; highly resistant to sunburn; ripens 1 week later than Payne. Tree: growth starts about the same time as Payne; starts bearing 1 year later than Payne; easier to shake than Payne; growth similar to Eureka, only more spreading and not as tall; more vigorous than Payne; pistillate bloom date about 3 days later than Payne; tends to bear most of crop on terminals and subterminals, about 10% of lateral buds fruitful; possibly more susceptible to frosted scale insect than other varieties of the Blackmer type.

Howard.—Orig. in Davis, Calif. by Harold I. Forde, Univ. of California, Davis. Introd.

in 1979. Plant patent 4405; assigned to Regents, Univ. of California, Davis. Pedro x UC 56-224; selected in 1968; tested as 64-182. Nut: resembles Payne; nearly round, shell fairly smooth, well-sealed; kernel 50% to 56% of nut, excellent quality. Tree: smaller than Chandler and Vina; harvest slightly ahead of Hartley; precocious, highly productive; Franquette and Cisco recommended as a pollenizers.

Howe.—Orig. in Brownsville, Ore., by Cecil B. Howe. Introd. about 1955. Chance seedling of Kirk; discovered about 1935. Nut: elongated; shell poorly sealed, relatively hard but not as hard as Franquette, which it resembles; kernels somewhat heavily veined but otherwise not objectionable with respect to color; ripens with Franquette. Tree: growth habit very different from Franquette, growing upright with a central leader, being very much higher (50 ft) than wide (30 ft) when 20 years old; leaf out is earlier than Franquette; reported to be tolerant to walnut blight caused by *Xanthomonas juglandis* (Pierce) Dows.

Idaho.—A protogynous selection from among the large bijou-type nuts. Idaho is said to have origniated in Parma, Idaho, and was introduced to California by E.T. Rice about 1947. Nut: large, but the kernel is of poor quality and its weight is only 35% of the nut weight. Yield is poor. Tree: leafs out with Payne and bears mostly on terminals; cold-hardy.

Illinois 3.—Orig. in Urbana, Ill., by A.S. Colby, Illinois. Agr. Expt. Sta. Introd. in 1940. Open-pollinated seedling of a Crath Carpathian; selected about 1940. Nut: small; quality good. Tree: productive at Bluffs, Ill., but not productive at Urbana.

Improved Franquette.—See Brown.

Jacobs.—Orig. in Elmore, Ohio, by Sylvester Shessler. Introd. about 1950. Chance seedling which arose from seed imported from Hanover, Germany, about 1873; parent tree planted on the Ernsthause farm; seedling tree from it was discovered about 1938 by Shessler on the Jacobs farm. Nut: averages 36 per lb; nearly round, slightly oblong, suture line slightly raised; kernel fills well, sweet; awarded fourth prize in 1950 in the Northern Nut Growers' Assn. contest, Norris, Tenn.

James.—Orig. in Scott County, Ill., by Royal Oakes, Bluffs, Ill. Introd. in 1957. Chance seedling of Carpathian type; discovered in 1952. Nut: kernel quality good. Tree: bears heavily.

Jensen.—Orig. in Hollister, Calif., by Ernest E. Jensen. Introd. in 1956. Open-pollinated seedling of Payne; discovered in the late 1930s. Nut: thick hull; quality good after hot weather; matures early; resembles Payne. Tree: blooms late; heavy producer; many lateral buds fruitful; not blight resistant.

Kentucky Giant.—Orig. in Jefferson County, Ky., by H.B. Briggs. Introd. in 1956 by Sterling Rouse, Florence, Ky. Seedling of Carpathian-type seed secured from the Wisconsin Hort. Society; first crop produced in 1943. Nut: Bijou type; very large; shell thin; matures early. Tree: blooms late; comes into bearing early, about 6 years of age; tolerant to low winter temperatures.

Kentucky Papershell.—Orig. in Erlanger, Kenton County, Ky., by Philip Taliaferro. Named, propagated, and introd. in 1953 by Sterling Rouse, Florence, Ky. Parentage unknown; from Carpathian introductions. Nut: shell very thin; 45 per lb; flavor rich and sweet. Tree: bears early; very vigorous; needs a pollenizer for best results; withstands cold.

King (*King's Crest*).—Orig. in Eagle Creek, Ore., by C.D. Bush. Introd. in 1926. Nut: Franquette type; size medium; shell attractive, well-sealed, symmetrical; ripens with Franquette. Tree: reported to pollenize Franquette.

King's Crest.—See King.

Lake.—Orig. in Bluffs, Ill., by Royal Oakes. Introd. about 1954. Open-pollinated seedling of a Crath Carpathian type; selected in 1950. Nut: large; quality very good; ripens in midseason. Tree: hardy; protogynous; not very productive.

Lewis.—Orig. in Vassar, Mich., by Dwight T. Lewis. Introd. in 1932. Three seeds secured from Hereford, England, by Thomas P. Lewis who planted them in Vassar; one tree lived; selected in 1912. Nut: large. Tree: bears every other year; yields well; cold resistant. Apparently no longer propagated.

Lindale.—See Conway Mayette.

Little York.—Orig. in Dayton, Ohio, by Arthur C. Orth. Introd. in 1964; assigned to Homer G. Bowser, Xenia, Ohio. Open-pollinated seedling of a Crath Carpathian type; seed planted in 1937; tested as Orth 6. Nut: medium small; flavor good. Tree: hardy.

Littlepage.—Orig. in Wassaic, N.Y., by William A. Benton and Gilbert L. Smith, Benton & Smith Nut Tree Nursery. Introd. in 1945. Plant patent 815, 14 Dec. 1948. Seedling of a Carpathian type; seed secured from Wisconsin Hort. Soc. in 1935 and planted that year; fruited in 1944; tested as Carpathian S-61. Nut: large; elliptical; shell light brown, thin, well-sealed, kernel very wrinkled, somewhat dark, medium oiliness, quality good but less so than other varieties; ripens early, about 15 Sept. at place of origin. Tree: very hardy; grows rapidly; bears heavily.

Lompoc.—Orig. in Davis, Calif., by E.F. Serr and H.I. Forde, California Agr. Expt. Sta. Introd. in 1968. Waterloo x Payne. Cross made in 1951; selected in 1957; tested as UC 52-48. Nut: large; pointed; shell seal good; kernel 54% of nut, quality good; 70% of those tested had light-colored kernels; matures midseason or later. Tree: productive; leafs out 5 days after Payne; 50% of lateral buds producing pistillate flowers; best adapted to cooler, semi-coastal regions; husk fly and blight control are needed where present. Named after a city in the Santa Ynez Valley.

Lyell.—Orig. in Spokane, Wash., by Mrs. Lyell J. Watt. Introd. in 1955. Ten seeds from Carpathian Mountains of Poland planted about 1935 by a Mrs. Garvin; one seedling given to Mrs. Watt. Nut: large; kernel flavor good.

Macks.—See Macomber.

Macomber (*Macks*).—Orig. in Escalon, Calif., by A.D. Macomber. Introd. in 1960. Parentage unknown; discovered in 1951 in an orchard of Waterloo. Nut: resembles Waterloo; harvested after Payne. Tree: foliage dense, leafs out 10 days after Payne.

Manregian.—Orig. in Chico, Calif., by U.S. Dept. of Agriculture. Plant Introd. Sta. Introd. and named in 1954 by Moses Adams, Salem, Ore., who received the tree from Chico in 1927; this tree is a seedling of P.I. 18256. Parentage unknown; seed of P.I. 18256 collected by Frank Meyer of the U.S. Dept. of Agriculture in 1906 in mountains north of Beijing, China. Nut: large; round; shell thin but rather hard, with a grayish cast; kernel quality excellent, dark in color. Tree: more hardy than standard varieties; promising as a source of vigorous rootstocks.

Marchetti.—Orig. in Stockton, Calif., by Giulio Marchetti. Introd. in 1953. Probably Payne x Eureka; selected about 1948. Nut: long; size medium, 6.4 g; oblong; kernel light colored; shell very well sealed; resembles Eureka. Tree: small to medium; high proportion (90%) of lateral buds on young trees produce pistillate flowers; production heavy; early bearing; leafing out about 6 days after Payne; requires vigorous rootstock and regular pruning; well adapted to hot, dry summer climate.

McDermid.—Orig. in Ontario, Canada, midway between Point Weller and St. Catherines on the Peter McDermid place. Introd. in 1929 into the U.S. by James Nielsen of Michigan State Univ., East Lansing. A third generation tree, being an open-pollinated seedling of a seedling of the original McDermid variety, which was introd. from Germany about 150 years ago (early 1800s) by Father McDermid at St. Catherines. Nut: large; shell rather thick, smooth, apex pointed; kernel sometimes astringent, in other districts, kernel quality very good. Tree: usually never bears heavily; apparently very susceptible to late spring frosts at bloom time, but in other locations, it bears well and appears to be hardy; vigorous grower.

McKinster.—Orig. in Columbus, Ohio, by Ray McKinster. Introd. in 1952. Seedling of Persian walnut from seed imported from Poland by Paul C. Crath. Fruit: flavor and quality high; crops large. Tree: well adapted in the Ohio climate; self-pollenizing.

Merkel (*Merkel N.W.*).—Orig. in Chelsea, Mich., by Henry Merkel. Introd. in 1964. Parentage unknown; selections of Rev. Crath from the Carpathian Mountains; selected in 1963. Nut: resembles French variety Mayette; shell thin; kernel averages 53% of nut; harvest mid-September. Tree: susceptible to husk maggot; yearly crop from 1963 to 1965 averaged 36 lbs of nuts.

Merkel N.W.—See Merkel.

Mesa.—Orig. in Las Cruces, N.M., by New Mexico State Univ. Introd. in 1980. Parentage unknown. Discovered in 1958. Nut: large; globose; well-sealed; resistant to sunburn; shell medium thick; kernel plump, light in color, excellent quality; harvests early. Tree: medium; spreading; vigor medium; hardiness to –10F.

Metcalfe.—Orig. in Webster, N.Y, by Mrs. Ward H. Metcalfe. Introd. in 1952. seedling of Carpathian type; seed planted in 1936. Nut: size medium, 42 to 45 per lb; shell thin, light amber; kernel white, sweet, quality very good, filling shell. Tree: hardy to –20F.; does not bear well, but pollination requirements not yet fully established; pollen shed early; very vigorous.

Meylan.—A French type originally from Meylan (Isère), France. Nut: seal is poor, yield is fair, kernel quality is good, but the percent kernel is low. Some say that the kernel lacks astringency and is therefore a good nut for people sensitive to walnuts. Tree: terminal bearing, protogynous, late leafing, and late in shedding pollen.

Midland.—Orig. in Davis, Calif., by E.F. Serr and H.I. Forde, California Agr. Expt. Sta. Introd. in 1968. Franquette x Payne; sibling of Vina. Cross made in 1948; selected in 1955; tested as UC 49-47. Nut: large, Payne type; shell seal good; kernel 50% of nut, quality high; 70% of those tested had light-colored kernels; matures in midseason. Tree: more vigorous than Vina and a little slower in coming into full bearing; does not require quite as heavy pruning as Vina when a young tree; productive; leafs out 12 days after Payne; 50% of lateral buds producing pistillate flowers; does not seem tolerant of very high heat and therefore probably not suited for the hottest parts of the interior valleys of California; extends harvest season; its adaptability to the middle section of the California Central Valley including Davis, Sacramento, and Stockton districts, suggested the name.

Morris.—Orig. in Amenia, Dutchess County, N.Y, by Benton & Smith Nut Tree Nursery. Introd. in 1948. Grown from seed imported from Poland in 1935; selected in 1946 or 1947. Nut: large; shell thin; kernel quality excellent, bright. Tree: productive. Won fourth prize in the 1949–50 contest of the Northern Nut Growers' Assn.

Moyer.—Orig. in Winston, Ore., by C.E. Schuster and C.E. Moyer. Introd. in 1958. Mayette seedling resulting from attempted cross with northern California black walnut (*Juglans hindsii*); cross made in 1938 by Schuster. Nut: Persian type; flavor excellent; resembles Franquette, but larger; drops well in advance of leaves. Tree: very upright; hardy, more so than Franquette and Mayette in Oregon; consistent but not a heavy bearer.

Myrtleford Jewell.—Orig. near Myrtleford, Victoria, Australia, by Valley Nut Groves. Introd. there and in the U.S. in 1945. Chance seedling; discovered in 1930. Nut: small; crack-out percentage low; a hardshell Mayette type, similar to Mayette and Poe. Tree: considered to be blight resistant in Australia, but has not shown resistance in California; leafs out late.

Nebraska.—Orig. in Lincoln, Neb., by George Brand. Introd. in 1979 by Nebraska Nut Growers' Assn. Seed originally came from Carpathian Mountains of Poland and was obtained in 1930s from the Wisconsin Hort. Soc. Nut: medium; round; shell thin, tight seam, smooth; kernel plump, 41% kernel to shell; harvest date 10 Oct. Tree: 35 ft with crown diam. 50 ft; very vigorous and hardy; productive. Named and introduced because of its vigor and use as germplasm for cold hardiness, has tolerated temperature of –30F.

Neyer.—Orig. in Ann Arbor, Mich., by Fred Neyer. Introd. in 1958. Seed imported from Paul C. Crath in Poland by the Wisconsin Hort. Soc.; planted in 1936; selected in 1954; tested as Neyer 1. Nut: round; kernel sweet, white, plump, very attractive; shell thin; ripens early but unevenly; resembles Hansen. A prize winner in the 1953 Michigan Nut Contest.

Neyer 2.—Orig. in Ann Arbor, Mich., by Fred Neyer. Introd. in 1958. Seeds imported from Paul C. Crath in Poland by the Wisconsin Hort. Soc.; planted in 1936; selected in 1957. Nut: shell fairly heavy; kernel white, not as sweet as Neyer; ripens very uniformly, midseason to late. A prize winner in the 1956 Michigan Nut Contest.

North Platte.—Orig. in Hamburg, Iowa, by Sam Diedrich, North Platte, Neb. Introd. in early 1980s. Named and introduced by Nebraska Nut Growers' Assn. for its use as germplasm for cold hardiness. Original tree planted as unknown seedling in 1962. Nut: quality acceptable. Tree: vigor moderate; productivity moderate; has tolerated winter temperatures of–30 to –32F.

Nu-Schafer.—Orig. in Clarkston, Wash., by H. Lynn Tuttle, Lynn Tuttle Nursery. Introd. in 1959. Schafer x self; selected in 1945. Carpathian type. Nut: slightly larger than Schafer; kernel averages 55% of nut; flavor rich, milder than Schafer, which it resembles; shell thin, closely sealed, relatively light colored; ripens in early fall. Tree: large; grows rapidly; produces large crops; self-fruitful; resistant to low winter temperatures.

Nugget.—Orig. in Yuba City, Calif., by C.E. Sullivan. Introd. in 1957. Plant patent 1475; 8 May 1956; assigned to Sierra Gold Nurseries, Yuba City. Open-pollinated seedling of Payne; selected in 1948. Nut: large; shell well-sealed, attractive; kernel 53% of nut, plump, cracks out easily, quality good; harvest in midseason, between Payne and Hartley; reported to have resistance to blight and sunburn. Tree: prolific; comes into bearing early; vigorous; leafs out late, about 2 weeks after Payne.

Olds.—Orig. in Wenatchee, Wash., by L.G. Olds, Introd. in 1958. Chance seedling; discovered in 1942. Nut: size medium; shell thin; kernel large, plump. Tree: very hardy, surviving the 1955 freeze with no damage; apparently self-fruitful.

Orth.—Orig. in Butler Township, Montgomery County, Ohio, by Arthur C. Orth. Introd. in 1953 by the Benton & Smith Nut Tree Nursery, Millerton, N.Y. Open-pollinated seedling of a Crath Carpathian type; seed planted in 1937; tested as Orth 1. Nut: medium large; oblong, pointed; cracks well. Tree: hardy.

Pedro.—Orig. in Davis, Calif., by E.F. Serr and H.I. Forde, California Agr. Expt. Sta. Introd. in 1968. Conway Mayette x Payne. Cross made in 1952; selected in 1958; tested as UC 53-113. Nut: large; shell heavy, seal fair; kernel 50% of nut; 70% of those tested had light-colored kernels; matures midseason to late. Tree: leafs out 18 days after Payne; consistently produces many catkins and sheds pollen over a long period; can be used as pollenizer for most early varieties such as Ashley, Lompoc, Marchetti, Serr, Gustine, and Vina; can be used as a main producing variety where moderately late leafing is desired; very high summer temperatures have injured shell and kernel quality some years in hotter districts. Name was suggested in keeping with Spanish names for varieties that are good pollenizers.

Pioneer.—Orig. in Davis, Calif., by E.F. Serr and H.I. Forde, California Agr. Expt. Sta. Introd. in 1968. Franquette x Payne; sibling of Vina. Cross made in 1950; selected in 1957; tested as UC 51-170. Nut: Payne shape; shell seal fair to good, more suitable for in-shell than for cracking; kernel 48% of nut; 70% of those tested had light-colored kernels; matures midseason. Tree: not as productive as Vina in the early bearing stage and does not require a heavy pruning; borne heavy crops in Davis when 8 to 10 years old; 40% of lateral buds producing pistillate flowers; leafs out 14 days after Payne; pollenizer needed; suitable for upper Sacramento Valley and similar situations. Name was selected because the variety seems able to produce with somewhat less cultural care, especially pruning, than more highly fruitful varieties.

Placentia.—One of the Santa Barbara softshells derived from the nuts bought at the wharf is San Francisco by Joseph Sexton about 1869. The origin of this sack of nuts is not known. The Placentia was the standard in-shell walnut until the Hartley came along. Nut: shell is thin but strong, with a good seal. Kernel color is poor. Tree: earlier leafing than Payne; susceptible to blight and frost damage; protogynous and terminal bearing.

Poe.—Orig. in Lake County, Calif. Nut: thick-shelled; kernel small. One gourmet food writer described Poe as less astringent than other walnuts and tasting slightly like butterscotch. Tree: midseason to late leafing; terminal bearing, protogynous, low yielding.

Rapier.—Orig. on the property of a Mr. Rapier, Texada Island, British Columbia,

Canada, by J.U. Gellatly, Gellatly Nut Nursery, Westbank, B. C. Introd. in 1931. Chance seedling; discovered about 1910. Nut: size medium; shape of Franquette; kernel sweet. Tree: very vigorous.

Roberts.—Orig. at the Roberts residence in Caseyville, Ill., by Louis J. Gerardi, Louis Gerardi Nursery, O'Fallon, Ill. Introd. in 1940. Chance seedling; discovered in 1939. Nut: resembles Broadview. Tree: bears at an early age; good crops; not hardy. Now obsolete.

Rodhouse.—Orig. at Pleasant Hill, Pike County, Ill., by Louis Rodhouse. Plant patent 3159; 9 May 1972; assigned to Stark Bros. Orchards & Nurseries Co., Louisiana, Mo. Parentage unknown, nut introduced from Poland. Nut: medium, slightly smaller than Lake, about 35 per lb; base round, point blunt; shell thin, cracks and releases easily; kernel round, flavor good, not bitter. Tree: large; moderate vigor; not injured at –20F; very productive, bears regularly; harvested first half of October in Pike County; self-fertile.

Russ.—See Breslau.

Rutherford Ehrhardt.—See Early Ehrhardt.

S-S Franquette.—Orig. in California. Introd. in 1952. Parentage unknown; tree propagated along with many standard Franquette trees, by W.E. Stuke, Stuke Nursery Co., Gridley, Calif.; trees planted by R.S. and W.M. Stuart, Stuart's Nursery, French Camp, Calif., in 1940; in 1951, this one tree observed as different and promising by Fred M. Charles, Farm Advisor, San Joaquin County, Stockton, Calif.; the name, signifying Stuke and Stuart, coined by E.F. Serr, Univ. of California, Davis. Nut: kernel quality high, averaging 90% light colored, tends to be smaller than Franquette; shell well-sealed; matures late, with Franquette; resembles Franquette in other characteristics. Tree: a typical Franquette in that it leafs out late in spring; differs in bearing habit from Franquette in that usually about 5% of lateral buds on the new growth are fruitful; promising for foothill areas of Lake County and those counties bordering the Sierra Nevada, Calif.; bears a little more heavily than Franquette.

Schafer.—Orig. in Yakima, Wash., by Wilhelm Schafer. Introd. in 1940 by H. Lynn Tuttle, Lynn Tuttle Nursery, Clarkson, Wash. Plant patent 494; 16 Dec. 1941; by Wilhelm Schafer and Marie Schafer; assignors to H. Lynn Tuttle. Seedling selected in 1937; grown from seed imported in 1920 into the U.S. by Mrs. Schafer from the northern foothills of the Carpathian Mountains, Province of Bukowina, Rumania. Nut: Carpathian type; large; shell thin, tightly sealed; high percentage of kernel, very fine quality, oily, rich; matures in a short growing season, in early fall. Tree: vigorous; tall, spreading; extremely hardy; bears annually and abundantly.

Scharsch Franquette.—Orig. in Ord Bend, Glenn County, Calif., by Fred (John) Scharsch. Introd. in 1962. Open-pollinated seedling of Franquette; discovered about 1956. Nut: medium size, averaging 32 mm in cheek diam.; shell seal good; high percentage of light-colored kernels; ripens late; otherwise resembles Franquette. Tree: starts bearing both pistillate and staminate flowers earlier in life than standard Franquette; up to 20% of lateral buds produce pistillate flowers; otherwise, growth characters similar to typical Franquette.

SeeandO.—Orig. in Wenatchee, Wash., by H.S. Simons, Columbia Okanogan Nursery Co. Introd. in 1934. Probably a seedling of Mayette; discovered in 1921. Nut: quality fine. Tree: hardy; bears well and regularly. Now more popular for home gardens than for commercial production.

Serr.—Orig. in Davis, Calif., by E.F. Serr and H.I. Forde, California Agr. Expt. Sta. Introd. in 1968. Payne x P.I. 159568; cross made in 1958; selected in 1963; tested as UC 59-129. Nut: large, Payne type; shell thin, seal fair to good, cracking quality good; kernel 59% of nut, quality good; 96% of those tested had light-colored kernels; highly resistant to sunburn; matures early to midseason. Tree: very vigorous when young and grows rapidly; leafs out 5 days after Payne; 50% of lateral buds producing pistillate flowers; well adapted to hotter parts of San Joaquin and Sacramento Valley, wherever late spring frosts are not a hazard. Susceptible to pistillate flower abscission. Named at the request of the Diamond Walnut Growers, in recognition of E.F. Serr's contribution on the walnut breeding program.

Sharkey (*Tribble*).—Orig. in California; scions secured by Millard Sharpe of Vacaville, Calif., from one of the Tribble brothers of Elk Grove, Calif. Introd. in 1955. Parentage unknown; discovered about 1925; tested by Sharpe and later by R.J. Button of Vacaville and E.F. Serr of the California Agr. Expt. Sta., Davis. Nut: very large; round; kernel percentage high, being 5% to 10% higher than most commercial varieties; kernel light colored and heavy (50% heavier than average varieties). Tree: very fruitful (many lateral buds produce pistillate flowers); early bearing. Named in memory of Millard "Sharkey" Sharpe (1884 to 1954).

Sinensis 5.—Nut: seal is good. Kernel weight is 50% of nut weight, but kernel quality is poor. Tree: leafs out five days after Payne and pistillate bloom is nine days earlier than Payne. Protogynous and bears nuts primarily on terminals.

Sinensis 7.—Nut: seal is good. Kernel weight is 55% of nut weight and kernel quality is fair. Tree: leafs out seven days after Payne and its pistillate boom is four days earlier than Payne. Protogynous and bears nuts primarily on terminals.

Somers.—Orig. in Perrinton, Mich., by Lee Somers. Introd. in 1962; approval given by Mich. Nut Growers' Assn. and Northern Nut Growers' Assn. Open-pollinated seedling of a Carpathian type; grown from seed distributed by the Wisconsin Hort. Soc. about 1937; from nuts collected in Poland by Paul C. Crath; selected in 1954; tested as Somers N2W4. Nut: averages 11.6 g; size medium, averaging 42.8 mm long, 34.9 mm wide at cheek, and 32.6 mm wide at suture; slightly elongated, flattened on ends; shell attractive, cracks easily; kernel color light, flavor sweet, averages 55.6% of nut, oil content averages 64.7%; ripens in early September, and is noted for this early maturity. Tree: consistent annual bearer.

Spurgeon (*Bruce, Spurgeon Special*).—Orig. in Vancouver, Wash., by John F. Spurgeon. Introd. about 1920. Considered to be a seedling of Franquette; discovered by Spurgeon in an orchard then owned by Edward Detrick; named and propagated by Spurgeon, then the owner of this orchard; orchard owned later by J.B. Bruce. Nut: large; shell very thin, with a good seal; kernel light brown, flavor good.

Spurgeon Special.—See Spurgeon.

Stoke.—See Crath 1.

Sturdy.—Orig. in Roanoke, Va., by Harvey F. Stoke. Introd. about 1938. Grown from seed imported from the Carpathian Mountains of Poland by Paul C. Crath. Nut: large; shell thin; kernel does not fill shell too well, tends to be too bitter for good quality; resembles Mayette, but larger. Tree: sturdy, very vigorous; extremely hardy; few pistillate flowers produced; unproductive. No longer being propagated.

Sunland.—Orig. at Davis, Calif., by Harold I. Forde, Univ. of California, Davis. Introd. in 1979. Plant patent 4389; assigned to Regents, Univ. of California, Davis. Lompoc x plant introd. 159568; selected in 1969; tested as 66-4. Nut: resembles Serr; long, oval, smooth, large; kernel 60% of nut, excellent quality. Tree: very vigorous; precocious; highly productive; harvests with Hartley. Chico, Tehama, or Hartley recommended as pollenizers.

Tehama.—Orig. in Davis, Calif., by E.F. Serr and H.I. Forde, California Agr. Expt. Sta. Introd. in 1968. Waterloo x Payne. Cross made in 1957; selected in 1963; tested as UC 58-11. Nut: Payne type; shell seal good, suitable for in-shell and cracking; kernel 53% of nut; 70% of those tested had light-colored kernels; matures in midseason. Tree: very productive; consistent pruning is needed to prevent overbearing in young trees; leafs out 18 days after Payne; pollenizer seems desirable to help set heavy crops, especially on young trees; 80% of lateral buds producing pistillate flowers; suitable for upper Sacramento Valley and other districts with similar frost and blight hazards. The name was suggested because of apparent suitability of this selection for the walnut districts of Tehama County.

Tribble.—See Sharkey.

Trinta.—Orig. in Gustine, Calif., by Manuel Trinta, Patterson, Calif. Introd. in 1963 by Sierra Gold Nurseries, Yuba City, Calif. Plant patent 2642; 7 June 1966; assigned to Sierra Gold Nurseries, Yuba City, Calif. Parentage unknown. Nut: medium large; averaging 31 in-shell per lb; long cylindrical; shell thick, light tan; kernel light beige,

48% to 50% of nut; flavor excellent; reported to be resistant to blight and sunburn. Tree: vigorous; extremely dense; tends to droop; moderately fast grower while young; well-adapted to different soils; resembles Waterloo, but ripens earlier and more productive, larger and more abundant foliage, blooming period and harvesting period approximately 2 weeks earlier.

Tulare.—Orig. at Davis, Calif., by H.I. Forde, G.H. McGranahan, R.G. Snyder, G.S. Sibbett, W. Reil, J. Hasey, and D.E. Ramos. Tehama x Serr; cross made in 1966; introd. in 1992 for use in hedgerow or high-density plantings. Nut: midseason maturity, similar to Hartley; almost round; 13.3 g/nut, 53.3% kernel, 75% light kernels. Tree: Upright growth habit; moderate vigor; good male/female bloom overlap. Precocious and productive; almost all terminals and 75% of lateral shoots are fruitful.

Valnur.—Orig. in Wenatchee, Wash., by Van Well Nursery. Introd. in 1956. Parentage unknown. Nut: large; long; shell thin; kernel plump. Tree: large; upright; vigor moderate; very hardy.

Van Buren.—Orig. in South Haven, Mich., by G.S. Tolles. Introd. in 1960; named in 1950. Open-pollinated seedling of a Carpathian type. Nut: appearance good; kernel flavor and quality very good. Tree: self-fruitful.

van Thoor.—Orig. in Limburg, The Netherlands, by C.J. Gerritsen, Inst. of Hort. Plant Breeding, Wageningen, The Netherlands. Introd. in 1954. Parentage unknown; selected in 1953. Nut: size intermediate to large; kernel taste and color satisfactory; rather difficult to shell, but kernel comes free easily. Tree: staminate flowers bloom early; female flowers bloom rather late.

Vina.—Orig. in Davis, Calif., by E.F. Serr and H.I. Forde, California Agr. Expt. Sta. Introd. in 1968. Franquette x Payne. Cross made in 1948; selected in 1955; tested as UC 49-49. Nut: pointed, somewhat similar in shape to Hartley but with a less flattened base; shell seal good; kernel 49% of nut, high quality; 70% of those tested had light-colored kernels; matures early to midseason. Tree: very productive and consistent bearer; leafs out 8 days after Payne; 80% of lateral buds producing pistillate flowers; tolerant of high summer temperature. Needs consistent pruning to prevent overbearing and lowering of quality. Named after the town of Vina.

Wallick 16 (*G.O.K. 16*).—Orig. in Peru, Ind., by Ford Wallick. Introd. in 1959. seedling of Carpathian type; planted in 1942; selected in 1953. Nut: medium to large; shell smooth, light colored; kernel cracks out about 50%, flavor good; ripens early. Tree: prolific; very hardy.

Waterloo.—Orig. in Stockton, Calif. Introd. about 1934. Discovered in a planting of Eureka seedlings; named by E.F. Serr, then Farm Advisor at Stockton, later Pomologist, Univ. of California at Davis. Nut: large; well-filled; good quality and color; matures 1 week later than Eureka. Tree: good producer; leafs out 2 weeks later than Eureka. Named after Waterloo Road in Stockton. This variety was used as one of the parents of Gustine, Lompoc, and Tehama.

Watt.—Orig. in Penticton, British Columbia, Canada, by J.U. Gellatly, Gellatly Nut Nursery, Westbank, British Columbia. Introd. in 1932. Parentage unknown; original tree grew on the Watt farm at Penticton; sold as a seedling tree to Watt by Gellatly. Nut: well sealed; shell thickness medium, easily cracked; kernel well flavored and fills out the shell. Tree: not as hardy as Broadview.

Weng.—Orig. in Harrison Township, Montgomery County, Ohio, by Lewis Weng. Introd. in 1953 by Benton & Smith Nut Tree Nursery, Millerton, N.Y. Open-pollinated seedling of a Crath Carpathian type; seed secured in 1936 from the Wisconsin State Hort. Soc. Nut: averages about 40 per lb; nearly round; ripens fairly early. Tree: hardy; productive; appears to be self-fruitful.

Rootstocks for Persian Walnut

Sullivan.—Orig. in Yuba City, Calif., by C.E. Sullivan. Introd. in 1960. Plant patent 1963; 9 Aug. 1960. Paradox walnut hybrid: northern California black walnut (*Juglans hindsii*) x Persian walnut (*J. regia*); discovered in 1952. Used as a clonal rootstock for Persian walnut; resistant to oak root fungus and root-lesion nematode (*Pratylenchus vulnus*); grows vigorously; propagated by cuttings or trench layering.

PERSIMMON

AMERICAN SPECIES
Diospyros virginiana

Buhrman.—Orig. in Pennsylvania as a chance seedling; named for a Mr. Buhrman on whose farm it was discovered. Introd. about 1950. Fruit: large; skin blushed red before maturing; flesh color and flavor poor; seeds few; ripens late. Tree: not as hardy, disease resistant, nor productive as Early Golden, Garretson, and Killen; twigs relatively slender, glabrous; probably 60-chromosome race.

Craggs (*Creggs*).—Orig. in Harrisburg, Ill., by a Mr. Craggs. selected in 1945; introd. in 1949. Parentage unknown. Fruit: large; somewhat acorn-shaped; skin often red-blushed, attractive; flesh translucent, light yellow, texture and flavor excellent; seeds few; ripens midseason, about 30 Sept. in Caseyville, Ill. Tree: sturdy growth; leaves glossy.

Creggs.—See Craggs.

Edris.—Orig. as a chance seedling on the property of Dr. Edris in 1920 in Oskaloosa, lowa. Introd. about 1925 by D.C. Snyder, Center Point lowa. Fruit: large; quality fine.

Ennis.—A chance seedling discovered about 1953 in Bedford, Ind., by Ray G. Ennis. Introd. in 1955. Fruit: medium-large; flesh rather firm, quality good, seedless; ripens early.

Ford.—An open-pollinated seedling discovered about 1918 in Rusk, Texas, by Frank B. Guinn. Introd. about 1925. Fruit: 1 1/2 inches in diam.; skin deep orange when mature; flavor good.

Garretson.—Orig. in Adams County, Pa.; originator unknown. Introd. about 1921 by the J.F. Jones Nursery, Lancaster, Pa. Probably an open-pollinated seedling of Early Golden; a 90-chromosome form discovered about 1920. Fruit: slightly smaller and more spherical than Early Golden; flesh a rich orange, very tender, highly flavored; matures during September and early October; moderately seeded, seeds small; matures well at Geneva, N.Y.; performs well on the Great Plains. Tree: very hardy; nearly identical with Early Golden. Garretson was considered in 1959 as the first choice variety for the northern states.

Illini Male.—Orig. in Urbana, Ill., by J.C. McDaniel, Illinois Agr. Expt. Sta. Incompletely sex-reversed bud mutation of Early Golden; discovered in 1952. Introd. in 1953. This variety is considered potentially valuable in breeding for fruit quality in the dioecious American persimmon species. While original branch has remained partially male through 1968, and other sex-reversed branchlets have appeared on the tree, all bud propagated trees reverted to typical Early Golden. (Similar transiently male branchlets have been seen in the Killen variety and in other Early Golden trees).

John Rick.—Orig. in Urbana, Ill., by Joseph C. McDaniel, Illinois Agr. Expt. Sta. Killen x an unknown, probably William; tested as 111. RR. and selected in 1958; introd. in 1963. Possesses 90 chromosomes; Fruit: one of largest and firmest among September-maturing American persimmons. The small, deeply inset calyx minimizes skin punctures. Tree requires pollination by a 90-chromosome staminate cultivar, e.g., William. Named in honor of a long-time persimmon experimenter of Reading, Pa.

Killen.—Orig. in Felton, Del., by the late Joseph Killen. Considered to be an open-pollinated seedling of Early Golden; selected about 1915; first propagated in Round Hill, Va., by J. Russell Smith, who probably named it. Introd. about 1930. Fruit: slightly larger, firmer and better flavored than Early Golden, which it resembles. Tree: occasionally bears staminate-flowering branchlets; needs a 90-chromosome pollinator for reliable fruiting. Killen is a promising parent in breeding.

Knowles.—See Owens.

Luke.—Orig. in Urbana, Ill., by Joseph C. McDaniel and Herbert C. Barrett, Illinois Agr. Expt. Sta. Open-pollinated male seedling of Killen; seed planted in 1951; selected in 1956; tested as seedling Illinois L. Introd. in 1959. Tree: useful as a pollinator among

American persimmons of 90-chromosome race; staminate flowers open early to fairly late; resistant to leafspot.

Meader.—Orig. in Rochester, N.H., by Elwyn M. Meader (Prof. retired, Univ. of New Hampshire). Open-pollinated seedling of Garretson, seed supplied by George L. Slate, Geneva, N.Y., selected in 1961 by Prof. Meader. Introd. in 1974 by the Gerardi Nursery, O'Fallon, Ill. Fruit: larger than Garretson when pollinated, skin orange color, thin; flesh firm, sweet, texture smooth and free of astringency when ripe, quality high; mostly parthenocarpic, but bears seeded fruit when pollinated; ripens early. Tree: hardy to –25F; highly productive; similar to Early Golden, Killen, and Garretson. Tree: sporadic staminate as with Garretson. Recommended for central and northern United States.

Mike.—Orig. in Urbana, Ill., by Joseph C. McDaniel and Herbert C. Barrett, Illinois Agr. Expt. Sta. Open-pollinated seedling of Killen; selected in 1955; tested as seedling Illinois MM, introd. in 1959. Fruit: none. Tree: male; useful as a pollinator for 90-chromosome race; fine twigs very pubescent.

Miles.—Orig. in Rutherford County, Tenn., by M.T. Miles. Chance seedling from the wild, discovered before 1900; grown locally until 1950. Introd. in 1952. Fruit: fair size; good quality; fewer seeds than usual, resembling Pennland, but larger. Tree: productive; strong grower.

Owens (*Knowles*).—Orig. tree some 100 years old discovered in Poseyville, Ind., on land owned by Oliver Knowles. Considered to be the largest native persimmon tree in the United States, 13 feet 2 inches in circumference 4 feet above ground, spread of about 60 feet, an estimated 85 feet tall; bears well, up to an estimated crop of 1 ton per year. First brought to public attention by W. Alfred Owens of Poseyville. Named and introduced in 1959 by W.B. Ward, Indiana Agr. Expt. Sta., Lafayette. Being propagated in considerable amounts since 1954. Fruit: size medium; oblate; flavor very fine; seedless; ripens in late September to early November. Tree: grafted trees slow to come into bearing, evidently self-fruitful.

Penland (*Pennland's Seedless*).—Orig. near Penland, N.C. Introd. about 1937 as a seedling growing in a cut-over forest; called to the attention of Spencer B. Chase, Tennessee Valley Authority, Division of Forestry Relations in 1935; named in 1935 by J.C. McDaniel; first grafted at Norris, Tenn., in 1937. Fruit: almost seedless. Variety apparently not extensively planted in North Carolina.

Pennland's Seedless.—See Penland.

Spencer.—Orig. near Harrisburg, Ill., by E.R. Spencer. Introd. in 1949. Fruit: quality fair; two to three seeds per fruit; early maturing, about 1 Sept. in Caseyville, Ill. Superseded by better varieties. No longer propagated .

William.—Orig. in Urbana, Ill., by J.C. McDaniel, Illinois Agr. Expt. Sta. Rootstock sucker discovered in 1951; introd. in 1952. Fruit: some small fruit in some years. Tree: sporadic pistillate; excellent pollinizer; blooms profusely over an extended season, producing scattered mediocre fruit; a good vigorous understock; produces many suckers.

ORIENTAL OR JAPANESE SPECIES
fs20, *Diospyros kaki*

Cultivars in this species are categorized into four classes: 1) pollination-variant, astringent (PCA); 2) pollination-variant, non-astringent (PCNA); 3) pollination-variant, astringent (PVA); and 4) pollination-variant, nonastringent (PVNA).

California Fuyu.—See Oku-Gosho.

California Maru.—(PVNA) Orig. in Japan. Introduced as PI 83790, erroneously labeled Jiro. Fruit is large, roundish; flesh crisp, sweet, and dark brown when seeded, similar to Zenji Maru. Bears well.

Chocolate.—See Tsuru-no-ko.

Dojo Hachiya.—See Hachiya.

Eureka.—(PCA) Fruit medium, round with reddish skin and a firm, light yellow flesh. Prolific bearer.

Fuji.—(PCA) Orig. in Japan. Fruit and tree identical but crops better than Hachiya in California. (isozyme patterns differ). Pistillate constant.

Fuyu (*Fuyugaki*).—(PCNA) Orig. in Gifu Prefecture, Japan, in 1902, probably a budsport of Ikura Gosho. Introduced as PI 72662 into Plant Introduction Gardens, Chico, Calif. Most widely planted in Japan and California. Fruit oblate, round when well-grown, with glossy orange-red skin.

Crop sets parthenocarpically but it improves when seeded. Matures in November, excellent quality and keeps well. Tree is vigorous but the branches tend to droop. Pistillate constant.

Fuyugaki.—See Fuyu.

Fuyuhana.—(PCNA) Hybrid of Fuyu x Hana Gosho selected in Brasil. Fruit characteristics similar to those of Fuyu and Jiro.

Giant Fuyu.—See Hana Fuyu.

Gosho (*Goshogaki*).—(PCNA) An old cultivar discovered in Japan. Fruit large with orange-yellow skin, flesh is firm, sweet, and slightly brown with dark dots. Flavor is inferior to that of Fuyu. Tree is vigorous and bears a good annual crop. Staminate sporadic.

Goshogaki.—See Gosho.

Hachiya (*Dojo Hachiya*).—(PVA) An old cultivar that originated in Japan in ancient times. It is the most popular PVA type in Japan and California. Very large fruit (250 g), oblong-conical, attractive, deep orange-red when ripe, often develops black splotches. Parthenocarpic fruit drops prematurely; sets better when seeded. Small region about seed turns brown-black. Eaten fresh, dried, or made into cookies and bread. Tree is vigorous and upright. Pistillate constant.

Hana Fuyu (*Giant Fuyu, Jumbu, Yotsudani*).—(PCNA) Orig. in Japan. Introd. in the U.S. as P.I. 83709 in 1930. Fruit large, roundish-oblate; with four distinct grooves, hence, the name Yotsudani (four valleys). Skin reddish-orange, attractive. Ripens late October and soften rapidly after harvest with a sweet but bland flavor. Pistillate constant. Tree vigorous, sets a light crop in some season.

Hiratanenashi.—(PVA) Orig. in Niizu, Niigata Prefecture, Japan in the 1600s. Widely planted in Japan because of its excellent flavor. Fruit, medium size (180 g), round to oblate, four-sided with four shallow grooves radiating from the apex; skin glossy, tough, and orange-red. Seeds abort; needs treatment to remove astringency.

Hyakume (*Koshu Hyakume*).—(PVNA) An old, large-fruited, Japanese cultivar. Requires pollination for the flesh to turn brown and edible; otherwise needs treatment with alcohol fumes or exposure to carbon dioxide or ethylene gases to remove astringency. Skin yellow with many "sugar" rings on distal end. Pistillate constant.

Ichikikei Jiro.—(PCNA) A budsport of Jiro very similar to Maekawa Jiro. The fruit matures early with good quality. Tree has low vigor and a compact growth habit. Pistillate constant.

Izu.—(PCNA) Developed and released by the Okitsu Branch, Fruit Tree Res. Sta. Japan, 1970. Progeny of Fuyu x (Oku-Gosho x Oku-Gosho). Fruit large, oblate, skin orange-red, flesh pale orange. Matures late September. Pistillate constant.

Jiro.—(PCNA) Orig. in Shizuoka, Prefecture, Japan. Selected by Jirokichi Matsumoto in the 1800s. Fruit: medium (200 g), oblate-truncate, squarish in cross section with four prominent longitudinal grooves. Skin yellow-orange, flesh yellow. Matures slightly earlier than Fuyu in late October in Japan. Pistillate constant.

Jumbu.—See Hana Fuyu.

Jumbu.—(PCNA) Orig. in Fullerton, Calif., by J.M. Alcorn. Introd. in 1938. Parentage unknown; discovered about 1930. Fruit: very large; misshapen; nonastringent. Probably identical with the Japanese variety Hana Fuyu. Tree: large; very vigorous.

Kaki Tipo.—(PCA, PCNA, PVNA, or PVA) Kaki tipo is an Italian name given to pistillate constant cultivars, such as the Hyakume, for export to northern Europe after treatment with carbon dioxide or ethylene gas.

Kawabata.—(PCNA) Orig. in Japan. Introd. in the U.S. as P.I. 71946 in 1927. Fruit: large; oblate, conical; skin red-orange; flesh yellow slightly speckled with brown, quality good, usually seeded; ripens late November. Some fruit have a constriction around the base like Tamopan.

Lantern.—(PCA) Orig. in Chu Li Chi Village of Hseangtong, China. Introd. in the U.S. as P.I. 91182 in 1931. Fruit: medium to small, round-oblate; skin orange-yellow; flesh yellow-orange, usually seeded, astringent until soft-ripe, quality poor to fair; ripens mid-November.

Lone Star.—(PCA) Orig. in Waco, Texas, by C. Falkner. A chance seedling discovered about 1915; introd. about 1920. Fruit: large; well-colored; ripens in late fall; resembles Eureka. Tree: very prolific.

Maekawa Jiro.—(PCNA) Probably a sport of Jiro that colors and matures 2 weeks earlier. Fruit large, oblate and squarish like Jiro, with orange-yellow skin. Apex of a large fruit may split. Pistillate constant.

Matsumoto Wase Fuyu.—(PCNA) Discovered in Kyoto Prefecture and released in 1935. Fruit mature 2 weeks earlier but are smaller and flatter than Fuyu. Calyx may separate in oversized fruit. Heavy set needs to be thinned or the tree may bear biennially. Pistillate constant.

Nishimura Wase.—(PVNA) Discovered and registered by Yazo Nishimura in 1960. Fruit medium to large, round to oblate, slightly squarish and matures in mid-September.

O Gosho.—(PCNA) Orig. in Japan. Introd. in the U.S. as P.I. 78487 in 1928. Fruit: size medium, conical to oblate; skin orange-yellow, thick, tough; flesh: light yellow with brown flecks, quality fair to good; seeds one or more; ripens mid-November.

Oku-Gosho (*California Fuyu*).—(PCNA) Orig. in Japan. Introduced as PI 26491 (Fuyu kaki). Fruit identical to Fuyu but matures slightly earlier. The tree is sporadically staminate and is used as a pollen parent.

Peiping.—(PCA) Orig. in Round Hill, Va., by the late Prof. J. Russell Smith, Swarthmore, Pa. Introd. about 1940 as scions collected north of Peking, near the Great Wall; imported by B.T. Galloway, U.S. Dept. of Agriculture, through arrangements made by Prof. Smith. Fruit: rounded, slightly pointed at the apex. Skin: strong orange; flesh: orange, quality good; seed content lower than most varieties. Tree: at maturity somewhat spreading, appears more hardy than most varieties.

Saijo.—(PCA) Orig. in Japan, principally for drying. Fruit small (150 g) oblong with four grooves, dull yellow skin. Softens rapidly on tree with excellent flavor, honey-sweet. Pistillate constant.

Square.—(PCA) Orig. in Chu Li Chi Village of Hseangtong, China. Introd. in the U.S. as P.I. 91184 in 1931. Fruit: small to medium, oblate; skin: reddish-orange; flesh orange, quality poor, usually seeded; ripens mid-November.

Suruga.—(PCNA) Developed and released by the Okitsu Branch, Tree Fruit Res. Sta. Japan, 1959. Progeny of Oku-Gosho x Oku-Gosho. Fruit: medium, oblate-conical, roundish in cross section. Calyx tends to separate from fruit. Bears heavily and matures late, mid-November, but keeps extremely well.

Tamopan.—(PCA) Orig. in China. Fruit very large with a pronounced constriction near the calyx. Fruit: large. Skin yellow, thick, and leathery, pulp soft and fibrous. Pistillate constant.

Tanenashi.—(PCA) Orig. in Japan. Fruit large, conical, and matures early. Sets well parthenocarpically but is susceptible to stem end cracks. The tree is moderately vigorous, tends to droop with cropping. Grown widely in southern states, especially Florida.

Twentieth Century.—(PCNA) Origin unknown. Received from the Yokohama Nursery Co., Yokohama, Japan, by the U.S. Dept. of Agriculture, Plant Introduction Sta., Chico, Calif., as P.I. 72960 in 1927. Twentieth Century (Nijisseki) not catalogued in Japan. The Plant Introduction Inventory describes it as "A new variety fruit, sweet and juicy." Fruit: size uniform, medium to large, oblate, quandrangular; cavity wide, depth medium; apex flattened, slightly depressed; skin medium orange-yellow, moderately thick and tough; flesh medium to reddish-orange with some dark specks, quality good to very good; seeds several, medium to large.

Triumph.—(PVNA) An important cultivar grown in Israel. Fruit is yellow-orange, pulp is brown and sweet when pollinated, but is yellow and astringent in parthenocarpic fruit.

Tsuru-no-ko (*Chocolate*).—(PVNA) Orig. in Japan. (Another cultivar exists with the same name or Tsurugaki, which is a PCA type.) Fruit: elongated, conical; skin: yellow-orange. Flesh dark brown when seeded. Chocolate is sporadically pistillate, thus an excellent pollinizer.

Yotsudani.—See Hana Fuyu.

Zenji Maru.—(PVNA) Discovered by a Buddhist priest in the 1200s. Mature fruit deep orange, flesh buttery texture, dark brown and sweet when seeded; parthenocarpic fruits astringent. Monoecious, producing an abundance of staminate flowers. Used as pollinizer. Tends to bear biennially if the tree is not pruned and thinned regularly.

PINEAPPLE

CO-2.—Plant patent 8863; 16. Aug. 1994. Developers are Calvin H. Oda of Honolulu, Hawaii, and David D.F. Williams of Fort Collins, Colo. Assignee: Del Monte Fresh Produce N.A., Coral Gables, Fla. A hybrid from a cross between two hybrid clones, 58-1184 and 59-443, both originated from Pineapple Research Institute of Hawaii. Plant: short, upright. Stem: length 24 to 32 cm in diam. at soil level, 22 to 26 cm at anthesis. Smooth, closely overlapping leaves, 5/13 phyllotaxy. Few small spines near leaf base. Fruit peduncle length 23 to 25 cm. Square-shouldered fruit with large, flat fruitlets, large crown with narrow semi-rigid leaves. Fruit: weight 1298 to 2592 g, average 2061 g, uniform pale-yellow flesh, 15 to 16.7 degree Brix, 0.42% to 0.91% TA, vitamin C at 30.8 to 55.5 mg/100 ml juice. Slightly more susceptible to *Phytophthora cinnamomi* and *P. parasitica* than Smooth Cayenne, but lower incidence of parasitic fruit diseases and resistance to internal browning. Produces 4% to 37% more than Smooth Cayenne.

Smooth Cayenne.—Origin obscure. The first recorded collection was made by Perrottet in 1819 from French Guiana as Maipourri. Five crowns were sent to Versailles, then to Great Britain. These were materials from which the pineapple industries in Australia, Jamaica, and Hawaii were originated. Smooth Cayenne is considered to be the best and the most widely cultivated cultivar because of its slicing quality. Plant: open-rosette, broad, short, with pale-green bases, spineless, except near tip. Relatively short fruit peduncle. Flower: lilac color. Fruit: large (0.5 to 4 kg), cylindrical, flat, and wide eyes and orangy-yellow skin when full-ripe. Flesh: pale yellow, nonfibrous, and more or less translucent, medium core; TSS 13.6%, TA 14.5%, vitamin C 20.3 mg/100 ml juice. Fruit qualities depend on clone, environment, time, and length of growth cycle and the amount of care provided. Very susceptible to wilt and nematodes. There are numerous clones of Smooth Cayenne, such as Champaka 153, Champaka 180, Hilo, 53-116, 59-656, and Cayenne Azores (F200) in Hawaii. G-25, G-32-33, and Cayenne de guinee in Guinea; Cayenne Guadeloupe, Sto Dominguo Cayenne in the West Indies; and Typhones 1, 2, 3, 4, and 5 in Taiwan.

Spanish Jewel.—Orig. in Honolulu, Hawaii, by Kenneth R. Kerns and Julius Lloyd Collins, Pineapple Research Inst. of Hawaii. Introd. in 1960. Plant patent 1993; 6 Dec. 1960; assigned to Pineapple Res. Inst. of Hawaii. Cayenne x Red Spanish; cross made in 1930; selected in 1934; tested as 10,388. Fruit: cylindrical, shoulders square; for a 5-lb. specimen, length was 7 1/6 inches and 6 1/8 inches in diam., exclusive of crown; eyes rather rectangular, flat; bracts yellow to salmon pink covering one-third of eye; overall eye color Spanish orange; more eyes than in Red Spanish; flesh white to primrose yellow, juicy, flavor resembles Red Spanish; general internal fruit characteristics resemble Red Spanish; ripens 20 to 25 days later than Cayenne. Plant: yield comparable to Cayenne, being above that of Red Spanish; extremely resistant to soilborne heart and root rot (*Phytophthora cinnamoni*), being a distinct advantage over Cayenne. Average degree Brix 13.5, acidity 0.82%.

PISTACHIO

Agency.—Syn. Aegina, Greek origin. Nut: large, reddish, very early maturity. Tree: very early bloom, medium size, modest yield, not commercial quality.

Ask.—Introduced from Israel, seedling from seed of Syrian origin (Hormaza et al., 1994. Econ. Bot. 48:349–358). Tree: early bloom, just after Chico. Prolific pollen producer.

Bronte.—Old cultivar from Sicily. Nut: many, green kernels, poor nut appearance, high percentage of nonsplits, good flavor, midseason maturity. Tree: large, not commercial quality.

Buenzle.—Orig. in northern California. Introd. in 1930. Chance seedling of *Pistacia vera*; discovered about 1902; tested at the U.S. Plant Introduction Sta., Chico, Calif. Nut: very small , being too small for commercial usage under present demands; inferior to Red Aleppo and Trabonella commercially and for home gardens. Tree: still some plantings in 1963, no known plantings 1996.

Chico.—Orig. in Chico, Calif., by the Plant Introduction Sta., U.S. Dept. of Agriculture. Introd. for trial in 1962 as P.I. 150646. Male seedling of *Pistacia vera* selected from seed introd. under P.I. 73396 from Aleppo, Syria; probably of species hybrid origin; tested as Chico 23. Tree: prolific pollen producer; blooms early; suitable pollinator for Bronte, Red Aleppo, Trabonella; blooms too early for Kerman, Lassen, and other later-blooming varieties; Chico and Peters are effective for good overlapping of bloom of the early varieties.

Damghan.—From Iran. Nut: very large, somewhat flattened, good percent splits. Tree: sporadic and low yield. Blooms late. Compatibility problems with some rootstocks. Not commercial quality.

Gazvin.—Introduced from Israel, from seed of Iranian origin (Hormaza et al., 1994). Tree: very early bloom, just after Chico. Prolific pollen producer.

Iran Large.—From Iran. Small tree. Nut: Very large, long nuts, inferior nut appearance, midseason maturity. Good percent splits. Tree: small tree, low yields, mid-season bloom. Not commercial quality.

Joley.—Seedling introduction from Damghan, Iran, selected at the Univ. of California, Davis, Calif., by Dr. J. Crane. Nut: good flavor, almond-shaped nuts similar to Trabonella or Bronte. Kernel color is greener than Kerman, but nut size is significantly smaller. High percentage of filled nuts and few blanks. However, at Davis, a high percentage of the nuts appear to be nonsplits (nonsplit shells) and the removal of the shell halves is not as easy as with Kerman. Early-midseason maturity, blooms and matures about 10 days earlier than Kerman. Tree: tree has good vigor. Few plantings. On a few young trees grown in Kern County, the yield has never been high.

Kastel.—Introd. from Israel, but probably originally from Iran. Nut: large, good yield. Nuts similar to Kerman, similar maturity to Kerman. Tree: large, vigorous, late bloom.

Kerman.—Orig. at the U.S. Plant Introduction Sta., Chico, Calif., by the New Crops Res. Branch, Agr. Res. Service, U.S. Dept. of Agriculture, Beltsville, Md. Introd. for trial in 1957; as P.l. 121776. Open-pollinated seedling of P.l. 86372 from Rafsinjan, Iran; *Pistacia vera*; selected in 1936. Nut: above average in size; shells split well, easily opened by hand; kernels commercially acceptable, size above average; shaken or knocked from tree readily when ripe. Medium-late maturity. Tree: vigorous; spreading; blooms late; produces heavy crops. The primary U.S. cultivar.

Lassen.—Orig. at U.S. Plant Introduction Sta., Chico, Calif., by New Crops Res. Branch, U.S. Dept. of Agriculture. Released for trail in 1962 as P.I. 150574. Female seedling selected in 1945 from seed introd. by W.E. Whitehouse, Plant Explorer, U.S. Dept. of Agriculture, as P.I. 86372 from Rafsinjan, Iran; from same seed lot as Kerman. Nut: size about average; shell splits well, easily opened by hand; easily knocked from tree; roasted quality good. Tree: yields above average; blooms late enough in most years to escape frost; pollinated by Peters.

Minasian.—Orig. in Orose, Calif., by J.A. Minasian. Introd. in 1930. Chance seedling of *Pistacia vera;* discovered about 1902; tested at the Plant Introduction Sta., U.S. Dept. of Agriculture, Chico, Calif. Nut: very small, being too small for commercial usage under present demands, inferior to Red Aleppo and Trabonella. Tree: still some plantings in 1963, no known plantings 1996.

Nazareth.—Introduced from Israel, from seed of Syrian origin (Hormaza et al., 1994). Tree: blooms later than Ask or Gazvin, but before Peters.

Peters.—Orig. in Fresno, Calif., by A.B. Peters. Introduction date unknown, but considered to be about 1930. Parentage unknown; *Pistacia vera;* discovered in the early 1900s; P.I. 246344. Male tree: good producer of pollen; anthesis coincides with stigma receptivity of Kerman, being a pollinator for that variety. Flowers over a relatively long period. Pollen is relatively durable and with good viability. Primary commercial pollinizer for Kerman.

Pontikis.—Orig. and tested at Dept. of Pomology, Agricultural College of Athens, Botanikos, Greece, by C.A. Pontikis. Aegenes open-pollinated. Fruit: large (25 x 16 mm); kernel weight 55% of fruit weight; oblong-ovate, with strong curing in dorsal suture near fruit stem; kernel oblong-ovate. External color yellow, with pink tint over one-third of surface; seedcoat purple over two-thirds of surface. Splitshell 90% to 98% every year, which is one-third higher than Aegenes. Percentage of blank nuts about 5% to 10%, same as Aegenes. Tree: medium size; semi-upright; and 5 x 6 m when grafted on *Pistacia terebinthus* seedling. Productivity same as Aegenes: 7.4 kg dry nuts per tree at 7 years, 16.2 at 11 years; biennial; flowering mid-April at Athens; fruit ripens 10–15 Sept. Pollination by male *P. vera* clones C and D. Adaptation to other growing regions unknown. Described in Pontikis, C.A. 1986. HortScience 21:1074.

Rashti.—Introduced from Israel, but probably originally from Iran. Nut: large, similar shape to Kerman. Excellent flavor. Matures very late in season. Tree: medium size, high yield, late bloom.

Red Aleppo.—Old cultivar from Syria or Turkey. Introd. into California by R.A. Fuller. May be seedling of Turkish or Syrian Red Aleppo. Nut: high percent splits, similar in quality to Kerman but with smaller nuts. Midseason maturity, before Kerman. Tree: medium size, medium yield, midseason bloom.

Sfax.—Old cultivar from Tunisia. Smallish tree. Nut: pale shell, large clusters with a tendency toward a high percent of blanks. Early maturity. Tree: early bloom, medium size, not commercial quality.

Sirora.—New Australian cultivar. Seedling of Red Aleppo. Nut: smaller than Kerman, green. Tree: late flowering, matures about 8 days before Kerman. (Maggs D.H. 1990. Fruit Var. J. 44:178–179.)

Trabonella.—Old cultivar from Sicily similar to Bronte. Nut: many, poor nut appearance but green kernels, high percentage of nonsplit nuts. Midseason maturity. Good flavor. Tree: large, midseason bloom, not commercial quality.

2-16.—Seedling selected from introduced seed from Azerbaijan (Hormaza et al., 1994). Tree: blooms 1 to 2 weeks before Peters. Good pollen producer.

2-18.—Seedling selection from introduced seed from Azerbaijan (Hormaza et al., 1994). Tree: blooms 1 to 2 weeks after Peters, some overlap. Good pollen producer, but less durable pollen than Peters (Polito, V.S. and J.G. Luza. 1988. J. Amer. Soc. Hort. Sci. 113:214–217).

PITANGA

Lolita.—Orig. in Carlsbad, Calif., by N.E. Westree. Introd. in 1957. Miscellaneous seedlings; discovered in 1950; tested as Westree 317. Fruit: averages 1 inch in diam.; skin black; moderate sugar content; used for jams, jellies, and ice cream flavoring. Tree: bears in January and June.

Nacha.—Orig. in Carlsbad, Calif., by N.E. Westree. Introd. in 1957. Misc. seedlings; discovered in 1950; tested as Westree 194. Fruit: large, up to 1 1/2 inches in diam.; skin brilliant red, attractive; flavor and texture good; moderate sugar content; used for jams, jellies, and ice cream flavoring. Tree: bears in January and June.

PLUM

Ace.—See Red Ace.

Acme.—Orig. in Saskatoon, Saskatchewan, Canada, by C.F. Patterson, Univ. of Saskatchewan. Introd. in 1960 for home gardens. Derived from *Prunus nigra* and *P. salicina* ; selected in 1948; tested as Sask. P58-105. Fruit: spherical, up to 2 inches under nonirrigated field conditions at Saskatoon; skin dark red; quality excellent; ripens in early September. Tree: hardy.

Akihime.—Chance seedling found in Aikawa, Sukawa, Yuzawa City, Akita prefecture, Japan, by Shoichiro Kojima; assigned to Nakajima Tenkoen Co. Plant patent 7858; 28 Apr. 1992. Fruit: round to ovoid with slightly pointed apex, very large, 150 to 200 g; skin purplish-red on yellow ground color; flesh: yellow, juicy, medium firmness, high sugar 13.7°Brix; matures with Kelsey. Blossom: small. Tree: vigorous; upright to spreading; low self-fertility requiring pollinator like Oishi Wase.

Albion.—Orig. in Geneva, N.Y., by Richard Wellington, New York State Agr. Expt. Sta. Introd. in 1929. Golden Drop x Grand Duke; cross made in 1908; seed germinated in 1909. Fruit: skin purplish-black; flesh a little coarse and stringy; stone clings tenaciously; ripens late; quality better than Grand Duke, which it resembles.

Alderman.—Orig. in Excelsior, Minn., by W.H. Alderman, Hort. Res. Center, Univ. of Minnesota. Introd. in 1985 by J.J. Luby, W.H. Alderman, S.T. Munson, D.K. Wildung, W.H. Gray, and E.E. Hoover. Burbank x Older; cross made in 1937; tested as MN416. Fruit: conic, large, 40 to 50 mm wide and 50 to 60 mm long; skin orange-red with little or no bloom; flesh: bright golden-yellow, sweet; quality excellent; clingstone. Ripens 1 week later than Superior or Toka. Tree: precocious; vigorous; round-headed, spreading; winter hardy; self-fruitful; pollinated by Toka and South Dakota.

Aleta Rose.—Bud sport of Simka found in Dinuba, Calif., by Loretta J., Phillip D., and Alfred W. Strohmaier. Plant patent 5231; 15 May 1984. Fruit: globose to cordate; medium to large, 57 mm; skin 90% to 100% red-purple; flesh yellow; stone semi-free. In central San Joaquin Valley, ripens 10 days before Simka, about 25 to 29 June. Blossom: medium, white, blooms late. Tree: vigorous; upright to spreading; partially self-fertile.

Algoma.—Orig. in Ottawa, Ont., Canada, by Canada Dept. Agr. Res. Sta. Introd. in 1937. Sandcherry (*Prunus besseyi*) x *P. salicina* 'Burbank', selected in 1935; tested as Ottawa 302. Fruit: good size and quality for a plum of this type; characteristics intermediate between parents. Tree: extremely hardy; bushy; early bearing.

Allred.—Orig. in Amity, Ark., by Ross Wolfe, Wolfe Nursery, Stephenville, Texas. Introd. in 1941. Parentage unknown; selection of myrobalan (*Prunus cerasifera*); discovered in 1939. Fruit: small, up to 1 1/4 inches in diam.; skin and flesh red, quality good, but very high in acidity; makes a delicious jelly. Tree: leaves and bark red; produces well; adapted over a wide area; a colorful ornamental, resembling Vesuvius.

Amador (*Erna*).—Orig. in Vacaville, Calif., by Millard Sharpe. Introd. about 1922. Open-pollinated seedling of Climax; selected about 1916. Fruit: Japanese type; very large; skin red, does not crack; flesh yellow, very sweet; semi-clingstone; resembles Climax, following it in ripening.

Amazon.—Orig. in Le Grand, Calif., by F.W. Anderson. Introd. in 1960. Plant patent 2043; 4 Apr. 1961; assigned to Reedley Nursery, Reedley, Calif. Open-pollinated seedling of Eldorado, selected in 1959. Fruit: larger than Eldorado, globose; flesh cream-colored, firm, meaty, melting, flavor delicate, vinous, eating quality good to best; skin deep blue shaded with dark red, giving an overall purple-blue appearance, thick to medium, with no tendency to crack; freestone; ripens several days before Nubiana and 1 week earlier than Laroda; shipping and keeping qualities good. Tree:

vigor medium; bears regularly; self-fruitful.

American Mirabelle.—Orig. in Geneva, N.Y., by Richard Wellington, New York State Agr. Expt. Sta. Introd. in 1925. *Prunus domestica* 'Imperial' x *P. insititia* 'Mirabelle'; cross made in 1911. Fruit: a small, sweet, golden yellow Damson; flavor very good; fresh and culinary purposes; season early; resembles Mirabelle in color but much larger. Tree: growth habit resembles *P. insititia*.

Anderson.—Orig. in Kingsburg, Calif., by Cleve Anderson. Introd. in 1964. Plant patent 2410; 9 June 1964. Chance seedling; Japanese type; discovered in 1959. Fruit: relatively small, averaging about 2 inches in diam. and 2 inches in length; cordate, symmetrical, suture relatively inconspicuous; skin green with a faint blush toward apex, medium thickness, tenacious to flesh, slight tendency to crack; flesh golden, firm, crisp, meaty, sweet, eating quality fair to good, semi-freestone; ripens late, about 5 weeks after Kelsey harvest season of a month (September) without becoming overripe on tree; continues firm in shipment or in cold storage; resembles Kelsey. Tree: size medium; upright; vigor medium; very productive; bears regularly; hardy.

Andys Pride.—Orig. in Le Grand, Calif., by F.W. Anderson, Merced, Calif. Introd. in 1965. Plant patent 2564; 19 Oct. 1965; assigned to Peach & Willow Farms, Fresno, Calif. Open-pollinated seedling of Eldorado; selected in 1962; tested as Anderson 17H970. Fruit: large to medium, uniform; averaging 2 1/16 inches in suture diam.; globose, symmetrical, ventral suture inconspicuous; skin yellow overspread with red, dark red at maturity, medium thick; flesh yellow, firm, meaty, few fibers, ripens evenly, subacid, flavor delicate; stone small, free; ripens 10 Sept., about 1 week after Grand Rosa; shipping quality very good; resembles Grand Rosa. Tree: large to medium; spreading; vigorous; very productive; bears regularly; requires heavy thinning.

An Gee.—Orig. in Fresno, Calif., by Ronald Paul. Plant patent 4024; 22 Mar. 1977. Discovered as a limb mutation on Casselman tree. Tree and fruit similar to Casselman, fruit is larger with lighter bloom and darker red color, more oblong; ripens 3 to 5 days earlier.

Angeleno.—Orig. in Fresno, Calif., by John M. Garabedian. Introd. in 1967. Plant patent 2747; 13 June 1967; assigned to Peach & Willow Farms, Fresno, Calif. Open-pollinated seedling of Queen Ann. Fruit: large to medium; roundish-oblate; skin deep purple-blue; flesh yellowish-amber, firm; stone free to semi-free; ripens 4 to 5 weeks later than Queen Ann. Tree: size medium; vigor medium; productive and regular bearer; leaf glands globose; flower small, waxy white.

Anna Sun.—Orig. in Del Rey, Calif., by Tom Chamberlin, H.P. Metzler & Sons. Introd. in 1986 . Parentage unknown. Plant patent 6077; 5 Jan. 1988; licensed exclusively to Agri Sun Nursery, Selma, Calif. Fruit resembles Kelsey; large; slightly elongated; skin medium green, smooth; flesh yellowish-green, slightly crisp; ripens 3 weeks after Kelsey. Tree: vigorous, productive; apparently self-fruitful.

Anoka.—Orig. in Excelsior, Minn., by the Univ. of Minnesota Fruit Breeding Farm. Introd. in 1922. *Prunus salicina* 'Burbank' x *P. americana* 'De Soto'; Cross made in 1913; selected in 1918; tested as Minn. 118. Fruit: medium to large; oval to slightly obovate; skin attractive red, thick, tough, easily removed from flesh; flesh yellow, firm, fine-grained, medium juicy, mildly subacid, quality fair; pit small; cling; ripens in late midseason. Tree: size medium; slightly drooping; vigorous; productive; hardy. Variety has been abandoned because of low quality.

AP-3.—Bud sport of Laroda found in Dinuba, Calif., by Alvin Peters. Plant patent 5074; 26 July 1983. Fruit: globose; medium size, 56 mm; skin red blush over orange-yellow ground color; flesh yellow; clingstone. In central San Joaquin Valley, ripens 6 to 7 weeks after Laroda, about 25 Aug. to 5 Sept. Blossom: medium, white. Tree: vigorous; semi-spreading.

Armelita.—Orig. near Reedley, Fresno County, Calif., by Dennis Surabian. Plant patent 4140; 1 Nov. 1977. Open-pollinated seedling of Laroda. Fruit: similar to Laroda, but final coloration of fruit occurs earlier and is much darker in color; freestone; globose; blue-violet skin with yellow, firm-textured flesh that assumes a brilliant red color in storage. Tree: size medium; vigor medium; regular bearer.

AU-Amber.—Orig. at Auburn, Ala., by J.D. Norton, Auburn Univ. Introd. in 1988. Parents are siblings of double backcross of Starcher No. 1 (a native selection at least partly *Prunus angustifolia*) to Methley; cross made in 1965; selected in 1975; tested as Methley 12-10. Fruit: medium; skin reddish-purple; flesh amber, moderately firm; high quality. Ripens 3 to 4 weeks before Santa Rosa, 1 week before Methley. Tree: vigorous on peach rootstock; spreading; resistant to bacterial spot and bacterial canker; tolerant of plum leaf scald; susceptible to black knot. Chilling requirement about 650 h. Adapted to the humid Southeast for local sales. Flowers large, self-infertile; blooms about 4 days before Santa Rosa.

AU-Cherry.—Orig. at Auburn, Ala., by J.D. Norton, Auburn Univ. Introd. in 1988. Parents are siblings of double backcross of Starcher No. 1 (a native selection at least partly *Prunus angustifolia*) to Methley; cross made in 1965; selected in 1975; tested as Methley C-50. Fruit: small; skin dark red; flesh dark red, moderately firm. good quality. Ripens with Santa Rosa and up to 10 days later. Resistant to bacterial spot, bacterial canker and black knot; tolerant of plum leaf scald. Chilling requirement 700 to 750 h. Adapted to the humid Southeast for local sales and backyard growers. Flowers large, self-infertile; blooms about with Santa Rosa

AU-Producer.—Orig. in Auburn, Ala., by J.D. Norton, Alabama Agr. Expt. Sta., Auburn Univ. Introduced in 1977. Bruce x Santa Rosa; cross made in 1963; selected in 1967; tested as Bruce 13-18. Fruit: round, 1 1/2 to 1 3/4 inches in diam.; firm, excellent eating quality, highly colored; resistant to bacterial canker, bacterial fruit spot, and bacterial leaf spot; has tolerance to black knot. Tree: large; spreading; vigorous; productive.

AU-Roadside.—Orig. in Auburn, Ala., by J.D. Norton. Alabama Agr. Expt. Sta., Auburn Univ. Introd. in 1983. Ozark Premier x Methley. Cross made in 1961; selected in 1963; tested as Ozark Premier F-2. Fruit: Japanese-type; larger than Methley and Bruce, same size as Ozark Premier; flesh not firm enough for shipping; turns dark red to purple upon ripening; pit medium, semi-cling; ripens 7 days before Ozark Premier. Tree: vigorous, precocious, productive; resistant to complex of bacterial diseases common to Southeastern states, also resistant to black-knot, brown rot, and plum leaf scald. Superior to Ozark Premier in yield, quality, disease resistance, and longevity.

AU-Rosa.—Orig. at Auburn, Ala, by J.D. Norton, Auburn Univ. Introd. in 1988. Parents are siblings of double backcross of Starcher No. I (a native selection at least partly *Prunus angustifolia*) to Methley; cross made in 1965; selected in 1975; tested as Santa Rosa A-S. Fruit: large; skin red; flesh amber, firm, high quality. Ripens a few days after Santa Rosa. Tree: vigorous and upright on peach rootstock; moderately resistant to bacterial spot and bacterial canker; tolerant of plum leaf scald and brown rot. Chilling requirement about 750 h. Better adapted to the humid Southeast than is *Santa Rosa*, which it replaces. Flowers large, self-infertile; blooms with Santa Rosa.

AU-Rubrum.—Orig. at Auburn, Ala., by J.D. Norton, G.E. Boyhan, D.A. Smith, and B.R. Abrahams, Dept. of Horticulture and Alabama Agr. Expt. Sta., Auburn Univ. Introd. in 1988. Bud mutation of Crimson; tested as Crimson M-1. Fruit: large, 60 mm in diameter, compared to 4.2 for Crimson; skin scarlet-red; flesh dark red, moderately firm, clingstone. Excellent eating quality; canned quality acceptable, but not equal to AU-Amber. At Auburn, fruit matures about 1 week after Methley, 5 weeks before Crimson; stores well at 0 C. Tree: more vigorous than Crimson; spreading; productive; dark green foliage. Chilling requirement about 700 h. Resistant to *Pseudomonas syringae*, *Xanthomonas pruni*, and *Apiosporina morobosa*; tolerant of plum leaf scald and brown rot. Flowers large, self-infertile; blooms 2 days before Santa Rosa.

Autumn Beaut.—Orig. in Modesto, Calif., by Chris, Gary, and Grant Zaiger and Leith Gardner. (Roysum open-pollinated) x Eldorado. Plant patent 8189; 30 Mar. 1993.

Fruit: globose to slightly oblate; large, 64 mm; skin maroon to blackish-blue; flesh light yellow, firm; clingstone; does not crack. In central California ripens 6 days before Roysum. Blossom: small to medium; white. Tree: vigorous; upright productive; self-infertile.

Autumn Bright.—See Autumn Brite.

Autumn Brite (*Autumn Bright*).—Orig. in Tulare, Calif., by Roy Sumruld. Introd. in 1972. Plant patent 3297; 2 June 1973, assigned to the Burchell Nursery, Modesto, Calif. Unknown parentage, discovered in 1967. Fruit: Japanese type; 1 to 1 1/2 inches in diam.; ovoid; skin red, blushed with purple, moderate bloom; flesh light yellow with red near skin, flavor mild, clingstone; ripens very late, 1–20 Oct. Tree: size medium; vigorous; moderately productive.

Autumn Giant.—Orig. in Modesto, Calif., by Floyd, Gary, and Grant Zaiger and Leith Gardner. Introd. in 1986. King David x Roysum. Plant patent 5624; 7 Jan. 1986; assigned to Zaiger's Genetics. Fruit: very large, 65 mm; globose; skin yellow ground color covered by red blush; flesh yellow, firm; flavor mild. Tree: vigorous; upright; productive.

Autumn Pride.—Orig. in Modesto, Calif., by Chris, Gary, and Grant Zaiger and Leith Gardner. Friar x (Mariposa x Ebony). Plant patent 7843; 31 Mar. 1992. Fruit: globose, slightly flattened; large, 64 to 67 mm; skin purple; flesh pale yellow; stone semi-free. In central California, ripens 7 to 10 days after Roysum, about the last week of September. Blossom: medium, white, blooms about 2 days after Santa Rosa. Tree: vigorous; upright; productive.

Autumn Queen.—Orig. in Parlier, Calif., by Seishi Iwanura. Plant patent 3850; 23 Mar. 1976. Fruit: medium, generally globose; skin deep crimson; flesh light yellow, firm, flavor delicate, sweet, slightly acid; excellent eating quality; freestone; ripens in early September, 3 weeks before Roysum, which it most nearly resembles; last picking extendible for 2 weeks because of high fruit quality. Tree: large; self-pollinating; vigorous; productive; regular bearing.

Beauty Junior.—Orig. in Santa Rosa, Calif., by Luther Burbank. Introd. in 1920. seedling of Beauty. Fruit: very large; round; flesh deep crimson; keeps well in cold storage; ripens in September and October.

Beaver.—Orig. in Valley River, Manitoba, Canada, by W.J. Boughen, Boughen Nurseries. Introd. in 1925. Open-pollinated seedling of Assiniboine; selected in 1925; sibling of Dandy. Fruit: large; skin yellow; ripens in midseason, after Dandy; quality good, for fresh and preserving uses; resembles Assiniboine. Tree: yields well.

Bee-Gee.—Orig. in Fresno, Calif., by John M. Garabedian. Introd. in 1969. Plant patent 2858; 21 Jan. 1969; assigned to Peach & Willow Farms, Fresno. Open-pollinated seedling of Eldorado. Fruit: large; generally globose, very symmetrical; resembles Laroda in form; skin straw-yellow, almost entirely overspread with red, shading to black; flesh yellow shading from a relatively deep red at the apex to a lighter red extending nearly to the shoulder; ripens 35 days after Eldorado. Tree: large; upright to spreading; vigorous; productivity medium, regular bearer.

Belgium Gage.—See Belgium Green Gage.

Belgium Green Gage (*Belgium Gage*).—Orig. in Deloraine, Manitoba, Canada, by Alidor Andries. Introd. in 1936. seedling *of Prunus domestica* secured in Belgium. Fruit: 1 1/4 inches in diam.; round; skin dark yellow, tender; flesh yellow, firm, fine, aromatic, flavor spicy and sweet, quality good; season early September. Tree: spreading; not hardy in upper fruit buds, but fruiting regularly on lower branches in Manitoba. Now obsolete.

Bella-Rosa.—Orig. in Ivanhoe, Calif., by John Luisi, Visalia, Calif. Introd. in 1964. Plant patent 2406; 26 May 1964. Bud mutation of Santa Rosa; discovered in 1955. Fruit: larger than parent, up to about 2 5/8 inches in diam.; uniformly globose to slightly ovoid; skin red, darkening as it becomes fully ripe, thick, tough, tenacious to flesh, bloom moderate; flesh yellowish-white with deep purplish-red next to skin at maturity, firm, crisp, moderately dry; stone size medium, clinging; stores and ships well; ripens 3 to 5 days before Santa Rosa. Tree: large, vigorous, upright; good productivity; bears regularly; blooms 2 to 7 days before Santa Rosa.

Betty Anne.—Orig. in Modesto, Calif., by

Chris, Gary, and Grant Zaiger and Leith Gardner. {[Friar x (Mariposa x Ebony)] x Autumn Giant}. Plant patent 8471; 23 Nov. 1993. Fruit: nearly globose, slightly flattened at base and on suture side; large, 58 to 64 mm; skin red with yellow ground color; flesh yellow-white, firm with mild flavor; clingstone. In central California, ripens the last of August. Blossom: medium, white. Tree: vigorous; upright; productive.

Bill Hengst October Gem.—Chance seedling found in Exeter, Calif., by William E. Hengst. Plant patent 7348; 9 Oct. 1990. Fruit: ovate; medium-large, 62 mm; dark red skin and flesh (Princeton); clingstone. In central San Joaquin Valley, ripens 4 weeks after Casselman. Blossom: small, white. Tree: medium vigor; upright; productive.

Billington (*Billington's Early*).—Orig. in Glen Eden, Auckland, New Zealand, on the property of a Mr. Billington. Introd. there about 1922; propagation rights acquired by Hayward Wright. Chance seedling; parentage unknown; selected about 1919. Fruit: size medium; round, flattened at stem end, apex round with slight depression, halves somewhat unequal; stem of medium length; skin smooth, dark purplish-red, with heavy white bloom; flesh dark red, firm, juicy, flavor and quality good, clingstone; holds well; season end of December into January in New Zealand. Tree: bearing habit very irregular.

Billington's Early.—See Billington.

Bird.—Orig. in Vacaville, Calif., by Millard Sharpe. Introd. in 1923. Parentage unknown; *Prunus domestica*. Fruit: obovate to oblong; skin dullish red; flesh very light amber yellow, sweet, mild, quality fair, freestone; ripens last week in August. No longer being propagated.

Blackamber.—Orig. in Fresno, Calif. Selected by J.H. Weinberger in cooperation with U.S. Dept. of Agriculture and Univ. of California at Kearney Hort. Field Sta., Parlier, Calif. Introd. in 1980 by D.W. Ramming. Friar x Queen Rosa. Cross made in 1970; selected in 1973; tested as K68-43. Fruit: oblate, averages 6.2 cm in diam.; skin black at full maturity; flesh light yellow with some diffusion of red from skin when overripe, quality good, firm; pit small; ripens end of June at Fresno, 1 week before Eldorado and 3 weeks before Friar. Tree: normal vigor, upright growth habit; flowers self-infertile, pollinizer required, blooms with and is intercompatible with Friar; recommended only for Japanese plum areas.

Black Beaut.—Orig. in Le Grand, Calif., by F.W. Anderson, Merced. Introd. in 1975. Plant patent 3617; 17 Sept. 1974; assigned to Reedley Nursery, Reedley, Calif. Open-pollinated seedling of Red Beaut, tested as 1S14B. Fruit: large to medium; globose to slightly oblate, uniform; skin deep purple; flesh yellow to red; semi-freestone; resembles Red Beaut except for skin color; ripens earlier than Santa Rosa, extends shipping season of Red Beaut type plum. Tree: large; upright to spreading; vigorous; productive.

Black Diamond™ (*Suplumeleven*).—Orig. in Fresno, Calif., by John H. Weinberger. Introd. in 1982. Plant patent 4902; 26 Oct. 1982; assigned to Superior Farming Co. Open-pollinated seedling of Angeleno (most probably pollinated by Midsummer). Fruit: large, 60 mm in diam.; round; skin black, smooth finish; inconspicuous suture, medium thick; flesh red, medium firm; semi-freestone; excellent eating quality; most nearly resembles Queen Ann. In central California, ripens 1 week before Queen Ann. Tree: medium vigor; very productive; average precocity; upright; midseason blossoming; Angeleno is suitable pollinizer.

Black Elephant Heart.—See Black Queen.

Black Flame™ (*Suplumfifteen*).—Orig. in Fresno, Calif., by John H. Weinberger. Introd. in 1985. Plant patent 5487; 4 June 1985; assigned to Superior Farming Co. Open-pollinated seedling of Simka; tested as 295-198. Fruit: large, 60 to 65 mm; globose; skin dark purple; flesh yellow, firm, clingstone. In central California, ripens about 3 weeks after Simka, which it most nearly resembles. Tree: very productive; regular bearer; medium vigorous; upright; late-blossoming; Black Torch is suitable pollinizer.

Black Gold.—Orig. in Selma, Calif., by Jim Mulligan. Plant patent 4643; 10 Feb. 1981. Introd. in 1980. Sport from Ace plum; selected in 1975. Fruit: skin black, resembles Damson plum, but larger; flesh golden; long storage life.

Black Gold™ (*Suplumtwelve*).—Orig. by John H. Weinberger; assigned to Superior Farming Co. Angeleno open-pollinated. Plant patent 4965; 14 Dec. 1982. Fruit: globose; medium size, 61 mm; skin dark purple; flesh red; stone semi-free. In central San Joaquin Valley, ripens with Eldorado, about 30 June to 8 July. Blossom: medium, white. Tree: medium vigor; spreading; very productive.

Black Jack.—Orig. in Modesto, Calif., by Floyd, Gary, and Grant Zaiger and Leith Gardner. Friar open-pollinated. Introd. in 1984. Plant patent 7896; 30 June 1992. Fruit: large, 67 mm in diam.; globose; skin blue; flesh yellow; semi-clingstone, nearly free; matures in mid- to late June in California. Tree: vigorous, upright; heavy producer; self-incompatible.

Black Jewel.—Orig. at Fresno, Calif., by J.M. Garabedian. Introd. in 1987. Plant patent 5905; 17 Mar. 1987. Chance seedling; Japanese type. Fruit medium-large, 63 mm wide and 60 mm long; skin dark maroon-red with moderately heavy blue-gray bloom; flesh light amber, melting to meaty texture; quality good to excellent; clingstone. In central California, ripens 8 Aug. Tree: productive; medium vigor; upright.

Black Kelsey.—Orig. near Fresno, Calif., by John M. Garabedian. Introd. in 1970. Plant patent 3154; 9 May 1972, assigned to Superior Farming Co., Bakersfield, Calif. Variety resulted from open-pollination of Kelsey, Queen Ann, or Mariposa. Fruit: large, uniform; shape like Kelsey; stem 1/2 inch in length; skin thick, tough, bitter, tenacious to flesh, smooth with excellent resistance to shriveling, dark purple, nearly black when fully ripe, bloom bluish gray, heavy, dots numerous, small yellowish; flesh brilliant red, mild, vinous, firm, fine, crisp to meaty; stone semi-free, generally large, eating quality very good. Tree: size medium; vigorous; substantially upright but spreading and open with density determined by pruning methods; hardy; productive; regular bearer; ripens with Kelsey.

Black Noble.—Orig. in Le Grand, Calif., by Norman and Glen Bradford. Red Beaut x unnamed seedling; cross made in 1983. Plant patent 7504; 23 Apr. 1991. Fruit: symmetrical, obovoid; medium size, 54 mm; skin very shiny, dark reddish-purple; flesh strong red, crisp, and extremely firm; flavor acidic and very sweet; juicy; clingstone. In central California, ripens between Black Beaut and Santa Rosa. Blossom: small; white; blooms with and pollinated by Wickson. Tree: vigorous; upright; self-unfruitful; productive.

Black Premium™ (*Suplumsixteen*).—Orig. by John H. Weinberger and Tim P. Sheehan, assigned to Sun World. Queen Rosa x Eldorado. Introd. in 1991. Plant patent 7574; 2 July 1991. Fruit: ovate with pointed blossom end; medium size, 60 mm; skin red-black; flesh yellow. In central San Joaquin Valley, ripens with Blackamber. Tree: medium vigor; semi-upright; regular bearing; medium productive.

Black Prince.—Orig. in Bedford, England, by Laxton Bros. Introd. there about 1930. Considered to be Kentish Black x Damson. Fruit: small to medium; skin bluish-black with a light blue bloom; flesh juicy, very acid, with Damson flavor; ripens in early August. Tree: vigorous; compact; blooms early; bears heavy crops annually. Not the Black Prince variety originated by R.E. Burton, Vacaville, Calif., and introduced before 1920.

Black Queen (*Black Elephant Heart*).—Orig. in Fresno, Calif., by John M. Garabedian. Introd. in 1964. Plant patent 2420; 7 July 1964; assigned to Peach & Willow Farms, Fresno. Open-pollinated seedling of Queen Ann. Fruit: large; some fruit slightly enlarged on one side of suture; skin smooth, black with blue-gray bloom; flesh bright red, turning dark red when overripe, firm; clingstone; hangs well on tree; ripens 5 days later than Santa Rosa, 3 weeks later than Duarte. Tree: size medium; vigor medium; productive and regular bearer; leaf glands globose; flower size medium, white.

Black Ruby.—Orig. in Byron, Ga., by J.M. Thompson. BY4-95 open-pollinated; seed planted in 1975. BY4-95 is a seedling of Queen Ann probably pollinated by Santa Rosa. Selected in 1978 and tested as BY7404-1. Released in 1980 by W.R. Okie. Fruit: round to oblate; medium size, 50 mm; skin reddish-black; flesh yellow, firm, good quality; freestone. In Georgia, ripens 1 week after Santa Rosa and Morris. Blossom: white with orange-red anthers; pollinated by Rubysweet, AU-Rubrum, and Methley. Tree: moderately vigorous and upright; productive; self-unfruitful. Moderately resistant to bacterial spot and bacterial canker, tolerant to plum leaf scald.

Black Torch™ (*Suplumthirteen*).—Orig. in Fresno, Calif., by John H. Weinberger. Introd. in 1984. Plant patent 5343; 20 Nov. 1984; assigned to Superior Farming Co. Open-pollinated seedling of an unnamed seedling, selected in 1979. Fruit: large, 60 mm; skin black; flesh cream, clingstone; round, smooth, uniform; attractive. In central California ripens 1 week after Santa Rosa; most nearly resembles Eldorado. Tree: vigor medium; productive; upright, round-topped.

Bluebell.—Orig. in Mountain Grove, Mo., by Paul H. Shepard, Missouri State Fruit Expt. Sta. Introd. in 1947. Stanley x President; selected in 1947; tested as PL-218; sibling of Bluefre and Radiance. Fruit: prune type; large, more so than Stanley; skin blue; flesh firm, yellow, flavor good, sweet, freestone; ripens about 14 Sept. shortly after Stanley, which it resembles. Tree: vigorous; upright; fruit held through a long harvesting period.

Bluefre.—Orig. in Mountain Grove, Mo., by Paul H. Shepard, Missouri State Fruit Expt. Sta. Introd. in 1947. Stanley x President; selected in 1947; tested as PL-228; sibling of Bluebell and Radiance. Fruit: large; skin blue; flesh firm, thick, greenish-yellow, flavor good, freestone; ripens latter part of August, just before Stanley; resembles President. Tree: shape resembles President; fruit hangs on tree in good condition for about 30 days after normal harvest time; very productive.

Blue Giant.—Orig. in Modesto, Calif., by Floyd, Gary, and Grant Zaiger and Leith Gardner. Laroda x Queen Ann. Introd. in 1988. Plant patent 6764; 25 Apr. 1989; assigned to Zaiger's Genetics. Fruit: very large, 63 to 70 mm in diam.; globose; skin blue; flesh yellow; clingstone; matures late June to early July in California. Tree: moderate vigor; semi-upright; very productive; self-incompatible.

Blue Knight.—Orig. in Modesto, Calif., by Chris, Gary, and Grant Zaiger and Leith Gardner. (Red Beaut open-pollinated) open-pollinated. Plant patent 5813; 2 Dec. 1986. Fruit: globose; medium size, larger than Red Beaut, 54 mm; skin blackish-blue; flesh yellow, firm; clingstone. In central California, ripens 10 days after Red Beaut, about 13 to 19 June. Blossom: medium, white. Tree: vigorous; upright; productive.

Bonnie.—Orig. in Mountain Grove, Mo., by Paul H. Shepard, Missouri State Fruit Expt. Sta. Introd. in 1947. Open-pollinated seedling of America; selected in 1947; tested as PL-129. Fruit: larger, redder, ripens earlier than America, which it resembles; firm; good shipper. Tree: vigorous; bud and bloom very hardy.

Bounty.—Orig. in Morden, Manitoba, Canada, by Canada Dept. Agr. Res. Sta. Introd. in 1939. Open-pollinated seedling of *Prunus nigra* 'Assiniboine'; selected in 1933; seed originally sent from Minnesota State Fruit Breeding Farm in 1922; tested as Morden 105. Fruit: medium to large; skin dark red; flesh orange yellow, tender, juicy, sweet; quality fair for dessert, good for canning; ripens third week of August. Tree: upright, spreading; very hardy; productive; suitable for northern conditions.

Bradgreen.—See Green Sweet™.

Brilliant.—Orig. in Mountain Grove, Mo., by Paul H. Shepard, Missouri State Fruit Expt. Sta. Introd. in 1947. Burbank x Methley; selected in 1947; tested as PL-174. Fruit: large; round; skin red with light blue bloom, smooth, attractive; flesh firm, yellow, sweet, flavor like Abundance; clingstone; ripens about mid-July. Tree: productive; vigorous; resembles Methley in growth habit; has winter bud hardiness.

Broken Heart.—Orig. at Visalia, Calif., by Frank A. Vieira. Introd. in 1988. Plant patent 6118; 1 Mar. 1988; assigned to L.C. and E.E. Huerta, Reedley, Calif. Sport of Elephant Heart. Fruit: medium, 58 mm wide and 58 mm long; broadly cordate; skin greenish yellow with heavy gray bloom; flesh yellowish-orange near the skin surface to lighter orange in and around the cavity, firm, meaty; quality excellent; semi-freestone; in central California, ripens second week of August. Tree: precocious; productive; vigorous.

Brooks Giant Italian.—See Brooks Italian.

Brooks Italian (*Brooks Giant Italian, Lafayette*).—Orig. in Lafayette, Ore., by B. Brooks, Lafayette Nursery Co. Introd. in 1946 by this company. Open-pollinated seedling of Italian Prune; discovered in 1942. Fruit: averages 50% larger than Italian Prune; oval; skin blue, attractive, moderately thick; flesh greenish-yellow, very firm, slightly tart, moderately juicy, sweeter than Italian; keeps and ships well; fair canning quality, good drying quality; ripens 1 week earlier than Italian Prune; similar in color and shape to Italian Prune. Tree: bears annually.

Bruce (*Bruce's Early*).—Orig. in Donley County, Texas, by A.L. Bruce. Introd. about 1921. [*Prunus salicina* (probably Abundance) x *P. angustifolia*]. Fruit: 1 1/2 inches or larger in diam.; quality fair; matures very early, being its primary outstanding fruit characteristic. Tree: bears heavily; may be propagated by hardwood cuttings, but trees on peach rootstock (among all kinds tested) appear to be best.

Bruce's Early.—See Bruce.

Burbank's Flaming Delicious.—See Flaming Delicious.

Burbank's Great Yellow.—See Great Yellow.

Burbank's June Redskin.—See June Redskin.

Burbank's Mammoth Cardinal.—See Mammoth Cardinal.

Burgundy.—Orig. in Dinuba, Calif., by Walter D. Krause, Reedley, Calif. Introd. in 1970. Open-pollinated seedling, probably Mariposa x Eldorado, discovered in 1970. Fruit: small, 1 1/2 inch diam.; round; skin and flesh dark maroon; flavor mellow sweet; ripens mid-July to mid-August, recommended for home gardens, good fresh and for preserves and canning, fruit too small for commercial market. Tree: upright, vase shaped; heavy productivity, requires thinning; self-fertile.

Burmosa.—Orig. in Winters, Calif., by Claron O. Hesse, California Agr. Expt. Sta., Davis. Introd. in 1952. Burbank x Formosa; cross made in 1940; selected in 1945; tested as Calif. 26-26. Fruit: Japanese; large, sizing better than does Beauty, with an average weight of 6.2 fruit per lb.; round-oval to round-oblate, slightly asymmetrical, halves equal or nearly 50; stem short, rather stout; skin greenish-yellow to light amber-yellow, bright, blushed medium red, to full red when completely ripe; bloom light grey, moderately heavy; dots numerous; flesh light or whitish to greenish-cream to light whitish-amber at full ripe, moderately sweet, mild, melting, juicy, very fine-textured; stone small, nearly completely free except at suture; very long storage; good holding characteristics both on and off the tree; reaches shipping maturity very early, about with Beauty, 15 days before Santa Rosa; resembles Formosa. Tree: productivity fair to good; vigorous; light to moderate chilling requirement; similar to Burbank; self-unfruitful, successfully pollinated by Becky Smith, Duarte, Flaming Delicious, Gaviota, Santa Rosa, Wickson.

Burton.—Orig. in Vacaville, Calif., by R.E. Burton. Introd. in 1926. Considered to be an open-pollinated seedling of Imperial; discovered in 1900; in 1915 some 160 trees grafted to this selection. Fruit: prune type; obovate, necked; skin purple; very large, averaging 20 to 30 dried prunes per lb.; suitable for fancy packs; some difficulty in dehydrating because of large size; resembles Giant.

Byrongold.—Orig. at Byron, Ga., by J.M. Thompson, U.S. Dept. of Agriculture S.E. Fruit & Tree Nut Res. Lab. Introd. in 1985. Open-pollinated seedling of BY68-87; BY68-87=[Gaviota open-pollinated x (Ozark Premier x *Prunus angustifolia*)]; seed collected in 1973; selected in 1976; tested as BY7401-5. Fruit: large; skin golden yellow; round; flesh yellow, firm, semi-freestone; good quality. Ripens 3 days after Ozark Premier 12 days after Santa Rosa. Tree: very vigorous on peach rootstock; spreading; very resistant to bacterial spot and bacterial canker; tolerant of plum leaf scald and brown rot. Flowers large, three or four per cluster; mostly self-incompatible; blooms about 3 days before Santa Rosa. Adapted to the humid Southeast.

California Gem.—See Del Norte.

Calita.—Orig. in Fresno, Calif., by U.S. Dept. of Agriculture and Istituto Sperimentale per la Frutticoltura, Rome, Italy. Introd. in 1984. Gaviota x Eldorado; cross made in

1963; selected by John H. Weinberger, U.S. Dept. of Agriculture, in cooperation with Univ. of California, Parlier; tested as FK5-37. Tested in Italy, California, Washington, and Georgia. Being introduced in Italy because of outstanding fruit quality, regular production, and firmness. Not recommended for California, Washington, or Georgia. Fruit: Japanese variety; 2 to 2 1/2 inches in diam.; round to slightly ovate and slightly asymmetrical; skin red-violet; flesh yellow, firm, softens slowly; clingstone; quality fair to excellent. Tree: unpright to spreading; vigorous; productive; flowers self-infertile, requiring a pollinizer; matures 10 to 12 days after Shiro and 4 to 6 days after Santa Rosa.

Campagne.—Orig. in Fresno, Calif., by Thomas A. Campagne, Kingsburg, Calif. Introd. in 1966. Plant patent 2674; 4 Oct. 1966. Bud mutation of Santa Rosa. Fruit: 2 inches in diam., uniform, round; stem length 1/4 to 3/8 of an inch, reducing fruit drop; skin deeper red than Santa Rosa, smooth, stronger than Santa Rosa; flesh deep yellow, juicy; aroma and texture similar to Santa Rosa; flavor more delicate than Santa Rosa; eating quality excellent; keeping quality very good; shipping quality good; freestone; without fibers; ripens evenly, mid-May, 1 to 2 weeks earlier than Burmosa, 3 to 4 weeks earlier than Santa Rosa; recommended for eating, shipping, and canning. Tree: similar to Santa Rosa; more productive than Santa Rosa; bears regularly; flower similar in size to Santa Rosa but light pink in color; good resistance to disease.

Caroline Harris.—Orig. in Bakersfield, Calif., by Caroline P. Harris. Plant patent 3823; 23 Dec. 1975; assigned to Plumwood Assn., Bakersfield, Calif. Parentage unknown. Fruit: large, globose; skin purple-red; flesh red, firm; flavor sweet, subacid; excellent eating quality, shipping and dessert use; stone small; clingstone; ripens late June. Tree: moderate to vigorous; spreading; productive.

Cascade.—See Krause.

Casselman.—Orig. in Exeter, Calif., by S.T. Casselman. Introd. in 1959 by Reedley Nursery, Reedley, Calif. Considered to be a bud mutation of Late Santa Rosa; discovered in 1956. Fruit: size and shape similar to parent; skin with more yellow flecking in ground color and lighter red over-color than in parent; ripens a few days later than Late Santa Rosa; shows no tendency to crack.

Casselman Special.—See Late Santa Rosa.

Castleton™.—Orig. in Geneva, N.Y., by R.L. Andersen, J.P. Watson, and K.G. Livermore. Valor x Iroquois; cross made in 1966; tested as NY 66.609.4; introd. in 1993. Fruit: 36 mm cheek to cheek, 38 mm at broad point of suture and 46 mm long; similar to Stanley in size but more round; skin dark purple with little waxy bloom, colors early. Ripens 2 to 4 days before Richards Early Italian. Excellent fresh fruit and processing quality. Nonfragmenting stone, which "pits" well in needle-type pitters; stone dimensions 25 x 17 x 9 mm, compared to Stanley's at 30 x 14 x 8 mm; suture cracks infrequent compared to Stanley. Tree: medium vigor, spreading; intermediate number of laterals; less prone to produce secondary lateral branches than Stanley; heavily spurred. Partially self-fertile; pollinized by Longjohn, Polly, and Stanley; consistently heavy yields.

Catalina.—Orig. in California by Walter Krause. Introd. in 1982. Angeleno open-pollinated. Japanese type. Fruit: large; round; skin dark purple with waxy bloom; flesh amber, very firm; outstanding flavor. In central California, ripens 7 to 10 days before Friar. Tree: very vigorous; very productive; not as hardy as Friar.

Celebration.—Orig. in Koelenhof, South Africa, by M.A. Oosthuizen and Jan Stadler, Fruit and Fruit Tech. Res. Inst., Stellenbosch. Introd. in 1989. An open-pollinated seedling of Songold (most probably pollinated by Eldorado). Fruit: size same as Sungold; round to oblate; skin black with heavy bloom; flesh deep yellow to orange, semi-melting, juicy; pit medium to large, clingstone; quality similar to Eldorado. Ripens 1 week later than Sungold. Tree: compact; upright, similar to Santa Rosa; self-unfruitful, pollinated by Harry Pickstone and with same bloom date.

Chadsey.—Orig. in Chilliwack, B.C., Canada, by Chester Chadsey. Introd. in 1925. Parentage unknown; European plum

(*Prunus domestica*); discovered about 1920. Fruit: large; freestone. Variety now obsolete.

Chilcott.—Orig. in Mandan, N.D., by W.P. Baird, Northern Great Plains Field Sta. Introd. in 1947. Open-pollinated seedling of *Prunus americana* secured from Edgeley, North Dakota; discovered in 1926. Fruit: large; roundish; skin red; clingstone; quality good for a native species; ripens in late August at place of origin; resembles Wastesa but larger. Tree: hardy; upright; productive; good pollinator.

Chinook.—Orig. in Mandan, N.D., by W.P. Baird, Northern Great Plains Field Sta. Introd. in 1957. Open-pollinated seedling of *Prunus nigra* 'Ojibwa'; selected in 1942; tested as Mandan 42-4. Fruit: large for a native plum, averaging about 20 per lb.; skin bright red, tender; flesh firm, juicy, sweet; flavor excellent; cooking tests rate high; usually ripens in late August. Tree: hardy; bears heavily annually; growth moderate, slightly spreading; most branches with wide-angled crotches. Recommended in the northern Great Plains area where a hardy native plum of superior quality and a good pollinator for other plums is desired.

C.K.C.—Orig. in Rosthern, Saskatchewan, Canada, by Seager Wheeler. Introd. in 1933. Parentage unknown. Fruit: size medium; skin red; flesh yellow, juicy; quality fair to good; season late August; resembles *Prunus nigra* 'Assiniboine'. Tree: upright, spreading; very productive; hardy.

Coin.—Orig. in Santa Rosa, Calif., by Luther Burbank. Introd. in 1927. Parentage not given but apparently a cross between Japanese and American *Prunus* spp. Fruit: round; skin yellow with large dots; flesh clear pale amber, sweet, mild, rich, freestone. No longer propagated.

Compact Friar.—Orig. in San Joaquin County, Calif., by Henry G. Wong. Irradiated mutation of Friar. Plant patent 8103; 19 Jan. 1993. Fruit: oblate; large, 43 to 46 mm; skin purplish-black, same color as Friar; flesh orange-yellow, firm, mild sweet. Ripens with Friar. Blossom: medium, white. Tree: medium vigor, dwarf, half the size of Friar; upright but not as much as Friar; self-unfruitful.

Crimson.—Orig. in Auburn, Ala., by J.D. Norton, Agr. Expt. Sta. Introd. in 1973. Bruce x Ozark Premier; cross made in 1961; tested as Bruce 12-114. Fruit: 1 1/2 to 1 3/4 inches in diam.; nearly round; skin crimson; flesh crimson, firm; quality excellent; clingstone; may be stored 3 weeks or more; ripens mid-July; recommended for local and distant market. Tree: medium; spreading; vigorous; excellent productivity; resistant to bacterial canker, fruit spot, leaf spot, black knot and brown rot; chilling requirement 850 h.

Crimson Nugget.—Orig. in Del Rey, Calif., by Frank T. Matoba. Red Beaut open-pollinated. Plant patent 8038; 27 June 1991. Fruit: ovate; small size; skin dark red. In central San Joaquin Valley, ripens 5 days before Red Beaut.

Crystal Red.—Orig. in Grimsby Beach, Ont., Canada, by A.W. Eickmeier. Introd. in 1943. Plant patent 560; 22 Dec. 1942. Parentage unknown, but of Japanese type; selected in 1940. Fruit: skin highly colored red, tough; flesh yellow, firm, juicy; flavor mild but sour next to seed; clingstone; matures 2 to 3 weeks earlier than Burbank, which it resembles; excellent shipper. Tree: regular and early bearer; strong grower; upright-spreading; hardy.

Daisy.—Orig. in Shenandoah, Iowa, by Oscar Forss. Introd. in the late 1920s. Chance seedling of a native species; selected in the early 1920s. Fruit: small; skin red; quality good, making excellent jelly and a fair sauce. Tree: exceedingly heavy production. Variety now obsolete.

Dandy.—Orig. in Valley River, Manitoba, Canada, by W.J. Boughen, Boughen Nurseries. Introd. in 1923. Open-pollinated seedling of *Prunus nigra* 'Assiniboine'; seed planted in 1920; selected in 1923. Fruit: quality very good, mild; yellow; for eating fresh or for culinary uses; ripens early. Tree: prolific bearer.

Del Norte (*California Gem*).—Orig. in Vacaville, Calif., by a Mr. Smith. Introd. in the 1930s. Chance seedling of a Japanese plum (*Prunus salicina*). Fruit: skin medium to dark red, bright, shiny; flesh amber, fairly firm; quality fair; ripens the latter part of August; tends to sunburn badly in area of origin. Now obsolete.

Del Rey Sun.—Orig. in Del Rey, Calif., by Thomas O. Chamberlin, Sr.; assigned to Metzler Investments and Metropolitan Life Insurance Co. Open-pollinated seedling of unknown origin. Plant patent 6991; 22 Aug. 1989. Fruit: round; medium, 54 to 64 mm; skin black; flesh yellow; freestone. In central San Joaquin Valley, ripens about 15 to 20 June. Blossom: white, one flower per bud. Tree: vigorous; upright; appears self-fertile.

Dolly.—Orig. in Modesto, Calif., by Chris, Gary, and Grant Zaiger and Leith Gardner. Red Beaut open-pollinated. Plant patent 6071; 29 Dec. 1987. Fruit: large size, 54 to 58 mm; ovate, elongated at apex; skin yellow; flesh yellow; clingstone. In central San Joaquin Valley, ripens 1 month after Red Beaut, about 6 to 11 July. Blossom: small to medium, white. Tree: vigorous; upright; self-unfruitful; production moderate.

Double x French Prune.—Sport of Petite. Fruit European type; large; skin dark blue; flavor rich, sweet. Very good for drying, long-distance shipping. Tree: productive; hardy. Partially self-fruitful.

Dropmore Blue.—Orig. in Dropmore, Manitoba, Canada, by F.L. Skinner, Dropmore Nurseries. Introd. in 1941. Parentage unknown; *Prunus nigra.* Fruit: 1 1/4 inches in diam., 1 1/2 inches long; skin purplish-red with a bluish bloom; flesh sweet, meaty; suitable for dessert and preserves; ripens in mid-September. Tree: vigorous; hardy; productive.

Ducie.—Orig. in Saskatoon, Saskatchewan, Canada, by C.F. Patterson, Univ. of Saskatchewan. Introd. in 1960 for home gardens. Derived from *Prunus nigra* and *P. salicina* ; selected in 1946; tested as Sask. P60-102. Fruit: up to 1 1/4 inches in length under nonirrigated orchard conditions in Saskatoon; skin red; flesh moderately firm; ripens during late September. Tree: hardy.

Durado.—Orig. in Parlier, Fresno County, Calif., by C.O. Hesse. Plant patent 3925; 22 June 1976; assigned to The Regents of the Univ. of California, Berkeley. 7a-31M x Burmosa; cross made in 1969; tested as 9.27-58. Fruit: medium; red to dark red skin color at maturity; green-amber flesh, becoming darker near the skin; freestone; excellent eating quality. Tree: vigorous; upright-spreading; outstanding characteristic of Durado is the early maturity of its fruit, 2 to 3 days earlier than Red Beaut.

Earliblue.—Orig. in Mountain Grove, Mo., by K.W. Hanson, Missouri State Fruit Expt. Sta. Introd. in 1972. Stanley x President; cross made in 1940; selected in 1947; tested as PL 237. Fruit: prune type; medium to large; long oval; skin purplish-blue, thin, tender smooth, nearly free; flesh greenish-yellow; texture fine, firm, crisp, juicy; flavor sweet; freestone; quality very good; ripens early, about mid-July; recommended for fresh market, drying, canning, juice. Tree: upright; very hardy; less productive than Stanley; flower self-fertile.

Earliqueen.—Orig. in Modesto, Calif., by Chris, Gary, and Grant Zaiger and Leith Gardner. Open-pollinated seedling of 61EC540 (unknown plum open-pollinated). Plant patent 8583; 8 Feb. 1994. Fruit: large size, 60 to 73 mm; globose; skin garnet-red to Bordeaux-red, becoming maroonish blue-black when overripe; flesh yellowish-white to pale yellow, firm; clingstone. In central San Joaquin Valley, ripens about 2 days after Red Beaut, about 25 May to 2 June. Blossom: medium, white. Tree: vigorous; upright.

Early Gar Rosa.—Orig. in Fresno, Calif., by John M. Garabedian, Peters and Garabedian, at the Peach and Willow Ranch. Introd. in 1964. Plant patent 2324; 3 Dec. 1963; assigned to Peach & Willow Farms, Fresno. Open-pollinated seedling of a bud mutation of Santa Rosa. Fruit: size medium, uniform; skin red nearly overspreading a bright yellow ground color, bloom light gray, medium texture, medium thickness; flesh yellow shading to orange-red near pit, firm, slightly acid, eating quality excellent; clingstone; size medium; ripens 10 days before Santa Rosa, which it resembles. Tree: large; slightly more than Santa Rosa; upright-spreading; very productive; bears regularly; blooms during mid- to late season.

Early Golden.—Orig. in Fonthill, Ont., Canada. Introd. before 1946. Chance seedling; Japanese type but may be myrobalan x Japanese plum (possibly Shiro or Burbank); first sold by Wellington & Davidson (now Fonthill Nurseries), Fonthill. Fruit:

treated commercially as a Japanese type; round; skin golden with high blush; not as large as Shiro, but ripens 10 to 14 days before that variety and just ahead of Wrights Early; flesh firm, golden, smooth, quality good; stone small, free. Tree: self-unfruitful, sets well with pollen of myrobalan; blooms very early; productive, tends to bear biennially; vigorous. Widely planted in Ontario.

Early Laxton.—Orig. in Bedford, England, by Laxton Bros. Introd. in England in 1916; introd. into the U.S. as P.I. 126891; first fruited at the Plant Introduction Sta., Chico, Calif., in 1923; received by the New York State Agr. Expt. Sta., Geneva, in 1925; first listed by the New York State Fruit Testing Cooperative Assn. in 1936. Parentage considered to be Jaune Hative x Early Orleans. Fruit: domestica type; rather small to medium; oval; skin yellow with pinkish blush; flesh juicy, golden, sweet; suitable for early dessert use and for cooking and canning; acceptable quality for its season, being extremely early. Tree: self-unfruitful; somewhat susceptible to bacterial canker; tends to drop fruit just before ripening time.

Early Santa Rosa.—See Star Rosa.

Eastman Special.—See Rosa Grande.

Ebony.—Orig. in Le Grand, Calif., by F.W. Anderson. Introd. in 1961. Plant patent 2049; 18 Apr. 1961; assigned to Reedley Nursery, Reedley, Calif. Open-pollinated seedling of Eldorado; selected in 1959. Fruit: larger than Eldorado; skin deep purple blue with slight gray bloom, tough, no tendency to crack; flesh yellow near pit, shading darker near skin, fibers few, firm, meaty, melting; eating quality good to best; semi-freestone; keeping and shipping qualities good. Tree: large, spreading; vigorous; bears regularly.

Ebony Sun.—Orig. in Del Rey, Calif., by Tom Chamberlin, H.P. Metzler & Sons. Introd. in 1986. Parentage unknown. Plant patent 6153; 19 Apr. 1988; licensed exclusively to Agri Sun Nursery, Selma, Calif. Fruit: large; slightly elongated; skin black, smooth; flesh yellow, crisp, firm. In central California, ripens mid-August. Tree: moderately vigorous, productive when provided good pollination.

Eclipse.—Orig. in Saskatoon, Saskatchewan, Canada, by C.F. Patterson, Univ. of Saskatchewan. Introd. in 1960 for home gardens. Derived from *Prunus nigra* and *P. salicina*; selected in 1954; tested as Sask. P55-30. Fruit: up to 1 1/4 inches long and 1 inch in diam. under nonirrigated orchard conditions at Saskatoon; skin dark red; flesh aromatic; quality very good; ripens in early September. Tree: hardy.

Edwards.—Orig. in Oakland, Calif., by Lawrence C. Edwards. Introd. in 1961. Plant patent 1213; 1 Sept. 1953; assigned to Columbia & Okanogan Nursery Co., Wenatchee, Wash. Chance seedling discovered about 1930. Fruit: European type (*Prunus domestica*); large; skin deep purplish-red; flesh greenish-amber, texture fine, melting, subacid to mild; semi-freestone; fleshy ring found about the stem base in the fruit cavity; ripens from the third week in July to second week in August. Tree: size medium; vigor medium; productive; suckers readily from the roots.

Eickmeier.—Orig. in Grimsby Beach, Ont., Canada, by A.W. Eickmeier & Sons. Introd. about 1930; imported into the U.S. in 1941 as P.I. 142470 from E.D. Smith & Sons, Winona, Ontario; not known if imported before this date. Parentage unknown; selected about 1924. Fruit: small , averaging 3. 4 cm in diam.; ovate; skin amber, medium heavy blush, light bloom; flesh yellow to amber, soft, sugar medium, acid medium to low, flavor mild, quality poor to fair, generally mediocre; stone size medium, cling; ripens very early, being noted primarily for this characteristic. Tree: medium yield; resembles Japanese plum (*Prunus salicina*). Apparently no longer being propagated.

El Doris.—See Late Santa Rosa.

Eldorosa (*Meldor*).—Orig. in Winters, Calif., by Claron O. Hesse, California Agr. Expt. Sta., Davis, and the U.S. Dept. of Agriculture. Not released or introduced by these agencies; named and shipped commercially by a grower. Gaviota x Santa Rosa; selected in 1945; tested as Calif. 16-26. Fruit: Santa Rosa type, but shorter, with greater transverse diam.; needs to be full color before harvest; resembles Santa Rosa.

Elephant Heart.—Orig. in Sebastopol, Calif., by Luther Burbank. Introd. in 1929 by Stark Bros. Nurseries & Orchards Co.,

Louisiana, Mo. Parentage unknown; Japanese type; selected about 1920. Fruit: very large; skin thick, mottled purple and green; flesh blood red, freestone. Tree: strong; hardy.

Elite.—Orig. in Saskatoon, Saskatchewan, Canada, by C.F. Patterson, Univ. of Saskatchewan. Introd. in 1960 for home gardens. Derived from *Prunus nigra* and *P. salicina* ; selected in 1948; tested as Sask. P58-107. Fruit: spherical, up to 2 inches under nonirrigated orchard conditions at Saskatoon; skin dark red; flesh juicy, quality very good; ripens in early September; resembles Acme and Prairie. Tree: vase-shape, low spreading; no trace of winter injury at Saskatoon.

Elliot.—Orig. in Excelsior, Minn., by the Univ.of Minnesota Fruit Breeding Farm. Introd. in 1920. *Prunus salicina* (probably 'Apple') x *P. americana* (unnamed seedling); first fruited and selected in 1911; tested as Minn. 8. Fruit: large; nearly round; skin attractive bright red, medium thick and tough, dots small and russeted, bloom medium; flesh yellow, firm, fine texture, tender, juicy, sweet, pleasant; quality good; pit medium size, smooth, free or nearly so; matures late. Tree: medium height; vigorous; very hardy; production heavy and reliable.

Ember.—Orig. in Excelsior, Minn., by the Univ. of Minnesota Fruit Breeding Farm. Introd. in 1936. Shiro x South Dakota 33; cross made in 1913; selected in 1918; tested as Minn. 83. Fruit: medium to large; skin yellow with red blush, bloom medium; flesh yellow, firm, fine-grained, tender, juicy, sweet, clingstone; high dessert and culinary quality; matures in early September; keeps well for 2 to 3 weeks after ripening. Tree: low, spreading; vigorous; hardy.

Emperor.—Orig. in Parlier, Calif., by J.F. Doyle and T.M. DeJong. Open-pollinated seedling of U.C. selection 11,15-27 = [open-pollinated seedling of U.C. selection 15-16 (= Burton x Yakima)]; cross made in 1985; tested as KAC5-6-38. Plant patent 8188; 30 Mar. 1993. Fruit: uniform; very large, 55 mm in diam., 71 mm long; oblong to oval; skin dark blue with waxy gray bloom; amber flesh; heat tolerant. Excellent fresh fruit; pit cling to semi-cling, 37–40 x 19 to 21 x 12–15 mm. Tree: very vigorous; large; upright to upright-spreading; very productive, regular bearer. Not highly self-fertile; successfully pollinized by French Prune; blooms 1 to 3 days before French Prune. Should be propagated on plum rootstock, preferably Marianna, not on peach.

Erliripe.—Orig. in Fresno, Calif., by Charles L. Vinyard. Introd. in 1965. Plant patent 2713; 14 Feb. 1967; assigned to Hervin Farms, Fresno. Bud mutation of Beauty; discovered in 1960; first fruited in 1964. Fruit: resembles Beauty being distinguished therefrom by its larger size, deeply set stem cavity, firmer flesh and ripening 3 to 4 days earlier. Tree: vigorous growth; productive; fruit sets more evenly along the branches for easier and more selective thinning than Beauty.

Erna.—See Amador.

Eugene.—Orig. in Elk Grove, Calif., by Claude Tribble. Introd. in 1944. Apparently an open-pollinated seedling of a Japanese type plum; selected in 1940. Fruit: large; quality good. Tree: very productive.

Eureka.—Orig. in Godfrey, Ill., by R.Z. Nolan. Introd. about 1942. Parentage unknown; discovered in 1937. Fruit: resembles Burbank. Not a synonym of Wickson.

Excell.—Orig. in Saskatoon, Saskatchewan, Canada, by C.F. Patterson, Univ. of Saskatchewan. Introd. in 1960 for home gardens. Derived from *Prunus nigra* and *P. salicina* ; selected in 1955; tested as Sask. P55-59. Fruit: up to 1 1/2 inches long under nonirrigated orchard conditions at Saskatoon; skin yellow, washed with red; flesh juicy, fairly firm; quality good; ripens in early September. Tree: hardy.

Explorer.—Orig. at Southeastern Fruit & Tree Nut Res. Lab., Byron, Ga., from seed of Queen Ann x Santa Rosa collected by John H. Weinberger near Fresno, Calif. Selected by Victor E. Prince in 1967; tested as BY 4-401 and evaluated by J.M. Thompson; introd. in 1980. Fruit: Japanese type; nearly round with diam. up to 2 1/4 inches, skin deep, lively purplish-black with heavy waxy bloom; flesh deep amber, flavor good to excellent; seeds small; semi-clingstone; ripens 115 days after full bloom, ripening as early as late June and as late as mid-July.

Tree: bears by third year; vigor moderate, young trees semi-upright, vase shape, which modifies with age into a spreading flat-top; flowers self-infertile, pollinizer necessary; moderate to good resistance to bacterial spot (*Xanthomonas pruni*); inconsistent production.

Fiebing (*Fiebing Prize, Phoebe*).—Orig. in Deephaven, Minn., by Charles Haralson. Introd. about 1929. Considered to be *Prunus salicina* 'Wickson' x (*P. americana* x *P. simonii*) 'Kaga'; selected about 1929. Fruit: large; roundish; skin dark red, thick, suture prominent; flesh yellow, firm, juicy, sweet, aromatic; quality very good; season early September. Tree: very hardy; vigorous; spreading; productive.

Fiebing Prize.—See Fiebing.

Fire Queen.—Orig. in Fresno, Calif., by John M. Garabedian. Introd. in 1965. Plant patent 2575; 16 Nov. 1965; assigned to Peach & Willow Farms, Fresno, Calif. Open-pollinated seedling of Santa Rosa. Fruit: large, 2 1/8 to 2 5/8 inches in diam.; globose to round; suture inconspicuous; skin greenish-yellow with bright red blush over most of surface; flesh light yellow, mottled red, juicy, firm, slightly acid, good sugar balance; eating quality good, clingstone; hangs well on tree; superior shipping quality; matures in early July. Tree: spreading habit; vigor medium; very productive.

First Beaut.—Mutation of Red Beaut found in Tulare County, Calif., by Charles H. Neufeld. Plant patent 7335; 25 Sept. 1990. Fruit: ovate; medium size, 46 mm; skin red-purple; flesh amber-yellow, firm, juicy; stone semi-cling. In central San Joaquin Valley, ripens 5 days before Red Beaut. Blossom: average size, white, pollen-sterile. Tree: vigorous; upright and broadly spreading.

Flaming Delicious (*Burbank's Flaming Delicious*).—Orig. in Sebastopol, Calif., by Luther Burbank. Introd. in 1934. Plant patent 14; 15 Apr. 1932; by Elizabeth Waters Burbank, assigned to Stark Bros. Nurseries & Orchards Co., Louisiana, Mo. Parentage unknown; a Japanese plum hybrid. Fruit: large, 2 1/4 inches in diam.; roundish oval to round cordate; skin deep carmine, shading to blackish red-purple, thin, waxy at maturity; flesh deep purplish-red, firm until mature when it becomes tender and melting, very juicy; flavor mild, subacid, aromatic, pleasing; pit freestone, small; ovate; ripens early. Tree: heavy bearer; hardy; leaf ovate, midrib and veins prominent, glands reniform.

Florida.—See Red Ace.

Fortune.—Orig. at Fresno, Calif., by D.W. Ramming, U.S. Dept. of Agriculture Hort. Crops Res . Lab. Introd. in 1988 . Laroda x B 65-11 (=Queen Ann x Late Santa Rosa). Cross made in 1967 and selected in 1971 by J.H. Weinberger; tested as C36-107. Fruit: large; skin bright red on yellow ground color; flesh yellow, firm, semi-clingstone; good quality. Ripens with Friar. Flower self-infertile; blooms with Santa Rosa. Released to provide an alternative to the standard black-skinned varieties.

Fresno Rosa.—Orig. in Fresno, Calif., by John M. Garabedian. Introd. in 1967. Plant patent 2783; 19 Dec. 1967; assigned to Peach & Willow Farms, Fresno. Open-pollinated seedling of Eldorado. Fruit: large; skin yellow-green partially overspread with a bright pinkish red; flesh yellow, clingstone; resembles Laroda, but ripening 10 days later. Tree: resembles Laroda; regular and productive bearer.

Friar.—Orig. in Fresno, Calif., by John H. Weinberger. Introd. in 1968. Gaviota x Nubiana. Cross made in 1957; tested as F34-29. Fruit: size equal to Nubiana; oblate, but more rounded than Nubiana; skin black when fully mature, rougher than Nubiana, practically free of growth cracks; flesh amber, firm, softens slowly after harvest, quality good; pit small, free; ripens end of July, several days later than Nubiana. Tree: very vigorous and productive; flowers self-infertile, pollinizer required; blooms with Santa Rosa; Santa Rosa, Late Santa Rosa, Laroda, Redroy, Redheart have been effective pollinizers; recommended only for Japanese-type plum growing areas; has not been thrifty at Fort Valley, Ga.

Frome Cross.—Orig. in Bristol, England, by G.T. Spinks, Univ. of Bristol, Long Ashton Res. Sta. Introd. in 1946. Coe's Golden Drop x Merryweather Damson; selected about 1938. Fruit: size medium; broad oval; skin dark blue with bloom; slight Damson flavor; culinary use; ripens in mid-late September; hangs well; resembles Merryweather Damson. Tree: produces good crops; self-fruitful.

Frontier.—Orig. in Fresno, Calif., by J.H. Weinberger, U.S. Dept. of Agriculture. Introd. in 1967. Mariposa x Laroda. Cross made in 1955; selected in 1960. Fruit: larger than Santa Rosa; round-ovate; skin blue-black like Laroda, more attractive than other red-fleshed varieties; flesh red, not as dark as Mariposa, firm, separates readily from pit; flavor good, similar to Mariposa; ripens 10 to 14 days after Santa Rosa, less than a week earlier than Eldorado; ripens slowly on tree; softens slowly after picking. Tree: vigorous; flowers are self-infertile, pollinizer is necessary; Santa Rosa, Redheart, and Redroy are effective pollinizers, blooming period overlapping with Frontier; recommended in Japanese plum growing areas.

Gar Beaut.—Bud sport of Nubiana found in Fresno, Calif., by John M. Garabedian. Plant patent 5137; 15 Nov. 1983. Fruit: oblate; small size, 47 mm; skin black, slightly bitter, slight tendency to crack on tip; flesh light cream to yellow, ripens tip first; stone semi-free. In central San Joaquin Valley, ripens 1 week before Red Beaut, about 14 to 21 May. Blossom: white; blooms 7 to 10 days before Red Beaut. Tree: vigorous; upright.

Gar-Belmont.—Orig. in Fresno, Calif., by John M. Garabedian. Open-pollinated seedling of unknown origin. Plant patent 7159; 20 Feb. 1990. Fruit: oblate; medium size, 57 to 59 mm; skin very dark red; flesh cream to pale yellow; stone semi-free. In central San Joaquin Valley, ripens 5 days after Early Gar-Rosa, about 1 to 10 June. Blossom: average, pink. Tree: average vigor; spreading; productive.

Gardner.—Orig. in Corvallis, Ore., by the Oregon Agr. Expt. Sta. Introd. in 1923. Petite d'Agen (French) x self; selected in 1923, Fruit: sugar content high; perhaps not superior to Petite d'Agen; inclined to split. Tree: heavy producer. Named after V.R. Gardner, a staff member of the above station from 1912–18. Variety believed to be no longer propagated.

Gar Jumbo.—Orig. at Fresno, Calif., by J.M. Garabedian. Introd. in 1987. Plant patent 5903; 17 Mar. 1987. Chance seedling, Japanese type. Fruit: large to very large, 60 mm wide and 58 mm long; globose, symmetrical, uniform; skin medium red to dark maroon to mahogany; red with blue-grey bloom; semi-freestone. In central California, ripens with Autumn Rosa. Tree: productive; vigorous.

Gar-Red.—Orig. at Fresno, Calif., by J.M. Garabedian. Chance seedling of unknown parentage. Plant patent 7076; 19 Dec. 1989. Fruit: large, 64 mm wide, 59 mm long; globose; skin pink; flesh light pink; semi-freestone; matures mid-August in Fresno. Tree: vigorous; upright; productive.

Gar-Rosa.—Orig. in Fresno, Calif., by John M. Garabedian. Introd. in 1956. Plant patent 1518; 18 Sept. 1956. seedling of an unnamed yellow-fleshed bud sport of the Santa Rosa variety, open-pollinated. Fruit: flesh medium red shading to amber, texture melting, flavor subacid, aroma distinct; eating quality of the best; clingstone; holds up well in cold storage; shipping quality good; resembles Santa Rosa but matures 8 to 10 days earlier, larger in size, and is more rounded in form than that variety. Tree: size medium; productive; uncertain bearer. A second plum was patented (#3164) under this name on 9 May 1972.

Geddes.—Orig. in Saskatoon, Saskatchewan, Canada, by C.F. Patterson, Univ. of Saskatchewan. Introd. in 1960 for home gardens. Derived from *Prunus nigra* and *P. salicina*; selected in 1947; tested as Sask. P58-102. Fruit: spherical, up to 1 1/2 inches under nonirrigated orchard conditions at Saskatoon; skin yellow, slightly blushed; quality good; ripens in late August. Tree: extremely hardy.

Gerrans Early French Prune.—Orig. in Meridian, Calif., by Lon Gerrans. Introd. in 1962 by Sierra Gold Nursery, Yuba City, Calif. Bud mutation of French Prune; discovered about 1955. Fruit: prune type; sugar content higher than parent; ripens about 7 to 10 days before French Prune, which it resembles.

Gilbert.—Orig. in Kristianstad, Sweden, by E.J. Olden, Balsgard Fruit Breeding Inst. Introd. in 1968. Ontario x Ruth Gerstetter. Cross made in 1952, selected in 1959. Fruit:

size medium; oval, a little compressed at top end, often with unequal halves; skin violet color, thick, tough; flesh light green-yellow, sweet, slightly acid; early ripening variety, 2 to 3 days later than Ruth Gerstetter or during first plum ripening week in Sweden; fruit drop easily when fully ripe, but can be picked as soon as they change color. Tree: size medium; spreading; vigor moderate; hardiness similar to Ontario; productivity moderate; tolerant to *Sclerotinia,* but a little susceptible to bacterial canker; recommended as a commercial variety in southern Sweden.

Gilman.—Orig. in Saskatoon, Saskatchewan, Canada, by C.F. Patterson, Univ. of Saskatchewan. Introd. in 1960 for home gardens. Derived from *Prunus nigra* and *P. salicina*; selected in 1944; tested as Sask. P58-132. Fruit: spherical, up to 1 1/4 inches under nonirrigated orchard conditions at Saskatoon; skin dark red; quality mild, good; ripens in early September; resembles Eclipse. Tree: hardy.

Gloucester.—Orig. in Ottawa, Ont., Canada, by M.B. Davis, Canada Dept. Agr., Res. Sta. Introd. in 1941. *Prunus salicina* 'Burbank' x *P. nigra.* Fruit: size medium; skin thick but tender, red; flesh yellow, juicy, sweet, quality good; season early September; culinary uses. Tree: very hardy; spreading; resembles *P. nigra* in most respects.

Goldbay.—Orig. in Wawanesa, Manitoba, Canada, by E. Robinson, Gaybird Nurseries. Introd. in 1943. F_2 open-pollinated seedling of *Prunus americana* 'Valley River'. Fruit: 1 inch in diam.; flesh yellow, quality good. Now obsolete.

Gold Dust.—Orig. in Hudson, N.Y., by Charles Collier of Bountiful Ridge Nurseries, Princess Anne, Md. Introd. in 1952 by this nursery. Parentage unknown; discovered in 1950. Fruit: Japanese type; size medium, but sometimes reaches 2 inches; skin golden yellow, similar to Shiro; flesh juicy, sweet, flavor typical of Green Gage, fine quality. Tree: vigorous, bears well and regularly in northeastern U.S., not so well in central areas of Atlantic states. Apparently no longer propagated.

Golden Beaut.—Orig. at Le Grand, Calif., by F.W. Anderson. Introd. in 1978. Plant patent 4316; 17 Oct. 1978; assigned to Reedley Nursery, Reedley, Calif. Cross between two unnamed varieties of plum trees. Fruit: as compared to Cassellman and Rosemary, clingstone fruit has less red interior color, but brighter yellow skin and flesh and is sweeter in all stages of maturity, harvest 2 weeks later than those of Cassellman and Rosemary. Tree: medium size; vigor medium; regular and productive bearer.

Golden Delicious.—Orig. in Santa Rosa, Calif., by Luther Burbank. Plant patent 18; 19 July 1932; by Elizabeth Waters Burbank, Executrix, Santa Rosa, Calif., assigned to Stark Bros. Nurseries & Orchards Co., Louisiana, Mo. Parentage unknown. Fruit: very large; cordate, unequal sides; skin smooth, glossy, golden with minute greenish dots, moderately thick; suture deep and distinctly marked; flesh deep chrome yellow, firm, melting, tender, quality excellent, freestone; stem short, stout. Tree: vigor moderate; bushy; stiff upright branches. No longer propagated. Included here because of plant patent.

Golden Globe.—Orig. in Modesto, Calif., by Floyd, Gary, and Grant Zaiger and Leith Gardner. Laroda x Queen Ann. Introd. in 1990. Plant patent 7474; 19 Mar. 1991; assigned to Zaiger's Genetics. Fruit: large, 63 mm in diam., 63 mm long; globose; skin yellow; flesh yellow; clingstone; matures late July to early August in California. Tree: vigorous; upright; productive; self-incompatible.

Golden La Crescent.—See La Crescent.

Golden Minnesota.—See La Crescent.

Golden Nectar.—Orig. in Exeter, Calif., by William E. Hengst. Introd. in 1987. Plant patent 4977; 25 Jan. 1983; assigned to L.E. Cooke Co., Visalia, Calif. Open-pollinated seedling of Mariposa. Fruit: very large; oblong, pointed, suture shallow but distinct; skin amber, waxy, thin; flesh amber, firm, sweet; flavor excellent. Pit very small. Ripens 3 weeks after Mariposa. Fruit stores well at room temperature and under refrigeration. Flowers small, white, between Mariposa and Casselman; self-fertile. Excellent for drying. Chilling requirement about 500 h.

Golden Rod.—Orig. in Excelsior, Minn., by Univ. of Minnesota Fruit Breeding Farm.

Introd. in 1923. Shiro x *Prunus americana* 'Howard Yellow'; cross made in 1913; selected in 1920; tested as Minn. 120. Fruit: large; roundish; suture broad, shallow; skin thin, tough, yellow, bloom whitish, dots medium size; flesh yellow, firm, moderately juicy, subacid to tart; quality fair; stone small, cling; ripens medium late; keeps very well in cold storage. Tree: very vigorous; upright-spreading. No longer recommended because it is a very shy bearer.

Golden Sugar.—Orig. in Sebastopol, Calif., by Luther Burbank. Introd. in 1931 by Stark Bros. Nurseries & Orchards Co., Louisiana, Mo. Parentage unknown; *Prunus domestica*; selected in 1927. Fruit: medium large; flesh clear yellow, very sweet, quality good, freestone. Propagation discontinued during the late 1930s.

Golden West (*Wilmore*).—Origin obscure, believed to have orig. in Wheat Ridge, Colo., by W.W. Wilmore, W.W. Wilmore Nurseries. Introd. in 1939. Open-pollinated seedling of a native plum; discovered in 1924. Fruit: large; flavor very good; resembles Waneta.

Gracious.—Orig. in Mandan, N.D., by W.P. Baird, Northern Great Plains Field Sta. Introd. in 1957. Open-pollinated seedling of Emerald; selected in 1941; probably the result of crossing with Red Wing, as leaf and fruit characteristics of Gracious indicate a combination of *Prunus americana* and *P. salicina*, tested as Mandan 41-15. Fruit: large, averaging about 12 per lb., skin an attractive mottled red with bloom, thin, tender; flesh yellow, medium firm, sweet, juicy, entirely freestone; rates very good in cooking tests; ripens in late August; more resistant to brown rot than Red Wing, which it resembles. Tree: very hardy, more so than Red Wing, moderately vigorous; slightly spreading. Recommended for planting in most sections of the Northern Great Plains, where a large plum of superior quality is wanted.

Grandoro.—Orig. in Le Grand, Calif., by F.W. Anderson, Merced, Calif. Introd. in 1964. Plant patent 2415; 30 June 1964; assigned to Kim Bros., Reedley Nursery, Reedley, Calif. Open-pollinated seedling of Eldorado; selected in 1958; tested as Anderson 17H562. Fruit: Japanese type; very large, averaging 2 1/2 inches in diam.; uniform; globose, symmetrical; skin deep purple-blue, thick, tough, displays no tendency to crack; flesh medium yellow, meaty, firm, flavor vinous, eating quality good to excellent; stone free, small; keeping and shipping qualities good; used for dessert, local and long-distance shipping; ripens 3 to 4 weeks after Eldorado, which it resembles. Tree: large; vigorous, spreading, open; productive; bears regularly; blooms in midseason about with Santa Rosa.

Grand Prize.—Orig. in Sebastopol, Calif., by Luther Burbank. Introd. in 1937 by Stark Bros. Nurseries & Orchards Co., Louisiana, Mo. Parentage unknown; selected in 1932. Fruit: very large, about twice the size of Italian Prune; skin purple; quality high; ripens about 15 Aug. Tree: size medium; spreading; vigor moderate.

Grand Rosa.—Orig. in Le Grand, Calif., by F.W. Anderson, Merced, Calif. Introd. in 1959. Plant patent 1756; 23 Sept. 1958; assigned to Reedley Nursery, Reedley, Calif. Open-pollinated seedling of Eldorado; selected in 1957. Fruit: large, averaging 2 to 2 3/8 inches in diam.; globose; skin red, shading to burgundy red, with a light purple bloom, being similar to Late Santa Rosa; flesh yellow, shading to red near skin; has the firmness and the good shipping and keeping qualities of Eldorado; free to semi-freestone; ripens about 2 weeks after Eldorado and about 1 week after Late Santa Rosa, thus extending the marketing season of Eldorado-type varieties; at harvest, few culls due to growth cracks and other defects. Tree: large; vigorous; productive and regular bearer.

Great Yellow (*Burbank's Great Yellow*).—Orig. in Sebastopol, Calif., by Luther Burbank. Introd. in 1931 by Stark Bros. Nurseries & Orchards Co., Louisiana, Mo. Plant patent 13; 5 Apr. 1932; by Elizabeth Waters Burbank, Executrix, Santa Rosa, Calif., assigned to Stark Bros. Nurseries & Orchards Co. Parentage unknown; selected about 1920. Fruit: large; roundish; skin clear yellow, thin; flesh clear yellow, fine texture; mild subacid flavor; ripens early. Tree: moderately vigorous; hardy.

Green Sweet™ (*Bradgreen*).—Orig. in Le Grand, Calif., by Norman and Glen Bradford. (Black Beaut open-pollinated) open-pollinated. Plant patent 8498; 14 Dec. 1993. Fruit: oblate; medium size, 60 mm; skin strong yellow-green; flesh light yellow-green, extremely firm and crisp; flavor very sweet with medium acidity, 18 to 22°Brix, very juicy; stores very well. In central California, ripens with Angeleno. Blossom: medium, white, blooms early; pollinated by Ambra. Tree: vigorous; spreading; very productive; self-unfruitful.

Grenville.—Orig. in Ottawa, Ont., Canada, by Canada Dept. Agr., Res. Sta. Introd. in 1941. *Prunus salicina* 'Burbank' x *P. nigra*; Japanese type; selected in 1932. Fruit: large; skin red; quality excellent; resembles Burbank. Tree: very hardy.

Gypsy Red.—Orig. in Reedley, Calif., by James W. Taylor; assigned to Ito Packing Co. Open-pollinated seedling of Elephant Heart. Plant patent 7119; 23 Jan. 1990. Fruit: round; medium size, 67 mm; skin reddish-purple with yellow-greenish ground color; flesh red, very sweet with less acid than Satsuma; clingstone. In central San Joaquin Valley, ripens about 5 to 15 Aug. Blossom: medium, white; anthers yellow. Tree: vigorous; upright; productive; self-infertile.

Hall.—Orig. in Geneva, N.Y., by Richard Wellington, New York State Agr. Expt. Sta. Introd. in 1923. Golden Drop x Grand Duke; cross made in 1908; European type. Fruit: large; skin reddish-purple, attractive; flesh yellowish-green, firm, meaty, tender, juicy, sweet, quality good, semi-clingstone; ripens second week in September; resembles Grand Duke.

Harlan Rosa.—See Late Santa Rosa.

Harry Pickstone.—Orig. in Stellenbosch, Cape Province, South Africa, by Fruit Tech. Res. Inst., Dept. of Agr. Tech. Serv. Introd. in 1973. Gaviota x (Methley x Wickson); cross made in 1959, selected in 1965, tested as 44/48. Fruit: medium large; roundish oval; skin dark red when fully ripe; stem medium long and medium thick; flesh golden yellow, flavor sweet, quality excellent; ripens second to third week of January, keeps well under dual temperature storage, 10 days at –0.5C and thereafter at 7.2C; shipping quality good; moderately tolerant to bacterial spot. Tree: somewhat spreading; vigorous; high productivity; tolerant to delayed foliation; leaves glossy and more tolerant to leaf spot disease than most Japanese varieties; flower white, self-fertile.

Havens 2B.—Orig. in Sutter, Calif., in the orchard of Fred Havens by L.H. Day and E.F. Serr, California Agr. Expt. Sta., Davis. Introd. in 1964. Chance seedling of Damson plum (*Prunus insititia*); discovered about 1930. Used as a clonal interstock because of its compatibility between Nonpareil almond (scion) and Marianna 2624 rootstock.

Heart of Gold.—Orig. in Santa Rosa, Calif., by Luther Burbank. Introd. about 1931 by Stark Bros. Nurseries & Orchards Co., Louisiana, Mo. Parentage unknown. Fruit: heart-shaped; skin yellow; flesh firm, juicy, sweet; season early. Tree: productive; very vigorous.

Hennepin.—Orig. in Excelsior, Minn., by Univ. of Minnesota Fruit Breeding Farm. Named and introd. in 1923. *Prunus salicina* Satsuma x *P. americana*; cross made in 1911, selected about 1918; tested as Minn. 132. Fruit: size medium; roundish oval; skin medium thick, rather tough, dark reddish-purple, bloom very heavy, dots many and conspicuous; flesh dark red, medium firm, medium juicy, sweet, slightly stringy, quality fair to good; stone broad, large, semi-cling; good preserving quality; season early. Tree: hardy; very productive; size medium; vigorous.

Hildreth.—Orig. at High Plains Grasslands, Res. Sta., Cheyenne, Wyo., by U.S. Dept. of Agriculture. Introd. in 1982. Open-pollinated seedling of Clyman. Selected in 1958: tested as 68-11. Fruit: round-oblong; 1 1/4 to 1 1/2 inches in diam.; flesh yellow, meaty, tender; skin thin, tender, greenish-yellow with rose or reddish blush; seeds small, freestone; ripens 90 days after full bloom; first week of September at Cheyenne. Tree: upright: vigor moderate; productive, thinning recommended; flowers self-fertile.

Hinchliff.—Orig. in Saskatoon, Saskatchewan, Canada, by C.F. Patterson, Univ. of Saskatchewan. Introd. in 1960 for home gardens. Derived from *Prunus nigra* and *P.*

salicina; selected in 1960; tested as Sask. P60-105. Fruit: spherical, up to 1 1/2 inches under nonirrigated field conditions at Saskatoon; skin yellowish, blushed red; quality good; ripens during the first 2 weeks in September. Tree: hardy.

Hoffman.—Orig. in Saskatoon, Saskatchewan, Canada, by C.F. Patterson, Univ. of Saskatchewan. Introd. in 1960 for home gardens. Derived from *Prunus nigra* and *P. salicina*; selected in 1955; tested as Sask P55-39. Fruit: spherical, up to 1 1/4 inches under nonirrigated orchard conditions at Saskatoon; skin yellow; flesh firm; quality good; season early September. Tree: hardy.

Hollywood.—Orig. in Modesto, Calif., by L.L. Brooks. Introd. in 1936. *Prunus pissardi* x *P. salicina* (direction of cross unknown); discovered about 1932; first commercially propagated by Ralph S. Moore, Visalia, Calif. Fruit: flesh red, quality good; ripens early, in Beauty season; hangs well on tree; used for jelly and canning; resembles Satsuma. Tree: resembles *P. pissardi.*

Homeside.—Orig. in Auburn, Ala., by J.D. Norton, Agr. Expt. Sta., Auburn Univ. Introd. in 1975. Methley x Ozark Premier; cross made in 1961; selected in 1963; tested as Methley A-39. Fruit: very large, 2 1/4 to 2 1/2 inches in diam.; nearly round; skin light red to orange; flesh cream, not firm enough for long distance shipping; quality excellent; recommended for home and roadside markets; ripens early July. Tree: large; vigorous; spreading; productive; resistant or tolerant to black knot, bacterial canker, bacterial fruit spot and bacterial leaf spot; chilling requirement 700 h.

Honey Dew Gage.—Orig. in Santa Rosa, Calif., by Luther Burbank. Introd. about 1931 by Stark Bros. Nurseries & Orchards Co., Louisiana, Mo. Parentage unknown. Fruit: Green Gage type; flesh sweet, juicy; ripens late. Tree: very productive; vigorous.

Honeymoon.—Orig. in Sebastopol, Calif., by Luther Burbank. Introd. in 1931 by Stark Bros. Nurseries & Orchards Co., Louisiana, Mo. Green Gage x an unknown variety; selected in 1920. Fruit: large; skin golden yellow; flesh quality high; ripens midseason; resembles known parent. No longer propagated.

Howard Miracle.—Orig. in Montebello, Calif., by Frederick H. Howard. Introd. in 1947. Plant patent 721; 31 Dec. 1946; assigned to Howard and Smith, Montebello, Calif. Parentage unknown; discovered about 1941. Fruit: Japanese type; large; skin yellow, blushed heavily with red at maturity, thin, tough; flesh yellow, firm, flavor very good and distinctive, juicy; freestone; matures late August in the vicinity of Montebello; ships well. Tree: very vigorous; needs cross-pollination.

Howard Sun.—Orig. in Del Rey, Calif., by Tom Chamberlin, H.P. Metzler & Sons. Introd. in 1986. Parentage unknown. Plant patent 6006; 8 Sept. 1987; licensed exclusively to Agri Sun Nursery, Selma, Calif. Fruit: resembles Friar; large; round; skin very dark blue to reddish-blue; flesh amber yellow, mild, sweet, excellent texture; ripens 1 week before Angeleno; excellent storage and shipping qualities. Tree: productive.

Improved Late Santa Rosa (*Late Santa Rosa Improved*).—Origin unknown; however, considered to be one of the two mutations discussed under Late Santa Rosa. Introd. in 1931. Fruit: tends not to crack along suture; otherwise, similar to Late Santa Rosa.

Improved Methley.—Orig. in Athens, Texas, by Athens Nursery Co. Introd. about 1950. Parentage unknown, probably Methley; discovered about 1915. Fruit: small to medium; oval; skin tough, yellow-green with red blush; flesh sweet, yellow amber; ripens 2 to 10 days after Methley. Tree: large, ascending, vigorous, hardy; productive.

Improved Satsuma.—See Mariposa; Satsuma Improved.

Iroquois.—Orig. in Geneva, N.Y., by John Watson, New York State Agr. Expt. Sta. Introd. in 1966. Italian Prune x Hall. Cross made in 1937; selected in 1949; tested as N.Y. 797. Fruit: size medium; more pointed than Stanley; skin deep blue, attractive; flesh greenish-yellow, firm, subacid, quality fair to good, freestone; ripens 1 week before Stanley; processed for baby food. Tree: size medium; upright, spreading; very productive; self-fruitful.

Ivanovka.—Orig. in Manchuria. Introd. by I.V. Mitchurin into Canada to the Canada

Dept. Agr., Res. Sta., Morden, Manitoba, in 1939. *Prunus salicina* (*P. triflora koreana*). Fruit: 1 1/2 inches in diam.; roundish; skin greenish-yellow, mostly covered with dull bluish-red, overlaid with medium bloom; flesh greenish-yellow to pink, somewhat stringy, sweet, sprightly; clingstone; quality good as dessert and canned; season late August. Tree: vigorous; upright, spreading; hardy; productive.

Jake's Best.—Orig. in Fresno, Calif., by John M. Garabedian. Open-pollinated seedling of unknown origin. Plant patent 7148; 13 Feb. 1990. Fruit: oblate, one cheek generally slightly larger than the other; medium size, 47 mm; skin dark red; flesh Cardinal; stone semi-free. In central San Joaquin Valley, ripens 2 weeks before Wuhl, about 27 May to 3 June. Blossom: average, white, blooming with Red Beaut. Tree: medium vigor; upright; productive.

JD-1.—Orig. in Modesto, Calif., by John D. Young; assigned to Young Family Trust. Purchased as budded ornamental plum tree of unknown origin. Plant patent 5926; 7 Apr. 1987. Growth closely resembles flowering Blireiana. Fruit: ovoid; small, larger than Blireiana, 54 mm; skin wine-red, glabrous; flesh deep red; freestone. Ripens 10 to 14 days before Hollywood. Leaves: a deeper bronze-purplish color than Blireiana. Blossom: single, white with pink tinge. Tree: vigorous; upright.

John-John.—Orig. in Fresno, Calif., by John M. Garabedian. Open-pollinated seedling of unknown origin. Plant patent 5906; 17 Mar. 1987. Fruit: globose to slightly oblate; one-half often larger apically; medium to large, 57 to 60 mm; skin purplish-red, bitter; flesh light amber-yellow; stone semi-free. In central San Joaquin Valley, ripens 7 to 10 days before Laroda, about 6 July. Blossom: medium, white. Tree: vigorous; upright spreading; very productive.

July Santa Rosa.—Orig. in Reedley, Calif., by Nick J. Friesen. Introd. in 1962. Bud mutation of Late Santa Rosa; discovered in 1959. Fruit: appearance of Late Santa Rosa; flesh firm, less acid than Santa Rosa; does not crack; ripens 7 to 10 days after Santa Rosa and before Late Santa Rosa; resembles Santa Rosa.

June Beaut.—Orig. in Le Grand, Calif., by Norman and Glen Bradford. Eldorado open-pollinated. Plant patent 5662; 18 Feb. 1986. Fruit: oblate; medium size, 51 mm; skin dark maroon, only slight tendency to crack; flesh greenish-yellow; clingstone. In central California, ripens 3 to 5 days after Eldorado. Blossom: medium, white, blooms 5 days after Eldorado. Tree: vigorous; upright; regular bearing; very productive; self-infertile.

June Redskin (*Burbank's June Redskin*).—Orig. in Sebastopol, Calif., by Luther Burbank. Introd. in 1934. Plant patent 12; 5 Apr. 1932; by Elizabeth Waters Burbank, Executrix, Santa Rosa, Calif.; assigned to Stark Bros. Nurseries & Orchards Co., Louisiana, Mo. Parentage unknown; selected about 1922. Fruit: medium to large; globular; skin deep crimson shading to purple, numerous conspicuous dots, thick, tough; flesh firm, yellow, acid to sweet; freestone; season early, 25 June at Sebastopol. Tree: size medium; upright to spreading.

Kota.—Orig. in Brookings, S.D., by N.E. Hansen, South Dakota Agr. Expt. Sta. Introd. in 1939. *Prunus americana* x *P. simonii*. Fruit: 1 3/4 inches in diam. and 1 1/2 inches from apex to stem end; oblate; skin vivid dark red, with large distinct yellow dots and white bloom; suture very wide and shallow; stem stout; flesh very firm and fragrant.

Krause (*Stanley Improved, Cascade*).—Orig. in Dinuba, Calif., by John Krause. Introd. in 1946 by Wilbur W. Wilhelm, Dinuba Nursery, Dinuba, Calif. Parentage unknown; tree came from a midwestern nursery; discovered in 1945. Fruit: large; appearance excellent; flesh firm, crisp; quality fair; ripens later than President. Tree: upright, being similar to President; self-unfruitful.

La Crescent (*Golden La Crescent, Golden Minnesota*).—Orig. in Excelsior, Minn., by Univ. of Minnesota. Fruit Breeding Farm. Introd. in 1923. (*Prunus americana* x *P. salicina*) Shiro x *P. americana* Howard Yellow; cross made in 1913; selected in 1919; tested as Minn. 109. Fruit: small; obliquely oval, apex rounded; skin thin, tender, easily removed, yellow, sometimes faintly blushed, bloom delicate, dots nearly absent; flesh yellow, juicy, very tender, melting, sweet, aromatic, suggestive of apricots; dessert quality good; stone small, free; ripens early. Tree: large; extremely vigorous; moderately productive; hardy.

Lafayette.—See Brooks Italian.

Lantz.—Orig. in Ames, Iowa, by H.L. Lantz, Iowa State College. Introd. in 1958. (Waneta x Moorpark) x Japex; seed planted in 1932; selected for testing about 1935–40; tested under the number Ames 200; the female seedling parent was produced by Harlow Rockhill of Conrad, Iowa; the pollen parent, a seedling of *Prunus salicina*, was selected by Luther Burbank. Fruit: round, uniform in size and form, about 1 to 1 1/2 inches in diam.; stem short; skin dark blue; flesh amber, tender, juicy, makes excellent sauce; clingstone; hangs well on tree; quite resistant to brown rot. Named after Professor H.L. Lantz.

Laroda.—Orig. in Winters, Calif., by Claron O. Hesse, California Agr. Expt. Sta., Davis, and Agr. Res. Serv., U.S. Dept. of Agriculture. Introd. in 1954. Gaviota x Santa Rosa; cross made in 1940; selected in 1943; tested as Calif. 15a-28. Fruit: Japanese-type; large; nearly round; skin deep reddish-purple, developing completely before shipping maturity; flesh light amber with some red next to skin at full tree maturity, firm, sweet; aroma mild; quality good; ships very well; excellent market acceptance; ripens end of July in California's foothill districts and a few days earlier in San Joaquin Valley plum areas. Tree: vigorous; numerous medium-sized spurs and buds; blooms with Santa Rosa; self-unfruitful, being pollinated by Santa Rosa and Red Rosa.

Late Duarte.—Orig. in Auburn, Calif., by Robert L. Holmes. Introd. in 1938. Bud mutation of Duarte; discovered before 1930. Fruit: skin deep red, mottled when full ripe; flesh medium pink; clingstone; ripening date for fresh-market shipment varies from end of regular Duarte to Kelsey and President. Tree: pollination required in some districts with Sharkey and Eldorado apparently as good pollinators.

Late President.—Orig. in Le Grand, Calif., by F.W. Anderson, Merced, Calif. Introd. in 1944. Plant patent 573; 16 Mar. 1943. Bud mutation of President; discovered in 1935. Fruit: ripens about 2 weeks later than President, otherwise identical with parental type.

Late Rosa.—See Late Santa Rosa.

Late Roza.—See Late Santa Rosa.

Late Santa Rosa (*Casselman Special, El Doris, Harlan Rosa, Late Rosa, Late Roza, Red Rosa, Red Roza, Red Santa Rosa*).—It appears that there are two mutations, indistinguishable from each other as concerns both fruit and tree characteristics except that the fruit cracks sometimes in one and apparently less frequently in the other. The northern California type: orig. in the Mount Pleasant district near Penryn, Calif., by Jack Bergtholdt. Introd. in 1931. Bud mutation of Santa Rosa; discovered in 1915; name of Late Rosa was given this variety by Harry Butler and F.E. Kayo. Fruit: matures about 1 month after Santa Rosa; in some years the fruit splits badly or has soft noses; otherwise, resembles parent. The southern California type: orig. in Yucaipa Valley, San Bernardino County, Calif., by S.A. Skinner who named it El Doris; A.H. Harlan, who purchased the orchard from Skinner, named it Harlan Rosa. Introd. in 1931. Bud mutation of Santa Rosa; discovered in 1924. Fruit: matures about 1 month after Santa Rosa; apparently does not crack or split as much as the northern California type; flesh firmer than other type, remains yellow to maturity, with little or no red next to skin.

Late Santa Rosa Improved.—See Improved Late Santa Rosa.

Late Tragedy.—Orig. in Placerville, Calif., by Charles Rayburn. Introd. in 1921; seedling of Tragedy; discovered in 1919. Fruit: skin deep blackish-purple; flesh bright whitish to yellowish green, sweet, mildly subacid, with considerable flavor, semifreestone; shipping quality good; ripens with Laroda, 15 to 17 days before Nubiana, and 20 to 23 days after Satsuma.

Laxton's Blue Tit.—Orig. in Bedford, England, by Laxton Bros. Introd. into the United States in 1938 as P.I. 127026. Considered to be Czar x Green Gage. Fruit: size medium; round; skin blue, with a heavy blue bloom; flesh very juicy; used for dessert purposes; ripens shortly after Czar. Tree: self-fruitful; bears well.

Laxton's Bountiful.—Orig. in Bedford, England, by Laxton Bros. Introd. about 1926. Considered to be Red Magnum Bonum x Victoria. Fruit: size medium; oval; skin pinkish-red with white dots; flesh firm; quality poor; ripens 7 to 10 days before Victoria. Tree: self-fruitful.

Laxton's Cropper.—Orig. in Bedford, England, by Laxton Bros. Introd. in 1928. Considered to be Victoria x Aylesbury. Fruit: slightly smaller than Victoria; oval; skin reddish-purple; used for cooking; ripens in mid-September; keeps well off tree. Tree: self-fruitful; bears well.

Laxton's Delicious.—Orig. in Bedford, England, by Laxton Bros. Introd. in 1920. Considered to be Coe's Golden Drop x Pond. Fruit: oblong; suture prominent; skin pink-red with a distinct lavender-blue bloom, with many white dots; flesh very juicy; flavor rich; pit small, clingstone; ripens in mid-September; keeps for many days after harvest. Tree: vigorous; branches brittle; self-unfruitful.

Laxton's Goldfinch.—Orig. in Bedford, England, by Laxton Bros. Introd. in 1935. Considered to be Early Transparent Gage x Jefferson. Fruit: round; suture deep; skin yellow; flesh very juicy, flavor rich, sweet; slightly clingstone; ripens in late August. Tree: very vigorous.

Laxton's Prosperity.—Orig. in Bedford, England, by Laxton Bros. Nursery. Introd. about 1920. Considered to be Grand Duke x Czar. Fruit: large; oval-oblong; skin purplish-blue; flesh pale yellow, freestone; ripens between Early Prolific and Victoria; a cooking plum.

Laxton's Supreme.—Orig. in Bedford, England, by Laxton Bros. Introd. in 1938 into the U.S. as P.I. 127033. Considered to be Denniston's Superb x Victoria; European plum (*Prunus domestica*). Fruit: resembles Denniston's Superb in size and shape; ripens early. Tree: leaf rather small; blooms in midseason.

Lehman.—Orig. in Cool, Calif., by William Lehman. Introd. in 1932. Parentage unknown; discovered in 1920. Fruit: Japanese type (*Prunus salicina*); skin straw colored with red cheek, very tough, tends to sunburn; flesh yellow, juicy.

Lewis.—See Prinlew.

Longjohn.—Orig. in Geneva, N.Y., by R.L. Andersen, J.P. Watson, and K.G. Livermore, New York State Agr. Expt. Sta. Iroquois x CA4A33L; selected in 1971; tested as NY 66.601.1; introd. in 1993. Fruit: large, 37.5 mm cheek to cheek, 40 mm at broadest point of suture, and 65 mm long; distinctive shape, pointed on both ends; skin dark purple-blue covered by waxy bloom. Flesh amber, smooth texture, similar in firmness to Stanley; freestone with excellent fresh and processed flavor and texture, sweet and aromatic. Ripens with Stanley, but larger and better quality. Tree: vigorous, moderately upright, with weeping laterals. Precocious and productive; consistent cropping; slightly self-fertile; effectively pollenized by Stanley, Polly, and Castleton.

Maglio.—Orig. in Nelson, B.C., Canada, by Carmine Maglio. Introd. in 1935. Plant patent 413; 23 July 1940. Parentage unknown; first fruited in 1925. Fruit: Japanese type; size medium; skin deep red, thin, tough, bitter; flesh yellow, juicy; flavor at pit unpleasant; hangs well on tree so picking season is long; good shipping qualities; matures about same season as Italian Prune, but cannot compete with Italian Prune for quality. Tree: upright, spreading; vigorous; not too winter hardy.

Majestic M.—Bud sport of Friar found in Orosi, Calif., by Merel R. Butts. Plant patent 6975; 8 Aug. 1989. Fruit: round; medium size, 56 to 59 mm; skin blackish-purple, slight tendency to crack; flesh orange-brown; freestone. In central San Joaquin Valley, ripens 2 to 3 weeks after Angeleno, about second to third week of October. Blossom: small, white. Tree: semi-spreading.

Mammoth Cardinal (*Burbank's Mammoth Cardinal*).—Orig. in Sebastopol, Calif., by Luther Burbank. Introd. in 1934. Plant patent 16; 10 May 1932; by Elizabeth Waters Burbank, Executrix, assigned to Stark Bros. Nurseries & Orchards Co., Louisiana, Mo. Parentage unknown; selected about 1919. Fruit: large; skin red, thick; flesh yellow; quality high; stone medium to large, semi-cling to clingstone; ships well; ripens 14 July at Sebastopol; similar if not identical with Formosa.

Mammoth Golden.—Apparently orig. in

Iowa. Introd. in 1938 when first listed in Sears, Roebuck & Co. catalog. Parentage unknown. Fruit: large; skin bright golden-yellow, blushed; quality good; ripens in July in the midwestern United States. Tree: small; needs a pollinator, being pollinated by Blue Damson, Italian Prune, Mt. Royal, Stanley.

Mandarin.—Orig. in Dropmore, Manitoba, Canada, by F.L. Skinner. Introd. in 1941 by the Manitoba Hardy Plant Nursery, Dropmore. Open-pollinated seedling of *Prunus salicina*; selected in 1940. Fruit: 1 to 1 1/4 inches in diam.; skin purple; flesh meaty, sweet, cooks well; ripens in August. Tree: hardy throughout western Canada.

Manet.—Orig. in Mandan, N.D., by W.P. Baird, U.S. Northern Great Plains Field Sta. Introd. in 1947. Open-pollinated seedling of *Prunus americana*. Fruit: large; roundish; skin an attractive red; flesh juicy, sweet. Tree: very hardy.

Margaret.—Orig. in Vacaville, Calif., by R.E. Burton. Introd. about 1922. Open-pollinated seedling of Splendor; discovered in 1916. Fruit: European type; skin dark blue, bloom whitish; flesh soft, sweet; freestone; for fresh eating, shipping, canning; dessert-type plum, not a drying prune; lighter in color and larger than Tragedy; ripens 10 to 12 days later than Tragedy.

Marianna 2623.—Orig. in Davis, Calif., by W.L. Howard, California Agr. Expt. Sta. Introd. about 1940. Open-pollinated seedling of Marianna, being a sibling of Marianna 2624; selected in 1926; Marianna is a cultivar of, apparently, *Prunus cerasifera* x *P. munsoniana* (?); however, some horticulturists consider Marianna as simply *P. cerasifera*. Fruit: not commercial. Used only as a clonal rootstock: resistant to root-knot nematode; not as resistant to oak root fungus as Marianna 2624; grows vigorously; propagated vegetatively by hardwood cuttings.

Marianna 2624.—Orig. in Davis, Calif., by W.L. Howard, California Agr. Expt. Sta. Introd. about 1940. Open-pollinated seedling of Marianna; selected in 1926. Marianna is a cultivar, apparently, of *Prunus cerasifera* x *P. munsoniana* (?); however, some horticulturists consider Marianna as just *P. cerasifera*. Fruit: not commercial. Used only as a rootstock: resistant to root-knot nematode; partially resistant to oak root fungus; grows vigorously; propagated vegetatively by hardwood cuttings.

Mariposa (*Improved Satsuma*).—Orig. in Pasadena, Calif., by Jennie Benedict Thompson. Introd. in 1935. Plant patent 111; 13 Nov. 1934; assigned to Armstrong Nurseries, Ontario, Calif. Chance seedling; selected in 1923. Fruit: large; round; skin maroon, thick, sweet; flesh maroon, firm, juicy, flavor sweet, freestone; sweeter than Satsuma, which it resembles. Tree: medium to large; vigorous; resistant to delayed foliation in southern California.

Marvel.—Orig. in Mountain Grove, Mo., by Paul H. Shepard, Missouri State Fruit Expt. Sta. Introd. in 1947. Burbank x Methley; selected in 1947; tested as PL-198. Fruit: large; round, conical; skin medium dark red, color better than either parent; flesh solid red, firm, sweet; clingstone; uniform in size and color; does not drop readily; ripens about 1 July. Tree: highly productive.

Mason Early Italian.—Orig. in Parker, Wash., by Don Mason. Introd. in 1969 by Carlton Nursery Co., Forest Grove, Ore. Bud mutation of Richards Early Italian. Discovered in 1965. Fruit: size medium; oblong, prune shape; skin blue; flesh amber; quality good; ripens 6 to 8 days earlier than parent tree; for commercial plantings and fresh fruit market. Tree: size medium; vigor good; hardy; productive.

Mathison Standard.—Orig. in Kingsburg, Calif., by Joe Mathison. Introd. in 1958 by Stribling's Nurseries, Merced, Calif.; European type. Fruit: larger than Standard; ripens 2 weeks before Standard, which it resembles. Tree: very productive.

McRobert.—Orig. in Gilbert Plains, Manitoba, Canada, by J.C. McRobert. Introd. in 1929. Open-pollinated seedling of *Prunus nigra*. Fruit: large; oval; skin greenish-yellow, washed dark red; flesh dark yellow, firm; quality fair as dessert or canned; season early September. Tree: upright to spreading; hardy; moderately vigorous; fairly productive.

Meldor.—See Eldorosa.

Mendota.—Orig. in Excelsior, Minn., by the Univ. of Minnesota Fruit Breeding Farm. Named and introd. in 1924. *Prunus salicina*

'Burbank' x *P. americana* 'Wolf'; cross made in 1908; selected about 1912; tested as Minn. 5; sibling of Mound, Redcoat, Red Wing, Tonka. Fruit: very large; long oval, apex pointed; skin medium thick, easily removed at maturity, attractive bright red, bloom thin; flesh light yellow, firm, meaty, tender, slightly fibrous, medium juicy, flavor mild, sweet; quality good; stone medium to large, freestone; ripens in late midseason. Tree: size medium; moderately vigorous; productive; moderately hardy.

Meredith.—Orig. in Von Ormy, Texas, by R.C. Aldridge, Aldridge Nursery. Introd. in 1946. Parentage unknown. Seedling discovered in 1940. Fruit: size medium; round; skin green with red cheek; flesh yellow, firm, crisp; shipping qualities excellent; ripens 10–15 June; fruit holds on tree 2 to 3 weeks after ripening. Tree: size medium; broad-shaped; vigorous; productive; hardy; recommended as a commercial or home variety.

Merton.—Orig. in Peach, Wash., by Lynn Tuttle. Introd. in 1950. Bud mutation of Italian Prune; discovered about 1935. Fruit: ripens 10 days before Italian Prune; has tendency to crack along suture some years and in some localities; resembles Italian Prune. Tree: vigorous; spreading, with wide-angle crotches; productive.

Midnight Sun.—Orig. in Del Rey, Calif., by Tom Chamberlin, H.P. Metzler & Sons. Introd. in 1986. Parentage unknown. Plant patent 6044; 27 Oct. 1987; assigned to Metropolitan Insurance Co.; licensed exclusively to Agri Sun Nursery, Selma, Calif. Fruit: large; nearly round; skin black, very smooth; flesh nearly white, firm; flavor mild with no hint of acidity. Good handling and storage characteristics. Tree: vigorous, productive.

Mid Red.—Orig. in Modesto, Calif., by Floyd, Gary, and Grant Zaiger and Leith Gardner. Introd. in 1986. Casselman limb sport x King David. Plant patent 5744; 13 May 1986; assigned to Zaiger's Genetics. Fruit: large, 65 mm; slightly elongated; skin red on yellow ground color; flesh yellow, firm; flavor mild subacid. Tree: vigorous; upright; productive.

Midrosas.—Bud sport of Late Santa Rosa found in Fresno, Calif., by John D. Pakchoian; assigned to Pakchoian Farms. Plant patent 5416; 12 Mar. 1985. Fruit: round; medium, 50 mm; skin reddish-purple, no suture stripe as in Late Santa Rosa; flesh yellow to yellow-red; clingstone. In central San Joaquin Valley, ripens 10 to 15 days before Late Santa Rosa, about 25 June. Blossom: medium, white; anthers yellow. Tree: vigorous; upright to upright-spreading.

Miller Early Sweet.—Orig. in Junction City, Ore., by Frank Miller. Introd. in 1934. Open-pollinated seedling of Italian Prune; discovered about 1925 or 1926. Fruit: slightly smaller than Italian Prune; flesh very sweet, nonacid, high-quality dried prune, but only fair for cooking purposes; ripens early. Tree: bears heavily. Called a prune rather than a plum, although it is treated commercially as a plum.

Milton Early Italian.—Orig. in Wenatchee, Wash., in the experimental plots of E.L. Reeves, U.S. Dept. of Agriculture. A bud sport of Fellenberg (Italian Prune) selected in 1943 by S.M. Zeller; tested as E-76; introd. by Q.B. Zielinski and J.A. Milbrath in 1954. Identical with Fellenberg, except flowering is 4 or 5 days earlier, fruit maturity is 10 to 14 days earlier, and productivity is somewhat less.

Mina.—Orig. in Morden, Manitoba, Canada, by Canada Dept. Agr., Res. Sta. Introd. in 1934. Open-pollinated seedling of *Prunus americana* 'Pembina'; tested as Morden 109. Fruit: size medium; oblong; skin deep red with light bloom; flesh yellow, fine, sweet, slight apricot flavor; good for cooking; season mid-August. Tree: upright; moderately vigorous; hardy; productive.

Minnesota Purple.—Orig. in Excelsior, Minn., by Univ. of Minnesota, Agr. Expt. Sta. Introd. in 1920s. Omaha x *P. cerasifera pissardi*; selected in 1920. Fruit: 3/4 inch in diam.; round; skin purplish-black, thin bloom, tender; flesh purplish-black, tender, moderately juicy, sour; quality poor; clingstone; ripens early August. Tree: small, bushy, spreading; not hardy in Minnesota. Variety obsolete.

Mohawk.—Orig. in Geneva, N.Y., by John Watson, New York Agr. Expt. Sta. Introd. in 1966. Italian Prune x Hall. Tested as N.Y. 795. Fruit: prune-type; larger than Stanley; skin deep blue, attractive; flesh yellowish-amber, firm, sweet; quality high; stone semi-free to free; ripens 2 weeks before Stanley and 1 week earlier than Iroquois. Tree: slightly larger and more upright than Iroquois; about as hardy as Stanley; moderately productive; not self-fruitful; desirable for fresh market and home fruit garden.

Monitor.—Orig. in Excelsior, Minn., by W.H. Alderman, Univ. of Minnesota Fruit Breeding Farm. Named and introd. in 1920. Burbank x a native American plum (*Prunus americana*); first fruited in 1918; tested as Minn. 70. Fruit: medium to large; roundish ovate; skin thick, dull bronze-red, russet dots very conspicuous, bloom light; flesh yellow, tender, very juicy, firm, sweet; quality good; clingstone; cracks in rainy weather; ripens in late midseason. Tree: vigorous; upright, spreading; very hardy; productive.

Monster.—Orig. in Sebastopol, Calif., by Luther Burbank. Introd. in 1941 by Stark Bros. Nurseries & Orchards Co., Louisiana, Mo. Parentage unknown; selected in 1936. Fruit: very large flesh red. Tree: productive. No longer propagated.

Mordel.—Orig. in Morden, Manitoba, Canada, by Canada Dept. Agr., Res. Sta. Introd. in 1930. Open-pollinated seedling of *P. nigra* 'Assiniboine'. Fruit: size medium; oblong; skin dull red with light bloom; flesh yellow, firm, juicy, mild; good canner; season last of August. Tree: upright; quite vigorous; hardy.

Morden 124.—Orig. in Morden, Manitoba, Canada, by Canada Dept. Agr., Res. Sta. Introd. for trial in 1941. (*Prunus besseyi* 'Brooks') x Sapa; selected in 1941. Fruit: nearly 1 inch in diam.; almost round; skin reddish-purple to purplish-black; flesh reddish-purple, fine-grained, mild, pleasant; season mid-August. Bush: small to medium; very hardy; form roundish; productive.

Morden 125.—Orig. in Morden, Manitoba, Canada, by Canada Dept. Agr., Res. Sta. Introd. in 1947. Open-pollinated seedling of *Prunus salicina* selected in 1936. Fruit: roundish; up to 1 1/4 inches in diam.; skin greenish becoming yellow when ripe; flesh meaty, texture smooth, flavor sweet, pleasant; quality good. Tree: medium tall; moderately vigorous; hardy.

Morris.—Orig. in College Station, Texas, by the Hort. Dept., Texas A&M Univ. Self-pollinated seedling of Methley; cross made in 1956; selected in 1961 by J.B. Storey, evaluated by H.H. Bowen; introd. in 1974; tested as P 1-13. Fruit: large; ovate; skin reddish-black; flesh red at maturity, firm, slightly dry, crisp, good flavor; ripens 1 week after Methley with or slightly ahead of Ozark Premier, about 10 June; has longer shelf life than Methley. Tree: upright; large; equal in vigor to Methley; chilling requirement more than 800 h; subject to spring frost injury; more resistant to brown rot and bacterial spot than Methley. Produces well in northeastern Texas and in southeastern United States.

Mound.—Orig. in Excelsior, Minn., by the Univ. of Minnesota Fruit Breeding Farm. Named and introd. in 1922. *Prunus salicina* 'Burbank' x *P. americana* 'Wolf'; cross made in 1908; tested as Minn. 50; sibling of Mendota, Redcoat, Red Wing, Tonka. Fruit: very large; ovate slightly compressed; skin fairly thin, slightly astringent, medium red over yellow, attractive, dots small, bloom thin; flesh yellow, firm, meaty, tender, moderately juicy, mildly sweet; quality fair; partly freestone; ripens in late midseason. Tree: vigorous; hardy; very productive; medium size.

Moyer Perfecto.—Orig. in Roseburg, Ore., by C.E. Moyer, C.E. Moyer Nurseries, Winston, Ore. Introd. about 1927. Chance seedling; discovered about 1925. Fruit: European type; large, averaging 1 1/2 inches in diam.; ovate; skin thick, blue, bloom heavy and whitish; flesh yellow, firm, juicy, somewhat coarse; quality good; semi-clingstone; a good drying prune; ripens about 30 Sept. at Corvallis, Ore., or about 7 to 10 days after Italian and somewhat after French Prune in California. Tree: vigorous; resembles Italian Prune; self-fruitful.

Mr. Paul.—Bud mutation of Queen Rosa found in Fresno, Calif., by Ronald Paul. Plant patent 5413; 12 Mar. 1985. Fruit: oblate; large, 65 mm; skin red to red-purple on yellow to yellow-green ground color; flesh very light yellow-amber; clingstone. In central San Joaquin Valley, ripens 1 month after Queen Rosa, about the third week of July. Blossom: late blooming. Tree: vigorous; very upright; productive.

Muir Red.—Orig. in Red Bluff, Calif., by Oliver C. Muir. Introd. in 1956. Parentage unknown; received in a shipment of trees from Oregon; European type; discovered in 1946. Fruit: prune type; large; skin reddish before full maturity; very similar to French when dried; resembles French Prune but larger.

Myrix.—Orig. in Woodland, Calif., by Jack A. Traylor, California Dept. Agr., Sacramento, Calif. Introd. in 1965. Open-pollinated seedling of myrobalan (*Prunus cerasifera*); selected in 1960; tested as FP 6-20. Used only as a rootstock for plums and prunes; used in growing virus indicator plants to provide seed eligible for the certification program of the California Dept. Agr.

Myrobalan B.—Original seedlings came from Barbier & Co., Orleans, France; selected at East Malling Res. Sta., East Malling, Maidstone, Kent, England. Introd. about 1920. Open-pollinated seedling of *Prunus cerasifera*, myrobalan; selected in 1914. A clonal rootstock; a vigorous rootstock for plums; resistant to bacterial canker.

Myrobalan 29C.—Orig. in Marysville, Calif., by Marion E. Gregory, Gregory Brothers Nursery Co., now of Brentwood, Calif. Introd. in 1920. Open-pollinated seedling of *Prunus cerasifera* selected in 1915; seed imported from France. Used only as a clonal rootstock; resistant to root-knot nematode; grows vigorously; propagated vegetatively by hardwood cuttings.

Noble French (*Noble Petite*).—Orig. in Myrtle Creek, Ore., by Noble Andrews. Introd. in 1924. Bud mutation of Imperial Epineuse; discovered about 1915. Fruit: prune type; large; oval to obovate; skin cerise, thin; flesh firm, amber, juicy, flavor pleasant and distinctive, soluble solids high, texture somewhat coarse; ripens about 12 Sept.; resembles Imperial Epineuse. Tree: hardy; grows rapidly.

Noble Petite.—See Noble French.

Norseman.—Orig. in Ottawa, Ont., Canada, by Canada Dept. Agr., Res. Sta. Introd. in 1963 by Percy H. Wright, Saskatoon, Saskatchewan, Canada; tested as Ottawa 273. Parentage unknown; open-pollinated seed from Russia. Japanese type; selected at Ottawa in 1935; propagation rights assigned to Wright about 1947. Fruit: small; quality very good, sweet. Tree: extremely hardy, apparently to –55F; may be propagated readily on hybrid sandcherry.

Norther.—Orig. in Morden, Manitoba, Canada, by Canada Dept. Agr., Res. Sta. Introd. in 1944. Open-pollinated seedling of *Prunus nigra* 'Assiniboine'; seed originally sent from Minnesota Agr. Expt. Sta., Fruit Breeding Farm, Excelsior, Minn.; seed planted in 1922; selected in 1933; tested as Morden M-101. Fruit: oval-ovate; 1 1/4 x 1 inches; skin thin, tender, bright red with light gray bloom; flesh yellowish, juicy, sweet; pleasant as a dessert fruit; fair for cooking; season very early (mid-August). Tree: moderately spreading; strong, sturdy; hardy; annually productive; suited to northern prairies. A good native-type plum of dessert quality.

Nubiana.—Orig. in Winters, Calif., by Claron O. Hesse, California Agr. Expt. Sta., Davis, and Agr. Res. Serv., U.S. Dept. of Agriculture. Introd. in 1954. Gaviota x Eldorado; cross made in 1940; first fruited in 1944; tested as Calif. 17-28. Fruit: Japanese; large, sizing well; flat, resembling Eldorado; skin deep reddish-blue, developing before shipping maturity; flesh light amber with no red, very firm, meaty, moderately sweet, quality fair; ships well; matures in late July in California, after Laroda and before President; resembles Eldorado. Tree: very vigorous; abundantly productive; self-fruitful; blooms with Santa Rosa.

October Sun.—Orig. in Del Rey, Calif., by Tom Chamberlin, H.P. Metzler & Sons. Introd. in 1990. Parentage unknown. Plant patent 7827; 17 Mar. 1992; licensed exclusively to Agri Sun Nursery, Selma, Calif. Fruit: resembles Casselman; medium to large; slightly elongated; skin red on yellow ground color; flesh yellow, juicy, sweet. In central California, ripens 10 days after Roysum. Tree: vigorous, productive.

Olean.—Orig. in Le Grand, Merced County, Calif., by F.W. Anderson. Plant patent 3991; 14 Dec. 1976; assigned to The Burchell Nursery Co., Modesto, Calif. President x Agen prune. Fruit: large to medium; skin thick to medium, deep blue; flesh crisp, firm texture; subacid; best eating quality;

free to semi-freestone; excellent keeping quality. Tree: medium; vigorous; productive; upright to spreading; high resistance to brown rot.

Olson.—Orig. near Lake Dauphin, Manitoba, Canada, by W.J. Boughen, Boughen Nurseries, Valley River, Man. Introd. in 1920. Selection in 1920 from a native plum, *Prunus nigra*. Fruit: skin tough, dark red; flesh very sweet; quality good; for dessert and preserves; ripens early. Tree: yields very well.

Oneida.—Orig. in Geneva, N.Y., by John Watson, New York Agr. Expt. Sta. Introd. in 1966. Albion x Italian Prune. Cross made in 1937; selected in 1949; tested as N.Y. 826. Fruit: prune-type; medium to large; roundish-oblong; skin dark reddish-black, almost purple; flesh greenish-amber, firm, sweet, quality good; stone usually free; ripens 1 Oct. at Geneva or with Albion, superior to Albion; good keeping quality. Tree: as large, or slightly larger than Stanley; slightly upright; good producer; self-fruitful; recommended for fresh fruit and processing, but does not fit into processing schedule based on earlier ripening varieties in New York area.

Ox-Heart.—Orig. in Mountain Grove, Mo., by Paul H. Shepard, Missouri State Fruit Expt. Sta. Introd. in 1947. Burbank x Methley; selected in 1947; tested as PL-179. Fruit: very large; conical, similar to Burbank, but larger and redder; flesh red, firm, flavor good, sweet; clingstone; ripens about 10 July. Tree: vigorous; resembles Methley in tree growth.

Ozark Premier (*Premier*).—Orig. in Mountain Grove, Mo., by Paul H. Shepard, Missouri State Fruit Expt. Sta. Introd. in 1946, Burbank x Methley; selected in 1943; tested as PL-167. Fruit: extremely large; nearly round; skin bright red, tough; suture shallow; flesh yellow, firm, fine-grained; juice tart; flavor good; pit small, cling; ripens 1 Aug., with Burbank, which it resembles. Tree: vigorous; productive.

Padre.—Orig. in Palo Alto, Calif., by W.F. Wight, U.S. Dept. of Agriculture. Introd. in 1938. Wickson x Santa Rosa; cross made in 1921 at the U.S. Plant Introduction Garden, Chico, Calif.; tree planted in 1923 in Palo Alto; tested as US 15-23. Fruit: Japanese type; 5.7 to 6.2 cm in diam.; somewhat oblate, very symmetrical; skin bright carmine red, dots yellowish-gray, very waxy, bloom thin grayish, fairly tough, easily peeled when fully ripe, suture line very indistinct; flesh yellow, streaked light red next to pit, attractive, firm, juicy, texture fine, flavor mild, very pleasing; quality excellent; well-adapted for shipping; ripens early, slightly later than Santa Rosa. Tree: vigorous; upright; leaf averages 4 to 5. 5 inches long; flower white.

Parkside.—Orig. in Edmonton, Alberta, Canada, by R. Simonet. Introd. in 1960 by A.J. Porter, Honeywood Nursery, Parkside, Saskatchewan, Canada. *Prunus salicina* x *P. nigra*; seedlings sent by Simonet to Porter; selected in 1955 by Porter. Fruit: up to 1 1/4 inches in diam., being of good size for this type; skin thin; flavor rich, good; pit very small; ripens in September. Tree: hardy; vigorous.

Parson.—Orig. in Forest Grove, Ore., by Peter R. Parson. Introd. in 1947. Plant pat. 872; 13 Sept. 1949. Parentage unknown; discovered about 1939. Fruit: prune type; higher sugar content and ripens earlier than Italian Prune, which it resembles. Tree: hardier and heavier bearing than Italian Prune.

Patterson Pride.—Orig. in Saskatoon, Saskatchewan, Canada, by C.F. Patterson, Univ. of Saskatchewan. Introd. in 1960 for home gardens. Derived from *Prunus nigra* and *P. salicina*; selected in 1949; tested as Sask. P58-104. Fruit: up to 1 3/4 inches in diam. under nonirrigated orchard conditions in Saskatoon; skin deep red; quality excellent; keeps well in ordinary storage; ripens in mid-September. Tree: low-growing; branches markedly recurving; yields well; hardy.

Pembina.—Orig. in Brookings, S.D., by N.E. Hansen, South Dakota Agr. Expt. Sta. Introd. about 1923. *Prunus salicina* 'Red June' x *P. nigra*; selected in 1917. Fruit: large; round-ovate; pointed; skin red, thick, heavy bluish bloom; flesh yellow, juicy, sweetish; quality good for dessert, only fair for canning; ripens last of August.

Perfection.—Orig. in Saskatoon, Saskatchewan, Canada, by C.F. Patterson, Univ. of Saskatchewan. Introd. in 1960 for home gardens. Derived from *Prunus nigra* and *P. salicina*; selected in 1954; tested as Sask. P54-32. Fruit: spherical, slightly flattened, up to 1 1/2 inches under nonirrigated orchard conditions at Saskatoon; skin dark red; flesh firm; quality excellent; ripens in late September; keeps well in ordinary storage. Tree: hardy.

Phoebe.—See Fiebing.

Pipestone.—Orig. in Excelsior, Minn., by Univ. of Minnesota Fruit Breeding Farm. Introd. for trial in 1928; introd. commercially in 1942. Burbank x (*Prunus salicina* x Wolf); cross made ln 1919; fruited in 1926; tested as Minn. 218. Fruit: large; skin solid deep red, thin, tough but peeling easily; flesh yellow, somewhat stringy, with excellent quality; for home and commercial use. Tree: vigorous; hardy; performance reliable.

Polly.—Orig. in Geneva, N.Y., by R.L. Andersen, J.P. Watson, and K.G. Livermore, New York State Agr. Expt. Sta. Oneida open-pollinated; selected in 1977; tested as NY 65.363.1; introd. in 1993. Fruit: large, kidney-shaped, flat (cheek to cheek dimensions less than across suture), 40 mm cheek to cheek, 50 mm at broad point of suture, 60 mm long. Skin purple-red with green ground color until fully ripe; flesh amber-green, medium firm, good retention of adequate firmness in storage; flavor mild, fruity. Ripens with Stanley. Tree: vigorous; spreading; self-thins, causing consistent cropping with large fruit. Self-fertile; pollinates most other European plums.

Prairie.—Orig. in Saskatoon, Saskatchewan, Canada, by C.F. Patterson, Univ. of Saskatchewan. Introd. in 1960 for home gardens. Derived from *Prunus nigra* and *P. salicina*; selected in 1948; tested as Sask. P58-106. Fruit: up to 2 inches long and 1 3/4 inches in diam. under nonirrigated orchard conditions at Saskatoon; skin dark red; quality good; somewhat freestone; ripens in early September; resembles Acme and Elite. Tree: size medium; vase shape, but not too spreading; trace of winter injury at Saskatoon.

Premier.—See Ozark Premier.

Premier.—Orig. in Merced, Calif., by Walter D. Krause, Stribling's Nurseries. Introd. in 1963. Plant patent 2496; 20 Apr. 1965; assigned to Stribling's Nurseries. Burmosa x Santa Rosa; selected in 1961; Japanese type; tested as Stribling 80-15061. Fruit : large, averaging a packed size of 4 x 4 to 3 x 4; shape similar to Santa Rosa; skin fully colored a deep rich red at shipping maturity, which is 4 to 5 days before Beauty; flesh light amber, firmer than Beauty and Burmosa, clingstone; resembles Santa Rosa. Tree: vigorous; self-fruitful; blooms with Burmosa and Mariposa; produces good crops.

Prima Black Plum 5-25.—Orig. in Sanger, Calif., by Michael R. Gerawan. Open-pollinated seedling of unknown origin. Plant patent 8067; 22 Dec. 1992. Fruit: round; medium size, 52 to 57 mm; skin black; flesh yellow to red under skin; clingstone. In central San Joaquin Valley, ripens 4 to 6 days after Red Beaut, about 1 June. Blossom: medium to large, white; anthers yellow with red-orange tinge; blooms early with Ambra. Tree: medium; upright to upright-spreading.

Prima Black Plum 8-15.—Orig. in Sanger, Calif., by Michael R. Gerawan. Open-pollinated seedling of unknown origin. Plant patent 8068; 22 Dec. 1992. Fruit: globose to slightly oblate with a pronounced point; large size, 66 to 71 mm; skin dark red (Catawba); flesh yellow (Nankeen); freestone. In central San Joaquin Valley, ripens 2 weeks after Friar, about the first week of August. Blossom: medium, white; anthers yellow. Tree: medium size; upright to spreading; productive.

Prima Red Plum 9-1.—Orig. in Sanger, Calif., by Michael R. Gerawan. Open-pollinated seedling of unknown origin. Plant patent 8057; 8 Dec. 1992. Fruit: oblate; large size, 73 to 80 mm; skin red; flesh cream; stone semi-free. In central San Joaquin Valley, ripens 7 to 10 days after Angeleno, about the second week of September. Blossom: medium, white; anthers yellow-orange with red margins; blooms early with Blackamber. Tree: upright to upright-spreading.

Prinlew (*Lewis*).—Orig. in Portland, Ore., by H.A. Lewis, Introd. in 1923. Italian Prune x Bradshaw; selected in 1918. Fruit: large; flesh firm, juicy, flavor good, texture fine; ships well.

Ptitsin.—Orig. in Harbin, Manchuria, by L.V.

Ptitsin. Received by Canada Dept. Agr., Expt. Sta., Morden, Manitoba, Canada, in 1939; received by Univ. of Minnesota, Agr. Expt. Sta., Excelsior, Minn. in 1945. *Prunus salicina* (*P. triflora koreana*). Fruit: 1 1/8 x 1 3/8 inches in diam.; truncate, unequal halves; skin yellow green with numerous white dots, thin bloom, tender; flesh yellow, moderately tender, moderately juicy, flavor good, aromatic; quality good; clingstone; ripens late August. Tree: low, spreading; productive; moderately hardy. Variety considered obsolete in Morden.

Ptitsin 5.—Orig. in Manchuria. Received by Canada Dept. Agr., Res. Sta., Morden, Manitoba, Canada, in 1939, from L.V. Ptitsin, Harbin, Manchuria. Introd. in Canada in 1939. *Prunus salicina* (*P. triflora koreana*). Fruit: similar to Ptitsin 9; medium to small; round; skin yellow, moderately heavy bloom; flesh yellow, smooth, very juicy, flavor excellent; freestone; ripens early. Tree: vigorous; not hardy in Minnesota.

Ptitsin 7.—Orig. in Harbin, Manchuria, by L.V. Ptitsin. Received by Canada Dept. Agr., Expt. Sta., Morden, Manitoba, Canada, in 1939; received by Univ. of Minnesota, Agr. Expt. Sta., Excelsior, Minn. in 1945. *Prunus salicina* (*P. triflora koreana*). Fruit: 1 1/8 x 1 3/8 inches in diam.; truncate, unequal halves; skin yellow green with numerous white dots, thin bloom, tender; flesh yellow, moderately tender, moderately juicy, flavor good, aromatic, quality good; clingstone; ripens late August. Tree: low, spreading; productive moderately hardy. Variety considered obsolete in Morden.

Ptitsin 9.—Orig. in Manchuria. Scions received from L.V. Ptitsin, Harbin, Manchuria, by Canada Dept. Agr., Res. Sta., Morden, Manitoba, Canada, in 1939. Introd. in Canada in 1939. *Prunus salicina* (*P. triflora koreana*). Fruit: medium, measuring 1 1/8 inches in diam.; nearly round; skin green becoming greenish-yellow when mature; flesh light greenish, firm, meaty, freestone; flavor very sweet, mild, pleasant, somewhat like Green Gage; season mid-August; quality very acceptable as dessert or preserves. Tree: bush type; rounded, spreading; hardy; productive; adaptable to prairie conditions.

Ptitsin 10.—Orig. in Manchuria. Scions received in 1939 from L.V. Ptitsin, Harbin, Manchuria, by Canada Dept. Agr., Res. Sta., Morden, Manitoba, Canada. Introd. in Canada in 1939. *Prunus salicina* (*P. triflora koreana*). Fruit: similar to Ptitsin 9 except fruit are larger and of slightly better quality. Tree: more vigorous than Ptitsin 9; not hardy in Minnesota.

Punian.—Bud sport of French Prune found in Yuba City, Calif., by Gurbachan Punian. Plant patent 7168; 27 Feb. 1990. All characteristics are the same as French Prune, except the fruit ripens 15 to 18 days earlier and the sugar content (30% to 31%) is higher at harvest. Fruit: ovate; medium size, 21 to 24 g fresh weight, 8.8 to 9.2 g dry weight; pronounced cavity; skin deep blue; flesh yellow to amber; stone medium to small, cling. Blossom: white, blooms with French Prune. Tree: vigorous; upright.

Purple.—Orig. in Auburn, Ala., by J.D. Norton, Agr. Expt. Sta. Introd. in 1973. Methley x Ozark Premier; cross made in 1961; tested as Methley 1143. Fruit: 1 3/4 inches in diam.; skin purple; flesh cream-colored, firm; quality good; semi-clingstone; may be stored 3 weeks or more; for local and distant markets; ripens about 20 July. Tree: medium; upright; vigorous; productivity excellent; resistant to bacterial canker, fruit spot, leaf spot and black knot; chilling requirement 850 h.

Purple Flame.—Orig. in Sebastopol, Calif., by Luther Burbank. Introd. in 1931 by Stark Bros. Nurseries & Orchards Co., Louisiana, Mo. Parentage unknown; selected about 1922. Fruit: flesh red, quality good. Tree: foliage red. Primarily for ornamental purposes.

Purple Heart.—See Purpleheart.

Purpleheart (*Purple Heart*).—Orig. in Enfield, N.H., by Forest L. Colby. Introd. in 1969 by Farmer Seed and Nursery Co., Faribault, Minn. Parentage unknown. Fruit: Japanese type; size medium; round; skin dull purplish red with heavy waxy bloom; flesh deep red, moderately firm, sweet, juicy; stone very small for size of plum; recommended for home gardens in northern areas; ripens August. Tree: small spreading; vigorous; very hardy; productive.

Purple King.—Orig. in Avondale, Auckland, New Zealand, on the property of a Mr. Quilard. Introd. there in 1921. Chance seedling of *Prunus salicina* type; discovered and propagated by Hayward Wright. Fruit: very large; round, symmetrical, apex slightly depressed; stem short, thick; skin smooth, purplish-red to dark purple with greenish dots and blue bloom, appearance excellent; flesh yellow, tinged red directly below skin, flavor fair; quality good; clingstone; season end of January into February in New Zealand. An important market variety.

Purple Majesty.—Orig. in Le Grand, Calif., by Norman and Glen Bradford. Red Beaut x unnamed seedling; cross made in 1982. Plant patent 7503; 23 Apr. 1991. Fruit: symmetrical, oblate; medium size, 60 mm; skin dark reddish-purple; flesh orange-yellow, very firm; flavor mild and very sweet, juicy; clingstone. In central California, ripens about with Santa Rosa. Blossom: small; white; blooms with and pollinated by Wickson. Tree: vigorous; upright; productive; self-unfruitful.

Queen Ann.—Orig. in Winters, Calif., by Claron O. Hesse, California Agr. Expt. Sta., Davis, and Agr. Res. Serv., U.S. Dept. of Agriculture. Introd. in 1954. Gaviota x Eldorado; cross made in 1940; first fruited in 1944; tested as Calif. 17-26. Fruit: large; somewhat heart-shaped, tendency to have a slightly irregular surface; skin develops a full deep mahogany before shipping maturity; flesh light amber, no red next to skin, firm, sweet, texture and quality very good, aroma not pronounced; holds in storage for rather long periods; ships very well; ripens late, mid-August in California, about with President. Tree: somewhat weak but better than Gaviota, appearing to be more vigorous on established peach or budded on Marianna 2624 rootstocks than when topworked to established Santa Rosa and Duarte trees; numerous flowers on medium-sized spurs; self-unfruitful, but satisfactorily pollinated by Wickson, Gaviota, Santa Rosa, Becky Smith, Redroy and Red Rosa.

Queen Rosa.—Orig. in Fresno, Calif., at U.S. Hort. Field Sta., by U.S. Dept. of Agriculture. Introd. in 1972. Queen Ann x Santa Rosa; cross made in 1961, selected in 1964 by John H. Weinberger. Fruit: Japanese type; large, round to slightly oblate; skin red, smooth, attractive; to avoid discoloration caused by handling and rolling, fruit should not be picked until shoulders well-colored; flesh light yellow, firm, softens slowly; good shipping quality; flavor similar to Queen Ann; clingstone; ripens late June or end of Santa Rosa season. Tree: vigorous; does not set heavy crop when young; flowers self-infertile, requires a pollinizer: Santa Rosa, Casselman, Simka, Laroda, Friar, or Redroy. Queen Rosa pollinates Laroda and Friar. Recommended only for Japanese plum growing areas.

Queensland Bellerosa.—Orig. in Applethorpe, Queensland, Australia, by B.L. Topp and D.M. Russell, Granite Belt Hort. Res. Sta. Introd. in 1988. Burbank x Santa Rosa. Cross made in 1979; selected in 1984; tested as GB 22-10. Fruit: medium to large, 48 mm; oblong with pronounced suture; skin 90% to 100% bright red at maturity, turning dark red with bloom on shelf, skin slightly bitter; flesh bright yellow tinged with red at stone and near skin; juicy; flavor aromatic. Ripens 11 days before Santa Rosa. Tree: vigorous; less upright than Santa Rosa; productive; flowers self-unfruitful; pollinated by Santa Rosa. Good levels of field resistance to bacterial spot (*Xanthomonas campestris* pv. *pruni*).

Queenston.—Orig. in Queenston, Ont., Canada, by Charles Lowrey. Released from bud restrictions in 1963. Chance seedling; discovered during 1930–35; first test trees planted in 1939. Fruit: European type; size medium; skin blue; ripens early, with or a few days before California Blue; freestone; drops less than California Blue; quality not as good as California Blue but acceptable in its ripening season; suitable for dessert but not for processing. Tree: productivity, vigor and the general tree itself are better than California Blue; vigorous, upright; productive annually.

Radiance.—Orig. in Mountain Grove, Mo., by Paul H. Shepard, Missouri State Fruit Expt. Sta. Introd. in 1947. Stanley x President; selected in 1947; tested as PL-219; sibling of Bluebell and Bluefre. Fruit: large; resembles Yellow Egg in size and shape; skin light yellow with pink blush, attractive; flesh yellow, firm, flavor good, sweet; freestone; ripens about 10 Sept. Tree: vigorous; product upright.

Radisson.—Orig. in Excelsior, Minn., by the Univ. of Minnesota Fruit Breeding Farm. Named and introd. in 1925. *Prunus salicina* x *P. americana*; cross made in 1907 or 1908; tested as Minn. 157. Fruit: large; round to slightly oval, pointed at apex; skin thick, strong, a rich attractive red, bloom heavy, dots conspicuous; flesh yellow, firm, tender, juicy, sweet, slightly fibrous near skin; quality very good; stone roundish oval, semi-freestone; ripens early; well-adapted for market purposes. Tree: upright, spreading; size medium.

Ranch 9-Golden.—Orig. at Fresno, Calif., by J.M. Garabedian. Chance seedling of unknown parentage. Plant patent 5912; 24 Mar. 1987. Fruit: small to medium, 52 mm wide, 47 mm long; broadly oval to oblate; skin very dark purplish-red; flesh light yellow; clingstone; matures September in Fresno. Tree: vigorous; upright; productive.

Ranch 9-Maroon.—Orig. in Fresno, Calif., by John M. Garabedian. Open-pollinated seedling of unknown origin. Plant patent 5902; 17 Mar. 1987. Fruit: oblate; medium, 60 mm; skin deep maroon; flesh deep red; stone semi-free. In central San Joaquin Valley, ripens with Roysum, about 8 Sept. Blossom: medium, white. Tree: medium; upright-spreading; productive.

Rancho Cinco.—Orig. at Fresno, Calif., by J.M. Garabedian. Chance seedling of unknown parentage. Plant patent 7075; 19 Dec. 1989. Fruit: medium to large, 54 mm wide, 60 mm long; oblong; skin dark red; flesh dark red; semi-freestone; matures early to mid-August in Fresno. Tree: vigorous; upright; productive.

Rancho Ocho.—Orig. at Fresno, Calif., by J.M. Garabedian. Chance seedling of unknown parentage. Plant patent 7082; 25 Dec. 1989. Fruit: large, 57 mm wide, 55 mm long; globose; skin dark red; flesh light red; semi-freestone; matures late June in Fresno. Tree: vigorous; upright; productive.

Raribank.—Orig. in New Brunswick, N.J., by J.H. Clark, New Jersey Agr. Expt. Sta. Introd. in 1949. Selection of wild beach plum (*Prunus maritima*); tested as N.J. 1. Fruit: skin purplish-red; quality good; freestone; good as jelly or canned; ripens during early September in New Brunswick. Tree: large; vigorous; very resistant to brown rot and Japanese beetle.

Ray's Hope.—Bud sport of Casselman found in Exeter, Calif., by Arthur R. Casselman. Plant patent 5650; 28 Jan. 1986. Fruit: ovoid to oblong; medium-small size, 48 mm; skin dark red on yellow ground color; flesh light yellow-amber, firm; clingstone. In central San Joaquin Valley, ripens 4 to 5 weeks after Casselman, about 10 Sept. to 3 Oct. Tree: vigorous; upright-spreading; very productive.

Red Ace (*Ace, Florida*).—Orig. in Sebastopol, Calif., by Luther Burbank. Introd. in 1931 by Stark Bros. Nurseries & Orchards Co., Louisiana, Mo.; first announced in print by Burbank in 1927. Parentage unknown; Japanese type; selected about 1923. Fruit: size medium; round-oval; skin crimson to dull red, bloom pale blue; flesh light crimson, sweet, delicious; quality excellent; semi-freestone to freestone; resembles Elephant Heart. Tree: very productive; spreading; hardy.

Red Beaut.—Orig. in Le Grand, Calif., by F.W. Anderson, Merced, Calif. Introd. in 1965. Plant patent 2539; 22 June 1956; assigned to Kim Bros., Reedley Nursery, Reedley, Calif. Open-pollinated early-ripening seedling of Eldorado x Burmosa; Japanese type; tested as Anderson 17H450. Fruit: size medium, averaging 2 inches long and 1 7/8 inches in suture diam.; globose; uniform; skin bright red when picking ripe, shading to reddish-purple when full ripe, thickness medium; flesh yellow when picking ripe, dark red when full ripe, crisp, melting, subacid, flavor delicate; stone medium to small, free; ships and keeps well; ripens very early, about 1 week before Beauty; eating quality superior; resembles Beauty. Tree: regular and productive bearer; size medium vigorous.

Redbud.—Orig. in Mountain Grove, Mo., by Paul H. Shepard, Missouri State Fruit Expt. Sta. Introd. in 1947. Burbank x Methley; selected in 1947; tested as PL-149. Fruit: size medium; skin red; flesh red, tart; primarily for culinary uses; ripens last week of July; resembles Methley. Tree: useful for both ornamental and fruiting purposes; purple foliage, red blossoms; very productive; new wood long, dark, streaked with red; extremely vigorous.

Red Cannon Ball.—Orig. in Santa Rosa, Calif., by Luther Burbank. Introd. about 1931 by Stark Bros. Nurseries & Orchards & Co., Louisiana, Mo. Parentage unknown. Fruit: large; round; skin shiny red, tough; flesh red, sweet, clingstone; ripens early. No longer propagated.

Redcoat.—Orig. in Excelsior, Minn., by Univ. of Minnesota Fruit Breeding Farm. Introd. in 1942. *Prunus salicina* 'Burbank' x *P. americana* 'Wolf'; cross made in 1908; tested as Minn. 17; sibling of Mendota, Mound, Red Wing, and Tonka. Fruit: skin crimson overlaid with a heavy bloom, thin, tender; flesh yellow, meaty, medium juicy, flavor sweet to mild, subacid, pleasant; freestone; season early, closely following Underwood; good canning and culinary qualities. Tree: productive; hardy; bears early.

Red Giant.—Orig. in Watervliet, Mich., by Walter Warsko. Introd. in 1942. Parentage unknown; chance seedling; discovered in 1940. Fruit: large; skin reddish; flesh greenish, flavor good; resembles Austrian prune. Tree: vigorous; productive; not widely planted.

Red Giant™ (*Suplumeighteen*).—Orig. in Fresno, Calif., by John H. Weinberger and Tim P. Sheehan; assigned to Sun World. Queen Rosa open-pollinated. Plant patent 7443; 12 Feb. 1991. Fruit: globose to oblate; large size, 60 mm; skin red purple; flesh red; stone semi-free. In central San Joaquin Valley, ripens 2 weeks before Queen Rosa. Blossom: average, white, early. Tree: average vigor; upright; productive.

Redglow.—Orig. in Excelsior, Minn., by the Univ. of Minnesota Fruit Breeding Farm. Introd. in 1949. *Prunus salicina* 'Burbank' x *P. munsoniana* 'Jewell'; selected in 1919 from seed planted in 1914; tested as Minn. 101. Fruit: large; skin dark red with heavy bloom, thick; flesh orange, juicy, tender, sweet, clingstone; ripens in midseason; resembles Tonka. Tree: hardy; productive.

Redheart.—Orig. in Winters, Calif., by Claron O. Hesse, California Agr. Expt. Sta., Davis. Introd. in 1952. Duarte x Wickson; cross made in 1940; selected in 1943; tested as Calif. 10-29. Fruit: medium to large; round-ovate to nearly cordate, smooth, regular; stem length medium; skin moderately tough, crisp, dull amber green, completely covered by a dull medium to darkish-red, sometimes slightly purplish; bloom moderately heavy, gray; dots small, numerous, conspicuous; flesh bright red over amber, typical blood plum, sweet, mild, aroma pronounced like Duarte, firm-melting, meaty, nearly crisp, fine-grained, excellent; stone size medium, nearly free; shipping and dessert use; season early, coming on market with, or slightly before first shipments of Duarte in San Joaquin Valley of California; resembles Duarte. Tree: upright-spreading; very vigorous; chilling requirement moderate; very productive; cross-fruitful with Becky Smith, Burbank, Del Norte, Elephant Heart, Flaming Delicious, Formosa, Gaviota, Red Rosa, Santa Rosa, Sharkey, Wickson; cross-unfruitful with Duarte and Eldorado; resembles Duarte in nearly all respects.

Red Noble.—Orig. in Le Grand, Calif., by Norman and Glen Bradford. Open-pollinated seedling of an unnamed seedling. Plant patent 7975; 15 Sept. 1992. Fruit: globose to slightly obovoid; medium size, 54 mm; skin dark purplish-red; flesh light yellow, firm; flavor very sweet with medium acidity; very juicy. In central California, ripens about 3 days before Red Beaut. Blossom: medium; white; blooms early; pollinated by Ambra. Tree: vigorous; upright; medium productive; self-unfruitful.

Red Nugget.—Orig. in Del Rey, Calif., by Frank T. Matoba. Red Beaut open-pollinated. Selected in 1985. Plant patent 7765; 14 Jan. 1992. Japanese type. Fruit: round to ovate, small, uniform, 44 to 46 mm diam. Skin medium red; flesh semi-cling; quality good. Ripens about 10 days before Red Beaut. Tree: medium; upright; productive.

Red Rosa.—See Late Santa Rosa.

Red Roza.—See Late Santa Rosa.

Red Santa Rosa.—See Late Santa Rosa.

Redroy.—Orig. in Winters, Calif., by Claron O. Hesse, California Agr. Expt. Sta., Davis, and Agr. Res. Serv., U.S. Dept. of Agriculture. Introd. in 1954. Gaviota x Elephant Heart; cross made in 1940; first fruited in 1944; tested as Calif. 15a-30. Fruit: Japanese;

size medium, sizes well; nearly round to slightly elongated; skin bright full red, unique color, appearance attractive; flesh deep, bright amber, firm, meaty, sweet; quality excellent; excellent shipping characteristics; ripens in mid-July in California, with Laroda. Tree: vigorous; produces abundant spurs and buds of medium size, self-fruitful; blooms with Santa Rosa; thin well for production of large fruit.

Redtalian.—Orig. in Winnipeg, Manitoba, Canada, by S. Juskow. Introd. in 1941. Parentage unknown. Fruit: large, 2 1/8 inches long by 1 1/2 inches in diam.; oblong and pointed; skin dark, medium bright, red under heavy blue bloom; flesh opaque to whitish, firm, flavor mild to being somewhat flat, quality fair, freestone; season mid- to late September. Tree: medium tall; upright; hardy; heavily foliaged. Now obsolete.

Red Wing.—Orig. in Excelsior, Minnesota, by the Univ. of Minnesota Fruit Breeding Farm. Named in 1920; introd. in 1921. *Prunus salicina* 'Burbank' x *P. americana* 'Wolf'; cross made in 1908; tested as Minn. 12; sibling of Mendota, Mound, Redcoat, Tonka. Fruit: large; roundish oblong, base and apex flattened; skin medium thick, easily removed, deep bright red, bloom heavy, russet dots numerous; flesh yellow, firm, moderately juicy, sweet; quality good; stone large, broad, free; ripens in mid-season. Tree: vigorous, spreading; productive; hardy; leaf large.

Reeves.—Orig. in Chilliwack, B.C., Canada, by A. Reeves. Introd. about 1940. Peach plum x self (probably). Fruit: *P. domestica;* medium large; oval; skin creamy pink blush on greenish-yellow cover; flesh light orange, juicy, flavor fair, slightly clingstone; ripens about 17 days later than Peach, about 26 Aug., otherwise Reeves is very similar to it. Tree: very productive; indistinguishable from peach cultivar.

Reine Red.—Orig. in Moscow, Idaho, by Leif Verner, Idaho Agr. Expt. Sta. Introd. in 1951. Bud mutation of Reine Claude; discovered in 1943. Fruit: skin red; high dessert quality of parent; resembles Reine Claude.

Richards Early Italian.—Orig. in Wapato, Wash., by E.C. Richards, May Nursery Co., Yakima, Wash. Introd. in 1935. Bud mutation of Italian Prune discovered in 1930. Fruit: prune type; slightly larger and sweeter than Italian Prune, which it resembles; ripens 3 days before Demaris and 2 weeks before Italian Prune; sometimes quality below that of Italian Prune. Tree: sometimes shows leaf disorders.

Rich Pride.—Orig. in Junction City, Ore., by Chris Rich. Introd. in 1946. Parentage unknown; selected about 1920. Fruit: French Prune-type; large; flesh fairly soft, very sweet, juicy, nearly freestone; especially good as fresh fruit; ripens very late, about 2 weeks after Italian; good drying qualities.

Rivers Delight.—Orig. in Greenville, S.C., by Daniel M. Rivers. Introd. in 1968. Plant patent 2785; 2 Jan. 1968. Parentage unknown; chance seedling found growing under a tree named Jap Gold; discovered in 1958; first fruited in 1961. Fruit: resembles Burbank in shape, being distinguished by firmness of flesh, medium heavy skin with deep red color and deep burgundy color of flesh; excellent shipping qualities; eating quality excellent, juicy; clingstone; ripens middle of July; tolerant to rot. Tree: large, vigorous; very productive; flowers large, white.

Roberts.—See Wisconsin.

Robinson.—Orig. in Saskatoon, Saskatchewan, Canada, by C.F. Patterson, Univ. of Saskatchewan. Introd. in 1960 for home gardens. Derived from *Prunus nigra* and *P. salicina;* selected in 1947; tested as Sask. P60-101. Fruit: up to 1 1/2 inches long under nonirrigated orchard conditions at Saskatoon; skin red; flesh firm; quality very good; ripens during the last 2 weeks of September. Tree: hardy.

Robusto.—Orig. in Byron, Ga., by Victor E. Prince, of U.S. Dept. of Agriculture. Introd. in 1980. (Queen Ann x Barstow) x (Ozark Premier x *Prunus angustifolia*). Cross made in 1968; selected in 1973 by J.M. Thompson; tested as BY 68-317. Fruit: Japanese-type; larger than Methley, smaller than Bruce; used as "green plum"; flesh firm for shipping but softens and turns bright red upon ripening, can be picked green, developing an attractive bright red while on shelf; pit small, clings tenaciously; ripens 10 days after Bruce, 5 days after Methley or late May or early June at Byron. Tree: vigorous, precocious, productive; resistant to complex of bacterial diseases common to Southeastern states; grows so robustly that heavy thinning and semiannual pruning recommended.

Rosa Grande (*Eastman Special, Super Rosa*).—Orig. in Ontario, Calif., by Arnold Abacherli. Introd. in 1954 by Matlin's Nursery, Ontario, Calif., and in 1957 by Stribling's Nurseries, Merced, Calif. Bud mutation of Santa Rosa; discovered in 1950. Fruit: large; skin deep red with gold specks; flesh red; pit small; ripens in Santa Rosa season. Tree: produces heavily; vigor greater than Santa Rosa.

Rosanna.—Orig. in Visalia, Calif., by Carl M. Lauritzen. Introd. in 1957. Bud mutation of Santa Rosa; discovered in 1955. Fruit: large; firm; ripens after Santa Rosa and before Late Santa Rosa; resembles Santa Rosa. Tree: vigorous.

Rosemary.—Orig. near Le Grand, Merced County, Calif., by F.W. Anderson. Plant patent 3797; 4 Nov. 1975; assigned to Burchell Nursery, Modesto, Calif. Open-pollinated seedling of Eldorado. Fruit: medium to large; symmetrical, globose; skin yellow, entirely overspread with red shading to darker red, many yellow flecks, medium thickness; flesh firm, crisp, meaty, subacid, yellow; semi-freestone; ripens 15 Sept. Tree: large; vigorous; upright to spreading; very productive; regular bearer.

Rose Zee.—Orig. in Modesto, Calif., by Floyd, Gary, and Grant Zaiger and Leith Gardner. Spring Beaut x Nubiana. Introd. in 1984. Plant patent 5287, 2 Oct. 1984. Fruit: medium size, 57 mm diam., 38 mm long; oblate; skin red; flesh yellow; clingstone. Matures in late May and early June in California. Tree: vigorous; upright; heavy production; self-incompatible.

Royal Diamond.—Orig. in Kingsburg, Calif., by Larry Kitahara; assigned to Kitahara Farms. Open-pollinated seedling of unknown origin. Plant patent 6756; 25 Apr. 1989. Fruit: oblate to round; large size, 59 mm; skin blue-black to dark blood-red; flesh light yellow amber; stone semi-free. In central San Joaquin Valley, ripens 3 to 4 weeks after Friar, about 17 to 27 Aug. Blossom: medium, white; anthers red-orange; late blooming, with Simka. Tree: medium vigor; very upright; productive.

Royal Garnet.—Orig. in Le Grand, Calif., by Fred Anderson. Red Beaut open-pollinated. Plant patent 4823; 9 Feb. 1982. Fruit: globose; medium size, 54 mm; skin deep purple; flesh yellow; clingstone. In central California, ripens 1 week after Red Beaut, about 24 to 30 May. Blooms with Red Beaut. Tree: spreading; productive.

Royal Purple.—Orig. in Santa Rosa, Calif., by Luther Burbank. Introd. about 1931 by Stark Bros. Nurseries & Orchards Co., Louisiana, Mo. Parentage unknown. Fruit: large, up to 2 1/2 inches in diam.; skin bluish-purple; flesh yellow, sweet, juicy, freestone; ships well. No longer propagated.

Royal Red.—Orig. in Le Grand, Calif., by F.W. Anderson, Merced, Calif. Introd. in 1974. Plant patent 3618; 17 Sept. 1974; assigned to Reedley Nursery, Reedley. Open-pollinated seedling of Red Beaut. Fruit: large to medium; globose; skin red, more attractive than Red Beaut; flesh dark greyish-yellow (bistre green); clingstone; ripens 3 to 5 days before Red Beaut. Tree: large; upright to spreading; vigorous; productive.

Royal Star.—Orig. in Kingsburg, Calif., by Lawrence K. and Kei Kitahara. Open-pollinated seedling of unknown origin. Plant patent 7192; 13 Mar. 1990. Fruit: ovate; large size, 56 mm; skin dark blue-black; flesh pink to red, firm, medium juicy; clingstone. In central San Joaquin Valley, ripens with Angeleno. Blossom: medium, white. Tree: vigorous; upright to spreading.

Royal Supreme.—Orig. in Merced, Calif., by Walter D. Krause, Stribling's Nurseries. Introd. in 1960. Open-pollinated seedling of Nubiana; Japanese type; selected in 1958. Fruit: large, fresh packs averaging 3 x 4; skin deep purple; flesh light amber, flavor fair to good, clingstone; ripens in early June, 3 days before Santa Rosa; resembles Nubiana and Eldorado. Tree: self-fruitful; pollinates other vars., including Queen Ann.

Royal Zee.—Orig. in Modesto, Calif., by Floyd, Gary, and Grant Zaiger and Leith Gardner. Introd. in 1985. Open-pollinated seedling of Royal Beaut. Plant patent 5486; 4 June 1985; assigned to Zaiger's Genetics. Tested as 42-GA-1200. Fruit: large, 60 to 65 mm; skin overcolor bluish red; globose;

flesh yellow, firm; mild flavor. Tree: medium vigor; upright; productive. For local and long-distance marketing.

Roysum.—Orig. in Tulare, Calif., by Roy Sumruld. Introd. in 1966 by Sierra Gold Nurseries, Yuba City, Calif. Plant patent 2619; 12 Apr. 1966. Bud mutation of Late Santa Rosa; discovered in 1955. Fruit: averages 2 1/4 inches in diam.; shape variable, oblong; suture shallow; skin reddish-blue, thick to medium, tenacious to flesh; bloom whitish and similar to parent; flesh uniform light yellow, firm, slightly acid, juicy; high quality; freestone; ripens evenly, very late, 2 months after Late Santa Rosa, in early October; keeps very well; resembles Late Santa Rosa. Tree: self-fruitful.

Rubysweet.—Orig. at Byron, Ga., by W.R. Okie, U.S. Dept. of Agriculture, S.E. Fruit & Tree Nut Res. Lab. Introd. in 1989. Mariposa x Morris. Cross made in 1968 by V.E. Prince; selected in 1973 by J.M. Thompson; tested as BY69-350. Fruit: large; skin reddish-bronze; round; flesh blood-red, firm; freestone; high-quality. Ripens between Santa Rosa and Ozark Premier. Tree: vigorous on peach rootstock; mostly upright; resistant to bacterial spot and bacterial canker; tolerant of plum leaf scald and brown rot. Adapted to the humid Southeast. Flowers large, two or three per cluster, self-incompatible; bloom about 2 days before Santa Rosa.

Ruth Gerstetter.—Orig. in Besigheim, Württemberg, by Adolf Gerstetter. Introd. in 1932. Czar x Bonne de Bry. Cross made in 1920. Fruit: size medium; skin thick and tough; flesh green-yellow, rather firm, subacid; freestone; ripens early, the end of July in Sweden. Tree: medium size; upright; vigor moderate; not hardy; unproductive; susceptible to bacterial canker.

Sage MacVerna.—See Stark MacVerna Delicious.

Salsa-Pride.—Orig. in Visalia, Calif., by John R. Salsa. Introd. in 1958. Plant patent 1695; 15 Apr. 1958. Parentage unknown, but considered to be an open-pollinated seedling of Late Santa Rosa; discovered in 1955. Fruit: Japanese type; about 3 inches long, 2 1/4 to 2 1/2 inches in diam.; colors similarly to Late Santa Rosa but about 4 to 5 days earlier; flavor different, pit smaller, and fruit flatter than that variety; flesh very firm, tinged with yellow and red; clingstone; does not split; matures 10–20 July; keeps very well. Tree: apparently identical to Late Santa Rosa.

Santa Rosa-Two.—Orig. at Fresno, Calif., by J.M. Garabedian. Introd. in 1987. Plant patent 5904; 17 Mar. 1987. Chance seedling. Fruit: Japanese type, medium to large, 60 mm wide and 59 mm long; broadly ovate, slightly asymmetrical; skin purplish-red with medium blue-gray bloom; flesh amber, firm to meaty; quality good to excellent; clingstone. Ripens in late June, 5 days before Santa Rosa. Tree: very productive; vigorous.

Saskatchewan.—Orig. in Saskatoon, Saskatchewan, Canada, by C.F. Patterson, Univ. of Saskatchewan. Introd. in 1960 for home gardens. Derived from *Prunus nigra* and *P. salicina* ; selected in 1953; tested as Sask. P53-13. Fruit: spherical, up to 1 1/4 inches under nonirrigated orchard conditions at Saskatoon; skin blushed; flesh moderately firm; quality good; ripens during the first half of September. Tree: hardy.

Satsuma Improved (*Improved Satsuma*).—Orig. apparently in Visalia, Calif., by Hathaway Nursery. Introd. about 1946. Parentage unknown; discovered about 1944. Fruit: larger than Satsuma; flesh firmer than Mariposa, flavor better than Ace; ripens in late June and early July in the San Joaquin Valley of California and before Satsuma, which it resembles. Tree: vigorous; consistent producer; blooms with Elephant Heart.

Scarlet Sun.—Orig. in Del Rey, Calif., by Tom Chamberlin, H.P. Metzler & Sons. Introd. in 1986. Bud mutation of Roysum Plum. Plant patent 6200; 14 June 1988; assigned to Metropolitan Insurance Co.; licensed exclusively to Agri Sun Nursery, Selma, Calif. Fruit: resembles Rosemary; large; round, slightly elongated; skin deep, bright red, slightly rough; flesh moderate orange-yellow; slightly tart; ripens 2 weeks after Rosemary. Excellent cold storage and shipping qualities. Tree: vigorous; productive; requires cross-pollination.

Segundo.—Orig. at Byron, Ga., by J.M. Thompson, U.S. Dept. of Agriculture S.E.

Fruit & Tree Nut Res. Lab. Introd. in 1985. (Queen Ann x Santa Rosa) x (Ozark Premier x *Prunus angustifolia*); seed collected in 1967; selected in 1973; tested as BY68-971. Fruit: medium; skin color changes from greenish-yellow to yellow red as it ripens; round; flesh yellowish-red, firm when green; clingstone; fair quality. Ripens 1 week after Santa Rosa, 10 days before Bruce. Tree: very vigorous on peach rootstock; spreading; very resistant to bacterial spot and bacterial canker; tolerant of plum leaf scald and brown rot. Adapted to the humid Southeast. Released for use as a green plum similar to Bruce. Flowers large, two per cluster; self-incompatible; blooms about 3 days before Santa Rosa.

Seneca.—Orig. in Geneva, N.Y., by N.Y. Agr. Expt. Sta. Introd. in 1972. Italian Prune x Prinlew; cross made in 1937, selected in 1949, tested as N.Y. 981. Fruit: large; oval; skin reddish-purple, attractive, slight cracking may occur during heavy rains at or near maturity; flesh amber, firm, fine-grained; quality good; ripens 1–10 Sept., extended ripening period, hangs well; recommended for home gardens and local market. Tree: slightly susceptible to brown rot.

Severn Cross.—Orig. in Bristol, England, by G.T. Spinks, Univ. of Bristol, Long Ashton Res. Sta. Introd. in 1932. Coe's Golden Drop x Giant; selected about 1927. Fruit: large; oval; skin golden with some red spots, slight bloom; flesh yellow, very juicy, sweet, flavor good, clingstone; dessert quality good; ripens in late September; hangs for a good period. Tree: upright; vigorous; self-fruitful. Royal Hort. Society Award of Merit, 1951.

Sharkey.—Orig. in Vacaville, Calif., by Millard Sharpe. Introd. in 1920. Open-pollinated seedling of Burbank. Fruit: round oval; skin light to medium greenish-yellow, with a dullish to deep red blush; flesh light amber, sweet, mild; quality fair; clingstone; ripens during first 2 weeks of August.

Sharron's.—Bud sport of Grand Rosa found in Sanger, Calif., by Giovanni B. Bubani; assigned to R.E.U.S., Pruner A.G.I, Pruner A.G.II, and Agri-Cal. Plant patent 6380; 8 Nov. 1988. Fruit: globose to ovate; large size, 58 to 61 mm; skin dark red, resistant to cracking; flesh yellow, resistant to heat damage; stone semi-free. In central San Joaquin Valley, ripens 4 days before Casselman, about 26 July to 18 Aug. Blossom: large, white; anthers light yellow; midseason. Tree: vigorous; upright.

Showtime.—Orig. in Modesto, Calif., by Eric Wuhl. Open-pollinated seedling of Santa Rosa. Plant patent 8037; 24 Nov. 1992. Fruit: size large, 58 to 62 mm; globose to ovate; skin red-purple; flesh yellow-red, firm; freestone. In central San Joaquin Valley, ripens with Santa Rosa, about 18 to 25 June. Blossom: small, white. Tree: upright.

Sierra.—Orig. in Ontario, Calif., by Herbert C. Swim, Armstrong Nurseries. Introd. in 1955. Plant patent 1482; 12 June 1956; by Herbert C. Swim; assigned to Armstrong Nurseries, Ontario, Calif. Inca x Santa Rosa; selected in 1949. Fruit: Japanese type; firmness and sweet flesh of Inca; roundness and coloring of Santa Rosa; flesh yellow; semi-freestone; good eating and keeping qualities; ripens 24 June–25 July. Tree: vigor of Santa Rosa; fairly self-fruitful at place of origin.

Simka (*Sim-Ka Rose*).—Orig. in Fowler, Calif., by Luke Kazarian. Introd. in 1959. Plant patent 1882; 1 Dec. 1959; by Coche D. Simonian and Luke Kazarian. Chance seedling; discovered in 1956. Fruit: large; uniform; suture very shallow; shape of Eldorado, but apex more pointed; skin ebony; flesh yellowish-white, sweet, texture very firm; freestone at full maturity; ripens from 15 July to 1 Aug. at place of origin; keeps well on tree until about 15 Aug. for local shipments; ships and stores well; resembles Nubiana. Tree: size medium; vigorous; regular productivity; partially self-thinning.

Sim-Ka Rose.—See Simka.

Songold.—Orig. in Stellenbosch, Cape Province, South Africa, by the Fruit & Food Tech. Res. Inst., Dept. of Agr. Tech. Serv. Introd. in 1972. Golden King x Wickson. Cross made in 1961; selected in 1968, tested as OR 5/14/76. Fruit: large, about 5 cm in diam.; roundish oval; skin greenish yellow at first ripening to maroon red; stem short, thick; flesh golden yellow; quality excellent, flavor sweet and pleasing; keeps well in storage even at 31F; ripens second to third week of February. Tree: upright, but less so than Santa Rosa; vigorous; productive; resistant to delayed foliation; leaves glossy, tendency to be slightly turned downward in width and more tolerant to leaf spot than most Japanese plum varieties; flowers white, numerous, self-incompatible, Santa Rosa recommended as pollinator. Recommended for areas prone to bacterial spot infection and as a shipping variety.

South Dakota.—Orig. in Brookings, S.D., by N.E. Hansen, South Dakota Agr. Expt. Sta. Introd. in 1949, by the Minnesota Fruit Breeding Farm, Excelsior, Minn., with the approval of the South Dakota Agr. Expt. Sta.; tested as S. Dak. 27. Parentage unknown but an American–Japanese hybrid; selected before 1907. Fruit: size medium; quality good; freestone. Tree: excellent pollinizer for American–Japanese hybrid varieties; very hardy; productive; excellent parent for breeding hardy, freestone plums.

Splendid.—Orig. in Faribault, Minn., by Brand Peony Farms. Introd. in 1930. Chance seedling. Fruit: American type; large, averaging 1 1/2 to 1 3/4 inches in diam.; nearly round; skin bright red with numerous white dots, thin; flesh amber, firm, meaty, flavor fine; pit small. Tree: rather upright.

Spring Beaut.—A bud sport of Red Beaut found by Chris Zaiger in Modesto, Calif. Plant patent 4394; 13 Mar. 1979. Fruit: size medium, 54 mm; globose to ovate; skin red; flesh yellow; clingstone. In central San Joaquin Valley, ripens 2 or 3 days before Red Beaut, about 1 to 8 June. Blossom: white, medium. Tree: vigorous; upright.

Stanley.—Orig. in Geneva, N.Y., by Richard Wellington, New York State Agr. Expt. Sta. Introd. in 1926. Agen x Grand Duke; cross made in 1912; seed germinated in 1913. Fruit: prune type; appearance attractive; ripens 1 week earlier than Italian Prune; better in fruit and tree than Italian Prune. Tree: more reliable in bearing and more adaptable than Italian Prune.

Stanley Improved.—See Krause.

Stark Blue Ribbon.—Orig. at Bothell, Snohomish County, Wash., by Susan A. Aldrich. Plant patent 4053; 10 May 1977; assigned to Stark Brothers Orchards & Nurseries Co., Louisiana, Mo. Discovered in 1940 as a seedling growing among Italian Prune trees. Fruit: closely resembles Italian Prune with the exception that it has increased foliage resistance to fungal leaf spot and bacterial leaf spot, including *Coccomyces* leaf spot; fruit is larger, ripens earlier and is of better flavor than Italian Prune. Tree: similar to Italian Prune but more spreading.

Starking Delicious.—Orig. in Oklahoma City, Okla., by Ruth S. Johnson. Introd. in 1951. Plant patent 1064; 1 Jan. 1952; assigned to Stark Bros. Nurseries & Orchards Co., Louisiana, Mo. Parentage unknown; discovered as a seedling in 1931. Fruit: large; skin dark red, tough, astringent; flesh blood red, firm, flavor good; clingstone; suture diam. about 2 inches; ripens the 1 Aug. Tree: very productive; highly resistant to bacterial canker and leaf spot.

Stark MacVerna Delicious (*Sage MacVerna*).—Orig. in Ohaupo, Hamilton, New Zealand, by Ernest W. Sage. Introd. in 1966 by Stark Bros. Nurseries & Orchards Co., Louisiana, Mo. Plant patent 2571; 16 Nov. 1965; assigned to Stark Bros. Nurseries & Orchards Co. Chance seedling of *Prunus salicina*, Japanese type; selected in 1938. Fruit: large, about 2 1/2 inches in diam.; broadly ovoid, sides unequal; suture inconspicuous; apex depressed; stem stout, about 3/4 inch long; skin reddish-purple, thin, tender, bitter; flesh reddish-purple, melting, coarse, sweet, aromatic, flavor delicate, juice abundant; quality high; stone small, semi-freestone; keeping and shipping qualities good; for dessert, market, culinary, canning purposes; ripens end of January at place of origin; resembles Maynard; resistant to bacterial leaf spot. Tree: large; vigorous; very productive; hardy; resembles Maynard and Purple King.

Star Rosa (*Early Santa Rosa*).—Orig. in Di Giorgio, Calif., by Elmer Stark, Di Giorgio Fruit Corp. Introd. in 1950. Plant patent 995; 28 Nov. 1950; assigned 15 July 1948, to the Di Giorgio Fruit Corp., San Francisco, Calif. Bud mutation of Santa Rosa, a spontaneous tetraploid chimera, the first reported in the diploid Japanese-type plums, derived in part from *Prunus salicina*. Fruit: larger, more flattened, coarser flesh than the parent; ripens 1 week earlier than Santa

Rosa, with or before Beauty; better keeping quality than parent, which it resembles. Tree: very vigorous; blooms 2 to 4 days after parent; apparently self-incompatible, satisfactory pollinators are Redheart, Inca, Myrobalan 5Q, possibly Elephant Heart, Nubiana, Beauty, Queen Ann.

Stevens.—Orig. in Valley River, Manitoba, Canada, by W.J. Boughen, Boughen Nurseries. Introd. about 1930; seedling of a native plum (*Prunus americana*). Fruit: size medium; skin yellow with a red blush; quality only fair, somewhat astringent; season late August. Tree: spreading; size medium; productive; very hardy.

Street.—Orig. in Chilliwack, B.C., Canada, by H.M. Eddie & Sons, Vancouver, B.C. Introd. about 1928 by the above concern. Chance seedling of a European plum (*Prunus domestica*); discovered in 1922. Fruit: large; round, tapering at stem end; flesh firm, juicy, flavor fine, freestone; ripens in early August in area of origin.

Summer Treat.—Orig. in Chico, Calif., by Elmer Nilsson; assigned to Nancy Nilsson. Unknown origin; purchased as a budded Santa Rosa tree. Plant patent 8693; 19 Apr. 1994. Fruit: round to oval; medium size, 49 to 54 mm; skin light apple-red to deep mahogany; flesh amber-yellow, firm, dense. Ripens 2 to 4 weeks after Santa Rosa. Can remain on the tree until September. Blossom: medium, white; anthers yellow-gold. Tree: vigorous; upright to upright-spreading; good production; self-fruitful.

Superb.—Orig. in Saskatoon, Saskatchewan, Canada, by C.F. Patterson, Univ. of Saskatchewan. Introd. in 1960 for home gardens. Derived from *Prunus nigra* and *P. salicina* ; selected in 1954; tested as Sask. P55-29. Fruit: up to 1 3/4 inches long and 1 1/4 inches in diam. under nonirrigated orchard conditions in Saskatoon; skin dark red; quality excellent; ripens early to midseason; keeps well in cold storage. Tree: hardy.

Superior.—Orig. in Excelsior, Minn., by Univ. of Minnesota Fruit Breeding Farm. Introd. in 1933. *Prunus salicina* 'Burbank' x (*P. americana* x *P. simonii*) 'Kaga'; selected in 1925; tested as Minn. 194. Fruit: large; conic; skin dark red with russet dots, with heavy, attractive bloom; flesh yellow, firm, dessert quality superior, juicy, flavor sprightly, slightly acid near skin; clingstone; ripens in late midseason. Tree: prolific bearer; vigorous; comes into bearing early.

Super Rosa.—See Rosa Grande.

Suplumeleven.—See Black Diamond™.

Suplumtwelve.—See Black Gold™.

Suplumthirteen.—See Black Torch™.

Suplumfourteen.—See Sweet Rosa™.

Suplumfifteen.—See Black Flame™.

Suplumsixteen.—See Black Premium™.

Suplumseventeen.—See Tokay Red™.

Suplumeighteen.—See Red Giant™.

Suplumtwenty.—Orig. in Madera County, Calif., by John H. Weinberger; assigned to Sun World. Suplumeleven x Queen Rosa. Plant patent 8363; 7 Sept. 1993. Fruit: oblate; 64 mm; skin red-black; flesh yellow; clingstone. In central California, ripens 2 weeks before Suplumeleven, about mid- to late June. Blossom: white, medium; anthers yellow with shadings of red. Tree: medium vigor; upright; very productive.

Supreme.—Orig. in Saskatoon, Saskatchewan, Canada, by C.F. Patterson, Univ. of Saskatchewan. Introd. in 1960 for home gardens. Derived from *Prunus nigra* and *P. salicina* ; selected in 1947; tested as Sask. P58-103. Fruit: up to 2 inches in diam. under nonirrigated orchard conditions at Saskatoon; skin blushed; quality excellent; ripens in late August. Tree: medium upright and only slightly spreading; moderately vigorous; no trace of winter injury at Saskatoon.

Sweet Delight.—Orig. in Reedley, Calif., by Jonathan P. Chakerian. Casselman x unknown pollen source. Plant patent 6824; 30 May 1989. Fruit: elliptical; medium to small size, 54 mm; skin dark red; flesh yellow; stone semi-free to cling. In central San Joaquin Valley, ripens 3 weeks before Roysum, about 8 to 20 Sept. with 19% to 21.5% sugar. Blossom: medium, white-yellow; blooms 7 to 8 days after Casselman. Tree: vigorous; upright; very productive; appearing self-fertile.

Sweet Gypsy.—Orig. in Reedley, Calif., by James W. Taylor; assigned to Ito Packing Co. Mariposa open-pollinated. Plant patent 6816; 23 May 1989. Fruit: round; medium to small size, 54 mm; skin reddish-purple;

flesh red, very sweet with pleasant acid; clingstone. In central San Joaquin Valley, ripens 1 week before Mariposa, about 28 June to 10 July. Blossom: medium, white; anthers yellow; blooms with Santa Rosa. Tree: vigorous; spreading; productive.

Sweetheart.—Orig . in Oregon City, Ore., by L.L. McGraw. Introd. in 1976; assigned to Men's Garden Club of Portland. Chance seedling, European type. Fruit flesh golden color, juicy, meaty, sweet, aromatic; quality good; freestone. Tree: precocious; productive; moderately vigorous; upright-spreading; probably self-fruitful.

Sweet Heart.—Orig. in Reedley, Calif., by Jonathan P. Chakerian. Casselman x unknown pollen source. Plant patent 5724; 22 Apr. 1986. Fruit: 54 to 56 mm; ovate; skin dark reddish-purple; no observed tendency to crack; flesh amber with red under the skin, firm; semi-freestone. Ripens with Casselman. Blossom: light pink. Tree: medium vigor.

Sweet Queen.—Bud sport of Queen Ann found in Kingsburg, Calif., by David C. Vart. Plant patent 8557; 25 Jan. 1994. Fruit: slightly elongated; large, 58 to 61 mm; skin bright green; flesh light yellow-green, firm, resistant to bruising, mild flavor; clingstone. In central San Joaquin Valley, ripens 1 to 5 July. Blossom: white. Tree: very vigorous.

Sweet Rosa™ (*Suplumfourteen*).—Orig. in Fresno, Calif., by John H. Weinberger. Introd. in 1984. Plant patent 5342; 20 Nov. 1984; assigned to Superior Farming Co. Open-pollinated seedling of Angeleno (most probably pollinated by Midsummer); tested as 193-65. Fruit: 5 to 10 mm larger than Santa Rosa, which it most nearly resembles; skin red to purple, with prominent lenticels; more ovate than Santa Rosa; flesh dark red; medium firm to soft; clingstone. Fruit attain very high sugar levels; eating quality excellent. Tree: medium vigor; upright; medium productive; average precocity; midseason blossoming.

Sweet Yosemite.—Orig. in Oakland, Calif., by Franklin B. Roche. Introd. in 1960. Plant patent 1867; 22 Sept. 1959; assigned to Stribling's Nurseries, Merced, Calif. Chance seedling; discovered in 1953. Fruit: large, averaging 2 1/4 inches in diam.; skin purplish-blue; flesh firm, yellow; juice sweet and abundant; pit small, freestone; ripens 6 to 10 days later than Santa Rosa and 6 to 10 days before Satsuma; resembles Eldorado. Tree: size medium; vigorous; productive and regular bearer.

Tawena.—Orig. in Brookings, S.D., by N.E. Hansen, South Dakota Agr. Expt. Sta. Introd. in 1924. *Prunus salicina* 'Apple' x *P. americana* 'Terry'; sibling of Kahinta and Waneta. Fruit: not as large as Waneta; round; canning quality better than Waneta. Tree: spreading; less hardy than Waneta; very productive.

Tecumseh.—Orig. in Brookings, S.D., by N.E. Hansen, South Dakota Agr. Expt. Sta. Introd. about 1923. Shiro x Surprise; cross made in 1918. Fruit: size medium; round ovate; skin bright red, bloom bluish; flesh yellow, firm, juicy, subacid; quality good; ripens in mid-August.

Teme Cross.—Orig. in Bristol, England, by G.T. Spinks, Univ. of Bristol, Long Ashton Res. Sta. Introd. in 1932. Open-pollinated seedling of Heron; selected about 1927. Fruit: size medium; oval; skin golden with red spots; flesh yellow, very juicy, sweet, good Gage flavor; dessert use; ripens in late August. Tree: vigorous and productive.

Thames Cross.—Orig. in Bristol, England, by G.T. Spinks, Univ. of Bristol, Long Ashton Res. Sta. Introd. in 1946. Coe's Golden Drop x Giant; selected about 1938. Fruit: large; oval; skin pale yellow; flesh very juicy, sweet; dessert use; ripens in early-mid-September. Tree: crops well; self-fruitful.

Tiorn Sladky.—Orig. in Russia. Received by Canada Dept. of Agr., Res. Sta., Morden, Manitoba in 1939; received by Univ. of Minnesota, Agr. Expt. Sta., Excelsior. 1943. *Prunus domestica*. Fruit: 7/8 x 1 inch in diam.; roundish; skin purple to blue; heavy bloom, tender; flesh green, meaty, dry, sweet; ripens early August. Tree: upright; vigorous; productive; not hardy; moderately susceptible to leaf spot. Variety considered obsolete at Manitoba.

Tokay Red™ (*Suplumseventeen*).—Orig. in Fresno, Calif., by John H. Weinberger and Tim P. Sheehan; assigned to Superior Farming Co. Queen Rosa open-pollinated. Plant patent 7355; 16 Oct. 1990. Fruit: globose to oblate; large, 60 mm; skin dark purple-black; flesh yellow, juicy, average to fine; clingstone. In central San Joaquin Valley, ripens with Black Beaut. Blossom: small, white, average bloom time. Tree: average vigor; spreading; productive.

Tonka.—Orig. in Excelsior, Minn., by the Univ. of Minnesota Fruit Breeding Farm. Named in 1920; introd. in 1921. *Prunus salicina* Burbank x *P. americana* Wolf; cross made in 1908; tested as Minn. 21; sibling of Mendota, Mound, Redcoat, Red Wing. Fruit: large; roundish; small apex pointed; skin medium thick, strong, easily removed from flesh when fully ripe, medium light red, attractive, bloom medium heavy, inconspicuous dots small and numerous; flesh yellow, firm, tender, juicy, sweet; good to very good; stone roundish, small, free when fully ripe; ripens in midseason; ships well. Tree: heavy and reliable producer; size medium; rather spreading; tends to be short-lived.

Trailblazer.—Orig. in Portland, Ore., by Mildred M. Smith. Introd. in 1955. Plant patent 1586; 2 Apr. 1957, by Smith; assigned to Samuel J. Rich, Rich & Sons Nursery, Hillsboro, Ore. Shiro x *Prunus cerasifera* var. *atropurpurea* (purple-leafed plum); cross made in 1947; selected in 1952. Fruit: skin cherry red; flesh apricot orange, slightly more acid than Shiro but sweeter than purple-leafed parent; ripens about 1 week later than Shiro. Tree: growth habit similar to Shiro; flowers white at anthesis, with prominent stamens; leaves when young red on both sides, at maturity dark green on upper surface, red on lower surface.

Tulare.—Orig. in Vacaville, Calif., by Millard Sharpe. Introd. about 1922. Open-pollinated seedling of Tragedy. Fruit: oval; skin purple; flesh greenish-yellow, quality fair to poor; freestone; ripens during last week of July into first week of August.

Twilite.—Orig. in Mountain Grove, Mo., by Paul H. Shepard, Missouri State Fruit Expt. Sta. Introd. in 1947. Open-pollinated seedling of Black Beauty; good probability that Santa Rosa was the pollen parent; considered to be a Japanese–American hybrid; tested as PL-97. Fruit: medium to large; skin red with light, blue bloom; flesh yellow tinged with red, fine-grained, juicy, sweet, flavor good, clingstone; very late ripening, about 10 Sept., remaining firm on tree for 3 to 4 weeks; ships well. Tree: fairly vigorous, resembling Japanese sp.; large crops produced.

Underwood.—Orig. in Excelsior, Minn., by the Univ. of Minnesota Fruit Breeding Farm. Named in 1920 and introd. in 1921. Shiro x *Prunus americana* Wyant; cross made in 1911; selected in 1916; tested as Minn. 91. Fruit: Japanese type; large; round conic to conic; skin dull red, but rich, satin-like, somewhat astringent; flesh golden yellow, juicy, melting, tender, slightly stringy, sweet, quality good; stone long, small, semi-cling; ripens early, in July, extends over a long season. Tree: vigorous; bears regularly and heavily; very hardy; pollinator required.

Valor.—Orig. in Vineland Station, Ont., Canada, by G.H. Dickson, G.W. Eaton, C.D. Whitty, and G. Tehrani, Ontario Dept. Agr., Hort. Res. Inst. Introd. in 1967. Imperial Epineuse x Grand Duke; selected in 1938; tested as V33021. Fruit: European; medium large, diam. 20 mm, length 41 mm; suture moderate; skin dark purple, speckled, attractive; flesh prune type, greenish-gold; semi-freestone; ripens 24 Sept. in Ontario, with Italian Prune or just ahead of it; recommended as fresh fruit variety. Tree: vigorous; productive; foliage dark.

Vanderbilt.—Orig. in Santa Rosa, Calif., by Luther Burbank. Introd. about 1931 by Stark Bros. Nurseries & Orchards Co., Louisiana, Mo. Parentage unknown. Fruit: large; egg-shaped; skin red; flesh yellow, firm, sweet, clingstone; matures in midseason. Tree: productive.

Vanier.—Orig. in Ottawa, Ont., by M.B. Davis, Div. of Hort., Central Expt. Farm. Introd. in 1983 and named as a replacement for Burbank by G. Tehrani, Ontario Ministry of Agr. & Food, Hort. Res. Inst. of Ontario, Vineland Sta. Burbank x Wickson; cross made in 1930s; tested as Ottawa 3010. Fruit: medium, round; 46 mm long and 49 mm wide; skin bright red; flesh orange-yellow, quality good; clingstone; ripens about 29 Aug., with Burbank and Ozark Premier. Tree: productive, precocious, vigorous with upright growth; self-unfruitful, pollinated by Burbank, Myrobalan B, Ozark Premier, and Santa Rosa. Only variety that will pollinate Ozark Premier in Ontario trials at HRIO.

Veeblue.—Orig. in Vineland Station, Ont., Canada, by G. Tehrani and G.H. Dickson, Ontario Ministry of Agr. and Food, Hort. Res. Inst. of Ontario. Imperial Epineuse x President; selected in 1938. Fruit: European plum medium (54 to 41 mm); skin blue; flesh orange, quality fair; semi-freestone; ripens 10 Sept., the same as Iroquois and Early Italian Prune. Tree: productive, precocious, semi-vigorous, and flat spreading; self-unfruitful, pollinated by Italian Prune, Verity, California Blue, Stanley, Iroquois, Vision, Valor, and Bluefre.

Verity.—Orig. in Vineland Station, Ont., Canada, by G.H. Dickson, G.W. Eaton, C.D. Whitty, and G. Tehrani, Ontario Dept. Agr., Hort. Res. Inst. Introd. in 1967. Imperial Epineuse x Grand Duke; selected in 1938; tested as V310110. Fruit: European; size medium, diam. 27 mm, length 36 mm; suture moderate; skin blue, attractive; flesh prune type with orange color; freestone; ripens 28 Sept. in Ontario, with Italian Prune; recommended for processing and fresh market. Tree: productive; hardy; flat spreading habit.

Victor Large French.—Orig. in Madrone, Calif., by Victor Locarnini, Jr., Morgan Hill, Calif. Introd. in 1970. Bud mutation of Agen (French) prune. discovered in 1957. *P. domestica*. Fruit: resembles Agen (French), distinguished by larger size; ripens 15 Aug.–15 Sept. Tree: size medium, spreading; vigor moderate; hardy; very productive and sets more regularly than French Prune; tolerant to mites. Recommended as a replacement for French Prune.

Victory.—Orig. in Vineland, Ont., Canada, by G. Tehrani and W. Lay, Hort. Res. Inst. of Ontario. Vision x Valor. Cross made in 1972; selected in 1980; tested as V72331; introd. in 1992. Fruit: European; large, 62 mm long, 52 mm wide; 59 g; heart-shaped. Skin violet-blue with waxy bloom, to black without waxy bloom. Flesh greenish-yellow, firm, juicy, fair to good flavor; semi-clingstone; medium ovate pit. Ripens with Bluebell, 4 or 5 days after Stanley. Tree: similar to Stanley; self-incompatible; Italian Prune, Stanley, Verity, Vision, and Voyageur are suitable pollinizers.

Viking Pride.—Bud sport of Durado found in Kingsburg, Calif., by Joseph W. Satterstrom. Plant patent 6712; 4 Apr. 1989. Fruit: oblate; medium to large, 49 to 60 mm; skin dark red; flesh yellow, firmer than Durado; clingstone. In central San Joaquin Valley, ripens 10 days after Durado, about 8 to 13 June. Tree: vigorous; upright.

Vision.—Orig. in Vineland Station, Ont., Canada, by G.H. Dickson, G.W. Eaton, C.D. Whitty, and G. Tehrani, Ontario Dept. Agr., Hort. Res. Inst. Introd. in 1967. Pacific x Albion; selected in 1944; tested as V37011. Fruit: European; large, diam. 32 mm, length 45 mm; suture moderate; skin blue; flesh prune type, yellow, quality very good, freestone; ripens with Reine Claude1 Oct., in Ontario; recommended as a fresh fruit variety. Tree: very vigorous; productive; spreading habit.

Voyageur.—Orig. in Vineland Station, Ont., by G. Tehrani, Hort. Res. Inst. of Ontario. Introd. in 1987. Open-pollinated seedling of Ruth Gerstetter; tested as V60322. Fruit: European; small, 39 mm wide and 44 mm long; elliptical; skin dark purple with bloom removed, violet purple with bloom intact; flesh yellow, firm, juicy; quality good; semi-clingstone. Ripens in late August, 1 week later than California Blue. Tree: productive; precocious; semi-vigorous, as Stanley. Self-fruitful; tends to overbear and benefits from fruit thinning; good pollinizer for Bluebell, Early Blue, Fellenberg, and Verity. Resistant to European red mite; low susceptibility to black knot and bacterial canker.

Vroege Tolse.—Orig. in Opheusden, The Netherlands, by Joh. van Dam, The Inst. of Hort. Plant Breeding, Wageningen, The Netherlands. Introd. in 1960. Parentage unknown; selected in 1958. Fruit: European; large; skin dark red to blue; quality very good; freestone; ripens very early. Tree: bears very well.

Waconia.—Orig. in Excelsior, Minn., by the Univ. of Minnesota Fruit Breeding Farm. Named and introd. in 1923. *Prunus salicina* 'Burbank' x *P. americana* 'Wolf'; cross made in 1908; selected in 1914; tested as Minn. 10. Fruit: size medium to large; long ovate conic, apex pointed; skin dark red, waxy, thick; flesh yellow, firm, moderately juicy, somewhat fibrous, mildly sweet; quality fair; stone large, cling; ripens in midseason.

Tree: uniformly productive; hardy. No longer propagated.

Wade.—Orig. in Clemson, S.C., by M.B. Hughes, Clemson Univ. Introd. in 1974. Beauty x Ozark Premier. Fruit: Japanese type; 1 3/4 to 2 1/4 inches in diam.; round; skin red; flesh yellow with red near skin, medium firm; ripens 11–23 June; resistant to cracking, brown rot, bacterial leaf spot and canker; self-infertile, pollinated by Methley or Bruce. Tree: large, upright, vigorous, productive.

Weatherspoon.—Orig. in Elgin, Ore., by H.H. Weatherspoon. Introd. about 1920. Parentage unknown. Fruit: European; large; skin moderately thick, tough; flesh firm, sweet, aromatic; good shipping quality; ripening 1 week ahead of Italian Prune, which it resembles. Tree: somewhat thorny, heavily branched; blooms heavily, but low productivity; not hardy; requires cross-pollination by other Italian-type varieties.

Weeping Santa Rosa.—Orig. in Tudor, Sutter County, Calif., by George R. McFeely, Oswald Nursery, Yuba City, Calif. Introd. in 1969. Plant patent 2938; 11 Nov. 1969. Bud mutation of Santa Rosa. Fruit: 1 5/8 to 2 inches in diam.; uniform, slightly oblate; skin yellow, substantially covered with red, many small yellow dots, blue-gray bloom; flesh pale golden yellow, juicy, flavor mild, firm, texture medium to fine, quality good; semi-freestone. Tree: resembles Santa Rosa, but characterized by a weeping habit of growth, with long slender limb; a good ornamental.

Westerner.—Orig. in Fresno, Calif., by John M. Garabedian. Open-pollinated seedling of unknown origin. Plant patent 5609; 24 Dec. 1985. Fruit: ovate to globular; large, 66 mm; skin dark red; flesh light creamy-white, becoming full red when ripe; stone semi-free. In central San Joaquin Valley, ripens with Laroda, about 15 July. Blossom: medium small, white; anthers yellow. Tree: vigorous; upright-spreading.

Wilbur.—Orig. in Valley River, Manitoba, Canada, by W.J. Boughen, Boughen Nurseries. Introd. in 1920. Selected in 1919 from seed of *Prunus americana*, secured from N.E. Hansen, South Dakota Agr. Expt. Sta., Brookings. Fruit: large; skin yellow with red splashes; flesh with no astringency; ripens after Mammoth.

Wilmore.—See Golden West.

Wilson River.—Orig. in Valley River, Manitoba, Canada, by W.J. Boughen, Boughen Nurseries. Introd. in 1920. Selection of a native plum. Fruit: large; skin red, attractive; flesh quality fair. Tree: very hardy; leaves large.

Winered.—Orig. in Dropmore, Manitoba, Canada, by F.L. Skinner, Skinner's Nursery. Introd. in 1955. *Prunus salicina* x (*P. besseyi* hybrid). Fruit: size medium; round; flesh dark purple; lacks the astringency of native American plums (*P. americana*); good cooked and for dessert purposes; resembles Damson when cooked. Tree: upright to 5 feet; yields well.

Winona.—Orig. in Excelsior, Minn., by the Univ. of Minnesota Fruit Breeding Farm. Named and introd. in 1922. *Prunus salicina* x *P. americana*; cross made in 1909 or 1910; tested as Minn. 30. Fruit: medium large; round; skin light red over greenish-yellow becoming fully red at maturity, easily removed from flesh, no astringency, dots small, bloom heavy; flesh yellow, very tender, fine-grained, juicy, deliciously sweet; quality very good; stone medium size, cling; ripens late. Tree: large; vigorous.

Wisconsin (*Roberts*).—Orig. in southern Oregon by F. Ross Brown, Albany, Ore. Introd. in 1939. Bud mutation of Italian Prune; discovered about 1909; budwood secured from the Oregon Agr. Expt. Sta. by R.H. Roberts, Wisconsin Agr. Expt. Sta., Madison; selected before 1937 by R.H. Roberts. Fruit: prune type; characteristics of Italian Prune. Tree: characteristics of Italian Prune except in Wisconsin it blooms several days after that variety; in Wisconsin blooms with Northwestern Greening apple; satisfactory crops produced in Midwest where Italian Prune is adapted and produces well.

Wuhl.—Orig. in Fresno, Calif., by Richard Peters and Eric Wuhl. Plant patent 5585; 3 Dec. 1985. Fruit: ovate; large size, 60 to 65 mm; skin purple; flesh red, little or no bitterness; stone semi-free, not prone to splitting like Black Beaut. In central San Joaquin Valley, ripens 3 days before Santa Rosa, about 15 June. Blossom: small, white. Tree: upright; productive; self-infertile.

Yakima.—Orig. in Bingen, Wash., by Theodore Suksdorf. Introd. about 1925 by

Washington Nurseries, Toppenish, Wash. Seedling of a 'Peach' plum; discovered about 1902; Washington Nurseries obtained buds of this tree in 1922. Fruit: very large; somewhat oblong; skin bright mahogany red; flesh yellow, tender, sweet, very firm; good quality; freestone. Tree: very shy bearer unless cross-pollinated with other European plums; vigorous; upright.

Ziegler.—Orig. in Carlton, Ore., by Charles Ziegler. Introd. in 1930. Open-pollinated seedling of Italian Prune (*Prunus domestica*); selected in 1919. Fruit: resembles Italian Prune, but is larger and ripens 10 days later.

not named.—Orig. by Luther Burbank, in Santa Rosa, Calif. Not introduced as still infected with virus. Plant patent 1041; 25 Sept. 1951; by Elizabeth Waters Burbank, Executrix, Santa Rosa, assigned to Stark Bros. Nurseries & Orchards Co., Louisiana, Mo. Fruit: medium large; skin thin, tough, plum-purple; flesh mild, sweet; freestone; ripens early. Tree: dwarf habit of growth; productive.

not named.—Orig. in Tulare, Calif., by Wilfred C. Swall. Not introd. commercially. Plant patent 2108; 5 Dec. 1961. Bud mutation of Santa Rosa. Fruit: ripens 3 weeks later than Santa Rosa; smaller than Santa Rosa and skin lighter in color, otherwise similar to that variety. Tree: similar to Santa Rosa.

not named.—Orig. in Tulare County, Calif., by R.L. Sumruld. Plant patent 3949; 14 Sept. 1976. Discovered as a sport limb growing from a Roysum tree. Fruit: similar to Roysum but ripens 16–20 Aug., about 12 days after Casselman and considerably earlier than Roysum. Tree: large; vigorous; upright; productive.

Rootstocks for Plums and Prunes

Citation.—See under Peaches. Reduces tree size by about 25%.

Eruni.—Orig. in Balsgard, Sweden, by E. Olden, Swedish Univ. of Agr. Introd. in 1987 by V. Trajokowski. Open-pollinated Pershore; selected in 1952; tested as BPR 32. European plums on Eruni make medium-sized, precocious tree. Compatible with wide range of *Prunus domestica* varieties. Readily propagated by layers or by hardwood cuttings . Root system is similar to that of Myrobalan B . Ungrafted tree is upright, semi-vigorous.

Ferciana.—See Ishtara®.

Fereley.—See Jaspi®.

GF 43.—Orig. at Pont-de-la-Maye, near Bordeaux, France, by R. Bernhard and Ch. Grasselly, INRA Res. Sta. Introd. in 1967. Open-pollinated seedling of French Prune; UPOV #708, assigned to Agri Obtention, Guyancourt, France. Clonal rootstock for hexaploid plums and for peaches. Very vigorous, comparable to peach seedling stocks; compatible with very broad spectrum of *P. domestica* varieties. Readily propagated by hardwood cuttings. Resistant to lime-induced chlorosis. More tolerant to waterlogging than peach seedling stocks. Resistant to crown gall; more resistant than Brompton to *Phytophthora*; susceptible to apple chlorotic leafspot virus. Winter hardiness doubtful. On an unbudded tree, flowers are white, self-fertile; fruit are blue with yellow-green flesh.

GF 557.—See under Peaches. Compatible with European and Japanese plum varieties.

GF 677.—See under Peaches. Incompatible with many varieties; compatible with Prune d'Ente, President, Imperial Epineuse, Fellenberg, and Coe's Golden Drop.

GF 1869.—See under Peaches; compatible with plums and prunes.

Hansen 536.—See under Almonds. Compatible with Japanese plums.

Hansen 2168.—See under Almonds. Compatible with Japanese plum varieties.

Ishtara® (*Ferciana, PPH2*).—Orig. in Pont-de-la-Maye, near Bordeaux, France, by R. Bernhard, INRA Fruit Res. Sta. Introd. in 1986 under the trademark Ishtara®. Belsiana plum x a natural plum–peach hybrid [F_1 (322 x 871)]; cross made in 1950; tested as PPH_2. Semi-dwarfing rootstock for almond, apricot, peach, plums, and prunes. Compatible with broad range of *P. domestica*, including Althan, Oullins, and California Blue. For some European plum cultivars, reduces tree size 15% to 20%; induces high productivity index and increased fruit size; scion foliage may be less intensely green than on other stocks. Fruit of Reine Claude and French Prune on this rootstock are larger than when grown on

Myrobolan B rootstocks. Nonsuckering. Propagated readily as hardwood or semi-hardwood cuttings. Sensitivity to winter waterlogging similar to that of peach seedling, greater than Marianna GF 8-1; anchorage may be weak in such conditions. Tolerance to *Armillaria rnellea* better than peach seedling stocks. Resistant to root-knot nematodes. As ungrafted tree, upright, precocious, well-anchored; partially female sterile; scarce fruit are intermediate between peach and plum, clingstone.

Jaspi® (*Fereley*).—Orig. at Pont-de-la-Maye, near Bordeaux, France, by R. Bernhard, INRA Fruit Res. Sta. Introd. in 1989; UPOV applied for. Methley x *Prunus spinosa* (2069 x 106)4. Compatible with European plums, including Reine Claude d'Althan, and most peaches and nectarines. Semi-dwarfing for French Prune, Reine Claude, and Mirabelle de Nancy. Induces earlier bearing and more efficient production than does Marianna GF 8-1. Anchorage excellent; no suckers are produced; good resistance to waterlogging; drought-tolerant; resistant to lime chlorosis. Readily propagated by hardwood cuttings and in micropropagation. In the nursery, liners are vigorous, produce no spines, and take the bud well. As an unbudded tree, vigorous, semi-upright, few branches, no spines; rust resistant.

Marianna GF 8-1.—Orig. at Pont-de-le-Maye, near Bordeaux, France, by R. Bernhard, INRA Fruit Res. Sta. Introd. in 1965. Open-pollinated seedling of diploid Marianna; GF8-1 is a triploid, probably derived by fertilization of an unreduced ovule by *P. cerasifera.* UPOV #698; assigned to Agri Obtention, Guyancourt, France. Very vigorous. Compatible with all varieties of European plums, although unions with California Blue and Althan Green Gage may be weak. Relatively shallow rooting; well-adapted to calcareous soils. More tolerant to root asphyxia than is myrobalan. Very sensitive to nematodes. Not very winter hardy. On an unbudded tree, flowers white; fruit small, 35 mm; round; skin red on yellow ground color; flesh yellow; clingstone; very vigorous, bushy.

Maridon.—A tetraploid Marianna type orig. in Republic of South Africa from open-pollinated Marianna seeds planted by N. Hurter and M.J. van Tonder, FFTRI, in the 1960s. Tested as no. 8-6. Released in 1990 by J.D. Stadler, G.F.A. Lotze, and J.H. Acker. Compatible with all Japanese plums grown in South Africa. No incompatibility problems have been observed with the several *P. domestica* varieties tested. Incompatible under peaches. Apricots on Maridon break at the graft union. Size of plum trees grafted on Maridon is similar to or slightly smaller than those on Marianna. Precocity and productivity are similar to Marianna. Plum varieties grafted on Maridon appear to be much less susceptible to bacterial canker than those on Marianna. Maridon is propagated commercially by hardwood cuttings; rooting success is similar to that of Marianna.

Mirabi®—Orig. at Pont-de-le-Maye, near Bordeaux, France, by Ch. Grasselly, INRA Fruit Res. Sta. Introd. in 1980. Open-pollinated seedling of P2032 (*Prunus cerasifera*), selected in 1957. A medium-vigor clonal rootstock. Waterlogging tolerance better than peach seedlings. As unbudded tree, flowers white, self-sterile; fruit clingstone.

Myran® (*Yumir, PPH$_1$*).—See under Peaches. Compatible with some, but not all, *P. domestica.*

Myrobolan GF 31.—See under Apricots. Compatible with most plums and prunes.

Pixy.—Orig. in Maidstone, Kent, U.K., by A.B. Beakbane. Introd. in 1975. Plant patent 4061; 14 June 1977. Open-pollinated *Prunus insititia*; seed collected in France in 1947; tested as E.340.4.6. Semi-dwarfing rootstock for European plums. Propagated by hardwood cuttings or layers. Trees on Pixy are precocious; productive, but fruit size smaller than usual; almost free from suckering. Very sensitive to drought. As unworked plant, resistant to bacterial canker and to silver leaf.

***PPH$_1$*.**—See Myran®.

***PPH$_2$*.**—See Ishtara®.

St. Julien GF 655-2.—See under Peaches. Compatible with most *Prunus domestica* varieties.

St. Julien X.—Selected in New Zealand from English seed of St. Julien (*P. insititia*). Semi-dwarfing for Japanese plums and peaches. Some suckers produced. Shoots are thorny. Propagated by hardwood cuttings. Induces somewhat early leaf fall of scion.

***Yumir*.**—See Myran®.

PLUMCOT

BY69-1637P.—Orig. at Byron, Ga., by J.M. Thompson, U.S. Dept. of Agriculture S.E. Fruit & Tree Nut Res. Lab. Released in 1985. Open-pollinated seedling of BY4-997 x seedling apricot; BY4-997 = Nubiana open-pollinated. Seed collected in 1968; selected in 1973. Fruit: large; skin dark purple-black, very light pubescence; round; flesh apricot colored, very firm, freestone, tart. Ripens 2 weeks after Santa Rosa. Tree: vigorous on peach rootstock; upright; only moderately productive. Moderately resistant to bacterial canker and bacterial spot; tolerant of plum leaf scald and brown rot. Adapted to the humid Southeast. Flowers large, pink, solitary, self-fertile; fertile with both plum and apricot. Released for breeding purposes because of tree health, fruit size, and fertility with both plum and apricot.

Flavor Delight.—Orig. in Modesto, Calif., by Floyd, Gary, and Grant Zaiger and Leith Gardner. Introd. as a home garden variety in 1989. Apricot x plumcot. Plant patent 7090; 2 Jan. 1990. Fruit: medium; skin dull yellow; flesh yellow, firm, freestone, with good sugar/acid balance; flavor delicate; high sugar, with 4°Brix higher than standard; cultivars; early maturing. Tree: vigorous; large; semi-upright; regular production.

Flavor Queen.—Orig. in Modesto, Calif., by Floyd, Gary, and Grant Zaiger and Leith Gardner. Introd. in 1990. Plant patent 7420; 15 Jan. 1991; assigned to Zaiger's Genetics. Fruit: large; globose; skin yellow; flesh yellow; high Brix. Matures late July in California. Tree: vigorous; semi-upright; productive; self incompatible.

Flavor Supreme.—Orig. at Modesto, Calif., by Floyd, Gary, and Grant Zaiger and L.M. Gardner. Introd. as a home garden variety in 1989. An interspecific plum x plumcot hybrid, from a cross between two open-pollinated seedlings of Red Beaut. Plant patent 6763, 25 Apr. 1989. Fruit: medium, 50 to 56 mm wide and 53 to 56 mm long; globose to slightly flattened at stem end; skin violet brown to purplish gray with moderate to heavy bloom; flesh deep red, firm, moderately juicy; quality excellent; clingstone; ripens 13–20 June. Suitable for long-distance shipping. Tree: precocious and productive; semi-vigorous; upright.

Orange.—Orig. in Sebastopol, Calif., by Luther Burbank. Introd. in 1931 by Stark Bros. Nurseries & Orchards Co., Louisiana, Mo. Parentage unknown; selected about 1922. Fruit: skin yellow with reddish-purple cheek, pubescent; flesh yellow, quality good, flavor subacid, plum-type. No longer propagated.

Purple.—Orig. in Sebastopol, Calif., by Luther Burbank. Introd. in 1931 by Stark Bros. Nurseries & Orchards Co., Louisiana, Mo. Parentage unknown; discovered about 1922. Fruit: skin purple, slightly pubescent, with heavy bloom; flesh deep purple, quality good, clingstone. No longer propagated.

Red Velvet.—Orig. in Le Grand, Calif., by Norman and Glen Bradford. Unknown plum by unknown apricot. Plant patent 7011; 5 Sept. 1989. Fruit: oblong, compressed transversely; medium size, 44 mm; skin dark purplish-red, pubescent; flesh brilliant yellow, medium firm, flavor acidic unless very ripe, juicy; clingstone. In central California, ripens about 5 days before Red Beaut. Blossom: medium, white; blooms early; pollinated by Castlebrite apricot. Tree: vigorous; spreading; light productivity; self-unfruitful.

Royal Velvet.—Orig. in Le Grand, Calif., by Norman and Glen Bradford. Unknown plum by unknown apricot. Plant patent 7431; 29 Jan. 1991. Fruit: globose; medium size, 53 mm; skin dark purplish-red, pubescent; flesh light orange-yellow, very firm; sweet and acidic even if under-ripe; juicy; clingstone. In central California, ripens about with Red Beaut. Blossom: medium, white; blooms early; pollinated by Castlebrite and other apricots. Tree: vigorous; upright; uncertain productivity; self-unfruitful.

POMEGRANATE

Early Wonderful.—Orig. near Visalia, Calif., by Leonard K. Wileman, Wileman Bros. & Elliott, Cutler, Calif. Plant patent 3520; 12 Mar. 1974; assigned to Orange County Nursery, Norwalk, Calif. Introd. in 1979. Bud mutation of Wonderful; discovered in 1972. Fruit: large, 2 1/4 to 3 1/4 inches in diam., round-oblate, uniform; skin deep red, attractive, smooth, glossy; flesh firm; ripens mid-September, 2 weeks before Wonderful; can be harvested in one picking. Tree: size medium, spreading, drooping, round topped; vigorous, hardy, productive; tolerant to most diseases and insects.

Foothill Early.—Orig. in Porterville, Calif., by R.J. Owen. Plant patent 3884; 11 May 1976. Bud mutation of Wonderful. Fruit: resembles Wonderful, distinguished by brighter red color of its blossoms, earlier coloring and smaller size of its fruit, and pulp that is less tart and has a mildly acid taste; ripens last week of August and first week of September or 2 weeks earlier than Wonderful and 2 weeks after Granada. Tree: identical to Wonderful.

Foothill Late.—Orig. in Porterville, Calif., by R.J. Owen. Plant patent 3883; 11 May 1976. Bud mutation of Wonderful. Fruit: resembles Wonderful, distinguished by brighter red color of its blossoms, earlier coloring, and larger size of its fruit, and pulp that is more tart and has an acid taste; ripens the last week in September and first week in October, or 2 weeks after Wonderful and 6 weeks later than the Granada variety. Tree: identical to Wonderful.

Granada.—Orig. in Lindsay, Calif., by Mitchell J. Slayman. Introd. in 1966. Plant patent 2618; 5 Apr. 1966. Bud mutation of Wonderful; discovered in 1960. Fruit: resembles Wonderful, distinguished by deeper red color of its blossoms and regular bearing of fruit that displays a red crown while in the green stage, smaller fruit size; ripens 1 month earlier than Wonderful, darker red color and less tart. Tree: identical to Wonderful.

Early Red.—Orig. in Orange Cove, Fresno County, Calif., by Benjamin Y. Kamada. Plant patent 2723; 7 Mar. 1967. Bud mutation of Wonderful. Fruit: resembles Wonderful, being distinguished by its lower ratio of acid to the total soluble solids at an earlier maturity date, more uniformly rounded form, a thinner, brighter, darker colored rind surface resistant to splitting; ripens 3 weeks earlier than Wonderful. Tree: vigorous; regularly productive.

Early Red 2.—Orig. in Orange Cove, Fresno Country, Calif., by Benjamin Y. Kamada. Plant patent 5685; 11 Mar. 1986. Bud mutation of Early Red. Fruit: resembles Early Red, being distinguished by early maturity date, slightly smaller fruit size and more uniformly red fruit; ripens 7 to 8 weeks before Wonderful. Tree: identical to Early Red.

Eversweet.—Orig. in Camarillo, Ventura Country, Calif., by Sassin J. Chater. Plant patent 5418; 19 Mar. 1985. Chance seedling from a Lebanese cultivar. Fruit: skin and flesh have less red color than Wonderful; flesh sweet even when immature; rind thick; two to three crops produced each year between mid-August and October. Tree: size medium, spreading, drooping.

Fleshman.—Orig. in Fallbrook, San Diego County, Calif., by Paul H. Thomson. Introd. in 1970. Chance seedling. Disc. in 1950s. Fruit: large, 3 to 4 inches in diam. round; skin pink; flesh sweet; seeds soft; ripens in September. Tree vigorous and productive.

Sweet.—Orig. unknown. Fruit: skin lighter in color than Wonderful, remains slightly greenish with red blush even when ripe; flesh pink, much sweeter than other cultivars. Tree: very ornamental, precocious, productive.

Trauernicht.—Orig. in Fort Worth, Texas, by Erich C. Trauernicht, Trauernicht Nursery Co. Introd. in 1936. Plant patent 184; 18 June 1936; assigned to Baker Bros. Co., Fort Worth. Chance seedling; discovered in 1934. Fruit: round; skin dark red at maturity, thin; very acid; seeds bright crimson or pink, numerous. Tree: blooms all summer; semi-dwarf, up to 4 feet tall; produces heavily; hardy; branches willowy, drooping; foliage small, graceful; very ornamental.

Wonderful.—Orig. In Florida. First propagated in California in 1896; now the leading commercial variety. Fruit: large, deep purple-red, glossy; rind medium thick, tough; flesh deep crimson in color, juicy, tart; seeds small and tender; ripens mid to late September. Tree: size medium, spreading, drooping, round topped, vigorous, hardy, productive.

PRUNE

Lembrun.—Orig. in Le Temple Sur Lot, France, by G.H. Odier, Lot-et-Garonne, France. Plant patent 3862; 13 Apr. 1976; assigned to Dave Wilson Nursery, Hughson, Calif. Bud mutation of Agen. Fruit: similar to Agen but has a smaller stone and a higher sugar content; ripens 28 Aug., 8 to 10 days earlier than Agen. Tree: similar to Agen but more upright and more compact; for culinary, canning, and drying use.

PUMMELO

Chandler.—Orig. in Riverside, Calif., by James W. Cameron and Robert K. Soost, California Citrus Res. Center and Agr. Expt. Sta. Introd. in 1961. Cross of two pummelo (*Citrus grandis* Osbeck) introductions from S.E. Asia, P.I. 10126 (C.E.S. 2240) x P.I. 10127 (C.E.S. 2241) 'Kao Pan'; cross made in 1945 by H.B. Frost and J.W. Lesley. Fruit: similar to a large grapefruit, 4 1/2 to 6 inches in diam.; globose to oblate, no neck, slightly bilaterally asymmetric; rind light yellow, 10 to 12 mm thick, smooth except for small, raised oil glands, sometimes lightly pubescent, albedo adherent to flesh, oil odor not prominent; core solid, 6 to 10 mm in diam.; 14 to 16 segments, membranes medium thick; pulp pale yellow with variable pink tinge, firm, tender, vescicles medium to large; juice pale yellow, pleasing but weakly aromatic, medium in soluble solids and acid, having moderately high ratio in December–Janauary; picking ripe in December but lasts until March; seeds numerous, 12 to 15 mm long, flattened Tree: vigorous, spreading, only few thorns, fruit borne single or in pairs; leaves large, broad, petioles broad. Named in honor of William Henry Chandler, Professor of Horticulture, Emeritus, Univ. of California, Los Angeles.

US 145.—Orig. in 1947 as a seedling of *Citrus grandis* 'Thong Dee'; selected in 1981 by H.C. Barrett, U.S. Dept. of Agriculture, Orlando, Fla. Fruit: large size, 100 mm, 1300 g; globose; rind yellow, smooth, 12 mm thick; flesh light yellow, moderately firm and juicy; soluble solids 12% to 15%; acids 1.05% to 1.13%; midseason. Tree: some cold tolerance; recovers well from freeze damage.

QUINCE

Chaenomeles sp.

Arthur Colby (*Chaenomeles*).—Orig. in Urbana, Ill., by Arthur S. Colby, Illinois Agr. Expt. Sta. Introd. in 1961. Derived from *Chaenomeles lagenaria* and *C. japonica*; from seed collected at Arnold Arboretum; planted in 1923; selected about 1930; tested as Ill. 9; selected for introduction by H.R. Kemmerer and J.C. McDaniel of the above Station. Fruit: large, being larger than the average of *C. lagenaria*; obovate; skin smooth, greenish-yellow; ripens late; useful for jelly. Plant: bush grows to about 5 feet; upright, spreading; production heavy when cross-pollinated; flowers very showy, bright red (geranium lake shade), to 2 inches in diam., appearing in midseason; foliage disease resistant; for ornamental purposes, also. Named in honor of Arthur S. Colby, Professor of Horticulture, Emeritus, Univ. of Illinois, Urbana.

Cydonia oblonga Miller

Apple.—See Orange.

Bereczcki.—See Bereczki.

Bereczki (*Bereczcki, Beretzki*).—A very old Serbian cultivar named after an eminent Hungarian pomologist, Professor Bereczki. Fruit: very large, pear shaped, golden yellow. Quality very good, tender when cooked. Tree: very vigorous, heavy yielding, precocious.

Beretzki.—See Bereczki.

Champion.—A standard quince cultivar since the late 1800s. Originated in the United States about 1870. Fruit: intermediate shape between apple and pear; large, up to 720 g (24 oz); skin yellowish-green; flesh white to yellow, almost as tender as apple; slightly astringent; ripens in midseason, generally October in northern latitudes; later than Orange; keeps very well. Tree: vigorous, larger and taller than Orange, cold hardy, precocious, very productive; may not mature in regions with short season.

Cooke's Jumbo (*Jumbo*).—Selected by Herb Kaprielian, Dinuba, Calif. One tree in his orchard of Van Deman quince consistently bore larger fruit than the other trees. Introduced by L.E. Cooke Nursery, Visalia, Calif., in 1972. Fruit: pyriform, large to very large, 12 to 15 cm in diam.; skin yellowish-green; flesh white; ripens in September and October.

Jumbo.—See Cooke's Jumbo.

Meech (*Meech's Prolific*).—Origin in Connecticut about 1850 where it was grown as Orange. May be astrain of Champion. Grown in Vineland, N.J., in the mid-1800s. Introduced in 1883 by Rev. W.W. Meech. Fruit: large, pyriform, up to 540g (18 oz), skin very fine-textured, bright yellow, exceedlingly fragrant, excellent flavor, ripens 2 weeks earlier than Champion. Tree: slow growing,heavy and annual bearing; precocious.

Meech's Prolific.—See Meech.

Monstruevx de Vrania.—See Vranja.

Orange (*Apple*).—Early ripening habit has made this the leading quince grown in the northeastern UnitedStates. Orange is a group rather than a distinct genotype. Its origin is uncertain. Many orange or apple-shaped quince came to be called "Orange" or "Apple," and they were thought to come true from seed, resulting in several strains. Fruit: large to very large, nearly round, faintly ribbed; skin bright golden-yellow to greenish-yellow, very pubescent; flesh orange-yellow, tender, fine texture, aromatic; early midseason. Quality very good when grown in the North where it ripens during cool weather. In NewJersey and the South, it often ripens when temperatures are high, resulting in inferior size, flavor, and color.Tree: vigorous, hardy, productive.

Pineapple.—One of the leading cultivars in California. Developed by Luther Burbank in Santa Rosa,Calif., and released in 1899. Fruit: large, roundish, resembles Orange, but smoother, more globular, lighter in color, and slightly larger; skin light golden, smooth; flesh white, pineapple-like flavor, slightly aromatic, tender when cooked.

Tree: ornamental, 3 to 8 m tall, cold hardy, tolerates wet soil, blooms late, flowers tinted pink, self-fertile, needs 200 to 300 h chilling, as productive as Orange.

Portugal.—Very old cultivar from Portugal. Fruit: very large, oblong-oval; skin almost orange when ripe, very wooly, flesh juicier than other types; turns fine purple or deep crimson when cooked; flavor very good; ripens very early, 10 days before Orange. Tree: vigorous, becoming large and spreading, not productive; late to begin bearing.

Rea (*Rea's Mammoth*).—A strain of Orange disseminated by Joseph Rea of Coxsackie, N.Y. in the 1800s. Fruit: larger than Orange, and ripens slightly later. Tree: not as vigorous or cold hardy as other cultivars.

Rea's Mammoth.—See Rea.

Rich.—Orig. in Hillsboro, Ore., by Charles Rich. Introduced in 1948. Parentage unknown; selected about 1940. Tree: uniform shape; less subject to dry rot than other varieties. No longer available.

Smyrna.—Brought from Smyrna, Turkey, by R.C. Roeding of Fresno, Calif., in 1897. Fruit: oblong to pyriform, large to very large, furrowed; skin lemon yellow with brown pubescence; flesh light yellow, fragrant, mild; ripens with Orange; excellent keeper, stores longer than other quince cultivars in refrigeration. Tree: small but vigorous; large, thick leaves; self-fruitful, needs 200 h chilling.

Van Deman.—Orig. in Santa Rosa, Calif., by Luther Burbank. Selected from among 700 crosses of Orange x Portugal. Introduced in 1891. Fruit: very large, oblong to pyriform; skin smooth, pale orange with little pubescence; flesh pale yellow, rather coarse, slightly astringent, aromatic, subacid, juicy, becomes deep red when cooked; core large, open. Tree: vigorous, prolific, hardier than other quince cultivars. Named in honor of pomologist H.E. Van Deman.

Vranja (*Monstruevx de Vrania*).—A Serbian cultivar, possibly the same as Bereczki. Orig. near Vranje (Vranja). Fruit: pyriform, matures late, October to November.

Raspberry

Algonquin.—Orig. in Vancouver, British Columbia, Canada, by H.A. Daubeny, Agr. Canada Res. Sta. and T.M. Sjulin, Washington State Univ., Puyallup, and released as germplasm BC 72-1-7; Haida x Canby; cross made in 1972; selected in 1975; introd. in 1989 by Hugh Daubeny and A. Dale, Hort. Res. Inst. Ontario, Simcoe. Fruit: medium size; bright, nondarkening red; may separate with difficulty from receptacle in some environments; medium firmness; some resistance to postharvest rot caused by *Rhizopus* spp. Plant: productive, upright; compact habit with medium length laterals, short internodes; primocanes numerous; glabrous, nonwaxy; relatively few spines; floricanes have basal cracking; winter hardy in the Pacific Northwest, Ontario, Nova Scotia, and Denmark; shows low chilling requirement in New Zealand. Homozygous for gene Ag_1 for resistance to the common strain of the North American aphid vector of the raspberry mosaic virus complex; slow to become infected by pollen transmission of raspberry bushy dwarf virus; moderately susceptible to cane *Botrytis*; some resistance to spur blight and to root rot.

Alkoopina.—A yellow-fruited, primocane-fruiting raspberry released for temperate Australia. Orig. in Toolangi, Australia, by G. McGregor, Inst. for Hort. Development, Victoria. Chilcotin x Amity; cross made in 1983 by H.A. Daubeny, Agriculture and Agri-Food Canada, Pacific Agr. Res. Centre, Vancouver, British Columbia; tested as A83-10-A1; introd. in 1992. Propagated under an agreement with the Australian Rubus Growers Assoc. Fruit: medium-large; yellow; drupelet coherance and firmness superior to most yellow raspberries; easy to harvest; somewhat susceptible to rain damage and sun scorch. Plant: yield equally distributed between floricane and primocane crop; spineless; canes similar to Willamette in rigidity; numerous primocanes.

Allegany.—Orig. in College Park, Md., by I.C. Haut, Agr. Ext. Serv. F_1 (Manteo selfed) x Dundee, tested as Md. 133-D, also may have been tested as Monocacy. Introd. in 1970. Fruit: large, black, firm; quality good; ripens a few days after Cumberland; for commercial or home use. Plant: very vigorous and productive. Resistant to cane spot.

Amethyst.—Orig. in Ames, Iowa by E.L. Denisen, Iowa State Univ. Robertson (black) x Cuthbert (red). Cross made in 1952, selected in 1956, tested as 50-5255. Introd. in 1968. Fruit: large, oval, glossy purple, lacks bloom which makes most purple varieties appear dusty, attractive; drupelets cohere to form a firm berry; ripens early July in central Iowa, picked over 2 to 3 weeks, fruit well exposed; processing use. Plant: 4 to 6 ft. when pinched, does not sucker; upright; very vigorous; hardy; very productive. Tolerant to cane spot.

Amity.—Orig. in Corvallis, Ore., by F.J. Lawrence, released jointly by the U.S. Dept. of Agriculture–Agr. Res. Serv., Oregon Agr. Expt. Sta., and Washington Agr. Res. Center. Selected in 1976. ORUS 1839 (Fallred ORUS 1347) x ORUS 1835 (PI 338908 x Heritage) (PI 338908 is East Malling selection 791/45 derived from Malling Landmark and *Rubus occidentalis*); tested as ORUS 2036; introd. in 1984. Fruit: medium size, uniform shape; very firm; fairly dark red; ease of separation from receptacle varies and can be rather difficult under cooler weather conditions; good fresh-market and processing qualities; fruit produced on primocanes in August; ripens up to 7 days earlier than Heritage in the Pacific Northwest. Plant: moderate yield, slightly less than Heritage over the entire season; primocanes vigorous, glabrous, nonwax green with purple overlay; scattered pale purple spines that are objectionable; fruiting laterals medium to short length. Some resistance to root rot; susceptible to infection by pollen transmission of raspberry bushy dwarf virus and to the common strain of the North American aphid vector of the raspberry mosaic virus complex.

Anita.—Orig. in Chapingo, Mexico, by J. Rodriguez at the Colegio de Postgraduados. Malling Exploit open-pollinated; selected in 1979; introd. in 1989. Fruit: medium to large size; symmetrical, conic; medium firm; light red, slightly less glossy than Malling Exploit; earlier season than Malling Exploit because of earlier budbreak. Plant: good yield; moderate vigor; easily propagated; estimated chilling requirement 250 units, probably the lowest requirement of any red raspberry variety.

Augustred.—Orig. in Durham, N.H., by A.F. Yeager and E.M. Meader, Agr. Ext. Serv. NH R7 [Durham x NH 102 (dwarf F_2 seedling x Taylor)] x N.Y.. 287 [N.Y. 18810 (Marcy x Indian Summer) x N.Y. 20990 (*Rubus strigosus* selection); cross made in mid-1950s; selected in 1960, tested as NH 64-3; introd. in 1973. Fruit: medium; round-elongated; bright, attractive red; good flavor, soft; ripens late July to August on primocanes. Plant: short, erect; primocanes often branched; precocious bearing; moderate productivity; needs nitrogen fertilization; winter hardy.

Autumn Bliss.—Orig. in East Malling, England, by E Keep, J.H. Parker, and V.H. Knight, Hort. Res. Intl. (East Malling Res. Sta.). Complex parentage involving *Rubus strigosus, R. arcticus, R. occidentalis,* and the red raspberry varieties Malling Landmark, Malling Promise, Lloyd George, Pyne's Royal, Burnetholm, and Norfolk Giant; cross made in 1974; selected in 1976; tested as EM 3676/78; introd. in 1984. Plant variety rights held by Plant Breeding Intl. Cambridge and Hort. Res. Intl. Fruit: large; oval-conic; slightly dark red; pleasant, rather mild flavor; produced on primocanes; earlier ripening than either Heritage or Zeva Herbsternte, overlapping with late floricane fruiting types; fairly easy separation from receptacle. Plant: productive; primocanes moderately numerous, glabrous, fairly erect, green with pink tinge; numerous light purple spines. Susceptible to infection by pollen transmission of raspberry bushy dwarf virus; has the gene A_{10} for resistance to all races of the European aphid vector of raspberry mosaic virus. Cane diseases and winter hardiness are not considered since primocanes are usually removed after fruiting.

Autumn Britten.—An early ripening, primocane-fruiting raspberry released for eastern Canada. Orig. in East Malling, Kent, U.K., by V.H. Knight, E. Keep, and J.H. Parker, Hort. Res. Intl. Sibling of Autumn Bliss, with complex derivation involving several red raspberry varieties and *Rubus arcticus, R. strigosus,* and *R. occidentalis*; tested as EM 3676/6; introd. in 1995. Plant variety rights held by Ontario Berry Growers Assn. and Hort. Res. Intl. Fruit: large; medium to dark red; firm; at times more uniform than Autumn Bliss; usually ripens a few days after Autumn Bliss; can be difficult to harvest in British Columbia but not in Ontario. Plant: yields similar to or slightly higher than Autumn Bliss; canes not as erect as Autumn Bliss; some spines. Susceptible to infection by pollen transmission of raspberry bushy dwarf virus; gene A_{10} gives resistance to the four known strains of the European aphid vector of the raspberry mosaic virus complex.

Autumn Cascade.—An early ripening, spineless primocane-fruiting raspberry that is easy to harvest. Orig. in East Malling, Kent, U.K., by V.H. Knight, E. Keep, and J.H. Parker, Hort. Res. Intl. Sibling of Autumn Cygnet with complex derivation involving several red raspberry varieties and *Rubus odoratus, R. arcticus, R. strigosus,* and *R. occidentalis*; tested as EM 4395/65; introd. in 1994. Plant variety rights held jointly by Specialist Fruit Plants and Hort. Res. Intl. Fruit: medium to large; dark red; firm; compared to Autumn Bliss, ripens 1 week later in Kent but a few days earlier in British Columbia; good flavor. Plant: yields less than Autumn Bliss; spineless. Susceptible to *Phytophthora*-incited root rot in greenhouse screening test; gene A10 gives resistance to the four known strains of the European aphid vector of the raspberry mosaic virus complex.

Autumn Cygnet.—A spineless, primocane-fruiting raspberry adapted to Tasmania, Australia. Orig. in East Malling, Kent, U.K., by V.H. Knight, E. Keep, and J.H. Parker, Hort. Res. Intl. Sibling of Autumn Cascade with complex derivation involving several red raspberry varieties and *Rubus odoratus*, *R. arcticus*, *R. strigosus*, and *R. occidentalis*; tested as Em 4395/75; introd. in 1994. Market rights held by Tasmanian Berry Gardens Pty., Battery Point. Fruit: medium size; glossy, medium-red color, lighter than Autumn Bliss; fairly soft, lacks good coherence; uneven shape; easier to harvest than either Autumn Bliss or Autumn Britten; ripens after Autumn Bliss and before Heritage in Australia but usually before Autumn Bliss in British Columbia; good flavor. Plant: good yield in Australia; canes nearly self-supporting; spineless. Gene A10 gives resistance to the four known strains of the European aphid vector of the raspberry mosaic virus complex.

Avon.—Orig. in Kentville, Nova Scotia, by D.L. Craig and L.E. Aalders, Agr. Canada, Res. Sta. Introd. in 1967. Malling Promise x Cuthbert, cross made in 1954, selected in 1957, tested as K57-12. Fruit: larger than Viking or Latham; conic; bright red, attractive, firm, slightly acid, moderately juicy, quality good; drupelets medium to small in size, numerous; separates easily from receptacle; for dessert and freezing; ripens with Viking. Plant: winter hardy; productive; yields well from basal as well as from mid- and terminal buds; primocanes hairy, spiny, free of wax, green in summer, partial purple overlay in fall; floricanes gray-brown, numerous; spines short, purple, not objectionable; secondary fruiting buds present, moderate lateral length.

Bababerry.—Orig. in Idyllwild, Calif., by Gertrude P. Millikan. Parentage unknown; introd. in 1979. U.S. plant patent 4732. Parentage unknown. Fruit: large, firm, sweet, good quality, flavor good. Plant: adapted to wide variety of soils and temperatures, where most other red raspberries will not flourish; probably has low chilling requirement.

Balder.—Orig. in Aas-NLH, Norway, by G. Redalen, Agricultural Univ. of Norway. Norna x Malling Jewel; cross made in 1975; selected in 1980; tested as H 4-10-07; introd. in 1988. Fruit: medium size; dark, dull red color; relatively soft; medium sweet; high acidity; separates readily from the receptacle; early, relatively concentrated season; processing use. Plant: productive; primocanes numerous, erect, and vigorous, with dark-purple spines largely restricted to the basal portions, slightly waxy and glabrous; floricanes erect and vigorous, light brownish gray; large leaves may hide fruit, making hand harvest difficult; very winter hardy; susceptible to cane spot.

Beskid.—Orig. in Brzezna, Poland, by J. Danek and Z. Pasiut, Expt. Sta. of the Res. Inst. of Pomology and Floriculture. EM 3650/3 x EM 2747/78; tested as 80083; introd. in 1991. Fruit: medium to large; conic with blunt tip; firm with small, closely adhered drupelets; glossy red; late ripening with prolonged season; low susceptibility to *Botrytis* rot; fresh-market and processing use. Plant: high yielding; winter hardy; vigorous with a medium number of slightly spreading primocanes; primocanes pubescent, ash green; floricanes brown, slightly waxed, covered with rudimental spines; fruiting laterals long, strongly attached, with numerous fruit uniformly dispersed from the top to about three-fourths of length. Moderately susceptible to cane *Botrytis* and spur blight; low susceptibility to the raspberry beetle; has both genes A_1 and A_{10} conferring resistance to all known strains of the European aphid vector of the raspberry mosaic virus complex.

Black Knight.—Orig. in Champaign, Ill., by Herschel L. Boll. Johnson Everbearing selfed; cross made in 1967. Introd. in 1973. Fruit: large, up to 1 inch in diam., hemispheric shape; black; firm; ripens early. Plant: everbearing, bears continuously from June until freezing weather, by bearing first on year-old canes, then on tips of new laterals produced from base of year-old canes and then new primocanes; highly productive and vigorous, produces up to 16 canes per plant per year, and grows up to 7 ft tall; propagates by tip-layers.

Black Magic.—See Johnson Everbearing.

Black Treasure.—Orig. in Champaiyn. Ill., by Herschel L. Boll. Johnson Everbearing, selfed in 1967; introd. in 1977. Fruit: large, up to 1 inch in diam.; hemispheric shape; black; firm; ripens early. Plant: fruits continuously on floricanes and then primocanes from June until freezing weather, highly productive; vigorous; propagated by tip-layers.

Bogong.—A high-yielding, primocane-fruiting raspberry with large fruit of excellent quality. Orig. in Toolangi, Australia, by G. McGregor, Inst. for Hort. Development, Victoria. Comox x Autumn Bliss; cross made in 1985; tested as A85-24-A7; introd. in 1994. Propagated under an agreement with the Australian Rubus Growers Assoc. Fruit: large; glossy, medium red; firmness similar to Heritage; uniform drupelets; easy to harvest; season similar to Heritage; fairly intense aromatic flavor. Plant: high-yielding; canes more branched than Heritage; basal spines.

Bountiful Giant.—Orig. in Champaign, Ill., by Herschel L. Boll. Johnson Everbearing selfed in 1967; introd. in 1979. Fruit: large, up to 1 inch in diam., hemispheric; black; firm; ripens early, about same time as Logan. Plant: fruits continuously on floricanes and then primocanes from June until freezing weather; highly productive, vigorous, produces up to 16 canes per plant, grows to 7 ft tall, propagated by tip-layers and produces medium number.

Brabant.—See Rusilva.

Brandywine.—Orig. in Geneva, N.Y., by D.K. Ourecky, New York Agr. Ext. Serv. New York 631 x Hilton; cross made in 1963; selected in 1966; tested as New York 905, introd. in 1976. Fruit: large, 5 to 6 g; round-conic; reddish-purple with moderate pubescence; tart, but of high quality; late season. Plant: very vigorous; productive; canes erect and strong, forming defined hills; propagated by tip-layering. Susceptible to verticillium wilt and the North American aphid vector of the raspberry mosaic virus complex.

Centennial.—Orig. in Puyallup, Wash., by P.P. Moore, T.M. Sjulin, B.H. Barritt, and H.A. Daubeny, Washington State Univ. Meeker x Skeena; cross made by H.A. Daubeny in 1974; selected by B.H. Barritt in 1978; tested as WSU 738, introd. in 1989. Fruit:large; long conic shape; bright, glossy, medium red; pleasant, mild flavor; medium firm; suitable for fresh market and processing; separates with some difficulty from receptacle and thus not suited to machine harvesting; ripening season similar to Meeker, several days later than Willamette and Skeena; relatively susceptible to pre- and postharvest rots. Plant: productive; more upright habit than Meeker, but less upright than Skeena; long, strong fruiting laterals; primocanes not numerous; vigorous; pubescent; numerous dark purple, moderately soft spines. Susceptible to spur blight, cane *Botrytis*, cane spot, infection by pollen transmission of raspberry bushy dwarf virus and the North American aphid vector of the raspberry mosaic virus complex; some resistance to root rot. Not winter hardy in British Columbia, or northern Washington.

Cherokee.—Orig. in Blacksburg, Va., by G.D. Oberle, Virginia Agr. Ext. Serv., Virginia Polytechnical Inst. & State Univ. Introd. in 1972. Hilton x VPI 5 (Taylor x St. Regis); cross made in 1958; selected in 1972; tested as VPI 13. Fruit: large; round conic to medium conic; medium to bright red; firm; seeds prominent; fruit well exposed; season extends from late June to mid July on floricanes and mid September to frost on primocanes. Plant: productive; good vigor; tolerant to fluctuating winter temperatures. Tolerant to root rot. Susceptible to the common strain of the North American aphid vector of the raspberry mocaic virus complex.

Chilcotin.—Orig. in Vancouver, British Columbia, Canada, by H.A. Daubeny, Agr. Canada Res. Sta. Sumner x Newburgh; cross made in 1963; selected in 1965; introd. in 1977. Fruit: medium size; conic, glossy, non-darkening red;medium firm; early to midseason ripening with long harvest season; ease of separation from receptable can vary; fresh market. Plant: productive; moderate vigor; canes moderately spiny; numerous primocanes; pubescent; moderately winter hardy in the Pacific Northwest. Susceptible to the North American aphid vector of the raspberry mosaic virus complex; resistant to the common strain of the raspberry bushy dwarf virus; resistant to cane *Botrytis*; only moderately susceptible to spur blight.

Chilliwack.—Orig. in Vancouver, British Columbia, Canada, by H.A. Daubeny, Agr.

Canada Res. Sta. BC 64-10-198 (Sumner x Carnival) x Skeena, cross made in 1974; selected in 1978; tested as BC 74-10-81, introd. in 1987. Fruit: large; conic; medium red, glossy; firm; drupelets numerous; high soluble solids; excellent fresh eating quality; suitable for processing; separates readily from receptacle and, thus, is well-suited to machine harvesting; ripens several days later than Skeena and Willamette; some resistance to pre- and postharvest rots. Plant: moderate yield; upright growth habit with long, strong fruiting laterals; primocanes numerous, erect, glabrous, and nonwaxy; purple spines largely restricted to basal portions; floricanes erect, reddish to yellow brown with basal cracking. Susceptible to cane *Botrytis,* cane spot and crown gall; some resistance to spur blight, root rot; slow to become infected by pollen transmission of raspberry bushy dwarf virus; resistant to the common strain of the North Amierican aphic cector of the raspberry mosaic virus complex. Winter hardy in the Pacific Northwest.

Clutha.—A high-yielding raspberry with easily harvested fruit adapted to medium-low chill conditions in New Zealand. Orig. in Vancouver, British Columbia, Canada, by H.A. Daubeny, Agr. and Agri-Food Canada, Pacific Agr. Res. Centre. Meeker x Skeena; cross made in 1974; selected in 1977; tested as BC 74-11-41; introd. into New Zealand in 1983 and released there in 1991 by H. Hall, Riwaka Res. Centre. Fruit: smaller than Willamette and Marcy; moderately glossy, medium-dark red; mid- to late season, ripening 3 to 4 days later and peaking 10 days later than Marcy; processing use. Plant: yield higher than Marcy and Skeena; moderate numbers of medium-strong, semi-upright canes; floricanes dark brown with few spines; strongly attached laterals with high fruit numbers; good budbreak in New Zealand. Moderately susceptible to cane spot in New Zealand; susceptible to cane *Botrytis*; susceptible to infection by pollen transmission of raspberry bushy dwarf virus, but slow to become infected in New Zealand, where it is under high infection pressure; resistant to the common strain of the North American aphid vector of the raspberry mosaic virus complex.

Colossus.—Orig. in Europe. and introd. in 1920 into the U.S. by John Scheepers, New York, N.Y. Tetraploid of unknown origin. Fruit: red; very large, soft, tart, with large drupelets; quality variable; matures late, fresh use and also good for canning. Plant: very vigorous; seems to be adapted to areas with long growing seasons; similar to but appears more rugose and more vigorous in plant habit than the tetraploids La France or Hailsham.

Comet.—Orig. in Ottawa, Ontario, Canada, by Canada Dept. Agr., Res. Sta. Ottawa x Madawaska; selected in 1943; tested as Ottawa 383; introd. in 1954. Fruit: large; red; conic; moderately firm, quality fair; matures 2 to 3 days before Latham. Plant: medium spines; tall, stout; hardy; productive.

Comox.—Orig. in Vancouver, British Columbia, Canada, by H.A. Daubeny, Agr. Canada Res. Sta. BC 64-9-81(Creston x Willamette) x Skeena; cross made in 1974; selected in 1978; tested as BC 74-12-42; introd. in 1987. Fruit:very large; conic; medium red; firm; drupelets numerous; good fresh eating quality; suitable for processing; fairly easy separation from receptacle and can be machine-harvested; ripens several days later than Willamette and Skeena; some resistance to pre- and postharvest rots. Plant very high yielder; vigorous and fairly upright growth habit, but laterals tend to droop because of large numbers of fruit; primocanes reasonably numerous, fairly upright, glabrous, nonwaxy; purple spines largely restricted to basal portions; floricanes fairly erect, yellow-brown with basal cracking. Susceptible to cane *Botrytis,* spur blight, cane spot, and root rot; slow to become infected by pollen transmission of raspberry bushy dwarf virus; resistant to the common strain of North American aphid vector of the raspberry mosaic virus complex; relatively winter hardy in the Pacific Northwest.

Creston.—Originator and place of origin uncertain but probably was a selection from the Agricultural Canada bredding program located in Ottawa. Tested as 151 in the Creston area of British Columbia; introduced in 1950. Fruit: small to medium size; medium red; conic; firm, fair flavor. Plant: moderately vigorous with strong and relatively smooth, upright canes. Resistant to the common strain of North American aphid vector of the raspberry mosaic virus complex; susceptible to infection by pollen transmission of raspberry bushy dwarf virus.

Crimson Cone.—Orig. in Urbana, Ill., by H.L. Boll, Illinois Agr. Expt. Sta. Latham x Milton; selected in 1949; tested as Ill. 45-113; introd. in 1955. Fruit: medium to large; conic; medium red; firm, good flavor; drupelets strongly coherent. Plant: produces many suckers; very productive; very tall; floricane fruiting.

Dinkum.—Orig. in Toolangi, Australia, by G. McGregor, Inst. for Hort. Development, Melbourne, Victoria. Autumn Bliss x Glen Moy; cross made in 1983 by D.L. Jennings, Scottish Crop Res. Inst.; selected by G. McGregor in 1985; tested as A-83-31-G5; introduced in 1992. Propagated under an agreement with the Australian *Rubus* Growers Assoc. Fruit: medium size; firm with medium size drupelets; medium dark red and slightly more glossy than Autumn Bliss; excellent flavor; easy to harvest; main production on primocanes; overwintered floricanes produce early summer fruit on lower portions; some reistance to postharvest *Botrytis* rot; ripens as much as 19 days earlier than Heritage and at the same time or 3 days earlier than Autumn Bliss; fresh market use. Plant: yield similar to Autumn Bliss but usually has more condensed ripening season and higher yield at each harvest; upright primocanes with relatively short fruiting laterals. Susceptible to root rot and to infection by pollen transmission of raspberry bushy dwarf virus.

Dixie.—Orig. in Raleigh, N.C., by Carlos F. Williams, North Carolina Agr. Expt. Sta. *Rubus biflorus* x Latham; cross made in 1928; introd. in 1938. Fruit: red; medium size; tart flavor. Plant: vigorous; productive; adapted to southeastern U.S.

Dormanred.—Orig. in State College, Miss., by J.P. Overcash, Mississippi Agr. Ext. Serv., Mississippi State Univ. *Rubus parvifolius* x Dorsett. Cross made in 1949, selected in 1951, tested as Miss. 158; introd. in 1972. Fruit: large; flattened globe; glossy medium red; juicy, medium firm, flavor mild acid; good for freezing; ripens mid-June. Plant: large; canes vigorous, productive; tolerant to high summer temperatures; adapted to Mississippi and adjacent states.

Double Delight.—A winter-hardy, relatively early ripening, primocane-fruiting raspberry with an attractive appearance. Orig. in Morden, Manitoba, Canada, by C. Davidson, Agr. and Agri-Food Canada, Morden Res. Centre. (Fall Red x primocane-fruiting type derived from *Rubus strigosus* from Cheyenne, Wyo.) x (Fall Red x Boyne); cross made in 1968; selected in 1978; tested as MRS#8114; introd. in 1993. Propagated under royalty agreement by Prairie Plant Systems, Saskatoon, Saskatchewan, Canada, and by Jefferson Nurseries, Portage la Prairie, Manitoba Fruit: small; medium bright red; attractive; firm with good drupelet coherence; conic with conspicuous round point; tendency for twin fruit to form from the same flower; flavor comparable to Red River; ripens before Heritage or Redwing, 10 days after Red River. Plant: yield similar to Red River; canes relatively long and thick; height 1 to 1.5 m; spines green, turning to red in fall; may need trellis; moderate primocane production; winter hardy to at least USDA zone 3.

Dundee.—Orig. in Geneva, N.Y., by Richard Wellington, New York State Agr. Expt. Sta. Smith 1 x Palmer; cross made in 1910; introd. in 1927. Fruit: large; glossy; black; small drupelets; attractive; moderately firm, mildly subacid, quality very good. Plant: tall; vigorous; productive.

Durham.—Orig. in Durham, N.H., by A.F. Yeager, New Hampshire Agr. Expt. Sta. Taylor x Nectar blackberry seedling is probably a parthenogenetic seed therefrom; selected in 1944; introd. in 1947. Fruit: medium size; red; firm; primocane fruiting, ripens 2 weeks before the fall crop of Indian Summer. Plant winter hardy; very vigorous.

Eallred.—Orig. in Durham, N.H., by A.F. Yeager and E.M. Meader, New Hampshire Agr. Expt. Sta. N.H. 7 x N.Y. 287 (a selection secured from George L. Slate, New York State Agr. Expt. Sta., Geneva); cross made by Yeager; selected in 1956 and tested as N.H. 56-2; introd. in 1964. Fruit: red; firm, good flavor; starts to ripen by mid-August at place of origin. Plant: produces crops continuously on highly productive branched primocanes until curtailed by frost; vigorous; upright; produces suckers freely; mowing of over-wintered canes recommended to eliminate hand pruning and as an aid to control diseases.

Earlibest.—Orig. in Hopkins, Minn., by Frank Lorenz. Latham o.p.; selected in 1953; introd. in 1960. Fruit: resembles Latham. Plant: canes large, spineless; vigorous; productive.

Earlired.—See Early Red.

Early Red (*Earlired*).—Orig. in South Haven, Mich., by Stanley Johnston, Mich. Agr. Expt. Sta. Lloyd George x Cuthbert; cross made in 1931; selected in 1934; introd. in 1951. Fruit: large; round-conic; light red when first ripe turning rather dark when fully ripe; easily separated from the receptacle; flavor good; cans well without crumbling; freezes as well as Latham; ripens early. Plant: upright, tall, vigorous; winter hardy; very productive; canes numerous, moderately stocky, quite thickly covered with stiff, sharp, reddish-brown spines.

Ebonee.—Orig. in Graettinger, Palo Alto County, Iowa, by Joseph Gabrielsen. Open-pollinated seedling of Cumberland; discovered in 1949; introd. in 1962. Plant patent 2101. Fruit: large; black; attractive; averages 5/8 inches long, 7/8 inches wide, maintaining size through season; firm, coherent, good sprightly flavor; ripens about 7 days after Cumberland, which it resembles; satisfactory for dessert, freezing, and canning. Plant: vigorous, very productive.

Evans.—Orig. in Geneva, N.Y., by Richard Wellington, New York State Agr. Expt. Sta. Watson Prolific x Honeysweet; cross made in 1921; introd. in 1936. Fruit: glossy black; attractive; moderately firm, high quality. Plant: canes easily broken by wind. Obsolete in New York, considered promising in southern California.

Fairview.—Orig. in Corvallis, Ore., by George F. Waldo, U.S. Dept. of Agriculture and Oregon Agr. Expt. Sta. U.S.-Ore. 782 x Washington; cross made in 1945; selected in 1950; tested as U.S.-Ore. 1022; introd. in 1961. Fruit: medium large, but smaller than Willamette; skin very bright medium red, attractive; firm, flavor very pleasing, more intense and with less acid than Willamette; especially suitable for frozen pack; not suitable for canning; ripens in midseason, beginning with Canby and Willamette and continuing for about 1 month. Plant: vigorous on well-drained soil, outliving Willamette on heavy poorly drained soil; productive, fruit produced on long fruiting laterals; winter hardy in the Pacific Northwest. Susceptible to the North American aphid vector of the raspberry mosaic virus complex.

Fallbrook.—Orig. in Brooks, Alberta, Canada, by S. Mahadeva, Alberta Special Crops and Hort. Res. Center. Selected from a population of open-pollinated seedlings from E.M. Meader, New Hampshire; introd. in 1984. Fruit: slightly smaller than Boyne; bright red; sweet flavor; ripens as early as late July and continues into mid-September, depending on season; home-garden use in southern Alberta. Plant: primocane fruiting; semi-erect canes.

Fallgold.—Orig. in Durham, N.H., by E.M. Meader, Agr. Ext. Serv. [NH 56-1 x (Taylor x *R. pungens oldhami*) F_2 o.p.]; selected in 1964; tested as NH 64-2; introd. in 1967. Fruit: similar to Fallred, but has yellow color; flavor good. Plant: vigorous and suckers freely, similar to Fallred; produces on both primocanes and floricanes; winter hardy.

Favorite.—A late-ripening primocane-fruiting raspberry with spineless canes. Orig. in Soings-en-Sologne, France by Jacques Marionnet. Autumn Bliss x Delmes. Protected by COV 64470. Fruit: large; medium red; long shape; good shelf life; good flavor; ripens one week later than Heritage in France. Plant: high yield; spine-free, erect, vigorous canes; few primocanes. Susceptible to raspberry bushy dwarf virus.

Fertodi zamatos.—Orig. in Hungary by L. Kollanyi, M. German, and M. Kovacs, Res. Sta. Fertod. Fertodi Hungaria x Canby; selected in 1971; tested as F.4; introd. in 1986. Fruit: medium to large size; attractive bright red; firm; flavor good; mid to late season, processing. Plant: very tall; winter hardy; productive.

Festival.—Orig. in Ottawa, Ontario, Canada,

by S.O. Fejer and L.P.S. Spangelo, Canada Dept. Agr. Res. Sta. Muskoka x Trent; cross made in 1948, selected in 1952 by L.P.S.Spangelo, tested as 48-26-02; introd. in 1972. Fruit: large, oval; medium red, glossy; firm, good cohesion, needs moderate picking force, more acid and less aromatic than Carnival; early to midseason ripening; suitable for jam, good freezing qualities. Plant: short canes, primocanes slender, erect, strong, with very few branches; very winter hardy; productive. Tolerant to leaf curl virus and spur blight; less susceptible to powdery mildew than Carnival; susceptible to the North American aphid vector of the raspberry mosaic virus complex.

Forever Amber.—Orig. in Monticello, Minn., by Marion Hagerstrom. Chance seedling discovered in 1968; introduced in 1973; U.S. plant patent 3717. Fruit: medium size, hemispheric shape, yellow to amber; firm; ripens midseason. Plant: highly productive and vigorous; winter hardy in Minnesota; canes arching.

Fraser.—Orig. in Saskatoon, Saskatchewan, Canada, by C.F. Patterson, Univ. of Sask. Chief x Viking; selected in 1957; introd. in 1960. Fruit: red; large; juicy, mild flavor, good quality; recommended for home gardens. Plant: canes short, stout; productivity good; extremely winter hardy.

Gaia.—Orig. in East Malling, England, by V.H. Knight and E. Keep, Hort. Res. Intl. From the cross of two East Malling selections of complex parentage, represents the sixth backcross from the Cumberland black raspberry (*Rubus occidentalis*); cross made in 1974; tested as EM 3655/48; introd. in 1992. Plant varieties rights held jointly by NSA Plants, East Malling, and Hort. Res. Intl. Fruit: large; firm; dark red; round-conic, slightly irregular in shape; easily removed from receptacle; later ripening than Leo and slightly earlier than Malling Augusta; good flavor fresh and in jam; canned and frozen products superior to Glen Clova but not as well-flavored as Leo. Plant: relatively high yielding; vigorous; good habit with tall, erect canes; well-presented fruit on moderately ascending lateral; less spiny than Leo or Malling Augusta. Relatively resistant to cane *Botrytis* and spur blight; moderately susceptible to cane spot, cane blight, and crown gall; may have some resistance to root rot. Has gene A_{10} conferring resistance to the four known strains of the European aphid vector of the raspberry mosaic virus complex; resistant to the common strain of raspberry bushy dwarf virus.

Galante.—A primocane-fruiting raspberry with spineless canes. Orig. in Soings-en-Sologne, France, by Jacques Marionnet. Autumn Bliss x Delmes. Protected by COV 64468. Fruit: large; medium red; long shape; good shelf life; good flavor; ripening season similar to Heritage in France. Plant: spine-free, very erect canes; few primocanes.

Garnet.—See Ruddy.

Gatineau.—Orig. in Ottawa, Ontario, Canada, by Canada Dept. Agr., Res. Sta. Lloyd George x Newman; selected in 1931; tested as Ottawa 276; introd. in 1943. Fruit: large; conic; dark red; moderately firm, quality fair; ripens very early. Plant: canes medium spiny; medium short; winter hardy; very productive. Very susceptible to anthracnose.

Gina.—Orig. in Chapingo, Mexico, by J. Rodriguez at the Colegio de Postgraduados. Malling Exploit o.p.; selected in 1979; introd. in 1989. Fruit: medium to large size, comparable to Malling Exploit; red; medium firm; symmetrical and conic; does not crumble when handled; earlier ripening than Malling Exploit. Plant: high yield; medium vigor; short internodes; has an estimated chilling requirement of 500 units.

Glen Ample.—A high-yielding raspberry with fruit especially well-adapted to fresh market in southern Britain. Orig. in Invergowrie, Scotland, by R.J. NcNicol and D.L. Jennings, Scottish Crop Res. Inst. Sibling of Glen Rosa, complex parentage includes Glen Prosen and Meeker; cross made in 1978; tested as SCRI 7815B8; introd. in 1994. Worldwide marketing rights held by NSA Plants, East Malling, Kent, U.K. Fruit: medium to large size; bright red; round-conic shape; firm with slight tendency to break at the collar in wet and cool conditions; easily removed from receptacle; adapted to machine harvest;

midseason ripening; excellent flavor; fresh market and processing. Plant: spine-free; upright, vigorous canes with long, upright laterals. Gene A_1 gives resistance to two strains of the European aphid vector of the raspberry mosaic virus complex.

Glen Clova.—Orig. in Dundee, Scotland, by D.L. Jennings, Scottish Hort. Res. Inst. Seedling of complex origin; year of crosses 1960; selected in 1963; tested as M9; introd. in 1969. Fruit: medium size; short conic; medium red, slightly downy; firm, good quality; for canning, freezing, and jam; ripens mid-July in Scotland. Plant: productive, erect; vigorous; hardy. Susceptible to powdery mildew; tolerant to cane spot.

Glen Garry.—Orig. in Invergowrie, Scotland, by D.L. Jennings and R.J. McNicol, Scottish Crop Res. Inst. Malling Delight x SCRI 7331/1 (SCRI 703/36 x Glen Prosen); cross made in 1975; tested as SCRI 7518E6; introd. in 1990, marketing rights held by R.W. Walpole, King's Lynn, Norfolk, U.K. Fruit: very large size due to presence of gene L_1, which is unstable; conic; firm; slightly pale color; excellent flavor; fresh-market use. Plant: high yielding; early to midseason; spine-free; moderate vigor; long, strong fruiting laterals. Plants not containing gene L_1 can be identified in vegetative stage by smaller stipules and less serrated leaves. Gene A_1 gives resistance to two strains of the European aphid vector of the raspberry mosaic virus complex; susceptible to raspberry bushy dwarf virus by graft inoculation and probably by pollen transmission.

Glen Lyon.—Orig. in Invergowrie, Scotland, by D.L. Jennings and R.J. McNicol, Scottish Crop Res. Inst. SCRI 7331/1 (SCRI 703/36 x Glen Prosen) x SCRI 7256/1 [SCRI 6820/35 (sibling of Glen Prosen) x Haida]; cross made in 1975; tested as SCRI 7515C5; introd. in 1991. Marketing rights held by NSA plants, East Malling, Kent, U.K. Fruit: medium to large; bright, glossy medium red; firm; easily removed from receptacle; early to midseason ripening; low sugar content; distinct acid flavor; good shelf life; fresh-market and processing uses. Plant: establishes rapidly, enabling it to produce higher initial yield than Glen Clova, otherwise yield similar to Glen Clova; easily managed growth habit; spine-free; moderate vigor; easy to harvest due to relatively short height and good fruit visibility. Resistant to cane *Botrytis* in Scotland, but susceptible in British Columbia; resistant to spur blight and leaf spot virus; has gene A_1 conferring resistance to two strains of the European aphid vector of the raspberry mosaic virus complex; susceptible to infection by pollen transmission of raspberry bushy dwarf virus.

Glen Magna.—A very high-yielding, late-ripening raspberry. Orig. in Invergowrie, Scotland, by R.J. McNicol and D.L. Jennings, Scottish Crop Res. Inst. Meeker x SCRI 7719B11; cross made in 1980; tested as SCRI 8032A3; introd. in 1994. Worldwide marketing rights held by NSA Plants, East Malling, Kent, U.K. Fruit: very large; deep red color; long conic shape; medium firm; excellent uniform appearance; removal from receptacle may not be sufficiently easy in some environments for machine harvest; excellent flavor with similarities to Meeker and Glen Moy; fresh market and processing, especially freezing. Plant: upright, vigorous canes with few spines. Gene A_1 gives resistance to two strains of the European aphid vector of the raspberry mosaic virus complex; gene Bu gives resistance to raspberry bushy dwarf virus.

Glen Moy.—Orig. in Scotland by D.L. Jennings, Scottish Crop Res. Inst. SCRI 688/12 x SCRI 6815/113 (parentage includes *Rubus occidentalis*, Glen Clova, Lloyd George, and Malling Landmark). Cross made in 1972; selected in 1976; tested as SCRI 7210/204; introd. in 1981. Marketing rights held by Plant Breeding Intl., Cambridge, U.K. Fruit: large; light to medium red; moderately firm; easy fruit abscission; ripens early. Plant: very productive, erect, vigorous, spine-free, pubescent canes, moderately winter hardy. Susceptible to cane midge and to root rot; gene A_1 gives resistance to two strains of the European aphid vector of the raspberry mosaic virus complex; gene Bu gives resistance to raspberry bushy dwarf virus.

Glen Prosen.—Orig. in Invergowrie, Scotland by D.L. Jennings, Scottish Crop Res. Inst. SCRI 6531/84 x SCRI 6549/1; cross made in 1968: selected in 1976; tested as 6820/54 (parentage includes Rubus occidentalis, Malling Jewel, Burnetholm, Lloyd George, Malling Landmark); cross made in 1968; slected in 1976; introd. in 1981. Marketing rights held by Plant Breeding Intl., Cambridge, U.K. Fruit: medium large; medium red; very firm; easy fruit abscission; mid to late season ripening; good flavor. Plant: productive, moderate vigor; erect, spine-free, subglabrous canes; moderately winter hardy. Susceptible to cane midge; susceptible to infection by pollen transmission of raspberry bushy dwarf virus; gene A1 gives resistance to two strains of the Ejuropean aphid vector of the raspberry mosaic virus complex.

Glen Rosa.—A raspberry with easily harvested fruit, well adapted to the cool growing conditions of Scotland. Orig. in Invergowrie, Scotland, by R.J. McNicol and D.L. Jennings, Scottish Crop Res. Inst. Sibling of Glen Ample; complex parentage includes Glen Prosen and Meeker; cross made in 1978; tested as SCRI 7815A12; introd. in 1994. Worldwide marketing rights held by NSA Plants, East Malling, Kent, U.K. Fruit: medium size; bright medium-red color; firm with good collar adhesion; easily removed from receptacle; adapted to machine harvest; midseason ripening; fairly good flavor; for processing, less well adapted to fresh market. Plant: production moderate; spine-free, relatively upright canes with medium vigor. Gene H gives resistance to spur blight and cane *Botrytis*; gene A_{10} gives resistance to the four known strains of the European aphid vector of the raspberry mosaic virus complex; gene Bu gives resistance to raspberry bushy dwarf virus.

Glen Shee.—A large-fruited raspberry probably best suited to home gardens and pick-your-own in Britain. Orig. in Invergowrie, Scotland, by R.J. McNicol and D.L. Jennings, Scottish Crop Res. Inst. Complex parentage with no named cultivar in its pedigree for four generations; cross made in 1980; tested as SCRI 8044C9; introd. in 1994. Worldwide marketing rights held by NSA Plants, East Malling, Kent, U.K. Fruit: slightly pale; firm; fleshy; slightly weak skin prone to wind rub; not adapted to machine harvest; moderate flavor. Plant: moderate yield; spine-free; canes relatively upright, strong, vigorous. Susceptible to midge blight; gene A_1 gives resistance to two strains of the European aphid vector of the raspberry mosaic virus complex; susceptible to infection by pollen transmission of raspberry bushy dwarf virus.

Glen Yarra.—A high-yielding raspberry adapted to temperate Australia. Orig. in Invergowrie, Scotland by D.L. Jennings, Scottish Crop Res. Inst. Complex parentage involving *Rubus occidentalis,* Malling Jewel, Burneholm, Lloyd George and Malling Landmark; cross made in 1968; tested as SCRI 6820/64; introd. in 1995 by G.R. McGregor, Inst. for Hort. Development, Melbourne, Victoria. Propagated under an agreement with the Australian Rubus Growers Assoc. Fruit: medium to large; firm; medium red color with downy appearance; midseason ripening. Plant: high yield; strong, erect spineless canes; few primocanes; low chilling requirement. Susceptible to root rot; reaction to raspberry bushy dwarf virus unknown.

Glencoe.—Orig. in Invergowrie, Scotland, by D.L. Jennings and R.J. McNicol, Scottish Crop Res. Inst. A derivative of the Munger black raspberry (*Rubus occidentalis*) crossed with Glen Prosen; tested as SCRI 53-14-6; introd. in 1989. Fruit: medium size; round-conic; dull, purple; very firm; intense flavor; easy to harvest; midseason ripening; good shelf life; specialized processing, fresh-market and home-garden uses. Plant: yield similar to Glen Clova in southern parts of Britain; vigorous, primocanes semi-erect, spine-free, and produced in moderate numbers; floricanes deep purple with conspicuous waxy bloom; fruiting laterals medium length and stiff. No exceptional susceptibilities identified, resistant to verticillium wilt. Not adapted to cold spring weather.

Golden Indian Summer.—See New York Times.

Goldenwest.—Orig. in Puyallup, Wash., by C.D. Schwartze and Myhre, Western Wash. Expt. Sta. Cuthbert x Lloyd George; selected in 1935; introduced in 1953. Fruit: thimble shaped; attractive yellow becoming pink when overripe; drupelets medium small, coherent, separating readily from the conic receptacle; medium firm, flavor good; resembles Washington, except in color; ripens late. Plant: vigorous; very productive; canes tall, nearly straight, bark light green becoming gray-brown when dormant; spines green, moderate in size and number; leaves large, rugose, medium green, fairly flat; fruiting laterals long; winter hardy. Resistant to the common strain of North American aphid vector of the raspberry mosaic virus complex; resistant to western yellow rust and powdery mildew.

Goldie (*Graton Gold*).—Orig. in Sonoma County, Calif., in 1987. Yellow-fruited, primocane-fruiting sport of Heritage. U.S. plant patent 7625; assigned to Nourse Farms, South Deerfield, Mass. Fruit: similar fruit firmness, flavor, and overall performance to Heritage; unique gold color is deeper yellow than other yellow-fruited raspberries. Plant: fruiting habit similar to Heritage; primocane production is more abundant than Heritage.

***Graton Gold*.**—See Goldie.

Haida.—Orig. in Agassiz, British Columbia, Canada, by H. A. Daubeny, Canada Dept. Agr., Res. Sta. Malling Promise x Creston; cross made in 1959, selected in 1962, tested as BC 201; introd. in 1973. Fruit: medium size; blunt conic; glossy medium red, sometimes separates with difficulty from receptacle; midseason ripening; fresh market use. Plant: productive; upright, compact habit with short to medium length laterals; numerous sturdy canes; primocanes rosy red; floricanes brown; very few spines; winter hardy. Resistant to the common strain of the North American aphid vector of the raspberry mosaic complex; resistant to both the common strain and a resistance-breaking strain of raspberry bushy dwarf virus; some resistance to spur blight and to root rot; moderately susceptible to cane *Botrytis*.

Haut.—Orig. in College Park, Md., by I.C. Haut, Univ. of Maryland. 37-22-11 [Manteo selfed (Manteo = Cumberland selfed)] x B16 (=Bristol selfed); introd. in 1987 by H.J. Swartz. Fruit: medium; spherical; black with moderate bloom; firm, cohesive; very sweet, aromatic; ripens over a longer period than most black raspberries. Moderately resistant to *Botrytis*. Plant: vigorous; freely branching; spiny. Susceptible to orange rust and to anthracnose; tobacco streak-infected plants show no loss of vigor or fertility.

Heritage.—Orig. in Geneva, N.Y., by L. Slate, New York State Agr. Expt. Sta. (Milton x Cuthbert) x Durham; cross made in 1958, selected in 1960; tested as N.Y. 696; introd. in 1969. Fruit: medium size; conic; red, attractive; firm, fair quality; relatively late ripening primocane crop. Plant: tall, 5 to 7 ft; very vigorous, winter hardy; suckers prolifically; primocanes erect, sturdy; leaves distinctive oblong shape; more long laterals are produced in Maryland than farther north where the growth is more compact. Susceptible to the North American aphid vector of the raspberry mosaic virus complex.

Hilton.—Orig. in Geneva, N.Y., by George L. Slate, New York State Agr. Expt. Sta. Newburgh x St. Walfried; cross made in 1945; selected in 1948; tested as N.Y. 183; introd. in 1965. Fruit: large; long conic; red; medium firm, thick, quality fair, cavity small; resembles Lloyd George; clings to plant until fully ripe; ripens midseason. Plant: very productive; vigorous; winter hardy.

Himbo Queen (*Rafzeter*).—Orig. in Rafz, Switzerland, by P. Hauenstein, Hauenstein AG, Nursery / Rosery, 8197 Rafz, Switzerland. Malling Exploit x Puyallup. Registered in Switzerland as no. 352547, international no. 512691. Fruit very large; conic; bright, nondarkening red; firm; aromatic; sweet flavor; separates readily from receptacle; good for fresh market, including long distance shipping; good for deep freezing. Plant: very productive; vigorous with thick canes; laterals very long, firmly attached; winter hardy; resistant to drought.. Minimum susceptibility to cane diseases.

Hollins.—Orig. in Watsonville, Calif., by S. Wilhelm. Sweetbriar x Reiter E4720; introd. in 1992. Plant patent 8027; assigned to Sweetbriar Development. Fruit: medium size; light red, glossy; separates easily from receptacle; floricane production about 1 May and primocane production about 20 July in central coastal California. Plant:

moderate vigor with production equally split between primocane and floricane crop; some resistance to powdery mildew and to late leaf rust.

Honeyking.—Orig. in Parkside, Saskatchewan, Canada, by A.J. Porter, Honeywood Nursery. Viking x a selection of *Rubus strigosus;* selected in 1945; introd. in 1948. Fruit: red; quality good for canning; fruit too small where standard sorts can be grown. Plant: productive; very winter hardy; yields well.

Honeyqueen.—Orig. in Carlos, Alberta, Canada, by R. Erskine. Open-pollinated seedling of Honeyking; discovered in 1964; introd. in 1966. Fruit: larger than Honeyking; yellow; softer than Amber. Plant: very winter hardy; shorter canes than Amber; productive.

Huron.—Orig. in Geneva, N.Y., by George L. Slate, New York State Agr. Expt. Sta. Rachel x Dundee; cross made in 1943; selected in 1947; tested as N.Y. 27189; introd. in 1965. Fruit: black; quality good; all around merit rather than any distinguishing characteristics; ripens midseason. Plant: vigorous; productive.

Indian Summer.—Orig. in Geneva, N.Y., by Richard Wellington, New York State Agr. Expt. Sta. N.Y. 1950 (Empire x Herbert) x Lloyd George; cross made in 1925; introd. in 1936. Fruit: red; aromatic; crumbles frequently; useful for jam; primocane fruiting variety. Plant: very productive; vigorous; resistant to the common strain of the North American aphid vector of the raspberry mosaic virus complex.

Iowa.—Orig. in Storm Lake, Iowa, by Wragg Nursery Co., Des Moines, Iowa. Parentage unknown; introduced in 1920. Fruit: large; dark red; quality good; ripens early. Plant: vigorous; winter hardy; productive.

Itasca.—Orig. in Excelsior, Minn., by E.T. Andersen, Univ. of Minnesota. Fruit Breeding Farm. Newburgh selfed made in 1937; tested as Minn. 399; introd. in 1965. Fruit: medium large, similar to Latham; round; orange-red, glossy; drupelets large; soft, juicy, flavor pleasant; hangs well; easily picked. Plant: vigorous; winter hardy; productive; strong fruiting laterals produced; cane smooth, nearly spine-free. Susceptible to anthracnose. Well adapted to the cool, rather humid areas of north-central Minnesota.

Jet.—Orig. in Urbana, Ill., by John L. Franklin. Chance seedling, discovered in 1968; introd. in 1973. U.S. plant patent 3720; assigned to Herschel L. Boll, Champaign, Ill. Fruit: large, up to 7/8 inch in diam., hemispheric shape, black, pubescent; firm; ripens early. Plant: highly productive and vigorous, producing up to 11 canes per plant per year, canes stocky and arching.

Jewel.—Orig. in Geneva, N.Y., by New York Agr. Expt. Sta. N.Y. 29773 (Bristol x Dundee) x Dundee. Cross made in 1954 by G.L. Slate, selected in 1957, tested as N.Y. 628; introd. in 1973. Fruit: large; black, glossy, slightly pubescent; coherent; firm, quality good; ripens midseason, concentrated ripening habit (two to three pickings). Plant: vigorous; erect; winter hardy; productive; slightly susceptible to powdery mildew.

Joan Squire.—A high-yielding, primocane-fruiting raspberry with good fruit quality. Orig. in Kent, U.K., by D.L. Jennings, Medway Fruits. Selected from parents of complex origin; tested as 89111E2; introd. in 1995. Worldwide marketing rights held by NSA Plants, East Malling, Kent. Fruit: size similar to Autumn Bliss, larger than Heritage; firm; cohesive; nearly as firm as Heritage; attractive medium red with some gloss, no tendency to develop blue tinge; in southern England, ripens 2 weeks later than Autumn Bliss, 2 weeks before Heritage; good shelf life; good flavor. Plant: more productive that either Autumn Bliss or Heritage; numerous vigorous, spine-free canes; spreading habit; canes need support as fruit ripens.

Joe Mello.—Orig. in Watsonville, Calif., by S. Wilhelm. Reiter 323 x Chilcotin; selected in 1983; introd. in 1988. U.S. plant patent 6493 assigned to Sweetbriar Co. Fruit: medium size, 2.5 to 4 g; very attractive, bright red with slight darkening after harvest. Plant: primocane fruiting; medium vigor; sturdy, erect canes.

John Robertson.—Orig. in Hot Springs, S.D., by John Robertson. Parentage unknown; discovered in 1930; introd. about 1935. Fruit: black; good quality. Plant: productive; winter hardy.

Johnson Everbearing (*Black Magic*).—Orig. in Davis, W.Va., by A.W. Johnson. Parentage unknown; discovered in 1934, supposedly as a wild plant; introd. in 1945. Fruit: black; large. Plant: primocane fruiting; resembles Cumberland.

Julia.—Orig. in East Malling, England, by V.H. Knight and E. Keep, Hort. Res. Intl. Result of cross of an East Malling selection of complex parentage and a sibling of Glen Prosen, fourth backcross from the Cumberland black raspberry (*Rubus occidentalis*); cross made in 1976; tested as EM 4301/9; introd. in 1992. Plant varieties rights held jointly by NSA Plants, East Malling, and Hort. Res. Intl. Fruit: large; uniform shape; moderately firm; medium red, slightly dull; easily removed from receptacle; midseason ripening; canned and frozen qualities superior to Glen Clova; produces good-flavored jam. Plant: high yielding; moderate number of long, fairly erect canes; canes moderately spiny; long laterals tend to droop but are well attached. Resistant to cane *Botrytis*, spur blight, and cane spot; moderately resistant to cane blight; susceptible to root rot; has gene A_{10} conferring resistance to the four known strains of the European aphid vector of the raspberry mosaic virus complex; susceptible to infection by pollen transmission of raspberry bushy dwarf virus.

Kelleriis 5.—Orig. in Kvistgaard, Denmark, by Svend Poulsen, D.T. Poulsen's Nurseries. Faistrup x Preussen; selected in 1944; introd. in 1956. Fruit: red; harvest time starts at the same time as Preussen but finishes 2 to 3 weeks later. Plant: productive.

Killarney.—Orig. in Morden, Manitoba, Canada, by C.R. Ure, Canada Dept. Agr., Res. Sta. Chief x Indian Summer; selected in 1949; tested as Morden 532; introd. in 1961. Fruit: red; somewhat larger than Latham; bright, medium red; very attractive; flavor sweet, sprightly, pleasant; good for dessert and as frozen or canned product. Plant: moderately tall; sturdy; canes thick; very winter hardy; very productive.

Kiwigold.—Orig. in Murrinsville, New Zealand, by C.E.H. and M.A. Thomas. Sport of Heritage discovered in 1984; introd. in 1988. U.S. plant patent pending; assigned to Nourse Farms, South Deerfield, Mass. Fruit: medium; conic; yellow; firm; excellent quality; produced on primocanes; ripening time similar to Heritage. Plant primocanes vigorous, numerous, erect, sturdy; more spines than Heritage; no red pigmentation.

Lake Geneva.—Orig. in Lake Geneva, Wis., by Phil T. Robers, Robers' Nursery. Considered to be Latham x Ranere (St. Regis); discovered in 1948; introd. in 1960. Fruit: red; larger and firmer than Latham, with similar color; flavor better, making a better frozen product, than Latham; juice deep red; ripens several days after Latham. Plant: moderately winter hardy; very productive; moderately vigorous; fruiting canes lack rigidity. Somewhat resistant to common raspberry diseases. Adapted to eastern and southern Wisconsin.

Latham.—Orig. in Excelsior, Minn., by Univ. of Minnesota Fruit Breeding Farm. King x Louden; cross made in 1908; selected in 1914; tested as Minn. 4; introd. in 1920. Fruit: small to medium size; round; bright attractive red; somwhat soft and tends to crumble, medium juicy, medium sweet, sprightly; season medium to late, ripening over a long period; good for canning and freezing. Plant: vigorous, productive; very winter hardy. Susceptible to the North American aphid vector of the raspberry mosaic virus complex; susceptible to powdery mildew.

Lawrence.—Orig. in Watsonville, Calif., by S. Wilhelm. Sweetbriar x Reiter 239; introd. in 1992. U.S. plant patent 8022 assigned to Sweetbriar Development. Fruit: medium to large size; sweet, relatively low acid flavor; separates easily from receptacle. Primocane production begins about 1 Aug., floricane production about 15 May in central coastal California. Plant: vigorous; produces about one-third of total crop on primocanes. Some resistance to powdery mildew and to late leaf rust.

Leo.—Orig. in East Malling, England, by E. Keep and J.H. Parker, Hort. Res. Intl. Fourth backcross derivative of the Cumberland black raspberry (*Rubus occidentalis*); pedigree includes red raspberries Malling Landmark, Lloyd George, Pyne's Royal, Norfolk Giant, and

Burnetholm; cross made in 1964; selected in 1969; tested as EM 1312/19; introd. in 1977. Plant variety rights held jointly by Plant Breeding Intl. and Hort. Res. Intl. Fruit: large; almost round; rather pubescent; late ripening; excellent fresh flavor; suitable for processing. Plant: relatively high yielding; moderate number of very tall canes; rather spreading habit; long, strong fruiting laterals. Resistant to cane *Botrytis* and spur blight; has genes A_1 and A_{10} conferring resistance to the four known races of the European aphid vector of the raspberry mosaic virus complex; susceptible to raspberry bushy dwarf virus by graft inoculation, but may have some resistance to infection by pollen transmission.

Lloyd George.—Orig. as a chance seedling in a wood in Dorsetshire, England. Parentage probably the native *Rubus idaeus*; introd. J.J. Kettle, Corfe Castle, Dorset; introd. into North America in 1923. Fruit: medium size, soft, dark, excellent flavor. Plant: moderately productive and vigorous. Resistant to the common strain of the North American aphid vector of the raspberry mosaic virus complex. Has been used extensively in North American breeding programs as a source of aphid resistance, conferred by gene Ag_1, and in both North American and European programs as a source of fruit quality, particularly flavor.

Lowden Purple.—Orig. in Ancaster, Ontario, Canada, by W. Lowden, Lowden's Better Plants & Seeds. Lowden x a very late unnamed black seedling; cross made in 1956; selected in 1959; introd. in 1975. U.S. plant patent 3634. Fruit: purple; medium size, long conic; cluster size medium to large; firm, seeds very small, flavor more like reds than purples or blacks, sweeter than most other purples, good for pick-your-own operations; ripens very late. Plant: 7 to 8 ft high; upright; canes purplish, stout; hardy; vigor moderate, uniform; productive; does not sucker; propagated by tip layering. Resistant to anthracnose.

Lowden.—Orig. in Ancaster, Ontario, Canada, by E.W. Lowden. Considered to be Bristol x Sodus; introd. in 1961; U.S. plant patent 2013. Fruit: large; dull purplish-black when fully ripe; juicy, sweet; canning quality good; ripens fairly late, averaging 17 July to 7 Aug. Plant: quite vigorous; upright; productive; cane color distinctive, ranging from dark red to bright maroon; spines about 1/8 inch long, borne at right angles to the cane.

Madawaska.—Orig. in Ottawa, Ontario, Canada, by Canada Dept. Agr., Res. Sta. Lloyd George x Newman; selected in 1931; tested as Ottawa 272; introd. in 1943. Fruit: medium large; medium dark red; moderately firm, some tendency to crumble, quality fair; ripens early; fair quality only and tends to be acid. Plant: canes medium spiny; medium short; winter hardy; productive.

Magnaberry.—Orig. in Ontario, Calif., by H.C. Swim, Armstrong Nurseries. F_3 from Nessberry x Sodus raspberry; selected in 1945; introd. in 1949; U.S. plant patent 1008. Fruit: black; large; nearly round; combines flavor of the raspberry and blackberry; excellent for jam or jelly with a distinct flavor and fine appearance; separates readily from receptacle; early ripening with long season. Plant: large; very vigorous; productive.

Malahat.—A high-yielding, early ripening fresh-market raspberry. Orig. in Vancouver British Columbia, Canada, by H.A. Daubeny and C. Kempler, Agriculture and Agri-Food Canada, Pacific Agr. Res. Centre. Meeker x BC/SCRI 7853/116 (selection has complex parentage, including Nootka, *Rubus occidentalis, R. phoenicolasius*); cross made by HAD in 1985; selected in 1987; tested as BC 85-5-24; introd. in 1996. Propagated under a royalty agreement by Sieglin Ent., Abbotsford, British Columbia Fruit: large; firm; conic; glossy medium red; easy to harvest; flavor not as sweet as Chilliwack but considered good; good shelf life; intermediate reaction to *Botrytis* postharvest rot; ripens earlier than any other high-quality Pacific Northwest variety. Plant: high yield; vigorous, fairly upright habit; adequate primocane numbers; winter hardiness appears similar to Meeker and less than Chilliwack. Relatively susceptible to spur blight; susceptible to cane spot; resistant to cane *Botrytis*; relatively susceptible to *Phytophthora* root rot in greenhouse screening test; susceptible to natural infection by raspberry bushy dwarf virus; resistant to the common strain of the North American aphid vector of the raspberry mosaic virus complex.

Malling Augusta.—Orig. in East Malling, England, by E. Keep and V.H. Knight, Hort. Res. Intl. Third backcross derivative of *Rubus cockburnianus* x Malling Joy; cross made in 1974; tested as EM 3650/6; introd. in 1989. Fruit: medium to large; firm; oval to conic; dark red, slightly dull; easily removed from receptacle; moderate acid flavor; late fresh-market use. Plant: moderate yield; moderate number of tall, rather spreading canes; deeply pigmented; glabrous with medium-sized spines; long, very floriferous laterals; fruit well presented; occasionally produces floriferous basal laterals that tend to ripen later than other laterals. Somewhat susceptible to cane *Botrytis* and to spur blight; some resistance to cane spot; has gene A_{10} conferring resistance to the four known races of the European aphid virus vector of the raspberry mosaic virus complex; appears to have resistance to infection by pollen transmission of the common strain of raspberry bushy dwarf virus but becomes infected by graft inoculation.

Malling Enterprise.—Orig. in East Malling, Kent, England, by N.H. Grubb, East Malling Res. Sta. Preussen x Malling 23/50; selected in 1932; tested as Malling 33/132 and as seedling E; introd. in 1943. Fruit: size medium; short conic; dark red, dull; drupelets cohere fairly well, varying in number and size; quality good; aroma strong; ripens fairly late; resembles Malling Jewel. Plant: canes tall, erect; leaves dull, grayish-green.

Malling Exploit.—Orig. in East Malling, England, by N.H. Grubb, East Malling Res. Sta. Newburgh x Malling 30/8 (Lloyd George x Pyne's Royal) selected in 1937; introd. in 1950. Fruit: large; firm, red; picks easily; ripens early; resembles Malling Promise. Plant: productive; adapted to heavy, poorly drained soil. Virus tolerance fair.

Malling Jewel.—Orig. in East Malling, England, by N.H. Grubb, East Malling Res. Sta.; Preussen x Malling 23/50 (Pyne's Royal selfed x Lloyd George selfed); selected in 1932; tested as Malling 33/54 and as Seedling J; introd. in 1950. Fruit: medium size; soft, dull, light red; easy to remove from receptacle; processing use. Plant: moderately productive, compact habit with moderate vigor; upright canes. Susceptible to all known strains of the European aphid vector of the raspberry mosaic virus complex but appears to have some tolerance to the complex itself.

Malling Joy.—Orig. in East Malling, England, by E. Keep, V.H. Knight, and J.H. Parker, Hort. Res. Intl. Parents are of complex origin; each is from a fourth backcross derivative of the Cumberland black raspberry (*Rubus occidentalis*) and also has Malling Promise, Malling Landmark, Burnetholm, Pyne's Royal, Lloyd George, and Norfolk Giant in its ancestry; cross made in 1968; selected in 1973; tested as EM 2467/123; introd. in 1980. Fruit: large, blunt conic, very firm; slightly dark and pubescent; fairly easily removed from receptacle; late ripening, up to 2 days earlier than Leo; good, slightly acid fresh flavor; postharvest susceptibility to fruit rots similar to Glen Clova and Malling Jewel; good quality for fresh market and jam making. Plant: very high yield due to large fruit size and high number of fruit per lateral; tall canes, rather spreading; laterals very long, strong, and well attached, held horizontally; spiny; fruit well exposed. Has genes A_1 and A_{10} conferring resistance to the four known strains of the European aphid vector of the raspberry mosaic virus complex; susceptible to raspberry bushy dwarf virus by graft inoculation and probably by pollen transmission.

Malling Landmark.—Orig. in East Malling, England, by N.H. Grubb, East Malling Res. Sta. Preussen x Baumforth A; selected in 1933; tested as Malling 34/42 and as Seedling L; introd. in 1943. Fruit: quality good; resembles Baumforth A. Plant: productive. Resistant to strains 1 and 3 of the European aphid vector of the raspberry mosaic virus complex.

Malling Notable.—Orig. in East Malling, England, by N.H. Grubb, East Malling Res. Sta. Preussen x Lloyd George; selected in 1932; tested as Malling 32/163 and as Seedling N; introd. in 1947. Fruit: large; round conic; light crimson; fairly firm, quality good. Plant: moderately strong growing; producing numerous erect canes.

Malling Promise.—Orig. in East Malling, England, by N.H. Grubb, East Malling Res. Sta. Newburgh x Malling 30/8 (Pyne's Royal selfed x Lloyd George selfed F_2); selected in 1937; tested as Malling 51/79; introd. in 1944. Fruit: good quality; ripens early; resembles Malling Exploit. Plant: productive. Virus tolerance fair.

Mandarin.—Orig. in Raleigh, N.C., by C.F. Williams, North Carolina Agr. Expt. Sta. (*Rubus parvifolius* x Taylor) x Newburgh; cross made in 1940; selected in 1943; introd. in 1955. Fruit: red; good quality, resembles Latham. Plant: adapted for areas south of where red raspberry varieties are usually grown; more resistant to septoria leaf spot and high temperatures than most varieties.

Manteo.—Orig. in Raleigh, N.C., by C.F. Williams, North Carolina Agr. Expt. Sta. Cumberland selfed, F_2 segregate from selfing in 1937; selected in 1939; introd. in 1953. Fruit: black; resembles Cumberland. Plant: survival in North Carolina better than any other black raspberry variety.

Marcy.—Orig. in Geneva, N.Y., by R. Wellington, New York State Agr. Expt. Sta. Lloyd George x Newman; seedling first fruited in 1928; introd. in 1936. Fruit: red; extremely large; soft, flavor mild; early ripening; quality good but too dark when fully ripe. Plant: vigorous; not winter hardy. Very susceptible to infection by pollen transmission of raspberry bushy dwarf virus.

Marion.—Orig. in Geneva, N.Y., by R. Wellington, New York State Agr. Expt. Sta. Bristol x N.Y. 2585 (Newman x Herbert); selected in 1930; introd. in 1937. Fruit: purple; large; moderately juicy, firm; tart, quality good; season late. Plant: resembles red raspberry in habit.

Marve.—A high-yielding, late-ripening raspberry adapted to western Europe. Orig. in Kapelle, The Netherlands, by J. Westveer. Probably an open-pollinated seedling of Jochems Roem; selected in 1973; introd. in 1990. Licensed by E.C. Meulbloc Boomkwekerij, 4434 AG Kwadendamme Zeeland, Netherlands. Fruit: large, bright medium to dark red; firm; conic shape; moderately easy to harvest; good flavor; ripens a few days later than Schonemann; some resistance to *Botrytis* rot. Plant: high yield; vigorous, strong canes with high density of spines; relatively few primocanes. Probably has some resistance to cane diseases and to *Phytophthora* root rot.

Mary Lewis.—Orig. in Holland, Mich., by E. De Roo. Considered to be Superlative x Ranere; selected in 1914; introd. in 1920. Fruit: large; pointed; crimson red; firm, good flavor; ripens early. Plant: productive; upright.

Matsqui.—Orig. in Agassiz, British Columbia, Canada, by H.A. Daubeny, Canada Dept. of Agr., Res. Sta. Sumner x Carnival. Selected in 1961, tested as 59-17-61 and BC 202; introd. in 1969. Fruit: medium size; firm; bright medium red, attractive, does not darken on picking; easily removed from the receptacle; early ripening; mild flavor; good for freezing; some tolerance to *Botrytis* rot. Plant: moderately productive; moderately vigorous; abundant primocane production; floricanes, brown with purple overlay, are smooth, sturdy, and upright; winter hardy. Some root rot tolerance; susceptible to the North American aphid vector of the raspberry mosaic virus complex.

Meco.—A fresh market raspberry adapted to France. Orig. in d'Angers, by V. Bellenot-Kapusta, Station D'Amelioration Des Especes Fruitieres Et Ornementales; Meeker x Rose de Cote d'Or; introd. in 1990. Fruit: medium size; soft; good flavor. Plant: very high yield. Tolerant to *Phytophthora* root rot in the field.

Meeker.—Orig. in Puyallup, Wash., by C.D. Schwartze, Western Washington Res. and Ext. Center, Washington State Univ. Willamette x Cuthbert; cross made in 1950; selected in 1953; tested as WSU 408; introd. in 1967. Fruit: medium size; medium red; firm, midseason ripening; good flavor; processing use. Plant: productive; tall; vigorous canes with pale pink spines; long laterals; only moderately winter hardy in the Pacific Northwest. Susceptible to infection by pollen transmission of raspberry bushy dwarf virus; susceptible to the North American aphid vector of the raspberry mosaic virus complex; susceptible to spur blight; resistant to cane *Botrytis*; some field resistance to root rot.

Milton.—Orig. in Geneva, N.Y., by G.L. Slate, New York State Agr. Expt. Sta. Lloyd George x Newburgh; cross made in 1927; introd. in 1942. Fruit: medium size; bright red; firm, good flavor, ripens late. Plant: vigorous; tall; productive; winter hardy.

Morris.—See Morrison.

Morrison (*Morris*).—Orig. in North Kingsville, Ohio, by F. Morris. Parentage unknown; discovered about 1925; introd. in 1942. Fruit: large; glossy, attractive black; firm, fair quality; ripens late; resembles Cumberland. Plant: productive.

Mortgage Lifter.—Orig. in Berrien County, Mich., by O.K. White. Parentage unknown; discovered in 1946 as a spontaneous seedling along a fence row; introd. in 1949. Fruit: medium size; black; sweet; early ripening, continuing to bear throughout the season; resembles both Cumberland and Logan.

Muskoka.—Orig. in Ottawa, Ontario, Canada, by Canada Dept. Agr., Res. Sta. Newman x Herbert; selected in 1924; tested as Ottawa 201; introd. in 1950. Fruit: large; deep red; firm. Plant: most valuable characteristic is its extreme hardiness on Canadian prairies; productive; canes smooth.

Mysore.—Orig. in India as a selection of a native species, *Rubus albescens* Roxb. Introd. into the U.S. in 1948. Achenes sent to the Univ. of Florida in 1947 by F.B. Harrington of Natal, South Africa; originally introd. from India; then selected plants were distributed to nurserymen in southern Florida. Fruit: black; covered with soft hairs; flavor mild, slightly sweet, flavor good; achenes small; similar to western native black raspberry; main crop ripens March to May in Florida. Plant: evergreen; shoots glaucous, with prominent spines; canes vigorous, growing 10 to 15 ft per season; adapted only in regions (subtropical) where freezing temperatures do not occur; apparently more disease resistant than most raspberries; flower to 1/2 inch in diam., petals purplish-pink, in terminal and axillary clusters of eight to many. Propagated by achenes, cuttings, or tip layering (the easiest method). Useful in breeding work and very suitable for home gardens.

New Hampshire.—Orig. in Durham, N.H., by A.F. Yeager, New Hampshire Agr. Expt. Sta. Open-pollinated F_2 seedling from an F_1 of Taylor x *Rubus chamaemorus*; introd. in 1955. Fruit: red; very firm; ripening in midseason; harvest season long and continuous. Plant: winter hardy in northern U.S.; productive.

New York Times (*Golden Indian Summer*).—Orig. in Otisville, Orange County, N.Y., by B. Altbayer. Bud mutation of Indian Summer; discovered in 1950; introd. in 1952. Fruit: large; yellow; good flavor; bears in summer and fall; picks easily; resembles Golden Queen. Plant: vigorous.

Newberry.—Orig. in Port Angeles, Wash., by F.A. Newberry. Bud mutation of either Cuthbert or Ranere; now considered by some to be a chance seedling, possibly of Lloyd George; introd. in 1951; U.S. plant patent 1031. Fruit: large; elongated; dark red; medium soft, good quality; highly flavored; drupelets large and numerous; matures in midseason; resembles Lloyd George. Plant: grows upright for a short distance and then bends to a horizontal position, easily trained.

Newburgh.—Orig. in Geneva, N.Y., by R. Wellington, New York State Agr. Expt. Sta. Newman x Herbert; cross made in 1922; introd. in 1929. Fruit: bright red, attractive; medium size; fairly firm; fair quality and flavor; keeping and shipping quality good; does not process well; early to midseason ripening. Plant: productive; winter hardy. Resistant to root rot; susceptible to the North American aphic vector of the raspberry mosaic virus complex; moderately susceptible to powdery mildew.

Newman (*Newman 23*).—Orig. in Ville La Salle, Que., Canada, by C.P. Newman. Parentage unknown; discovered about 1918; introd. in 1924. Fruit: red; large; flavor mild; midseason. Plant: susceptible to mosaic.

Newman 23.—See Newman.

Nootka.—Orig. in Vancouver, British Columbia, Canada, by H.A. Daubeny, Agr. Canada Res. Sta. Carnival x Willamette; cross made in 1964; selected in 1966; tested as BC 64-6-142; introd. in 1978. Fruit: round to conic, firm, medium to dark red; separates readily from receptacle, resistant to pre- and postharvest rot, long harvest season; drupelets numerous and medium; well adapted to machine harvest. Plant: productive; vigorous; canes fairly upright; relatively long laterals; winter hardy in the Pacific Northwest. Susceptible to spur blight; moderately susceptible to cane *Botrytis*; some resistance to root rot; resistant to the common strain of the North American aphid vector of the raspberry mosaic virus complex.

Nordic.—Orig. in St. Paul, Minn. by J. Luby, E. Hoover, D. Bedford, S. Munson, W. Gray, D. Wilding, and C. Stushnoff, Dept. of Hort. Sci. and Landscape Architecture, Univ.of Minnesota. Boyne x Fall Red; cross made in 1969; selected in 1972; tested as MN 603; introd. in 1987. Fruit: produced on floricanes and on primocanes; most similar to Boyne, but lighter color and superior firmness and flavor; frozen product similar to Boyne for appearance, texture, and flavor; primocane fruit ripens 2 to 3 weeks later than Heritage. Plant: high yield from floricanes, low yield from primocanes; primocanes green, sparse purple spines largely restricted to basal portions; primocane production moderate with frequent branching; winter hardy. Reaction to spur blight similar to Boyne; less susceptible to cane spot than Boyne, susceptible to tomato ring spot virus; susceptible to infection from pollen transmission of raspberry bushy dwarf virus.

Norna.—Orig. at the Njos Res. Sta., Norway. Preussen x Lloyd George; cross made in 1944; introd. in 1964. Fruit: large; round; dark red with medium gloss; medium firm; well exposed and easy to harvest; large drupelets; suited for preserving; flavor mild, not sweet; early to medium-early ripening. Plant: moderate number of canes; erect; very vigorous; easy to manage; canes dull, not waxy, some pubescence; moderate number of spines; winter hardy. Susceptible to cane *Botrytis*.

Nova.—Orig. in Kentville, Nova Scotia, by D.L. Craig. Res. Sta., Agr. Can. Southland x Boyne; selected in 1970; introd. in 1981. Fruity: medium size; firm; good quality; somewhat acid; bright red, attractive; ripens midseason. Plant: very winter hardy; primocanes sturdy. upright, only a few spines; produces long laterals with fruit exposed to outside of row; productive.

OAC Regal.—Orig. in Vineland, Ontario, Canada, by L. Ricketson, Ontario Ministry of Agr. and Food. Latham x Milton; cross made in 1968; tested as GU 74; introd. in 1991 by J.A. Sullivan, Dept. Hort. Sci., Univ. of Guelph. Fruit: medium size, larger than Boyne; medium red; medium firm; conic; separates readily from receptacle; season 4 or 5 days later than Boyne; fresh-market use. Plant: productive; upright habit; numerous, vigorous primocanes with purple overlay; relatively few spines; winter hardy in Ontario and the Montreal area of Quebec. Susceptible to anthracnose.

OAC Regency.—Orig. in Vineland, Ontario, Canada, by L. Ricketson, Ontario Ministry of Agr. and Food. Creston x Muskoka; cross made in 1968; tested as GU 75; introd. in 1991 by J.A. Sullivan, Dept. of Hort. Sci., Univ. of Guelph. Fruit: medium size, larger and firmer than Boyne; conic; separates readily from receptacle; season 1 or 2 days later than Boyne; fresh-market use. Plant: more productive than OAC Regal and Boyne; moderate number of vigorous primocanes; relatively few spines; very winter hardy in Ontario and the Montreal area of Quebec. Susceptible to cane disease.

Ott's Pennridge.—Orig. in Ottsville, Pa., by H. Lloyd Ott. Chance seedling of Latham; discovered in 1957; introd. in 1965. Fruit: larger than Latham and ripens 10 days later; red; flavor good. Plant: vigorous, averages 1 to 2 ft taller than Latham. Plant: very productive.

Ottawa.—Orig. in Ottawa, Ontario, by Canada Dept. Agr., Res. Sta. Viking x (Logan blackberry x Ranere); selected in 1931; tested as Ottawa 275; introd. in 1943. Fruit: medium size; firm; red; good shipper; resembles Viking; ripens with Newburgh. Plant: unproductive. Very susceptible to powdery mildew.

Park Black.—Orig. in Des Moines, Iowa, by C. Park. Open-pollinated seedling of Cumberland; discovered in 1929; introd. in 1950 by the Earl Ferris Nursery, Hampton, Iowa. Fruit: long and large; uniform size; black; good flavor. Plant: winter hardy; vigorous; productive.

Pathfinder.—Orig. in Cheyenne, Wyo., by G.S. Howard, Agr. Res. Sta., U.S. Dept. of Agriculture. Augustred x Cheyenne 62-17 (*Rubus strigosus*); cross made in 1964; selected in 1967; tested as 67-10; introd. in 1976. Cheyenne 62-17 is a third generation selection from indigenous *Rubus strigiosus* near Greybull, Wyo. Fruit: up to 22 mm in diam., hemispheric; medium to deep red; medium firm, somewhat less flavorful than standard varieties; primocane fruit ripens first week of August; fruiting season lasts about 6 weeks or until heavy frost occurs. Plant: vigorous; produces numerous canes; medium to high yield; winter hardy.

Perron's Red.—Orig. in Quebec, Canada, by W.H. Perron and Co. Zeva Herbstemte x Pathfinder; selected in 1977; introd. in 1987. Fruit: large; conic; medium red, slightly more glossy than Heritage; less firm than Heritage; produced on primocanes; season similar to Heritage in British Columbia. Plant: vigorous; numerous canes, fairly erect, glabrous, light green with some red overlay, purple spines. Susceptible to the North American aphid vector of raspberry mosaic virus complex.

Pocahontas.—Orig. in Blacksburg, Va., by G.D. Oberle, Virginia Agr. Ext. Serv., Virginia Polytechnical Inst. & State Univ. Hilton x VPI 5 (Taylor x St. Regis). Cross made in 1958, selected in 1964, tested as VPI 14; introd. in 1972. Fruit: red; large; medium conic to long conic; medium red with dusky overcast, attractive; firm, small drupelets, texture fine, flavor aromatic; ripens 25 June to 15 July, somewhat hidden. Plant: larger than average; upright; very vigorous; winter hardiness above average, tolerant to fluctuating winter temperatures; very productive. Tolerant to anthracnose, cane blight, and leaf spot.

Polana.—Orig. in Brzezna, Poland, by Jan Danek and Zofia Pasiut, Expt. Sta. of the Res. Inst. of Pomology and Floriculture. Heritage x Zeva Herbsternte; tested as 81221; introd. in 1991. Fruit: medium size; compact, wide-conic; ripens earlier than Heritage; aromatic and tasty but quality deteriorates in late season. Plant: primocane fruiting; more productive than Heritage; canes numerous, uniform medium height, erect, covered with wax, pubescent; spines scarce; primocane fruit borne on laterals distributed from the tip to mid-cane; some fruit also produced from laterals on the lower part of the canes in summer cropping season. Low to medium susceptibility to diseases and pests. In laboratory tests showed some resistance to the European aphid vector of the raspberry mosaic virus complex.

Polaris.—Orig. in Roseau, Minn., by L. Renner. Chance seedling; introd. in 1972. Fruit: red; characteristics similar to Latham, ripening 10 days earlier. Plant: short cane; prolific plant maker; winter hardy throughout Minnesota.

Potomac.—Orig. in Glenn Dale, Md., by G.F. Waldo, U.S. Dept. of Agriculture. Farmer (black) x Newman (red); selected in 1924; introd. in 1933. Fruit: purple; medium size; firm, good flavor; suited for canning and preserving. Plant: vigorous and productive; resistant to fungus disease.

Prestige.—Orig. in Durham, N.H., by A.F. Yeager and E.M. Meader, New Hampshire Agr. Expt. Sta. Second generation seedling of 'Taylor' x *Rubus pugens oldhami*; introd. in 1979. Tested as NH TXP15; selected in 1964; introd. in 1979. Fruit: medium size, bright red, tart, pleasant flavor, ripens early; large drupelets; easily separated from distinctive yellow or orange receptacle when fully ripe. Plant: winter hardy, vigorous, productive, suckers sparingly, propagated by regeneration from roots left in ground when dug or transplanted.

Princess.—An early-ripening primocane-fruiting raspberry. Orig. in Soings-en-Sologne, France by Jacques Marionnet. Autumn Bliss x Meeker. Protected by COV 64469. Fruit: medium size; medium red; length and width similar; good flavor; early ripening season similar to Autumn Bliss in France. Plant: high yield; very few spines; moderate primocane numbers.

Purple Autumn.—Orig. in Urbana, Ill., by A.S. Colby, Ill. Agr. Expt. Sta. Bristol (black) x Indian Summer (red); cross made in 1938; selected in 1940; introd. in 1953. Fruit: large; medium purple; firm, tart, good flavor; easily picked; primocane fruit matures later than the September variety; floricane fruit earlier than Sodus and Marion. Plant: productivity medium to high; easily propagated by leaf–bud cuttings.

Puyallup.—Orig. in Puyallup, Wash., by C.D. Schwartze and A.S. Myhre, Washington State Univ. Washington x Taylor; selected in 1943; introd. in 1953. Fruit: mild, sweet flavor, medium red; medium firm; midseason ripening; fresh and processing use. Plant: productive on well-drained soil; witer hardy in the Pacific Northwest; canes vigorous, medium to large, straight, fruiting laterals medium long and strong. Susceptible to powdery mildew, root rot, and to the North American aphid vector of the raspberry mosaic virus complex.

Qualicum.—A high-yielding, midseason raspberry producing large, firm fruit with a good shelf life. Orig. in Vancouver, British Columbia, Canada, by H.A. Daubeny, Agr. and Agri-Food Canada, Pacific Agr. Res. Centre. Glen Moy x Chilliwack; cross made in 1983; selected in 1986; tested as BC 83-16-50; introd in 1995. Propagated under a royalty agreement by Sieglin Ent., Abbotsford, British Columbia. Fruit: large; conic; medium red, slightly less glossy than Chilliwack; easily removed from the receptacle; can be machine-harvested; relatively high soluble solids; pleasant fresh flavor; some resistance to pre- and postharvest fruit rots; fresh market and processing, including individual quick freeze. Plant: vigorous, upright habit; primocanes fairly numerous, light rosy-red color with dark purple spines mostly at base; floricanes light brown with gray overlay and basal cracking; long, strongly attached laterals tend to droop; fruit spaced well apart along laterals; winter hardiness appears adequate in the Pacific Northwest. Some resistance to spur blight; resistant to powdery mildew; susceptible to cane *Botrytis* and cane spot; susceptible to *Phytophthora*-incited root rot in greenhouse screening test; susceptible to infection by pollen transmission of raspberry bushy dwarf virus; resistant to the common strain of the North American aphid vector of the raspberry mosaic virus complex.

Rafzeter.—See Himbo Queen.

Rakaia.—A midseason raspberry adapted to cooler conditions in New Zealand. Orig. in Lincoln, New Zealand, by I.K. Lewis, DSIR Crop Research Division. Marcy x Fairview; cross made in 1979; tested as E6; introd in 1986. Fruit: larger than Marcy; dark red; conic ovate shape; easy to harvest; ripens 12 to 16 days after Marcy. Plant: yields less than Marcy or Skeena under warmer conditions but higher than either under cooler conditions; small numbers of medium-strength, semi-upright canes; floricanes dark brown with prominent basal spines of moderate length; strongly attached laterals bear 20 to 30 fruit. Very susceptible to raspberry bud moth; may have resistance to infection by pollen transmission of raspberry bushy dwarf virus.

Red River.—Orig. in Morden, Manitoba, by C. Davidson, Agr. Canada Res. Sta. (Fall Red x primocane-fruiting type derived from *Rubus strigosus* from Cheyenne, Wyo.) x (Fall Red x Boyne); cross made in 1968; selected in 1978; tested as MRS#8008; introd. in 1993. Propagated under a royalty agreement by Jefferies Nurseries, Portage la Prairie, Manitoba. Fruit: intermediate size maintained during main part of harvest season; glossy medium red; hemispheric; good drupelet coherence; high soluble solids; low pH; sweet and tart, good flavor; earlier ripening than either Redwing or Heritage; fresh-market use. Plant: primocane-fruiting; short, stout canes with sparse short spines; trellis support may be required to support heavy fruit loads; moderate production of canes; winter hardy, suitable for harsh prairie climate. Susceptible to anthracnose but can be controlled by an annual mowing system.

Redbrook.—Orig. in Brooks, Alberta, Canada, by S. Mahadeva, Alberta Special Crops and Horticultural Res. Center. Chief x Boyne; tested as BR14; introd. in 1984. Fruit: large, bright red; sweet, with typical raspberry flavor. Plant: high yield on sturdy canes; winter hardy.

Redman.—Orig. in Parkside, Saskatchewan, Canada, by A.J. Porter, Honeywood Nursery. Seedling of purple raspberry x (Viking x a native wild selection); selected about 1952; introd. in 1961. Fruit: slightly smaller than Viking, red; firm, good quality; early season, ripening most of its crop during the first 2 weeks of the season; fresh and processing use. Plant: more winter hardy than Honeyking, but not as productive.

Redwing.—Orig. in St. Paul, Minn., by J. Luby, E. Hoover, D. Bedford, S. Munson, W. Gray, D. Wilding, and C. Stushnoff, Dept. of Hort. Sci. and Landscape Architecture, Univ. of Minnesota. Heritage x Fall Red; cross made in 1969; selected in 1972; tested as MN 629; introd. in 1987. Fruit: size and color similar to Heritage in Minnesota, but color lighter and brighter in British Columbia; firmness slightly less than Heritage; flavor similar or slightly inferior to Heritage; produced on primocanes in August; ripens 10 to 14 days earlier than Heritage in Minnesota; season similar to Heritage in British Columbia Plant: yields usually lower than Heritage; primocanes vigorous; moderate numbers; stout; green with light red overlay; moderately dense; light red to pink canes; short, purple spines. Susceptible to the North American aphid vector of the raspberry mosaic virus complex; susceptible to cane spot.

Reveille.—Orig. in College Park, Md., by I.C. Haut, Maryland Agr. Expt. Sta. [VPI 1 (Indian Summer x Sunrise)] x September; introd. in 1966. Fruit: large; bright red; good quality, but too soft for commercial use; for home or roadside sales; ripens early. Plant: excellent resistance to fluctuating winter temperatures.

Rideau.—Orig. in Ottawa, Ontario, Canada, by Canada Dept. Agr., Res. Sta. Lloyd George x Newman; selected in 1931; tested as Ottawa 262; introd. in 1943. Fruit: small; red; conic; bright and attractive color; firm, good quality; good shipper; good for processing. Plant: medium short; winter hardy; productive. Susceptible to powdery mildew. Resistant to the common strain of the North American aphid vector of the raspberry mosaic virus complex.

Romy.—Orig. in Meyrin, near Geneva, Switzerland, by R. Romy. Seedling of Lloyd George. Introd. in 1954. Plant patent 1258; assigned to W. Hauenstein Sohne, Rafz (Zurich), Switzerland. Fruit: medium large; red; juicy, soft, flavor subacid, aromatic, quality good; shipping quality fair. Plant: primocane fruiting; canes relatively erect; habit of growth low growing; produces many shoots throughout the growing season; productive.

Rossana.— Orig. at Univ. of Torino, Italy, by R. Paglietti. Open-pollinated Malling Promise; tested as MP 45. Fruit: medium size, larger than Heritage; bright red; conic; good flavor; not as firm as Heritage; ripens 1 week later than Heritage in Italy and 3 to 4 weeks later than Heritage in Minnesota; local and distant fresh markets. Plant: primocane fruiting; productive; fruit spaced well apart along laterals and well-exposed; overwintered canes not winter hardy in Minnesota; hardy to USDA zone 4. Some resistance to rot.

Royalty.—Orig. in Geneva, N.Y., by J.C. Sanford and D.K. Ourecky, Cornell Univ. N.Y. 253 (Cumberland x Newburgh) x N.Y. 17861 (Newburgh x Indian Summer). Cross made in 1954; selected in 1957; tested as N.Y. 632; introd. in 1982. U.S. plant patent 5405. Fruit: very large; purple; good flavor; sweet. Plant: productive; widely adapted; winter hardy. Resistant to the North American aphid vector of the raspberry mosaic virus complex, to the aphid vector of raspberry leaf curl, and to budworm.

Ruby (*Watson*).—Orig. in Geneva, N.Y., by J. Sanford, K. Maloney, and J. Reich. Heritage x Titan. Cross made in 1980; selected in 1982; tested as NY114; introd. in 1988. U.S. plant patent 7067; assigned to Nourse Farms, South Deerfield, Mass. Fruit large; produced on primocanes; ripens several days later than Heritage in New York; conic; bright medium red; flavor rated similar to or slightly inferior to that of Heritage; ease of separation from receptacle can vary; susceptible to pre- and postharvest fruit rot. Plant: yields similar to Heritage; vigorous; numerous primocanes; somewhat spreading; green with some purple overlay; sharp, light purple spines distributed uniformly and sparsely. Susceptible to root rot, crown gall, and the North American aphid vector of the raspberry mosaic virus complex; in California, reported resistant to yellow rust and powdery mildew.

Rucami.—Orig. by R. Bauer at the Max-Planck-Inst., Koln. '4a' x Andenken an Paul Camenzind; cross made in 1960; selected in 1976 in Breitbrunn/Chiemsee; introd. in 1979 by Ahrens and Sieberz, D-5200 Siegburg-Seligenthal. Fruit: large; oval-conic with large drupelets; medium to dark red; medium firm; aromatic flavor with high acidity; ripens midseason; fresh-market and deep-freezing use. Plant: moderately productive; vigorous; thick spiny canes with long, strong fruiting laterals; primocanes numerous. Resistant to the most common race of the European aphid vector of the raspberry mosaic virus complex and to cane and root diseases. Somewhat susceptible to wet and windy weather so needs windbreak.

Rucanta.—See Rutrago

Ruddy (*Garnet*).—Orig. in Fargo, N.D., by A.F. Yeager, North Dakota Agr. Expt. Sta. Selected F_2 seedling (Plum Farmer x Latham); selected in 1933; introd. in 1937. Fruit: purple; medium size; flavored like a black raspberry; fresh use and for jams and jellies. Plant: productive; reproduces by suckers as do red varieties; winter hardy; resistant to red spider mite.

Rumilo.—See Rumiloba.

Rumiloba (*Rumilo*).—Orig. by R. Bauer at the Max-Planck-Inst., Koln. Promiloy [Malling Promise x (Lloyd George x Preussen)] x '4a'; cross made in 1960, selected in 1976 in Breitbrunn/Chiemsee; introd. in 1985 by Haberli, CH-9315 Neukirch-Egnach. Fruit: large; broad oval-conic; drupelets numerous, large; attractive medium red; easy to harvest; very aromatic flavor; late ripening season; moderately susceptible to preharvest *Botrytis* rot; good shelf life; fresh-market and freezing use. Plant: medium to high yield; very vigorous; upright habit; moderately spiny; number of primocanes high. Resistant to the most common race of the European aphid vector of the raspberry mosaic virus complex, and to cane and root diseases.

Rusilva (*Brabant*).—Orig. by R. Bauer at the Max-Planck-Inst., Koln. Geloy [Geneva 17861 x (Lloyd George x Preussen)] x '4a'; cross made in 1960; selected in 1976 in Breitbrunn/Chiemsee; introd. in 1987 by Haberli, CH-9315 Neukirch-Egnach. Fruit: round; light to medium red; firm; rather sour; aromatic; ripens midseason; well displayed on the plant and easy to pick; moderately susceptible to preharvest *Botrytis* rot; fresh-market use. Plant: productive; moderate vigor; nearly spine-free; thin canes with long to medium length laterals; primocanes numerous; tends to develop flowers and fruit in late autumn. Resistant to the most common race of the European aphid vector of the raspberry mosaic virus complex, and to cane and root diseases.

Rutrago (*Rucanta*).—Orig. by R. Bauer at the Max-Planck-Inst., Koln. '4a' x Tragilo [Geneva 16134 x (Lloyd George x Preussen)]; cross made in 1960; selected in 1976 in Breitbrunn/Chiemsee; introd. in 1979 by Ahrens and Sieberz, D-5200 Siegburn-Seligenthal. Fruit: medium size; round conic; firm; medium-red color; very aromatic with high sugar and acid components; late ripening season; easy to pick; fresh-market and freezing uses. Plant: productive; vigorous; very strong, erect-growing canes; medium length to long basal laterals; medium number of spiny primocanes. Resistant to the most common race of the European aphid vector of the raspberry mosaic virus complex, and to cane and root diseases.

Scepter.—Orig. in College Park, Md., by I.C. Haut, Maryland Agr. Expt. Sta. September x Durham; introd. in 1966. Fruit: large; medium red; moderately soft; resembles September; ripens 10 days earlier than September. Plant: very vigorous; highly tolerant to fluctuating winter temperature injury; primocane fruiting.

Selwyn.—An early ripening raspberry adapted to New Zealand. Orig. in Nelson, New Zealand, by H. Hall, Riwaka Res. Center. Marcy x Malling Delight; cross made in 1979 by H.A. Daubeny, Agr. and Agri-Food Canada, Pacific Agr. Res. Centre, Vancouver, British Columbia; Canada, seedlings grown in 1984 and selection made in Lincoln, New Zealand, in 1987; tested as F36; introd. in 1992. Fruit: medium size; medium red; soft; conic ovate shape; difficult to harvest; season similar to Marcy. Plant: slow to establish with small to medium number of upright canes; floricanes medium red to brown with bloom and basal spines; strongly attached, medium-length laterals with 10 to 20 fruit each. May have resistance to infection by pollen transmission of raspberry bushy dwarf virus.

Sensation.—Orig. in Champaiyn, Ill., by H.L. Boll. Indian Summer selfed. Cross made in 1967; introd. in 1974. Berry: large, up to 1 1/2 inch; long conic; medium red; firm, high in flavor; ripens early. Plant: produces on both primocanes and floricanes; first crop larger than second; low to medium productivity; medium vigor, growing up to 4 ft tall; producing medium number of suckers.

Sentinel.—Orig. in College Park, Md., by I.C. Haut, Maryland Agr. Expt. Sta. Sunrise x Milton; introd. in 1966. Fruit: medium light red; firm; resembles Milton; ripens late to midseason. Plant: excessive suckering; some resistance to injury from fluctuating winter temperatures.

Sentry.—Orig. in College Park, Md., by I. C. Haut, Maryland Agr. Expt. Sta. Sunrise x Taylor; introd. in 1966. Fruit: medium red; firm, high quality; resembles Taylor; ripens early to midseason. Plant: some resistance to injury from fluctuating winter temperatures under mid-Atlantic conditions.

September.—Orig. in Geneva, N.Y., by G.L. Slate, New York State Agr. Expt. Sta. Marcy x Ranere; cross made in 1934; selected for second test in 1939; tested as N.Y. 17438; introd. in 1947. Fruit: medium size; bright red, attractive; firm; does not crumble; quality of the floricane fruit is only fair, but the primocane fruit is considered good quality; floricane fruit is as early as that of Indian Summer and June and about 5 days earlier than Newburgh; fall fruit matures 2 to 4 weeks before that of Indian Summer; tends to cling to the receptacle. Plant: vigorous; winter hardy; productive.

Shuttleworth.—Orig. in Silver Creek, N.Y., by A.L. Shuttleworth. Parentage unknown; introd. in 1933. Fruit: large, glossy black; firm, superior quality; ripens very early. Plant: canes strong; productive.

Skeena.—Orig. in Vancouver, British Columbia, Canada, by H.A. Daubeny, Agr. Canada Res. Sta. Creston x SCRI 6010/52 (derived from Burnetholm and Malling Jewel); cross made in 1965; selected in 1967; tested as BC 65-2-16; introd. in 1937. Fruit: medium size; firm; medium red; conic; easy separation from receptacle; adapted to fresh market and frozen retail packs. Plant: productive; strong; upright; nonpubescent canes with few spines; strongly attached fruiting laterals; good primocane production; winter hardy in the Pacific Northwest. Resistant to the common strain of the North American aphid vector of the raspberry mosaic virus complex; some resistance to spur blight; susceptible to root rot and to infection by pollen tranmission of raspberry bushy dwarf virus.

Sodus.—Orig. in Geneva, N.Y., by G.L. Slate, New York State Agr. Expt. Sta. Dundee (black raspberry) x Newburgh (red raspberry); cross made in 1927; introd. in 1935. Fruit: large; firm, purple; sprightly, good quality. Plant: vigorous; productive; winter hardy; more resistant to drought than Columbian. Susceptible to verticillium wilt.

Somo.—Orig. in Mountain Grove, Mo., by P.H. Shepard, Missouri State Fruit Expt. Sta. Parentage unknown; tested as Mo. R-101; introd. in 1956. Fruit: black; average size; nearly round; attractive; quality very good fresh or frozen; firm, clings well to a small receptacle, but readily picked when ripe; does not crumble when handled; ripens early, before Cumberland and Kansas; somewhat smaller as season advances due partly to high production and long harvest season. Plant: productive; vigorous and tall; roots readily at tips.

Sonoma.—Orig. in Glen Ellen, Calif., by E.E. Roach, Log Cabin Nursery. (Ranere x Latham) x (Cuthbert x Latham); selected in 1947; introd. in 1951. Fruit: large. Plant: primocane fruiting; able to withstand and bear fruit in hot summer and low humidity; produces to mid-December.

Souris.—A hardy, floricane-fruiting raspberry with better flavor than Boyne. Orig. in Morden, Manitoba, Canada, by C. Davidson, Agr. and Agri-Food Canada, Morden Res. Center. Asker x Boyne; cross made in mid-1960s; tested as MRS#1x8; introd. in 1993. Propagated under a royalty agreement by Jefferies Nurseries, Portage la Prairie, Manitoba. Fruit: small; hemispheric shape; appearance similar to Boyne but with larger drupelets; season similar to Boyne; fresh market. Plant: yield greater than or equal to Boyne; canes have fewer spines than Boyne; self-supporting under test conditions; more primocanes produced than Boyne. Winter hardy to USDA zone 2.

Southland.—Orig. in Raleigh, N.C., by C.F. Williams, Agr. Expt. Sta. and Crops Res. Div., U.S. Dept. of Agriculture. N.C. 237 x Md. S420-5, a complex hybrid involving several raspberry species; cross made in 1950; selected in 1953; introd. in 1968. Fruit: medium size; symmetric and conic; light red, does not darken on maturity; firm; does not crumble; flavor slightly acid with good dessert quality, but color may be too light for frozen pack; ripens early, 2 to 3 days after Sunrise, also has substantial primocane fruit in early or mid-August. Plant: moderate vigor; multiplies freely, thrives with high soil fertility and good drainage; under these conditions, plants need to be maintained in thin narrow rows or grown in the hill system; winter hardy. Highly resistant to leaf spot, powdery mildew, and cane spot.

Sparkle.—Orig. in Escondido, Calif., by W. Ferguson. Parentage unknown; selected in 1950; introd. in 1953. Fruit: bright, light red; firm, good quality.

Sparkling Gem.—Orig. in Champaign, Ill., by H.L. Boll. Selected in 1971 from a single plant of *Rubus phoenicolasius* of unknown origin; introd. in 1973. Fruit: medium size. hemispheric, bright medium red, leaving inside of sepals and receptacle a beautiful orange when picked, glossy, attractive; flesh slightly soft, sweet if allowed to mature on plant; ripens late in Michigan, about 20 July. Plant: low to medium productivity; vigorous, producing up to 22 canes per plant per year up to 7/8 inch in diam. and 18 ft in length; canes arching, covered with bright red hairs and a medium number of short orange prickles; leaves whitish tomentose underneath. Resistant to leaf spot.

Starking Black Giant.—Orig. in Sterling, Ill., by E.F. Stouffer, Stark Bros. Nurseries & Orchards Co., Louisiana, Mo. Open-pollinated seedling of Cumberland; selected in 1947; introd. in 1955. Plant patent 1494. Fruit: black; resembles Bristol but larger, better flavored, and with smaller achenes; early ripener. Plant: extremely vigorous and productive; heat and drought resistant.

Stonehurst.—Orig. in Santa Clara Valley, Calif., by J.M. Reiter and E.V. Goldsmith. Parentage unknown; selected in 1973; introd. in 1979. U.S. plant patant 4485. Fruit: medium size; skin rosy red to pomegranate purple; round; sometimes irregular; partially hidden by leaves; easy separation from receptacle; ripens on primocanes from early August until mid-September and on floricanes during last week of May and the first 2 weeks of June; can produce continuously from May until December. Plant: productive, with two thirds on primocanes and one third on floricanes; primocanes upright and rigid, nonbranching, light green with soft, reddish-brown spines at base; floricanes light brown and branching. Some root rot resistance.

Success.—Orig. in Durham, N.H., by A.F. Yeager and E.M. Meader, New Hampshire Agr. Expt. Sta. Morrison x New Hampshire 100 (a red raspberry selection); tested as N.H. P-4; introd. in 1956. Fruit: purple; medium to large; highly flavored, sweet; good for fresh, frozen, and canned uses; ripens in midseason. Plant: productive; winter hardy. Apparently resistant to spur blight.

Summit.—Orig. in Corvallis, Ore., by F.J. Lawrence, U.S. Dept. of Agriculture–Agr. Res. Serv. and Oregon State Univ. and Washington State Univ. ORUS 1838 (Fallred x ORUS 1347) x ORUS 1842 (NY 600 x OR 1347) (ORUS 1347 is a very late primocane fruiting selection with firm, attractive fruit with a pedigree involving Newburgh, Washington, and Willamette.); selected in 1976; tested as ORUS 2033; introd. in 1989. Fruit: small to medium; medium red; firm; round; produced on primocanes in August; ripens 10 to 14 days earlier than Heritage in the Pacific Northwest; reasonably easy separation from receptacle; good fresh-market and processing qualities. Plant: high yield; primocanes vigorous, nonpubescent, waxy; a few scattered purple spines; fruiting laterals strong and medium long. Resistant to root rot; susceptible to infection by pollen transmission of raspberry bushy dwarf virus and to the North American aphid vector of the raspberry mosaic virus complex.

Sumner.—Orig. in Puyallup, Wash., by C.D. Schwartze and Arthur S. Myhre,

Washingon State Univ. Washington x Tahoma; selected in 1938; introd. in 1956. Fruit: medium size; medium red; firm; sweet flavor; midseason ripening; fresh and processing use. Plant: productive; numerous canes of medium height and upright habit; winter hardy in Pacific Northwest. Resistant to powdery mildew and to western yellow rust; susceptible to the North American aphid vector of the raspberry mosaic virus complex.

Sunrise.—Orig. in Glenn Dale, Md., by G.M. Darrow, U.S. Dept. of Agriculture. Latham x Ranere; selected in 1923; introd. in 1939. Fruit: red; fine texture; quality better than Latham; does not crumble; matures very early, 2 days before Ranere and nearly 2 weeks before Latham; picks easily. Plant: very winter hardy; resistant to leaf spot and can spot.

Sweetbriar.—Orig. in Santa Clara Valley, Calif., by J.M. Reiter and E.V. Goldsmith. Parentage unknown; selected in 1973; introd. in 1979; U.S. plant patent 4486. Fruit: medium size; rose to rose red; firmer than Willamette, softer than Heritage; conic and symmetric; individual drupelets sometimes irregular in size; well-exposed; easy separation from receptacle; Watsonville, Calif. production peaks on primocanes during September and into early October and on floricanes during June and first week in July; can be produced continuously from May until December. Plant: productive; with 65% of the crop on primocanes and 35% on floricanes; primocanes upright, rigid, nonbranching, light green with soft light brown spines at base; floricanes light brown and branching.

Tahoma.—Orig. in Puyallup, Wash., by C.D. Schwartze, Washington State Univ. Latham x Lloyd George; selected in 1935; introd. in 1938. Fruit: medium size, conic to thimble-shape; bright medium red, attractive; drupelets medium to small, usually coherent; medium firm, flavor good but sour. Plant: winter hardy in Pacific Northwest; canes moderate in height, slender, and straight; laterals short to medium length; becomes dormant early. Resistant to western yellow rust and to root rot.

Taylor.—Orig. in Geneva, N.Y., by R. Wellington. Newman x Lloyd George; cross made in 1925; introd. in 1935. Fruit: bright, attractive red; good quality; does not crumble; easily removed from receptacle; withstands handling; ripens with Latham. Plant: vigorous; productive; tall. Susceptible to the North American aphid vector of the raspberry mosaic virus complex.

Tennessee Autumn.—Orig. in Knoxville, Tenn., by B. D. Drain, Tennessee Agr. Expt. Sta. Tenn 181 (Van Fleet x Latham) x Lloyd George; tested as Tenn. Y944; introd. in 1940. Plant patent 512. Fruit: large, to 1 inches diam., hemispheric; dark red; fair flavor, tart; ripens main crop on primocanes in September and October, some fruit produced in May to June. Plant: vigorous; productive; upright habit; suckers profusely; spiny. Adapted to mild winter climate.

Tennessee Luscious.—Orig. in Knoxville, Tenn., by B.D. Drain, Tennessee Agr. Expt. Sta. Lloyd George x Tenn. VVF 169 (Van Fleet x Viking); tested as Tenn. X37; introd. in 1944. U.S. plant patent 653. Fruit: large; conic; glossy medium red; firm, subacid, good flavor; drupelets strongly coherent; resembles Cuthbert; season late. Plant: upright; vigorous; multiplies slowly from suckers; canes purple in late summer.

Tennessee Prolific.—Orig. in Knoxville, Tenn., by B.D. Drain, Tennessee Agr. Expt. Sta. Lloyd George x Tennessee seedling VVF 169; cross made in 1935; introd. in 1948. Fruit: firm and handles well; good freezing quality; ripens about the same season as Latham; resembles Tennessee Luscious. Plant: adapted to Tennessee and nearby areas.

Thames.—Orig. in Ottawa, Ontario, Canada, by Canada Dept. Agr., Res. Sta. Lloyd George x Newman; selected in 1931; tested as Ottawa 271; introd. in 1952. Fruit: bright red; holds color well after picking; good for freezing; matures very late. Plant: canes tall, strong, branched; unproductive.

Titan.—Orig. in Geneva, N.Y., by J.C. Sanford, D.K. Ourecky, and J.E. Reich, New York State Agr. Expt. Sta., Cornell Univ. Hilton x N.Y. 598 (Newburgh x St. Walfried); cross made in 1963; selected in 1966; tested as N.Y. 883; introd. in 1985. U.S. plant patent 5404. Fruit: very large; bright red; long conic; large cavity; firm in New York, but soft in British Columbia; pleasant, mild flavor, early ripening; can be difficult to remove from receptacle and thus not suited to machine harvesting; susceptible to pre- and postharvest rots. Plant: unusually high yield potential; moderate vigor; primocanes usually not numerous; stout; green; glabrous; spines small, basal, green, with darkened tips; floricanes dark brown; some cracking; winter hardy. Susceptible to cane diseases in New York; in British Columbia, shows some resistance to spur blight and moderate susceptibility to cane *Botrytis* and cane spot; susceptible to pollen transmission of raspberry bushy dwarf and tobacco streak viruses; susceptible to tomato ringspot virus and to root rot; susceptible to the North American aphid vector of the raspberry mosaic virus complex and to the aphid vector of leaf curl virus and to the viruses themselves.

Toms.—Orig. in Nevada, Iowa, by L.E. Toms, Inter-State Nurseries, Hamburg, Iowa. Chief x Plum Farmer; discovered in 1927 or 1928; introd. in 1948. U.S. plant patent 779. Fruit: medium size, 3/4 inch in diam.; bright red; good quality; matures during July. Plant very disease resistant; bears well over a long period; fruit laterals long.

Trailblazer.—Orig. in Cheyenne, Wyo., by G.S. Howard, U.S. Dept. of Agriculture. Augustred x Cheyenne 63-16 (*Rubus strigosus*); cross made in 1964; selected in 1965; tested as 69-14; introd. in 1976. Fruit: hemispheric to conic, up to 25 mm in diam.; medium to deep red; medium firm; somewhat less flavorful than standard varieties; primocane fruit ripens first week of August and season lasts about 6 weeks or until heavy frost occurs. Plant: vigorous; produces numerous canes: medium to high yield; winter hardy.

Trent.—Orig. in Ottawa, Ontario, Canada, by Canada Dept. Agr., Res. Sta. Newman x Lloyd George; selected in 1931; tested as Ottawa 264; introd. in 1943. Fruit: small; medium dark red; firm, tends to crumble, fair quality; early ripening. Plant: productive; canes short; strong; winter hardy. Susceptible to powdery mildew.

Tulameen.—Orig. in Vancouver, British Columbia, by H.A. Daubeny and A. Anderson, Agr. Canada Res. Sta. Nootka x Glen Prosen; cross made in 1980; selected in 1983; tested as BC 80-28-53; introd. in 1989. Fruit: very large; long-conic; medium red, glossy; firm; drupelets numerous; separates readily from receptacle and can be machine-harvested; starts to ripen with Meeker but has a longer season that overlaps early primocane-fruiting cultivars; relatively susceptible to preharvest botrytis rot, some resistance to postharvest rot; good flavor; fresh-market and processing uses, including individual quick freeze. Plant: very high yield; vigorous; fairly upright habit; primocanes reasonably numerous, nonpubescent, green with an irregularly distributed purple overlay concentrated at the base; sharp, purple spines largely restricted to basal portions; floricanes sturdy, fairly upright, grayish-yellow with general basal cracking; fruiting laterals strongly attached, relatively long, with fruit well spaced; long pedicels aid in giving excellent exposure for hand-harvesting; moderately winter hardy in the Pacific Northwest. Susceptible to spur blight, moderately susceptible to cane *Botrytis* and cane spot; moderately susceptible to root rot; slow to become infected by pollen transmission of raspberry bushy dwarf virus; resistant to the common strain of the North American aphid vector of the raspberry mosaic virus complex and to powdery mildew.

Tweed.—Orig. in Ottawa, Ontario, Canada, by Canada Dept. Agr., Res. Sta. Newman x Lloyd George; selected in 1931; tested as Ottawa 263; introd. in 1946. Fruit: dark red; small; moderately firm, fair quality; early ripening, about 1 day before Trent. Plant: short; hardy; unproductive. Susceptible to mildew.

Van Fleet.—Orig. in Chico, Calif., by W. Van Fleet, U.S. Dept. of Agriculture. *Rubus kuntleanus* x Cuthbert; cross made in 1910 at Chico; seed from this cross planted in 1911 in Glenn Dale, Md.; seed of *R. kuntzeanus* collected in 1907 in China by H.E. Wilson, then planted and fruited in Chico; introd. in 1924. Fruit: red; medium size; too soft for commercial use, subacid, pleasant, dessert quality fair to good;

drupelets small; ripens late, with a season of about 5 weeks. Plant: very vigorous; productive; adapted to warm climates with low chill hours.

Vene.—Orig. in Stjordal, Norway, by R. Nestby, Kvithamar Res. Sta. Veten x Newburgh; cross made in 1970; selected in 1975; introd. in 1987. Fruit: small; round; glossy; dark red; softer than Veten but with little tendency to crumble; ripens early. Some resistance to preharvest *Botrytis* rot; local fresh-market and processing uses. Plant: young primocanes green, pubescent; mature primocanes very spiny; winter hardy. Some resistance to cane botrytis; moderately susceptible to spur blight.

Veten.—Orig. in Njos, Norway, State Expt. Sta. Asker x Lloyd George; introd. in 1961. Fruit: large; conic; dark red, medium bright to dull; large drupelets, soft but not crumbly; early to midseason ripening; acidic flavor; processing use. Plant: productive; medium number of erect canes with moderate growth; canes dull, not waxy, light green with a little red in autumn and red-brown in winter; cane tips not pubescent; spines few, except on young canes; easy to manage; moderately winter hardy. Susceptible to cane *Botrytis*.

Viking.—Orig. in Vineland, Ontario, Canada, by Ontario Dept. Agr., Hort. Res. Inst. Cuthbert x Marlboro; cross made in 1914; selected in 1918; introd. in 1924. Fruit: medium to small; bright red; fairly soft; ripens midseason. Plant: moderatelly productive; canes relatively smooth; strong; very tall; winter hardy Susceptible to powdery mildew; very susceptible to both orange and late yellow rust.

Waiau.—A high-yielding raspberry adapted to medium-low chill New Zealand conditions. Orig. in Nelson, New Zealand, by H. Hall, Riwaka Res. Center. Fairview x Marcy; cross made in 1979 by H.A. Daubeny, Agr. and Agri-Food Canada, Pacific Agr. Res. Centre, Vancouver, British Columbia, Canada; seedlings grown in 1984 and selection made in Lincoln, New Zealand, in 1987; tested as R61; introd. in 1990. Fruit: large; light-medium red and moderately glossy; firm; moderately difficult to harvest; mid- to late season, ripening 6 to 8 days after Marcy. Plant: large numbers of semi-upright canes that arch into row; floricanes light brown with gray bloom and basal spines; moderate budbreak; medium-length, strongly attached laterals with 30 to 40 fruit each; leafy habit hides some fruit. Susceptible to root rot and to infection by pollen transmission of raspberry bushy dwarf virus.

Washington.—Orig. in Puyallup, Wash., by C.D. Schwartze, Washington State Univ. Cuthbert x Lloyd George; selected in 1935; introd. in 1938. Fruit: medium size; bright medium red; sweet flavor, fresh and processing use; midseason ripening. Plant: moderately productive; moderately vigorous; strong laterals. Apparently resistant to the common strain of the North American aphid vector of the raspberry mosaic virus complex; susceptible to western yellow rust and to root rot.

Watson.—See Ruby.

Wawi.—A fresh market raspberry adapted to France. Orig. in d'Angers, by V. Bellenot–Kapusta, Station D'Amerlioration Des Especes Fruitierest Et Ornementales. Washington x Willamette; introd. in 1990. Fruit: medium to small size; firm; good flavor; attractive. Plant: high yield. Tolerant to *Phytophthora* root rot in the field.

Webster.—Orig. in Geneva, N.Y., by Richard Wellington, N.Y. State Agr. Expt. Sta. Smith 1 (black raspberry) x a purple raspberry seedling; cross made in 1910; introd. in 1926. Fruit: medium size; dark, dull purple; firm, tart, fair quality; medium juicy, drupelets numerous; late season. Plant: slow growing; not very productive.

Willamette.—Orig. in Corvallis, Ore., by G.F. Waldo, U.S. Dept. of Agriculture. Newburgh x Lloyd George; selected in 1936; introd. in 1943. Fruit: medium size; dark red; nearly round; moderately firm, slightly acid, quality good for all forms of processing; early ripening. Plant: moderately productive; moderately vigorous canes; strong laterals; spiny; abundant primocanes; moderately winter heardy in the Pacific Northwest. Susceptible to the North American aphid vector of the raspbery mosaic virus complex; susceptible to root rot and to spur blight; resistant to the common strain of raspberry bushy dwarf virus; resistant to cane *Botrytis* and cane spot.

Zenith.—Orig. in Arslev, Denmark, by A. Thuesen, Res. Centre for Hort. Pyne's Royal x Spangsbierg 1307/49; tested as Spangsbjerg 2051/67; introd. in 1985. Fruit: medium size; medium to dark red, glossy; short conic; soft to moderately firm; easily removed from the receptacle; no tendency to crumble; fairly late ripening with a long season; fresh and processing uses. Plant: primocanes numerous; canes erect, very long, relatively large diam.; density of spines sparse to medium; fruiting laterals long.

Sandcherry

Advance.—Orig. in Rosthern, Saskatchewan, Canada, by Seager Wheeler. Introd. in 1929. Chance seedling. Fruit: medium to large; flavor mild; season early August. Bush: medium; productive. Now obsolete.

Amber.—Orig. in Brookings, S.D., by N.E. Hansen, Agr. Ext. Serv. Introd. in 1944. Selection of *Prunus besseyi* of South Dakota. Fruit: 3/4 inch in diam.; skin golden yellow; flesh quality very good; ripens in July, earlier than Oahe.

Black Beauty.—Orig. in Charleswood, Manitoba, Canada, by G.F. Chipman of Winnipeg, Man. Introd. in 1938. A selected sandcherry; tested as Manitoba 842. Fruit: roundish; 3/4 inch in diam.; skin thin, shiny black; flesh firm, juicy, sweet, quality fair to good; season early August; resembles Mando or Manmoor. Bush: upright; productive; hardy.

Brooks.—Orig. in Brooks, Alberta, Canada, by A. Griffin, Canada Dept. Agr., Res. Sta. Introd. in 1934. A superior seedling of the native sandcherry (*Prunus pumila*); tested as Brooks 1. Fruit: very large for this species, often 1 inch in diam.; roundish-oval; skin purplish-black, thick, tough; flesh greenish-yellow, firm, juicy, tender, mild for a sandcherry; cans well; ripens during early August. Bush: upright, spreading; vigorous; hardy; productive; makes an ideal low-growing ornamental.

Dakota Ruby.—See Ruby.

Honeywood.—Orig. in Parkside, Saskatchewan, Canada, by A.J. Porter, Honeywood Nursery. Introd. in 1956. Open-pollinated seedling of Sapa; selected in 1953. Fruit: large, averaging 3/4 to 7/8 inch in diam. Bush: spreading; leaf glossy, dark green, turning coppery bronze in fall; resistant to mildew; cross-pollination necessary, with its own seedlings being satisfactory. Also, considered to be an ornamental plant.

Leafland.—Orig. in Carlos, Alberta, Canada, by Robert Erskine. Introd. in 1957. Parentage unknown. Discovered about 1950. Fruit: to 3/4 inch in diam.; globose; flavor mild and sweet. Bush: vigorous hardy, very productive.

Mando.—Orig. in Morden, Manitoba, Canada, by Canada Dept. Agr., Res. Sta. Introd. in 1931. Selection of *Prunus besseyi*. Fruit: round, 3/4 inch in diam.; skin purplish-black; flesh greenish, firm, flavor mild; good canner; season early August. Bush: upright; size medium; vigorous; productive.

Manmoor.—Orig. in Morden, Manitoba, Canada, by Canada Dept. Agr., Res. Sta. Introd. in 1929. Selection of *Prunus besseyi*. Fruit: round, 3/4 inch in diam.; skin purplish-black; flesh greenish, firm, flavor mild; good canner; season early August. Bush: upright; size medium; vigorous; productive.

Manorette.—Orig. in Moose Range, Saskatchewan, Canada, by Percy M. Wright, Carrot Valley Nursery, Saskatoon, Sask. Introd. in 1952. Manor cherry plum x common sandcherry o.p. Cross made about 1948, selected about 1951. Fruit: small; round, flesh good flavor, not bitter; ripens 25 Aug.; recommended for eating, pie making or any cherry use. Bush: 5 ft tall, vigorous; productive; hardy to –60F.

Oahe.—Orig. in Brookings, S.D., by N.E. Hansen, South Dakota Agr. Expt. Sta. Introd. in 1937. Seedling of *Prunus besseyi*. Fruit: small, averages 5/8 inch in diam.; round; skin purplish-black; flesh greenish, juicy, tender, sweet, mild; quality good; ripens in early August. Bush: size medium, spreading; productive.

Ruby (*Dakota Ruby*).—Orig. in Brookings, S.D., by S.A. McCrory, South Dakota Agr. Expt. Sta. Introd. in 1953. Open-pollinated seedling of *Prunus besseyi;* selected about 1948. Fruit: flesh dark red; tannic, very suitable for jelly, jam, and sauce, may be eaten fresh. Bush: spreading, 3 to 5 ft tall; ornamental.

Teepee.—Orig. in Brookings, S.D., by N.E. Hansen, Agr. Ext. Serv. Introd. in 1937. Seedling of Sioux. Fruit: 5/8 inch in diam.;

skin black; flesh quality good; ripens late July. Bush: upright, spreading, productive.

Wampum.—Orig. in Brookings, S.D., by N.E. Hansen, Agr. Ext. Serv. Introd. in 1937. Seedling of native bush sandcherry. Fruit: 3/4 inch in diam.; skin black; flesh quality excellent; ripens 1 Aug. Bush: 4 ft tall; upright, open; productive.

Sapodilla

Brown Sugar.—Orig. in Homestead, Fla., by Florida Subtrop. Expt. Sta. Introd. in 1948. Parentage unknown; seed planted in 1936; selected about 1945 by Robert Newcomb of Homestead. Fruit: medium small, with diam. averaging 2 1/2 to 2 3/4 inches and length 2 to 2 1/2 inches; nearly round with smooth apex; skin light, scurfy brown; flesh pale brown, fragrant, juicy, very sweet and rich, texture slightly granular, quality very good. Tree: tall, bushy; prolific.

Prolific.—Orig. in Homestead, Fla., by Florida Subtrop. Expt. Sta. Introd. in 1951. Open-pollinated seedling; seed planted in 1936. Fruit: 2 1/2 to 3 1/2 inches long and nearly as wide; skin scurfy, brown, becoming nearly smooth as fruit reaches maturity; flesh light tan, mildly fragrant, texture smooth, flavor sweet; quality good. Tree: precocious and prolific in fruiting; only moderately susceptible to rust.

Russell.—Orig. in Islamorada, Fla., by R.H. Fitzpatrick, Homestead, Fla. Introd. in 1935. Chance seedling discovered in 1935; selected, named and first propagated by Fitzpatrick. Fruit: 3 to 4 inches long, 2 3/4 to 3 3/4 inches broad; skin scurfy brown; flesh pinkish-tan, shading to greenish-tan under the skin, mildly fragrant, texture somewhat granular, flavor rich, sweet. Tree: slower to bear and produces less fruit than does Prolific.

Tikal.—An early season selection introd. at Tropical Res. and Education Center, Univ. of Florida, Homestead, in 1987. Grown from open-pollinated seed collected in Mexico in 1949. Selected by C.W. Campbell, S.E. Malo, and J. Popenoe. Fruit: large, 80 to 325 g, averaging 120 g; skin light brown, smooth texture. Flesh light tan, smooth texture, excellent flavor. Fruit is less susceptible to Caribbean fruit fly than are seedlings. Season from late December to March. Tree: medium vigor, forming full canopy to make a large tree; regularly productive. Tree is well adapted to warm subtropical climate; tolerant to periods of drought; requires adequate water to crop well.

SASKATOON

Most *Amelanchier* cultivars grown for fruit production belong to the species *A. alnifolia*, unless otherwise noted. Similar quality fruit from selections of closely related and naturally hybridizing species are treated as saskatoons here, in the scientific literature, and in commerce. Cultivars are self-fertile unless otherwise noted. All validly introduced fruit cultivars have been included. Cultivar descriptions are based on limited published data, unpublished research data, and comments by originators or their successors; scientifically acceptable comparative data are not yet available. U.S. Dept. of Agriculture hardiness zones are used. The Department of Horticulture Science, University of Saskatchewan, is the International Registration Authority for the genus *Amelanchier*.

Altaglow—Orig. in Red Deer River valley, Alberta (51°N), by A. Griffin prior to 1923; wild plant transplanted to Alberta Horticultural Research Station, Brooks. Initial selection and distribution for testing, 1928. Tested at Agriculture Canada Research Station, Beaverlodge (BRS), Alberta as Brooks White. Selected as an ornamental by J.A. Wallace, BRS, 1946. Introd. by P.D. Hargrave, Brooks, Alberta, in 1964. Fruit up to 16 mm diam., nearly spherical, white, easily bruised; typically 5–9 per cluster, clusters tight, even ripening, very uneven in size; flavor bland but sweet. Apparently self-sterile. Shrub to 7 m high; tall and erect habit, 3 m spread at maturity; moderate to good suckering close to crown, crown expanding indefinitely; crown long-lived, 40+ years. Hardy to zone 1. Introduced as an ornamental for its habit in large landscapes and for the oddity of its white fruit, but mainly for its splendid fall foliage colors.

Bluff—Orig. near Buffalo Lake, Sexsmith, Alberta (55°N) by P. Student; wild clump discovered on his farm in 1946. Selected in 1975 by J.G.N. Davidson and K.T. (Student) Davidson, introd. in 1990. Fruit up to 13 mm diam., nearly spherical, blue-black with light bloom; typically 7–13 per cluster, cluster tight, even size, and exceptionally even ripening; good, well-balanced flavor, moderately tangy; holds its flavor when cooked better than other cultivars; seeds few and small. Shrub to 5 m high; initially upright, tends to retain tall, erect habit, 2.5 m spread at maturity unless pruned; moderate to good and close suckering near crown; crown long-lived, 50+ years. Hardy to zone 1. May have some resistance to leaf diseases. Currently being evaluated in a comprehensive cultivar trial.

Buffalo—Orig. near Buffalo Lake, Sexsmith, Alberta (55°N) by A. Student; wild clump discovered on his farm in 1925. Selected in 1980 from suckers transplanted in 1975 by J.G.N. Davidson and K.T. (Student) Davidson, introd. in 1990. Fruit up to 14 mm diam., obovate to nearly spherical, blue-black with slight bloom; typically 7–13 per cluster, cluster fairly loose, fairly even ripening; excellent flavor with very good balance between tanginess and sweetness, best fresh but also cooks, cans, and jams well. Shrub to 4 m high; initially upright, to spreading at maturity, 5 m spread; moderate suckering near crown, crown expands similarly to Pembina; crown long-lived, 70+ years. Hardy to zone 1. Currently being evaluated in a comprehensive cultivar trial.

Forestburg—Orig. near Forestburg, Alberta (52°30′N); wild plant discovered by A. Nixon on his farm. Transplanted to Agriculture Canada Research Station, Beaverlodge (BRS), Alberta, in 1948; tested as B.E.F. 0003. Selected by J.A. Wallace, BRS, introd. in 1963. Fruit up to 16 mm diam., nearly spherical, blue-black with bloom; typically 7–11 per cluster, clusters very tight, fairly even ripening, later than Smoky; flavor mild, quite sweet, juicy; pH 4.2. Shrub to 4 m high; initially upright to arching-spreading, 5 m spread at maturity; moderate to light suckering near crown, crown expands slowly; crown long-lived, 40+ years. Hardy to zone 2. Heavy producer of large fruits. More drought-tolerant than the cultivars of Alberta origin. Currently being evaluated in a comprehensive cultivar trial.

Honeywood—Orig. near Parkside, Saskatchewan (53°N) by A.J. Porter. Seedling selection from wild plant discovered by A.J. Porter ca. 1955 near his Honeywood Nursery, introd. by him in 1973. Flowers 4–8 days later than other cultivars, and ripens somewhat later also. Fruit up to 16 mm diam., basically flattened to spherical, blue-black with little bloom; typically 9–15 per cluster, cluster fairly tight, fairly even ripening; excellent full and tangy flavor; pH 3.7–3.9; seeds large. Shrub to 5 m high; initially upright to arching-spreading, 4 m spread at maturity; sparse suckering near crown, crown expands slowly like Pembina. May have some resistance to leaf spot and powdery mildew. Very productive and precociously fruitful. Currently being evaluated in a comprehensive cultivar trial.

Lee #3—Orig. near Barrhead, Alberta (54°N), by L. Lee in 1989; cross between Pembina and Northline(?). Introd. by K. Pruski, Alberta Agriculture, Crop Diversification Centre–North, Edmonton, Alberta, in 1994. Fruit up to 16 mm diam., intense flavor and bouquet; few-seeded. Shrub to 3 m high, compact, spreads slowly.

Martin—Orig. in Langham, Saskatchewan (52°N), by D. Martin (D. Martin Nursery); introd. by D. Martin in 1990. Seedling selection from Thiessen for large fruit size and more uniform ripening, otherwise similar to Thiessen. Currently being evaluated in a comprehensive cultivar trial.

Moonlake—Orig. near Saskatoon, Saskatchewan (52°N), by G. Krahn (Lakeshore Tree Farms); wild plant discovered by Moon Lake; introduced by G. Krahn in 1974. Fruit up to 16 mm diam., obovate to nearly spherical, blue-black with light bloom; typically 6–10 per cluster, cluster open; flavor mild, sweet, good; pH 3.9. Shrub to 3 m high; initially upright to arching-spreading, 3 m spread at maturity; suckering moderate. Productivity moderate, somewhat sporadic. May have some resistance to powdery mildew. Currently being evaluated in a comprehensive cultivar trial.

Nelson—Orig. near Bradwell, Saskatchewan (52°N); wild plant discovered by S.H. Nelson in 1974. Introd. by R. St-Pierre, Univ. of Saskatchewan, Saskatoon, in 1992. Flowers 3–7 days later than other cultivars. Fruit up to 13 mm diam., nearly spherical, blue- black with little bloom, few seeds; typically 6–12 fruit per cluster, cluster compact, ripening somewhat uneven; good tangy flavor. Shrub, compact, to 1.5 m high; suckering moderate. Possibly some resistance to saskatoon-juniper rust (*Gymnosporangium* spp.). Currently being evaluated in a comprehensive cultivar trial.

Northline—Orig. near Beaverlodge, Alberta (55°N), by J.A. Wallace (Beaverlodge Nursery) in 1958; introd. by him in 1965. Fruit up to 16 mm diam., obovate to nearly spherical, blue-black with bloom, very firm; typically 7–13 per cluster, fairly even ripening; excellent full flavor, similar in quality to Pembina, fairly sweet; pH 3.8-3.9; resistant to cracking. Shrub to 4 m high; initially upright to arching-spreading, 6 m spread at maturity; suckering quite freely near crown, crown expands indefinitely; crown long-lived, 50+ years. Hardy to zone 1. Very productive, exceeds Smoky at some locations. Occupies an increasing proportion of commercial hectarage in Canada, the third largest in 1993. Currently being evaluated in a comprehensive cultivar trial.

Parkhill—Orig. in Michigan; wild plant selection. Introd. by Parkhill Nursery, Bismarck, North Dakota in 1974. Species uncertain; recent genetic fingerprinting studies suggest a hybrid of *A. stolonifera* with *A. alnifolia*. Fruit up to 13 mm diam., obovate to nearly spherical, blue-black with bloom; typically 7–11 per cluster, cluster fairly open, fairly even ripening; flavor mild and bland; pH 4.1. Shrub to 1.5 m high; initially upright to spreading, 2.5 m spread at maturity; low to moderate suckering, crown expands slowly. Hardy to zone 2. Susceptible to powdery mildew. Currently being evaluated in a comprehensive cultivar trial.

Pembina—Orig. near Barrhead, Alberta (54°N), by J.A. Wallace, Agriculture Canada, Beaverlodge, Alberta; wild plant discovered in Pembina River valley ca. 1928; tested as Barrhead No. 1 and as B.E.F. 3501; reselected 1950; introd. in 1952. Fruit up to 14 mm diam., obovate to nearly spherical, blue-black with bloom; typically 9–13 per cluster, fairly even ripening; excellent, full, tangy flavor, fairly sweet; pH 4.1. Shrub to 5 m high; initially upright to upright-spreading, 5 m spread at maturity; moderate to sparse suckering near crown, crown expands more slowly than Smoky; crown long-lived, 70+ years. Hardy to zone 1. Possible susceptibility to woolly elm aphid. Nearly as productive as Smoky. The full-flavor standard against which other cultivars are judged. Currently being evaluated in a comprehensive cultivar trial.

Regent—Orig. near Regent, North Dakota (46°N). Open-pollinated seedling introd. in 1977 by J. Candrian, Farmer Seed and Nursery Co., Faribault, Minnesota. Species uncertain; recent genetic fingerprinting studies suggest a hybrid of *A. stolonifera* with *A. alnifolia.* Fruit up to 13 mm diam., ovoid to nearly spherical, blue-black with bloom; typically 7–11 per cluster, cluster loose and open; flavor somewhat plum-like, mild, sweet and somewhat bland; pH 4.1–4.5; few and small seeds. Shrub to 2 m high; initially upright, 2 m spread at maturity; suckering low to moderate. Hardy to zone 3, marginal in 2. Precociously fruitful. Possible resistance to leaf spot and saskatoon-juniper rust. It is also used as an ornamental, having attractive fall foliage colors. Currently being evaluated in a comprehensive cultivar trial.

Smoky—Orig. near Beaverlodge, Alberta (55°N), by W.D. Albright; wild plant discovered on the Beaverlodge Research Station (in the Smoky River drainage basin), transplanted in 1918; tested as Selection no. 9 (1928), and B.E.F. 3502 (1935), reselected 1950. Introd. in 1952 by J.A. Wallace, Agriculture Canada, Beaverlodge, Alberta. Fruit up to 14 mm diam., spherical, blue-black with bloom; typically 7–11 per cluster, uneven ripening; good, mild flavor, sweetest cultivar with highest sugar/acid ratio so far; pH 4.1–4.5; large and many seeds. Shrub to 4.5 m high; initially upright to arching-spreading, 6 m spread at maturity; freely suckering near crown, crown expands indefinitely; crown long-lived, 70+ years. Hardy to zone 1. Most productive commercial cultivar so far, yielding up to 6,000 kg/ha. The cultivar that enabled commercial production to start in the prairies; in 1990 it occupied about 85% of the hectarage in Canada. Currently being evaluated in a comprehensive cultivar trial.

Success—Orig. in Pennsylvania mountains (ca. 41°N). Selection made before 1868 from seedlings of wild plants. Acquired 1873 by H.E. Van Deman, Kansas, and introduced by him in 1878. Van Deman sold more than 10,000 plants by 1888. Species uncertain; recent genetic fingerprinting studies suggest a hybrid of *A. stolonifera* with *A. alnifolia.* Fruit up to 14 mm diam., obovate to nearly spherical, blue-black with bloom; typically 7–11 per cluster, cluster loose, ripens slowly, fruit held firmly; flavor good but mild, somewhat apple-like, quite sweet; pH 4.0; seeds large. Shrub to 2 m high; initially upright to upright-spreading, 2 m spread at maturity; moderate suckering near crown. Hardy to zone 3, marginal in 2. Susceptible to powdery mildew. In one study it made the best fruit leather of 9 cultivars. By far the oldest surviving cultivar. Currently being evaluated in a comprehensive cultivar trial.

Thiessen—Orig. west of Hepburn, Saskatchewan (52°N), near the North Saskatchewan River; wild plant discovered in 1906 by Maria Loewen and transplanted to her parents' farm near Debenham. Years later, removed to their farm near Langham, Saskatchewan. Obtained from this farm and introduced by G. Krahn (Lakeshore Tree Farms, Saskatoon, Saskatchewan) in 1976. Flowers a few days earlier than other cultivars. Fruit up to 17 mm diam., nearly spherical, blue-black with slight bloom; typically 6–12 per cluster, cluster fairly loose, uneven ripening; excellent flavor, fresh and juicy. Shrub to 5 m high; initially upright, but tends to sprawl from an early age, eventually becoming a large bush, up to 6 m spread at maturity; moderate to good suckering near crown, crown expands indefinitely; crown long-lived, 70+ years. Hardy to zone 1. Some resistance to powdery mildew. It is much favored for U-pick orchards. It had the second largest commercial hectarage in Canada in 1993. Winner of the Canadian Society for Horticultural Science Outstanding Cultivar Award in 1994. Currently being evaluated in a comprehensive cultivar trial.

STRAWBERRY

Acadia.—Orig. in Kentville, Nova Scotia, Canada, by D.L. Craig and L.E. Aalders, Canada Dept. Agr., Res. Sta. Introd. in 1965. Redcrop x Sparkle; selected in 1959. Fruit: size medium; conical, uniform; skin medium red, glossy; calyx attractive; achenes even with surface or slightly sunken; flesh medium red, moderately firm, fine-grained, tender, no core, slightly acid, flavor excellent; resembles Sparkle; ripens 2 to 3 days before Sparkle, in midseason to late; suitable fresh and for freezing. Plant: medium to large; upright; extremely vigorous; yields well; numerous runners produced; free of Latent A virus.

Adirondack.—Orig. in Glens Falls, N.Y.., by George A. Webster. Introd. in 1948. Fairfax x Howard 17; selected in 1940. Fruit: large, conic; flesh firm, red throughout, aromatic; calyx very showy. Plant: vigorous; deep rooted; heavy bearer over a long season. In quality ranks with Fairfax; in yield with Howard 17.

Adonis.—See Midland.

Agassiz.—Orig. in Agassiz, British Columbia, Canada, by Canada Dept. Agr., Res. Sta. Introd. in 1957. Pathfinder x British Sovereign; cross made in 1947; selected in 1949 by T.H. Anstey and J.A. Freeman. Fruit: very large, uniform, holding its size well during entire season; skin bright red; flesh as firm as British Sovereign; flavor slightly acid; ripens at place of origin about with British Sovereign; freezing quality rates high. Plant: vigorous; productive; yields higher than British Sovereign; some drought resistance; some red stele resistance; not everbearing.

Alamo.—Orig. in Winter Haven, Texas, by Texas Agr. Expt. Sta. Introd. in 1937. Blakemore x Ettersburg 80?; selected in 1935; tested as Tex. 21. Fruit: large; flavor mild. Plant: yields heavily; large; low plant producer.

Alaska Pioneer.—Orig. in College, Alaska, by Arvo Kallio, Alaska Agr. Expt. Sta. and U.S. Dept. of Agr. Introd. in 1968. [Senga Sengana x Alaska 292 (*Fragaria virginiana*); cross made in 1958; selected in 1960; tested as Alaska 6344. Fruit: medium to small; regular, conic; skin red, attractive, tender; flesh red, firm; flavor pleasant, rather acid, characteristic wild aroma; good for dessert, canning, freezing and jam; calyx medium, medium dark green, easily removed; achenes small, slightly sunken; ripens middle of July to early August. Plant: extremely hardy; vigorous; runners produced abundantly; leaves medium dark, not glossy; tolerant to drought, low temperature fungi, and leaf spot; under certain conditions a slight susceptibility to mildew; adaptable to a wide range of soil types including silt loam, sandy loam, acid and alkaline soils low in organic matter; a home garden variety.

Albritton.—Orig. in Raleigh, N.C., by E.B. Morrow, North Carolina Agr. Expt. Sta. and George M. Darrow, U.S. Dept. of Agriculture. Introd. in 1951. Southland selfed (N.C. 1065) x Massey selfed (N.C. 1053); raised in 1945; selected in 1946; tested as N.C. 1369. Fruit: medium to large, about the size of Massey; conic, uniform; skin tough, vivid red, turning to a deeper red when fully ripe, glossy, holds this appearance well; flesh very firm, firmer than Massey, red to center, subacid, flavor good to excellent; achenes borne on surface; ripening season about same as Massey in eastern North Carolina; well adapted for freezing. Plant: vigorous; productive, being better than for other varieties in eastern North Carolina; resistant to leaf spot; makes runners freely.

Alcatraz.—Orig. in Ettersburg, Calif., by Albert F. Etter. Introd. about 1930. Red Sugar x Kalicene; sibling of Golden Gate. Fruit: large but short, conical; skin light red; flesh light pink, firm, flavor fair.

Aliso.—Orig. in Davis, Calif., by R.S. Bringhurst and Victor Voth, Agr. Expt. Sta. Introd. in 1967. Cal. 52.16-12 x self (Cal. 39.117-4 x 39.96-18); selected in 1958 at Univ. of California So. Coast Field Sta., Santa Ana; tested as 56S115-11. Fruit: large;

blunt conic, tends to button off (incomplete tip pollination) on early fruit; skin red with bright finish, durable; achenes large and exposed; flesh light color, fairly firm; flavor fair to good; yield high, starts early and persists. Plant: size medium; foliage light green; flower stocks long; stolon production light; susceptible to verticillium wilt. Recommended for winter planting in central and southern coastal California, the source of planting stocks should be from nurseries at high elevation.

All Season (*Lovett's All Season*).—Orig. in Little Silver, N.J., by John A. Kemp. Introd. in 1923 by Lovett's Nursery, Little Silver. One of Van Fleet's hybrids x Superb; selected in 1917; considered by some to be Progressive x Early Jersey Giant. Fruit: large; skin bright red; resembles Big Joe in appearance. Plant: vigorous; everbearing. Variety now obsolete.

Allstar.—Orig. at Wye Inst. on the Eastern Shore of Maryland by Maryland Agr. Expt. Sta. College Park, Md., and U.S. Dept. of Agriculture. Introd. in 1981. US 4419 (MdUS 2905 x MdUS 2650) x MdUS 3184 (NC 1768 x Surecrop). Cross made in 1969; selected in 1971 by D.H. Scott and A.D. Draper; tested as MdUS 4429. Berry: large, firm, symmetrical, attractive external and internal color; flavor pleasant; ripens with Guardian and Scott. Plant: vigorous, productive; runners freely; resistant to five races of red stele root rot; moderately resistant to venicillium wilt; foliage resistant to leaf scorch and powdery mildew.

Amazing.—Orig. in Champaign, Ill., by Herschel L. Boll. Introd. in 1973. Plant patent 3714; 6 May 1975. [(Canall x Twentieth Century) x Twentieth Century] x Ogallala. Cross made in 1967. Berry: large at beginning of season, and small at end of season, wedge to conic to blunt conic, medium to medium-dark red; flesh medium red; calyx large, most sepals three-toothed; ripens early, about same time as Sunrise. Plant: highly productive, vigorous, produces medium number of runners; few leaves have four-leaflets.

Ambrosia.—See Ambrosia Late.

Ambrosia Late (*Ambrosia*).—Orig. in Platteville, Wis., by A.E. Lang, Dodgeville, Wis. Introd. in 1932. Parentage unknown; discovered in 1928. Fruit: ships well. Plant: yields well; foliage heavy. No longer propagated.

American Sweetheart.—Orig. in Champaign, Ill., by Herschel L. Boll. Introd. in 1973. Plant patent 3713; 6 May 1975. Bud mutation of Armore; discovered in 1959. Berry: medium to large, conic, medium red, glossy, evenly colored, attractive; seeds raised; flesh medium light red; very firm, pedicels long and slender; calyx small to medium; ripens midseason, about with Catskill. Plant; everbearing, runners usually produce flower buds before roots; highly productive, bears from June until freezing; vigorous; produces few runners; leaves slightly glossy to glossy.

Anaheim.—Orig. at Univ. of California, South Coast Res. and Ext. Center, near Irvine, by V. Voth and R.S. Bringhurst. Irvine x Cal 85.92-602; cross made in 1988; selected in 1989; tested as Cal 88.66-610; introd. in 1992. U.S. plant patent pending. Compared to Chandler, fruit slightly smaller; firmer; external color lighter, more orange, and less glossy; internal color lighter; less aromatic. Achenes yellow to light red and slightly extruded; very good flavor; fresh-market, processing, and home-garden uses. Plant: short-day type producing fruit over an extended period in arid, subtropical climates; production pattern similar to Chandler, except Anaheim is somewhat later in cool Mediterranean climates, such as central California, and yield is slightly less. More vigorous and more erect habit than Chandler; moderately resistant to common leaf spot and powdery mildew; equal or greater tolerance to two-spotted spider mite than Chandler; tolerant to viruses occurring in California.

Annapolis.—Orig. in Kentville, Nova Scotia, Canada, by D.L. Craig, A.R. Jamieson, K.A. Sanford, and N.L. Nickerson, Agr. Canada Res. Sta. K74-5 (Micmac x Raritan) x Earliglow; cross made in 1977; selected by D.L. Craig in 1978; tested as K78-6; introd. in 1984. Fruit: large primaries, medium-size secondaries; medium firm; light- to medium-red external and internal color; primaries conic, secondaries globose to globose-conic; primaries have reflexed calyx, secondaries and later have repressed calyx; calyx moderately difficult to remove; early flowering and ripening; fresh-market use. Plant: vigorous; abundant but not excessive runners; small crowns; medium-length scapes. Susceptible to powdery mildew; moderately resistant to verticillium wilt; highly resistant to races A-4, A-6, and A-7 of the red stele causal organism, *Phytophthora fragariae*; moderate yield where red stele is not present and high yield where it is; winter hardy in Atlantic Canada.

Apollo.—Orig. in Willard, N.C., by North Carolina Agr. Expt. Sta., Raleigh, and the U.S. Dept. of Agriculture, Beltsville, Md. Introd. in 1970. NC 1759 x NC 1729. Cross made in 1956, selected in 1958, tested as NC 2841 by G.J. Galletta, North Carolina Agr. Expt. Sta. and D.H. Scott, the U.S. Dept. of Agriculture, Beltsville. Fruit: about 40% larger than Albritton; symmetrical; skin glossy, attractive, deep scarlet; flesh sweet, flavor good; ripens late, slightly after Albritton. Plant: size medium; vigor medium; runner production medium; susceptible to red stele; tolerant to leaf spot, powdery mildew, and leaf scorch; performs well in matted row or hills. Recommended for commercial plantings in southeastern and southcentral U.S.

Aptos.—Orig. at Univ. of California, Wolfskill Expt. Orchard, Winters by R.S. Bringhurst and Victor Voth, Agr. Expt. Sta., Univ. of California, Davis. Introd. in 1979. Plant patent pending; to be assigned to Regents, Univ. of California, Davis. Tufts x 65.63-601 (Sequoia x 58.45-11). Third backcross generation derivative from a male *Fragaria virginiana glauca* collected in the Wasatch Mountains near Salt Lake City, Utah. Selected 1971 to 1972. Fruit: flesh flavor and quality equal or superior to those of currently grown California short-day cultivars; high in ascorbic acid; skin darker, more red than that of Tioga, tends to be dull when overripe. Plant: smaller than Tioga; leaves same color as those of Tioga; less susceptible to verticillium wilt than Tioga and Tufts. Recommended for home gardens as day-neutral (can be made to produce fruit 3 months after planting). Named for city of Aptos, Santa Cruz County, Calif.

Arapahoe.—Orig. in Cheyenne, Wyo., by LeRoy Powers, E.D. Krouch, and D.H. Scott, U.S. Dept. of Agriculture, Hort. Field Sta. Introd. in 1954. (Early Cheyenne 1 x Cheyenne 430-2) x (Wayzata x Cheyenne 3). Fruit: medium to large; skin bright, glossy red, tender; flesh very firm, red to center, subacid; very good dessert quality; suitable for freezing. Plant: everbearing; hardy; drought tolerant; extremely vigorous; good runner production; recommended for Colorado and Wyoming.

Arch Red.—Orig. in Brigham City, Utah, by Leland L. Larsen, Larsen Expt. Farm & Nursery. Introd. in 1956. Lindalicious x Maytime; selected in 1947. Fruit: size above average; quality good; matures early; resembles Maytime. Plant: a good plant maker.

ArKing.—Orig. in Bold Knob, Ark., by J.N. Moore, Univ. of Ark., Fayetteville. Introd. in 1981. Cardinal x Ark. 543 1 (MdUS 3082 x Delite). Cross made in 1971; selected in 1974; tested as Ark. 6224. Fruit: large, blunt conic; skin medium red, glossy, attractive; flesh very firm, good aroma and flavor; ripens late. Plant; vigorous, very productive; tolerant to leaf spot, leaf scorch, powdery mildew, five eastern races of red stele.

Armore (*Red Cluster*).—Orig. in Columbia, Mo., by H.G. Swartout, Missouri Agr. Expt. Sta. Introd. in 1950. Blakemore x Aroma; selected in 1940. Fruit: medium to large, continuous throughout season; short wedge to blunt conic, irregular; skin red; flesh firm, light red, extremely sweet; ships very well; ripens in late midseason; resembles Aroma in southwest Missouri. Plant: very productive; sets and matures nearly all of its flowers; susceptible to leaf spot and mildew; produces runners freely; grows best in heavy silt loam.

Arrowhead.—Orig. in Excelsior, Minn., by Univ. of Minnesota Fruit Breeding Farm. Introd. in 1946. Duluth x Senator Dunlap selected in 1929; tested as Minn. 1118. Fruit: size medium; attractive; flesh red, very firm, medium to poor quality, subacid; often hollow centers; firm enough to stand shipping; ripens early. Plant: hardy; vigorous; forms many runners; flowers perfect.

Athena.—Orig. near Comacchio (Ferrara), Italy, by M. Leis and D. Masacchi,

Consorzio Italiano Vivaisti. Cross made in 1985; plant patent 37-NV/89; introd. in 1989. Flower: abundant pollen production. Fruit: large; firm; bright red external and red internal color; elongated conic and slightly flattened; good flavor. Ripens 2 days after Pajaro; good shelf life; fresh-market use. Plant: medium vigor; upright habit; withstands sudden temperature changes; resistant to common diseases; adapted to southern Italy.

Atlas.—Orig. in Willard, N.C., by North Carolina Agr. Expt. Sta., Raleigh, and the U.S. Dept. of Agriculture, Beltsville, Md. Introd. in 1970. NC 1759 x Albritton. Cross made in 1956, selected in 1959, tested as NC 2934 by G.J. Galletta, North Carolina Agr. Expt. Sta. and D.H. Scott, the U.S. Dept. of Agriculture, Beltsville. Fruit: larger than Albritton; skin medium red, glossy; flesh firm as Earlibelle, flavor subacid; ripens midseason between Earlibelle and Albritton. Plant: very large, vigorous; runner production moderate; more productive than Albritton; susceptible to red-stele root rot and powdery mildew; tolerant to leaf scorch and leaf spot. Recommended for commercial planting in the southeastern and southcentral U.S. on fertile moist soils.

Auchincruive 6.—Orig. in Auchincruive, Ayr, Scotland, by R.D. Reid, Dept. Agr. for Scotland. Introd. about 1941. Open-pollinated seedling of Scotland 52, which is probably Frith. One of the first introductions in Europe resistant to red core disease (*Phytophthora fragariae*); only of local interest and now completely obsolete; supplied to breeders in Canada as a source of genes; not grown or recognized as a variety in Europe now. Superseded in 1947 by Auchincruive Climax.

Auchincruive Climax.—Orig. in Auchincruive, Ayr, Scotland, by Robert D. Reid, Dept. of Agr. for Scotland. Introd. in 1947 in Great Britain; sent to the U.S. Dept. of Agriculture, Beltsville, Md., in 1945 for trial purposes and usage in the breeding program of that department. Seedling TD8 x Aberdeen. Fruit: quality high in Europe; suitable for all purposes. Plant: high yielding; resistant to red stele disease. Variety was widely grown in Europe, Australia, and New Zealand, but now completely ruined by June yellows and has gone out of cultivation. Under U.S. conditions, variety was too soft for commercial use and was never propagated.

Autumn Surprise.—Orig. in Champaign, Ill., by Herschel L. Boll. Introd. in 1973. Plant patent 3715; 6 May 1975. [(Canall x Twentieth Century) x Twentieth Century] x Ogallala. Berry: large, up to 1 3/4 inches in diam., wedge to conic, medium dark red; flesh medium red; fruit stems thick; calyx large, most sepals three-toothed; ripens early, about with Sunrise. Plant: everbearing, first crop slightly larger than second; highly productive, vigorous, produces medium number of runners for an everbearer; leaves slightly dull.

Badgerbelle.—Orig. in Sturgeon Bay, Wis., by F.A. Gilbert, Peninsular Branch Expt. Sta., Univ. of Wisconsin. Introd. in 1967. Robinson x Jerseybelle; selected in 1959; tested as Wis. 5827. Fruit: large, 28 to 30 berries per lb; globose conic; skin medium red, glossy, attractive; flesh medium red, moderately firm, mildly subacid, quality fair; calyx large; fresh fruit and freezing; ripens late with Jerseybelle. Plant: vigorous; prolific plant producer, runners root easily; consistently productive; leaf deep green, glossy, medium large, pubescent; tolerant to leaf spot, moderately susceptible to leaf scorch, resistance to red stele and verticillium wilt undetermined.

Badgerglo.—Orig. in Peninsular Expt. Sta., Sturgeon Bay, Wis., by Franklin A. Gilbert, Univ. of Wisconsin, Madison. Introd. in 1972. Sparkle x Stelemaster, cross made in 1959, selected in 1960, tested as Wisc. 599. Fruit: large, uniformly conic; skin bright red, very attractive, glossy; seeds yellow, slightly raised; caps adhere very tightly; flesh medium red, firm, flavor fair-good; good freezing quality; ripens late mid-season. Plant: large; vigorous; productive; susceptible to red stele; tolerant to leaf spot and scorch.

Beacon.—Orig. in Geneva, N.Y., by Richard Wellington, New York State Agr. Expt. Sta. Introd. in 1923. President x Marshall; cross made in 1910. Fruit: ripens with Dunlap. Plant: unproductive and blossoms susceptible to frost. Now obsolete.

Beaute Rouge.—See Beauty Red.

Beauty Red (*Beaute Rouge*).—Orig. in Grapevine, Texas, by F.D. Fry. Introd. in

1967. Parentage unknown, selected in 1967. Fruit: large; heart-shaped; skin medium red; flesh pale to red, core hollow, flavor tart; ripens a week after Blakemore; recommended as fresh fruit. Plant: large, bears heavily; vigorous; runner production high; blossoms more susceptible to freezes than Blakemore or Dixieland.

Beaver.—Orig. in Eau Claire, Wis., by C.H. Beaver. Introd. in 1925. Considered to be Dr. Burrill x Howard 17; selected about 1921. Fruit: quality only fair; ships very well. Plant: hardy; very productive; well adapted to light, sandy soils. Not to be confused with the Beaver variety of unknown parentage found about 1901 by J.F. Beaver, Dayton, Ohio and introduced commercially in 1906.

Belle Rouge.—A high-yielding strawberry adapted to forcing under plastic and for open culture in cooler regions of Japan. Orig. at the Morioka Branch, National Res. Inst. of Vegetables, Ornamental Plants and Tea, Japan; introd. in 1989. Fruit: large; glossy red; conic; mid- to late season; firm flesh and skin; good shelf life; adapted to long-distance shipping. Plant: productive.

Bellmar.—Orig. in Glenn Dale, Md., by George M. Darrow, U.S. Dept. of Agriculture. Introd. in 1932. Missionary x Howard 17; cross made in 1923; selected in 1925; tested as USDA 655. Fruit: large; long conic, broad shoulder; cap reflexed, very large; skin bright red, extremely glossy; achenes yellow or reddish, even with surface; flesh light red, firm, tender, rarely with a cavity, juicy, subacid, aromatic; dessert and shipping quality very high, also general market use; ripens with Howard 17 in Maryland, earlier in North Carolina. Plant: high yielder; very vigorous; produces runners freely; flower perfect.

Belrubi.—Orig. in Station d'Amelioration des Plantes Maraicheres, INRA, Montfavet (Vaucluse), France, by G. Risser. Commercially introd. in 1970. Pocahontas x Redcoat, cross made in 1961, selected in 1962. Fruit: large; conical; flesh red, high quality, firm; ripens late-midseason. Plant: large; upright; high vigor; very high productivity; variety susceptible to mildew (both plant and fruit) and *Botrytis*. Very good adaptability to soils with high pH; outstanding fruit quality.

Benton.—Orig. in Corvallis, Ore., by the U.S. Dept. of Agriculture and Oregon and Washington Agr. Expt. Stations. Introd. in 1975. OR-US 2414 x Vale, cross made in 1958 and selected in 1960 by George F. Waldo, tested as OR-US 2785. Fruit: large, holds size well through season; conic to round conic; skin medium bright red, medium-tender, slightly raised yellow seeds; flesh red, medium-firm, fresh flavor excellent; calyx more easily removed than Northwest and less so than Hood; ripens mid-season to late; a fresh market and home garden variety, as frozen product similar to Hood and not as good as Northwest because of lighter color and lower texture rating though flavor good to very good, satisfactory preserves. Plant: vigorous; generally upright, possibility of mechanical harvest, although fruit may be too soft; produces runners freely, averages yield 6.8 tons per acre at Corvallis; greater winter hardiness than Northwest and Hood but not as good as Cheam or Shuksan; more field tolerance of red stele than Northwest and equal to Hood, more resistance to powdery mildew than Northwest and equal to Hood, not as field tolerant of viruses as Northwest but better than Hood. Recommended for southern Washington, Willamette Valley, and southern Oregon.

Berri-Supreme (*Super Giant*).—Orig. in Butler, Utah, by T. Rasuga, Sandy, Utah. Introd. in 1935. Considered to be a bud mutation or seedling of Superb; discovered in 1930 in a planting of that variety. Fruit: unusually large for being an everbearer. Plant: everbearing. No longer propagated.

Black.—See Fairfax.

Black Hornet.—See Fairfax.

Black Lee.—Orig. in Brigham City, Utah, by Leland L. Larsen, Larsen Expt. Farm & Nursery. Introd. in 1954. Golden Gate x Lindalicious; selected in 1940. Fruit: medium to large; skin dark maroon, high gloss; good for dessert use and preserving; resembles Golden Gate.

Blakemore.—Orig. in Glenn Dale, Md., by George M. Darrow and George F. Waldo, U.S. Dept. of Agriculture. Introd. in 1930. Missionary x Howard 17; cross made in 1923 by Geo. M. Darrow; selected in 1925;

tested as U.S.D.A. 659; preserving tests by C.P. Lathrop, National Preservers' Assn.; field and shipping tests by Charles Dearing, North Carolina Agr. Expt. Sta. Fruit: size medium; blunt conic, shoulder broad; skin bright light red, slightly glossy, very attractive, fairly tough; hull reflexed, size medium, easily hulled; achenes fairly small, yellow or greenish-yellow, even with surface; flesh firm, light red, not darkening on holding, rarely with a cavity, juicy, acid, high pectin content, very good flavor and dessert quality; very good for preserving; easily picked; ripens somewhat before Missionary and about 2 weeks before Howard 17. Plant: superior vigor, somewhat more so than Missionary; runners produced freely; flower perfect; resistant to leaf spot, verticillium wilt, leaf scorch; well-adapted to hill culture; adapted to southern states; very productive; grown on a wide range of soil types. Named for Marcus Blakemore, first president of the National Preservers' Assn.

Blaze.—Orig. in Durham, N.H., by L.P. Latimer, New Hampshire Agr. Expt. Sta. Introd. in 1952. (Simcoe x Catskill) x (Tupper x Fairfax); selected in 1949; tested as N.H. 288. Fruit: skin very bright, glowing red, appearance very attractive; flavor pleasant; extremely uniform in color, size, shape; excellent for freezing for home use although it has a white center; retains brilliant color on removal from freezer; resembles Tupper but with much brighter color and firmer texture; ripens late. Plant: flowers perfect; high yields.

Bliss.—Orig. in Geneva, N.Y., by Richard Wellington, New York State Agr. Expt. Sta. Introd. in 1923. Chesapeake x Atkins Continuity; seed borne in 1911. Fruit: ripens late; quality excellent. Plant: withstands drought well but lacks vigor and productivity. Variety now obsolete.

Blomidon.—Orig. in Kentville, Nova Scotia, Canada, by D.L. Craig, A.R. Jamieson, and K.A. Sanford, Agr. Canada Res. Sta. K72-4 x [Micmac x (Guardsman x Tioga)]; cross made in 1975; selected by D.L. Craig in 1976; tested as K76-3; introd. in 1984. Fruit: large to very large primaries and medium-size secondaries; firm; glossy medium- to deep red external color except white under calyx; medium-red internal color with white fibro-vascular bundles; globose conic to short wedge shape; calyx moderately difficult to remove; tart flavor. Ripens mid- to late season; fresh-market use. Plant: moderate to high yielding; vigorous; abundant but not excessive runners; medium-length scapes; moderately resistant to susceptible to powdery mildew and to verticillium wilt; moderately resistant to botrytis fruit rot and to A-6, the most common race of the red stele causal organism, *Phvtophthora fragariae*, in Atlantic Canada; susceptible to June yellows.

Bogota.—Orig. in Wageningen, Holland, by Inst. for Hort. Plant Breeding. Introd. in 1978. Zb.53.116 x Tago. Selected in 1971. Licensed in 1977. Berry: large, irregularly conical; skin orange-red, moderately glossy, seeds deep set; flesh firm, pink, moderately juicy, subacid; late ripening, 5 days after Tago. Plant: productive, erect, broad; leaf large, irregularly waved; moderately susceptible to verticillium wilt, mildew, and *Phytophthora fragariae*.

Boonemore.—Orig. in northern Boone County, Ark., by H.W. Smith. Introd. in the 1930s. Open-pollinated seedling of New Home. Fruit: large; quality good. Plant: a very strong grower; self-sterile. Apparently no longer propagated.

Boquet.—Orig. in Geneva, N.Y., by Richard Wellington, New York State Agr. Expt. Sta. Introd. in 1923. Chesapeake x Pan American; cross made in 1911. Fruit: large; skin light red; quality good but not as good as Bliss; ripens in midseason. Plant: runner production scanty; flowers perfect. Obsolete, being superseded by more recent varieties superior in vigor and productivity.

Borden.—Orig. in Ottawa, Ontario, Canada, by Canada Dept. Agr., Res. Sta. Introd. in 1936. (Nor J x Parsons Beauty) x (*Fragaria* sp. x Jessie); selected in 1934. Fruit: large; skin bright scarlet; flesh very firm, below medium quality. Plant: very productive; light runnering; subject to drought. Not grown commercially.

Bountiful.—Orig. in Corvallis, Ore., by F.J. Lawrence, M.M. Stahler, P.P. Moore, L.W. Martin, G.W. Varseveld, and W.A. Sheets. Linn x Totem; selected in 1972; tested as ORUS 4688; introd. in 1993 jointly by U.S.

Dept. of Agriculture–Agr. Res. Serv., Oregon State Univ., and Washington State Univ. Fruit: same size or slightly smaller than Totem; firm; conic; uniform bright red external and internal color; nonreflexed calyx easy to remove; late season with concentrated ripening; higher susceptibility to preharvest botrytis fruit rot than Totem; processing use. Plant: high yielding; vigorous; abundant runners; low spreading habit; some flowers exposed above leaf canopy. Probably as tolerant as Totem to the aphid-borne virus complex in the Pacific Northwest; relatively resistant to powdery mildew; more susceptible than Totem to a composite of races of the red stele causal organism.

Bounty.—Orig. in Kentville, Nova Scotia, by D.L. Craig and L.E. Aalders, Canada Dept. Agr., Res. Sta. Introd. in 1972. Jerseybelle x Senga Sengana; cross made in 1963, selected in 1964, tested as K64-436. Fruit: size medium, globose conic; skin glossy, medium red; achenes yellow, level with surface; flesh medium red, medium firm, tender, flavor good; calyx separates easily; ripens late, 7 to 30 July; recommended for fresh fruit and processing. Plant: large; vigorous; productive; tolerant to leaf spot, mildew (in the Atlantic region); susceptible to mildew (British Columbia) and verticillium wilt at Vineland, Ontario, slightly susceptible to leaf scorch. Tested and suited to Atlantic Canada (New Brunswick, Prince Edward Island, and Nova Scotia).

Branford.—Orig. in New Haven, Conn., by D.F. Jones, Connecticut Agr. Expt. Sta. Introd. in 1939. (Kalicene x Howard 17) x (Progressive x Howard 17); first selected in 1923; tested as Conn. 431. Fruit: size and shape even; color light and bright, being excellent in frozen product and in preserves; ripens early June. Plant: less vigorous than most varieties, needing special conditions to produce satisfactory crops.

Brightmore.—Orig. in Corvallis, Ore., by George M. Darrow and George F. Waldo, U.S. Dept. of Agriculture. Introd. in 1942. Blakemore x Ore. 154 (Ettersburg 121 x Marshall); cross made in 1932 by Darrow; selected in 1934; tested as Ore. 1084. Fruit: size medium, tending to be small in late season; conic to long conic or necked, uniform; skin very glossy, slightly darker than Blakemore, lighter red than Marshall; achenes yellow, only slightly sunken; flesh medium red, firm, acid medium, flavor good; ripens with Marshall; quality excellent for frozen pack, very good for preserves, fresh-fruit market; resembles Blakemore. Plant: fairly resistant to virus diseases; vigorous; productive except on the very heaviest soils; susceptible to red stele root disease; foliage medium light green; produces many runners.

Brighton.—Orig. at Univ. of California Woltskill Expt. Orchard, Winters by R.S. Bringhurst and Victor Voth, Agr. Expt. Sta., Univ. of Calif., Davis. Introd. in 1979. Plant patented to Regents, Univ. of California, Davis. 65.65-601 (complex parentage) x Tufts. Third backcross generation derivative from a male *Fragaria virginiana glauca* collected in the Wasatch Mountains near Salt Lake City, Utah. Selected 1971 to 1972. Fruit: flesh flavor and quality not as good as those of Aptos or Hecker; skin dark reddish-orange, glossy shellac-like sheen. Plant: smaller than Tioga; leaves lighter and more yellow than those of Tioga; less susceptible to verticillium wilt than Tioga and Tufts. Recommended for home gardens as day-neutral (can be made to produce fruit 3 months after planting). Named for Brighton Valley of Big Cottonwood Canyon, Utah, source of wild strawberry *Fragaria virginiana glauca* in parentage.

Brilliant.—See Gem.

Bristol.—Orig. in New Haven, Conn., by D.F. Jones, Connecticut Agr. Expt. Sta. Introd. in 1939. (Chesapeake x Marshall) x (Progressive x Howard 17); selected in 1923. Fruit: attractive; quality high; resembles Chesapeake; ripens early June.

British Sovereign (*Victoria*).—Orig. in Coghlan, British Columbia, Canada, by K. Kinoshita and O. Kamachi. Introd. in 1923. Parentage unknown; six plants discovered in a field of Campbell (Paxton) variety in 1920. Fruit: medium to large; shape very uniform; skin medium to dark red right to the core; flesh mildly subacid with rich flavor; excellent for jam, good for canning, and satisfactory for fresh market and freezing. Plant: hardy; deep-rooted and moderately vigorous; susceptible to red stele, mildew, and fruit rots but shows some tolerance to root rots. Was the standard variety in British Columbia for many years, but now not grown.

Brunes Marvel.—Orig. in Pequot Lakes, Minn., by Charles Brunes. Introd. in 1942. Assumed to be Progressive x Mastodon seedling; selected in 1935. Fruit: sugar content high, acid low, recommended as nonacidic; resembles Progressive.

Burgess.—Orig. in Alabama. Introd. in 1930. Considered to be a chance seedling of Aroma; discovered about 1922. Fruit: skin dark red, appearance good; nearly as firm as Aroma; flavor poor. Plant: very vigorous; about equal in production to Aroma. No longer propagated.

Burgundy.—Orig. in Excelsior, Minn., by Univ. of Minnesota. Fruit Breeding Farm. Introd. in 1943. Easypicker x Duluth; selected in 1929; tested as Minn. 1192. Fruit: large; roundish cordate, regular; skin dark red, evenly colored, rather glossy, somewhat pubescent; achenes medium few, raised; flesh very dark red, firm, meaty; flavor aromatic, subacid; quality very good; ripens late. Plant: hardy; vigorous, upstanding; very productive; June-bearing; flowers pistillate, late, being 2 weeks after Howard 17.

Cacak 45/7.—Orig. in Cacak, Yugoslavia, by P.D. Misic, Ing. V.Z. Bugarcic, and M.B. Tesic, Fruit Res. Inst. Introd. in 1967. Yugoslavia plant patent 187/1; 14 Jan. 1967; by Dept. of Agr. Forestry of Yugoslavia; assigned to Fruit Res. Inst., Cacak. Talisman x Macherauchs Frühernte (107). Selected in 1962. Fruit: large; skin red; flesh firm, aromatic, dessert quality; calyx small and separates from the flesh with slightly greater difficulty than with the parents; picking season short, ripens early. Plant: vigor and productivity superior to parents.

Cal Parker.—Orig. in South Coast Field Sta. Santa Ana, Calif., by Victor Voth and R.S. Bringhurst, Agr. Expt. Sta., Univ. of California, Davis. Introd. in 1983. Plant patent pending; to be assigned to Regents, Univ. of California, Davis. Douglas x 71.98-604 [Tufts x 69.7-101 (parent of Pajaro)]. Cross made in 1977; selected in 1979; tested as Cal 77.30-605 and later as C3. Fruit: about the size of Douglas; long conic to long flat wedge; skin slightly darker than Tioga, glossy, and attractive; flesh very firm, more durable than Tioga, Tufts, or Aiko; flavor good; ripens later than Tufts, Tioga, and Douglas; recommended for fresh market and processing; a short-day type. Plant: semi-erect; runner production better than Tioga; low chilling requirement, similar to Tioga or Tufts; self-fertile; performed well in winter and summer plantings in south and central coastal California. Named for the late Bob Parker of Lassen Canyon Nursery, Redding, Calif.

Caledonia.—Orig. in Geneva, N.Y., by Richard Wellington, New York Agr. Expt. Sta. Introd. in 1931. Marshall x Howard 17; cross made in 1923. Fruit: large; attractive; flesh red, firm, quality good; made an excellent preserve for which purpose it was introduced. Plant: numerous plant maker; vigorous; productive; eventually proved too susceptible to leaf spot and is out of cultivation.

Calypso.—Orig. in East Malling, England, by D. Simpson, Hort. Res. Intl. Rapella x Selva; selected in 1986; tested as EMR 28; introd. in 1992. Fruit: firm; primaries can be irregular; good flavor; quality better than standard Rapella; day-neutral with similar season as Rapella; fresh-market use. Plant: smaller than Rapella; abundant runners; moderately resistant to verticillium wilt; susceptible to powdery mildew, but symptoms largely confined to leaves, seldom found on fruit; susceptible to two-spotted spider mite.

Camarosa.—Orig. at Univ. of California, South Coast Res. and Ext. Center, near Irvine, by V. Voth and R.S. Bringhurst. Douglas x Cal 85.218-605; cross made in 1988; selected in 1989; tested as Cal 88.24-603; introd. in 1992. U.S. plant patent pending. Fruit: larger and firmer than Chandler; very flat conic; external color similar to Chandler, glossy; internal color darker than Chandler; achenes light to dark red, even with surface or slightly indented; very good flavor, less aromatic than Chandler; fresh-market, processing, and home-garden uses. Plant: short-day type, producing fruit over an extended period in arid, subtropical climates; production pattern similar to Chandler, but produces more early season fruit; high yielding; more vigorous than Chandler. Moderately susceptible to common leaf spot; relatively resistant to powdery mildew; equal or greater tolerance to two-spotted mite than Chandler; tolerant to viruses occurring in California.

Cambridge Aristocrat.—Orig. in Cambridge, England, by D. Boyes, Hort. Res. Sta., School of Agr., Univ. of Cambridge. Introd. about 1952. A variety from the U.S. x Early Cambridge; selected in 1944; tested as Cambridge 641. Fruit: long, conical; flavor rich. Plant: red core resistant.

Cambridge Early Pine.—Orig. in Cambridge, England, by D. Boyes, Hort. Res. Sta., School of Agr., Univ. of Cambridge. Introd. in 1947. *Fragaria chiloensis* type seedling x Blakemore; selected in 1944; tested as Cambridge 448. Fruit: round conical; very distinct pine flavor; ripens very early; quite distinct, not resembling any variety. Plant: yield poor. Not widely grown.

Cambridge Epicure.—Orig. in Cambridge, England, by D. Boyes, Hort. Res. Sta., School of Agr., Univ. of Cambridge. Introd. in 1947. Early Cambridge x (Euresko x Ettersburg 121); selected in 1944; tested as Cambridge 143. Fruit: small; flavor good; ripens very early; resembles Euresko. Plant: yield small. Not widely grown.

Cambridge Favourite.—Orig. in Cambridge, England, by D. Boyes, Hort. Res. Sta., School of Agr., Univ. of Cambridge. Introd. in 1947. *Fragaria chiloensis* type seedling x Blakemore; selected in 1944; tested as Cambridge 422. Fruit: short, conical; appearance good; ripens early, with a long picking season; very good for canning; ships well; distinct, not resembling other varieties; widely grown for dessert and canning.

Cambridge Forerunner.—Orig. in Cambridge, England, by D. Boyes, Hort. Res. Sta., School of Agr., Univ. of Cambridge. Introd. in 1947. Early Cambridge x Bellmar; selected in 1941; tested as Cambridge 54. Fruit: large; flesh soft; early for cloching; resembles Early Cambridge. No longer propagated.

Cambridge Late Pine.—Orig. in Cambridge, England, by D. Boyes, Hort. Res. Sta., School of Agr., Univ. of Cambridge. Introd. in 1947. *Fragaria chiloensis* type seedling x Fairfax; selected in 1944; tested as Cambridge 490. Fruit: skin very dark; flavor good; resembles Fairfax. Plant: grown in home gardens only, because fruit is very dark.

Cambridge Premier.—Orig. in Cambridge, England, by D. Boyes, Hort. Res. Sta., School of Agr., Univ. of Cambridge. Introd. in 1947. (Royal Sovereign x Blakemore) x Euresko seedling; selected in 1941; tested as Cambridge 276. Fruit: appearance good; resembles Royal Sovereign. Plant: a mildew-resistant Royal Sovereign. Not widely grown.

Cambridge Prizewinner.—Orig. in Cambridge, England, by D. Boyes, Hort. Res. Sta., School of Agr., Univ. of Cambridge. Introd. in 1947. Early Cambridge x a U.S. variety; selected in 1941; tested as Cambridge 134. Fruit: large; attractive; resembles Royal Sovereign; ripens early. Variety widely grown.

Cambridge Rearguard.—Orig. in Cambridge, England, by D. Boyes, Hort. Res. Sta., School of Agr., Univ. of Cambridge. Introd. about 1952. *Fragaria chiloensis* type seedling x a U.S. variety; selected in 1944; tested as Cambridge 503. Fruit: large; flesh dark, flavor poor; distinct, not resembling other varieties; ripens late. Plant no longer grown.

Cambridge Regent.—Orig. in Cambridge, England, by D. Boyes, Hort. Res. Sta., School of Agr., Univ. of Cambridge. Introd. in 1947. A U.S. variety x Royal Sovereign; selected in 1944; tested as Cambridge 679. Fruit: flesh rather dark; resembles Royal Sovereign, but dark; ripens very early. Variety not widely grown.

Cambridge Rival.—Orig. in Cambridge, England, by D. Boyes, Hort. Res. Sta., School of Agr., Univ. of Cambridge. Introd. in 1948. Dorsett x Early Cambridge; selected in 1941; tested as Cambridge 632. Fruit: medium large; conical or wedge; skin shiny, bright crimson; flesh firm; ripens early to second early season. Plant: resistant to red stele.

Cambridge Sentry.—Orig. in Cambridge, England, by D. Boyes, Hort. Res. Sta., School of Agr., Univ. of Cambridge. Introd. in 1946. Dorsett x Early Cambridge; selected in 1941; tested as Cambridge 614.

Fruit: long and pointed; skin dark crimson; flesh very firm, red; ripens midseason to late. Plant: vigorous; resistant to red stele; high yield.

Cambridge Vigour.—Orig. in Cambridge, England, by D. Boyes, Hort. Res. Sta., School of Agr., Univ. of Cambridge. Introd. in about 1947 [US-3378 (Aberdeen x Fairfax) x Early Cambridge]. Fruit: large; skin attractive, light red; flesh firm, flavor good; ripens early to midseason. Plant: moderately productive; tolerant to red stele; susceptible to verticillium wilt and drought. Parent of Crusader introd. in 1966.

Camden.—Orig. in Geneva, N.Y., by Richard Wellington, New York State Agr. Expt. Sta. Introd. in 1931. Marshall x Howard 17; cross made in 1923. Fruit: large, holding up well throughout season; skin glossy, attractive; flavor mild, quality fairly good; when overripe, color tends to darken. Variety now obsolete.

Campbell (*Riverview*).—Orig. in Davis, Calif., by R.E. Baker and Victor Voth, California Agr. Expt. Sta. Introd. in 1949. Cal. 177.21 [USDA 543 x Cal. 68.24 (USDA 634 x New York 4626)] x Cal. 103.22 [Cal. A.08 (USDA 253 x F_2 USDA 253) x Nich Ohmer]; tested as Cal. 467-1. Fruit: size medium to medium large; conic; skin medium red; flesh medium firm, juicy, bright red; achenes flush with the surface and yellow; dessert quality equal to Tahoe, Shasta, or Sierra; ripens from late April to early May until late fall; shipping quality good. Plant: upright; similar to Tahoe; produces an abundance of runners; only slightly more resistant to verticillium wilt than Lassen; tolerant to yellows but susceptible to mildew and injury from redspider.

Canall.—Orig. in Brigham City, Utah, by L.L. Larsen, Larsen Expt. Farm. Introd. in 1950. Catskill x Golden Gate; selected in 1947. Fruit: large; skin dark red, nearly black; highly flavored; good dessert quality. Plant: heavy cropper.

Candy Red.—Orig. in Grandview, Ind., by Alben B. Cook. Introd. in 1979 by Herschel L. Boll, Champaign, Ill. Ozark Beauty x Tennessee Shipper. Cross made in 1965. Berry: large, mostly long conic; skin medium dark red; glossy, attractive; flesh firm; ripens late. Plant: highly productive, vigorous and large; produces many runners.

Canoga.—Orig. in Geneva, N.Y., by Donald K. Ourecky, New York State Agr. Expt. Sta. Introd. in 1979. NY 1123 (Senga sengana x Midland) x Holiday. Cross made in 1969; selected in 1971; tested as NY 1362. Fruit: large, wedge conic; skin intense red brick color, tough; seeds yellow, even with surface; flesh dark red, very firm; ripens late midseason. Plant: vigorous, upright, very productive; runner production moderate.

Capitola.—Orig. at Univ. of California, Wolfskill Experimental Orchards, near Davis, by R.S. Bringhurst and V. Voth. Cal 75.121-101 x Parker; cross made in 1983; selected in 1984; tested as Cal 83.93-6 and later as CN 93; introd. in 1991. U.S. plant patent 7615, 6 Aug. 1991. Fruit: as large or larger than Selva but less firm; external and internal color similar to Chandler; symmetrical medium conic; achenes bright yellow to slightly red; pleasant subacid flavor as good as or better than Selva; fresh-market and processing uses. Plant: more strongly day-neutral than Selva but not as strongly as Hecker; both mother and daughter plants tend to flower and fruit; vigorous; runner production equal or better than Selva; scape more erect than Selva. Quite susceptible to common leaf spot; moderately susceptible to two-spotted spider mite; very tolerant to viruses occurring in California, including mild yellow edge.

Cardinal.—Orig. in Fayetteville, Ark., by J.N. Moore, H.L. Bowden, and W.A. Sistrunk, Ark. Agr. Ext. Serv. Introd. in 1974. Earlibelle x Ark. 5063; cross made in 1967 by J.N. Moore; selected in 1969; tested as Ark. 5744. Fruit: large; slightly necked; skin very glossy, dry, attractive; flesh medium red throughout, firm, quality good both fresh and frozen; calyx easily removed in processing; ascorbic acid averages 50 mg/100 g fresh weight; ripens midseason. Plant: vigorous; very productive; prolific plant producer; high yield of usable fruit in one picking by mechanical harvest; resistant to leaf spot, leaf scorch, and powdery mildew; tolerant of two-spotted spider mite; may have wide adaptation in eastern U.S., recommended for trial from Maryland and southward and west to Texas.

Carlsbad.—Orig. at Univ. of California, South Coast Res. and Ext. Center, near Irvine, by V. Voth and R.S. Bringhurst. Irvine x Cal 85.218-605; cross made in 1988; selected in 1989; tested as Cal 88.70-613; introd.in 1992. U.S. plant patent pending. Fruit: large; firmer than Chandler; very flat conic primary, blocky conic secondary; less glossy, with lighter external and internal color than Chandler; achenes yellow to dark red and even to or slightly extruded from the fruit surface. Very good flavor, somewhat less aromatic than Chandler; fresh-market, processing, and home-garden uses. Plant: short day, produces fruit over an extended period for arid, subtropical climates; production pattern similar to Chandler but produces greater quantities of early season fruit; more vigorous than Chandler. Moderately resistant to common leaf spot and powdery mildew; equal or greater tolerance to two-spotted mite than Chandler; tolerant to viruses occurring in California.

Carver Bell.—Orig. in Carver, Mass., by James H. Jennings. Introd. in 1951. Parentage unknown; discovered in 1949. Fruit: large; shape varies, being bell-shaped in early season but pointed later; skin pink; flesh not too firm, flavor very good; resembles Empire; matures early, with Earlidawn. Plant: hardy; vigorous, withstands dry weather very well; susceptible to leaf spot but considered tolerant of virus.

Cascade.—Orig. in Puyallup, Wash., by C.D. Schwartze, Western Washington Expt. Sta. Introd. in 1960. Shasta x Northwest; selected in 1949. Fruit: size medium; rounded or blunt; skin dark red, glossy; flesh very firm, sweet, rich flavor; superior texture and flavor for processing (freezing, preserves, and ice cream); ripens in midseason, with Northwest, which it resembles. Plant: adapted to medium light, well-drained soils; vigorous; productive; hardy in western Washington; somewhat resistant to virus diseases, more so than Marshall; susceptible to mildew and red stele diseases; blooms with Northwest.

Cato.—Orig. in Geneva, N.Y., by Richard Wellington, New York State Agr. Expt. Sta. Introd. in 1931. Marshall x Howard 17; cross made in 1923; sibling of Clermont and Culver. Fruit: large, retaining good size during the season; roundish wedge conic, apex pointed; skin dark red, glossy, attractive, bruises very easily; achenes slightly raised; flesh dark red, moderately firm, subacid, juicy, quality very good; ripens about 4 days after Howard 17, between Clermont and Culver. Plant: vigorous, productive; superior to Marshall; flower perfect, size medium. A home garden variety; apparently now superseded by other varieties.

Catskill.—Orig. in Geneva, N.Y., by Richard Wellington, New York State Agr. Expt. Sta. Introd. in 1934. Marshall x Howard 17; cross made in 1923; selected in 1925. Fruit: large; long conic, slightly irregular; skin glossy, bright crimson, attractive; flesh light red, moderately firm, apparently too soft for shipping, very satisfactory for freezing, mildly subacid, dessert quality excellent; ripens in midseason. Plant: very vigorous and productive; fairly resistant to verticillium wilt, powdery mildew, leaf scorch and leaf spot, but sensitive to virus diseases; runners produced freely; adapted to wide range of soil types.

Cavalier.—Orig. in Ottawa, Ontario, Canada, by Canada Dept. Agr., Res. Sta. Introd. in 1957. Valentine x Sparkle; cross made in 1948; selected in 1950; tested as Ottawa 481. Fruit: large, but tends to become smaller later in season; skin attractive, bright red, glossy; flesh lighter red than Grenadier and Guardsman but darker than Redcoat, acidity not high, flavor fine, firmer than Catskill, Dunlap, and Howard 17; good for dessert, processing, and freezing; ripens before Howard 17. Plant: virus-free; significantly outyielded Howard 17 during first 6 days of harvest, with total yield equal to or higher than that of Howard 17 and Dunlap; produces runners freely; vigorous but not as tall as Redcoat; somewhat susceptible to leaf spot, leaf scorch, and powdery mildew; highly resistant to verticillium wilt.

Cavendish.—Orig. in Kentville, Nova Scotia, Canada, by D.L. Craig, Canada Res. Sta Introd. in 1990. Glooscap x Annapolis; cross made in 1981, selected in 1983; tested as K83-4. Fruit: medium to very large; surface deep glossy red, medium firm; flesh firm, red except white beneath calyx near core; flavor sweet, less tart than Honeoye, melon-like; tolerant to gray mold. Ripens midseason. Plants: medium vigor; runner well, but not excessively; crown size small to medium; very productive. Tolerant or resistant to verticillium wilt; resistant to stele races A4, A-6, and A-7; tolerant to powdery mildew; susceptible to green petal virus.

Chambly.—Orig. in L'Acadie, Quebec, by S. Khanizadeh, D. Buszard, M. Lareau, and D. Bagnara, Agr. Canada, Res. Sta., St-Jean-sur-Richelieu, and Dept. of Plant Science, Macdonald College, Ste.-Anne-de-Bellevue, Quebec. Sparkle x Honeoye; cross made in 1982; selected in 1984; tested as SJ84187-3; introd. in 1990. Fruit: medium size; firm; conic shape with white, raised neck; glossy, deep red external and solid red internal color; reflexed calyx relatively difficult to remove; flavor similar to Bounty and Sparkle; fresh-market and processing uses. Plant: high yielding; low vigor with sparse appearance; scapes long and thick, erect during flowering, semi-erect as fruit matures. Probably resistant to powdery mildew, leaf scorch, and leaf blight; resistant to some races of the red stele causal organism; high level of winter hardiness; partially tolerant to herbicide terbacil.

Chandler.—Orig. in Davis, Calif. by R.S. Bringhurst and Victor Voth, Agr. Expt. Sta., Univ. of California. Introd. in 1983. Plant patented (plant patent 4481), assigned to Regents, Univ. of California, Davis. Douglas x Cal 72.361-105 = C55. Cross made in 1977; selected in 1979; tested as Cal 77.32-103 and C24. Fruit: larger than Tioga; long conic to long flat wedge; skin glossy and attractive, resembles Pajaro; flesh same color as the skin throughout, firm; higher ascorbic acid content than Tioga, Tufts. or Douglas, but lower than Pajaro and Aiko; flavor excellent; recommended for fresh market and processing; short-day type; ripens early. Plant: semi-erect, about same size as Tioga: self-fertile; performed well in winter and summer plantings in south and central coastal California. Runner production excellent, better than Tioga. Named for the late William Henry Chandler, former Chairman of the Dept. of Pomology, Univ. of California, Davis.

Chaska.—Orig. in Excelsior, Minn., by Univ. of Minnesota Fruit Breeding Farm. Introd. in 1922. (Dunlap x Pocomoke) x Brandywine; tested as Minn. 801. Fruit: large; roundish to slightly conic; skin medium dark red, glossy; flesh dark red, firm, juicy, slightly subacid, aromatic; quality good; good canner; midseason. Plant: vigorous; productive; leaves large, moderately dark green; petioles long. No longer propagated.

Cheam.—Orig. in Agassiz, British Columbia, Canada, by Hugh A. Daubeny, Canada Dept. of Agr., Res. Sta. Introd. in 1968. Siletz x Puget Beauty. Selected in 1960, tested as 59-22-38 and BC3. Fruit: medium to large; uniform and conical; skin glossy, medium red; flesh medium red, moderately firm; ripens a few days earlier than Northwest. Plant: vigorous, with long petioles and tendency to sprawl; runner production profuse; very hardy; productive; tolerant to most races of *Phytophthora fragariae*; susceptible to fruit rot; recommended as a replacement for Siletz in the Pacific Northwest.

Cheyenne 1.—See Early Cheyenne.

Cheyenne 2.—Orig. in Cheyenne, Wyo., by LeRoy Powers, U.S. Dept. of Agriculture, Hort. Field Sta. Introd. in 1942. F_1 *Fragaria ovalis* (P.I. 36979) x Fairfax; tested as Station 956. Fruit: larger than Early Cheyenne 1; long conic, skin bright red; flesh red, soft, sweet, aromatic; ripens 10 to 12 days later than Early Cheyenne 1. Plants: vigorous; prolific as Dorsett and Fairfax, produces large number of runners; very hardy in Rocky Mountain and Great Plains regions. Recommended for home gardens.

Cheyenne 3.—Orig. in Cheyenne, Wyo., by LeRoy Powers, U.S. Dept. of Agriculture, Hort. Field Sta. Introd. in 1942. Fairfax x [Fairfax x *Fragaria ovalis* (P.I. 36979)]; tested as Station 1187. Fruit: resembles Fairfax closely in shape, size, firmness, appearance, and quality; ripens 10 days to 2 weeks after Cheyenne 2 or a month later than Early Cheyenne 1. Plant: hardy in the Rocky Mountain and Great Plains regions.

Chief Bemidji.—Orig. in Port Hope Township, Beltrami County, Minn., by Rufus Jacobs, Bemidji, Minn. Introd. in 1960. Plant patent 2147; 22 May 1962; assigned to Farmer Seed & Nursery Co., Faribault,

Minn. Chance seedling discovered in row of Montana Progressive seedlings. Fruit: large; rounded wedge-shaped; skin bright red; flesh firm, sweet; resembles Armore. Plant: everbearing; extremely hardy; vigorous; produces 15 to 20 large runners per plant; roots thicker and longer than Montana Progressive; does well on poor soil.

Churchill.—See Louise.

Citation.—Orig. in Lexington, Ky., by Carl E. Chaplin, Kentucky Agr. Expt. Sta. Introd. in 1964. Fairland x Tennessee Shipper; cross made in 1951; selected in 1953; tested as Ky. 12-51-10. Fruit: medium large, maintaining this size well throughout the season; conic, more uniform than many varieties; skin a very bright medium red; achenes flush with surface; flesh bright red; superior shipping and freezing qualities; ripens in midseason, about 1 or 2 days after Pocahontas, which it resembles. Plant: very vigorous; productive; makes a good matted row; moderately susceptible to leaf spot.

Clare.—Orig. in Ames, Iowa, by E.L. Denison, Iowa State Univ. Iowa 3-7165 (Spotlight x Sunrise) x Iowa 3145 (Surecrop x Sunrise); cross made in 1977; tested as Iowa 7-75065; introd. in 1989. Fruit: large; firm; concentrated midseason ripening; calyx slightly reflexed. Plant: vigorous; fairly numerous runners; petioles sturdy, medium thick; crown solid with seven to nine fruit buds each.

Claribel.—Orig. in Ottawa, Ontario, Canada, by Canada Dept. of Agr., Res. Sta. Introd. in 1935. Ettersburg 121 x Cassandra; selected in 1933. Fruit: short fruit stalks; flesh very firm, quality poor; ripens very late. Plant: flowers imperfect. No longer propagated.

Clermont.—Orig. in Geneva, N.Y., by Richard Wellington, New York State Agr. Expt. Sta. Named in 1929; introd. in 1931. Marshall x Howard 17; cross made in 1923; selected in 1925; sibling of Cato and Culver. Fruit: very large, holding up well throughout season; conic, regular, apex pointed; hull green, attractive, size medium; skin bright red, very glossy, does not bruise easily; achenes slightly raised; flesh medium red, moderately firm, juicy, slightly hollow at center, mildly subacid, pleasant flavor; quality good; excellent for preserving as well as dessert; ripens early, 2 or 3 days after Howard 17. Plant: large; good yields; somewhat susceptible to leaf spot; flower perfect.

Coleman's President.—Orig. in Bellingham, Wash., by Frank Coleman. Introd. about 1938. Open-pollinated seedling of Marshall; discovered about 1935. Fruit: large declining to small; round conic; skin rather dull medium red; calyx adherent; flesh red, fairly firm, flavor good; ripens early midseason. Plant: larger than Marshall, more upright; good runner maker; leaves bluish-green on strong petioles. Variety believed to be obsolete.

Columbia.—Orig. in Puyallup, Wash., by C.D. Schwartze, Western Washington Expt. Sta. Introd. in 1960. Wash. 157 x Wash. 175; selected in 1948. Fruit: medium to large, averaging slightly smaller than Northwest; irregularly conic; skin bright red, soft, easily bruised; hull large, attractive; flesh red, being similar in flavor, color, and processing qualities to Northwest; needs special care in harvesting and handling; ripens 5 to 7 days after Northwest, which it resembles. Plant: very vigorous and productive; hardy in western Washington; similar to Siletz in resistance to red stele and root rots; resistant to mildew and fruit rot (*Botrytis*); somewhat resistant to virus diseases; blooms 5 to 7 days after Northwest; yield approaches Northwest under good cultivation practices.

Comet.—Orig. in Fayetteville, Ark., by J.N. Moore, H.L. Bowden, and W.A. Sistrunk, Ark. Agr. Expt. Sta. Introd. in 1975. Earlibelle x Ark. 5063; cross made in 1967; selected in 1969; tested as Ark. 5734. Fruit; medium large; conic; symmetrical; skin firm, glossy, bright red; flesh firm, uniform red, good flavor. Ripens early; recommended for use in fresh market and processing; good shipping and processing qualities. Plant: very vigorous; prolific; good productivity; resistant to leaf spot, leaf scorch, and powdery mildew; susceptible to red stele; tolerant to spider mite.

Cornwallis.—Orig. in Kentville, Nova Scotia, Canada, by D.L. Craig, A.R. Jamieson, K.A. Sandford, and N.L. Nickerson, Agr. Canada Res. Sta. Earliglow x Kent; cross made in 1977; selected by D.L.

Craig in 1978; tested as K78-6; introd. in 1984. Fruit: medium size; medium firm; uniform; deep red external and internal color; short conic; calyx slightly reflexed, easy to medium-difficult to remove; good flavor; fresh-market and processing use. Plant: vigorous; abundant but not excessive runners; long scape becomes prostrate as fruit matures. Moderately resistant to powdery mildew and to verticillium wilt; highly resistant to races A-4, A-6, and A-7 of the red stele causal organism, *Phythophthora fragariae*; moderate yield where red stele is not present and high yield where it is present; hardy in Atlantic Canada.

Cortina.—A high-yielding, fresh-market strawberry producing firm fruit. Orig. near Comacchio (Ferrara), Italy, by M. Leis, D. Musacchi, A. Martinelli, Consorzio Italiano Vivaisti. Addie x Parajo; cross made in 1984; introd. in 1992. Fruit: medium to large size; intense bright red external and red to orange internal color; some hollowness; heart shape with pointed apex; large calyx easily removed; achenes beneath surface; season similar to Addie; good flavor with medium sweetness and considerable acidity. Plant: medium vigor; moderate runner production; moderately upright habit; flowers early and at level of leaf canopy.

Corvallis.—Orig. in Oregon by C.E. Schuster, Oregon Agr. Expt. Sta. Introd. in 1930. Plant patent 60; 18 Apr. 1933; dedicated by assignment to the People of the United States of America. Ettersburg 121 x Marshall; selected in 1922; tested as O.S.C. 12. Fruit: size medium; conic to oblate, no neck; skin bright crimson; flesh firm; excellent for canning and freezing; ripens after Marshall; resembles Ettersburg. Plant: vigorous; upright; productive.

Cream.—Orig. in Ettersburg, Calif., by Albert F. Etter (1872 to 1950). Introd. in 1920. Ettersburg 330 x Trebla. Fruit: medium to small; conic; skin yellow cream; flesh medium firm, very juicy, subacid, quality fair; ripens late. Plant: small; productive. Apparently no longer propagated.

Crimson Flash.—See Midland.

Crimson Glow.—Orig. in New Brunswick, N.J., by J.H. Clark, New Jersey Agr. Expt. Sta. Introd. under restrictions in 1940; introd. and named in 1943. Fairfax x N.J. 51 (Pearl x Aberdeen); selected in 1933; tested as N.J. 311. Fruit: medium to large; round conic; skin smooth, glossy, dark red; flesh sweet, very good, quality high; picking easy since stem breaks back of calyx. Plant: vigorous; principally for home gardens.

Crimson King.—Orig. by Marion Hagerstrom, Monticello, Minn. Unknown origin with Fairfax in derivation; introd. in 1984. U.S. plant patent 4413, assigned to Stark Bros., Louisiana, Mo. Fruit: large; soft to medium firm; oval; glossy red external color and uniform red internal color; semisweet flavor; early ripening; home-garden and pick-your-own uses. Plant: very vigorous; very winter hardy; relatively susceptible to leaf spot and leaf scorch.

Crusader.—Orig. in Auchincruive, Ayr, Scotland, by Robert D. Reid, Scottish Hort. Res. Inst. (West of Scotland Unit). Introd. in 1966. Auchincruive seedling 50X5 [seedling 6M17 (Temple selfed) x seedling 11 (Aberdeen open)] x Cambridge Vigour; cross made in 1957; tested as Breeder's no. 57E28 and Brogdale A.45. Fruit: large to medium; bluntly conical with rounded apex; skin light to dark red, bright but darkens after picking, shiny, slightly pitted; flesh white in center suffused with pink towards exterior, fairly firm, crisp; flavor sweet, very pleasant; ripens slightly earlier than Redgauntlet. Plant: large to very large; vigorous, upright; leaflets usually broader than long; flowers not liable to frost damage; resistant to red core (*Phytophthora fragariae*), susceptible to mildew in dry areas.

Cuesta.—Orig. at Univ. of California, Wolfskill Experimental Orchards, near Davis, by R.S. Bringhurst and V. Voth. Seascape x Cal 83.25-2 (Fern x Parker); cross made in 1987; selected in 1988; tested as Cal 87.109-3; introd. in 1992. U.S. plant patent pending. Fruit: larger than Chandler; firmness similar to Chandler; long conic; compared to Chandler, external color slightly darker and less glossy and internal color darker; achenes light to dark red, even with surface or slightly indented. Very good flavor, slightly less aromatic than Chandler; fresh-market, processing, and home-garden uses. Plant: short-day type producing fruit over an extended period in arid, subtropical climates; production pattern similar to Chandler, except it produces larger and more late-season fruit; higher yield than Chandler; more erect and open habit than Chandler. Moderately susceptible to common leaf spot and powdery mildew; tolerant to two-spotted spider mite; tolerant to viruses occurring in California.

Culver.—Orig. in Geneva, N.Y., by Richard Wellington, New York State Agr. Expt. Sta. Introd. in 1931. Marshall x Howard 17; cross made in 1923; selected in 1925; sibling of Cato and Clermont. Fruit: large to very large; quite uniform, conic to wedge-conic, apex pointed; skin dark glossy red, attractive, coloring evenly, bruises slightly; hull very large, flat, bright green, achenes sunken; flesh dark red, juicy, tender, sprightly; very well adapted for preserves, for nearby markets and home use; ripens in early midseason, 5 days after Howard 17. Plant: large, vigorous, very tall; moderately productive; usually resistant to leaf spot; flower perfect, size medium.

Cummerbund.—See Fairfax.

Cupertino.—Orig. in Davis, Calif., by R.E. Baker and Victor Voth, California Agr. Expt. Sta. Introd. in 1949. Cal. 611.4 (168.3 x 186.1) x Cal. 468.4 (177.21 x 135.18); derived from Nich Ohmer, Banner, Fendalcino, USDA 253, USDA 634, and New York 4626; tested as Cal. 829.9. Fruit: skin exceptionally glossy; flavor good, high quality, aromatic; production peaks follow other University varieties (e.g., Shasta, Tahoe) and thus fills in slack picking periods. Plant: virus tolerant.

Curtuno.—A fresh-market strawberry producing uniform-appearing fruit, adapted to protected culture. Orig. in Tudela-Narvarra, Spain, by J.M.A. Lopez. Parents were seedlings obtained from open pollination among 32 varieties; tested as 89-33-018; introd. in 1992. Assignee, Plantas de Navarra, S.A. Valtierra, Spain. Plant patent 8632. Fruit: somewhat similar to Chandler; large primaries; firm; orange-red, glossy external; medium-red internal color; nearly conic; achenes below surface; reflexed calyx; medium sweet and low acid. Plant: medium vigor; erect inflorescence above foliage; medium chilling requirement.

Cyclone.—Orig. in Ames, Iowa, by Ervin L. Denisen, Iowa Agr. Expt. Sta. Introd. in 1959. Iowa 1-3713 (Rockhill selfed) x Iowa 68-3702 (Beaver x Dorsett); cross made in 1950; fruited for the first time in 1952; selected in 1952; tested as Iowa 1-35. Fruit: large; slightly necked so that hull is removed easily; flesh brilliant red, juicy, flavor excellent; June bearing; freezes well; designed for home gardens and local markets as it is not firm enough for distant shipping. Plant: high yielding; vigorous; produces many runners; hardy.

Dabreak.—Orig. in Baton Rouge, La., by P.L. Hawthorne, La. Agr. Expt. Sta. Introd. in 1961. Headliner x La. 3-1169 (complex hybrid including Klonmore, Suwanee, Konvoy, and Marion Bell); cross made in 1954; selected in 1956; tested as La. 5-325. Fruit: medium to large; long conic to wedge; skin glossy, bright red, attractive; hull dark green, large; achenes small, somewhat sunken; flesh firm, juicy, medium red, quality good; shipping quality very good; for fresh market, processing, and dessert; season early to late spring, being early March through early June in Baton Rouge. Plant: vigorous, runner production medium to good; productive; flower perfect; leaf diseases must be controlled.

Daddy Leman.—See Leman.

Darrow.—Orig. in Beltsville, Md., by U.S. Dept. of Agriculture and Maryland Agr. Expt. Sta. Introd. in 1974. Md-US 2713 (Redglow x Surecrop) x Md-US 2787 (Md-US 2359 x Md-US 2101). Cross made in 1963, seedlings grown in Salisbury, Md. in 1964, selected in 1965 by D.H. Scott, I.C. Haut, and A.D. Draper. Fruit: medium-large; secondary berries symmetrical conic; skin tough, deep red, glossy, flesh very firm, handles well, uniform red, flavor good, not as tart as Surecrop; good for frozen pack; ripens 2 to 3 days later than Earlidawn, 3 to 5 days before Surecrop. Plant: very productive; medium vigor; produces moderate number of runners; tolerant to five races of red stele root rot, moderately tolerant to verticillium wilt, leaf scorch, mildew and leaf spot. Recommended for areas where red stele root rot may be a problem and where early ripening is desired. Named in honor of George

M. Darrow, former Head of Small Fruit Investigations, U.S. Dept. of Agriculture.

Daybreak.—Orig. in Glenn Dale, Md., by George F. Waldo, U.S. Dept. of Agriculture, and E.B. Morrow, North Carolina Agr. Expt. Sta., Raleigh. Introd. in 1939. Missionary x Fairfax; selected in 1932. Fruit: large; matures early. Plant: very vigorous; large crops. No longer propagated because of presence of virus disease.

Deep Red.—Orig. in Monticello, Minn., by Marion Hagerstrom. Introd. in 1976 by Herschel L. Boll, Champaign, Ill. Probably Midland x Red Rich. Cross made in 1966. Berry: large, up to 2 1/4 inches in diam., wedge to conic, medium dark red; flesh medium red; ripens late. Plant: highly productive, vigorous, hardy; produces many runners.

Deephaven.—Orig. in Excelsior, Minn., by Univ. of Minnesota Fruit Breeding Farm. Introd. in 1922. Parentage unknown; selected in 1919; tested as Minn. 41. Fruit: large; roundish or oblate conic; skin medium red, glossy; flesh medium red, juicy, subacid, quality good; early for spring crop, fall crop medium early. Plant: everbearing; vigorous; productive. No longer propagated.

Deet.—See Robinson.

Delite.—Orig. in Carbondale, Ill., by U.S. Dept. of Agriculture and Southern Illinois Univ. Introd. in 1974. Albritton x MDUS 2650. Cross made in 1958, selected in 1961 by R.C. Blake, tested as SIUS 169. Fruit: medium-large, cone shaped; skin bright red, glossy; flesh pink, moderately firm, flavor slightly acid, aroma and quality good; calyxes medium size; ripens 5 to 7 days later than Surecrop. Plant vigorous, produces runners freely; productive; tolerant to at least five races of red stele root rot, verticillium wilt, leaf spot and leaf blight. Recommended for the south-central U.S. where red stele root rot and verticillium wilt are problems.

Delmarvel.—A disease-resistant strawberry with excellent fruit quality; adapted to mid-Atlantic and adjacent states. Earliglow x Atlas; cross made by A.D. Draper in 1977; selected in 1979 by G.J. Galletta and A.D. Draper; tested as MDUS 4923; introd. in 1994 by U.S. Dept. of Agriculture, Beltsville, Md.; Ohio Agr. Res. and Development Center, Ohio State Univ.; and New Jersey Agr. Expt. Sta., Rutgers Univ. Fruit: large; firm; symmetrical; good color; early midseason ripening; some rot resistance; withstands handling, adapted to shipping; good flavor with outstanding aroma. Plant: productive in matted rows or hill culture in either light or heavy soils; vigorous; abundant runners. Resistant to five races of red stele root rot causal organism; resistant to most leaf and stem diseases.

Dixieland.—Orig. in Beltsville, Md., by George M. Darrow and D.H. Scott, U.S. Dept. of Agriculture. Introd. in 1953, by the U.S. Dept. of Agriculture and the North Carolina Agr. Expt. Sta. Tennessee Shipper x Midland; cross made in 1946; selected in 1948; tested as US-3758. Fruit: medium to large; long blunt conic, uniform; skin strong, vivid red, slightly deeper red than Blakemore, glossy, tough; flesh bright red, very firm, about as firm as Albritton and more so than Blakemore, flavor tart at first, becoming sweet when fully ripe, good; ripening season similar to that of Blakemore; very good for freezing and preserving. Plant: vigorous and productive; some plants show yellow variegation; yields better than most other varieties from Washington, D.C., to southern North Carolina and west to Arkansas, being promising throughout this area; susceptible to red stele, leaf scorch, and verticillium wilt; especially promising as an early variety to be grown with Albritton as a later-maturing variety; produces runners freely.

Donna.—Orig. at the Turner & Growers Mangere Res. Sta. near Auckland, New Zealand, by I.K. Lewis and M.W. Hall. Fern seedling x Douglas; cross made in 1984; selected in 1985; tested as 84-1-38 and later as T-30; introd. in 1993. Flower: self-fertile, with ample pollen production throughout the season and, consequently, little chance of malformed fruit. Fruit: size similar to Tioga but varies as the season advances; very firm; glossy; external color slightly darker than Tioga, internal color similar with lighter ring around the core; medium cordate; solid to slightly hollow; achenes bright yellow, position slightly raised above surface; good flavor; late and offseason production for fresh-market use. Plant: strongly day-neutral, similar to Hecker and Brighton; low chilling requirement; thick peduncles remain semi-erect until fruit ripens; runner production less than Selva.

Donner.—Orig. in Wheatland, Calif., by Harold E. Thomas and Earl V. Goldsmith, Calif. Agr. Expt. Sta., Berkeley. Introd. in 1945. Calif. 145.52 [Calif. BH-14 (mixed crosses) x Redheart] x Calif. 222 [Calif. 66.2 (USDA 634 x Banner) x Calif. 7.20 (Blakemore x Nich Ohmer)]; cross made in 1936; selected in 1938; tested as Calif. 579.4. Fruit: medium large to large, maintaining size during the season; long conic, moderately irregular; skin medium red, attractive, glossy; achenes yellow, small, flush with surface; flesh red, soft to medium firm, flavor excellent, subacid; dessert quality very good; shipping quality fair; hull size medium, varies from clasping to reflexed; ripens early; flower stem never long. Plant: bears a good spring crop, fall production light; highly susceptible to verticillium wilt; vigorous; good runner production. Grown in Japan; apparently no longer grown in California. Named for a celebrated lake in California.

Dorsett.—Orig. in Glenn Dale, Md., by George M. Darrow, U.S. Dept. of Agriculture. Introd. in 1933. Royal Sovereign x Howard 17; selected in 1925. Fruit: appearance attractive; dessert quality high; ripens early. Plant: very vigorous; productive; tolerant to leaf spot.

Douglas.—Orig. at Watsonville, Calif., by R.S. Bringhurst and Victor Voth, Agr. Expt. Sta., Univ. of Calif., Davis. Introd. in 1979. Plant patented, assigned to Regents, Univ. of California, Davis. 64.57-108 (Tioga x Sequoia) x Tufts. Fruit: larger than that of Tioga; flesh quality and flavor equal or superior to those of Aiko, Tioga, and Tufts; skin resembles that of Tufts, but redder. Plant: larger than Tioga; leaflets exceptionally large, less green than those of Tioga; highly productive; as susceptible to verticillium wilt as Tioga, Tufts, and Aiko; highly susceptible to red stele (*Phytophthora fragariae*). Recommended for commercial production in southern California as an early, high-quality, winter-planting variety and as an improved summer-cropping variety for the central coast. Named for the late Malcom B. Douglas, manager of California Strawberry Advisory Board.

Dresden.—Orig. in Geneva, N.Y., by George L. Slate, New York State Agr. Expt. Sta. Introd. in 1939. Beacon x Howard 17; cross made in 1929; selected in 1931. Fruit: skin light red, smooth; flesh quality fair; ripens early; maintains size throughout the season; few or no nubbins. Plant: best adapted to northern tier of states and Canada; very productive; crown appears to lack hardiness. Variety now obsolete.

Driscoll E11.—Orig. in Shasta County, Calif., by H.A. Johnson, Jr. and H.E. Thomas. Plant patent 3001; 17 Nov. 1970; assigned to Driscoll Strawberry Associates, Watsonville, Calif. The Strawberry Inst. of California. Selection L88.101 x Goldsmith. Fruit: large, dropping in size during the season; irregular short wedge; large, conspicuous flower, anthers produce an abundance of pollen even under wet, cold, windy conditions; scarlet red surface and flesh color; skin firm enough for commercial shipment to eastern markets; dessert quality medium, strong acid taste with low sugar. Plant: medium; vigorous; everbearing.

Driscoll E17.—Orig. in Shasta County, Calif., by H.A. Johnson, Jr. Plant patent 3274; 19 Dec. 1972; assigned to Driscoll Strawberry Associates, Watsonville. Strawberry Inst. of California (Selection L44.1 x Selection E2). Fruit: main crop produces berries large in size; medium to short wedge; flesh and epidermis firm, calyx large; seeds exerted; good shipping quality; dessert quality good, sweeter than Goldsmith. Plant: everbearer; vigorous with ample chilling before planting; less tolerant to two-spotted mite than Goldsmith and same susceptibility to mildew as Goldsmith, thrip injury may cause bronzing of fruit.

Driscoll E18.—Orig. in Watsonville, Calif., by H.A. Johnson, Jr. Introd. in 1971. Plant patent 3529; 19 Mar. 1974; assigned to Driscoll Strawberry Associates, Watsonville. Driscoll E12 x Goldsmith; cross made in 1964; selected in 1966; tested as X31.103 and Driscoll E18. Fruit: 35 mm long, wedge shaped; skin scarlet, firm; flesh scarlet, becoming peach red at center; ripens in May and June. Plant; everbearer; small to medium; semi-open; productivity good for an everbearer; very susceptible to two-spotted mite. Included here because variety has been patented.

Driscoll E20.—Orig. in Shasta County, Calif., by H.A. Johnson, Jr. Plant patent 4020; 1 Mar. 1977; assigned to Driscoll Strawberry Associates, Watsonville, Calif. Driscoll Associates Selection E3 x Driscoll Associates Selection NM55.5. Fruit: main crop medium to large; shape varying from long to short wedge, some lobed; dessert quality medium Plant: everbearing; medium, becoming larger during the fruiting season; vigorous with ample chilling; extensive root system more susceptible to *Botrytis* than E18, greater tolerance to two-spotted mite, *Mycosphaerella* leaf spot, and powdery mildew than E18.

Driscoll E21.—Orig. in Shasta County, Calif., by H.A. Johnson, Jr. Plant patent 3982; 16 Nov. 1976; assigned to Driscoll Strawberry Associates, Watsonville, Calif. Driscoll Associates E3 x Driscol Associates F5. Fruit: small to medium crown crop; main crop medium to large; medium wedge to medium conic to globose conic; surface color carmine; dessert quality good. Plant: medium to large; everbearer; vigorous; extensive root system; susceptible to two-spotted mite, but less susceptible to injury caused by mite; moderately susceptible to mildew.

Driscoll G5.—Orig. in Shasta County, Calif., by H.A. Johnson, Jr. Plant patent 3275; 19 Dec. 1972; assigned to Driscoll Strawberry Associates, Watsonville, Calif. The Strawberry Inst. of California (Selection L44.1 x Selection E2). Fruit: large; high quality when plants are vigorous; medium wedge to medium conic; epidermis and flesh firm with high gloss; medium to good dessert quality, sweeter than Goldsmith. Plant: small to medium; vigorous if given ample chilling before planting; equally susceptible to two-spotted mite as Goldsmith; tolerant to mildew.

Driscoll G6.—Orig. in Watsonville, Calif., by H.A. Johnson, Jr. Introd. in 1972. Plant patent 3473; 12 Feb. 1973; assigned to Driscoll Strawberry Associates, Watsonville, Calif. Driscoll E5 x Driscoll B5; cross made in 1960; selected in 1964; tested as T57.11A and Driscoll G6. Fruit: large short wedge to globose; skin scarlet; seeds medium to large, apricot yellow, become dark when exposed to the sun; flesh scarlet, firm with medium core; consistent size and productivity over full and fruiting season; summer planted, ripens in April; recommended for central coastal and southern California. Plant: medium for summer planted variety; vigor good in heavy, well drained soil; productivity good, especially during spring and late fall; susceptible to two-spotted mite and vascular collapse if grown as winter planted. Included here because variety has been patented.

Driscoll G9.—Orig. in Shasta County, Calif., by H.A. Johnson, Jr. Plant patent 4022; 8 Mar. 1977; assigned to Driscoll Strawberry Associates, Watsonville, Calif. Driscoll F5 x Driscoll E17. Fruit: crown crop is large in size, fruit from subsequent crops are medium to large; irregular to long wedge; skin and flesh firm; good dessert quality. Plant: medium; vigorous only if given ample chilling before being planted; extensive root system; in comparison to Heidi, more susceptible to two-spotted mite, less susceptible to powdery mildew and more resistant to *Botrytis*.

Driscoll H4.—Orig. in Shasta County, Calif., by H.A. Johnson, Jr. Plant patent 3987; 14 Dec. 1976; assigned to Driscoll Strawberry Associates, Watsonville, Calif. Driscoll Y5 x Driscoll D9. Fruit: first crop of berries produced are medium to large, subsequent crops have berries dropping in size; medium conic to medium wedge; flesh firm, but lacks moisture with low sweetness; poor to medium dessert quality; high ascorbic acid content of fresh fruit. Plant: medium to large; vigorous; extensive root system; equal to G6 in susceptibility to two-spotted mite and mildew; more susceptible to *Mycosphaerella* leaf spot.

Dry Weather.—Orig. in Fargo, N.D., by A. F. Yeager, North Dakota Agr. Expt. Sta. Introd. in 1930. Americus x Howard 17; selected in 1925. Fruit: quality very good but somewhat soft; primarily for home gardens. Plant: drought resistant; everbearing.

Duluth.—Orig. in Excelsior, Minn., by Univ. of Minnesota Fruit Breeding Farm. Introd.

in 1920. Pan American x Dunlap; cross made in 1909; selected in 1913; tested as Minn. 1017. Fruit: large; roundish; dark red; firm; center tends to be hollow; quality good; ripens in midseason. Plant: adapted to heavy soils; vigorous; everbearer; productive; susceptible to leaf-spot diseases. No longer recommended.

Earle Felten.—Orig. in Pennsauken Township, Camden County, N.J., by Oscar Earle Felten, East Merchantville, N.J. Introd. in 1951. Plant patent 1033; 28 Aug. 1951. Felten seedling 3001 (inbred sib-crosses of Marvel x Howard 17) x Mildred Felten (Felten seedling 3001 x Fairfax); selected in 1938. Fruit: large; mostly conic; skin bright red, covered with many bright yellow achenes giving a golden cast; flesh tart, firm, juicy; resembles Mildred Felten. Plant: upright; productive; self-pollinizing.

Earlee.—Orig. in Brigham City, Utah, by Leland L. Larsen, Larsen Expt. Farm Nursery. Introd. in 1949. Premier x unnamed seedling; selected in 1937. Fruit: size medium. Plant: heavy producer; a spring and fall cropper in Utah.

Earlibelle.—Orig. in Willard, N.C., by the North Carolina Agr. Expt. Sta. and the Crops Res. Div., U.S. Dept. of Agriculture. Introd. in 1964. Albritton x Md-US 2101; selected in 1956 by E.B. Morrow, North Carolina Agr. Expt. Sta., and D.H. Scott, U.S. Dept. of Agriculture; extensively tested in North Carolina from 1960 to 1964 by G.J. Galletta and J.M. O'Neal, North Carolina Agr. Expt. Sta.; tested as N.C. 2486. Fruit: medium large, averaging 62 berries per lb, with 66 for Albritton; long conic, uniform; skin bright red, turning deep red at maturity, uniform, very glossy; flesh a uniform bright red, firm, flavor tart, good; ripens before Albritton in eastern North Carolina; yields well, averaging 25% greater than Albritton. Plant: very vigorous; comparatively small crowns; runners produced so freely, they must be thinned for best fruiting performance; not productive in Beltsville, Md. and northward, productive in southern Virginia and southward; leaf gray-green, not glossy, medium size, glabrous, resistant to leaf spot and leaf scorch. Recommended for North Carolina and southward as an early commercial variety because of its attractive appearance, firmness and productivity to complement the later-maturing Albritton variety.

Earlidawn.—Orig. in Beltsville, Md., by D.H. Scott and George M. Darrow, U.S. Dept. of Agriculture. Introd. in 1956. Midland x Tennessee Shipper; raised in 1947; selected in 1948; tested as US3939. Fruit: large, maintaining large average size throughout the picking season; short conic, fairly uniform shape; skin attractive, vivid red, somewhat deeper than that of Blakemore, glossy, moderately tough; flesh bright red, medium firm, firmer than that of Howard 17, Catskill, and Sparkle, about as firm as Pocahontas and less firm than Blakemore; flavor tart, good; dessert quality good; when frozen, rated high for color, texture, flavor; achenes slightly sunken; matures very early, about 5 days before Blakemore and Howard 17, escaping frost even better than Howard 17 because its short flower stems with flowers are well covered by leaves. Plant: vigorous; hardy; produces fewer runners than most other varieties except Midland; very productive, especially in wide matted rows, with yields above those of most other varieties; shown no yellow variegation; resistant to leaf spot and leaf scorch; susceptible to red stele root-rot and verticillium wilt; adapted in New England, from Pennsylvania to North Carolina and westward to Indiana; most promising in Michigan, Ohio, Pennsylvania; succeeded by Earliglow.

Earliglow.—Orig. in Salisbury, Md., by Agr. Res. Serv., U.S. Dept. of Agriculture, and Maryland Agr. Expt. Sta. Introd. in 1975. MdUS 2359 (Fairland x Midland) x MdUS 2713 (Redglow x Surecrop), cross made in 1964, selected in 1966 by D.H. Scott, A.D. Draper, and I.C. Haut. Fruit: size medium; symmetrical conic; skin deep red, tough, glossy; flesh red, uniform, firm, handles well, flavor sweet; ripens 2 to 3 days after Earlidawn; very good for frozen pack, promising for U-pick operations. Plant: vigorous; prolific runner producer; resistant to five races of red-stele root rot, moderately resistant to verticillium wilt, leaf scorch and leaf spot, partially resistant to powdery mildew; has been productive and fruit attractive in narrow, matted beds in

Maryland, Pennsylvania, New Jersey, Ohio, Michigan, and Missouri. Recommended for soils where red stele may be a problem and where early ripening is desired.

EarliMiss.—Orig. in Mississippi State, Miss., by J.P. Overcash, Hort. Dept. Mississippi State Univ. Introd. in 1975. Albritton x Tennessee Shipper; cross made in 1956; selected in 1958; tested as MS 58-10. Fruit: large, medium conic; skin glossy, red; flesh light red; ripens mid to late April; good for home use or local market. Plant: excellent vigor; productivity very good; tolerant to common local leaf diseases (Leaf spot, leaf scorch, leaf blight). Excellent plant production and survival during hot, dry summers in Mississippi. Heavy yields under Mississippi conditions.

Earlimore.—Orig. in Excelsior, Minn., by A.N. Wilcox and T.S. Weir, Univ. of Minnesota Fruit Breeding Farm. Introd. in 1958. Minn. SY5305-46 (Campbell x self) x Howard 17; cross made in 1940; selected in 1941; distributed for trial in 1950; processing tests by J.D. Winter and Shirley Trantanella; tested as Minn. 1636. Fruit: medium large, being maintained through a relatively long picking season; rounded conic; hull medium small; achenes small, protruding slightly above surface of skin; skin attractive bright red; flesh firm, juicy; flavor pleasant, aromatic, very good, slightly acid; most suitable for fresh dessert use, fair to good for freezing, satisfactory for shipping moderate distances; June bearing, ripening very early. Plant: high yields; very vigorous; hardy; highly resistant to leaf spot and leaf scorch; makes a wide row, well spaced; flower perfect, blooms in midseason.

Early Cambridge.—Orig. in Cambridge, England, by Sir Rowland Biffen and D. Boyes, Hort. Res. Sta., School of Agr., Univ. of Cambridge. Introd. in 1937. Royal Sovereign x Ekey; selected before 1932. Fruit: quality not as good as other varieties; satisfactory for jam; resembles Royal Sovereign. Plant: extremely resistant to red core; heavier yields than Royal Sovereign; can be forced under cloches.

Early Cheyenne 1 (*Cheyenne* 1).—Orig. in Cheyenne, Wyo., by LeRoy Powers, U.S. Dept. of Agriculture, Hort. Field Sta. Introd. in 1942. F_1 Dorsett x *Fragaria ovalis* (P.I. 361261); tested at Station 911. Fruit: size medium; conic; skin bright red; flesh red, soft, subacid, aromatic, flavor good, quality good; ripens very early. Plant: extremely hardy in the Rocky Mountain and Great Plains regions; makes runners freely; productive. Recommended for home gardens.

Early Davis.—Orig. in Canadian Maritime Province. Introd. about 1925; considered to have been introduced in the U.S. by Sunnyside Fruit Farm, Athol, Mass. Parentage unknown. Fruit: ripens 10 to 14 days before Howard 17. Now obsolete.

Early Midway. (*Midway 1*).—Origin unknown, believed to have originated in selections from U.S. Dept. of Agriculture Expt. Sta. in Beltsville, Md. Previously mixed and sold as Midway, but ripens 5 days earlier; fruit tends to be lighter and brighter in color, calyx wraps around the berry more and foliage lighter in color than Midway; otherwise very similar to Midway.

Easypicker.—Orig. in Excelsior, Minn., by Univ. Minn. Fruit Breeding Farm. Introd. in 1922. Dunlap x Crescent; tested as Minn. 775. Fruit: medium to large; roundish to roundish conic; skin rich medium red; flesh firm, juicy, fine-grained, rich dark red throughout, sweet, mild, aromatic, quality good; midseason. Plant: vigorous; productive; leaves medium dark green; petioles medium in length and thickness; flowers pistillate (imperfect), small to medium, calyx small.

Eden.—Orig. in Geneva, N.Y., by George L. Slate, New York State Expt. Sta. Introd. in 1951. Dresden x Fairfax; cross made in 1940; tested as N.Y. 24248. Fruit: wedge-conic to conic; skin attractive, glossy, tough, medium to slightly dark red; flesh firm, red to center, tart, fair to good; achenes yellow, even with surface of the berry; outstanding for processing, freezing, preserving, and canning; flavor rather tart for a good dessert berry; first picking with or slightly before Sparkle. Plant: vigorous, healthy, propagates rapidly, one of the more productive varieties.

Eleanor Roosevelt.—Orig. in Glenn Dale, Md., by George F. Waldo, U.S. Dept. of Agriculture, and E.B. Morrow, North Carolina

Agr. Expt. Sta., Raleigh. Introd. in 1939. Bellmar x Fairfax; selected in 1933. Fruit: immense; skin very dark and dull; flesh firm, deep red; season long; shipping quality good. Plant: very productive.

Elgin.—Orig. in Ottawa, Ontario, Canada Dept. Agr., Res. Sta. Introd. in 1944. Ettersburg 214 x Wm. Belt; selected in 1940; tested as Ottawa 271. Fruit: resembles Ettersburg 214; quality medium; ripens very late. Plant: susceptible to leaf spot; flowers perfect. Virus-free plants not available. Not grown commercially.

Elinor.—Orig. in Ettersburg, Calif., by Albert F. Etter. Introd. in 1933. Duchesne x Trebla. Fruit: large; wedge shaped; flesh soft, flavor good, pinkish, acidity low; dessert quality very good; achenes sunken; matures late; pedicels very long. Plant: yields well. Primarily for home gardens.

Elista.—Orig. in Elst, The Netherlands, by the Inst. of Hort. Plant Breeding, Wageningen, The Netherlands. Introd. in 1964. (Jucunda x self) x US 3763; selected in 1955; tested as IVT. 5573. Fruit: large to fairly large; roundish-conical, symmetrical; achenes sunken; flesh bright red, flavor good with a somewhat unusual aroma; difficult to cap; for the fresh market; ripens late. Plant: productive; fairly vigorous; open-type growth habit.

Elsanta.—Orig. in Wageningen, Holland, by Inst. for Hort. Plant Breeding. Introd. in 1981. Gorella x Holiday. License applied for. Selected in 1975. Berry: large; rounded conical; skin red, glossy, sometimes insufficiently set at top; flesh firm, orange, juicy, taste good; keeping quality good; late ripening. Plant: highly productive; long, firm trusses; susceptible to verticillium wilt and *Phytophthora fragariae, Botrytis cinerea*, mildew, and *Phytophthora cactorum*.

Elvira.—Orig. in Wageningen,. Holland, by Inst. for Hort. Plant Breeding. Introd. in 1977. Gorella x Vola (Bowa x Empire); selected in 1967, licensed in 1977. Berry: large. conical: skin orange-red, glossy; flesh firm and juicy, flavor pleasant and brisk; early ripening, with Gorella. Plant: productive, low-spreading open growth; strongly ramified inflorescences that slant outward; susceptible to mildew, moderately susceptible to *Botrytis* rot, verticillium wilt, and *Phytophthora fragariae*. Introd. primarily for production in frames and coldhouses.

Empire All Red (*Pitt, Saito*).—Orig. in New Westminster, British Columbia, Canada, by P. Saito. Introd. in 1934. A single, different plant selected from a shipment of an unknown variety from California about 1932. Plant: some tolerance to wet soil conditions; everbearing characteristics; not recommended for commercial plantings. Virtually extinct except for a small planting in the Mission area of British Columbia.

Empire State.—Orig. in Jewell, N.Y., by Raymond and Clayton Garlick. Introd. in 1932; assigned to R.M. Kellogg Co., Three Rivers, Mich. Considered to be Glen Mary x Aroma; first fruited in 1926. Description made by George L. Slate, New York State Agr. Expt. Sta., Geneva. Fruit: size medium; skin dull, dark red, unattractive, coloring unevenly; flesh medium firm, quality poor. Plant: vigor medium; dwarfed; very susceptible to mildew. Now obsolete.

Empire.—Orig. in Geneva, N.Y., by George L. Slate, New York State Agr. Expt. Sta. Introd. in 1951. Dresden x Sparkle; cross made in 1940. Fruit: large, uniform; skin color light, bright, very attractive, maintained throughout the season; light flesh color is a fault; very good dessert quality; ripens midseason to late, 4 to 6 days after Howard 17. Plant: very productive; susceptible to leaf spot; promising for northeastern areas, but too soft for Maryland and southward.

Enzed Levin (Orion).—Orig. at the Turner & Growers Mangere Res. Sta. near Auckland, New Zealand, by I.K. Lewis and M.W. Hall. 84-1-38 x 84-8-443 (these selections involve Hecker, Parajo, Fern, and Douglas in their derivations); cross made in 1986; selected in 1987; tested as 86-19-25 and later as T-42; introd. in 1993. Flower: self-fertile; usually ample pollen production, except in early part of season or after prolonged low light intensity conditions—therefore, possibility of some malformed fruit. Fruit: large; firm; glossy; external and internal color lighter than Tioga, with lighter ring in the core region; long, smooth conic with primaries frequently wedged; some hollow centers; achenes bright yellow, at surface or slightly indented; early season; very good flavor; good shelf life; fresh-market use. Plant: short-day type; peduncles semi-erect until fruit ripens; very good runner production.

Erie.—Orig. in Geneva, N.Y., by George L. Slate, New York State Agr. Expt. Sta. Introd. in 1951. Sparkle x Howard 17; cross made in 1940. Fruit: very large, large average size being maintained throughout the season; attractive appearance; flesh fairly soft, quality fair; for local fresh-fruit market; ripens late, about with Sparkle or 1 week after Howard 17. Plant: very productive.

Essex.—Orig. in Geneva, N.Y., by George L. Slate, New York State Agr. Expt. Sta. Introd. in 1951. Howard 17 x Deutsch Evern; cross made in 1926. Fruit: ripens 5 days earlier than Howard 17; better in quality than Howard 17; useful for the home garden, as too small and too soft for commercial use. Plant: vigorous, productive.

Esteem.—Orig. in Platteville, Wis., by Albin E. Lang, Dodgeville, Wis. Introd. in 1954, by the Fieldhouse Fruit Farm, Dodgeville, Wis. Plant patent 1408; 2 Aug. 1955. Ambrosia Late x Gem; cross made in 1947; selected in 1951. Fruit: large; flavor good; produces throughout growing season. Plant: highly productive; vigorous.

Ettersburg 450.—Orig. in Ettersburg, Calif., by Albert F. Etter (1872-1950). Introd. in 1920. Parentage unknown; tested as E-450. Fruit: rather small; skin tough, red, attractive; flesh solid; flavor excellent, sweet, rich red to center, very good canning quality. Plant: not very vigorous.

Ever Berry.—An ever-bearing strawberry adapted to cooler regions of Japan. Orig. at the Morioka Branch of the National Res. Inst. of Vegetables, Ornamental Plants and Tea, Japan; introd. in 1992. Fruit: medium size; relatively soft; glossy red-orange; conic; uniform. Plant: high-yielding; intermediate vigor and runner production; high chilling requirement.

Evermore.—Orig. in Excelsior, Minn., by Univ. of Minnesota Fruit Breeding Farm. Introd. in 1945. Duluth x Senator Dunlap; selected in 1929; tested as Minn. 1166. Fruit: large to medium; conic; attractive; flesh firm, quality fair. Plant: everbearer; productive; hardy; drought resistant; produces runners quite freely.

Evita.—Orig. in Faversham, Kent, England, by Peter Vinson, Commercial Fruit Plants, Brooker Farm, New Church, Romney Marsh, Kent. Chandler x (Gorella x Brighton); cross made in 1988; selected in 1989 by Peter Vinson and Simon Warren; tested as 89-A24; introd. in 1993; owned by Commercial Fruit Plants. Fruit: large to medium size; conic to wedge shape; firm; glossy; medium-red external and uniform red internal color; achenes level with surface; large, reflexed calyx; flavor good with slight acidity. Day-neutral with ripening season from July to October in southeastern England; fresh-market use. Plant: high yielding; open habit; few runners produced; flowers above canopy and fruit well displayed; some resistance to powdery mildew.

Fairfax (*Black*, *Black Hornet*, *Cummerbund*, *Grandview*, all are apparently similar to or identical to Fairfax).—Orig. in Glenn Dale, Md., by George M. Darrow, U.S. Dept. of Agriculture. Introd. in 1933. Royal Sovereign x Howard 17; cross made in 1923; selected in 1925. Fruit: size medium; wedge to short blunt conic; skin quite bright red, turning dark if not picked and marketed promptly when first ripening; flesh deep red, mildly subacid; excellent dessert quality; ripens in early midseason. Plant: runner production moderate; resistant to leaf spots and leaf scorch, sensitive to virus diseases; especially productive when late-season runners are removed; grown from southern New England to Maryland and westward to Kansas.

Fairland.—Orig. in Beltsville, Md., by George M. Darrow, U.S. Dept. of Agriculture. Introd. in 1947. Aberdeen x Fairfax; selected in 1938. Fruit: size medium; skin bright red; flesh soft but firmer and with more flavor than either Howard 17 or Catskill; ripens early. Plant: resistant to red stele; susceptible to leaf spot; very productive; adapted to northeastern U.S.

Fairmore.—Orig. in Glenn Dale, Md., by George M. Darrow, U.S. Dept. of Agriculture, and E. B. Morrow, North Carolina Agr. Expt. Sta., Raleigh. Introd. in 1939. Blakemore x Fairfax; selected in 1934. Fruit: flesh very firm, high flavor in southern states; shipping quality good; season early. Plant: vigorous; resistant to leaf spot. No longer propagated because of presence of virus disease.

Fairpeake.—Orig. in Glenn Dale, Md., by George F. Waldo and George M. Darrow, U.S. Dept. of Agriculture Introd. in 1944. Chesapeake x Fairfax, sibling of Starbright and Redstar; cross made in 1931 by Waldo; selected in 1933; tested as USDA 2119. Fruit: medium to large; wedge to short blunt conic; hull clasping, sometimes reflexed, size medium; skin bright medium red, uniform, glossy, attractive; achenes yellow, often reddish, small, raised above surface; flesh light red to the center, aromatic, firm, juicy, sweet, highly flavored; dessert quality excellent; ripens late, with Aroma and Gandy in Maryland, later than Starbright, before Redstar. Plant: vigorous; productive, more so than Starbright; runner production fair; resistant to leaf spot and leaf scorch.

Fantasy.—Orig. in Champaign, Ill., by Herschel L. Boll. Introd. in 1976. [(Canall x Twentieth Century) x Twentieth Century] x Ogallala. Cross made in 1967. Berry; large, mostly wedge shape, medium dark red; flesh medium red; calyx large; ripens late. Plant: highly productive, vigorous; produces medium number of runners.

Fendalcino.—Orig. in Ettersburg, Calif., by Albert F. Etter (1872 to 1950). Introd. in 1920. Fendall x Ettersburg 121. Fruit: large; skin bright red; flesh moderately firm, quality good; ripens early. Plant: productive; resistant to mildew. Apparently no longer propagated.

Fern.—Orig. in Davis, Calif., by R.S. Bringhurst and Victor Voth, Agr. Expt. Sta., Univ. of California. Introd. in 1983. Plant patent assigned to Regents, Univ. of California, Davis. Tufts x 69.62-103. Cross made in 1971; selected in 1972: tested as Cal 72.8-105 and CN12. Fruit: about as large as Tioga; medium conic to flat wedge; internally from solid throughout to slightly hollow; skin is slightly darker than Pajaro; flesh same color as skin with slightly lighter ring around the core; flesh as firm and durable as Tioga, Tufts, or Aiko; flavor excellent; early maturity; day-neutral variety, weather permitting can be made to fruit year-round, commencing about 3 months after planting; recommended for fresh-market, processing, and commercial usage. Plant: semi-erect; more spreading than Hecker and about same size, but only 75% as large as standard Tioga; runner production good, comparable to Hecker; low chilling requirement, similar to Hecker; self-fertile. Named for the late Fern Miller who was prominent in community affairs of Davis, Calif.

Fletcher.—Orig. in Geneva, N.Y., by George L. Slate, New York State Agr. Expt. Sta. Introd. in 1959. Midland x Suwannee; cross made in 1949; selected in 1951; tested as N.Y. 423. Fruit: large at first of season but average medium size during the season; conic, regular, apex pointed; skin medium red, glossy, attractive, bruising slightly but much firmer than Catskill, about as firm as Sparkle; flesh red, juicy, center solid, flavor subacid, quality very good; achenes numerous, slightly sunken; excellent for freezing, slightly better frozen than Sparkle, Eden, and Frontenac; ripens late, about 3 days after Sparkle. Plant: vigorous; tall; very productive; makes runners freely; little or no leaf spot, leaf scorch and mildew; flower perfect, large. Named in honor of S.W. Fletcher, Dean and Director of Research, School of Agr., now Dean Emeritus, Pennsylvania State Univ.

Florida Belle.—Orig. at Agr. Res. Center, Dover, Fla.. Sequoia x Earlibelle; cross made in 1969; tested as selection 69-736. Introd. in 1975. Fruit: blunt, conic; deep-red color, although at times has a white shoulder around the calyx. Plant: open; upright; smaller than Tioga; resistant to anthracnose (*Collectotrichum fragariae* Brooks); plant shape and size facilitates pesticide coverage of foliage and fruit.

Florida Ninety.—Orig. in Lakeland, Fla., by A.N. Brooks, Strawberry Lab., Florida Agr. Expt. Sta. Introd. in 1952. Open-pollinated seedling of Missionary; selected in 1948. Fruit: large; shape unique, being long, pointed to crocodile nose; skin bright red; flavor excellent; very well suited for fresh market; ripens extremely early, November and December. Plant: yields very well; very vigorous; rest period short; produces numerous runners; very subject to leaf spots and leaf scorch.

Fort Laramie.—Orig. in Cheyenne, Wyo., by U.S. Dept. of Agriculture. Introd. in 1973. The parentage of this variety is Geneva x S.65122 (Earlidawn x Chief Bemidji); cross made in 1966, selected in 1968 by Gene S. Howard and James P. Huck, U.S. Dept. of Agriculture, tested as 67105-41. Fruit: large; round-conic; skin bright scarlet-red, yellow seeds at surface level; flesh pink to scarlet, firm in areas with cool nights, aromatic, sweet. Plant: everbearing, will bloom, fruit and produce runners simultaneously, vigor excellent; abundant runner production; very productive; hardy, –30F; self-fertile; some susceptibility to mildew in Maryland and Oregon.

Fortune.—Orig. in Geneva, N.Y., by George L. Slate and John Watson, New York State Agr. Expt. Sta. Named and introd. in 1961. USDA 2827 (Dorsett x USDA 367) x self; cross made in 1948; selected in 1950; tested as N.Y. 386. Fruit: large; skin light bright red, glossy, very attractive; flesh red, tender, juicy, slightly hollow, subacid, as firm as Empire and firmer than Catskill; flavor superior; not suitable for freezing; hull medium size, adhering firmly; achenes numerous, even to slightly raised; ripens in early midseason. Plant: leaves dark green with faint bluish cast; very vigorous, tall; productive; produces runners freely; flower perfect, medium size. Recommended for home gardens and local markets.

Fresno.—Orig. in Torrey Pines, Calif., by R.S. Bringhurst and Victor Voth, California Agr. Expt. Sta., Davis. Introd. in 1961. Lassen x Calif. 42.8-16; selected in 1955; tested as Calif. 53.9-10 (including Sierra, Blakemore, Narcissa, Nich Ohmer, N.Y. 4626, U.S. 543, and U.S. 634); sibling of Torrey. Fruit: large, uniform; long conic, being wedge shaped at certain times; hull separates easily from fruit; skin attractive, tough, lighter color than Lassen, does not darken readily; achenes yellow, medium size, flush with fruit surface; flesh very firm, quality good; dessert quality superior to Lassen; ripens with Lassen and later than Wiltguard and Torrey; resembles Lassen. Plant: very vigorous; many runners produced; tolerant to salinity; highly susceptible to verticillium wilt; more susceptible to virus than Lassen; general Calif. adaptation including warm-winter areas; performs well in southern California, the Central Valley, Santa Maria, and Santa Clara Valley; resembles Lassen.

Frontenac.—Orig. in Geneva, N.Y., by George L. Slate, New York State Agr. Expt. Sta. Introd. in 1959. Erie (Sparkle x Howard) x (Fairfax x Dresden); cross made in 1945; selected in 1947; tested as N.Y. 96; named in 1959. Fruit: large; conic to wedge, uniform; skin medium red, attractive, glossy, fairly tough; flesh firm, red, juicy, flavor moderately acid, center slightly hollow, quality good; achenes numerous, slightly raised; ripens very late, 5 days after Sparkle; fully as good for freezing as Eden and Sparkle. Plant: very large, tall; vigorous; bears heavy crops; apparently suffers from drought in some seasons; low plant production, but the fruiting row is usually adequate; good for the hill system; flower perfect, large.

Frost Proof.—Orig. in Mankato, Minn., by F.M. Schwab. Introd. about 1936. Chance seedling; discovered about 1935. Fruit: large; flesh very firm; matures very late. Plant: frost resistant in late spring.

Fukuba.—Orig. in Shinyuku Garden, Tokyo, Japan, by I. Fukuba. Introd. about 1902; imported into U.S. in 1956, as P.I. 231088. Open-pollinated seedling of General Chanzy; cross made from 1897 to 1901. Fruit: very large, 2 1/2 to 3 1/2 inches long, 1 to 1 1/2 inches in diam., 15 to 22 g, in Japan being achieved by allowing only one berry per cluster to grow; thin, long wedge-conic; skin bright red, attractive; achenes inconspicuous; flesh delicious, very tender, remarkable aroma. Plant: in Japan forced on sunny slopes in cold frames.

Fulton.—Orig. in Geneva, N.Y., by George L. Slate, New York State Agr. Expt. Sta. Introd. in 1959. Starbright x Pathfinder; cross made in 1947; selected in 1949; tested as N.Y. 233. Fruit: size medium; short conic, uniform, apex pointed; skin glossy, attractive, medium red, colors evenly, tough; achenes numerous, raised, somewhat seedy in appearance; flesh very firm, medium red, flavor subacid, juicy, center solid, quality fair; hull small, clasping, adhering tightly; satisfactory for freezing; ripens in midseason, just after Catskill. Plant:

tall; vigorous; productive; numerous slender runners produced; free from leaf diseases; flower perfect, size medium.

Gala.—Orig. in Geneva, N.Y., by George L. Slate and Don Ourecky, New York State Agr. Expt. Sta. Introd. in 1966. Midland x Suwanee; tested as N.Y. 418; cross made in 1949; selected in 1951. Fruit: large; skin dark, slightly rough; flesh firm; quality good; ripens 1 or 2 days later than Earlidawn, firmer than Earlidawn. Plant: vigorous; productive; makes a good fruiting row without crowding; late frosts may injure the flowers, not recommended for frosty areas.

Gardena.—A high-yielding, fresh-market strawberry. Orig. near Comacchio (Ferrara), Italy, by M. Leis, D. Musacchi, A. Martinelli, Consorzio Italiano Vivaisti. Addie x Parajo; cross made in 1984; introd. in 1992. Fruit: medium to large size; deep red external and red to orange internal color; firm; some hollowness; mostly uniform heart shape with pointed apex; large calyx that is difficult to remove; achenes beneath surface. Ripens 2 days before Addie; medium-sweet flavor with considerable acidity. Plant: medium vigor; sparse runner production; upright habit; flowers at leaf canopy level.

Garnet.—Orig. in Geneva, N.Y., by George L. Slate, New York State Agr. Expt. Sta. Introd. in 1965. Midland x Suwannee; selected in 1951; tested as N.Y. 430. Fruit: large; skin attractive, medium red; flesh moderately firm, quality good; ripens late. Plant: vigorous; very productive.

Gem (*Superfection, Brilliant, Gem Everbearing, Giant Gem*).—Orig. in Farwell, Mich., by Frank J. Keplinger, Heart-O-Michigan Farms Nurseries. Introd. in 1933. Bud mutation of Champion Everbearing (Lupke); discovered in 1928. Fruit: small; short wedge to oblate, irregular; skin deep red, glossy; flesh pale red, acid; dessert quality good. Plant: everbearing; quite productive; susceptible to leaf spots; resistant to leaf scorch.

Gem Everbearing.—See Gem.

Gemzata.—Orig. in Bristol, Ind., by Victor Judson. Introd. in 1934. Parentage unknown; discovered in 1932. Fruit: large; resembles Gem and Wayzata. Plant: produces well; good plant maker.

Geneva.—Orig. in Geneva, N.Y., by George L. Slate and John Watson, New York State Agr. Expt. Sta. Named and introd. in 1961. N.Y. 316 (Streamliner x Fairfax) x Red Rich [Wayzata (Rockhill) x Fairfax]; cross made in 1952; selected in 1954; tested as N.Y. 547. Fruit: large, size somewhat smaller later in season; short wedge conic to round conic; skin dark red, glossy, attractive, tender; flesh red, medium firm, subacid, juicy, superior in quality; hull medium size, adheres firmly; ripens in late midseason, in June with lull period before recropping in fall, and continuing until frost. Plant: everbearing; vigorous; productive; height medium; runners thick with long internodes, production satisfactory under very good growing conditions when blossoms are removed; hill culture preferred; flower perfect, medium size. Recommended for home gardens and local use.

Gerida.—Orig. in Neukirch, Switzerland, by G. Spiegler, Haberli AG, a private nursery. Elsanta x Elvira; cross made in 1983; introd. in 1990; registered for Plant Variety Rights in most European countries. Fruit: large to medium; firm; conic, very regular shape; glossy, bright red. Excellent flavor; extended shelf life; relatively late ripening season; easy to harvest; fresh-market use. Plant: high yielding; vigorous, open habit, less erect than Elsanta; inflorescence beneath the leaf canopy. Some resistance to powdery mildew; high resistance to crown rot; can be grown on heavier soils.

Giant Gem.—See Gem.

Gilbert.—Orig. in Sturyeon Bay, Wis., by Franklin A. Gilbert. Peninsular Expt. Sta., Univ. of Wisconsin. Introd. in 1982. Cyclone x Badgerglo. Selected in 1971; tested as Wl 7028. Fruit: large: few first fruit are double or wedge-shaped. others conic: skin bright red, glossy, attractive: flesh medium firm, quality fair, good interior color; ripens early midseason. Plant: very productive. long harvest season, runners root easily; appears to be winter hardy; resistant to scorch.

Glamour.—Orig. in Ancaster, Ontario, Canada, by Edward Lowden. Introd. in 1965. Parentage unknown; discovered in 1955. Fruit: larger than Sparkle; skin shiny light red, very attractive; flesh light red, being of a distinctive coloration, good quality; ripens late, about with or slightly after Sparkle. Plant: heavy yielder; grows well in heavy soil; flower pistillate, being pollinated by Sparkle or Robinson.

Glen.—Orig. in Miami, Manitoba, Canada, by William Oakes, Glenelm Nursery. Introd. in 1944 (an English seedling x Minnesota 4) x Marvel; selected in 1939. Fruit: size medium; skin medium red, glossy; flesh firm, quality fair to good; does not sunscald. Plant: very hardy; stems upright; vigorous; very productive.

Glenalice.—Orig. in Miami, Manitoba, Canada, by William Oakes, Glenelm Nursery. Introd. in 1946. (Gem x Rockhill) x Simcoe; selected in 1943. Fruit: size medium; long, pointed; skin dark red, glossy; flesh deep red with no white streaks. Plant: hardy; yields very heavily; produces few runners.

Glenciss.—Orig. in Miami, Manitoba, Canada, by William Oakes, Glenelm Nursery. Introd. in 1947. Gem x Rockhill; selected in 1944. Fruit: sweeter than Gem, which it resembles. Plant: everbearing; does not produce as many runners as Gem.

Glenda.—Orig. in Miami, Manitoba, Canada, by William Oakes, Glenelm Nursery. Introd. in 1958. Glenheart x Elgin; selected in 1955. Fruit: very large; attractive; ripens late. Plant: large; vigorous; very hardy.

Glenheart.—Orig. in Miami, Manitoba, Canada, by William Oakes. Introd. about 1941. Probable parentage: Gem x Rockhill. Fruit: large; skin bright medium red, very attractive coloring and shape; flesh firm, sweet, pleasant. Plant: strong, vigorous; strong plant maker; very hardy; productive; moderately susceptible to leaf spot; perfect flowers.

Glenman.—Orig. in Miami, Manitoba, Canada, by William Oakes, Glenelm Nursery. Introd. in 1959. Glen x Elgin; selected in 1953. Fruit: ripens late; stem long; resembles Glen Mary. Plant: vigorous; hardy; very productive.

Glenmore.—Orig. in Miami, Manitoba, Canada, by Wm. Oakes, Glenelm Nursery. Introd. in 1942. (Minnesota 4 x Glenelm 2) x (seedling x Marvel); selected in 1934. Fruit: flesh firm; is easily picked; for dessert and canning. Plant: hardy; bears well.

Glenrich.—Orig. in Miami, Manitoba, Canada, by William Oakes, Glenelm Nursery. Introd. in 1958. Glenciss x Rancho Santa Ana Ornamental Hybrid 25; selected in 1954. Fruit: flesh firm, deep red; somewhat rough. Plant: everbearing; yields heavily; hardy; vigorous; healthy.

Glooscap.—Orig. in Kentville, Nova Scotia by D.L. Craig and G.W Bishop, Res. Sta., Agr. Can. Introd. in 1983. Micmac x Bounty: selected in 1974; tested as K74-12. Fruit: medium; uniform, globose conic; flesh medium red, medium firm, tender core, good flavor, hulls easily; skin medium to dark red, glossy. Plant: runners freely; vigorous; very productive, tolerant to common foliage diseases; roots susceptible to red stele. Recommended for fresh market and processing.

Gold Heart.—Orig. in Mankato, Minn., by F.M. Schwab. Introd. in 1936. Parentage unknown; discovered in 1935. Fruit: somewhat less firm than Howard 17. Variety now obsolete.

Golden Gate.—Orig. in Ettersburg, Calif., by Albert F. Etter (1872 to 1950). Introd. probably in the late 1920s. Red Sugar x Kalicene. Fruit: large; flesh firm, flavor good, very dark red.

Goldsmith (*Institute Z5A*).—Orig. in Waterford, Calif., by Earl V. Goldsmith and Harold E. Thomas, Strawberry Inst. of California, Morgan Hill. Introd. in 1958. Plant patent 1735; 29 July 1958; by Earl V. Goldsmith, Margaret S. Goldsmith and Harold E. Thomas; Margaret S. Goldsmith assignor to the Strawberry Inst. of California; distribution restricted. Two Univ. of California selections were crossed: Cal. 638.45 x Cal. 469.5; a complex hybrid derived from U.S. 543, U.S. 634, New York 4626, Nich Ohmer, Marshall, Banner, and Fendalcino; selected in 1950. Fruit: larger than Shasta, with a higher surface gloss; shape long wedge; flesh moderately red, flavor mildly subacid; variety for local and long-distance shipment; good for freezing; its long fruiting season normally starts later than Shasta. Plant: small; moderately vigorous.

Gorella.—Orig. in Wageningen, Holland, by the Inst. of Hort. Plant Breeding. Introd.

in 1960. Juspa x US 3763; selected in 1955. Fruit: very large, retaining good size to end of season; conical; skin bright red, glossy; flesh red, firm, flavor good; ripens fairly early. Plant: growth erect; leaves tough, sometimes 4 to 5 foliate; trusses very firm.

Governor Simcoe.—Orig. in Vineland Sta., Ontario, Canada, by C.L. Ricketson. Holiday x Guardian; cross made in 1972; selected in 1974; tested as V7236R-3; introd. in 1985 by A. Dale, Hort. Res. Inst. of Ontario. Fruit: large; firm; bright, medium-red exterior and pale interior color; moderately sweet and aromatic flavor. Ripens mid- to late season; good shelf life; fresh-market use. Plant: high yielding; vigorous; abundant runners but not excessive; grows best on warm, sandy soil. Moderately susceptible to verticillium wilt and leaf scorch; susceptible to powdery mildew.

Grand Champion.—Orig. in New Albany, Ind., by W.E. Tellis. Introd. in 1930. Considered to be a bud mutation of Aroma because of resemblance to it; discovered in 1925. Fruit: large; round; skin light to dark red, much like Robinson, but less brilliant; flesh soft, especially in years of ample rain; resembles Aroma, but easily distinguishable from it. Plant: very vigorous when not infected with leaf spot; prolific; highly susceptible to leaf spot. Now discarded in favor of Blakemore and other varieties.

Grandview.—See Fairfax.

Great Bay.—Orig. in Durham, N.H., by L.P. Latimer, New Hampshire Agr. Expt. Sta. Introd. in 1949. Simcoe x Catskill; selected in 1943. Fruit: berries remain large throughout a long midseason; flesh red; processes well as frozen product. Plant: vigorous; drought resistant; crown and root system very large; heavy yield.

Green Mountain.—Orig. in Putney, Vt., by George D. Aiken, The Aiken Nurseries. Introd. in 1935. Plant patent 112; 20 Nov. 1934. Howard 17 x Superb; first crop produced in 1927. Fruit: averages 1 3/4 inches long; conical; hull curled, easily removed; skin dark red; achenes large, yellow; flesh red throughout, flavor good; first ripens in midseason, about 1 week later than Howard 17, fruiting during summer, with fall berries ripening 15 Sept. to 15 Oct. Plant: everbearing; vigorous; produces runners freely; very productive under favorable conditions; leaf light green, medium size, tips rounded; flower perfect.

Grenadier.—Orig. in Ottawa, Ontario, Canada, by Canada Dept. Agr., Res. Sta. Introd. in 1957. Valentine x Fairfax; cross made in 1948; selected in 1950; tested as Ottawa 483. Fruit: large, usually larger than Howard 17; skin glossy, darker than Cavalier but with acceptable appearance; flesh firm, darker than Cavalier and Redcoat, acidity low, flavor acceptable, quality higher than that of Cavalier, Guardsman and Redcoat; very good for dessert, good for processing frozen and as preserves; ripens in early midseason. Plant: virus free; has outyielded Howard 17 during the first 6 days of harvest season; yield equal to or greater than that of Howard 17 and Dunlap; plant vigor and runner production similar to Cavalier; somewhat susceptible to leaf spot and powdery mildew.

Guardian.—Orig. in Salisbury, Md., by Md. Agr. Expt. Sta. and Crops Res. Div., U.S. Dept. of Agriculture. Introd. in 1969. NC 1768 x Surecrop. Cross made in 1958; selected in 1960 by I. C. Haut and D.H. Scott; tested as Md-US 3079. Fruit: large; primary berries have an irregular conic shape with a rather rough outline; secondary and later berries are symmetrical and smooth with slight necks; skin light red, glossy; flesh firm, light red, which is not attractive in frozen pack; ripens 4 to 5 days later than Surecrop. Plant: vigor moderate, producing moderate number of runners, resistant to five races of red stele root rot and verticillium wilt; leaves resistant to leaf scorch and mildew; moderate susceptibility to leaf spot; has yielded satisfactorily in narrow, matted rows in Maryland, New Jersey, Ohio, Michigan, southern Illinois, and Missouri; fruit have been unattractive in North Carolina and Arkansas.

Guardsman.—Orig. in Ottawa, Ontario, Canada, by Canada Dept. Agr., Res. Sta. Introd. in 1957. Claribel x Sparkle; cross made in 1948; selected in 1950; tested as Ottawa 484. Fruit: medium to large; skin bright red, glossy, very attractive; flesh firm, darker red than that of Redcoat, turning red before full ripe, so if not picked carefully it may be high in acidity; ripens late—the primary reason for its introduction; good for freezing and Jam. Plant: virus free; very high yields, as well as or better than Dunlap and Howard 17; vigorous; produces ample runners; young leaflets conspicuously light green; resistant to some races of red stele, susceptible to verticillium wilt.

Hagerstrom Red-Rich.—See Red Rich.

Hagerstrom's Everbearing.—See Red Rich.

Headliner.—Orig. in Baton Rouge, La., by P.L. Hawthorne, J.C. Miller, W.F. Wilson, Jr., N.L. Horn, and W.A. Sistrunk, Louisiana Agr. Expt. Sta. Introd. in 1957. Seedling L-7-27 [Suwanee x L-6-30-45-7 (Klonmore selfed)] x seedling L-7-42 (Suwanee x Konvoy); cross made in 1949; first grown from seed in 1950; produced first fruit in 1951; tested as La. L-0-188. Fruit: medium to large, 30% to 50% larger than those of older varieties grown in Louisiana; blunt conic; skin medium bright red, glossy, attractive for fresh market; flesh medium light red, juicy, mildly subacid, firm, quality good; "achenes" medium small, slightly sunken; processing superior to leading commercial varieties now being grown in area of introduction; begins fruiting early in the season and bears until late in the spring or early summer, being early March through early June in Baton Rouge; picking qualities excellent; shipping qualities good. Plant: vigorous; very productive; produces runners freely; susceptible to bird eye spot; resistant to leaf spot; flower perfect.

Hebron.—Orig. in New Haven, Conn., by D.F. Jones, Connecticut Agr, Expt. Sta. Introd. in 1939. Inbred Chesapeake x (F| cross of inbred Progressive x Howard 17). Fruit: skin light, bright; calyces green on maturing; flesh soft and tart; resembles Chesapeake; ripens late.

Hecker.—Orig. at Univ. of California Wolfskill Expt. Orchard, Winters by R.S. Bringhurst and Victor Voth, Agr. Expt. Sta. Univ. of California, Davis. Introd. in 1979. Plant patent assigned to Regents, Univ. of California, Davis. 65.65-601 (complex parentage) x 66.96-101 [self x 42.8-16 (male parent of Tioga)]. Third backcross generation derivative from a male *Fragaria virginiana glauca* collected in the Wasatch Mountains near Salt Lake City, Utah. Fruit: flesh flavor and quality equal or superior to those of currently grown California short-day cultivars; high in ascorbic acid; skin dark reddish-orange. Plant: small; leaves same color as those of Tioga; less susceptible to verticillium wilt than Tioga and Tufts. Recommended for home gardens as day-neutral (can be made to produce fruit 3 months after planting). Named for Hecker Pass, Santa Cruz County, Calif.

Heidi.—Orig. in Watsonville, Calif., by Harold A. Johnson, Jr., Driscoll Strawberry Assn. Introd. in 1971. Plant patent 3123; 18 Apr. 1972, assigned to Driscoll Strawberry Assn. Driscoll E3 x Driscoll NM55.5. Cross made in 1963, selected in 1965, tested as G3. Fruit: large; medium conic to medium wedge; skin dark red when fully ripe, glossy; flesh scarlet, light red at core, firm, high flavor; seeds yellow, darken when exposed to full sun; ripens April. Plant: medium large, semi-open; vigorous; consistently productive from spring to fall, both first and second year; susceptible to mildew and calyx rot; tolerant to the two-spotted mite and high levels of soil salts; recommended for central coast area of California.

Hera.—Orig. near Comacchio (Ferrara), Italy, by M. Leis and D. Musacchi, Consorzio Italiano Vivaisti. Cross made in 1984; introd. in 1989. Italian plant patent 37-NV/89. Flower: abundant pollen production. Fruit: large; firm; bright red external and internal color; regular, conic; good flavor. Ripening season 4 days before Pajaro; excellent shelf life; adapted to shipping. Plant: medium vigor with upright growth habit; scape upright with fruit above leaf canopy; adapted to southern Italy.

Hilcrest.—Orig. in Sherldan, Ore., by F.J. Barton. Introd. about 1923. Marshall (Oregon) x Dunlap. Fruit: round; ripens evenly. Plant: fairly productive. Apparently commercial propagation stopped about 1935.

Hinomine.—A high-yielding, early ripening strawberry adapted to forcing under plastic in both warm and cool regions of Japan. Orig. at the Kurume Branch of the National Res. Inst. of Vegetables, Ornamental Plants and Tea, Japan; introd. in 1989. Fruit: large; firm; glossy deep scarlet. Plant: productive; establishes quickly. Resistant to powdery mildew and fusarium wilt.

Holiday.—Orig. in Geneva, N.Y., by D.K. Ourecky, Agr. Expt. Sta. Introd. in 1972. Raritan x New York 844; cross made in 1963, selected in 1965, tested as New York 1144. Fruit: large; oblate; skin bright red, attractive, tough; seeds sunken; flesh light red, very firm, flavor distinct; pedicel brittle; ripens 4 to 5 days after Earlidawn, concentrated ripening habit; good freezing quality. Plant: large; free running; vigorous; more productive than Sunrise; foliage light green; less susceptible to fruit rot than other varieties.

Honeoye.—Orig. in Geneva, N.Y., by Donald K. Ourecky, New York State Agr. Expt. Sta. Introd. in 1979. Vibrant x Holiday. Cross made in 1970; selected in 1972; tested as N.Y. 1409. Fruit: large, conic; skin bright red, glossy; flesh average firmness; ripens early midseason. Plant; very vigorous, upright, very productive; good runner production.

Honey Lump.—Orig. in Pequot Lakes, Minn., by Charles C. Brunes. Introd. in 1962. Plant patent 1989; 22 Nov. 1960; assigned to Farmers Seed & Nursery Co., Fairbault, Minn. Considered to be an open-pollinated seedling of Marvel; discovered. in 1955. Fruit: larger than Marvel, with average length of 1 5/8 inches; wedge shape varying to flat oval; skin an intense red, very glossy; flesh red, firm, juicy, very sweet, nonacid; dessert quality extremely good, good for freezing and canning; stem unusually long; ripens early; resembles Armore. Plant: bisexual; everbearing.

Hood.—Orig, in Corvallis, Ore., by U.S. Dept. of Agriculture and Oregon Agr. Expt. Sta. Introd. in 1965. US-Ore. 2315 x Puget Beauty; cross made in 1955 and selected in 1951 by G.F. Waldo, U.S. Dept. of Agriculture; tested in Corvallis and the North Willamette Expt. Sta., Aurora, Ore., by Waldo and R.M. Bullock during 1958 to 1964. Fruit: borne well above soil on strong, upright clusters; berry large, round conic; skin bright medium red, glossy, attractive; flesh light red, medium firm, pleasantly subacid flavor; ripens in midseason, harvest season similar to Northwest; preserves and jams equal or superior to those made from Marshall and other varieties now used; satisfactory for local fresh-fruit markets. Plant: yields of 6 tons per acre have been obtained; large; moderate number of runners produced; leaf large, mildew resistant; susceptible to red stele but out-produces Northwest on red stele–infested soil.

Howard 25.—See Howard's Supreme.

Howard's Supreme (*Howard* 25).—Orig. in Belchertown, Mass., by E.C. Howard. Introd. in 1931. Plant patent 71; 25 July 1933; assigned to E.W. Townsend and Sons, Salisbury, Md. Howard seedling 103 x Howard 17; selected in 1909. Fruit: skin dark rich red; flesh juicy, firm, texture, quality good; good shipper. Plant: vigorous, healthy; high production; flowers imperfect, which has taken it out of production commercially.

Howe.—Orig. in Ottawa, Ontario, Canada, by Canada Dept. Agr., Res. Sta. Introd. in 1936. (Red Sugar x Howard 17) x (Willard x Santiago de Chile); selected in 1934. Fruit: large; appearance attractive; frequently rough; quality medium; ripens midseason. Plant: flowers imperfect. Obsolete.

Improved Clark.—See Ulrich.

Indra.—Orig. in Alnarp, Sweden, by Emil Johansson, State Hort. Res. Sta. Introd. in Sweden in 1951; in the U.S. in 1953. A pistillate variety from the U.S. called Southland (in Sweden) x Alnarp Luna; cross made in 1936; selected in 1939; tested as Alnarp 235. Fruit: conical; skin bright red; flesh light red, juicy, sweet, with a very pleasant flavor; ripens medium early; more suitable for dessert than for preservation; later-ripening fruit too small. Plant: very vigorous, producing high yields. Indra is the name of a Hindu god in one of August Strindberg's plays.

Institute D4.—Orig. in Big Bend, Calif., by Harold A. Johnson, Jr., and Harold E. Thomas, Morgan Hill, Calif. Introd. in 1964. Plant patent 2613; 22 Mar. 1966; assigned to the Strawberry Inst. of Calif., Morgan Hill. Univ. of Calif. 544.3 x Strawberry Inst. Calif. C52.3. Fruit: medium to large; conic to globose conic, varying to wedge or rounded wedge; skin color lighter than Goldsmith or Lassen, glossy surface persistent; flesh white, firm, flavor mild and subacid, quality good; core hollow; ripens very early in frost free localities when winter planted. Plant: upright; open crown; grows well in heavy soils and those of moderate alkalinity or saline content; inflorescence on long stem extending above foliage level; robust growth; leaves heavier and thicker than Lassen; surface undulated and wrinkled; vigorous and dense; very productive during first year when winter set; susceptible to verticillium wilt and two-spotted mite; resistance to red stele not determined.

Institute E2.—Orig. in Big Bend, Calif., by Harold A. Johnson, Jr. and Harold E. Thomas, Morgan Hill, Calif. Introd. in 1964. Plant patent 2611; 22 Mar. 1966; assigned to the Strawberry Inst. of California, Morgan Hill. A cross between two Strawberry Inst. of California seedlings, E101.10 x C52.3. Fruit: large; long conic to long wedge; skin red, glossy, tough, with some longitudinal furrows; irregular ripening producing blotches; flesh firm, but juicy, flavor mild; resembles Goldsmith; tendency to split primaries in the summer crop. Plant: vigor and size medium; open crowned, with characteristic prostrate habit; leaf small to medium low and parallel to ground; leaflets with margins cupped downward, smooth surface; runner production poor to medium; susceptible to verticillium wilt.

Institute E5.—Orig. in Big Bend, Calif., by Harold A. Johnson, Jr. and Harold E. Thomas, Morgan Hill, Calif. Introd. in 1964. Plant patent 2612; 22 Mar. 1966; assigned to the Strawberry Inst. of California. Strawberry Inst. of California selections C52.3 x F121.7. Fruit: medium to medium large; conic to wedge shape; skin red, mostly glossy but may lighten in spots, furrowed giving the fruit an irregular appearance; flesh firm, very light color, flavor mild; resembles Goldsmith. Plant: size and vigor medium; resembles Goldsmith; in comparison to other everbearing varieties produces early crops and high yield; inflorescence often with three equal secondary peduncles; tendency to produce simultaneous ripening of the primary and the secondary fruit on a given inflorescence; susceptible to a type of crown degeneration common to California strawberry fields, susceptible to verticillium wilt and mildly susceptible to Rhizoctonia root rot; two-spotted mites affect the plant seriously and thrips have caused flower blasting.

Institute X2.—Orig. in Waterford, Calif., by Earl V. Goldsmith and Harold E. Thomas, Strawberry Inst. of California, Morgan Hill. Introd. in 1958 by this institute; restricted distribution. Plant patent 1737; 29 July 1958; Earl V. Goldsmith, Margaret S. Goldsmith and Harold E. Thomas; Margaret S. Goldsmith assignor to the Strawberry Inst. of California. Two Univ. of California selections were crossed: Cal. 638.45 x Cal. 794.19; a complex hybrid derived from Banner, Fendalcino, U.S. 253, U.S. 634, New York 2626, Nich Ohmer, and Blakemore; selected in 1950. Fruit: large; skin bright, glossy; these characteristics make a superior packaged product for interstate and local shipments; flesh moderately red; good for frozen trade; ripens normally with Shasta, but somewhat more continuously during summer than Shasta; flavor mildly subacid; dessert quality about equal to Shasta. Plant: tall vigorous; bears well the first year in cool areas; high yielding.

Institute Z4.—Orig. in Waterford, Calif., by Earl V. Goldsmith and Harold E. Thomas, Strawberry Inst. of California, Morgan Hill. Introd. in 1958. Plant patent 1736; 29 July 1958; by Earl V. Goldsmith, Margaret S. Goldsmith and Harold E. Thomas; Margaret S. Goldsmith assignor to the Strawberry Inst. of California; restricted distribution. Two Univ. of California selections were crossed: Cal. 800.13 x Cal. 814.6; a complex hybrid derived from U.S. 634, U.S. 543, U.S. 253, Ruby, Nich Ohmer, Banner, Fendalcino, New York 4626, and Redheart; selected in 1950. Fruit: medium to large; round conic to long conic; skin bright, glossy; achenes medium sized, slightly embedded to even with surface of fruit; flavor mildly subacid. Plant: very rugged; comparatively tolerant to virus diseases prevailing in California; ability to grow well under poor soil conditions.

Institute Z5A.—See Goldsmith.

Irvine.—Orig. at Univ. of California, South Coast Res. and Ext. Center, near Irvine, by

V. Voth and R.S. Bringhurst. Douglas x Muir; cross made in 1982; selected in 1983; tested as Cal 82.14-603 and later as CN14; introd. in 1988. U.S. plant patent 7172, 2 Feb. 1990. Fruit: size usually less than Selva, variable; very firm; medium conic but can be flat or wedge-shaped and may have hollow center; bright red, lighter than Selva; flavor usually better than Selva; fresh-market use. Plant: more strongly day-neutral than Selva; lower chilling requirement than Selva; much more erect growth habit than Selva. Probably susceptible to verticillium wilt. Moderately susceptible to common leaf spot; quite susceptible to anthracnose and to two-spotted spider mite; tolerant to viruses occurring in California.

Isabella.—Orig. in Bologna, Italy, by Enrico Baldini, Inst. di Coltivazioni Arboree, University, Bologna. Introd. in 1971. The final patent number is 891782 (Italy), 26 Dec. 1969, assigned to Zanzi Nursery, Ferrara, Italy. Pocohontas x Surprise des Halles. Cross made in 1963, selected in 1969. Fruit: large; conic to long wedge; skin red; flesh pink, firm, juicy, flavor excellent, aromatic; recommended for fresh market and freezing; ripens about 5 days before Gorella. Plant: highly vigorous; hardy; productivity high; easily capped.

Jerseybelle.—Orig. in New Brunswick, N.J., by F.A. Gilbert, New Jersey Agr. Expt. Sta. Introd. in 1955. N.J. 953 [(Lupton x Aberdeen) x Fairfax] x N.J. 925 (Pathfinder x Fairfax); selected in 1948; tested as N.J. 7A. Fruit: very large; blunt conic; skin very attractive, glossy, medium red, tough; calyx large, remains an attractive green, double on first berries; flesh fairly firm, but too soft for distant shipping, flavor mild, quality good fresh, white streaks remain at core until fully ripe; ripens in late midseason. Plant: large; moderately productive; vigorous; produces fewer runner plants than Sparkle; susceptible to leaf spot, leaf scorch, red stele, verticillium wilt; best adapted to New Jersey, Pennsylvania, and northward,

Jewel.—Orig. in Geneva, N.Y., by J.C. Sanford, D.K. Ourec, and J.E. Reich, New York State Agr. Expt. Sta. NY1221 {[(Tennesee Shipper x Fairfax) x Geneva] x Senga Sengana} x Holiday; cross made in 1969; selected in 1971; tested as NYl324. Plant patent 5897; assigned to Cornell Research Foundation. Fruit: large, firm, attractive; glossy, bright red; wedge-conic; very flavorful; good for fresh or processed use. Ripens in late midseason. Plants: runner moderately; not quite as productive as Honeoye; hardy, drought-tolerant; open habit with dark green foliage; susceptible to red stele and to Verticillium root rots.

Jim.—Orig. in Ottawa, Ontario, Canada, by Canada Dept. Agr., Res. Sta. Introd. in 1932. Open-pollinated seedling of Willard; selected in 1931. Fruit: large; too pale in color; flesh soft, flavor good; ripens midseason. Obsolete.

Joan.—Orig. in Beltsville, Md., by George M. Darrow, U.S. Dept. of Agriculture. Introd. in 1933 by Reith Brothers Nursery, Sawyer, Mich. Ananas de Guemene x Matodon; first fruited in 1931. Fruit: large; flesh sweet, highly flavored. Plant: average yield; flower imperfect. Apparently now obsolete.

John.—Orig. in Ottawa, Ontario, Canada, by Canada Dept. Agr., Res. Sta. Introd. in 1933. Open-pollinated seedling of Willard; selected in 1931. Fruit: fairly attractive; quality good; ripens midseason. Obsolete.

Joyce.—Orig. in Salisbury, Md., by a Mr. Bassett. Introd. in 1932. Parentage unknown; discovered in 1932. Fruit: large. Variety now obsolete.

Jubilee.—Orig. in Parkside, Saskatchewan, Canada, by A.J. Porter, Honeywood Nursery. Introd. in 1955. Louise x Northerner; selected in 1952. Fruit: slightly smaller than Parkland; quality reasonably good; resembles Louise. Plant: extremely hardy; productive.

Julymorn.—Orig. in New Brunswick, N.J., by J.H. Clark, New Jersey Agr. Expt. Sta. Introd. under restriction in 1938; introd. and named in 1943. Redheart x N.J. 5 (Mastodon x Howard 17); selected in 1932; tested as N.J. 225. Fruit: skin dark red; flesh red throughout, very firm, tart, good for processing; resembles Redheart; ripens very late. Plant: size medium; good runner production; blooms late.

Jumbo.—Orig. in Thornhill, Ontario, Canada, by Matthew James Johnson. Introd. about 1950 by Kellogg Nursery Co., Three Rivers, Mich. Beaderarena x Howard

17; sibling of Nectarena; selected about 1940. Fruit: skin dark red; sometimes rough and irregular in shape; flesh fairly firm; ripens in midseason; fine for freezing, quality fair for commercial processing and for the home garden. Apparently this variety is neither of the Jumbo varieties given in *The Small Fruits of New York* by U.P. Hedrick, 1925.

June Rockhill.—Orig. in Conrad, Iowa, by Harlow Rockhill. Introd. in 1947 by the Wheelock Wilson Nursery, Marshalltown, Iowa. Plant patent 854; 19 July 1949; assigned to the estate of the late Harlow Rockhill. Progressive x Early Jersey giant; selected in 1932. Fruit: identical with Rockhill 26, except that June Rockhill fruits in June. Plant: prolific plant producer.

Jupiter.—Orig. in Salisbury, Md., by Oliver C. Cordrey, Hannah E. Cordrey, and Ernest W. Townsend. Introd. in 1931 by Ernest W. Townsend. Plant patent 46; 29 Nov. 1932; assigned to Ernest W. Townsend. Howard 17 x Gandy; selected in 1925. Fruit: size similar to Blakemore, but somewhat elongated, similar to the last half of a heavy crop of Howard 17; flesh firm, quality poor; ripens early. Plant: very productive. Apparently identical with Ridgeley as grown in Maryland.

Juspa.—Orig. in Elst, The Netherlands, by H.G. Kronenberg and L.M. Wassenaar, Inst. for Hort. Plant Breeding, Wageningen, The Netherlands. Introd. under license in 1957. Jucunda x Sparkle; selected in 1949; tested as I.V.T. 667. Fruit: size medium; short conic; skin clear red; flesh light red, flavor fair; ripens in midseason; some tendency of the first fruit to make nubbins; good capping; resembles Jucunda. Plant: compact; productive; flowers hidden under leaves; leaves rather susceptible to mildew and leaf scorch; makes runners freely.

Kalicene.—Orig. in Ettersburg, Calif., by Albert F. Etter (1872 to 1950). Introd. in 1920. Ettersburg 216 x Trebla or Ettersburg 121 x Trebla. Fruit: medium to small; conic; skin medium red, glossy; flesh firm, red throughout, moderately juicy, subacid; quality fair; ripens early. Plant: productive; vigorous. Apparently no longer propagated.

Kanner King.—Orig. in Sawyer, Mich., by Bert W. Keith and Bud H. Keith. Introd. in 1932. Plant patent 26; 30 Aug. 1932. Chance seedling of Glen Mary; discovered in 1926. Fruit: medium to large, uniform; conic, like Dunlap; skin bright red, somewhat glossy, flesh bright red throughout, firm, tender, subacid; fresh and canning quality very good; achenes bright yellow, even with surface; average to long picking season, ripens in spring; hull size medium, easily removed. Plant: upright to spreading; leaflets slightly crimped; healthy, hardy; produces runners freely.

Kardinal King.—See Robinson.

Karmyn.—Orig. in Pretoria, South Africa, by E.P. Evans, Hort. Res. Inst. Introd. in 1966. (Senga 752 x 1/67/22) x Festival; selected in 1962. Fruit: medium to large; skin darker ruby red than Robyn, durable; achenes large and exposed; flesh very dark red and evenly colored; flavor good throughout production period (June to December). Plant: size medium; vigorous grower; produces runners less freely than Robyn; tolerant to both leaf spot and leaf scorch; for home garden and local market.

Kasuga.—Orig. in Sandy, Utah, by Taijiro Kasuga. Introd. in 1955. Plant patent 1264; 2 Mar. 1954. Starbright x Twentieth Century; cross made in 1943; selected in 1953. Fruit: uniform shape; skin glossy; flesh very firm; holds shape and color well; excellent freezing quality; resembles Shasta but is more conic in shape and of better quality. Plant: taller and more vigorous than Twentieth Century; tolerant of high salt and alkali conditions; flowers perfect.

Kellogg Beauty (*New Beauty*).—Orig. in Parkville, Mich., by Dan Sidler. Introd. in 1927 by R.M. Kellogg Co., Three Rivers, Mich. Parentage unknown; discovered in 1923. Fruit: ripens late. Plant: bisexual. Propagation discontinued in 1946.

Kendall (*Nil*).—Orig. in Narara, New South Wales, by Hort. Res. Sta., Dept. of Agr., New South Wales. Introd. in 1961. Open-pollinated seedling of Richmond Red. Fruit: size medium; blunt conic; skin considered too dark for market requirements, tough; flesh firm, flavor good; recommended for Coastal regions of New South Wales; ripens September to December. Plant: size medium; rounded; vigor, hardiness and productivity moderate; susceptible to mildew; some tolerance to simple virus infection and common leaf disease; variety has been superseded by Torrey.

Kent.—Orig. in Kentville, Nova Scotia by D.L. Craig, L.E. Aalders, and G.W. Bishop, Res. Sta. Agr. Can. Introd. in 1981. K68-58 (Redgauntlet x Tioga) x Raritan. Cross made in 1973; selected in 1974; tested as K74-10. Berry: large; uniform conic; skin medium red, glossy: flesh red throughout, firm, quality very good; ripens midseason, after Redcoat, before Micmac. Plant: productive, abundant runners, vigorous, hardy, slightly susceptible to mildew.

King.—Orig. in Ottawa, Ontario, Canada, by Canada Dept. Agr., Res. Sta. Introd. in 1934. (Delecto x Cassandra) x self. Fruit: large; irregularly wedged; neck well-defined; calyx large; skin dark red; flesh medium dark; excellent for canning and jam; dessert quality very good; firm enough to ship well; ripens medium early, shortly after Howard 17; in some districts fruit tends to bruise slightly on the surface. Plant: flowers perfect; grown only to a limited extent in the Province of Quebec.

Klonmore.—Orig. in Baton Rouge, La., by Julian C. Miller, Louisiana Agr. Expt. Sta. Introd. in 1940. Blakemore x Klondike; cross made in 1933; selected in 1935. Fruit: small; blunt conic; skin a light bright red, achenes yellow; flesh medium firm, subacid; matures over a long season; ships well; freezes well; dessert quality fair to medium; resembles Blakemore. Plant: vigorous; produces runners freely; resistant to leaf spot; susceptible to leaf scorch. Apparently now superseded by other varieties (e.g., Headliner and Dabreak).

Konvoy.—Orig. in Baton Rouge, La., by Julian C. Miller, Louisiana Agr. Expt. Sta. Introd. in 1942. Fairmore x Klondike; selected in 1938. Fruit: resembles Klondike; not as firm as some of the better shipping varieties; not recommended for shipping but as a home garden variety; ripens slightly earlier than Klondike and has larger fruit. Plant: resistant to leaf spot and scorch; productive.

Korona.—Orig. in Wageningen, Holland, by Inst. for Hort. Plant Breeding. Introd. in 1978. Tamella x Induka. Licensed in 1977. Berry: large, conical; skin dark red; flesh red, quality good, firm, juicy, aroma and taste good; ripens midseason. Plant: productive, erect, open; leaf dark green, four to five foliolate; inflorescences abundant; moderately susceptible to Botrytis cinerea and vertitillium wilt, mildew, and *Phytophthora fragariae.*

Kristina.—Orig. in Kristianstad, Sweden, by Balsgård Fruit Breeding Inst. Introd. in 1968. Senga x Valentine. Cross made in 1957, selected in 1960, tested as F2608. Fruit: size medium; round or short conical; skin bright red, shiny; flesh red, firm, acid and aromatic; freezing quality good; ripens about 20 to 25 June in Sweden, 10 to 12 days earlier than Senga. Plant: size medium, low bush shape; vigor moderate; hardy; productive; seems to be tolerant to *Botrytis* and mildew.

Laguna.—Orig. at Univ. of California, South Coast Res. and Ext. Center, near Irvine, by V. Voth and R.S. Bringhurst. Irvine x Cal 85.92-602; cross made in 1988; selected in 1989; tested as Cal 88.66-616; introd. in 1992. U.S. plant patent pending. Fruit: size similar to Chandler; firmer than Chandler; symmetrical, conic; compared to Chandler, external color lighter, more orange, and less glossy, and internal color lighter; achenes yellow to light red and slightly indented; very good flavor, slightly less aromatic than Chandler; fresh-market, processing, and home-garden uses. Plant: short-day type producing fruit over an extended period in arid, subtropical climates; production pattern similar to Chandler, but production persists later in season in cool Mediterranean climates; adapted to central California, as well as southern California; plant larger, more erect and vigorous than Chandler. Moderately resistant to common leaf spot and powdery mildew; equal or greater tolerance to two-spotted spider mite than Chandler; tolerant to viruses occurring in California.

Lambada.—A high-quality, early ripening strawberry adapted in The Netherlands to both forcing under plastic tunnels and outdoor cultivation. Orig. at CPRO-DLO in Wageningen, Netherlands. (Sivetta x Holiday) x (Karina x Primella); introd. in 1992. Fruit: large; bright red; relatively firm, does not bruise readily; conic with exposed neck; uniform; reflexed calyx. Ripens early, 2 or 3 days before Gorella; considerable rot resistance; excellent flavor. Plant: moderate productivity; vigorous; stiff petioles; long scapes. Probably resistant to powdery mildew and verticillium wilt; some resistance to red stele and crown rot; moderately susceptible to leaf spot.

Lassen.—Orig. in Wheatland, Calif., by Harold E. Thomas and Earl V. Goldsmith, Calif. Agr. Expt. Sta., Berkeley. Introd. in 1945. Calif. 21.9 [Blakemore x Calif. Z9 (Banner x Fendalcino)] x Calif. 161.1 [Nich Ohmer x Calif. 86.6 (USDA 634 x Calif. Zll (Banner x Fendalcino)]; cross made in 1936; selected in 1938; tested as Calif. 544.2. Fruit: large, holding well through the season; normally blunt conic or rounded, varying to short wedge; skin light red, glossy, good market appearance, tender; flesh soft, juicy, medium red; slightly hollow; no aroma; dessert quality fair, shipping quality fair; hull size medium, separating easily from fruit. Plant: particularly well-adapted to southern California. because of a relatively low chilling requirement and tolerance to salinity; production high, with three seasonal peaks per crop season, from April to November; long-lived, but used on an annual planting basis only; resistant to virus diseases; very susceptible to verticillium wilt, moderately susceptible to yellows disease; nubbin berries produced in some seasons during summer crop. Named for a mountain peak in northern California.

Lateglow.—Orig. in Beltsville, Md., by D.H. Scott, the U.S. Dept. of Agriculture. Introd. in 1987. Tamella x MDUS 3184 (NCUS 1768 x Surecrop); cross made in 1974; selected at Wye Inst., near Queenstown, Md., in 1976 by A.D. Draper and P. Rosati; tested as MDUS 4839. Fruit: large to very large; uniform in symmetry; skin firm, deep glossy scarlet; flesh soft medium, medium red; sweet, juicy, aromatic; tolerant to gray mold and leather rot. Ripens after Allstar, with Delite and Rosanne. Plants: runner freely; medium size; marketable yields very high, like those of Lester. Very resistant to most races of red stele; resistant to verticillium wilt; tolerant to leaf spot and leaf scorch; susceptible to anthracnose leaf blight, and powdery mildew. Flowers late with Allstar and Guardian.

Latestar.—Orig. in Beltsville, MD by G.J. Galletta, J.L. Maas, and J.M. Enns, U.S. Dept. of Agr. Intro. in 1995. Lateglow x Allstar; cross made in 1978; selected in 1980; tested as MDUS 5084. Fruit: large; attractive; firm skin and flesh; glossy red skin; light red flesh color; pleasant, mild, slighly acidic fresh flavor. Plant: resistant to five races of red stele, most of the leaf and stem diseases and fruit rots but susceptible to leaf blight; vigorous, runners well, flowers and fruits later than Allstar, Lateglow, or Jewel; very productive; grows well in matted row or raised beds, planted in spring or summer.

Lee Teague.—Orig. in Logan, Iowa, by Lee Roy Teague. Introd. in 1959. Plant patent 1827; 28 Apr. 1959; assigned to the originator and his heirs. Parentage unknown; discovered in 1956. Fruit: size medium; form flattened; skin deep, dark red; flesh firm, no hard core, flavor pleasing, subacid, dark red throughout; ripens evenly. Plant: everbearer; large and vigorous; runners numerous but easily controlled; productive; flower perfect.

Leman (*Daddy Leman*).—Orig. in Clark County, Ohio. Introd. in 1930 by W.N. Scarff & Sons Nursery, New Carlisle, Ohio. Chance seedling; discovered in 1920. Fruit: large; ripens late, but more evenly than Gandy, which it resembles. Plant: vigorous; foliage dark green. Variety now obsolete.

Lester.—Orig. in Beltsville, Md., by D.H. Scott, the U.S. Dept. of Agriculture. Introd. 1984. Raritan x MDUS 3413 {[(Tennessee Shipper x Maytime) Stelemaster] x Earlidawn}; cross made in 1969; selected at Wye Inst. near Queenstown, Md. in 1971; tested as MDUS 4359. Fruit: medium large, very attractive, symmetrical, uniform deep red; flesh pink, firm, mild flavor; tolerant to gray mold and leather rot. Plants: vigorous medium size; medium-heavy production; runner freely to form medium-dense fruiting beds. Resistant to seven races of red stele; susceptible to verticillium wilt; tolerant or resistant to powdery mildew, leaf scorch, and leaf blight; susceptible to anthracnose.

Lincoln.—Orig. at the Turners & Growers Mangere Res. Sta. near Auckland, New Zealand, by I.K. Lewis and M.W. Hall.

Donna x 85-24-1 (derived from Cruz, Pajaro, Douglas, and Holiday); cross made in 1987; selected in 1988; tested as 87-2-129 and later as T-69; introd. in 1993. Flower: self-fertile, with ample pollen throughout season and, consequently, little chance of malformed fruit. Fruit: large; as firm as Chandler; glossy; external and internal color slightly darker than Pajaro; bright yellow achenes at surface or slightly indented; conic; solid, seldom hollow. Midseason ripening; excellent flavor; fresh-market and processing uses. Plant: weak day-neutral with lower chilling requirement than Selva; yields comparable to Chandler; semi-erect habit; peduncles remain semi-erect until fruit ripens; runner production comparable to Selva.

Lindalicious.—Orig. in Brigham City, Utah, by L.L. Larsen, Larsen Expt. Farm. Introd. in 1948. Fairfax x unnamed seedling; selected in 1937. Fruit: medium to large; solid red; turns red on inside before completely colored on outside; no core; no green tip; tough skin; high freezing and shipping quality. Plant: everbearing in some localities; heavy producer.

Linn.—Orig. by the Oregon Agr. Ext. Serv., Washington Agr. Expt. Sta. and Agr. Res. Service, U.S. Dept. of Agriculture. Introd. in 1976; MD-US 3184 x OR-US 2414; cross made in 1967; selected in 1969; tested as OR-US 4005 in Oregon and Washington. Fruit: medium to large, conic, rarely rough or grooved; color bright red, glossy, good red internal color; flesh very firm; skin medium tender with raised yellow seeds; ripens 1 to 2 days after Northwest. Plant: vigorous; semi-upright foliage; runners freely; good field tolerance to red stele root rot; superior to Northwest in mildew resistance and superior to Hood for virus tolerance. Introduced because hand-harvesting is facilitated by the almost complete exposure of the fruit and firmness and crop concentration facilitates machine harvesting.

Locke Lake Ruby.—Orig. in Clearwater, Minn., by Bertha Rosa Zum Brunnen, Locke Lake Nursery, Monticello, Minn. Introd. in 1963. Plant patent 2429; 28 July 1964. Open-pollinated seedling of Red Rich. Fruit: fairly large, up to 1 1/2 inches long; cordate; skin glossy, medium dark red; achenes relatively small, protruding somewhat above the surface; flesh quite firm, red with a whitish core area, relatively fine-textured, quite juicy, flavor sweet, nonacid; ships well; suitable for canning and freezing. Plant: everbearing; bisexual; quite hardy; very productive; apparently resistant to leaf spot; crown medium-heavy; medium number of runner plants produced, some bearing the first season.

Louise (*Churchill*).—Orig. in Ottawa, Ontario, Canada, by Canada Dept. Agr., Res. Sta. Named and introd. in 1942. Ettersburg 80 x self; selected in 1933. Fruit: medium to large; very attractive; only fairly firm, not outstanding for shipping; good for dessert; ripens late; resembles Ettersburg 80. Plant: flowers imperfect, very susceptible to leaf scorch and leaf spot; virus-free stocks not available. Not widely grown.

Lovett's All Season.—See All Season.

Lowridge.—Orig. in Matteson, Ill., by Edna Graham, Mahopac, N.Y. Introd. in 1971. Plant patent 3049; 20 Apr. 1971. Parentage complex, including native Illinois wild strawberry stock crossed with Dunlap, Fairfax, Gem, Mastodon, Superfection, and Red Rich. Selected in 1949. Fruit: large to medium; conical, blunted point; skin medium red, shiny; seeds bright yellow, flush with surface; flesh medium red, juicy, sweet; recommended for the home garden, fresh fruit, freezing and jam; not a shipping variety; its unique characteristic is the successive ripening habit, from first spring fruiting until heavy frost (about May to October). Plant; small; extremely hardy; prolific in runner production, producing berries on runner plants as soon as they begin to root and sometimes without a separate root system; not drought resistant.

Luscious Red.—Orig. in Ortonville, Minn., by Louis H. Block. Introd. in 1962. Plant patent 2230; 26 Feb. 1963. Chance seedling; discovered in 1956. Fruit: long conic, neck medium to distinct; skin attractive, bright, medium red; achenes project slightly above surface; flesh medium red, pink around core, flavor mild, subacid, tender, medium juicy, almost sweet, mildly aromatic; quality good to very good. Plant: everbearing; vigorous; production heavy; very hardy; flowers form early on runner plants.

Lustre.—Orig. in East Merchantville, N.J., by Oscar Earle Felten. Introd. in 1949. Plant patent 929; 28 Mar. 1950; assigned to O.E. Felten, East Merchantville, N. J. and W.T. Wilkins, Moorestown, N.J. seedling 3001 (Howard 17 x Marvel) x seedling 3700 (seedling 3001 x Fairfax); selected in 1945. Fruit: large; skin dark red, fairly tough; flesh firm, flavor mild; achenes in position to protect berry in shipment; resembles Sparkle. Plant: stems hold berries off ground; vigorous.

Macherauchs Frühernte (*Müncheberger Frühe*).—Orig. in Muncheberg, Mark, Germany, by Friedrich Gruber, at the Kaiser Wilhelm Institut by the Planck Institut f. Züchtungsforschung, Kohn-Vogelsang, Germany. Introd. in 1951 in Germany and somewhat later in the U.S. and Canada. German patent 623151; assigned to O. Macherauch Gr. Minmelage. Amerikanischer Samling II (Geneva) x Deutsch Evern; selected in 1936. Fruit: size medium; flesh medium firm, light to medium red; resembles Deutsch Evern; ripens early; easily picked. Plant: high yielding ability; resistant to late frosts (flowers protected under leaves); subject to mildew; tolerant of poor soils and fall drought.

Mackenzie.—Orig. in Ottawa, Ontario, Canada, by Canada Dept. Agr., Res. Sta. Introd. in 1941. Excelsior x Howard 17; selected in 1934. Fruit: about same season as Howard 17, which it resembles; quality good. Plant: flowers perfect; more productive under certain conditions than Howard 17. Not grown commercially. No virus-free plants available.

Madame Lefeber.—See Madame Levebvre.

Madame Lefebvre (*Madame lefeber, President Roosevelt*).—Orig. in France. Introd. into England about 1920 by R. Hendriksen, Reewyk, The Netherlands. Introd. into Canada in the early 1930s. Fruit: ripens very early; quality poor.

Madame Lefebvre•M49.—A special reselected virus-tested clone of Madame Lefebvre, developed in 1949 by the East Malling Res. Sta., England, and officially accepted there for certification purposes. Fruit: size medium; spherical conical, blunt; skin dark red; flesh moderately firm; ripens early. Plant: medium large; spreading open habit, vigor medium. Variety has been largely replaced by other varieties such as Cambridge Prizewinner.

Mae.—Orig. in Carroll County, Md., by E.H. Devilbliss. Introd. in 1931 by Jesse E. Stoner, Westminster Nurseries, Westminster, Md. Parentage unknown; selected in 1927. Fruit: large; ships well; ripens about 4 days later than Howard 17. Plant: vigorous; flower imperfect. Variety apparently obsolete.

Maine 55.—Orig. in Monmouth, Maine, at the Highmoor Farm of the Maine Agr. Expt. Sta. Introd. in 1952. Howard 17 x Aberdeen; selected in 1940. Fruit: quality high, both fresh and frozen. Plant: hardy; highly resistant to Aberdeen-type of red stele. No longer propagated.

Majestic.—Orig. in Jonesville, Va., by E.L. Russell. Introd. in 1940. Plant patent 345; 24 Oct. 1939; assigned to E.L. Russell. Seedling of Howard 17; discovered in 1933. Fruit: large; flesh firm, red; ripens in midseason; good shipping quality; tends to darken color when fully ripe. Plant: large; very productive; sets many plants; resistant to drought and leaf spot; sturdy and upright stems; more productive and a little more tart than Blakemore or Howard 17.

Marion Bell.—Orig. in Baton Rouge, La., by Julian C. Miller, Louisiana Agr. Expt. Sta. Introd. in 1946. Fairmore x self; selected in 1942. Fruit: skin bright; resembles Klonmore; flavor excellent; matures over a long period; good shipper. Plant: resistant to leaf spot and scorch.

Marlate.—Orig. in College Park, Md., by I.C. Haut, Univ. of Maryland. Introd. in 1969. Jerseybelle x [(Klonmore x Midland) x (Tennessee Shipper x Midland)]. Cross made in 1956, selected in 1959, tested as Md 88E. Fruit: large; very symmetrical, conic; skin uniform bright red, tough; flesh moderately light red, firm; most valuable characteristics: holds size exceptionally well, very firm, easily picked, tough skin; ripens late, between Jerseybelle and Vesper. Plant: medium size; vigor good; productive. Recommended for mid-Atlantic region and northward.

Marmion.—Orig. at Scottish Hort. Res. Inst., Auchincruive, Ayr. Crusader x Redgauntlet; cross made in 1960. Fruit: conical to round; medium to large; orange-red skin with pale pink flesh; slightly soft; calyx medium to small, clasping fruit. Plant: dense with partially upright and vigorous growth; red-core resistance; only slightly susceptible to mildew and Botrytis.

Marmolada.—Orig. near Comacchio (Ferrara), Italy, by M. Leis and D. Musacchi, Consorzio Italiano Vivaisti. Cross made in 1985; Gorella x Salvi 15 or selection no. 8 (possibly the Driscoll variety, Heidi); introd. in 1989. Italian plant patent 36-NV/89. Fruit: large; very firm; attractive; bright red external and internal color; uniform trunk conic shape; midseason ripening with long harvest period; sweet flavor. Moderately resistant to rhizoctonia and anthracnose rots; tolerant to botrytis rot; excellent shelf life; fresh-market use. Plant: very high yielding; compact; abundant runners; erect scape; tolerant to angular leaf spot and verticillium wilt; adapted to northern Italy.

Marvel.—Orig. in North Lake, Wis., by Percy Schuckhardt. Introd. in 1922 by R.M. Kellogg Co., Three Rivers, Mich. Senator Dunlap x Warfield; discovered in 1915. Fruit: large; slightly necked; skin bright red; flesh dark red, firm; quality very good. Plant: vigorous; very productive. Propagation discontinued in 1934.

Massey.—Orig. in Glenn Dale, Md., by George M. Darrow, U.S. Dept. of Agriculture, and E.B. Morrow, North Carolina Agr. Expt. Sta., Raleigh. Introd. in 1940. US 634 (Royal Sovereign x Howard 17) x Blakemore; selected in 1934. Fruit: very large; extremely attractive; dessert quality high; good shipping quality; ripens late. Plant: very vigorous; moderately productive; tolerant to leaf scorch and leaf spot. Adapted to coastal plain areas of North and South Carolina and Virginia.

Masterpiece.—Orig. in Shenandoah, Iowa, by Henry Field, Henry Field Seed Nursery Co. Introd. in 1931 by this company. Mastodon is one of the parents; selected about 1927. Fruit: very large; skin dark red; flavor good; better than Mastodon. Variety now obsolete.

Mastodon.—Orig. in Peru, Ind., by George Voer. Introd. in 1921. Superb x Kellogg; selected in 1917. Fruit: large, somewhat variable and irregular; quality good; colors somewhat unevenly. Plant: very vigorous, productive, superior to Progressive and Superb; everbearing, crops borne in spring and autumn; moderate runner production; flower perfect.

Maytime.—Orig. in Glenn Dale, Md., by George M. Darrow, U.S. Dept. of Agriculture. Introd. in 1941. Missionary x Fairfax; cross made in 1931 by George F. Waldo; selected in 1933; tested as USDA 2267. Fruit: large in first part of season, later becoming medium sized; wedge or conic, uniform; skin rich red, not turning dark on holding, glossy, very attractive; achenes yellow, even with surface; flesh rich red throughout, juicy, subacid, aromatic, dessert quality high; shipping quality very good; ripens early, 4 days before Blakemore; firmer, earlier, more highly flavored than Howard 17. Plant: runner plants produced freely; leaf free from scorch; productive; flower clusters borne beneath leaves.

McClintock.—Orig. in Knoxville, Tenn., by Tennessee Agr. Expt. Sta. Introd. in 1932. Aroma x self; selected in 1924; tested as Tenn. 6. Fruit: appearance more attractive and larger yield than Klondike, which it resembles.

Melody.—Orig. in Invergowrie, Scotland, by R.J. McNicol, Scottish Hort. Res. Inst. SCRI 66M1 (highly red-stele-resistant third generation derivative of *Fragaria virginia*) x Senga Sengana; cross made by H.J. Gooding at Auchincruive in 1971; tested as SCRI 71WC64; introd. in 1992; entered for Plant Breeders' Rights. Fruit: small to medium size; firm; easy calyx removal; good external and internal color; ripens midseason; processing use. Plant: yield same or slightly better than Cambridge Favourite; runners well. Highly field resistant to red stele; moderately resistant to verticillium wilt in laboratory tests.

Merrimack.—Orig. in Durham, N.H., by E.M. Meader. Introd. in 1957. N.H. 179 (Tupper x Fairfax) x Fairpeake; cross made in 1949; tested as N.H. F-13. Fruit: large; flavor sweet; ripens late. Plant: free from harmful virus diseases; vigorous, with heavy crowns; produces runners freely; primarily for the New England states.

Merton Herald.—Orig. in Bayfordbury, Hertford, Herts, England, by Hedley Williams, John Innes Inst. Introd. there in 1965. Tennessee Beauty x Royal Sovereign; selected in 1953. Fruit and plant: earlier and more mildew resistant than Royal Sovereign, yet retains the superb quality.

Merton Princess.—Orig. in Bayfordbury, Hertford, Herts, England, by the John Innes Hort. Institution. Introd. in 1957. Royal Sovereign x Merton 31 (possibly Northstar); selected in 1952. Fruit: appearance very good. Plant: yields heavily; resistant to mildew.

Merton Ruby.—Orig. in Bayfordbury, Hertford, Herts, England, by Hedley Williams, John Innes Inst. Introd. in 1965. USDA 3378 x Early Cambridge; cross made in 1950; selected in 1953; tested as J.I. 663. Fruit: medium to large; short conical or conical; skin light red; flesh firm, light red; very distinctive; ripens midseason to late, with many of the fruit ripening at one time. Plant: vigorous; upright; yields very heavily.

Micmac.—Orig. in Kentville. Nova Scotia by D.L. Craig, L.E. Aalders and G.W. Bishop, Res. Sta. Agr. Can. Introd. in 1976. Tioga x K61-87 (Guardsman S_1). Cross made in 1967; selected in 1968: tested as K68-108. Fruit: large; globose conic, slightly necked, uniform; skin light to medium red, glossy; flesh firm. red throughout, mild, subacid, aromatic; flavor good; achenes yellow, level with surface; calyx removal somewhat difficult, pedicel brittle, picks easily; ripens mid to late, 3 to 4 days after Redcoat. Plant: productive; vigorous, upright petioles: leaves large, adequate runner production, slightly susceptible to powdery mildew; hardiness equal to those of Redcoat and Bounty.

Midland (*Adonis, Crimson Flash*).—Orig. in Glenn Dale, Md., by George F. Waldo and George M. Darrow, U.S. Dept. of Agriculture. Introd. in 1944. Howard 17 x Redheart; cross made in 1929; selected in 1931; tested as USDA 1812. Fruit: very large; blunt wedge to short conic, relatively uniform; skin dark red, glossy, slightly darker than Dorsett; hull reflexed, size medium; achenes yellow to red, even with surface or slightly raised; flesh deep red, moderately firm, tender, juicy, mildly subacid, highly flavored, aromatic; excellent dessert quality; freezes very well; season early, with Howard 17; shipping quality good. Plant: productive; vigorous; runners produced freely; leaf large, resistant to leaf spot and leaf scorch; sensitive to virus diseases; popular from southern New England to Virginia and west to Iowa and Kansas.

Midway.—Orig. in Salisbury, Md., by U.S. Dept. of Agriculture, Beltsville, Md., and the Maryland Agr. Expt. Sta., College Park. Released for propagation in 1959; introd. in 1960. Dixieland x Temple; cross made in 1951; selected in 1953; tested as Md-US-2389 from Massachusetts to Virginia and westward, performing satisfactorily in Maryland, New Jersey, Pennsylvania, Indiana, Ohio, Massachusetts, Missouri, and eastern Washington. Fruit: medium to large, maintaining good size during the picking season; shape irregular, medium long conic to blunt conic; skin uniform deep red, tough, glossy; flesh firm, flavor subacid with good dessert quality, satisfactory in frozen pack; ripens in midseason, with Temple and Northwest. Plant: moderately vigorous; produces many runners; resistant to common race of red stele and should be tested as a red-stele-resistant replacement for Fairland and Temple; leaves somewhat susceptible to leaf scorch, leaf spot, verticillium wilt; productive as grown in narrow matted rows in Beltsville and Salisbury.

Midway 1.—See Fairfax.

Mildred Felten.—Orig. in East Merchantville, N.J., by Oscar Earle Felten. Introd. in 1947. Plant patent 761; 16 Sept. 1947. Fairfax x seedling 3001 (Howard 17 x Marvel); selected in 1934. Fruit: larger, more tart, lighter colored than Fairfax, which it resembles, picking season longer than Fairfax and later berries hold up in size much better. Plant: larger, heavier bearer than Fairfax.

Millay Everbearer.—Orig. in Bowdoinham, Maine, by Ruth Millay, Portland, Maine. Introd. in 1972. Plant patent 3153; 9 May 1972. Bud mutation of Red Rich, discovered in 1953. Fruit and plant: shows a general resemblance to Red Rich, differentiated by: superior fruit flavor; longer, straighter, stiffer, and more runners up to 6 ft in length before freezing; larger plant; more continuous everbearing habit; more productive. Compared to Ozark Beauty: more productive, seeds less prominent and skin darker red. Compared to Geneva: more productive for longer period and skin darker.

Minnehaha.—Orig. in Excelsior, Minn., by the Univ. of Minnesota. Fruit Breeding Farm. Introd. in 1920. Minnesota x Abington; cross made in 1911; tested as Minn. 935. Fruit: large; wedge shaped; skin medium red, producing green tips in early pickings; flesh light color, firm; ships very well; late season; poor canning variety. Plant: vigorous; hardy; productive; leaves large, dark green; resistant to decay. No longer propagated.

Minnesota.—Orig. in Excelsior, Minn., by the Univ. of Minnesota Fruit Breeding Farm. Introd. in 1920. Dunlap x Pocomoke; cross made in 1907; tested as Minn. 3. Fruit: medium to large; conic; skin bright, glossy, medium-dark red, attractive; flesh red, firm, fine-grained, juicy, sweet, aromatic, quality good; early mid-season. Plant: productive; subject to mosaic disease (June yellows).

Miss Utah.—Orig. in Sandy, Utah, by T. Kasuga. Introd. in 1955. May have originated in the breeding program of the California Agr. Expt. Sta. as selection Calif. 1002.5; this hybrid did not appear to do well in California but had done much better in Utah; selected in 1953 by Kasuga as worthy of naming and introducing. Fruit: long, conic, extremely uniform; flesh very firm, high dessert and freezing quality.

Mohawk.—An early ripening, fresh-market strawberry adapted to southwestern Canada and the northeastern United States. Orig. in Beltsville, Md. MDUS 4587 x Earliglow; cross made in 1979 by G.J. Galletta; selected in 1981; tested as MDUS 5122; introd. in 1994 by G.J. Galletta, J.L. Maas, J.M. Enns, and A.D. Draper, U.S. Dept. of Agriculture, Beltsville; A. Dale, Hort. Res. Inst. of Ontario; and H.J. Swartz, Dept. of Hort., Univ. of Maryland. Fruit: medium large; irregular short conic; glossy, deep scarlet exterior, pink interior; flesh juicy but firm texture, skin firm; achenes yellow and recessed; calyx clasping to partially reflexed. Ripens 2 or 3 days before Earliglow in Maryland and before Veestar in Ontario; good shelf life; flavor mild, sweet, and subacidic. Plant: vigorous; medium to tall; dense crowns with many leaves; petioles erect and thick; abundant runners; yields similar to Earliglow in Maryland and less than Veestar in Ontario. Resistant or tolerant to powdery mildew, leaf scorch, leaf blight. Resistant to at least six races of the red stele causal organism.

Molalla.—Orig. in Corvallis, Ore., by George F. Walto, U.S. Dept. of Agriculture and Oregon Agr. Expt. Sta. Introd. in 1961. US-Ore. 2136 x US-Ore. 2242; cross made in 1952; selected in 1954; tested as US-Ore. 2416; named in 1961. Fruit: size medium but larger than Siletz, becoming smaller toward end of season, becoming more uniform as season progresses; flesh a uniform rich red throughout, very firm, subacid; flavor and appearance excellent for frozen pack; ripens in midseason, about 1 week after Marshall; resembles Northwest. Plant: vigorous; foliage dark green; produces many runners; slightly susceptible to mildew; appears resistant to red stele disease, similar to Columbia or Siletz; appears tolerant of virus diseases; yields have ranged from 5 to 8 tons per acre.

Monmouth.—Orig. in Monmouth, Maine, at the Highmoor Farm of the Maine Agr. Expt. Sta. Introd. in 1952. Aberdeen x Howard 17; selected in 1940. Fruit: soft but freezes well; resembles Senator Dunlap. Plant: hardy; highly resistant to Aberdeen-type red stele; productive.

Montana Progressive.—Orig. in Bozeman, Mont., by H.E. Morris and M.M. Afanasiev, Montana Agr. Expt. Sta. Introd. in 1942. Open-pollinated seedling of Progressive; selected in 1935. Fruit: medium to large; globose to globose conic to blunt wedge, slight neck on some fruit; skin dark to bright red; achenes red, even with surface; calyx medium large, reflexed; flesh tender, pink to reddish, subacid, firm, aromatic; dessert and shipping quality good. Plant: medium to large; somewhat spreading; everbearing; resistant to yellows and leaf spot; spring crop fair to good; fall crop heavy; flowers perfect.

Montrose.—Orig. at Scottish Hort. Res. Inst., Auchincruive, Ayr. Crusader x Redgauntlet; cross made in 1960. Fruit: large; round, orange-red; soft texture although acceptable for shipping; midseason to late ripening; recommended for dessert use when grown on redcore-free land where its productivity can be fully exploited. Plant: moderately susceptible to *Botrytis* and red core and only slightly susceptible to mildew.

Mr. Big.—Orig. in Carpenter, Ohio, by Truman A. Penn. Introd. in 1953. Considered to be an open-pollinated seedling of Robinson. Fruit: extremely large; softer than Robinson. Its chief value is as a novelty.

Mrak.—Orig. at Univ. of California, Wolfskill Experimental Orchards, near Davis, by R.S. Bringhurst and V. Voth. Cal 69.141-101 x Aiko; cross made in 1975; selected in 1976; tested as Cal 75.34-105 and later as CN 27; introd. in 1987. U.S. plant patent 6579, 31 Jan. 1989. Fruit: compared to Fern, smaller and less intense red, with similar firmness; glossy; symmetrical, medium to long conic; slightly hollow center; flavor not quite as good as Fern; fresh-market use. Plant: strongly day-neutral with relatively high chilling requirement; low-growing and smaller than Fern; runner production good; yield comparable to Fern. Susceptible to powdery mildew and common leaf spot; resistant to verticillium wilt; less susceptible to two-spotted spider mite than Fern; tolerant to viruses occurring in California.

Muir.—Orig. at Univ. of California, Wolfskill Experimental Orchards, near Davis, by R.S. Bringhurst and V. Voth. Cal 70.3-117 x Cal 71.98-605; cross made in 1975; selected in 1976; tested as Cal 75.7-105 and later as CN 17; introd. in 1987. U.S. plant patent 6558. Fruit: almost as large as Selva but less firm; bright red exterior; medium to long conic, slightly bulbous, somewhat flat to wedge shape; center may be hollow; flavor as good as or better than Selva; more seedy than Selva; fresh-market use. Plant: more strongly day-neutral than Selva; higher chilling requirement than short-day types; more erect habit than Selva; very good runner production; productivity similar to Selva. Moderately susceptible to common leaf spot; highly susceptible to verticillium wilt; less susceptible to two-spotted spider mite than Selva; tolerant to viruses occurring in California.

Müncheberger Frühe.—See Macherauchs Frühernte.

Nanaimo.—A strawberry producing fruit with high soluble solids, suited to both fresh and processing markets. Orig. in Vancouver, British Columbia, Canada, by H.A. Daubeny, Agr. and Agri-Food Canada, Pacific Agr. Res. Centre. Totem x Korona. Cross made in 1986; selected in 1988; tested as BC 86-33-2; introd. in 1995. Fruit: medium to large; bright medium-red external and internal color; medium firm; mostly wedge shape; calyx easily removed; achenes slightly raised; midseason ripening; reactions to pre- and postharvest rot similar to Totem; excellent fresh flavor. Plant: productive; good vigor; relatively upright habit; adequate runner production; as winter hardy as Totem. Slightly more susceptible to powdery mildew than Totem; tolerant to the aphid-borne virus complex in the Pacific Northwest; reaction to red stele unknown.

Nancy Lee.—Orig. in Princess Anne, Md., by Homer S. Kemp, Bountiful Ridge Nurseries. Introd. in 1934. Chance seedling; discovered in 1930 in nursery trial plots. Fruit: resembles Howard 17 but a week earlier in ripening; high flavor. Plant: comparatively disease resistant; bears heavily. Variety apparently now obsolete.

Narabelle.—Orig. in Narara, New South Wales, by Hort. Res. Sta., Dept. of Agr., New South Wales. Introd. in 1966. Kendan x unnamed seedling (N56/7). Cross made in 1961, selected in 1963. Fruit: size medium; conic; skin considered too soft for commercial production, attractive; flesh flavor fair. Plant: size medium; upright; vigor and hardiness fair; productivity moderate; slightly susceptible to leaf spot (*Mycosphaerella fragariae*). Doubtful the variety is now being grown.

Naratoga.—Orig. in Gosford, New South Wales, Australia, by J. Cox, Gosford Hort. Res. Sta., New South Wales Dept. Agr. Introd. in 1975. Cambridge Favourite x Tioga; cross made in 1969, selected in 1970.

Fruit: large; slightly wedge shaped; skin firm, gloss medium; flesh firm; a dessert variety; ripens about 3 weeks before Torrey. Plant: small; dense, low growth habit; vigor moderate; productivity high, about 40 oz per plant; moderately resistant to powdery mildew, susceptible to leaf spot.

Narcissa.—Orig. in Glenn Dale, Md., by George M. Darrow, U.S. Dept. of Agriculture Introd. in 1933. Royal Sovereign x Howard 17; selected in 1925. Fruit: high dessert quality; ripens early; resistant to fruit rots in the Northwest. Plant: high yields.

Nared.—Orig. in Narara, New South Wales, by Hort. Res. Sta., Dept. of Agr., New South Wales. Introd. in 1964. Kendan x unnamed seedling (N56/7). Cross made in 1962, selected in 1964. Fruit: size medium; regular conic; skin tough, highly attractive; flesh firm, flavor good; recommended for fresh market; ripens September to December. Plant: size medium; low, rounded; vigor, hardiness and productivity moderate; tolerant to common leaf diseases; seldom planted because of high susceptibility to leaf blight (*Dendrophoma obscurans*).

Nectarena.—Orig. in Thornhill, Ontario, Canada, by Matthew James Johnson. Introd. in 1947. Plant patent 780; 20 Jan. 1948; assigned to the R.M. Kellogg Company, Three Rivers, Mich. Beaderarena x Howard 17; selected about 1940. Fruit: flavor sweet and mild, similar to Beaderarena. Plant: vigor and productiveness similar to that of Howard 17.

Nedspride.—Orig. in Tehama County, Calif., by H.A. Johnson, Jr. and H.E. Thomas. Plant patent 2796; 13 Feb. 1968; assigned to Driscoll Strawberry Associates, Watsonville, Calif. (The Strawberry Inst. of California Selection C72.26 = Cal. 677.10 x Cal. 752.46) x Goldsmith. Fruit: large, dropping in size during the season, primary berries irregular wedge to conic, secondary and tertiary berries globose conic to conic; smooth, firm with high gloss; surface color carmine; firm and juicy. Plant: smaller and more open than Lassen; dense; extensive root system; tolerant to verticillium wilt, susceptible to red stele, two-spotted mite, and powdery mildew.

New Beauty.—See Kellogg Beauty.

New Victoria.—Orig. in Saanichton, Vancouver Island, British Columbia, Canada, by Canada Dept. Agr., Res. Sta. Introd. in 1930. Royal Sovereign x Magoon; selected in 1926. Fruit: large; conical; skin bright red; achenes prominent; flesh firm. Plant: vigorous; productive; flower perfect. By 1937 the yields were poor, probably because of virus infection; propagation discontinued.

New Washington.—See Washington.

Nil.—See Kendall.

Nisqually.—Orig. in Puyallup, Wash., by C.D. Schwartze, Western Washington Res. and Ext. Center, Washington State Univ. Introd. in 1967. WSU 615 everbearing x WSU 592 noneverbearing; selected in 1957; cross first made in 1942; tested as WSU 916. Fruit: resembles Rockhill; conic, size variable; appearance and quality good; produces moderate crop in June becoming more productive July through September; for gardens and local market. Plant: everbearing; flower buds initiated in short and long days; produces good runners; very productive; tolerant to mildew. Named after a western Washington Indian tribe.

Nokomis.—Orig. in Excelsior, Minn., by Univ. of Minnesota Fruit Breeding Farm. Introd. in 1922. Dunlap x Abington; selected in 1918; tested as Minn. 489. Fruit: large; roundish conic; skin light to medium scarlet; flesh light pinkish-red, juicy, mild, sweet; flavor good; recommended for home gardens, fruit too soft for shipment and not attractive canned; midseason. Plant: vigorous; healthy; leaves medium light green; petioles long; drought resistant; productive. No longer propagated.

Northeaster.—An early ripening, red stele-resistant strawberry adapted to the northeastern United States. Orig. in Beltsville, Md. MDUS 4380 x Holiday; cross made by D.H. Scott in 1974; selected in 1976 by A.D. Draper and P. Rosati; tested as MDUS 4787; introd. in 1994 by U.S. Dept. of Agriculture, Beltsville, and New Jersey Agr. Expt. Sta., Rutgers Univ., and Ohio Agr. Expt. Sta., Ohio State Univ. Fruit: large; firm; ripening season similar to Earliglow; recommended for fresh market and processed frozen; strong flavor and aroma, similar to Holiday. Plant: yields on sandy soil similar to Earliglow, on heavier soils yields more than Earliglow; less vigorous than Earliglow on lighter soils and more vigorous on heavier soils. Resistant to five races of red stele causal organism; susceptible to powdery mildew.

Northerner.—Orig. in Parkside, Saskatchewan, Canada, by A.J. Porter, Honeywood Nursery. Introd. in 1955. Sparta x a native wild strawberry; selected in 1940. Fruit: medium to small; characteristics intermediate between the parents. Plant: extremely hardy; everbearing; very productive; vigorous; makes few runners.

Northland.—Orig. in Excelsior, Minn., by A.N. Wilcox, Univ. of Minnesota, St. Paul. Introd. in 1982. Burgundy x Premier. Selected in 1948; tested as Minn. 1868. Fruit: large to medium; conic; flesh glossy, bright red, quality good, flavor mild, soft; skin glossy, bright red, very attractive; ripens midseason. Plant: hardy; vigorous, produces runners freely; moderately susceptible to leaf spot.

Northstar.—Orig. in Glenn Dale, Md., by George F. Waldo and George M. Darrow, U.S. Dept. of Agriculture. Introd. in 1939. Howard 17 x Redheart; cross made in 1928; selected in 1930; tested as USDA 1425. Fruit: large; short, blunt conic to wedge shape; hull size medium, clasping to reflexed; skin light to medium red, glossy, exceptionally attractive; achenes small, yellow, slightly raised to even with surface; flesh light red, juicy, firm, tender, flavor tart, aromatic, usually without a cavity; dessert, shipping, and processing quality very good; ripens evenly, early to midseason, about with or slightly before Fairfax; suitable for commercial and home garden use. Plant: very large, in a thin matted row; produces relatively few runners; leaf large, deeply serrated, free from spot and scorch.

Northwest.—Orig. in Puyallup, Wash., by C.D. Schwartze, Western Wash. Expt. Sta. Introd. in 1949. Brightmore x a complex hybrid derived from Narcissa, Wilson and one of Etter's cultivars; selected in 1943; tested as Wash. 220. Fruit: large in early pickings, medium size toward end of season; conic to long conic, uniform; skin bright crimson, glossy; flesh light red throughout, firm, subacid, well flavored when sugared; very good for fresh market and commercial freezing, good for canning; ripens midseason to late, 5 to 6 days after Marshall, largely replacing Marshall in Washington and Oregon. Plant: very productive; single crop produced each year, no tendency toward everbearing habit; vigorous; adapted to matted row or hill system of planting; somewhat resistant to virus diseases; susceptible to cyclamen mites, red stele, leaf spot and root rots; grows best on medium light, well-drained, irrigated soil; blooms 5 to 6 days after Marshall. Recommended for areas that can no longer grow Marshall because of yellows virus disease.

Notley.—Orig. apparently in Chatsworth, Calif. Introd. about 1931. Parentage unknown, but presumably one of Albert Etter's varieties was one parent. Fruit: very large; skin bright red, glossy; flesh firm, flavor good, sweet; ripens fairly early. Plant: everbearing. Now obsolete.

Nyoho.—A high-yielding strawberry adapted to forcing under plastic in cool regions of Japan. Orig. at the Tochigi Agr. Expt. Sta., Japan, by F. Honda. Kei 210 x Reiko; cross made in 1979; introd. in 1984. Fruit: relatively small; glossy scarlet color; conic; aromatic. Plant: vigorous; relatively low chilling requirement. Relatively susceptible to powdery mildew; susceptible to fusarium wilt and anthracnose.

Oak Hill Special.—Orig. in New Buffalo, Mich., by Edw. L. Lubke, Oak Hill Nursery. Introd. about 1933. Open-pollinated seedling of Beauty. Fruit: large; round; flesh very sweet. Plant: vigorous, tall.

Ogallala.—Orig. in North Platte, Neb., by Glen Viehmeyer, North Platte Expt. Sta., and in Cheyenne, Wyo., by LeRoy Powers, U.S. Hort. Field Sta. Introd. in 1958. Several varieties as well as the wild Rocky Mountain species, *Fragaria ovalis*, entered into the parentage: (Rockhill x Cheyenne 3) x (Midland x Cheyenne 2); cross made in Winter 1945–46; seedlings planted at North Platte in 1947; tested as NP 47222-1. Fruit: medium to large, carrying a good size through the season; skin rich, dark red, too tender for good shipping; achenes slightly depressed; flesh bright red, firm, tender, sweet, high quality, aroma mild; slight bitterness in overripe fruit during hot weather; processes very well; season starts in late May or early June in North Platte to early November; fruiting continuous (everbearing) except for 3 weeks immediately after the spring crop; fruiting scapes short. Plant: foliage dense, dark green, leathery; very vigorous; highly resistant to leaf spot; extremely hardy; yields high; appears to be well adapted to the northern Great Plains region. Named after a Sioux Indian tribe.

Oka.—Orig. in L'Acadie, Quebec, Canada by S. Khanizadeh, M. Lareau, D. Buszard, and D. Bagnara, Agr. Canada Res. Sta., St-Jeanne-sur-Richelieu, Quebec, and Dept. of Plant Science, Macdonald College, Ste.-Anne-de-Bellevue, Quebec K75-13 (K71-8 x Micmac) x Honeoye; cross made in 1982; selected in 1983; tested as SJ83184-3; introd. in 1992; plant patent pending. Fruit: large; wedge shape; moderately firm; moderately glossy; medium-red external and light-red internal color. Flavor similar to Chambly; midseason ripening; fresh-market use. Plant: high yielding; low vigor with many runners; peduncles erect during flowering, semi-erect as fruit matures. Probably resistant to powdery mildew and leaf scorch; some susceptibility to leaf spot; resistant to some races of the red stele causal organism; high level of winter hardiness; partially tolerant to herbicide terbacil.

Olga Petrova.—Orig. in Guelph, Ontario, Canada, by C.B. Stevenson. Introd. about 1925. Considered to be a seedling of Howard 17. Fruit: small to medium; skin dull scarlet; achenes sunken; flesh bright scarlet, quality poor, too soft for commercial use. Plant: produces runners easily. Variety apparently obsolete.

Olympus.—Orig. in Puyallup, Wash., by Bruce H. Barritt and C.D. Schwartze, Western Washington Res. and Ext. Center, Washington State Univ. Introd. in 1973 Molalla x Columbia; cross made in 1960; selected in 1963; tested as WSU 1142. Fruit: medium large, similar to Northwest but smaller than Hood; short conic; skin bright red, tender; seeds sunken; flesh bright red throughout, flesh firmness equal to Northwest; low susceptibility to *Botrytis* fruit rot; calyx very easily removed, reflexed; ripens with Northwest; very high yield of marketable fruit, recommended for processing industry of Pacific Northwest. Plant: extremely vigorous; prolific producer of branch crowns; shy runner producer; resistant to several races of red stele root rot; virus tolerant; foliage susceptible to mildew.

Orion.—See Enzed Levin.

Orland.—Orig. in Monmouth, Maine, at the Highmoor Farm of the Maine Agr. Expt. Sta. Introd. in 1952. Aberdeen x Howard 17; selected in 1940. Fruit: attractive; resembles Howard 17. Plant: productive; carries Aberdeen-type red stele resistance.

Oshkosh.—Orig. in Oshkosh, Wis., by H.C. Christensen. Introd. about 1935. Considered to be a chance seedling Fruit: skin dark red; flesh firm; ripens several days before Dunlap. Variety now obsolete.

Oso Grande.—Orig. at Univ. of California, Wolfskill Experimental Orchards, near Davis, by R.S. Bringhurst and V. Voth. Parker x Cal 77.3-603; cross made in 1981; selected in 1982; tested as Cal 81.43-605; introd. in 1987; U.S. plant patent 6578, 31 Jan. 1989. Fruit: very large; firmer than Chandler but not as firm as Parker; bright red exterior and lighter red interior than Chandler; blocky to wedge shape, medium conic; center may be hollow; very good, mildly subacid flavor; fresh-market and processing uses. Plant: short-day type producing fruit over an extended period in arid, subtropical climates; production pattern similar to Chandler; yield as much or more than Chandler; very good runner production. Susceptible to common leaf spot; susceptibility to two-spotted mite similar to Chandler; tolerant to viruses occurring in California.

Ourown.—Orig. in Aurora, Waushara County, Wis., by C. Olson. Plant patent 3814; 9 Dec. 1975. Gem x Temple. Fruit: medium to large, ridged and irregular; red near epidermis, pink at core, cavity large; high dessert quality, excellent for freezing. Plant: everbearing; medium to large; vigorous; hardy; large rugged root system; runners light green to red and vigorous.

Ozark Beauty.—Orig. in Westfork, Ark., by J.B. Winn, Winn's Berry Farms. Introd. in 1955. Red Rich x Twentieth Century; selected in 1949. Fruit: long-necked; skin glossy; flavor very mild, sweet; resembles Twentieth Century. Plant: blooms late; very productive; prolific plant maker.

Pajaro.—Orig. at Watsonville, Calif., by R.S. Bringhurst and Victor Voth, Agr. Expt. Sta., Univ. of California, Davis. Introd. in 1979. Plant patent assigned to Regents, Univ. of California, Davis. Sequoia x 63.7-101 (complex parentage). Fruit: larger than that of Tioga, well-shaped; flesh flavor and quality equal or superior to those of Aiko, Tioga, and Tufts, firm, more ascorbic acid than that of Tufts but less than that of Aiko. Plant: smaller than Tioga; leaves less yellow than those of Tioga; as susceptible to red stele (*Phytophyhora fragariae*). Recommended for commercial production in southern California as an early, high-quality, winter-planting variety and as an improved summer-cropping variety for the central coast. Named for Pajaro Valley, Santa Cruz County, Calif.

Pam Ann.—Orig. in Aitkin, Minn., by Walter Honeyager. Introd. in 1965. Parentage unknown but considered to be an open-pollinated seedling of Superfection or Streamliner; discovered in 1960. Fruit: flesh dark red; keeps well; good for freezing and jam. Plant: very hardy; everbearing.

Pandora.—Orig. in East Malling, England, by D. Simpson at Hort. Res. Intl. (Von Humboldt x Redstar) x Merton Dawn; cross made in 1979; introd. in 1989. Flower: pistillate; requires pollinizer; midseason- and late-flowering varieties must be used; no more than four to six rows of Pandora should be planted to one row of pollinizer variety. Fruit: large; relatively firm; regular conic shape; attractive, glossy, orange-red external color; pleasant flavor with succulent, juicy texture. Fresh-market use; in southeastern England, Pandora has unique late fruiting season providing a good fresh-market overlap with day-neutral and everbearing varieties. Moderately resistant to preharvest botrytis rot. Plant: high yielding; very vigorous; moderately resistant to verticillium wilt; highly resistant to powdery mildew; susceptible to leaf spot.

Parish Perfect.—Orig. in Platteville, Grant County, Wis., by Edith Parish. Introd. in 1954. Plant patent 1442; 3 Jan. 1956; licensed to Moundview Nursery, Platteville, Wis. and to Judson Nursery, Bristol, Ind. Parentage unknown; discovered in 1950. Fruit: large; skin medium dark red; flesh nonfibrous; achenes indented in flesh; ripens mid-June to early July. Plant: resembles Dunlap.

Parker.—Orig. in Parkside, Saskatchewan, Canada, by A.V. Franklin. Introd. in 1953. Blakemore x Sparta; selected in 1948. Fruit: medium to large; stems short; flavor very good; resembles Sparta but not as firm. Plant: everbearer; makes runners quite freely for an everbearer; needs little or no winter protection in area of origination.

Parkland.—Orig. in Parkside, Saskatchewan, Canada, by A.J. Porter, Honeywood Nursery. Introd. in 1953. F_2 (Sparta x a native selection); selected in 1950. Fruit: large; flesh red, flavor fair to good; resembles Glenmore; good summer crop in late June to early July and ripens in late August with a good fall crop. Plant: everbearer; very hardy; vigorous; produces well.

Pathfinder.—Orig. in New Brunswick, N.J., by J.H. Clark, New Jersey Agr. Expt. Sta. Introd. in 1938. Howard 17 x Aberdeen; tested as N.J. 35. Fruit: medium to large; skin bright red; brisk subacid, quality good; ripens 3 days after Howard 17. Plant: very resistant to Aberteen strain of red stele in New Jersey.

Paymaster.—See Sparkle.

Pegasus.—Orig. in East Malling, England, by D. Simpson at Hort. Res. Intl. Redgauntlet x Gorella; tested as ES 608; introd. in 1990. Fruit: large; regular conic; softer than Elsanta; glossy; midseason ripening; can become dark; fresh-market use. Plant: yield similar to Elsanta; moderately resistant to verticillium wilt.

Pennholm.—Orig. in Carpenter, Ohio, by Truman A. Penn. Introd. in 1953. Open-pollinated seedling of Twentieth Century; selected in 1951. Plant: everbearer; grows farther south than any known ever bearing variety; resembles Twentieth Century.

Phelps.—Orig. in Durham, N.H., by L.P. Latimer, New Hampshire Agr. Expt. Sta.

Introd. in 1954. King x (Tupper x Fairfax); selected in 1949; tested as N.H. 243. Fruit: skin attractive, medium dark red; flesh red, firm, excellent favor; ripens 2 to 4 days later than Howard 17, season long; resembles Fairfax. Plant: vigorous runner formation; yields very well.

Pineapple.—Orig. in St. Louis, Mich., by Carl Baney of Riverside Gardens. Introd. in 1935. Considered to be a bud mutation of Dunlap; discovered in 1930. Fruit: skin bright, lighter red than Dunlap; ships and keeps well; resembles Dunlap.

Pitt.—See Empire All Red.

Pixie.—Orig. in Parksite, Saskatchewan, Canada, by A.J. Porter, Honeywood Nursery. Introd. in 1947. Open-pollinated seedling of August Beauty; selected in 1943. Fruit: small to medium; medium long conic; skin deep red; quality very good; for dessert and canning. Plant: everbearer; yields heavily; vigorous; spring and fall crops. Variety believed to be obsolete due to winter kill.

Plentiful.—Orig. in Urbana, Ill., by A.S. Colby, Illinois Agr. Expt. Sta. Introd. in 1953. Redstar x Pathfinder; cross made in 1941; selected in 1945; tested as Ill. 41-11; sib of Vermilion. Fruit: very large; skin medium to light red; flesh medium firm, flavor good; for local, fresh market, being too soft for distant shipment; matures in late midseason. Plant: resistant to red stele, leaf spot, and leaf scorch; produces many runners; very high yielder.

Pocahontas.—Orig. in Beltsville, Md.,by the U.S. Dept. of Agriculture and the Virginia Truck Expt. Sta. Introd. in 1953. Tennessee Shipper x Midland; cross made in 1946; selected in 1948; tested as US-3745. Fruit: large, maintaining better size throughout the picking season than Blakemore; blunt conic, fairly uniform; skin medium to deep red, glossy, attractive, tough; achenes slightly sunken; flesh red, medium firm, about as firm, as Blakemore, very tart; dessert quality good; very good for freezing; ripens early, about 1 week after Blakemore. Plant: vigorous; very productive; makes runners freely; has shown no yellow variegation, susceptible to red stele root rot and verticillium wilt; resistant to leaf scorch, partially so to leaf spot in the south; introduced for the northern Blakemore area, from Norfolk, Va., to southern New England and to Missouri.

Porter's Pride.—Orig. in Parkside, Saskatchewan, Canada, by A.J. Porter, Honeywood Nursery. Introd. in 1960. Open-pollinated seedling of F_2 (Sparta x wild species); selected about 1952. Fruit: large; flavor good; sufficiently firm for local market but not for distant shipping. Plant: hardy; vigorous; resistant to mildew; productive under prairie conditions.

Potagold.—Orig. in Plympton, Mass., by E.A. Wright. Introd. in 1956. Chance seedling; discovered in 1949. Fruit: large; quality very good; matures late, 3 days before Redstar. Plant: large, hardy, healthy; yields well.

Prairie Belle.—Orig. in Kipling, Saskatchewan, Canada, by Mrs. E.W. Carey. Introd. in 1934. Parentage unknown. Fruit: size medium; short conic to wedge; flesh firm, light red, meaty, quality fair; season same as Dunlap. Plant: strong grower; prolific plant maker; yields well; appears drought resistant.

Precoce di Romagna.—Orig. in Bologna, Italy, by Enrico Baldini, Inst. di Coltivazioni Arboree, University, Bologna. Introd. in 1971. The final patent number is 891783 (Italy), 24 Dec. 1969, assigned to Zanzi Nursery, Ferlala, Italy. Pocahontas x Blakemore. Cross made in 1963, selected in 1969. Fruit: size medium; conic to long conic; skin deep red, seeds set superficially; flesh dark pink, firm flavor good, sweet, subacid; recommended for fresh market; ripens about 1 week before Gorella. Plant: vigor moderate; productivity high; prostrate growing habit, easily capped.

Prelude.—Orig. in Raleigh, N.C., by G.J. Galletta in cooperation with North Carolina Agr. Res. Serv., Dept. of Hort. Sci. and U.S. Dept. of Agriculture. Introd. in 1980. Titan x NC2967. Selected in Willard, N.C., in 1967 by Galletta, D.H. Scott, and A.D. Draper; tested 1969 to 1979 as NC 3419 by Galletta, J.R. Ballington. Scott, and Draper in North Carolina and Maryland. Berry: slightly larger than Apollo, attractive; flesh medium firm, excellent flavor, high quality; ripens very early, ahead of Earlibelle; tolerant to resistant to leaf scorch, leaf spot, and powdery mildew; susceptible to anthracnose and tolerant to resistant to mite infestation; best adapted to southern Coastal Plain and Piedmont areas; adaptable to matted row or hill culture.

President Roosevelt.—See Madame Lefebvre.

Primetime.—Orig. in Beltsville, Md., by G.J. Galletta, J.L. Maas, and J.M Enns, U.S. Dept. of Agriculture. Introd. in 1995. MDUS 4377 [Sunrise x MDUS 3082 (sibling of Redchief and Guardian)] x Earliglow; cross made in 1978; selected in 1980; tested as MDUS5069. Fruit: very large; attractive and symmetrical; medium in skin firmness and toughness; bright scarlet skin, pink-fleshed; pleasant, mild, lightly aromatic fresh flavor. Plant: resistant to 5 races of red stele, tolerant to most of the leaf and stem diseases and fruit rots except powdery mildew; vigorous and runners well; flowers and fruits in midseason with or slightly ahead of Redchief and Lester; grows well in matted rows or hill culture, on light or heavy soils, tolerates heat.

Prince Albert (*Volunteer*).—Orig. in Parkside, Saskatchewan, Canada, by A.J. Porter, Honeywood Nursery. Introd. in 1952. Open-pollinated seedling of Easy Picker; selected in 1940. Fruit: resembles Easy Picker; skin thin, bruises easily; flesh quality only fair; ripens late June to early July. Plant: hardy; high productiveness under northern prairie conditions. Variety believed to be obsolete.

Protem.—Orig. in Beaverlodge, Ala., Canada, by R.E. Harris, Canada Dept. Agr., Res. Sta. Introd. in 1964. Glenheart x Cheyenne 2; selected in 1957. Fruit: uniform shape; resembles Red Rich. Plant: hardy.

Providence.—Orig. in Long Ashton, England, by D. Wilson. Introd in 1984; Gorella x ES 318 (Cambridge Favourite x NY 844). Fruit: medium size; firm; bright red exterior and mid-red interior; conic; calyx very easily removed; flavor superior to Cambridge Favourite; ripens midseason; excellent shelf life, about twice that of Cambridge Favourite; fresh-market use; excellent jam product. Plant: vigorous; erect growth habit; fruit moderately well displayed on medium-length scapes.

Puget Beauty.—Orig. in Puyallup, Wash., by C.D. Schwartze and Arthur S. Myhre, Western Washington Expt. Sta. Introd. in 1956. US-Ore. 1765 (includes Marshall, Ettersburg 121, Howard 17, Royal Severeign) x Sparkle (Aberdeen x Fairfax); cross made in 1946; selected in 1948; tested as Wash. 518. Fruit: large; conic to long conic, sometimes slightly necked; hull large; skin light crimson, glossy, exceptionally attractive; achenes small, yellow, level with surface; flesh highly flavored, not as firm as Northwest, texture resembles Marshall, light bright red, very sweet, aromatic; quality fine for dessert; produces excellent, aromatic preserves and acceptable frozen products; ripens about with Marshall; easily picked; resembles Sparkle. Plant: large to very large under favorable growing conditions; upright; leaf large, leathery, dark green; runner production moderate; hardier than Northwest; less productive than Northwest; very resistant to powdery mildew; in some locations susceptible to red stele root disease; adapted to moderately heavy, poorly drained soils; blooms after Marshall. Named after the Puget Sound country where it was developed.

Puget Sound Wonder.—See Washington.

Quinault.—Orig. in Puyallup, Wash., by C.D. Schwartze, Western Wash. Res. and Ext. Center, Washington State Univ. Introd. in 1967. WSU 901 (sib. of Nisqually) x Puget Beauty. Cross made in 1958; selected in 1959; tested as WSU 1105. Fruit: very large; conic; skin attractive; flesh soft, color and flavor good; resembles Puget Sound; good for northern gardens; produces moderate crop in June, becoming more productive July through September. Plant: everbearing; flower buds initiated in short and long days; produces good runners. Named after a western Washington Indian tribe.

Radiance.—Orig. at Cheyenne, Wyo., by LeRoy Powers, E.D. Krouch, and D.H. Scott, U.S. Dept. of Agriculture, Hort. Field Sta. Introd. in 1954. Montana x (Early Cheyenne 1 x Cheyenne 430-2); Cheyenne 430-2 is (*Fragaria ovalis* x Fairfax) x Fairfax. Fruit: size medium skin deep red; flesh firm; suitable for freezing. Plant: everbearing; hardy at Cheyenne; vigorous; recommended for Colorado and Wyoming.

Rainier.—Orig. in Puyallup, Wash., by B.H. Barritt and C.D. Schwartze, West Washington Res. Ext. Center, Washington State Univ. Introd. in 1972. WSU 685 (Northwest x Sierra) x Columbia. Cross made in 1962 by C.D. Schwartze, selected in 1965, tested as WSU 1232, sister seedling of Shuksan. Fruit: large throughout season; wedge-conic; skin bright red, slightly sunken yellow seeds; flesh firmer than Northwest, bright red throughout; calyx large with narrow pointed sepals, appressed and difficult to remove, not suited for mechanical calyx removal; ripens late, 2 to 3 days later than Northwest and Shuksan; makes outstanding frozen slice pack and preserves, also suited for fresh-fruit market, with good shelf life. Plant: vigorous; productivity slightly lower than Shuksan and Northwest, similar to Hood, less hardy than Shuksan, more so than Northwest; petioles long and twisting, giving illusion of wilting; produces fewer runners than Hood, Shuksan, or Northwest; tolerant to powdery mildew and red-stele root rot (not all races); more susceptible to *Botrytis* fruit rot than Shuksan but not as susceptible as Northwest; recommended for processing and fresh market in western Washington, Oregon and southwestern British Columbia.

Ralph.—Orig. in Ottawa, Ontario, Canada, by Canada Dept. Agr., Res. Sta. Introd. in 1931. Superb x Westney. Fruit: size medium; conical; quality good; ripens early. Plant: yields consistently; crops large; fairly drought resistant; flowers perfect. No longer propagated.

Ranger.—Orig. in Winter Haven, Texas, by the Texas Agr. Expt. Sta. Introd. in 1937. Texas x Missionary; selected in 1935; tested as Texas. 68. Fruit: large; often has cavities, colors evenly; quality comparable to Klondike; ripens early; before Klondike. Plant: foliage covers flowers and fruits well; heavy plant producer; adapted to southern U.S. conditions; vigorous.

Raritan.—Orig. in New Brunswick, N.J., by L. Fredric Hough, New Jersey Agr. Expt. Sta. Introd. in 1968. Redglow x Jerseybelle. Cross made in 1955; selected in 1957; tested as N.J. 857. Fruit: large; skin glossy, bright red, with yellow achenes partially embedded, calyx large, showy and stays green; as attractive as Jerseybelle; flesh firm, quality very good, color acceptable for fresh market but not for processing; ripens midseason. Plant: produces runners freely, making a good matted row; productive; susceptible to red stele root rot and verticillium wilt.

Red Cluster.—See Armore.

Red Giant.—Orig. in Monticello, Minn., by Marion Hagerstrom. Introd. in 1968. Plant patent 2884; 6 May 1969; assigned to Stark Bros. Nurseries & Orchards Co., Louisiana, Mo. Red Rich x Midland. Cross made in 1959, selected in 1961. Fruit: large; wedge shaped; skin glossy, bright red, attractive; achenes depressed; flesh red, firm, texture fine, juicy, flavor excellent; ripens midseason 15 to 25 June; for fresh fruit or processing. Plant: large; vigorous; hardy; productive.

Red-Glo.—See Red Rich.

Red Gold.—Orig. in Osage, Iowa, by Clark E. Gardner, Gardner Nursery Co. Introd. about 1925. Unknown seedling x Howard 17. Fruit: large; conical, shape irregular; skin bright red; flesh firm, red flavor mild; ripens early about 7 days before Howard 17. Plant: large, vigorous. Variety now obsolete.

Red Rich (*Red-Glo, Hagerstrom Red-Rich, Hagerstrom's Everbearing*).—Orig. in Enfield, Minn., by Marion Hagerstrom. Introd. in 1949. Plant patent 993; 28 Nov. 1950; assignor of one-half to Carl R. Hagerstrom, Enfield. Wayzata x Fairfax; selected in 1940. Fruit: large, larger than Wayzata but smaller than Fairfax; conical wedge, contour irregular; skin an attractive rich red, glossy; achenes conspicuous, flush with surface; flesh dark red, firm, juicy, mildly acid with a sweet, rich, tart flavor; good for dessert; keeps well; freezing quality good; resembles Fairfax. Plant: everbearer; runners produced freely, but in some areas (e.g., eastern Canada) few develop; flower stems thick, generally upright; productive; hardy; resistant to leaf spot, leaf scorch; adapted to northern states.

Red Sugar.—Orig. in Ettersburg, Calif., by Albert F. Etter (1872 to 1950). Introd. about 1924. Open-pollinated seedling of

Ettersburg 80. Fruit: excellent canning quality. Used primarily as a breeding stock.

Red Titan.—Orig. in Stanton, Montcalm County, Mich., by Floyd Hartman. Introd. in 1961 by Judson Nurseries, Bristol, Ind. Plant patent 1848; 23 June 1959; assigned to John H. Rhoades and Evelyn E. Rhoades, Lansing, Mich. A natural cross of Howard 17 x Robinson or a mutation of Howard 17; discovered in 1953. Fruit: large; elongated, conical; skin dark red, glossy; flesh red, core very small, high flavor but tart; ripens in early midseason, with Howard 17; good for processing, freezing, fresh market. Plant: vigorous; leaflets darker green and broader than Howard 17; blooms less freely than Howard 17; good runner production. No longer being propagated.

Redchief.—Orig. in Salisbury, Md., by Maryland Agr. Expt. Sta., College Park, and U.S. Dept. of Agriculture, Crops Res. Div., Beltsville. Introd. in 1968. N.C. 1768 x Surecrop; cross made in 1958; selected in 1960 by I.C. Haut and D.H. Scott of the above stations; tested as Md-US 3068. Fruit: medium to large; cone shaped; skin deep red, glossy; flesh firm, flavor subacid; good dessert quality maintained in frozen products; secondary and later berries have slight necks; calyx is easily removed; ripens 2 to 3 days later than Surecrop. Plant: vigor medium; runners produced freely, but not so abundantly as Surecrop; tolerant to five races of red stele root rot and intermediate in resistance to verticillium wilt; leaves tolerant to leaf scorch and mildew, but susceptible to leaf spot; productive in Maryland, New Jersey, Massachusetts, Michigan, Ohio, Missouri, and southern Illinois; unproductive and unattractive in North Carolina and Arkansas; recommended where red stele root rot may be a problem and where Surecrop is not productive.

Redcoat.—Orig. in Ottawa, Ontario, Canada, by Canada Dept. Agr., Res. Sta. Introd. in 1957. Sparkle x Valentine; cross made in 1948; selected in 1950; tested as Ottawa 487. Fruit: good commercial size throughout the harvest season; skin glossy, light red, highly attractive; flesh firm, flavor acceptable, color rather light; acceptable for freezing, good for jam; ripens in early midseason, after Cavalier. Plant: vigorous, tall; runners produced freely; virus-free; produces high yields; susceptible to verticillium wilt and leaf spot, fairly resistant to powdery mildew.

Redcrest.—Orig. in Corvallis, Ore., by F.J. Lawrence; released jointly by U.S. Dept. of Agriculture–Agr. Res. Service, Oregon State Univ., and Washington State Univ. Linn x Totem; selected in 1976; tested as ORUS 4930; introd. in 1990. Fruit: slightly smaller than Totem; very firm but not tough; uniform conic; bright, uniform red external and internal color; calyx easy to remove; late, concentrated ripening season; acidic flavor; processing use. Plant: relatively high yielding; fairly vigorous. Moderately susceptible to powdery mildew and leaf blight; intermediate level of field resistance to red stele; less tolerant than Totem and Benton and more tolerant than Hood to the aphid-borne virus complex in the Pacific Northwest.

Redcrop.—Orig. in New Brunswick, N.J., by J.H. Clark, New Jersey Agr. Expt. Sta. Introd. in 1949. Aberdeen x Fairfax; tested as N.J. 377. Fruit: large; flesh medium firm, red throughout, quality good; very good for frozen pack; attractive calyx. Plant: vigorous; productive; resistant to Aberdeen strain of red stele in New Jersey.

Redgauntlet.—Orig. in Auchincruive, Ayr, Scotland, by Robert D. Reid, Scottish Hort. Res. Inst. (West of Scotland Unit). Introd. in 1957. New Jersey 1051 x Auchincruive Climax; cross made in 1946; tested as 6J27 and Auchincruive 25. Fruit: medium to large or very large, maintaining size throughout season; round conical, symmetrical but occasionally misshapen; skin scarlet to crimson, darkening if foliage scarce, not easily bruised; achenes yellow to red, even and regular, sunken but not pitted; plug small and soft, remains in fruit when calyx removed; little or no cavity; flesh white suffused red, juicy, subacid; ships well; ripens 1 week earlier than Talisman or Climax. Plant: vigorous; upright to spreading; open habit; foliage medium to dark green; resistant to some races of red core root rot (*Phytophthora fragariae*).

Redgem.—Orig. in Corvallis, Ore., by F.J. Lawrence, M.M. Stahler, P.P. Moore, L.W. Martin, G.W. Varseveld, and W.A. Sheets;

released jointly by U.S. Dept. of Agriculture–Agr. Res. Service, Oregon State Univ., and Washington State Univ. Benton x ORUS 3596 (Earlibelle x ORUS 2853); selected in 1972; tested as ORUS 4459; introd. in 1993. Fruit: smaller and less firm than Totem; uniform, blunt conic; bright, uniform red external and light-red internal color; calyx nonreflexed and easy to remove; mid-late ripening season; slightly more susceptible to preharvest botrytis rot than Totem; good flavor; local fresh-market and processing, especially individual quick-frozen, uses; suited to machine harvesting. Plant: yield similar to Totem or slightly less; vigorous; abundant runners; peduncle moderately erect; flowers remain in canopy. Moderately susceptible to powdery mildew; resistance to red stele similar to Totem; some tolerance to the aphid-borne virus complex in the Pacific Northwest.

Redglow.—Orig. in Beltsville, Md., by D.H. Scott and George M. Darrow, U.S. Dept. of Agriculture. Introd. in 1956. Fairland x Tennessee Shipper; raised in 1948; selected in 1949; tested as US-4194. Fruit: large, maintaining large average size during the picking season; shape uniform; skin vivid red, deeper than that of Blakemore, very glossy, handsome; yellow achenes about even with surface; flesh a uniform red, moderately firm, more so than Fairland and Sparkle but less so than Stelemaster; flavor subacid, very good; freezing quality very good; matures 3 to 4 days later than Midland and about with Blakemore and Howard 17. Plant: very vigorous, producing runners freely; susceptible to leaf spot; resistant to the common race of red stele root-rot fungus; moderately productive, being most productive in narrow matted rows; tested from Pennsylvania to North Carolina and westward to Indiana and Kentucky. Suggested for trial plantings in Maryland and the middle Atlantic coast states as an early red-stele-resistant variety of about the Blakemore and Howard 17 season.

Redheart.—Orig. in Glenn Dale, Md., by George M. Darrow, U.S. Dept. of Agriculture Introd. in 1932. Portia x Euresko; cross made in 1923; selected in 1925; tested as USDA 632. Fruit: large to medium; long, conic to blunt conic, irregular; hull large, reflexed; skin dark red, firm; achenes yellow to red, raised above surface; flesh rich red, very firm, juicy, rarely with cavity, acid, aromatic; dessert quality very high, canning and freezing quality very good; season fairly early, before Ettersburg 121 in the Pacific Northwest, after Howard 17 in the northern states. Plant: productive; quite vigorous; foliage relatively susceptible to leaf scorch; flower perfect.

Redlands Horizon.—A fresh-market strawberry for the coastal southeastern region of Queensland, Australia. Orig. Queensland Dept. of Primary Industries Hort. Centre, Redlands Res. Sta., by M. Herrington, P. Brown, and S. Prytz. Earlisweet x Pajaro; cross made in 1985; tested as 24/86; introd. in 1993. Fruit: large size in early season to medium later; glossy red external and paler red internal color; conic to globose shape; early to midseason ripening; pleasant, slightly acidic flavor. Plant: small; flat, open habit allows good spray penetration; fruit well-displayed; sparse runner production. Tolerance to leaf blight similar to Earlisweet.

Redstar.—Orig. in Glenn Dale, Md., by George M. Darrow and George F. Waldo, U.S. Dept. of Agriculture. Introd. in 1940. Chesapeake x Fairfax; cross made in 1931 by Waldo; selected in 1933; tested as USDA 2124. Fruit: very large; short conic with wide shoulder, sometimes ridged; hull reflexed to clasping, medium to very large; skin bright medium red, not darkening, very glossy, tough, nearly as attractive as Starbright; achenes yellow, raised above surface; flesh light bright red, firm, juicy, rather tart until quite ripe; tender, aromatic; dessert quality very good; shipping quality good; ripens late, later than Gandy, Chesapeake, and Lupton; a commercial and home garden variety; resembles Fairfax. Plant: very large; quite productive; produces runners freely; leaf deep green, glossy, cupped, free from leaf spot and leaf scorch; tolerant to virus diseases; flower perfect; one of the better late varieties grown from Maryland to southern New England and west to Missouri and Iowa.

Redwing.—Orig. in New Brunswick, N.J., by J.H. Clark, New Jersey Agr. Expt. Sta.

Introd. under restrictions in 1940; introd. and named in 1943. N.J. 46 (Pearl x Aberdeen) x Fairfax; selected in 1933; tested as N.J. 347. Fruit: very large; skin light red; flesh firm; good for frozen pack; ripens late but before Julymorn. Plant: very large.

Regina.—Orig. in Müncheberg, Mark, Germany, at the Kaiser Wilhelm Institut by the Max Planck Institut f. Züchtungsforschung (Friedrich Gruber), Köln-Vogelsang. Introd. in 1951 in Germany and somewhat later in the U.S. and Canada. German patent 645 300; assigned to Christian Fey, Meckheim. Amerikanischer Samling II (Geneva) x Deutsch Evern; selected in 1936. Fruit: size medium; skin glossy, light to medium red; ripens early, about 2 days before Deutsch Evern; achenes only slightly depressed; broad; flesh firm, pink to light red, slightly acid, aromatic; not useful as processed fruit but excellent as fresh fruit. Plant: vigor medium; habit spreading; foliage yellow-green, petioles and young leaves pubescent; some tolerance to virus; susceptible to mildew.

Rena.—Orig. in Ettersburg, Calif., by Albert F. Etter (1872 to 1950). Introd. in 1920. Wild beach strawberry x Ettersburg 84. Fruit: small; roundish; skin glossy; flesh whitish, juicy, moderately firm, sweet; quality fair; ripens late. Now obsolete.

Repita.—Orig. in Wageningen, Holland, by the Inst. of Hort. Plant Breeding. Introd. in 1960. Ada Herzberg x Auchincruive Climax; selected in 1956. Fruit: size medium; conical; skin bright red; flesh fairly firm, flavor good, juicy. Plant: everbearing; moderately vigorous; inflorescences fairly long; flowers abundant; rather susceptible to *Diplocarpon earliana*.

Revada.—Orig. in Wageningen, Holland, by the Inst. of Hort. Plant Breeding. Introd. in 1960. Auchincruive Climax x Ada Herzberg; selected in 1956. Fruit: size medium; conical; skin light red; acid mild. Plant: everbearing; vigorous, spreading; healthy; good cropper; sensitive to *Botrytis cineria*; inflorescences borne just above leaves.

Reward.—Orig. in Vancouver, British Columbia, Canada, by D.M. McPherson. Introd. in 1939. Chance seedling; selected in 1935; registered with the Canadian Hort. Council under no. 199. Fruit: large; skin dark red; fine flavor; exceptionally easy to hull; ships well. Plant: heavy cropper; produces many runners. Apparently no longer in existence.

Rhapsody.—A high-yielding, mid-late-season strawberry adapted to Northern England and Scotland. Orig. in Invergowrie, Scotland, by R.J. McNicol, Scottish Hort. Res. Inst. SCRI 61G51 (Talisman x Cambridge Vigor) x Cambridge Favorite; cross made by H.J. Gooding at Auchincruive in 1969; tested as SCRI 69EW30; introd. in 1988; entered for Plant Breeders' Rights. Fruit: medium to large size; firm; glossy red exterior and solid red interior color, tip tends to remain white; conic; calyx removal moderately difficult, tends to plug; juicy; good flavor; late fresh market. Plant: consistently higher yielding than Cambridge Favourite; erect habit with medium to strong vigor; pale-green crinkled foliage with interveinal lighter green areas; moderate runner production. Field resistance to red stele; moderately resistant to verticillium wilt.

Ridgely.—Orig. in Greenwood, Del., by W.S. Todd. Introd. in 1928. Chance seedling. Fruit: large; shape uniform; skin dark glossy red; flesh firm, quality good; ripens about 3 days later than Howard 17; resembles Gandy. Plant: vigorous; productive.

Riogrande.—Orig. in Winter Haven, Texas, by the Texas Agr. Expt. Sta. Introd. in 1927. Blakemore x Ettersburg 80?; selected in 1935; tested as Tex. 15. Fruit: flavor good; season early; resembles Blakemore. Plant: good plant producer, adapted to southern U.S. winters.

Riverview.—See Campbell.

Robinson (*Kardinal King, Deet, Scarlet Beauty*).—Orig. in Sister Lakes, Mich., by John C. Haley. Introd. about 1940. Howard 17 x Washington; cross made in 1932; selected in 1934. Fruit: medium to large; conic; skin red, showy; flesh light red, soft, flavor mild; not adapted to freezing; replaced by firmer, better flavored varieties Plant: small; vigorous; very productive; makes runners very freely; partially resistant to leaf spots, susceptible to leaf scorch, tolerant to virus diseases.

Robyn.—Orig. in Pretoria, South Africa, by E.P. Evans, Hort. Res. Inst. Introd. in 1966 [Senga 752 x 1/67/22] x Festival; selected in 1962. Fruit: large; skin dark ruby-red with bright finish, durable; achenes large and exposed; flesh dark red and evenly colored; flavor pleasant during cooler months of June to August, but fairly acid during summer months; ripens June to mid-December. Plant: size medium; moderately vigorous, prolific runner production; tolerant to leaf spot and slightly susceptible to leaf scorch; for home garden and local market.

Rock of Ages.—Orig. in Pennsauken Township, Camden County, N.J., by Oscar Earle Felten, East Merchantville, N. J. Introd. in 1951. Plant patent 1005; 27 Feb. 1951. The result of inbreeding of unnamed Felten seedling; selected in 1941. Fruit: large; skin bright red; flesh firm, juicy, subacid; resembles Howard 17 and Dorsett; ripens the first week in June at place of origin. Plant: vigorous; upright; productive; self-pollinating; flowers perfect.

Rockhill (Wayzata, Wazata).—Orig. in Conrad, Iowa, by Harlow Rockhill. Introd. in 1923 by the R.M. Kellogg Co., Three Rivers, Mich. Early Jersey Giant x Progressive; cross made in 1916; selected in 1918; tested as Rockhill 26. Fruit: large; globose conic to wedge shaped, irregular; skin red to dark red; flesh firm, subacid, quality very good; ships well. Plant: large; vigorous; everbearer; makes very few runners; may be propagated by crown division; foliage dark green, healthy; grown in Oregon, Minnesota, Iowa and other northern states.

Rosanne.—Orig. in Raleigh, N.C., by North Carolina Agr. Res. Serv., Dept. of Hort. Sci., and U.S. Dept. of Agriculture. Introd. in 1980. Self-pollination of hybrid selection NC 3140 in 1967 by G.J. Galletta; selected in Willard, N.C., in 1969; tested as NC 3645 in North Carolina and Maryland from 1971 to 1979. Berry: large as Apollo,as firm and tough as Earlibelle; symmetrical, pale colored until it receives ample sunlight and heat; rot resistant; flavor unique with rose-like aroma; ripens very late, with Catskill in Coastal Plains and with Tennessee Beauty in mountains. Plant: vigorous; runner moderate to freely; yields as well as Apollo from first or second year, beds grown in either hill or matted rows; resistant to leaf spot, leaf scorch, powdery mildew, anthracnose and mites.

Rossella.—Orig. in Bologna, Italy, by Enrico Baldini, Inst. di Coltivazioni Arboree, University, Bologna. Introd. in 1971. The final patent number is 891780 (Italy), 24 Dec. 1969, assigned to Zanzi Nursery, Ferlala, Italy. Open-pollinated seedling of Paymaster. Selected in 1969. Fruit: size medium; oblate, regular; skin shiny red, seeds raised, flesh firm, bright red, juicy, subacid; recommended for fresh market; ripens about 5 days before Gorella. Plant: moderately vigorous, hardy; productive; long, upstanding runners, making capping difficult.

Royalty.—Orig. in Grandview, Ind., by Alben B. Cook. Introd. in 1978 by Herschel L. Boll, Champaign, Ill. Ozark Beauty x Tennessee Shipper. Cross made in 1965. Berry; large, mostly long wedge; skin medium dark red, glossy, attractive; flesh firm; ripens late. Plant: highly productive, vigorous and large; produces many runners.

Saanich Belle.—Orig. in Saanichton, British Columbia, Canada, by Canada Dept. Agr., Res. Sta. Introd. in 1934. Magoon x Royal Sovereign; selected in 1930. Plant: heavy cropper; suited to heavy soils but lacked quality for commercial production. No longer propagated.

Saito.—See Empire All Red.

Salinas.—Orig. in Davis, Calif., by R.S. Bringhurst and Victor Voth, Agr. Expt. Sta. Introd. in 1967. Cal 37.2-11 x Cal 42.17-18; selected in 1959 in Davis; tested as 57.4-1. Fruit: large; conic primaries tend to furcate; skin red with bright finish, durable; achenes small and flush with surface; flesh light to bright red, firm, flavor excellent. Plant: size medium; productive; foliage dark green, leaves up-cupped; flower stocks long; runner production profuse; resistant to most strains of verticillium wilt; susceptible to bacterial collapse (*Xanthomonas fragaria*). Recommended for winter and summer planting in central coastal California.

Santana.—Orig. in South Coast Field Sta., Santa Ana, Calif. by Victor Voth and R.S. Bringhurst, Agr. Expt. Sta., Univ. of California, Davis. Introd. in 1983. Plant patent assigned to Regents, Univ. of California, Davis. Cal. 72.309-601 [64.57-108 (Sequoia x Tioga) x Tufts] x Cal 72.359-605 (64.57-108 x Tufts) Cross made in 1977; first fruited and selected in 1979; tested as Cal 77.36-604 and C6. Fruit: as large as Tioga; medium to short conic; skin slightly darker than Tioga, glossy. and attractive; flesh about the same color as skin with a lighter ring around the core, firm and durable as Tioga, Tufts, or Aiko; internally slightly hollow; earliest fruiting of all California short-day type cultivars including Douglas. Plant: erect; about the same size as Tioga; self-fertile; suitable for fresh market and processing. Named after the town of Santa Ana, Calif.

Scarff.—Orig. in New Carlisle, Ohio, by W.N. Scarff, Scarff & Sons Nursery. Introd. in 1924. Chance seedling; selected about 1915 to 1917. Fruit: large; quality high; mildly subacid; resembles Robinson in size and quality. Variety now obsolete.

Scarlet Beauty.—See Robinson.

Schwab.—Orig. in Mankato, Minn., by F.M. Schwab. Introd. in 1933. Parentage unknown; discovered in 1930. Fruit: quality good; keeps well; resembles Dunlap. Plant: bears well.

Scotland.—Orig. in Vineland Sta., Ontario, Canada, by C.L. Ricketson. Guardian x V6747R-6 (Veestar x NY 844); cross made in 1972; selected in 1974; tested as V7251-1; introd. in 1991 by A. Dale, Hort. Res. Inst. of Ontario. Fruit: large; very firm; bright, medium-pale external and pale internal color; flavor moderately acidic; ripens as late as Bounty; can be decapped more effectively than Midway on CML decapper; can be machine-harvested; processing use. Plant: moderately productive; vigorous; adequate runner production; resistant to leaf scorch.

Scott.—Orig. at Wye Inst., Eastern Shore of Maryland, by D.H. Scott and A.D. Draper, U.S. Dept. of Agriculture and Maryland Agr. Expt. Sta., College Park. Introd. 1979 by Galletta et al. Sunrise x Tioga. Cross made in 1969; selected in 1971; tested MdUS 4376. Berry: attractive; large; symmetrical, short wedge to conic; skin tough, resists bruising, well-colored; flesh red, firm, flavor mild, freezes well; ripens with Guardian. Plant: medium large; vigorous; hardy; prolific runner ability, makes better plant bed than Guardian or Delite; more productive than Guardian and equal to Delite; resistant to five races of red stele, powdery mildew, and leaf scorch; partially resistant to verticillium wilt; adapted to Massachusetts, Michigan, North Carolina, and Missouri. Named for Donald H. Scott, former head of Small Fruit Investigations of U.S. Dept. of Agriculture.

Seascape.—Orig. at Univ. of California, Wolfskill Experimental Orchards, near Davis, by R.S. Bringhurst and V. Voth. Cross made in 1983; selected in 1984; tested as Cal 83.49-1 and later as advanced selection CN 49; introd in 1991. U.S. plant patent 7614, 6 Aug. 1991. Fruit: usually as large or larger than Selva and almost as firm; medium to occasionally long conic; bright red external and internal color; achenes bright yellow to slightly reddish, and flush to slightly exerted from surface; good flavor, superior to Selva; fresh-market, processing, and home-garden uses. Plant: compared to Selva, somewhat less strongly day-neutral; less productive; slightly lower chilling requirement; equal or better runner production; similarly erect habit. Moderately susceptible to common leaf spot; somewhat susceptible to two-spotted spider mite; tolerant to viruses occurring in California.

Secord.—Orig. in Vineland Sta., Ontario, Canada, by C.L. Ricketson. Guardian x Holiday; cross made in 1972; selected in 1974; tested as V7236-3; introd. in 1985 by A. Dale, Hort. Res. Inst. of Ontario. Fruit: large; firm; bright, medium-red external and relatively pale internal color; flavor moderately sweet, aromatic; ripens midseason; fresh-market use. Plant: high yielding; moderately vigorous; moderate runner production; resistant to verticillium wilt. Variety recalled from cultivation.

Selekta—Orig. in Roodeplaat, Pretoria, South Africa, by Edward P. Evans, Hort. Res. Inst. Introd. in 1972. Earlibelle x Torrey. Cross made in 1967, selected in 1968, tested as 68-41. Fruit: very large; wedge shaped; skin attractive, bright red; flesh bright red with white near center, flavor good during cool months, but best during summer months, winter crop is mostly hollow; keeping quality good; recommended for fresh fruit and processing; ripens June to December. Plant: large to very large; vigorous; upright; abundant runner production; center leaflet slightly longer than broad with acute leaf angle; slightly susceptible to *Diplocarpon earliana*.

Selkirk.—Orig. in Vineland Sta., Ontario, Canada, by C.L. Ricketson. Earlibelle x Holiday; cross made in 1972; selected in 1974; tested as V7210-5; introd. in 1991 by A. Dale, Hort. Res. Inst. of Ontario. Fruit: moderately large to large; very firm; very bright, medium-red external and moderately pale internal color; flavor moderately sweet; ripens early to midseason; processing use, especially as frozen product. Plant: moderate yield; vigorous; adequate runner production. Moderately resistant to leaf spot; susceptible to leaf scorch and verticillium wilt; very susceptible to powdery mildew.

Sella.—A high-yielding, fresh-market strawberry with firm fruit. Orig. near Comacchio (Ferrara), Italy, by M. Leis, D. Musacchi, A. Martinelli, Consorzio Italiano Vivaisti. Addie x Pajaro. Cross made in 1984; introd. in 1992. Fruit: large; intense bright red external and red to orange internal color; some hollowness; heart shape with pointed apex; large calyx; achenes beneath surface; ripens 4 days after Addie; good flavor with medium sweetness and acidity. Plant: vigorous; moderate runner production; moderately upright growth habit; flowers at level of leaf canopy.

Selva.—Orig. in Davis, Calif. by R.S. Bringhurst and Victor Voth, Agr. Expt. Sta., Univ. of California. Introd. in 1983. Plant patent assigned to Regents, Univ. of California, Davis. Cal 70.3-117 (sister of Brighton) x Cal. 71.98-605 (Tufts x 63.7-101 (parent of Pajaro). Cross made in 1975; first fruited in 1976; selected in 1977: tested as Cal 75.71-107 and CN18. Fruit: larger than Tioga: medium to long conic with some flat and wedge-shaped; internally somewhat hollow; skin darker than Tioga, bright, and attractive: flesh same color as skin with somewhat lighter ring around the core, firmer than Tioga, Tufts, Aiko, Pajaro, or any of the other day-neutrals. Plant: semierect, larger than Hecker: for winter and summer plantings as well as at times when short-day plants will only produce runners; fruits about 3 months after planting. Named for La Selva Beach near Santa Cruz, Calif.

Seneca.—Orig. in Geneva, N.Y., by K. Maloney, D.K. Ourecky, J.D. Reich, and J.C Sanford, New York State Agr. Expt. Sta. NY 1261 (Redcoat x NY 844) x Holiday; cross made in 1974; selected in 1976; tested as NY 1529; introd. in 1992. Plant patent applied for; assigned to Cornell Res. Foundation. Fruit: large; blunt primaries and globose secondaries; very firm; medium-red external and light-red translucent internal color; pleasant aromatic flavor; late midseason ripening; fresh market and processing for jam and freezing. Plant: high yield; good vigor; abundant runners; susceptible to red stele, black root rot, and verticillium wilt; winter hardy in northeastern United States.

Senga Sengana.—Orig. in Hamburg, Germany, by Reinhold von Sengbusch. Introd. in Europe in 1954 and in the U.S. in 1960 as P.I. 264680. German patents 920279 and 947271; assigned to the originator. Markee x Sieger; selected in 1942. Fruit: large, maintained into later pickings; skin red, attractive; flesh firmness medium; freezing quality good; calyx comes free easily, frequently remaining on plant. Plant: very high yielding; healthy; runners begin to form in late August. Imported by the Northwestern Washington Expt. Sta., Washington State Univ., Mount Vernon.

Sentinel.—Orig. in Raleigh, N.C., by North Carolina Agr. Res. Serv., Dept. of Hort. Sci. and U.S. Dept. of Agriculture. Introd. in 1980. Titan x NC 2967. Cross made in 1965 by G.J. Galletta; selected in Willard, N.C., in 1967; tested as NC 3418 in North Carolina and Maryland, from 1969 to 1979. Berry: as large, firm, well-colored, and flavorful as Titan, but more attractive, more symmetrical and tougher skinned; tolerant of field rot; matures late with Apollo in Coastal Plain and with Tennessee Beauty in mountains. Plant: large and vigorous; runners moderate; resistant to leaf spot and leaf scorch; tolerant to powdery mildew; susceptible to anthracnose and mites;

should be grown in thin matted rows in Coastal Plain, but can be grown in hills or rows in mountains; has tendency to yield less on second-year than on first-year beds; produces slightly less than Apollo in Coastal Plain, but as well as Apollo and Tennessee Beauty in mountains.

Sequoia.—Orig. in Univ. of California Field Sta., Santa Ana, by Royce Bringhurst and Victor Voth, California Agr. Expt. Sta. Plant patent 3178; 16 May 1972; assigned to the Regents of the Univ. of California, Berkeley. Introd. in 1968. Hybrid between Cal 52.16-15 (a sister of the Wiltguard variety and the only parent of the Aliso variety) and Cal 51sl-1 (selected from a first generation selfed population from the Lassen variety); cross made in 1956; selected in 1958; tested as 56.124.12. Fruit: exceptionally large; long conic; skin dark red, tender, smooth, attractive; must be harvested at frequent intervals during hot periods or finish will become dull; flesh soft, requiring extra care in handling and shipping; flavor excellent; begins to bear in January, hitting peak production in March. Plant: erect; vigorous; attractive and distinctly different in appearance from other Univ. of California varieties; low chilling requirement; plants runner readily; susceptible to verticillium wilt, although less susceptible than Lassen or Shasta; recommended for winter planting in the south coastal and central coastal production areas. Only high elevation nursery plants should be used. Designated as a new strawberry variety of the Univ. of California to commemorate the centennial year.

Serenta.—An ever-bearing strawberry that flowers in southern England continuously from late April until frost. Orig. in Buckinghamshire by J.R. Ellis. 82/12-10 x Pink Panda; introd. in 1993. Assigned to K.M. Muir, Essex, England; U.S. plant patent 8801. Flowers: dark pink; abundant. Fruit: small; glossy light red external, near white to pinkish-red interior; firm skin but soft flesh; elliptical; achenes at surface; slightly dry and sweet with musk flavor; average keeping quality. Plant: vigorous; abundant runners. Resistant to powdery mildew.

Settler.—Orig. in Vineland Sta., Ontario, Canada, by C.L. Ricketson. Guardian x Holiday; cross made in 1972; selected in 1974; tested as V7236-3; introd. in 1986 by A. Dale, Hort. Res. Inst. of Ontario. Fruit: large; flesh firm but skin weak; bright, medium-red external and pale internal color; flavor moderately sweet and aromatic; early to midseason ripening; local fresh-market use. Plant: high yielding; good vigor. Moderately resistant to verticillium wilt and leaf scorch; susceptible to powdery mildew; more widely adapted than Governor Simcoe.

Sharon.—Orig. in Dodgeville, Wis., by the Fieldhouse Fruit Farm. Introd. in 1954 by the Fieldhouse Fruit Farm. Wis. 537 (Corvallis x Howard 17); cross made in 1936 by R.H. Roberts, Wisconsin Agr. Expt. Sta.; selection made in 1950. Fruit: large; skin dark red, appearance very fine; ships well. Plant: yields very well; good plant maker.

Shasta.—Orig. in Davis, Calif., by Harold E. Thomas and Earl V. Goldsmith, California Agr. Expt. Sta., Berkeley. Introd. in 1945. Calif. 67.5 (Nich Ohmer x USDA 634) x Calif. 177.21 [USDA 543 x Calif. 68.24 (USDA 634 x New York 4626)]; N.Y. 4626 = Marshall x Howard 17; cross made in 1935; selected in 1937; tested as Calif. 403.8. Fruit: large, producing good sizes throughout the harvest season; blunt conic; skin attractive, medium tough, bright red, glossy; achenes greenish-yellow, flush with surface; flesh red, firm, flavor good, medium juicy, mildly subacid, processing quality good; ripens in midseason, and continues in production through the summer in the central coast area of California. Plant: productive; fairly vigorous; runner production medium; foliage dark bluish-green; best adapted in central coast area of California; fair resistance to mildew and virus diseases; susceptible to verticillium wilt.

Shelton.—Orig. in New Haven, Conn., by D.F. Jones, Connecticut Agr. Expt. Sta. (Chesapeake x Marshall) x (F_1 cross of inbred Progressive x Howard 17). Fruit: color attractive, retained after picking; flesh soft; resembles Chesapeake; subject to sunscald; ripens early June. Plant: foliage healthy.

Shortcake.—Orig. in Champaign, Ill., by Herschel L. Boll. Introd. in 1978. [(Canall x

Twentieth Century) x Twentieth Century] x Ogallala. Cross made in 1967. Berry: medium to large, long conic to conic; skin medium red. slightly glossy, attractive; flesh firm; ripens midseason about with Catskill. Plant: everbearing; highly productive, vigorous; produces medium number of runners.

Shuksan.—Orig. in Puyallup, Wash., by B.H. Barritt and C.D. Schwartze, Western Washington Res. & Ext. Center. Introd. in 1970. The parentage of this variety is (Northwest x Sierra) x Columbia. Cross made in 1962, selected in 1965, tested as WSU 1239. Fruit: large, broad wedge to conic; skin bright red, glossy, abrasion resistant; flesh firm, bright red throughout; for fresh market and processing; ripens mid-season with Northwest. Plant: large; vigorous; very hardy; productivity equal to Northwest; tolerant to red-stele races A-1 and A-3, virus and Botrytis.

Shuswap.—Orig. in Vancouver, British Columbia, by H.A. Daubeny, P.P. Moore, T.M. Sjulin, F.J. Lawrence, and B.H. Barritt, Agr. Canada Res. Sta. Holiday x WSU 1651 [Olympus x OR-US 3965 (Earlibelle x Hood)]; cross made in 1976 by B.H. Barritt; selected in 1977 by H.A. Daubeny; tested as BC 76-7-20; introd. in 1991. Fruit: large; firm; globose conic; glossy; well-maintained, light-red external and light- to pale-red internal color; smooth surface with distinctive white area around reflexed calyx; yellow achenes at surface or slightly sunken; calyx relatively difficult to remove; mild, subacid flavor, sweeter under higher temperatures and light intensities; midseason ripening; some resistance to both pre- and postharvest botrytis rots; has prolonged shelf life for fresh-market use. Plant: high yielding; vigorous; abundant runners; flowers at or below the leaf canopy with erect to semi-erect peduncles; winter hardiness level similar to Totem and less than Sumas. Relatively tolerant to aphid-borne virus complex in the Pacific Northwest; slightly susceptible to leaf spot and powdery mildew; some tolerance to two-spotted spider mite; appears susceptible to most races of the red stele causal organism.

Sierra.—Orig. in Davis, Calif., by Harold E. Thomas and Earl V. Goldsmith, California Agr. Expt. Sta., Berkeley. Introd. in 1945. Nich Ohmer x Calif. 177.21 [(USDA 543 x Calif. 68.24 (USDA 634 x New York 4626)]; cross made in 1935; selected in 1937; tested as Calif. 537.5. Fruit: flowering stem long; primary berries large, size smaller as later berries mature; conic, sometimes wedge conic; skin medium red, glossy, very attractive, medium firm; seed position varies from raised to flush with surface; flesh medium firm, a good red, aroma slight, good dessert quality; white tips in some seasons on mature berries; ships well; hull medium large, reflexed, separates easily from fruit; ripens late. Plant: total production medium but with a period of low yields in summer along coastal area of California; moderately resistant to yellows disease and virus diseases; resistant to verticillium wilt; best adapted to foothill area of California. Named for the Sierra Nevada range. Variety no longer propagated.

Siletz.—Orig. in Corvallis, Ore., by George F. Waldo, U.S. Dept. of Agriculture and the Oregon Agr. Expt. Sta. Introd. in 1955. U.S.-Ore. 2012 x U.S.-Ore. 1816; selected in 1949; tested as U.S.-Ore. 2172. Fruit: medium to small unless plants well spaced; round conic; flavor very good, especially frozen (main usage); flesh dark red, soft; midseason. Plant: resistant to red stele, verticillium wilt, mildew, leaf spot; apparently tolerant to virus diseases; very vigorous; very productive, expected to outyield Marshall in irrigated areas where red stele is a problem; adapted to the Pacific Northwest.

Silva.—Orig. in Sebastopol, Calif., by William E. Silva. Introd. in 1950. Plant patent 890; 1 Nov. 1949. Rockhill x (Gem x Wayzata); selected in 1945. Fruit: large; skin red; flesh red, firm, sweet. Plant: produces no runners; everbearer; productive.

Simcoe.—Orig. in Ottawa, Ontario, Canada, by Canada Dept. Agr., Res. Sta. Introd. in 1936. (Red Sugar x Howard 17) x (Delecto x Cassandra); selected in 1934. Fruit: large; attractive; loses size after two or three pickings; quality medium; ripens midseason. Flower imperfect. Obsolete.

Sioux.—Orig. in Cheyenne, Wyo., by LeRoy Powers, U.S. Hort. Field Sta. and by Glenn Veihmeyer, North Platte Expt. Sta., North

Platte, Neb.. Introd. in 1948. (Fairfax x *Fragaria ovalis*) x Fairfax. Fruit: good quality. Plant: extreme hardiness; very productive.

Snow King.—Orig. in Allegan County, Mich., by the Burgess Seed Plant Co., Galesburg, Mich. Introd. in 1922. seedling of a wild species; discovered in 1918. Fruit: strictly a novelty; size medium; nearly round; skin and flesh white; flesh very sweet; wild strawberry flavor; because of color, birds not attracted.

Solana.—Orig. in Davis, Calif., by R.S. Bringhurst and Victor Voth, California Agr. Expt. Sta. Introd. in 1957. Cal 177-19 x Cal 103-22; cross made in 1935; parentage includes Nich Ohmer, Royal Sovereign, Howard 17 and Klondike; selected in 1937; test as Cal 35.93-11. Fruit: large to medium; skin glossy, bright red slightly pubescent, yellow achenes flush with surface; flesh uniformly red, aromatic, moderately firm, juicy flavor subacid; quality high. Plant: adapted to southern California conditions; tolerant of warm winters and virus; good tolerance to salinity; characteristically has four to five leaflets instead of the normal three. Named for town of Solana Beach, where much of testing was done.

Sonjana.—Orig. in Stuttgart, Germany, by Reinhold Hummel. Introd. in Germany in 1958. Plant patent 1691; 18 Mar. 1958. Result of crossing seedlings derived from the cross Holstein x Sona Horstmann. Fruit: size medium; skin red; flesh whitish; aroma very good; keeps well; everbearing, maturing 25 May to 15 Oct. at place of origin but prime condition is end of May. Plant: produces vigorous runners, which may be espaliered or trained on a support, thus is termed a "climber," blooms during the period 20 April to 5 June.

Soquel.—Orig. in Davis, Calif., by R.S. Bringhurst and Victor Voth, Agr. Expt. Sta., Univ. of California. Introd. in 1983. Plant patent assigned to Regents, Univ. of California, Davis. Cruz x Aiko. Cross made in 1972; first fruited and selected in 1974. Tested as Cal 72.330-107 and C95. Fruit: larger than Tioga or Aiko; medium conic. generally symmetrical: internally hollow; skin about the same color as Aiko, but more glossy and attractive: flesh lighter than skin, particularly during the early part of the season, firm and durable as Tioga, Tufts, or Aiko: achenes are bright yellow, flush with surface in contrast to those of Aiko, which tend to be slightly embedded. Plant: low to semi-erect, slightly larger than Aiko; self-fertile. Matures earlier than Aiko but fruits as long; has performed well in winter and summer plantings in central coastal California. Named for the town of Soquel near Santa Cruz, Calif.

Southland.—Orig. in Washington, D.C., by George M. Darrow, U.S. Dept. of Agriculture. Introd. in 1932. Ettersburg 80 x Howard 17 (Premier); cross made in 1920; selected in 1922 in Glenn Dale, Md.; tested as USDA 44. Fruit: large; globose, sometimes slightly conic; hull clasping, very large; skin bright red, glossy; achenes slightly sunken; flesh light red, tender, sweet, juicy, rarely with a cavity, not as firm as Blakemore and Bellmar, mildly subacid, aromatic; dessert quality very high; shipping quality fair; ripens early, with Missionary. Plant: heavy early crop; vigorous; blooms early; flower perfect; adapted to the south.

Sparkle (*Paymaster*).—Orig. in New Brunswick, N.J., by J.H. Clark, New Jersey Agr. Expt. Sta. Introd. in 1942 under restrictions; named and released commercially in 1943. Fairfax x Aberdeen; selected in 1933; tested as N.J. 312. Fruit: size medium; short blunt conic to oblate; skin rather dark red, glossy, attractive; flesh soft, mildly subacid, quality excellent; good for dessert and frozen pack; ripens in midseason to late. Plant: productive to very productive; last flowers on cluster usually set; moderately susceptible to verticillium wilt, leaf scorch, leaf spot, but resistant to powdery mildew; resistant to one type of red stele disease; susceptible to virus diseases; produces many runners; grown in the northeastern states to Wisconsin.

Sparta.—Orig. in Parkside, Saskatchewan, Canada, by A.J. Porter, Honeywood Nursery. Introd. there in 1941; introd. in the U.S. in 1949 as P.I. 181700. Gem x Fairfax; selected in 1936. Fruit: medium to large; conic; skin bright red; flesh very firm, quality very good, flavor fair to good; keeps well; ripens after Gem. Plant: large; vigorous; very resistant to leaf spot and mildew; everbearing; variable in making runners; often less productive than Gem.

Spring Beauty.—Orig. in Champaign, Ill., by Herschel L. Boll. Introd. in 1976. [(Canall x Twentieth Century) x Twentieth Century] x Ogallala. Cross made in 1967. Berry: large, mostly wedge in shape; skin light red; seeds indented; flesh slightly soft; calyx large and frilly, making fruit attractive; ripens early about with Sunrise. Plant highly productive; vigorous; producing many runners.

Spring Giant.—Orig. in Grandview, Ind., by Alben B. Cook. Introd. in 1973. Plant patent 3716; 6 May 1975; assigned to Herschel L. Boll, Champaign, Ill. (Canall x Twentieth Century) x Twentieth Century. Berry: verylarge, up to 2 1/2 inches in diam., wedge to irregular in shape; skin medium dark red and slightly purplish; flesh firm, sweet, highly flavorful and unusual, no noticeable core; fruit stems thick; flowers pistillate; ripens late, about 2 to 3 days after Jerseybelle. Plant: low to medium in productivity; extremely vigorous, large and tall; produces medium number of runners but makes a wide row as internodes are thick and long; leaves large, up to 9 inches in width, light green to yellowish-green, slightly glossy to glossy, rugose.

St. Williams.—Orig. in Vineland Sta., Ontario, Canada, by C.L. Ricketson. Guardsman x V6747R-6 (Veestar x NY 844); cross made in 1972; selected in 1974; tested as V7261-3; introd. in 1991 by A. Dale, Hort. Res. Inst. of Ontario. Fruit: moderately large; very firm flesh and firm skin; bright, medium-red external and red internal color; flavor moderately sweet; ripens mid- to late season; processing use, particularly as frozen product. Plant: high yielding; vigorous; adequate runners. Resistant to leaf scorch, leaf spot, and powdery mildew; moderately resistant to verticillium wilt.

Starbright.—Orig. in Glenn Dale, Md., by George M. Darrow and George F. Waldo, U.S. Dept. of Agriculture. Introd. in 1940. Chesapeake x Fairfax; cross made in 1931 by Waldo; selected in 1933; tested as USDA 2120. Fruit: large; blunt conic to blunt wedge, very uniform; skin very glossy, tough, bright medium red, not turning dark, very attractive; hull usually clasping, size medium; achenes yellow, slightly raised above surface; flesh light red, juicy, tender, firm, subacid, dessert quality good, aromatic but not quite as much aroma as Dorsett; shipping quality exceptional, superior to most varieties; fairly long season, about with Catskill in Maryland. Plant: fairly productive; vigorous; susceptible to hot weather; fairly freely produces runner plants; leaf large, unusually free of leaf spot and leaf scorch; flower perfect.

Starkrimson.—Orig. in Jefferson County, Mo., by Clarence H. Christ. Introd. in 1956. Midland x Fairpeake; selected in 1949. Plant patent 1365; 22 Mar. 1955; assigned to Stark Bros. Nurseries & Orchards Co., Louisiana, Mo. Fruit: large; round; skin dark red; flesh very firm, red, quality very good; ripens 6 days later than Bellmar. Plant: extremely vigorous; productive.

Startler.—Orig. in New Westminster, British Columbia, Canada, by A. Keir. Introd. in 1930. Parentage unknown. Fruit: large; skin dark red. Plant: produces too few runners. Variety out of commercial production in Canada.

Startyme.—A high-yielding, late-ripening, fresh-market strawberry with firm fruit. Orig. at Hort. Res. Inst. of Ontario, Simcoe, Canada, and Univ. of Guelph by A. Dale, A.A. Vandenberg, W.D. Evans, and J.A. Sullivan. V7705-3 (derived from Holiday, Midway, and Redglow) x GU66Q50 (derived from Catskill, Redcoat, Sparkle, and Guardsman); cross made by A. Dale and W.D. Evans in 1984; selected by A. Dale in 1986; tested as225C1; introd. in 1995. Canadian Plant Breeders Rights granted in 1994. Fruit: large; bright, glossy external and slightly pale internal color; short cone shape; slightly softer than Governor Simcoe; uniform; good shelf life. Ripens late, with Governor Simcoe; calyx easy to remove; good flavor; fresh and processing. Plant: vigorous; winter hardy. Low susceptibility to leaf spot and leaf scorch; susceptible to powdery mildew. Tolerant to herbicide sinbar.

Stelemaster.—Orig. in Beltsville, Md., by D.H. Scott and George M. Darrow, U.S. Dept. Agr. and in Salisbury, Md., by W.F. Jeffers, Maryland Agr. Expt. Sta. Introd. in

1954. Fairland x Md. 683; first selected in 1950; seedlings. screened for resistance to red stele at Beltsville and resistant plants grown at Salisbury for evaluation of fruit. Fruit: size medium, generally uniform and smooth; surface medium red, glossy; flesh light red, firm, juicy, flavor mildly subacid; ripens about the same time as Blakemore and 5 to 7 days earlier than Temple; not satisfactory for frozen pack. Plant: vigorous, productive; leaves resistant to scorch but somewhat subject to leaf spot; its chief merit is multiple resistance to red stele.

Stoplight.—Orig. in Ames, Iowa, by E.L. Denisen, Iowa State Univ. Introd. in 1971. 142-5909 (Florida Ninety x Cyclone) x 6-5908 6-5908 is a two generation self of Dunlap. Cross made in 1962, selected in 1964, tested as 8-6214. Fruit: attractive; large; broad conic; skin bright red, shiny; flesh red throughout, firm to medium firm, no noticeable core or stringiness; flavor excellent; ripens mid-season (5 to 15 June in Iowa); harvest concentrated in two to three pickings, possibility of mechanical harvest; excellent for freezing and jam. Plant: large, erect; vigorous; very hardy; extremely productive; good plant maker; tolerant of leaf spot and leaf scorch.

Strafford.—Orig. in Durham, N.H., by L.P. Latimer, New Hampshire Agr. Expt. Sta. Introd. in 1954. (Simcoe x Catskill) x (Tupper x Fairfax); selected in 1949; tested as N.H. 266. Fruit: large; flesh dark red, flavor good, tends to be slightly soft in cloudy weather; ripens 10 to 14 days later than Howard 17. Plant: perfect flowered; crowns very large; yields heavily.

Strain 7.—Orig. in McFarland, Calif., by Walter J. Cordua, A.J. Shrier & Sons. Introd. in 1951. Strain of Cupertino; selected in 1951. Fruit: resembles Cupertino. Plant: vigorous, yet low enough to shade berries; some resistance to red spider; produces heavy crop in spring under hot San Joaquin Valley climate; fall crop moderate; spring crop about equal to that of Lassen.

Streamliner.—Orig. in Lostine, Ore., by Roy C. Edgmand. Introd. in 1944. Parentage unknown; discovered in 1938. Fruit: flesh firm; good keeper for home use. Plant: hardy; heavy spring cropper; everbearer; makes runners freely.

Sultan.—Orig. in Platteville, Wis., by Albin E. Lang. Introd. in 1956. Plant patent 1745; 26 Aug. 1958. Esteem x self; cross made in 1951; selected in 1953. Fruit: skin dark red; tendency toward white under the calyx and on the inside; flesh relatively firm, flavor rich, mildly acid; core relatively small. Plant: runners hardy, freely produced; everbearer.

Sumas.—Orig. in Vancouver, British Columbia, Canada, by H.A. Daubeny, Agr. Canada, Res. Sta. Introd. in 1986. Cheam x Tioga; cross made in 1969; tested as BC 69-5-34. Fruit: large; bright, does not darken; slightly aromatic; suitable for fresh use and processing; ripens ahead of Totem and Benton; tolerant to fruit rots. Plants: high yields of marketable fruit; winter-hardy. Resistant to red stele races A1 and A-4; tolerant to two-spotted mites and powdery mildew; susceptible to leaf spot.

Sumner.—Orig. in Raleigh, N.C. by North Carolina Agr. Res. Serv. Dept. of Hort. Sci. and U.S. Dept. of Agriculture. Introd. in 1980. Titan x MdUS 2856. Cross made in 1965 by G.J. Galletta; selected in Willard, N.C., in 1967; tested in North Carolina and Maryland from 1969 to 1979 as NC 3469. Berry: large, comparable to Apollo; flesh firm, like Earlibelle; flavor good; ripens early midseason in Coastal Plains and late season in mountains; susceptible to rotting in spring. Plant: runner moderately, occasionally freely; resistant to leaf spot, leaf scorch, and mites; variable in response to powdery mildew; can be grown in hills or matted rows; excellent yields in second year; culture should be avoided in areas of Coastal Plain where anthracnose is prevalent. Named for Jesse W. Sumner of Rocky Mount, formerly superintendant of several outlying research stations for the North Carolina Dept. of Agriculture.

Sunburst.—Orig. in Grandview, Ind., by Alben B. Cook. Introd. in 1973. Plant patent 3718; 13 May 1975; assigned to Herschel L. Boll, Champaign, Ill.[(Canall x Twentieth Century) x self] x Streamliner. Cross made in 1964. Berry: large, conic; skin orange-red, slightly glossy; flesh white to light orange-red, sweet; core soft, easily removed as calyx is removed; calyx sunken; ripens midseason, about with Catskill. Plant:

everbearing; productive, bears from June until freeze; vigorous; produces medium number of runners; leaves glossy.

Sunrise.—Orig. in Salisbury, Md., by U.S. Dept. of Agriculture, Beltsville, Md., and Maryland Agr. Expt. Sta., College Park. Introd. in 1964. US 4152 x Stelemaster; cross made in 1952, seedlings. grown at Salisbury, Md. in 1953, selected in 1954 by I C. Haut, G.M. Darrow and D.H. Scott; tested as Md-US 2601. Fruit: size medium; conic, symmetrical, slight neck on secondary and later berries; skin light, bright red that does not darken, glossy; flesh light pink, firm, subacid; dessert quality good but flesh too pale for good frozen pack appearance; ripens early, about with Dixieland. Plant: very vigorous; runners produced freely; resistant to three races of red stele fungus and to verticillium wilt; foliage resistant to leaf scorch and mildew, very susceptible to leaf spot; moderate to very productive in narrow matted rows; most satisfactorily grown in southern midwest U.S.

Sunset.—Orig. at the Univ. of California, Wolfskill Experimental Orchards, near Davis, by R.S. Bringhurst and V. Voth. Cal 75.121-101 x Cal 81.16-604 (Cal 71.98-605 x Chandler); cross made in 1985; selected in 1986; tested as 85.22-1; introd. in 1992. U.S. plant patent pending. Fruit: larger but less firm than Selva; flat conic to heart-shaped; both external and internal color similar to Selva but more glossy; achenes vary from yellow to light red, slightly indented; flavor substantially better than Selva; fresh-market, processing, and home-garden use. Plant: moderate to weak day-neutral, somewhat comparable to Selva; production pattern is slightly later than Selva; often higher yielding than Selva; more vigorous and erect than Selva. Moderately susceptible to common leaf spot; highly susceptible to powdery mildew; less susceptible to two-spotted spider mite than Selva; tolerant to viruses occurring in California.

Superfection.—See Gem.

Super Giant.—See Berri-Supreme.

Surecrop.—Orig. in Salisbury, Md., by D.H. Scott and George M. Darrow, U.S. Dept. of Agriculture, Beltsville, Md., and W.F. Jeffers, Maryland Agr. Expt. Sta., College Park. Introd. in 1956. Fairland x Maryland-US-1972; raised in 1950; selected in 1951 tested as Md.-US-2233. Fruit: large, maintaining this large size during the picking season; round conic, irregular; skin attractive, glossy, tough, light bright red that becomes a rich red but does not turn dark; achenes yellow, even with the surface; flesh light red, firm; flavor tart, good; rated as good for dessert, satisfactory for frozen pack; matures in midseason, about with Fairland and Temple. Plant: large; very vigorous; produces many runners; drought resistant; resistant to several races of red stele, fairly resistant to verticillium wilt; very productive in narrow matted rows. Tested from Pennsylvania and New York to southern Virginia and westward to Ohio; most useful in red-stele infested soils because of resistance to more than one race of red stele; leaves resistant to mildew, leaf scorch, and leaf spot.

Suwannee.—Orig. in Glenn Dale, Md., by George F. Waldo, U.S. Dept. of Agriculture. Introd. in 1945. Missionary x Howard 17; cross made in 1931; selected in 1933; tested as U.S. 2183. Fruit: high flavor and quality under adverse weather conditions when most varieties are quite deficient in these qualities. Plant: resembles Southland in adaptation, uses and limitations; also, adapted to home gardens.

Sweet Abundance.—Orig. in Monticello, Minn., by Marion Hagerstrom. Introd. in 1979 by Herschel L. Boll, Champaign, Ill. (Red Giant x Luscious Red) x Red Rich. Cross made in 1966. Berry; medium to large, mostly conic; skin medium light red, some have seedy tips; stem long and thick; ripens midseason, about with Catskill. Plant; everbearing; highly productive, especially in autumn; many berries per cluster; vigorous and tall; hardy; produces many runners.

Sweet Charlie.—An early ripening strawberry adapted to the Mid-Atlantic and southern United States. Orig. in Dover, Fla., by C.K. Chandler, C.M. Howard, and E.E. Albregts, Agr. Res. and Education Center, Univ. of Florida. Fl 80-456 x Pajaro; selected by C.M. Howard in 1981; tested as Fl 8S-4925; introd. in 1994. U.S. plant patent 8729. Fruit: large; orange-red external and orange streaked with white internal color; primaries wedge-shaped and secondaries wedge to conic; moderately firm; greenish-yellow achenes slightly recessed; easy to harvest; good flavor. Plant: productive; vigorous; abundant runner production. Tolerant to anthracnose.

Sweet Sue.—See Wisconsin 261.

Sweet Surprise.—Orig. in Grandview, Ind., by Alben B. Cook. Introd. in 1977 by Herschel L. Boll, Champaign, Ill. Ozark Beauty x Tennessee Shipper. Cross made in 1965. Berry: large, wedge to conic; skin medium red; flesh firm; ripens midseason about same time as Catskill. Plant: highly productive, vigorous, produces many runners; leaves large, up to 8 inches in width.

Sweetheart.—Orig. in Parkside, Saskatchewan, Canada, by A.J. Porter. Introd. in 1947. Open-pollinated seedling of August Beauty; selected in 1943. Fruit: highly flavored, somewhat soft for shipping. Plant: everbearer.

Symphony.—A fresh-market, red-stele-resistant strawberry. Orig. in Invergowrie, Scotland, by R.J. McNicol, Scottish Crop Res. Inst. Rhapsody x Holiday; cross made in 1979; tested as SCRI 69EW30; introd. in 1994. Worldwide marketing rights held by Commercial Fruit Plants, Kent, U.K. Fruit: large; firm; glossy, red-crimson exterior; conic to long conic; difficult to cap; mid-late season ripening 10 days after Elsanta; excellent shelf life; juicy; excellent flavor. Plant: high-yielding; abundant runner production. Has escaped crown rot and verticillium wilt despite exposure.

Tahoe.—Orig. in Wheatland, Calif., by Harold E. Thomas and Earl V. Goldsmith, California Agr. Expt. Sta., Berkeley. Introd. in 1945. Calif. 144.21 (Narcissa x Nich Ohmer) x Calif. 143.32 [Narcissa x Calif. BH-14 (mixed crosses)]; cross made in 1936; selected in 1938; tested as Calif. 567.6. Fruit: medium large; long conic, symmetrical; skin medium red, moderately glossy, moderately tender; achenes flush with surface, size medium; flesh light red, moderately firm and juicy; slight hollow; very little aroma; dessert quality medium good, shipping quality good; ripens late, starting with Sierra. Plant: consistently high producer in coastal area, particularly in the Salinas district; long lived; excellent producer of runners; stems strong, upright, high branching; fairly resistant to verticillium wilt and to yellows disease; some nubbin berries produced during summer. Name is taken from the largest lake in California.

Talisman.—Orig. in Auchincruive, Ayr, Scotland, by Robert D. Reid, Scottish Hort. Res. Inst. (West of Scotland Unit). Introd. in 1955. New Jersey 1051 x Auchincruive Climax; cross made in 1946. Fruit: resembles Auchincruive Climax. Plant: highly resistant to red core root rot (*Phytophthora fragariae*) but susceptible to certain specialized physiological races.

Tangi.—Orig. at Baton Rouge, La., by P.L. Hawthorne, W.L. Poillion, D.W. Newsom, W.F. Wilson, Jr., N.L. Horn, J.C. Taylor, J.E. Love, and R.J. Constantin, Louisiana Agr. Ext. Serv. Introd. in 1973. L-59-2012 x L-59-1938; cross made in 1962; tested as L63-377. Fruit: four to 11 per peduncle, medium to large; primary fruit long, conical to wedged, slightly furrowed, others conical to long conical; skin red, glossy; seeds small, slightly sunken, bright yellow to reddish; flesh red throughout, firm; ships well, market quality equal or superior to other Louisiana varieties; ripens March to June. Plant: vigorous; productivity very good, outyields Headliner and Dabreak; over 3000 pints per acre; good plant producer; fruit resistant to weather damage; resistant to leaf spot and leaf scorch, highly susceptible to crown rot and powdery mildew; frequent irrigation important during fruiting season for highest yield. Recommended for southern strawberry growing areas.

Tango.—Orig. in East Malling, England, by D. Simpson, Hort. Res. Intl. Rapella x Selva; selected in 1986; introd. in 1993. Fruit: firmer than Rapella; shape and appearance good; flavor acidic; fresh-market use. Plant: high yielding; strongly day-neutral with heaviest production in late July and first half of August in southeastern England; very few runners. Resistant to verticillium wilt and some races of the red stele causal organism; susceptible to powdery mildew and two-spotted spider mite.

Tarda.—See Vicoda.

Tardiva di Romagna.—Orig. in Bologna, Italy, by Enrico Baldini, Inst. di

Coltivazioni Arboree, University, Bologna. Introd. in 1971. Covered by Italian plant patent 891781 (Italy); 24 Dec. 1969; assigned to Zanzi Nursery, Ferlara, Italy, Macherauch Spaternte x Cambridge Favourite. Cross made in 1963, selected in 1969. Fruit:, large, conic to short wedge; skin deep red, with raised seeds; flesh firm, juicy, subacid, excellent flavor, very aromatic; recommended for fresh market and freezing; ripens about 10 days after Gorella. Plant: very vigorous; productivity high; hardy; easily capped.

Templar.—Orig. in Auchincruive, Scotland, by R.D. Reid, Scottish Hort. Res. Inst.. Introd. in 1964. Seedling Auchincruive 11 x Cambridge Vigour; cross made in 1957; breeder's number 57C33, tested in National Fruit Trials (Brogdale) as A.38. Fruit: medium large to very large through the season; globose conic to short conic, very symmetrical; skin light red turning to dark red when fully ripe, shiny; achenes small, yellow or reddish, not deeply pitted; flesh whitish suffused light red, cavity lacking; texture firm, juicy, fine, flavor rich, slightly acid; calyx not large, sepals medium to short, in most fruit slightly reflexed, giving fleshy neck that separates easily; maturity late, extending 3 to 4 weeks, similar to Talisman. Plant: medium to large; foliage dense, compact, upright; with many vigorous stolons; flower truss short, simple, covered by leaf canopy, bearing six to eight fruit; peduncle short, thick, strong, hairs slight, short and outright; leaf medium to large, medium to dark green, slightly shiny, sometimes mottled; lamina flat to convex; leaflets about as broad as long, base rounded, downward cupped, serrations sharp, medium depth, pubescence silky on upper surface; petiole with base tinged pink, furrow shallow. Selected for resistance to three races of red core root rot (*Phytophthora fragariae*).

Temple.—Orig. in Beltsville, Md., by George M. Darrow, U.S. Dept. of Agriculture, and W.F. Jeffers, Maryland Agr. Expt. Sta. Introd. in 1943. Aberdeen x Fairfax; selected in 1939. Fruit: large; earlier, firmer and more resistant to leaf scorch disease than Aberdeen; brighter red than Fairfax. Plant: very free plant maker; highly resistant to red-stele root disease.

Temptation.—Orig. in Champaign, Ill., by H.L. Boll. Plant patent 3798; 4 Nov. 1975. [(Canall x Twentieth Century) x Twentieth Century] x Ogallala; cross made in 1967. Introduced in 1974. Fruit: large; blunt wedge to round wedge; slightly glossy to glossy, firm and attractive in appearance; medium dark red color; flesh medium red, sweet and slightly above average in flavor; ripens 26 May. Calyx large. Plant: produces about one quart fruit per plant; ripens early, about the same time as Sunrise; vigorous and tall; produces a medium number of runners, making a moderately wide row; not especially resistant to any diseases; does well even on poor soil; few leaves have 4 to 5 leaflets.

Tendersweet.—Orig. in Parkside, Saskatchewan, Canada, by A.J. Porter, Honeywood Nursery. Introd. in 1961, by T. & T. Seeds, Winnipeg, Manitoba Northerner x Elgin; selected in 1956. Fruit: large; appearance very attractive; quality medium; ripens late, between season of summer bearers and the fall crop of everbearers; resembles Elgin.

Tennessean.—Orig. in Jackson, Tenn., by Brooks D. Drain, West Tennesee Expt. Sta. Introd. in 1950. Tenn. 230 x Tenn. 586; both selections are comprised of complex hybrids of several varieties (e.g., Klondike, Aroma, Missionary, Howard 17, Cumberland, Pickerproff, and Hoffman in Tenn. 586, and Missionary, Crescent, Howard 17, Haverland, Belmont, Sharpless, Cyclone, Green Prolific, etc.); the cross that produced Tennessean was made in 1942 by J.P. Overcash; selected in 1945; tested as Tenn. 965. Fruit: size medium, somewhat larger than Blakemore; long conic, necked, smooth; hull large, reflexed; skin medium red, colors evenly, glossy; achenes numerous, medium size, sunken; flesh light red, whitish toward center, core usually hollow, firm, juicy, subacid, aromatic, sprightly; quality good; ripens early; resembles Tennessee Supreme. Plant: vigorous; good production; produces many runners; flower perfect.

Tennessee Beauty.—Orig. in Knoxville, Tenn., by Brooks D. Drain, Tennessee Agr. Expt. Sta. Introd. in 1942. Plant patent 629; 6 June 1944; Brooks D. Drain, assignor to the Univ. of Tennessee Res. Corp., Knoxville. Missionary x Howard 17; cross made in 1933; selected in 1935; tested as Tenn. 263. Fruit: large, size retained well throughout picking season; shape regular, round conic to wedge, apex pointed; skin attractive, glossy, medium to deep red, colors evenly; hull large, medium green; achenes sunken; flesh medium red, whitish toward core, firm, subacid, well flavored; dessert and shipping quality good; freezes well; ripens in midseason with about a 3-week season; resembles Aroma but with better quality. Plant: healthy; production very good; vigorous; medium in height; somewhat susceptible to leaf spot; resistant to leaf scorch, tolerant to virus diseases; flower perfect, blooms fairly early; runners produced freely.

Tennessee Shipper.—Orig. in Knoxville, Tenn., by Brooks D. Drain and Louis A. Fister, Tennessee Agr. Expt. Sta. Introd. in 1941. Plant patent 570; 2 Mar. 1943; Brooks D. Drain, assignor to the Univ. Tennesee Research Corp., Knoxville. Missionary x Blakemore; selected in 1935; tested as Tenn. 148. Fruit: size uniform, medium, but retains good size throughout the picking season, larger than Blakemore; conic to round conic, necked; stems long, strong, making for easy picking; hull medium light green; skin smooth, glossy, light red, very attractive; achenes sunken in flesh; flesh medium red, whitish toward core, very firm, medium juicy, subacid, sprightly; dessert quality good, ships well, medium flavor in frozen pack, desirable for preserves; ripens early, first of May at Jackson, Tenn., length of season about 1 month; resembles Blakemore. Plant: healthy; vigor medium, low growing; heavy bearer; produces few to many runners but fewer than Blakemore; less winter injury than other varieties grown in the same area; flower perfect, blooms medium early; resembles Blakemore.

Tennessee Supreme.—Orig. in Knoxville, Tenn., by Brooks D. Drain, Tennesee Agr. Expt. Sta. Introd. in 1940. Plant patent 502; 3 Feb. 1942; Brooks D. Drain, assignor to the Univ. Tennesee Research Corp., Knoxville. Missionary x Howard 17; selected in 1935; tested as Tenn. 260. Fruit: medium to large; round conic to wedge; skin glossy, evenly colored medium red; flesh red, solid, juicy, medium firm, subacid, flavor fine; selected for frozen pack; season early; resembles Howard 17. Plant: prolific bearer; develops large numbers of runner plants; adapted south of Howard 17.

Tillikum.—Orig. in Puyallup, Wash. by Bruce H. Barritt, Western Wash. Res. & Ext. Center. Introd. in 1983. Totem x Cal. 65.65-601. Selected in 1976; tested as WSU 1727E. Fruit: small to medium: conic; skin glossy, dark red; flesh bright red, firmness average, very good acidic flavor. Plant: vigorous, productive; everbearing; tolerant to aphid-transmitted viruses.

Tioga.—Orig. in Davis, Calif., by R.S. Bringhurst and Victor Voth, California Agr. Expt. Sta. Introd. in 1963. Lassen (25% Blakemore, 25% Nich Ohmer, 19% Banner, 19 Fendalcino, 12% U.S. 634) x Cal 42. 8-16, a complex hybrid involving Sierra as the female parent (37% Nich Ohmer, 19% U.S. 543, 16% U.S. 634, 16% New York 4626, 6% Blakemore, and 6% Narcissa); Tioga is a sib of Fresno and Torrey; cross made in 1953; selected in 1955; tested as Cal 53.9-2. Fruit: average size 12 to 14 g; long conic to wedge; skin medium red, glossy, tough; seed yellow, flush with surface; flesh medium red, exceptionally firm, flavor good, soluble solids 7% to 10%, acidity medium high; shipping quality excellent; production very high, comparable to Lassen, higher than Fresno, Torrey, Solana, and Shasta; ripens early, about with Fresno, later than Torrey and Lassen. Plant: similar to Lassen color and general conformation but more vigorous; highly susceptible to verticillium wilt; moderate resistance to virus (occasionally manifests crinkle symptoms); fairly tolerant to salinity (similar to Lassen); adaptable to all growing areas of California, particularly with summer planting.

Titan.—Orig. in Raleigh, N.C., by Agr. Expt. Sta. and U.S. Dept. of Agriculture. Introd. in 1971. NC 1767 x Albritton. Cross made in 1956, selected in 1958 in Willard, N.C., by D.H. Scott and F.E. Correll, tested as NC 2840 by G.J. Galletta and D.H. Scott from 1960 to 1969. Fruit 80% to 90% larger than Earlibelle or Albritton, but not as symmetrical; skin medium dark red; flesh as firm as

Earlibelle and as pleasant flavored as Albritton; ships well; ripens with Earlibelle, recommended for local market, freezing and home gardens in the mid-South. Plant: large, vigorous; very productive, out-yields Earlibelle; moderate runner production; leaves resistant to leaf spot and leaf scorch; susceptible to powdery mildew and red stele root rot; tested from Massachusetts to North Carolina and westward to Arkansas, has proven superior only in the Coastal Plain and Piedmont sections of North Carolina and Bald Knob, Ark.

Torrey.—Orig. in Torrey Pines, Calif., by R.S. Bringhurst and Victor Voth, California Agr. Expt. Sta., Davis. Introd. in 1961. Lassen x Cal 42.8-16; selected in 1955; tested as Cal 53.9-9; sibling of Fresno. Fruit: large, uniform; wedge shaped; skin darker red than Fresno, about like Lassen; hulls easily; flesh very firm, quality good; ripens before Lassen and after Wiltguard; resembles Lassen. Plant: vigorous; produces numerous runners; leaflet deeply incised, lobes rather irregular, giving wrinkled appearance to leaf blade; tolerant to warm winters, having a short rest period; performs best in southern California; susceptible to verticillium wilt; fair tolerance to salinity; virus tolerant.

Totem.—Orig. in Agassiz, British Columbia, Canada, by Hugh A. Daubeny, Canada Dept. of Agr. Res. Sta., Agassiz. Introd. in 1971. Puget Beauty x Northwest. Cross made in 1962, selected in 1963, tested as 62-65-51 and BC 18. Fruit: slightly larger than Northwest; conical, but not as elongated as Northwest; skin dark red, achenes at surface or slightly sunken; flesh medium to dark red, flavor good, firmer than Northwest; for fresh market and freezing; ripens 2 to 3 days after Northwest. Plant: vigorous, petioles upright, thick; first year runner production adequate, but less than Northwest, second year heavy; hardier, more productive and more tolerant to fruit rot than Northwest. Released as a possible replacement for Northwest.

Toyonoka.—The leading strawberry in Japan, adapted to forcing under plastic in warmer regions, producing from November to May. Orig. at the Kurume Branch, Natl. Res. Inst. of Vegetables, Ornamental Plants and Tea, Japan, by F. Honda. Harunoka pollen parent; introd. in 1983. Fruit: large; firm; glossy scarlet; globular shape; good shelf life; juicy, sweet flavor. Plant: productive; low chilling requirement. Susceptible to powdery mildew; intermediate susceptibility to fusarium wilt.

Trailblazer.—Orig. in Glen Ellen, Calif., by E.E. Roach, Log Cabin Nursery. Introd. in 1951. (Gem x a native wild strawberry) x Streamliner; selected in 1949. Fruit: large; red to core; aromatic; extremely early. Plant: very prolific.

Tribute.—Orig. in Beltsville, Md., by D.H. Scott and A.D. Draper in cooperation with Maryland Agr. Expt. Sta., College Park, Md., and U.S. Dept. of Agriculture. Introd. in 1981. EB 18 (MdUS 3082 x Cal 65.65-601) x MdUS 4258 (MdUS 2713 x MdUS 3364). Selected in 1975, tested as EB 60. Selection EB 18 inherited the everbearing habit from California selection 65.65.601 obtained from R.S. Bringhurst, Univ. of California, Davis. The everbearing character of the California selection descended from clone of *Fragaria virginiana glauca* (= *F. ovalis*) collected in the Wasatch mountains of Utah by Bringhurst. Berry: medium to small, according to season, symmetrical, red, attractive; flesh firm; flavor sprightly. Plant: everbearer, bears in spring, again in mid- to late summer and continues until fall; resistant to five races of red stele root rot and powdery mildew; partially resistant to verticillium wilt and leaf scorch; runners less freely than June-bearers, but produces more freely than the standard everbearing cultivars; recommended for soils where red stele root rot may be a problem.

Tristar.—Orig. in Beltsville, Md., by D.H. Scott and A.D. Draper in cooperation with Maryland Agr. Expt. Sta., College Park, Md., and U.S. Dept. of Agriculture. Introd. in 1981. EB 18(MdUS 3082 x Cal 65.65-601) x MdUS 4258 (MdUS 2713 x MdUS 3364). Selected in 1975, tested as (EB 62). Selection EB 18 inherited the everbearing habit from California selection 65.65-601 obtained from R.S. Bringhurst of Univ. of California, Davis. The everbearing character of the California selection descended from clone of *Fragaria virginiana glauca* (= *F. ovalis*) collected in the Wasatch mountains of Utah by Bringhurst. Berry: medium to small, according to season, symmetrical, red, attractive; flesh firm; flavor good, sweet and aromatic. Plant: everbearer, bears in spring, again in mid to late summer, and continues until fall; resistant to five races of red stele root rot and powdery mildew; partially resistant to verticillium wilt and leaf scorch; runners less freely than June-bearers, but produces more freely than the standard everbearing cultivars; recommended for soils where red stele root rot may be a problem.

Trumpeter.—Orig. in Excelsior, Minn., by A.N. Wilcox, Univ. of Minnesota. Fruit Breeding Farm. Introd. in 1960. Burgundy x SYS 3672179 (seedling of Howard 17 selfed); cross made in 1944; selected in 1946; tested as Minn. SY 445. Fruit: large; rounded conic to cordate, not necked, relatively uniform; skin bright red, attractive; flesh an intense red throughout, very firm, juicy slightly acid, aromatic, flavor very good; freezing quality superior; June bearing, ripening relatively late; hull medium to large, bright green; achenes medium to small, bright yellow, raised slightly above surface; stores well. Plant: very tall, very vigorous; foliage bright, light green; hardy; apparently highly resistant to root rot, leaf spot and scorch; flower perfect; crop yields higher than Howard 17 and Dunlap; performs well on several soil types.

Tufts.—Orig. at South Coast Field Sta., Santa Ana, Calif., by R.S. Bringhurst and Victor Voth, Agr. Ext. Serv., Univ. of Calif., Davis. Introd. in 1972. Plant patent 3561; 4 June 1974; assigned to the Regents of the Univ. of California. Cal 46.5-1 x Tioga. Cross made in 1963 selected 1965, tested as Cal 63. 120-11. Fruit: large; long, flat conic skin bright red, tough; flesh orange-red, firm; ripens earlier than Tioga but harvest peaks less strongly and remains in production longer. Plant: medium size; erect, high canopy; vigorous; low chilling requirement; productivity comparable to Tioga, susceptible to verticillium wilt, less susceptible to leaf spot than Tioga; recommended throughout California, including winter planting in southern California.

Tupper.—Orig. in Ottawa, Ontario, Canada, by Canada Dept. Agr., Res. Sta. Introd. in 1942. Ettersburg 214 x Cassandra; selected in 1934. Fruit: late; valuable for home use and local markets; resembles Ettersburg 214. Plant: flowers imperfect; very productive.

Tustin.—Orig. in South Coast Field Sta., Santa Ana, Calif. by Victor Voth and R.S. Bringhurst. Agr. Expt. Sta., Univ. of California, Davis. Introd. in 1983. Plant patent assigned to Regents, Univ. of California, Davis. Douglas x Cal 72.361-105 = C55 (plant patent 4481). Cross made in 1977; first fruited and selected in 1979; tested as Cal 77.32-606 and C5. Fruit: larger than Tioga; blunt medium conic with some wedgy; internally slightly hollow; skin similar to Tufts, glossy, and attractive; flesh about same color as skin with a light ring around the core, firm and durable as that of Douglas, softer than Tufts; flavor excellent; for fresh market and processing. Plant: semi-compact, about 75% to 85% as large as Tioga: self-fertile; early fruiting, about with Tufts or Tioga, later than Douglas; has performed well in winter and summer plantings in south and central coastal California. Named for the town of Tustin, Calif.

Twentieth Century (*Utah Centennial*).—Orig. in Cottonwood Height, Utah, by Taiiro Kasuga. Introd. in 1932. Berri-Supreme x Rockhill; selected in 1926. Fruit: highly flavored; resembles Rockhill; ships well. Plant: yields heavily; heavy plant maker.

Tyee.—Orig. in Vancouver and Abbotsford, British Columbia, by Hugh A. Daubeny of Agr. Canada, Res. Sta. Vancouver. Introd. in 1980. Totem x Olympus. Cross made in 1970; selected in 1971; tested as BC 70-17-12. Berry: medium size; skin medium red, glossy; flesh firm, quality good fresh or frozen; resistant to *Botrytis*; adapted to machine harvest. Plant: vigorous; abundant runner production; tolerant to verticillium wilt; resistant to red stele and powdery mildew, and some resistance to two-spotted spider mite; winter hardy. Variety succumbed to June yellows.

Ulrich (*Improved Clark*).—Orig. in Mill Creek Canyon, The Dalles, Ore., by Julius Ulrich. Introd. in 1925. Chance seedling; discovered in 1920 near a planting of Gold

Dollar and Clark Seedling; possibly a hybrid of Gold Dollar with wild, low-growing native strawberry. Fruit: almost spherical, being similar to the wild fruit; a good canning and freezing berry; stem short. Plant: foliage dark green; very vigorous; appears resistant to strawberry mite in Hood River Valley, Ore. Now obsolete, as Marshall supplanted it.

Ulrich Imperial.—Orig. in Rochester, Minn., by Arnold E. Ulrich. Introd. in 1960. Plant patent 2085; 5 Sept. 1961. Burgundy x Howard 17; selected in 1949. Fruit: large to very large; round conical, axially curved; skin bright red, glossy; achenes flush with surface; flesh dark red, solid, small core, flavor slightly tart, quality very good; excellent for freezing; ripens about 3 days after Howard 17, with a season of 10 June to 3 July; early fruit often green tipped. Plant: very hardy; runners freely produced; more productive than either parent; foliage above average in resistance to common diseases.

Uncle Cap.—Orig. in Sawyer, Mich., by Bert Keith, Keith Bros. Nursery. Introd. in 1933. Chance seedling of Mastodon; first fruited in 1928; selected in 1930. Fruit: large; flesh white centered, quality good but rather soft. Variety now obsolete, but is the known parent of Waybright.

Universal Res.—Orig. in Champaign, Ill., by Herschel L. Boll. Introd. in 1978. [(Canall x Twentieth Century) x Twentieth Century] x Ogallala. Cross made in 1967. Berry: large, up to 2 1/4 inches in diam., wedge to conic: skin medium red; flesh slightly soft; calyx large; ripens midseason, about with Catskill. Plant: highly productive, vigorous; produces many runners.

Utah Centennial.—See Twentieth Century.

Utah Shipper.—Orig. in Brigham City, Utah, by L.L. Larsen, Larsen Experimental Farm. Introd. for trial in 1950. Lindalicious x Catskill; selected in 1947. Fruit: large to extra large; skin medium red; flesh firm; good shipper; resembles Lindalicious, except color is lighter. Plant: heavy cropper.

Vale.—Orig. in Corvallis, Ore., by G.F. Waldo, Oregon Agr. Expt. Sta. Introd. in 1966. US-Ore. 2012 x US-Ore. 2100; cross made in 1950; selected in 1952; tested as US-Ore. 2331. Fruit: large, becoming smaller toward end of season; conic, slightly necked and separates easily from the calyx; skin medium dark red, not bright; flesh red, medium firm in eastern Oregon but too soft for commercial use in western Oregon; flavor pleasing, mildly acid. Plants: hardy; produce an abundance of leaves but only a moderate number of runners; productive; flower stems medium to long, 7 to 15 blossoms per cluster; resistant to red stele, but not tolerant of virus disease, somewhat tolerant of salinity; for home garden and local market; recommended for eastern Oregon, eastern Washington and western Idaho. Named for the town of Vale.

Valentine.—Orig. in Vineland, Ontario, Canada, by Ontario Dept. Agr., Hort. Res. Inst. Introd. there in 1941; tested as Ontario 2701-14; introd. in the U.S. in 1950 as P.I. 189193 Howard 17 x Vanguard cross made in 1927; selected in 1930. Fruit: dark red; size medium, flesh firm, quality good; ripens very early; resembles Parsons. Exceptionally valuable in breeding.

Vandyke.—Orig. in Vineland, Ontario, Canada, by Ontario Dept. Agr., Hort. Res. Inst. Introd. in 1928. Open-pollinated seedling of Ontario seedling 1467 (Dunlap seedling x Early Ozark); selected in 1922 from 1919 breeding. Fruit: matures early; resembles Dunlap. Variety now lost because of susceptibility to yellows.

Vanguard.—Orig. in Vineland, Ontario, Canada, by Ontario Dept. Agr., Hort. Res. Inst. Introd. in 1924. Pocomoke x Early Ozark selected in 1915 from 1913 breeding. Fruit: matures early; resembles Early Ozark. Obsolete.

Vanrouge.—Orig. in Vineland, Ontario, Canada, by Ontario Dept. Agr., Hort. Res. Inst. Introd. in 1938. Ontario seedling 180115 [Admiral x Ontario seedling 1563 (Dunlap x Early Ozark)] x Bliss; selected in 1933 from 1930 breeding. Fruit: flesh deep red throughout; excellent for frozen pack. Obsolete.

Veestar.—Orig. in Vineland Station, Ontario, Canada, by C.L. Ricketson, Ontario Dept. Agr., Hort. Res. Inst. Introd. in 1967. Valentine x Sparkle; selected in 1960. Fruit: size medium; conic, uniform; skin medium red, glossy, smooth, attractive; flesh light to medium red, slightly pale center, moderately firm, inconspicuous core, slightly acid, excellent flavor; achenes even with surface or slightly sunken; ripens early, about 1 day after Earlidawn; a fresh-market variety. Plant: upright; size medium; vigor medium; leaflets slightly cupped upward; runner production moderate; susceptible to leaf scorch; moderately resistant to verticillium wilt; productive.

Vermilion.—Orig. in Urbana, Ill., by A.S. Colby, Ill. Agr. Expt. Sta. Introd. in 1950. Redstar x Pathfinder; cross made in 1941; selected in 1943; sibling of Plentiful. Fruit: large; short blunt conic, irregular; skin bright vermilion red, attractive; achenes yellow; flesh pale red, subacid; high dessert quality; ripens in midseason with Howard 17; resembles Pathfinder. Plant: resistant to red-stele root rot, leaf spot, leaf blight, and leaf scorch; vigorous; productive; runners produced freely; some stocks troubled by variegation.

Vesper.—Orig. in New Brunswick, N.J., by J.N. Moore and L.F. Hough, New Jersey Agr. Expt. Sta. Introd. in 1962. Utah Shipper x Jerseybelle; cross made in 1955; selected in 1957; tested as N.J. 157. Fruit: very large, with first pickings averaging up to 1 1/2 inches in diam.; skin medium tough, glossy, very attractive; achenes yellow and prominent; taste and texture equal to Catskill, with quality surpassing that of Jerseybelle, which it resembles; ripens very late, 3 to 5 days after Jerseybelle or about 15 June in New Brunswick; requires more frequent harvesting than some other varieties because it ripens rapidly; not satisfactory for freezing. Plant: vigorous; recommended for central New Jersey and northward; not resistant to disease; yields higher than average; virus free.

Vibrant.—Orig. in Vineland Station, Ontario, Canada, by C.L. Ricketson, Ontario Dept. Agr., Hort. Res. Inst. Introd. in 1967. Sparkle x Valentine; selected in 1960. Fruit: size medium; conic; slight tendency to split primaries; skin medium dark red, glossy, smooth, attractive; flesh bright medium red, firm, slightly acid, flavor fair; achenes even with surface or slightly sunken; ripens early midseason about 1 day ahead of Redcoat; a fresh-market and processing variety; quality of jam, and canned and frozen berries is excellent; calyx difficult to remove by machines. Plant: upright; size medium; vigor medium; leaflets slightly cupped upward; runner production moderate; moderately susceptible to verticillium wilt; productive.

Vicoda (*Tarda*).—A high-yielding, extremely late-ripening strawberry. Orig. in Belgium; propagated by Vissers Aardbeienplanten, America, Nds. Parentage unknown; introd. in 1989. Fruit: large; soft with tendency to bruise; dull, light red; tip remains white; long conic, uniform except for primaries. Ripens 30 days after Gorella; long harvest period; susceptible to rot; fair flavor. Plant: moderately vigorous; many crowns and several scapes per crown; scapes short and flowers at canopy level. Susceptible to powdery mildew, leaf spot, and all major root diseases.

Victoria.—See British Sovereign.

Viking.—Orig. in St. Paul, Minn., by Carl Helgeson. Introd. in 1967. Unknown seedling, discovered in 1964, selected in 1965. Fruit: 1 1/2 x 1 1/4 inches in diam.; elongated; skin deep red, flesh deep red, ripens mid-June. Plant: 16 to 19 inches high, mound shaped, holds fruit off ground, sturdy; vigor good; hardy; productivity equal to Dunlap or Jumbo.

Vista.—Orig. at Watsonville, Calif., by R.S. Bringhurst and Victor Voth, Agr. Expt. Sta., Univ. of California, Davis. Introd. in 1979. Plant patent assigned to Regents, Univ. of California, Davis. Albritton x Tufts. Fruit: larger than Tioga; flesh flavor and quality equal or superior to those of Aiko, Tioga, and Tufts; more ascorbic acid than that of Tufts but less than that of Aiko; skin same color as that of Tioga, dark reddish-orange. Plant: larger than Tioga; leaflets lighter and less yellow than those of Tioga; highly productive; as susceptible to verticillium wilt as Tioga, Tufts, and Aiko; highly susceptible to red stele (*Phytophthora fragariae*). Recommended for commercial production in southern California as an early, high-quality, winter planting variety and as an improved summer-cropping variety for the central coast. Named for the city of Vista, San Diego County, Calif.

Vola.—Orig. in Zaltbommel, The Netherlands, by the Hort. Expt. Sta. of the V.V.T.B.

Introd. in 1961; assigned plant breeder's right. Bowa x Empire; selected in 1955. Fruit: medium large; long conical; skin bright orange-red, shiny; flesh orange red, fairly firm, juicy, flavor good; ripens in midseason; good for fresh market but not for processing. Plant: vigorous; productive; flowers self-fertile.

Volunteer.—See Prince Albert.

Von Humboldt.—Orig. in Ettersburg, Calif., by A.F. Etter. Introd. in 1930. (Rose Ettersburg x *Fragaria chiloensis*) x self. Fruit: quality very good. Plant: vigorous; foliage heavy, evergreen. Now obsolete.

Washington (*New Washington, Puget Sound Wonder*).—Orig. in the state of Washington by John Nordheim. Introd. in 1930 by R.M. Kellogg Co., Three Rivers, Mich. Parentage unknown. Fruit: large; skin deep, glossy red; flesh very sweet; cans well; ripens very late, late June and early July. Plant: bisexual. Cultivar now obsolete.

Waybright.—Orig. in Veradale, Wash., by Harry R. Waybright. Introd. in 1935. Uncle Capp x an unknown seedling; selected in 1933. Fruit: size medium; skin dark red; resembles known parent.

Wayzata.—See Rockhill.

Wazata.—See Rockhill.

Welch Late.—Orig. in Shenandoah, Iowa, by J.C. Welch, Welch Nursery. Introd. about 1934. Chance seedling; discovered in 1929. Fruit: large; flesh firm, flavor good; ripens late, about 2 weeks after Dunlap. No longer propagated.

What a Flavor.—Orig. in Harrison Hot Springs, British Columbia, Canada, by Victor Macdonald. Introd. in 1921. Discovered about 1918 in a wood lot as a naturalized seedling. Fruit: extremely large; flesh soft, sweet, flavor excellent; resembles British Sovereign.

Wild Flavor.—Orig. in Roseau, Minn., by Louise Wenner. Introd. in 1958. Open-pollinated seedling of Dunlap; discovered in 1935. Fruit: flavor excellent.

Wiltguard.—Orig. in Davis, Calif., by R.S. Bringhurst and Victor Voth, California Agr. Expt. Sta. Introd. in 1961. Cal 39. 117-4 (sibling of Cupertino)x Cal 39.96-18; selected in 1954; tested as Cal 52.16-7. Fruit: fairly large; blunt conic to conic; skin bright red, does not darken, attractive finish; achenes bright yellow, small, flush with skin surface; flesh red, not darkening, highly flavored, very good dessert quality; tends to bear quite a number of buttontype split fruit, more so toward end of season; ripens very early, before Lassen, Fresno, and Torrey; resembles Cupertino. Plant: vigorous, semi-dense; leaf size medium, dark green, leaflets uncupped; productive; produces numerous runners; resistant to verticillium wilt, but less so than Sierra; begins production very early, especially in the central valleys of California; considered to be excellent for home gardens in California.

Wine Dark.—Orig. in Falmouth, Mass., by John Peterson. Introd. in 1953. Parentage unknown; selected about 1950 by Peterson from seedlings received from the Wayside Gardens Co., Mentor, Ohio. Fruit: small; skin very dark; flesh soft, quality poor; matures late, about 2 weeks after Howard 17. Flower: pistillate, requiring cross-pollination, blooms very late.

Winona.—Orig. in St. Paul, MN by J. Luby and D. Wildung of Univ. of Minnesota, and G.J. Galletta of U.S. Dept. of Agr. Intro. in 1996. Plant patent applied for by Regents of Univ. of Minnesota. Earliglow x MNUS 52 (Lateglow x MDUS 4616); cross made in 1985; selected in 1987; tested as MNUS 210. Fruit: large; firm flesh and medium to firm skin; scarlet skin color; flesh medium orange-red to pink at center; blunt wedge to blunt conic; large, showy, and partially reflexed calyx; achenes yellow and slightly raised; fresh flavor well balanced between sugars and acids, slighly aromatic. Plant: hardy; very productive; resistant to 5 races of red stele and tolerant or resistant to balck root rot, powdery mildew, leaf spot, and leaf scorch; late maturing with Lateglow; large-crowned with thick runners; adapted to matted row and continental climate.

Wisconsin 214.—Orig. in Madison, Wis., by R.H. Roberts, Wisconsin Agr. Expt. Sta. Introd. in 1949. Corvallis x Wisconsin 7128 [78 (Beaver x Premier) x 134 (Vanguard x Premier)]; selected in 1943. Fruit: very attractive; freezes well. Plant: vigorous; production varies with season.

Wisconsin 261 (*Sweet Sue*).—Orig. in Madison, Wis., by R.H. Roberts, Wisconsin Agr. Expt. Sta. Introd. in 1949. Its complicated parentage contains such varieties as Corvallis, Beaver, Howard 17, and Vanguard. Fruit: flesh soft; shipping quality poor; picking season short. Plant: highly productive. This selection apparently promoted by a commercial nursery that named it Sweet Sue.

Wisconsin Queen.—Orig. in Madison, Wis., by R.H. Roberts, Wisconsin Agr. Expt. Sta. Introd. in 1959. Wis. 214 x Wis. 6-2; selected in 1950; tested as Wis. 50-144; a selection from original crosses of Corvallis x Premier; Premier parent used was not identical to Howard 17, according to Roberts. Fruit: medium to large; skin bright red, attractive; flesh orange red; good table and superior freezing qualities. Plant: hardy; produces few late season runners; vigorous; average susceptibility to leaf spot and leaf scorch.

Wonderbearing.—Orig. in Randolph, Wis., by the J.W. Jung Seed Co. Introd. in 1926. Selected in 1924 from a field of Dunlap. Fruit: resembles Dunlap. Plant: similar to Dunlap but more vigorous; requires rich soil and ample moisture. Variety now obsolete.

Wray Red.—Orig. in White Salmon, Wash., by E.P. Wray. Introd. about 1930. Plant patent 101; 7 Aug. 1934. (Americus x Bederarena) x Chesapeake; selected about 1920. Fruit: large; skin medium dark red, rough; flesh firm; good canning qualities; ripens early. Plant: medium height; very productive in eastern Washington; flowers perfect.

Wright.—Orig. in Ottawa, Ontario, Canada, by Canada Dept. Agr., Res. Sta. Introd. in 1936. (Portia x Ettersburg 512) x (Santiago de Chile x self); selected in 1934. Fruit: large; flesh sweet, quality medium, flavor flat; ripens midseason. Plant: foliage leathery and thick; production poor. Obsolete.

Wyona.—Orig. in Bedford, Va., by A.S. Johnson. Introd. about 1923. Perhaps an open-pollinated seedling of Gandy; selected before 1922. Fruit: flesh firm; quality good for a subacid berry; long keeper; late maturity. No longer planted.

Xenion.—Orig. in Spangsbjerg, Esberg, Denmark, by the Danish State Res. Sta. Introd. in 1959. Deutsch Evern x Valentine; selected in 1950. Fruit: medium to small; dessert quality high; ripens extremely early. Plant: good for forcing; resistant to powdery mildew.

Ydun.—Orig. in Spangsbjerg, Denmark, by the Danish State Res. Sta. Introd. in Denmark in 1948, in the U.S. in 1949 as P.I. 173993, and in Canada in 1952. (Spate von Leopoldshall x Deutsch Evern) x Culver; selected in 1938. Fruit: not quite firm enough for shipping. Plant: very high yielding.

Yolo.—Orig. at Univ. of California, Wolfskill Experimental Orchards, near Davis, by R.S. Bringhurst and V. Voth. Cal 69.141-101 x Cal 71.98-605; cross made in 1975; selected in 1976; tested as Cal 75.54-115 and later as CN 75; introd. in 1987. Fruit: size comparable to Fern and smaller than Selva; less firm than Selva and firmer than Fern; less red and more orange than Fern, glossy; symmetrical, medium to short conic with occasional wedge; centers may be somewhat hollow; good flavor, comparable or superior to Fern; fresh-market use. Plant: strongly day-neutral with high chilling requirement; compact habit; yield similar to Fern. Relatively resistant to powdery mildew; resistant to verticillium wilt; moderately susceptible to common leaf spot; relatively susceptible to two-spotted spider mite; tolerant to viruses occurring in California.

Yorkshire.—Orig. in North Collings, N.Y, by C.J. Loretto, Sunnyhill Nursery. Introd. in 1950. Parentage unknown; discovered in 1945. Fruit: large; conical; flesh firm, sweet; ships well. Plant: hardy; withstands wet soils.

Zenith.—Orig. in Cascades, Quebec, Canada, by Wesley MacDiarmid. Introd. in 1930. Open-pollinated seedling of Dunlap; discovered in 1921. Fruit: resembles known parent, except the fruit is larger and without the neck of that variety.

not named.—Orig. in Watsonville, Calif., by Harold A. Johnson, Jr., and Harold E. Thomas. Not introduced, included here because of plant patent 3042; 13 Apr. 1971; assigned to Driscoll Strawberry Assn., Watsonville. F121.7 x Goldsmith, selected in 1963, tested as Driscoll-E7. Variety's most valuable characteristics: high fruit flavor and productivity. Not introd. because fruit not firm enough and size not consistently large.

not named.—Orig. in Watsonville, Calif., by Harold A. Johnson, Jr., and Harold E. Thomas. Not introd., included here because of plant patent 3043, 13 Apr. 1971, assigned to Driscoll Strawberry Assn., Watsonville. F121.7 x Goldsmith, selected in 1963, tested as Driscoll-E6. Variety's most valuable characteristics: high fruit flavor and productivity. Not introd. because fruit not firm or large enough.

not named.—Orig. in Pequot Lakes, Minn., by Charles C. Brunes. Plant patent 2407; 2 June 1964. Chance seedling. Fruit: 1 1/4 x 1 3/8 inches in diam.; short conical, symmetrical; skin medium dark red, glossy; flesh medium dark red, interior flesh marked with characteristic white zone that is oval shaped and tapering to apex, very firm, medium juicy, texture medium, very sweet, nonacid; good dessert quality, excellent for canning and freezing; ripens 15 June to 10 Oct. Plant: everbearing; vigorous; hardy; prolific and rapid production; bisexual; adapted to many types of soils and weather conditions. Included here because of plant patent.

TANGELO

Broward.—Orig. in Eustis, Fla., by F.W. Savage, U.S. Dept. of Agriculture. Introd. in 1939. Duncan grapefruit x Dancy mandarin; selected in 1912. Fruit: medium to large; quality good; ripens during December and January at Eustis.

Clement.—Orig. in Eustis, Fla., by W.T. Swingle, U.S. Dept. of Agriculture. Introd. in 1931. Bowen grapefruit x Clementine mandarin; cross made in 1914. Fruit: shape irregular; rind light yellow, thick, pebbled, wrinkled, unattractive; peels fairly well; pulp smooth, mild, flavor tart in California; few monoembryonic seeds; resembles Alamoen grapefruit. Of no economic value.

Gill.—Orig. in Lakeland, Polk County, Fla., by Marion E. Gill, Winter Haven, Fla. Introd. in 1962. Plant patent 2148; 22 May 1962; assigned to Philip's Pride, Winter Haven. King mandarin orange x Duncan grapefruit; seed planted in 1951; selected in 1958. Fruit: large, 4 1/4 to 5 1/4 inches in diam.; appearance of a grapefruit; rind thickness medium, easily peeled; flavor excellent, no bitterness; juice abundant; flesh color resembles mandarin orange; segments easily separated; suitable fresh, canned, frozen; seeds medium large, averaging 15 per fruit; matures in midseason to early late. Tree: hardy; crops large; resembles Valencia orange.

Kauffman Early.—See K-Early.

K-Early (*Kauffman Early, Sunrise*).—Orig. in Eustis, Fla., at U.S. Dept. of Agriculture. Citrus Breeding Sta. Introd. in 1949 by John H. Kauffman, Sr., Grand Island Nurseries, Eustis. Hybrid of grapefruit and Dancy mandarin. Fruit: approximately shape of Orlando tangelo but slightly larger; internal quality poor, with low solids; external quality good; rind smooth, almost yellow, colors early; flavor good, but rag is bitter like the grapefruit parent; shipping quality good; ripens 15 Sept. to 1 Oct. Tree: very vigorous, reaching full size 2 years earlier than Orlando tangelo; dense; tolerant to *xyloporosis*, but known to carry this virus; regular producer; cross-pollination appears to make variety more productive.

Lake.—See Orlando.

Minneola.—Orig. in Eustis, Fla., by E.M. Savage, U.S. Dept. of Agriculture. Introd. in 1931.Duncan grapefruit x Dancy mandarin; cross made in 1911 by W. T. Swingle, U.S. Dept. of Agriculture; selected in 1912; sibling of Orlando and Seminole. Fruit: large; oblate-obovate, necked; rind deep reddish-orange, very attractive, nearly smooth, moderately adherent; flesh dark orange, flavor and aroma delicious, very juicy; seeds about 10; season December through April depending on location. Tree: vigorous; productive when cross-pollinated; susceptible to citrus scab and alternaria.

Nova.—Orig. in Orlando, Fla., by John M. Bellows, Philip C. Reece, C. J. Hearn, and F. E. Gardner, U.S. Hort. Field Sta., U.S. Dept. of Agriculture. Introd. in 1964. Clementine mandarin x Orlando tangelo; *Citrus reticulata* x (*C. paradisi* x *C.reticulata*) cross made in 1942 by Bellows; first fruited in 1950; selected and tested in Fla. by Reece, Hearn and Gardner as Orlando 426-9-22; classed as a tangelo although tree shows strong mandarin characteristics, the fruit is same size and shape as Orlando. Fruit: size medium; rind orange, smooth, very attractive, easily removed even though close fitting to pulp; flesh orange, firm, juicy, flavor more pleasant and sweeter than Orlando, total soluble solids range from 10% in late September to 13% or more in December in Florida, usually 11 segments; seedy in mixed plantings; ripens early, before Orlando, in October, reaching its prime in November in Florida. Tree: as hardy as Orlando under Fla. conditions.

Orlando (*Lake*).—Orig. in Eustis, Fla., by E.M. Savage, U.S. Dept. of Agriculture. Introd. in 1931. Duncan grapefruit x Dancy mandarin; cross made in 1911 by W. T. Swingle, U.S. Dept. of Agriculture; selected in 1912; sibling of Minneola and Seminole. Fruit: medium-large, 2 3/4 inches in diam.; slightly flat; rind deep orange, attractive, thin, adheres firmly to pulp, slightly pebbled; flesh pale orange, very juicy, mildly sweet, fine flavor; seeds about 10; matures early, ripening in November through March depending on location; little resemblance to either parent, thus sometimes known on the market as Orlando orange. Tree: vigorous; productive; resistant to citrus scab; requires cross-pollination, with Temple or Dancy being satisfactory. One of the best tangelo varieties.

Page.—Orig. in Orlando, Fla. by Philip C. Reece and Frank E. Gardner, U.S. Hort. Field Sta. Introd. in 1963. Minneola tangelo x Clementine (Algerian tangerine); although fruit resembles an orange, it is considered as a tangelo hybrid; tested as Orlando 421-17-8. Fruit: size medium; round; rind deep orange to scarlet, attractive, easily peeled; flesh color dark orange, flavor excellent, sweet, rich, solids reach 12% to 13% in November with 1% acids; quality excellent; used primarily for local fresh market in Florida; higher quality if grown on sweet orange or sour orange rootstocks; fruit from trees on Rough lemon rootstock tends to be low in solids and juice volume; seeds numerous in mixed plantings, averaging 18; matures early, in October, reaching prime condition in November. Tree: moderately vigorous; nearly thornless; branches upright, spreading under heavy fruit load; vigorous; best suited to areas where climatic conditions are similar to those of Florida; similar in hardiness to Robinson, Lee, and Osceola mandarins.

Pearl.—Orig. in Riverside, Calif., by H.B. Frost, California Citrus Expt. Sta. Introd. in 1940. Imperial grapefruit x Willow Leaf mandarin orange; selected in 1929. Fruit: size medium small; slightly flat; rind yellow, nearly smooth, thin; unattractive; flesh juicy, firm, mild, sweet, aroma unique, quality very good; acid and sugar similar to sweet orange; seeds many; ripens medium early; loses quality if left on tree past maturity. Tree: vigorous; spreading, drooping; somewhat alternate bearing.

San Jacinto.—Orig. in Indio, Calif., by Walter T. Swingle, U.S. Dept. of Agriculture. Introd. in 1931. F_2 hybrid seedling of unknown tangelo; resembles Orlando; from the same parentage as Thornton. Fruit: size medium; slightly necked to round; rind yellow-orange, nearly smooth, peels fairly well; quality fair; tart to sweet; ripens mid-season December to March; more resistant to handling than Thornton; variable in size and quality. Tree: vigorous; productive.

Seminole.—Orig. in Eustis, Fla., by E.M. Savage, U.S. Dept. of Agriculture. Introd. in 1931. Duncan grapefruit x Dancy mandarin; cross made in 1911 by W.T. Swingle, U.S. Dept. of Agriculture; selected in 1912; sibling of Minneola and Orlando. Fruit: size medium; slightly flat; rind red-orange, attractive, slightly pebbled, peels fairly well; flesh very juicy, flavor and aroma good, somewhat tart at times; seeds many, may averages 22 per fruit; season late, February through April depending on location; ships well; resembles Minneola. Tree: fairly vigorous and productive; leaves unifoliate, 1 3/4 to 2 1/2 inches long, boat-shaped; less susceptible to scab than Minneola.

Sunrise.—See K-Early.

Webber.—Orig. in Eustis, Fla., by E.M. Savage, U.S. Dept. of Agriculture. Introd. in 1932. Duncan grapefruit x Dancy mandarin; selected in 1909. Fruit: size medium, flat; skin smooth, thin, orange colored; flavor fine, very juicy; ripens December to February. Named for H.J. Webber, who deemed it one of the best of the tangelos.

Wekiwa.—Orig. in Eustis, Fla., by E.M. Savage, U.S. Dept. of Agriculture. Introd. in 1931. Duncan grapefruit seedling 37 x Sampson tangelo; selected in 1909. Fruit: medium small; rind pale yellow, smooth; flesh pink, sweet, unpleasant when overripe; seeds few; ripens December and January. Tree: lacking in vigor; productive.

Yalaha.—Orig. in Eustis, Fla., by F.W. Savage, U.S. Dept. of Agriculture. Introd. in 1931. Duncan grapefruit x Dancy mandarin; cross made in 1911. Fruit: matures late; characteristics very similar to those of Minneola, Sampson, and Thornton; chief objections are large number of seeds and open core. Never achieved much popularity.

TANGOR

Altoona.—See Umatilla.

Dweet.—Orig. in Riverside, Calif., by Howard B. Frost and James W. Cameron, California. Citrus Expt. Sta. Introd. in 1950. Maltese Oval (Mediterranean Sweet) sweet orange x Dancy mandarin; selected in 1930. Fruit: size similar to Valencia orange, fairly uniform; short obovoid or short pyriform, usually with a very short, often irregular, neck; skin often somewhat pebbled; rind yellow-orange, overripe fruit may become greenish, medium to rather thin, not easily peeled, slightly puffed when overripe; oil glands very numerous; nine to 11 flesh segments, separating with some difficulty, pulp vesicles large, stout, very tender, extremely juicy, juice very easily extracted, sugar medium, juice aroma pronounced, flavor tart to sweet, good to very good; seeds small, plump, 12 to 19 per fruit; drops somewhat easily; ripens February or March to May or June or even later at Riverside. Tree: large; vigorous; Dweet was coined from the names Dancy and Mediterranean Sweet.

Honey Murcotte.—See Murcott.

Mency.—Orig. in Riverside, Calif., by Howard B. Frost, Calif. Citrus Expt. Sta. Introd. in the early 1920s. Citrus *sinensis* cultivar Maltese Oval (Mediterranean Sweet) x C. *reticulata* cultivar Dancy; selected about 1915. Fruit: size medium; round to slightly necked; rind slightly pebbled, red-orange, peels fairly well; soft; seeds many; drops easily; ripens in coastal-interior of Calif. during January to February, in desert area of California during October to November; usually rather sour at Riverside, therefore seems best on Rough lemon stock; sunburns badly under extreme temperatures. Tree: vigorous; very productive but yields lightly under extreme temperatures.

Murcott (*Murcott Honey Orange, Murcott Mandarin, Honey Murcott, Smith Tangerine*).—Origin somewhat obscure. Believed that parent tree was sent for trial from U.S. Dept. of Agriculture nursery at Little River (Miami), Fla., to R.D. Hoyt of Safety Harbor, Fla. A neighbor, Charles Murcott Smith, obtained a bud of this tree from Hoyt for propagation. Introd. about 1922; variety propagated on a small scale by several nurseries under various names but not coming into commercial importance until about 1952. Thought to be C. *reticulata* x C. *sinensis* (mandarin orange x sweet orange). Fruit: size medium; round to flat; rind texture pebbled, pale orange color, thin, not easily peeled as peel adheres fairly tightly to segments; pulp quality superior, very deep orange, firm, juice rich and sweet; seeds many; matures late, March to April in coastal California and February to March in Florida or California desert area; ships well. Tree: bush, with rather stiff branches; fruit borne mostly near the ends of branches; subject to sun scald and scab; productive, tendency toward alternate bearing.

Murcott Honey Orange.—See Murcott.

Murcott Mandarin.—See Murcott.

Ruddy.—Orig. in Riverside, Calif., by Howard B. Frost, Calif. Agr. Expt. Sta. Introd. in the early 1920s. *Citrus reticulata* cultivar King x C. *sinensis* cultivar Maltese Oval (Mediterranean Sweet); mandarin x sweet orange; selected about 1915. Fruit: fairly large; flesh acid, aroma of juice distinctive; rind suitable for candied peel; entire fruit suitable for marmalade. No longer propagated.

Smith Tangerine.—See Murcott.

Sue-Linda Temple.—Orig. at Merritt Island, Fla., by R.L. Schlernitzaur. Orig. in 1940 as a single tree within a commercial planting of Temple tangor trees. Fruit: internal fruit characteristics differ from Temple by having fewer seeds, earlier maturity, a deeper orange flesh and juice color, and a greater juice content; external fruit characteristics, color, size, and shape are similar to the standard Temple. Tree: several feet taller than Temple.

Thoro Seedless Temple.—Orig. in Miami, Fla. by A. Thoro. Plant patent 4161; 6 Dec. 1977. Apparently a mutation of the Temple orange. Fruit: resembles Temple but is seedless. Tree: somewhat drooping; medium; good vigor; blooms continuously.

Umatilla (*Altoona, Oranguma*).—Orig. in Eustis, Fla., by E.M. Savage, U.S. Dept. of Agriculture. Introd. in 1931. Satsuma mandarin x Ruby orange; selected in 1912. Fruit: large; size and shape of King mandarin; rind red-orange, smooth; flesh very juicy, rich, tender, sweeter than Seminole tangelo, usually 10 segments; zero to 10 seeds; a very good late variety, ripening January to March; resembles varieties of the tangelo group. Tree: slow growth with spreading open top; leaves oval, long, pointed, almost wingless petioles; resembles Satsuma.

Tung

Cooter.—Orig. in Brooker, Fla., by U.S. Dept. of Agriculture. Introd. in 1948. Parentage unknown; original tree was planted in 1931 in the orchard of H.W. Bennett near Brooker, Fla.; tested as USDA F-4. Fruit: size medium; spherical, sometimes slightly oblate; borne in small clusters; oil content intermediate; matures early. Tree: productivity very high; moderately cold resistant.

Folsom.—Orig. in Isabel, La., by the U.S. Dept. of Agriculture. Introd. in 1953. Parentage unknown; found in Fall 1938 in an orchard planted about 1928 near Isabel, by the Great Southern Lumber Co. Selected as a promising variety because of its large size, heavy crop, previous fruiting, and good tree form. Seedling progenies of it were grown and planted along with many others in a test orchard near Folsom, La., in Spring 1940; tested as USDA L-47. Fruit: size normally large; borne on short to medium long stems, usually in clusters; slightly flattened; sometimes have a rather pronounced neck at the stem and have a pronounced tip at the apex; normally develop a bright purple color as they approach maturity; at 15% moisture content, they generally contain 21% oil or slightly more; mature as late or later than La Crosse. Tree: highly productive; sometimes overbears even as a young tree and then tends to bear biennially; resistant to cold, especially from early fall freezes. First-generation seedlings come true to type from seed, but it is not uncommon to have 10% or more off-type seedlings from second-generation seed. Folsom and its first-generation seedlings are recommended for trial throughout the Tung Belt.

Gahl.—Orig. in Isabel, La., by U.S. Dept. of Agriculture. Introd. in 1948. Parentage unknown; original tree located in an orchard planted about 1928 by the Great Southern Lumber Co. near Isabel, La.; tested as USDA L-51. Fruit: large; borne singly or in small clusters; almost spherical except for tip of apex; percentage of kernel medium; oil content medium high; matures early. Tree: productivity very high; resistant to cold injury.

Isabel.—Orig. in Isabel, La., by U.S. Dept. of Agriculture. Introd. in 1948. Parentage unknown; original tree located in an orchard planted about 1928 by the Great Southern Lumber Co., near Isabel, La.; tested as USDA L-2. Fruit: large; borne singly or in small clusters; rather prominently ridged at sutures; kernel percentage high; oil content very high; matures early. Tree: productivity very high; moderately resistant to cold injury.

La Crosse.—Orig. in Brooker, Fla., by U.S. Dept. of Agriculture. Introd. in 1948. Parentage unknown; original tree planted in 1931 in the orchard of H.W. Bennett, near Brooker, Fla.; tested as USDA F-99. Fruit: small; length shorter than breadth; flattened at apical and basal ends; cluster type; matures late; oil content high. Tree: productivity very high; top moderately dense with fruiting wood throughout.

Lamont.—Orig. in Lamont, Fla., by U.S. Dept. of Agriculture. Introd. in 1948. Parentage unknown; original tree located in the Chase orchard near Lamont and planted about 1931; tested as USDA F-542. Fruit: size medium to large; slightly flattened at apical end; frequently one fruit in each cluster is small and abnormal in shape; cluster medium type; percentage of kernel high; oil content high; matures in midseason. Tree: productivity very high; moderately resistant to cold injury.

Lampton.—Orig. in Leon County, Fla., by the U.S. Dept. of Agriculture. Introd. in 1950. Open-pollinated seedling of L-14; selected in 1943; tested as USDA F-578. Fruit: kernel percentage high; oil content 22% to 23% on a whole-fruit basis at 15% moisture content; resembles Isabel. Tree: highly productive; vigorous growth.

White Sapote

Chapman.—Orig. in Lomita, Calif. Introd. in 1923 by Armstrong Nurseries, Ontario, Calif. Parentage unknown; selected about 1920. Fruit: large, averaging 4 inches in diam.; round slightly ribbed; skin yellowish-green; flesh creamy white, melting, flavor very good; ripens September to January.

Collins Seedless.—Orig. in southern California. Introd. in 1924 by Armstrong Nurseries, Ontario, Calif. Parentage unknown; selected about 1920. Fruit: grows in clusters; quality very good; very little core; no seeds; ripens October to January. Tree: vigorous; large crops.

Dade.—Orig. in Homestead, Fla., by the Florida Subtrop. Expt. Sta. Introd. in 1943. Open-pollinated seedling, selected in 1939. Fruit: flavor fair; no bitter element; shape and size uniform. Tree: bears well.

Flournoy.—Orig. on the Flournoy ranch, El Cajon, Calif., by James H. Macpherson, Williams & Macpherson Subtrop. Nursery, Encinitas, Calif. Introd. in 1930. Open-pollinated seedling of Wilson; selected in 1928. Fruit: quality fine; resembles Wilson. Tree: bears heavily; hardy.

Galloway.—Orig. in Orange, Calif. Introd. in 1924 by Armstrong Nurseries, Ontario, Calif. A seedling nursery tree sold by this nursery to A.E. Zapf, Orange, Calif.; selected about 1920. Fruit: size medium; shape flattened; flavor rich, sweet, aromatic, pleasing; ripens December to May. Tree: strong, vigorous; bears at an early age; produces good crops; a good ornamental shade tree.

Hammond.—Orig. in Ventura County, Calif. Introd. in 1925 by Armstrong Nurseries, Ontario, Calif. A Mexican seedling; selected about 1920. Fruit: size medium, averaging 1/2 to 3/4 lb; almost round, slightly oblong; quality excellent; ripens January to March. Tree: vigorous.

Lenparken.—Orig. in San Diego, Calif., by Mrs. Oscar Kendall. Introd. in 1945. Chance seedling Fruit: large, up to 18 oz, skin thin, yellow green; flavor good. Apparently no longer being planted.

Maltby.—Orig. in Carlsbad, Calif., by Guy Maltby. Introd. about 1933. Parentage unknown; discovered about 1930. Fruit: large, weighing up to 1 lb; skin too thin for commercial shipping, protuberances prominent; flavor poor; seeds numerous. Tree: heavy bearer.

Max Golden.—Orig. in Carlsbad, Calif., by Fred Ramsey. Introd. about 1932. Parentage unknown; wooly leaf white sapote (*Casimiroa tetrameria*). Fruit: large; flesh dark yellow, strong aromatic flavor; tends to have few seed; superior to fruit of other varieties of this species. No longer being planted commercially.

May.—Orig. in San Diego, Calif., by L.S. Whitmoyer. Introd. in 1945. Parentage unknown. Fruit: sugar content high; matures in September and October.

Pike.—Orig. in Santa Barbara, Calif., by the U.S. Dept. of Agriculture. Introd. in the late 1920s. Parentage unknown; grown from seed secured from the U.S. Dept. of Agriculture; planted on the Pike ranch. Fruit: large; quality good; skin thin, fragile; matures in fall. Tree: considered productive.

Suebelle (*Hubbell*).—Orig. in Encinitas, Calif., by Susan Hubbell, Rancho Leucadia Nursery. Introd. in 1931 as Hubbell by this Nursery; later introd. as Suebelle by Mr. C. Tanner. Fruit: smooth; round; 2 to 6 oz; flesh custard-like, melting; no ginger taste in skin. Tree: everbearing.

Vinmar.—Orig. in Buena Vista, Calif., by Martin Hankins. Introd. in 1943. Parentage unknown. Fruit: quality good; keeps well; sugar content averages 25%.

Wilson.—Orig. in Monrovia, Calif., by W.C. Wilson. Introd. in 1927 by Armstrong Nurseries, Ontario, Calif. Fruit: somewhat flattened; quality good; matures in fall. Tree: produces well.

Wood.—Orig. in Encinitas, Calif., by Winfield M. Wood. Introd. in 1943. Parentage unknown; discovered in 1939. Fruit: large; skin contains practically no bitterness; flesh sweet, mild; ripens in early summer. Tree: bears well.

Conversion factors for U.S. and metric units.[z]

To convert column 1 into column 2, multiply by:	Column 1	Column 2	To convert column 2 into column 1, multiply by:
	Length		
0.621	kilometer (km)	mile (mi)	1.609
3.281	meter (m)	foot (ft)	0.305
0.394	centimeter (cm)	inch	2.54
	Area		
0.386	square kilometer (km^2)	square mile (mi^2)	2.59
247.1	square kilometer (km^2)	acre	0.00405
2.471	hectare (ha)	acre	0.405
	Volume		
0.00973	cubic meter (m^3)	acre-inch	102.8
35.32	cubic meter (m^3)	cubic foot (ft^3)	0.0283
0.0353	liter	cubic foot (ft^3)	28.32
0.0284	liter	bushel (bu)	35.24
1.057	liter	quart (qt)	0.0946
	Force		
0.102	newton (N)	kilograms of force	9.8
	Mass		
1.102	metric ton (t)	ton	0.9072
2.205	kilogram (kg)	pound (lb)	0.454
0.035	gram (g)	ounce (oz)	28.35
	Pressure		
14.22	kilogram per square meter	pound per square inch (psi)	0.0703
14.50	bar	pound per square inch (psi)	0.06895
0.9869	bar	atmosphere (atm)	1.013
0.9678	kilogram per square centimeter	atmosphere (atm)	1.033
14.70	atmosphere (atm)	pound per square inch (psi)	0.06805
0.01	kilopascals (kPa)	bar	100
	Yield or Rate		
0.446	metric ton per hectare	ton per acre	2.24
0.892	kilogram per hectare	pound per acre	1.12
0.107	liter per hectare	gallon per acre	9.354
	Water measurement		
8.108	hectare-meter	acre-foot	0.1233
97.29	hectare-meter	acre-inch	0.01028
0.08101	hectare-centimeter	acre-foot	12.33
0.973	hectare-centimeter	acre-inch	1.028
0.00973	cubic meter (m^3)	acre-inch	102.8
0.981	hectare centimeter per hour	cubic foot per second	1.0194
440.3	hectare centimeter per hour	gallon per minute	0.00227
0.00981	cubic meter per hour	cubic foot per second	101.94
4.403	cubic meter per hour	gallon per minute	0.227
	Temperature		
(9/5)(°C) + 32	Celsius	Fahrenheit	(5/9)(°F – 32)

[z]Taken from ASHS Publications Manual, 1991. Amer. Soc. for Hort. Sci., Alexandria, Va.